P9-AGA-417

The Cumulative Standardized Normal Distribution (Continued)

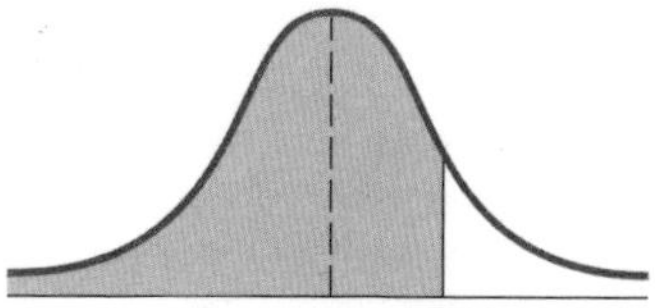

Entry represents area under the stadardized normal distribution from −∞ to Z

Z	0.00	0.01	0.02	0.03	0.04	0.05	0.06	0.07	0.08	0.09
0.0	0.5000	0.5040	0.5080	0.5120	0.5160	0.5199	0.5239	0.5279	0.5319	0.5359
0.1	0.5398	0.5438	0.5478	0.5517	0.5557	0.5596	0.5636	0.5675	0.5714	0.5753
0.2	0.5793	0.5832	0.5871	0.5910	0.5948	0.5987	0.6026	0.6064	0.6103	0.6141
0.3	0.6179	0.6217	0.6255	0.6293	0.6331	0.6368	0.6406	0.6443	0.6480	0.6517
0.4	0.6554	0.6591	0.6628	0.6664	0.6700	0.6736	0.6772	0.6808	0.6844	0.6879
0.5	0.6915	0.6950	0.6985	0.7019	0.7054	0.7088	0.7123	0.7157	0.7190	0.7224
0.6	0.7257	0.7291	0.7324	0.7357	0.7389	0.7422	0.7454	0.7486	0.7518	0.7549
0.7	0.7580	0.7612	0.7642	0.7673	0.7704	0.7734	0.7764	0.7794	0.7823	0.7852
0.8	0.7881	0.7910	0.7939	0.7967	0.7995	0.8023	0.8051	0.8078	0.8106	0.8133
0.9	0.8159	0.8186	0.8212	0.8238	0.8264	0.8289	0.8315	0.8340	0.8365	0.8389
1.0	0.8413	0.8438	0.8461	0.8485	0.8508	0.8531	0.8554	0.8577	0.8599	0.8621
1.1	0.8643	0.8665	0.8686	0.8708	0.8729	0.8749	0.8770	0.8790	0.8810	0.8830
1.2	0.8849	0.8869	0.8888	0.8907	0.8925	0.8944	0.8962	0.8980	0.8997	0.9015
1.3	0.9032	0.9049	0.9066	0.9082	0.9099	0.9115	0.9131	0.9147	0.9162	0.9177
1.4	0.9192	0.9207	0.9222	0.9236	0.9251	0.9265	0.9279	0.9292	0.9306	0.9319
1.5	0.9332	0.9345	0.9357	0.9370	0.9382	0.9394	0.9406	0.9418	0.9429	0.9441
1.6	0.9452	0.9463	0.9474	0.9484	0.9495	0.9505	0.9515	0.9525	0.9535	0.9545
1.7	0.9554	0.9564	0.9573	0.9582	0.9591	0.9599	0.9608	0.9616	0.9625	0.9633
1.8	0.9641	0.9649	0.9656	0.9664	0.9671	0.9678	0.9686	0.9693	0.9699	0.9706
1.9	0.9713	0.9719	0.9726	0.9732	0.9738	0.9744	0.9750	0.9756	0.9761	0.9767
2.0	0.9772	0.9778	0.9783	0.9788	0.9793	0.9798	0.9803	0.9808	0.9812	0.9817
2.1	0.9821	0.9826	0.9830	0.9834	0.9838	0.9842	0.9846	0.9850	0.9854	0.9857
2.2	0.9861	0.9864	0.9868	0.9871	0.9875	0.9878	0.9881	0.9884	0.9887	0.9890
2.3	0.9893	0.9896	0.9898	0.9901	0.9904	0.9906	0.9909	0.9911	0.9913	0.9916
2.4	0.9918	0.9920	0.9922	0.9925	0.9927	0.9929	0.9931	0.9932	0.9934	0.9936
2.5	0.9938	0.9940	0.9941	0.9943	0.9945	0.9946	0.9948	0.9949	0.9951	0.9952
2.6	0.9953	0.9955	0.9956	0.9957	0.9959	0.9960	0.9961	0.9962	0.9963	0.9964
2.7	0.9965	0.9966	0.9967	0.9968	0.9969	0.9970	0.9971	0.9972	0.9973	0.9974
2.8	0.9974	0.9975	0.9976	0.9977	0.9977	0.9978	0.9979	0.9979	0.9980	0.9981
2.9	0.9981	0.9982	0.9982	0.9983	0.9984	0.9984	0.9985	0.9985	0.9986	0.9986
3.0	0.99865	0.99869	0.99874	0.99878	0.99882	0.99886	0.99889	0.99893	0.99897	0.99900
3.1	0.99903	0.99906	0.99910	0.99913	0.99916	0.99918	0.99921	0.99924	0.99926	0.99929
3.2	0.99931	0.99934	0.99936	0.99938	0.99940	0.99942	0.99944	0.99946	0.99948	0.99950
3.3	0.99952	0.99953	0.99955	0.99957	0.99958	0.99960	0.99961	0.99962	0.99964	0.99965
3.4	0.99966	0.99968	0.99969	0.99970	0.99971	0.99972	0.99973	0.99974	0.99975	0.99976
3.5	0.99977	0.99978	0.99978	0.99979	0.99980	0.99981	0.99981	0.99982	0.99983	0.99983
3.6	0.99984	0.99985	0.99985	0.99986	0.99986	0.99987	0.99987	0.99988	0.99988	0.99989
3.7	0.99989	0.99990	0.99990	0.99990	0.99991	0.99991	0.99992	0.99992	0.99992	0.99992
3.8	0.99993	0.99993	0.99993	0.99994	0.99994	0.99994	0.99994	0.99995	0.99995	0.99995
3.9	0.99995	0.99995	0.99996	0.99996	0.99996	0.99996	0.99996	0.99996	0.99997	0.99997
4.0	0.99996832									
4.5	0.99999660									
5.0	0.99999971									
5.5	0.99999998									
6.0	0.99999999									

A Roadmap for Selecting a Statistical Method

Type of Analysis	TYPE OF DATA: Numerical	TYPE OF DATA: Categorical
Describing a Group or Several Groups	Ordered array, stem-and-leaf display, frequency distribution, relative frequency distribution, percentage distribution, cumulative percentage distribution, histogram, polygon, cumulative percentage polygon **(Sections 2.2 and 2.3)** Mean, median, mode, quartiles, geometric mean, range, interquartile range, standard deviation, variance, coefficient of variation, box-and-whisker plot **(Sections 3.1–3.3)** Index numbers **(Section 16.9)**	Summary table, bar chart, pie chart, Pareto diagram **(Section 2.1)**
Inference about One Group	Confidence Interval Estimate for the Mean **(Sections 8.1 and 8.2)** *Z* Test for the Mean **(Section 9.2)** *t* Test for the Mean **(Section 9.4)** Chi-Square Test for a Variance or Standard Deviation **(Section 12.5)**	Confidence Interval Estimate for the Proportion **(Section 8.3)** *Z* Test of Hypothesis for the Proportion **(Section 9.5)**
Comparing Two Groups	Tests for the Difference in the Means of Two Independent Populations **(Section 10.1)** Wilcoxon Rank Sum Test **(Section 12.7)** Paired *t* Test **(Section 10.2)** Wilcoxon Signed Rank Test **(Section 12.8)**	*Z* Test for the Difference between Two Proportions **(Section 10.3)** Chi-Square Test for the Difference between Two Proportions **(Section 12.1)** McNemar Test for Two Related Samples **(Section 12.4)**
Comparing More than Two Groups	One-Way Analysis of Variance **(Section 11.1)** Kruskal-Wallis Test **(Section 12.9)** Randomized Block Design **(Section 11.2)** Friedman Test **(Section 12.10)** Two-Way Analysis of Variance **(Section 11.3)**	Chi-Square Test for Differences among More than Two Proportions **(Section 12.2)**
Analyzing the Relationship between Two Variables	Scatter Diagram, Time Series Plot (**Section 2.5**) Covariance, Coefficient of Correlation **(Section 3.5)** Simple Linear Regression **(Chapter 13)** *t* Test of Correlation **(Section 13.7)**	Contingency Table, Side-by-Side Bar Chart **(Section 2.4)** Chi-Square Test of Independence **(Section 12.3)**
Analyzing the Relationship between Two or More Variables	Multiple Regression **(Chapters 14 and 15)** Time Series Forecasting **(Chapter 16)**	Logistic Regression **(Section 14.7)**

HAVE YOU THOUGHT ABOUT *Customizing* THIS BOOK?

Customizing

THE PRENTICE HALL JUST-IN-TIME PROGRAM IN DECISION SCIENCE

You can combine chapters from this book with chapters from any of the Prentice Hall titles listed on the following page to create a text tailored to your specific course needs. You can add your own material or cases from our extensive case collection. By taking a few minutes to look at what is sitting on your bookshelf and the content available on our Web site, you can create your ideal textbook.

The Just-In-Time program offers:

- **Quality of Material to Choose From**—In addition to the books listed, you also have the option to include any of the cases from Prentice Hall Custom Business Resources, which gives you access to cases (and teaching notes where available) from Darden, Harvard, Ivey, NACRA, and Thunderbird. Most cases can be viewed online at our Web site.
- **Flexibility**—Choose only the material you want,either from one title or several titles (plus cases) and sequence it in whatever way you wish.
- **Instructional Support**—You have access to the text-specific CD-ROM that accompanies the traditional textbook and desk copies of your JIT book.
- **Outside Materials**—There is also the option to include up to 20% of the text from materials outside of Prentice Hall Custom Business Resources.
- **Cost Savings**—Students pay only for material you choose. The base price is $6.00, plus $2.00 for case material, plus $.09 per page. The text can be shrink-wrapped with other Pearson textbooks for a 10% discount. Outside material is priced at $.10 per page plus permission fees.
- **Quality of Finished Product**—Custom cover and title page—including your name, school, department, course title, and section number. Paperback, perfect bound, black-and-white printed text. Customized table of contents. Sequential pagination throughout the text.

Visit our Web site at www.prenhall.com/custombusiness and create your custom text on our bookbuildsite or download order forms online.

THE PRENTICE HALL Just-In-Time program

Just-In-Time

PEARSON Prentice Hall

PEARSON Custom Publishing

YOU CAN CUSTOMIZE YOUR TEXTBOOK WITH CHAPTERS FROM ANY OF THE FOLLOWING PRENTICE HALL TITLES: *

BUSINESS STATISTICS

- Berenson/Levine/Krehbiel, BASIC BUSINESS STATISTICS, 10/e
- Groebner/Shannon/Fry/Smith, BUSINESS STATISTICS, 6/e
- Levine/Stephan/Krehbiel/Berenson, STATISTICS FOR MANAGERS USING MICROSOFT EXCEL, 4/e
- Levine/Krehbiel/Berenson, BUSINESS STATISTICS: A FIRST COURSE, 4/e
- Newbold/Carlson/Thorne, STATISTICS FOR BUSINESS AND ECONOMICS, 5/e
- Shannon/Groebner/Fry/Smith, A COURSE IN BUSINESS STATISTICS, 3/e

PRODUCTION/OPERATIONS MANAGEMENT

- Anupindi/Chopra/Deshmukh/Van Mieghem/Zemel, MANAGING BUSINESS PROCESS FLOWS
- Handfield/Nichols, Jr., SUPPLY CHAIN MANAGEMENT
- Haksever/Render/Russell/Murdick, SERVICE MANAGEMENT AND OPERATIONS, 2/e
- Hanna/Newman, INTEGRATED OPERATIONS MANAGEMENT
- Heineke/Meile, GAMES AND EXERCISES IN OPERATIONS MANAGEMENT
- Heizer/Render, OPERATIONS MANAGEMENT, 7/e
- Krajewski/Ritzman, OPERATIONS MANAGEMENT, 7/e
- Latona/Nathan, CASES AND READINGS IN POM
- Russell/Taylor, OPERATIONS MANAGEMENT, 4/e
- Schmenner, PLANT AND SERVICE TOURS IN OPERATIONS MANAGEMENT, 5/e
- Nicholas, PROJECT MANAGEMENT, 2/e

MANAGEMENT SCIENCE/SPREADSHEET MODELING

- Eppen/Gould, INTRODUCTORY MANAGEMENT SCIENCE, 5/e
- Moore/Weatherford, DECISION MODELING WITH MICROSOFT EXCEL, 6/e
- Render/Stair/Hanna, QUANTITATIVE ANALYSIS FOR MANAGEMENT, 8/e
- Render/Stair/Balakrishnan, MANAGERIAL DECISION MODELING WITH SPREADSHEETS
- Render/Stair, CASES AND READINGS IN MANAGEMENT SCIENCE
- Taylor, INTRODUCTION TO MANAGEMENT SCIENCE, 8/e

For more information, or to speak to a customer service representative, contact us at 1-800-777-6872.

www.prenhall.com/custombusiness

* Selection of titles on the JIT program is subject to change

BASIC BUSINESS STATISTICS

TENTH EDITION

Concepts and Applications

BASIC BUSINESS STATISTICS

TENTH EDITION

Concepts and Applications

MARK L. BERENSON
Department of Information and Decision Sciences
School of Business, Montclair State University

DAVID M. LEVINE
Department of Statistics and Computer Information Systems
Zicklin School of Business, Baruch College, City University of New York

TIMOTHY C. KREHBIEL
Department of Decision Sciences and Management Information Systems
Richard T. Farmer School of Business, Miami University

PEARSON
Prentice Hall

Upper Saddle River, New Jersey 07458

Library of Congress Cataloging-in-Publication Data
Berenson, Mark L.
Basic business statistics: concepts and applications / Mark L. Berenson, David M. Levine, Timothy C. Krehbiel.—10th ed.
p. cm.
Includes bibliographical references and index.
ISBN 0-13-153686-9
1. Commercial statistics. 2. Statistics. I. Levine, David M.– II. Krehbiel, Timothy C. III. Title.
HF1017.B38 2005
519.5—dc22 2004060020

Executive Editor: Mark Pfaltzgraff
Editorial Director: Jeff Shelstad
Managing Editor (Editorial): Alana Bradley
Senior Editorial Assistant: Jane Avery
Media Project Manager: Nancy Welcher
Marketing Manager: Debbie Clare
Marketing Assistant: Joanna Sabella
Managing Editor (Production): Cynthia Regan
Production Editor: Anne Graydon
Permissions Supervisor: Charles Morris
Production Manager: Arnold Vila
Design Manager: Maria Lange
Designer: Steve Frim
Director, Image Resource Center: Melinda Reo
Manager, Rights and Permissions: Zina Arabia
Manager, Visual Research: Beth Brenzel
Manager, Cover Visual Research & Permissions: Karen Sanatar
Manager, Print Production: Christy Mahon
Composition/Full-Service Project Management: GGS Book Services, Atlantic Highlands
Printer/Binder: Courier

Photo Credits: Chapter 1, page 2, Corbis/Stock Market; Chapter 2, page 22, Photo Researchers, Inc.; Chapter 3, page 72, PhotoEdit; Chapter 4, page 126, PhotoEdit; Chapter 7, page 228, ImageState/International Stock Photography Ltd.; Chapter 8, page 260, PhotoEdit; Chapter 9, page 300; PhotoEdit; Chapter 10, page 346, Stock Boston; Chapter 11, page 396, Stock Boston; Chapter 12, page 446, Stock Boston; Chapter 13, page 514, Getty Images, Inc.—Photodisc; Chapter 14, page 570, Sharyn Levine; Chapter 15, page 620, Stock Boston; Chapter 16, page 658, Photo Researchers, Inc.; Chapter 17, page 726, Corbis/Stock Market; Chapter 18, page 752, PhotoEdit.

Credits and acknowledgments borrowed from other sources and reproduced, with permission, in this textbook appear on appropriate page within text.
Microsoft Excel and Windows are registered trademarks of Microsoft Corporation in the U.S.A. and other countries. Screen shots and icons reprinted with permission from the Microsoft Corporation. This book is not sponsored or endorsed by or affiliated with Microsoft Corporation.
MINITAB is a registered trademark of the Minitab, Inc., at
3081 Enterprise Drive
State College, PA 16801 USA
ph: 814.238.3280 fax: 814.238.4383
e-mail: **info@minitab.com**
URL: **http://www.minitab.com**

SPSS® is a registered trademark of the SPSS®, Inc., at
233 South Wacker Drive, 11th Floor
Chicago, IL 60606-6412 USA
ph: 312.651.3000 fax: 312.651.3668
URL: **http://www.spss.com**
Six Sigma® Management is a registered trademark of Motorola Corporation.

Copyright © 2006, 2004, 2002, 1999, 1996, 1992 by Pearson Education, Inc., Upper Saddle River, New Jersey, 07458. All rights reserved. Printed in the United States of America. This publication is protected by Copyright and permission should be obtained from the publisher prior to any prohibited reproduction, storage in a retrieval system, or transmission in any form or by any means, electronic, mechanical, photocopying, recording, or likewise. For information regarding permission(s), write to: Rights and Permissions Department.

Pearson Prentice Hall™ is a trademark of Pearson Education, Inc.
Pearson® is a registered trademark of Pearson plc
Prentice Hall® is a registered trademark of Pearson Education, Inc.

Pearson Education LTD.
Pearson Education Australia PTY, Limited
Pearson Education Singapore, Pte. Ltd
Pearson Education North Asia Ltd
Pearson Education, Canada, Ltd
Pearson Educación de Mexico, S.A. de C.V.
Pearson Education—Japan
Pearson Education Malaysia, Pte. Ltd

PEARSON
Prentice Hall

10 9 8 7 6 5 4 3 2 1
ISBN 0-13-153686-9

To our wives,
Rhoda B., Marilyn L., and Patti K.

and to our children,
Kathy, Lori, Sharyn, Ed, Rudy, and Rhonda

Brief Contents

CD-ROM TOPICS

CONTENTS

10 TWO-SAMPLE TESTS 345

11 ANALYSIS OF VARIANCE 395

PREFACE

Educational Philosophy

In our many years of teaching business statistics, we have continually searched for ways to improve the teaching of these courses. Our active participation in a series of Making Statistics More Effective in Schools and Business, Decision Sciences Institute, and American Statistical Association conferences as well as the reality of serving a diverse group of students at large universities has shaped our vision for teaching these courses. Over the years, our vision has come to include these key principles:

1. Students need to be shown the relevance of statistics.
   - Students need a frame of reference when learning statistics, especially when statistics is not their major. That frame of reference for business students should be the functional areas of business—that is, accounting, economics and finance, information systems, management, and marketing. Each statistical topic needs to be presented in an applied context related to at least one of these functional areas.
   - The focus in teaching each topic should be on its application in business, the interpretation of results, the presentation of assumptions, the evaluation of the assumptions, and the discussion of what should be done if the assumptions are violated.
2. Students need to be familiar with the software used in the business world.
   - Integrating spreadsheet or statistical software into all aspects of an introductory statistics course allows the course to focus on interpretation of results instead of computations.
   - Introductory business statistics courses should recognize that in business, spreadsheet software is typically available on a decision-maker's desktop (and sometimes statistical software is as well).
3. Students need to be given sufficient guidance on using software.
   - Textbooks should provide enough instructions so that students can effectively use the software integrated with the study of statistics, without having the software instruction dominate the course.
4. Students need ample practice in order to understand how statistics is used in business.
   - Both classroom examples and homework exercises should involve actual or realistic data as much as possible.
   - Students should work with data sets, both small and large, and be encouraged to look beyond the statistical analysis of data to the interpretation of results in a managerial context.

New to This Edition

This new tenth edition of *Basic Business Statistics: Concepts and Applications* has been improved in a number of important areas.

Improved accessibility for students

- Every chapter in the text has undergone major rewriting and the text now uses a more active, conversational writing style that students will appreciate. Sentences have been shortened and simplified.
- The text now includes many more examples from everyday life. Such examples include those on online shopping (Chapter 2), time to get ready in the morning (Chapter 3), and waiting time at a fast-food restaurant (Chapter 9).
- Many problems have been simplified so that they contain no more than four parts.
- Key formulas are now included at the end of each chapter.
- Worked-out solutions to Self-Test Questions are provided in the back of the text.
- A roadmap for selecting the proper statistical method is included at the front of the text to help students select the proper technique and to make connections between topics.
- Many new applied examples and exercises with data from *The Wall Street Journal*, *USA Today*, *Consumer Reports*, and other sources have been added to the text.

Enhanced software instruction

- End-of-chapter Microsoft Excel appendixes now discuss how to use standard Excel worksheets to perform most statistical analyses. Instructors and students who wish to avoid using add-ins will find these new instructions immediately useful. (Those who choose to use the PHStat2 Excel add-in will find that all explanations of PHStat2 commands have been placed together in a new Appendix G for easy reference.)
- Many of the standard Excel worksheets discussed in the Excel appendixes are included as in-chapter illustrations. Each illustration (see example below) include a listing of all cell formulas contained in the worksheet. (PHStat2 users will also find these illustrations informative as they are consistent with the worksheets that PHStat2 produces for you.)

	A	B	
1	**Estimate for the Mean Sales Invoice Amount**		
2			
3	**Data**		
4	**Sample Standard Deviation**	**28.95**	
5	**Sample Mean**	**110.27**	
6	**Sample Size**	**100**	
7	**Confidence Level**	**95%**	
8			
9	Intermediate Calculations		
10	Standard Error of the Mean	2.8950	**=B4/SQRT(B6)**
11	Degrees of Freedom	99	**=B6 - 1**
12	*t* Value	1.9842	**=TINV(1-B7,B11)**
13	Interval Half Width	5.7443	**=B12 * B10**
14			
15	**Confidence Interval**		
16	**Interval Lower Limit**	**104.53**	**=B5 - B13**
17	**Interval Upper Limit**	**116.01**	**=B5 + B13**

- Updated version of PHStat2—PHStat2 version 2.5, the newest version of Prentice Hall's add-in for Microsoft Excel is bundled free with this text. This updated version includes enhancements such as multiple regression with independent variables in noncontiguous columns, improved stem-and-leaf displays and box-and-whisker plots, the Z test for the difference in two means, Levene's test for the homogeneity of variance, and the Marascuilo multiple comparisons procedure for proportions. (Support for PHStat2, including free updates when available, can be found at **www.prenhall.com/phstat**.)
- Use of Minitab Version 14, the latest version of the Minitab statistical software—All Minitab output in the text and all Minitab appendixes are from Minitab Version 14, the latest version of the Minitab statistical software.

Reorganization of the hypothesis-testing chapters

- All tests involving the normal and t distribution are covered in Chapters 9 and 10 *prior to* coverage of the F test.
- The Analysis of Variance is covered in Chapter 11.
- All chi-square tests are covered in Chapter 12 along with nonparametric tests.

More comprehensive coverage of topics

- The sample covariance, the normal approximation to the binomial distribution, the power of the test, the McNemar test, and the chi-square goodness-of-fit test are now included in the text.

Chapter-by-Chapter Changes in the Tenth Edition

Each chapter has a new opening page that shows the sections and subsections for the chapter.

- *Chapter 1* has rewritten sections 1.1, 1.2, and 1.3. The sections on survey sampling have been moved to Chapter 7.

- *Chapter 2* has a new data set concerning mutual fund returns for 1999–2003. Graphs for categorical variables are discussed prior to graphs for numerical variables. All graphs for one variable are discussed before any graphs for two variables. The examples within the chapter refer to online shopping and the cost of restaurant meals in addition to mutual fund returns.
- *Chapter 3* has a new data set concerning mutual fund returns for 1999–2003. The examples within the chapter refer to the time to get ready in the morning as well as mutual fund returns. Z scores for detecting outliers are now included. The sample covariance is now included along with the coefficient of correlation.
- *Chapter 4* now includes additional examples.
- *Chapter 5* covers the Poisson distribution prior to the hypergeometric distribution.
- *Chapter 6* has a simplified section on the normal probability plot and a section on the Normal Approximation to the Binomial Distribution.
- *Chapter 7* now includes Types of Survey Sampling Methods and Survey Worthiness.
- *Chapter 8* has numerous new problems.
- *Chapter 9* uses a simpler, six-step method to perform hypothesis tests using the critical value approach, a straightforward five-step method to perform hypothesis tests using the p-value, and includes a section on the power of the test. The section on the χ^2 test for a variance has been moved to Chapter 12.
- *Chapter 10* is reorganized so that two sample tests for means and proportions precede the F test for the difference between the variances. The Wilcoxon rank sum test and the Wilcoxon signed ranks test have been moved to Chapter 12.
- *Chapter 11* no longer includes the Kruskal-Wallis test or the Friedman rank test, which have been moved to Chapter 12.
- *Chapter 12* now includes the McNemar test, the χ^2 test for a variance, the χ^2 test for goodness of fit, the Wilcoxon rank sum test, the Wilcoxon signed ranks test, the Kruskal-Wallis test, and the Friedman rank test.
- *Chapter 13* now includes computations for the regression coefficients and sum of squares in chapter examples.
- *Chapter 14* now covers R^2, adjusted R^2, and the overall F test prior to residual analysis.
- *Chapter 15* provides additional coverage of stepwise regression.
- *Chapter 16* has new and updated data sets for the in-chapter examples.
- *Chapter 18* has more coverage of Six Sigma Management.

Hallmark Features

We have continued many of the traditions of past editions. We've highlighted some of those features below.

- **"Using Statistics" business scenarios**—Each chapter begins with a "Using Statistics" example that shows how statistics is used in accounting, finance, management, or marketing. Each scenario is used throughout the chapter to provide an applied context for the concepts.

USING STATISTICS

Comparing the Performance of Mutual Funds

Among the many investment choices available today, mutual funds, a market basket of a portfolio of securities, are a common choice for those thinking about their retirement. If you decided to purchase mutual funds for your retirement account, how would you go about making a reasonable choice among the many funds available today?

You first would want to know the strategies of the professionals who manage the funds. Do they invest in high-risk securities or do they make more conservative choices? Does the fund specialize in a certain sized company, one whose outstanding stock totals a large amount (large cap) or one that is quite small (small cap)? Does the fund charge management fees that reduce the percentage return earned by an investor? And, of course, you would want to know how well the fund performed in the past.

All of this is a lot of data to review if you consider several dozen or more mutual funds. How could you "get your hands around" such data and explore it in a comprehensible manner?

- **Emphasis on data analysis and interpretation of computer output**—We believe that the use of computer software is an integral part of learning statistics. Our focus emphasizes analyzing data by interpreting the output from Microsoft Excel and Minitab, while reducing emphasis on doing computations. Therefore, we have included more computer output and integrated this output into the fabric of the text. For example, in the coverage of tables and charts in Chapter 2, the focus is on the interpretation of various charts, not on their construction by hand. In our coverage of hypothesis testing in Chapters 9 through 12, extensive computer output has been included so that focus can be placed on the *p*-value approach. In our coverage of simple linear regression in Chapter 13, we assume that Microsoft Excel or Minitab will be used. Thus, the focus is on the interpretation of the output, not on hand calculations.
- **Pedagogical aides** such as an active writing style, boxed numbered equations, set-off examples to provide reinforcement for learning concepts, problems divided into Learning the Basics and Applying the Concepts, and key terms are included.
- **End-of-chapter appendices** using standard Microsoft Excel and Minitab Version 14, with illustrations, provide easy-to-follow instructions. PHStat2 instructions are included in Appendix G. SPSS appendices are included on the CD-ROM that accompanies this text.
- **Answers** to most of the even-numbered exercises are provided in an appendix at the end of the book.
- **PHStat2,** a supplemental add-in program for Microsoft Excel, that enhances the statistical capabilities of Microsoft Excel and executes for you the low-level menu selection and worksheet entry tasks associated with implementing statistical analysis in Excel is included on the student CD-ROMs. When combined with Microsoft Excel's own Data Analysis ToolPak add-in, virtually all statistical methods taught in an introductory statistics course can be illustrated using Microsoft Excel.
- **Web cases**—A chapter-ending Web case is included for each of the first 17 chapters. By visiting Web sites related to the companies and researching the issues raised in the "Using Statistics" scenarios that start each chapter, students learn to identify misuses of statistical information. The Web cases require students to sift through claims and assorted information in order to discover the data most relevant to the case. Students then determine whether the conclusions and claims are supported by the data. (Instructional tips for using the Web cases and solutions to the Web cases are included in the Instructor's Solutions Manual.)
- **Case Studies and Team Projects**—Detailed case studies are included at the end of numerous chapters. The *Springville Herald* case is included at the end of virtually all chapters as an integrating theme. A Team Project relating to mutual funds is included at the end of many chapters as an integrating theme.
- **Visual Explorations**—a Microsoft Excel workbook bundled free with this text—allows students to interactively explore important statistical concepts in descriptive statistics, probability, the normal distribution, and regression analysis. For example, in descriptive statistics, students observe the effect of changes in the data on the mean, median, quartiles, and standard deviation. In sampling distributions, students use simulation to explore the effect of sample size on a sampling distribution. With the normal distribution, students get to see the effect of changes in the mean and standard deviation on the areas under the normal curve. In regression analysis, students have the opportunity of fitting a line and observing how changes in the slope and intercept affect the goodness of fit. (Visual Explorations requires a Microsoft Excel security setting of Medium.)

Supplement Package

The supplement package that accompanies this text includes the following:

- **Instructor's Solution Manual**—This manual includes teaching tips for each chapter, extra detail in the problem solutions, and many Excel and Minitab solutions.
- **Student Solutions Manual**—This manual provides detailed solutions to virtually all the even-numbered exercises.
- **Test Item File**—The Test Item File contains true/false, multiple choice, fill-in, and problem-solving questions based on the definitions, concepts, and ideas developed in each chapter of the text.

- **TestGen testing software**—The printed test bank is designed for use with the TestGen test-generating software. This computerized package allows instructors to custom design, save, and generate classroom tests. The test program permits instructors to edit, add, or delete questions from the test banks; edit existing graphics and create new graphics; analyze test results; and organize a database of tests and student results. This software allows for greater flexibility and ease of use. It provides many options for organizing and displaying tests, along with a search and sort feature. The program is available both on the Instructor's CD-ROM and on the Prentice Hall online catalog for download.
- **Instructor's Resource Center**—The Instructor's Resource Center contains the electronic files for the complete Instructor's Solutions Manual (MS Word), the Test Item File (MS Word), the computerized Test Item File (MS Word), TestGen, and PowerPoint presentations.
- **Course and Homework Management Tools**
  - **Prentice Hall's OneKey** offers the best teaching and learning resources all in one place. OneKey for *Basic Business Statistics, 10e*, is all you need to plan and administer your course, and is all your students need for anytime, anywhere access to your course materials. Conveniently organized by textbook chapter, the compiled resources include: links to quizzes, PowerPoint presentations, data files, links to Web cases, PHStat2 download, Visual Explorations download, Student Solutions Manual, as well as additional instructor resources.
  - **WebCT and Blackboard**—With a local installation of either course management system, Prentice Hall provides content designed especially for this textbook to create a complete course suite, tightly integrated with the system's course management tools.
  - **PH GradeAssist**—This online homework and assessment system allows the instructor to assign problems for student practice, homework, or quizzes. The problems, taken directly from the text, are algorithmically generated, so each student gets a slightly different problem with a different answer. This feature allows students multiple attempts for more practice and improved competency. PH GradeAssist grades the results and can export them to Microsoft Excel worksheets.
- **Companion Web site**—This site contains
  - An online study guide with true/false, multiple choice, and essay questions designed to test student's comprehension of chapter topics.
  - PowerPoint presentation files with chapter outlines and key formulas.
  - Student data files for text problems in Excel, Minitab, and SPSS.
- **Student version of Minitab**—For a reasonable additional cost, a student version of Minitab Version 14 can be packaged with this text. Please contact your Prentice Hall Sales Representative for ordering information.
- **Student version of SPSS**—For a reasonable additional cost, a student version of SPSS 12 can be packaged with the text. Please contact your Prentice Hall Sales Representative for ordering information.
- **Text Web site**—The text has a home page on the World Wide Web at **www.prenhall.com/berenson**. This site provides many resources for both faculty members and students.

  PHStat2 has a home page on the World Wide Web at **www.prenhall.com/phstat**.

  An index page for the supporting material for all the Web cases included in the text can be found at **www.prenhall.com/Springville/Springvillecc.htm**.

Acknowledgments

We are extremely grateful to the many organizations and companies that allowed us to use their data in developing problems and examples throughout the text. We would like to thank *The New York Times*, Consumers Union (publishers of *Consumer Reports*), Mergent's Investor Service (publishers of *Mergent's Handbook of Common Stocks*), and CEEPress.

In addition, we would like to thank the Biometrika Trustees, American Cyanimid Company, the Rand Corporation, the American Society for Testing and Materials (for their kind permission to publish various tables in Appendix E), and the American Statistical Association (for its permission to publish diagrams from the *American Statistician*). Finally, we are grateful to Professors George A. Johnson and Joanne Tokle of Idaho State University and Ed Conn, Mountain States Potato Company, for their kind permission to incorporate parts of their work as our Mountain States Potato Company case in Chapter 15.

A Note of Thanks

We would like to thank Randy Craig, Salem State University; Mark Eakin, University of Texas–Arlington; Kathy Ernstberger, Indiana University–Southeast; Kimberley Killmer Hollister, Montclair State University; C. P. Kartha, University of Michigan, Flint; Robert Lemke, Lake Forest College; Ram Misra, Montclair State University; Prashant Palvia, University of North Carolina, Greensboro; Susan Pariseau, Merrimack College; Brock Williams, Texas Tech University; Frederick Wiseman, Northeastern University; Reginald Worthley, University of Hawaii, Manoa; and Charles Zimmerman, Robert Morris College, for their comments that have made this a better book.

We would especially like to thank Debbie Clare, Mark Pfaltzgraff, Jeff Shelstad, Alana Bradley, Anne Graydon, Cynthia Regan, Nancy Welcher, and Jane Avery of the editorial, marketing, and production teams at Prentice Hall. It has been our privilege to work with Tom Tucker on this project and many previous ones. As Tom now moves on to a new career, we will greatly miss his insight, encouragement, and dedication. Thank you, Tom, and good luck!

We would like to thank our statistical readers and accuracy checkers Annie Puciloski, Stonehill College, and James Zimmer, Chattanooga State University, for their diligence in checking our work; Robie Grant for her proofreading; Julie Kennedy for her copyediting; and Sandra Krausman of GGS Book Services, Atlantic Highlands, for her work in the production of this text.

We are extremely grateful for the love and support given to us by our families. Our parents Nat and Ethel Berenson, Reuben and Lee Levine, Marvin Krehbiel, and Roberta Reed, have blessed us with a lifetime of encouragement. Finally, we would like to thank our wives and children for their patience, understanding, love, and assistance in making this book a reality. It is to them that we dedicate this book.

Concluding Remarks

We have gone to great lengths to make this text both pedagogically sound and error-free. If you have any suggestions or require clarification about any of the material, or if you find any errors, please contact us at **David_Levine@BARUCH.CUNY.EDU** or **KREHBITC@MUOHIO.EDU**. Include the phrase "BBS version 10" in the subject line of your e-mail. For more information about using PHStat2, see Appendixes F and G, and the PHStat2 readme file on the CD-ROM packaged with this book.

Mark L. Berenson

David M. Levine

Timothy C. Krehbiel

BASIC BUSINESS STATISTICS

TENTH EDITION

Concepts and Applications

CHAPTER 1

Introduction and Data Collection

LEARNING OBJECTIVES

In this chapter, you learn:

- How statistics is used in business
- The sources of data used in business
- The types of data used in business

USING STATISTICS

Good Tunes—Part I

Good Tunes, a privately held online retailer of home entertainment systems, seeks to expand its business by opening several stores. To get the financing necessary to underwrite this expansion, Good Tunes needs to apply for loans at local area banks. The managers of the firm agree to develop an electronic slide show that will explain their business and state the facts that will convince the bankers to loan Good Tunes the money it needs. You have been asked to assist in the process of preparing the slide show. What facts would you include? How would you present those facts?

Every day you use news and information sources to gather the facts that you need to lead your life. You might listen to a weather forecast to decide what clothes to wear, and if you live in a large city, you might listen to a commuter report to learn about the best route for traveling to your job or school.

Your personal likes and dislikes shape some of your decisions, too. In spite of hearing bad reviews of a motion picture that suggest you skip seeing it, you might decide to go anyhow just because you happen to like a particular actor who appears in that film.

Likewise, every day business managers have to make decisions. Although managers sometimes resort to "gut instincts" to make some decisions (this is more formally known as unstructured decision making), they more typically make decisions that are directly influenced by hard facts. As a business student, you cannot really learn how to make unstructured decisions, as such decisions require instincts and insights that require years of experience to form. You can learn, though, the procedures and methods that will help you make better decisions that are based on hard facts. When you begin focusing on the procedures and methods involved in the collecting, presenting, and summarizing of a set of data, or forming conclusions about that data, you have discovered statistics.

In the Good Tunes scenario, you should proceed with the reasonable assumption that the bankers seek to make a decision based on the hard facts you help present, and not on other factors, such as whims or personal likes or dislikes. Presenting the wrong information or the correct information in the wrong fashion could lead the bankers to make a bad business decision, which could jeopardize the future of Good Tunes. You need to know something about statistics to provide the hard facts that are needed, and to know something about statistics, you first need to know the basic concepts of statistics.

1.1 BASIC CONCEPTS OF STATISTICS

Statistics is the branch of mathematics that examines ways to process and analyze data. Statistics provides procedures to collect and transform data in ways that are useful to business decision-makers. To understand anything about statistics, you need to first understand the meaning of a variable.

VARIABLES

Variables are characteristics of items or individuals.

Examples of variables are your gender, your major field of study, the amount of money you have in your wallet, and the amount of time it takes you to get ready to go to school in the morning. The key aspect of the word *variable* is the idea that items differ and people differ. The person next to you may be male rather than female, may be majoring in a different field of study than you, almost certainly has a different amount of money in their wallet, and undoubtedly takes a different amount of time to get ready in the morning than you do. You should distinguish between a variable, such as gender, and its *value* for an individual observation (e.g., "male").

All variables should have an **operational definition**, a universally accepted meaning that is clear to all associated with an analysis. Without operational definitions, confusion can occur. A famous example of such confusion that illustrates the importance of operational definitions relates to the 2000 U.S. presidential election and the disputed ballots in the state of Florida (Jackie Calmes and Edward P. Foldessy, "In Election Review, Bush Wins with No Supreme Court Help," *The Wall Street Journal*, November 12, 2001, A1, A14). A review of 175,010 Florida ballots that were rejected for either no presidential votes or votes for two or more candidates was conducted with the help of the National Opinion Research Center of the University of Chicago. Nine standards or operational definitions were used to evaluate these ballots. The nine standards led to different results. Three of the standards (including one pursued by Al Gore) led to margins of victory for George Bush that ranged from 225 to 493 votes. Six of the standards (including one pursued by George Bush) led to margins of victory for Al Gore that ranged from 42 to 171 votes.

Now that variables have been defined, you need to understand the meaning of population, sample, parameter, and statistic.

POPULATION

A **population** consists of all of the members of a group about which you want to draw a conclusion.

SAMPLE

A **sample** is the portion of the population selected for analysis.

PARAMETER

A **parameter** is a numerical measure that describes a characteristic of a population.

STATISTIC

A **statistic** is a numerical measure that describes a characteristic of a sample.

Examples of populations are all the full-time students at a college, all the registered voters in New York, and all the people who went shopping at the local mall this weekend. Samples could be selected from each of the three populations mentioned above. Examples include 10 full-time students selected for a focus group, 500 registered voters in New York who were contacted via telephone for a political poll, and 30 mall shoppers who were asked to complete a customer satisfaction survey. In each case, the people in the sample represent a portion or subset of the people comprising the population.

The average amount spent by all the people who went shopping at the local mall this weekend is a parameter. Information from all the shoppers in the entire population is needed to compute this parameter. The average amount spent by the 30 shoppers completing the customer satisfaction survey is a statistic. Information from only 30 people who went to the local mall this weekend is used in calculating the statistic.

Statistics, itself, is divided into two branches, both of which are applicable to managing businesses. **Descriptive statistics** focuses on collecting, summarizing, and presenting a set of data. **Inferential statistics** uses sample data to draw conclusions about a population.

Descriptive statistics has its roots in the recordkeeping needs of large political and social organizations. For example, every decade since 1790, the United States has conducted a census that collects and summarizes data about its citizens. Through the years, the U.S. Census Bureau has been one of the many groups that have refined the methods of descriptive statistics. The foundation of inferential statistics is based on the mathematics of probability theory. Inferential methods use sample data to calculate statistics that provide estimates of the characteristics of the entire population.

Today, applications of statistical methods can be found in different areas of business. Accounting uses statistical methods to select samples for auditing purposes and to understand the cost drivers in cost accounting. Finance uses statistical methods to choose between alternative portfolio investments and to track trends in financial measures over time. Management uses statistical methods to improve the quality of the products manufactured or the services delivered by an organization. Marketing uses statistical methods to estimate the proportion of customers who prefer one product over another and why they do, and to draw conclusions about what advertising strategy might be most useful in increasing sales of a product.

1.2 THE GROWTH OF STATISTICS AND INFORMATION TECHNOLOGY

During the past century, statistics has played an important role in spurring the use of information technology and, in turn, such technology has spurred the wider use of statistics. At the beginning of the twentieth century, the expanding data-handling requirements associated with the federal census led directly to the development of tabulating machines that were the forerunners of today's business computer systems. Statisticians such as Pearson, Fisher, Gosset, Neyman, Wald, and Tukey established the techniques of modern inferential statistics in response to the need to analyze large sets of population data that had become increasingly costly, time-consuming, and cumbersome to collect. The development of early computer systems permitted others to develop computer programs to ease the calculational and data-processing burdens imposed by those techniques. These first programs, in turn, allowed greater use of statistical methods by business decision-makers, and this greater use, as well as more recent advances in information technology, have completed the cycle by spurring the development of even more sophisticated statistical methods.

Today, when you hear of retailers investing in a "customer-relationship management system" or a packaged goods producer engaging in "data mining" to uncover consumer preferences, you should realize that statistical techniques form the foundations of such cutting-edge applications of information technology. Even though cutting-edge applications might require custom programming, for many years businesses have had access to **statistical packages**, such as Minitab and SPSS, that are standardized sets of programs that help managers use a wide range of statistical techniques by automating the data processing and calculations these techniques require. Whereas such packages were once available only in corporate computing centers, the increasing power and connectivity of personal computers have brought the statistical power of these packages to the desktop, where they have joined such familiar tools as word processing, worksheet, and Web browser programs.

The leasing and training costs associated with statistical packages have led many to consider using some of the graphical and statistical functions of Microsoft Excel. However, you need to be aware of concerns that many statisticians have about the accuracy and completeness of the statistical results that Excel produces. Unfortunately, some investigators have determined that certain Microsoft Excel statistical capabilities contain flaws that can lead to invalid results, especially when the data sets used are very large or have unusual statistical properties (see reference 3). Clearly, when you use Microsoft Excel, you must be careful about the data and the analysis you are undertaking. Whether this complication outweighs the benefits of Excel's attractive features is still an unanswered question in business today.

1.3 HOW THIS TEXT IS ORGANIZED

The primary goal of this text is helping you learn to understand how the methods of statistics can be used in decision-making processes. For business students, this understanding includes the following objectives:

- To properly present and describe business data and information
- To draw conclusions about large populations based solely on information collected from samples
- To make reliable forecasts about business trends
- To improve business processes

This text uses these four objectives listed above as its organizing principle. Figure 1.1 shows how each chapter relates to these objectives. You will explore the methods involved in the collection, presentation, and description of information in the remaining portion of this chapter and

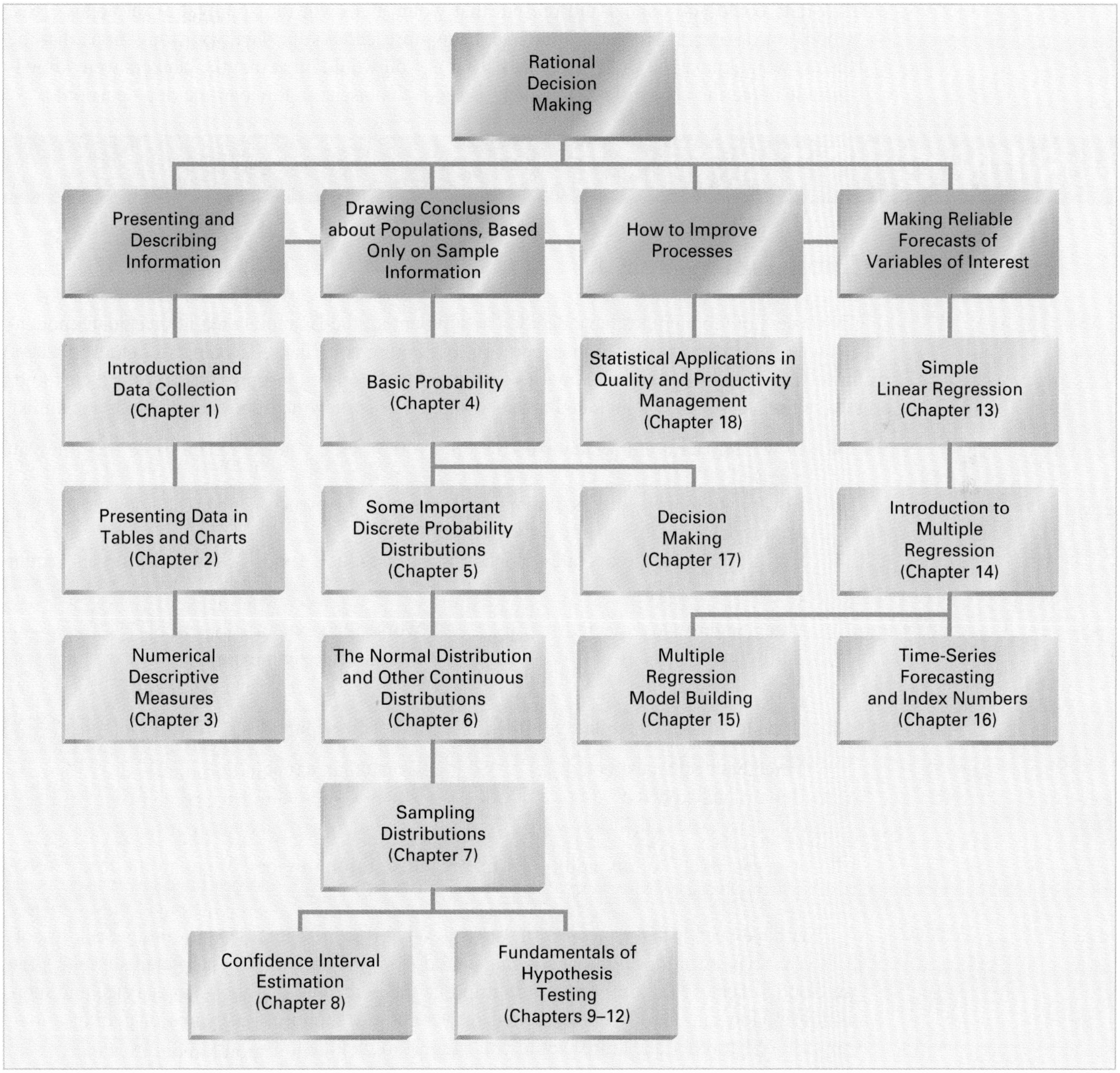

FIGURE 1.1 Structure Chart for This Text

Chapters 2 and 3. In Chapters 4 through 7 you will learn about the basic concepts of probability, and the binomial, normal, and other distributions to better understand, in Chapters 8 through 12, how you can draw conclusions about large populations based solely on information from samples. In Chapters 13–16 you will focus on regression analysis and time series modeling that can be used to make forecasts. In Chapter 18 you will learn methods for improving business processes.

Because learning in context enhances comprehension, each chapter begins with a "Using Statistics" scenario, such as the "Good Tunes—Part I" on page 2, that identifies a business problem in which statistics can be applied to change data into the useful information required for a rational decision. Questions raised in the scenarios lead to answers in the form of statistical methods presented in subsequent sections of the text. By thinking about these questions, you will gain an appreciation of how business managers are using statistics today to solve problems and improve the quality of their products and services.

For the "Good Tunes—Part I" scenario, selecting what to present is just as important as selecting the proper method for presentation and summarization. In this case, presumably the bankers themselves would demand some of the data, the "financials" of the business. But what other data could you collect and present that would help win the approval of the loans? (See "Good Tunes—Part II" below.) Of course, having presented your data, you would hope that the bankers would make the right inferences. That is, you would hope that the bankers were knowledgeable about the appropriate statistical methods that assist in the loan-making decision!

USING STATISTICS

Good Tunes—Part II

The owners of Good Tunes have decided to supplement the financial data in their loan application with data concerning customer perceptions about Good Tunes. To help assess these perceptions, Good Tunes has been asking its customers to complete and promptly return a customer satisfaction survey that is included in every order. The survey includes the following questions:

- How many days did it take from the time you ordered your merchandise to the time you received it? ________
- How much money (in U.S. dollars) do you expect to spend on stereo and consumer electronics equipment in the next twelve months? ________
- How do you rate the overall service provided by Good Tunes with respect to your recent purchase?

Much better than expected	☐	Worse than expected	☐
Better than expected	☐	Much worse than expected	☐
About as expected	☐		

- How do you rate the quality of the items you recently purchased from Good Tunes?

Much better than expected	☐	Worse than expected	☐
Better than expected	☐	Much worse than expected	☐
About as expected	☐		

- Are you likely to buy additional merchandise through Good Tunes in the next twelve months? Yes ________ No ________

You have been asked to review the survey. What type of data does the survey seek to collect? What type of information can be generated from the data of the completed survey? How can Good Tunes use that information to improve the perceived quality of the service and merchandise? How can Good Tunes use that information to increase its chance of getting a loan approval? What other questions would you suggest to include in the survey?

1.4 COLLECTING DATA

Managing a business effectively requires collecting the appropriate data. In most instances, the data are measurements acquired from items in a sample. The samples are chosen from populations in such a manner that the sample is as representative of the population as possible. The most common technique to ensure proper representation is to use a random sample. (See Chapter 7 for a detailed discussion of sampling techniques.)

Many different types of circumstances require the collection of data:

- A marketing research analyst needs to assess the effectiveness of a new television advertisement.
- A pharmaceutical manufacturer needs to determine whether a new drug is more effective than those currently in use.
- An operations manager wants to monitor a manufacturing process to find out whether the quality of product being produced is conforming to company standards.
- An auditor wants to review the financial transactions of a company in order to determine whether or not the company is in compliance with generally accepted accounting principles.
- A potential investor wants to determine which firms within which industries are likely to have accelerated growth in a period of economic recovery.

Identifying Sources of Data

Identifying the most appropriate source of data is a critical aspect of statistical analysis. If biases, ambiguities, or other types of errors flaw the data being collected, even the most sophisticated statistical methods will not produce accurate information. Four important sources of data are:

- Data distributed by an organization or an individual
- A designed experiment
- A survey
- An observational study

Data sources are classified as being either **primary sources** or **secondary sources**. When the data collector is the one using the data for analysis, the source is primary. When one organization or individual has compiled the data that are used by another organization or individual, the source is secondary.

Organizations and individuals that collect and publish data typically use that data as a primary source and then let others use it as a secondary source. For example, the United States federal government collects and distributes data in this way for both public and private purposes. The Bureau of Labor Statistics collects data on employment as well as distributing the monthly *Consumer Price Index*. The Census Bureau oversees a variety of ongoing surveys regarding population, housing, and manufacturing and undertakes special studies on topics such as crime, travel, and health care.

Market research firms and trade associations also distribute data pertaining to specific industries or markets. Investment services such as Mergent's provide financial data on a company-by-company basis. Syndicated services such as A. C. Nielsen provide clients with data enabling the comparison of client products with those of their competitors. Daily newspapers are filled with numerical information regarding stock prices, weather conditions, and sports statistics.

As listed above, conducting an experiment is another important data collection source. For example, to test the effectiveness of laundry detergent, an experimenter determines which brands in the study are more effective in cleaning soiled clothes by actually washing dirty laundry instead of asking customers which brand they believe to be more effective. Proper experimental designs are usually the subject matter of more advanced texts, because they often involve sophisticated statistical procedures. However, some fundamental experimental design concepts will be considered in Chapters 10 and 11.

Conducting a survey is a third important data source. Here the people being surveyed are asked questions about their beliefs, attitudes, behaviors, and other characteristics. Responses are then edited, coded, and tabulated for analysis.

Conducting an observational study is the fourth important data source. In such a study, a researcher observes the behavior directly, usually in its natural setting. Observational studies take many forms in business. One example is the **focus group**, a market research tool that is used for eliciting unstructured responses to open-ended questions. In a focus group, a moderator leads the discussion, and all the participants respond to the questions asked. Other, more structured types of studies involve group dynamics and consensus building and use various organizational behavior tools such as brainstorming, the Delphi technique, and the nominal-group method. Observational study techniques are also used in situations in which enhancing teamwork, or improving the quality of products and service are management goals.

1.5 TYPES OF DATA

Data are the observed values of variables, for example, the responses to a survey. Statisticians develop surveys to deal with a variety of different variables. As illustrated in Figure 1.2, there are two types of variables—categorical and numerical.

FIGURE 1.2
Types of Variables

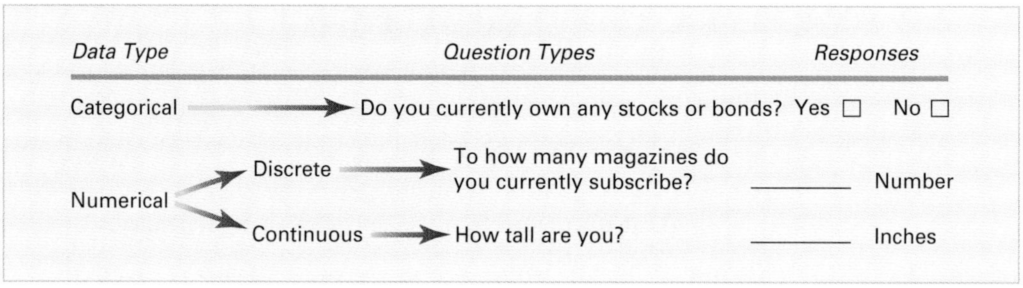

Categorical variables yield categorical responses, such as yes or no answers. An example is the response to the question "Do you currently own any stocks or bonds?" because it is limited to a simple yes or no answer. Another example is the response to the question on the Good Tunes survey (presented on page 6), "Are you likely to buy additional merchandise through Good Tunes in the next 12 months?" Categorical variables can also yield more than two possible responses. For example, "Which day of the week are you most likely to eat dinner in a restaurant?"

Numerical variables yield numerical responses such as your height in inches. Other examples are how much money you expect to spend on stereo equipment in the next 12 months (from the Good Tunes customer satisfaction survey) or the response to the question "To how many magazines do you currently subscribe?" There are two types of numerical variables: discrete and continuous.

Discrete variables produce numerical responses that arise from a counting process. "The number of magazines subscribed to" is an example of a discrete numerical variable, because the response is one of a finite number of integers. You subscribe to zero, one, two, and so on, magazines.

Continuous variables produce numerical responses that arise from a measuring process. Your height is an example of a continuous numerical variable, because the response takes on any value within a continuum or interval, depending on the precision of the measuring instrument. For example, your height may be 67 inches, $67\frac{1}{4}$ inches, $67\frac{7}{32}$ inches, or $67\frac{58}{250}$ inches, depending on the precision of the available instruments.

No two persons are exactly the same height and the more precise the measuring device used, the greater the likelihood of detecting differences between their heights. However, most measuring devices are not sophisticated enough to detect small differences. Hence, *tied observations* are often found in experimental or survey data even though the variable is truly continuous, and theoretically all values of a continuous variable are different.

Levels of Measurement and Types of Measurement Scales

Data are also described in terms of their level of measurement. There are four widely recognized levels of measurement: nominal, ordinal, interval, and ratio scales.

Nominal and Ordinal Scales Data from a categorical variable are measured on a nominal scale or on an ordinal scale. A **nominal scale** (Figure 1.3) classifies data into various distinct categories in which no ranking is implied. In the Good Tunes customer satisfaction survey, the answer to the question "Are you likely to buy additional merchandise through Good Tunes in the next 12 months?" is an example of a nominally scaled variable, as is your favorite soft drink, your political party affiliation, and your gender. Nominal scaling is the weakest form of measurement because you cannot specify any ranking across the various categories.

Categorical Variable	*Categories*
Personal Computer Ownership	Yes ☐ No ☐
Type of Stocks Owned	Growth ☐ Value ☐ Other ☐ None ☐
Internet Provider	Microsoft Network ☐ AOL ☐ Other ☐

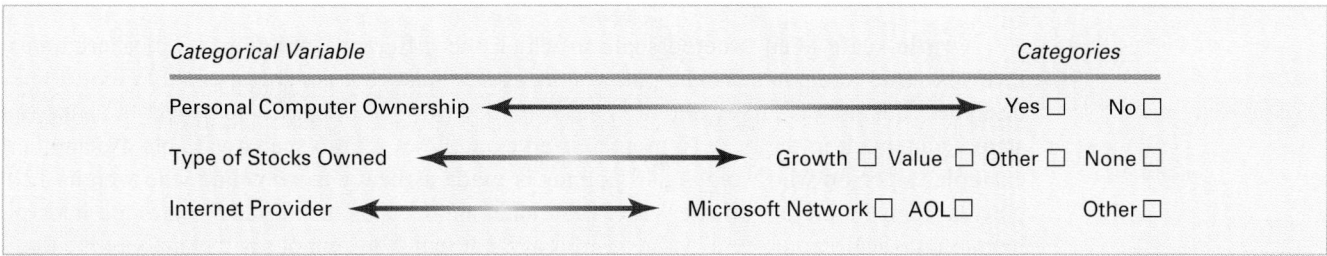

FIGURE 1.3 Examples of Nominal Scaling

An **ordinal scale** classifies data into distinct categories in which ranking is implied. In the Good Tunes survey, the answers to the question "How do you rate the overall service provided by Good Tunes with respect to your recent purchase?" represent an ordinal scaled variable because the responses "much better than expected, better than expected, about as expected, worse than expected, and much worse than expected" are ranked in order of satisfaction level. Figure 1.4 lists other examples of ordinal scaled variables.

Categorical Variable	*Ordered Categories*
Student class designation	Freshman Sophomore Junior Senior
Product satisfaction	Very Unsatisfied Fairly Unsatisfied Neutral Fairly Satisfied Very Satisfied
Faculty rank	Professor Associate Professor Assistant Professor Instructor
Standard & Poor's bond ratings	AAA AA A BBB BB B CCC CC C DDD DD D
Student grades	A B C D F

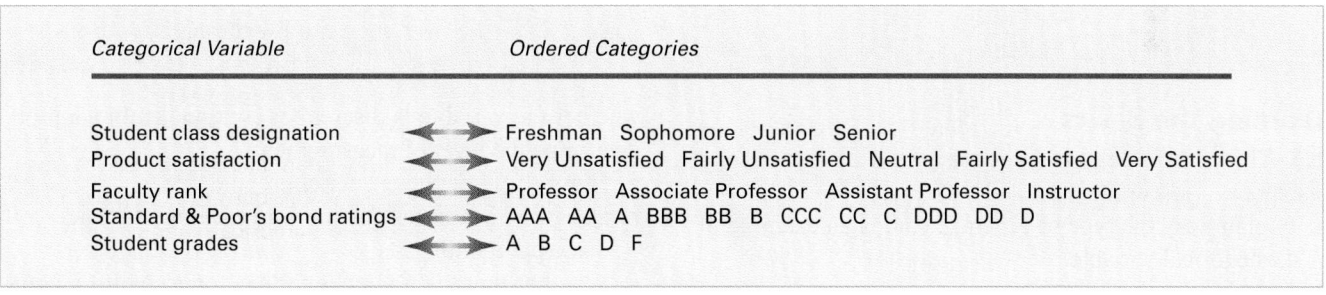

FIGURE 1.4 Examples of Ordinal Scaling

Ordinal scaling is a stronger form of measurement than nominal scaling because an observed value classified into one category possesses more of a property than does an observed value classified into another category. However, ordinal scaling is still a relatively weak form of measurement because the scale does not account for the amount of the differences *between* the categories. The ordering implies only *which* category is "greater," "better," or "more preferred"—not by *how much*.

Interval and Ratio Scales Data from a numerical variable are measured on an interval or ratio scale. An **interval scale** (Figure 1.5) is an ordered scale in which the difference between measurements is a meaningful quantity but does not involve a true zero point. For example, a noontime temperature reading of 67 degrees Fahrenheit is 2 degrees warmer than a noontime reading of 65 degrees. In addition, the 2 degrees Fahrenheit difference in the noontime temperature readings is the same if the two noontime temperature readings were 74 and 76 degrees Fahrenheit, since the difference has the same meaning anywhere on the scale.

FIGURE 1.5
Examples of Interval and Ratio Scales

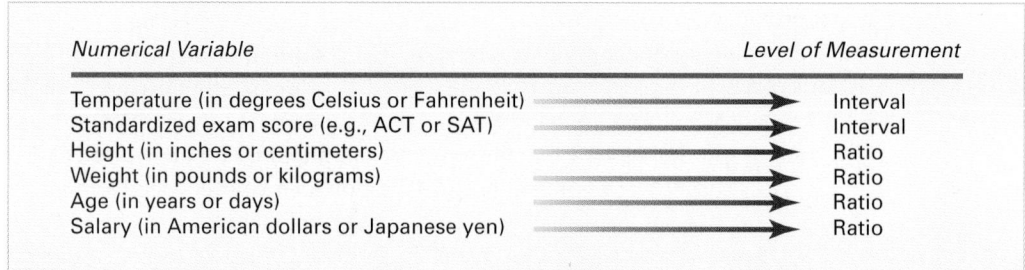

Numerical Variable		*Level of Measurement*
Temperature (in degrees Celsius or Fahrenheit)	→	Interval
Standardized exam score (e.g., ACT or SAT)	→	Interval
Height (in inches or centimeters)	→	Ratio
Weight (in pounds or kilograms)	→	Ratio
Age (in years or days)	→	Ratio
Salary (in American dollars or Japanese yen)	→	Ratio

A **ratio scale** is an ordered scale in which the difference between the measurements involves a true zero point as in height, weight, age, or salary measurements. In the Good Tunes customer satisfaction survey, the amount of money (in U.S. dollars) you expect to spend on stereo equipment in the next 12 months is an example of a ratio scaled variable. As another example, a person who weighs 240 pounds is twice as heavy as someone who weighs 120 pounds. Temperature is a trickier case: Fahrenheit and Celsius (centigrade) scales are interval but not ratio scales; the "zero" value is arbitrary, not real. You cannot say that a noontime temperature reading of 4 degrees Fahrenheit is twice as hot as 2 degrees Fahrenheit. But a Kelvin temperature reading, in which zero degrees means no molecular motion, is ratio scaled. In contrast, the Fahrenheit and Celsius scales use arbitrarily selected zero-degree beginning points.

Data measured on an interval scale or on a ratio scale constitute the highest levels of measurement. They are stronger forms of measurement than an ordinal scale, because you can determine not only which observed value is the largest but also by how much.

PROBLEMS FOR SECTION 1.5

Learning the Basics

1.1 Three different beverages are sold at a fast-food restaurant—soft drinks, tea, and coffee.

a. Explain why the type of beverage sold is an example of a categorical variable.

b. Explain why the type of beverage sold is an example of a nominally scaled variable.

1.2 Soft drinks are sold in three sizes in a fast-food restaurant—small, medium, and large. Explain why the size of the soft drink is an example of an ordinal scaled variable.

1.3 Suppose that you measure the time it takes to download an MP3 file from the Internet.

a. Explain why the download time is a numerical variable.

b. Explain why the download time is a ratio scaled variable.

Applying the Concepts

SELF Test **1.4** For each of the following variables, determine whether the variable is categorical or numerical. If the variable is numerical, determine whether the variable is discrete or continuous. In addition, determine the level of measurement.

a. Number of telephones per household

b. Length (in minutes) of the longest long-distance call made per month

c. Whether there is a telephone line connected to a computer modem in the household

d. Whether there is a fax machine in the household

PH Grade ASSIST **1.5** The following information is collected from students upon exiting the campus bookstore during the first week of classes:

a. Amount of time spent shopping in the bookstore

b. Number of textbooks purchased

c. Academic major

d. Gender

Classify each of these variables as categorical or numerical. If the variable is numerical, determine whether the variable is discrete or continuous. In addition, determine the level of measurement.

PH Grade ASSIST **1.6** For each of the following variables, determine whether the variable is categorical or numerical. If the variable is numerical, determine whether the variable is discrete or continuous. In addition, determine the level of measurement.

a. Name of Internet provider
b. Amount of time spent surfing the Internet per week
c. Number of e-mails received in a week
d. Number of online purchases made in a month

1.7 For each of the following variables, determine whether the variable is categorical or numerical. If the variable is numerical, determine whether the variable is discrete or continuous. In addition, determine the level of measurement.

a. Amount of money spent on clothing in the last month
b. Favorite department store
c. Most likely time period during which shopping for clothing takes place (weekday, weeknight, or weekend)
d. Number of pairs of winter gloves owned

1.8 Suppose the following information is collected from Robert Keeler on his application for a home mortgage loan at the Metro County Savings and Loan Association:

a. Monthly Payments: $1,427
b. Number of Jobs in Past 10 Years: 1
c. Annual Family Salary Income: $86,000
d. Marital Status: Married

Classify each of the responses by type of data and level of measurement.

1.9 One of the variables most often included in surveys is income. Sometimes the question is phrased "What is your income (in thousands of dollars)?" In other surveys, the respondent is asked to "Place an X in the circle corresponding to your income level" and given a number of ranges to choose from.

a. In the first format, explain why income might be considered either discrete or continuous.
b. Which of these two formats would you prefer to use if you were conducting a survey? Why?
c. Which of these two formats would likely bring you a greater rate of response? Why?

1.10 If two students score a 90 on the same examination, what arguments could be used to show that the underlying variable—test score—is continuous?

1.11 The director of market research at a large department store chain wanted to conduct a survey throughout a metropolitan area to determine the amount of time working women spend shopping for clothing in a typical month.

a. Describe both the population and the sample of interest, and indicate the type of data the director might wish to collect.
b. Develop a first draft of the questionnaire needed in (a) by writing a series of three categorical questions and three numerical questions that you feel would be appropriate for this survey.

SUMMARY

In this chapter you have studied data collection and the various types of data used in business. In the "Using Statistics" scenario you were asked to review the customer survey used by the Good Tunes Company (see page 6). The first two questions shown will produce numerical data and the last three will produce categorical data. The responses to the first question (number of days) are discrete, and the responses to the second question (amount of money spent) are continuous. After the data have been collected, they must be organized and prepared in order to make various analyses. In the next two chapters, tables and charts and a variety of descriptive numerical measures that are useful for data analysis are developed.

KEY TERMS

CHAPTER REVIEW PROBLEMS

Checking Your Understanding

1.12 What is the difference between a sample and a population?

1.13 What is the difference between a statistic and a parameter?

1.14 What is the difference between descriptive and inferential statistics?

1.15 What is the difference between a categorical and a numerical variable?

1.16 What is the difference between a discrete and a continuous variable?

1.17 What is an operational definition and why is it so important?

1.18 What are the four types of measurement scales?

Applying the Concepts

1.19 The Data and Story Library **lib.stat.cmu.edu/DASL** is an online library of data files and stories that illustrate the use of basic statistical methods. The stories are classified by method and by topic. Go to this site and click on **List all topics**. Pick a story and summarize how statistics were used in the story.

1.20 Go to the official Microsoft Excel Web site **www.microsoft.com/office/excel**. Explain how you think Microsoft Excel could be useful in the field of statistics.

1.21 Go to the official Minitab Web site **www.minitab.com**. Explain how you think Minitab could be useful in the field of statistics.

1.22 Go to the official SPSS Web site **www.spss.com**. Explain how you think SPSS could be useful in the field of statistics.

1.23 The Gallup organization releases the results of recent polls at its Web site **www.gallup.com**. Go to this site and click on an article of interest to you in the "Top Stories" section.

a. Give an example of a categorical variable found in the article.
b. Give an example of a numerical variable found in the article.
c. Is the variable you selected in (b) discrete or continuous?

1.24 The U.S. Census Bureau **www.census.gov** site contains survey information on people, business, geography, and other topics. Go to the site and click on **Housing** in the "People" section. Then click on **American Housing Survey**.

a. Briefly describe the American Housing Survey.
b. Give an example of a categorical variable found in this survey.
c. Give an example of a numerical variable found in this survey.
d. Is the variable you selected in (c) discrete or continuous?

1.25 On the U.S. Census Bureau **www.census. gov** site, click on **Survey of Business Owners** in the "Business" section and read the description of The Survey of Business Owners and Self-Employed Persons (SBO). Click on **SBO-1** in the "Forms and Instructions" section to view the actual survey form used.

a. Give an example of a categorical random variable found in this survey.
b. Give an example of a numerical random variable found in this survey.
c. Is the variable you selected in (b) discrete or continuous?

1.26 In a report based on U.S. Transportation Department statistics, the budget carrier JetBlue was number 1 in quality among all U.S. airlines in 2003. JetBlue had the second-best on-time performance, arriving on time 86% of the time. Also, JetBlue customers filed fewer complaints than all other airlines but one ("JetBlue ranked No. 1 Airline, Report Says," USAToday.com, April 5, 2004).

a. Which of the four types of data sources listed in Section 1.4 on page 7 do you think were used in this study?
b. Name a categorical variable discussed in this article.
c. Name a numerical variable discussed in this article.

1.27 According to a Goldman Sachs survey, only about 4% of U.S. households bank online. A survey by Cyber Dialogue investigated reasons people quit online banking after trying it. A partial listing of the results of the Cyber Dialogue survey are given below ("USA Snapshots," *USA Today*, February 21, 2000, A1).

Why Did You Quit Online Banking?

Too complicated or time-consuming	27%
Unhappy with customer service	25%
No need/not interested	20%
Concerns about security or fraud	11%
Too costly	11%
Concerns about privacy	5%

a. Describe the population for the Goldman Sachs survey.
b. Describe the population for the Cyber Dialogue survey.
c. Is a response to the question "Why did you quit online banking?" categorical or numerical?
d. Twenty-seven percent of respondents indicated that online banking was too complicated or too time-consuming. Is this a parameter or a statistic?

1.28 A manufacturer of cat food was planning to survey households in the United States to determine purchasing habits of cat owners. Among the questions to be included are those that relate to

1. where cat food is primarily purchased.
2. whether dry or moist cat food is purchased.
3. the number of cats living in the household.
4. whether or not the cat is pedigreed.

a. Describe the population.
b. For each of the four items listed, indicate whether the variable is categorical or numerical. If numerical, is it discrete or continuous?
c. Develop five categorical questions for the survey.
d. Develop five numerical questions for the survey.

INTRODUCTION TO THE WEB CASES

LEARNING FROM THE WEB CASES IN THIS TEXT

People use statistical techniques to help communicate and present important information to others both inside and outside their businesses. And every day, people misuse these techniques:

- A sales manager working with an "easy-to-use" charting program chooses an inappropriate chart that obscures data relationships.
- The editor of an annual report presents a chart of revenues with an abridged *Y*-axis that creates the false impression of greatly rising revenues.
- An analyst generates meaningless statistics about a set of categorical data using analyses designed for numerical data.

Although much of the misuse of statistics is unintentional, you need to be able to identify all such misuses in order to be an informed manager. The primary goal of the Web Cases throughout this text is to help you develop this type of skill.

Web Cases ask you to visit the Web sites that are related to the companies and issues raised in the "Using Statistics" scenario that starts each chapter or a Web page that supports the continuing story of the *Springville Herald*, a small-city daily newspaper. You review internal documents as well as publicly stated claims, seeking to identify and correct the misuses of statistics. Unlike a traditional text case study, but much like real-world situations, not all of the information you encounter will be relevant to your task, and you may occasionally discover conflicting information that you need to resolve before continuing with the case.

To assist your learning, the Web Case for each chapter begins with the learning objective and a synopsis of the scenario under consideration. You will be directed to a specific Web site or Web page and given a set of questions that will guide your exploration. If you prefer, you can also explore the Web pages for the cases by linking to the Springville Chamber of Commerce page **www.prenhall.com/ Springville/SpringvilleCC.htm**.

Complementing the Web Case in most chapters is a traditional case study exercise in which you are asked to apply your knowledge of statistics to a problem being faced by the management of the *Springville Herald*.

To illustrate how to use a Web Case, link to the Web site for Good Tunes **www.prenhall.com/Springville/Good_Tunes.htm**, the online retailer mentioned in the "Using Statistics" scenarios for this chapter. Recall that the privately held Good Tunes is seeking financing to expand its business by opening retail locations. Since it is in management's interest to show that Good Tunes is a thriving business, it is not too surprising to discover the "our best sales year ever" claim in the "Good Times at Good Tunes" entry at the top of their home page.

The claim is also a hyperlink, so click on "our best sales year ever" to display the page that supports the claim. How would you support such a claim? With a table of numbers? A chart? Remarks attributed to a knowledgeable source? Good Tunes has used a chart to present "two years ago" and "latest twelve month's" sales data by category. Are there any problems with the choices made on this Web page? *Absolutely*!

First, note that there are no scales for the symbols used, so it is impossible to know what the actual sales volumes are. In fact, as you will learn in section 2.6, charts that incorporate symbols in this way are considered examples of *chartjunk* and would never be used by people seeking to properly use graphs.

This important point aside, another question that arises is whether the sales data represent the number of units sold or something else. The use of the symbols creates the impression that unit sales data are being presented. If the data are unit sales, does such data best support the claim being made—or would something else, such as dollar volumes—be a better indicator of sales at Good Tunes?

Then there are those curious chart labels. "Latest twelve months" is ambiguous—it could include months from the current year as well as months from one year ago and therefore may not be an equivalent time period to "two years ago." Since the business was established in 1997, and the claim being made is "best sales year ever," why hasn't management included sales figures for *every* year?

Is Good Tunes management hiding something or are they just unaware of the proper use of statistics? Either way, they have failed to properly communicate a vital aspect of their "story."

In subsequent Web Cases, you are asked to provide this type of analysis, using the open-ended questions presented in this text as guidance. Not all the cases are as straightforward as this sample and some cases include perfectly appropriate applications of statistics.

REFERENCES

1. Kendall, M. G., and R. L. Plackett, eds., *Studies in the History of Statistics and Probability*, vol. 2 (London: Charles W. Griffin, 1977).
2. Kirk, R. E., ed., *Statistical Issues: A Reader for the Behavioral Sciences* (Monterey, CA: Brooks/Cole, 1972).
3. McCullough, B. D., and B. Wilson, "On the accuracy of statistical procedures in Microsoft Excel 97," *Computational Statistics and Data Analysis*, 31 (1999), 27–37.
4. *Microsoft Excel 2003* (Redmond, WA: Microsoft Corporation, 2002).
5. *Minitab Release 14* (State College, PA: Minitab, Inc., 2004).
6. Pearson, E. S., ed., *The History of Statistics in the Seventeenth and Eighteenth Centuries* (New York: Macmillan, 1978).
7. Pearson, E. S., and M. G. Kendall, eds., *Studies in the History of Statistics and Probability* (Darien, CT: Hafner, 1970).
8. *SPSS® Base 12.0 Brief Guide* (Upper Saddle River, NJ: Prentice Hall, 2003).

Appendix 1 Introduction to Using Statistical Programs

ABOUT THIS APPENDIX

Section A1.1 Read this section if you are unfamiliar with the basics of Microsoft Windows operations that are required in order to use Microsoft Excel, Minitab, or SPSS effectively.

Section A1.2 Read this section only if you plan to use Microsoft Excel with this text.

Section A1.3 Read this section only if you are using Minitab with this text.

Section A1.4 Read this CD-ROM section only if you are using SPSS with this text.

Note: Throughout this appendix and all other appendices in this text, the symbol ➔ is used to represent a sequence of menu selections. For example, the instruction "select **File ➔ Open** " means to first select the **File** menu choice and then select the **Open** choice from the submenu that appears.

A1.1 USING WINDOWS

Using the Mouse

In Microsoft Windows, you frequently use a mouse or other pointing device to make choices by pointing to an onscreen object and pressing a mouse button. By convention, Windows expects pointing devices to contain two buttons, one designated as the primary button, the other as the secondary button. You can move your mouse and press and release the mouse buttons in the following ways:

Click or **Select:** Move the mouse over an object and press the primary button.

Drag: Move the mouse over an object. Then while pressing and holding down the primary button, move the mouse pointer somewhere else on the screen and release the primary button. Dragging either moves objects to another part of the screen or allows you to select multiple items.

Double-click: Move the mouse over an object and click the primary button twice in rapid succession.
Right-click: Move the mouse over an object and click the secondary button.

By default, Microsoft Windows defines the left mouse button as the primary button and the right button as the secondary button (this gives rise to the phrase "right-click"), but you can swap the definitions by selecting the Windows Mouse Control Panel icon.

Opening Programs

You can choose one of two ways to directly open a program such as Microsoft Excel, Minitab, or SPSS for use. These ways are:

- **Program icon click:** Double-click the Windows desktop icon representing the program (in some Windows versions, you may need only to single click the icon).
- **Start Menu selection:** Press the **Windows key** (or click the onscreen Start button) and select the **Programs** or **All Programs** choice. From the menu list that appears, select the entry for the program. If the program is listed on a submenu, you will have to first select the submenu, and then select the program.

Experienced Microsoft Windows users may know other ways to open a program. You can, of course, use these other ways if you prefer.

Program Windows

Microsoft Windows gets its name because every time you open a program, an onscreen rectangular frame or "window" opens as well. Inside this window, you interact with the program and many interactions lead to the opening of additional windows into which you make entries and selections.

The opening window of most programs contains these common elements:

A **title bar** at the top of the window that identifies the program and any file in use.
Resize buttons on the right side of the title bar area that affect the displayed size of the window.
A **Close program button** on the right margin of the title bar that allows you to quickly end your use of a program.
A **menu bar**, a horizontal list of words below the title bar that contains command choices on one or more menus.
One or more **tool bars** that contain buttons that are command shortcuts.

Figure A1.1 shows these common elements for the opening windows in Microsoft Excel 2003, Minitab Release 14, and SPSS Student Version 12. Note the windows for these programs additionally have a worksheet area composed of rows and columns that you use for data entry.

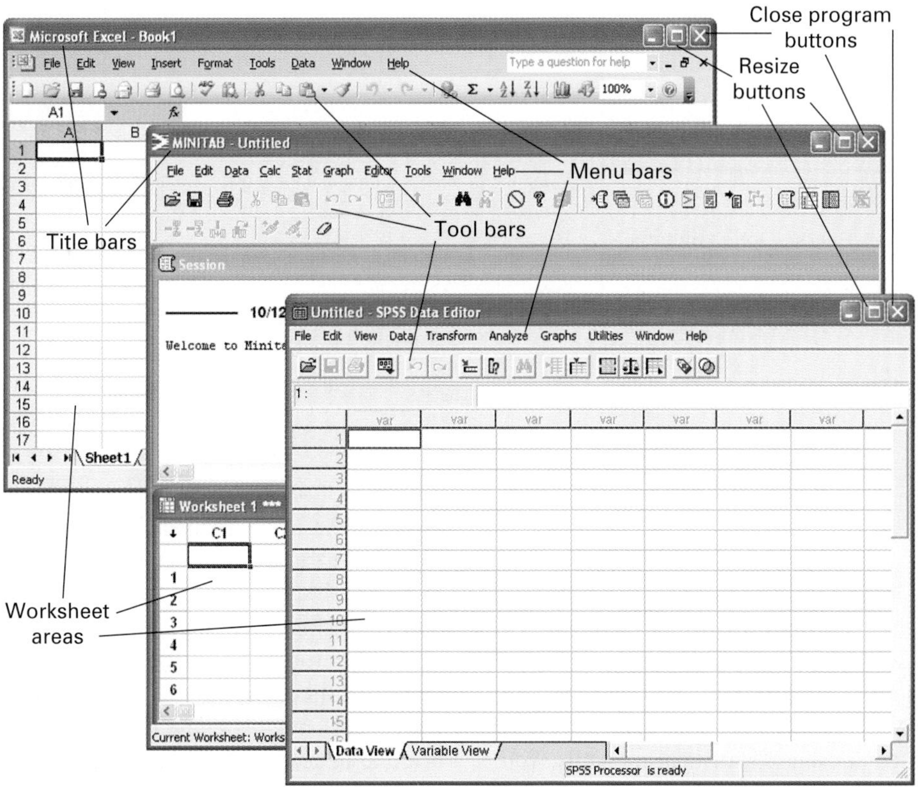

FIGURE A1.1 Program Windows for Microsoft Excel, Minitab, and SPSS

Dialog Boxes

Many entries and selections you make as you use a program trigger the display of additional windows known as dialog boxes. The Microsoft Excel 2003 dialog boxes to open or print a file (see Figure A1.2) contain the following elements commonly found in all dialog boxes:

Question mark help button: Clicking this button allows you to then click on an element of the dialog box to display a help message about that element.
Drop-down list box: Displays a list of choices when you click the drop-down button that appears at the right edge of the box.
List box: Displays a list of choices. Sometimes includes **scroll buttons** or a **slider** if the list of choices is larger than the size the box can display.
Text box: Provides a space into which you can type an entry. These boxes are sometimes combined with either a drop-down list or **spinner buttons** (seen in the Pages From box in Figure A1.2) that provide alternative ways to specifying an entry.
Check box: Provides a set of choices in which you can make zero, one, or more than one choice (compare with option buttons).
Icons: Allows you direct access to other places in your Windows system where files may be stored.
Option buttons: Provides a set of mutually exclusive choices in which only one choice can be selected at a time.
Command buttons: Causes the program to take some action that usually closes the current dialog box and triggers the display of an additional dialog box. An **OK button** causes the program to take an action using the current values and settings of the dialog box. A **Cancel button** closes a dialog box and cancels the operation associated with the dialog box.

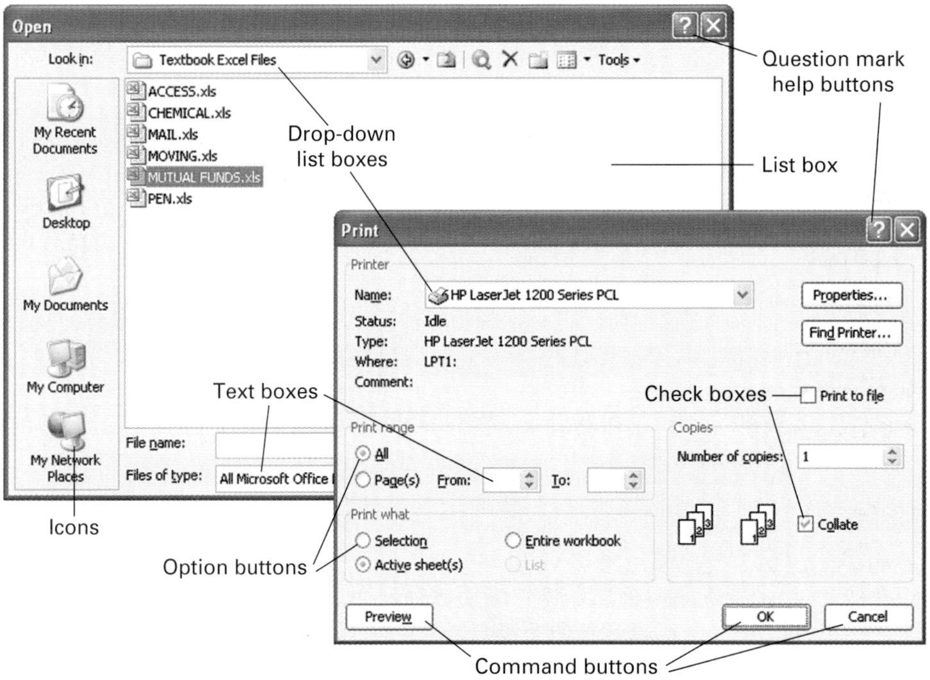

FIGURE A1.2 Common Dialog Box Elements

A1.2 INTRODUCTION TO MICROSOFT EXCEL

Microsoft Excel Overview

Microsoft Excel is the electronic worksheet program of Microsoft Office. Although not a specialized statistical program, Excel contains basic statistical functions and includes Data Analysis Tool Pak procedures that you can use to perform selected advanced statistical methods. You can also install the Prentice Hall PHStat2 add-in (included with the CD-ROM packaged with this text) to extend and enhance the Data Analysis Tool Pak that Microsoft Excel contains. (You do not need to use PHStat2 in order to use Microsoft Excel with this text, although using PHStat2 will simplify using Excel for statistical analysis.)

In Microsoft Excel, you create or open and save files that are called **workbooks**. Workbooks are collections of worksheets and related items such as charts that contain the original data as well as the calculations and results associated with one or more analyses. Because of its widespread distribution, Microsoft Excel is a convenient program to use, but some statisticians express concern about its lack of fully reliable and accurate results for some statistical procedures. Although Microsoft has improved many statistical functions starting with Excel 2003, you should be cautious about using Microsoft Excel to perform analyses on data other than the data used in this text. (If you plan to install PHStat2, make sure to first read Appendix F and any PHStat2 read-me file that may appear on the included CD-ROM.)

Using Microsoft Excel Worksheets

In Microsoft Excel, you enter data into worksheets that are organized as lettered columns and numbered rows. Typically, you enter the data for each variable in a separate column, using the row 1 cell for a variable label and each subsequent row for a single observation. You should follow the good practice of entering only one set of data per worksheet.

To refer to a specific entry, or cell, you use a *Sheetname!ColumnRow* notation. For example, Data!A2 refers to the cell in column A and row 2 in the Data worksheet. To refer to a specific group or **range** of cells, you use a *Sheetname!Upperleftcell:Lowerrightcell* notation. For example, Data!A2:B11 refers to the 20 cells that are in rows 2 though 11 in columns A and B of the Data worksheet.

Each Microsoft Excel worksheet has its own name. Automatically, Microsoft Excel names worksheets in the form of **Sheet1**, **Sheet2**, and so on. You should rename your worksheets, giving them more self-descriptive names, by double-clicking the sheet tabs that appear at the bottom of each sheet, typing a new name, and pressing the Enter key.

Using Formulas in Excel Worksheets

Formulas are worksheet cell entries that perform a calculation or some other task. You enter formulas by typing the equal sign symbol (=) followed by some combination of mathematical or other data-processing operations.

For simple formulas, you use the symbols +, –, *, /, and ^ for the operations addition, subtraction, multiplication, division, and exponentiation (a number raised to a power), respectively. For example, the formula =Data!B2 + Data!B3 + Data!B4 + Data!B5 adds the contents of the cells B2, B3, B4, and B5 of the Data worksheet and displays the sum as the value in the cell containing the formula. You can also use Microsoft Excel *functions* in formulas to simplify formulas. For example, the formula =SUM(Data!B2:B5) that uses the Excel SUM() function is a shorter equivalent to the formula of the previous sentence. You can also use cell or cell range references that do not contain the *Sheetname*! part, such as B2 or B2:B5. Such references *always* refer to the worksheet in which the formula has been entered.

Formulas allow you to create generalized solutions and give Excel its distinctive ability to automatically recalculate results when you change the values of the supporting data. Typically, when you use a worksheet, you see only the results of any formulas entered, not the formulas themselves. However, for your reference, many illustrations of Microsoft Excel worksheets in this text also show the underlying formulas adjacent to the results they produce. When using Excel, you can select **Tools ➔ Options** and in the **View** tab of the **Options** dialog box that appears, select the **Formulas** check box, and click the **OK** button to see onscreen the formulas themselves and not their results. To restore the original view, uncheck the **Formulas** check box.

Using the Microsoft Excel Chart Wizard

You use the Microsoft Excel Chart Wizard to generate a wide variety of charts. The Chart Wizard is one of several Microsoft Office **wizards**, sets of linked dialog boxes that guide you, step-by-step, through the task of creating something. To use the Chart Wizard, you first select **Insert ➔ Chart**. Then you make selections and enter information about the properties of the chart as you step through the dialog boxes by clicking a Next button. Clicking the **Finish** button in the last dialog box ends the wizard and creates the chart. At any point, you can cancel the operation of the wizard by clicking a **Cancel** button or move to a previous dialog box by clicking **Back.**

The Chart Wizards of the various versions of Microsoft Excel differ slightly. For Microsoft Excel 2003, the four-step wizard (see Figure A1.3 on page 18) requires you to do the following:

Step 1: Choose the chart type.

Step 2: Enter the workbook locations of the source data for the values to be plotted and the source data for chart labeling information (if any).

Step 3: Specify the formatting and labeling options for the chart. (See further comments below.)

Step 4: Choose the workbook location of the chart. You will always create a better-scaled chart if you choose the "as new sheet" instead of the "as object in (a worksheet)" option.

You can change these settings after the chart has been produced by right-clicking on the chart and making the appropriate selection from the shortcut menu that appears. For example, to reconsider the settings associated with the Step 3 dialog box, you would select **Chart Options** from the shortcut menu.

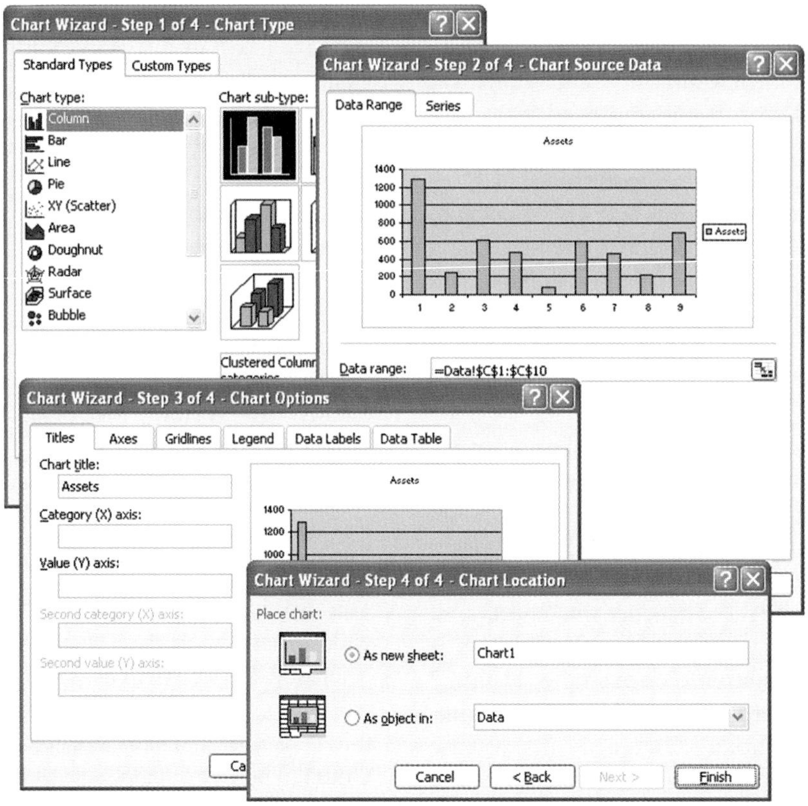

FIGURE A1.3 Microsoft Excel Chart Wizard Dialog Boxes

The automatic settings in the Step 3 dialog box create imperfectly designed charts. When you are using the Chart Wizard to generate charts for problems and examples from this text, in the dialog box you should select the tabs listed below (see Figure A1.4) and apply the following instructions (if a tab does not appear in the Step 3 dialog box for a particular chart type, ignore the instruction for the tab):

- Select the **Titles** tab and enter a title and axis labels, if appropriate.
- Select the **Axes** tab and then select both the **(X) axis** and **(Y) axis** check boxes. Also select the **Automatic** option button under the (X) axis check box.
- Select the **Gridlines** tab and deselect (uncheck) all the choices under the (X) axis heading and under the (Y) axis heading.
- Select the **Legend** tab and deselect (uncheck) the **Show legend** check box.
- Select the **Data Labels** tab and in the Data labels group, select the **None** option button.

FIGURE A1.4 Chart Wizard Step 3 Dialog Box Tabs

Opening and Saving Workbooks

You open workbooks to use data and results that have been created by you or others at an earlier time. To open a Microsoft Excel workbook, first select **File → Open**. In the Open dialog box that appears (see Figure A1.5), you select the file to be opened and then click the **OK** button. If you cannot find your file, you may need to do one or more of the following:

- Use the scroll bars or the slider, if present, to scroll through the entire list of files.
- Select the correct folder from the **Look in** drop-down list at the top of the dialog box.
- Change the **Files of type** value from the drop-down list at the bottom of the dialog box. You should select **Text Files** from the list to see any text files; to list every file in the folder, select **All Files**.

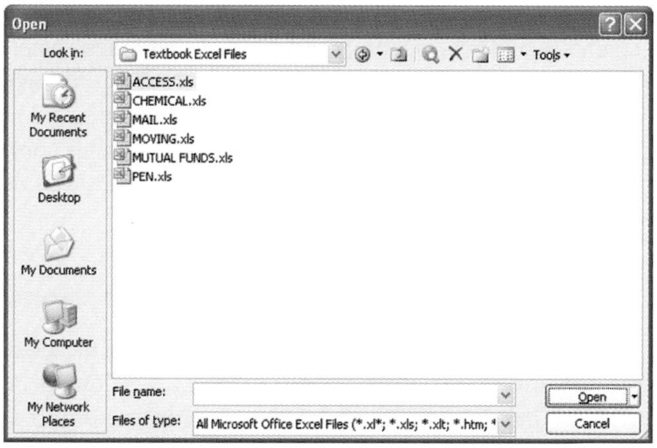

FIGURE A1.5 Microsoft Excel Open Dialog Box

To save a workbook, select **File → Save As** to display the Save As dialog box, which is similar to the Open dialog box. Enter (or edit) the name of the file in the **File name** box and click the **OK** button. If applicable, you can also do the following:

- Change to another folder by selecting that folder from the **Save in** drop-down list.
- Change the **Save as type** value to something other than the default choice, **Microsoft Excel Workbook**. "**Text (Tab delimited)**" or "**CSV (Comma delimited)**" are two file types sometimes used to share Excel data with other programs.

After saving your work, you should consider saving your file a second time, using a different name, in order to create a backup copy of your work. Files opened from nonwritable disks, such as the CD-ROM packaged with this text, cannot be saved to their original folders.

Printing Workbooks

To quickly print Excel worksheets you can select **File → Print** and then click the **OK** button in the Print dialog box. However, except for the simplest worksheets, you will probably want to preview the printed output and make any necessary adjustments before actually printing. To do this, select the worksheet to be printed and then select **File → Print Preview**. If the preview contains formatting errors, click the **Close** button, make the changes necessary, and reselect **File → Print Preview**. When you are satisfied with the results, click the **Print** button in the Print Preview window and then click the **OK** button in the Print dialog box.

The Print dialog box (see Figure A1.2 on page 16) contains settings to select the printer to be used, what parts of the workbook to print (the active worksheet is the default), and the number of copies to produce (1 is the default). If you need to change these settings, change them before clicking the **OK** button.

After printing, you should verify the contents of your printout. Most printing failures will trigger the display of an error message that you can use to figure out the source of the failure. You can customize your printouts by selecting **File → Page Setup** (or clicking the Setup button in the Print Preview window) and making the appropriate entries in the Page Setup dialog box (not shown) before printing your worksheets.

A1.3 INTRODUCTION TO MINITAB

Minitab Overview

Minitab is a statistical program that initially evolved from efforts at the Pennsylvania State University to improve the teaching of statistics. Today, while still used in many schools, Minitab has become a commercial product that is used in large corporations worldwide including Ford Motor Company, 3M, and GE.

In Minitab, you create and open **projects** to store all of your data and results. A **session**, or log of activities, a **Project Manager** that summarizes the project contents, and any worksheets or graphs used are the components that form a project. Project components are displayed in separate windows *inside* the Minitab application window. By default, you will see only the session and one worksheet window when you begin a new project in Minitab. (You can bring any window to the front by selecting the window in the Minitab Windows menu.) You can open and save an entire project or, as is done in this text, open and save worksheets. Minitab's accuracy, availability for many different types of computer systems, and commercial acceptance makes this program a great tool for learning statistics.

Using Minitab Worksheets

You enter data in a Minitab worksheet so that each variable is assigned to a column. Minitab worksheets are organized as numbered rows and columns numbered in the form *Cn* in which C1 is the first column. You enter variable labels in a special unnumbered row that precedes row 1. Unlike worksheets in programs such as Microsoft Excel, Minitab worksheets do not accept formulas and do not automatically recalculate themselves when you change the values of the supporting data.

By default, Minitab names open worksheets serially in the form of Worksheet1, Worksheet2, and so on. Better names are ones that reflect the content of the worksheets, such as Funds for a worksheet that contains mutual funds data. To give a sheet a descriptive name, open the Project Manager window, right-click the icon for the worksheet, and select **Rename** from the shortcut menu and type in the new name.

Opening and Saving Worksheets and Other Components

You open worksheets to use data that have been created by you or others at an earlier time. To open a Minitab worksheet, first select **File → Open Worksheet**. In the Open Worksheet dialog box that appears (see Figure A1.6), you select the file to be opened and then click the **OK** button. If you cannot find your file, you may need to do one or more of the following:

- Use the scroll bars or the slider, if present, to scroll through the entire list of files.
- Select the correct folder from the **Look in** drop-down list at the top of the dialog box.
- Change the **Files of type** value from the drop-down list at the bottom of the dialog box. You should select **Text Files** from the list to see any text files; to list every file in the folder, select **All Files**.

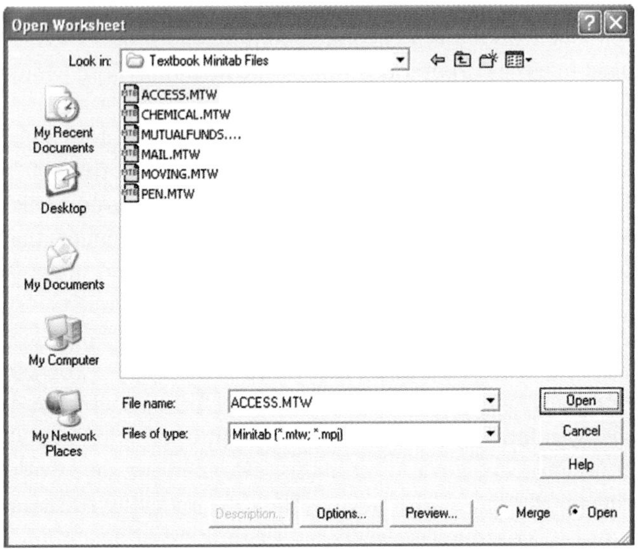

FIGURE A1.6 Open Worksheet Dialog Box

To open a Minitab Project that can include the session, worksheets, and graphs, select **File → Open Project**.

To save a worksheet, select **File → Save Current Worksheet As** to display the Save Worksheet As dialog box, which is similar to the Open Worksheet As dialog box. Enter (or edit) the name of the file in the **File name** box and click the **OK** button. If applicable, you can also do the following:

- Change to another folder by selecting that folder from the **Save in** drop-down list.
- Change the **Save as type** value to something other than the default choice, **Minitab**. "**Minitab Portable** " or an earlier version of **Minitab**, such as "**Minitab 13**" are commonly chosen alternatives.

After saving your work, you should consider saving your file a second time, using a different name, in order to create a backup copy of your work. Files opened from non-writable disks, such as the CD-ROM packaged with this text, cannot be saved to their original folders.

To save a Minitab Project, select the similar **File → Save Project As**. The Save Project As dialog box contains an **Options** button that displays a dialog box in which you can select which project parts other than worksheets will be saved.

Individual graphs and the session can also be saved separately by first selecting their windows and then selecting the similar **File → Save Graph As** or **File → Save Session As**, as appropriate. Minitab graphs can be saved in either a Minitab graph format or any one of several common graphics formats, and Session files can be saved as simple or formatted text files.

Printing Worksheets, Graphs, and Sessions

To print a specific worksheet, graph, or session, first select the window of the worksheet, graph, or session to be printed. Then select **File → Print *object***, where ***object*** is either **Worksheet**, **Graph**, or **Session Window**, depending on the window you selected.

If you are printing a graph or a session window, you will then see the Print dialog box. If you are printing a worksheet, you will first see a Data Window Print Options dialog box (Figure A1.7) that allows you to select formatting options for your printout (the default selections should be fine for most of your printouts). Click the **OK** button in that dialog box to proceed to the Print dialog box.

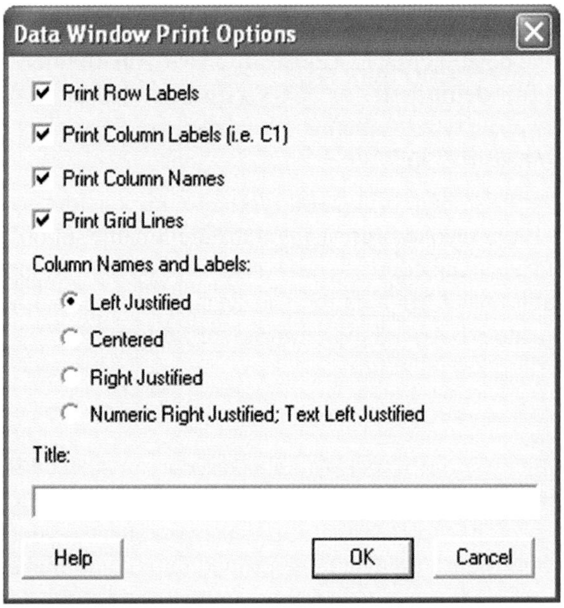

FIGURE A1.7 Data Window Print Options Dialog Box

The Print dialog box contains settings to select the printer to be used, what pages to print, and the number of copies to produce (1 is the default). If you need to change these settings, change them before clicking the **OK** button to produce your printout.

After printing, you should verify the contents of your printout. Most printing failures will trigger the display of onscreen information that you can use to figure out the source of the failure. You can change the paper size or paper orientation of your printout by selecting **File → Page Setup** and making the appropriate choices and clicking the **OK** button.

CHAPTER 2

Presenting Data in Tables and Charts

LEARNING OBJECTIVES

In this chapter, you learn:

- To develop tables and charts for categorical data
- To develop tables and charts for numerical data
- The principles of properly presenting graphs

USING STATISTICS

Comparing the Performance of Mutual Funds

Among the many investment choices available today, mutual funds, a market basket of a portfolio of securities, is a common choice for those thinking about their retirement. If you decided to purchase mutual funds for your retirement account, how would you go about making a reasonable choice among the many funds available today?

You first would want to know the different categories of mutual funds available. You would want to know the strategies of the professionals who manage the funds. Do they invest in high-risk securities or do they make more conservative choices? Does the fund specialize in a certain sized company, one whose outstanding stock totals a large amount (large cap) or one that is quite small (small cap)? Does the fund charge management fees that reduce the percentage return earned by an investor? And, of course, you would want to know how well the fund performed in the past.

All of this is a lot of data to review if you consider several dozen or more mutual funds. How could you "get your hands around" such data and explore it in a comprehensible manner?

One of the ways you can answer the questions raised in the "Using Statistics" scenario is by studying past data on the performance of mutual funds. The data relating to a sample of 121 mutual funds are in the **MUTUALFUNDS2004** file. As an investor you would want to examine both categorical and numerical variables. Did mutual funds with a growth objective have lower returns than mutual funds with a value objective? Do growth funds tend to be riskier investments than value funds? Reading this chapter will help you to select and develop proper tables and charts to try to answer these and other questions.

2.1 TABLES AND CHARTS FOR CATEGORICAL DATA

When you have categorical data, you tally responses into categories and then present the frequency or percentage in each category in tables and charts.

The Summary Table

The **summary table** indicates the frequency, amount, or percentage of items in a set of categories so that you can see differences between categories. A summary table lists the categories in one column and the frequency, amount, or percentage in a different column or columns. Table 2.1 illustrates a summary table based on a recent survey that asked why people shop for holiday gifts online (USA Today Snapshots, "Convenience, Shipping Make Online Appealing," *USA Today*, December 24, 2003, A1). In Table 2.1 the most common reasons for shopping online are free shipping and convenience followed by comparison shopping. Very few respondents shopped online because of larger selection or speed.

TABLE 2.1
Reasons for Shopping for Holiday Gifts Online

Reason	Percentage (%)
Comparison shopping	23
Convenience	33
Free shipping	34
Larger selection	6
Speed	4
Total	100%

EXAMPLE 2.1

SUMMARY TABLE OF LEVELS OF RISK OF MUTUAL FUNDS

The 121 mutual funds that are part of the "Using Statistics" scenario (see page 22) are classified according to the risk level of the mutual funds, categorized as low, average, and high. Construct a summary table of the mutual funds categorized by risk.

SOLUTION

Most of the mutual funds are either low risk or average risk (104 or approximately 85%). Very few of the mutual funds are high-risk funds (14%).

TABLE 2.2
Frequency and Percentage Summary Table Pertaining to Risk Level for 121 Mutual Funds

Fund Risk Level	Number of Funds	Percentage of Funds %
Low	58	47.93
Average	46	38.02
High	17	14.05
Total	121	100.00

The Bar Chart

In a **bar chart**, a bar shows each category, the length of which represents the amount, frequency or percentage of values falling into a category. Figure 2.1 displays the bar chart for the reasons for shopping for holiday gifts online presented in Table 2.1.

FIGURE 2.1
Microsoft Excel Bar Chart of the Reasons for Shopping for Holiday Gifts Online

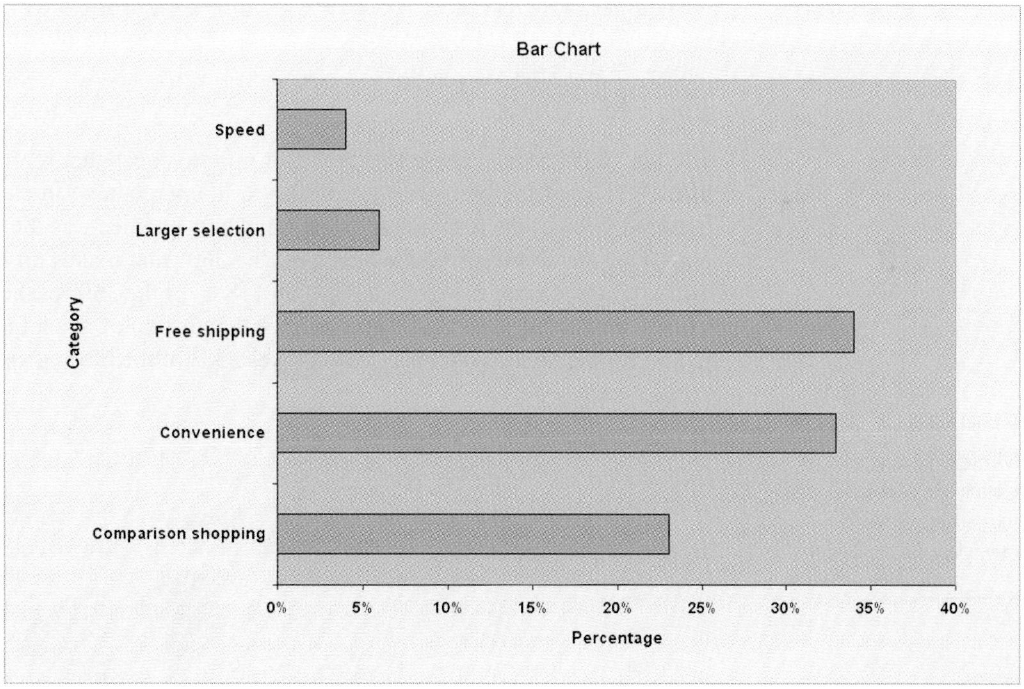

Bar charts allow you to compare percentages in the different categories. In Figure 2.1 the most common reasons for shopping online are free shipping and convenience followed by comparison shopping. Very few respondents shopped online because of larger selection or speed.

EXAMPLE 2.2

BAR CHART OF LEVELS OF RISK OF MUTUAL FUNDS

Construct a bar chart for the levels of risk of mutual funds (based on information in Table 2.2) and interpret the results.

SOLUTION

Most of the mutual funds are either low risk or average risk (more than 100 or 85%). Very few of the mutual funds are high risk funds (fewer than 20 or 15%).

FIGURE 2.2
Microsoft Excel Bar Chart of the Levels of Risk of Mutual Funds

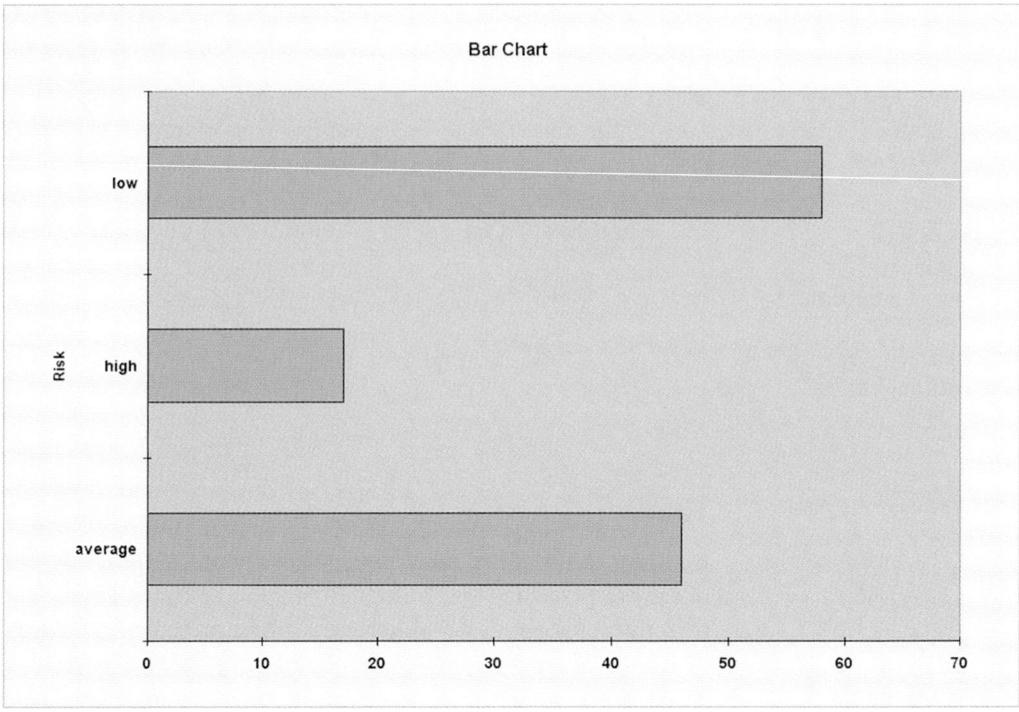

The Pie Chart

The **pie chart** takes a circle and breaks it up into slices that represent the categories. The size of each slice of the pie varies according to the percentage in each category. In Table 2.1 for example, 33% of the respondents stated that convenience was the main reason for online shopping. Thus, in constructing the pie chart, the 360° that makes up a circle is multiplied by 0.33, resulting in a slice of the pie that takes up 118.8° of the 360° of the circle. From Figure 2.3, you can see that the pie chart lets you visualize the portion of the entire pie that is in each category. In this figure, the convenience reason takes 33% of the pie and speed takes only 4%.

FIGURE 2.3
Microsoft Excel Pie Chart of the Reasons for Shopping for Holiday Gifts Online

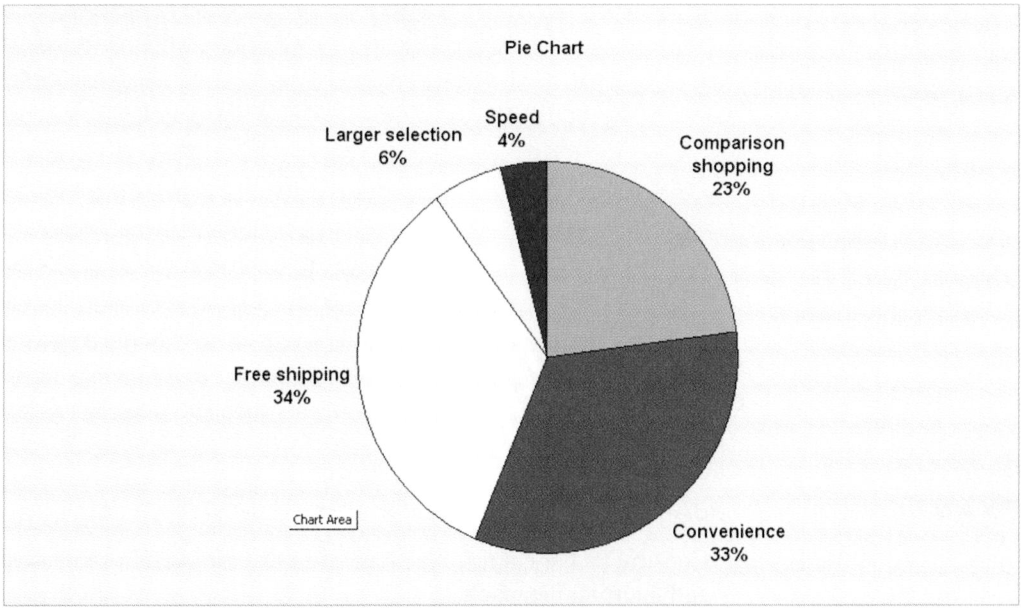

Which chart should you use? The selection of a particular chart often depends on your intention. If a comparison of categories is most important, you should use a bar chart. If observing the portion of the whole that is in a particular category is most important, you should use a pie chart.

EXAMPLE 2.3

PIE CHART OF LEVELS OF RISK OF MUTUAL FUNDS

Construct a pie chart for the levels of risk of mutual funds (see Table 2.2 on page 23) and interpret the results.

SOLUTION

(See Figure 2.4.) Most of the mutual funds are either low risk or average risk (more than 85%). Very few of the mutual funds are high-risk funds (less than 15%).

FIGURE 2.4

Microsoft Excel Pie Chart of the Levels of Risk of Mutual Funds

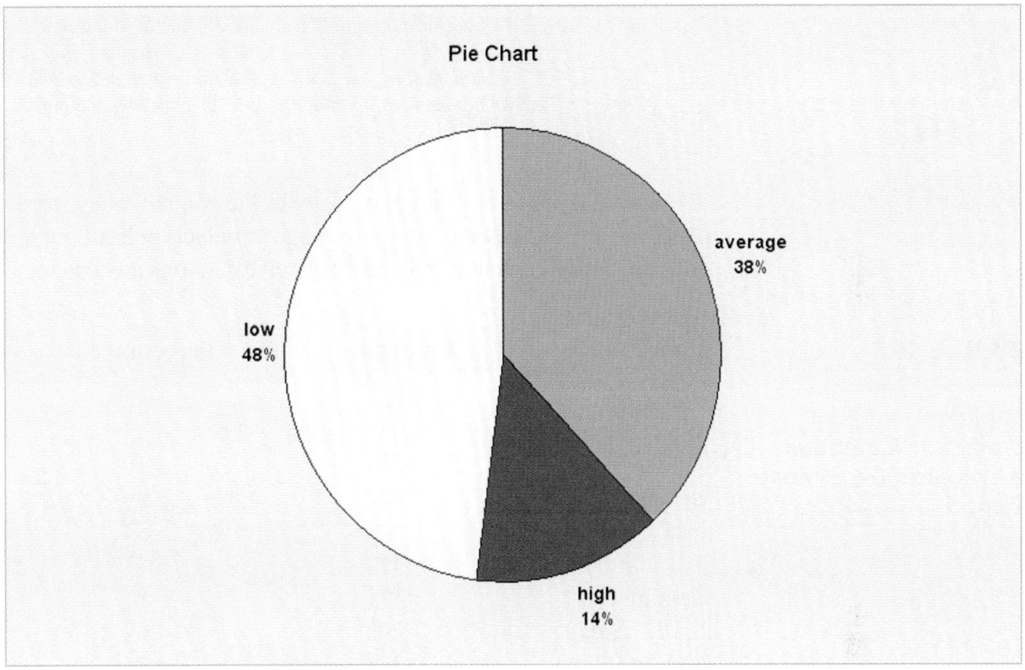

The Pareto Diagram

In a **Pareto diagram** the categorized responses are plotted in the descending order according to their frequencies and are combined with a cumulative percentage line on the same chart. The Pareto diagram can identify situations where the Pareto principle occurs.

PARETO PRINCIPLE

The **Pareto principle** exists when the majority of items in a set of data occur in a small number of categories, and the few remaining observations are spread out over a large number of categories. These two groups are often referred to as the "vital few" and the "trivial many."

The Pareto diagram has the ability to separate the "vital few" from the "trivial many," enabling you to focus on the important categories. In situations where the data involved consist of defective or nonconforming items, the Pareto diagram becomes a powerful tool for prioritizing improvement efforts.

Table 2.3 presents data from a large injection-molding company that manufactures plastic molded components used in computer keyboards, washing machines, automobiles, and television sets. The data presented in Table 2.3 consist of all computer keyboards with defects produced during a 3-month period. **KEYBOARD**

TABLE 2.3

Summary Table of Causes of Defects in Computer Keyboards in a 3-Month Period

Cause	Frequency	Percentage
Black Spot	413	6.53
Damage	1,039	16.43
Jetting	258	4.08
Pin mark	834	13.19
Scratches	442	6.99
Shot mold	275	4.35
Silver streak	413	6.53
Sink mark	371	5.87
Spray mark	292	4.62
Warpage	1,987	31.42
Total	6,324	100.01*

*Result differs slightly from 100.00 due to rounding.
Source: *U. H. Acharya and C. Mahesh, "Winning Back the Customer's Confidence: A Case Study on the Application of Design of Experiments to an Injection-Molding Process,"* Quality Engineering, *11, 1999, 357–363.*

Table 2.4 presents a summary table for the computer keyboard defects data in which the categories are ordered based on the percentage of defects present (rather than arranged alphabetically). The cumulative percentages for the ordered categories are also included as part of the table.

TABLE 2.4

Ordered Summary Table of Causes of Defects in Computer Keyboards in a 3-Month Period

Cause	Frequency	Percentage	Cumulative Percentage
Warpage	1,987	31.42	31.42
Damage	1,039	16.43	47.85
Pin mark	834	13.19	61.04
Scratches	442	6.99	68.03
Black spot	413	6.53	74.56
Silver streak	413	6.53	81.09
Sink mark	371	5.87	86.96
Spray mark	292	4.62	91.58
Shot mold	275	4.35	95.93
Jetting	258	4.08	100.00
Total	6,324	100.01*	

*Result differs slightly from 100.00 due to rounding.

In Table 2.4, the first category listed is warpage (with 31.42% of the defects), followed by damage (with 16.43%), followed by pin mark (with 13.19%). The two most frequently occurring categories—warpage and damage—account for 47.85% of the defects; the three most frequently occurring categories—warpage, damage, and pin mark—account for 61.04% of the defects, and so on. Figure 2.5 is a Pareto diagram based on the results displayed in tabular form in Table 2.4.

Figure 2.5 presents the bars vertically along with a cumulative percentage line. The cumulative line is plotted at the midpoint of each bar at a height equal to the cumulative percentage. If you follow the line you see that these first three categories account for about 60% of the corrections. Since the categories in the Pareto diagram are ordered by the frequency of occurrences, decision makers can see where to concentrate efforts to improve the process. Attempts to reduce defects due to warpage, damage, and pin marks should produce the greatest payoff. Then efforts can be made to reduce scratches and black spots.

In order for a summary table to include all categories, even those with few defects, in some situations you need to include a category labeled *Other* or *Miscellaneous*. In these situations, the bar representing these categories is placed to the right of the other bars.

FIGURE 2.5
Microsoft Excel Pareto Diagram for the Keyboard Defects Data

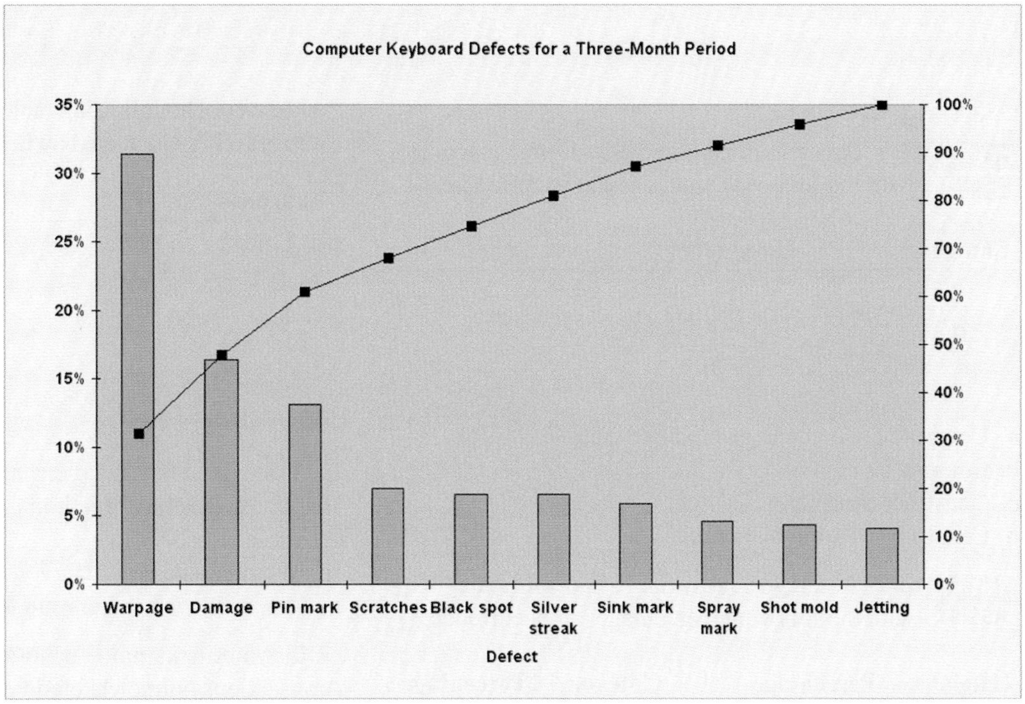

EXAMPLE 2.4 PARETO DIAGRAM OF REASONS FOR SHOPPING FOR HOLIDAY GIFTS ONLINE

Construct a Pareto diagram of the reasons for shopping online (see Table 2.1 on page 22).

SOLUTION
In Figure 2.6, free shipping and convenience account for 67% of the reasons for online shopping, and free shipping, convenience, and comparison shopping account for 90% of the reasons for online shopping.

FIGURE 2.6
Minitab Pareto Diagram of the Reasons for Shopping for Holiday Gifts Online

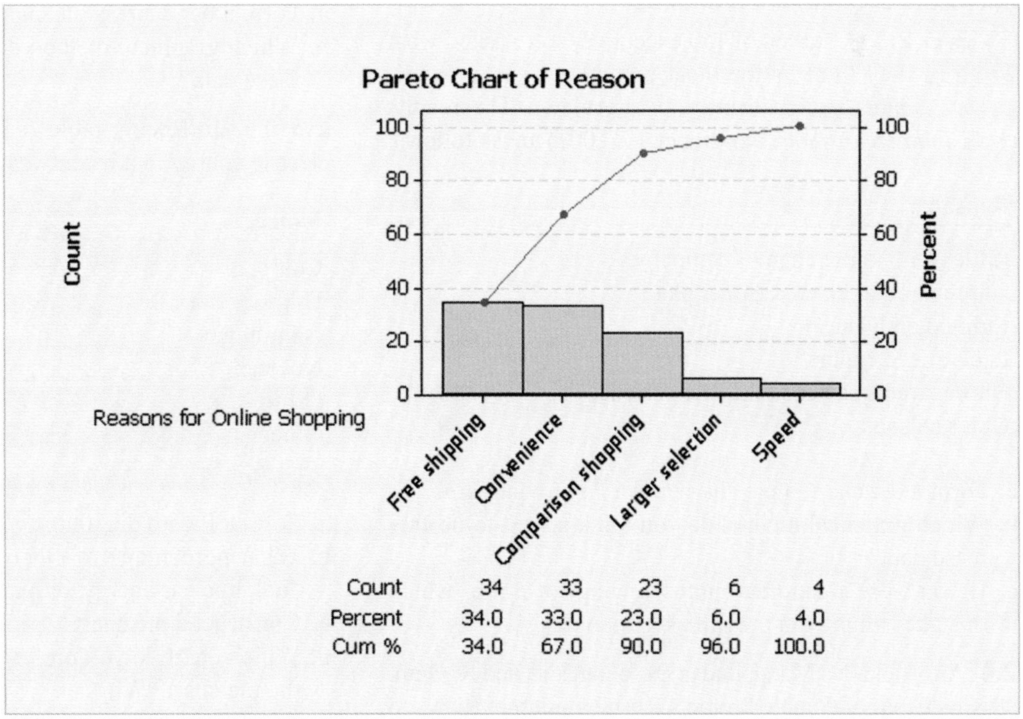

PROBLEMS FOR SECTION 2.1

Learning the Basics

PH Grade ASSIST **2.1** A categorical variable had three categories with the following frequency of occurrence:

Category	Frequency
A	13
B	28
C	9

a. Compute the percentage of values in each category.
b. Construct a bar chart.
c. Construct a pie chart.
d. Construct a Pareto diagram.

PH Grade ASSIST **2.2** A categorical variable had four categories with the following percentages of occurrence:

Category	Percentage	Category	Percentage
A	12	C	35
B	29	D	24

a. Construct a bar chart.
b. Construct a pie chart.
c. Construct a Pareto diagram.

Applying the Concepts

You can solve problems 2.3–2.10 manually or by using Microsoft Excel, Minitab or SPSS.

SELF Test **2.3** A survey of 150 executives were asked what they think is the most common mistake candidates make during job interviews. The results (*USA Today Snapshots*, November 19, 2001) are as follows:

Reason	%
Little or no knowledge of company	44
Unprepared to discuss career plans	23
Limited enthusiasm	16
Lack of eye contact	5
Unprepared to discuss skills/experience	3
Other reasons	9

a. Form a bar chart, a pie chart, and a Pareto diagram.
b. Which graphical method do you think is best to portray these data?
c. If you were a candidate interviewing for a job, which mistakes might you try hardest to avoid?

2.4 An article (M. Mangalindan, N. Wingfield, and R. Guth, "Rising Clout of Google Prompts Rush by Internet Rivals to Adapt," *The Wall Street Journal*, July 16, 2003, A1, A6) discussed the wide influence that Google had on the World Wide Web. The following table provides the market share of Web searches conducted by U.S. Internet users in May 2003.

Source	Percentage (%)
Ask Jeeves	3
AOL Time Warner	19
Google	32
MSN-Microsoft	15
Yahoo	25
Other	6

a. Form a bar chart, a pie chart, and a Pareto diagram.
b. Which graphical method do you think is best to portray these data?
c. What conclusions can you reach concerning the market share for Web searches in May 2003?

2.5 Americans paid for more than 50 billion dollars in transactions online via credit card in 2000 (Byron Acohido, "Microsoft, Banks Battle to Control Your e-info," *USA Today*, August 13, 2001, 1B–2B). These transactions were broken down as follows:

Credit Card	Amount ($billions)	%
American Express	8.04	15.6
Discover	1.97	3.8
MasterCard	15.57	30.2
Visa	25.96	50.4

a. Form a bar chart, a pie chart, and a Pareto diagram.
b. Which graphical method do you think is best to portray these data?

2.6 The following table represents the U.S. sources of electric energy in a recent year:

Source	%
Coal	51
Hydropower	6
Natural gas	16
Nuclear	21
Oil	3
Other	3

Source: U.S. Department of Energy.

a. Form a Pareto diagram.
b. What percentage of electricity is derived from either coal, nuclear energy, or natural gas?
c. Construct a pie chart.
d. Which chart do you prefer to use, the Pareto diagram or the pie chart? Why?

2.7 An article (P. Kitchen, "Retirement Plan: To Keep Working," *Newsday*, September 24, 2003) discussed the results of a sample of 2,001 Americans ages 50 to 70 who were employed full-time or part-time. The following table represents their plans for retirement.

Plans	Percentage (%)
Not work for pay at all	29
Start own business	10
Work full-time	7
Work part-time	46
Don't know	3
Other	5

a. Form a bar chart and a pie chart.
b. Which graphical method do you think is best to portray these data?

2.8 Junk e-mail or spam has become a serious problem for productivity (J. Hopkins, "Spam Blaster Does Job for Merrill," *USA Today*, January 7, 2004, 7B). The following table concerning company use of anti-spam software is based on a survey of tech executives.

Company Use of Anti-Spam Software	Percentage (%)
Have software for some users	12
Have software for all users	59
Plan for software in the next 12 months	20
Don't plan for software	9

a. Form a bar chart and a pie chart.
b. Which graphical method do you think is best to portray these data?

2.9 A network analyst has recorded the root cause of network crashes during the past six months.

Reason for Failure	Frequency
Physical connection	1
Power failure	3
Server software	29
Server hardware	2
Server out of memory	32
Inadequate bandwidth	1

a. Form a Pareto diagram.
b. Discuss the "vital few" and "trivial many" reasons for the root causes of network crashes.

2.10 The following data represent complaints about hotel rooms.

Reason	Number
Room dirty	32
Room not stocked	17
Room not ready	12
Room too noisy	10
Room needs maintenance	17
Room has too few beds	9
Room doesn't have promised features	7
No special accommodations	2

a. Form a Pareto diagram.
b. What reasons for complaints do you think the hotel should focus on if it wants to reduce the number of complaints? Explain.

2.2 ORGANIZING NUMERICAL DATA

When the number of data values is large, you can organize numerical data into an ordered array or a stem-and-leaf display to help you to understand the information you have. Suppose that you decide to undertake a study that compares the cost for a restaurant meal in a major city to the cost of a similar meal in the suburbs outside the city. Table 2.5 shows the data for 50 city restaurants and 50 suburban restaurants. RESTRATE The data are not arranged in order from lowest to highest. This arrangement makes it difficult to draw conclusions about the price of meals in the two geographical areas.

TABLE 2.5
Price per Person at 50 City Restaurants and 50 Suburban Restaurants

City									
50	38	43	56	51	36	25	33	41	44
34	39	49	37	40	50	50	35	22	45
44	38	14	44	51	27	44	39	50	35
31	34	48	48	30	42	26	35	32	63
36	38	53	23	39	45	37	31	39	53
Suburban									
37	37	29	38	37	38	39	29	36	38
44	27	24	34	44	23	30	32	25	29
43	31	26	34	23	41	32	30	28	33
26	51	26	48	39	55	24	38	31	30
51	30	27	38	26	28	33	38	32	25

The Ordered Array

An **ordered array** is a sequence of the data in rank order from the smallest value to the largest value. Table 2.6 contains ordered arrays for the price of meals at city restaurants and suburban restaurants. From Table 2.6 you can see that the price of a meal at the city restaurants is between \$14 and \$63, and the price of a meal at the suburban restaurants is between \$23 and \$55.

TABLE 2.6
Ordered Array of Price per Person at 50 City Restaurants and 50 Suburban Restaurants

RESTRATE

City									
14	22	23	25	26	27	30	31	31	32
33	34	34	35	35	35	36	36	37	37
38	38	38	39	39	39	39	40	41	42
43	44	44	44	44	45	45	48	48	49
50	50	50	50	51	51	53	53	56	63
Suburban									
23	23	24	24	25	25	26	26	26	26
27	27	28	28	29	29	29	30	30	30
30	31	31	32	32	32	33	33	34	34
36	37	37	37	38	38	38	38	38	38
39	39	41	43	44	44	48	51	51	55

The Stem-and-Leaf Display

The **stem-and-leaf display** organizes data into groups (called stems) such that the values within each group (the leaves) branch out to the right on each row. The resulting display allows you to see how the data are distributed and where concentrations of data exist. To see how a stem-and-leaf display is constructed, suppose that 15 students from your class eat lunch at a fast-food restaurant. The following data are the amounts spent for lunch.

5.35 4.75 4.30 5.47 4.85 6.62 3.54 4.87 6.26 5.48 7.27 8.45 6.05 4.76 5.91

To form the stem-and-leaf display you put the values in ascending order first. Then you use the units column as the stems and round the decimals (the leaves) to one decimal place.

3	5
4	83998
5	4559
6	631
7	3
8	5

The first value of 5.35 is rounded to 5.4. Its stem (row) is 5 and its leaf is 4. The second value of 4.75 is rounded to 4.8. Its stem (row) is 4 and its leaf is 8.

EXAMPLE 2.5 STEM-AND-LEAF DISPLAY OF THE 2003 RETURN OF MUTUAL FUNDS

In the "Using Statistics" scenario, you are interested in studying the 2003 return of mutual funds MUTUALFUNDS2004. Construct a stem-and-leaf display.

SOLUTION

From Figure 2.7, you can conclude that:

- The lowest 2003 return was 14%.
- The highest 2003 return was 78%.
- The 2003 returns were concentrated between 25% and 50%.
- Only four mutual funds had returns below 20% and only two mutual funds had returns above 70%.

FIGURE 2.7
Minitab Stem-and-Leaf Display of 2003 Returns

Stem-and-Leaf Display: Return 2003

```
Stem-and-leaf of Return 2003  N  = 121
Leaf Unit = 1.0

   1   1  4
   4   1  589
   7   2  033
  22   2  566777777889999
  33   3  00111123344
  55   3  5555555666777777788899
 (17)  4  00000011222233444
  49   4  555666777778888999
  31   5  0000123344
  21   5  56789
  16   6  11122234
   8   6  666778
   2   7  1
   1   7  8
```

PROBLEMS FOR SECTION 2.2

Learning the Basics

PH Grade ASSIST **2.11** Form an ordered array given the following data from a sample of $n = 7$ midterm exam scores in accounting:

68 94 63 75 71 88 64

PH Grade ASSIST **2.12** Form a stem-and-leaf display given the following data from a sample of $n = 7$ midterm exam scores in finance:

80 54 69 98 93 53 74

PH Grade ASSIST **2.13** Form an ordered array given the following data from a sample of $n = 7$ midterm exam scores in marketing:

88 78 78 73 91 78 85

PH Grade ASSIST **2.14** Form an ordered array given the following stem-and-leaf display from a sample of $n = 7$ midterm exam scores in information systems:

5	0
6	
7	446
8	19
9	2

Applying the Concepts

PH Grade ASSIST **2.15** The following is a stem-and-leaf display representing the amount of gasoline purchased in gallons (with leaves in tenths of gallons) for a sample of 25 cars that use a particular service station on the New Jersey Turnpike:

9	147
10	02238
11	125566777
12	223489
13	02

a. Place the data into an ordered array.
b. Which of these two displays seems to provide more information? Discuss.
c. What amount of gasoline (in gallons) is most likely to be purchased?
d. Is there a concentration of the purchase amounts in the center of the distribution?

PH Grade ASSIST **2.16** The following data represent the bounced check fee in dollars for a sample of 23 banks for direct-deposit customers who maintain a $100 balance. **BANKCOST1**

26 28 20 20 21 22 25 25 18 25 15 20
18 20 25 25 22 30 30 30 15 20 29

Source: "The New Face of Banking," Copyright © 2000 by Consumers Union of U.S., Inc., Yonkers, NY 10703–1057. Adapted with permission from Consumer Reports, *June 2000.*

a. Place the data into an ordered array.
b. Set up a stem-and-leaf display for these data.
c. Which of these two displays seems to provide more information? Discuss.
d. Around what value, if any, are the bounced check fees concentrated? Explain.

PH Grade ASSIST **2.17** The following data represent the monthly service fee in dollars if a customer's account falls below the minimum required balance for a sample of 26 banks for direct-deposit customers who maintain a $1,500 balance. BANKCOST2

12	8	5	5	6	6	10	10	9	7	10	7	7
5	0	10	6	9	12	0	5	10	8	5	5	9

Source: "The New Face of Banking," Copyright © 2000 by Consumers Union of U.S. Inc., Yonkers, NY 10703–1057. Adapted with permission from Consumer Reports, *June 2000.*

a. Place the data into an ordered array.
b. Set up a stem-and-leaf display for these data.
c. Which of these two displays seems to provide more information? Discuss.
d. Around what value, if any, are the monthly service fees concentrated? Explain.

SELF Test **2.18** The following data represent the total fat for burgers and chicken items from a sample of fast-food chains. FASTFOOD

BURGERS
19 31 34 35 39 39 43

CHICKEN
7 9 15 16 16 18 22 25 27 33 39

Source: "Quick Bites," Copyright © 2001 by Consumers Union of U.S., Inc., Yonkers, NY 10703–1057. Adapted with permission from Consumer Reports, March 2001.

a. Place the data for burgers and chicken into two ordered arrays.
b. Construct stem-and-leaf displays for burgers and chicken.
c. Does the ordered array or the stem-and-leaf display provide more information? Discuss.
d. Compare the burgers and chicken items in terms of the total fat. What conclusions can you make?

2.19 The following data represent the daily average hotel cost and rental car cost for 20 U.S. cities during a week in October 2003. HOTEL-CAR

City	Hotel	Cars
San Francisco	205	47
Los Angeles	179	41
Seattle	185	49
Phoenix	210	38
Denver	128	32
Dallas	145	48
Houston	177	49
Minneapolis	117	41
Chicago	221	56
St. Louis	159	41
New Orleans	205	50
Detroit	128	32
Cleveland	165	34
Atlanta	180	46
Orlando	198	41
Miami	158	40
Pittsburgh	132	39
Boston	283	67
New York	269	69
Washington D.C.	204	40

Source: Extracted from The Wall Street Journal, *October 10, 2003, W4.*

a. Place the data for hotel cost and rental car cost into two ordered arrays.
b. Construct stem-and-leaf displays for hotel cost and rental car cost.
c. Does the ordered array or the stem-and-leaf display provide more information? Discuss.
d. Around what value, if any, are the hotel cost and rental car cost concentrated? Explain.

2.3 TABLES AND CHARTS FOR NUMERICAL DATA

When you have a data set that contains a large number of values, reaching conclusions from an ordered array or stem-and-leaf is often difficult. You need to use tables and charts in such circumstances. There are many different tables and charts to visually present numerical data. These include the frequency and percentage distributions, histogram, polygon, and cumulative percentage polygon (ogive).

The Frequency Distribution

The frequency distribution helps you draw conclusions from a large set of data.

A **frequency distribution** is a summary table in which the data are arranged into numerically ordered class groupings.

In constructing a frequency distribution, you must give attention to selecting the appropriate *number* of **class groupings** for the table, determining a suitable *width* of a class grouping, and establishing the *boundaries* of each class grouping to avoid overlapping.

The number of class groupings you use depends on the number of values in the data. Larger numbers of values allow for a larger number of class groups. In general, the frequency distribution should have at least five class groupings but no more than 15. Having too few or too many class groupings provides little new information.

When developing the frequency distribution, you define each class grouping by class intervals of equal width. To determine the **width of a class interval**, you divide the **range** (the highest value–the lowest value) of the data by the number of class groupings desired.

DETERMINING THE WIDTH OF A CLASS INTERVAL

$$\text{Width of interval} = \frac{\text{range}}{\text{number of desired class groupings}} \quad \textbf{(2.1)}$$

The city restaurant data consist of a sample of 50 restaurants. With this sample size, 10 class groupings are acceptable. From the ordered array in Table 2.6 on page 30, the range of the data is \$63 – \$14 = \$49. Using Equation (2.1), you approximate the width of the class interval as follows:

$$\text{Width of interval} = \frac{49}{10} = 4.9$$

You should choose an interval width that simplifies reading and interpretation. Therefore, instead of using an interval width of \$4.90, you should select an interval width of \$5.00.

To construct the frequency distribution table, you should establish clearly defined **class boundaries** for each class grouping so that the values can be properly tallied into the classes. You place each value in one and only one class. You must avoid overlapping of classes.

Because you have set the width of each class interval for the restaurant meal cost data at \$5, you need to establish the boundaries of the various class groupings so as to include the entire range of values. Whenever possible, you should choose these boundaries to simplify reading and interpretation. Thus, for the city restaurants, since the cost ranges from \$14 to \$63, the first class interval ranges from \$10 to less than \$15, the second from \$15 to less than \$20, and so on, until they have been tallied into 11 classes. Each class has an interval width of \$5, without overlapping. The center of each class, the **class midpoint**, is halfway between the lower boundary of the class and the upper boundary of the class. Thus, the class midpoint for the class from \$10 to under \$15 is \$12.5, the class midpoint for the class from \$15 to under \$20 is \$17.5, and so on. Table 2.7 is a frequency distribution of the cost per meal for the 50 city restaurants and for the 50 suburban restaurants.

TABLE 2.7
Frequency Distribution of the Cost per Meal for 50 City Restaurants and 50 Suburban Restaurants

Cost per Meal (\$)	City Frequency	Suburban Frequency
10 but less than \$15	1	0
15 but less than \$20	0	0
20 but less than \$25	2	4
25 but less than \$30	3	13
30 but less than \$35	7	13
35 but less than \$40	14	12
40 but less than \$45	8	4
45 but less than \$50	5	1
50 but less than \$55	8	2
55 but less than \$60	1	1
60 but less than \$65	1	0
Total	50	50

The summary table allows you to draw conclusions about the major characteristics of the data. For example, Table 2.7 shows that the cost of meals at city restaurants are concentrated between \$30 and \$55 as compared to the cost of meals at suburban restaurants, which are clustered between \$25 and \$40.

If the data set does not contain many values, one set of class boundaries may provide a different picture than another set. For example, for the restaurant cost data, using a class-interval width of 4.0 instead of 5.0 (as was used in Table 2.7) may cause shifts in the way in which the values distribute among the classes.

You can also get shifts in data concentration when you choose different lower and upper class boundaries. Fortunately, as the sample size increases, alterations in the selection of class boundaries affect the concentration of data less and less.

EXAMPLE 2.6

FREQUENCY DISTRIBUTION OF THE 2003 RETURN FOR GROWTH AND VALUE MUTUAL FUNDS

In the "Using Statistics" scenario, you are interested in comparing the 2003 return of growth and value mutual funds. MUTUALFUNDS2004 Construct frequency distributions for the growth funds and for the value funds.

SOLUTION

The 2003 percentage returns of the growth funds are highly concentrated between 30 and 50, with some concentration between 20 and 30 (see Table 2.8). The 2003 percentage returns of the value funds are concentrated between 30 and 50, with some also between 20 and 30 and 50 and 70. You should not directly compare the frequencies of the growth funds since there are 49 growth funds and 72 value funds in the sample.

TABLE 2.8

Frequency Distribution of the 2003 Return for Growth and Value Mutual Funds

2003 Percentage Return	Growth Frequency	Value Frequency
10 but less than 20	2	2
20 but less than 30	9	9
30 but less than 40	13	20
40 but less than 50	15	20
50 but less than 60	5	10
60 but less than 70	5	9
70 but less than 80	0	2
Total	49	72

The Relative Frequency Distribution and the Percentage Distribution

Because you usually want to know the proportion or the percentage of the total that is in each group, the relative frequency distribution or the percentage distribution is preferred to the frequency distribution. When you are comparing two or more groups that differ in their sample sizes, you must use either a relative frequency distribution or a percentage distribution.

You form the **relative frequency distribution** by dividing the frequencies in each class of the frequency distribution (see Table 2.7 on page 33) by the total number of values. You form the **percentage distribution** by multiplying each relative frequency by 100%. Thus, the relative frequency of meals at city restaurants that cost between \$30 and \$35 is 7 divided by 50 or 0.14, and the percentage is 14%. Table 2.9 presents the relative frequency distribution and percentage distribution of the cost of restaurant meals at city and suburban restaurants.

From Table 2.9, you conclude that meals cost more at city restaurants than at suburban restaurants—16% of the meals at city restaurants cost between \$40 and \$45 as compared to 8% of the suburban restaurants; 16% of the meals at city restaurants cost between \$50 and \$55 as compared to 4% of the suburban restaurants; while only 6% of the meals at city restaurants cost between \$25 and \$30 as compared to 26% of the suburban restaurants.

TABLE 2.9
Relative Frequency Distribution and Percentage Distribution of the Cost of Restaurant Meals at City and Suburban Restaurants

	City		Suburban	
Cost per Meal ($)	Relative frequency	Percentage	Relative frequency	Percentage
10 but less than $15	0.02	2.0	0.00	0.0
15 but less than $20	0.00	0.0	0.00	0.0
20 but less than $25	0.04	4.0	0.08	8.0
25 but less than $30	0.06	6.0	0.26	26.0
30 but less than $35	0.14	14.0	0.26	26.0
35 but less than $40	0.28	28.0	0.24	24.0
40 but less than $45	0.16	16.0	0.08	8.0
45 but less than $50	0.10	10.0	0.02	2.0
50 but less than $55	0.16	16.0	0.04	4.0
55 but less than $60	0.02	2.0	0.02	2.0
60 but less than $65	0.02	2.0	0.00	0.0
Total	1.00	100.0	1.00	100.0

EXAMPLE 2.7

RELATIVE FREQUENCY DISTRIBUTION AND PERCENTAGE DISTRIBUTION OF THE 2003 RETURN FOR GROWTH AND VALUE MUTUAL FUNDS

In the "Using Statistics" scenario, you are interested in comparing the 2003 return for growth and value mutual funds. **MUTUALFUNDS2004** Construct relative frequency distributions and percentage distributions for the growth funds and for the value funds.

SOLUTION
You conclude (see Table 2.10) that the 2003 return for the growth funds is slightly lower than for the value funds and that 18.37% of growth funds have returns between 20 and 30 as compared to 12.5% of the value funds. Slightly more of the value funds have higher returns (between 50 and 60, and 60 and 70) than the growth funds.

TABLE 2.10
Relative Frequency Distribution and Percentage Distribution of the 2003 Return for Growth and Value Mutual Funds

	Growth		Value	
Annual Percentage Return in 2003	Proportion	Percentage	Proportion	Percentage
10 but less than 20	0.0408	4.08	0.0278	2.78
20 but less than 30	0.1837	18.37	0.1250	12.50
30 but less than 40	0.2653	26.53	0.2778	27.78
40 but less than 50	0.3061	30.61	0.2778	27.78
50 but less than 60	0.1020	10.20	0.1389	13.89
60 but less than 70	0.1020	10.20	0.1250	12.50
70 but less than 80	0.0000	0.00	0.0278	2.78
Total	1.0000	100.0	1.0000	100.0

The Cumulative Distribution

The **cumulative percentage distribution** provides a way of presenting information about the percentage of values that are less than a certain value. For example, you might want to know what percentage of the city restaurant meals cost less than $20, less than $30, less than $50, and so on. The percentage distribution is used to form the cumulative percentage distribution. From Table 2.12, 0.00% of the meals cost less than $10, 2% cost less than $15, 2% also cost less than $20 (since none of the meals cost between $15 and $20), 6% (2% + 4%) cost less than $25, and so on, until all 100% of the meals cost less than $65. Table 2.11 illustrates how to develop the cumulative percentage distribution for the cost of meals at city restaurants.

TABLE 2.11

Developing the Cumulative Percentage Distribution for the Cost of Meals at City Restaurants

Cost per Meal ($)	Percentage	Percentage of Funds Less than Lower Boundary of Class Interval
10 but less than $15	2	0
15 but less than $20	0	2
20 but less than $25	4	2 = 2 + 0
25 but less than $30	6	6 = 2 + 0 + 4
30 but less than $35	14	12 = 2 + 0 + 4 + 6
35 but less than $40	28	26 = 2 + 0 + 4 + 6 + 14
40 but less than $45	16	54 = 2 + 0 + 4 + 6 + 14 + 28
45 but less than $50	10	70 = 2 + 0 + 4 + 6 + 14 + 28 + 16
50 but less than $55	16	80 = 2 + 0 + 4 + 6 + 14 + 28 + 16 + 10
55 but less than $60	2	96 = 2 + 0 + 4 + 6 + 14 + 28 + 16 + 10 + 16
60 but less than $65	2	98 = 2 + 0 + 4 + 6 + 14 + 28 + 16 + 10 + 16 + 2
$65 but less than $70	0	100 = 2 + 0 + 4 + 6 + 14 + 28 + 16 + 10 + 16 + 2 + 2

Table 2.12 summarizes the cumulative percentages of the cost of city and suburban restaurant meals. The cumulative distribution clearly shows that the cost of meals is lower in suburban restaurants than in city restaurants—34% of the suburban restaurants cost less than $30 as compared to only 12% of the city restaurants; 60% of the suburban restaurants cost less than $35 as compared to only 26% of the city restaurants; 84% of the suburban restaurants cost less than $40 as compared to only 54% of the city restaurants.

TABLE 2.12

Cumulative Percentage Distributions of the Cost of City and Suburban Restaurant Meals

Cost ($)	City Percentage of Restaurants Less than Indicated Value	Suburban Percentage of Restaurants Less than Indicated Value
10	0	0
15	2	0
20	2	0
25	6	8
30	12	34
35	26	60
40	54	84
45	70	92
50	80	94
55	96	98
60	98	100
65	100	100

EXAMPLE 2.8

CUMULATIVE PERCENTAGE DISTRIBUTION OF THE 2003 RETURN FOR GROWTH AND VALUE MUTUAL FUNDS

In the "Using Statistics" scenario, you are interested in comparing the annual return in 2003 of growth and value mutual funds. MUTUALFUNDS2004 Construct cumulative percentage distributions for the growth funds and for the value funds.

SOLUTION

The cumulative distribution in Table 2.13 indicates that slightly more of the growth funds have lower returns than the value funds—22.45% of the growth funds have returns below 30 as compared to 15.28% of the value funds; 48.98% of the growth funds have returns below 40 as compared to 43.06% of the value funds; 79.59% of the growth funds have returns below 50 as compared to 70.83% of the value funds.

TABLE 2.13
Cumulative Percentage Distributions of the 2003 Return for Growth and Value Funds

Annual Return	Growth Fund Percentage Less than Indicated Value	Value Fund Percentage Less than Indicated Value
10	0.00	0.00
20	4.08	2.78
30	22.45	15.28
40	48.98	43.06
50	79.59	70.83
60	89.80	84.72
70	100.00	97.22
80	100.00	100.00

The Histogram

The **histogram** is a bar chart for grouped numerical data in which the frequencies or percentages of each group of numerical data are represented as individual bars. In a histogram, there are no gaps between adjacent bars as there are in a bar chart of categorical data. You display the variable of interest along the horizontal (X) axis. The vertical (Y) axis represents either the frequency or the percentage of values per class interval.

Figure 2.8 displays the Minitab frequency histogram for the cost of restaurant meals at city restaurants. The histogram indicates that the cost of restaurant meals at city restaurants is concentrated between approximately \$30 and \$50. Very few meals cost less than \$20 or more than \$60.

FIGURE 2.8
Minitab Histogram for the Cost of Restaurant Meals at City Restaurants

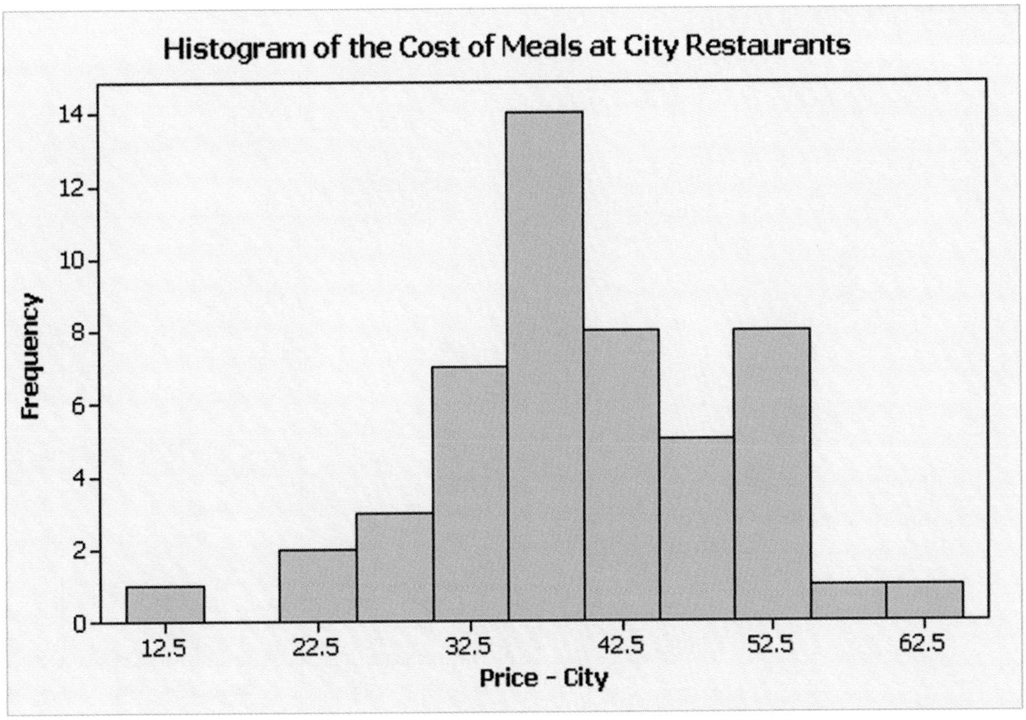

EXAMPLE 2.9 HISTOGRAM OF THE 2003 RETURN FOR GROWTH AND VALUE MUTUAL FUNDS

In the "Using Statistics" scenario, you are interested in comparing the 2003 return of growth and value mutual funds. MUTUALFUNDS2004 Construct histograms for the growth funds and for the value funds.

SOLUTION
Figure 2.9 shows that the distribution of the growth funds has more lower returns as compared to the value funds, which have more higher returns.

FIGURE 2.9A

Histogram of 2003 Percentage Return (Panel A—Growth Funds and Panel B—Value Funds)

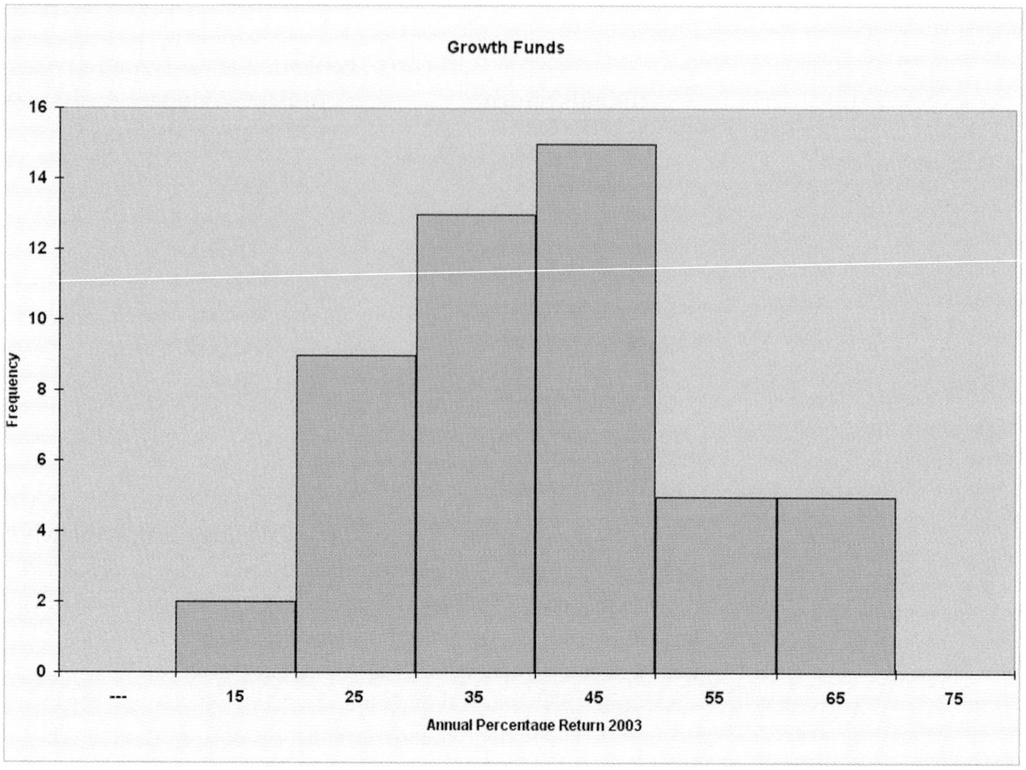

FIGURE 2.9B

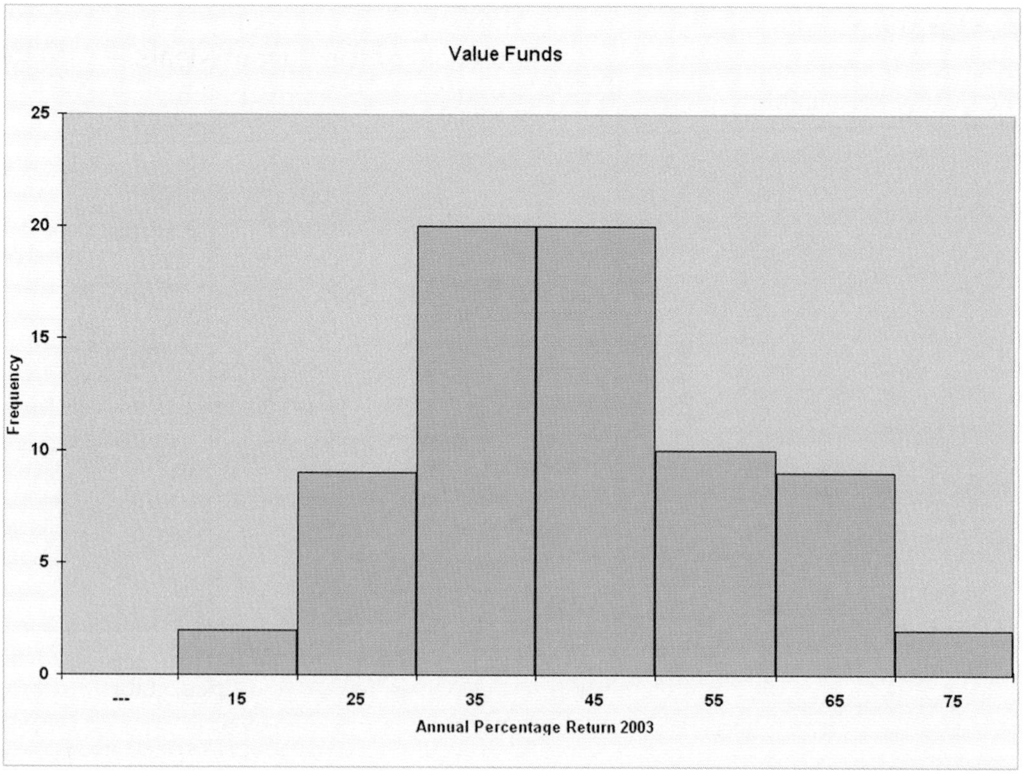

The Polygon

Constructing multiple histograms on the same graph when comparing two or more sets of data is difficult and confusing. Superimposing the vertical bars of one histogram on another histogram causes difficulty in interpretation. When there are two or more groups, you should use a percentage polygon.

PERCENTAGE POLYGON

The **percentage polygon** is formed by having the midpoint of each class represent the data in that class and then connecting the sequence of midpoints at their respective class percentages.

Figure 2.10 displays the percentage polygons for the cost of restaurant meals for city and suburban restaurants. The polygon for the suburban restaurants is concentrated to the left of (corresponding to lower cost) the polygon for city restaurants. The highest percentages of cost for the suburban restaurants are for class midpoints of $27.50 and $32.50, while the highest percentages of cost for the city restaurants are for a class midpoint of $37.50.

FIGURE 2.10

Percentage Polygons for the Cost of Restaurant Meals for City and Suburban Restaurants

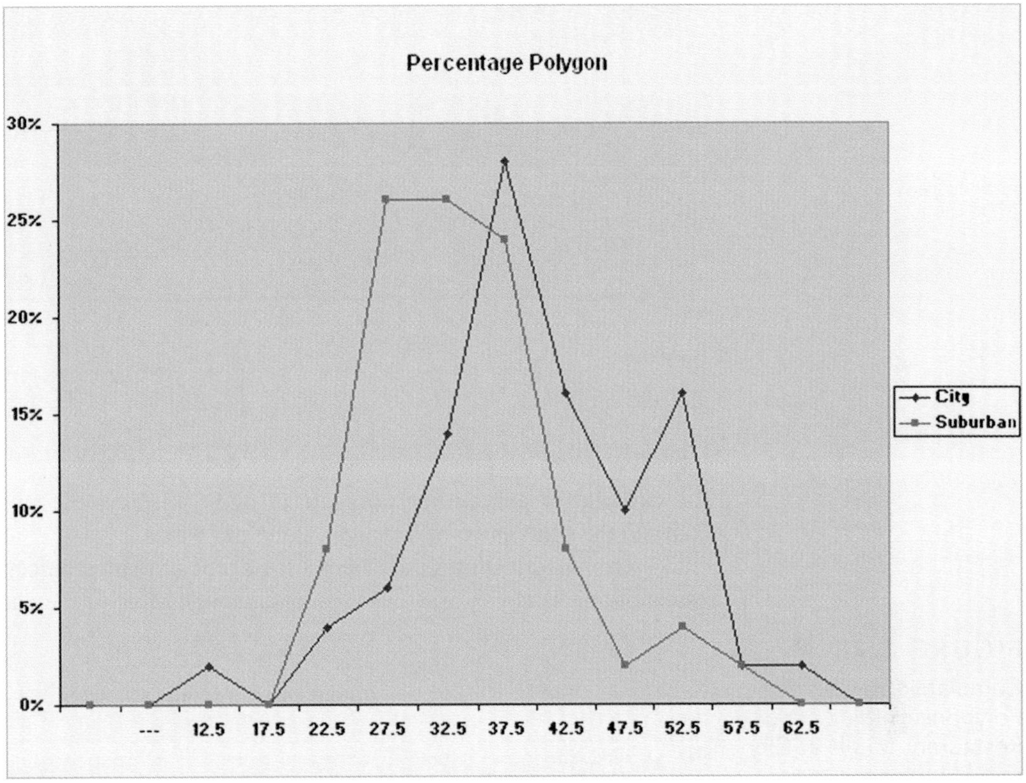

The polygons in Figure 2.10 have points whose values on the X axis represent the midpoint of the class interval. For example, look at the points plotted on the X axis at 22.5 ($22.50). The point for the suburban restaurants (the higher one) represents the fact that 8% of these restaurants have meal costs between $20 and $25. The point for the city restaurants (the lower one) represents the fact that 4% of these restaurants have meal costs between $20 and $25.

When you construct polygons or histograms, the vertical (Y) axis should show the true zero or "origin" so as not to distort the character of the data. The horizontal (X) axis does not need to specify the zero point for the variable of interest, although the range of the variable should constitute the major portion of the axis.

EXAMPLE 2.10 PERCENTAGE POLYGONS OF THE 2003 RETURN FOR GROWTH AND VALUE MUTUAL FUNDS

In the "Using Statistics" scenario, you are interested in comparing the 2003 return of growth and value mutual funds. MUTUALFUNDS2004 Construct percentage polygons for the growth funds and for the value funds.

SOLUTION

Figure 2.11 shows that the distribution of the growth funds has more lower annual returns as compared to the value funds, which have more higher returns.

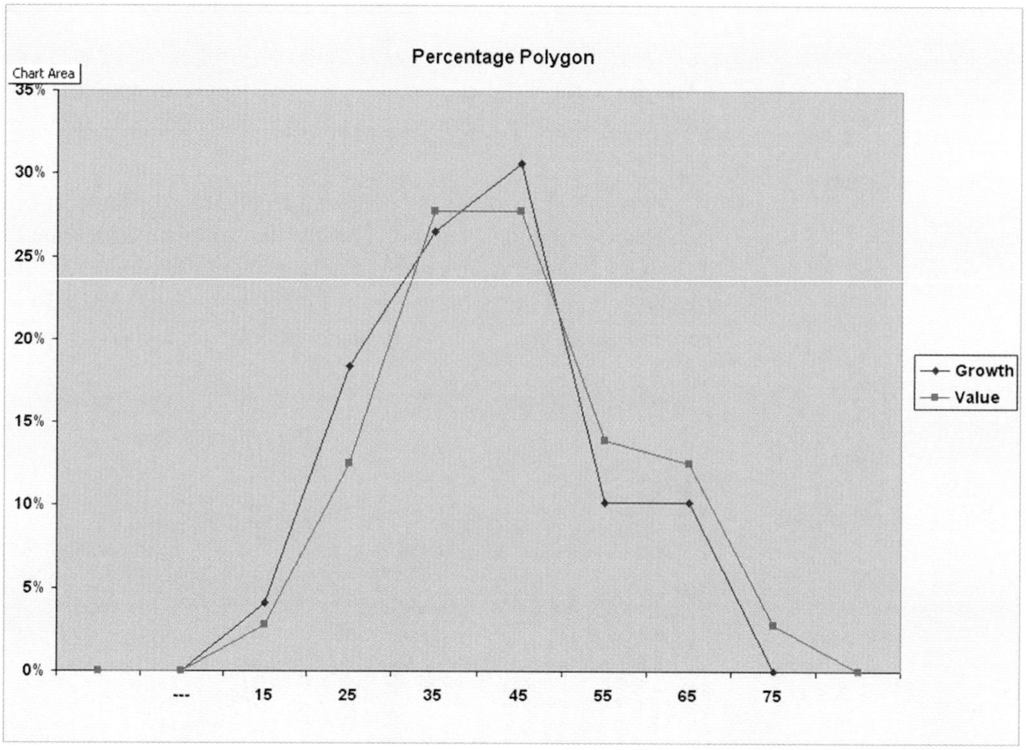

FIGURE 2.11
Percentage Polygons of 2003 Percentage Return

The Cumulative Percentage Polygon (Ogive)

The **cumulative percentage polygon**, or **ogive**, displays the variable of interest along the *X* axis, and the cumulative percentages along the *Y* axis.

Figure 2.12 illustrates the Microsoft Excel cumulative percentage polygons of the cost of restaurant meals at city and suburban restaurants. Most of the curve for the city restaurants is

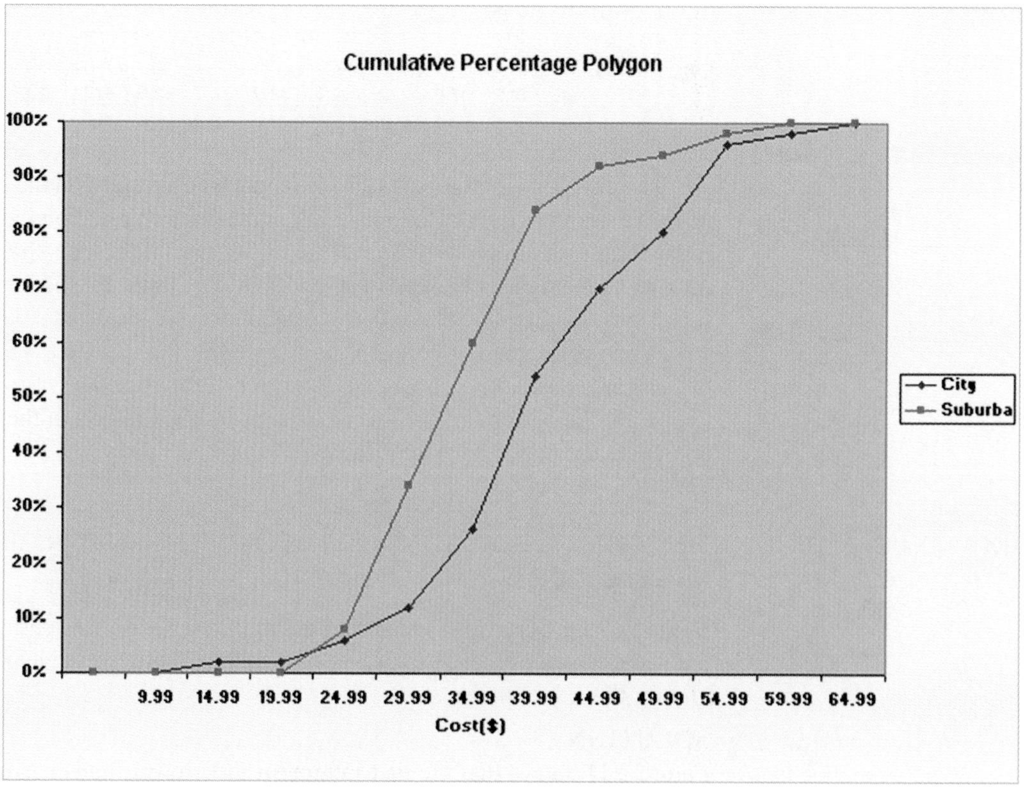

FIGURE 2.12
Cumulative Percentage Polygons of the Cost of Restaurant Meals at City and Suburban Restaurants

located to the right of the curve for the suburban restaurants. This indicates that the city restaurants have fewer meals that cost below a particular value. For example, 12% of the city restaurant meals cost less than $30 as compared to 34% of the suburban restaurant meals.

EXAMPLE 2.11

CUMULATIVE PERCENTAGE POLYGONS OF THE 2003 RETURN FOR GROWTH AND VALUE MUTUAL FUNDS

In the "Using Statistics" scenario, you are interested in comparing the 2003 return of growth and value mutual funds. MUTUALFUNDS2004 Construct cumulative percentage polygons for the growth funds and for the value funds.

SOLUTION

Figure 2.13 illustrates the Microsoft Excel cumulative percentage polygons of the 2003 percentage return for growth and value funds. The curve for the value funds is located slightly to the right of the curve for the growth funds. This indicates that the value funds have fewer returns below a particular value. For example, 70.83% of the value funds have returns less than 50 as compared to 79.59% of the growth funds.

FIGURE 2.13
Cumulative Percentage Polygons of 2003 Percentage Return

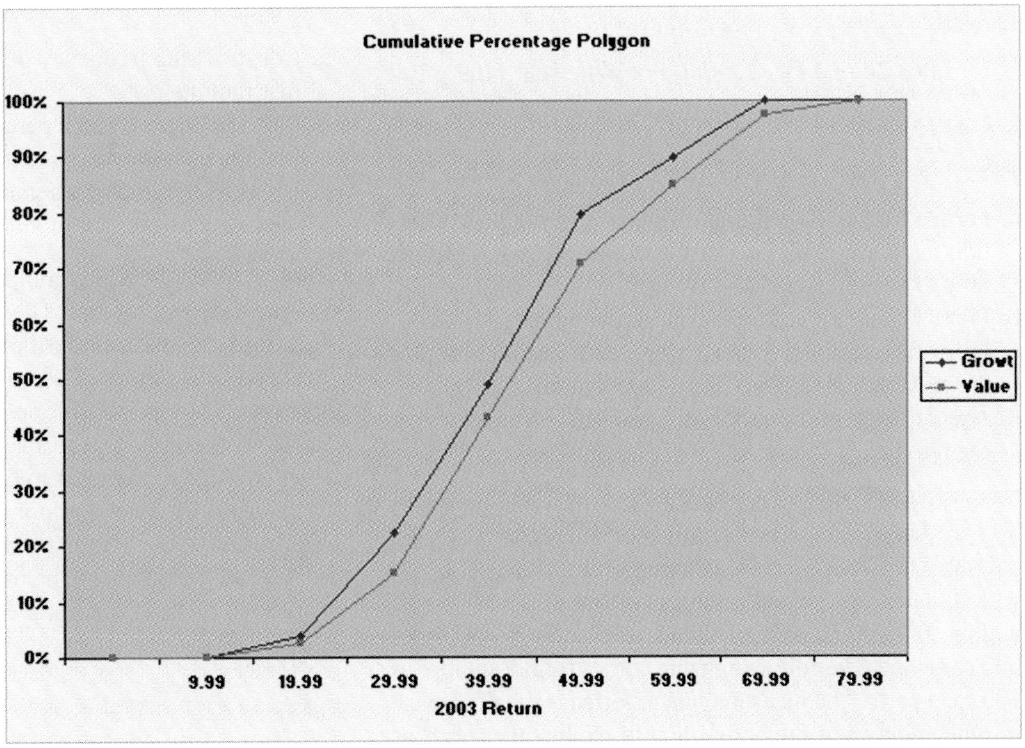

PROBLEMS FOR SECTION 2.3

Learning the Basics

PH Grade ASSIST **2.20** The values for a set of data vary from 11.6 to 97.8.

a. If these values are grouped into classes, indicate the class boundaries.
b. What class-interval width did you choose?
c. What are the nine class midpoints?

PH Grade ASSIST **2.21** In constructing an ogive (i.e., a cumulative percentage polygon) pertaining to the GMAT scores from a sample of 50 applicants to an MBA program, previous data indicated that none of the applicants scored below 450. The frequency distribution was formed by choosing class intervals 450 to 499, 500 to 549, and so on, with the last class grouping being 700 to 749. If two applicants scored in the interval 450 to 499 and 16 applicants scored in the interval 500 to 549:

a. What percentage of applicants scored below 500?
b. What percentage of applicants scored between 500 and 549?
c. What percentage of applicants scored below 550?
d. What percentage of applicants scored below 750?

Applying the Concepts

You can solve problems 2.22–2.27 manually or by using Microsoft Excel, Minitab, or SPSS.

PH Grade ASSIST **2.22** The data displayed below represent the cost of electricity during July 2004 for a random sample of 50 one-bedroom apartments in a large city. UTILITY

Raw Data on Utility Charges ($)

96	171	202	178	147	102	153	197	127	82
157	185	90	116	172	111	148	213	130	165
141	149	206	175	123	128	144	168	109	167
95	163	150	154	130	143	187	166	139	149
108	119	183	151	114	135	191	137	129	158

a. Form a frequency distribution and a percentage distribution that have class intervals with the upper class limits $99, $119, and so on.
b. Plot a histogram and a percentage polygon.
c. Form the cumulative percentage distribution and plot the ogive (cumulative percentage polygon).
d. Around what amount does the monthly electricity cost seem to be concentrated?

SELF Test **2.23** One operation of a mill is to cut pieces of steel into parts that will later be used as the frame for front seats in an automobile. The steel is cut with a diamond saw and requires the resulting parts to be within ±0.005 inch of the length specified by the automobile company. The following table comes from a sample of 100 steel parts. The measurement reported is the difference in inches between the actual length of the steel part, as measured by a laser measurement device, and the specified length of the steel part. For example, the first value, −0.002, represents a steel part that is 0.002 inch shorter than the specified length. STEEL

−0.002	0.002	0.0005	−0.0015	−0.001
0.0005	0.001	0.001	−0.0005	−0.001
0.0025	0.001	0.0005	−0.0015	0.0005
0.001	0.001	0.001	−0.0005	−0.0025
0.002	−0.002	0.0025	−0.0005	0.0025
0.001	−0.003	0.001	−0.001	0.002
0.005	−0.0015	0	−0.0015	0.0025
−0.002	−0.0005	−0.0025	0.0025	−0.002
0	0	−0.001	0.001	0
0.001	−0.0025	0.0035	0.0005	−0.0005
−0.0025	−0.003	0	0	−0.001
−0.003	−0.001	−0.003	0.002	0
0.001	0.002	−0.002	−0.0005	−0.002
−0.0005	−0.001	−0.001	0.0005	0
0	0	−0.0015	0.0005	0
−0.003	0.003	−0.0015	0	0.002
−0.001	0.0015	−0.002	−0.0005	−0.003
0.0005	0	0.001	0.002	−0.0005
0.0025	0	−0.0025	0.001	−0.002
−0.0025	−0.0025	−0.0005	−0.0015	−0.002

a. Construct the frequency distribution and the percentage distribution.
b. Plot a histogram and a percentage polygon.
c. Plot the cumulative percentage polygon.
d. Is the steel mill doing a good job in meeting the requirements set by the automobile company? Explain.

2.24 A manufacturing company produces steel housings for electrical equipment. The main component part of the housing is a steel trough that is made out of a 14-gauge steel coil. It is produced using a 250-ton progressive punch press with a wipe-down operation putting two 90-degree forms in the flat steel to make the trough. The distance from one side of the form to the other is critical because of weatherproofing in outdoor applications. The company requires that the width of the trough be between 8.31 inches and 8.61 inches. The following are the widths of the troughs in inches for a sample of $n = 49$. TROUGH

8.312	8.343	8.317	8.383	8.348	8.410	8.351	8.373
8.481	8.422	8.476	8.382	8.484	8.403	8.414	8.419
8.385	8.465	8.498	8.447	8.436	8.413	8.489	8.414
8.481	8.415	8.479	8.429	8.458	8.462	8.460	8.444
8.429	8.460	8.412	8.420	8.410	8.405	8.323	8.420
8.396	8.447	8.405	8.439	8.411	8.427	8.420	8.498
8.409							

a. Construct the frequency distribution and the percentage distribution.
b. Plot a histogram and a percentage polygon.
c. Plot the cumulative percentage polygon.
d. What can you conclude about the number of troughs that will meet the company's requirements of troughs being between 8.31 and 8.61 inches wide?

2.25 The manufacturing company in problem 2.24 also produces electric insulators. If the insulators break when in use, a short circuit is likely to occur. To test the strength of the insulators, destructive testing in high-powered labs is carried out to determine how much *force* is required to break the insulators. Force is measured by observing how many pounds must be applied to the insulator before it breaks. The strength of 30 insulators tested are as follows. **FORCE**

1,870	1,728	1,656	1,610	1,634	1,784	1,522	1,696
1,592	1,662	1,866	1,764	1,734	1,662	1,734	1,774
1,550	1,756	1,762	1,866	1,820	1,744	1,788	1,688
1,810	1,752	1,680	1,810	1,652	1,736		

a. Construct the frequency distribution and the percentage distribution.
b. Plot a histogram and a percentage polygon.
c. Plot the cumulative percentage polygon.
d. What can you conclude about the strength of the insulators if the company requires a force measurement of at least 1,500 pounds before breaking?

2.26 The ordered arrays in the accompanying table deal with the life (in hours) of a sample of forty 100-watt lightbulbs produced by manufacturer A and a sample of forty 100-watt lightbulbs produced by manufacturer B. **BULBS**

Manufacturer A					Manufacturer B				
684	697	720	773	821	819	836	888	897	903
831	835	848	852	852	907	912	918	942	943
859	860	868	870	876	952	959	962	986	992
893	899	905	909	911	994	1,004	1,005	1,007	1,015
922	924	926	926	938	1,016	1,018	1,020	1,022	1,034
939	943	946	954	971	1,038	1,072	1,077	1,077	1,082
972	977	984	1,005	1,014	1,096	1,100	1,113	1,113	1,116
1,016	1,041	1,052	1,080	1,093	1,153	1,154	1,174	1,188	1,230

a. Form the frequency distribution and percentage distribution for each manufacturer using the following class-interval widths for each distribution:
(1) Manufacturer A: 650 but less than 750, 750 but less than 850, and so on.
(2) Manufacturer B: 750 but less than 850, 850 but less than 950, and so on.
b. Plot the percentage histograms on separate graphs and plot the percentage polygons on one graph.
c. Form the cumulative percentage distributions and plot the ogives on one graph.
d. Which manufacturer has bulbs with a longer life— manufacturer A or manufacturer B? Explain.

2.27 The following data represent the amount of soft drink in a sample of 50 2-liter bottles. **DRINK**

2.109	2.086	2.066	2.075	2.065	2.057	2.052	2.044
2.036	2.038	2.031	2.029	2.025	2.029	2.023	2.020
2.015	2.014	2.013	2.014	2.012	2.012	2.012	2.010
2.005	2.003	1.999	1.996	1.997	1.992	1.994	1.986
1.984	1.981	1.973	1.975	1.971	1.969	1.966	1.967
1.963	1.957	1.951	1.951	1.947	1.941	1.941	1.938
1.908	1.894						

a. Construct the frequency distribution and the percentage distribution.
b. Plot a histogram and a percentage polygon.
c. Form the cumulative percentage distribution and plot the cumulative percentage polygon.
d. On the basis of the results of (a) through (c), does the amount of soft drink filled in the bottles concentrate around specific values?

2.4 CROSS TABULATIONS

The study of patterns that may exist between two or more categorical variables is common in business.

The Contingency Table

A **cross-classification** (or **contingency**) **table** presents the results of two categorical variables. The joint responses are classified so that the categories of one variable are located in the rows and the categories of the other variable is located in the columns. The values located at the intersections of the rows and columns are called **cells**. Depending on the type of contingency table constructed, the cells for each row-column combination contain either the frequency, the percentage of the overall total, the percentage of the row total, or the percentage of the column total.

Suppose that in the "Using Statistics" scenario, you want to examine whether or not there is any pattern or relationship between level of risk and the objective of the mutual fund (growth versus value). Table 2.14 summarizes this information for all 121 mutual funds.

TABLE 2.14
Contingency Table Displaying Fund Objective and Fund Risk

	RISK LEVEL			
OBJECTIVE	**High**	**Average**	**Low**	**Total**
Growth	14	23	12	49
Value	3	23	46	72
Total	17	46	58	121

You construct this contingency table by tallying the joint responses for each of the 121 mutual funds with respect to objective and risk into one of the six possible cells in the table. Thus, the first fund listed (AFBA Five Star USA Global Institutional) is classified as a growth fund with an average risk. Thus, you tally this joint response into the cell that is the intersection of the first row and second column. The remaining 120 joint responses are recorded in a similar manner. Each cell contains the frequency for the row-column combination.

In order to further explore any possible pattern or relationship between objective and fund risk, you can construct contingency tables based on percentages. You first convert these results into percentages based on the following three totals:

1. The overall total (i.e., the 121 mutual funds)
2. The row totals (i.e., 49 growth funds and 72 value funds)
3. The column totals (i.e., the three levels of risk)

Tables 2.15, 2.16, and 2.17 summarize these percentages.

TABLE 2.15
Contingency Table Displaying Fund Objective and Fund Risk Based on Percentage of Overall Total

	RISK LEVEL			
OBJECTIVE	**High**	**Average**	**Low**	**Total**
Growth	11.57	19.01	9.92	40.50
Value	2.48	19.01	38.02	59.50
Total	14.05	38.02	47.93	100.00

TABLE 2.16
Contingency Table Displaying Fund Objective and Fund Risk Based on Percentage of Row Total

	RISK LEVEL			
OBJECTIVE	**High**	**Average**	**Low**	**Total**
Growth	28.57	46.94	24.49	100.00
Value	4.17	31.94	63.89	100.00
Total	14.05	38.02	47.93	100.00

TABLE 2.17
Contingency Table Displaying Fund Objective and Fund Risk Based on Percentage of Column Total

	RISK LEVEL			
OBJECTIVE	**High**	**Average**	**Low**	**Total**
Growth	82.35	50.00	20.69	40.50
Value	17.65	50.00	79.31	59.50
Total	100.00	100.00	100.00	100.00

Table 2.15 shows that 14.05% of the mutual funds sampled are high risk, 40.5% are growth funds, and 11.57% are high-risk funds that are growth funds. Table 2.16 shows that 28.57% of the growth funds are high risk and 24.49% are low risk. Table 2.17 shows that 82.35% of the high-risk funds, and only 20.69% of the low-risk funds are growth funds. The tables reveal that growth funds are more likely to be high risk while value funds are more likely to be low risk.

The Side-by-Side Bar Chart

A useful way to visually display the results of cross-classification data is by constructing a **side-by-side bar chart**. Figure 2.14, which uses the data from Table 2.14, is a Microsoft Excel side-by-side bar chart that compares the three fund risk levels based on their objective. An examination of Figure 2.14 reveals results consistent with those of Tables 2.15, 2.16, and 2.17. Growth funds are more likely to be high risk, while value funds are more likely to be low risk.

FIGURE 2.14
Microsoft Excel Side-by-Side Bar Chart for Fund Objective and Risk

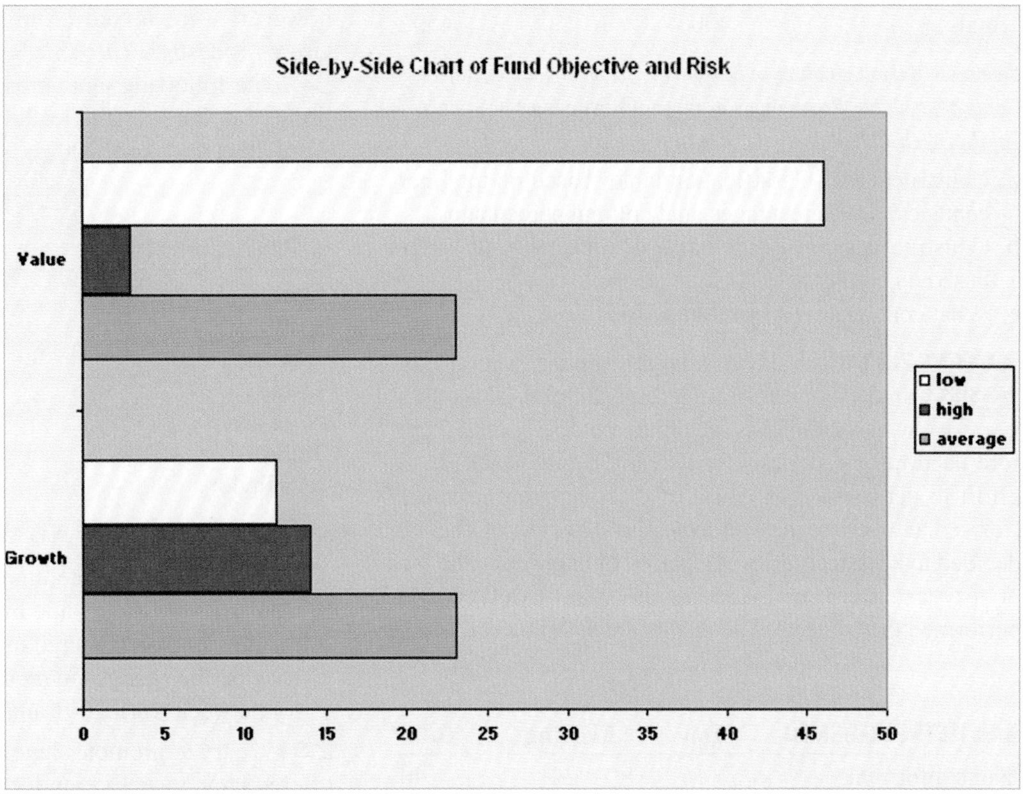

PROBLEMS FOR SECTION 2.4

Learning the Basics

PH Grade ASSIST **2.28** The following data represent the responses to two questions asked in a survey of 40 college students majoring in business—What is your gender? (Male = M; Female = F) and What is your major? (Accountancy = A; Computer Information Systems = C; Marketing = M):

Gender: M M M F M F F M F M F M M M M F F M F F

Major: A C C M A C A A C C A A A M C M A A A C

Gender: M M M M F M F F M M F M M M M F M F M M

Major: C C A A M M C A A A C C A A A A C C A C

a. Tally the data into a contingency table where the two rows represent the gender categories and the three columns represent the academic-major categories.

b. Form cross-classification tables based on percentages of all 40 student responses, based on row percentages, and based on column percentages.

c. Using the results from (a), construct a side-by-side bar chart of gender based on student major.

PH Grade ASSIST **2.29** Given the following cross-classification table, construct a side-by-side bar chart comparing *A* and *B* for each of the three-column categories on the vertical axis.

	1	2	3	Total
A	20	40	40	100
B	80	80	40	200

Applying the Concepts

PH Grade ASSIST SELF Test **2.30** The results of a study made as part of a yield improvement effort at a semiconductor manufacturing facility provided defect data for a sample of 450 wafers. The following table presents a summary of the responses to two questions: Was a particle found on the die that produced the wafer? and Is the wafer good or bad?

	CONDITION OF DIE		
QUALITY OF WAFER	**No Particles**	**Particles**	**Totals**
Good	320	14	334
Bad	80	36	116
Totals	400	50	450

Source: S. W. Hall, Analysis of Defectivity of Semiconductor Wafers by Contingency Table, Proceedings Institute of Environmental Sciences, *Vol. 1 (1994), 177–183.*

a. Construct cross-classification tables based on total percentages, row percentages, and column percentages.
b. Construct a side-by-side bar chart of quality of wafers based on condition of die.
c. What conclusions do you draw from these analyses?

PH Grade ASSIST **2.31** Each day at a large hospital, several hundred laboratory tests are performed. The rate at which these tests are done improperly (and therefore need to be redone) seems steady, at about 4%. In an effort to get to the root cause of these nonconformances (tests that need to be redone), the director of the lab decided to keep records over a period of one week. The laboratory tests were subdivided by the shift of workers who performed the lab tests. The results are as follows:

	SHIFT		
LAB TESTS PERFORMED	**Day**	**Evening**	**Total**
Nonconforming	16	24	40
Conforming	654	306	960
Total	670	330	1,000

a. Construct cross-classification tables based on total percentages, row percentages, and column percentages.
b. Which type of percentage—row, column, or total—do you think is most informative for these data? Explain.
c. What conclusions concerning the pattern of nonconforming laboratory tests can the laboratory director reach?

2.32 A sample of 500 shoppers was selected in a large metropolitan area to determine various information concerning consumer behavior. Among the questions asked was "Do you enjoy shopping for clothing?" The results are summarized in the following cross-classification table:

	GENDER		
ENJOY SHOPPING FOR CLOTHING	**Male**	**Female**	**Total**
Yes	136	224	360
No	104	36	140
Total	240	260	500

a. Construct cross-classification tables based on total percentages, row percentages, and column percentages.
b. Construct a side-by-side bar chart of enjoy shopping for clothing based on gender.
c. What conclusions do you draw from these analyses?

2.33 Retail sales in the United States for April 2002 were slightly higher than April 2001. Discounters such as Wal-Mart, Costco, Target, and Dollar General all had increased sales of 9% or more. Retail sales in the apparel industry, however, were mixed (Ann Zimmerman, "Retail Sales Grow Modestly," *The Wall Street Journal*, May 10, 2002, B4). The following table presents total retail sales in millions of dollars for the leading apparel companies during April 2001 and April 2002.

	TOTAL SALES IN MILLIONS $$$	
APPAREL COMPANY	**April 01**	**April 02**
Gap	1,159.0	962.0
TJX	781.7	899.0
Limited	596.5	620.4
Kohl's	544.9	678.9
Nordstrom	402.6	418.3
Talbots	139.9	130.1
AnnTaylor	114.2	124.8

Source: Extracted from The Wall Street Journal.

a. Construct a table of column percentages.
b. Construct a side-by-side bar chart to visually highlight the information gathered in (a).
c. Discuss the changes in retail sales for the apparel industry between April 2001 and April 2002.

2.34 To try to promote sluggish sales in 2003, automakers offered large incentives in the form of cash rebates to customers purchasing new cars. For example, buyers of new Lincolns received an average cash rebate of $4,086. In spite of these rebates, the American automakers still lost global market share to international competition.

	CASH REBATES (IN DOLLARS)	
BRAND	**2001**	**2003**
Buick	1,939	3,655
Chevrolet	1,654	3,231
Chrysler	1,835	2,832
Ford	1,334	2,752
Lincoln	2,449	4,086

Source: K. Lundegaard and S. Freeman, "Detroit's Challenge: Weaning Buyers from Years of Deals," The Wall Street Journal, *January 6, 2004, A1.*

a. Construct a side-by-side bar chart for the five brands.
b. Discuss the changes in the size of cash rebates from 2001 to 2003.

2.35 U.S. car sales rose 3.3% in January 2004 compared to January 2003. Japanese automakers experienced a much larger increase. The following table contains the sales of cars and light trucks from some big automakers during January 2003 and 2004.

CAR MAKER	SALES OF NEW CARS AND LIGHT TRUCKS 2003	2004
Nissan	55,213	72,164
Honda	89,993	90,173
Toyota	119,376	143,729
Chrysler	144,826	162,205
Ford	242,068	229,238
GM	291,254	296,788

Source: Adapted from S. Freeman and J. B. White, "U.S. Car Sales Rose 3.3% in January," The Wall Street Journal, *February 4, 2004, A2.*

a. Construct a side-by-side bar chart for the six brands.
b. Discuss the changes in the sales of new cars and light trucks in January 2004 compared to January 2003.

2.5 SCATTER DIAGRAMS AND TIME-SERIES PLOTS

The Scatter Diagram

When analyzing a single numerical variable such as the cost of a restaurant meal or the 2003 return, you use the histogram, the polygon, and the cumulative percentage polygon developed in section 2.3. You use the **scatter diagram** to examine possible relationships between two numerical variables. You plot one variable on the horizontal *X* axis and the other variable on the vertical *Y* axis. For example, a marketing analyst could study the effectiveness of advertising by comparing weekly sales volumes and weekly advertising expenditures. Or, a human resources director interested in the salary structure of the company could compare the employees' years of experience with their current salaries.

To demonstrate a scatter diagram, you can study the relationship between the expense ratio and 2003 return. For each mutual fund, you plot the expense ratio on the horizontal *X* axis, and the 2003 return on the vertical *Y* axis. Figure 2.15 represents Microsoft Excel output for these two variables.

FIGURE 2.15
Microsoft Excel Scatter Diagram of Expense Ratio and 2003 Return

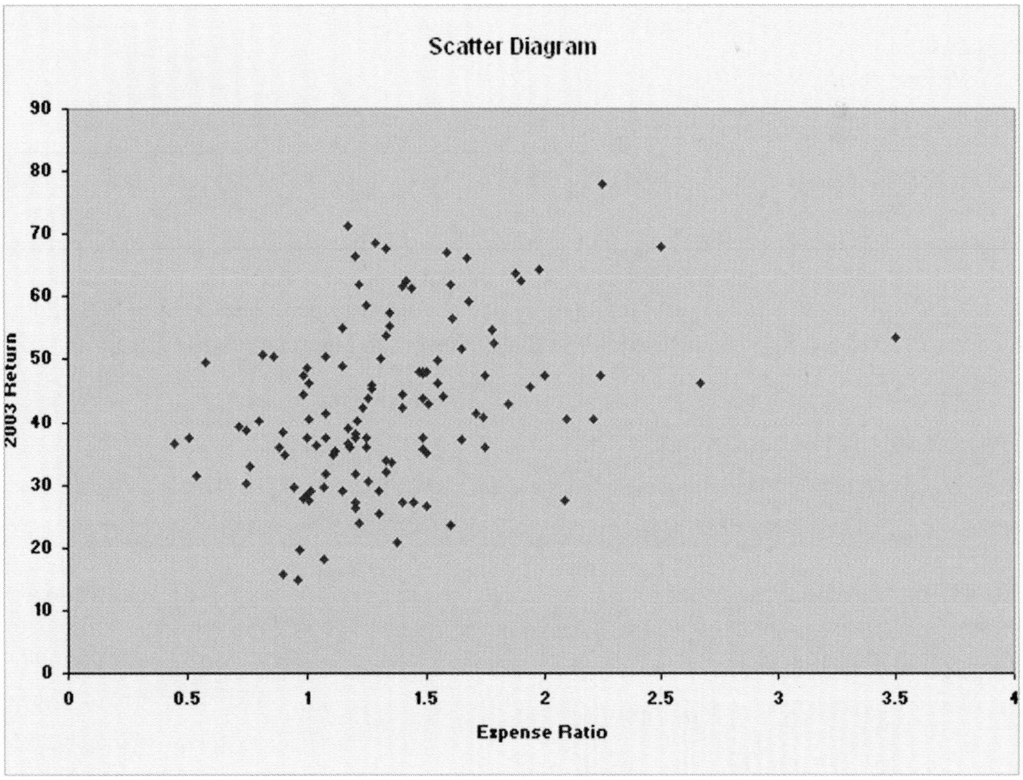

Although there is a great deal of variation in the expense ratio and 2003 return of the mutual funds, there appears to be an increasing (positive) relationship between the expense ratio and 2003 return. In other words, funds that have a low expense ratio have a low 2003 return. Other pairs of variables may have a decreasing (negative) relationship in which one variable decreases as the other increases. The scatter diagram is revisited in Chapter 3 when the coefficient of correlation and the covariance are studied, and in Chapter 13 when regression analysis is developed.

The Time-Series Plot

The **time-series plot** is used to study patterns in the values of a variable over time. Each value is plotted as a point in two dimensions. A time-series plot displays the time period on the horizontal X axis and the variable of interest on the vertical Y axis.

Figure 2.16 is a time-series plot of the monthly mortgage payment (in 2002 dollars) from 1988 to 2002. HOUSESNY

FIGURE 2.16
Microsoft Excel Time-Series Plot of the Monthly Mortgage Payment in 2002 Dollars from 1988 to 2002

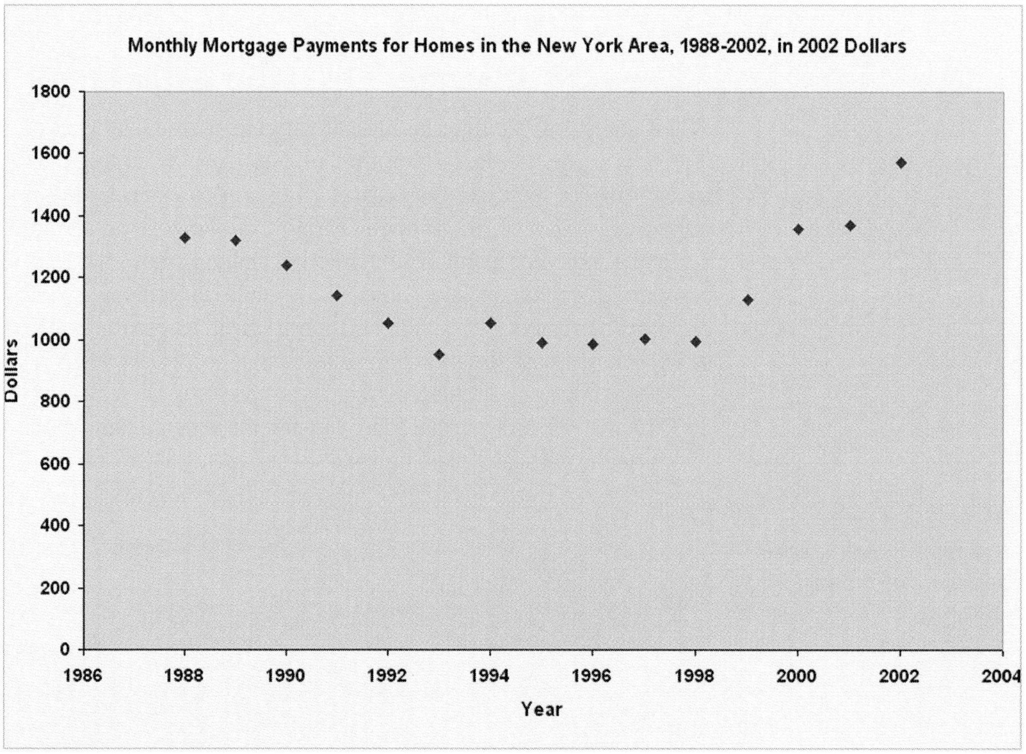

Monthly mortgage payments (when considered in 2002 dollars) dropped in the late 1980s and early 1990s, only to level off. They started rising again since 2000.

PROBLEMS FOR SECTION 2.5

Learning the Basics

PH Grade ASSIST **2.36** The following is a set of data from a sample of $n = 11$ items.

X	7	5	8	3	6	10	12	4	9	15	18
Y	21	15	24	9	18	30	36	12	27	45	54

a. Plot the scatter diagram.
b. Is there a relationship between X and Y? Explain.

PH Grade ASSIST **2.37** The following is a series of real annual sales (in millions of constant 1995 dollars) over an 11-year period (1992 to 2002).

Year	1992	1993	1994	1995	1996	1997	1998	1999	2000	2001	2002
Sales	13.0	17.0	19.0	20.0	20.5	20.5	20.5	20.0	19.0	17.0	13.0

a. Construct a time-series plot.
b. Does there appear to be any change in real annual sales over time? Explain.

Applying the Concepts

You can solve problems 2.38–2.45 manually or by using Microsoft Excel, Minitab or SPSS.

2.38 The following data represent the approximate retail price (in $) and the energy cost per year (in $) of 15 refrigerators. **REFRIGERATOR**

Model	Price	Energy Cost
Maytag MTB1956GE	825	36
Kenmore7118	750	43
Maytag MTB2156GE	850	39
Kenmore Elite	1000	38
Amana ART2107B	800	38
GE GTS18KCM	600	40
Kenmore 7198	750	35
Frigidaire Gallery GLHT216TA	680	38
Kenmore 7285	680	40
Whirlpool Gold GR9SHKXK	940	37
Frigidaire Gallery GLRT216TA	680	40
GE GTS22KCM	650	44
Whirlpool ETF1TTXK	800	43
Whirlpool Gold GR2SHXK	1050	40
Frigidaire FRT18P5A	510	40

Source: "Refrigerators," Copyright 2002 by Consumers Union of U.S., Inc., Yonkers, NY 10703–1057. Adapted with permission from Consumer Reports, *August 2002, 26.*

a. Construct a scatter diagram with energy cost on the *X* axis and price on the *Y* axis.
b. Does there appear to be a relationship between price and energy cost? If so, is the relationship positive or negative?
c. Would you expect the higher-priced refrigerators to have the greatest energy efficiency? Is this borne out by the data?

2.39 The following data **SECURITY** represent the turnover rate of pre-boarding screeners at airports in 1998 and 1999 and the security violations detected per million passengers.

City	Turnover	Violations
St. Louis	416	11.9
Atlanta	375	7.3
Houston	237	10.6
Boston	207	22.9
Chicago	200	6.5
Denver	193	15.2
Dallas	156	18.2
Baltimore	155	21.7
Seattle/Tacoma	140	31.5
San Francisco	110	20.7
Orlando	100	9.9
Washington–Dulles	90	14.8
Log Angeles	88	25.1
Detroit	79	13.5
San Juan	70	10.3
Miami	64	13.1
New York–JFK	53	30.1
Washington–Reagan	47	31.8
Honolulu	37	14.9

Source: Alan B. Krueger, "A Small Dose of Common Sense Would Help Congress Break the Gridlock over Airport Security," The New York Times, *November 15, 2001, C2.*

a. Construct a scatter diagram with turnover rate of pre-boarding screeners on the *X* axis and security violations detected on the *Y* axis.
b. What conclusions can you reach about the relationship between the turnover rate of pre-boarding screeners and the security violations detected?

SELF Test **2.40** The following data **CELLPHONE** represents the digital-mode talk time in hours and the battery capacity in milliampere-hours of cellphones.

Talk Time	Battery Capacity	Talk Time	Battery Capacity
4.50	800	1.50	450
4.00	1500	2.25	900
3.00	1300	2.25	900
2.00	1550	3.25	900
2.75	900	2.25	700
1.75	875	2.25	800
1.75	750	2.50	800
2.25	1100	2.25	900
1.75	850	2.00	900

Source: "Service Shortcomings," Copyright 2002 by Consumers Union of U.S., Inc., Yonkers, NY 10703–1057. Adapted with permission from Consumer Reports, *February 2002, 25.*

a. Construct a scatter diagram with battery capacity on the *X* axis and digital-mode talk time on the *Y* axis.
b. What conclusions can you reach about the relationship between the battery capacity and the digital-mode talk time?
c. You would expect the cellphones with higher battery capacity to have a higher talk time. Is this borne out by the data?

2.41 The following data **BATTERIES2** represents the price and cold-cranking amps (which denotes the starting current the battery can deliver) of automobile batteries.

Name	Price ($)	Cca
NAPA Legend Professional Line 7575	60	630
Exide Nascar Select 75-84N	80	630
DieHard Weatherhandler 30375 (South)	60	525
DieHard Weatherhandler 30075 (North)	60	650
EverStart 75-5	30	525
Duralast 75-D	50	650
Interstate Mega-Tron MT-75	80	650
EverStart 75-2	60	650
ACDelco Maintenance free 75A-72	80	650
Motorcraft Premier Silver Series BXT-75	80	700
DieHard Gold 33165 (South)	80	700
EverStart Extreme 65-2N (North)	60	850
ACDelco Maintenance Free 65-84	92	850
Exide 65-60	85	850
EverStart Extreme 65-2 (South)	60	675
DieHard Gold 33065 (North)	80	900
Duralast Gold 34DT-DGS (South)	70	800
Duralast Gold 34DT-DGN (North)	70	900
Interstate Mega-Tron Plus MTP-78DT	96	800
Optima Red Top 34/78-1050	140	750
ACDelco Professional 78DT-7YR	80	850
EverStart High Power DT-3	40	630
DieHard Weatherhandler 30034 (North)	60	540
DieHard Weatherhandler 30334 (South)	60	525

Source: "Leading the Charge," Copyright 2001 by Consumers Union of U.S., Inc., Yonkers, NY 10703–1057. Adapted with permission from Consumer Reports, *October 2001, 25.*

a. Set up a scatter diagram with cold-cranking amps on the *X* axis and price on the *Y* axis.
b. What conclusions can you reach about the relationship between the cold-cranking amps and price?
c. You would expect the batteries with higher cold-cranking amps to have a higher price. Is this borne out by the data?

2.42 The U.S. Bureau of Labor Statistics compiles data on a wide variety of workforce issues. The following table gives the monthly seasonally adjusted civilian unemployment rate for the United States from 1998 to 2003. UERATE

Seasonally Adjusted U.S. Unemployment Rate (in %)

Month	1998	1999	2000	2001	2002	2003
January	4.7	4.3	4.0	4.2	5.6	5.9
February	4.6	4.4	4.1	4.2	5.6	6.0
March	4.7	4.2	4.0	4.3	5.7	5.9
April	4.3	4.3	4.0	4.5	5.9	6.1
May	4.4	4.2	4.1	4.4	5.8	6.2
June	4.5	4.3	4.0	4.5	5.8	6.4
July	4.5	4.3	4.0	4.5	5.8	6.3
August	4.5	4.2	4.1	4.9	5.8	6.2
September	4.5	4.2	3.9	4.9	5.7	6.2
October	4.5	4.1	3.9	5.4	5.8	6.1
November	4.4	4.1	4.0	5.6	5.9	6.1
December	4.4	4.1	4.0	5.8	6.0	5.8

Source: U.S. Bureau of Labor Statistics.

a. Construct a time-series plot of the U.S. unemployment rate.
b. Does there appear to be any pattern?

2.43 The following data DRINK represent the amount of soft drink filled in a sample of 50 consecutive 2-liter bottles. The results are listed horizontally in the order of being filled.

2.109	2.086	2.066	2.075	2.065	2.057	2.052	2.044	2.036	2.038
2.031	2.029	2.025	2.029	2.023	2.020	2.015	2.014	2.013	2.014
2.012	2.012	2.012	2.010	2.005	2.003	1.999	1.996	1.997	1.992
1.994	1.986	1.984	1.981	1.973	1.975	1.971	1.969	1.966	1.967
1.963	1.957	1.951	1.951	1.947	1.941	1.941	1.938	1.908	1.894

a. Construct a time-series plot for the amount of soft drink on the *Y* axis and the bottles number (going consecutively from 1 to 50) on the *X* axis.
b. What pattern, if any, is present in the data?
c. If you had to make a prediction of the amount of soft drink filled in the next bottle, what would you predict?
d. Based on the results of (a) through (c), explain why it is important to construct a time-series plot and not just a histogram as was done in problem 2.27 on page 43.

2.44 The data in the following table represent the number of U.S. households actively using online banking and/or online bill payment from 1995 to 2003. ONLINEBANKING

Year	Number of Households (millions)
1995	0.6
1996	2.5
1997	4.5
1998	7.0
1999	10.5
2000	15.5
2001	22.0
2002	28.0
2003	33.0

Source: Extracted from R. J. Dalton, "In the Mainstream," Newsday, *February 8, 2004, F6–F7.*

a. Construct a time-series plot for the number of U.S. households actively using online banking and/or online bill payment.
b. What pattern, if any, is present in the data?
c. If you had to make a prediction of the number of U.S. households actively using online banking and/or online bill payment in 2004, what would you predict?

2.45 The data in the following table represent the average number of television viewers (excluding local broadcasts) per game (in millions) for the National Football League (NFL), the National Basketball Association (NBA), Major League Baseball (MLB), and the National Hockey League (NHL). SPORTSTV

Year	NFL	NBA	MLB	NHL
1995	19.6	10.6	15.9	3.6
1996	18.5	10.2	9.8	3.2
1997	17.4	10.8	10.4	2.4
1998	18.1	7.8	9.4	2.6
1999	18.3	7.2	10.0	3.3
2000	17.0	6.7	7.7	2.8
2001	16.9	6.8	9.8	3.1
2002	18.6	5.8	8.9	2.6

Source: Extracted from S. Fatsis, "Salaries, Promos, and Flying Solo," The Wall Street Journal, *February 9, 2004, R.4.*

a. For each of the four sports, construct a time-series plot.
b. What pattern, if any, is present in the data?
c. If you had to make a prediction of the number of viewers for each sport in 2003, what would you predict?

2.6 MISUSING GRAPHS AND ETHICAL ISSUES

Good graphical displays reveal what the data are conveying. Unfortunately many graphs presented in newspapers and magazines as well as graphs that can be developed using the Chart Wizard of Microsoft Excel either are incorrect, misleading, or are so unnecessarily complicated that they never should be used. To illustrate the misuse of graphs, the first graph presented is one that was printed in *Time* magazine as part of an article on increasing exports of wine from Australia to the United States.

FIGURE 2.17

Figure 2.17 "Improper" Display of Australian Wine Exports to the United States in Millions of Gallons

Source: *Adapted from S. Watterson, "Liquid Gold—Australians Are Changing the World of Wine. Even the French Seem Grateful,"* Time, *November 22, 1999, 68.*

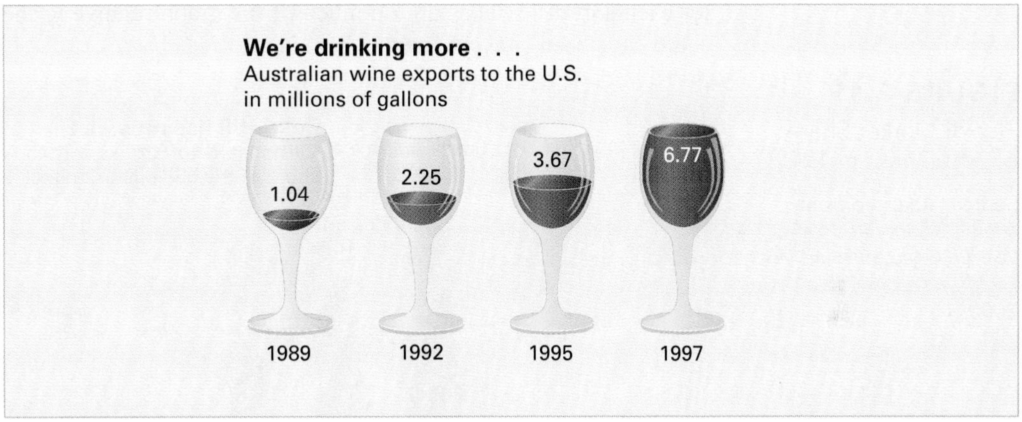

In Figure 2.17, the wineglass icon representing the 6.77 million gallons for 1997 does not appear to be almost twice the size of the wineglass icon representing the 3.67 million gallons for 1995, nor does the wineglass icon representing the 2.25 million gallons for 1992 appear to be twice the size of the wineglass icon representing the 1.04 million gallons for 1989. Part of the reason for this is that the three-dimensional wineglass icon is used to represent the two dimensions of exports and time. Although the wineglass presentation may catch the eye, the data should be presented in a summary table or a time-series plot.

In addition to the type of distortion created by the wineglass icons in the *Time* magazine graph displayed in Figure 2.17, improper use of the vertical and horizontal axes leads to distortions. Figure 2.18 on page 52 presents another graph used in the same *Time* magazine article.

There are several problems in the graph. First, there is no zero point on the vertical axis. Second, the acreage of 135,326 for 1949 to 1950 is plotted above the acreage of 150,300 for 1969 to 1970. Third, it is not obvious that the difference between 1979 to 1980 and 1997 to 1998 (71,569 acres) is approximately three and a half times the difference between 1979–1980 and 1969–1970 (21,775 acres). Fourth, there are no scale values on the horizontal axis. Years are plotted next to the acreage totals, not on the horizontal axis. Fifth, the values for the time dimension are not properly spaced along the horizontal axis. The value for 1979–1980 is much closer to 1990 than it is to 1969–1970.

FIGURE 2.18

"Improper" Display of Amount of Land Planted with Grapes for the Wine Industry

Source: *Adapted from S. Watterson, "Liquid Gold—Australians Are Changing the World of Wine. Even the French Seem Grateful,"* Time, *November 22, 1999, 68–69.*

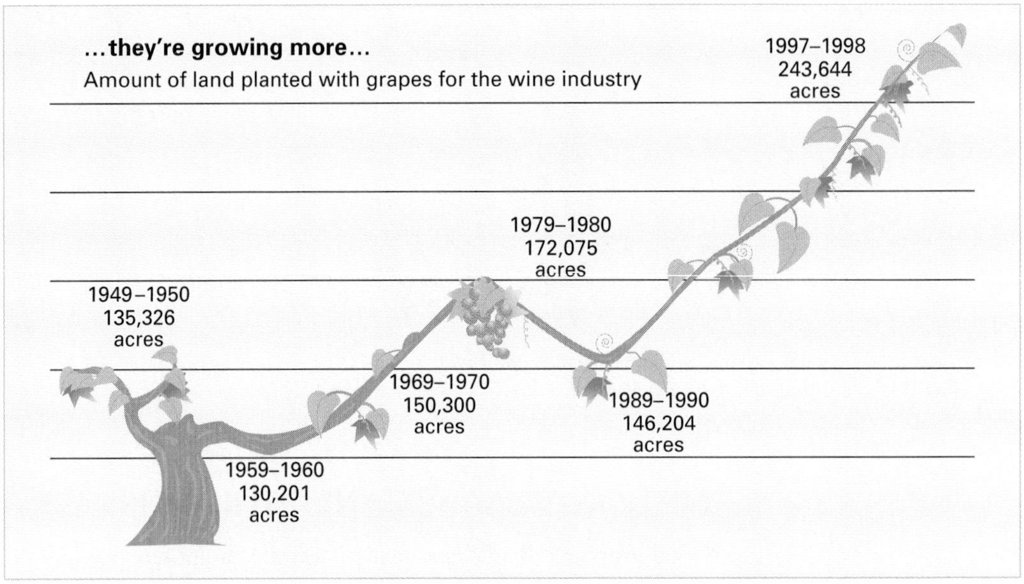

Other types of eye-catching displays that you typically see in magazines and newspapers often include information that is not necessary and just add excessive clutter. Figure 2.19 represents one such display. The graph in Figure 2.19 shows those products with the largest market share for soft drinks in 1999. The graph suffers from too much clutter although it is designed to show the differences in market share among the soft drinks. The display of the fizz for each soft drink takes up too much of the graph relative to the data. The same information could have been conveyed with a bar chart or pie chart.

FIGURE 2.19

Plot of Market Share of Soft Drinks in 1999

Source: *Adapted from Anne B. Carey and Sam Ward, "Coke Still Has Most Fizz,"* USA Today, *May 10, 2000, 1B.*

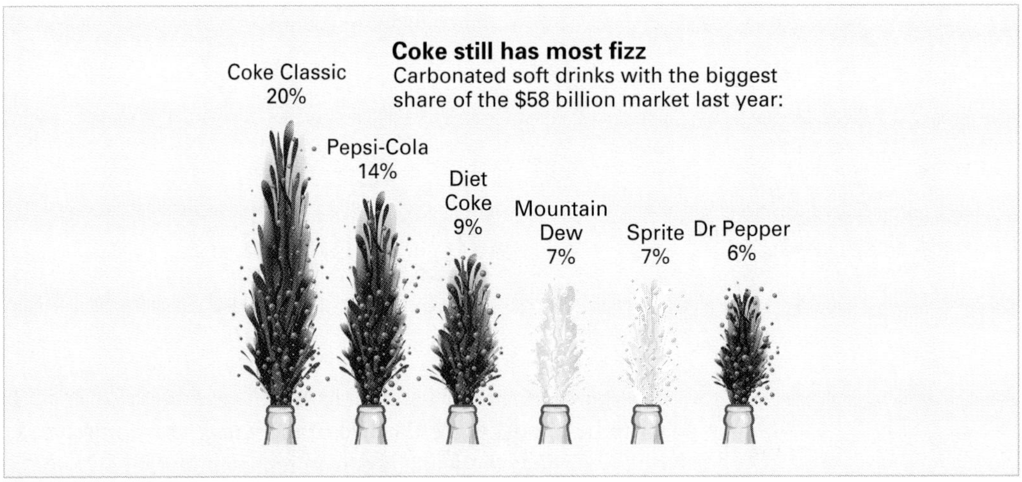

Some guidelines for developing good graphs are as follows:

- The graph should not distort the data.
- The graph should not contain unnecessary adornments (sometimes referred to as **chartjunk**).
- Any two-dimensional graph should contain a scale for each axis.
- The scale on the vertical axis should begin at zero.
- All axes should be properly labeled.
- The graph should contain a title.
- The simplest possible graph should be used for a given set of data.

One of the biggest sources of improper graphs is the Chart Wizard of Microsoft Excel. Figure 2.20 represents the Step 1 dialog box of the Chart Wizard. You can select from among column, bar, line, pie, and area charts as well as more complicated types such as doughnut, radar, surface, bubble, stock, cylinder, cone, and pyramid charts. These more complicated charts should rarely be used since they are harder to interpret than the simpler charts covered in this chapter.

FIGURE 2.20
Step 1 Dialog Box of the Microsoft Excel Chart Wizard

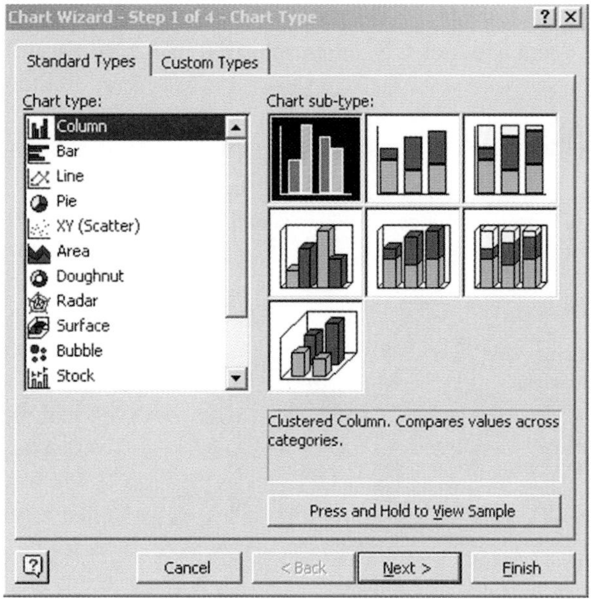

Most instances of misleading graphs are the result of people not being aware of the guidelines for creating good graphs. Ethical issues arise, however, when charts are constructed to purposely mislead the reader. In either case, you must use extreme caution when trying to draw conclusions from graphs that deviate from the guidelines given in this chapter.

PROBLEMS FOR SECTION 2.6

Applying the Concepts

2.46 (Student Project) Bring a chart to class from a newspaper or magazine that you believe to be a poorly drawn representation of a numerical variable. Be prepared to submit the chart to the instructor with comments as to why you believe it is inappropriate. Do you believe that the intent of the chart is to purposely mislead the reader? Also, be prepared to present and comment on this in class.

2.47 (Student Project) Bring a chart to class from a newspaper or magazine that you believe to be a poorly drawn representation of a categorical variable. Be prepared to submit the chart to the instructor with comments as to why you consider it inappropriate. Do you believe that the intent of the chart is to purposely mislead the reader? Also, be prepared to present and comment on this in class.

2.48 (Student Project) Bring a chart to class from a newspaper or magazine that you believe to contain too many unnecessary adornments (i.e., chartjunk) that may cloud the message given by the data. Be prepared to submit the chart to the instructor with comments about why you think it is inappropriate. Also, be prepared to present and comment on this in class.

2.49 The following visual display contains an overembellished chart that appeared in *USA Today* dealing with the number of deaths from lightning strikes in the United States.

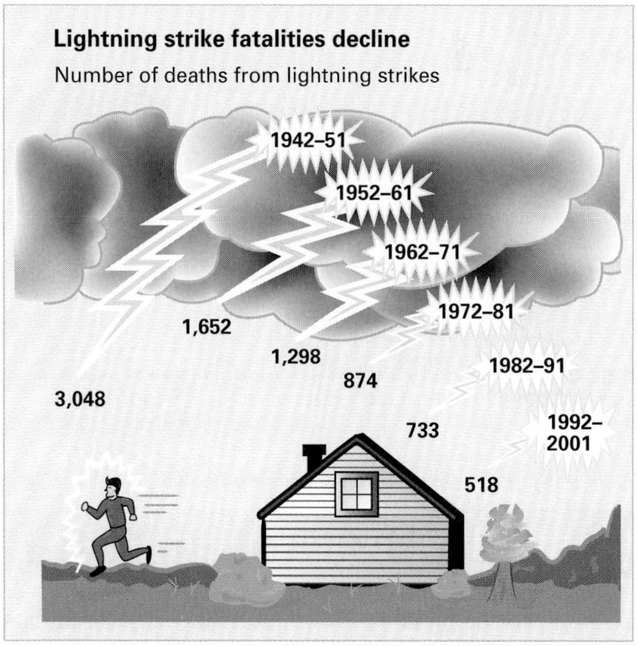

a. Describe at least one good feature of this visual display.
b. Describe at least one bad feature of this visual display.
c. Redraw the graph, using the guidelines given on page 52.

2.50 The following visual display concerning the relative size of police departments in major U.S. cities appeared in *USA Today:*

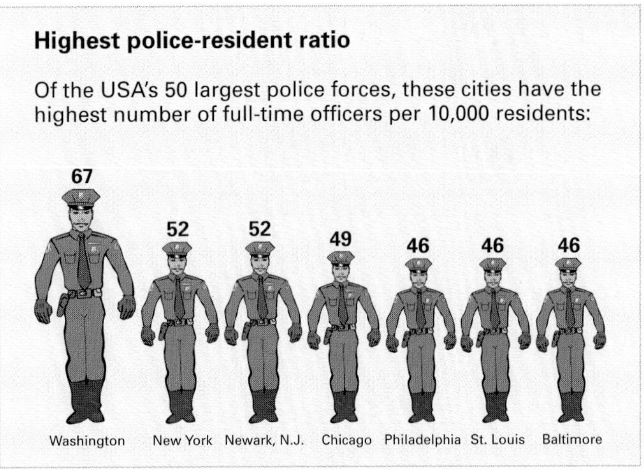

Source: Adapted from USA Today, *February 2000.*

a. Indicate a feature of this chart that violates the principles of good graphs.

b. Set up an alternative graph for the data provided in this figure.

2.51 The following visual display concerning where the United States gets its electricity appeared in *USA Today:*

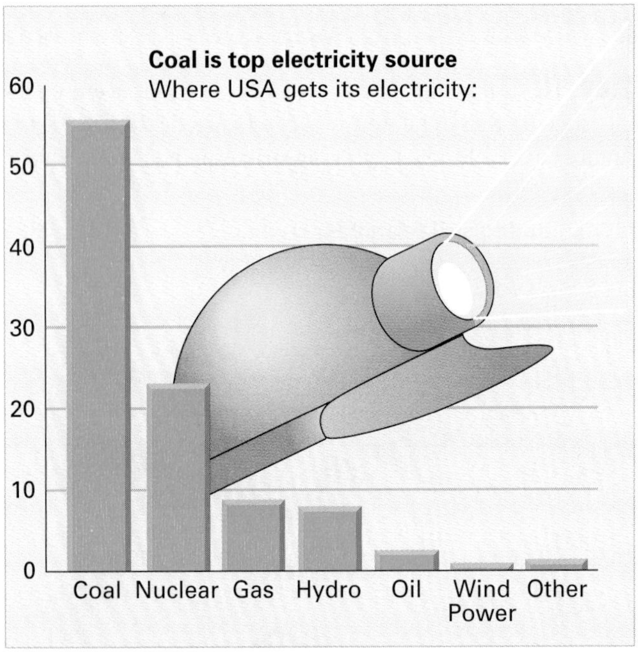

Source: Adapted from USA Today, *January 30, 2002.*

a. Describe at least one good feature of this visual display.

b. Describe at least one bad feature of this visual display.

c. Redraw the graph using the guidelines given on page 52.

2.52 An article in *The New York Times* (Donna Rosato, "Worried about the Numbers? How about the Charts?" *The New York Times*, September 15, 2002, Business 7) reported on research done on annual reports of corporations by Professor Deanna Oxender Burgess of Florida Gulf Coast University. Professor Burgess found that even slight distortions in a chart changed readers' perception of the information. The article displayed sales information from the annual report of Zale Corporation and showed how results were exaggerated.

Go to the World Wide Web or the library and study the most recent annual report of a selected corporation. Find at least one chart in the report that you think needs improvement and develop an improved chart. Explain why you believe the improved chart is better than the one included in the annual report.

2.53 Figures 2.1, 2.3, and 2.6 consist of a bar chart, a pie chart, and a Pareto diagram for the online shopping data.

a. Use the Chart Wizard of Microsoft Excel to develop a doughnut chart, a cone chart, and a pyramid chart for the online shopping data.

b. Which graphs do you prefer—the bar chart, the pie chart, and the Pareto diagram or the doughnut chart, a cone chart, and a pyramid chart? Explain.

2.54 Figures 2.2 and 2.4 consist of a bar chart and a pie chart for the risk level for the mutual fund data. **MUTUALFUNDS2004**

a. Use the Chart Wizard of Microsoft Excel to develop a doughnut chart, a cone chart, and a pyramid chart for the risk level of the mutual funds.

b. Which graphs do you prefer—the bar chart and pie chart or the doughnut chart, a cone chart, and a pyramid chart? Explain.

SUMMARY

As you can see in Table 2.18, this chapter discussed data presentation. You have used various tables and charts to draw conclusions about online shopping, the cost of restaurant meals in a city and its suburbs, and the set of mutual funds that were first introduced in the "Using Statistics" scenario at the beginning of the chapter. Now that you have studied tables and charts, in Chapter 3 you will learn about a variety of numerical descriptive measures useful for data analysis and interpretation.

TABLE 2.18
Roadmap for Selecting Tables and Charts

	Type of Data	
Type of Analysis	**Numerical**	**Categorical**
Tabulating, organizing, and graphically presenting the values of a variable	Ordered array, stem-and-leaf display, frequency distribution, relative frequency distribution, percentage distribution, cumulative percentage distribution, histogram, polygon, cumulative percentage polygon **(sections 2.2 and 2.3)**	Summary table, bar chart, pie chart, Pareto diagram **(section 2.1)**
Graphically presenting the relationship between two variables	Scatter diagram, time-series plot **(section 2.5)**	Contingency table, side-by-side bar chart **(section 2.4)**

KEY TERMS

bar chart 23
cells 43
chartjunk 52
class boundaries 33
class groupings 33
class midpoint 33
contingency table 43
cross-classification table 43
cumulative percentage distribution 35
cumulative percentage polygon 40
frequency distribution 32
histogram 37
ogive (cumulative percentage polygon) 40
ordered array 30
Pareto diagram 25
Pareto principle 25
percentage distribution 34
percentage polygon 39
pie chart 24
range 33
relative frequency distribution 34
scatter diagram 47
side-by-side bar chart 45
stem-and-leaf display 30
summary table 22
time-series plot 48
width of class interval 33

CHAPTER REVIEW PROBLEMS

Checking Your Understanding

2.55 How do histograms and polygons differ with respect to their construction and use?

2.56 Why would you construct a summary table?

2.57 What are the advantages and/or disadvantages of using a bar chart, a pie chart, or a Pareto diagram?

2.58 Compare and contrast the bar chart for categorical data with the histogram for numerical data.

2.59 What is the difference between a time-series plot and a scatter diagram?

2.60 Why is it said that the main feature of the Pareto diagram is its ability to separate the "vital few" from the "trivial many"? Discuss.

2.61 What percentage breakdowns can help you interpret the results found in a cross-classification table?

Applying the Concepts

You can solve problems 2.62–2.74 manually or by using Microsoft Excel, Minitab, or SPSS. We recommend that you use Microsoft Excel, Minitab, or SPSS to solve problems 2.75–2.85.

2.62 The data on the top of page 56 represent the breakdown of the price of a new college textbook.

Revenue Categories	Percentage	
Publisher	64.8	
Manufacturing costs		32.3
Marketing and promotion		15.4
Administrative costs and taxes		10.0
After-tax profit		7.1
Bookstore	22.4	
Employee salaries and benefits		11.3
Operations		6.6
Pretax profit		4.5
Author	11.6	
Freight	1.2	

Source: Extracted from T. Lewin, "When Books Break the Bank," The New York Times, *September 16, 2003, B1, B4.*

a. Using the four categories of publisher, bookstore, author, and freight, construct a bar chart, a pie chart, and a Pareto diagram.
b. Using the four subcategories of publisher and three subcategories of bookstore along with the author and freight categories, construct a Pareto diagram.
c. Based on the results of (a) and (b), what conclusions can you reach concerning who gets the revenue from the sales of new college textbooks? Do any of these results surprise you? Explain.

2.63 The following data represent the market share for the repair of cars and light trucks in 1992 and 2002.

Source	1992 Percentage	2002 Percentage
Foreign specialists	3.9	6.0
Parts stores with service bays	7.3	6.4
Repair specialists	12.7	16.2
Service stations, garages	39.1	29.5
Tire stores	8.1	8.9
Vehicle dealers	21.6	26.6
Others	7.3	6.4

Source: Extracted from A. Frangos, " Corner Garages Battle Dealers to Fix Your Car," The Wall Street Journal, *June 3, 2003, B1, B4.*

a. For each year, construct a bar chart, a pie chart, and a Pareto diagram.
b. Construct a side-by-side bar chart of the market share in 1992 and in 2002.
c. Based on the results of (a) and (b), what changes in market share have occurred between 1992 and 2002?

2.64 The following data represent how consumers made in-store payments in 1999, 2001, and 2003.

Type of Payment	1999 Percentage	2001 Percentage	2003 Percentage
Cash	39	33	32
Check	18	18	15
Debit	21	26	31
Credit	22	21	21
Other	0	2	1

Source: Extracted from M. Ingebretsen and M. Ballinger, "Charge It," The Wall Street Journal, *February 9, 2004, R2.*

a. Construct a side-by-side bar chart of the types of payment in 1999, 2001, and 2003.
b. Based on the results of (a), what changes in the types of payment have occurred in 1999, 2001, and 2003?

2.65 The following data represent the per-capita consumption of beverages (in gallons) sold at retail stores in 1998, 2000, and 2002.

Type of Drink	1998 Consumption	2000 Consumption	2002 Consumption
Bottled water	2.5	4.1	6.7
Dairy/other	0.3	0.3	0.3
Juice drinks	3.1	3.7	4.0
Soft drinks	54.0	53.0	52.5
Sports drinks	1.9	2.2	2.5
Tea	1.9	2.0	1.9
Total	63.7	65.3	67.9

Source: Extracted from T. Howard, "Coke, Pepsi Sales Up, but Core Colas Flat," USA Today, *July 21, 2003, 3B.*

a. For each year, form a percentage summary table for the types of drinks.
b. For each year, construct a bar chart, a pie chart, and a Pareto diagram.
c. Construct a side-by-side bar chart of the market share of the types of drinks in 1998, 2000, and 2002.
d. Based on the results of (a) through (c), what changes in market share have occurred between 1998 and 2002?

2.66 Brazil is the second-largest coffee consuming country in the world. Unlike most major markets where a handful of corporations dominate the coffee roasting and selling market, over 2,000 small roasters are active in Brazil. The Sara Lee Corporation has become the leading coffee retailer in Brazil by acquiring several Brazilian coffee roasters (Miriam Jordan, "Sara Lee Wants to Percolate through All of Brazil," *The Wall Street Journal*, May 8, 2002, A14). Consumption by the seven largest coffee-consuming nations and a breakdown of the market leaders in Brazil are given on page 57.

Coffee Consumption in Major Markets in 2000

Country	Consumption (in Millions of 60-Kg Bags)
United States	18.6
Brazil	12.8
Germany	9.2
Japan	6.7
France	5.4
Netherlands	1.8
Finland	0.9

Source: Extracted from The Wall Street Journal.

Leading Coffee Brands in Brazil

Brand	Market Share
Sara Lee owned brands	27.6%
Nescafe	6.1%
Tres Coracoes	4.8%
Melitta	4.0%
All Others	57.5%

Source: Extracted from The Wall Street Journal.

a. Construct a graph for the data concerning the major coffee-consuming countries. Which type of graph is most appropriate? Explain.

b. Construct a graph for the data concerning the market share of coffee in Brazil. Which type of graph is most appropriate? Explain.

2.67 The following data represent proven conventional oil reserves in billions of barrels, subdivided by region and country.

Region and Country	Proven Conventional Reserves (Billions of Barrels)	
North America	54.8	
Mexico		28.3
U.S.		21.8
Canada		4.7
Central and South America	95.2	
Venezuela		76.9
Brazil		8.1
Other Central and South America		10.2
Western Europe	17.2	
Norway		9.5
Britain		5.0
Other Western Europe		2.7
Africa	74.9	
Libya		29.5
Nigeria		22.5
Algeria		9.2
Angola		5.4
Other Africa		8.3
Middle East	683.6	
Saudi Arabia		259.2
Iraq		112.5
United Arab Emirates		97.8
Kuwait		94.0
Iran		89.7
Qatar		13.2
Oman		5.5
Other Middle East		11.7
Far East and Oceania	44.0	
China		24.0
Indonesia		5.0
India		4.7
Other Far East and Oceania		10.3
Eastern Europe and Former USSR	59.0	
Russia		48.6
Kazakhstan		5.4
Other Eastern Europe and Former USSR		5.0

Source: United States Department of Energy.

Using the set of countries:

a. Form a bar chart, a pie chart, and a Pareto diagram.

Using the set of regions:

b. Form a bar chart, a pie chart, and a Pareto diagram.

c. Which graphical method do you think is best to portray these data?

d. Based on the results of (a) and (b), what conclusions can you make concerning the proven conventional oil reserves for the different countries and regions?

2.68 In the aftermath of the attacks of September 11, 2001, statisticians at the National Center for Health Statistics became more concerned with their ability to track and classify victims of terrorism (E. Weinstein, "Tracking Terror's Rising Toll," *The Wall Street Journal*, January 25, 2002, A13). The following data represent deaths due to terrorism on U.S. soil from 1990 to 2001 and also the deaths in the United States in 2000 due to various causes.

Year	Deaths Due to Terrorism in the United States
1990	0
1991	0
1992	0
1993	6
1994	1
1995	169
1996	2
1997	0
1998	1
1999	3
2000	0
2001	2,717

Cause	Deaths in Thousands
Smoke and fire	3.3
Accidental drowning	3.3
Alcohol-induced deaths	18.5
Alzheimer's disease	49.0
Assault by firearms	10.4
Assault by non-firearms	5.7
Asthma	4.4
Cancer	551.8
Strokes and related diseases	166.0
Emphysema	16.9
Diabetes	68.7
Heart diseases	710.0
Falls	12.0
HIV	14.4
Influenza and pneumonia	67.0
Injuries at work	5.3
Motor vehicle accidents	41.8
Suicide	28.3
Drug-related deaths	15.9

Source: Federal Bureau of Criminal Justice Statistics, National Center for Health Statistics, National Highway Transportation Safety Administration, Department of Defense.

a. Construct a time-series plot of deaths due to terrorism on U.S. soil. Is there any pattern to the deaths due to terrorism on U.S. soil between 1990 and 2001?

For the deaths in thousands due to different causes:

b. Construct a bar chart, a pie chart, and a Pareto diagram.

c. Which graphical method do you think is best to portray these data?

d. Based on the results of (c), what conclusions can you make concerning the deaths in the United States in 2000 due to various causes?

2.69 The owner of a restaurant serving Continental-style entrées was interested in studying patterns of patron demand for the Friday to Sunday weekend time period. Records were maintained that indicated the number of entrées ordered for each type. The data were as follows:

Type of Entrée	Number Served
Beef	187
Chicken	103
Duck	25
Fish	122
Pasta	63
Shellfish	74
Veal	26

a. Form a percentage summary table for the types of entrées ordered.

b. Construct a bar chart, a pie chart, and a Pareto diagram for the types of entrées ordered.

c. Do you prefer a Pareto diagram or a pie chart for these data? Why?

d. What conclusions can the restaurant owner draw concerning demand for different types of entrées?

2.70 Suppose that the owner of the restaurant in problem 2.69 was also interested in studying the demand for dessert during the same time period. She decided that two other variables, along with whether a dessert was ordered, were to be studied: the gender of the individual, and whether a beef entrée was ordered. The results are as follows:

	GENDER		
DESSERT ORDERED	Male	Female	Total
Yes	96	40	136
No	224	240	464
Total	320	280	600

	BEEF ENTRÉE		
DESSERT ORDERED	Yes	No	Total
Yes	71	65	136
No	116	348	464
Total	187	413	600

For each of the two cross-classification tables:

a. Construct a table of row percentages, column percentages, and total percentages.

b. Which type of percentage (row, column, or total) do you think is most informative for each gender? for beef entrée? Explain.

c. What conclusions concerning the pattern of dessert ordering can the owner of the restaurant reach?

2.71 An article in *The New York Times* (William McNulty and Hugh K. Truslow, "How It Looked Inside the Booth," *The New York Times*, November 6, 2002) provided the following data on the method for recording votes in 1980, 2000, and 2002, broken down by percentage of counties in the United States using each method and percentage of registered voters using each method. The results are as follows:

	PERCENTAGE OF COUNTIES USING		
METHOD	1980	2000	2002
Punch cards	18.5	18.5	15.5
Lever machines	36.7	14.4	10.6
Paper ballots	40.7	11.9	10.5
Optical scan	0.8	41.5	43.0
Electronic	0.2	9.3	16.3
Mixed	3.1	4.4	4.1

METHOD	PERCENTAGE OF REGISTERED VOTERS USING 1980	2000	2002
Punch cards	31.7	31.4	22.6
Lever machines	42.9	17.4	15.5
Paper ballots	10.5	1.5	1.3
Optical scan	2.1	30.8	31.8
Electronic	0.7	12.2	19.6
Mixed	12.0	6.7	9.3

a. Set up separate pie charts for each year for the percentage of counties and the percentage of registered voters using the various methods.
b. Set up side-by-side bar charts by year for the percentage of counties and the percentage of registered voters using the various methods.
c. Which type of graphical display is more helpful in depicting the data? Explain.
d. What differences are there in the results for the counties and the registered voters?

2.72 In summer 2000, a growing number of warranty claims on Firestone tires sold on Ford SUVs prompted Firestone and Ford to issue a major recall. An analysis of warranty-claims data helped identify which models to recall. A breakdown of 2,504 warranty claims based on tire size is given in the following table.

Tire Size	Warranty Claims
23575R15	2,030
311050R15	137
30950R15	82
23570R16	81
331250R15	58
25570R16	54
Others	62

Source: Extracted from Robert L. Simison, "Ford Steps Up Recall without Firestone," The Wall Street Journal, *August 14, 2000, A3.*

The 2,030 warranty claims for the 23575R15 tires can be categorized into ATX models and Wilderness models. The type of incident leading to a warranty claim, by model type, is summarized in the following table.

Incident	ATX Model Warranty Claims	Wilderness Warranty Claims
Tread separation	1,365	59
Blow out	77	41
Other/unknown	422	66
Total	1,864	166

Source: Extracted from Robert L. Simison, "Ford Steps Up Recall without Firestone," The Wall Street Journal, *August 14, 2000, A3.*

a. Construct a Pareto diagram for the number of warranty claims by tire size. What tire size accounts for most of the claims?
b. Construct a pie chart to display the percentage of the total number of warranty claims for the 23575R15 tires that come from the ATX model and Wilderness model. Interpret the chart.
c. Construct a Pareto diagram for the type of incident causing the warranty claim for the ATX model. Does a certain type of incident account for most of the claims?
d. Construct a Pareto diagram for the type of incident causing the warranty claim for the Wilderness model. Does a certain type of incident account for most of the claims?

2.73 One of the major measures of the quality of service provided by any organization is the speed with which the organization responds to customer complaints. A large family-held department store selling furniture and flooring including carpet had undergone a major expansion in the past several years. In particular, the flooring department had expanded from 2 installation crews to an installation supervisor, a measurer, and 15 installation crews. During a recent year the company got 50 complaints concerning carpet installation. The following data represent the number of days between the receipt of the complaint and the resolution of the complaint.
FURNITURE

54	5	35	137	31	27	152	2	123	81	74	27
11	19	126	110	110	29	61	35	94	31	26	5
12	4	165	32	29	28	29	26	25	1	14	13
13	10	5	27	4	52	30	22	36	26	20	23
33	68										

a. Form the frequency distribution and the percentage distribution.
b. Plot the histogram and the percentage polygon.
c. Form the cumulative percentage distribution and plot the ogive (cumulative percentage polygon).
d. On the basis of the results of (a) through (c), if you had to tell the president of the company how long a customer should expect to wait to have a complaint resolved, what would you say? Explain.

2.74 The data in the file PIZZA represent the cost of a slice in dollars, number of calories per slice, and amount of fat in grams per slice for a sample of 36 pizza products.

Source: "Frozen Pizza on the Rise," Copyright © 2002 by Consumers Union of U.S., Inc., Yonkers, NY 10703–1057. Adapted with permission from Consumer Reports, *January 2002, 40–41.*

a. Construct frequency distributions and percentage distributions for fat, cost, and calories.
b. Plot histograms and percentage polygons for fat, cost, and calories.

c. Form cumulative percentage distributions and plot ogives (cumulative percentage polygons) for fat, cost, and calories.
d. Construct scatter diagrams of cost and calories, cost and fat, and calories and fat.
e. Based on (a) through (d), what conclusions can you reach about the cost, fat, and calories of these pizza products?

2.75 An article in *Quality Engineering* examined the viscosity (resistance to flow) of a chemical product produced in batches. Assume that the viscosity of the chemical needs to be between 13 and 18 to meet company specifications. The data for 120 batches are in the data file CHEMICAL

Source: D. S. Holmes and A. E. Mergen, "Parabolic Control Limits for the Exponentially Weighted Moving Average Control Charts," Quality Engineering, *Vol. 4(1992), 487–495.*

a. Develop the ordered array.
b. Construct a frequency distribution and a percentage distribution.
c. Plot the percentage histogram.
d. What percentage of the batches is within company specifications?

2.76 Studies conducted by a manufacturer of "Boston" and "Vermont" asphalt shingles have shown product weight to be a major factor in the customer's perception of quality. Moreover, the weight represents the amount of raw materials being used and is therefore very important to the company from a cost standpoint. The last stage of the assembly line packages the shingles before the packages are placed on wooden pallets. Once a pallet is full (a pallet for most brands holds 16 squares of shingles), it is weighed, and the measurement is recorded. The company expects pallets of their "Boston" brand-name shingles to weigh at least 3,050 pounds but less than 3,260 pounds. For the company's "Vermont" brand-name shingles, pallets should weigh at least 3,600 pounds but less than 3,800. The data file PALLET contains the weights (in pounds) from a sample of 368 pallets of "Boston" shingles and 330 pallets of "Vermont" shingles.
a. For the "Boston" shingles, form a frequency distribution and a percentage distribution having eight class intervals, using 3,015, 3,050, 3,085, 3,120, 3,155, 3,190, 3,225, 3,260, and 3,295 as the class boundaries.
b. For the "Vermont" shingles, form a frequency distribution and a percentage distribution having seven class intervals, using 3,550, 3,600, 3,650, 3,700, 3,750, 3,800, 3,850, and 3,900 as the class boundaries.
c. Form histograms for the "Boston" shingles and for the "Vermont" shingles.
d. Comment on the distribution of pallet weights for the "Boston" and "Vermont" shingles. Be sure to identify the percentage of pallets that are underweight and overweight.

2.77 Do marketing promotions, such as bobble-head giveaways, increase attendance at Major League Baseball games? An article in *Sport Marketing Quarterly* reported on the effectiveness of marketing promotions (T. C. Boyd, and T. C. Krehbiel, "Promotion Timing in Major League Baseball and the Stacking Effects of Factors that Increase Game Attractiveness," *Sport Marketing Quarterly*, Vol. 12, (2003), 173–184). The data file ROYALS includes the following variables for the Kansas City Royals during the 2002 baseball season:

GAME = Home games in the order they were played
ATTENDANCE = Paid attendance for the game
PROMOTION 1 = If a promotion was held; 0 = if no promotion was held

a. Construct a percentage histogram for the attendance variable. Interpret the histogram.
b. Construct a percentage polygon for the attendance variable. Interpret the polygon.
c. Which graphical display do you prefer, the one in (a) or (b)? Explain.
d. Construct a graphical display containing two percentage polygons for attendance—one for the 43 games with a promotion and the second for the 37 games without a promotion. Compare the two attendance distributions.

2.78 The data in the file PROTEIN indicate fat and cholesterol information concerning popular protein foods (fresh red meats, poultry, and fish).

Source: United States Department of Agriculture.

For the data relating to the number of calories and the amount of cholesterol for the popular protein foods:
a. Construct the frequency distribution and the percentage distribution.
b. Plot the histogram and the percentage polygon.
c. Form the cumulative percentage distribution and plot the cumulative percentage polygon.
d. What conclusions can you draw from these analyses?

2.79 Suppose that you wish to study characteristics of the model year 2002 automobiles in terms of the following variables: horsepower, miles per gallon, length, width, turning circle requirement, weight, and cargo volume. AUTO2002

Source: "The 2002 Cars," Copyright © 2002 by Consumers Union of U.S., Inc., Yonkers, NY 10703–1057. Adapted with permission from Consumer Reports, *April 2002, 22–71.*

For each of these variables:
a. Construct the frequency distribution and the percentage distribution.
b. Plot the histogram and the percentage polygon.
c. Form the cumulative percentage distribution and plot the cumulative percentage polygon.
d. What conclusions can you draw concerning the 2002 automobiles?

2.80 Referring to the characteristics on model year 2002 automobiles AUTO2002 in problem 2.79,
a. Form a cross classification table of type of drive with type of gasoline.

b. Plot a side-by-side bar chart of type of drive with type of gasoline.
c. Based on the results of (a) and (b), does there appear to be a relationship between type of drive with type of gasoline?

2.81 The data in the file STATES represent the results of the American Community Survey, a sampling of households taken in all states during the 2000 U.S. Census. For each of the variables of average travel-to-work time in minutes, percentage of homes with eight or more rooms, median household income, and percentage of mortgage-paying homeowners whose housing costs exceed 30% of income:
a. Form the frequency distribution and the percentage distribution.
b. Plot the histogram and the percentage polygon.
c. Construct the cumulative percentage distribution and plot the cumulative percentage polygon.
d. What conclusions about these four variables can you make based on the results of (a) through (c)?

2.82 The economics of baseball has caused a great deal of controversy with owners arguing that they are losing money, players arguing that owners are making money, and fans complaining about how expensive it is to attend a game and watch games on cable television. In addition to data related to team statistics for the 2001 season, the file BB2001 contains team-by-team statistics on ticket prices; the fan cost index; regular season gate receipts; local television, radio, and cable receipts; all other operating revenue; player compensation and benefits; national and other local expenses; and income from baseball operations. For each of these variables,
a. Form the frequency distribution and the percentage distribution.
b. Plot the histogram and the percentage polygon.
c. Construct the cumulative percentage distribution and plot the cumulative percentage polygon.
d. Construct a scatter diagram to predict the number of wins on the *Y* axis from the player compensation and benefits on the *X* axis. What conclusions can you reach from this scatter diagram?
e. What conclusions about these variables can you reach based on the results of (a) through (c)?

2.83 The data in the file AIRCLEANERS represent the price, yearly energy cost, and yearly filter cost of room air cleaners.
a. Construct a scatter diagram with price on the *Y* axis and energy cost on the *X* axis.
b. Construct a scatter diagram with price on the *Y* axis and filter cost on the *X* axis.
c. What conclusions about the relationship of energy cost and filter cost to the price of the air cleaners can you make?

Source: "Portable Room Air Cleaners," Copyright © 2002 by Consumers Union of U.S., Inc., Yonkers, NY 10703–1057. Adapted with permission from Consumer Reports, *February 2002, 47.*

2.84 The data in the file PRINTERS represent the price, text speed, text cost, color photo time, and color photo cost of computer printers.
a. Construct scatter diagrams with price and text speed, price and text cost, price and color photo time, and price and color photo cost.
b. Based on the results of (a), do you think that any of the other variables might be useful in predicting printer price? Explain.

Source: "Printers," Copyright 2002 by Consumers Union of U.S., Inc., Yonkers, NY 10703–1057. Adapted with permission from Consumer Reports, *March 2002, 51.*

2.85 The S&P 500 Index tracks the overall movement of the stock market by considering the stock price of 500 large corporations. The data file STOCKS2003 contains weekly data for this index as well as the weekly closing stock price for three companies during 2003. The variables included are:
WEEK—Week ending on date given
S & P—Weekly closing value for the S&P 500 Index
SEARS—Weekly closing stock price for Sears
TARGET—Weekly closing stock price for the Target
SARA LEE—Weekly closing stock price for the Sara Lee

Source: ***finance.yahoo.com***

a. Construct a time-series plot for the weekly closing values of the S&P 500 index, Sears, Roebuck and Company, Target Corporation, and Sara Lee.
b. Explain any patterns present in the plots.
c. Write a short summary of your findings.

2.86 (Class Project) Let each student in the class respond to the question "Which carbonated soft drink do you most prefer?" so that the teacher can tally the results into a summary table.
a. Convert the data to percentages and construct a Pareto diagram.
b. Analyze the findings.

2.87 (Class Project) Let each student in the class be cross-classified on the basis of gender (male, female) and current employment status (yes, no) so that the teacher can tally the results.
a. Construct a table with either row or column percentages, depending on which you think is more informative.
b. What would you conclude from this study?
c. What other variables would you want to know regarding employment in order to enhance your findings?

Report Writing Exercises

2.88 Referring to the results from problem 2.76 on page 60 concerning the weight of "Boston" and "Vermont" shingles, write a report that evaluates whether the weight of the pallets of the two types of shingles are what the company expects. Be sure to incorporate tables and charts into the report.

2.89 Referring to the results from problem 2.72 on page 59 concerning the warranty claims on Firestone tires, write a report that evaluates warranty claims on Firestone tires sold on Ford SUVs. Be sure to incorporate tables and charts into the report.

TEAM PROJECT

The data file **MUTUALFUNDS2004** contains information regarding 12 variables from a sample of 121 mutual funds. The variables are:

Fund—The name of the mutual fund
Category—Type of stocks comprising the mutual fund—small cap, mid cap, large cap
Objective—Objective of stocks comprising the mutual fund—growth or value
Assets—In millions of dollars
Fees—Sales charges (no or yes)
Expense ratio—Ratio of expenses to net assets in percentage
2003 Return—Twelve-month return in 2003
Three-year return—Annualized return 2001–2003
Five-year return—Annualized return 1999–2003
Risk—Risk-of-loss factor of the mutual fund classified as low, average, or high
Best quarter—Best quarterly performance 1999–2003
Worst quarter—Worst quarterly performance 1999–2003

2.90 For the expense ratio variable:
a. Plot a histogram.
b. Plot percentage polygons of the expense ratio for mutual funds that have fees and mutual funds that do not have fees on the same graph.
c. What conclusions about the expense ratio can you reach based on the results of (a) and (b)?

2.91 For the variable containing five-year annualized return from 1999 to 2003:
a. Plot a histogram.
b. Plot percentage polygons of the five-year annualized return from 1999 to 2003 for growth mutual funds and value mutual funds on the same graph.
c. What conclusions about the five-year annualized return from 1999 to 2003 can you reach based on the results of (a) and (b)?

2.92 For the variable containing three-year annualized return from 2001 to 2003:
a. Plot a histogram.
b. Plot percentage polygons of the three-year annualized return from 2001 to 2003 for growth mutual funds and value mutual funds on the same graph.
c. What conclusions about the three-year annualized return from 2001 to 2003 can you reach based on the results of (a) and (b)?

RUNNING CASE MANAGING THE *SPRINGVILLE HERALD*

Advertising fees are an important source of revenue for any newspaper. In an attempt to boost these revenues and to minimize costly errors, the management of the *Herald* has established a task force charged with improving customer service in the advertising department. Review the task force's data collection (open **Ad_Errors.htm** in the Springville HeraldCase folder on the CD-ROM that accompanies this text or link to **www.prenhall.com/HeraldCase/Ad_Errors.htm**) and identify the data that are important in describing the customer service problems. For each set of data you identify, construct the graphical presentation you think is most appropriate for the data and explain your choice. Also, suggest what other information concerning the different types of errors would be useful to examine. Offer possible courses of action for either the task force or management to take that would support the goal of improving customer service.

WEB CASE

In the "Using Statistics" scenario, you were asked to gather information that would help you make wise investment choices. Sources for such information include brokerage firms and investment counselors. Apply your knowledge about the proper use of tables and charts in this Web Case about the claims of foresight and excellence by a Springville investment service.

Visit the StockTout Investing Service Web site **www.prenhall.com/Springville/StockToutHome.htm**. Review their investment claims and supporting data and then answer the following.

1. How does the presentation of the general information about StockTout on its home page affect your perception of their business?
2. Is their claim about having more winners than losers a fair and accurate reflection about the quality of their investment service? If you do not think that the claim is a fair and accurate one, provide an alternate presentation that you think is fair and accurate.
3. StockTout's "Big Eight" mutual funds are part of the sample found in the MUTUALFUNDS2004 file. Is there any other relevant data from that file that could have been included in the Big Eight table? How would that new data alter your perception of StockTout's claims?
4. StockTout is proud that all "Big Eight" funds have gained in value over the past five years. Do you agree that they should be proud of their selections? Why or why not?

REFERENCES

1. Huff, D., *How to Lie with Statistics* (New York: Norton, 1954).
2. *Microsoft Excel 2003* (Redmond, WA: Microsoft Corporation, 2002).
3. *Minitab for Windows Version* 14 (State College; PA: Minitab Inc., 2004).
4. *SPSS ® Base 12.0 Brief Guide* (Upper Saddle River, NJ: Prentice Hall, 2003).
5. Tufte, E. R., *Envisioning Information* (Cheshire, CT: Graphics Press, 1990).
6. Tufte, E. R., *The Visual Display of Quantitative Information*, 2nd ed (Cheshire, CT: Graphics Press, 2002).
7. Tufte, E. R., *Visual Explanations* (Cheshire, CT: Graphics Press, 1997).
8. Wainer, H., *Visual Revelations: Graphical Tales of Fate and Deception from Napoleon Bonaparte to Ross Perot* (New York: Copernicus/Springer-Verlag, 1997).

Appendix 2 Using Software for Tables and Charts

A2.1 MICROSOFT EXCEL

You can use Microsoft Excel to create many of these tables and charts discussed in this chapter. If you have not already read Appendix 1.2, "Introduction to Microsoft Excel," on page 16, you should do so now.

Summary Tables

Use the PivotTable Wizard to generate a summary table. If you are not familiar with PivotTables, first read the "Using the PivotTable Wizard" (see Appendix F). To generate a summary table similar to Table 2.2 on page 23, open the **MUTUALFUNDS2004.xls** workbook to the **Data** worksheet. Select **Data → PivotTable and PivotChart Report** (**Data → PivotTable Report** in Microsoft Excel 97) and make these entries in the PivotTable Wizard dialog boxes:

Step 1: Select the **Microsoft Excel list or database** option and the **PivotTable** option (if it appears) and click **Next**.

Step 2: Enter **J1:J122** as the **Range** and click **Next**.

Step 3: Select the **New worksheet** option and click the **Layout** button.

In the Layout dialog box, first drag a copy of the **Risk** label to the **ROW** area. Then drag a second copy of the **Risk** label to the **DATA** area, which changes the label to **Count of Risk**. Click **OK** to return to the main Step 3 dialog box and click the **Options** button to continue.

In the PivotTable Options dialog box, enter a self-descriptive name for the table in the **Name** edit box and **0** in the **For empty cells, show** edit box. Click **OK** to return to the main Step 3 dialog box.

Click **Finish** in the main Step 3 dialog box to produce the PivotTable.

Rename the new worksheet with a self-descriptive name. (You can close any floating toolbars or windows that appear over the PivotTable to better view your table.)

To add a percentage column, enter **Percentage** in cell **C4** of the new worksheet and enter the formula **=B5/B$8** in cell **C5**. Copy this formula down through cell C7. Format the cell range **C5:C7** for percentage display. Adjust the number of decimals displayed and adjust the column C width if you would like to produce a table similar to Figure A2.1.

Count of Risk		
Risk ▾	Total	Percentage
average	46	38%
high	17	14%
low	58	48%
Grand Total	121	

FIGURE A2.1 Completed Summary Table

OR See section G.1 (**One-Way Tables & Charts**) if you want PHStat2 to produce a summary table for you.

Bar or Pie Chart

Use the Microsoft Excel Chart Wizard to generate a bar or pie chart. If you are not familiar with this wizard, first read "Using the Chart Wizard" (see page 17). First create a PivotTable summary table. With the table onscreen, click a cell outside the table, select **Insert → Chart**, and make these entries in the Chart Wizard dialog boxes:

Step 1: Click either **Bar** (for a bar chart) or **Pie** from the **Standard Types Chart type** box and leave the first **Chart sub-type** selected. Click **Next**.

Step 2: With the blinking cursor in the **Data range** box, click the PivotTable to have Excel fill in the address of the PivotTable for you. Click **Next**.

Step 3: Select the formatting and labeling chart options for the chart. (See "Using the Microsoft Excel Chart Wizard" on page 17 for suggestions.) Click **next**.

Step 4: Select **As new sheet** and click **Finish**.

If field buttons appear on the chart, right-click any field button and select **Hide PivotChart Field Buttons** from the shortcut menu.

OR See section G.1 (**One-Way Tables & Charts**) if you want PHStat2 to produce a bar or pie chart for you.

Pareto Diagram

See section G.1 (**One-Way Tables & Charts**) if you want PHStat2 to produce a Pareto diagram as a Microsoft Excel chart. (There are no Microsoft Excel commands that directly produce a Pareto diagram.)

Ordered Array

Organize your worksheet so that each variable appears in its own column, enter a variable column heading in row 1, and enter the values for the variable starting in row 2. (This is the format of the Excel files included on the CD-ROM packaged with this text.) Select **Data → Sort**. In the Sort dialog box, select the variable to be sorted from the **Sort by** drop-down list. Select either the first **Ascending** or **Descending** option button and leave the **Header row** option button selected and click **OK**.

Stem-and-Leaf Display

See section G.2 (**Stem-and-Leaf Display**) if you want PHStat2 to produce a stem-and-leaf display as a Microsoft Excel chart. (There are no Microsoft Excel commands that directly produce stem-and leaf displays.)

Frequency Distributions and Histograms

Use the Data Analysis ToolPak to create frequency distributions and histograms. Open to the worksheet containing the data you want to summarize. Select **Tools → Data Analysis**. From the list that appears in the Data Analysis dialog box, select **Histogram** and click **OK**. In the Histogram dialog box (see Figure A2.2), enter the cell range of the data in the **Input Range**. Then select **Labels** if you are using data that are arranged like the data in the Excel files on the CD-ROM packaged with this text. Finish by selecting **Chart Output** and clicking **OK**. (See section G.3 (**Histograms & Polygons**) for an explanation of the **Bin Range**.)

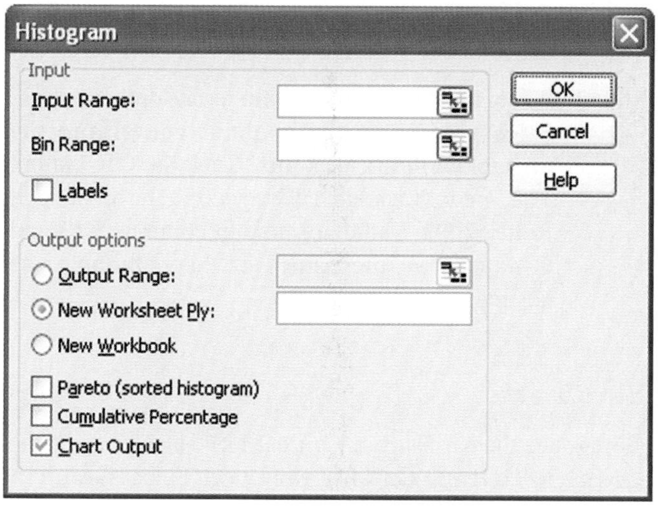

FIGURE A2.2 Data Analysis Histogram Dialog Box

The frequency distribution and histogram will appear together on a separate worksheet.

OR See section G.3 (**Histogram & Polygons**) if you want PHStat2 to produce a frequency distribution and a histogram for you.

Percentage and Cumulative Percentage Polygons

See section G.3 (**Histogram & Polygons**) if you want PHStat2 to produce percentage and cumulative percentage polygons as Microsoft Excel charts. (There are no Microsoft Excel commands that directly produce percentage and cumulative percentage polygons.)

Cross-Classification Table and Side-by-Side Chart

Use the PivotTable and Chart Wizards to create a two-way cross-classification table and a side-by-side chart. To create a cross-classification table similar to Table 2.14 on page 44, open the **MUTUALFUNDS2004.XLS** workbook to the **Data** worksheet. Select **Data ➔ PivotTable and PivotChart Report** (**Data ➔ PivotTable Report** in Microsoft Excel 97) and make these entries in the PivotTable Wizard dialog boxes:

Step 1: Select the **Microsoft Excel list or database** option and the **PivotTable** option (if it appears) and click **Next**.

Step 2: Enter **C1:J122** as the **Range** and click **Next**.

Step 3: Select the **New worksheet** option and click the **Layout** button.

In the Layout dialog box, first drag a copy of the **Objective** label to the **ROW** area. Then drag a second copy of the **Objective** label to the **DATA** area, which changes the label to **Count of Objective**. Drag a copy of the **Risk** label to the **COLUMN** area. Click **OK** to return to the main Step 3 dialog box and click the **Options** button to continue.

In the PivotTable Options dialog box, enter a self-descriptive name for the table in the **Name** edit box and **0** in the **For empty cells, show** edit box. Click **OK** to return to the main Step 3 dialog box.

Click **Finish** in the main Step 3 dialog box to produce the PivotTable.

To create a side-by-side chart, click a cell outside the two-way table, select **Insert ➔ Chart**, and make these entries in the Chart Wizard dialog boxes:

Step 1: Click **Bar** from the **Standard Types Chart type** box and leave the first **Chart sub-type** selected. Click **Next**.

Step 2: With the blinking cursor in the **Data range** box, click the PivotTable to have Excel fill in the address of the PivotTable for you. Click **Next**.

Step 3: Select the formatting and labeling chart options for the chart. (See "Using the Microsoft Excel Chart Wizard" on page 17 for suggestions.) Click **Next**.

Step 4: Select **As new sheet** and click **Finish**.

If field buttons appear on the chart, right-click any button and select **Hide PivotChart Field Buttons** from the shortcut menu.

OR See section G.4 (**Two-Way Tables & Charts**) if you want PHStat2 to produce a two-way summary table and a side-by-side chart for you.

Scatter Diagram

Use the Chart Wizard to generate a scatter diagram. To create a scatter diagram similar to Figure 2.15 on page 47, open the **MUTUALFUNDS2004.XLS** workbook to the **Data** worksheet. Select **Insert ➔ Chart**, and make these entries in the Chart Wizard dialog boxes:

Step 1: Click **XY (Scatter)** from the **Standard Types Chart type** box and leave the first **Chart sub-type** selected. Click **Next**.

Step 2: Enter **F1:G122** in the **Data range** box, select the **Columns** option, and click **Next**.

Step 3: Select the formatting and labeling chart options for the chart. (See "Using the Microsoft Excel Chart Wizard" on page 17 for suggestions.) Click **Next**.

Step 4: Select **As new sheet** and click **Finish**.

Be aware that the Chart Wizard always assumes that the first column of the data range (column F in this example) contains the *X* variable data. If you have a sheet in which the Y variable data appears first, you will need to first rearrange your columns (or copy them out of order to a new sheet) before using the Chart Wizard.

A2.2 MINITAB

You can use Minitab to create many of the tables and charts discussed in this chapter. If you have not already read Appendix 1.3, "Introduction to Minitab," on page 19, you should do so now.

Unstacking Data

Data are usually arranged so that all of the values of a variable are stacked vertically down a column. In many cases, you need to separately analyze different subgroups in terms of a numerical variable of interest. For example, in the mutual fund data, you may want to separately analyze the 2003 percentage return of the growth funds and the value funds. This can be accomplished by unstacking the 2003 percentage return variable so that the 2003 percentage returns for the growth funds are located in one column and the 2003 percentage returns for the value funds are located in a different column.

To accomplish this task, open the MUTUALFUNDS2004.MTW worksheet. **Select Data → Unstack Columns**. Then do the following:

Step 1: In the Unstack Columns dialog box (see Figure A2.3), enter **C7** or **Return 2003** in the **Unstack the data in:** edit box.

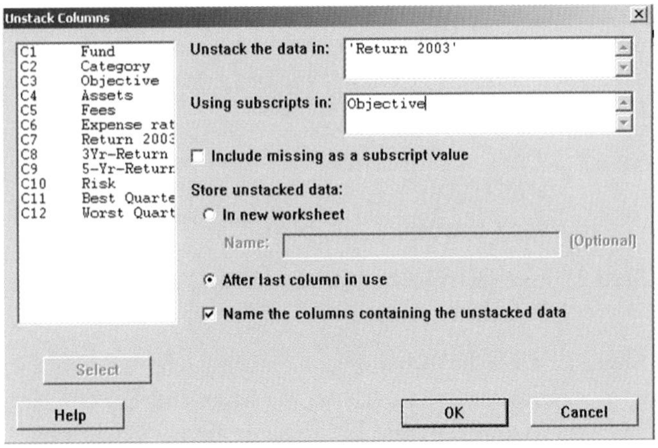

FIGURE A2.3 Minitab Unstack Columns Dialog Box

Step 2: Enter **C3** or **Objective** in the **Using Subscripts in:** edit box.

Step 3: Select the **After last column in use** option button. Select the **Name the columns containing the unstacked data** check box. Click the **OK** button. The new variables Return2003_Growth and Return2003_Value are now in columns C13 and C14. Change the names of these variables as desired.

Bar Chart

To produce the bar chart in Figure 2.1 on page 23, open the **ONLINESHOPPING.MTW** worksheet. Select **Graph → Bar Chart**. Then do the following:

Step 1: In the Bar Charts dialog box (see Figure A2.4), in the Bars represent: drop-down list box, select **Values from a table** since the frequencies in each category are provided. (If you are using raw data such as from the **MUTUALFUNDS2004.MTW** worksheet, select Counts of unique values in the bars represent dialog box.) Select the **Simple** graph box. Click the **OK** button.

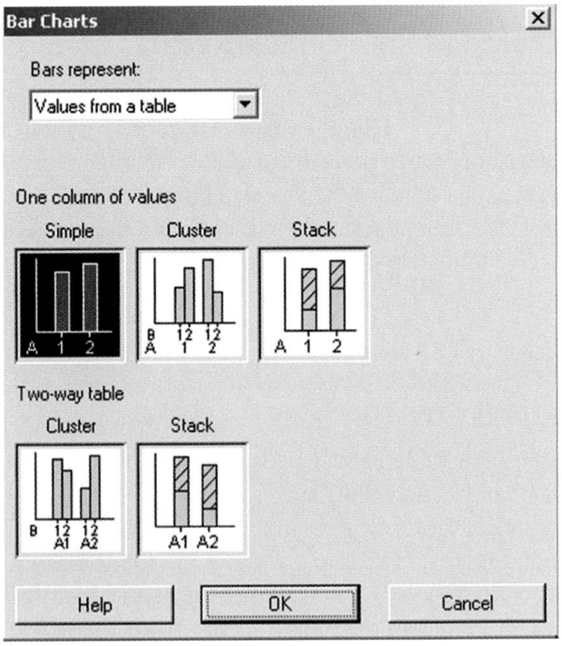

FIGURE A2.4 Minitab Bar Charts Dialog Box

Step 2: In the Bar Chart—Values from a table, One column of values, Simple dialog box (see Figure A2.5), enter **C2** or **Percentage (%)** in the Graph Variables: edit box. Enter **C1** or **Reason** in the Categorical variable: edit box. Click the **OK** button.

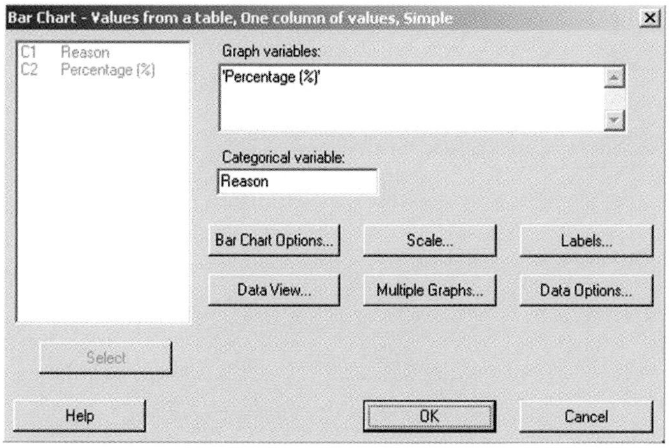

FIGURE A2.5 Minitab Bar Chart—Values from a Table, One Column of Values, Simple Dialog Box

To select colors for the bars and borders in the bar chart:

Step 1: Right-click on any of the bars of the bar chart.

Step 2: Select **Edit Bars**.

Step 3: In the Attributes tab of the Edit Bars dialog box, enter selections for Fill Pattern and Border and Fill Lines.

Pie Chart

To produce a pie chart similar to Figure 2.4 on page 25, open the **MUTUALFUNDS2004.MTW** worksheet. Select **Graph → Pie Chart**. Then do the following:

Step 1: In the Pie Chart dialog box (see Figure A2.6), select the **Chart raw data** option button since you are using the raw data from the worksheet. (If you are using the frequencies in each category such as from the **ONLINE SHOPPING.MTW** worksheet, select the **Chart values from a table** option button). Enter **C10** or **Risk** in the categorical variables: edit box.

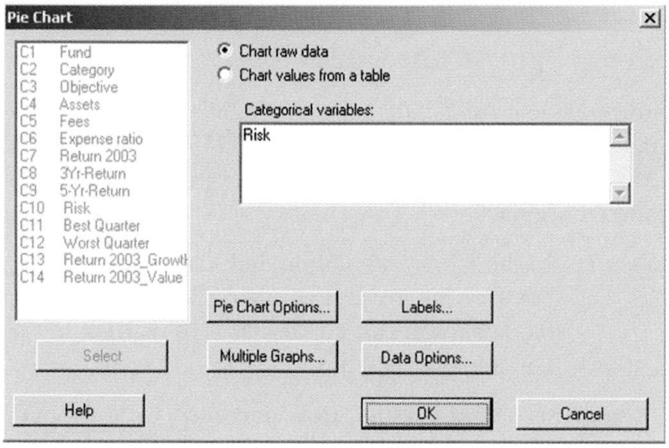

FIGURE A2.6 Minitab Pie Charts Dialog Box

Step 2: Select the **Labels** button. In the Pie Chart—Labels dialog box (see Figure A2.7), select the **Slice Labels** tab. Then select the **Category name** and **Percent** check boxes. Click the **OK** button to return to the Pie Chart dialog box. Click the **OK** button.

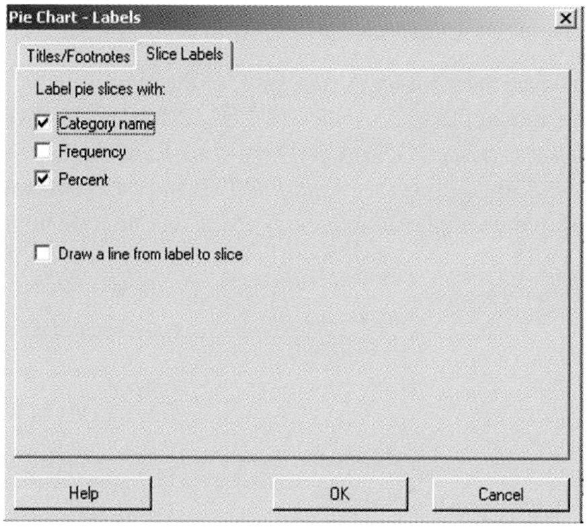

FIGURE A2.7 Minitab Pie Charts—Labels Dialog Box

Pareto Diagram

To produce the Pareto diagram of Figure 2.6 on page 27, open the **KEYBOARD.MTW** worksheet. This data set contains the causes of the defects in column C1 and the frequency of defects in column C2. Select **Stat → Quality Tools → Pareto Chart**. In the Pareto Chart dialog box (see Figure A2.8)

Step 1: Select the **Chart defects table** option button.

Step 2: In the Labels in: edit box, enter **C1** or **Cause**.

Step 3: In the Frequencies in: edit box, enter **C2** or **Frequency**.

Step 4: In the Combine defects after the first edit box, enter **99.9**.

Step 5: Click the **OK** button.

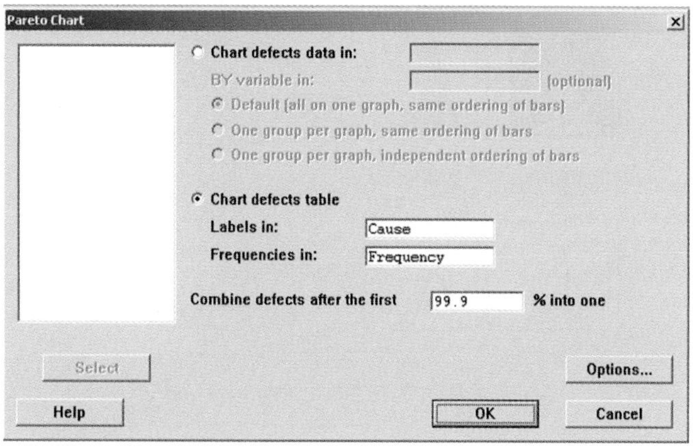

FIGURE A2.8 Minitab Pareto Chart Dialog Box

If the variable of interest was located in a single column and is in raw form with each row indicating a type of error, you would select the Charts defects data in: option button and you would enter the appropriate column number or variable name in the Chart defects data in: edit box.

Stem-and-Leaf Display

To produce the stem-and-leaf display of the returns in 2003 for all the mutual funds, open the **MUTUALFUNDS2004.MTW** worksheet. Select **Graph ➔ Stem-and-Leaf**. In the Stem-and-Leaf dialog box (see Figure A2.9), enter **C7** or **'Return 2003'** in the Graph variables: edit box. Click the **OK** button.

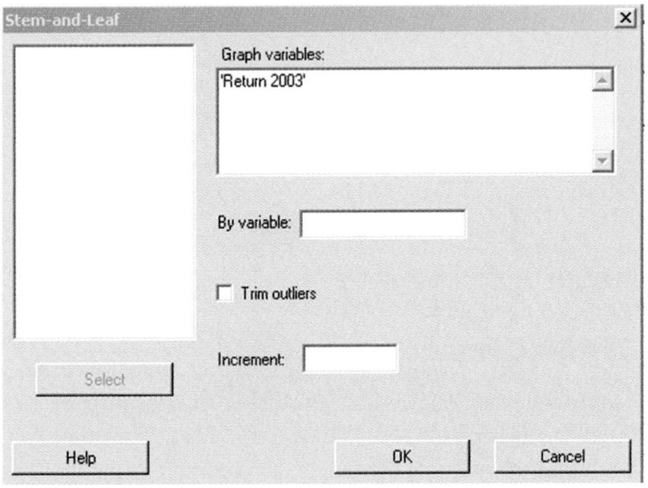

FIGURE A2.9 Minitab Stem-and-Leaf Dialog Box

Histogram

To produce the histogram of the returns in 2003 for all the mutual funds, open the **MUTUALFUNDS2004.MTW** worksheet. Select **Graph ➔ Histogram**.

Step 1: In the Histograms dialog box (see Figure A2.10), select **Simple**. Click the **OK** button.

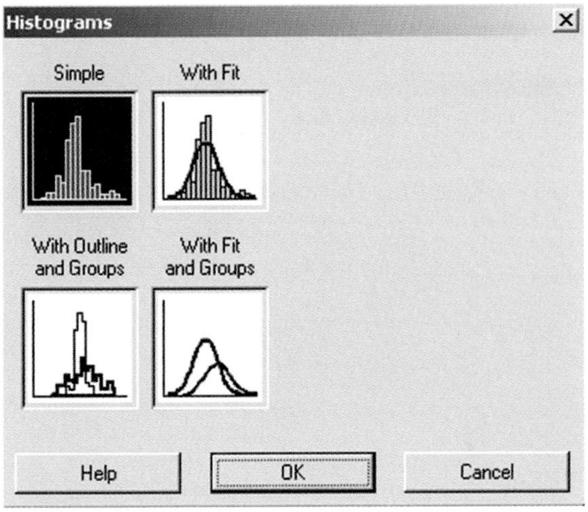

FIGURE A2.10 Minitab Histograms Dialog Box

Step 2: In the Histogram—Simple dialog box (see Figure A2.11), enter **C7** or **Return2003** in the Graph variables: edit box. Click the **OK** button.

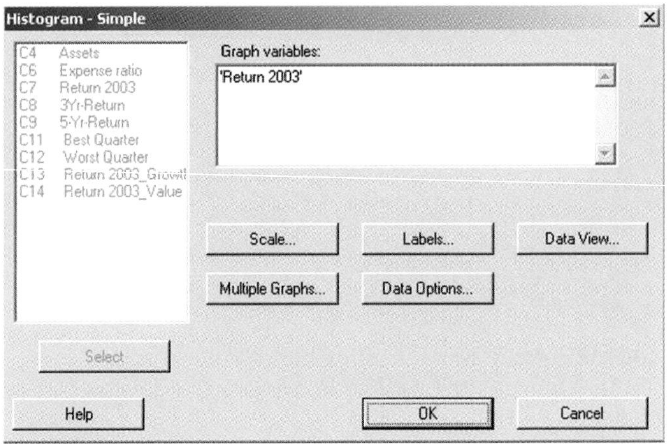

FIGURE A2.11 Minitab Histogram—Simple Dialog Box

To select colors for the bars and borders in the bar chart,

Step 1: Right-click on any of the bars of the bar chart.

Step 2: Select **Edit Bars**.

Step 3: In the Attributes tab of the Edit Bars dialog box, enter selections for Fill Pattern and Border and Fill Lines.

Step 4: To define your own class groupings, select the **Binning** tab. Select the **Midpoint** option button to specify midpoints or the **Cutpoints** option button to specify class limits. Select the **Midpoint/Cutpoint positions**: option button. Enter the set of values in the edit box.

If you wish to create separate histograms for the growth and value funds similar to Figure 2.9 on page 38, you first need to unstack the data (see page 66) and create separate variables for the 2003 return for the growth and value funds. Then you can separately create histograms for each of the two groups.

Cross-Tabulation Table

To produce cross-tabulation tables similar to Tables 2.14 through 2.17 on page 44, open the **MUTUALFUNDS2004.MTW** worksheet. Select **Stat ➔ Tables ➔ Cross Tabulation and Chi-Square**.

Step 1: In the Cross Tabulation and Chi-Square dialog box (see Figure A2.12), enter **C3** or **Objective** in the For rows: edit box. Enter **C10** or **Risk** in the For columns: edit box.

Step 2: Select the **Counts**, **Row percents**, **Column percents**, and **Total percents** check boxes. Click the **OK** button.

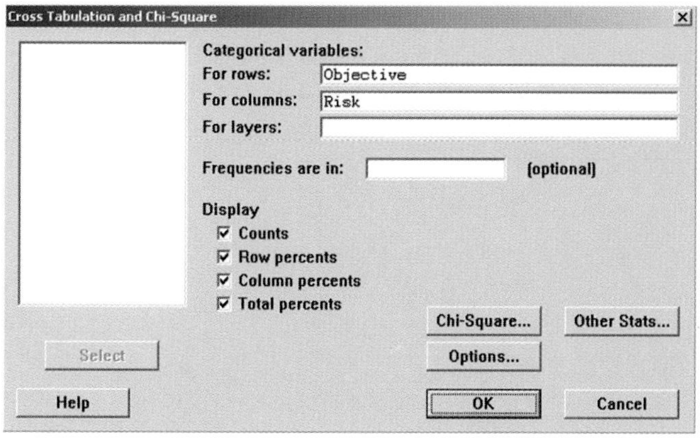

FIGURE A2.12 Minitab Cross-Tabulation and Chi-Square Dialog Box

Side-by-Side Bar Chart

To produce a side-by-side bar chart similar to Figure 2.14 on page 45, open the **MUTUALFUNDS2004.MTW** worksheet. Select **Graph → Bar Chart**

Step 1: In the Bar Charts dialog box (see Figure A2.4 on page 66), in the Bars represent: drop-down list box, select **Counts of unique values** since you are using raw data. Select the **Cluster** graph. Click the **OK** button.

Step 2: In the Bar Chart—Counts of unique values, Cluster dialog box (see Figure A2.13), enter **C3** or **Objective** and **C10** or **Risk** in the Categorical variables: edit box. Click the **OK** button.

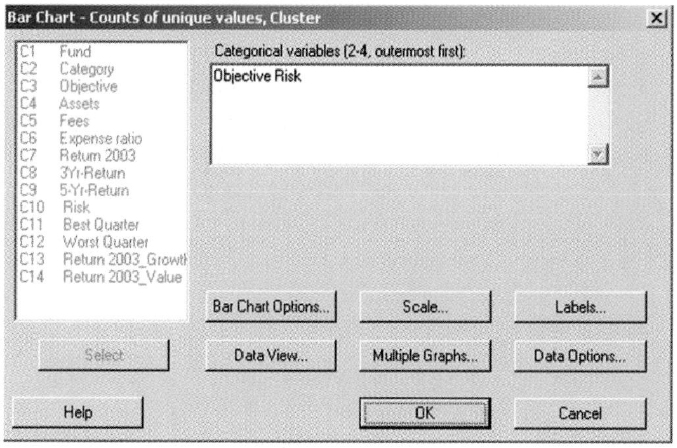

FIGURE A2.13 Minitab Bar Chart—Counts of Unique Values, Cluster Dialog Box

Scatter Diagram and Time-Series Plot

To produce the scatter diagram of the expense ratio and 2003 return for all the mutual funds (see Figure 2.15 on page 47), open the **MUTUALFUNDS2004.MTW** worksheet. Select **Graph → Scatterplot**.

Step 1: In the Scatterplots dialog box (see Figure A2.14), select **Simple**. Click the **OK** button.

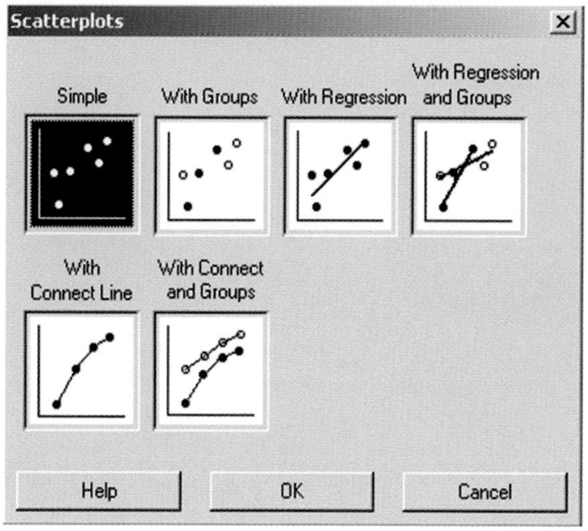

FIGURE A2.14 Minitab Scatterplots Dialog Box

Step 2: In the Scatterplot—Simple dialog box (see Figure A2.15), enter **C7** or **'Return 2003'** in the Y variables: edit box in row 1. Enter **C6** or **'Expense ratio'** in the X variables: edit box in row 1. Click the **OK** button.

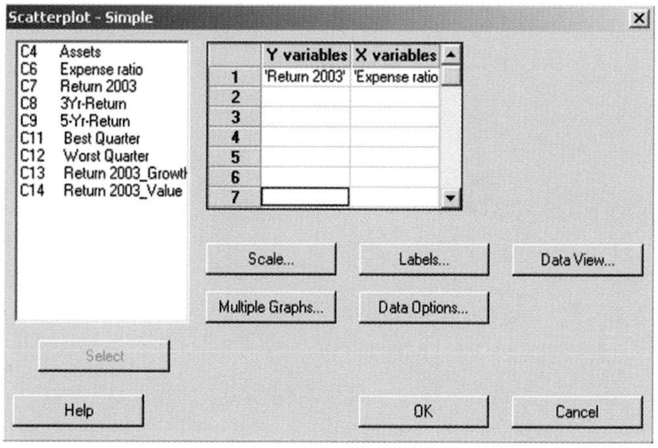

FIGURE A2.15 Minitab Scatterplot—Simple Dialog Box

To create a time-series plot, place the time period on the *X* axis and the variable of interest on the *Y* axis.

CHAPTER 3

Numerical Descriptive Measures

LEARNING OBJECTIVES

In this chapter, you learn:

- To describe the properties of central tendency, variation, and shape in numerical data
- To calculate descriptive summary measures for a population
- To construct and interpret a box-and-whisker plot
- To describe the covariance and the coefficient of correlation

USING STATISTICS

Evaluating the Performance of Mutual Funds

Return to the study of mutual funds introduced in Chapter 2. You want to decide which types of mutual funds to invest in. In the last chapter you learned how to *present* data in tables and charts. However, when dealing with numerical data, such as the return on investments in mutual funds in 2003, you also need to summarize the data, and ask statistical questions. What is the central tendency for returns of the various funds? For example, what is the mean return in 2003 for the low-risk, average-risk, and high-risk mutual funds? How much variability is present in the returns? Are the returns for high-risk funds more variable than for average-risk funds or low-risk funds? How can you use this information when deciding what mutual funds to invest in?

For numerical variables, you need more than just the visual picture of what a variable looks like than you get from the graphs discussed in Chapter 2. For example, for the 2003 returns, you would like to determine not only whether the riskier funds had a higher 2003 return, but whether they also had greater variation, and how the returns for each risk group were distributed. You also want to examine whether there is a relationship between the expense ratio and the 2003 return. Reading this chapter will allow you to learn about some of the methods to measure:

- **central tendency**, the extent to which all of the data values group around a central value
- **variation**, the amount of dispersion or scattering of values away from a central point
- **shape**, the pattern of the distribution of values from the lowest value to the highest value

You will also learn about the covariance and the coefficient of correlation that help measure the strength of the association between two numerical variables.

3.1 MEASURES OF CENTRAL TENDENCY, VARIATION, AND SHAPE

You can characterize any set of data by measuring its central tendency, variation, and shape. Most sets of data show a distinct central tendency to group around a central point. When people talk about an "average value" or the "middle value" or the most popular or frequent value, they are talking informally about the mean, median, and mode, three measures of central tendency.

Variation measures the **spread** or **dispersion** of values in a data set. One simple measure of variation is the range, the difference between the highest and lowest value. More commonly used in statistics are the standard deviation and variance, two measures explained later in this section. The shape of a data set represents a pattern of all the values from the lowest to highest value. As you will learn later in this section, many data sets have a pattern that looks approximately like a bell, with a peak of values somewhere in the middle.

The Mean

The **arithmetic mean** (typically referred to as the **mean**) is the most common measure of central tendency. The mean is the only common measure in which all the values play an equal role. The mean serves as a "balance point" in a set of data (like the fulcrum on a seesaw). You calculate the mean by adding together all the values in a data set and then dividing that sum by the *number* of values in the data set.

The symbol $\bar{X}$, called *X bar*, is used to represent the mean of a sample. For a sample containing *n* values, the equation for the mean of a sample, is written as

$$\bar{X} = \frac{\text{sum of the values}}{\text{number of values}}$$

Using the series $X_1, X_2, \ldots, X_n$ to represent the set of *n* values and *n* to represent the number of values, the equation becomes:

$$\bar{X} = \frac{X_1 + X_2 + \cdots + X_n}{n}$$

By using summation notation (discussed fully in Appendix B), you replace the numerator $X_1 + X_2 + \cdots + X_n$ by the term $\sum_{i=1}^{n} X_i$ that means sum all the X_i values from the first *X* value, X_1, to the last *X* value, X_n, to form Equation (3.1), a formal definition of the sample mean.

SAMPLE MEAN

The **sample mean** is the sum of the values divided by the number of values.

$$\bar{X} = \frac{\sum_{i=1}^{n} X_i}{n} \qquad \textbf{(3.1)}$$

where $\bar{X}$ = sample mean

n = number of values or sample size

X_i = *i*th value of the variable *X*

$\sum_{i=1}^{n} X_i$ = summation of all X_i values in the sample

Because all the values play an equal role, a mean will be greatly affected by any value that is greatly different from the others in the data set. When you have such extreme values, you should avoid using the mean.

The mean can suggest what is a "typical" or central value for a data set. For example, if you knew the typical time it takes you to get ready in the morning, you might be able to better plan your morning and minimize any excessive lateness (or earliness) going to your destination. Suppose you define the time to get ready as the time in minutes (rounded to the nearest minute) from when you get out of bed to when you leave your home. You collect the times shown below for 10 consecutive work days:

Day:	1	2	3	4	5	6	7	8	9	10
Time (minutes):	39	29	43	52	39	44	40	31	44	35

TIMES

The mean time is 39.6 minutes, computed as follows:

$$\bar{X} = \frac{\text{sum of the values}}{\text{number of values}}$$

$$\bar{X} = \frac{\sum_{i=1}^{n} X_i}{n}$$

$$\bar{X} = \frac{39 + 29 + 43 + 52 + 39 + 44 + 40 + 31 + 44 + 35}{10}$$

$$\bar{X} = \frac{396}{10} = 39.6$$

Even though no one day in the sample actually had the value 39.6 minutes, allotting about 40 minutes to get ready would be a good rule for planning your mornings, but only because the 10 days does not contain extreme values.

Contrast this to the case in which the value on day four was 102 minutes instead of 52 minutes. This extreme value would cause the mean to rise to 44.6 minutes as follows:

$$\bar{X} = \frac{\text{sum of the values}}{\text{number of values}}$$

$$\bar{X} = \frac{\sum_{i=1}^{n} X_i}{n}$$

$$\bar{X} = \frac{446}{10} = 44.6$$

The one extreme value has increased the mean by more than 10% from 39.6 to 44.6 minutes. In contrast to the original mean that was in the "middle," greater than 5 of the get-ready times (and less than the 5 other times), the new mean is greater than 9 of the 10 get-ready times. The extreme value has caused the mean to be a poor measure of central tendency.

EXAMPLE 3.1

THE MEAN 2003 RETURN FOR SMALL CAP MUTUAL FUNDS WITH HIGH RISK

The 121 mutual funds that are part of the "Using Statistics" scenario (see page 72) are classified according to the risk level of the mutual funds (low, average, and high) and type (small cap, mid cap, and large cap). Compute the mean 2003 return for the small cap mutual funds with high risk.

SOLUTION The mean 2003 return for the small cap mutual funds with high risk (**MUTUALFUNDS2004**) is 51.53, calculated as follows:

$$\bar{X} = \frac{\text{sum of the values}}{\text{number of values}}$$

$$= \frac{\sum_{i=1}^{n} X_i}{n}$$

$$= \frac{463.8}{9} = 51.53$$

The ordered array for the nine small cap mutual funds with high risk is:

37.3 39.2 44.2 44.5 53.8 56.6 59.3 62.4 66.5

Four of these returns are below the mean of 51.53 and five of these returns are above the mean.

The Median

The **median** is the value that splits a ranked set of data into two equal parts. The median is not affected by extreme values, so you can use the median when extreme values are present.

The median is the middle value in a set of data that has been ordered from lowest to highest value.

To calculate the median for a set of data, you first rank the values from smallest to largest. Then use Equation (3.2) to compute the rank of the value that is the median.

MEDIAN

50% of the values are smaller than the median and 50% of the values are larger than the median.

$$\text{Median} = \frac{n+1}{2} \text{ ranked value} \qquad \textbf{(3.2)}$$

You compute the median value by following one of two rules:

- ***Rule 1*** If there are an *odd* number of values in the data set, the median is the middle ranked value.
- ***Rule 2*** If there are an *even* number of values in the data set, then the median is the *average* of the two middle ranked values.

To compute the median for the sample of 10 times to get ready in the morning, you rank the daily times as follows:

Ranked values:

29 31 35 39 39 40 43 44 44 52

Ranks:

1 2 3 4 5 6 7 8 9 10

↑

Median = 39.5

Because the result of dividing $n + 1$ by 2 is $(10 + 1)/2 = 5.5$ for this sample of 10, you must use Rule 2 and average the fifth and sixth ranked values, 39 and 40. Therefore, the median is 39.5. The median of 39.5 means that for half of the days, the time to get ready is less than or equal to 39.5 minutes, and for half of the days the time to get ready is greater than or equal to 39.5 minutes. The median time to get ready of 39.5 minutes is very close to the mean time to get ready of 39.6 minutes.

EXAMPLE 3.2

COMPUTING THE MEDIAN FROM AN ODD-SIZED SAMPLE

The 121 mutual funds that are part of the "Using Statistics" scenario (see page 72) are classified according to the risk level of the mutual funds (low, average, and high) and type (small cap, mid cap, and large cap). Compute the median 2003 return for the nine small cap mutual funds with high risk. **MUTUALFUNDS2004**

SOLUTION Because the result of dividing $n + 1$ by 2 is $(9 + 1)/2 = 5$ for this sample of nine, using Rule 1, the median is the fifth ranked value. The percentage return in 2003 for the nine small cap mutual funds with high risk are ranked from the smallest to the largest:

Ranked values:

37.3 39.2 44.2 44.5 53.8 56.6 59.3 62.4 66.5

Ranks:

1 2 3 4 5 6 7 8 9

↑

Median

The median return is 53.8. Half the small cap high-risk mutual funds have returns equal to or below 53.8 and half have returns equal to or above 53.8.

The Mode

The **mode** is the value in a set of data that appears most frequently. Like the median and unlike the mean, extreme values do not affect the mode. You should use the mode only for descriptive purposes as it is more variable from sample to sample than either the mean or the median. Often there is no mode or there are several modes in a set of data. For example, consider the time to get ready data shown below.

29 31 35 39 39 40 43 44 44 52

There are two modes, 39 minutes and 44 minutes, since each of these values occurs twice.

EXAMPLE 3.3 COMPUTING THE MODE

A systems manager in charge of a company's network keeps track of the number of server failures that occur in a day. Compute the mode for the following data that represents the number of server failures in a day for the past two weeks.

1 3 0 3 26 2 7 4 0 2 3 3 6 3

SOLUTION The ordered array for these data is

0 0 1 2 2 3 3 3 3 3 4 6 7 26

Because 3 appears five times, more times than any other value, the mode is 3. Thus, the systems manager can say that the most common occurrence is having three server failures in a day. For this data set, the median is also equal to 3 while the mean is equal to 4.5. The extreme value 26 is an outlier. For these data, the median and the mode better measure central tendency than the mean.

A set of data will have no mode if none of the values is "most typical." Example 3.4 presents a data set with no mode.

EXAMPLE 3.4 DATA WITH NO MODE

Compute the mode for the 2003 return for the small cap mutual funds with high risk. MUTUALFUNDS2004

SOLUTION The ordered array for these data is

37.3 39.2 44.2 44.5 53.8 56.6 59.3 62.4 66.5

These data have no mode. None of the values is most typical because each value appears once.

Quartiles

Quartiles split a set of data into four equal parts—the **first quartile Q_1** divides the smallest 25.0% of the values from the other 75.0% that are larger. The **second quartile Q_2** is the median—50.0% of the values are smaller than the median and 50.0% are larger. The **third quartile Q_3** divides the smallest 75.0% of the values from the largest 25.0%. Equations (3.3) and (3.4) define the first and third quartiles.[1]

[1] *The Q_1, median, and Q_3 are also the 25th, 50th, and 75th percentile, respectively. Equations (3.2), (3.3), and (3.4) can be expressed generally in terms of finding percentiles: (p * 100)th percentile = p * (n + 1) ranked value.*

FIRST QUARTILE Q_1

25.0% of the values are smaller than Q_1, the first quartile, and 75.0% are larger than the first quartile Q_1.

$$Q_1 = \frac{n+1}{4} \text{ ranked value} \tag{3.3}$$

THIRD QUARTILE Q_3

75.0% of the values are smaller than the third quartile Q_3, and 25.0% are larger than the third quartile Q_3.

$$Q_3 = \frac{3(n+1)}{4} \text{ ranked value} \tag{3.4}$$

Use the following rules to calculate the quartiles:

- ***Rule 1*** If the result is a whole number, then the quartile is equal to that ranked value. For example, if the sample size $n = 7$, the first quartile Q_1 is equal to the $(7 + 1)/4$ = second ranked value.
- ***Rule 2*** If the result is a fractional half (2.5, 4.5, etc.), then the quartile is equal to the average of the corresponding ranked values. For example, if the sample size $n = 9$, the first quartile Q_1 is equal to the $(9 + 1)/4 = 2.5$ ranked value, halfway between the second ranked value and the third ranked value.
- ***Rule 3*** If the result is neither a whole number nor a fractional half, you round the result to the nearest integer and select that ranked value. For example, if the sample size $n = 10$, the first quartile Q_1 is equal to the $(10 + 1)/4 = 2.75$ ranked value. Round 2.75 to 3 and use the third ranked value.

To illustrate the computation of the quartiles for the time-to-get-ready data, rank the data from smallest to largest.

Ranked values:

29 31 35 39 39 40 43 44 44 52

Ranks:

1 2 3 4 5 6 7 8 9 10

The first quartile is the $(n + 1)/4 = (10 + 1)/4 = 2.75$ ranked value. Using the third rule for quartiles, you round up to the third ranked value. The third ranked value for the get-ready time data is 35 minutes. You interpret the first quartile of 35 to mean that on 25% of the days the time to get ready is less than or equal to 35 minutes, and on 75% of the days, the time to get ready is greater than or equal to 35 minutes.

The third quartile is the $3(n + 1)/4 = 3(10 + 1)/4 = 8.25$ ranked value. Using the third rule for quartiles, you round this down to the eighth ranked value. The eighth ranked value for the get-ready time data is 44 minutes. You interpret this to mean that on 75% of the days, the time to get ready is less than or equal to 44 minutes, and on 25% of the days, the time to get ready is greater than or equal to 44 minutes.

EXAMPLE 3.5 COMPUTING THE QUARTILES

The 121 mutual funds that are part of the "Using Statistics" scenario (see page 72) are classified according to the risk level of the mutual funds (low, average, and high) and type (small cap, mid cap, and large cap). Compute the first quartile (Q_1) and third quartile (Q_3) 2003 return for the small cap mutual funds with high risk. **MUTUALFUNDS2004**

SOLUTION Ranked from smallest to largest, the percentage return in 2003 for the nine small cap mutual funds with high risk is:

Ranked value:	37.3	39.2	44.2	44.5	53.8	56.6	59.3	62.4	66.5
Ranks:	1	2	3	4	5	6	7	8	9

For these data

$$Q_1 = \frac{(n+1)}{4} \text{ ranked value}$$

$$= \frac{9+1}{4} \text{ ranked value} = 2.5 \text{ ranked value}$$

Therefore, using the second rule, Q_1 is the 2.5 ranked value, halfway between the second ranked value and the third ranked value. Since the second ranked value is 39.2 and the third ranked value is 44.2, the first quartile Q_1 is halfway between 39.2 and 44.2. Thus,

$$Q_1 = \frac{39.2 + 44.2}{2} = 41.7$$

To find the third quartile Q_3

$$Q_3 = \frac{3(n+1)}{4} \text{ ranked value}$$

$$= \frac{3(9+1)}{4} \text{ ranked value} = 7.5 \text{ ranked value}$$

Therefore, using the second rule, Q_3 is the 7.5 ranked value, halfway between the seventh ranked value and the eighth ranked value. Since the seventh ranked value is 59.3 and the eighth ranked value is 62.4, the third quartile Q_3 is halfway between 59.3 and 62.4. Thus,

$$Q_3 = \frac{59.3 + 62.4}{2} = 60.85$$

The first quartile of 41.7 indicates that 25% of the returns in 2003 for small cap high-risk funds are below or equal to 41.7 and 75% are greater than or equal to 41.7. The third quartile of 60.85 indicates that 75% of the returns in 2003 for small cap high-risk funds are below or equal to 60.85 and 25% are greater than or equal to 60.85.

The Geometric Mean

The geometric mean and the geometric rate of return measure the status of an investment over time. The **geometric mean** measures the rate of change of a variable over time. Equation (3.5) defines the geometric mean.

GEOMETRIC MEAN

The geometric mean is the *n*th root of the product of *n* values

$$\bar{X}_G = (X_1 \times X_2 \times \cdots \times X_n)^{1/n} \quad \textbf{(3.5)}$$

Equation (3.6) defines the geometric mean rate of return.

GEOMETRIC MEAN RATE OF RETURN

$$\bar{R}_G = [(1 + R_1) \times (1 + R_2) \times \cdots \times (1 + R_n)]^{1/n} - 1 \quad \textbf{(3.6)}$$

where R_i is the rate of return in time period i

To illustrate using these measures, consider an investment of \$100,000 that declined to a value of \$50,000 at the end of year 1 and then rebounded back to its original \$100,000 value at the end of year 2. The rate of return for this investment for the two-year period is 0, because the starting and ending value of the investment is unchanged. However, the arithmetic mean of the yearly rates of return of this investment is

$$\bar{X} = \frac{(-0.50) + (1.00)}{2} = 0.25 \text{ or } 25\%$$

since the rate of return for year 1 is

$$R_1 = \left(\frac{50,000 - 100,000}{100,000}\right) = -0.50 \text{ or } -50\%$$

and the rate of return for year 2 is

$$R_2 = \left(\frac{100,000 - 50,000}{50,000}\right) = 1.00 \text{ or } 100\%$$

Using Equation (3.6), the geometric mean rate of return for the two years, is

$$\begin{aligned}\bar{R}_G &= [(1 + R_1) \times (1 + R_2)]^{1/n} - 1 \\ &= [(1 + (-0.50)) \times (1 + (1.0))]^{1/2} - 1 \\ &= [(0.50) \times (2.0)]^{1/2} - 1 \\ &= [1.0]^{1/2} - 1 \\ &= 1 - 1 = 0\end{aligned}$$

Thus, the geometric mean rate of return more accurately reflects the (zero) change in the value of the investment for the two-year period than does the arithmetic mean.

EXAMPLE 3.6

COMPUTING THE GEOMETRIC MEAN RATE OF RETURN

The percentage change in the NASDAQ Composite Index was –31.53% in 2002 and +50.01% in 2003. Compute the geometric rate of return.

SOLUTION Using Equation (3.6), the geometric mean rate of return in the NASDAQ Composite Index for the two years is

$$\begin{aligned}\overline{R}_G &= [(1 + R_1) \times (1 + R_2)]^{1/n} - 1\\ &= [(1 + (-0.3153)) \times (1 + (0.5001))]^{1/2} - 1\\ &= [(0.6847) \times (1.5001)]^{1/2} - 1\\ &= [1.0271]^{1/2} - 1\\ &= 1.0135 - 1 = 0.0135\end{aligned}$$

The geometric rate of return in the NASDAQ Composite Index for the two years is 1.35%.

The Range

The **range** is the simplest numerical descriptive measure of variation in a set of data.

RANGE

The range is equal to the largest value minus the smallest value.

$$\text{Range} = X_{\text{largest}} - X_{\text{smallest}} \quad \textbf{(3.7)}$$

To determine the range of the times to get ready, you rank the data from smallest to largest:

29 31 35 39 39 40 43 44 44 52

Using Equation (3.7), the range is 52 – 29 = 23 minutes. The range of 23 minutes indicates that the largest difference between any two days in the time to get ready in the morning is 23 minutes.

EXAMPLE 3.7

COMPUTING THE RANGE IN THE 2003 RETURN OF SMALL CAP HIGH-RISK MUTUAL FUNDS

The 121 mutual funds that are part of the "Using Statistics" scenario (see page 72) are classified according to the risk level of the mutual funds (low, average, and high) and type (small cap, mid cap, and large cap). Compute the range of the 2003 return for the small cap mutual funds with high risk. **MUTUALFUNDS2004**

SOLUTION Ranked from the smallest to the largest, the 2003 return for the nine small cap mutual funds with high risk is:

37.3 39.2 44.2 44.5 53.8 56.6 59.3 62.4 66.5

Therefore, using Equation (3.7), the range = 66.5 – 37.3 = 29.2.

The largest difference between any two returns for the small cap mutual funds with high risk is 29.2.

The range measures the *total spread* in the set of data. Although the range is a simple measure of total variation in the data, it does not take into account *how* the data are distributed between the smallest and largest values. In other words, the range does not indicate if the values are evenly distributed throughout the data set, clustered near the middle, or clustered near one or both extremes. Thus, using the range as a measure of variation when at least one value is an extreme value is misleading.

The Interquartile Range

The **interquartile range** (also called **midspread**) is the difference between the *third* and *first quartiles* in a set of data.

INTERQUARTILE RANGE

The interquartile range is the difference between the third quartile and the first quartile.

$$\text{Interquartile range} = Q_3 - Q_1 \qquad \textbf{(3.8)}$$

The interquartile range measures the spread in the middle 50% of the data; therefore, it is not influenced by extreme values. To determine the interquartile range of the times to get ready

29 31 35 39 39 40 43 44 44 52

you use Equation (3.8) and the earlier results on page 78, $Q_1 = 35$ and $Q_3 = 44$.

$$\text{Interquartile range} = 44 - 35 = 9 \text{ minutes}$$

Therefore, the interquartile range in the time to get ready is 9 minutes. The interval 35 to 44 is often referred to as the *middle fifty*.

EXAMPLE 3.8

COMPUTING THE INTERQUARTILE RANGE FOR THE 2003 RETURN OF SMALL CAP HIGH-RISK MUTUAL FUNDS

The 121 mutual funds that are part of the "Using Statistics" scenario (see page 72) are classified according to the risk level of the mutual funds (low, average, and high) and type (small cap, mid cap, and large cap). Compute the interquartile range of the 2003 return for the small cap mutual funds with high risk. **MUTUALFUNDS2004**

SOLUTION Ranked from smallest to largest, the 2003 return for the nine small cap mutual funds with high risk is:

37.3 39.2 44.2 44.5 53.8 56.6 59.3 62.4 66.5

Using Equation (3.8) and the earlier results on page 78, $Q_1 = 41.7$ and $Q_3 = 60.85$.

$$\text{Interquartile range} = 60.85 - 41.7 = 19.15$$

Therefore, the interquartile range in the 2003 return is 19.15.

Because the interquartile range does not consider any value smaller than Q_1 or larger than Q_3, it cannot be affected by extreme values. Summary measures such as the median, Q_1, Q_3, and the interquartile range, which cannot be influenced by extreme values, are called **resistant measures**.

The Variance and the Standard Deviation

Although the range and the interquartile range are measures of variation, they do not take into consideration *how* the values distribute or cluster between the extremes. Two commonly used measures of variation that take into account how all the values in the data are distributed are the **variance** and the **standard deviation**. These statistics measure the "average" scatter around the mean—how larger values fluctuate above it and how smaller values distribute below it.

A simple measure of variation around the mean might take the difference between each value and the mean and then sum these differences. However, if you did that, you would find that because the mean is the balance point in a set of data, for *every* set of data these differences would sum to zero. One measure of variation that would differ from data set to data set would *square* the difference between each value and the mean and then sum these squared differences. In statistics, this quantity is called a **sum of squares** (or **SS**). This sum is then divided by the number of values minus 1 (for sample data) to get the sample variance (S^2). The square root of the sample variance is the sample standard deviation (S).

Because the sum of squares are a sum of squared differences that by the rules of arithmetic will always be nonnegative, *neither the variance nor the standard deviation can ever be negative*. For most sets of data, the variance and standard deviation will be a positive value, although both of these statistics will be zero if there is no variation at all in a set of data and each value in the sample is the same.

For a sample containing n values, $X_1, X_2, X_3, \ldots, X_n$, the sample variance (given by the symbol S^2) is

$$S^2 = \frac{(X_1 - \bar{X})^2 + (X_2 - \bar{X})^2 + \cdots + (X_n - \bar{X})^2}{n - 1}$$

Equation (3.9) expresses the equation using summation notation.

SAMPLE VARIANCE

The **sample variance** is the sum of the squared differences around the mean divided by the sample size minus one.

$$S^2 = \frac{\sum_{i=1}^{n}(X_i - \bar{X})^2}{n - 1} \tag{3.9}$$

where $\bar{X}$ = mean

n = sample size

X_i = ith value of the variable X

$\sum_{i=1}^{n}(X_i - \bar{X})^2$ = summation of all the squared differences between the X_i values and $\bar{X}$

SAMPLE STANDARD DEVIATION

The **sample standard deviation** is the square root of the sum of the squared differences around the mean divided by the sample size minus one.

$$S = \sqrt{S^2} = \sqrt{\frac{\sum_{i=1}^{n}(X_i - \bar{X})^2}{n - 1}} \tag{3.10}$$

If the denominator were n instead of $n - 1$, Equation (3.9) [and the inner term in Equation (3.10)] would calculate the average of the squared differences around the mean. However, $n - 1$ is used because of certain desirable mathematical properties possessed by the statistic S^2 that make it appropriate for statistical inference (which will be discussed in Chapter 7). As the sample size increases, the difference between dividing by n or $n - 1$ becomes smaller and smaller.

You will most likely use the sample standard deviation as your measure of variation [defined in Equation (3.10)]. Unlike the sample variance, which is a squared quantity, the standard deviation is always a number that is in the same units as the original sample data. The standard deviation helps you to know how a set of data clusters or distributes around its mean. For almost all sets of data, the majority of the observed values lie within an interval of plus and minus one standard deviation above and below the mean. Therefore, knowledge of the mean and the standard deviation usually helps define where at least the majority of the data values are clustering.

To hand-calculate the sample variance S^2 and the sample standard deviation S:

Step 1: Compute the difference between each value and the mean.

Step 2: Square each difference.

Step 3: Add the squared differences.

Step 4: Divide this total by $n - 1$ to get the sample variance.

Step 5: Take the square root of the sample variance to get the sample standard deviation.

Table 3.1 shows the first four steps for calculating the variance and standard deviation for the getting ready times data with a mean ($\bar{X}$) equal to 39.6 (see page 74 for the calculation of the mean). The second column of Table 3.1 shows Step 1. The third column of Table 3.1 shows Step 2. The sum of the squared differences (Step 3) is shown at the bottom of Table 3.1. This total is then divided by $10 - 1 = 9$ to compute the variance (Step 4).

TABLE 3.1

Computing the Variance of the Getting Ready Times

$\bar{X} = 39.6$

Time (X)	***Step 1:*** $(X_i - \bar{X})$	***Step 2:*** $(X_i - \bar{X})^2$
39	−0.60	0.36
29	−10.60	112.36
43	3.40	11.56
52	12.40	153.76
39	−0.60	0.36
44	4.40	19.36
40	0.40	0.16
31	−8.60	73.96
44	4.40	19.36
35	−4.60	21.16

	Step 3: **Sum:**	***Step 4:*** **Divide by (n − 1):**
	412.40	45.82

You can also calculate the variance by substituting values for the terms in Equation (3.9):

$$S^2 = \frac{\sum_{i=1}^{n}(X_i - \bar{X})^2}{n-1}$$

$$= \frac{(39-39.6)^2 + (29-39.6)^2 + \cdots + (35-39.6)^2}{10-1}$$

$$= \frac{412.4}{9}$$

$$= 45.82$$

Because the variance is in squared units (in squared minutes for these data), to compute the standard deviation you take the square root of the variance. Using Equation (3.10) on page 82, the sample standard deviation S is

$$S = \sqrt{S^2} = \sqrt{\frac{\sum_{i=1}^{n}(X_i - \bar{X})^2}{n-1}} = \sqrt{45.82} = 6.77$$

This indicates that the get-ready times in this sample are clustering within 6.77 minutes around the mean of 39.6 minutes (i.e., clustering between $\bar{X} - 1S = 32.83$ and $\bar{X} + 1S = 46.37$). In fact, 7 out of 10 get-ready times lie within this interval.

Using the second column of Table 3.1, you can also calculate the sum of the differences between each value and the mean to be zero. For any set of data, this sum will always be zero:

$$\sum_{i=1}^{n}(X_i - \bar{X}) = 0 \text{ for all sets of data}$$

This property is one of the reasons that the mean is used as the most common measure of central tendency.

EXAMPLE 3.9

COMPUTING THE VARIANCE AND STANDARD DEVIATION OF THE 2003 RETURN OF SMALL CAP HIGH-RISK MUTUAL FUNDS

The 121 mutual funds that are part of the "Using Statistics" scenario (see page 72) are classified according to the risk level of the mutual funds (low, average, and high) and type (small cap, mid cap, and large cap). Compute the variance and standard deviation of the 2003 return for the small cap mutual funds with high risk. **MUTUALFUNDS2004**

SOLUTION Table 3.2 illustrates the computation of the variance and standard deviation for the return in 2003 for the small cap mutual funds with high risk. Using Equation (3.9) on page 82

$$S^2 = \frac{\sum_{i=1}^{n}(X_i - \bar{X})^2}{n-1}$$

$$= \frac{(44.5-51.53)^2 + (39.2-51.53)^2 + \cdots + (66.5-51.53)^2}{9-1}$$

$$= \frac{891.16}{8}$$

$$= 111.395$$

TABLE 3.2

Computing the Variance of the 2003 Return for the Small Cap Mutual Funds with High Risk

$\bar{X} = 51.5333$

Return 2003	***Step 1:*** $(X_i - \bar{X})$	***Step 2:*** $(X_i - \bar{X})^2$
44.5	−7.0333	49.4678
39.2	−12.3333	152.1111
62.4	10.8667	118.0844
59.3	7.7667	60.3211
56.6	5.0667	25.6711
53.8	2.2667	5.1378
37.3	−14.2333	202.5878
44.2	−7.3333	53.7778
66.5	14.9667	224.0011

	Step 3: **Sum:**	***Step 4:*** **Divide by $(n - 1)$:**
	891.16	111.395

Using Equation (3.10) on page 82, the sample standard deviation S is

$$S = \sqrt{S^2} = \sqrt{\frac{\sum_{i=1}^{n}(X_i - \bar{X})^2}{n-1}} = \sqrt{111.395} = 10.55$$

The standard deviation of 10.55 indicates that the 2003 returns for the small cap mutual funds with high risk are clustering within 10.55 around the mean of 51.53 (i.e., clustering between $\bar{X} - 1S = 40.98$ and $\bar{X} + 1S = 62.08$). In fact, 55.6% (5 out of 9) of the 2003 returns lie within this interval.

The following summarizes the characteristics of the range, interquartile range, variance, and standard deviation.

- The more spread out, or dispersed, the data are, the larger the range, interquartile range, variance, and standard deviation.
- The more concentrated, or homogeneous the data are, the smaller the range, interquartile range, variance, and standard deviation.
- If the values are all the same (so that there is no variation in the data), the range, interquartile range, variance, and standard deviation will all equal zero.
- None of the measures of variation (the range, interquartile range, standard deviation, and variance) can *ever* be negative.

The Coefficient of Variation

Unlike the previous measures of variation presented, the **coefficient of variation** is a *relative measure* of variation that is always expressed as a percentage rather than in terms of the units of the particular data. The coefficient of variation, denoted by the symbol *CV*, measures the scatter in the data relative to the mean.

COEFFICIENT OF VARIATION

The coefficient of variation is equal to the standard deviation divided by the mean, multiplied by 100%.

$$CV = \left(\frac{S}{\overline{X}}\right)100\% \quad \textbf{(3.11)}$$

where S = sample standard deviation

$\overline{X}$ = sample mean

For the sample of 10 get-ready times, since $\overline{X} = 39.6$ and $S = 6.77$, the coefficient of variation is

$$CV = \left(\frac{S}{\overline{X}}\right)100\% = \left(\frac{6.77}{39.6}\right)100\% = 17.10\%$$

For the get-ready times, the standard deviation is 17.1% of the size of the mean.

You will find the coefficient of variation very useful when comparing two or more sets of data that are measured in different units as Example 3.10 illustrates.

EXAMPLE 3.10

COMPARING TWO COEFFICIENTS OF VARIATION WHEN TWO VARIABLES HAVE DIFFERENT UNITS OF MEASUREMENT

The operations manager of a package delivery service is deciding on whether to purchase a new fleet of trucks. When packages are stored in the trucks in preparation for delivery, you need to consider two major constraints—the weight (in pounds) and the volume (in cubic feet) for each item.

The operations manager samples 200 packages, and finds that the mean weight is 26.0 pounds, with a standard deviation of 3.9 pounds, and the mean volume is 8.8 cubic feet, with a standard deviation of 2.2 cubic feet. How can the operations manager compare the variation of the weight and the volume?

SOLUTION Because the measurement units differ for the weight and volume constraints, the operations manager should compare the relative variability in the two types of measurements.

For weight, the coefficient of variation is

$$CV_W = \left(\frac{3.9}{26.0}\right)100\% = 15\%$$

For volume, the coefficient of variation is

$$CV_V = \left(\frac{2.2}{8.8}\right)100\% = 25.0\%$$

Thus, relative to the mean, the package volume is much more variable than the package weight.

Z Scores

An **extreme value** or **outlier** is a value located far away from the mean. *Z* scores are useful in identifying outliers. The larger the *Z* score, the farther the distance from the value to the mean. The ***Z* score** is the difference between the value and the mean, divided by the standard deviation.

Z SCORES

$$Z = \frac{X - \bar{X}}{S} \tag{3.12}$$

For the time to get ready in the morning data, the mean is 39.6 minutes and the standard deviation is 6.77 minutes. The time to get ready on the first day is 39.0 minutes. You compute the Z score for day 1 from

$$\begin{aligned} Z &= \frac{X - \bar{X}}{S} \\ &= \frac{39.0 - 39.6}{6.77} \\ &= -0.09 \end{aligned}$$

Table 3.3 shows the Z scores for all 10 days. The largest Z score is 1.83 for day 4 on which the time to get ready was 52 minutes. The lowest Z score was −1.57 for day 2 on which the time to get ready was 29 minutes. As a general rule, a Z score is considered an outlier if it is less than −3.0 or greater than +3.0. None of the times met that criterion to be considered outliers.

TABLE 3.3
Z Scores for the 10 Get-Ready Times

	Time (X)	Z Score
	39	−0.09
	29	−1.57
	43	0.50
	52	1.83
	39	−0.09
	44	0.65
	40	0.06
	31	−1.27
	44	0.65
	35	−0.68
Mean	39.6	
Standard deviation	6.77	

EXAMPLE 3.11

COMPUTING THE Z SCORES OF THE 2003 RETURN OF SMALL CAP HIGH-RISK MUTUAL FUNDS

The 121 mutual funds that are part of the "Using Statistics" scenario (see page 72) are classified according to the risk level of the mutual funds (low, average, and high) and type (small cap, mid cap, and large cap). Compute the Z scores of the 2003 return for the small cap mutual funds with high risk. **MUTUALFUNDS2004**

SOLUTION Table 3.4 illustrates the Z scores of the 2003 return for the small cap mutual funds with high risk. The largest Z score is 1.42 for a percentage return of 66.5. The lowest Z score is −1.35 for a percentage return of 37.3. As a general rule, a Z score is considered an outlier if it is less than -3.0 or greater than +3.0. None of the percentage returns met that criterion to be considered outliers.

TABLE 3.4

Z Scores of the 2003 Return for the Small Cap Mutual Funds with High Risk

	Return 2003	Z Scores
	44.5	−0.67
	39.2	−1.17
	62.4	1.03
	59.3	0.74
	56.6	0.48
	53.8	0.21
	37.3	−1.35
	44.2	−0.69
	66.5	1.42
Mean	51.53	
Standard Deviation	10.55	

Shape

A third important property that describes a set of numerical data is shape. Shape is the pattern of the distribution of data values throughout the entire range of all the values. A distribution will either be **symmetrical**, when low and high values balance each other out, or **skewed**, not symmetrical and showing an imbalance of low values or high values.

Shape influences the relationship of the mean to the median in the following ways:

- Mean < median; negative or left-skewed
- Mean = median; symmetric or zero skewness
- Mean > median; positive or right-skewed

Figure 3.1 depicts three data sets, each with a different shape.

FIGURE 3.1

A Comparison of Three Data Sets Differing in Shape

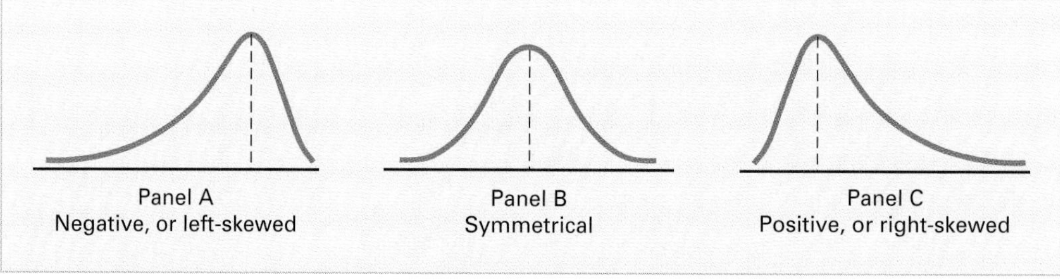

The data in panel A are negative, or **left-skewed**. In this panel, most of the values are in the upper portion of the distribution. There is a long tail and distortion to the left that is caused by some extremely small values. These extremely small values pull the mean downward so that the mean is less than the median.

The data in panel B are symmetrical. Each half of the curve is a mirror image of the other half of the curve. The low and high values on the scale balance, and the mean equals the median.

The data in panel C are positive, or **right-skewed**. In this panel, most of the values are in the lower portion of the distribution. There is a long tail on the right of the distribution and a distortion to the right that is caused by some extremely large values. These extremely large values pull the mean upward so that the mean is greater than the median.

Visual Explorations: Exploring Descriptive Statistics

Use the Visual Explorations Descriptive Statistics procedure to see the effect of changing data values on measures of central tendency, variation, and shape. Open the **Visual Explorations.xla** macro workbook and select **VisualExplorations ➔ Descriptive Statistics** from the Microsoft Excel menu bar. Read the instructions in the popup box (see illustration on page 89) and click **OK** to examine a dot scale diagram for the sample of 10 get-ready times used throughout this chapter.

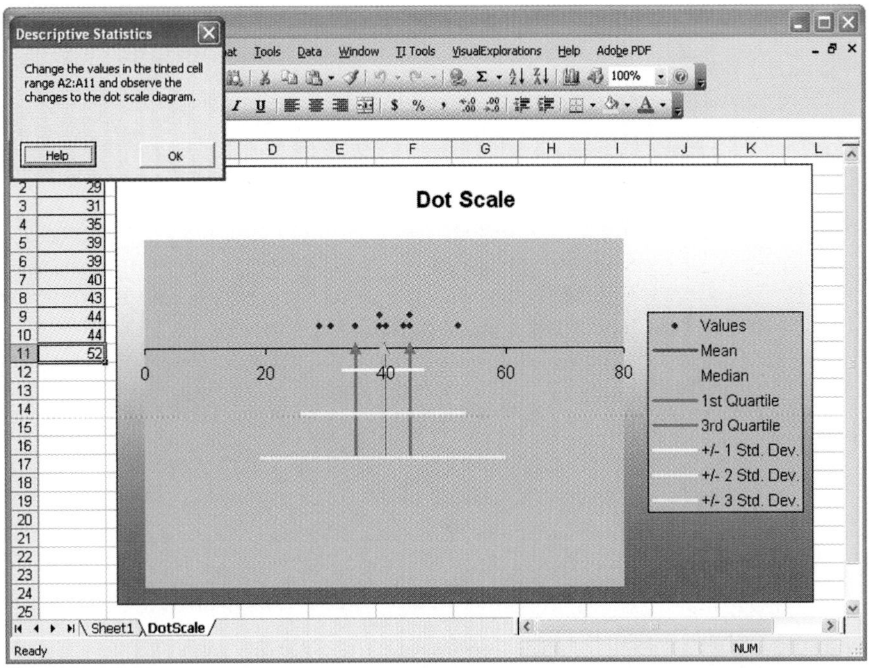

Experiment by entering an extreme value such as 10 minutes into one of the tinted cells of column A. Which measures are affected by this change? Which ones are not? You can flip between the "before" and "above" diagrams by repeatedly pressing **Crtl-Z** (undo) followed by **Crtl-Y** (redo) to help see the changes the extreme value caused in the diagram.

Microsoft Excel Descriptive Statistics Output

The Microsoft Excel Data Analysis ToolPak generates the mean, median, mode, standard deviation, variance, range, minimum, maximum, and count (sample size) on a single worksheet, all of which have been discussed in this section. In addition, Excel computes the standard error, along with statistics for kurtosis and skewness. The *standard error* is the standard deviation divided by the square root of the sample size and will be discussed in Chapter 7. *Skewness* measures the lack of symmetry in the data and is based on a statistic that is a function of the *cubed* differences around the mean. A skewness value of zero indicates a symmetric distribution. *Kurtosis* measures the relative concentration of values in the center of the distribution as compared with the tails and is based on the differences around the mean raised to the fourth power. This measure is not discussed in this text (see reference 2).

From Figure 3.2, the Excel descriptive statistics output for the 2003 return of the funds based on risk level, there appears to be slight differences in the 2003 percentage return for the

FIGURE 3.2
Microsoft Excel Descriptive Statistics of the 2003 Returns Based on Risk Level

	A	B	C	D
1	Descriptive Statistics of 2003 Return by Risk			
2		Low	Average	High
3	Mean	41.36207	42.96304	45.99412
4	Standard Error	1.631596	2.045318	3.117886
5	Median	40.25	41	44.5
6	Mode	35.8	37.5	#N/A
7	Standard Deviation	12.42586	13.87202	12.85537
8	Sample Variance	154.402	192.433	165.2606
9	Kurtosis	-0.09275	-0.33478	0.711316
10	Skewness	0.358427	0.586116	-0.55254
11	Range	55.4	58.3	51.6
12	Minimum	15.8	19.7	14.9
13	Maximum	71.2	78	66.5
14	Sum	2399	1976.3	781.9
15	Count	58	46	17

three risk levels. High-risk funds had a slightly higher mean and median than did low-risk and average-risk funds. There was very little difference in the standard deviations of the three groups.

Minitab Descriptive Statistics Output

For descriptive statistics, Minitab computes the sample size (labeled as N), the mean, median, standard deviation (labeled StDev), minimum, maximum, coefficient of variation (labeled CoefVar), first and third quartiles, range, and interquartile range (labeled IQR), all of which have been discussed in this section.

From Figure 3.3, the Minitab descriptive statistics output for the 2003 return of the funds based on risk level, there appears to be slight differences in the 2003 percentage return for the three risk levels. High-risk funds had a slightly higher mean, median, and quartiles than did low-risk and average-risk funds. There was very little difference in the standard deviations or interquartile ranges of the three groups.

FIGURE 3.3
Minitab Descriptive Statistics of the 2003 Returns Based on Risk Level

Results for: MUTUALFUNDS2004.MTW

Descriptive Statistics: Return 2003

Variable	Risk	N	Mean	StDev	CoefVar	Minimum	Q1	Median	Q3
Return 2003	average	46	42.96	13.87	32.29	19.70	32.63	41.00	49.83
	high	17	45.99	12.86	27.95	14.90	38.25	44.50	57.05
	low	58	41.36	12.43	30.04	15.80	31.23	40.25	48.95

Variable	Risk	Maximum	Range	IQR
Return 2003	average	78.00	58.30	17.20
	high	66.50	51.60	18.80
	low	71.20	55.40	17.73

PROBLEMS FOR SECTION 3.1

Learning the Basics

PH Grade ASSIST **3.1** The following is a set of data from a sample of $n = 5$:

7 4 9 8 2

a. Compute the mean, median, and mode.
b. Compute the range, interquartile range, variance, standard deviation, and coefficient of variation.
c. Compute the Z scores. Are there any outliers?
d. Describe the shape of the data set.

PH Grade ASSIST **3.2** The following is a set of data from a sample of $n = 6$:

7 4 9 7 3 12

a. Compute the mean, median, and mode.
b. Compute the range, interquartile range, variance, standard deviation, and coefficient of variation.
c. Compute the Z scores. Are there any outliers?
d. Describe the shape of the data set.

PH Grade ASSIST **3.3** The following set of data is from a sample of $n = 7$:

12 7 4 9 0 7 3

a. Compute the mean, median, and mode.
b. Compute the range, interquartile range, variance, standard deviation, and coefficient of variation.
c. Describe the shape of the data set.

PH Grade ASSIST **3.4** The following is a set of data from a sample of $n = 5$:

7 −5 −8 7 9

a. Compute the mean, median, and mode.
b. Compute the range, interquartile range, variance, standard deviation, and coefficient of variation.
c. Describe the shape of the data set.

PH Grade ASSIST **3.5** Suppose that the rate of return for a particular stock during the past two years was 10% and 30%. Compute the geometric mean rate of return. (*Note:* A rate of return of 10% is recorded as 0.10 and a rate of return of 30% is recorded as 0.30.)

Applying the Concepts

Problems 3.6–3.20 can be solved manually or by using Microsoft Excel, Minitab, or SPSS.

PH Grade ASSIST **3.6** The operations manager of a plant that manufactures tires wants to compare the actual inner diameter of two grades of tires, each of which is expected to be 575 millimeters. A sample of five tires of each grade was selected, and the results representing the inner diameters of the tires, ranked from smallest to largest, are as follows:

Grade *X*					Grade *Y*				
568	570	575	578	584	573	574	575	577	578

a. For each of the two grades of tires, compute the mean, median, and standard deviation.
b. Which grade of tire is providing better quality? Explain.
c. What would be the effect on your answers in (a) and (b) if the last value for grade *Y* were 588 instead of 578? Explain.

SELF Test **3.7** The following data represent the total fat for burgers and chicken items from a sample of fast-food chains. FASTFOOD

Burgers

19 31 34 35 39 39 43

Chicken

7 9 15 16 16 18 22 25 27 33 39

Source: Extracted from "Quick Bites," Copyright © 2001 by Consumers Union of U.S., Inc., Yonkers, NY 10703–1057. Adapted with permission from Consumer Reports, *March 2001, 46.*

For the burgers and chicken items separately:
a. Compute the mean, median, first quartile, and third quartile.
b. Compute the variance, standard deviation, range, interquartile range, and coefficient of variation.
c. Are the data skewed? If so, how?
d. Based on the results of (a) through (c), what conclusions can you reach concerning the differences in total fat of burgers and chicken items?

3.8 The median price of a home in December 2003 rose to \$173,200, an increase of 6.7% from December 2002. For the full year, sales hit a record 6.1 million homes (James R. Hagerty, "Housing Prices Continue to Rise," *The Wall Street Journal*, January 27, 2004, D1).
a. Describe the shape of the distribution of the price of homes sold.
b. Why do you think the article reports the median home price and not the mean home price?

3.9 In the 2002–2003 academic year, many public universities in the United States raised tuition and fees due to a decrease in state subsidies (Mary Beth Marklein, "Public Universities Raise Tuition, Fees—and Ire," *USA Today*, August 8, 2002, 1A–2A). The following represents the change in the cost of tuition, a shared dormitory room, and the most popular meal plan between the 2001–2002 academic year and the 2002–2003 academic year for a sample of 10 public universities. COLLEGECOST

University	Change in Cost ($)
University of California, Berkeley	1,589
University of Georgia, Athens	593
University of Illinois, Urbana–Champaign	1,223
Kansas State University, Manhattan	869
University of Maine, Orono	423
University of Mississippi, Oxford	1,720
University of New Hampshire, Durham	708
Ohio State University, Columbus	1,425
University of South Carolina, Columbia	922
Utah State University, Logan	308

a. Compute the mean, median, first quartile, and third quartile.
b. Compute the variance, standard deviation, range, interquartile range, coefficient of variation, and *Z* scores.
c. Are the data skewed? If so, how?
d. Based on the results of (a) through (c), what conclusions can you reach concerning the change in costs between the 2001–2002 and 2002–2003 academic years?

3.10 The following data COFFEEDRINK represent the calories and fat (in grams) of 16-ounce iced coffee drinks at Dunkin' Donuts and Starbucks.

Product	Calories	Fat
Dunkin' Donuts Iced Mocha Swirl latte (whole milk)	240	8.0
Starbucks Coffee Frappuccino blended coffee	260	3.5
Dunkin' Donuts Coffee Coolatta (cream)	350	22.0
Starbucks Iced Coffee Mocha Expresso (whole milk and whipped cream)	350	20.0
Starbucks Mocha Frappuccino blended coffee (whipped cream)	420	16.0
Starbucks Chocolate Brownie Frappuccino blended coffee (whipped cream)	510	22.0
Starbucks Chocolate Frappuccino Blended Crème (whipped cream)	530	19.0

Source: Extracted from "Coffee as Candy at Dunkin' Donuts and Starbucks," Copyright © 2004 by Consumers Union of U.S., Inc., Yonkers, NY 10703–1057. Adapted with permission from Consumer Reports, *June 2004, 9.*

For each variable (calories and fat),

a. Compute the mean, median, first quartile, and third quartile.
b. Compute the variance, standard deviation, range, interquartile range, coefficient of variation, and Z scores. Are there any outliers? Explain.
c. Are the data skewed? If so, how?
d. Based on the results of (a) through (c), what conclusions can you reach concerning the calories and fat in iced coffee drinks at Dunkin' Donuts and Starbucks?

3.11 The following data represent the daily hotel cost and rental car cost for 20 U. S. cities during a week in October 2003. **HOTEL-CAR**

City	Hotel	Cars
San Francisco	205	47
Los Angeles	179	41
Seattle	185	49
Phoenix	210	38
Denver	128	32
Dallas	145	48
Houston	177	49
Minneapolis	117	41
Chicago	221	56
St. Louis	159	41
New Orleans	205	50
Detroit	128	32
Cleveland	165	34
Atlanta	180	46
Orlando	198	41
Miami	158	40
Pittsburgh	132	39
Boston	283	67
New York	269	69
Washington, D.C.	204	40

Source: Extracted from The Wall Street Journal, *October 10, 2003, W4.*

For each variable (hotel cost and rental car cost),

a. Compute the mean, median, first quartile, and third quartile.
b. Compute the variance, standard deviation, range, interquartile range, coefficient of variation, and Z scores. Are there any outliers? Explain.
c. Are the data skewed? If so, how?
d. Based on the results of (a) through (c), what conclusions can you reach concerning the daily cost of a hotel and rental car?

3.12 The cost of 14 models of 3-megapixel digital cameras at a camera specialty store during 2003 was as follows. **CAMERA**

340 450 450 280 220 340 290
370 400 310 340 430 270 380

a. Compute the mean, median, first quartile, and third quartile.
b. Compute the variance, standard deviation, range, interquartile range, coefficient of variation, and Z scores. Are there any outliers? Explain.
c. Are the data skewed? If so, how?
d. Based on the results of (a) through (c), what conclusions can you reach concerning the price of 3-megapixel digital cameras at a camera specialty store during 2003?

3.13 A software development and consulting firm located in the Phoenix metropolitan area develops software for supply chain management systems using systematic software reuse. Instead of starting from scratch when writing and developing new custom software systems, the firm uses a database of reusable components totaling more than 2,000,000 lines of code collected from 10 years of continuous reuse effort. Eight analysts at the firm were asked to estimate the reuse rate when developing a new software system. The following data are given as a percentage of the total code written for a software system that is part of the reuse database. **REUSE**

50 62.5 37.5 75.0 45.0 47.5 15.0 25.0

Source: M. A. Rothenberger, and K. J. Dooley, "A Performance Measure for Software Reuse Projects," Decision Sciences, *30(Fall 1999), 1131–1153.*

a. Compute the mean, median, and mode.
b. Compute the range, variance, and standard deviation.
c. Interpret the summary measures calculated in (a) and (b).

3.14 A manufacturer of flashlight batteries took a sample of 13 batteries from a day's production and used them continuously until they were drained. The numbers of hours they were used until failure were: **BATTERIES**

342 426 317 545 264 451
1,049 631 512 266 492 562 298

a. Compute the mean, median, and mode. Looking at the distribution of times to failure, which measures of location do you think are most appropriate and which least appropriate to use for these data? Why?
b. Calculate the range, variance, and standard deviation.
c. What would you advise if the manufacturer wanted to be able to say in advertisements that these batteries "should last 400 hours"? (*Note:* There is no right answer to this question; the point is to consider how to make such a statement precise.)
d. Suppose that the first value was 1,342 instead of 342. Repeat (a) through (c), using this value. Comment on the difference in the results.

3.15 A bank branch located in a commercial district of a city has developed an improved process for serving customers during the noon to 1:00 P.M. lunch period. The waiting time in minutes (defined as the time the customer enters the line to when he or she reaches the teller window) of all customers during this hour is recorded over a period of one week. A random sample of 15 customers is selected, and the results are as follows: **BANK1**

4.21 5.55 3.02 5.13 4.77 2.34 3.54
3.20 4.50 6.10 0.38 5.12 6.46 6.19 3.79

a. Compute the mean, median, first quartile, and third quartile.
b. Compute the variance, standard deviation, range, interquartile range, coefficient of variation, and Z scores. Are there any outliers? Explain.
c. Are the data skewed? If so, how?
d. As a customer walks into the branch office during the lunch hour, she asks the branch manager how long she can expect to wait. The branch manager replies, "Almost certainly less than five minutes." On the basis of the results of (a) and (b), evaluate the accuracy of this statement.

3.16 Suppose that another branch, located in a residential area, is also concerned with the noon to 1 P.M. lunch hour. The waiting time in minutes (defined as the time the customer enters the line to the time he or she reaches the teller window) of all customers during these hours is recorded over a period of one week. A random sample of 15 customers is selected, and the results are as follows: **BANK2**

9.66 5.90 8.02 5.79 8.73 3.82 8.01
8.35 10.49 6.68 5.64 4.08 6.17 9.91 5.47

a. Compute the mean, median, first quartile, and third quartile.
b. Compute the variance, standard deviation, range, interquartile range, and coefficient of variation. Are there any outliers? Explain.
c. Are the data skewed? If so, how?
d. As a customer walks into the branch office during the lunch hour, he asks the branch manager how long he can expect to wait. The branch manager replies, "Almost certainly less than five minutes." On the basis of the results of (a) and (b), evaluate the accuracy of this statement.

SELF Test **3.17** China is the fastest-growing market for passenger car sales and fourth biggest after the United States, Japan, and Germany. Passenger car sales increased 61% in 2002 and 55% in 2003 (Peter Wonacott, "A Fear Amid China's Car Boom," *The Wall Street Journal*, February 2, 2004, A17). Compute the geometric mean rate of increase. (*Hint:* Denote an increase of 61% as $R_1 = 0.61$.)

3.18 The time period from 2000 to 2003 saw a great deal of volatility in the value of stocks. The data in the following table **STOCKRETURN** represent the total rate of return of the Dow Jones Industrial Index, the Standard & Poor's 500, the Russell 2000 Index, and the Wilshire 5000 Index from 2000 to 2003.

Year	DJIA	SP500	Russell2000	Wilshire5000
2003	25.30	26.40	45.40	29.40
2002	−15.01	−22.10	−21.58	−20.90
2001	−5.44	−11.90	−1.03	−10.97
2000	−6.20	−9.10	−3.02	−10.89

Source: Extracted from The Wall Street Journal, *January 2, 2004.*

a. Calculate the geometric rate of return for the Dow Jones Industrial Index, the Standard & Poor's 500, the Russell 2000 Index, and the Wilshire 5000 Index.
b. What conclusions can you reach concerning the geometric rates of return of the four stock indexes?
c. Compare the results of (b) to those of problems 3.19 (b) and 3.20 (b).

3.19 The time period from 2000 to 2003 saw a great deal of volatility in the value of investments. The data in the following table **BANKRETURN** represent the total rate of return of the one-year certificate of deposit, the 30-month certificate of deposit, and the money market deposit from 2000 to 2003.

Year	One Year	30 Month	Money Market
2003	1.20	1.76	0.61
2002	1.98	2.74	1.02
2001	3.60	3.97	1.73
2000	5.46	5.64	2.09

Source: Extracted from The Wall Street Journal, *January 2, 2004.*

a. Calculate the geometric rate of return for the one-year certificate of deposit, the 30-month certificate of deposit, and the money market deposit.
b. What conclusions can you reach concerning the geometric rates of return of the three deposits?
c. Compare the results of (b) to those of problems 3.18 (b) and 3.20 (b).

3.20 The time period from 2000 to 2003 saw a great deal of volatility in the value of metals. The data in the following table **METALRETURN** represent the total rate of return for platinum, gold, and silver from 2000 to 2003.

Year	Platinum	Gold	Silver
2003	34.2	19.5	24.0
2002	24.5	24.5	5.5
2001	−21.3	1.2	-3.0
2000	−23.3	1.8	-5.9

Source: Extracted from The Wall Street Journal, *January 2, 2004.*

a. Calculate the geometric rate of return for platinum, gold, and silver.
b. What conclusions can you reach concerning the geometric rates of return of the three metals?
c. Compare the results of (b) to those of problems 3.18 (b) and 3.19 (b).

3.2 NUMERICAL DESCRIPTIVE MEASURES FOR A POPULATION

Section 3.1 presented various *statistics* that described the properties of central tendency, variation, and shape for a *sample*. If your data set represents numerical measurements for an entire *population*, you need to calculate and interpret *parameters*, summary measures for a population. In this section, you will learn about three descriptive population parameters, the population mean, population variance, and population standard deviation.

To help illustrate these parameters, first review Table 3.5 that contains the five biggest bond funds (in terms of total assets) as of March 1, 2004. The 52-week return for each of these funds is also listed. LARGEST BONDS.

TABLE 3.5
2003 Return for the Population Consisting of the Five Largest Bond Funds

Bond Fund	52-Week Return (in %)
Vanguard GNMA	3.8
Vanguard Total Bond Index	6.5
Pimco Total Return Admin	7.0
Pimco Total Return Instl	7.3
America Bond Fund	12.9

Source: Extracted from The Wall Street Journal, *March 25, 2004, C2.*

The Population Mean

The **population mean** is represented by the symbol μ, the Greek lowercase letter *mu*. Equation (3.13) defines the population mean.

POPULATION MEAN

The population mean is the sum of the values in the population divided by the population size N.

$$\mu = \frac{\sum_{i=1}^{N} X_i}{N} \tag{3.13}$$

where μ = population mean

X_i = ith value of the variable X

$\sum_{i=1}^{N} X_i$ = summation of all X_i values in the population

To compute the mean return for the population of bond funds given in Table 3.5, use Equation (3.13),

$$\mu = \frac{\sum_{i=1}^{N} X_i}{N} = \frac{3.8 + 6.5 + 7.0 + 7.3 + 12.9}{5} = \frac{37.5}{5} = 7.5$$

Thus, the mean 2003 return for these bond funds is 7.5%.

The Population Variance and Standard Deviation

The **population variance** and the **population standard deviation** measure variation in a population. Like the related sample statistics, the population standard deviation is the square root of the population variance. The symbol σ^2, the Greek lowercase letter *sigma* squared, represents the population variance and the symbol σ, the Greek lowercase letter *sigma*, represents the population standard deviation. Equations (3.14) and (3.15) define these parameters. The denominators for the right-side terms in these equations use N and not the $(n - 1)$ term that is used in the equations for the sample variance and standard deviation [see Equations (3.9) and (3.10) on page 82].

POPULATION VARIANCE

The population variance is the sum of the squared differences around the population mean divided by the population size N.

$$\sigma^2 = \frac{\sum_{i=1}^{N}(X_i - \mu)^2}{N} \qquad \textbf{(3.14)}$$

where μ = population mean

X_i = ith value of the variable X

$\sum_{i=1}^{N}(X_i - \mu)^2$ = summation of all the squared differences between the X_i values and μ

POPULATION STANDARD DEVIATION

$$\sigma = \sqrt{\frac{\sum_{i=1}^{N}(X_i - \mu)^2}{N}} \qquad \textbf{(3.15)}$$

To compute the population variance for the data of Table 3.5 on page 94, you use Equation (3.14),

$$\begin{aligned}\sigma^2 &= \frac{\sum_{i=1}^{N}(X_i - \mu)^2}{N} \\ &= \frac{(3.8 - 7.5)^2 + (6.5 - 7.5)^2 + (7.0 - 7.5)^2 + (7.3 - 7.5)^2 + (12.9 - 7.5)^2}{5} \\ &= \frac{13.69 + 1.00 + 0.25 + 0.04 + 29.16}{5} \\ &= \frac{44.14}{5} = 8.828\end{aligned}$$

Thus, the variance of the returns is 8.828 squared percentage return. The squared units make the variance hard to interpret. You should use the standard deviation that uses the original units of the data (percentage return). From Equation (3.15),

$$\sigma = \sqrt{\sigma^2} = \sqrt{\frac{\sum_{i=1}^{N}(X_i - \mu)^2}{N}} = \sqrt{8.828} = 2.97$$

Therefore, the typical 2003 return differs from the mean of 7.5 by approximately 2.97. This large amount of variation suggests that these large bond funds produce results that differ greatly.

The Empirical Rule

In most data sets, a large portion of the values tend to cluster somewhat near the median. In right-skewed data sets, this clustering occurs to the left of the mean, that is, at a value less than the mean. In left-skewed data sets, the values tend to cluster to the right of the mean, that is, at a value greater than the mean. In symmetrical data sets, where the median and mean are the same, the values often tend to cluster around the median and mean producing a bell-shaped distribution. You can use the **empirical rule** to examine the variability in bell-shaped distributions:

- Approximately 68% of the values are within a distance of ±1 standard deviation from the mean.
- Approximately 95% of the values are within a distance of ±2 standard deviations from the mean.
- Approximately 99.7% are within a distance of ±3 standard deviations from the mean.

The empirical rule helps you measure how the values distribute above and below the mean. This can help you to identify outliers when analyzing a set of numerical data. The empirical rule implies that for bell-shaped distributions only about one out of 20 values will be beyond two standard deviations from the mean in either direction. As a general rule, you can consider values not found in the interval $\mu \pm 2\sigma$ as potential outliers. The rule also implies that only about three in 1,000 will be beyond three standard deviations from the mean. Therefore, values not found in the interval $\mu \pm 3\sigma$ are almost always considered outliers. For heavily skewed data sets, or those not appearing bell-shaped for any other reason, the Chebyshev rule discussed on page 97 should be applied instead of the empirical rule.

EXAMPLE 3.12 USING THE EMPIRICAL RULE

A population of 12-ounce cans of cola is known to have a mean fill-weight of 12.06 ounces and a standard deviation of 0.02. The population is also known to be bell-shaped. Describe the distribution of fill-weights. Is it very likely that a can will contain less than 12 ounces of cola?

SOLUTION

$$\mu \pm \sigma = 12.06 \pm 0.02 = (12.04, 12.08)$$
$$\mu \pm 2\sigma = 12.06 \pm 2(0.02) = (12.02, 12.10)$$
$$\mu \pm 3\sigma = 12.06 \pm 3(0.02) = (12.00, 12.12)$$

Using the empirical rule, approximately 68% of the cans will contain between 12.04 and 12.08 ounces, approximately 95% will contain between 12.02 and 12.10 ounces, and approximately 99.7% will contain between 12.00 and 12.12 ounces. Therefore, it is highly unlikely that a can will contain less than 12 ounces.

The Chebyshev Rule

The **Chebyshev rule** (reference 1) states that for any data set, regardless of shape, the percentage of values that are found within distances of k standard deviations from the mean must be at least

$$(1 - 1/k^2) \times 100\%$$

You can use this rule for any value of k greater than 1. Consider $k = 2$. The Chebyshev rule states that at least $[1 - (1/2)^2] \times 100\% = 75\%$ of the values must be found within ±2 standard deviations of the mean.

The Chebyshev rule is very general and applies to any type of distribution. The rule indicates *at least* what percentage of the values fall within a given distance from the mean. However, if the data set is approximately bell-shaped, the empirical rule will more accurately reflect the greater concentration of data close to the mean. Table 3.6 compares the Chebyshev and empirical rules.

TABLE 3.6
How Data Vary Around the Mean

	% of Values Found in Intervals Around the Mean	
Interval	**Chebyshev (for any distribution)**	**Empirical Rule (bell-shaped distribution)**
$(\mu - \sigma, \mu + \sigma)$	At least 0%	Approximately 68%
$(\mu - 2\sigma, \mu + 2\sigma)$	At least 75%	Approximately 95%
$(\mu - 3\sigma, \mu + 3\sigma)$	At least 88.89%	Approximately 99.7%

EXAMPLE 3.13 USING THE CHEBYSHEV RULE

As in Example 3.12, a population of 12-ounce cans of cola is known to have a mean fill-weight of 12.06 ounces and a standard deviation of 0.02. However, the shape of the population is unknown and you cannot assume that it is bell-shaped. Describe the distribution of fill-weights. Is it very likely that a can will contain less than 12 ounces of cola?

SOLUTION

$$\mu \pm \sigma = 12.06 \pm 0.02 = (12.04, 12.08)$$
$$\mu \pm 2\sigma = 12.06 \pm 2(0.02) = (12.02, 12.10)$$
$$\mu \pm 3\sigma = 12.06 \pm 3(0.02) = (12.00, 12.12)$$

Because the distribution may be skewed, you cannot use the empirical rule. Using the Chebyshev rule, you cannot say anything about the percentage of cans containing between 12.04 and 12.08 ounces. You can state that at least 75% of the cans will contain between 12.02 and 12.10 ounces, and at least 88.89% will contain between 12.00 and 12.12 ounces. Therefore, between 0 and 11.11% of the cans contain less than 12 ounces.

You can use these two rules for understanding how data are distributed around the mean when you have sample data. In each case, use the value you calculated for $\bar{X}$ in place of μ and the value you calculated for S in place of σ. The results you compute using the sample statistics are approximations since you used sample statistics ($\bar{X}$, S) and not population parameters (μ, σ).

PROBLEMS FOR SECTION 3.2

Learning the Basics

PH Grade ASSIST **3.21** The following is a set of data for a population with $N = 10$:

7 5 11 8 3 6 2 1 9 8

a. Compute the population mean.
b. Compute the population standard deviation.

PH Grade ASSIST **3.22** The following is a set of data for a population with $N = 10$:

7 5 6 6 6 4 8 6 9 3

a. Compute the population mean.
b. Compute the population standard deviation.

Applying the Concepts

SELF Test **3.23** The following data represent the quarterly sales tax receipts (in thousands of dollars) submitted to the comptroller of the Village of Fair Lake for the period ending March 2004 by all 50 business establishments in that locale: TAX

10.3	11.1	9.6	9.0	14.5
13.0	6.7	11.0	8.4	10.3
13.0	11.2	7.3	5.3	12.5
8.0	11.8	8.7	10.6	9.5
11.1	10.2	11.1	9.9	9.8
11.6	15.1	12.5	6.5	7.5
10.0	12.9	9.2	10.0	12.8
12.5	9.3	10.4	12.7	10.5
9.3	11.5	10.7	11.6	7.8
10.5	7.6	10.1	8.9	8.6

a. Compute the mean, variance, and standard deviation for this population.
b. What proportion of these businesses have quarterly sales tax receipts within ±1, ±2, or ±3 standard deviations of the mean?
c. Compare and contrast your findings with what would be expected on the basis of the empirical rule. Are you surprised at the results in (b)?

PH Grade ASSIST **3.24** Consider a population of 1,024 mutual funds that primarily invested in large companies. You determined that μ, the mean one-year total percentage return achieved by all the funds, is 8.20 and that σ, the standard deviation, is 2.75. In addition, suppose you determined that the range in the one-year total returns is from −2.0 to 17.1 and that the quartiles are, respectively, 5.5 (Q_1) and 10.5 (Q_3). According to the empirical rule, what percentage of these funds is expected to be

a. within ±1 standard deviation of the mean?
b. within ±2 standard deviations of the mean?
c. According to the Chebyshev rule, what percentage of these funds are expected to be within ±1, ±2, or ±3 standard deviations of the mean?
d. According to the Chebyshev rule, at least 93.75% of these funds are expected to have one-year total returns between what two amounts?

3.25 The following table ASSETS represents the assets in billions of dollars of the five largest bond funds.

Bond Fund	Assets (Billions $)
Vanguard GNMA	19.5
Vanguard Total Bond Mkt. Index	16.8
Bond Fund of America A	13.7
Franklin Calif. Tax-Free Inc. A	12.8
Vanguard Short-Term Corp.	10.9

a. Compute the mean for this population of the five largest bond funds. Interpret this parameter.
b. Compute the variance and standard deviation for this population. Interpret these parameters.
c. Is there a lot of variability in the assets of the bond funds?

3.26 The data in the file ENERGY contains the per capita energy consumption in kilowatt hours for each of the 50 states and the District of Columbia during 1999.
a. Compute the mean, variance, and standard deviation for the population.
b. What proportion of these states has average per capita energy consumption within ±1 standard deviation of the mean, within ±2 standard deviations of the mean, and within ±3 standard deviations of the mean?
c. Compare and contrast your findings versus what would be expected based on the empirical rule. Are you surprised at the results in (b)?
d. Do (a) through (c) with the District of Columbia removed. How have the results changed?

3.27 The data in the file DOWRETURN give the 10-year annualized return (1994–2003) for the 30 companies in the Dow Jones Industrials.
a. Compute the mean for this population. Interpret this number.
b. Compute the variance and standard deviation for this population. Interpret the standard deviation.
c. Use the empirical rule or the Chebyshev rule, whichever is appropriate, to further explain the variation in this data set.
d. Using the results in (c), are there any outliers? Explain.

3.3 COMPUTING NUMERICAL DESCRIPTIVE MEASURES FROM A FREQUENCY DISTRIBUTION

Sometimes you have only a frequency distribution, not the raw data. When this occurs, you can compute approximations to the mean and the standard deviation.

When you have data from a sample that has been summarized into a frequency distribution, you can compute an approximation of the mean by assuming that all values within each class interval are located at the midpoint of the class.

APPROXIMATING THE MEAN FROM A FREQUENCY DISTRIBUTION

$$\bar{X} = \frac{\sum_{j=1}^{c} m_j f_j}{n} \tag{3.16}$$

where $\bar{X}$ = sample mean

n = number of values or sample size

c = number of classes in the frequency distribution

m_j = midpoint of the jth class

f_j = numbers of values in the jth class

To calculate the standard deviation from a frequency distribution, you assume that all values within each class interval are located at the midpoint of the class.

APPROXIMATING THE STANDARD DEVIATION FROM A FREQUENCY DISTRIBUTION

$$S = \sqrt{\frac{\sum_{j=1}^{c} (m_j - \bar{X})^2 f_j}{n-1}} \tag{3.17}$$

Example 3.14 illustrates the computation of the mean and the standard deviation from a frequency distribution.

EXAMPLE 3.14 APPROXIMATING THE MEAN AND STANDARD DEVIATION FROM A FREQUENCY DISTRIBUTION

Consider the frequency distribution of the 2003 return of growth funds (Table 3.7). Compute the mean and standard deviation.

TABLE 3.7
Frequency Distribution of the 2003 Return for Growth Mutual Funds

Annual Percentage 2003 Return	Frequency
10 but less than 20	2
20 but less than 30	9
30 but less than 40	13
40 but less than 50	15
50 but less than 60	5
60 but less than 70	5
Total	49

SOLUTION The computations that you need to calculate the approximations of the mean and standard deviation of the 2003 return for growth mutual funds are summarized in Table 3.8.

Percentage Return	Number of Funds(f_j)	Midpoint(m_j)	$m_j f_j$	$(m_j - \bar{X})$	$(m_j - \bar{X})^2$	$(m_j - \bar{X})^2 f_j$
10 but less than 20	2	15	30	−25.51	650.7601	1,301.5202
20 but less than 30	9	25	225	−15.51	240.5601	2,165.0409
30 but less than 40	13	35	455	−5.51	30.3601	394.6813
40 but less than 50	15	45	675	4.49	20.1601	302.4015
50 but less than 60	5	55	275	14.49	209.9601	1,049.8005
60 but less than 70	5	65	325	24.49	599.7601	2,998.8005
Total	49		1,985			8,212.2449

TABLE 3.8 Computations Needed to Calculate the Approximations of the Mean and Standard Deviation of the 2003 Return for Growth Mutual Funds

Using Equations (3.16) and (3.17) on page 99,

$$\bar{X} = \frac{\sum_{j=1}^{c} m_j f_j}{n}$$

$$\bar{X} = \frac{1,985.0}{49} = 40.51$$

and

$$S = \sqrt{\frac{\sum_{j=1}^{c} (m_j - \bar{X})^2 f_j}{n - 1}}$$

$$S = \sqrt{\frac{8,212.2449}{49 - 1}}$$

$$= \sqrt{171.08843}$$

$$= 13.08$$

PROBLEMS FOR SECTION 3.3

Learning the Basics

3.28 Given the following frequency distribution for $n = 100$:

Class Intervals	Frequency
0—Under 10	10
10—Under 20	20
20—Under 30	40
30—Under 40	20
40—Under 50	10
	100

Approximate

a. the mean.

b. the standard deviation.

3.29 Given the following frequency distribution for $n = 100$:

Class Intervals	Frequency
0—Under 10	40
10—Under 20	25
20—Under 30	15
30—Under 40	15
40—Under 50	5
	100

Approximate

a. the mean.

b. the standard deviation.

Applying the Concepts

3.30 A wholesale appliance distributing firm wished to study its accounts receivable for two successive months. Two independent samples of 50 accounts were selected for each of the two months. The results are summarized in the following table:

Frequency Distributions for Accounts Receivable

Amount	March Frequency	April Frequency
\$0 to under \$2,000	6	10
\$2,000 to under \$4,000	13	14
\$4,000 to under \$6,000	17	13
\$6,000 to under \$8,000	10	10
\$8,000 to under \$10,000	4	0
\$10,000 to under \$12,000	0	3
Total	50	50

For each month, approximate the

a. mean.
b. standard deviation.
c. On the basis of (a) and (b), do you think the mean and the standard deviation of the accounts receivable have changed substantially from March to April? Explain.

3.31 The following table contains the cumulative frequency distributions and cumulative percentage distributions of braking distance (in feet) at 80 miles per hour for a sample of 25 U.S.-manufactured automobile models and for a sample of 72 foreign-made automobile models in a recent year:

Braking Distance (in Ft)	U.S.-Made Automobile Models "Less Than" Indicated Values		Foreign-Made Automobile Models "Less Than" Indicated Values	
	Number	Percentage	Number	Percentage
210	0	0.0	0	0.0
220	1	4.0	1	1.4
230	2	8.0	4	5.6
240	3	12.0	19	26.4
250	4	16.0	32	44.4
260	8	32.0	54	75.0
270	11	44.0	61	84.7
280	17	68.0	68	94.4
290	21	84.0	68	94.4
300	23	92.0	70	97.2
310	25	100.0	71	98.6
320	25	100.0	72	100.0

(continued)

For U.S.- and foreign-made automobiles

a. Construct a frequency distribution for each group.
b. On the basis of the results of (a), approximate the mean of the braking distance.
c. On the basis of the results of (a), approximate the standard deviation of the braking distance.
d. On the basis of the results of (b) and (c), do U.S.- and foreign-made automobiles seem to differ in their braking distance? Explain.

3.32 The following data represent the distribution of the ages of employees within two different divisions of a publishing company.

Age of Employees (Years)	A Frequency	B Frequency
20—Under 30	8	15
30—Under 40	17	32
40—Under 50	11	20
50—Under 60	8	4
60—Under 70	2	0

For each of the two divisions (A and B), approximate the

a. mean.
b. standard deviation.
c. On the basis of the results of (a) and (b), do you think there are differences in the age distribution between the two divisions? Explain.

3.4 EXPLORATORY DATA ANALYSIS

Section 3.1 discussed sample statistics for numerical data that are measures of central tendency, variation, and shape. Another way of describing numerical data is through exploratory data analysis that includes the five-number summary and the box-and-whisker plot (references 5 and 6).

The Five-Number Summary

A **five-number summary** that consists of

$$X_{\text{smallest}} \quad Q_1 \quad \text{Median} \quad Q_3 \quad X_{\text{largest}}$$

provides a way to determine the shape of the distribution. Table 3.9 explains how the relationships among the "five numbers" allows you to recognize the shape of a data set.

TABLE 3.9
Relationships among the Five-Number Summary and the Type of Distribution

	Type of Distribution		
Comparison	**Left-Skewed**	**Symmetric**	**Right-Skewed**
Distance from X_{smallest} to the median versus the distance from the median to X_{largest}.	The distance from X_{smallest} to the median is greater than the distance from the median to X_{largest}.	Both distances are the same.	The distance from X_{smallest} to the median is less than the distance from the median to X_{largest}.
Distance from X_{smallest} to Q_1 versus the distance from Q_3 to X_{largest}.	The distance from X_{smallest} to Q_1 is greater than the distance from Q_3 to X_{largest}.	Both distances are the same.	The distance from X_{smallest} to Q_1 is less than the distance from Q_3 to X_{largest}.
Distance from Q_1 to the median versus the distance from the median to Q_3.	The distance from Q_1 to the median is greater than the distance from the median to Q_3.	Both distances are the same.	The distance from Q_1 to the median is less than the distance from the median to Q_3.

For the sample of 10 get-ready times, the smallest value is 29 minutes and the largest value is 52 minutes (see pages 75 and 77). Calculations done previously in section 3.1 show that the median = 39.5, the first quartile = 35, and the third quartile = 44. Therefore, the five-number summary is

$$29 \quad 35 \quad 39.5 \quad 44 \quad 52$$

The distance from X_{smallest} to the median ($39.5 - 29 = 10.5$) is slightly less than the distance from the median to X_{largest} ($52 - 39.5 = 12.5$). The distance from X_{smallest} to Q_1 ($35 - 29 = 6$) is slightly less than the distance from Q_3 to X_{largest} ($52 - 44 = 8$). Therefore, the get-ready times are slightly right-skewed.

EXAMPLE 3.15 COMPUTING THE FIVE-NUMBER SUMMARY OF THE 2003 PERCENTAGE RETURN OF SMALL CAP HIGH-RISK MUTUAL FUNDS

The 121 mutual funds that are part of the "Using Statistics" scenario (see page 72) are classified according to the risk level of the mutual funds (low, average, and high) and type (small cap, mid cap, and large cap). Compute the five-number summary of the 2003 return for the small cap mutual funds with high risk. **MUTUALFUNDS2004**

SOLUTION From previous computations for the 2003 return for the small cap mutual funds with high risk (see pages 76 and 78), the median = 53.8, the first quartile = 41.7, and the third quartile = 60.85. In addition, the smallest value in the data set is 37.3 and the largest value is 66.5. Therefore, the five-number summary is

37.3 41.7 53.8 60.85 66.5

The distance from the median to $X_{largest}$ (66.5 – 53.8 = 12.7) is less than the distance from $X_{smallest}$ to the median (53.8 – 37.3 = 16.5). This indicates left skewness. The distance from $X_{smallest}$ to Q_1 (41.7 – 37.3 = 4.4) is slightly less than the distance from Q_3 to $X_{largest}$ (66.5 – 60.85 = 5.65). This indicates slight right-skewness. Therefore, the results are inconsistent.

The Box-and-Whisker Plot

A **box-and-whisker plot** provides a graphical representation of the data based on the five-number summary. Figure 3.4 illustrates the box-and-whisker plot for the get-ready times.

FIGURE 3.4

Box-and-Whisker Plot of the Time to Get Ready

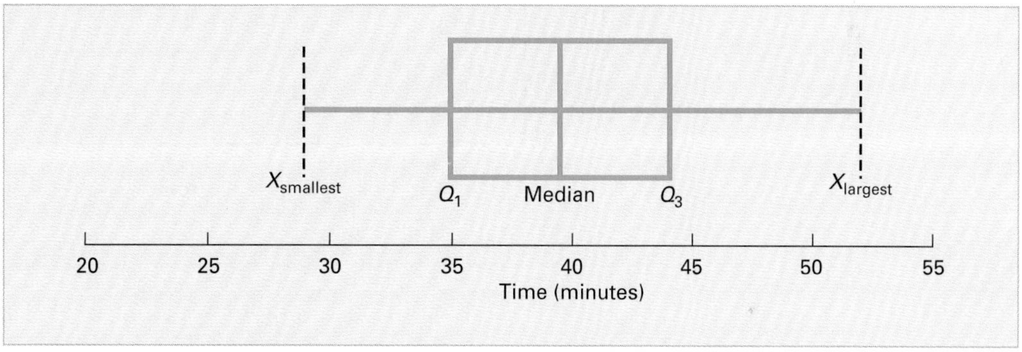

The vertical line drawn within the box represents the median. The vertical line at the left side of the box represents the location of Q_1 and the vertical line at the right side of the box represents the location of Q_3. Thus, the box contains the middle 50% of the values in the distribution. The lower 25% of the data are represented by a line (i.e., a *whisker*) connecting the left side of the box to the location of the smallest value, $X_{smallest}$. Similarly, the upper 25% of the data are represented by a whisker connecting the right side of the box to $X_{largest}$.

The box-and-whisker plot of the get-ready times in Figure 3.4 indicates very slight right-skewness since the distance between the median and the highest value is slightly more than the distance between the lowest value and the median. The right whisker is slightly longer than the left whisker.

EXAMPLE 3.16 THE BOX-AND-WHISKER PLOT OF THE 2003 PERCENTAGE RETURN OF LOW-RISK, AVERAGE RISK, AND HIGH-RISK MUTUAL FUNDS

The 121 mutual funds that are part of the "Using Statistics" scenario (see page 72) are classified according to the risk level of the mutual funds (low, average, and high) and type (small cap, mid cap, and large cap). Construct the box-and-whisker plot of the 2003 return for low-risk, average-risk, and high-risk mutual funds. **MUTUALFUNDS2004**

SOLUTION Figure 3.5 is the Minitab box-and-whisker plot of the 2003 return for low-risk, average-risk, and high-risk mutual funds. Minitab displays the box-and-whisker plot vertically from bottom (low) to top (high). The asterisk (*) for the average-risk fund represents the presence of outlier values.[2] The median percentage return and the quartiles are higher for the high-risk funds than for the low-risk and average-risk funds. The average-risk funds are right-skewed due to the extremely large return of one fund (78). The high-risk funds appear left-skewed because of the long lower whisker, but the median return is closer to the first quartile than to the third quartile. The low-risk funds appear to be slightly right-skewed since the upper whisker is longer than the lower whisker.

[2]If there are outliers, the whiskers in the Minitab box-and-whisker plot extend to 1.5 times the interquartile range beyond the quartiles or to the highest value.

FIGURE 3.5
Minitab Box-and-Whisker Plot of the 2003 Return for Low-Risk, Average-Risk, and High-Risk Mutual Funds

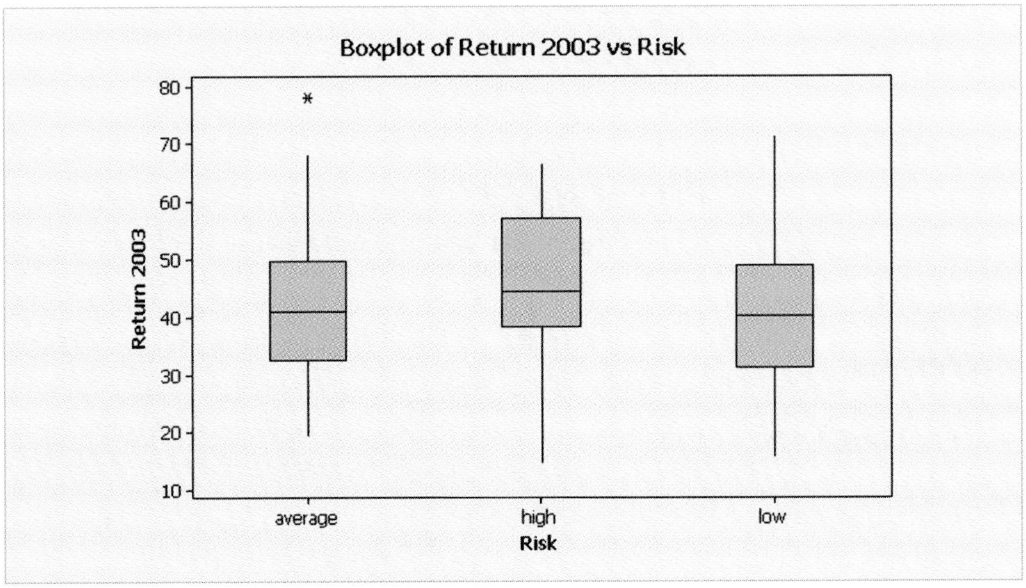

Figure 3.6 demonstrates the relationship between the box-and-whisker plot and the polygon for four different types of distributions. (*Note:* The area under each polygon is split into quartiles corresponding to the five-number summary for the box-and-whisker plot.)

FIGURE 3.6
Box-and-Whisker Plots and Corresponding Polygons for Four Distributions

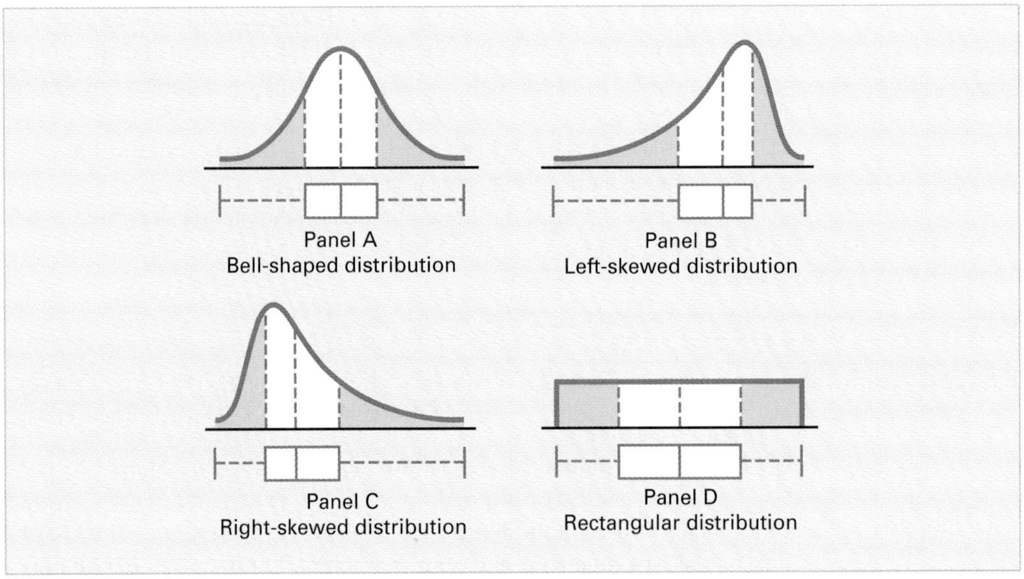

Panels A and D of Figure 3.6 are symmetric. In these distributions, the mean and median are equal. In addition, the length of the left whisker is equal to the length of the right whisker, and the median line divides the box in half.

Panel B of Figure 3.6 is left-skewed. The few small values distort the mean toward the left tail. For this left-skewed distribution, the skewness indicates that there is a heavy clustering of values at the high end of the scale (i.e., the right side); 75% of all values are found between the left edge of the box (Q_1) and the end of the right whisker ($X_{largest}$). Therefore, the long left whisker contains the smallest 25% of the values, demonstrating the distortion from symmetry in this data set.

Panel C of Figure 3.6 is right-skewed. The concentration of values is on the low end of the scale (i.e., the left side of the box-and-whisker plot). Here, 75% of all data values are found between the beginning of the left whisker ($X_{smallest}$) and the right edge of the box (Q_3), and the remaining 25% of the values are dispersed along the long right whisker at the upper end of the scale.

PROBLEMS FOR SECTION 3.4

Learning the Basics

PH Grade ASSIST **3.33** The following is a set of data from a sample of $n = 5$:

7 4 9 8 2

a. List the five-number summary.
b. Construct the box-and-whisker plot and describe the shape.
c. Compare your answer in (b) with that from problem 3.1(d) on page 90. Discuss.

PH Grade ASSIST **3.34** The following is a set of data from a sample of $n = 6$:

7 4 9 7 3 12

a. List the five-number summary.
b. Construct the box-and-whisker plot and describe the shape.
c. Compare your answer in (b) with that from problem 3.2(d) on page 90. Discuss.

PH Grade ASSIST **3.35** The following is a set of data from a sample of $n = 7$:

12 7 4 9 0 7 3

a. List the five-number summary.
b. Construct the box-and-whisker plot and describe the shape.
c. Compare your answer in (b) with that from problem 3.3(c) on page 90. Discuss.

3.36 The following is a set of data from a sample of $n = 5$:

7 −5 −8 7 9

a. List the five-number summary.
b. Construct the box-and-whisker plot and describe the shape.
c. Compare your answer in (b) with that from problem 3.4(c) on page 90. Discuss.

Applying the Concepts

Problems 3.37–3.42 can be solved manually or by using Microsoft Excel, Minitab, or SPSS.

PH Grade ASSIST SELF Test **3.37** A manufacturer of flashlight batteries took a sample of 13 batteries from a day's production and used them continuously until they were drained. The number of hours until failure are in the file. BATTERIES

342 426 317 545 264 451
1,049 631 512 266 492 562 298

a. List the five-number summary.
b. Construct the box-and-whisker plot and describe the shape.

3.38 In the 2002–2003 academic year, many public universities in the United States raised tuition and fees due to a decrease in state subsidies (Mary Beth Marklein, "Public Universities Raise Tuition, Fees—and Ire," *USA Today*, August 8, 2002, 1A–2A). The following represents the change in the cost of tuition, a shared dormitory room, and the most popular meal plan from the 2001–2002 academic year to the 2002–2003 academic year for a sample of 10 public universities. COLLEGECOST.XLS

University	Change in Cost ($)
University of California, Berkeley	1,589
University of Georgia, Athens	593
University of Illinois, Urbana–Champaign	1,223
Kansas State University, Manhattan	869
University of Maine, Orono	423
University of Mississippi, Oxford	1,720
University of New Hampshire, Durham	708
Ohio State University, Columbus	1,425
University of South Carolina, Columbia	922
Utah State University, Logan	308

a. List the five-number summary.
b. Construct the box-and-whisker plot and describe the shape.

3.39 A software development and consulting firm located in the Phoenix metropolitan area develops software for supply chain management systems using systematic software reuse. Instead of starting from scratch when writing and developing new custom software systems, the firm uses a database of reusable components totaling more than 2,000,000 lines of code collected from 10 years of continuous reuse effort. Eight analysts at the firm were asked to estimate the reuse rate when developing a new software system. The following data are given as a percentage of the total code written for a software system that is part of the reuse database. **REUSE**

50 62.5 37.5 75.0 45.0 47.5 15.0 25.0

Source: M. A. Rothenberger, and K. J. Dooley, "A Performance Measure for Software Reuse Projects," Decision Sciences, *30(Fall 1999), 1131–1153.*

a. List the five-number summary.
b. Construct the box-and-whisker plot and describe the shape of the data.

3.40 The following data represent the bounced check fee (in dollars) for a sample of 23 banks for direct-deposit customers who maintain a $100 balance and the monthly service fee (in dollars) for direct-deposit customers if their accounts fall below the minimum required balance of $1500 for a sample of 26 banks. **BANKCOST1 BANKCOST2**

26 28 20 20 21 22 25 25 18 25 15 20 18 20 25 25 22 30 30 30 15 20 29

12 8 5 5 6 6 10 10 9 7 10 7 7 5 0 10 6 9 12 0 5 10 8 5 5 9

Source: Extracted from "The New Face of Banking," Copyright © 2000 by Consumers Union of U.S., Inc., Yonkers, NY 10703–1057. Adapted with permission from Consumer Reports, *June 2000.*

a. List the five-number summary of the bounced check fee and of the monthly service fee.
b. Construct the box-and-whisker plot of the bounced check fee and the monthly service fee.
c. What similarities and differences are there in the distributions for the bounced check fee and the monthly service fee?

3.41 The following data represent the total fat for burgers and chicken items from a sample of fast-food chains. **FASTFOOD**

Burgers

19 31 34 35 39 39 43

Chicken

7 9 15 16 16 18 22 25 27 33 39

Source: Extracted from "Quick Bites," Copyright © 2001 by Consumers Union of U.S., Inc., Yonkers, NY 10703–1057. Adapted with permission from Consumer Reports, *March 2001, 46.*

a. List the five-number summary for the burgers and for the chicken items.
b. Construct the box-and-whisker plot for the burgers and the chicken items, and describe the shape of the distribution for the burgers and chicken items.
c. What similarities and differences are there in the distributions for the burgers and the chicken items?

3.42 A bank branch located in a commercial district of a city has developed an improved process for serving customers during the noon to 1:00 P.M. lunch period. The waiting time in minutes (operationally defined as the time the customer enters the line to the time he or she reaches the teller window) of all customers during this hour is recorded over a period of one week. A random sample of 15 customers is selected, and the results are as follows: **BANK1**

4.21 5.55 3.02 5.13 4.77 2.34 3.54
3.20 4.50 6.10 0.38 5.12 6.46 6.19 3.79

Another branch, located in a residential area, is also concerned with the noon to 1 P.M. lunch hour. The waiting time in minutes (defined as the time the customer enters the line until he or she reaches the teller window) of all customers during these hours is recorded over a period of one week. A random sample of 15 customers is selected, and the results are as follows: **BANK2**

9.66 5.90 8.02 5.79 8.73 3.82 8.01
8.35 10.49 6.68 5.64 4.08 6.17 9.91 5.47

a. List the five-number summary of the waiting time at the two bank branches.
b. Construct the box-and-whisker plot and describe the shape of the distribution of the two bank branches.
c. What similarities and differences are there in the distribution of the waiting time at the two bank branches?

3.5 THE COVARIANCE AND THE COEFFICIENT OF CORRELATION

In section 2.5, you used scatter diagrams to visually examine the relationship between two numerical variables. In this section, the covariance and the coefficient of correlation that measure the strength of the relationship between two numerical variables are discussed.

The Covariance

The **covariance** measures the strength of the linear relationship between two numerical variables (X and Y). Equation (3.18) defines the **sample covariance** and Example 3.17 illustrates its use.

THE SAMPLE COVARIANCE

$$\text{cov}(X,Y) = \frac{\sum_{i=1}^{n}(X_i - \bar{X})(Y_i - \bar{Y})}{n-1} \tag{3.18}$$

EXAMPLE 3.17

COMPUTING THE SAMPLE COVARIANCE

Consider the expense ratio and the 2003 return for the small cap high-risk funds. Compute the sample covariance.

SOLUTION Table 3.10 presents the expense ratio and 2003 return for the small cap high-risk funds and Figure 3.7 contains a Microsoft Excel worksheet that calculates the covariance for these data. The Calculations area of Figure 3.7 breaks down Equation (3.18) into a set of smaller calculations. From cell C17, or by using Equation (3.18) directly, the covariance is 1.19738.

$$\text{cov}(X,Y) = \frac{9.579}{9-1}$$
$$= 1.19738$$

TABLE 3.10
Expense Ratio and 2003 Return for the Small Cap High-Risk Funds

Expense Ratio	2003 Return
1.25	37.3
0.72	39.2
1.57	44.2
1.40	44.5
1.33	53.8
1.61	56.6
1.68	59.3
1.42	62.4
1.20	66.5

FIGURE 3.7
Microsoft Excel Worksheet for the Covariance between Expense Ratio and 2003 Return for the Small Cap High-Risk Funds

	A	B	C	
1	Expense ratio (X)	Return 2003 (Y)	(X-XBar)(Y-YBar)	
2	1.25	37.3	1.47078	=(A2 - C13) * (B2 - C14)
3	0.72	39.2	7.81111	=(A3 - C13) * (B3 - C14)
4	1.57	44.2	-1.58889	=(A4 - C13) * (B4 - C14)
5	1.4	44.5	-0.32822	=(A5 - C13)* (B5 - C14)
6	1.33	53.8	-0.05289	=(A6 - C13) * (B6 - C14)
7	1.61	56.6	1.30044	=(A7 - C13) * (B7 - C14)
8	1.68	59.3	2.53711	=(A8 - C13) * (B8 - C14)
9	1.42	62.4	0.72444	=(A9 - C13) * (B9 - C14)
10	1.2	66.5	-2.29489	=(A10 - C13) * (B10 - C14)
11				
12		Calculations		
13		XBar	1.353333333	=AVERAGE(A2:A10)
14		YBar	51.53333333	=AVERAGE(B2:B10)
15		*n*-1	8	=COUNT(A2:A10) - 1
16		Sum	9.57900	=SUM(C2:C10)
17		Covariance	1.19738	=C16/C15

The covariance has a major flaw as a measure of the linear relationship between two numerical variables. Since the covariance can have any value, you are unable to determine the relative strength of the relationship. To better determine the relative strength of the relationship, you need to compute the coefficient of correlation.

The Coefficient of Correlation

The **coefficient of correlation** measures the relative strength of a linear relationship between two numerical variables. The values of the coefficient of correlation range from −1 for a perfect negative correlation to +1 for a perfect positive correlation. *Perfect* means that if the points were plotted in a scatter diagram, all the points could be connected with a straight line. When dealing with population data for two numerical variables, the Greek letter ρ is used as the symbol for the coefficient of correlation. Figure 3.8 illustrates three different types of association between two variables.

FIGURE 3.8
Types of Association between Variables

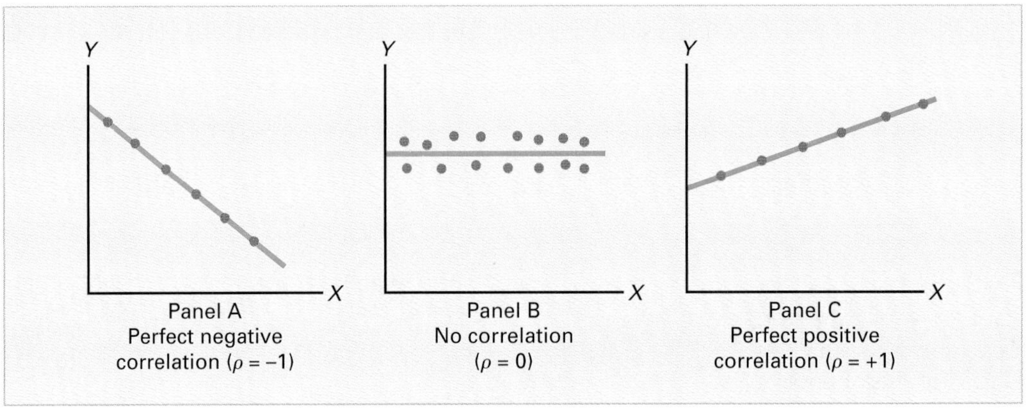

In panel A of Figure 3.8 there is a perfect negative linear relationship between X and Y. Thus, the coefficient of correlation ρ equals −1, and when X increases, Y decreases in a perfectly predictable manner. Panel B shows a situation in which there is no relationship between X and Y. In this case, the coefficient of correlation ρ equals 0, and as X increases, there is no tendency for Y to increase or decrease. Panel C illustrates a perfect positive relationship where ρ equals +1. In this case, Y increases in a perfectly predictable manner when X increases.

When you have sample data, the sample coefficient of correlation r is calculated. When using sample data, you are unlikely to have a sample coefficient of exactly +1, 0, or −1. Figure 3.9 on page 109 presents scatter diagrams along with their respective sample coefficients of correlation r for six data sets, each of which contains 100 values of X and Y.

In panel A, the coefficient of correlation r is −0.9. You can see that for small values of X there is a very strong tendency for Y to be large. Likewise, the large values of X tend to be paired with small values of Y. The data do not all fall on a straight line, so the association between X and Y cannot be described as *perfect*. The data in panel B have a coefficient of correlation equal to −0.6, and the small values of X tend to be paired with large values of Y. The linear relationship between X and Y in panel B is not as strong as in panel A. Thus, the coefficient of correlation in panel B is not as negative as in panel A. In panel C the linear relationship between X and Y is very weak, $r = -0.3$, and there is only a slight tendency for the small values of X to be paired with the larger values of Y. Panels D through F depict data sets that have positive coefficients of correlation because small values of X tend to be paired with small values of Y, and the large values of X tend to be associated with large values of Y.

In the discussion of Figure 3.9, the relationships were deliberately described as *tendencies* and not as *cause-and-effect*. This wording was used on purpose. Correlation alone cannot prove

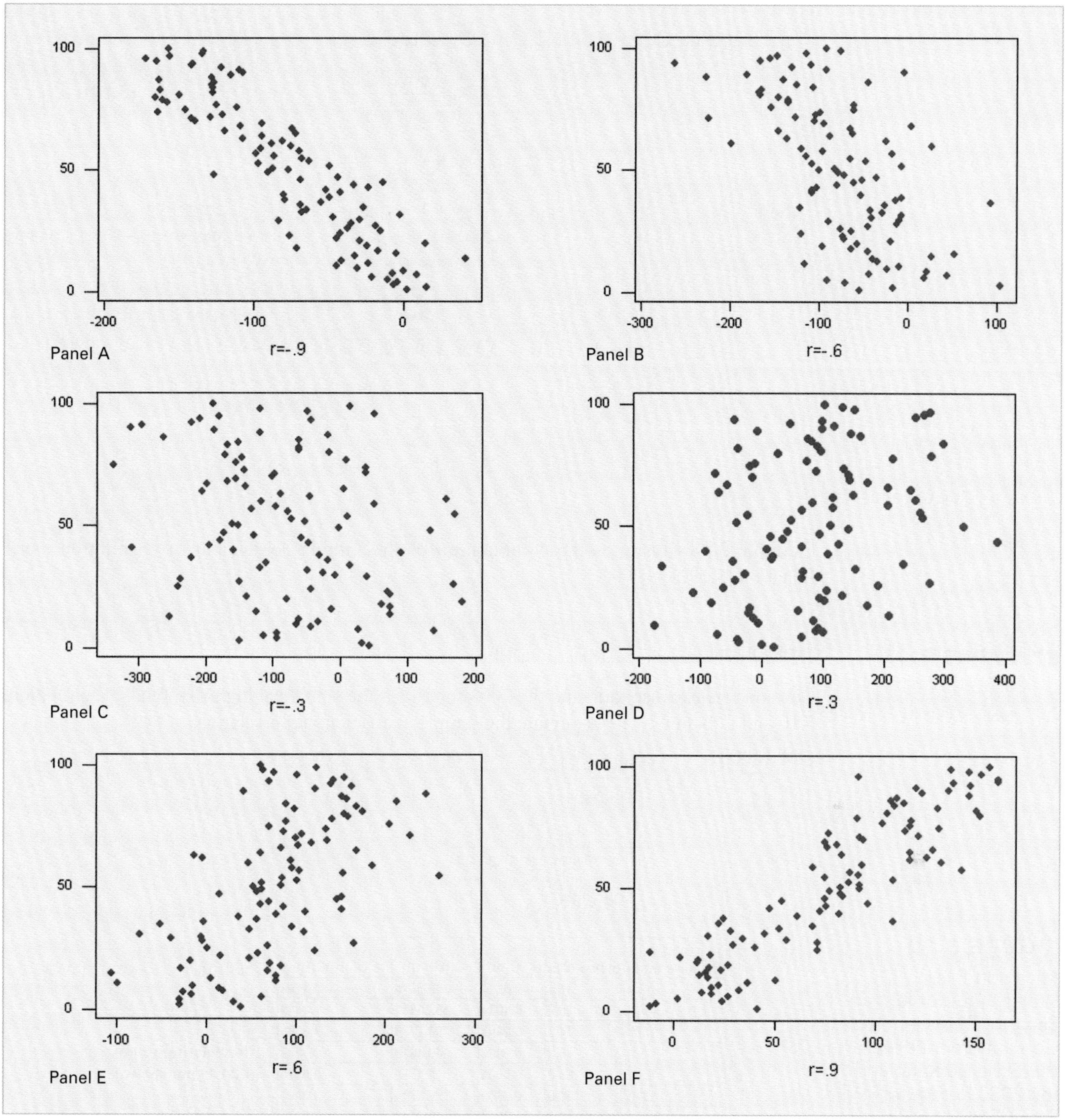

FIGURE 3.9 Six Scatter Diagrams Created from Minitab and Their Sample Coefficients of Correlation r

that there is a causation effect, that is, that the change in the value of one variable *caused* the change in the other variable. A strong correlation can be produced simply by chance, by the effect of a third variable not considered in the calculation of the correlation, or by a cause-and-effect relationship. You would need to perform additional analysis to determine which of these three situations actually produced the correlation. Therefore, you can say that causation implies correlation, but correlation alone does not imply causation.

Equation (3.19) defines the **sample coefficient of correlation** r and Example 3.18 illustrates its use.

THE SAMPLE COEFFICIENT OF CORRELATION

$$r = \frac{\text{cov}(X,Y)}{S_X S_Y} \tag{3.19}$$

where
$$\text{cov}(X,Y) = \frac{\sum_{i=1}^{n}(X_i - \bar{X})(Y_i - \bar{Y})}{n-1}$$

$$S_X = \sqrt{\frac{\sum_{i=1}^{n}(X_i - \bar{X})^2}{n-1}}$$

$$S_Y = \sqrt{\frac{\sum_{i=1}^{n}(Y_i - \bar{Y})^2}{n-1}}$$

Example 3.18 illustrates the computation of the sample coefficient of correlation using Equation (3.19).

EXAMPLE 3.18

COMPUTING THE SAMPLE COEFFICIENT OF CORRELATION

Consider the expense ratio and the 2003 return for the small cap high-risk funds. From Figure 3.10 and Equation (3.19), compute the sample coefficient of correlation.

SOLUTION

$$r = \frac{\text{cov}(X,Y)}{S_X S_Y}$$

$$= \frac{1.19738}{(0.287663)(10.554383)}$$

$$= 0.3943786$$

FIGURE 3.10

Microsoft Excel Worksheet for the Sample Coefficient of Correlation r between the Expense Ratio and the 2003 Return for Small Cap High-Risk Funds

	A	B	C	D	E	
1	Expense ratio	Return 2003	(X-XBar)²	(Y-YBar)²	(X-XBar)(Y-YBar)	
2	1.25	37.3	0.0107	202.5878	1.4708	
3	0.72	39.2	0.4011	152.1111	7.8111	
4	1.57	44.2	0.0469	53.7778	-1.5889	
5	1.4	44.5	0.0022	49.4678	-0.3282	(formulas for range
6	1.33	53.8	0.0005	5.1378	-0.0529	C2:E11 not shown)
7	1.61	56.6	0.0659	25.6711	1.3004	
8	1.68	59.3	0.1067	60.3211	2.5371	
9	1.42	62.4	0.0044	118.0844	0.7244	
10	1.2	66.5	0.0235	224.0011	-2.2949	
11		Sums:	0.662	891.16	9.5790	
12						
13				Calculations		
14				XBar	1.353333333	=AVERAGE(A2:A10)
15				YBar	51.53333333	=AVERAGE(B2:B10)
16				*n*-1	8	=COUNT(A2:A10) - 1
17				Covariance	1.19738	=E11/E16
18				S_X	0.287662997	=SQRT(C11/E16)
19				S_Y	10.55438298	=SQRT(D11/E16)
20				*r*	0.394378596	=CORREL(A2:A10,B2:B10) or =E17/(E18 * E19)

The expense ratio and the 2003 return for the small cap high-risk funds are positively correlated. Those mutual funds with the lowest expense ratios tend to be associated with the lowest 2003 returns. Those mutual funds with the highest expense ratios tend to be associated with the highest 2003 returns. This relationship is fairly weak, as indicated by a coefficient of correlation, $r = 0.394$.

You cannot assume that having a low expense ratio caused the low 2003 return. You can only say that this is what tended to happen in the sample. As with all investments, past performance does not guarantee future performance.

In summary, the coefficient of correlation indicates the linear relationship, or association, between two numerical variables. When the coefficient of correlation gets closer to +1 or −1, the linear relationship between the two variables is stronger. When the coefficient of correlation is near 0, little or no linear relationship exists. The sign of the coefficient of correlation indicates whether the data are positively correlated (i.e., the larger values of X are typically paired with the larger values of Y) or negatively correlated (i.e., the larger values of X are typically paired with the smaller values of Y). The existence of a strong correlation does not imply a causation effect. It only indicates the tendencies present in the data.

PROBLEMS FOR SECTION 3.5

Learning the Basics

3.43 The following is a set of data from a sample of $n = 11$ items:

X	7	5	8	3	6	10	12	4	9	15	18
Y	21	15	24	9	18	30	36	12	27	45	54

a. Compute the covariance.
b. Compute the coefficient of correlation.
c. How strong is the relationship between X and Y? Explain.

Applying the Concepts

Problems 3.44–3.49 can be solved manually or by using Microsoft Excel, Minitab, or SPSS.

3.44 A recent article (J. Clements, "Why Investors Should Put up to 30% of Their Stock Portfolio in Foreign Funds," *The Wall Street Journal*, November 26, 2003, D1) that discussed investment in foreign stocks stated that the coefficient of correlation between the return on investment of U.S. stocks and International Large Cap stocks was 0.80, U.S. stocks and International Small Cap stocks was 0.53, U.S. stocks and International Bonds was 0.03, U.S. stocks and Emerging market stocks was 0.71, U.S. stocks and Emerging market debt was 0.58.

a. What conclusions about the strength of the relationship between the return on investment of U. S. stocks and these five other types of investments can you make?
b. Compare the results of (a) to those of problem 3.45 (a).

3.45 A recent article (J. Clements, "Why Investors Should Put up to 30% of Their Stock Portfolio in Foreign Funds," *The Wall Street Journal*, November 26, 2003, D1) that discussed investment in foreign bonds stated that the coefficient of correlation between the return on investment of U.S. bonds and International Large Cap stocks was −0.13, U.S. bonds and International Small Cap stocks was −0.18, U.S. bonds and International Bonds was 0.48, U.S. bonds and Emerging market stocks was −0.20, U.S. bonds and Emerging market debt was 0.10.

a. What conclusions about the strength of the relationship between the return on investment of U.S. bonds and these five other types of investments can you make?
b. Compare the results of (a) to those of problem 3.44 (a).

3.46 The following data COFFEEDRINK represent the calories and fat (in grams) of 16-ounce iced coffee drinks at Dunkin' Donuts and Starbucks:

Product	Calories	Fat
Dunkin' Donuts Iced Mocha Swirl latte (whole milk)	240	8.0
Starbucks Coffee Frappuccino blended coffee	260	3.5
Dunkin' Donuts Coffee Coolatta (cream)	350	22.0
Starbucks Iced Coffee Mocha Expresso (whole milk and whipped cream)	350	20.0
Starbucks Mocha Frappuccino blended coffee (whipped cream)	420	16.0
Starbucks Chocolate Brownie Frappuccino blended coffee (whipped cream)	510	22.0
Starbucks Chocolate Frappuccino Blended Crème (whipped cream)	530	19.0

Source: Extracted from "Coffee as Candy at Dunkin' Donuts and Starbucks," Copyright © 2004 by Consumers Union of U.S., Inc., Yonkers, NY 10703–1057. Adapted with permission from Consumer Reports, *June 2004, 9.*

a. Compute the sample covariance.
b. Compute the coefficient of correlation.
c. Which do you think is more valuable in expressing the relationship between calories and fat—the covariance or the coefficient of correlation? Explain.
d. What conclusions can you reach about the relationship between calories and fat?

3.47 The following data represent the value of exports and imports in 2001 for various countries: EXPIMP

Country	Exports	Imports
European Union	874.1	912.8
United States	730.8	1180.2
Japan	403.5	349.1
China	266.2	243.6
Canada	259.9	227.2
Hong Kong	191.1	202.0
Mexico	158.5	176.2
South Korea	150.4	141.1
Taiwan	122.5	107.3
Singapore	121.8	116.0

Source: *Extracted from N. King and S. Miller, "Post-Iraq Influence of U.S. Faces Test at New Trade Talks,"* The Wall Street Journal, *September 9, 2003, A1.*

a. Compute the covariance.
b. Compute the coefficient of correlation.
c. Which do you think is more valuable in expressing the relationship between exports and imports—the covariance or the coefficient of correlation? Explain.
d. What conclusions can you reach about the relationship between exports and imports.

SELF Test **3.48** The following data SECURITY represent the turnover rate of pre-boarding screeners at airports in 1998–1999 and the security violations detected per million passengers.

City	Turnover	Violations
St. Louis	416	11.9
Atlanta	375	7.3
Houston	237	10.6
Boston	207	22.9
Chicago	200	6.5
Denver	193	15.2
Dallas	156	18.2
Baltimore	155	21.7
Seattle/Tacoma	140	31.5
San Francisco	110	20.7
Orlando	100	9.9
Washington–Dulles	90	14.8
Los Angeles	88	25.1
Detroit	79	13.5
San Juan	70	10.3
Miami	64	13.1
New York–JFK	53	30.1
Washington–Reagan	47	31.8
Honolulu	37	14.9

Source: *Extracted from Alan B. Krueger, "A Small Dose of Common Sense Would Help Congress Break the Gridlock over Airport Security,"* The New York Times, *November 15, 2001, C2.*

a. Compute the covariance.
b. Compute the coefficient of correlation.
c. What conclusions can you reach about the relationship between the turnover rate of pre-boarding screeners and the security violations detected?

3.49 The following data CELLPHONE represent the digital-mode talk time in hours and the battery capacity in milliampere-hours of cellphones.

Talk Time	Battery Capacity	Talk Time	Battery Capacity
4.50	800	1.50	450
4.00	1500	2.25	900
3.00	1300	2.25	900
2.00	1550	3.25	900
2.75	900	2.25	700
1.75	875	2.25	800
1.75	750	2.50	800
2.25	1100	2.25	900
1.75	850	2.00	900

Source: Extracted from "Service Shortcomings," Copyright 2002 by Consumers Union of U.S., Inc., Yonkers, NY 10703–1057. Adapted with permission from Consumer Reports, *February 2002, 25.*

a. Compute the covariance.
b. Compute the coefficient of correlation.
c. What conclusions can you reach about the relationship between the battery capacity and the digital-mode talk time?
d. You would expect cellphones with higher battery capacity to have a higher talk time. Is this borne out by the data?

3.6 PITFALLS IN NUMERICAL DESCRIPTIVE MEASURES AND ETHICAL ISSUES

In this chapter you studied how a set of numerical data can be characterized by various statistics that measure the properties of central tendency, variation, and shape. Your next step is analysis and interpretation of the calculated statistics. Your analysis is *objective*; your interpretation is *subjective*. You must avoid errors that may arise either in the objectivity of your analysis or in the subjectivity of your interpretation.

The analysis of the mutual funds based on risk level is *objective* and reveals several impartial findings. Objectivity in data analysis means reporting the most appropriate numerical descriptive measures for a given data set. Now that you have read the chapter and have become familiar with various numerical descriptive measures and their strengths and weaknesses, how should you proceed with the objective analysis? Because the data distribute in a slightly asymmetrical manner, shouldn't you report the median in addition to the mean? Doesn't the standard deviation provide more information about the property of variation than the range? Should you describe the data set as right-skewed?

On the other hand, data interpretation is *subjective*. Different people form different conclusions when interpreting the analytical findings. Everyone sees the world from different perspectives. Thus, because data interpretation is subjective, you must do it in a fair, neutral, and clear manner.

Ethical Issues

Ethical issues are vitally important to all data analysis. As a daily consumer of information, you need to question what you read in newspapers and magazines, what you hear on the radio or television, and what you see on the World Wide Web. Over time, much skepticism has been expressed about the purpose, the focus, and the objectivity of published studies. Perhaps no comment on this topic is more telling than a quip often attributed to the famous nineteenth-century British statesman Benjamin Disraeli: "There are three kinds of lies: lies, damned lies, and statistics."

Ethical considerations arise when you are deciding what results to include in a report. You should document both good and bad results. In addition, when making oral presentations and presenting written reports, you need to give results in a fair, objective, and neutral manner. Unethical behavior occurs when you willfully choose an inappropriate summary measure (e.g., the mean for a very skewed set of data) to distort the facts in order to support a particular position. In addition, unethical behavior occurs when you selectively fail to report pertinent findings because it would be detrimental to the support of a particular position.

SUMMARY

This chapter was about numerical descriptive measures. In this and the previous chapter, you studied descriptive statistics—how data are presented in tables and charts, and then summarized, described, analyzed, and interpreted. When dealing with the mutual fund data, you were able to present useful information through the use of pie charts, histograms, and other graphical methods. You explored characteristics of past performance such as central tendency, variability, and shape using numerical descriptive measures such as the mean, median, quartiles, range, standard deviation, and coefficient of correlation. Table 3.11 provides a list of the numerical descriptive measures covered in this chapter.

In the next chapter, the basic principles of probability are presented in order to bridge the gap between the subject of descriptive statistics and the subject of inferential statistics.

TABLE 3.11
Summary of Numerical Descriptive Measures

Type of Analysis	Numerical Data
Describing central tendency, variation, and shape of a numerical variable	Mean, median, mode, quartiles, geometric mean, range, interquartile range, standard deviation, variance, coefficient of variation, *Z* scores, box-and-whisker plot **(sections 3.1–3.4)**
Describing the relationship between two numerical variables	Covariance, coefficient of correlation **(section 3.5)**

KEY FORMULAS

Sample Mean

$$\bar{X} = \frac{\sum_{i=1}^{n} X_i}{n} \quad \textbf{(3.1)}$$

Median

$$\text{Median} = \frac{n+1}{2} \text{ ranked value} \quad \textbf{(3.2)}$$

First Quartile Q_1

$$Q_1 = \frac{n+1}{4} \text{ ranked value} \quad \textbf{(3.3)}$$

Third Quartile Q_3

$$Q_3 = \frac{3(n+1)}{4} \text{ ranked value} \quad \textbf{(3.4)}$$

Geometric Mean

$$\bar{X}_G = (X_1 \times X_2 \times \cdots \times X_n)^{1/n} \quad \textbf{(3.5)}$$

Geometric Mean Rate of Return

$$\bar{R}_G = [(1 + R_1) \times (1 + R_2) \times \cdots \times (1 + R_n)]^{1/n} - 1 \quad \textbf{(3.6)}$$

Range

$$\text{Range} = X_{\text{largest}} - X_{\text{smallest}} \quad \textbf{(3.7)}$$

Interquartile Range

$$\text{Interquartile range} = Q_3 - Q_1 \quad \textbf{(3.8)}$$

Sample Variance

$$S^2 = \frac{\sum_{i=1}^{n} (X_i - \bar{X})^2}{n-1} \quad \textbf{(3.9)}$$

Sample Standard Deviation

$$S = \sqrt{S^2} = \sqrt{\frac{\sum_{i=1}^{n} (X_i - \bar{X})^2}{n-1}} \quad \textbf{(3.10)}$$

Coefficient of Variation

$$CV = \left(\frac{S}{\bar{X}}\right) 100\% \quad \textbf{(3.11)}$$

***Z* Scores**

$$Z = \frac{X - \bar{X}}{S} \quad \textbf{(3.12)}$$

Population Mean

$$\mu = \frac{\sum_{i=1}^{N} X_i}{N} \quad \textbf{(3.13)}$$

Population Variance

$$\sigma^2 = \frac{\sum_{i=1}^{N} (X_i - \mu)^2}{N} \quad \textbf{(3.14)}$$

Population Standard Deviation

$$\sigma = \sqrt{\frac{\sum_{i=1}^{N} (X_i - \mu)^2}{N}} \quad \textbf{(3.15)}$$

Approximating the Mean from a Frequency Distribution

$$\bar{X} = \frac{\sum_{j=1}^{c} m_j f_j}{n} \quad \textbf{(3.16)}$$

Approximating the Standard Deviation from a Frequency Distribution

$$S = \sqrt{\frac{\sum_{j=1}^{c} (m_j - \bar{X})^2 f_j}{n-1}} \quad \textbf{(3.17)}$$

Sample Covariance

$$\text{cov}(X, Y) = \frac{\sum_{i=1}^{n} (X_i - \bar{X})(Y_i - \bar{Y})}{n-1} \quad \textbf{(3.18)}$$

Sample Coefficient of Correlation

$$r = \frac{\text{cov}(X, Y)}{S_X S_Y} \quad \textbf{(3.19)}$$

KEY TERMS

CHAPTER REVIEW PROBLEMS

Checking Your Understanding

3.50 What are the properties of a set of numerical data?

3.51 What is meant by the property of central tendency?

3.52 What are the differences among the mean, median, and mode, and what are the advantages and disadvantages of each?

3.53 How do you interpret the first quartile, median, and third quartile?

3.54 What is meant by the property of variation?

3.55 What does the Z score measure?

3.56 What are the differences among the various measures of variation such as the range, interquartile range, variance, standard deviation, and coefficient of variation, and what are the advantages and disadvantages of each?

3.57 How does the empirical rule help explain the ways in which the values in a set of numerical data cluster and distribute?

3.58 How do the empirical rule and the Chebychev rule differ?

3.59 What is meant by the property of shape?

3.60 How do the covariance and the coefficient of correlation differ?

Applying the Concepts

You can solve problems 3.61–3.67 manually or by using Microsoft Excel, Minitab, or SPSS. We recommend that you solve problems 3.68–3.86 using Microsoft Excel, Minitab, or SPSS.

3.61 A quality characteristic of interest for a tea-bag-filling process is the weight of the tea in the individual bags. If the bags are underfilled, two problems arise. First, customers may not be able to brew the tea to be as strong as they wish. Second, the company may be in violation of the truth-in-labeling laws. For this product, the label weight on the package indicates that, on average, there are 5.5 grams of tea in a bag. If the mean amount of tea in a bag exceeds the label weight, the company is giving away product. Getting an exact amount of tea in a bag is problematic because of variation in the temperature and humidity inside the factory, differences in the density of the tea, and the extremely fast filling operation of the machine (approximately 170 bags a minute). The following table provides the weight in grams of a sample of 50 tea bags produced in one hour by a single machine. TEABAGS

5.65	5.44	5.42	5.40	5.53	5.34	5.54	5.45	5.52	5.41
5.57	5.40	5.53	5.54	5.55	5.62	5.56	5.46	5.44	5.51
5.47	5.40	5.47	5.61	5.53	5.32	5.67	5.29	5.49	5.55
5.77	5.57	5.42	5.58	5.58	5.50	5.32	5.50	5.53	5.58
5.61	5.45	5.44	5.25	5.56	5.63	5.50	5.57	5.67	5.36

a. Compute the mean, median, first quartile, and third quartile.
b. Compute the range, interquartile range, variance, standard deviation, and coefficient of variation.
c. Interpret the measures of central tendency and variation within the context of this problem. Why should the company producing the tea bags be concerned about the central tendency and variation?
d. Construct a box-and-whisker plot. Are the data skewed? If so, how?
e. Is the company meeting the requirement set forth on the label that, on average, there are 5.5 grams of tea in a bag? If you were in charge of this process, what changes, if any, would you try to make concerning the distribution of weights in the individual bags?

3.62 In New York State, savings banks are permitted to sell a form of life insurance called Savings Bank Life Insurance (SBLI). The approval process consists of underwriting, which includes a review of the application, a medical information bureau check, possible requests for additional medical information and medical exams, and a policy compilation stage during which the policy pages are generated and sent to the bank for delivery. The ability to deliver approved policies to customers in a timely manner is critical to the profitability of this service to the bank. During a period of one month, a random sample of 27 approved policies was selected and the following total processing time in days was recorded: INSURANCE

73	19	16	64	28	28	31	90	60	56	31	56	22	18
45	48	17	17	17	91	92	63	50	51	69	16	17	

a. Compute the mean, median, first quartile, and third quartile.
b. Compute the range, interquartile range, variance, standard deviation, and coefficient of variation.
c. Construct a box-and-whisker plot. Are the data skewed? If so, how?
d. What would you tell a customer who enters the bank to purchase this type of insurance policy and asks how long the approval process takes?

3.63 One of the major measures of the quality of service provided by any organization is the speed with which it responds to customer complaints. A large family-held department store selling furniture and flooring, including carpet, had undergone a major expansion in the past several years. In particular, the flooring department had expanded from 2 installation crews to an installation supervisor, a measurer, and 15 installation crews. A sample of 50 complaints concerning carpet installation was selected during a recent year. The following data represent the number of days between the receipt of the complaint and the resolution of the complaint. FURNITURE

54	5	35	137	31	27	152	2	123	81	74	27
11	19	126	110	110	29	61	35	94	31	26	5
12	4	165	32	29	28	29	26	25	1	14	13
13	10	5	27	4	52	30	22	36	26	20	23
33	68										

a. Compute the mean, median, first quartile, and third quartile.
b. Compute the range, interquartile range, variance, standard deviation, and coefficient of variation.
c. Construct a box-and-whisker plot. Are the data skewed? If so, how?
d. On the basis of the results of (a) through (c), if you had to tell the president of the company how long a customer should expect to wait to have a complaint resolved, what would you say? Explain.

3.64 A manufacturing company produces steel housings for electrical equipment. The main component part of the housing is a steel trough that is made out of a 14-gauge steel coil. It is produced using a 250-ton progressive punch press with a wipe-down operation putting two 90-degree forms in the flat steel to make the trough. The distance from one side of the form to the other is critical because of weatherproofing in outdoor applications. The company requires that the width of the trough be between 8.31 inches and 8.61 inches. The following are the widths of the troughs in inches for a sample of $n = 49$. TROUGH

8.312	8.343	8.317	8.383	8.348	8.410	8.351	8.373	8.481	8.422
8.476	8.382	8.484	8.403	8.414	8.419	8.385	8.465	8.498	8.447
8.436	8.413	8.489	8.414	8.481	8.415	8.479	8.429	8.458	8.462
8.460	8.444	8.429	8.460	8.412	8.420	8.410	8.405	8.323	8.420
8.396	8.447	8.405	8.439	8.411	8.427	8.420	8.498	8.409	

a. Calculate the mean, median, range, and standard deviation for the width. Interpret these measures of central tendency and variability.
b. List the five-number summary.
c. Construct a box-and-whisker plot and describe the shape.
d. What can you conclude about the number of troughs that will meet the company's requirements of troughs being between 8.31 and 8.61 inches wide?

3.65 The manufacturing company in problem 3.64 also produces electric insulators. If the insulators break when in use, a short-circuit is likely to occur. To test the strength of the insulators, destructive testing is carried out to determine how much *force* is required to break the insulators. Force is measured by observing how many pounds must be applied to the insulator before it breaks. The data from 30 insulators from this experiment are as follows: FORCE

1,870	1,728	1,656	1,610	1,634	1,784	1,522	1,696	1,592	1,662
1,866	1,764	1,734	1,662	1,734	1,774	1,550	1,756	1,762	1,866
1,820	1,744	1,788	1,688	1,810	1,752	1,680	1,810	1,652	1,736

a. Calculate the mean, median, range, and standard deviation for the force variable.
b. Interpret the measures of central tendency and variability in (a).
c. Construct a box-and-whisker plot and describe the shape.
d. What can you conclude about the strength of the insulators if the company requires a force measurement of at least 1,500 pounds?

3.66 Problems with a telephone line that prevent a customer from receiving or making calls are disconcerting to both the customer and the telephone company. The following data represent samples of 20 problems reported to two different offices of a telephone company and the time to clear these problems (in minutes) from the customers' lines: PHONE

Central Office I Time to Clear Problems (minutes)

1.48	1.75	0.78	2.85	0.52	1.60	4.15	3.97	1.48	3.10
1.02	0.53	0.93	1.60	0.80	1.05	6.32	3.93	5.45	0.97

Central Office II Time to Clear Problems (minutes)

7.55	3.75	0.10	1.10	0.60	0.52	3.30	2.10	0.58	4.02
3.75	0.65	1.92	0.60	1.53	4.23	0.08	1.48	1.65	0.72

For each of the two central office locations:
a. Compute the mean, median, first quartile, and third quartile.
b. Compute the range, interquartile range, variance, standard deviation, and coefficient of variation.
c. Construct a side-by-side box-and-whisker plot. Are the data skewed? If so, how?
d. On the basis of the results of (a) through (c), are there any differences between the two central offices? Explain.

3.67 In many manufacturing processes the term "work-in-process" (often abbreviated WIP) is used. In a book manufacturing plant the WIP represents the time it takes for sheets from a press to be folded, gathered, sewn, tipped on end sheets, and bound. The following data represent samples of 20 books at each of two production plants and the processing time (operationally defined as the time in days from when the books came off the press to when they were packed in cartons) for these jobs. WIP

Plant A

5.62	5.29	16.25	10.92	11.46	21.62	8.45	8.58	5.41	11.42
11.62	7.29	7.50	7.96	4.42	10.50	7.58	9.29	7.54	8.92

Plant B

9.54	11.46	16.62	12.62	25.75	15.41	14.29	13.13	13.71	10.04
5.75	12.46	9.17	13.21	6.00	2.33	14.25	5.37	6.25	9.71

For each of the two plants:

a. Compute the mean, median, first quartile, and third quartile.

b. Compute the range, interquartile range, variance, standard deviation, and coefficient of variation.

c. Construct a side-by-side box-and-whisker plot. Are the data skewed? If so, how?

d. On the basis of the results of (a) through (c), are there any differences between the two plants? Explain.

3.68 The data contained in the file CEREALS consists of the cost in dollars per ounce, calories, fiber in grams, and sugar in grams for 33 breakfast cereals.

Source: Extracted from Copyright 1999 by Consumers Union of U.S., Inc., Yonkers, NY 10703–1057. Adapted with permission from Consumer Reports, *October 1999, 33–34.*

For each variable:

a. Compute the mean, median, first quartile, and third quartile.

b. Compute the range, interquartile range, variance, standard deviation, and coefficient of variation.

c. Construct a box-and-whisker plot. Are the data skewed? If so, how?

d. What conclusions can you reach concerning the cost per ounce in cents, calories, fiber in grams, and the sugar in grams for the 33 breakfast cereals?

3.69 State budget cuts forced a rise in tuition at public universities during the 2003–2004 academic year. The data in the file TUITION include the difference in tuition between 2002–2003 and 2003–2004 for in-state students and out-of-state students.

a. Compute the mean, median, first quartile, and third quartile for the difference in tuition between 2002–2003 and 2003–2004 for in-state students and out-of-state students.

b. Compute the range, interquartile range, variance, standard deviation, and coefficient of variation for the difference in tuition between 2002–2003 and 2003–2004 for in-state students and out-of-state students.

c. Construct a box-and-whisker plot of the difference in tuition between 2002–2003 and 2003–2004 for in-state students and out-of-state students. Are the data skewed? If so, how?

d. What conclusions can you reach concerning the difference in tuition between 2002–2003 and 2003–2004 for in-state students and out-of-state students?

3.70 Do marketing promotions, such as bobble-head give-aways, increase attendance at Major League Baseball games? An article in *Sport Marketing Quarterly* reported on the effectiveness of marketing promotions (T. C. Boyd and T. C. Krehbiel, "Promotion Timing in Major League Baseball and the Stacking Effects of Factors that Increase Game Attractiveness," *Sport Marketing Quarterly*, 12(2003), 173–183). The data file ROYALS includes the following variables for the Kansas City Royals during the 2002 baseball season:

GAME = Home games in the order they were played
ATTENDANCE = Paid attendance for the game
PROMOTION—Y = a promotion was held; N = no promotion was held

a. Calculate the mean and standard deviation of attendance for the 43 games where promotions were held and for the 37 games without promotions.

b. Construct a five-number summary for the 43 games where promotions were held and for the 37 games without promotions.

c. Construct a graphical display containing two box-and-whisker plots; one for the 43 games where promotions were held and one for the 37 games without promotions.

d. Discuss the results of (a) through (c) and comment on the effectiveness of promotions at Royals' games during the 2002 season.

3.71 The data contained in the file PETFOOD2 consist of the cost per serving, cups per can, protein in grams, and fat in grams for 97 varieties of dry and canned dog and cat food.

Source: Extracted from Copyright 1998 by Consumers Union of U.S., Inc., Yonkers, NY 10703–1057. Adapted with permission from Consumer Reports, *February 1998, 18–19.*

For the four types of food (dry dog food, canned dog food, dry cat food and canned cat food), for the variables of cost per serving, protein in grams, and fat in grams:

a. Compute the mean, median, first quartile, and third quartile.

b. Compute the range, interquartile range, variance, standard deviation, and coefficient of variation.

c. Construct a side-by-side box-and-whisker plot for the four types (dry dog food, canned dog food, dry cat food, and canned cat food). Are the data for any of the types of food skewed? If so, how?

d. What conclusions can you reach concerning any differences among the four types (dry dog food, canned dog food, dry cat food, and canned cat food)?

3.72 The manufacturer of Boston and Vermont asphalt shingles provide their customers with a 20-year warranty on most of their products. To determine whether a shingle will last as long as the warranty period, accelerated-life testing is conducted at the manufacturing plant. Accelerated-life testing exposes the shingle to the stresses it would be subject to in a lifetime of normal use in a laboratory setting via an experiment that takes only a few minutes to conduct. In this test, a shingle is repeatedly scraped with a brush for a short period of time and the amount of shingle granules that are removed by the brushing is weighed (in grams). Shingles that experience low amounts of granule loss are expected to last longer in normal use than shingles that experience high amounts of granule loss. In this situation, a shingle should experience no more than 0.8 grams of granule loss if it is expected to last the length of the warranty period. The data file **GRANULE** contains a sample of 170 measurements made on the company's Boston shingles, and 140 measurements made on Vermont shingles.

a. List the five-number summary for the Boston shingles and for the Vermont shingles.
b. Construct side-by-side box-and-whisker plots for the two brands of shingles and describe the shapes of the distributions.
c. Comment on the shingles' ability to achieve a granule loss of 0.8 grams or less.

3.73 The data in the file **STATES** represent the results of the American Community Survey, a sampling of 700,000 households taken in each state during the 2000 U.S. Census. For each of the variables of average travel-to-work time in minutes, percentage of homes with eight or more rooms, median household income, and percentage of mortgage-paying homeowners whose housing costs exceed 30% of income:

a. Compute the mean, median, first quartile, and third quartile.
b. Compute the range, interquartile range, variance, standard deviation, and coefficient of variation.
c. Construct a box-and-whisker plot. Are the data skewed? If so, how?
d. What conclusions can you reach concerning the mean travel-to-work time in minutes, percentage of homes with eight or more rooms, median household income, and percentage of mortgage-paying homeowners whose housing costs exceed 30% of income?

3.74 The economics of baseball has caused a great deal of controversy with owners arguing that they are losing money, players arguing that owners are making money, and fans complaining about how expensive it is to attend a game and watch games on cable television. In addition to data related to team statistics for the 2001 season, the file **BB2001** contains team-by-team statistics on ticket prices; the fan cost index; regular season gate receipts; local television, radio, and cable receipts; all other operating revenue, player compensation and benefits; national and other local expenses; and income from baseball operations. For each of these variables,

a. Compute the mean, median, first quartile, and third quartile.
b. Compute the range, interquartile range, variance, standard deviation, and coefficient of variation.
c. Construct a box-and-whisker plot. Are the data skewed? If so, how?
d. Compute the correlation between the number of wins and player compensation and benefits. How strong is the relationship between these two variables?
e. What conclusions can you reach concerning the regular season gate receipts; local television, radio, and cable receipts; all other operating revenue; player compensation and benefits; national and other local expenses; and income from baseball operations?

3.75 The data in the file **AIRCLEANERS** represent the price, yearly filter energy cost, and yearly filter cost of room air cleaners.

a. Compute the coefficient of correlation between price and energy cost.
b. Compute the coefficient of correlation between price and filter cost.
c. What conclusions about the relationship of energy cost and filter cost to the price of the air cleaners can you make?

Source: Extracted from "Portable Room Air Cleaners," Copyright © 2002 by Consumers Union of U.S., Inc., Yonkers, NY 10703–1057. Adapted with permission from Consumer Reports, *February 2002, 47.*

3.76 The data in the file **PRINTERS** represent the price, text speed, text cost, color photo time, and color photo cost of computer printers.

a. Compute the coefficient of correlation between price and each of the following: text speed, text cost, color photo time, and color photo cost.
b. Based on the results of (a), do you think that any of the other variables might be useful in predicting printer price? Explain.

Source: Extracted from "Printers," Copyright © 2002 by Consumers Union of U.S., Inc., Yonkers, NY 10703–1057. Adapted with permission from Consumer Reports, *March 2002, 51.*

3.77 You want to study characteristics of the model year 2002 automobiles in terms of the following variables: miles per gallon, length, width, turning circle requirement, weight, and luggage capacity. **AUTO2002**

Source: Extracted from "The 2002 Cars," Copyright © 2002 by Consumers Union of U.S., Inc., Yonkers, NY 10703–1057. Adapted with permission from Consumer Reports, *April 2002.*

For each of these variables:

a. Compute the mean, median, first quartile, and third quartile.

b. Compute the range, interquartile range, variance, standard deviation, and coefficient of variation.

c. Construct a box-and-whisker plot. Are the data skewed? If so, how?

d. What conclusions can you draw concerning the 2002 automobiles?

3.78 Refer to the data of problem 3.77. You want to compare sports utility vehicles (SUVs) with non-SUVs in terms of miles per gallon, length, width, turning circle requirement, weight, and luggage capacity. For SUVs and non-SUVs, for each of these variables:

a. Compute the mean, median, first quartile, and third quartile.

b. Compute the range, interquartile range, variance, standard deviation, and coefficient of variation.

c. Construct a side-by-side box-and-whisker plot. Are the data skewed? If so, how?

d. What conclusions can you reach concerning differences between SUVs and non-SUVs?

3.79 Zagat's publishes restaurant ratings for various locations in the United States. The data file RESTRATE contains the Zagat rating for food, decor, service, and the price per person for a sample of 50 restaurants located in New York City, and 50 restaurants located on Long Island.

Source: Extracted from Zagat Survey 2002 New York City Restaurants and Zagat Survey 2002 Long Island Restaurants.

For New York City and Long Island restaurants, for the variables of food rating, decor rating, service rating, and price per person:

a. Compute the mean, median, first quartile, and third quartile.

b. Compute the range, interquartile range, variance, standard deviation, and coefficient of variation.

c. Construct a side-by-side box-and-whisker plot for the New York City and Long Island restaurants. Are the data for any of the variables skewed? If so, how?

d. What conclusions can you reach concerning differences between New York City and Long Island restaurants?

3.80 As an illustration of the misuse of statistics, an article by Glenn Kramon ("Coaxing the Stanford Elephant to Dance," *The New York Times* Sunday Business Section, November 11, 1990) implied that costs at Stanford Medical Center had been driven up higher than at competing institutions because the former was more likely than other organizations to treat indigent, Medicare, Medicaid, sicker, and more complex patients. The chart below was provided to compare the average 1989 to 1990 hospital charges for three medical procedures (coronary bypass, simple birth, and hip replacement) at three competing institutions (El Camino, Sequoia, and Stanford).

Suppose you were working in a medical center. Your CEO knows you are currently taking a course in statistics and calls you in to discuss this. She tells you that the article was presented in a discussion group setting as part of a meeting of regional area medical center CEOs last night and that one of them mentioned that this chart was totally meaningless and asked her opinion. She now requests that you prepare her response. You smile, take a deep breath, and reply . . .

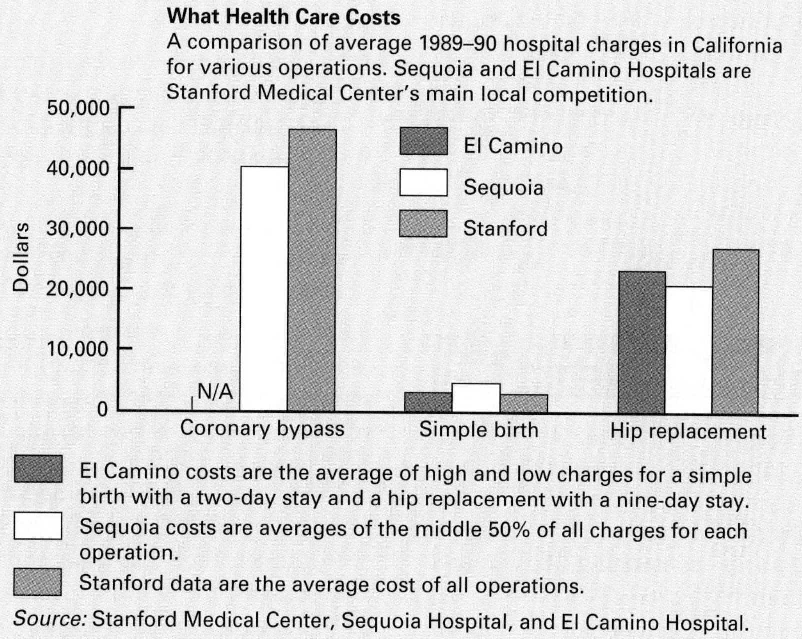

Source: Stanford Medical Center, Sequoia Hospital, and El Camino Hospital.

3.81 You are planning to study for your statistics examination with a group of classmates, one of whom you particularly want to impress. This individual has volunteered to use Microsoft Excel, Minitab, or SPSS to get the needed summary information, tables, and charts for a data set containing several numerical and categorical variables assigned by the instructor for study purposes. This person comes over to you with the printout and exclaims, "I've got it all—the means, the medians, the standard deviations, the box-and-whisker plots, the pie charts—for all our variables. The problem is, some of the output looks weird—like the box-and-whisker plots for gender and for major and the pie charts for grade point index and for height. Also, I can't understand why Professor Krehbiel said we can't get the descriptive stats for some of the variables—I got it for everything! See, the mean for height is 68.23, the mean for grade point index is 2.76, the mean for gender is 1.50, the mean for major is 4.33." What is your reply?

Report Writing Exercises

3.82 The data found in the data file BEER represent the price of a six-pack of 12-ounce bottles, the calories per 12 fluid ounces, the percent of alcohol content per 12 fluid ounces, the type of beer (craft lagers, craft ales, imported lagers, regular and ice beers, and light and nonalcoholic beers), and the country of origin (U.S. versus imported) for each of the 69 beers that were sampled.

Your task is to write a report based on a complete descriptive evaluation of each of the numerical variables—price, calories, and alcoholic content—regardless of type of product or origin. Then perform a similar evaluation comparing each of these numerical variables based on type of product—craft lagers, craft ales, imported lagers, regular and ice beers, and light or nonalcoholic beers. In addition, perform a similar evaluation comparing and contrasting each of these numerical variables based on the origins of the beers—those brewed in the United States versus those that were imported. Appended to your report should be all appropriate tables, charts, and numerical descriptive measures.

Source: Extracted from "Beers," Copyright © 1996 by Consumers Union of U.S., Inc., Yonkers, NY 10703–1057. Adapted with permission from Consumer Reports, *June 1996.*

TEAM PROJECTS

The data file MUTUALFUNDS2004 contains information regarding 12 variables from a sample of 121 mutual funds. The variables are:

Fund—The name of the mutual fund
Category—Type of stocks comprising the mutual fund—small cap, mid cap, large cap
Objective—Objective of stocks comprising the mutual fund—growth or value
Assets—In millions of dollars
Fees—Sales charges (no or yes)
Expense ratio—ratio of expenses to net assets in percentage
2003 Return—Twelve-month return in 2003
Three-year return—Annualized return 2001–2003
Five-year return—Annualized return 1999–2003
Risk—Risk-of-loss factor of the mutual fund classified as low, average, or high
Best quarter—Best quarterly performance 1999–2003
Worst quarter—Worst quarterly performance 1999–2003

3.83 For expense ratio in percentage, 2003 Return, three-year return, and five-year return,

a. Compute the mean, median, first quartile, and third quartile.
b. Compute the range, interquartile range, variance, standard deviation, and coefficient of variation.
c. Construct a box-and-whisker plot. Are the data skewed? If so, how?
d. What conclusions can you reach concerning these variables?

3.84 You wish to compare mutual funds that have fees to those that do not have fees. For each of these two groups, for the variables expense ratio in percentage, 2003 Return, three-year return, and five-year return,

a. Compute the mean, median, first quartile, and third quartile.
b. Compute the range, interquartile range, variance, standard deviation, and coefficient of variation.
c. Construct a box-and-whisker plot. Are the data skewed? If so, how?
d. What conclusions can you reach about differences between mutual funds that have fees and those that do not have fees?

3.85 You wish to compare mutual funds that have a growth objective to those that have value objective. For each of these two groups, for the variables expense ratio in percentage, 2003 Return, three-year return, and five-year return,

a. Compute the mean, median, first quartile, and third quartile.
b. Compute the range, interquartile range, variance, standard deviation, and coefficient of variation.
c. Construct a box-and-whisker plot. Are the data skewed? If so, how?
d. What conclusions can you reach about differences between mutual funds that have a growth objective to those that have value objective?

3.86 You wish to compare small cap, mid cap, and large cap mutual funds. For each of these three groups, for the variables expense ratio in percentage, 2003 Return, three-year return, and five-year return,

a. Compute the mean, median, first quartile, and third quartile.
b. Compute the range, interquartile range, variance, standard deviation, and coefficient of variation.
c. Construct a box-and-whisker plot. Are the data skewed? If so, how?
d. What conclusions can you reach about differences between small cap, mid cap, and large cap mutual funds?

RUNNING CASE MANAGING THE *SPRINGVILLE HERALD*

For what variable in the Chapter 2 Managing the *Springville Herald* case (see page 62) are numerical descriptive measures needed? For the variable you identify:

1. Compute the appropriate numerical descriptive measures, and generate a box-and-whisker plot.
2. Identify another graphical display that might be useful and construct it. What conclusions can you form from that plot that cannot be made from the box-and-whisker plot?

Summarize your findings in a report that can be included with the task force's study.

WEB CASE

Apply your knowledge about the proper use of numerical descriptive measures in this continuing Web Case from Chapter 2.

Visit the StockTout Investing Service Web site **www.prenhall.com/Springville/StockToutHome.htm** a second time and reexamine their supporting data and then answer the following:

1. Reexamine the data you inspected when working on the Web Case for Chapter 2. Can descriptive measures be computed for any variables? How would such summary statistics support StockTout's claims? How would those summary statistics affect your perception of StockTout's record?
2. Evaluate the methods StockTout used to summarize the results of its customer survey **www.prenhall.com/Springville/ST_Survey.htm**. Is there anything you would do differently to summarize these results?
3. Note that the last question of the survey has fewer responses. What factors may have limited the number of responses to that question?

REFERENCES

1. Kendall, M. G., and A. Stuart, *The Advanced Theory of Statistics*, vol. 1 (London: Charles W. Griffin, 1958).
2. *Microsoft Excel 2003* (Redmond, WA: Microsoft Corporation, 2002).
3. *Minitab Version 14* (State College, PA: Minitab Inc., 2004).
4. *SPSS Base 12.0 Brief Guide* (Upper Saddle River, NJ: Prentice Hall, 2003).
5. Tukey, J., *Exploratory Data Analysis* (Reading, MA: Addison-Wesley, 1977).
6. Velleman, P. F., and D. C. Hoaglin, *Applications, Basics, and Computing of Exploratory Data Analysis* (Boston, MA: Duxbury Press, 1981).

Appendix 3 Using Software for Descriptive Statistics

A3.1 MICROSOFT EXCEL

For Descriptive Statistics

Use the Data Analysis ToolPak. Open to the worksheet containing the data you want to summarize. Select **Tools → Data Analysis**. From the list that appears in the Data Analysis dialog box, select **Descriptive Statistics** and click **OK**. In the Descriptive Statistics dialog box (see Figure A3.1), enter the cell range of the data in the **Input Range** box. Choose the **Columns** option and **Labels in First Row** if you are using data that are arranged like the data in the Excel files on the CD-ROM packaged with this text. Finish by selecting **New Worksheet Ply**, **Summary statistics**, **Kth Largest**, and **Kth Smallest**, and clicking **OK**. Results appear on a separate worksheet.

OR you can use any of these sample statistics worksheet functions in your own formulas including AVERAGE (for mean), MEDIAN, MODE, QUARTILE, STDEV, VAR, MIN, MAX, SUM, COUNT, LARGE, or SMALL.

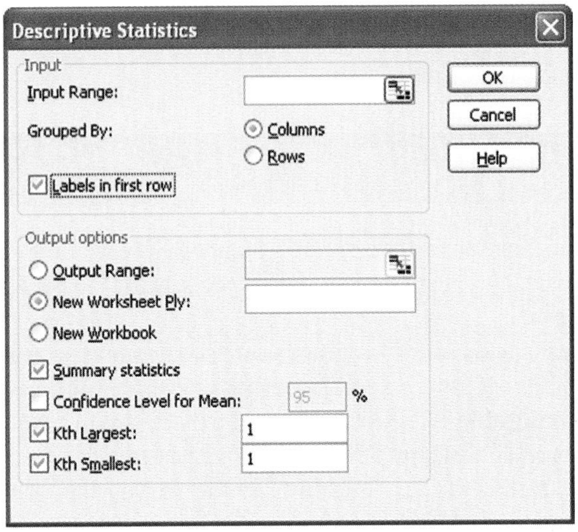

FIGURE A3.1 Data Analysis Descriptive Statistics Dialog Box

To enter one of these functions into a worksheet, select an empty cell and then select **Insert → Function.** In the Function dialog box, select **Statistical** from the drop-down list and then scroll to and select the function you want to use. Click **OK**. In the Function Arguments dialog box, enter the cell range of the data to be summarized and click **OK**. (For LARGE and SMALL, enter **1** as the **K** value; and for QUARTILE, enter either **1** or **3** as the **Quart** value, for either first or third quartile.) In versions of Microsoft Excel earlier than Excel 2003, you may encounter errors in results when using the QUARTILE function.

For Box-and-Whisker Plot

See section G.5 (**Box-and-Whisker Plot)** if you want PHStat2 to produce a box-and-whisker plot as a Microsoft Excel chart. (There are no Microsoft Excel commands that directly produce box-and-whisker plots.)

For Covariance

Open the **Covariance.xls** Excel file, shown in Figure 3.7 on page 107. Follow the onscreen instructions for modifying the table area if you want to use this worksheet with other pairs of variables. Note in Figure 3.7 that cell C15 contains a formula that uses the COUNT function. This allows Excel to automatically update the value of n when the size of the table area is changed, and ensures that the $n - 1$ term is always correct.

For Coefficient of Correlation

Open the **Correlation.xls** Excel file, shown in Figure 3.10 on page 110. Follow the onscreen instructions for modifying the table area if you want to use this worksheet with other pairs of variables. Note in Figure 3.10 that cell E16 contains a formula that uses the COUNT function. This allows Excel to automatically update the value of n when the size of the table area is changed and ensures that the $n - 1$ term is always correct.

The worksheet uses the CORREL function to calculate the coefficient of correlation. As shown in Figure 3.10, the formula **=E17/(E18 * E19)** could also be used in this particular worksheet to calculate the statistic, since the covariance S_X and S_Y already appear in the worksheet.

A3.2 MINITAB

Computing Descriptive Statistics

To produce descriptive statistics for the 2003 return for different risk levels shown in Figure 3.3 on page 90, open the **MUTUALFUNDS2004.MTW** worksheet. Select **Stat → Basic Statistics → Display Descriptive Statistics**.

Step 1: In the Display Descriptive Statistics dialog box (see Figure A3.2), enter **C7** or **'Return 2003'** in the Variables: edit box. Enter **C10** or **Risk** in the By variables (optional): edit box.

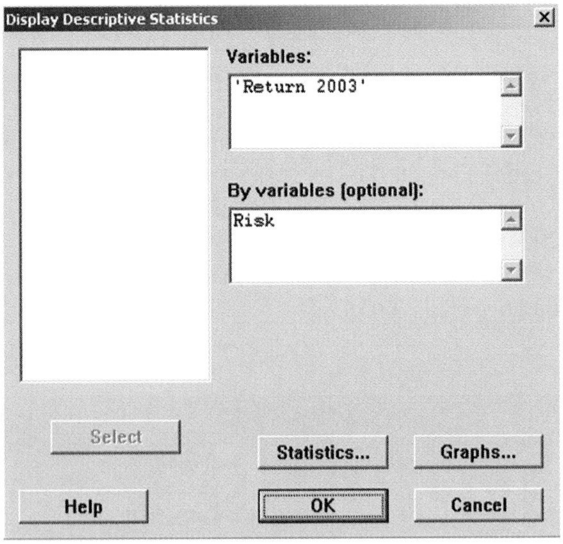

FIGURE A3.2 Minitab Display Descriptive Statistics Dialog Box

Step 2: Select the **Statistics** button. In the Display Descriptive Statistics—Statistics dialog box (see Figure A3.3), select the **Mean**, **Standard deviation**, **Coefficient of variation**, **First quartile**, **Median**, **Third quartile**, **Interquartile range**, **Minimum**, **Maximum**, **Range**, and **N total** (the sample size) check boxes. Click the **OK** button to return to the

Display Descriptive Statistics dialog box. Click the **OK** button again to compute the descriptive statistics.

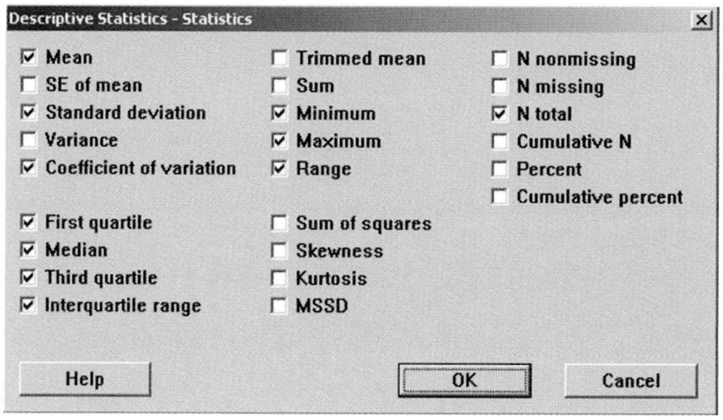

FIGURE A3.3 Minitab Display Descriptive Statistics—Statistics Dialog Box

Using Minitab to Create a Box-and-Whisker Plot

To create a box-and-whisker plot for the 2003 return for different risk levels shown in Figure 3.5 on page 104, open the **MUTUALFUNDS2004.MTW** worksheet. Select **Graph ➔ Boxplot.**

Step 1: In the Boxplots dialog box (see Figure A3.4), select the **One Y With Groups** choice. (If you want to create a box-and-whisker plot for one group, select the **One Y Simple** choice.) Click the **OK** button.

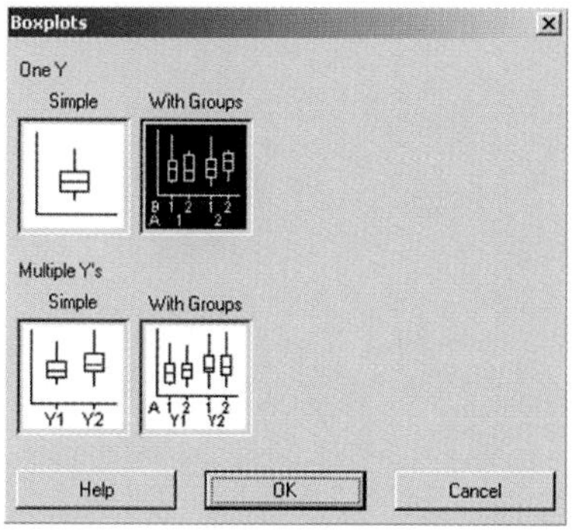

FIGURE A3.4 Minitab Boxplots Dialog Box

Step 2: In the Boxplot—One Y, With Groups dialog box (see Figure A3.5), enter **C7** or **'Return 2003'** in the Graph variables: edit box. Enter **C10** or **Risk** in the Categorical variables edit box. Click the **OK** button.

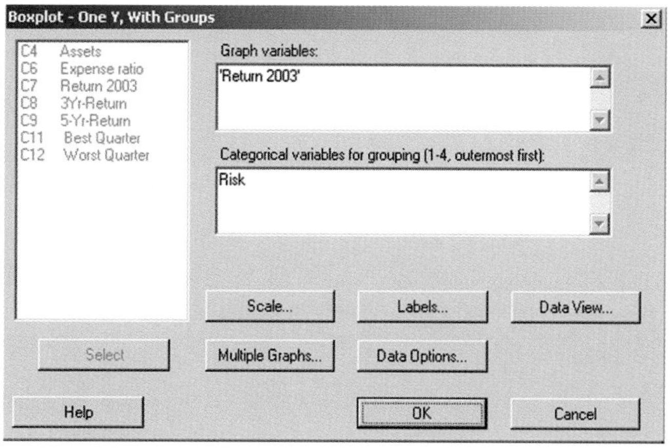

FIGURE A3.5 Minitab Boxplots—One Y, With Groups Dialog Box

The output will be similar to Figure 3.5 on page 104.

Calculating a Coefficient of Correlation

To compute the coefficient of correlation for the expense ratio and the 2003 return for *all* the mutual funds, open the **MUTUALFUNDS2004.MTW** worksheet. Select **Stat ➔ Basic Statistics ➔ Correlation**. In the Correlation dialog box (see Figure A3.6), enter **C6** or **'Expense ratio'** and **C7** or **'Return 2003'**. Click the **OK** button.

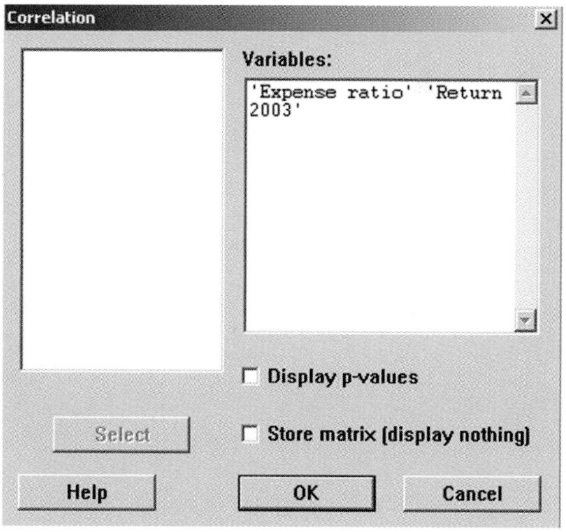

FIGURE A3.6 Minitab Correlation Dialog Box

CHAPTER 4

Basic Probability

LEARNING OBJECTIVES

In this chapter, you learn:

- Basic probability concepts
- Conditional probability
- To use Bayes' theorem to revise probabilities
- Various counting rules

USING STATISTICS

The Consumer Electronics Company

You are the marketing manager for the Consumer Electronics Company. You are analyzing the survey results of 1,000 households concerning their intentions to purchase a big-screen television set (defined as 31 inches or larger) in the next 12 months. Investigations of this type are known as intent to purchase studies. As a follow-up, you will survey the same households 12 months later to see whether they actually purchased the television set. In addition, for those who did purchase a big-screen television set, you are interested in whether they purchased a high-definition television (HDTV) set, whether they also purchased a DVD player in the last 12 months, and whether they were satisfied with their purchase of the big-screen television set. Some of the questions you would like to answer include the following:

- What is the probability that a household is planning to purchase a big-screen television set in the next year?
- What is the probability that the household will actually purchase a big-screen television set?
- What is the probability that a household is planning to purchase a big-screen television set and actually purchases the television set?
- Given that the household is planning to purchase a big-screen television set, what is the probability that the purchase is made?
- Does knowledge of whether the household *plans* to purchase the television set change the likelihood of predicting whether the household *will* purchase the television set?
- What is the probability that a household that purchases a big-screen television set will purchase an HDTV?
- What is the probability that a household that purchases a big-screen television set will also purchase a DVD player?
- What is the probability that a household that purchases a big-screen television set will be satisfied with their purchase?

Answers to these questions and others can help you develop future sales and marketing strategies. For example, should marketing campaigns for your big-screen television sets target those customers indicating intent to purchase? Are those individuals purchasing big-screen television sets easily persuaded to buy a higher-priced HDTV and/or a DVD player?

The principles of probability help bridge the worlds of descriptive statistics and inferential statistics. Reading this chapter will help you learn about different types of probabilities and how to revise probabilities in light of new information. These topics are the foundation for the probability distribution, the concept of mathematical expectation, and the binomial, hypergeometric, and Poisson distributions (topics that will be covered in Chapter 5).

4.1 BASIC PROBABILITY CONCEPTS

What is meant by the word *probability*? A **probability** is the numeric value representing the chance, likelihood, or possibility a particular event will occur, such as the price of a stock increasing, a rainy day, a nonconforming unit of production, or the outcome five in one toss of a die. In all these instances, the probability attached is a proportion or fraction whose value ranges between 0 and 1 inclusively. An event that has no chance of occurring (i.e., the **impossible event**) has a probability of 0. An event that is sure to occur (i.e., the **certain event**) has a probability of 1. There are three approaches to the subject of probability:

- *a priori* classical probability
- empirical classical probability
- subjective probability

In ***a priori*** **classical probability**, the probability of success is based on prior knowledge of the process involved. In the simplest case, where each outcome is equally likely, the chance of occurrence of the event is defined in Equation (4.1).

PROBABILITY OF OCCURRENCE

$$\text{Probability of occurrence} = \frac{X}{T} \quad \textbf{(4.1)}$$

where X = number of ways in which the event occurs
T = total number of possible outcomes

Consider a standard deck of cards that has 26 red cards and 26 black cards. The probability of selecting a black card is 26/52 = 0.50 since there are $X = 26$ black cards and $T = 52$ total cards. What does this probability mean? If each card is replaced after it is drawn, does it mean that one out of the next two cards selected will be black? No, because you cannot say for certain what will happen on the next several selections. However, you can say that in the long run, if this selection process is continually repeated, the proportion of black cards selected will approach 0.50.

EXAMPLE 4.1

FINDING *A PRIORI* PROBABILITIES

A standard six-sided die has six faces. Each face of the die contains either one, two, three, four, five, or six dots. If you roll a die, what is the probability you will get a face with five dots?

SOLUTION Each face is equally likely to occur. Since there are six faces, the probability of getting a face with five dots is $\frac{1}{6}$.

The above examples use the *a priori* classical probability approach because the number of ways the event occurs and the total number of possible outcomes are known from the composition of the deck of cards or the faces of the die.

In the **empirical classical probability** approach, the outcomes are based on observed data, not on prior knowledge of a process. Examples of this type of probability are the proportion of individuals in the "Using Statistics" scenario who actually purchase a television, the proportion of registered voters who prefer a certain political candidate, or the proportion of students who have a part-time job. For example, if you take a survey of students and 60% state that they have a part-time job, then there is a 0.60 probability that an individual student has a part-time job.

The third approach to probability, **subjective probability**, differs from the other two approaches because subjective probability differs from person to person. For example, the development team for a new product may assign a probability of 0.6 to the chance of success for the product while the president of the company is less optimistic and assigns a probability of 0.3. The assignment of subjective probabilities to various outcomes is usually based on a combination of an individual's past experience, personal opinion, and analysis of a particular situation. Subjective probability is especially useful in making decisions in situations in which you cannot use *a priori* classical probability or empirical classical probability.

Events and Sample Spaces

The basic elements of probability theory are the individual outcomes of a variable under study. You need the following definitions to understand probabilities.

Each possible outcome of a variable is referred to as an **event**.

A **simple event** is described by a single characteristic.

For example, when you toss a coin, the two possible outcomes are heads and tails. Each of these represents a simple event. When you roll a standard six-sided die in which the six faces of the die contain either one, two, three, four, five, or six dots, there are six possible simple events. An event can be any one of these simple events, a set of them, or a subset of all of them. For example, the event of an *even number of dots* consists of three simple events (i.e., two, four, or six dots).

A **joint event** is an event that has two or more characteristics.

Getting two heads on the toss of two coins is an example of a joint event since it consists of heads on the toss of the first coin and heads on the toss of the second coin.

The **complement** of event A (given the symbol A') includes all events that are not part of A.

The complement of a head is a tail since that is the only event that is not a head. The complement of face five is not getting face five. Not getting face five consists of getting face one, two, three, four, or six.

The collection of all the possible events is called the **sample space**.

The sample space for tossing a coin consists of heads and tails. The sample space when rolling a die consists of one, two, three, four, five, and six dots.

EXAMPLE 4.2 EVENTS AND SAMPLE SPACES

The "Using Statistics" scenario on page 126 concerns The Consumer Electronics Company. Table 4.1 presents the results of the sample of 1,000 households in terms of purchase behavior for big-screen television sets.

TABLE 4.1
Purchase Behavior for Big-Screen Television Sets

	ACTUALLY PURCHASED		
PLANNED TO PURCHASE	**Yes**	**No**	**Total**
Yes	200	50	250
No	100	650	750
Total	300	700	1,000

What is the sample space? Give examples of simple events and joint events.

SOLUTION The sample space consists of the 1,000 respondents. Simple events are "planned to purchase," "did not plan to purchase," "purchase," and "did not purchase." The complement of the event "planned to purchase" is "did not plan to purchase." The event "planned to purchase and actually purchased" is a joint event because the respondent must plan to purchase the television set *and* actually purchase it.

Contingency Tables and Venn Diagrams

There are several ways to present a sample space. Table 4.1 uses a **table of cross-classifications** to present a sample space. This table is also called a **contingency table** (see section 2.3). You get the values in the cells of the table by subdividing the sample space of 1,000 households according to whether someone planned to purchase and actually purchased the big-screen television set. For example, 200 of the respondents planned to purchase a big-screen television set and subsequently did purchase the big-screen television set.

A **Venn diagram** is a second way to present a sample space. This diagram graphically represents the various events as "unions" and "intersections" of circles. Figure 4.1 presents a typical Venn diagram for a two-variable situation, with each variable having only two events (A and A', B and B'). The circle on the left (the red one) represents all events that are part of A. The circle on the right (the yellow one) represents all events that are part of B. The area contained within circle A and circle B (center area) is the intersection of A and B (written as $A \cap B$), since it is part of A and also part of B. The total area of the two circles is the union of A and B (written as $A \cup B$) and contains all outcomes that are just part of event A, just part of event B, or part of both A and B. The area in the diagram outside of $A \cup B$ contains outcomes that are neither part of A nor part of B.

You must define A and B in order to develop a Venn diagram. You can define either event as A or B, as long as you are consistent in evaluating the various events. For the consumer electronics example, you can define the events as follows:

A = planned to purchase	B = actually purchased
A' = did not plan to purchase	B' = did not actually purchase

In drawing the Venn diagram (see Figure 4.2), you must determine the value of the intersection of A and B in order to divide the sample space into its parts. $A \cap B$ consists of all 200 households who planned to purchase and actually purchased a big-screen television set. The remainder of event A (planned to purchase) consists of the 50 households who planned to purchase a big-screen television set but did not actually purchase one. The remainder of event B (actually purchased) consists of the 100 households who did not plan to purchase a big-screen television set but actually purchased one. The remaining 650 households represent those who neither planned to purchase nor actually purchased a big-screen television set.

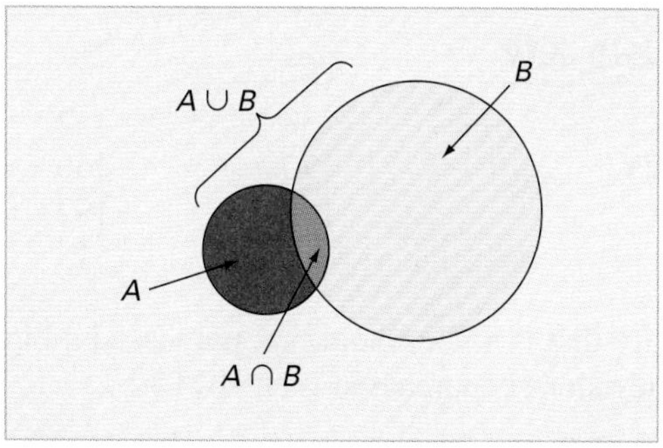

FIGURE 4.1
Venn Diagram for Events A and B

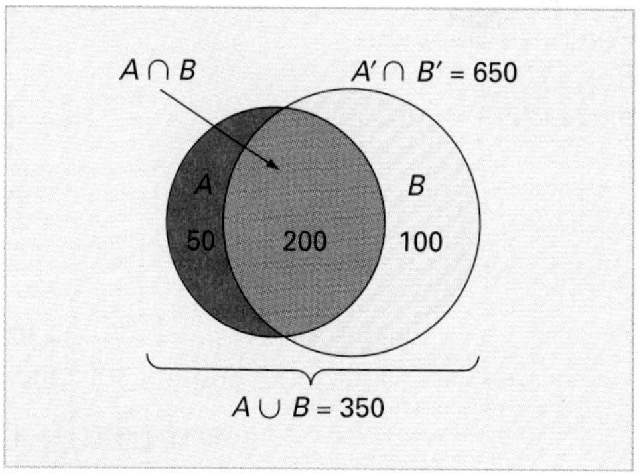

FIGURE 4.2
Venn Diagram for the Consumer Electronics Example

Simple (Marginal) Probability

Now you can answer some of the questions posed in the "Using Statistics" scenario. Since the results are based on data collected in a survey (see Table 4.1 on page 128), you can use the empirical classical probability approach.

As stated previously, the most fundamental rule for probabilities is that they range in value from 0 to 1. An impossible event has a probability of 0, and an event that is certain to occur has a probability of 1.

Simple probability refers to the probability of occurrence of a simple event, $P(A)$. A simple probability in the "Using Statistics" scenario is the probability of planning to purchase a big-screen television set. How can you determine the probability of selecting a household that planned to purchase a big-screen television set? Using Equation (4.1) on page 127:

$$\text{Probability of occurrence} = \frac{X}{T}$$

$$P(\text{planned to purchase}) = \frac{\text{number who planned to purchase}}{\text{total number of households}}$$

$$= \frac{250}{1{,}000} = 0.25$$

Thus, there is a 0.25 (or 25%) chance that a household planned to purchase a big-screen television set.

Simple probability is also called **marginal probability**, because you can compute the total number of successes (those who planned to purchase) from the appropriate margin of the contingency table (see Table 4.1 on page 128). Example 4.3 illustrates another application of simple probability.

EXAMPLE 4.3

COMPUTING THE PROBABILITY THAT THE BIG-SCREEN TELEVISION SET PURCHASED WILL BE AN HDTV

In the "Using Statistics" follow-up survey, additional questions were asked of the 300 households that actually purchased a big-screen television set. Table 4.2 indicates the consumers' responses to whether the television set purchased was an HDTV and whether they also purchased a DVD player in the last 12 months.

TABLE 4.2

Purchase Behavior Regarding HDTVs and DVD Players

	PURCHASED DVD		
PURCHASED HDTV	**Yes**	**No**	**Total**
HDTV	38	42	80
Not HDTV	70	150	220
Total	108	192	300

Find the probability that if a household that purchased a big-screen television set is randomly selected, the television set purchased is an HDTV.

SOLUTION Using the following definitions:

A = purchased an HDTV — B = purchased a DVD player

A' = did not purchase an HDTV — B' = did not purchase a DVD player

$$P(\text{HDTV}) = \frac{\text{number of HDTV television sets}}{\text{total number of television sets}}$$

$$= \frac{80}{300} = 0.267$$

There is a 26.7% chance that a randomly selected big-screen television set purchase is an HDTV.

Joint Probability

Marginal probability refers to the probability of occurrence of simple events. **Joint probability** refers to the probability of an occurrence involving two or more events. An example of joint probability is the probability that you will get heads on the first toss of a coin and heads on the second toss of a coin.

Referring to Table 4.1 on page 128, those individuals who planned to purchase and actually purchased a big-screen television set consist only of the outcomes in the single cell "yes—planned to purchase *and* yes—actually purchased." Because this group consists of 200 households, the probability of picking a household that planned to purchase *and* actually purchased a big-screen television set is

$$P(\text{planned to purchase } \textit{and} \text{ actually purchased}) = \frac{\text{planned to purchase } \textit{and} \text{ actually purchased}}{\text{total number of respondents}}$$

$$= \frac{200}{1{,}000} = 0.20$$

Example 4.4 also demonstrates how to determine joint probability.

EXAMPLE 4.4 DETERMINING THE JOINT PROBABILITY THAT A BIG-SCREEN TELEVISION SET CUSTOMER PURCHASED AN HDTV AND A DVD PLAYER

In Table 4.2 on page 130, the purchases are cross-classified as HDTV or not HDTV and whether or not the household purchased a DVD player. Find the probability that a randomly selected household that purchased a big-screen television set also purchased an HDTV and a DVD player.

SOLUTION Using Equation (4.1) on page 127,

$$P(\text{HDTV } \textit{and} \text{ DVD player}) = \frac{\text{number that purchased an HDTV } \textit{and} \text{ a DVD player}}{\text{total number of big-screen television set purchasers}}$$

$$= \frac{38}{300} = 0.127$$

Therefore, you have a 12.7% chance that a randomly selected household that purchased a big-screen television set purchased an HDTV and a DVD player.

You can view the marginal probability of a particular event using the concept of joint probability just discussed. The marginal probability of an event consists of a set of joint probabilities. For example, if B consists of two events, B_1 and B_2, then $P(A)$, the probability of event A, consists of the joint probability of event A occurring with event B_1 and the joint probability of event A occurring with event B_2. Use Equation (4.2) to compute marginal probabilities.

MARGINAL PROBABILITY

$$P(A) = P(A \textit{ and } B_1) + P(A \textit{ and } B_2) + \ldots + P(A \textit{ and } B_k) \quad \textbf{(4.2)}$$

where $B_1, B_2, \ldots, B_k$ are k mutually exclusive and collectively exhaustive events.

Mutually exclusive events and collectively exhaustive events are defined as follows.

Two events are **mutually exclusive** if both the events cannot occur simultaneously.

Heads and tails in a coin toss are mutually exclusive events. The result of a coin toss cannot simultaneously be a head and a tail.

A set of events is **collectively exhaustive** if one of the events must occur.

Heads and tails in a coin toss are collectively exhaustive events. One of them must occur. If heads does not occur, tails must occur. If tails does not occur, heads must occur.

Being male and being female are mutually exclusive and collectively exhaustive events. No one is both (they are mutually exclusive), and everyone is one or the other (they are collectively exhaustive).

You can use Equation (4.2) to compute the marginal probability of planned to purchase a big-screen television set.

$$\begin{aligned} P(\text{planned to purchase}) &= P(\text{planned to purchase } \textit{and} \text{ purchased}) \\ &\quad + P(\text{planned to purchase } \textit{and} \text{ did not purchase}) \\ &= \frac{200}{1{,}000} + \frac{50}{1{,}000} \\ &= \frac{250}{1{,}000} = 0.25 \end{aligned}$$

You will get the same result if you add the number of outcomes that make up the simple event "planned to purchase."

General Addition Rule

The general addition rule allows you to find the probability of event "*A or B*." This rule considers the occurrence of either event *A* or event *B* or both *A* and *B*. How can you determine the probability that a household planned to purchase *or* actually purchased a big-screen television set? The event "planned to purchase *or* actually purchased" includes all households who planned to purchase and all households who actually purchased the big-screen television set. You examine each cell of the contingency table (Table 4.1 on page 128) to determine whether it is part of this event. From Table 4.1, the cell "planned to purchase *and* did not actually purchase" is part of the event, because it includes respondents who planned to purchase. The cell "did not plan to purchase *and* actually purchased" is included because it contains respondents who actually purchased. Finally, the cell "planned to purchase *and* actually purchased" has both characteristics of interest. Therefore, the probability of planned to purchase *or* actually purchased is:

$$\begin{aligned} P(\text{planned to purchase } \textit{or} \text{ actually purchased}) &= P(\text{planned to purchase } \textit{and} \text{ did not actually} \\ &\quad \text{purchase}) + P(\text{did not plan to purchase } \textit{and} \\ &\quad \text{actually purchased}) + P(\text{planned to purchase} \\ &\quad \textit{and} \text{ actually purchased}) \\ &= \frac{50}{1{,}000} + \frac{100}{1{,}000} + \frac{200}{1{,}000} = \frac{350}{1{,}000} = 0.35 \end{aligned}$$

Often you will find it easier to determine $P(A \text{ or } B)$, the probability of the event *A or B*, by using the **general addition rule** defined in Equation (4.3).

GENERAL ADDITION RULE

The probability of A or B is equal to the probability of A plus the probability of B minus the probability of A and B.

$$P(A \text{ or } B) = P(A) + P(B) - P(A \text{ and } B) \quad \textbf{(4.3)}$$

Applying this equation to the previous example produces the following result:

$$\begin{aligned} P(\text{planned to purchase } \textit{or} \text{ actually purchased}) &= P(\text{planned to purchase}) + P(\text{actually purchased}) \\ &\quad - P(\text{planned to purchase } \textit{and} \text{ actually purchased}) \\ &= \frac{250}{1{,}000} + \frac{300}{1{,}000} - \frac{200}{1{,}000} \\ &= \frac{350}{1{,}000} = 0.35 \end{aligned}$$

The general addition rule consists of taking the probability of A and adding it to the probability of B, and then subtracting the joint event of A *and* B from this total because the joint event has already been included both in computing the probability of A and the probability of B. Referring to Table 4.1 on page 128, if the outcomes of the event "planned to purchase" are added to those of the event "actually purchased," the joint event "planned to purchase *and* actually purchased" has been included in each of these simple events. Therefore, because this joint event has been double-counted, you must subtract it to provide the correct result. Example 4.5 illustrates another application of the general addition rule.

EXAMPLE 4.5

USING THE GENERAL ADDITION RULE FOR THE HOUSEHOLDS THAT PURCHASED BIG-SCREEN TELEVISION SETS

In Example 4.3 on page 130, the purchases were cross-classified as an HDTV or not HDTV and whether or not the household purchased a DVD player. Find the probability that among households that purchased a big-screen television set, that they purchased an HDTV or a DVD player.

SOLUTION Using Equation (4.3) above,

$$\begin{aligned} P(\text{HDTV } \textit{or} \text{ DVD player}) &= P(\text{HDTV}) + P(\text{DVD player}) - P(\text{HDTV } \textit{and} \text{ DVD player}) \\ &= \frac{80}{300} + \frac{108}{300} - \frac{38}{300} \\ &= \frac{150}{300} = 0.50 \end{aligned}$$

Therefore, you have a 50.0% chance that a randomly selected household that purchased a big-screen television set purchased an HDTV or a DVD player.

PROBLEMS FOR SECTION 4.1

Learning the Basics

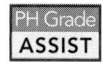

PH Grade ASSIST **4.1** Two coins are tossed.

a. Give an example of a simple event.

b. Give an example of a joint event.

c. What is the complement of a head on the first toss?

4.2 An urn contains 12 red balls and 8 white balls. One ball is to be selected from the urn.

a. Give an example of a simple event.

b. What is the complement of a red ball?

PH Grade ASSIST **4.3** Given the following contingency table:

	B	B'
A	10	20
A'	20	40

What is the probability of
a. event A?
b. event A'?
c. event A *and* B?
d. event A *or* B?

PH Grade ASSIST **4.4** Given the following contingency table:

	B	B'
A	10	30
A'	25	35

What is the probability of
a. event A'?
b. event A *and* B?
c. event A' *and* B'?
d. event A' *or* B'?

Applying the Concepts

PH Grade ASSIST **4.5** For each of the following, indicate whether the type of probability involved is an example of *a priori* classical probability, empirical classical probability, or subjective probability.
a. The next toss of a fair coin will land on heads.
b. Italy will win soccer's World Cup the next time the competition is held.
c. The sum of the faces of two dice will be 7.
d. The train taking a commuter to work will be more than 10 minutes late.

4.6 For each of the following, state whether the events created are mutually exclusive and collectively exhaustive. If they are not mutually exclusive and collectively exhaustive, either reword the categories to make them mutually exclusive and collectively exhaustive or explain why this would not be useful.
a. Registered voters in the United States were asked whether they registered as Republicans or Democrats.
b. Respondents were classified by type of car he or she drives: American, European, Japanese, or none.
c. People were asked, "Do you currently live in (i) an apartment or (ii) a house?"
d. A product was classified as defective or not defective.

4.7 The probability of each of the following events is zero. For each, state why.
a. A voter in the United States who is registered as a Republican and a Democrat
b. A product that is defective and not defective
c. An automobile that is a Ford and a Toyota

SELF Test **4.8** A U.S. Census American Housing Survey studied how U.S. homeowners get to work ("How People Get to Work," *USA Today* Snapshots, February 25, 2003, 1A). Suppose that the survey consisted of a sample of 1,000 homeowners and 1,000 renters.

Drives to Work	Homeowner	Renter	Total
Yes	824	681	1,505
No	176	319	495
Total	1,000	1,000	2,000

a. Give an example of a simple event.
b. Give an example of a joint event.
c. What is the complement of "drives to work"?
d. Why is "drives to work and is a homeowner" a joint event?

4.9 Referring to the contingency table in problem 4.8, if a respondent is selected at random, what is the probability that he or she
a. drives to work?
b. drives to work and is a homeowner?
c. drives to work or is a homeowner?
d. Explain the difference in the results in (b) and (c).

4.10 A yield improvement study at a semiconductor manufacturing facility provided defect data for a sample of 450 wafers. The following table presents a summary of the responses to two questions: "Was a particle found on the die that produced the wafer?" and "Is the wafer good or bad?"

	CONDITION OF DIE		
QUALITY OF WAFER	No Particles	Particles	Totals
Good	320	14	334
Bad	80	36	116
Totals	400	50	450

Source: S. W. Hall, Analysis of Defectivity of Semiconductor Wafers by Contingency Table, Proceedings Institute of Environmental Sciences, *Vol. 1 (1994), 177–183.*

a. Give an example of a simple event.
b. Give an example of a joint event.
c. What is the complement of a good wafer?
d. Why is a "good wafer" and a die "with particles" a joint event?

4.11 Referring to the contingency table in problem 4.10, if a wafer is selected at random, what is the probability that
a. it was produced from a die with no particles?
b. it is a bad wafer and was produced from a die with no particles?

c. it is a bad wafer or was produced from a die with particles?
d. Explain the difference in the results in (b) and (c).

PH Grade ASSIST **4.12** Are large companies less likely to offer board members stock options than small- to mid-sized companies? A survey conducted by the Segal Company of New York found that in a sample of 189 large companies, 40 offered stock options to their board members as part of their non-cash compensation packages. For small- to mid-sized companies, 43 of the 180 surveyed indicated that they offer stock options as part of their non-cash compensation packages to their board members (Kemba J. Dunham, "The Jungle: Focus on Recruitment, Pay and Getting Ahead," *The Wall Street Journal*, August 21, 2001, B6). Construct a contingency table or a Venn diagram to evaluate the probabilities. If a company is selected at random, what is the probability that the company
a. offered stock options to their board members?
b. is small- to mid-sized and did not offer stock options to their board members?
c. is small- to mid-sized or offered stock options to their board members?
d. Explain the difference in the results in (b) and (c).

4.13 Are whites more likely to claim bias? A survey conducted by Barry Goldman ("White Fight: A Researcher Finds Whites Are More Likely to Claim Bias," *The Wall Street Journal*, Work Week, April 10, 2001, A1) found that of 56 white workers terminated, 29 claimed bias. Of 407 black workers terminated, 126 claimed bias. Construct a contingency table or a Venn diagram to evaluate the probabilities. If a worker is selected at random, what is the probability that he or she
a. claimed bias?
b. is black and did not claim bias?
c. is black or claimed bias?
d. Explain the difference in the results in (b) and (c).

4.14 A sample of 500 respondents was selected in a large metropolitan area to study consumer behavior. Among the questions asked was "Do you enjoy shopping for clothing?" Of 240 males, 136 answered yes. Of 260 females, 224 answered yes. Construct a contingency table or a Venn diagram to evaluate the probabilities. What is the probability that a respondent chosen at random
a. enjoys shopping for clothing?
b. is a female and enjoys shopping for clothing?
c. is a female or enjoys shopping for clothing?
d. is a male or a female?

4.15 Each year, ratings are compiled concerning the performance of new cars during the first 90 days of use. Suppose that the cars have been categorized according to whether the car needs warranty-related repair (yes or no) and the country in which the company manufacturing the car is based (United States or not United States). Based on the data collected, the probability that the new car needs a warranty repair is 0.04, the probability that the car is manufactured by a U.S.-based company is 0.60, and the probability that the new car needs a warranty repair *and* was manufactured by a U.S.-based company is 0.025. Construct a contingency table or a Venn diagram to evaluate the probabilities of a warranty-related repair. What is the probability that a new car selected at random
a. needs a warranty-related repair?
b. needs a warranty repair and is manufactured by a company based in the United States?
c. needs a warranty repair or was manufactured by a U.S.-based company?
d. needs a warranty repair or was not manufactured by a U.S.-based company?

4.2 CONDITIONAL PROBABILITY

Computing Conditional Probabilities

Each example in section 4.1 involved finding the probability of an event when sampling from the entire sample space. How do you determine the probability of an event if certain information about the events involved are already known?

Conditional probability refers to the probability of event A, given information about the occurrence of another event B.

CONDITIONAL PROBABILITY

The probability of A given B is equal to the probability of A *and* B divided by the probability of B

$$P(A|B) = \frac{P(A \text{ and } B)}{P(B)} \quad \textbf{(4.4a)}$$

The probability of B given A is equal to the probability of A *and* B divided by the probability of A

$$P(B|A) = \frac{P(A \text{ and } B)}{P(A)} \quad \textbf{(4.4b)}$$

where

$P(A \text{ and } B)$ = joint probability of A *and* B

$P(A)$ = marginal probability of A

$P(B)$ = marginal probability of B

Referring to the "Using Statistics" scenario involving the purchase of big-screen television sets, suppose you were told that a household planned to purchase a big-screen television set. Now, what is the probability that the household actually purchased the television set? In this example the objective is to find P (actual purchase | planned to purchase). Here you are given the information that the household planned to purchase the big-screen television set. Therefore, the sample space does not consist of all 1,000 households in the survey. It consists of only those households that planned to purchase the big-screen television set. Of 250 such households, 200 actually purchased the big-screen television set. Therefore, (see Table 4.1 on page 128 or Figure 4.2 on page 129) the probability that a household actually purchased the big-screen television set given that he or she planned to purchase is

$$P(\text{actually purchased}|\text{planned to purchase}) = \frac{\text{planned to purchase } \textit{and} \text{ actually purchased}}{\text{planned to purchase}}$$

$$= \frac{200}{250} = 0.80$$

You can also use Equation (4.4b) to compute this result.

$$P(B|A) = \frac{P(A \text{ and } B)}{P(A)}$$

where

event A = planned to purchase

event B = actually purchased

Then

$$P(\text{actually purchased}|\text{planned to purchase}) = \frac{200/1{,}000}{250/1{,}000}$$

$$= \frac{200}{250} = 0.80$$

Example 4.6 further illustrates conditional probability.

EXAMPLE 4.6

FINDING A CONDITIONAL PROBABILITY CONCERNING THE HOUSEHOLDS THAT ACTUALLY PURCHASED A BIG-SCREEN TELEVISION SET

Table 4.2 on page 130 is a contingency table for whether the household purchased an HDTV and a DVD player. Of the households that purchased an HDTV, what is the probability that they also purchased a DVD player?

SOLUTION Because you know that the household purchased an HDTV, the sample space is reduced to 80 households. Of these 80 households, 38 also purchased a DVD player. Therefore, the probability that a household purchased a DVD player, given that the household purchased an HDTV, is:

$$P(\text{purchased DVD player} \mid \text{purchased HDTV}) = \frac{\text{number purchasing HDTV } \textit{and} \text{ DVD player}}{\text{number purchasing HDTV}}$$

$$= \frac{38}{80} = 0.475$$

If you use Equation (4.4a) on page 135:

A = purchased DVD player $\qquad$ B = purchased HDTV

then

$$P(A \mid B) = \frac{P(A \textit{ and } B)}{P(B)} = \frac{38/300}{80/300} = 0.475$$

Therefore, given that the household purchased an HDTV, there is a 47.5% chance that the household also purchased a DVD player. You can compare this conditional probability to the marginal probability of purchasing a DVD player, which is 108/300 = 0.36, or 36%. These results tell you that households that purchased an HDTV are more likely to purchase a DVD player than are households that purchased a big-screen television set that is not an HDTV.

Decision Trees

In Table 4.1 on page 128 households are classified according to whether they planned to purchase and whether they actually purchased a big-screen television set. A **decision tree** is an alternative to the contingency table. Figure 4.3 represents the decision tree for this example.

FIGURE 4.3

Decision Tree for the Consumer Electronics Example

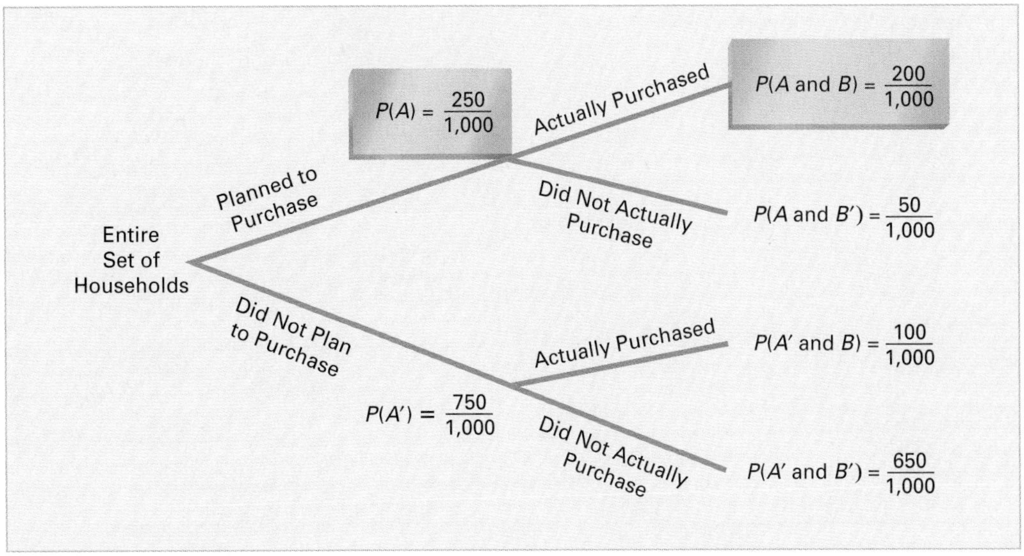

In Figure 4.3, beginning at the left with the entire set of households, there are two "branches" for whether or not the household planned to purchase a big-screen television set. Each of these branches has two subbranches, corresponding to whether the household actually purchased or did not actually purchase the big-screen television set. The probabilities at the end of the initial branches represent the marginal probabilities of A and A'. The probabilities at the end of each of the four subbranches represent the joint probability for each combination of events A and B. You compute the conditional probability by dividing the joint probability by the appropriate marginal probability.

For example, to compute the probability that the household actually purchased given that the household planned to purchase the big-screen television set, take P (planned to purchase *and* actually purchased) and divide by P (planned to purchase). From Figure 4.3

$$P(\text{actually purchased} \mid \text{planned to purchase}) = \frac{200/1,000}{250/1,000}$$

$$= \frac{200}{250} = 0.80$$

Example 4.7 illustrates how to construct a decision tree.

EXAMPLE 4.7

FORMING THE DECISION TREE FOR THE HOUSEHOLDS THAT PURCHASED BIG-SCREEN TELEVISION SETS

Using the cross-classified data in Table 4.2 on page 130, construct the decision tree. Use the decision tree to find the probability that a household purchased a DVD player, given that the household purchased an HDTV.

SOLUTION The decision tree for purchased a DVD player and an HDTV is displayed in Figure 4.4. Using Equation (4.4b) on page 136 and the following definitions:

$$A = \text{purchased HDTV} \qquad B = \text{purchased DVD player}$$

$$P(B \mid A) = \frac{P(A \text{ and } B)}{P(A)} = \frac{38/300}{80/300} = 0.475$$

FIGURE 4.4

Decision Tree for Purchased a DVD Player and an HDTV

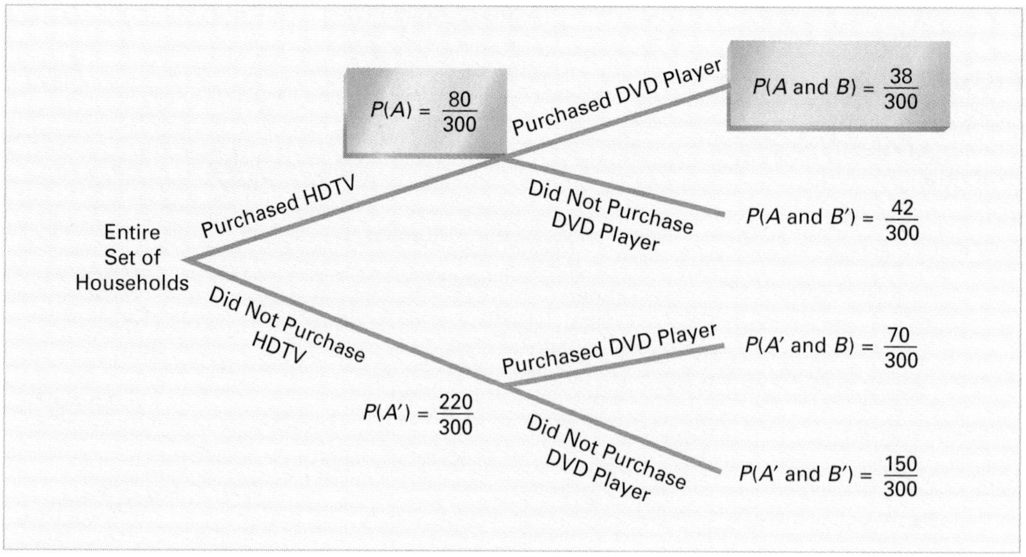

Statistical Independence

In the example concerning the purchase of big-screen television sets, the conditional probability is 200/250 = 0.80 that the selected household actually purchased the big-screen television set, given that the household planned to purchase. The simple probability of selecting a household that actually purchased is 300/1,000 = 0.30. This result shows that the prior knowledge that the household planned to purchase affected the probability that the household actually purchased the television set. In other words, the outcome of one event is *dependent* on the outcome of a second event.

When the outcome of one event does *not* affect the probability of occurrence of another event, the events are said to be statistically independent. **Statistical independence** can be determined by using Equation (4.5).

STATISTICAL INDEPENDENCE

Two events A and B are statistically independent if and only if

$$P(A \mid B) = P(A) \quad \textbf{(4.5)}$$

where $P(A \mid B)$ = conditional probability of A given B

$P(A)$ = marginal probability of A

Example 4.8 demonstrates the use of Equation (4.5).

EXAMPLE 4.8

DETERMINING STATISTICAL INDEPENDENCE

In the follow-up survey of the 300 households that actually purchased big-screen television sets, the households were asked if they were satisfied with their purchase. Table 4.3 cross-classifies the responses to the satisfaction question with their responses to whether the television set was an HDTV.

TABLE 4.3
Satisfaction with Purchase of Big-Screen Television Sets

	SATISFIED WITH PURCHASE?		
TYPE OF TELEVISION	**Yes**	**No**	**Total**
HDTV	64	16	80
Not HDTV	176	44	220
Total	240	60	300

Determine whether being satisfied with the purchase and type of television set purchased are statistically independent.

SOLUTION For these data,

$$P(\text{satisfied} \mid \text{HDTV}) = \frac{64/300}{80/300} = \frac{64}{80} = 0.80$$

which is equal to

$$P(\text{satisfied}) = \frac{240}{300} = 0.80$$

Thus, being satisfied with the purchase and type of television set purchased are statistically independent. Knowledge of one event does not affect the probability of the other event.

Multiplication Rules

By manipulating the formula for conditional probability, you can determine the joint probability $P(A \text{ and } B)$ from the conditional probability of an event. The **general multiplication rule** is derived using Equation (4.4a) on page 135,

$$P(A|B) = \frac{P(A \text{ and } B)}{P(B)}$$

and solving for the joint probability $P(A \text{ and } B)$.

GENERAL MULTIPLICATION RULE

The probability of A and B is equal to the probability of A given B times the probability of B.

$$P(A \text{ and } B) = P(A \mid B)P(B) \qquad \textbf{(4.6)}$$

Example 4.9 demonstrates the use of the general multiplication rule.

EXAMPLE 4.9

USING THE MULTIPLICATION RULE

Consider the 80 households that purchased an HDTV. In Table 4.3 on page 139 you see that 64 households are satisfied with their purchase and 16 households are dissatisfied. Suppose two households are randomly selected from the 80 customers. Find the probability that both households are satisfied with their purchase.

SOLUTION Here you can use the multiplication rule in the following way. If:

A = second household selected is satisfied

B = first household selected is satisfied

then, using Equation (4.6)

$$P(A \text{ and } B) = P(A \mid B)P(B)$$

The probability that the first household is satisfied with the purchase is 64/80. However, the probability that the second household is also satisfied with the purchase depends on the result of the first selection. If the first household is not returned to the sample after the satisfaction level is determined (sampling without replacement), then the number of households remaining will be 79. If the first household is satisfied, the probability that the second is also satisfied is 63/79, because 63 satisfied households remain in the sample. Therefore,

$$P(A \text{ and } B) = \left(\frac{63}{79}\right)\left(\frac{64}{80}\right) = 0.6380$$

There is a 63.80% chance that both of the households sampled will be satisfied with their purchase.

The **multiplication rule for independent events** is derived by substituting $P(A)$ for $P(A \mid B)$ in Equation (4.6).

MULTIPLICATION RULE FOR INDEPENDENT EVENTS

If A and B are statistically independent, the probability of A *and* B is equal to the probability of A times the probability of B.

$$P(A \text{ and } B) = P(A)P(B) \qquad \textbf{(4.7)}$$

If this rule holds for two events, A and B, then A and B are statistically independent. Therefore, there are two ways to determine statistical independence.

1. Events A and B are statistically independent if and only if $P(A \mid B) = P(A)$.
2. Events A and B are statistically independent if and only if $P(A \text{ and } B) = P(A)P(B)$.

Marginal Probability Using the General Multiplication Rule

In section 4.1 marginal probability was defined using Equation (4.2) on page 131. You can state the formula for marginal probability using the general multiplication rule. If

$$P(A) = P(A \text{ and } B_1) + P(A \text{ and } B_2) + \cdots + P(A \text{ and } B_k)$$

then, using the general multiplication rule, Equation (4.8) defines the marginal probability.

MARGINAL PROBABILITY USING THE GENERAL MULTIPLICATION RULE

$$P(A) = P(A \mid B_1)P(B_1) + P(A \mid B_2)P(B_2) + \cdots + P(A \mid B_k)P(B_k) \quad \textbf{(4.8)}$$

where $B_1, B_2, \ldots, B_k$ are the k mutually exclusive and collectively exhaustive events.

To illustrate this equation, refer to Table 4.1 on page 128. Using Equation (4.8), the probability of planning to purchase is:

$$P(A) = P(A \mid B_1)P(B_1) + P(A \mid B_2)P(B_2)$$

where

$P(A)$ = probability of "planned to purchase"

$P(B_1)$ = probability of "actually purchased"

$P(B_2)$ = probability of "did not actually purchase"

$$P(A) = \left(\frac{200}{300}\right)\left(\frac{300}{1{,}000}\right) + \left(\frac{50}{700}\right)\left(\frac{700}{1{,}000}\right)$$

$$= \frac{200}{1{,}000} + \frac{50}{1{,}000} = \frac{250}{1{,}000} = 0.25$$

PROBLEMS FOR SECTION 4.2

Learning the Basics

PH Grade ASSIST

4.16 Given the following contingency table:

	B	B'
A	10	20
A'	20	40

What is the probability of

a. $A \mid B$?
b. $A \mid B'$?
c. $A' \mid B'$?
d. Are events A and B statistically independent?

4.17 Given the following contingency table:

	B	B'
A	10	30
A'	25	35

What is the probability of

a. $A \mid B$?
b. $A' \mid B'$?
c. $A \mid B'$?
d. Are events A and B statistically independent?

PH Grade ASSIST **4.18** If $P(A \text{ and } B) = 0.4$ and $P(B) = 0.8$, find $P(A \mid B)$.

PH Grade ASSIST **4.19** If $P(A) = 0.7$ and $P(B) = 0.6$, and if A and B are statistically independent, find $P(A \text{ and } B)$.

PH Grade ASSIST **4.20** If $P(A) = 0.3$ and $P(B) = 0.4$, and if $P(A \text{ and } B) = 0.2$, are A and B statistically independent?

Applying the Concepts

SELF Test **4.21** A U.S. Census American Housing Survey studied how U.S. homeowners get to work ("How People Get to Work," *USA Today* Snapshots, February 25, 2003, 1A). Suppose that the survey consisted of a sample of 1,000 homeowners and 1,000 renters.

Drives to Work	Homeowner	Renter	Total
Yes	824	681	1,505
No	176	319	495
Total	1,000	1,000	2,000

a. Given that the respondent drives to work, what then is the probability that he or she is a homeowner?
b. Given that the respondent is a homeowner, what then is the probability that he or she drives to work?
c. Explain the difference in the results in (a) and (b).
d. Are the two events, driving to work and whether the respondent is a homeowner or a renter, statistically independent?

4.22 A yield improvement study at a semiconductor manufacturing facility provided defect data for a sample of 450 wafers. The following table presents a summary of the responses to two questions: "Were particles found on the die that produced the wafer?" and "Is the wafer good or bad?"

	CONDITION OF DIE		
QUALITY OF WAFER	No Particles	Particles	Totals
Good	320	14	334
Bad	80	36	116
Totals	400	50	450

Source: S. W. Hall, Analysis of Defectivity of Semiconductor Wafers by Contingency Table, Proceedings Institute of Environmental Sciences, *Vol. 1 (1994), 177–183.*

a. Suppose you know that a wafer is bad. What then is the probability that it was produced from a die that had particles?
b. Suppose you know that a wafer is good. What then is the probability that it was produced from a die that had particles?
c. Are the two events, a good wafer and a die with no particle, statistically independent? Explain.

PH Grade ASSIST **4.23** Are large companies less likely to offer board members stock options than small- to mid-sized companies? A survey conducted by the Segal Company of New York found that in a sample of 189 large companies, 40 offered stock options to their board members as part of their non-cash compensation packages. For small- to mid-sized companies, 43 of the 180 surveyed indicated that they offer stock options as part of their non-cash compensation packages to their board members (Kemba J. Dunham, "The Jungle: Focus on Recruitment, Pay and Getting Ahead," *The Wall Street Journal*, August 21, 2001, B6).

a. Given that a company is large, what then is the probability that the company offered stock options to their board members?
b. Given that a company is small- to mid-sized, what then is the probability that the company offered stock options to their board members?
c. Is the size of the company statistically independent of whether stock options are offered to their board members? Explain.

4.24 Are whites more likely to claim bias? A survey conducted by Barry Goldman ("White Fight: A Researcher Finds Whites Are More Likely to Claim Bias," *The Wall Street Journal*, Work Week, April 10, 2001, A1) found that of 56 white workers terminated, 29 claimed bias. Of 407 black workers terminated, 126 claimed bias.

a. Given that a worker is white, what then is the probability that the worker has claimed bias?
b. Given that a worker has claimed bias, what then is the probability that the worker is white?
c. Explain the difference in the results in (a) and (b).
d. Are the two events, "being white" and "claiming bias," statistically independent? Explain.

4.25 A sample of 500 respondents was selected in a large metropolitan area to study consumer behavior with the following results:

	GENDER		
ENJOYS SHOPPING FOR CLOTHING	Male	Female	Total
Yes	136	224	360
No	104	36	140
Total	240	260	500

a. Suppose the respondent chosen is a female. What, then, is the probability that she does not enjoy shopping for clothing?
b. Suppose the respondent chosen enjoys shopping for clothing. What, then, is the probability that the individual is a male?
c. Are enjoying shopping for clothing and the gender of the individual statistically independent? Explain.

4.26 Each year, ratings are compiled concerning the performance of new cars during the first 90 days of use. Suppose that the cars have been categorized according to whether or not the car needs warranty-related repair (yes or no) and the country in which the company manufacturing the car is based (United States or not United States). Based on the data collected, the probability that the new car needs a warranty repair is 0.04, the probability that the car is manufactured by a U.S.-based company is 0.60, and the probability that the new car needs a warranty repair *and* was manufactured by a U.S.-based company is 0.025.

a. Suppose you know that a company based in the United States manufactured the car. What, then, is the probability that the car needs a warranty repair?

b. Suppose you know that a company based in the United States did not manufacture the car. What, then, is the probability that the car needs a warranty repair?

c. Are need for a warranty repair and location of the company manufacturing the car statistically independent?

4.27 In 34 of the 54 years from 1950 to 2003, the S&P 500 finished higher after the first 5 days of trading. In 29 of those 34 years the S&P 500 finished higher for the year. Is a good first week a good omen for the upcoming year? The following table gives the first-week and annual performance over this 54-year period.

	S & P 500'S ANNUAL PERFORMANCE	
FIRST WEEK	Higher	Lower
Higher	29	5
Lower	10	10

Source: Adapted from Aaron Luchetti, "Stocks Enjoy a Good First Week," The Wall Street Journal, *January 12, 2004, C1.*

a. If a year is selected at random, what is the probability that the S&P finished higher for the year?

b. Given that the S&P 500 finished higher after the first five days of trading, what then is the probability that it finished higher for the year?

c. Are the two events, first-week performance and annual performance, statistically independent? Explain.

d. In 2004, the S&P 500 was up 0.9% after the first 5 days. Look up the 2004 annual performance of the S&P 500 at **finance.yahoo.com**. Comment on the results.

4.28 A standard deck of cards is being used to play a game. There are four suits (hearts, diamonds, clubs, and spades), each having 13 faces (ace, 2, 3, 4, 5, 6, 7, 8, 9, 10, jack, queen, and king), making a total of 52 cards. This complete deck is thoroughly mixed, and you will receive the first two cards from the deck without replacement.

a. What is the probability that both cards are queens?

b. What is the probability that the first card is a 10 and the second card is a 5 or 6?

c. If you were sampling with replacement, what would be the answer in (a)?

d. In the game of blackjack, the picture cards (jack, queen, king) count as 10 points and the ace counts as either 1 or 11 points. All other cards are counted at their face value. Blackjack is achieved if your two cards total 21 points. What is the probability of getting blackjack in this problem?

PH Grade ASSIST **4.29** A box of nine golf gloves contains two left-handed gloves and seven right-handed gloves.

a. If two gloves are randomly selected from the box without replacement, what is the probability that both gloves selected will be right-handed?

b. If two gloves are randomly selected from the box without replacement, what is the probability there will be one right-handed glove and one left-handed glove selected?

c. If three gloves are selected with replacement, what is the probability that all three will be left-handed?

d. If you were sampling with replacement, what would be the answers to (a) and (b)?

4.3 BAYES' THEOREM

Bayes' theorem is used to revise previously calculated probabilities when you have new information. Developed by the Rev. Thomas Bayes in the eighteenth century (see reference 1), Bayes' theorem is an extension of what you previously learned about conditional probability.

You can apply Bayes' theorem to the following situation. The Consumer Electronics Company is considering marketing a new model of television set. In the past, 40% of the television sets introduced by the company have been successful and 60% have been unsuccessful. Before introducing the television set to the marketplace, the marketing research department conducts an extensive study and releases a report, either favorable or unfavorable. In the past, 80% of the successful television sets had received a favorable market research report and 30% of the unsuccessful television sets had received a favorable report. For the new model of television set under consideration, the marketing research department has issued a favorable report. What is the probability that the television set will be successful?

Bayes' theorem is developed from the definition of conditional probability. To find the conditional probability of B given A, consider Equation (4.4b) [originally presented on page 136 and given below]:

$$P(B|A) = \frac{P(A \text{ and } B)}{P(A)} = \frac{(P(A|B)P(B)}{P(A)}$$

Bayes' theorem is derived by substituting Equation (4.8) on page 141 for $P(A)$ in the above equation.

BAYES' THEOREM

$$P(B_i|A) = \frac{P(A|B_i)P(B_i)}{P(A|B_1)P(B_1) + P(A|B_2)P(B_2) + \cdots + P(A|B_k)P(B_k)} \quad \textbf{(4.9)}$$

where B_i is the ith event out of k mutually exclusive and collectively exhaustive events.

To use Equation (4.9) for the television marketing example, let

event S = successful television set   event F = favorable report

event S' = unsuccessful television set   event F' = unfavorable report

and

$$P(S) = 0.40 \qquad P(F|S) = 0.80$$

$$P(S') = 0.60 \qquad P(F|S') = 0.30$$

Then, using Equation (4.9),

$$\begin{aligned} P(S|F) &= \frac{P(F|S)P(S)}{P(F|S)P(S) + P(F|S')P(S')} \\ &= \frac{(0.80)(0.40)}{(0.80)(0.40) + (0.30)(0.60)} \\ &= \frac{0.32}{0.32 + 0.18} = \frac{0.32}{0.50} \\ &= 0.64 \end{aligned}$$

The probability of a successful television set, given that a favorable report was received, is 0.64. Thus, the probability of an unsuccessful television set, given that a favorable report was received, is 1 – 0.64 = 0.36. Table 4.4 summarizes the computation of the probabilities and Figure 4.5 presents the decision tree.

TABLE 4.4 Bayes' Theorem Calculations for the Television-Marketing Example

Event S_i	Prior Probability $P(S_i)$	Conditional Probability $P(F \mid S_i)$	Joint Probability $P(F \mid S_i)P(S_i)$	Revised Probability $P(S_i \mid F)$
S = successful television set	0.40	0.80	0.32	0.32/0.50 = 0.64 = $P(S \mid F)$
S' = unsuccessful television set	0.60	0.30	0.18	0.18/0.50 = 0.36 = $P(S' \mid F)$
			0.50	

FIGURE 4.5
Decision Tree for Marketing a New Television Set

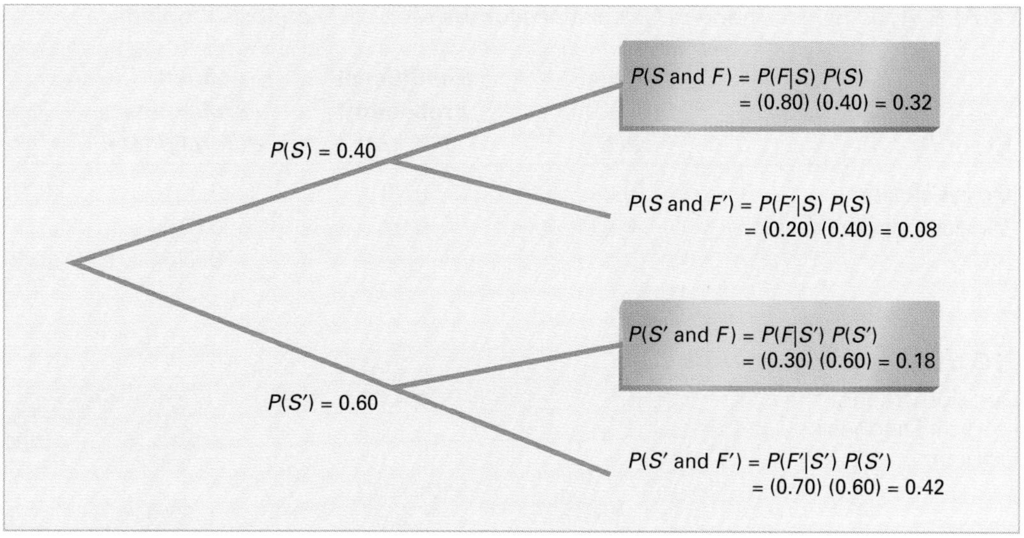

Example 4.10 applies Bayes' theorem to a medical diagnosis problem.

EXAMPLE 4.10

USING BAYES' THEOREM IN A MEDICAL DIAGNOSIS PROBLEM

The probability that a person has a certain disease is 0.03. Medical diagnostic tests are available to determine whether the person actually has the disease. If the disease is actually present, the probability that the medical diagnostic test will give a positive result (indicating that the disease is present) is 0.90. If the disease is not actually present, the probability of a positive test result (indicating that the disease is present) is 0.02. Suppose that the medical diagnostic test has given a positive result (indicating that the disease is present). What is the probability that the disease is actually present? What is the probability of a positive test result?

SOLUTION

Let event D = has disease; event T = test is positive

event D' = does not have disease; event T' = test is negative

and

$$P(D) = 0.03 \qquad P(T \mid D) = 0.90$$

$$P(D') = 0.97 \qquad P(T \mid D') = 0.02$$

Using Equation (4.9) on page 144,

$$\begin{aligned} P(D|T) &= \frac{P(T|D)P(D)}{P(T|D)P(D) + P(T|D')P(D')} \\ &= \frac{(0.90)(0.03)}{(0.90)(0.03) + (0.02)(0.97)} \\ &= \frac{0.0270}{0.0270 + 0.0194} = \frac{0.0270}{0.0464} \\ &= 0.582 \end{aligned}$$

The probability that the disease is actually present given a positive result has occurred (indicating that the disease is present) is 0.582. Table 4.5 summarizes the computation of the probabilities and Figure 4.6 presents the decision tree.

TABLE 4.5 Bayes' Theorem Calculations for the Medical Diagnosis Problem

Event D_i	Prior Probability $P(D_i)$	Conditional Probability $P(T \mid D_i)$	Joint Probability $P(T \mid D_i)P(D_i)$	Revised Probability $P(D_i \mid T)$
D = has disease	0.03	0.90	0.0270	$0.0270/0.0464 = 0.582 = P(D \mid T)$
D' = does not have disease	0.97	0.02	0.0194	$0.0194/0.0464 = 0.418 = P(D' \mid T)$
			0.0464	

FIGURE 4.6
Decision Tree for the Medical Diagnosis Problem

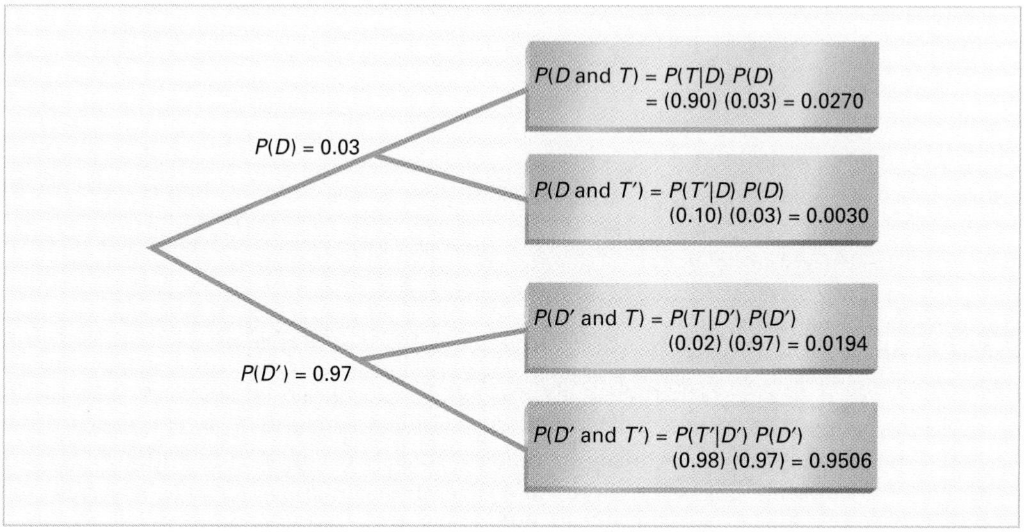

The denominator in Bayes' theorem represents $P(T)$, the probability of a positive test result, which in this case is 0.0464 or 4.64%.

PROBLEMS FOR SECTION 4.3

Learning the Basics

PH Grade ASSIST **4.30** If $P(B) = 0.05$, $P(A \mid B) = 0.80$, $P(B') = 0.95$, and $P(A \mid B') = 0.40$, find $P(B \mid A)$.

PH Grade ASSIST **4.31** If $P(B) = 0.30$, $P(A \mid B) = 0.60$, $P(B') = 0.70$, and $P(A \mid B') = 0.50$, find $P(B \mid A)$.

Applying the Concepts

4.32 In Example 4.10 on page 145, suppose that the probability that a medical diagnostic test will give a positive result if the disease is not present is reduced from 0.02 to 0.01. Given this information,

a. If the medical diagnostic test has given a positive result (indicating the disease is present), what is the probability that the disease is actually present?

b. If the medical diagnostic test has given a negative result (indicating that the disease is not present), what is the probability that the disease is not present?

PH Grade ASSIST SELF Test **4.33** An advertising executive is studying television viewing habits of married men and women during prime-time hours. On the basis of past viewing records, the executive has determined that during prime time, husbands are watching television 60% of the time. When the husband is watching television, 40% of the time the wife is also watching. When the husband is not watching television, 30% of the time the wife is watching television. Find the probability that

a. if the wife is watching television, the husband is also watching television.

b. the wife is watching television in prime time.

PH Grade ASSIST **4.34** Olive Construction Company is determining whether it should submit a bid for a new shopping center. In the past, Olive's main competitor, Base Construction Company, has submitted bids

70% of the time. If Base Construction Company does not bid on a job, the probability that Olive Construction Company will get the job is 0.50. If Base Construction Company bids on a job, the probability that Olive Construction Company will get the job is 0.25.

a. If Olive Construction Company gets the job, what is the probability that Base Construction Company did not bid?

b. What is the probability that Olive Construction Company will get the job?

4.35 Laid-off workers who become entrepreneurs because they cannot find meaningful employment with another company are known as *entrepreneurs by necessity. The Wall Street Journal* reports that these entrepreneurs by necessity are less likely to grow into large businesses than are *entrepreneurs by choice* (Jeff Bailey, "Desire—More Than Need—Builds a Business," *The Wall Street Journal*, May 21, 2001, B4). This article states that 89% of the entrepreneurs in the United States are entrepreneurs by choice and 11% are entrepreneurs by necessity. Only 2% of entrepreneurs by necessity expect their new business to employ 20 or more people within five years, while 14% of entrepreneurs by choice expect to employ at least 20 people within five years.

a. If an entrepreneur is selected at random and that individual expects that their new business will employ 20 or more people within five years, what then is the probability that this individual is an entrepreneur by choice?

b. Discuss several possible reasons why entrepreneurs by choice are more likely to believe that they will grow their business.

4.36 The editor of a textbook publishing company is trying to decide whether to publish a proposed business statistics textbook. Information on previous textbooks published indicates that 10% are huge successes, 20% are modest successes, 40% break even, and 30% are losers. However, before a publishing decision is made, the book will be reviewed. In the past, 99% of the huge successes received favorable reviews, 70% of the moderate successes received favorable reviews, 40% of the break-even books received favorable reviews, and 20% of the losers received favorable reviews.

a. If the proposed text receives a favorable review, how should the editor revise the probabilities of the various outcomes to take this information into account?

b. What proportion of textbooks receives favorable reviews?

4.37 A municipal bond service has three rating categories (*A*, *B*, and *C*). Suppose that in the past year, of the municipal bonds issued throughout the United States, 70% were rated *A*, 20% were rated *B*, and 10% were rated *C*. Of the municipal bonds rated *A*, 50% were issued by cities, 40% by suburbs, and 10% by rural areas. Of the municipal bonds rated *B*, 60% were issued by cities, 20% by suburbs, and 20% by rural areas. Of the municipal bonds rated *C*, 90% were issued by cities, 5% by suburbs, and 5% by rural areas.

a. If a new municipal bond is to be issued by a city, what is the probability that it will receive an A rating?

b. What proportion of municipal bonds is issued by cities?

c. What proportion of municipal bonds is issued by suburbs?

4.4 COUNTING RULES

In Equation (4.1) on page 127, the probability of occurrence of an outcome was defined as the number of ways the outcome occurs, divided by the total number of possible outcomes. In many instances, there are a large number of possible outcomes and it is difficult to determine the exact number. In these circumstances, rules for counting the number of possible outcomes have been developed. In this section five different counting rules are presented.

COUNTING RULE 1

If any one of k different mutually exclusive and collectively exhaustive events can occur on each of n trials, the number of possible outcomes is equal to

$$k^n \quad \textbf{(4.10)}$$

EXAMPLE 4.11 COUNTING RULE 1

Suppose you toss a coin five times. What is the number of different possible outcomes (the sequences of heads and tails)?

SOLUTION If you toss a coin (having two sides) five times, using Equation (4.10), the number of outcomes is $2^5 = 2 \times 2 \times 2 \times 2 \times 2 = 32$.

EXAMPLE 4.12 ROLLING A DIE TWICE

Suppose you roll a die twice. How many different possible outcomes can occur?

SOLUTION If a die (having six sides) is rolled twice, using Equation (4.10) the number of different outcomes is $6^2 = 36$.

The second counting rule is a more general version of the first, and allows for the number of possible events to differ from trial to trial.

COUNTING RULE 2

If there are k_1 events on the first trial, k_2 events on the second trial, . . . , and k_n events on the nth trial, then the number of possible outcomes is

$$(k_1)(k_2)\ldots(k_n) \quad \textbf{(4.11)}$$

EXAMPLE 4.13 COUNTING RULE 2

A state motor vehicle department would like to know how many license plate numbers are available if the license plates consist of three letters followed by three numbers.

SOLUTION Using Equation (4.11), if a license plate consists of three letters followed by three numbers (0 through 9), the total number of possible outcomes is (26)(26)(26)(10)(10)(10) = 17,576,000.

EXAMPLE 4.14 DETERMINING THE NUMBER OF DIFFERENT DINNERS

A restaurant menu has a price-fixed complete dinner that consists of an appetizer, entrée, beverage, and dessert. You have a choice of five appetizers, ten entrées, three beverages, and six desserts. Determine the total number of possible dinners.

SOLUTION Using Equation (4.11), the total number of possible dinners is (5)(10)(3)(6) = 900.

The third counting rule involves the computation of the number of ways that a set of items can be arranged in order.

COUNTING RULE 3

The number of ways that all n items can be arranged in order is

$$n! = (n)(n-1)\ldots(1)$$

where $n!$ is called *n factorial* and 0! is defined as 1.

EXAMPLE 4.15 COUNTING RULE 3

If a set of six textbooks is to be placed on a shelf, in how many ways can the six books be arranged?

SOLUTION To begin, you must realize that any of the six books could occupy the first position on the shelf. Once the first position is filled, there are five books to choose from in filling the second. You continue this assignment procedure until all the positions are occupied. The number of ways that you can arrange six books is

$$n! = 6! = (6)(5)(4)(3)(2)(1) = 720$$

In many instances you need to know the number of ways in which a subset of the entire group of items can be arranged in *order*. Each possible arrangement is called a **permutation**.

COUNTING RULE 4

Permutations: The number of ways of arranging X objects selected from n objects in order is

$$_{n}P_{X} = \frac{n!}{(n-X)!} \qquad \textbf{(4.13)}$$

EXAMPLE 4.16

COUNTING RULE 4

Modifying Example 4.15, if you have six textbooks, but there is room for only four books on the shelf, in how many ways can you arrange these books on the shelf?

SOLUTION Using Equation (4.13), the number of ordered arrangements of four books selected from six books is equal to

$$_{n}P_{X} = \frac{n!}{(n-X)!} = \frac{6!}{(6-4)!} = \frac{(6)(5)(4)(3)(2)(1)}{(2)(1)} = 360$$

In many situations you are not interested in the *order* of the outcomes, but only in the number of ways that X items can be selected from n items, *irrespective of order*. This rule is called the rule of **combinations**.

COUNTING RULE 5

Combinations: The number of ways of selecting X objects from n objects, irrespective of order, is equal to

$$_{n}C_{X} = \frac{n!}{X!(n-X)!} \qquad \textbf{(4.14)}$$

Comparing this rule to the previous one, you see that it differs only in the inclusion of a term $X!$ in the denominator. When permutations were used, all of the arrangements of the X objects are distinguishable. With combinations, the $X!$ possible arrangements of objects are irrelevant.

EXAMPLE 4.17

COUNTING RULE 5

Modifying Example 4.16, if the order of the books on the shelf is irrelevant, in how many ways can you arrange these books on the shelf?

SOLUTION Using Equation (4.14), the number of combinations of four books selected from six books is equal to

$$_{n}C_{X} = \frac{n!}{X!(n-X)!} = \frac{6!}{4!(6-4)!} = \frac{(6)(5)(4)(3)(2)(1)}{(4)(3)(2)(1)(2)(1)} = 15$$

PROBLEMS FOR SECTION 4.4

Applying the Concepts

SELF Test **4.38** If there are ten multiple-choice questions on an exam, each having three possible answers, how many different sequences of correct answers are there?

4.39 A lock on a bank vault consists of three dials, each with 30 positions. In order for the vault to open, each of the three dials must be in the correct position.

a. How many different possible "dial combinations" are there for this lock?

b. What is the probability that if you randomly select a position on each dial, you will be able to open the bank vault?

c. Explain why "dial combinations" are not mathematical combinations expressed by Equation (4.14).

4.40 **a.** If a coin is tossed seven times, how many different outcomes are possible?

b. If a die is tossed seven times, how many different outcomes are possible?

c. Discuss the differences in your answers to (a) and (b).

SELF Test **4.41** A particular brand of women's jeans is available in seven different sizes, three different colors, and three different styles. How many different jeans does the store manager need to order to have one pair of each type?

4.42 You would like to make a salad that consists of lettuce, tomato, cucumber, and sprouts. You go to the supermarket intending to purchase one type of each of these ingredients. You discover that there are eight types of lettuce, four types of tomatoes, three types of cucumbers, and three types of sprouts for sale at the supermarket. How many different salads do you have to choose from?

SELF Test **4.43** If each letter is used once, how many different four-letter "words" can be made from the letters E, L, O, and V?

4.44 In Major League Baseball, there are five teams in the Western Division of the National League: Arizona, Los Angeles, San Francisco, San Diego, and Colorado. How many different orders of finish are there for these five teams? Do you believe that all these orders are equally likely? Discuss.

4.45 Referring to problem 4.44, how many different orders of finish are possible for the first four positions?

4.46 A gardener has six rows available in his vegetable garden to place tomatoes, eggplant, peppers, cucumbers, beans, and lettuce. Each vegetable will be allowed one and only one row. How many ways are there to position these vegetables in his garden?

SELF Test **4.47** The Big Triple at the local racetrack consists of picking the correct order of finish of the first three horses in the ninth race. If there are 12 horses entered in today's ninth race, how many Big Triple outcomes are there?

4.48 The Quinella at the local racetrack consists of picking the horses that will place first and second in a race *irrespective* of order. If eight horses are entered in a race, how many Quinella combinations are there?

SELF Test **4.49** A student has seven books that she would like to place in an attaché case. However, only four books can fit into the attaché case. Regardless of the arrangement, how many ways are there of placing four books into the attaché case?

4.50 A daily lottery is conducted in which two winning numbers are selected out of 100 numbers. How many different combinations of winning numbers are possible?

4.51 A reading list for a course contains 20 articles. How many ways are there to choose three articles from this list?

4.5 ETHICAL ISSUES AND PROBABILITY

Ethical issues can arise when any statements relating to probability are presented to the public, particularly when these statements are part of an advertising campaign for a product or service. Unfortunately, many people are not comfortable with numerical concepts (see reference 3) and tend to misinterpret the meaning of the probability. In some instances, the misinterpretation is not intentional, but in other cases, advertisements may unethically try to mislead potential customers.

One example of a potentially unethical application of probability relates to advertisements for state lotteries. When purchasing a lottery ticket, the customer selects a set of numbers (such as 6) from a larger list of numbers (such as 54). Although virtually all participants know that they are unlikely to win the lottery, they also have very little idea of how unlikely it is for them to select all 6 winning numbers from the list of 54 numbers. They have even less of an idea of the probability of winning a consolation prize by selecting either 4 or 5 winning numbers.

Given this background, you might consider a recent commercial for a state lottery that stated, "We won't stop until we have made everyone a millionaire," to be deceptive and possibly unethical. Given the fact that the lottery brings millions of dollars into the state treasury, the state is never going to stop running it, although in our lifetime no one can be sure of becoming a millionaire by winning the lottery.

Another example of a potentially unethical application of probability relates to an investment newsletter promising a 90% probability of a 20% annual return on investment. To make the claim in the newsletter an ethical one, the investment service needs to (a) explain the basis on which this probability estimate rests, (b) provide the probability statement in another format such as 9 chances in 10, and (c) explain what happens to the investment in the 10% of the cases in which a 20% return is not achieved (e.g., Is the entire investment lost?).

PROBLEMS FOR SECTION 4.5

Applying the Concepts

4.52 Write an advertisement for the state lottery that ethically describes the probability of winning.

4.53 Write an advertisement for the investment newsletter that ethically states the probability of a 20% return.

SUMMARY

This chapter developed concepts concerning basic probability, conditional probability, Bayes' theorem, and counting rules. In the next chapter, important discrete probability distributions such as the binomial, hypergeometric, and Poisson distributions will be developed.

KEY FORMULAS

Probability of Occurrence

$$\text{Probability of occurrence} = \frac{X}{T} \quad \textbf{(4.1)}$$

Marginal Probability

$$P(A) = P(A \text{ and } B_1) + P(A \text{ and } B_2) + \cdots + P(A \text{ and } B_k) \quad \textbf{(4.2)}$$

General Addition Rule

$$P(A \text{ or } B) = P(A) + P(B) - P(A \text{ and } B) \quad \textbf{(4.3)}$$

Conditional Probability

$$P(A|B) = \frac{P(A \text{ and } B)}{P(B)} \quad \textbf{(4.4a)}$$

$$P(B|A) = \frac{P(A \text{ and } B)}{P(A)} \quad \textbf{(4.4b)}$$

Statistical Independence

$$P(A \mid B) = P(A) \quad \textbf{(4.5)}$$

General Multiplication Rule

$$P(A \text{ and } B) = P(A \mid B)P(B) \quad \textbf{(4.6)}$$

Multiplication Rule for Independent Events

$$P(A \text{ and } B) = P(A)P(B) \quad \textbf{(4.7)}$$

Marginal Probability Using the General Multiplication Rule

$$P(A) = P(A \mid B_1)P(B_1) + P(A \mid B_2)P(B_2) + \cdots + P(A \mid B_k)P(B_k) \quad \textbf{(4.8)}$$

Bayes' Theorem

$$P(B_i|A) = \frac{P(A|B_i)P(B_i)}{P(A|B_1)P(B_1) + P(A|B_2)P(B_2) + \cdots + P(A|B_k)P(B_k)} \quad \textbf{(4.9)}$$

Counting Rule 1

$$k^n \quad \textbf{(4.10)}$$

Counting Rule 2

$$(k_1)(k_2)\ldots(k_n) \quad \textbf{(4.11)}$$

Factorials

$$n! = (n)(n-1)\ldots(1) \quad \textbf{(4.12)}$$

Permutations

$${}_nP_X = \frac{n!}{(n-X)!} \quad \textbf{(4.13)}$$

Combinations

$${}_nC_X = \frac{n!}{X!(n-X)!} \quad \textbf{(4.14)}$$

KEY TERMS

CHAPTER REVIEW PROBLEMS

Checking Your Understanding

4.54 What are the differences between *a priori* classical probability, empirical classical probability, and subjective probability?

4.55 What is the difference between a simple event and a joint event?

4.56 How can you use the addition rule to find the probability of occurrence of event *A or B*?

4.57 What is the difference between mutually exclusive events and collectively exhaustive events?

4.58 How does conditional probability relate to the concept of statistical independence?

4.59 How does the multiplication rule differ for events that are and are not independent?

4.60 How can you use Bayes' theorem to revise probabilities in light of new information?

4.61 What is the difference between a permutation and a combination?

Applying the Concepts

4.62 A soft-drink bottling company maintains records concerning the number of unacceptable bottles of soft drink from the filling and capping machines. Based on past data, the probability that a bottle came from machine I and was nonconforming is 0.01 and the probability that a bottle came from machine II and was nonconforming is 0.025. Half the bottles are filled on machine I and the other half are filled on machine II. If a filled bottle of soft drink is selected at random, what is the probability that

a. it is a nonconforming bottle?
b. it was filled on machine I *and* is a conforming bottle?
c. it was filled on machine I *or* is a conforming bottle?
d. Suppose you know that the bottle was produced on machine I. What is the probability that it is nonconforming?
e. Suppose you know that the bottle is nonconforming. What is the probability that it was produced on machine I?
f. Explain the difference in the answers to (d) and (e).

(*Hint:* Construct a 2 × 2 contingency table or a Venn diagram to evaluate the probabilities.)

4.63 A survey asked workers which aspects of his or her job are extremely important. The results in percentages are as follows:

	Is Aspect Extremely Important?	
Aspect of Job	**Men**	**Women**
Good relationship with boss	63%	77%
Up-to-date equipment	59	69
Resources to do the job	55	74
Easy commute	48	60
Flexible hours at work	40	53
Able to work at home	21	34

Source: "Snapshot," USA Today, *May 15, 2000.*

Suppose the survey was based on the responses of 500 men and 500 women. Construct a contingency table for the different responses concerning each aspect of the job. If a respondent is chosen at random, what is the probability that

a. he or she feels that a good relationship with the boss is an important aspect of the job?
b. he or she feels that an easy commute is an important aspect of the job?
c. the person is a male and feels that a good relationship with the boss is an important aspect of the job?
d. the person is a female and feels that having flexible hours is an important aspect of the job?
e. Given that the person feels that having a good relationship with the boss is an important aspect of the job, what is the probability that the person is a male?

f. Are any of the things that workers say are extremely important aspects of a job statistically independent of the gender of the respondent? Explain.

4.64 Many companies use Web sites to conduct business transactions such as taking orders or performing financial exchanges. These sites are referred to as transactional public Web sites. An analysis of 490 firms listed in the Fortune 500 identified firms by their level of sales and whether or not the firm had a transactional public Web site (D. Young, and J. Benamati, "A Cross-Industry Analysis of Large Firm Transactional Public Web Sites," *Mid-American Journal of Business*, 19(2004), 37–46). The results of this analysis are given in the following table.

	TRANSACTIONAL PUBLIC WEB SITE	
SALES (IN BILLIONS OF DOLLARS)	**Yes**	**No**
Greater than $10 billion	71	88
Up to $10 billion	99	232

a. Give an example of a simple event and a joint event.

b. What is the probability that a firm in the Fortune 500 has a transactional public Web site?

c. What is the probability that a firm in the Fortune 500 has sales in excess of ten billion dollars and a transactional Web site?

d. Are the events *sales in excess of ten billion dollars* and *has a transactional public Web site* independent? Explain.

4.65 The owner of a restaurant serving Continental-style entrées was interested in studying ordering patterns of patrons for the Friday to Sunday weekend time period. Records were maintained that indicated the demand for dessert during the same time period. The owner decided to study two other variables along with whether a dessert was ordered: the gender of the individual and whether a beef entrée was ordered. The results are as follows:

	GENDER		
DESSERT ORDERED	**Male**	**Female**	**Total**
Yes	96	40	136
No	224	240	464
Total	320	280	600

	BEEF ENTRÉE		
DESSERT ORDERED	**Yes**	**No**	**Total**
Yes	71	65	136
No	116	348	464
Total	187	413	600

A waiter approaches a table to take an order. What is the probability that the first customer to order at the table

a. orders a dessert?

b. orders a dessert *or* a beef entrée?

c. is a female *and* does not order a dessert?

d. is a female *or* does not order a dessert?

e. Suppose the first person that the waiter takes the dessert order from is a female. What is the probability that she does not order dessert?

f. Are gender and ordering dessert statistically independent?

g. Is ordering a beef entrée statistically independent of whether the person orders dessert?

4.66 Unsolicited commercial e-mail messages containing product advertisements, commonly referred to as spam, are routinely deleted before being read by more than 80% of all e-mail users. Furthermore, a small percentage of those reading the spam actually follow through and purchase items. Yet, many companies use these unsolicited e-mail advertisements because of the extremely low cost involved. Movies Unlimited, a mail-order video and DVD business in Philadelphia, is one of the more successful companies in terms of generating sales through this form of e-marketing. Ed Weiss, general manager of Movies Unlimited, estimates that somewhere from 15% to 20% of their e-mail recipients read the advertisements. Moreover, approximately 15% of those who read the advertisements place an order (Stacy Forster, "E-Marketers Look to Polish Spam's Rusty Image," *The Wall Street Journal*, May 22, 2002, D2).

a. Using Mr. Weiss's lower estimate that the probability a recipient will read the advertisement is 0.15, what is the probability that a recipient will read the advertisement and place an order?

b. Movies Unlimited uses a 175,000-customer database to send e-mail advertisements. If an e-mail advertisement is sent to everyone in its customer database, how many customers do you expect will read the advertisement and place an order?

c. If the probability a recipient will read the advertisement is 0.20, what is the probability that a recipient will read the advertisement and place an order?

d. What is your answer to (b) assuming the probability that a recipient will read the advertisement is 0.20?

PH Grade ASSIST **4.67** In February 2002, the Argentine peso lost 70% of its value compared to the United States dollar. This devaluation drastically raised the price of imported products. According to a survey conducted by AC Nielsen in April 2002, 68% of the consumers in Argentina were buying fewer products than before the devaluation, 24% were buying the same number of products, and 8% were buying more products. Furthermore, in a trend toward purchasing less-expensive brands, 88% indicated that they had changed the brands they purchased (Michelle Wallin, "Argentines Hone Art of Shopping in a Crisis," *The Wall Street Journal*, May 28, 2002, A15). Suppose the following complete set of results were reported.

	NUMBER OF PRODUCTS PURCHASED			
BRANDS PURCHASED	**Fewer**	**Same**	**More**	**Total**
Same	10	14	24	48
Changed	262	82	8	352
Total	272	96	32	400

What is the probability that a consumer selected at random:

a. purchased fewer products than before?

b. purchased the same number or more products than before?

c. purchased fewer products and changed brands?

d. Given that a consumer changed the brands they purchased, what then is the probability that the consumer purchased fewer products than before?

e. Compare the results from (a) with (d).

PH Grade ASSIST **4.68** Sport utility vehicles (SUVs), vans, and pickups are generally considered to be more prone to roll over than cars. In 1997, 24.0% of all highway fatalities involved a rollover; 15.8% of all fatalities in 1997 involved SUVs, vans, and pickups, given that the fatality involved a rollover. Given that a rollover was not involved, 5.6% of all fatalities involved SUVs, vans, and pickups (Anna Wilde Mathews, "Ford Ranger, Chevy Tracker Tilt in Test," *The Wall Street Journal*, July 14, 1999, A2). Consider the following definitions:

A = fatality involved an SUV, van, or pickup

B = fatality involved a rollover

a. Use Bayes' theorem to find the probability that the fatality involved a rollover, given that the fatality involved an SUV, van, or pickup.

b. Compare the result in (a) to the probability that the fatality involved a rollover, and comment on whether SUVs, vans, and pickups are generally more prone to rollover accidents.

PH Grade ASSIST **4.69** Enzyme-linked immunosorbent assays (ELISA) is the most common type of screening test for detecting the HIV virus. A positive result from an ELISA indicates that the HIV virus is present. For most populations, ELISA has a high degree of sensitivity (to detect infection) and specificity (to detect noninfection). (See HIVInsite, at **HIVInsite.ucsf.edu/**.) Suppose that the probability a person is infected with the HIV virus for a certain population is 0.015. If the HIV virus is actually present, the probability that the ELISA test will give a positive result is 0.995. If the HIV virus is not actually present, the probability of a positive result from an ELISA is 0.01. If the ELISA has given a positive result, use Bayes' theorem to find the probability that the HIV virus is actually present.

WEB CASE

Apply your knowledge about contingency tables and the proper application of simple and joint probabilities in this continuing Web Case from Chapter 3.

Visit the StockTout Guaranteed Investment Package Web page **www.prenhall.com/Springville/ST_Guaranteed.htm**. Read the claims and examine the supporting data. Then answer the following:

1. How accurate is the claim of the probability of success for StockTout's Guaranteed Investment Package? In what ways is the claim misleading? How would you calculate and state the probability of having an annual rate of return not less than 15%?
2. What mistake was made in reporting the 7% probability claim? Using the table found on the "Winning Probabilities" Web page **ST_Guaranteed3.htm**, compute the proper probabilities for the group of investors.
3. Are there any probability calculations that would be appropriate for rating an investment service? Why, or why not?

REFERENCES

1. Kirk, R. L., ed., *Statistical Issues: A Reader for the Behavioral Sciences* (Belmont, CA: Wadsworth, 1972).
2. *Microsoft Excel 2003* (Redmond, WA: Microsoft Corp., 2002).
3. Paulos, J. A., *Innumeracy* (New York: Hill and Wang, 1988).

Appendix 4 Using Software for Basic Probability

A4.1 USING MICROSOFT EXCEL

For Basic Probabilities

Open the **Probabilities.xls** file. This worksheet already contains the entries for Table 4.2 on page 130. To adapt this worksheet to other problems, change the entries in the tinted cells in rows 3 through 6.

OR if using PHStat2, select **PHStat → Probability & Prob. Distributions → Simple & Joint Probabilities** to generate a worksheet into which you enter probability data in the empty tinted cells in rows 3 through 6.

For Bayes' Theorem

Open the **Bayes.xls** file. This worksheet already contains the entries for the Table 4.4 on page 144. To adapt this worksheet to other problems, change the entries for the prior and conditional probabilities in the tinted cell range B5:C6.

CHAPTER 5

Some Important Discrete Probability Distributions

LEARNING OBJECTIVES

In this chapter, you learn:

- The properties of a probability distribution
- To compute the expected value and variance of a probability distribution
- To calculate the covariance and understand its use in finance
- To compute probabilities from binomial, hypergeometric, and Poisson distributions
- How to use the binomial, hypergeometric, and Poisson distributions to solve business problems

USING STATISTICS

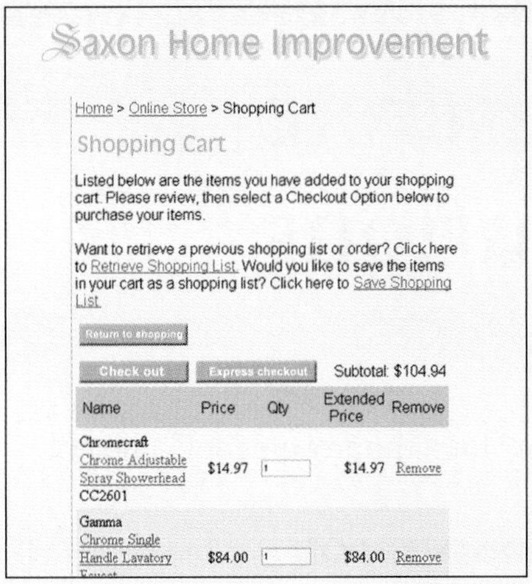

The Accounting Information System of the Saxon Home Improvement Company

Accounting information systems collect, process, store, transform, and distribute financial information to decision-makers both internal and external to a business organization (see reference 5). These systems continuously audit accounting information, looking for errors or incomplete or improbable information. For example, when customers of the Saxon Home Improvement Company submit an online order, the order forms are reviewed by the company's accounting information system for possible mistakes. Any questionable invoices are *tagged* and included in a daily *exceptions report*. Recent data collected by the company show that the likelihood is 0.10 that an order form will be tagged. Saxon would like to determine the likelihood of finding a certain number of tagged forms in a sample of a specific size. For example, what would be the likelihood that none of the order forms are tagged in a sample of four forms? One of the order forms is tagged?

How could the Saxon Home Improvement Company determine the solution to this type of probability problem? One way is to use a model, or small-scale representation, that approximates the process. By using such an approximation, Saxon managers could make inferences about the actual order process. Although model-building is a difficult task for some endeavors, in this case the Saxon managers can use *probability distributions*, mathematical models suited for solving the type of probability problems the managers are facing. Reading this chapter will help you learn about characteristics of a probability distribution and how to specifically apply the binomial, Poisson, and hypergeometric distributions to business problems.

5.1 THE PROBABILITY DISTRIBUTION FOR A DISCRETE RANDOM VARIABLE

In section 1.5, a *numerical variable* was defined as a variable that yielded numerical responses such as the number of magazines you subscribe to or your height in inches. Numerical variables are classified as *discrete* or *continuous*. Continuous numerical variables produce outcomes that come from a measuring process; for example, your height. Discrete numerical variables produce outcomes that come from a counting process, such as the number of magazines you subscribe to. This chapter deals with probability distributions that represent discrete numerical variables.

A **probability distribution for a discrete random variable** is a mutually exclusive listing of all possible numerical outcomes for that random variable such that a particular probability of occurrence is associated with each outcome.

For example, Table 5.1 gives the distribution of the number of mortgages approved per week at the local branch office of a bank. The listing in Table 5.1 is collectively exhaustive because all possible outcomes are included. Thus, the probabilities must sum to 1. Figure 5.1 is a graphical representation of Table 5.1.

TABLE 5.1
Probability Distribution of the Number of Home Mortgages Approved per Week

Home Mortgages Approved per Week	Probability
0	0.10
1	0.10
2	0.20
3	0.30
4	0.15
5	0.10
6	0.05

FIGURE 5.1
Probability Distribution of the Number of Home Mortgages Approved per Week

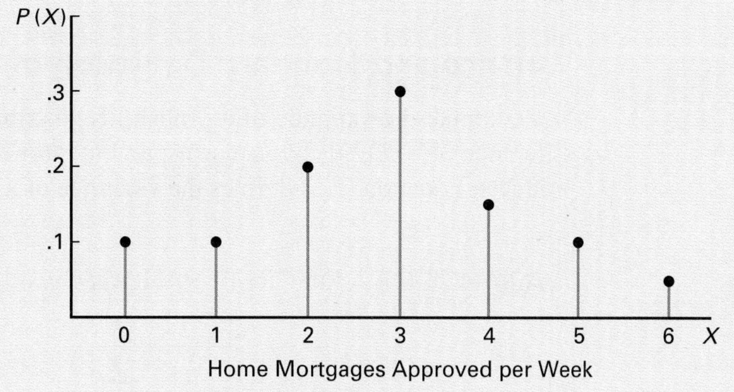

Expected Value of a Discrete Random Variable

The mean μ of a probability distribution is the *expected value* of its random variable. To calculate the expected value, you multiply each possible outcome X by its corresponding probability $P(X)$ and then sum these products.

EXPECTED VALUE μ OF A DISCRETE RANDOM VARIABLE

$$\mu = E(X) = \sum_{i=1}^{N} X_i P(X_i) \qquad \textbf{(5.1)}$$

where X_i = the ith outcome of the discrete random variable X
$P(X_i)$ = probability of occurrence of the ith outcome of X

For the probability distribution of the number of home mortgages approved per week (Table 5.1), the expected value is computed in Table 5.2 using Equation (5.1).

TABLE 5.2
Computing the Expected Value of the Number of Home Mortgages Approved per Week

Home Mortgages Approved per Week (X_i)	Probability $P(X_i)$	$X_i P(X_i)$
0	0.10	(0)(0.10) = 0.0
1	0.10	(1)(0.10) = 0.1
2	0.20	(2)(0.20) = 0.4
3	0.30	(3)(0.30) = 0.9
4	0.15	(4)(0.15) = 0.6
5	0.10	(5)(0.10) = 0.5
6	0.05	(6)(0.05) = 0.3
	1.00	$\mu = E(X) = 2.8$

$$\mu = E(X) = \sum_{i=1}^{N} X_i P(X_i)$$

$$= (0)(0.1) + (1)(0.1) + (2)(0.2) + (3)(0.3) + (4)(0.15) + (5)(0.1) + (6)(0.05)$$

$$= 0 + 0.1 + 0.4 + 0.9 + 0.6 + 0.5 + 0.3$$

$$= 2.8$$

The expected value of 2.8 for the number of mortgages approved is not "literally meaningful" because the actual number of mortgages approved in a given week must be an integer value. The expected value represents the *mean* number of mortgages approved per week.

Variance and Standard Deviation of a Discrete Random Variable

You compute the variance of a probability distribution by multiplying each possible squared difference $[X_i - E(X)]^2$ by its corresponding probability $P(X_i)$ and then summing the resulting products. Equation (5.2) defines the **variance of a discrete random variable**.

VARIANCE OF A DISCRETE RANDOM VARIABLE

$$\sigma^2 = \sum_{i=1}^{N} [X_i - E(X)]^2 P(X_i) \qquad \textbf{(5.2)}$$

where X_i = the ith outcome of the discrete random variable X
$P(X_i)$ = probability of occurrence of the ith outcome of X

Equation (5.3) defines the **standard deviation of a discrete random variable**.

STANDARD DEVIATION OF A DISCRETE RANDOM VARIABLE

$$\sigma = \sqrt{\sigma^2} = \sqrt{\sum_{i=1}^{N} [X_i - E(X)]^2 P(X_i)} \qquad \textbf{(5.3)}$$

In Table 5.3, the variance and the standard deviation of the number of home mortgages approved per week are computed using Equations (5.2) and (5.3):

$$\sigma^2 = \sum_{i=1}^{N} [X_i - E(X)]^2 P(X_i)$$

$$= (0 - 2.8)^2(0.10) + (1 - 2.8)^2(0.10) + (2 - 2.8)^2(0.20) + (3 - 2.8)^2(0.30) + (4 - 2.8)^2(0.15) + (5 - 2.8)^2(0.10) + (6 - 2.8)^2(0.05)$$

$$= 0.784 + 0.324 + 0.128 + 0.012 + 0.216 + 0.484 + 0.512$$

$$= 2.46$$

and

$$\sigma = \sqrt{\sigma^2} = \sqrt{2.46} = 1.57$$

Thus, the mean number of mortgages approved per week is 2.8, the variance is 2.46, and the standard deviation is 1.57.

TABLE 5.3
Computing the Variance and Standard Deviation of the Number of Home Mortgages Approved per Week

Home Mortgages Approved per Week (X_i)	Probability $P(X_i)$	$X_iP(X_i)$	$[X_i - E(X)]^2P(X_i)$
0	0.10	(0)(0.10) = 0.0	$(0 - 2.8)^2(0.10) = 0.784$
1	0.10	(1)(0.10) = 0.1	$(1 - 2.8)^2(0.10) = 0.324$
2	0.20	(2)(0.20) = 0.4	$(2 - 2.8)^2(0.20) = 0.128$
3	0.30	(3)(0.30) = 0.9	$(3 - 2.8)^2(0.30) = 0.012$
4	0.15	(4)(0.15) = 0.6	$(4 - 2.8)^2(0.15) = 0.216$
5	0.10	(5)(0.10) = 0.5	$(5 - 2.8)^2(0.10) = 0.484$
6	0.05	(6)(0.05) = 0.3	$(6 - 2.8)^2(0.05) = 0.512$

$$\sigma^2 = \sum_{i=1}^{N}[X_i - E(X)]^2 P(X_i) = 2.46$$

$$\sigma = 1.57$$

PROBLEMS FOR SECTION 5.1

Learning the Basics

PH Grade ASSIST **5.1** Given the following probability distributions:

Distribution A		Distribution B	
X	$P(X)$	X	$P(X)$
0	0.50	0	0.05
1	0.20	1	0.10
2	0.15	2	0.15
3	0.10	3	0.20
4	0.05	4	0.50

a. Compute the expected value for each distribution.
b. Compute the standard deviation for each distribution.
c. Compare and contrast the results of distributions *A* and *B*.

PH Grade ASSIST **5.2** Given the following probability distributions:

Distribution C		Distribution D	
X	$P(X)$	X	$P(X)$
0	0.20	0	0.10
1	0.20	1	0.20
2	0.20	2	0.40
3	0.20	3	0.20
4	0.20	4	0.10

a. Compute the expected value for each distribution.
b. Compute the standard deviation for each distribution.
c. Compare and contrast the results of distributions *C* and *D*.

Applying the Concepts

PH Grade ASSIST **5.3** Using the company records for the past 500 working days, the manager of Konig Motors, a suburban automobile dealership, has summarized the number of cars sold per day into the following table:

Number of Cars Sold per Day	Frequency of Occurrence
0	40
1	100
2	142
3	66
4	36
5	30
6	26
7	20
8	16
9	14
10	8
11	2
Total	500

a. Form the probability distribution for the number of cars sold per day.
b. Compute the mean or expected number of cars sold per day.
c. Compute the standard deviation.

5.4 The following table contains the probability distribution for the number of traffic accidents daily in a small city.

Number of Accidents Daily (X)	$P(X)$
0	0.10
1	0.20
2	0.45
3	0.15
4	0.05
5	0.05

a. Compute the mean or expected number of accidents per day.
b. Compute the standard deviation.

SELF Test **5.5** The manager of a large computer network has developed the following probability distribution of the number of interruptions per day:

Interruptions (X)	**$P(X)$**
0	0.32
1	0.35
2	0.18
3	0.08
4	0.04
5	0.02
6	0.01

a. Compute the mean or expected number of interruptions per day.
b. Compute the standard deviation.

5.6 In the carnival game Under-or-Over-Seven, a pair of fair dice is rolled once, and the resulting sum determines whether the player wins or loses his or her bet. For example, the player can bet \$1.00 that the sum will be under 7—that is, 2, 3, 4, 5, or 6. For this bet the player will lose \$1.00 if the outcome equals or exceeds 7 or will win \$1.00 if the result is under 7. Similarly, the player can bet \$1.00 that the sum will be over 7—that is, 8, 9, 10, 11, or 12. Here the player wins \$1.00 if the result is over 7 but loses \$1.00 if the result is 7 or under. A third method of play is to bet \$1.00 on the outcome 7. For this bet the player will win \$4.00 if the result of the roll is 7 and lose \$1.00 otherwise.
a. Construct the probability distribution representing the different outcomes that are possible for a \$1.00 bet on being under 7.
b. Construct the probability distribution representing the different outcomes that are possible for a \$1.00 bet on being over 7.
c. Construct the probability distribution representing the different outcomes that are possible for a \$1.00 bet on 7.
d. Show that the expected long-run profit (or loss) to the player is the same, no matter which method of play is used.

5.2 COVARIANCE AND ITS APPLICATION IN FINANCE

In section 5.1 the expected value, variance, and standard deviation of a discrete random variable of a probability distribution were discussed. In this section the covariance between two variables is introduced and applied to portfolio management, a topic of great interest to financial analysts.

The Covariance

The **covariance** σ_{XY} is a measure of the strength of the relationship between two discrete random variables X and Y. A positive covariance indicates a postive relationship. A negative covariance indicates a negative relationship. A covariance of 0 indicates that the two variables are independent. Equation (5.4) defines the covariance.

COVARIANCE

$$\sigma_{XY} = \sum_{i=1}^{N} [X_i - E(X)][Y_i - E(Y)]P(X_iY_i) \qquad \textbf{(5.4)}$$

where X = discrete variable X

X_i = ith outcome of X

$P(X_iY_i)$ = probability of occurrence of the ith outcome of X and the ith outcome of Y

Y = discrete variable Y

Y_i = ith outcome of Y

$i = 1, 2, \ldots, N$

To illustrate the covariance, suppose that you are deciding between two alternative investments for the coming year. The first investment is a mutual fund that consists of the stocks that make up the Dow Jones Industrial Average. The second investment is a mutual fund that is expected to perform best when economic conditions are weak. Your estimate of the returns for each investment (per \$1,000 investment) under three economic conditions, each with a given probability of occurrence, is summarized in Table 5.4.

TABLE 5.4
Estimated Returns for Each Investment under Three Economic Conditions

		Investment	
$P(X_iY_i)$	Economic Condition	Dow Jones Fund	Weak-Economy Fund
0.2	Recession	–\$100	+\$200
0.5	Stable economy	+100	+50
0.3	Expanding economy	+250	–100

The expected value and standard deviation for each investment and the covariance of the two investments are computed as follows:

$$\text{Let } X = \text{Dow Jones fund, and } Y = \text{weak-economy fund}$$

$$E(X) = \mu_X = (-100)(0.2) + (100)(0.5) + (250)(0.3) = \$105$$

$$E(Y) = \mu_Y = (+200)(0.2) + (50)(0.5) + (-100)(0.3) = \$35$$

$$Var(X) = \sigma_X^2 = (-100 - 105)^2(0.2) + (100 - 105)^2(0.5) + (250 - 105)^2(0.3)$$

$$= 14{,}725$$

$$\sigma_X = \$121.35$$

$$Var(Y) = \sigma_Y^2 = (200 - 35)^2(0.2) + (50 - 35)^2(0.5) + (-100 - 35)^2(0.3)$$

$$= 11{,}025$$

$$\sigma_Y = \$105.00$$

$$\sigma_{XY} = (-100 - 105)(200 - 35)(0.2) + (100 - 105)(50 - 35)(0.5) + (250 - 105)(-100 - 35)(0.3)$$

$$= -6{,}765 - 37.5 - 5{,}782.5$$

$$= -12{,}675$$

Thus, the Dow Jones fund has a higher expected value (i.e., larger expected return) than the weak-economy fund but has a higher standard deviation (i.e., more risk). The covariance of −12,675 between the two investments indicates a negative relationship in which the two investments are varying in the *opposite* direction. Therefore, when the return on one investment is high, the return on the other is typically low.

The Expected Value, Variance, and Standard Deviation of the Sum of Two Random Variables

Equation (5.4) defined the covariance between two variables *X* and *Y*. Now, the **expected value**, **variance**, and **standard deviation of the sum of two random variables** are defined.

EXPECTED VALUE OF THE SUM OF TWO RANDOM VARIABLES

The expected value of the sum of two random variables is equal to the sum of the expected values.

$$E(X + Y) = E(X) + E(Y) \tag{5.5}$$

VARIANCE OF THE SUM OF TWO RANDOM VARIABLES

The variance of the sum of two random variables is equal to the sum of the variances plus twice the covariance.

$$Var(X + Y) = \sigma_{X+Y}^2 = \sigma_X^2 + \sigma_Y^2 + 2\sigma_{XY} \tag{5.6}$$

STANDARD DEVIATION OF THE SUM OF TWO RANDOM VARIABLES

The standard deviation is the square root of the variance.

$$\sigma_{X+Y} = \sqrt{\sigma^2_{X+Y}} \quad \textbf{(5.7)}$$

To illustrate the expected value, variance, and standard deviation of the sum of two random variables, consider the two investments previously discussed. If X = Dow Jones fund and Y = weak-economy fund, using Equations (5.5), (5.6), and (5.7),

$$\begin{aligned} E(X+Y) &= E(X) + E(Y) = 105 + 35 = \$140 \\ \sigma^2_{X+Y} &= \sigma^2_X + \sigma^2_Y + 2\sigma_{XY} \\ &= 14{,}725 + 11{,}025 + (2)(-12{,}675) \\ &= 400 \\ \sigma_{X+Y} &= \$20 \end{aligned}$$

The expected return of the sum of the Dow Jones fund and the weak-economy fund is \$140 with a standard deviation of \$20. The standard deviation of the sum of the two investments is much less than the standard deviation of either single investment because there is a large negative covariance between the investments.

Portfolio Expected Return and Portfolio Risk

Now that the covariance and the expected return and standard deviation of the sum of two random variables have been defined, these concepts can be applied to the study of a group of assets referred to as a **portfolio**. Investors combine assets into portfolios to reduce their risk (see references 1 and 2). Often the objective is to maximize the return while minimizing the risk. For such portfolios, rather than studying the sum of two random variables, each investment is weighted by the proportion of assets assigned to that investment. Equations (5.8) and (5.9) define **portfolio expected return** and **portfolio risk**.

PORTFOLIO EXPECTED RETURN

The portfolio expected return for a two-asset investment is equal to the weight assigned to asset X multiplied by the expected return of asset X plus the weight assigned to asset Y multiplied by the expected return of asset Y.

$$E(P) = wE(X) + (1 - w)E(Y) \quad \textbf{(5.8)}$$

where $E(P)$ = portfolio expected return

w = portion of the portfolio value assigned to asset X

$(1 - w)$ = portion of the portfolio assigned to asset Y

$E(X)$ = expected return of asset X

$E(Y)$ = expected return of asset Y

PORTFOLIO RISK

$$\sigma_p = \sqrt{w^2\sigma^2_X + (1-w)^2\sigma^2_Y + 2w(1-w)\sigma_{XY}} \quad \textbf{(5.9)}$$

In the previous example, you evaluated the expected return and risk of two different investments, a Dow Jones fund and a weak-economy fund. You also computed the covariance of the two investments. Now suppose that you wish to form a portfolio of these two investments that consists of an equal investment in each of these two funds. To compute the portfolio expected return and the portfolio risk, using Equations (5.8) and (5.9), with $w = 0.50$, $E(X) = \$105$, $E(Y) = \$35$, $\sigma_X^2 = 14{,}725$, $\sigma_Y^2 = 11{,}025$, and $\sigma_{XY} = -12{,}675$

$$E(P) = (0.5)(105) + (1 - 0.5)(35) = \$70$$

$$\sigma_p = \sqrt{(0.5)^2(14{,}725) + (1 - 0.5)^2(11{,}025) + 2(0.5)(1 - 0.5)(-12{,}675)}$$

$$= \sqrt{100} = \$10$$

Thus, the portfolio has an expected return of \$70 for each \$1,000 invested (a return of 7%) and has a portfolio risk of \$10. The portfolio risk here is small because there is a large negative covariance between the two investments. The fact that each investment performs best under different circumstances has reduced the overall risk of the portfolio.

PROBLEMS FOR SECTION 5.2

Learning the Basics

PH Grade ASSIST **5.7** Given the following probability distributions for variables X and Y:

$P(X_iY_i)$	X	Y
0.4	100	200
0.6	200	100

Compute

a. $E(X)$ and $E(Y)$

b. σ_X and σ_Y

c. σ_{XY}

d. $E(X + Y)$

PH Grade ASSIST **5.8** Given the following probability distributions for variables X and Y:

$P(X_iY_i)$	X	Y
0.2	−100	50
0.4	50	30
0.3	200	20
0.1	300	20

Compute

a. $E(X)$ and $E(Y)$

b. σ_X and σ_Y

c. σ_{XY}

d. $E(X + Y)$

PH Grade ASSIST **5.9** Two investments X and Y have the following characteristics:

$$E(X) = \$50, E(Y) = \$100, \sigma_X^2 = 9{,}000,$$
$$\sigma_Y^2 = 15{,}000, \text{ and } \sigma_{XY} = 7{,}500$$

If the weight assigned to investment X of portfolio assets is 0.4, compute the

a. portfolio expected return.

b. portfolio risk.

Applying the Concepts

5.10 The process of being served at a bank consists of two independent parts—the time waiting in line and the time it takes to be served by the teller. Suppose that the time waiting in line has an expected value of 4 minutes with a standard deviation of 1.2 minutes and the time it takes to be served by the teller has an expected value of 5.5 minutes with a standard deviation of 1.5 minutes. Compute the

a. expected value of the total time it takes to be served at the bank.

b. standard deviation of the total time it takes to be served at the bank.

PH Grade ASSIST **5.11** In the example at the top of this page, half the portfolio assets are invested in the Dow Jones fund and half in a weak-economy fund. Recalculate the portfolio expected return and the portfolio risk if

a. 30% are invested in the Dow Jones fund and 70% in a weak-economy fund.

b. 70% are invested in the Dow Jones fund and 30% in a weak-economy fund.

c. Which of the three investment strategies (30%, 50%, or 70% in the Dow Jones fund) would you recommend? Why?

SELF Test **5.12** You are trying to develop a strategy for investing in two different stocks. The anticipated annual return for a $1,000 investment in each stock has the following probability distribution:

	Returns	
Probability	**Stock *X***	**Stock *Y***
0.1	−$100	$50
0.3	0	150
0.3	80	−20
0.3	150	−100

Compute the

a. expected return for stock *X* and for stock *Y*.
b. standard deviation for stock *X* and for stock *Y*.
c. covariance of stock *X* and stock *Y*.
d. Would you invest in stock *X* or stock *Y*? Explain.

5.13 Suppose that in problem 5.12 you wanted to create a portfolio that consists of stock *X* and stock *Y*. Compute the portfolio expected return and portfolio risk for each of the following percentages invested in stock *X*:

a. 30%
b. 50%
c. 70%
d. On the basis of the results of (a) through (c), which portfolio would you recommend? Explain.

5.14 You are trying to develop a strategy for investing in two different stocks. The anticipated annual return for a $1,000 investment in each stock has the following probability distribution:

	Returns	
Probability	**Stock *X***	**Stock *Y***
0.1	−$50	−$100
0.3	20	50
0.4	100	130
0.2	150	200

Compute the

a. expected return for stock *X* and for stock *Y*.
b. standard deviation for stock *X* and for stock *Y*.
c. covariance of stock *X* and stock *Y*.
d. Would you invest in stock *X* or stock *Y*? Explain.

5.15 Suppose that in problem 5.14 you wanted to create a portfolio that consists of stock *X* and stock *Y*. Compute the portfolio expected return and portfolio risk for each of the following percentages invested in stock *X*:

a. 30%
b. 50%
c. 70%
d. On the basis of the results of (a) through (c), which portfolio would you recommend? Explain.

5.16 You are trying to set up a portfolio that consists of a corporate bond fund and a common stock fund. The following information about the annual return (per $1,000) of each of these investments under different economic conditions is available along with the probability that each of these economic conditions will occur.

Probability	**State of the Economy**	**Corporate Bond Fund**	**Common Stock Fund**
0.10	Recession	−$30	−$150
0.15	Stagnation	50	−20
0.35	Slow growth	90	120
0.30	Moderate growth	100	160
0.10	High growth	110	250

Compute the

a. expected return for the corporate bond fund and for the common stock fund.
b. standard deviation for the corporate bond fund and for the common stock fund.
c. covariance of the corporate bond fund and the common stock fund.
d. Would you invest in the corporate bond fund or the common stock fund? Explain.

5.17 Suppose that in problem 5.16 you wanted to create a portfolio that consists of a corporate bond fund and a common stock fund. Compute the portfolio expected return and portfolio risk for each of the following percentages invested in a corporate bond fund:

a. 30%
b. 50%
c. 70%
d. On the basis of the results of (a) through (c), which portfolio would you recommend? Explain.

5.3 BINOMIAL DISTRIBUTION

The next three sections use mathematical models to solve business problems.

A **mathematical model** is a mathematical expression representing a variable of interest.

When a mathematical expression is available, you can compute the exact probability of occurrence of any particular outcome of the random variable.

The **binomial probability distribution** is one of the most useful mathematical models. You use the binomial distribution when the discrete random variable of interest is the number of successes in a sample of n observations. The binomial distribution has four essential properties:

- The sample consists of a fixed number of observations, n.
- Each observation is classified into one of two mutually exclusive and collectively exhaustive categories, usually called *success* and *failure*.
- The probability of an observation being classified as success, p, is constant from observation to observation. Thus, the probability of an observation being classified as failure, $1 - p$, is constant over all observations.
- The outcome (i.e., success or failure) of any observation is independent of the outcome of any other observation. To ensure independence, the observations can be randomly selected either from an *infinite population without replacement* or from a *finite population with replacement*.

Returning to the "Using Statistics" scenario presented on page 158 concerning the accounting information system, suppose *success* is defined as a tagged order form and *failure* as any other outcome. You are interested in the number of tagged order forms in a given sample of orders.

What results can occur? If the sample contains four orders, there could be none, one, two, three, or four tagged order forms. The binomial random variable, the number of tagged order forms, cannot take on any other value because the number of tagged order forms cannot be more than the sample size n and cannot be less than zero. Therefore, the binomial random variable has a range from 0 to n.

Suppose that you observe the following result in a sample of four orders:

First Order	Second Order	Third Order	Fourth Order
Tagged	Tagged	Not Tagged	Tagged

What is the probability of having three successes (tagged order forms) in a sample of four orders in this particular sequence? Because the historical probability of a tagged order is 0.10, the probability that each order occurs in the sequence is

First Order	Second Order	Third Order	Fourth Order
$p = 0.10$	$p = 0.10$	$1 - p = 0.90$	$p = 0.10$

Each outcome is independent of the others because the order forms were selected from an extremely large or practically infinite population without replacement. Therefore, the probability of having this particular sequence is

$$\begin{aligned} pp(1-p)p &= p^3(1-p)^1 \\ &= (0.10)(0.10)(0.10)(0.90) \\ &= (0.10)^3(0.90)^1 \\ &= 0.0009 \end{aligned}$$

This result indicates only the probability of three tagged order forms (successes) out of a sample of four order forms in a *specific sequence*. To find the number of ways of selecting X objects out of n objects *irrespective of sequence*, you use the **rule of combinations** given in Equation (5.10).

COMBINATIONS

The number of combinations of selecting X objects out of n objects is given by

$$_nC_X = \frac{n!}{X!(n-X)!} \tag{5.10}$$

where $n! = (n)(n-1). \ldots (1)$ is called n factorial. By definition, $0! = 1$.

With $n = 4$ and $X = 3$, there are

$$_nC_X = \frac{n!}{X!(n-X)!} = \frac{4!}{3!(4-3)!} = \frac{4 \times 3 \times 2 \times 1}{(3 \times 2 \times 1)(1)} = 4$$

such sequences. The four possible sequences are:

Sequence 1 = *tagged, tagged, tagged, not tagged* with probability

$$ppp(1-p) = p^3(1-p)^1 = 0.0009$$

Sequence 2 = *tagged, tagged, not tagged, tagged* with probability

$$pp\,(1-p)p = p^3(1-p)^1 = 0.0009$$

Sequence 3 = *tagged, not tagged, tagged, tagged* with probability

$$p\,(1-p)pp = p^3(1-p)^1 = 0.0009$$

Sequence 4 = *not tagged, tagged, tagged, tagged* with probability

$$(1-p)ppp = p^3(1-p)^1 = 0.0009$$

Therefore, the probability of three tagged order forms is equal to

$$\begin{gathered}(\text{number of possible sequences}) \times (\text{probability of a particular sequence}) \\ = (4) \times (0.0009) = 0.0036\end{gathered}$$

You can make a similar, intuitive derivation for the other possible outcomes of the random variable—zero, one, two, and four tagged order forms. However, as n, the sample size, gets large, the computations involved in using this intuitive approach become time consuming. A mathematical model provides a general formula for computing any binomial probability. Equation (5.11) is the mathematical model representing the binomial probability distribution for computing the number of successes (X), given the values of n and p.

BINOMIAL PROBABILITY DISTRIBUTION

$$P(X) = \frac{n!}{X!(n-X)!}p^X(1-p)^{n-X} \tag{5.11}$$

where $P(X)$ = probability of X successes given n and p

n = number of observations

p = probability of success

$1 - p$ = probability of failure

X = number of successes in the sample ($X = 0, 1, 2, \ldots, n$)

Equation (5.11) restates what you had intuitively derived. The binomial random variable X can have any integer value X from 0 through n. In Equation (5.11) the product

$$p^X(1-p)^{n-X}$$

indicates the probability of exactly X successes out of n observations in a *particular sequence*. The term

$$\frac{n!}{X!(n-X)!}$$

indicates *how many combinations* of the X successes out of n observations are possible. Hence, given the number of observations n and the probability of success p, the probability of X successes is:

$$\begin{aligned} P(X) &= (\text{number of possible sequences}) \times (\text{probability of a particular sequence}) \\ &= \frac{n!}{X!(n-X)!}p^X(1-p)^{n-X} \end{aligned}$$

Example 5.1 illustrates the use of Equation (5.11).

EXAMPLE 5.1

DETERMINING $P(X = 3)$, GIVEN $n = 4$ AND $p = 0.1$

If the likelihood of a tagged order form is 0.1, what is the probability that there are three tagged order forms in the sample of four?

SOLUTION Using Equation (5.11), the probability of three tagged orders from a sample of four is

$$\begin{aligned} P(X=3) &= \frac{4!}{3!(4-3)!}(0.1)^3(1-0.1)^{4-3} \\ &= \frac{4!}{3!(4-3)!}(0.1)^3(0.9)^1 \\ &= 4(0.1)(0.1)(0.1)(0.9) = 0.0036 \end{aligned}$$

Examples 5.2 and 5.3 show the computations for other values of X.

EXAMPLE 5.2

DETERMINING $P(X \geq 3)$, GIVEN $n = 4$ AND $p = 0.1$

If the likelihood of a tagged order form is 0.1, what is the probability that there are three or more (i.e., at least three) tagged order forms in the sample of four?

SOLUTION In Example 5.1 you found that the probability of *exactly* three tagged order forms from a sample of four is 0.0036. To compute the probability of *at least* three tagged order forms, you need to add the probability of three tagged order forms to the probability of four tagged order forms. The probability of four tagged order forms is

$$\begin{aligned} P(X=4) &= \frac{4!}{4!(4-4)!}(0.1)^4(1-0.1)^{4-4} \\ &= \frac{4!}{4!(0)!}(0.1)^4(0.9)^0 \\ &= 1(0.1)(0.1)(0.1)(0.1) = 0.0001 \end{aligned}$$

Thus, the probability of at least three tagged order forms is

$$P(X \geq 3) = P(X = 3) + P(X = 4)$$
$$= 0.0036 + 0.0001$$
$$= 0.0037$$

There is a 0.37% chance that there will be at least three tagged order forms in a sample of four.

EXAMPLE 5.3

DETERMINING $P(X < 3)$, GIVEN $N = 4$ AND $P = 0.1$

If the likelihood of a tagged order form is 0.1, what is the probability that there are fewer than three tagged order forms in the sample of four?

SOLUTION The probability that there are fewer than three tagged order forms is

$$P(X < 3) = P(X = 0) + P(X = 1) + P(X = 2)$$

Use Equation (5.11) on page 168 to compute each of these probabilities,

$$P(X = 0) = \frac{4!}{0!(4-0)!}(0.1)^0(1-0.1)^{4-0} = 0.6561$$

$$P(X = 1) = \frac{4!}{1!(4-1)!}(0.1)^1(1-0.1)^{4-1} = 0.2916$$

$$P(X = 2) = \frac{4!}{2!(4-2)!}(0.1)^2(1-0.1)^{4-2} = 0.0486$$

Therefore, $P(X < 3) = 0.6561 + 0.2916 + 0.0486 = 0.9963$.

$P(X < 3)$ could also be calculated from its complement, $P(X \geq 3)$ as follows:

$$P(X < 3) = 1 - P(X \geq 3)$$
$$= 1 - 0.0037 = 0.9963$$

Computations such as those in Example 5.3 can become tedious, especially as n gets large. To avoid computational drudgery, you can find many binomial probabilities directly from Table E.6, a portion of which is reproduced in Table 5.5. Table E.6 provides binomial probabilities for $X = 0, 1, 2, \ldots, n$, for various selected combinations of n and p. For example, to find the probability of exactly two successes in a sample of four when the probability of success is 0.1, first find $n = 4$ and then you look in the row $X = 2$ and column $p = 0.10$. The result is 0.0486.

TABLE 5.5

Finding a Binomial Probability for $n = 4$, $X = 2$, and $p = 0.1$

		p			
n	X	**0.01**	**0.02**	....	**0.10**
4	0	0.9606	0.9224	....	0.6561
	1	0.0388	0.0753	....	0.2916
	2	0.0006	0.0023	....	0.0486
	3	0.0000	0.0000	....	0.0036
	4	0.0000	0.0000	....	0.0001

Source: Table E.6.

You can also compute the binomial probabilities given in Table E.6 by using Microsoft Excel or Minitab. Figure 5.2 presents a Microsoft Excel worksheet for computing binomial probabilities, and Figure 5.3 illustrates Minitab output.

FIGURE 5.2
Microsoft Excel Worksheet for Computing Binomial Probabilities

	A	B	
1	**Tagged Orders**		
2			
3	**Data**		
4	**Sample size**	**4**	
5	**Probability of success**	**0.1**	
6			
7	**Statistics**		
8	**Mean**	**0.4**	=B4 * B5
9	**Variance**	**0.36**	=B8 * (1 - B5)
10	**Standard deviation**	**0.6**	=SQRT(B9)
11			
12	**Binomial Probabilities Table**		
13	X	P(X)	
14	0	0.6561	=BINOMDIST(A14, B4, B5, FALSE)
15	1	0.2916	=BINOMDIST(A15, B4, B5, FALSE)
16	2	0.0486	=BINOMDIST(A16, B4, B5, FALSE)
17	3	0.0036	=BINOMDIST(A17, B4, B5, FALSE)
18	4	0.0001	=BINOMDIST(A18, B4, B5, FALSE)

FIGURE 5.3
Minitab Binomial Distribution Calculations for $n = 4$ and $p = 0.1$

```
Binomial with n = 4 and p = 0.1

x   P( X = x )
0       0.6561
1       0.2916
2       0.0486
3       0.0036
4       0.0001
```

The shape of a binomial probability distribution depends on the values of n and p. Whenever $p = 0.5$, the binomial distribution is symmetrical, regardless of how large or small the value of n. When $p \neq 0.5$, the distribution is skewed. The closer p is to 0.5 and the larger the number of observations n, the less skewed the distribution becomes. For example, the distribution of the number of tagged order forms is highly skewed to the right because $p = 0.1$ and $n = 4$ (see Figure 5.4).

FIGURE 5.4
Microsoft Excel Histogram of the Binomial Probability Distribution with $n = 4$ and $p = 0.1$

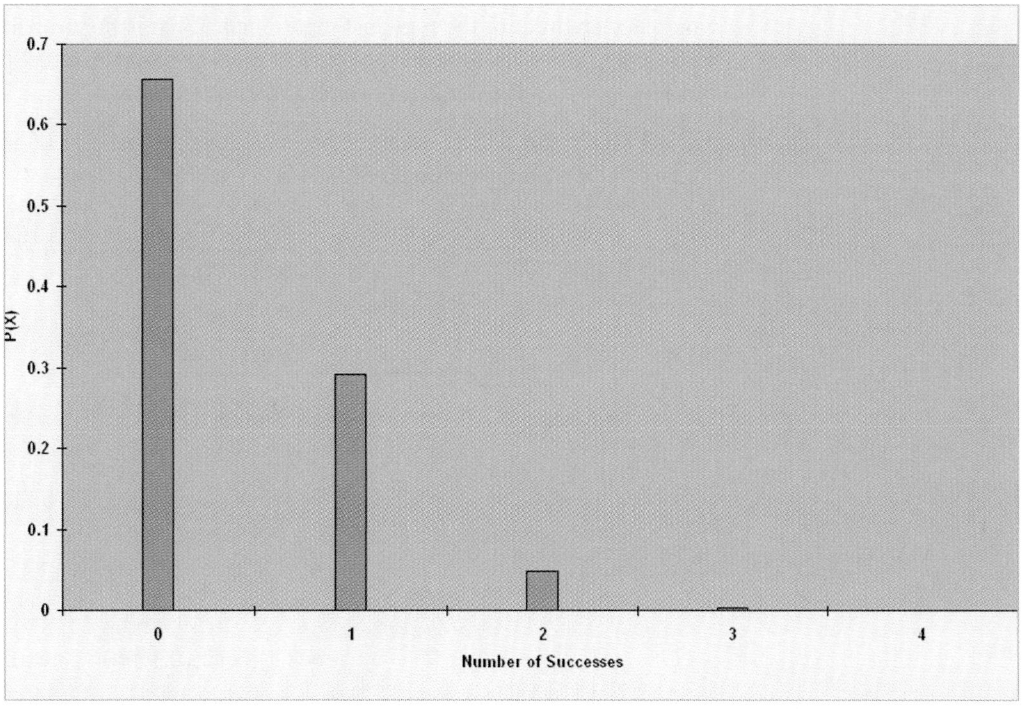

The mean of the binomial distribution is equal to the product of n and p. Instead of using Equation (5.1) on page 159 to compute the mean of the probability distribution, you use Equation (5.12) to compute the mean for variables that follow the binomial distribution.

THE MEAN OF THE BINOMIAL DISTRIBUTION

The mean μ of the binomial distribution is equal to the sample size n multiplied by the probability of success p.

$$\mu = E(X) = np \qquad \textbf{(5.12)}$$

On the average, over the long run, you theoretically expect $\mu = E(X) = np = (4)(0.1) = 0.4$ tagged order form in a sample of four orders.

The standard deviation of the binomial distribution is calculated using Equation (5.13).

THE STANDARD DEVIATION OF THE BINOMIAL DISTRIBUTION

$$\sigma = \sqrt{\sigma^2} = \sqrt{Var(X)} = \sqrt{np(1-p)} \qquad \textbf{(5.13)}$$

The standard deviation of the number of tagged order forms is

$$\sigma = \sqrt{4(0.1)(0.9)} = 0.60$$

This is the same result that you would calculate if you used Equation (5.3) on page 160.

EXAMPLE 5.4

COMPUTING BINOMIAL PROBABILITIES

Accuracy in taking orders at the drive-thru window is an important feature for fast-food chains. Each month *QSR Magazine* **www.qsrmagazine.com** publishes the results of its surveys. Accuracy is measured as the percentage of orders consisting of a main item, side item, and drink (but omitting one standard item such as a pickle) that are filled correctly. In a recent month, the percentage of correct orders of this type filled at Burger King was approximately 88%. Suppose that you and two friends go to the drive-thru window at Burger King and each of you place an order of the type just mentioned. What is the probability that all three orders will be filled accurately? None of the three? At least two of the three? What is the average and standard deviation of the number of orders filled accurately?

SOLUTION Since there are three orders and the probability of an accurate order is 88%, $n = 3$ and $p = 0.88$. Using Equations (5.11), (5.12), and (5.13)

$$\begin{aligned} P(X = 3) &= \frac{3!}{3!(3-3)!}(0.88)^3(1-0.88)^{3-3} \\ &= \frac{3!}{3!(3-3)!}(0.88)^3(0.12)^0 \\ &= 1(0.88)(0.88)(0.88)(1) = 0.6815 \end{aligned}$$

$$\begin{aligned} P(X = 0) &= \frac{3!}{0!(3-0)!}(0.88)^0(1-0.88)^{3-0} \\ &= \frac{3!}{0!(3-0)!}(0.88)^0(0.12)^3 \\ &= 1(1)(0.12)(0.12)(0.12) = 0.0017 \end{aligned}$$

$$P(X = 2) = \frac{3!}{2!(3-2)!}(0.88)^2(1-0.88)^{3-2}$$

$$= \frac{3!}{2!(3-2)!}(0.88)^2(0.12)^1$$

$$= 3(0.88)(0.88)(0.12) = 0.2788$$

$$P(X \geq 2) = P(X = 2) + P(X = 3)$$

$$= 0.2788 + 0.6815$$

$$= 0.9603$$

$$\mu = E(X) = np = 3(0.88) = 2.64$$

$$\sigma = \sqrt{\sigma^2} = \sqrt{Var(X)} = \sqrt{np(1-p)}$$

$$= \sqrt{3(0.88)(0.12)}$$

$$= \sqrt{0.3168} = 0.563$$

The probability that all three orders are filled accurately is 0.6815 or 68.15%. The probability that none of the orders are filled accurately is 0.0017 or 0.17%. The probability that at least two orders are filled accurately is 0.9603 or 96.03%. The mean number of accurate orders filled in a sample of three orders is 2.64 and the standard deviation is 0.563.

In this section the binomial distribution has been developed. The binomial distribution plays an even more important role when it is used in statistical inference problems involving the estimation or testing of hypotheses about proportions (as will be discussed in Chapters 8 and 9).

PROBLEMS FOR SECTION 5.3

Problems 5.18–5.26 can be solved manually or by using Microsoft Excel or Minitab. We recommend that you solve problems 5.27–5.29 using Microsoft Excel or Minitab.

Learning the Basics

PH Grade ASSIST **5.18** Determine the following:

a. For $n = 4$ and $p = 0.12$, what is $P(X = 0)$?
b. For $n = 10$ and $p = 0.40$, what is $P(X = 9)$?
c. For $n = 10$ and $p = 0.50$, what is $P(X = 8)$?
d. For $n = 6$ and $p = 0.83$, what is $P(X = 5)$?

5.19 If $n = 5$ and $p = 0.40$, what is the probability that

a. $X = 4$
b. $X \leq 3$
c. $X < 2$
d. $X > 1$

PH Grade ASSIST **5.20** Determine the mean and standard deviation of the random variable X in each of the following binomial distributions:

a. $n = 4$ and $p = 0.10$
b. $n = 4$ and $p = 0.40$
c. $n = 5$ and $p = 0.80$
d. $n = 3$ and $p = 0.50$

Applying the Concepts

5.21 The increase or decrease in the price of a stock between the beginning and the end of a trading day is assumed to be an equally likely random event. What is the probability that a stock will show an increase in its closing price on five consecutive days?

5.22 Sixty percent of Americans read their employment contract, including the fine print ("Snapshots," **usatoday.com**,

January 20, 2004). Assume that the number of employees who read every word of their contract can be modeled using the binomial distribution. For a group of five employees, what is the probability that

a. all five will have read every word of their contract?
b. at least three will have read every word of their contract?
c. less than two will have read every word of their contract?
d. What are your answers in (a) through (c) if the probability is 0.80 that an employee reads every word of their contract?

PH Grade ASSIST **5.23** A student is taking a multiple-choice exam in which each question has four choices. Assuming that she has no knowledge of the correct answers to any of the questions, she has decided on a strategy in which she will place four balls (marked *A*, *B*, *C*, and *D*) into a box. She randomly selects one ball for each question and replaces the ball in the box. The marking on the ball will determine her answer to the question. There are five multiple-choice questions on the exam. What is the probability that she will get

a. five questions correct?
b. at least four questions correct?
c. no questions correct?
d. no more than two questions correct?

SELF Test **5.24** In Example 5.4 on page 172, you and two friends decided to go to Burger King. Instead, suppose that you went to McDonald's, which last month filled 90% of the orders accurately. What is the probability that

a. all three orders will be filled accurately?
b. none of the three will be filled accurately?
c. at least two of the three will be filled accurately?
d. What is the mean and standard deviation of the number of orders filled accurately?

5.25 The commission rate commercial airlines pay travel agents has been declining for several years. In an attempt by travel agencies to raise revenue, many agencies are now charging their customers a ticket fee, typically between $10 and $25. According to the American Society of Ticket Agents, about 90% of travel agents charge customers fees when purchasing an airline ticket (Kortney Stringer, "American Air Fees for Travel Agents to Be Cut Again," *The Wall Street Journal*, August 20, 2001, B2).

a. Is the 90% figure quoted by the American Society of Ticket Agents best classified as *a priori* classical probability, empirical classical probability, or subjective probability?
b. You select a random sample of 10 travel agencies. Assume that the number of the 10 travel agencies charging a ticket fee is distributed as a binomial random variable. What are the mean and standard deviation of this distribution?
c. What assumptions are necessary in (b)?

5.26 Referring to problem 5.25, compute the probability that of the 10 travel agencies:

a. zero charge ticket fees?
b. exactly one charges ticket fees?
c. two or fewer charge ticket fees?
d. three or more charge ticket fees?

PH Grade ASSIST **5.27** When a customer places an order with Rudy's On-Line Office Supplies, a computerized accounting information system (AIS) automatically checks to see if the customer has exceeded his or her credit limit. Past records indicate that the probability of customers exceeding their credit limit is 0.05. Suppose that, on a given day, 20 customers place orders. Assume that the number of customers that the AIS detects as having exceeded their credit limit is distributed as a binomial random variable.

a. What are the mean and standard deviation of the number of customers exceeding their credit limits?
b. What is the probability that zero customers will exceed their limit?
c. What is the probability that one customer will exceed his or her limit?
d. What is the probability that two or more customers will exceed their limits?

5.28 The broadcast networks introduce new television shows each fall. In an attempt to get viewers interested in the new shows, television commercials are aired during the summer as part of a promotional campaign conducted prior to the fall debuts. The networks then conduct surveys to see what percentage of the viewing public is *aware* of the shows. According to network sources, in the fall of 2001, 68% of viewers aged 18–49 were aware of the new series *Criminal Intent*, whereas only 24% of viewers aged 18–49 were aware of *Inside Schwartz* (Joe Flint, "Viewers Awareness of New Shows Rises," *The Wall Street Journal*, August 20, 2001, B7).

a. Are the 68% and 24% figures quoted by networks best classified as a priori classical probability, empirical classical probability, or subjective probability?

Suppose that a random sample of 20 viewers aged 18–49 is selected. What is the probability that:

b. less than five of the viewers are aware of *Criminal Intent*?
c. 10 or more are aware of *Criminal Intent*?
d. all 20 are aware of *Criminal Intent*?

5.29 Referring to problem 5.28, consider another sample of 20 viewers aged 18–49. For the new show *Inside Schwartz*, what is the probability that:

a. less than five of the viewers are aware of *Inside Schwartz*?
b. 10 or more are aware of *Inside Schwartz*?
c. all 20 are aware of *Inside Schwartz*?
d. Compare the results of (a) through (c) to those for *Criminal Intent* in problem 5.28 (b) through (d).

5.4 POISSON DISTRIBUTION

Many studies are based on counts of the times a particular event occurs in a given *area of opportunity*. An **area of opportunity** is a continuous unit or interval of time, volume, or such area in which more than one occurrence of an event can occur. Examples are the surface defects on a new refrigerator, the number of network failures in a day, or the number of fleas on the body of a dog. When you have an area of opportunity such as these, you can use the **Poisson distribution** to calculate probabilities if:

- You are interested in counting the number of times a particular event occurs in a given area of opportunity. The area of opportunity is defined by time, length, surface area, and so forth.
- The probability that an event occurs in a given area of opportunity is the same for all of the areas of opportunity.
- The number of events that occur in one area of opportunity is independent of the number of events that occur in any other area of opportunity.
- The probability that two or more events will occur in an area of opportunity approaches zero as the area of opportunity becomes smaller.

Consider the number of customers arriving during the lunch hour at a bank located in the central business district in a large city. You are interested in the number of customers that arrive each minute. Does this situation match the four properties of the Poisson distribution given above? First, the *event* of interest is a customer arriving and the *given area of opportunity* is defined as a 1-minute interval. Will zero customers arrive, one customer arrive, two customers arrive, and so on? Second, it is reasonable to assume that the probability that a customer arrives during a 1-minute interval is the same as the probability for all the other 1-minute intervals. Third, the arrival of one customer in any 1-minute interval has no effect on (i.e., is statistically independent of) the arrival of any other customer in any other 1-minute interval. Finally, the probability that two or more customers will arrive in a given time period approaches zero as the time interval becomes small. For example, the probability is virtually zero that two customers will arrive in a time interval with a width of 1/100th of a second. Thus, you can use the Poisson distribution to determine probabilities involving the number of customers arriving at the bank in a 1-minute time interval during the lunch hour.

The Poisson distribution has one parameter, called λ (the Greek lowercase letter *lambda*), which is the mean or expected number of events per unit. The variance of a Poisson distribution is also equal to λ, and the standard deviation is equal to $\sqrt{\lambda}$. The number of events X of the Poisson random variable ranges from 0 to infinity.

Equation (5.14) presents the mathematical expression for the Poisson distribution for computing the probability of X events, given that λ events are expected.

POISSON PROBABILITY DISTRIBUTION

$$P(X) = \frac{e^{-\lambda}\lambda^X}{X!} \quad \textbf{(5.14)}$$

where $P(X)$ = the probability of X events in an area of opportunity

λ = expected number of events

e = mathematical constant approximated by 2.71828

X = number of events

To demonstrate the Poisson distribution, suppose that the mean number of customers who arrive per minute at the bank during the noon to 1 P.M. hour is equal to 3.0. What is the probability that in a given minute exactly two customers will arrive? And what is the probability that more than two customers will arrive in a given minute?

Using Equation (5.14) and $\lambda = 3$, the probability that in a given minute exactly two customers will arrive is

$$P(X = 2) = \frac{e^{-3.0}(3.0)^2}{2!} = \frac{9}{(2.71828)^3(2)} = 0.2240$$

To determine the probability that in any given minute more than two customers will arrive

$$P(X > 2) = P(X = 3) + P(X = 4) + \ldots + P(X = \infty)$$

Because all the probabilities in a probability distribution must sum to 1, the terms on the right side of the equation $P(X > 2)$ also represent the complement of the probability that X is less than or equal to 2 [i.e., $1 - P(X \leq 2)$]. Thus,

$$P(X > 2) = 1 - P(X \leq 2) = 1 - [P(X = 0) + P(X = 1) + P(X = 2)]$$

Now, using Equation (5.14),

$$P(X > 2) = 1 - \left[\frac{e^{-3.0}(3.0)^0}{0!} + \frac{e^{-3.0}(3.0)^1}{1!} + \frac{e^{-3.0}(3.0)^2}{2!}\right]$$

$$= 1 - [0.0498 + 0.1494 + 0.2240]$$

$$= 1 - 0.4232 = 0.5768$$

Thus, there is a 57.68% chance that more than two customers will arrive in the same minute.

To avoid computational drudgery involved in these computations, you can find Poisson probabilities directly from Table E.7, a portion of which is reproduced in Table 5.6. Table E.7 provides the probabilities that the Poisson random variable takes on values of $X = 0, 1, 2, \ldots$, for selected values of the parameter λ. To find the probability that exactly two customers will arrive in a given minute when the mean number of customers arriving is 3.0 per minute, you can read the probability corresponding to the row $X = 2$ and column $\lambda = 3.0$ from the table. The result is 0.2240, as demonstrated in Table 5.6.

TABLE 5.6

Computing a Poisson Probability for $\lambda = 3$

	λ			
X	**2.1**	**2.2**	....	**3.0**
0	.1225	.1108	....	.0498
1	.2572	.2438	....	.1494
2	.2700	.2681	....	.2240
3	.1890	.1966	....	.2240
4	.0992	.1082	....	.1680
5	.0417	.0476	....	.1008
6	.0146	.0174	....	.0504
7	.0044	.0055	....	.0216
8	.0011	.0015	....	.0081
9	.0003	.0004	....	.0027
10	.0001	.0001	....	.0008
11	.0000	.0000	....	.0002
12	.0000	.0000	....	.0001

Source: Table E.7.

You can also compute the Poisson probabilities given in Table E.7 by using Microsoft Excel or Minitab. Figure 5.5 presents a Microsoft Excel worksheet for the Poisson distribution with $\lambda = 3$. Figure 5.6 illustrates Minitab output.

FIGURE 5.5
Microsoft Excel Worksheet for Computing Poisson Probabilities

	A	B	C	D	E	
1	Customer Arrivals Analysis					
2						
3	Data					
4	Average/Expected number of successes:				3	
5						
6	Poisson Probabilities Table					
7	X	P(X)				
8	0	0.049787				=POISSON(A8, E4, FALSE)
9	1	0.149361				=POISSON(A9, E4, FALSE)
10	2	0.224042				=POISSON(A10, E4, FALSE)
11	3	0.224042				=POISSON(A11, E4, FALSE)
12	4	0.168031				=POISSON(A12, E4, FALSE)
13	5	0.100819				=POISSON(A13, E4, FALSE)
14	6	0.050409				=POISSON(A14, E4, FALSE)
15	7	0.021604				=POISSON(A15, E4, FALSE)
16	8	0.008102				=POISSON(A16, E4, FALSE)
17	9	0.002701				=POISSON(A17, E4, FALSE)
18	10	0.000810				=POISSON(A18, E4, FALSE)
19	11	0.000221				=POISSON(A19, E4, FALSE)
20	12	0.000055				=POISSON(A20, E4, FALSE)
21	13	0.000013				=POISSON(A21, E4, FALSE)
22	14	0.000003				=POISSON(A22, E4, FALSE)
23	15	0.000001				=POISSON(A23, E4, FALSE)
24	16	0.000000				=POISSON(A24, E4, FALSE)
25	17	0.000000				=POISSON(A25, E4, FALSE)
26	18	0.000000				=POISSON(A26, E4, FALSE)
27	19	0.000000				=POISSON(A27, E4, FALSE)
28	20	0.000000				=POISSON(A28, E4, FALSE)

FIGURE 5.6
Minitab Poisson Distribution Calculations for $\lambda = 3$

Poisson with mean = 3

x	P(X = x)
0	0.049787
1	0.149361
2	0.224042
3	0.224042
4	0.168031
5	0.100819
6	0.050409
7	0.021604
8	0.008102
9	0.002701
10	0.000810
11	0.000221
12	0.000055
13	0.000013
14	0.000003
15	0.000001

EXAMPLE 5.5 COMPUTING POISSON PROBABILITIES

The number of faults per month that arise in the gearboxes of buses is known to follow a Poisson distribution with a mean of 2.5 faults per month. What is the probability that in a given month no faults are found? At least one fault is found?

SOLUTION Using Equation (5.14) on page 175 with $\lambda = 2.5$ (or using Table E.7, Microsoft Excel, or Minitab), the probability that in a given month no faults are found is

$$P(X=0) = \frac{e^{-2.5}(2.5)^0}{0!} = \frac{1}{(2.71828)^{2.5}(1)} = 0.0821$$

$$\begin{aligned} P(X \geq 1) &= 1 - P(X=0) \\ &= 1 - 0.0821 \\ &= 0.9179 \end{aligned}$$

The probability that there will be no faults in a given month is 0.0821. The probability that there will be at least one fault is 0.9179.

PROBLEMS FOR SECTION 5.4

Learning the Basics

PH Grade ASSIST **5.30** Assume a Poisson distribution.

a. If $\lambda = 2.5$, find $P(X = 2)$.
b. If $\lambda = 8.0$, find $P(X = 8)$.
c. If $\lambda = 0.5$, find $P(X = 1)$.
d. If $\lambda = 3.7$, find $P(X = 0)$.

PH Grade ASSIST **5.31** Assume a Poisson distribution.

a. If $\lambda = 2.0$, find $P(X \geq 2)$.
b. If $\lambda = 8.0$, find $P(X \geq 3)$.
c. If $\lambda = 0.5$, find $P(X \leq 1)$.
d. If $\lambda = 4.0$, find $P(X \geq 1)$.
e. If $\lambda = 5.0$, find $P(X \leq 3)$.

5.32 Assume a Poisson distribution with $\lambda = 5.0$. What is the probability that

a. $X = 1$.
b. $X < 1$.
c. $X > 1$.
d. $X \leq 1$.

Applying the Concepts

Problems 5.33– 5.43 can be solved manually or by using Microsoft Excel or Minitab.

5.33 Assume that the number of network errors experienced in a day on a local area network (LAN) is distributed as a Poisson random variable. The mean number of network errors experienced in a day is 2.4. What is the probability that in any given day

a. zero network errors will occur?
b. exactly one network error will occur?
c. two or more network errors will occur?
d. fewer than three network errors will occur?

SELF Test **5.34** The quality control manager of Marilyn's Cookies is inspecting a batch of chocolate-chip cookies that has just been baked. If the production process is in control, the mean number of chip parts per cookie is 6.0. What is the probability that in any particular cookie being inspected

a. fewer than five chip parts will be found?
b. exactly five chip parts will be found?
c. five or more chip parts will be found?
d. either four or five chip parts will be found?

5.35 Refer to problem 5.34. How many cookies in a batch of 100 should the manager expect to discard if company policy requires that all chocolate-chip cookies sold must have at least four chocolate-chip parts?

5.36 The U.S. Department of Transportation maintains statistics for mishandled bags per 1,000 passengers. In 2003 Jet Blue had 3.21 mishandled bags per 1,000 passengers. What is the probability that in the next 1,000 passengers Jet Blue will have

a. no mishandled bags?
b. at least one mishandled bag?
c. at least two mishandled bags?
d. Compare the results in (a) through (c) to those of Delta in problem 5.37 (a) through (c).

5.37 The U.S. Department of Transportation maintains statistics for mishandled bags per 1,000 passengers. In 2003 Delta had 3.84 mishandled bags per 1,000 passengers. What is the probability that in the next 1,000 passengers Delta will have

a. no mishandled bags?
b. at least one mishandled bag?
c. at least two mishandled bags?
d. Compare the results in (a) through (c) to those of Jet Blue in problem 5.36 (a) through (c).

PH Grade ASSIST **5.38** Based on past experience, it is assumed that the number of flaws per foot in rolls of grade 2 paper follows a Poisson distribution with a mean of 1 flaw per 5 feet of paper (0.2 flaw per foot). What is the probability that in a

a. 1-foot roll there will be at least 2 flaws?
b. 12-foot roll there will be at least 1 flaw?
c. 50-foot roll there will be between 5 and 15 (inclusive) flaws?

5.39 J.D. Power & Associates calculates and publishes various statistics concerning car quality. The Initial Quality score measures the number of problems per new car sold. For 2003 model cars, the Lexus was the top brand with 1.63 problems per car. Korea's Kia came in last with 5.09 problems per car (L. Hawkins, "Finding a Car That's Built to Last?" *The Wall Street Journal*, July 9, 2003, D1, D5). Let the random variable X be equal to the number of problems with a newly purchased Lexus.

a. What assumptions must be made in order for X to be distributed as a Poisson random variable? Are these assumptions reasonable?

Making the assumptions as in (a), if you purchased a 2003 Lexus, what is the probability that the new car will have:

b. zero problems?
c. two or fewer problems?
d. Give an operational definition for "problem." Why is the operational definition important in interpreting the Initial Quality score?

5.40 Refer to problem 5.39. If you purchased a 2003 Kia, what is the probability that the new car will have:

a. zero problems?
b. two or fewer problems?
c. Compare your answers in (a) and (b) to those for the Lexus in problem 5.39 (b) and (c).

5.41 In 2004, both Lexus and Kia improved their performance (D. Hakim, "Hyundai Near Top of a Quality Ranking," *The New York Times*, April 29, 2004, C.8). Lexus had 0.87 problems per car and Korea's Kia had 1.53 problems per car. If you purchased a 2004 Lexus, what is the probability that the new car will have:

a. zero problems?
b. two or fewer problems?
c. compare your answers in (a) and (b) to those of the 2003 Lexus in problem 5.39 (b) and (c).

5.42 Refer to problem 5.41. If you purchased a 2004 Kia, what is the probability that the new car will have:

a. zero problems?
b. two or fewer problems?
c. Compare your answers in (a) and (b) to those for the 2003 Kia in problem 5.40 (a) and (b).

5.43 A toll-free phone number is available from 9 A.M. to 9 P.M. for your customers to register a complaint with a product purchased from your company. Past history indicates that an average 0.4 calls are received per minute.

a. What properties must be true about the situation described above in order to use the Poisson distribution to calculate probabilities concerning the number of phone calls received in a 1-minute period?

Assuming that this situation matches the properties you discussed in (a), what then is the probability that during a 1- minute period:

b. zero phone calls will be received?
c. three or more phone calls will be received?
d. What is the maximum number of phone calls that will be received in a 1-minute period 99.99% of the time?

5.5 HYPERGEOMETRIC DISTRIBUTION

Both the binomial distribution and the **hypergeometric distribution** are concerned with the number of successes in a sample containing n observations. These two discrete probability distributions differ in the way in which you get the data. For the binomial distribution, the sample data are drawn *with* replacement from a *finite* population or *without* replacement from an *infinite* population. Thus, the probability of success p is constant over all observations and the outcome of any particular observation is independent of any other. For the hypergeometric distribution, the sample data are drawn *without* replacement from a *finite* population. Thus, the outcome of one observation is dependent on the outcomes of the previous observations.

Consider a population of size N. Let A represent the total number of successes in the population. The hypergeometric distribution is then used to find the probability of X successes in a sample of size n selected without replacement. Equation (5.15) presents a mathematical expression of the hypergeometric distribution for finding X successes, given a knowledge of n, N, and A.

HYPERGEOMETRIC DISTRIBUTION

$$P(X) = \frac{\binom{A}{X}\binom{N-A}{n-X}}{\binom{N}{n}} \quad \textbf{(5.15)}$$

where $P(X)$ = the probability of X successes, given knowledge of n, N, and A

n = sample size

N = population size

A = number of successes in the population

$N - A$ = number of failures in the population

X = number of successes in the sample

$\binom{A}{X} = {}_AC_X$ (see p. 168)

The number of successes in the sample, represented by X, cannot be greater than the number of successes in the population A or the sample size n. Thus, the range of the hypergeometric random variable is limited to the sample size or to the number of successes in the population, whichever is smaller.

The Mean

Equation (5.16) defines the mean of the hypergeometric distribution.

THE MEAN OF THE HYPERGEOMETRIC DISTRIBUTION

$$\mu = E(X) = \frac{nA}{N} \quad \textbf{(5.16)}$$

The Standard Deviation

Equation (5.17) defines the standard deviation of the hypergeometric distribution.

THE STANDARD DEVIATION OF THE HYPERGEOMETRIC DISTRIBUTION

$$\sigma = \sqrt{\frac{nA(N-A)}{N^2}} \cdot \sqrt{\frac{N-n}{N-1}} \quad \textbf{(5.17)}$$

In Equation (5.17), the expression $\sqrt{\frac{N-n}{N-1}}$ is a **finite population correction factor** that results from sampling without replacement from a finite population.

To illustrate the hypergeometric distribution, suppose that you are forming a team of 8 executives from different departments within your company. Your company has a total of 30 executives, and 10 of these people are from the finance department. If members of the team are to be selected at random, what is the probability that the team will contain 2 executives from the finance department? Here the population of $N = 30$ executives within the company is finite. In addition, $A = 10$ are from the finance department. A team of $n = 8$ members is to be selected.

Using Equation (5.15),

$$P(X = 2) = \frac{\binom{10}{2}\binom{20}{6}}{\binom{30}{8}}$$

$$= \frac{\dfrac{10!}{2!(8)!} \times \dfrac{(20)!}{(6)!(14)!}}{\dfrac{30!}{8!(22)!}}$$

$$= 0.298$$

Thus, the probability that the team will contain two members from the finance department is 0.298, or 29.8%.

Such computations can become tedious, especially as N gets large. However, you can compute the probabilities by using Microsoft Excel or Minitab. Figure 5.7 presents a Microsoft Excel worksheet and Figure 5.8 presents Minitab output for the team-formation example. Note that the number of executives from the finance department (i.e., the number of successes in the sample) can be equal to 0, 1, 2, . . . , 8.

FIGURE 5.7

Microsoft Excel Worksheet for the Team Member Example

	A	B	
1	Team Formation Analysis		
2			
3	Data		
4	Sample size	8	
5	No. of successes in population	10	
6	Population size	30	
7			
8	Hypergeometric Probabilities Table		
9	X	P(X)	
10	0	0.0215	=HYPGEOMDIST(A10, B4, B5, B6)
11	1	0.1324	=HYPGEOMDIST(A11, B4, B5, B6)
12	2	0.2980	=HYPGEOMDIST(A12, B4, B5, B6)
13	3	0.3179	=HYPGEOMDIST(A13, B4, B5, B6)
14	4	0.1738	=HYPGEOMDIST(A14, B4, B5, B6)
15	5	0.0491	=HYPGEOMDIST(A15, B4, B5, B6)
16	6	0.0068	=HYPGEOMDIST(A16, B4, B5, B6)
17	7	0.0004	=HYPGEOMDIST(A17, B4, B5, B6)
18	8	0.0000	=HYPGEOMDIST(A18, B4, B5, B6)

FIGURE 5.8

Minitab Output for the Team Member Example

```
Hypergeometric with N = 30, M = 10, and n = 8

x  P( X = x )
0    0.021523
1    0.132447
2    0.298005
3    0.317872
4    0.173836
5    0.049083
6    0.006817
7    0.000410
8    0.000008
```

PROBLEMS FOR SECTION 5.5

Learning the Basics

PH Grade ASSIST **5.44** Determine the following:

a. If $n = 4$, $N = 10$, and $A = 5$, find $P(X = 3)$.

b. If $n = 4$, $N = 6$, and $A = 3$, find $P(X = 1)$.

c. If $n = 5$, $N = 12$, and $A = 3$, find $P(X = 0)$.

d. If $n = 3$, $N = 10$, and $A = 3$, find $P(X = 3)$.

PH Grade ASSIST **5.45** Referring to problem 5.44, compute the mean and standard deviation for the hypergeometric distributions described in (a) through (d).

Applying the Concepts

Problems 5.46–5.50 can be solved manually or by using Microsoft Excel or Minitab.

PH Grade ASSIST SELF Test **5.46** An auditor for the Internal Revenue Service is selecting a sample of 6 tax returns for an audit. If 2 or more of these returns are "improper," the entire population of 100 tax returns will be audited. What is the probability that the entire population will be audited if the true number of improper returns in the population is

a. 25?

b. 30?

c. 5?

d. 10?

e. Discuss the differences in your results depending on the true number of improper returns in the population.

5.47 The dean of a business school wishes to form an executive committee of 5 from among the 40 tenured faculty members at the school. The selection is to be random, and at the school there are 8 tenured faculty members in accounting. What is the probability that the committee will contain

a. none of them?

b. at least 1 of them?

c. not more than 1 of them?

d. What is your answer to (a) if the committee consisted of 7 members?

5.48 From an inventory of 48 cars being shipped to local automobile dealers, 12 have had defective radios installed. What is the probability that if eight cars are shipped to a particular dealership

a. all eight will have defective radios?

b. none will have defective radios?

c. at least 1 will have a defective radio?

d. What would be your answers to (a) through (c) if 6 cars have had defective radios installed?

5.49 A state lottery is conducted in which 6 winning numbers are selected from a total of 54 numbers. What is the probability that if 6 numbers are randomly selected,

a. all 6 numbers will be winning numbers?

b. 5 numbers will be winning numbers?

c. none of the numbers will be winning numbers?

d. What are your answers to (a) through (c) if the 6 winning numbers were selected from a total of 40 numbers?

5.50 In a shipment of 15 hard disks, 5 are defective. If 4 of the disks are inspected, what is the probability that

a. exactly 1 is defective?

b. at least 1 is defective?

c. no more than 2 are defective?

d. What is the mean number of defective hard disks that you would expect to find in the sample of 4 hard disks?

5.6 (*CD-ROM TOPIC*) USING THE POISSON DISTRIBUTION TO APPROXIMATE THE BINOMIAL DISTRIBUTION

section 5.6.pdf

Under certain circumstances the Poisson distribution can be used to approximate the binomial distribution. To study this topic, go to the **section 5.6.pdf** file located on the CD-ROM that accompanies this book.

SUMMARY

In this chapter you studied mathematical expectation, the covariance, and the development and application of the binomial, Poisson, and hypergeometric distributions. In the "Using Statistics" scenario, you learned how to calculate probabilities from the binomial distribution concerning the observation of tagged invoices in the accounting information system used by the Saxon Home Improvement Company. In the following chapter, important continuous distributions such as the normal distribution will be developed.

To help decide what probability distribution to use for a particular situation, you need to ask the following questions:

- Is there a fixed number of observations n, each of which is classified as success or failure; or is there an area of opportunity? If there is a fixed number of observations n, each of which is classified as success or failure, you use the binomial or hypergeometric distribution. If there is an area of opportunity, you use the Poisson distribution.
- In deciding whether to use the binomial or hypergeometric distribution, is the probability of success constant over all trials? If yes, you can use the binomial distribution. If no, you can use the hypergeometric distribution.

KEY FORMULAS

Expected Value μ of a Discrete Random Variable

$$\mu = E(X) = \sum_{i=1}^{N} X_i P(X_i) \quad \textbf{(5.1)}$$

Variance of a Discrete Random Variable

$$\sigma^2 = \sum_{i=1}^{N} [X_i - E(X)]^2 P(X_i) \quad \textbf{(5.2)}$$

Standard Deviation of a Discrete Random Variable

$$\sigma = \sqrt{\sigma^2} = \sqrt{\sum_{i=1}^{N} [X_i - E(X)]^2 P(X_i)} \quad \textbf{(5.3)}$$

Covariance

$$\sigma_{XY} = \sum_{i=1}^{N} [X_i - E(X)][Y_i - E(Y)]P(X_iY_i) \quad \textbf{(5.4)}$$

Expected Value of the Sum of Two Random Variables

$$E(X + Y) = E(X) + E(Y) \quad \textbf{(5.5)}$$

Variance of the Sum of Two Random Variables

$$Var(X + Y) = \sigma^2_{X+Y} = \sigma^2_X + \sigma^2_Y + 2\sigma_{XY} \quad \textbf{(5.6)}$$

Standard Deviation of the Sum of Two Random Variables

$$\sigma_{X+Y} = \sqrt{\sigma^2_{X+Y}} \quad \textbf{(5.7)}$$

Portfolio Expected Return

$$E(P) = wE(X) + (1 - w)E(Y) \quad \textbf{(5.8)}$$

Portfolio Risk

$$\sigma_p = \sqrt{w^2\sigma^2_X + (1 - w)^2\sigma^2_Y + 2w(1 - w)\sigma_{XY}} \quad \textbf{(5.9)}$$

Combinations

$${}_nC_X = \frac{n!}{X!(n - X)!} \quad \textbf{(5.10)}$$

Binomial Distribution

$$P(X) = \frac{n!}{X!(n - X)!} p^X (1 - p)^{n-X} \quad \textbf{(5.11)}$$

The Mean of the Binomial Distribution

$$\mu = E(X) = np \quad \textbf{(5.12)}$$

The Standard Deviation of the Binomial Distribution

$$\sigma = \sqrt{\sigma^2} = \sqrt{Var(X)} = \sqrt{np(1 - p)} \quad \textbf{(5.13)}$$

Poisson Distribution

$$P(X) = \frac{e^{-\lambda}\lambda^X}{X!} \quad \textbf{(5.14)}$$

Hypergeometric Distribution

$$P(X) = \frac{\binom{A}{X}\binom{N - A}{n - X}}{\binom{N}{n}} \quad \textbf{(5.15)}$$

The Mean of the Hypergeometric Distribution

$$\mu = E(X) = \frac{nA}{N} \quad \textbf{(5.16)}$$

The Standard Deviation of the Hypergeometric Distribution

$$\sigma = \sqrt{\frac{nA(N - A)}{N^2}} \cdot \sqrt{\frac{N - n}{N - 1}} \quad \textbf{(5.17)}$$

KEY TERMS

CHAPTER REVIEW PROBLEMS

Checking Your Understanding

5.51 What is the meaning of the expected value of a probability distribution?

5.52 What are the four properties of a situation that must be present in order to use the binomial distribution?

5.53 What are the four properties of a situation that must be present in order to use the Poisson distribution?

5.54 When do you use the hypergeometric distribution instead of the binomial distribution?

Applying the Concepts

Problems 5.55–5.72 can be solved manually or by using Microsoft Excel or Minitab.

5.55 Event insurance allows promoters of sporting and entertainment events to protect themselves from financial losses due to uncontrollable circumstances such as rain-outs. For example, each spring Cincinnati's Downtown Council puts on the Taste of Cincinnati. This is a rainy time of year in Cincinnati, and the chance of receiving an inch or more of rain during a spring weekend is about one out of four. An article in the *Cincinnati Enquirer*, by Jim Knippenberg ("Chicken Pox Means 3 Dog Night Remedy," *Cincinnati Enquirer*, May 28, 1997, E1), gave the details for an insurance policy purchased by the Downtown Council. The policy would pay \$100,000 if it rained more than an inch during the weekend festival. The cost of the policy was reported to be \$6,500.

a. Determine whether or not you believe that these dollar amounts are correct. (*Hint:* Calculate the expected value of the profit to be made by the insurance company.)

b. Assume that the dollar amounts are correct. Is this policy a good deal for Cincinnati's Downtown Council?

5.56 From 1872 to 2000, stock prices have risen in 74% of the years (Mark Hulbert, "The Stock Market Must Rise in 2002? Think Again," *The New York Times*, December 6, 2001, Business, 6). Based on this information, and assuming a binomial distribution, what do you think the probability is that the stock market will rise

a. next year?

b. the year after next?

c. in four of the next five years?

d. in none of the next five years?

e. For this situation, what assumption of the binomial distribution might not be valid?

5.57 The mean cost of a phone call handled by an automated customer-service system is \$0.45. The mean cost of a phone call passed on to a "live" operator is \$5.50. However, as more and more companies have implemented automated systems, customer annoyance with such systems has grown. Many customers are quick to leave the automated system when given an option such as "Press zero to talk to a customer-service representative." According to the Center for Client Retention, 40% of all callers to automated customer-service systems will automatically opt to go to a live operator when given the chance (Jane Spencer, "In Search of the Operator," *The Wall Street Journal*, May 8, 2002, D1).

If 10 independent callers contact an automated customer-service system, what is the probability

a. zero will automatically opt to talk to a live operator?

b. exactly one will automatically opt to talk to a live operator?

c. two or fewer will automatically opt to talk to a live operator?

d. all ten will automatically opt to talk to a live operator?

e. If all ten automatically opt to talk to a live operator, do you think that the 40% figure given in the article applies to this particular system? Explain.

5.58 One theory concerning the Dow Jones Industrial Average is that it is likely to increase during U.S. presidential election years. From 1964 through 2000, the Dow Jones Industrial Average has increased in eight of the ten U.S. presidential election years. Assuming that this indicator is a random event with no predictive value, you would expect that the indicator would be correct 50% of the time. What is the probability of the Dow Jones Industrial Average increasing in eight or more of the ten U.S. presidential election years if the true probability of an increase in the Dow Jones Industrial Average is

a. 0.50?

b. 0.70?

c. 0.90?

d. Based on the results of (a) through (c), what do you think is the probability that the Dow Jones Industrial Average will increase in a U.S. presidential election year?

5.59 Priority Mail is the United States Postal Service's alternative to commercial express mail companies like Federal Express. An article in the *Wall Street Journal* presents some interesting conclusions comparing Priority Mail shipments with the much cheaper first-class shipments (Rick Brooks, "New Data Reveal 'Priority Mail' Is Slower Than a Stamp," *The Wall Street Journal*, May 29, 2002, D1). When comparing shipments intended for delivery in 3 days, first-class deliveries failed to deliver on-time 19% of the time, while Priority Mail failed 33% of the time. Note that at the time of the article, first-class deliveries started as low as \$0.34 and Priority Mail started at \$3.50.

If 10 items are to be shipped first-class to 10 different destinations claimed to be in a 3-day delivery location, what is the probability that

a. zero items will take more than 3 days?
b. exactly one will take more than 3 days?
c. two or more will take more than 3 days?
d. What are the mean and the standard deviation of the probability distribution?

5.60 Refer to problem 5.59. If the shipments are made using Priority Mail, what is the probability that

a. zero items will take more than 3 days?
b. exactly one will take more than 3 days?
c. two or more will take more than 3 days?
d. What are the mean and the standard deviation of the probability distribution?
e. Compare the results of (a) through (c) to those of problem 5.59 (a) through (c).

5.61 Cinema advertising is increasing. Normally 60 to 90 seconds long, these advertisements are longer and more extravagant, and tend to have more captive audiences than television advertisements. Thus, it is not surprising that the recall rates for viewers of cinema advertisements are higher than for television advertisements. According to survey research conducted by the ComQUEST division of BBM Bureau of Measurement in Toronto, the probability a viewer will remember a cinema advertisement is 0.74, whereas the probability a viewer will remember a 30-second television advertisement is 0.37 (Nate Hendley, "Cinema Advertising Comes of Age," *Marketing Magazine*, May 6, 2002, 16).

a. Is the 0.74 probability reported by the BBM Bureau of Measurement best classified as *a priori* classical probability, empirical classical probability, or subjective probability?
b. Suppose that 10 viewers of a cinema advertisement are randomly sampled. Consider the random variable defined by the number of viewers that recall the advertisement. What assumptions must be made in order to assume that this random variable is distributed as a binomial random variable?
c. Assuming that the number of viewers that recall the cinema advertisement is a binomial random variable, what are the mean and standard deviation of this distribution?
d. Based on your answer to (c), if none of the viewers can recall the ad, what can be inferred about the 0.74 probability given in the article?

5.62 Refer to problem 5.61. Compute the probability that of the 10 viewers:

a. exactly zero can recall the advertisement.
b. all 10 can recall the advertisement.
c. more than half can recall the advertisement.
d. eight or more can recall the advertisement.

5.63 Refer to problem 5.61. For a television advertisement using the given probability of recall, 0.37, compute the probability that of the 10 viewers:

a. exactly zero can recall the advertisement.
b. all 10 can recall the advertisement.
c. more than half can recall the advertisement.
d. eight or more can recall the advertisement.
e. Compare the results of (a) through (d) to those of problem 5.62 (a) through (d).

5.64 In a survey conducted by the Council for Marketing and Opinion Research (CMOR), a national nonprofit research industry trade group based in Cincinnati, 1,628 of 3,700 adults contacted in the United States refuse to participate in phone surveys (Steve Jarvis, "CMOR Finds Survey Refusal Rate Still Rising," *Marketing News*, February 4, 2002, 4). Suppose that you are to randomly call 10 adults in the United States and ask them to participate in a phone survey. Using the results of the CMOR study, what is the probability:

a. all 10 will refuse?
b. exactly 5 will refuse?
c. at least 5 will refuse?
d. less than 5 will refuse?
e. less than 5 will agree to be surveyed?
f. What is the expected number of people that will refuse to participate? Explain the practical meaning of this number.

5.65 Credit card companies are increasing their revenues by raising the late fees charged to their customers. According to a study by **cardweb.com**, late fees represent the third largest revenue source for card companies after interest charges and payments from the merchants who accept their cards. In the last year, 58% of all credit card customers had to pay a late fee (Ron Lieber, "Credit-Card Firms Collect Record Levels of Late Fees," *The Wall Street Journal*, May 21, 2002, D1).

If a random sample of 20 credit card holders is selected, what is the probability that

a. zero had to pay a late fee?
b. no more than 5 had to pay a late fee?
c. more than 10 had to pay a late fee?
d. What assumptions did you have to make to answer (a) through (c)?

5.66 For e-commerce merchants, getting a customer to visit a Web site isn't enough. Merchants must also persuade online shoppers to spend money by completing a purchase. Experts at Andersen Consulting estimated that 88% of Web shoppers abandon their virtual shopping carts before completing their transaction (Rebecca Quick, "The Lessons Learned," *The Wall Street Journal*, April 17, 2000, R6). Consider a sample of 20 customers who visit an e-commerce Web site, and assume that the probability a customer will leave the site before completing their transaction is 0.88. Use the binomial model to answer the following questions:

a. What is the expected value, or mean, of the binomial distribution?
b. What is the standard deviation of the binomial distribution?
c. What is the probability that all 20 of the customers will leave the site without completing a transaction?
d. What is the probability that 18 or more of the customers will leave the site without completing a transaction?
e. What is the probability that 15 or more of the customers will leave the site without completing a transaction?

5.67 Refer to problem 5.66. If the Web site is enhanced so that only 70% of the customers will leave the site without completing their transaction,

a. what is the expected value, or mean, of the binomial distribution?
b. what is the standard deviation of the binomial distribution?
c. what is the probability that all 20 of the customers will leave the site without completing a transaction?
d. what is the probability that 18 or more of the customers will leave the site without completing a transaction?
e. what is the probability that 15 or more of the customers will leave the site without completing a transaction?
f. Compare the results of (a) through (e) to those of problem 5.66 (a) through (e).

5.68 One theory concerning the Standard & Poor's 500 index is that if it increases during the first five trading days of the year, it is likely to increase during the entire year. From 1950 through 2003, the Standard & Poor's 500 index had these early gains in 34 years. In 29 of these 34 years, the Standard & Poor's 500 index increased. Assuming that this indicator is a random event with no predictive value, you would expect that the indicator would be correct 50% of the time. What is the probability of the Standard & Poor's 500 index increasing in 29 or more years if the true probability of an increase in the Standard & Poor's 500 index is

a. 0.50?
b. 0.70?
c. 0.90?
d. Based on the results of (a) through (c), what do you think is the probability that the Standard & Poor's 500 index will increase if there is an early gain in the first five trading days of the year? Explain.

5.69 Spurious correlation refers to the apparent relationship between variables that either have no true relationship or are related to other variables that have not been measured. One widely publicized stock market indicator in the United States that is an example of spurious correlation is the relationship between the winner of the National Football League Superbowl and the performance of the Dow Jones Industrial Average in that year. The indicator states that when a team representing the National Football Conference wins the Superbowl, the Dow Jones Industrial Average will increase in that year. When a team representing the American Football Conference wins the Superbowl, the Dow Jones Industrial Average will decline in that year. During the time period 1967 to 2003, a 37-year period, the indicator has been correct 31 out of 37 times. Assuming that this indicator is a random event with no predictive value, you would expect that the indicator would be correct 50% of the time.

a. What is the probability that the indicator would be correct 31 or more times in 37 years?
b. What does this tell you about the usefulness of this indicator?

5.70 Worldwide golf ball sales total more than $1 billion annually. One reason for such a large number of golf ball purchases is that golfers lose them at a rate of 4.5 per 18-hole round ("Snapshots," **usatoday.com**, January 29, 2004). Assume that the number of golf balls lost in an 18-hole round is distributed as a Poisson random variable.

a. What assumptions need to be made so that the number of golf balls lost in an 18-hole round is distributed as a Poisson random variable?

Making the assumptions given in (a), what is the probability that

b. 0 balls will be lost in an 18-hole round?
c. 5 or fewer balls will be lost in an 18-hole round?
d. 6 or more balls will be lost in an 18-hole round?

5.71 A study of the homepages for Fortune 500 companies reports that the mean number of bad links per homepage is 0.4 and the mean number of spelling errors per homepage is 0.16 (Nabil Tamimi, Murii Rajan, and Rose Sebastianella, "Benchmarking the Home Pages of 'Fortune' 500 Companies," *Quality Progress*, July 2000). Use the Poisson distribution to find the probability that a randomly selected homepage will contain

a. exactly 0 bad links.
b. 5 or more bad links.
c. exactly 0 spelling errors.
d. 10 or more spelling errors.

5.72 Mega Millions is one of the most popular lottery games in the United States. Participating states in Mega Millions are Georgia, Illinois, Maryland, Massachusetts, Michigan, New Jersey, New York, Ohio, and Virginia. Rules for playing and the list of prizes are given below ("Win Megamoney Playing Ohio's Biggest Jackpot Game," Ohio Lottery Headquarters, 2002):

Rules:

- Select five numbers from a pool of numbers from 1 to 52, and one Mega Ball number from a second pool of numbers from 1 to 52.
- Each wager costs $1.

Prizes:

- Match all five numbers + Mega Ball—win jackpot (minimum of $10,000,000)
- Match all five numbers—win $175,000
- Match four numbers + Mega Ball—win $5,000
- Match four numbers—win $150
- Match three numbers + Mega Ball—win $150
- Match two numbers + Mega Ball—win $10
- Match three numbers—win $7
- Match one number + Mega Ball—win $3
- Match Mega Ball—win $2

Find the probability of winning

a. the jackpot.
b. the $175,000 prize. Note that this requires matching all five numbers but not matching the Mega Ball.
c. $5,000.
d. $150.
e. $10.
f. $7.
g. $3.
h. $2.
i. nothing.
j. All stores selling Mega Millions tickets are required to have a brochure that gives complete game rules and probabilities of winning each prize (the probability of having a losing ticket is not given). The slogan for all lottery games in the state of Ohio is "Play Responsibly. Odds Are, You'll Have Fun." Do you think Ohio's slogan and the requirement of making available complete game rules and probabilities of winning is an ethical approach to running the lottery system?

RUNNING CASE
MANAGING THE *SPRINGVILLE HERALD*

The *Herald* marketing department is seeking to increase home-delivery sales through an aggressive direct-marketing campaign that includes mailings, discount coupons, and telephone solicitations. Feedback from these efforts indicates that getting their newspapers delivered early in the morning is a very important factor for both prospective as well as existing subscribers.

After several brainstorming sessions, a team consisting of members from the marketing and circulation departments decided that guaranteeing newspaper delivery by a specific time could be an important selling point in retaining and getting new subscribers. The team concluded that the *Herald* should offer a guarantee that customers will receive their newspapers by a certain time or else that day's issue is free.

To assist the team in setting a guaranteed delivery time, Al Leslie, the research director, noted that the circulation department had the data that would show the percentage of newspapers yet undelivered every quarter-hour from 6 A.M. to 8 A.M. Jan Shapiro remembered that customers were asked on their subscription forms at what time would they be looking for their copy of the *Herald* to be delivered. These data were subsequently combined and posted on an internal *Herald* Web page (see **Circulation_Data.htm** in the HeraldCase folder on the CD-ROM that accompanies this text or link to **www.prenhall.com/HeraldCase/Circulation_Data.htm**).

EXERCISES

Review the internal data and propose a reasonable time (to the nearest quarter-hour) to guarantee delivery. To help explore the effects of your choice, calculate the following probabilities:

SH5.1 If a sample of 50 customers is selected on a given day, what is the probability, given your selected delivery time, that:

a. fewer than 3 customers would receive a free newspaper?
b. 2, 3, or 4 customers would receive a free newspaper?
c. more than 5 customers would receive a free newspaper?

SH5.2 Consider the effects of improving the newspaper delivery process so that the percentage of newspapers that go undelivered by your guaranteed delivery time decreases by 2%. If a sample of 50 customers is selected on a given day, what is the probability, given your selected delivery time (and the delivery improvement), that:

a. fewer than 3 customers would receive a free newspaper?
b. 2, 3, or 4 customers would receive a free newspaper?
c. more than 5 customers would receive a free newspaper?

WEB CASE

Apply your knowledge about expected value and the covariance in this continuing Web Case from Chapters 3 and 4.

Visit the StockTout Bulls and Bears Web page **www.prenhall.com/Springville/ST_BullsandBears.htm**, read the claims, and examine the supporting data. Then, answer the following:

1. Are there any "catches" about the claims the Web site makes for the rate of return of Happy Bull and Worried Bear Funds?
2. What subjective data influence the rate-of-return analyses of these funds? Could StockTout be accused of making false and misleading statements? Why, or why not?
3. The expected-return analysis seems to show that the Worried Bear Fund has a greater expected return than the Happy Bull Fund. Should a rational investor then never invest in the Happy Bull Fund? Why, or why not?

REFERENCES

1. Bernstein, P. L., *Against the Gods: The Remarkable Story of Risk* (New York: Wiley, 1996).
2. Emery, D. R., and J. D. Finnerty, *Corporate Financial Management*, 2nd ed. (Upper Saddle River, NJ: Prentice Hall, 2000).
3. Kirk, R. L., ed., *Statistical Issues: A Reader for the Behavioral Sciences* (Belmont, CA: Wadsworth, 1972).
4. Levine, D. M., P. Ramsey, and R. Smidt, *Applied Statistics for Engineers and Scientists Using Microsoft Excel and Minitab* (Upper Saddle River, NJ: Prentice Hall, 2001).
5. Mescove, S. A., M. G. Simkin, and A. Barganoff, *Core Concepts of Accounting Information Systems*, 7th ed. (New York: John Wiley, 2001).
6. *Microsoft Excel 2003* (Redmond, WA: Microsoft Corp., 2002).
7. *Minitab for Windows Version 14* (State College, PA: Minitab Inc., 2004).

Appendix 5 Using Software for Discrete Probability Distributions

A5.1 MICROSOFT EXCEL

For Expected Value of a Discrete Random Variable

Open the **Expected Value.xls** file. This worksheet already contains the entries for the Table 5.1 home mortgage approval example on page 159 and uses the SUM and SQRT (square root) functions to calculate statistics. To adapt this worksheet to other problems:

- If you have more or less than seven outcomes, first add or delete table rows by selecting the cell range **A5:E5** and then either **Insert → Cells** or **Edit → Delete**. (If a box of options appears, choose **Shift cells down** if adding cells or choose **Shift cells up** if deleting.)
- If adding rows, copy the formulas in cell range C4:E4 down through the new table rows.
- Enter a corrected list of X values in column A starting with **1** in cell A5.
- Enter the new values for P(X) in column B.

For Portfolio Expected Return and Portfolio Risk

Open the **Portfolio.xls** file. This worksheet already contains the entries for the investment data of Table 5.4 on page 163. To adapt this worksheet to other problems:

- If you have more or less than three outcomes, first add or delete table rows by selecting **row 5** and then either **Insert → Rows** or **Edit → Delete**.
- If adding rows, copy the formulas in cell range F4:J4 down through the new table rows.

- Enter the new probability and outcome values and the new **Weight Assigned to X** values in the tinted cells near the top of the worksheet.

OR See section G.6 (**Covariance and Portfolio Analysis**) if you want PHStat2 to produce this worksheet for you.

For Binomial Probabilities

Open the **Binomial.xls** Excel file, shown in Figure 5.2 on page 171. This worksheet already contains the entries for the tagged orders example used in section 5.3. The worksheet uses the BINOMDIST function to calculate binomial probabilities (see section G.7 for additional information).

To adapt this worksheet to other problems:

- If you have more or less than three outcomes, first add or delete table rows by selecting **row 15** and then either **Insert → Rows** or **Edit → Delete**.
- If adding rows, copy the entries in cell range A14:B14 down through the entire table to update the table.
- Enter the new sample size and probability of success values in cells B4 and B5.

OR See section G.7 (**Binomial**) if you want PHStat2 to produce this worksheet for you.

For Poisson Probabilities

Open the **Poisson.xls** Excel file, shown in Figure 5.5 on page 177. This worksheet already contains the entries for the bank arrivals example used in section 5.4. The worksheet uses the POISSON function to calculate Poisson probabilities (see section G.8 for additional information).

To adapt this worksheet to other problems:

- If you need more than 20 successes, first add table rows by selecting **row 9** and then **Insert → Rows**. Next, copy the entries in cell range A8:B8 down through the entire table to update the table.
- Enter the new **Average/Expected number of successes** value in cell E4.

OR See section G.8 (**Poisson**) if you want PHStat2 to produce this worksheet for you.

For Hypergeometric Probabilities

Open the **Hypergeometric.xls** Excel file, shown in Figure 5.7 on page 181. This worksheet already contains the entries for the team formation example used in section 5.5. The worksheet uses the HYPGEOMDIST function to calculate hypergeometric probabilities (see section G.9 for additional information).

To adapt this worksheet to other problems:

- If your sample size is more or less than 8, first add or delete table rows by selecting **row 11** and then either **Insert → Rows** or **Edit → Delete**.
- If adding rows, copy the entries in cell range A10:B10 down through the entire table to update the table.
- Enter the new values for **Sample size, No. of successes in population**, and **Population size** in the cell range B4:B6.

OR See section G.9 (**Hypergeometric**) if you want PHStat2 to produce this worksheet for you.

A5.2 MINITAB

You can use Minitab to find probabilities from the binomial, Poisson, and hypergeometric distributions.

Using Minitab to Calculate Binomial Probabilities

To illustrate the use of Minitab, consider the accounting information system discussed in section 5.3. To compute the results of Figure 5.3 on page 171,

1. Enter the values **0, 1, 2, 3**, and **4** in rows 1 to 5 of column C1.
2. Select **Calc → Probability Distributions → Binomial** to compute binomial probabilities. In the Binomial Distribution dialog box (see Figure A5.1), select the **Probability** option button to compute the probabilities of X successes for all values of X. In the Number of trials: edit box enter the sample size of 4. In the Probability of success: edit box enter **.10**. Select the **Input column:** option button and enter **C1** in the edit box. Click the **OK** button.

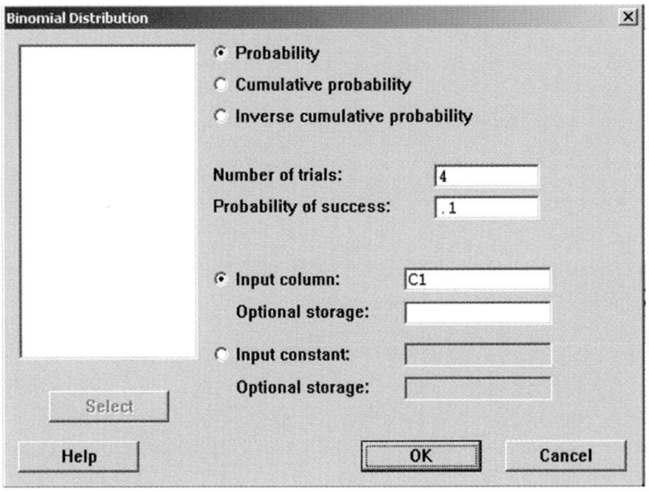

FIGURE A5.1 Minitab Binomial Distribution Dialog Box

Using Minitab to Calculate Poisson Probabilities

To illustrate how to calculate Poisson probabilities using Minitab, return to the bank customer arrival example of section 5.4. To compute the result shown in Figure 5.6 on page 177,

1. Enter the values **0** through **15** in rows 1 to 16 of column C1.
2. Select **Calc → Probability Distributions → Poisson** to compute Poisson probabilities. In the Poisson Distribution dialog box (see Figure A5.2 below), select the **Probability** option button to compute the exact probabilities of X successes for all values of X. In the Mean: edit box enter the λ value of **3**. Select the **Input column**: option button and enter **C1** in the edit box. Click the **OK** button.

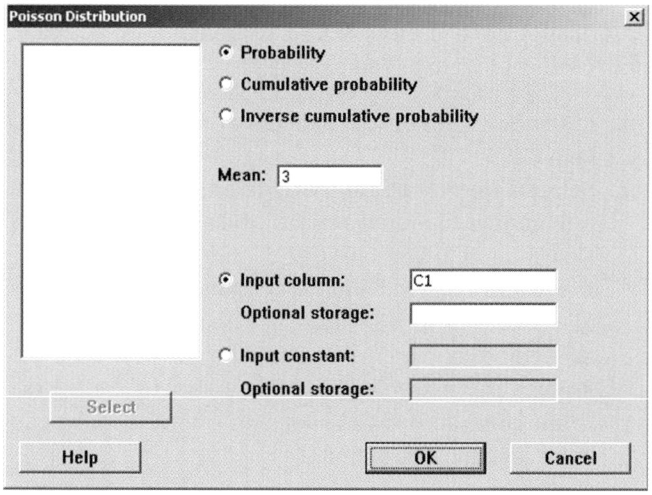

FIGURE A5.2 Minitab Poisson Distribution Dialog Box

Using Minitab to Calculate Hypergeometric Probabilities

To illustrate how to calculate hypergeometric probabilities, return to the team-formation example of section 5.5. You computed the probability of exactly 2 members from the finance department in a sample of 8, from a population of 30 executives where 10 are from the finance department, to be 0.2980. To compute this result, enter the values **0, 1, 2, 3, 4, 5, 6, 7,** and **8** in rows 1 to 9 of column C1. Select **Calc → Probability Distributions → Hypergeometric** to compute hypergeometric probabilities. In the Hypergeometric Distribution dialog box (see Figure A5.3), select the **Probability** option button. In the Population size (N): edit box enter the population size of **30**. In the Successes in population (M): edit box enter the value of **10**. In the Sample size (n): edit box, enter the sample size of **8**. Select the **Input column**: option button, and enter **C1** in its edit box. Click the **OK** button.

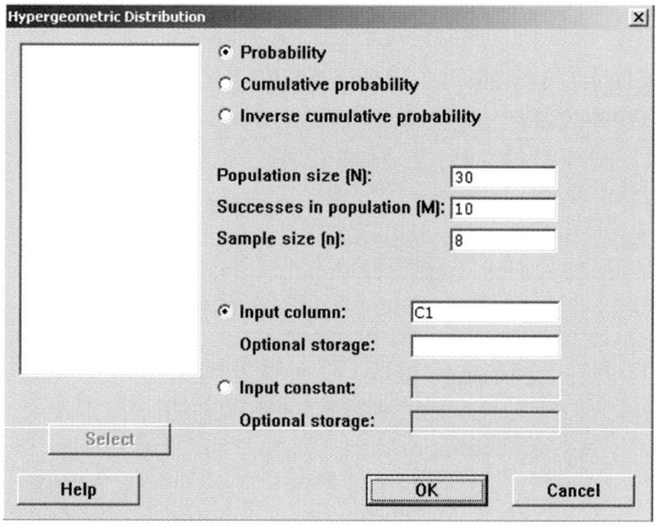

FIGURE A5.3 Minitab Hypergeometric Distribution Dialog Box

CHAPTER 6

The Normal Distribution and Other Continuous Distributions

USING STATISTICS: Download Time for a Web Site Homepage

LEARNING OBJECTIVES

In this chapter, you learn:

- To compute probabilities from the normal distribution
- To use the normal probability plot to determine whether a set of data is approximately normally distributed
- To compute probabilities from the uniform distribution
- To compute probabilities from the exponential distribution
- To compute probabilities from the normal distribution to approximate probabilities from the binomial distribution

USING STATISTICS

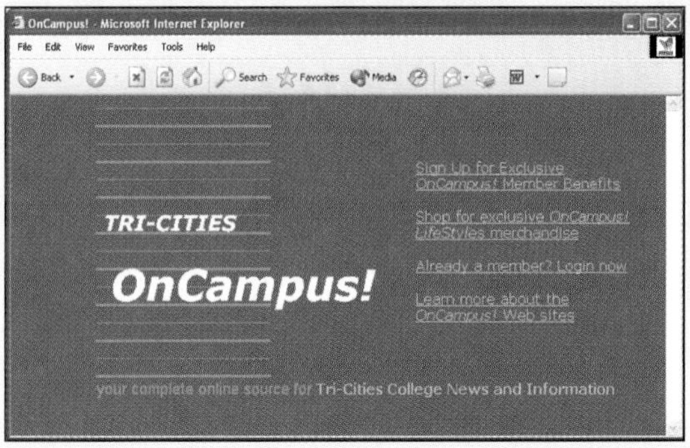

Download Time for a Web Site Homepage

You are a designer for the OnCampus! Web site that targets college students. To attract and retain users, you need to make sure that the homepage downloads quickly. Both the design of the homepage and the load on the company's Web server affect the download time. To check how fast the homepage loads, you open a Web browser on a PC at the corporate offices of OnCampus! and measure the download time, the number of seconds that passes from first linking to the Web site until the homepage is fully displayed.

Past data indicate that the mean download time is 7 seconds and that the standard deviation is 2 seconds. Approximately two-thirds of the download times are between 5 and 9 seconds, and about 95% of the download times are between 3 and 11 seconds. In other words, the download times are distributed as a bell-shaped curve with a clustering around the mean of 7 seconds. How could you use this information to answer questions about the download times of the current homepage?

In the last chapter, Saxon Home Improvement Company managers wanted to be able to solve problems about the number of occurrences of a certain type of outcome in a given sample size. As an OnCampus! Web designer, you face a different task, one that involves a continuous measurement because a download time could be any value and not just a whole number. How then can you answer questions about this *continuous numerical variable* such as:

- What proportion of the homepage downloads take more than 10 seconds?
- How many seconds elapse before 10% of the downloads are complete?
- How many seconds elapse before 99% of the downloads are complete?
- How would redesigning the homepage to download faster affect the answers to the above questions?

As in the previous chapter, you can use a probability distribution as a model. Reading this chapter will help you learn about characteristics of a continuous probability distribution and how to use the normal, uniform, and exponential distributions to solve business problems.

6.1 CONTINUOUS PROBABILITY DISTRIBUTIONS

A **continuous probability density function** is the mathematical expression that defines the distribution of the values for a continuous random variable. Figure 6.1 graphically displays the three continuous probability density functions discussed in this chapter. Panel A depicts a normal distribution. The normal distribution is symmetric and bell-shaped, implying that most values tend to cluster around the mean, which, due to its symmetric shape, is equal to the median. Although the values in a normal distribution can range from negative infinity to positive infinity, the shape of the distribution makes it very unlikely that extremely large or extremely small values will occur. Panel B depicts a uniform distribution where the probability of occurrence of a value is equally likely to occur anywhere in the range between the smallest value a and the largest value b. Sometimes referred to as the rectangular distribution, the uniform distribution is symmetric and therefore the mean equals the median. An exponential distribution is illustrated in panel C. This distribution is skewed to the right, making the mean larger than the median. The range for an exponential distribution is zero to positive infinity but its shape makes the occurrence of extremely large values unlikely.

FIGURE 6.1
Three Continuous Distributions

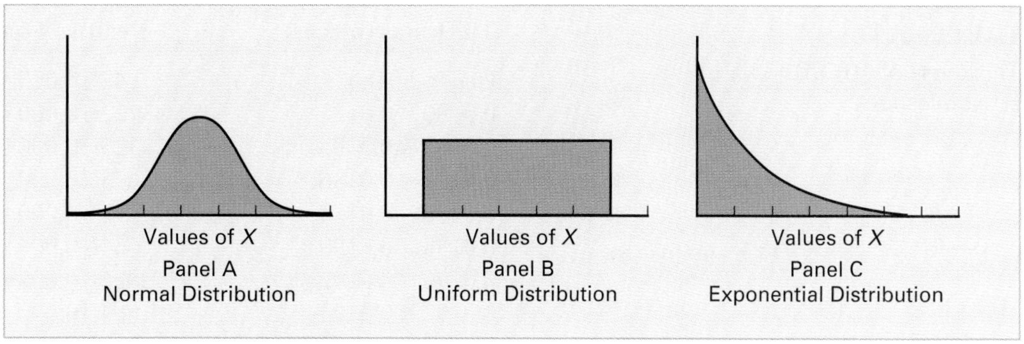

6.2 THE NORMAL DISTRIBUTION

The **normal distribution** (sometimes referred to as the *Gaussian* distribution) is the most common continuous distribution used in statistics. The normal distribution is vitally important in statistics for three main reasons:

- Numerous continuous variables common in the business world have distributions that closely resemble the normal distribution.
- The normal distribution can be used to approximate various discrete probability distributions.
- The normal distribution provides the basis for *classical statistical inference* because of its relationship to the *central limit theorem* (which is discussed in section 7.2).

The normal distribution is represented by the classic bell-shape depicted in panel A of Figure 6.1. In the normal distribution, you can calculate the probability that various values occur within certain ranges or intervals. However, the *exact* probability of a *particular value* from a continuous distribution such as the normal distribution is zero. This property distinguishes continuous variables, which are measured, from discrete variables, which are counted. As an example, time (in seconds) is measured and not counted. Therefore you can determine the probability that the download time for a home page on a Web browser is between 7 and 10 seconds or the probability that the download time is between 8 and 9 seconds or the probability that the download time is between 7.99 and 8.01 seconds. However, the probability that the download time is *exactly* 8 seconds is zero.

The normal distribution has several important theoretical properties:

- It is bell-shaped (and thus symmetrical) in its appearance.
- Its measures of central tendency (mean, median, and mode) are all identical.
- Its "middle fifty" is equal to 1.33 standard deviations. This means that the interquartile range is contained within an interval of two-thirds of a standard deviation below the mean to two-thirds of a standard deviation above the mean.
- Its associated random variable has an infinite range ($-\infty < X < \infty$).

In practice, many variables have distributions that closely resemble the theoretical properties of the normal distribution. The data in Table 6.1 represent the thickness (in inches) of 10,000 brass washers manufactured by a large company. The continuous variable of interest, thickness, can be approximated by the normal distribution. The measurements of the thickness of the 10,000 brass washers cluster in the interval 0.0190 to 0.0192 inch and distribute symmetrically around that grouping, forming a "bell-shaped" pattern. As demonstrated in Table 6.1, if the nonoverlapping (*mutually exclusive*) listing contains all possible class intervals (is *collectively exhaustive*), the probabilities will sum to 1. Such a probability distribution is a relative frequency distribution, as described in section 2.3, where, except for the two open-ended classes, the midpoint of every other class interval represents the data in that interval.

TABLE 6.1
Thickness of 10,000 Brass Washers

Thickness (inches)	Relative Frequency
Under 0.0180	48/10,000 = 0.0048
0.0180 < 0.0182	122/10,000 = 0.0122
0.0182 < 0.0184	325/10,000 = 0.0325
0.0184 < 0.0186	695/10,000 = 0.0695
0.0186 < 0.0188	1,198/10,000 = 0.1198
0.0188 < 0.0190	1,664/10,000 = 0.1664
0.0190 < 0.0192	1,896/10,000 = 0.1896
0.0192 < 0.0194	1,664/10,000 = 0.1664
0.0194 < 0.0196	1,198/10,000 = 0.1198
0.0196 < 0.0198	695/10,000 = 0.0695
0.0198 < 0.0200	325/10,000 = 0.0325
0.0200 < 0.0202	122/10,000 = 0.0122
0.0202 or above	48/10,000 = 0.0048
Total	1.0000

Figure 6.2 depicts the relative frequency histogram and polygon for the distribution of the thickness of 10,000 brass washers. For these data, the first three theoretical properties of the normal distribution are approximately satisfied; however, the fourth does not hold. The random variable of interest, thickness, cannot possibly be zero or below, and a washer cannot be so thick that it becomes unusable. From Table 6.1 above you see that only 48 out of every 10,000 brass washers manufactured are expected to have a thickness of 0.0202 inch or more, whereas an equal number are expected to have a thickness under 0.0180 inch. Thus, the chance of randomly getting a washer so thin or so thick is 0.0048 + 0.0048 = 0.0096—or less than 1 in 100.

FIGURE 6.2
Relative Frequency Histogram and Polygon of the Thickness of 10,000 Brass Washers

Source: *Data are taken from Table 6.1.*

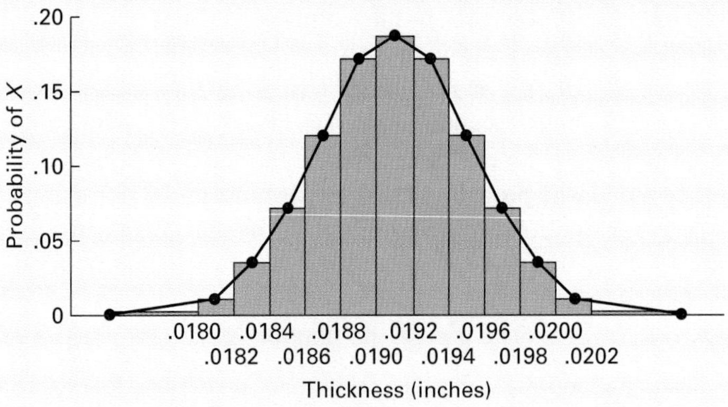

The mathematical expression representing a continuous probability density function is denoted by the symbol $f(X)$. For the normal distribution, the **normal probability density function** is given in Equation (6.1).

THE NORMAL PROBABILITY DENSITY FUNCTION

$$f(X) = \frac{1}{\sqrt{2\pi}\sigma} e^{-(1/2)[(X-\mu)/\sigma]^2} \quad \textbf{(6.1)}$$

where e is the mathematical constant approximated by 2.71828
π is the mathematical constant approximated by 3.14159
μ is the mean
σ is the standard deviation
X is any value of the continuous variable, where $(-\infty < X < \infty)$

Because e and π are mathematical constants, the probabilities of the random variable X are dependent only on the two parameters of the normal distribution—the mean μ and the standard deviation σ. Every time you specify a *particular combination* of μ and σ, a *different* normal probability distribution is generated. Figure 6.3 illustrates three different normal distributions. Distributions A and B have the same mean (μ) but have different standard deviations. Distributions A and C have the same standard deviation (σ) but have different means. Distributions B and C depict two normal probability density functions that differ with respect to both μ and σ.

FIGURE 6.3
Three Normal Distributions

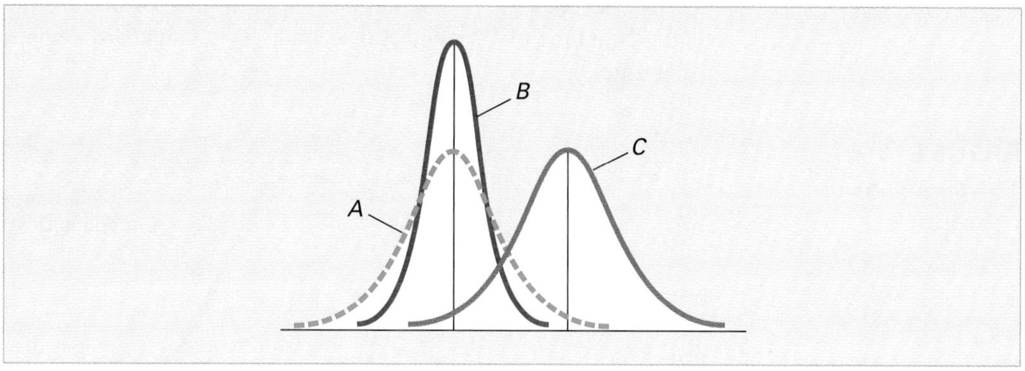

The mathematical expression in Equation (6.1) is computationally tedious and requires integral calculus. Fortunately, normal probability tables are available and you can avoid these complicated computations. The first step in finding normal probabilities is to use the **transformation formula**, given in Equation (6.2), to convert any normal random variable X to a **standardized normal random variable** Z.

THE TRANSFORMATION FORMULA

The Z value is equal to the difference between X and the mean μ, divided by the standard deviation σ.

$$Z = \frac{X - \mu}{\sigma} \quad \textbf{(6.2)}$$

Although the original data for the random variable X had mean μ and standard deviation σ, the standardized random variable Z will always have mean $\mu = 0$ and standard deviation $\sigma = 1$.

By substituting $\mu = 0$ and $\sigma = 1$ in Equation (6.1), the probability density function of a standardized normal variable Z is given in Equation (6.3).

THE STANDARDIZED NORMAL PROBABILITY DENSITY FUNCTION

$$f(Z) = \frac{1}{\sqrt{2\pi}} e^{-(1/2)Z^2} \quad \textbf{(6.3)}$$

Any set of normally distributed values can be converted to its standardized form. Then you can determine the desired probabilities using Table E.2, the **cumulative standardized normal distribution**.

To see how the transformation formula is applied and the results used to find probabilities from Table E.2[1], recall from the "Using Statistics" scenario on page 192 that past data indicate that the time to download the Web page is normally distributed with a mean $\mu = 7$ seconds and a standard deviation $\sigma = 2$ seconds. From Figure 6.4, you see that every measurement X has a corresponding standardized measurement Z computed from the transformation formula [Equation (6.2)]. Therefore, a download time of 9 seconds is equivalent to 1 standardized unit (i.e., 1 standard deviation above the mean) because

$$Z = \frac{9 - 7}{2} = +1$$

[1] *This text uses Table E.2, the Cumulative Standardized Normal table. To use the Standardized Normal table, see Table E.14 and section 6.1a, "Using the Standardized Normal Distribution Table," on the CD-ROM.*

A download time of 1 second is equivalent to 3 standardized units (3 standard deviations) below the mean because

$$Z = \frac{1-7}{2} = -3$$

Thus, the standard deviation is the unit of measurement. In other words, a time of 9 seconds is 2 seconds (i.e., 1 standard deviation) higher, or *slower*, than the mean time of 7 seconds. Similarly, a time of 1 second is 6 seconds (i.e., 3 standard deviations) lower, or *faster*, than the mean time.

FIGURE 6.4
Transformation of Scales

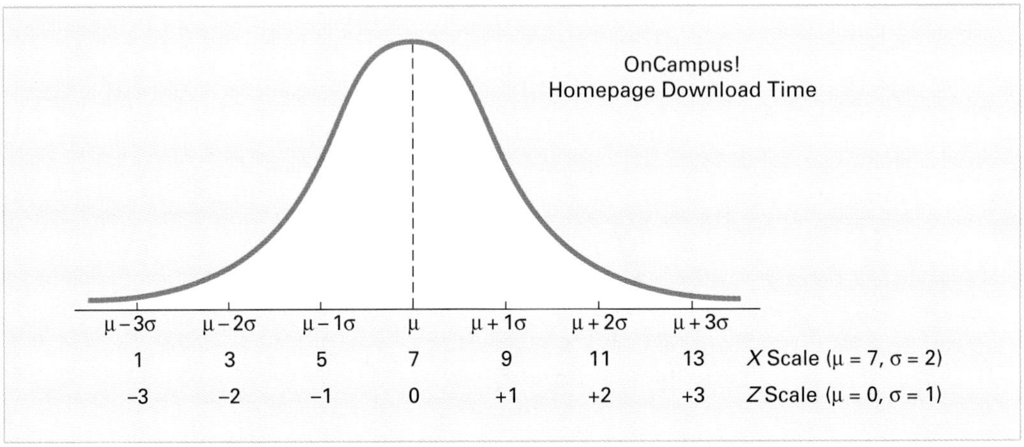

To further illustrate the transformation formula, suppose that the homepage of another Web site has a download time that is normally distributed with a mean $\mu = 4$ seconds and a standard deviation $\sigma = 1$ second. This distribution is illustrated in Figure 6.5.

FIGURE 6.5
A Different Transformation of Scales

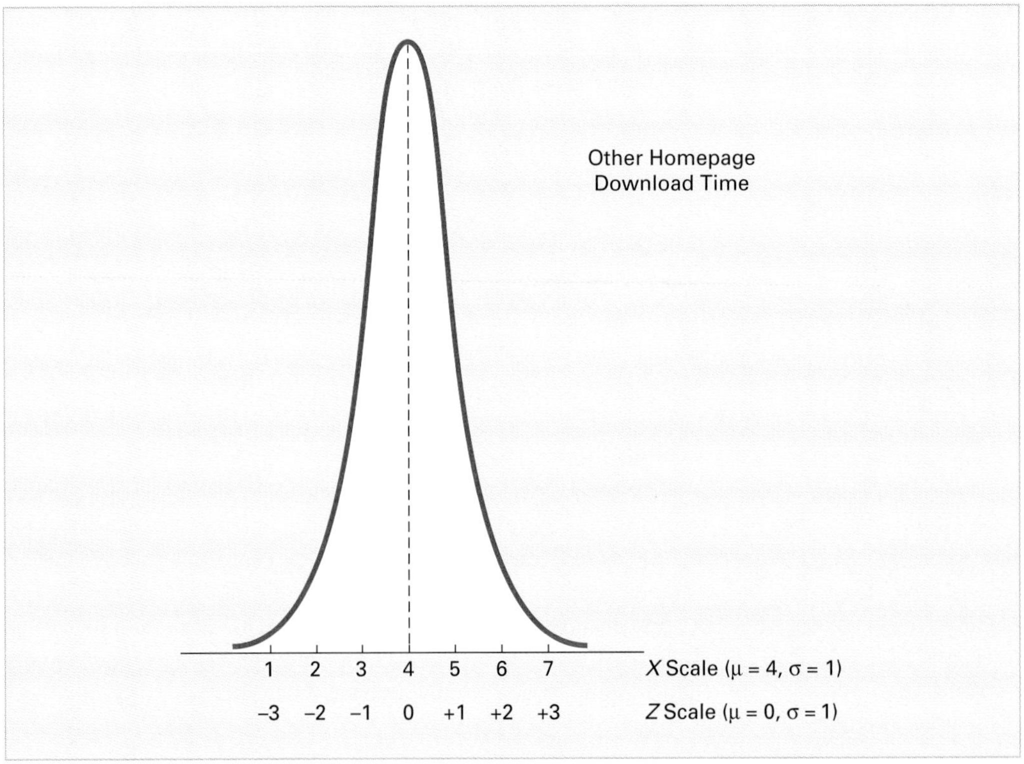

Comparing these results with those of the OnCampus! Web site, you see that a download time of 5 seconds is 1 standard deviation above the mean download time because

$$Z = \frac{5-4}{1} = +1$$

A time of 1 second is 3 standard deviations below the mean download time because

$$Z = \frac{1-4}{1} = -3$$

The two bell-shaped curves in Figures 6.4 and 6.5 show the relative frequency polygons of the normal distributions representing the download time (in seconds) for the two Web sites. Because the download times represent the entire population, the *probabilities* or proportion of area under the entire curve must add to 1.

Suppose you wanted to find the probability that the download time for the OnCampus! site is less than 9 seconds. First, you use Equation (6.2) on page 195 to transform $X = 9$ to standardized Z units. Because $X = 9$ is one standard deviation above the mean, $Z = +1.00$. Next you use Table E.2 to find the cumulative area under the normal curve calculated less than (i.e., to the left of) $Z = +1.00$. To read the probability or area under the curve less than $Z = +1.00$, scan down the Z column from Table E.2 until you locate the Z value of interest (in 10ths) in the Z row for 1.0. Next read across this row until you intersect the column that contains the 100ths place of the Z value. Therefore, in the body of the table, the tabulated probability for $Z = 1.00$ corresponds to the intersection of the row $Z = 1.0$ with the column $Z = .00$, as shown in Table 6.2, which is extracted from Table E.2. This probability is 0.8413. As illustrated in Figure 6.6, there is an 84.13% chance that the download time will be less than 9 seconds.

TABLE 6.2
Finding a Cumulative Area under the Normal Curve

Z	.00	.01	.02	.03	.04	.05	.06	.07	.08	.09
0.0	.5000	.5040	.5080	.5120	.5160	.5199	.5239	.5279	.5319	.5359
0.1	.5398	.5438	.5478	.5517	.5557	.5596	.5636	.5675	.5714	.5753
0.2	.5793	.5832	.5871	.5910	.5948	.5987	.6026	.6064	.6103	.6141
0.3	.6179	.6217	.6255	.6293	.6331	.6368	.6406	.6443	.6480	.6517
0.4	.6554	.6591	.6628	.6664	.6700	.6736	.6772	.6808	.6844	.6879
0.5	.6915	.6950	.6985	.7019	.7054	.7088	.7123	.7157	.7190	.7224
0.6	.7257	.7291	.7324	.7357	.7389	.7422	.7454	.7486	.7518	.7549
0.7	.7580	.7612	.7642	.7673	.7704	.7734	.7764	.7794	.7823	.7852
0.8	.7881	.7910	.7939	.7967	.7995	.8023	.8051	.8078	.8106	.8133
0.9	.8159	.8186	.8212	.8238	.8264	.8289	.8315	.8340	.8365	.8389
1.0	.8413	.8438	.8461	.8485	.8508	.8531	.8554	.8577	.8599	.8621

Source: Extracted from Table E.2.

FIGURE 6.6
Determining the Area Less Than Z from a Cumulative Standardized Normal Distribution

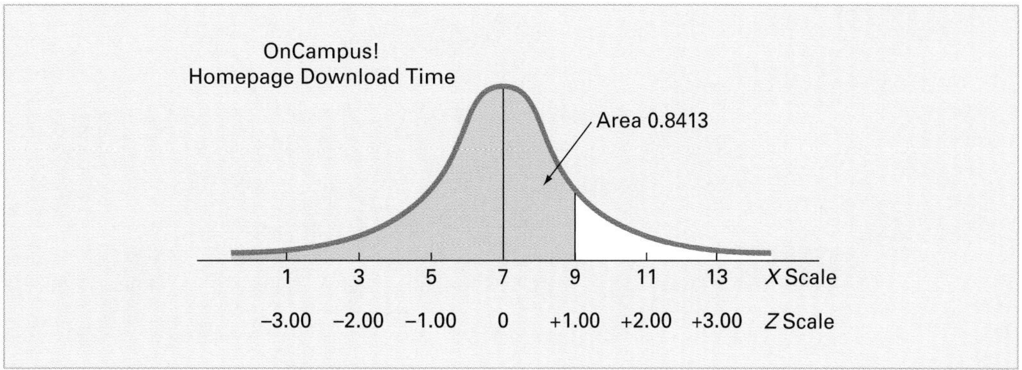

However, for the other homepage, from Figure 6.5 on page 196, you see that a time of 5 seconds is 1 standardized unit above the mean time of 4 seconds. Thus, the probability that the download time will be less than 5 seconds is also 0.8413. Figure 6.7 shows that regardless of the value of the mean μ and standard deviation σ of a normally distributed variable, Equation (6.2) can transform the problem to Z values.

FIGURE 6.7
Demonstrating a Transformation of Scales for Corresponding Cumulative Portions under Two Normal Curves

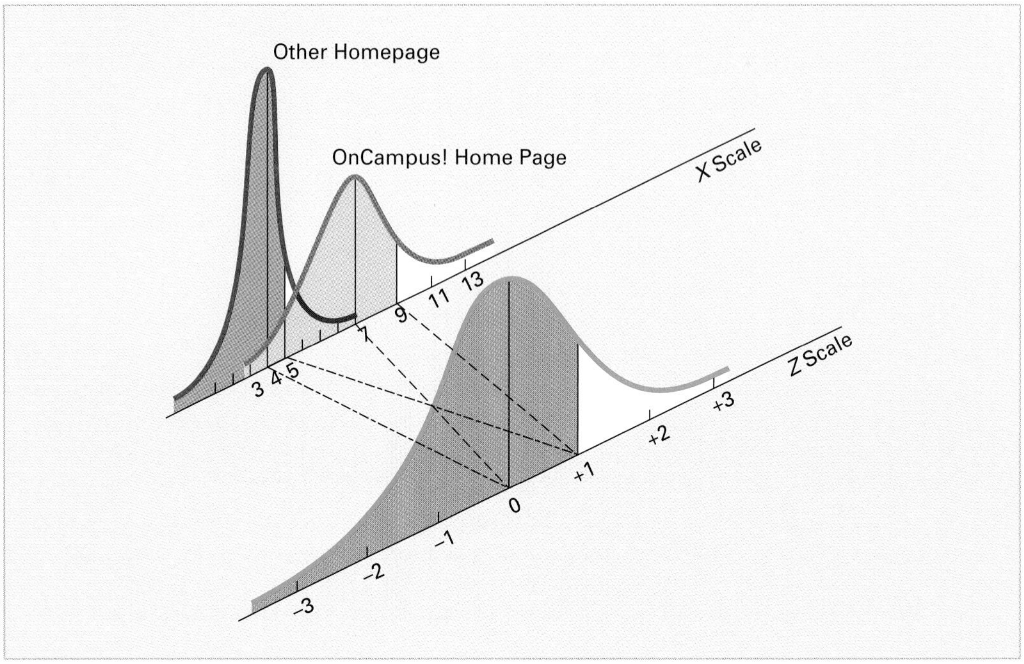

Now that you have learned to use Table E.2 with Equation (6.2), you can answer many questions related to the OnCampus! homepage using the normal distribution.

EXAMPLE 6.1

FINDING $P(X > 9)$

What is the probability that the download time will be more than 9 seconds?

SOLUTION The probability that the download time will be less than 9 seconds is 0.8413 (see Figure 6.6 on page 197). Thus, the probability that the download time will be more than 9 seconds is the *complement* of less than 9 seconds, 1 – 0.8413 = 0.1587. Figure 6.8 illustrates this result.

FIGURE 6.8
Finding P(X > 9)

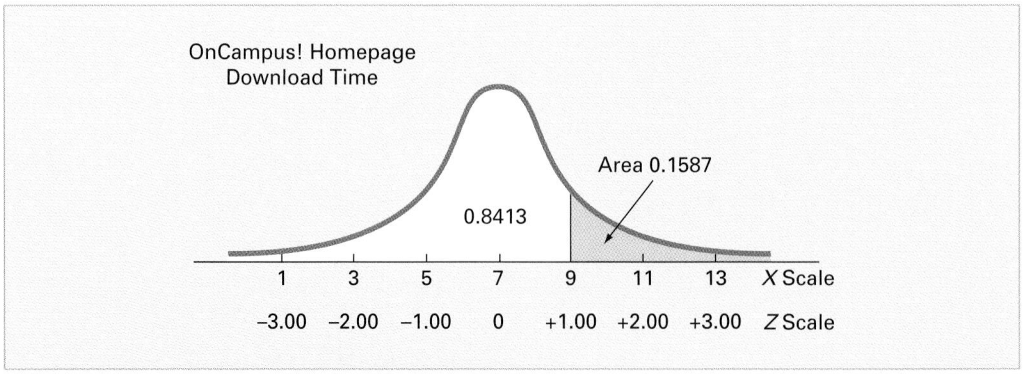

EXAMPLE 6.2 FINDING $P(7 < X < 9)$

What is the probability that the download time will be between 7 and 9 seconds?

SOLUTION From Figure 6.6 on page 197, you already determined that the probability that a download time will be less than 9 seconds is 0.8413. Now you must determine the probability that the download time will be under 7 seconds and subtract this from the probability that the download time is under 9 seconds. This is shown in Figure 6.9.

FIGURE 6.9
Finding $P(7 < X < 9)$

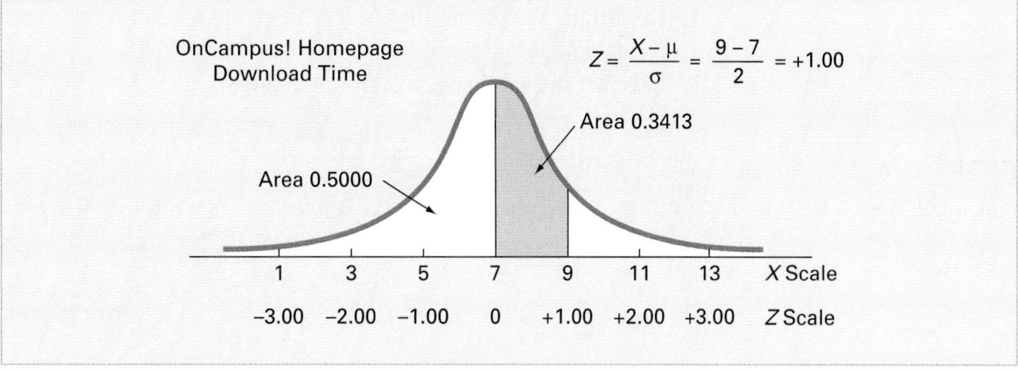

Using Equation (6.2) on page 195

$$Z = \frac{7-7}{2} = 0.00$$

Using Table E.2, the area under the normal curve less than the mean of $Z = 0.00$ is 0.5000. Hence, the area under the curve between $Z = 0.00$ and $Z = 1.00$ is $0.8413 - 0.5000 = 0.3413$.

EXAMPLE 6.3 FINDING $P(X < 7$ OR $X > 9)$

What is the probability that the download time is under 7 seconds or over 9 seconds?

SOLUTION From Figure 6.9, the probability that the download time is between 7 and 9 seconds is 0.3413. The probability that the download time is under 7 seconds or over 9 seconds is its complement, $1 - 0.3413 = 0.6587$.

Another way to view this problem, however, is to separately calculate both the probability of a download time of less than 7 seconds and the probability of a download time of over 9 seconds and to then add these two probabilities together to compute the desired result. This result is depicted in Figure 6.10. Because the mean and median are the same for normally distributed data, 50% of download times are under 7 seconds. From Example 6.1, the probability of a download time of over 9 seconds is 0.1587. Hence, the probability that a download time is under 7 or over 9 seconds, $P(X < 7 \text{ or } X > 9)$, is $0.5000 + 0.1587 = 0.6587$.

FIGURE 6.10
Finding $P(X < 7$ or $X > 9)$

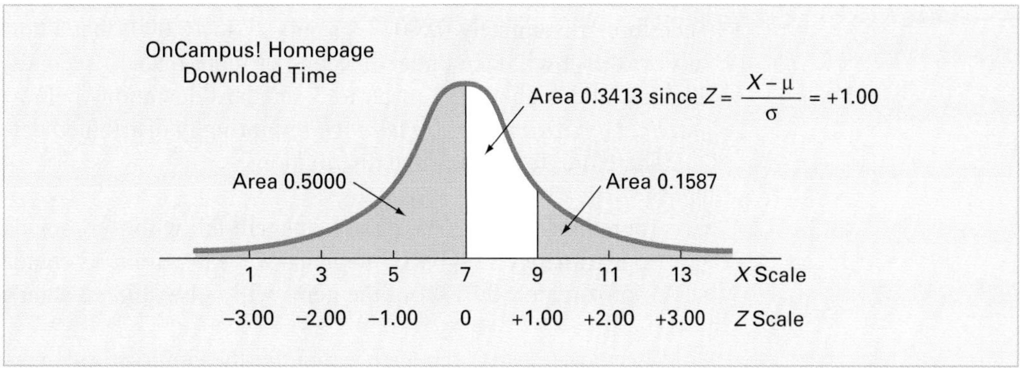

EXAMPLE 6.4 FINDING $P(5 < X < 9)$

What is the probability that the download time will be between 5 and 9 seconds, that is, $P(5 < X < 9)$?

SOLUTION In Figure 6.11, you can see that the area of interest is located between two values, 5 and 9. Because Table E.2 permits you only to find probabilities less than a particular value of interest, use the following three steps to find the desired probability:

1. Determine the probability of less than 9 seconds.
2. Determine the probability of less than 5 seconds.
3. Subtract the smaller result from the larger.

FIGURE 6.11
Finding $P(5 < X < 9)$

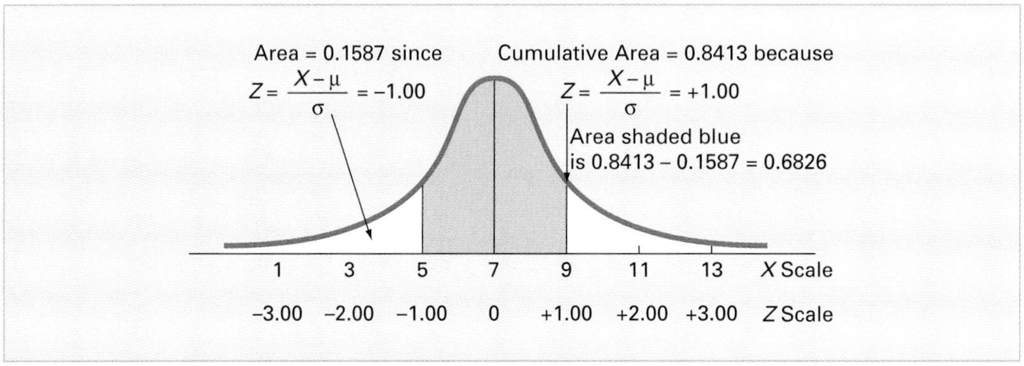

For this example, you have already completed step 1. The area under the normal curve less than 9 seconds is 0.8413. To find the area under the normal curve less than 5 seconds (step 2),

$$Z = \frac{5 - 7}{2} = -1.00$$

Using Table E.2, you look up $Z = -1.00$ and find 0.1587. From step 3, the probability that the download time will be between 5 and 9 seconds is $0.8413 - 0.1587 = 0.6826$, as displayed in Figure 6.11.

The result of Example 6.4 is important and allows you to generalize the findings. For any normal distribution there is a 0.6826 chance that a randomly selected item will fall within ±1 standard deviation of the mean. From Figure 6.12, slightly more than 95% of the items will fall within ±2 standard deviations of the mean. Thus, 95.44% of the download times are between 3 and 11 seconds. From Figure 6.13, 99.73% of the items will fall within ±3 standard deviations above or below the mean. Thus, 99.73% of the download times are between 1 and 13 seconds. Therefore, it is unlikely (0.0027, or only 27 in 10,000) that a download time will be so fast or so slow that it will take under 1 second or more than 13 seconds. This is why 6σ (i.e., 3 standard deviations above the mean to 3 standard deviations below the mean) is often used as a *practical approximation of the range* for normally distributed data.

Therefore, for any normal distribution:

- Approximately 68.26% of the items will fall within ±1 standard deviations of the mean.
- Approximately 95.44% of the items will fall within ±2 standard deviations of the mean.
- Approximately 99.73% of the items will fall within ±3 standard deviations of the mean.

The above result is the justification for the empirical rule presented on page 96. The closer a data set follows the normal distribution, the more accurate the empirical rule is.

FIGURE 6.12
Finding $P(3 < X < 11)$

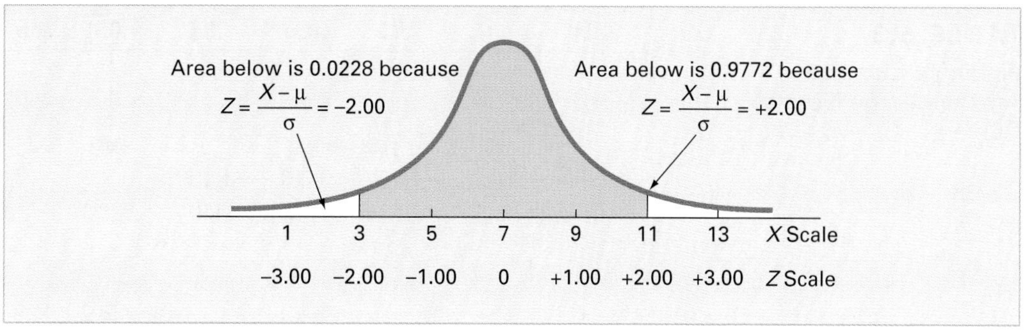

FIGURE 6.13
Finding $P(1 < X < 13)$

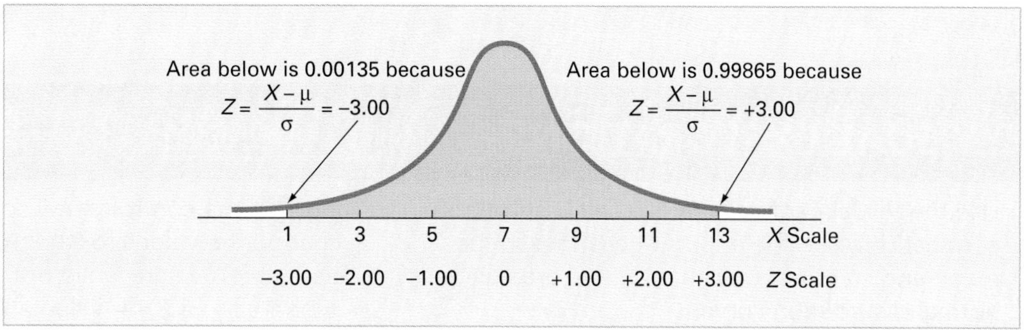

EXAMPLE 6.5

FINDING $P(X < 3.5)$

What is the probability that a download time will be under 3.5 seconds?

SOLUTION To calculate the probability that a download time will be under 3.5 seconds, you need to examine the shaded lower left-tail region of Figure 6.14.

FIGURE 6.14
Finding $P(X < 3.5)$

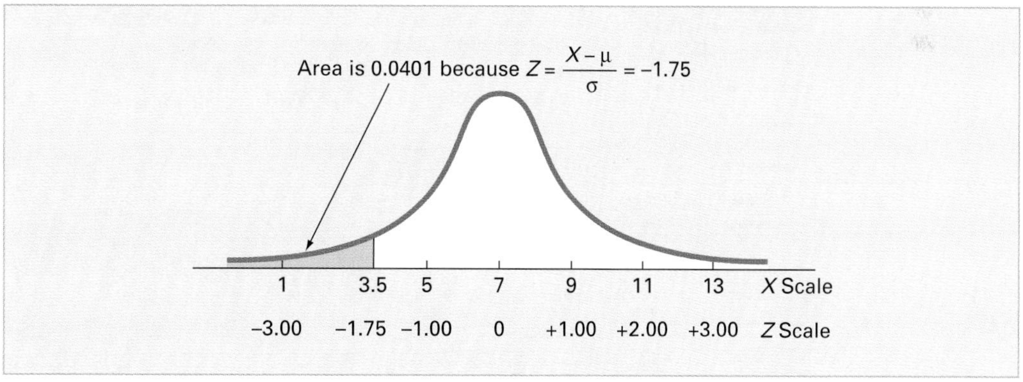

To determine the area under the curve below 3.5 seconds, first calculate

$$Z = \frac{X - \mu}{\sigma} = \frac{3.5 - 7}{2} = -1.75$$

Look up the Z value of -1.75 in Table E.2 by matching the appropriate Z row (-1.7) with the appropriate Z column (.05) as shown in Table 6.3 (which is extracted from Table E.2). The resulting probability or area under the curve less than -1.75 standard deviations below the mean is 0.0401.

TABLE 6.3
Finding a Cumulative Area under the Normal Curve

Z	.00	.01	.02	.03	.04	.05	.06	.07	.08	.09
.	.	.	.	.	.	.	.	.	.	.
.	.	.	.	.	.	.	.	.	.	.
.	.	.	.	.	.	.	.	.	.	.
−1.7	.0446	.0436	.0427	.0418	.0409	.0401	.0392	.0384	.0375	.0367
−1.6	.0548	.0537	.0526	.0516	.0505	.0495	.0485	.0475	.0465	.0455

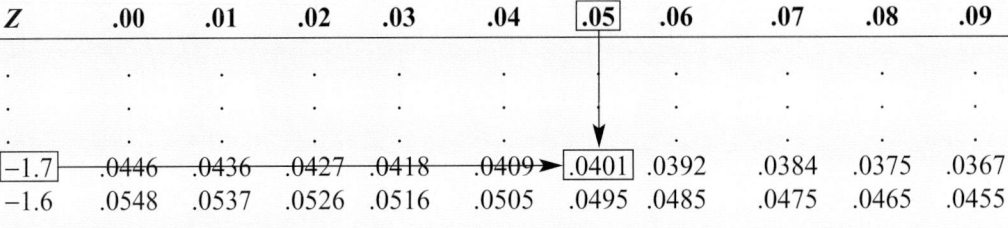

Source: Extracted from Table E.2.

VISUAL EXPLORATIONS Exploring the Normal Distribution

Use the Visual Explorations Normal Distribution command to see the effects of changes in the mean and standard deviation on the area under a normal distribution curve.

Open the **Visual Explorations.xla** file and select **VisualExplorations → Normal Distribution** from the Microsoft Excel menu bar. You will see a normal curve for the "Using Statistics" homepage download example and a floating control panel that allows you to adjust the shape of the curve and the shaded area under the curve (see illustration below).

Use the control panel spinner buttons to change the values for the mean, standard deviation, and *X* value, while noting their effects on the probability of *X* <= value and the corresponding shaded area under the curve (see illustration below). If you prefer, you can select the *Z* Values option button to see the normal curve labeled with *Z* Values.

Click the Reset button to reset the control panel values or click Help for additional information about the problem. Click Finish when you are done exploring.

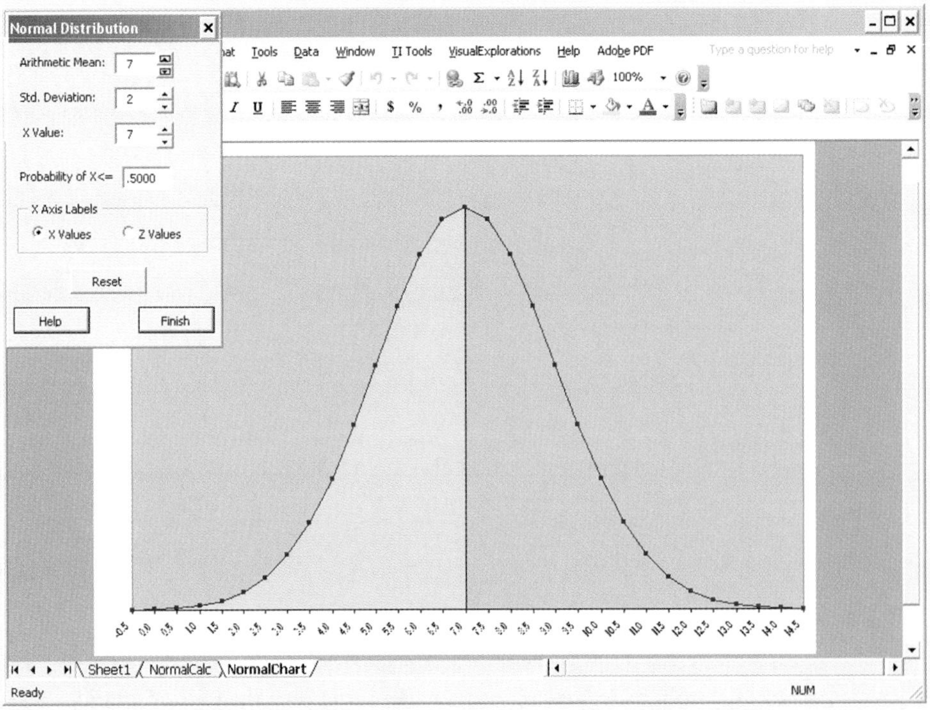

Examples 6.1 through 6.5 require you to use the normal tables to find an area under the normal curve that corresponds to a specific X value. There are many circumstances when you want to do the opposite. Examples 6.6 and 6.7 illustrate how to find the X value that corresponds to a specific area.

EXAMPLE 6.6

FINDING THE X VALUE FOR A CUMULATIVE PROBABILITY OF 0.10

How much time (in seconds) will elapse before 10% of the downloads are complete?

SOLUTION Because 10% of the homepages are expected to download in under X seconds, the area under the normal curve less than this Z value is 0.1000. Using the body of Table E.2, you search for the area or probability of 0.1000. The closest result is 0.1003, as shown in Table 6.4 (which is extracted from Table E.2).

TABLE 6.4
Finding a Z Value Corresponding to a Particular Cumulative Area (0.10) under the Normal Curve

Z	.00	.01	.02	.03	.04	.05	.06	.07	.08	.09
.	.	.	.	.	.	.	.	.	.	.
.	.	.	.	.	.	.	.	.	.	.
.	.	.	.	.	.	.	.	.	.	.
−1.5	.0668	.0655	.0643	.0630	.0618	.0606	.0594	.0582	.0571	.0559
−1.4	.0808	.0793	.0778	.0764	.0749	.0735	.0721	.0708	.0694	.0681
−1.3	.0968	.0951	.0934	.0918	.0901	.0885	.0869	.0853	.0838	.0823
−1.2	.1151	.1131	.1112	.1093	.1075	.0156	.0138	.1020	.1003	.0985

Source: Extracted from Table E.2.

Working from this area to the margins of the table, the Z value corresponding to the particular Z row (−1.2) and Z column (.08) is −1.28 (see Figure 6.15).

FIGURE 6.15
Finding Z to Determine X

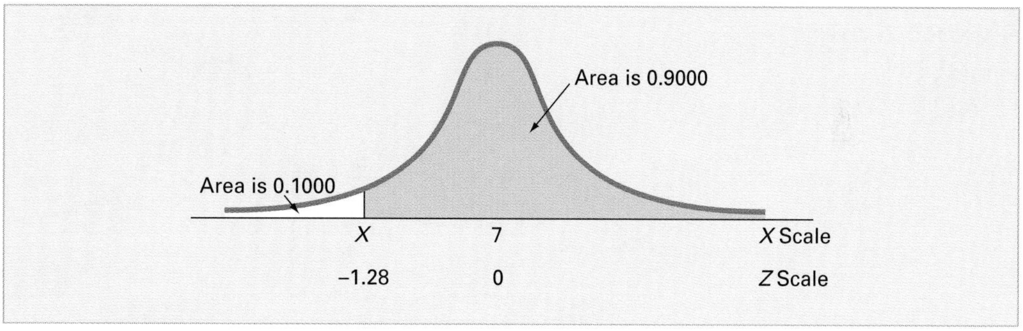

Once you find Z, use the transformation formula Equation (6.2) on page 195 to determine the X value as follows. Let

$$Z = \frac{X - \mu}{\sigma}$$

then

$$X = \mu + Z\sigma$$

Substituting $\mu = 7$, $\sigma = 2$, and $Z = -1.28$,

$$X = 7 + (-1.28)(2) = 4.44 \text{ seconds}$$

Thus, 10% of the download times are 4.44 seconds or less.

Equation (6.4) is used for finding an X value.

FINDING AN X VALUE ASSOCIATED WITH KNOWN PROBABILITY

The X value is equal to the mean μ plus the product of the Z value and the standard deviation σ.

$$X = \mu + Z\sigma \tag{6.4}$$

To find a *particular* value associated with a known probability, follow these steps.

1. Sketch the normal curve, and then place the values for the means on the respective X and Z scales.
2. Find the cumulative area less than X.
3. Shade the area of interest.
4. Using Table E.2, determine the Z value corresponding to the area under the normal curve less than X.
5. Using Equation (6.4), solve for X:

$$X = \mu + Z\sigma$$

EXAMPLE 6.7

FINDING THE X VALUES THAT INCLUDE 95% OF THE DOWNLOAD TIMES

What are the lower and upper values of X, located symmetrically around the mean, that include 95% of the download times?

SOLUTION First, you need to find the lower value of X (called X_L). Then you find the upper value of X (called X_U). Since 95% of the values are between X_L and X_U, and X_L and X_U are equal distance from the mean, 2.5% of the values are below X_L (see Figure 6.16).

FIGURE 6.16
Finding Z to Determine X_L

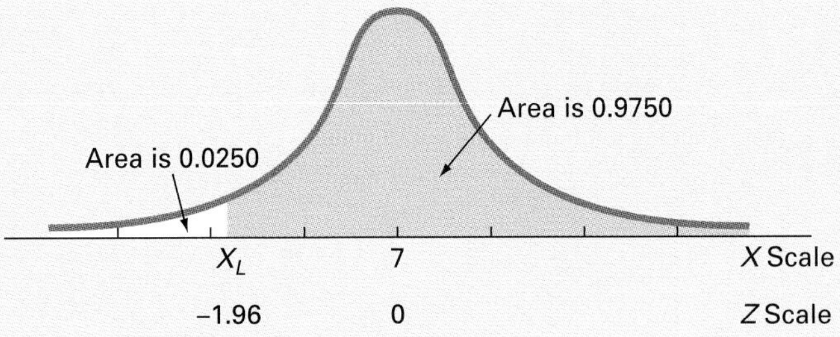

Although X_L is not known, you can find the corresponding Z because the area under the normal curve less than this Z is 0.0250. Using the body of Table 6.5, you search for the probability 0.0250.

TABLE 6.5
Finding a Z Value Corresponding to a Cumulative Area of 0.025 under the Normal Curve

Z	.00	.01	.02	.03	.04	.05	.06	.07	.08	.09
.	.	.	.	.	.	.	.	.	.	.
.	.	.	.	.	.	.	.	.	.	.
.	.	.	.	.	.	.	.	.	.	.
−2.0	.0228	.0222	.0217	.0212	.0207	.0202	.0197	.0192	.0188	.0183
−1.9	.0287	.0281	.0274	.0268	.0262	.0256	.0250	.0244	.0239	.0233
−1.8	.0359	.0351	.0344	.0336	.0329	.0232	.0314	.0307	.0301	.0294

Source: Extracted from Table E.2.

Working from the body of the table to the margins of the table, you see that the Z value corresponding to the particular Z row (−1.9) and Z column (.06) is −1.96.

Once you find Z, the final step is to use Equation (6.4) on page 204 as follows,

$$\begin{aligned} X &= \mu + Z\sigma \\ &= 7 + (-1.96)(2) \\ &= 7 - 3.92 \\ &= 3.08 \text{ seconds} \end{aligned}$$

You use a similar process to find X_U. Since only 2.5% of the homepage downloads take longer than X_U seconds, 97.5% of the homepage downloads take less than X_U seconds. From the symmetry of the normal distribution, the desired Z value as shown in Figure 6.17 is +1.96 (because Z lies to the right of the standardized mean of 0). You can also extract this Z value from Table 6.6. Note that 0.975 is the area under the normal curve less than the Z value of +1.96.

FIGURE 6.17
Finding Z to Determine X_U

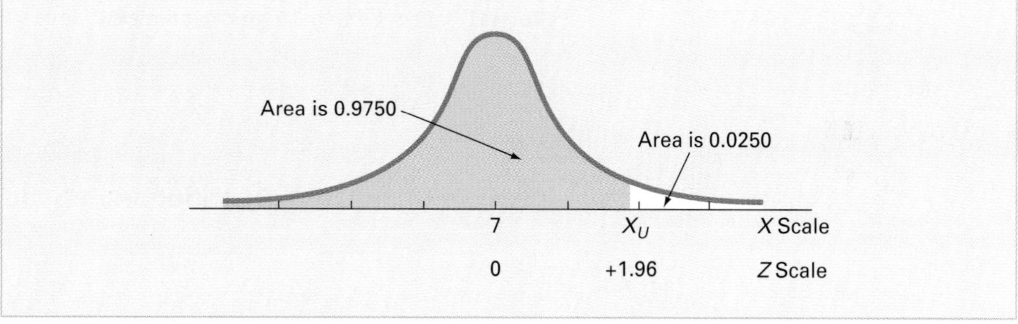

TABLE 6.6
Finding a Z Value Corresponding to a Cumulative Area of 0.975 under the Normal Curve

Z	.00	.01	.02	.03	.04	.05	.06	.07	.08	.09
.	.	.	.	.	.	.	.	.	.	.
.	.	.	.	.	.	.	.	.	.	.
.	.	.	.	.	.	.	.	.	.	.
+1.8	.9641	.9649	.9656	.9664	.9671	.9678	.9686	.9693	.9699	.9706
+1.9	.9713	.9719	.9726	.9732	.9738	.9744	.9750	.9756	.9761	.9767
+2.0	.9772	.9778	.9783	.9788	.9793	.9798	.9803	.9808	.9812	.9817

Source: Extracted from Table E.2.

Therefore, using Equation (6.4) on page 204,

$$\begin{aligned} X &= \mu + Z\sigma \\ &= 7 + (+1.96)(2) \\ &= 7 + 3.92 \\ &= 10.92 \text{ seconds} \end{aligned}$$

Therefore, 95% of the download times are between 3.08 and 10.92 seconds.

You can also use Microsoft Excel or Minitab to compute normal probabilities. Figure 6.18 illustrates a Microsoft Excel worksheet for Examples 6.5 and 6.6, and Figure 6.19 illustrates Minitab output for Examples 6.1 and 6.6.

FIGURE 6.18
Microsoft Excel Worksheet for Computing Normal Probabilities

	A	B	
1	Normal Probabilities		
2			
3	Common Data		
4	Mean	7	
5	Standard Deviation	2	
6			
7	Probability for X <=		
8	X Value	3.5	
9	Z Value	-1.75	=STANDARDIZE(B8, B4, B5)
10	P(X<=3.5)	0.0401	=NORMDIST(B8, B4, B5, TRUE)
11			
12	Find X and Z Given Cum. Pctage.		
13	Cumulative Percentage	10.00%	
14	Z Value	-1.2816	=NORMSINV(B13)
15	X Value	4.4369	=NORMINV(B13, B4, B5)

FIGURE 6.19
Minitab Normal Probabilities

Cumulative Distribution Function

```
Normal with mean = 7 and standard deviation = 2

x  P( X <= x )
9     0.841345
```

Inverse Cumulative Distribution Function

```
Normal with mean = 7 and standard deviation = 2

P( X <= x )         x
        0.1   4.43690
```

PROBLEMS FOR SECTION 6.2

Learning the Basics

PH Grade ASSIST **6.1** Given a standardized normal distribution (with a mean of 0 and a standard deviation of 1, as in Table E.2), what is the probability that

a. Z is less than 1.57?
b. Z is greater than 1.84?
c. Z is between 1.57 and 1.84?
d. Z is less than 1.57 or greater than 1.84?

6.2 Given a standardized normal distribution (with a mean of 0 and a standard deviation of 1, as in Table E.2), what is the probability that

a. Z is between -1.57 and 1.84?
b. Z is less than -1.57 or greater than 1.84?
c. What is the value of Z if only 2.5% of all possible Z values are larger?
d. Between what two values of Z (symmetrically distributed around the mean) will 68.26% of all possible Z values be contained?

PH Grade ASSIST **6.3** Given a standardized normal distribution (with a mean of 0 and a standard deviation of 1 as in Table E.2), what is the probability that

a. Z is less than 1.08?
b. Z is greater than -0.21?
c. Z is less than -0.21 or greater than the mean?
d. Z is less than -0.21 or greater than 1.08?

6.4 Given a standardized normal distribution (with a mean of 0 and a standard deviation of 1 as in Table E.2), determine the following probabilities:

a. $P(Z > 1.08)$
b. $P(Z < -0.21)$
c. $P(-1.96 < Z < -0.21)$
d. What is the value of Z if only 15.87% of all possible Z values are larger?

PH Grade ASSIST **6.5** Given a normal distribution with $\mu = 100$ and $\sigma = 10$, what is the probability that

a. $X > 75$?
b. $X < 70$?
c. $X < 80$ or $X > 110$?
d. 80% of the values are between what two X values (symmetrically distributed around the mean)?

PH Grade ASSIST **6.6** Given a normal distribution with $\mu = 50$ and $\sigma = 4$, what is the probability that
a. $X > 43$?
b. $X < 42$?
c. 5% of the values are less than what X value?
d. 60% of the values are between what two X values (symmetrically distributed around the mean)?

Applying the Concepts

6.7 During 2001, 61.3% of U.S. households purchased ground coffee and spent an average of \$36.16 on ground coffee during the year ("Annual Product Preference Study," *Progressive Grocer*, May 1, 2002, 31). Consider the annual ground coffee expenditures for households purchasing ground coffee, assuming that these expenditures are approximately distributed as a normal random variable with a mean of \$36.16 and a standard deviation of \$10.00.
a. Find the probability that a household spent less than \$25.00.
b. Find the probability that a household spent more than \$50.00.
c. What proportion of the households spent between \$30.00 and \$40.00?
d. 99% of the households spent less than what amount?

PH Grade ASSIST SELF Test **6.8** Toby's Trucking Company determined that on an annual basis the distance traveled per truck is normally distributed with a mean of 50.0 thousand miles and a standard deviation of 12.0 thousand miles.
a. What proportion of trucks can be expected to travel between 34.0 and 50.0 thousand miles in the year?
b. What percentage of trucks can be expected to travel either below 30.0 or above 60.0 thousand miles in the year?
c. How many miles will be traveled by at least 80% of the trucks?
d. What are your answers to (a) through (c) if the standard deviation is 10.0 thousand miles?

PH Grade ASSIST **6.9** The breaking strength of plastic bags used for packaging produce is normally distributed with a mean of 5 pounds per square inch and a standard deviation of 1.5 pounds per square inch. What proportion of the bags have a breaking strength of
a. less than 3.17 pounds per square inch?
b. at least 3.6 pounds per square inch?
c. between 5 and 5.5 pounds per square inch?
d. Between what two values symmetrically distributed around the mean will 95% of the breaking strengths fall?

6.10 A set of final examination grades in an introductory statistics course is normally distributed with a mean of 73 and a standard deviation of 8.
a. What is the probability of getting a grade of 91 or less on this exam?
b. What is the probability that a student scored between 65 and 89?
c. The probability is 5% that a student taking the test scores higher than what grade?
d. If the professor grades on a curve (gives A's to the top 10% of the class regardless of the score), are you better off with a grade of 81 on this exam or a grade of 68 on a different exam where the mean is 62 and the standard deviation is 3? Show your answer statistically and explain.

6.11 A statistical analysis of 1,000 long-distance telephone calls made from the headquarters of the Bricks and Clicks Computer Corporation indicates that the length of these calls is normally distributed with $\mu = 240$ seconds and $\sigma = 40$ seconds.
a. What is the probability that a call lasted less than 180 seconds?
b. What is the probability that a particular call lasted between 180 and 300 seconds?
c. What is the probability that a call lasted between 110 and 180 seconds?
d. What is the length of a particular call if only 1% of all calls are shorter?

6.12 The number of shares traded daily on the New York Stock Exchange (NYSE) is referred to as the *volume* of trading. On April 23, 2004, 1.395 billion shares of stock were traded ("NYSE Volume," *The Wall Street Journal*, April 26, 2004, C2). This volume of trading is near the mean volume for the NYSE. Assume that the number of shares traded on the NYSE is a normal random variable with a mean of 1.4 billion and a standard deviation of 0.15 billion. For a randomly selected day, what is the probability that the volume of trading on the NYSE is:
a. below 1.7 billion?
b. below 1.25 billion?
c. below 1.0 billion?
d. above 1.0 billion?

6.13 Many manufacturing problems involve the accurate matching of machine parts such as shafts that fit into a valve hole. A particular design requires a shaft with a diameter of 22.000 mm., but shafts with diameters between 21.900 mm. and 22.010 mm. are acceptable. Suppose that the manufacturing process yields shafts with diameters normally distributed with a mean of 22.002 mm. and a standard deviation of 0.005 mm. For this process, what is
a. the proportion of shafts with a diameter between 21.90 mm. and 22.00 mm.?
b. the probability a shaft is acceptable?
c. the diameter that will be exceeded by only 2% of the shafts?
d. What would be your answers in (a) through (c) if the standard deviation of the shaft diameters were 0.004 mm.?

6.3 EVALUATING NORMALITY

As discussed in section 6.2, many continuous variables used in business closely resemble a normal distribution. However, many important variables cannot even be approximated by the normal distribution. This section presents two approaches for evaluating whether a set of data can be approximated by the normal distribution:

1. Compare the data set's characteristics with the properties of the normal distribution.
2. Construct a normal probability plot.

Evaluating the Properties

The normal distribution has several important theoretical properties:

- It is symmetrical, thus the mean and median are equal.
- It is bell-shaped, thus the empirical rule applies.
- The interquartile range equals 1.33 standard deviations.
- The range is infinite.

In actual practice, some continuous variables may have characteristics that approximate these theoretical properties. However, many continuous variables are neither normally distributed nor approximately normally distributed. For such variables, the descriptive characteristics of the data do not match well with the properties of a normal distribution. One approach to check for normality is by comparing the actual data characteristics with the corresponding properties from an underlying normal distribution, as follows.

- Construct charts and observe their appearance. For small- or moderate-sized data sets, construct a stem-and-leaf display or a box-and-whisker plot. For large data sets, construct the frequency distribution and plot the histogram or polygon.
- Compute descriptive numerical measures and compare the characteristics of the data with the theoretical properties of the normal distribution. Compare the mean and median. Is the interquartile range approximately 1.33 times the standard deviation? Is the range approximately 6 times the standard deviation?
- Evaluate how the values in the data are distributed. Determine whether approximately two-thirds of the values lie between the mean ±1 standard deviation. Determine whether approximately four-fifths of the values lie between the mean ±1.28 standard deviations. Determine whether approximately 19 out of every 20 values lie between the mean ±2 standard deviations.

Do the mutual fund returns in 2003 discussed in Chapters 2 and 3 contain the properties of the normal distribution? Figure 6.20 displays descriptive statistics for these data and Figure 6.21 presents a box-and-whisker plot.

FIGURE 6.20

Microsoft Excel Descriptive Statistics for the 2003 Returns of Mutual Funds

	A	B
1	*Return 2003*	
2		
3	Mean	42.6215
4	Standard Error	1.1852
5	Median	40.8
6	Mode	37.5
7	Standard Deviation	13.0369
8	Sample Variance	169.9609
9	Kurtosis	-0.2828
10	Skewness	0.3500
11	Range	63.1
12	Minimum	14.9
13	Maximum	78
14	Sum	5157.2
15	Count	121
16	Largest(1)	78
17	Smallest(1)	14.9

FIGURE 6.21

Minitab Box-and-Whisker Plot for the 2003 Returns of Mutual Funds

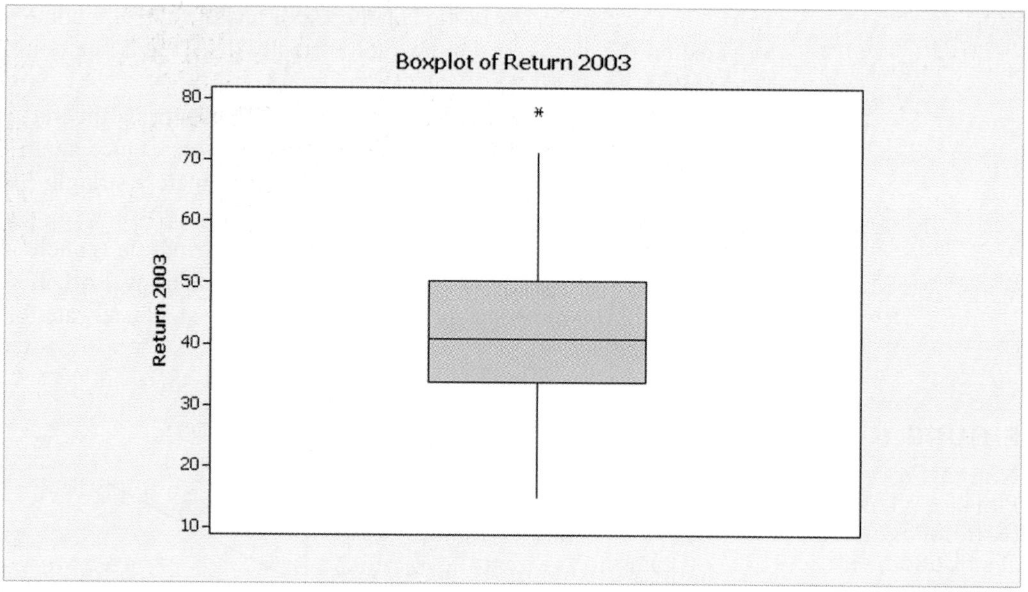

From these figures, you can make the following statements:

1. The mean of 42.62 is slightly higher than the median of 40.8.
2. The box-and-whisker plot appears slightly right-skewed with an outlier at 78.
3. The interquartile range of 16.45 is approximately 1.33 standard deviations.
4. The range of 63.1 is equal to 4.84 standard deviations.
5. 65.6% of the returns are within ± 1 standard deviation of the mean.
6. 78.7% of the returns are within ± 1.28 standard deviations of the mean.

Based on these statements and the criteria given above, you can conclude that the returns for 2003 are approximately normally distributed. However, statements 1 and 2 indicate that the 2003 returns are slightly right-skewed.

Constructing the Normal Probability Plot

A **normal probability plot** is a graphical approach for evaluating whether data are normally distributed. One common approach is called the **quantile–quantile plot**. In this method, you transform each ordered value to a Z score, and then plot the data values versus the Z scores. For example, if you have a sample of $n = 19$, the Z value for the smallest value corresponds to a cumulative area of $\frac{1}{n+1} = \frac{1}{19+1} = \frac{1}{20} = 0.05$. The Z value for a cumulative area of 0.05 (from Table E.2) is −1.65. Table 6.7 illustrates the entire set of Z values for a sample of $n = 19$.

TABLE 6.7

Ordered Values and Corresponding Z Values for a Sample of $n = 19$

Ordered Value	Z Value	Ordered Value	Z Value
1	−1.65	11	0.13
2	−1.28	12	0.25
3	−1.04	13	0.39
4	−0.84	14	0.52
5	−0.67	15	0.67
6	−0.52	16	0.84
7	−0.39	17	1.04
8	−0.25	18	1.28
9	−0.13	19	1.65
10	0.00		

The Z values are plotted on the X axis and the corresponding values of the variable are plotted on the Y axis. If the data are normally distributed, the points will plot along an approximately straight line.

A second approach (used by Minitab) transforms the vertical Y axis in a more complicated fashion that is beyond the scope of this text. Once again, if the data are normally distributed, the points will plot along an approximately straight line. Figure 6.22 illustrates the typical shape of normal probability plots for a left-skewed distribution (panel A), a normal distribution (panel B), and a right-skewed distribution (panel C). If the data are left-skewed, the curve will rise more rapidly at first, and then level off. If the data are right-skewed, the data will rise more slowly at first, and then rise at a faster rate for higher values of the variable being plotted.

FIGURE 6.22

Normal Probability Plots for a Left-Skewed Distribution, a Normal Distribution, and a Right-Skewed Distribution

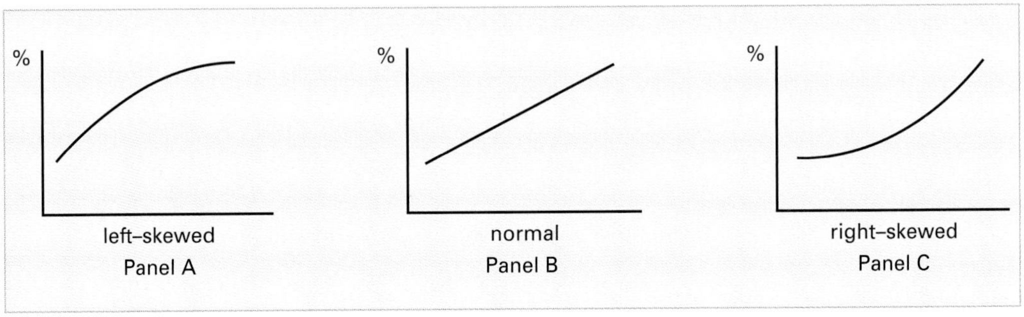

Figure 6.23 shows a Microsoft Excel quantile–quantile normal probability plot and Figure 6.24 displays a Minitab normal probability plot for the 2003 returns.

FIGURE 6.23

Microsoft Excel Normal Probability Plot for 2003 Returns

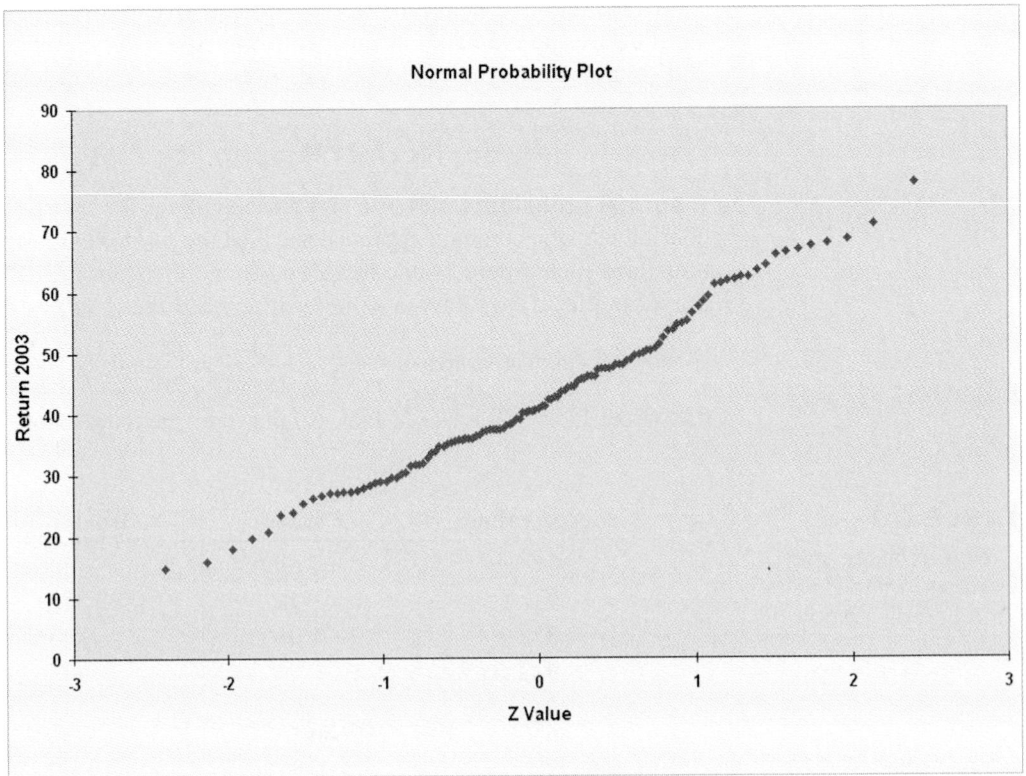

Figures 6.23 or 6.24 show that the normal probability plot of the 2003 returns approximate a straight line. You can conclude that the 2003 returns are approximately normally distributed.

FIGURE 6.24
Minitab Normal Probability Plot for 2003 Returns

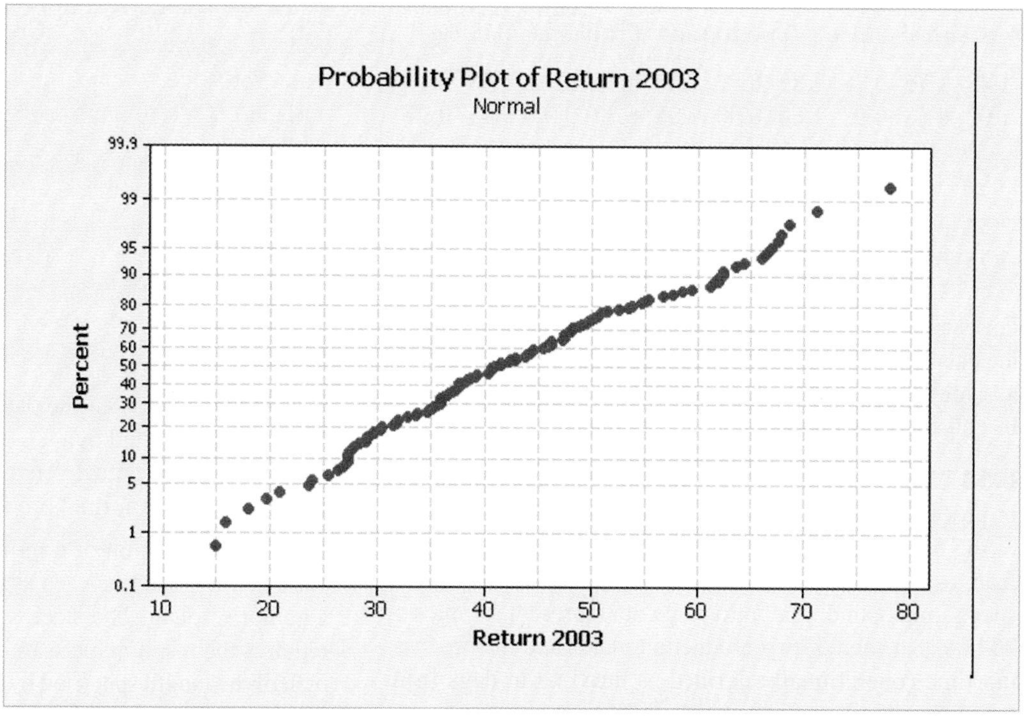

PROBLEMS FOR SECTION 6.3

Learning the Basics

PH Grade ASSIST **6.14** Show that for a sample of $n = 39$, the smallest and largest Z values are -1.96 and $+1.96$, and the middle (i.e., 20th) Z value is 0.00.

6.15 For a sample of $n = 6$, list the six Z values.

Applying the Concepts

You can solve problems 6.16–6.19 manually or by using Microsoft Excel, Minitab, or SPSS. We recommend that you solve Problems 6.20–6.22 using Microsoft Excel, Minitab, or SPSS.

SELF Test **6.16** The daily hotel rate for 25 cities in March 2004 was as follows:

City	Hotel Rate
Anaheim	95.26
Atlanta	78.91
Boston	112.92
Chicago	96.90
Dallas	77.43
Denver	74.22
Detroit	77.71
Houston	76.26
Los Angeles	95.78
Miami	140.61
Minneapolis	78.64
Nashville	74.61
New Orleans	121.59
New York	167.43
Norfolk	62.88
Oahu Island	119.76
Orlando	98.57
Philadelphia	95.02
Phoenix	123.19
San Diego	110.23
San Francisco	123.51
Seattle	95.09
St. Louis	74.68
Tampa	97.08
Washington	123.27

HOTEL-PRICE

Source: Extracted from USA Today, *April 27, 2004, 5B.*

Decide whether or not the data appear to be approximately normally distributed by:

a. evaluating the actual versus theoretical properties.
b. constructing a normal probability plot.

6.17 A problem with a telephone line that prevents a customer from receiving or making calls is disconcerting to both the customer and the telephone company. The data at the top of page 212 represent two samples of 20 problems reported to two different offices of a telephone company. The time to clear these problems from the customers' lines is recorded in minutes. **PHONE**

Central Office I Time to Clear Problems (minutes)

1.48 1.75 0.78 2.85 0.52 1.60 4.15 3.97 1.48 3.10
1.02 0.53 0.93 1.60 0.80 1.05 6.32 3.93 5.45 0.97

Central Office II Time to Clear Problems (minutes)

7.55 3.75 0.10 1.10 0.60 0.52 3.30 2.10 0.58 4.02
3.75 0.65 1.92 0.60 1.53 4.23 0.08 1.48 1.65 0.72

For each of the two central office locations, decide whether the data appear to be approximately normally distributed by:
a. evaluating the actual versus theoretical properties.
b. constructing a normal probability plot.

6.18 Many manufacturing processes use the term work-in-process (often abbreviated as WIP). In a book manufacturing plant, the WIP represents the time it takes for sheets from a press to be folded, gathered, sewn, tipped on end sheets, and bound. The following data represent samples of 20 books at each of two production plants and the processing time (operationally defined as the time in days from when the books came off the press to when they were packed in cartons) for these jobs: WIP

Plant A

15.62 5.29 16.25 10.92 11.46 21.62 8.45 8.58 5.41 11.42
11.62 7.29 17.50 17.96 14.42 10.50 7.58 9.29 7.54 18.92

Plant B

9.54 11.46 16.62 12.62 25.75 15.41 14.29 13.13 13.71 10.04
5.75 12.46 19.17 13.21 16.00 12.33 14.25 15.37 16.25 19.71

For each of the two plants, decide whether or not the data appear to be approximately normally distributed by:
a. evaluating the actual versus theoretical properties.
b. constructing a normal probability plot.

6.19 Credit scores are three-digit numbers used by lenders when evaluating your credit worthiness. The scores for residents of twenty metropolitan areas are as follows:

City	Credit Score
Atlanta	670
Boston	705
Chicago	680
Cleveland	690
Dallas	653
Denver	675
Detroit	675
Houston	655
Los Angeles	667
Miami	672
Minneapolis	707
New York	688
Orlando	671
Philadelphia	688
Phoenix	660
Sacramento	676
San Francisco	686
Seattle	691
Tampa	675
Washington	693

CREDITSCORE

Decide whether or not the data appear to be approximately normally distributed by:
a. evaluating the actual versus theoretical properties.
b. constructing a normal probability plot.

6.20 One operation of a mill is to cut pieces of steel into parts that will later be used as the frame for front seats in an automotive plant. The steel is cut with a diamond saw and requires the resulting parts to be within plus or minus 0.005 inch of the length specified by the automobile company. The data come from a sample of 100 steel parts. STEEL The measurement reported is the difference in inches between the actual length of the steel part, as measured by a laser measurement device, and the specified length of the steel part.

Decide whether or not the data appear to be approximately normally distributed by:
a. evaluating the actual versus theoretical properties.
b. constructing a normal probability plot.

6.21 In a rubber-edge manufacturing factory, the raw rubber is compounded through a kneading machine and then cut into thin ribbon strips. The strips are loaded onto mold machines and thermocast into the desired shapes of rubber edges. The weights (in grams) of a sample of rubber edges were as follows: RUBBER

8.63 8.59 8.63 8.67 8.64 8.57 8.53 8.59 8.66 8.54
8.65 8.61 8.67 8.65 8.64 8.64 8.51 8.61 8.65 8.62
8.57 8.60 8.54 8.69 8.52 8.63 8.72 8.58 8.66 8.66
8.57 8.66 8.62 8.66 8.69 8.57 8.58 8.65 8.68 8.56
8.54 8.65 8.65 8.62 8.66 8.61 8.64 8.73 8.62 8.60
8.69 8.50 8.58 8.63 8.66 8.59 8.69 8.70 8.54 8.62
8.63 8.61 8.65 8.59 8.61 8.56 8.64 8.65 8.67 8.61
8.64 8.61 8.67 8.65 8.55 8.71 8.75 8.56 8.62 8.66

Source: W. L. Pearn and K. S. Chen, "A Practical Implementation of the Process Capability Index Cpk," Quality Engineering, *1997, 9, 721–737.*

Decide whether or not the data appear to be approximately normally distributed by:
a. evaluating the actual versus theoretical properties.
b. constructing a normal probability plot.

6.22 The following data represent the electricity cost in dollars during the month of July 2004 for a random sample of 50 two-bedroom apartments in a large city: UTILITY

96	171	202	178	147	102	153	197	127	82
157	185	90	116	172	111	148	213	130	165
141	149	206	175	123	128	144	168	109	167
95	163	150	154	130	143	187	166	139	149
108	119	183	151	114	135	191	137	129	158

Decide whether the data appear to be approximately normally distributed by:

a. evaluating the actual versus theoretical properties.

b. constructing a normal probability plot.

6.4 THE UNIFORM DISTRIBUTION

In the **uniform distribution**, a value has the same probability of occurrence anywhere in the range between the smallest value a and the largest value b. Because of its shape, the uniform distribution is sometimes called the **rectangular distribution** (see panel B of Figure 6.1 on page 193). Equation (6.5) defines the continuous probability density function for the uniform distribution.

THE UNIFORM DISTRIBUTION

$$f(X) = \frac{1}{b - a} \text{ if } a \leq X \leq b \text{ and 0 elsewhere} \qquad \textbf{(6.5)}$$

where

a = the minimum value of X

b = the maximum value of X

Equation (6.6) defines the mean of the uniform distribution.

THE MEAN OF THE UNIFORM DISTRIBUTION

$$\mu = \frac{a + b}{2} \qquad \textbf{(6.6)}$$

Equation (6.7) defines the variance and standard deviation of the uniform distribution.

THE VARIANCE AND STANDARD DEVIATION OF THE UNIFORM DISTRIBUTION

$$\sigma^2 = \frac{(b - a)^2}{12} \qquad \textbf{(6.7a)}$$

$$\sigma = \sqrt{\frac{(b - a)^2}{12}} \qquad \textbf{(6.7b)}$$

One of the most common uses of the uniform distribution is in the selection of random numbers. When you use simple random sampling (see section 7.4), each value is assumed to come from a uniform distribution that has a minimum value of 0, and a maximum value of 1.

Figure 6.25 illustrates the uniform distribution with $a = 0$, and $b = 1$. The total area under the rectangle is equal to the base (1.0) times the height (1.0). Thus, the resulting area of 1.0 satisfies the requirement that the area under any probability density function equals 1.0. In such a distribution, what is the probability of getting a random number between 0.10 and 0.30? The area between 0.10 and 0.30, depicted in Figure 6.26, is equal to the base (which is $0.30 - 0.10 = 0.20$) times the height (1.0). Therefore,

$$P(0.10 < X < 0.30) = (\text{Base})(\text{Height}) = (0.20)(1.0) = 0.20$$

FIGURE 6.25
Probability Density Function for a Uniform Distribution with $a = 0$, and $b = 1$

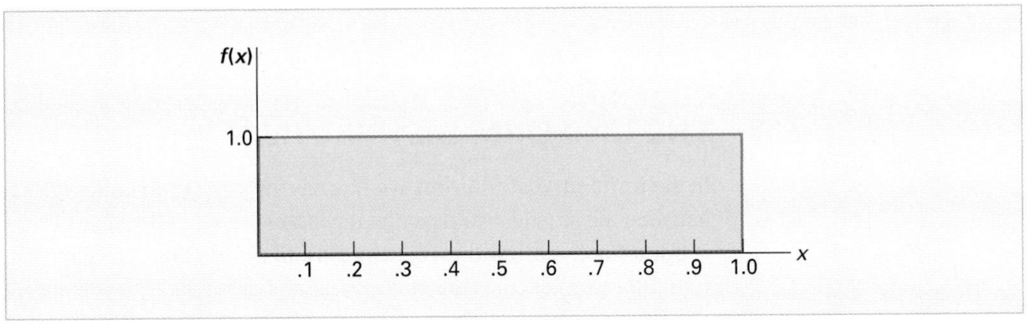

FIGURE 6.26
Finding $P(0.10 < X < 0.30)$ for a Uniform Distribution with $a = 0$, and $b = 1$

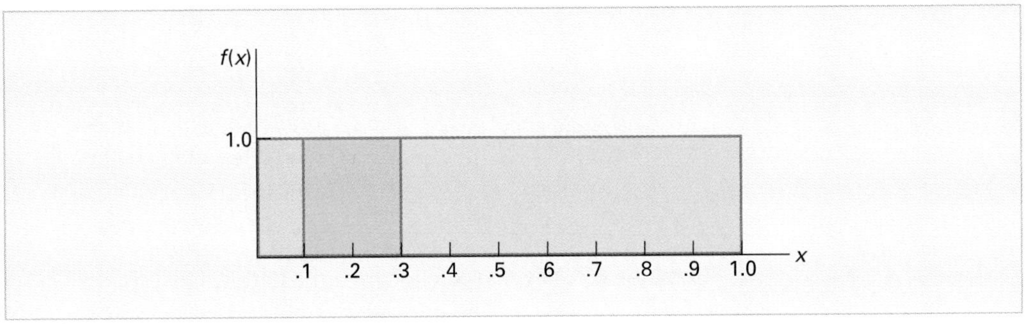

From Equations (6.6) and (6.7), the mean and standard deviation of the uniform distribution for $a = 0$ and $b = 1$ are computed as follows:

$$\mu = \frac{a+b}{2}$$

$$= \frac{0+1}{2} = 0.5$$

and

$$\sigma^2 = \frac{(b-a)^2}{12}$$

$$= \frac{(1-0)^2}{12}$$

$$= \frac{1}{12} = 0.0833$$

$$\sigma = \sqrt{0.0833} = 0.2887$$

Thus, the mean is 0.5 and the standard deviation is 0.2887.

PROBLEMS FOR SECTION 6.4

Learning the Basics

PH Grade ASSIST **6.23** Suppose you sample one value from a uniform distribution with $a = 0$ and $b = 10$. What is the probability of getting a value:

a. between 5 and 7?
b. between 2 and 3?
c. What is the mean?
d. What is the standard deviation?

Applying the Concepts

PH Grade ASSIST SELF Test **6.24** The time between arrivals of customers at a bank during the noon to 1 P.M. hour has a uniform distribution over an interval from 0 to 120 seconds. What is the probability that the time between the arrival of two customers will be:

a. less than 20 seconds?
b. between 10 and 30 seconds?
c. more than 35 seconds?
d. What is the mean and standard deviation of the time between arrivals?

6.25 At an ocean-side nuclear power plant, seawater is used as part of the cooling system. This system raises the temperature of the water that is discharged back into the ocean. The amount that the water temperature is raised has a uniform distribution over the interval from 10 to 25° C.

a. What is the probability that the temperature will increase less than 20° C?
b. What is the probability that the temperature will increase between 20 and 22° C?
c. A temperature increase of more than 18° C is considered to be potentially dangerous to the environment. What is the probability that at any point of time, the temperature increase is potentially dangerous?
d. What is the mean and standard deviation of the temperature increase?

PH Grade ASSIST **6.26** The time of failure for a continuous operation monitoring device of air quality has a uniform distribution over a 24-hour day.

a. If a failure occurs on a day when daylight is between 5:55 A.M. and 7:38 P.M., what is the probability that the failure will occur during daylight hours?
b. If the device is in secondary mode from 10 P.M. to 5 A.M., what is the probability that if a failure occurs, it will happen during secondary mode?
c. If the device has a self-checking computer chip that determines if the device is operational every hour on the hour, what is the probability that a failure will be detected within ten minutes of its occurrence?
d. If the device has a self-checking computer chip that determines if the device is operational every hour on the hour, what is the probability that it will take at least 40 minutes to detect that a failure has occurred?

6.27 The scheduled commuting time on the Long Island Railroad from Glen Cove to New York is 65 minutes. Suppose that the actual commuting time is uniformly distributed between 64 and 74 minutes. What is the probability that the commuting time will be:

a. less than 70 minutes?
b. between 65 and 70 minutes?
c. greater than 65 minutes?
d. What is the mean and standard deviation of the commuting time?

6.5 THE EXPONENTIAL DISTRIBUTION

The **exponential distribution** is a continuous distribution that is right-skewed and ranges from zero to positive infinity (see panel C of Figure 6.1 on page 193). The exponential distribution is widely used in waiting line (or queuing) theory to model the length of time between arrivals in processes such as customers at a bank's ATM, clients in a fast-food restaurant, patients entering a hospital emergency room, and hits on a Web site.

The exponential distribution is defined by a single parameter, its mean λ, the mean number of arrivals per unit of time. The value $1/\lambda$ is equal to the mean time between arrivals. For example, if the mean number of arrivals in a minute is $\lambda = 4$, then the mean time between arrivals is $1/\lambda = 0.25$ minutes or 15 seconds. Equation (6.8) defines the probability that the length of time before the next arrival is less than X.

EXPONENTIAL DISTRIBUTION

$$P(\text{arrival time} < X) = 1 - e^{-\lambda X} \quad \textbf{(6.8)}$$

where

e = the mathematical constant approximated by 2.71828
λ = the mean number of arrivals per unit
X = any value of the continuous variable where $0 < X < \infty$

To illustrate the exponential distribution, suppose that customers arrive at a bank's ATM at the rate of 20 per hour. If a customer has just arrived, what is the probability that the next customer will arrive within 6 minutes (i.e., 0.1 hour)?

For this example, $\lambda = 20$ and $X = 0.1$. Using Equation (6.8),

$$
\begin{aligned}
P(\text{arrival time} < 0.1) &= 1 - e^{-20(0.1)} \\
&= 1 - e^{-2} \\
&= 1 - 0.1353 = 0.8647
\end{aligned}
$$

Thus, the probability that a customer will arrive within 6 minutes is 0.8647, or 86.47%.

You can also use Microsoft Excel and Minitab to compute this probability. Figure 6.27 illustrates a Microsoft Excel worksheet, while Figure 6.28 illustrates Minitab output.

FIGURE 6.27
Microsoft Excel Worksheet for Finding Exponential Probabilities (Mean = λ)

	A	B	
1	Exponential Probability		
2			
3	Data		
4	Mean	20	
5	X Value	0.1	
6			
7	Results		
8	P(<=X)	0.8647	=EXPONDIST(B5, B4, TRUE)

FIGURE 6.28
Minitab Exponential Probabilities (Mean = $1/\lambda$)

Cumulative Distribution Function

```
Exponential with mean = 0.05

  x   P( X <= x )
0.1      0.864665
```

EXAMPLE 6.8 COMPUTING EXPONENTIAL PROBABILITIES

In the ATM example, what is the probability that the next customer will arrive within 3 minutes (i.e., 0.05 hour)?

SOLUTION For this example, $\lambda = 20$ and $X = 0.05$. Using Equation (6.8),

$$
\begin{aligned}
P(\text{arrival time} < 0.05) &= 1 - e^{-20(0.05)} \\
&= 1 - e^{-1} \\
&= 1 - 0.3679 = 0.6321
\end{aligned}
$$

Thus, the probability that a customer will arrive within 3 minutes is 0.6321, or 63.21%.

PROBLEMS FOR SECTION 6.5

Learning the Basics

6.28 Given an exponential distribution with a mean of $\lambda = 10$, what is the probability that the arrival time is
a. less than $X = 0.1$?
b. greater than $X = 0.1$?
c. between $X = 0.1$ and $X = 0.2$?
d. less than $X = 0.1$ or greater than $X = 0.2$?

6.29 Given an exponential distribution with a mean of $\lambda = 30$, what is the probability that the arrival time is
a. less than $X = 0.1$?
b. greater than $X = 0.1$?
c. between $X = 0.1$ and $X = 0.2$?
d. less than $X = 0.1$ or greater than $X = 0.2$?

6.30 Given an exponential distribution with a mean of $\lambda = 20$, what is the probability that the arrival time is
a. less than $X = 4$?
b. greater than $X = 0.4$?
c. between $X = 0.4$ and $X = 0.5$?
d. less than $X = 0.4$ or greater than $X = 0.5$?

Applying the Concepts

PH Grade ASSIST SELF Test **6.31** Autos arrive at a tollbooth located at the entrance to a bridge at the rate of 50 per minute during the 5:00–6:00 P.M. hour. If an auto has just arrived,
a. what is the probability that the next auto arrives within 3 seconds (0.05 minute)?
b. what is the probability that the next auto arrives within 1 second (0.0167 minute)?
c. What are your answers to (a) and (b) if the rate of arrival of autos were 60 per minute?
d. What are your answers to (a) and (b) if the rate of arrival of autos were 30 per minute?

6.32 Customers arrive at the drive-up window of a fast-food restaurant at the rate of 2 per minute during the lunch hour.
a. What is the probability that the next customer will arrive within 1 minute?
b. What is the probability that the next customer will arrive within 5 minutes?
c. During the dinner time period, the arrival rate is 1 per minute. What are your answers to (a) and (b) for this period?

6.33 Telephone calls arrive at the information desk of a large computer software company at the rate of 15 per hour.
a. What is the probability that the next call will arrive within 3 minutes (0.05 hour)?
b. What is the probability that the next call will arrive within 15 minutes (0.25 hour)?
c. Suppose the company has just introduced an updated version of one of its software programs, and telephone calls are now arriving at the rate of 25 per hour. Given this information, redo (a) and (b).

6.34 An on-the-job injury occurs once every 10 days on average at an automobile plant. What is the probability that the next on-the-job injury will occur within
a. 10 days?
b. 5 days?
c. 1 day?

PH Grade ASSIST **6.35** The time between unplanned shutdowns of a power plant has an exponential distribution with a mean of 20 days. Find the probability that the time between two unplanned shutdowns is:
a. less than 14 days.
b. more than 21 days.
c. less than 7 days.

PH Grade ASSIST **6.36** Golfers arrive at the starter's booth of a public golf course at the rate of 8 per hour during the Monday-to-Friday midweek period. If a golfer has just arrived,
a. what is the probability that the next golfer arrives within 15 minutes (0.25 hour)?
b. what is the probability that the next golfer arrives within 3 minutes (0.05 hour)?
c. The actual arrival rate on Fridays is 15 per hour. What are your answers to (a) and (b) on Fridays?

6.37 Trafficweb.org claims that it can deliver 10,000 hits to your Web site in the next 60 days for only $21.95 (**www.trafficWeb.org**, April 26, 2004). If this amount of Web site traffic is experienced, then the time between hits has as a mean of 8.64 minutes (or 0.116 per minute). Assume that your Web site does get 10,000 hits in the next 60 days, and that the time between hits has an exponential distribution. What is the probability that the time between two hits is:
a. less than 5 minutes?
b. less than 10 minutes?
c. more than 15 minutes?
d. Do you think that it is reasonable to assume that the time between hits has an exponential distribution?

6.6 THE NORMAL APPROXIMATION TO THE BINOMIAL DISTRIBUTION

In the earlier sections of this chapter you learned about the normal probability distribution. In this section you will learn how to use the normal distribution to approximate the binomial distribution.

Need for a Correction for Continuity Adjustment

There are two major reasons why you need to use a correction for continuity adjustment. First, discrete random variables that follow the binomial distribution can take on only specified (integer) values, while a continuous random variable (such as the normal) can take on any values within a continuum or interval. When using the normal distribution to approximate the binomial distribution, you can get more accurate approximations of the probabilities if you use a correction for continuity adjustment.

Second with a continuous distribution (such as the normal), the probability of getting a specific value of a random variable is zero. However, when the normal distribution is used to approximate a discrete distribution, you can use a correction for continuity adjustment to get the approximate probability of a specific value of the binomial distribution.

Consider an experiment in which you toss a fair coin 10 times. Suppose you want to compute the probability of getting *exactly* 4 heads. Whereas a discrete random variable can have only a specified value (such as 4), a continuous random variable used to approximate it could take on any values within an interval around that specified value, as demonstrated on the accompanying scale:

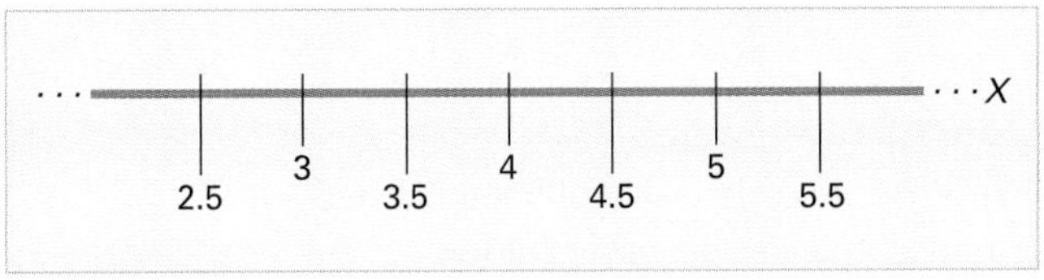

The correction for continuity adjustment requires adding or subtracting 0.5 from the value or values of the discrete random variable X as needed. To use the normal distribution to approximate the probability of getting *exactly* 4 heads (i.e., $X = 4$), you need to find the area under the normal curve from $X = 3.5$ to $X = 4.5$, the lower and upper boundaries of 4. To determine the approximate probability of getting *at least* 4 heads, you find the area under the normal curve greater than or equal to $X = 3.5$, since 3.5 is the lower boundary of X. Similarly, to determine the approximate probability of getting *at most* 4 heads, you find the area under the normal curve equal to or less than $X = 4.5$ since 4.5 is the upper boundary of X.

When using the normal distribution to approximate discrete probability distributions, semantics are important. To determine the approximate probability of getting *fewer* than 4 heads, you find the area under the normal curve less than or equal to $X = 3.5$. To determine the approximate probability of getting *more than* 4 heads, you find the area under the normal curve greater than or equal to $X = 4.5$. To determine the approximate probability of getting 4 *through* 7 heads, you find the area under the normal curve from $X = 3.5$ to $X = 7.5$.

Approximating the Binomial Distribution

In section 5.3 you learned that the binomial distribution is symmetric (like the normal distribution) whenever $p = 0.5$. When $p \neq 0.5$, the binomial distribution is not symmetric. However, the closer p is to 0.5 and the larger the sample size n, the more symmetric the distribution becomes.

On the other hand, the larger the sample size, the more tedious it is to compute the exact probabilities of success by using Equation (5.11) on page 168. Fortunately, whenever the sample size is large, you can use the normal distribution to approximate the exact probabilities of success.

As a general rule, you can use the normal distribution to approximate the binomial distribution whenever np and $n(1-p)$ are both at least 5. From section 5.3, you know that the mean of the binomial distribution is

$$\mu = np$$

and the standard deviation of the binomial distribution is

$$\sigma = \sqrt{np(1-p)}$$

Substituting these results into the transformation formula [Equation (6.2) on page 195],

$$Z = \frac{X - \mu}{\sigma} = \frac{X - np}{\sqrt{np(1-p)}}$$

so that, for large enough n, the random variable Z is approximately normally distributed. Hence, you find approximate probabilities corresponding to the values of the discrete random variable X, by using Equation (6.9).

NORMAL APPROXIMATION TO THE BINOMIAL DISTRIBUTION

$$Z = \frac{X_a - np}{\sqrt{np(1-p)}} \quad \textbf{(6.9)}$$

where $\mu = np$, mean of the binomial distribution

$\sigma = \sqrt{np(1-p)}$, standard deviation of the binomial distribution

X_a = adjusted number of successes for the discrete random variable X, such that $X_a = X - 0.5$ or $X_a = X + 0.5$ as appropriate

EXAMPLE 6.9

USING THE NORMAL DISTRIBUTION TO APPROXIMATE THE BINOMIAL DISTRIBUTION

You select a random sample of n = 1,600 tires from an ongoing production process in which 8% of all such tires produced are defective. What is the probability that 150 or fewer tires will be defective?

SOLUTION Since both np = 1,600(.08) = 128 and $n(1-p)$ = 1,600(0.92) = 1,472 are greater than 5, you can use the normal distribution to approximate the binomial. Here X_a, the adjusted number of successes, is 150.5.

$$Z \cong \frac{X_a - np}{\sqrt{np(1-p)}} = \frac{150.5 - 128}{\sqrt{(1{,}600)(0.08)(0.92)}} = \frac{22.5}{10.85} = +2.07$$

Using Table E.2, the area under the curve to the left of Z = +2.07 is 0.9808 (see Figure 6.29).

FIGURE 6.29
Approximating the Binomial Distribution

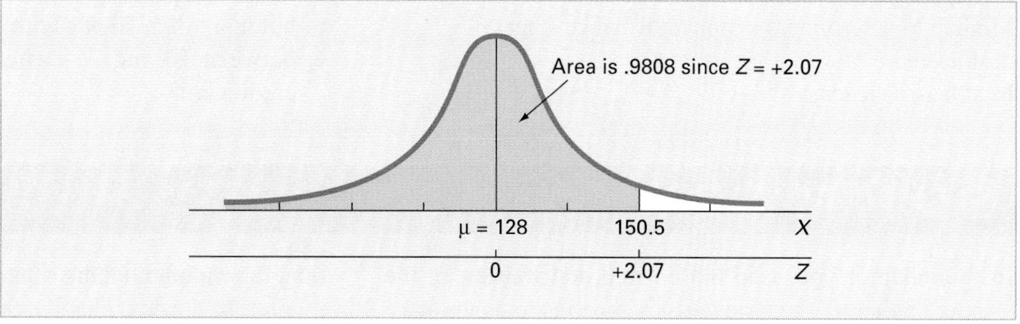

Computing a Probability Approximation for an Individual Value

Suppose that you want to approximate the probability of getting *exactly* 150 defective tires. The correction for continuity defines the integer value of interest to range from one-half unit below it to one-half unit above it. Therefore, you define the probability of getting 150 defective tires as the area (under the normal curve) between 149.5 and 150.5. Using Equation (6.9), you approximate the probability as follows:

$$Z = \frac{150.5 - 128}{\sqrt{(1{,}600)(0.08)(0.92)}} = \frac{22.5}{10.85} = +2.07$$

and

$$Z = \frac{149.5 - 128}{\sqrt{(1{,}600)(0.08)(0.92)}} = \frac{21.5}{10.85} = +1.98$$

From Table E.2, the area under the normal curve less than $X = 150.5$ ($Z = +2.07$) is 0.9808 and the area under the curve less than $X = 149.5$ ($Z = +1.98$) is 0.9761. Thus, the approximate probability of getting 150 defective tires is the difference in the two areas, 0.0047.

PROBLEMS FOR SECTION 6.6

Learning the Basics

6.38 For $n = 100$ and $p = 0.20$, use the normal distribution to approximate the probability that

a. $X = 25$.
b. $X > 25$.
c. $X \leq 25$.
d. $X < 25$.

6.39 For $n = 100$ and $p = 0.40$, use the normal distribution to approximate the probability that

a. $X = 40$.
b. $X > 40$.
c. $X \leq 40$.
d. $X < 40$.

Applying the Concepts

6.40 Consider an experiment in which you toss a fair coin 10 times and count the number of heads. Use Equation (5.11) on page 168 or Table E.6, or Microsoft Excel or Minitab to determine the probability of getting:

a. 4 heads.
b. at least 4 heads.
c. 4 through 7 heads.
d. Use the normal approximation to the binomial distribution to approximate the probabilities in (a) through (c).

6.41 For overseas flights, an airline has three different choices on its dessert menu—ice cream, apple pie, and chocolate cake. Based on past experience, the airline feels that each dessert is equally likely to be chosen. If a random sample of 90 passengers is selected, what is the *approximate* probability that:

a. at least 20 will choose ice cream for dessert?
b. exactly 20 will choose ice cream for dessert?
c. less than 20 will choose ice cream for dessert?

6.42 Based upon past experience, 40% of all customers at Miller's Automotive Service Station pay for their purchases with a credit card. If a random sample of 200 customers is selected, what is the *approximate* probability that:

a. at least 75 pay with a credit card?
b. not more than 70 pay with a credit card?
c. between 70 and 75 customers, inclusive, pay with a credit card?

SUMMARY

In this chapter you used the normal distribution in the "Using Statistics" scenario to study the time to download a Web page. In addition, you studied the uniform distribution, the exponential distribution, and the normal probability plot. In the next chapter, the normal distribution is used in developing the subject of statistical inference.

KEY FORMULAS

The Normal Probability Density Function

$$f(X) = \frac{1}{\sqrt{2\pi}\sigma} e^{-(1/2)[(X-\mu)/\sigma]^2} \quad \textbf{(6.1)}$$

Finding a Z Value

$$Z = \frac{X - \mu}{\sigma} \quad \textbf{(6.2)}$$

The Standardized Normal Probability Density Function

$$f(Z) = \frac{1}{\sqrt{2\pi}} e^{-(1/2)Z^2} \quad \textbf{(6.3)}$$

Finding an X Value

$$X = \mu + Z\sigma \quad \textbf{(6.4)}$$

The Uniform Distribution

$$f(X) = \frac{1}{b - a} \text{ if } a \le X \le b \text{ and 0 elsewhere} \quad \textbf{(6.5)}$$

Mean of the Uniform Distribution

$$\mu = \frac{a + b}{2} \quad \textbf{(6.6)}$$

Variance and Standard Deviation of the Uniform Distribution

$$\sigma^2 = \frac{(b - a)^2}{12} \quad \textbf{(6.7a)}$$

$$\sigma = \sqrt{\frac{(b - a)^2}{12}} \quad \textbf{(6.7b)}$$

Exponential Distribution

$$P(\text{arrival time} < X) = 1 - e^{-\lambda X} \quad \textbf{(6.8)}$$

Normal Approximation to the Binomial Distribution

$$Z \cong \frac{X_a - np}{\sqrt{np(1 - p)}} \quad \textbf{(6.9)}$$

KEY TERMS

continuous probability density function 192
cumulative standardized normal distribution 195
exponential distribution 215
normal distribution 193
normal probability density function 194
normal probability plot 209
quantile-quantile plot 209
rectangular distribution 213
standardized normal random variable 195
transformation formula 195
uniform distribution 213

CHAPTER REVIEW PROBLEMS

Checking Your Understanding

6.43 Why is it that only one normal distribution table such as Table E.2 is needed to find any probability under the normal curve?

6.44 How do you find the area between two values under the normal curve?

6.45 How do you find the X value that corresponds to a given percentile of the normal distribution?

6.46 What are some of the distinguishing properties of a normal distribution?

6.47 How can you use the normal probability plot to evaluate whether a set of data is normally distributed?

6.48 Under what circumstances can you use the exponential distribution?

6.49 Why is a correction for continuity adjustment needed when approximating a binomial probability with a normal distribution?

6.50 When can you use the normal distribution to approximate the binomial distribution?

Applying the Concepts

6.51 An industrial sewing machine uses ball bearings that are targeted to have a diameter of 0.75 inch. The lower and upper specification limits under which the ball bearing can operate are 0.74 inch and 0.76 inch, respectively. Past experience has indicated that the actual diameter of the ball bearings is approximately normally distributed with a mean of 0.753 inch and a standard deviation of 0.004 inch. What is the probability that a ball bearing is

a. between the target and the actual mean?
b. between the lower specification limit and the target?
c. above the upper specification limit?
d. below the lower specification limit?
e. 93% of the diameters are greater than what value?

6.52 The fill amount of soft drink bottles is normally distributed with a mean of 2.0 liters and a standard deviation of 0.05 liter. Bottles that contain less than 95% of the listed net content (1.90 liters in this case) can make the manufacturer subject to penalty by the state office of consumer affairs. Bottles that have a net content above 2.10 liters may cause excess spillage upon opening. What proportion of the bottles will contain

a. between 1.90 and 2.0 liters?
b. between 1.90 and 2.10 liters?
c. below 1.90 liters or above 2.10 liters?
d. 99% of the bottles contain at least how much soft drink?
e. 99% of the bottles contain an amount that is between which two values (symmetrically distributed) around the mean?

6.53 In an effort to reduce the number of bottles that contain less than 1.90 liters, the bottler in problem 6.52 sets the filling machine so that the mean is 2.02 liters. Under these circumstances, what are your answers in (a) through (e)?

6.54 An orange juice producer buys all his oranges from a large orange grove. The amount of juice squeezed from each of these oranges is approximately normally distributed with a mean of 4.70 ounces and a standard deviation of 0.40 ounce.

a. What is the probability that a randomly selected orange will contain between 4.70 and 5.00 ounces?
b. What is the probability that a randomly selected orange will contain between 5.00 and 5.50 ounces?
c. 77% of the oranges will contain at least how many ounces of juice?
d. 80% of the oranges are between what two values (in ounces) symmetrically distributed around the population mean?

6.55 According to *Investment Digest* ("Diversification and the Risk/Reward Relationship," Winter 1994, 1–3), the mean of the annual return for common stocks from 1926 to 1992 was 12.4%, and the standard deviation of the annual return was 20.6%. The article claims that the distribution of annual returns for common stocks is approximately bell-shaped and symmetric. Assume that the distribution is normally distributed with the mean and standard deviation given above. Find the probability that the return for common stocks will be

a. greater than 0%.
b. greater than 10%.
c. greater than 20%.
d. less than −10%.

6.56 During the same 67-year time span mentioned in problem 6.55, the mean of the annual return for long-term government bonds was 5.2%, and the standard deviation was 8.6%. The article claims that the distribution of annual returns for long-term government bonds is approximately bell-shaped and symmetric. Assume that the distribution is normally distributed with the mean and standard deviation given above. Find the probability that the return for long-term government bonds will be

a. greater than 0%.
b. greater than 10%.
c. greater than 20%.
d. less than −10%.
e. Discuss the differences between common stocks and long-term government bonds.

6.57 *The Wall Street Journal* reported that almost all the major stock market indexes had posted strong gains in the last 12 months ("What's Hot . . . and Not," *The Wall Street Journal*, April 26, 2004, C3). The one-year return for the S&P 500, a group of 500 very large companies, was approximately +27%. The one-year return in the Russell 2000, a group of 2000 small companies, was approximately +52%. Historically, the one-year returns are approximately normal. The standard deviation in the S&P 500 returns is approximately 20%, and in the Russell 2000 the standard deviation is approximately 35%.

a. What is the probability that a stock in the S&P 500 gained 30% or more in the last year? Gained 60% or more in the last year?
b. What is the probability that a stock in the S&P 500 lost money in the last year? Lost 30% or more?
c. Repeat (a) and (b) for a stock in the Russell 2000.
d. Write a short summary on your findings. Be sure to include a discussion of the risks associated with a large standard deviation.

6.58 The *New York Times* reported (Laurie J. Flynn, "Tax Surfing," *The New York Times*, March 25, 2002, C10) that the mean time to download the homepage from the Internal Revenue Service Web site **www.irs.gov** is 0.8 seconds. Suppose that the download time is normally distributed with a standard deviation of 0.2 seconds. What is the probability that a download time is

a. less than 1 second?
b. between 0.5 and 1.5 seconds?
c. above 0.5 second?
d. 99% of the download times are above how many seconds?
e. 95% of the download times are between what two values symmetrically distributed around the mean?

6.59 The same article mentioned in problem 6.58 also reported that the mean download time for the H&R Block Web site **www.hrblock.com** is 2.5 seconds. Suppose that the download time is normally distributed with a standard deviation of 0.5 seconds. What is the probability that a download time is

a. less than 1 second?
b. between 0.5 and 1.5 seconds?
c. above 0.5 second?
d. 99% of the download times are above how many seconds?
e. Compare the results for the IRS site computed in problem 6.58 to those of the H&R Block site.

6.60 (**Class Project**) According to Burton G. Malkiel, the daily changes in the closing price of stock follow a *random walk*—that is, these daily events are independent of each other and move upward or downward in a random manner—and can be approximated by a normal distribution. To test this theory, use either a newspaper or the Internet to select one company traded on the New York Stock Exchange, one company traded on the American Stock Exchange, and one company traded "over the counter" (i.e., on the NASDAQ national market) and then do the following:

1. Record the daily closing stock price of each of these companies for 6 consecutive weeks (so that you have 30 values per company).
2. Record the daily changes in the closing stock price of each of these companies for 6 consecutive weeks (so that you have 30 values per company).

For each of your six data sets, decide whether the data are approximately normally distributed by

a. examining the stem-and-leaf display, histogram or polygon and the box-and-whisker plot.
b. evaluating the actual versus theoretical properties.
c. constructing a normal probability plot.
d. Discuss the results of (a), (b), and (c). What can you now say about your three stocks with respect to daily closing prices and daily changes in closing prices? Which, if any, of the data sets are approximately normally distributed?

Note: The random-walk theory pertains to the daily *changes* in the closing stock price, not the daily closing stock price.

TEAM PROJECT

The data file MUTUALFUNDS2004 contains information regarding 12 variables from a sample of 121 mutual funds. The variables are:

Fund—The name of the mutual fund.
Category—Type of stocks comprising the mutual fund—small cap, mid cap, large cap
Objective—Objective of stocks comprising the mutual fund—growth or value
Assets—In millions of dollars
Fees—Sales charges (no or yes)
Expense ratio—Ratio of expenses to net assets in percentage
2003 Return—Twelve-month return in 2003
Three-year return—Annualized return 2001–2003
Five-year return—Annualized return 1999–2003
Risk—Risk-of-loss factor of the mutual fund classified as low, average, or high
Best quarter—Best quarterly performance 1999–2003
Worst quarter—Worst quarterly performance 1999–2003

6.61 Consider the variables expense ratio, three-year annualized return, and five-year annualized return. For each of these variables decide whether the data are approximately normally distributed by

a. evaluating the actual versus theoretical properties.
b. constructing a normal probability plot.

RUNNING CASE MANAGING THE *SPRINGVILLE HERALD*

The production department of the newspaper has embarked on a quality improvement effort. Its first project relates to the blackness of the newspaper print. Each day a determination needs to be made concerning how "black" the newspaper is printed. Blackness is measured on a standard scale in which the target value is 1.0. Data collected over the past year indicate that the blackness is normally distributed with a mean of 1.005 and a standard deviation of 0.10.

Each day, one spot on the first newspaper printed is chosen and the blackness of the spot is measured. The blackness of the newspaper is considered acceptable if the blackness of the spot is between 0.95 and 1.05.

EXERCISE

SH6.1 Assuming that the distribution has not changed from what it was in the past year, what is the probability that the blackness of the spot is:

a. less than 1.0?
b. between 0.95 and 1.0?
c. between 1.0 and 1.05?
d. less than 0.95 or greater than 1.05?

SH6.2 The objective of the production team is to reduce the probability that the blackness is below 0.95 or above 1.05. Would it be better off focusing on process improvement that lowered the mean to the target value of 1.0 or on process improvement that reduced the standard deviation to 0.075? Explain.

WEB CASE

Apply your knowledge about the normal distribution in this Web Case that extends the "Using Statistics" scenario from this chapter.

To satisfy concerns of potential advertisers, the management of OnCampus! has undertaken a research project to learn the amount of time viewers linger at their Web sites (known as the "stickiness" of viewers). The marketing department has collected data and has made some claims based on the assertion that the data follow a normal distribution. These data and conclusions can be found in a report located on the internal Web page **www.prenhall.com/Springville/OC_MarketingSurvey.htm**.

Read this marketing report and then answer the following:

1. Can the collected data be approximated by the normal distribution?
2. Review and evaluate the conclusions made by the OnCampus! marketing department. Which conclusions are correct? Which ones are incorrect?
3. If OnCampus! can improve the mean time by five minutes, how would the probabilities change?

REFERENCES

1. Cochran, W. G., *Sampling Techniques*, 3rd ed. (New York: Wiley, 1977).
2. Gunter, B., "Q-Q Plots," *Quality Progress* (February 1994), 81–86.
3. Levine, D. M., P. Ramsey, and R. Smidt, *Applied Statistics for Engineers and Scientists Using Microsoft Excel and Minitab* (Upper Saddle River, NJ: Prentice Hall, 2001).
4. Marascuilo, L. A., and M. McSweeney, *Nonparametric and Distribution-Free Methods for the Social Sciences* (Monterey, CA: Brooks/Cole, 1977).
5. *Microsoft Excel 2003* (Redmond, WA: Microsoft Corp., 2003).
6. *Minitab for Windows Version 14* (State College, PA: Minitab Inc., 2004).

Appendix 6 Using Software for Continuous Probability Distributions

A6.1 MICROSOFT EXCEL

For Normal Probabilities

Open the **Normal.xls** file, shown in Figure 6.18 on page 206. This worksheet solves the problems for Examples 6.5 and 6.6. Open the similar **Normal Expanded.xls** file to solve problems similar to Examples 6.2 and 6.3 as well as Examples 6.5 and 6.6. Both worksheets use the STANDARDIZE, NORMDIST, NORMSINV, and NORMINV functions to calculate normal probabilities and related values (see section G.10 for additional information).

To adapt this worksheet to other problems, change the **Mean, Standard Deviation, X Value**, and **Cumulative Percentage Values** in the aqua-tinted cells.

Or see section G.10 **(Normal)** if you want PHStat2 to produce a worksheet for you.

For Normal Probability Plot

There are no Microsoft Excel commands that directly produce a Normal Probability Plot. If you want PHStat2 to produce a Normal Probability Plot, see section G. 11.

For Exponential Probability

Open the **Exponential.xls** file, shown in Figure 6.27 on page 216. This worksheet already contains the entries for the bank ATM problem of section 6.5. The worksheet uses the EXPONDIST function to calculate exponential probabilities (see section G.12 for additional information). To adapt this worksheet for other problems, change the **Mean** and **X Value** values in cells B4 and B5.

Or see section G.12 **(Exponential)** if you want PHStat2 to produce this worksheet for you.

A6.2 MINITAB

Using Minitab to Compute Normal Probabilities

You can use Minitab instead of Table E.2 to compute normal probabilities. To find the probability that a download time is less than 9 seconds with $\mu = 7$ and $\sigma = 2$:

1. Enter **9** in the first row of column C1 of a blank worksheet.
2. Select **Calc ➔ Probability Distributions ➔ Normal**.
3. In the Normal Distribution dialog box (see Figure A6.1) select the **Cumulative probability** option button. Enter **7** in the Mean: edit box and **2** in the Standard deviation: edit box. Select the **Input column**: option button and enter **C1** in its edit box. Click the **OK** button. You will get the output displayed in the top portion of Figure 6.19 on page 206.

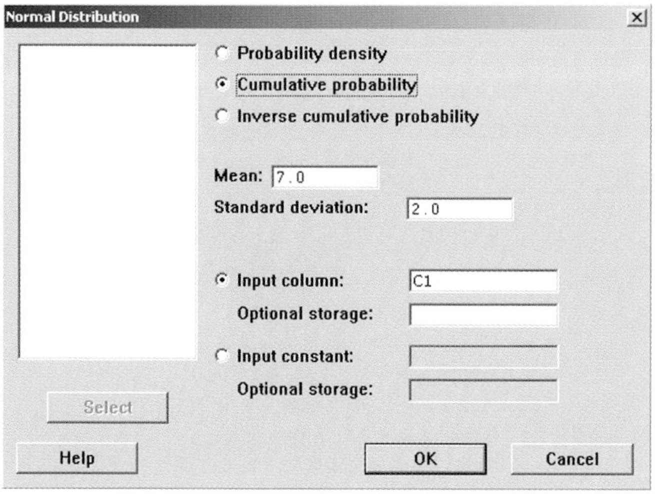

FIGURE A6.1 Minitab Normal Distribution Dialog Box

To find the Z value corresponding to a cumulative area of 0.10,

1. Enter **.10** in row 1 of column C2.
2. Select **Calc ➔ Probability Distributions ➔ Normal**.
3. Select the **Inverse cumulative probability** option button. Enter **7** in the Mean: edit box and **2** in the Standard Deviation: edit box.
4. Select the **Input Column** option button and enter **C2** in the edit box. Click the **OK** button.

You will get the output displayed in the bottom portion of Figure 6.19 on page 206.

Using Minitab for a Normal Probability Plot

To construct a normal probability plot in Minitab for the 2003 returns of mutual funds, open the **MUTUALFUNDS2004.MTW** worksheet. Then,

1. Select **Graph ➔ Probability Plot**.
2. In the Probability Plots dialog box, select **Single**. Click the **OK** button.
3. In the Probability Plot-Single dialog box (see Figure A6.2), in the Graph variables: edit box, enter **C7** or **'Return 2003'**.

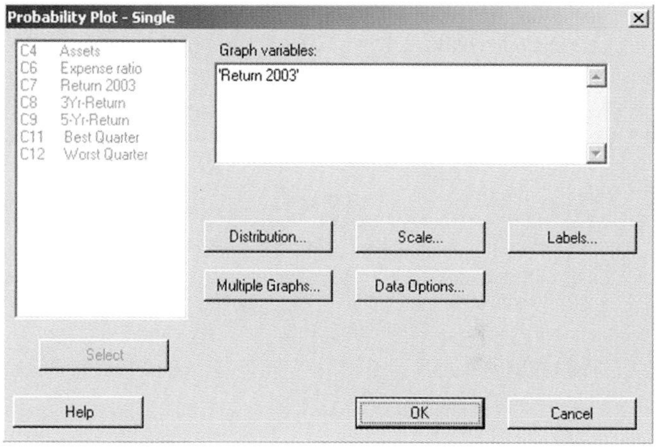

FIGURE A6.2 Minitab Probability Plot-Single Dialog Box

4. Click the **Distribution** button. In the Probability Plot - Distribution dialog box (see Figure A6.3), select **Normal** in the Distribution drop-down list box. Click the **OK** button to return to the Probability Plot - Single dialog box. Click the **OK** button.

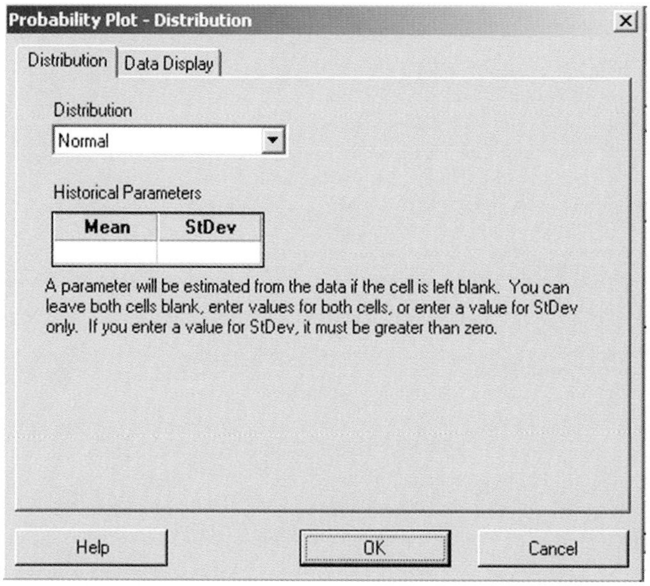

FIGURE A6.3 Minitab Probability Plot-Distribution Dialog Box

Using Minitab to Compute Exponential Probabilities

To compute exponential probabilities for the ATM customer-arrival example using Minitab,

1. Enter **.1** in column C1 of a blank worksheet.
2. Select **Calc → Probability Distributions → Exponential**.
3. In the Exponential Distribution dialog box (see Figure A6.4), select the **Cumulative probability** option button. In the Scale: edit box, enter **.05**. This entry is made because Minitab defines the mean as the mean time *between* arrivals, $1/\lambda = 1/20 = 0.05$, not the mean number of arrivals, $\lambda = 20$. Select the **Input column:** option button and enter **C1** in the Input column: edit box. Click the **OK** button.

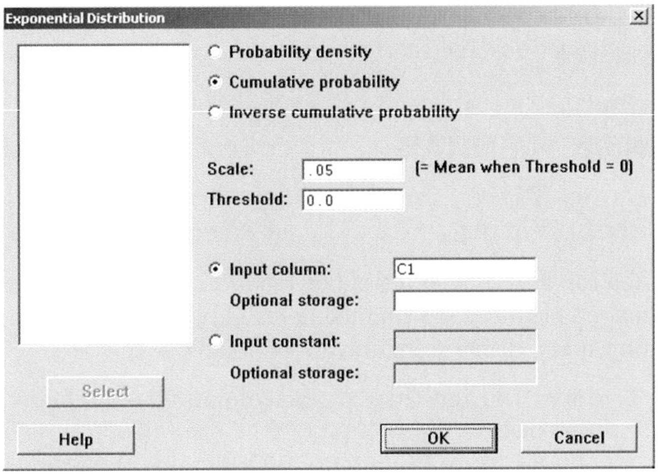

FIGURE A6.4 Minitab Exponential Distribution Dialog Box

CHAPTER 7

Sampling Distributions

LEARNING OBJECTIVES

In this chapter, you learn:

- The concept of the sampling distribution
- To compute probabilities related to the sample mean and the sample proportion
- The importance of the Central Limit Theorem
- To distinguish between different survey sampling methods

USING STATISTICS

Cereal-Fill Packaging Process

At Oxford Cereal Company plants, thousands of boxes of cereal are filled during each eight-hour shift. As the operations manager of a plant, you are in charge of monitoring the amount of cereal contained in the boxes. Boxes are supposed to contain a mean of 368 grams of cereal as indicated on the package label. Because of the speed of the process, the cereal weight varies from box to box, causing some boxes to be underfilled and some overfilled. If the process is not working properly, the mean weight in the boxes could vary too much from the label weight of 368 grams to be acceptable. Because weighing every single box is too time-consuming, costly, and inefficient, you must take a sample of boxes and make a decision regarding the probability that the cereal-filling process is working properly. Each time you select a sample of boxes and weigh the individual boxes, you calculate a sample mean $\bar{X}$. You need to determine the probability that such an $\bar{X}$ could have been randomly drawn from a population whose population mean is 368 grams. Based on this assessment, you will have to decide whether to maintain, alter, or shut down the process.

In the last chapter, you used the normal distribution to study the distribution of download times for the OnCampus! Web site. In this chapter, you need to make a decision about the cereal-filling process based on a *sample* of cereal boxes. You will learn about *sampling distributions* and how to use them to solve business problems. As in the previous chapter, the normal distribution is used to calculate probabilities.

7.1 SAMPLING DISTRIBUTIONS

In many applications you want to make statistical inferences, that is, to use statistics calculated from samples to estimate the values of population parameters. In this chapter you will learn about the sample mean, a statistic used to estimate a population mean (a parameter). You will also learn about the sample proportion, a statistic used to estimate the population proportion (a parameter). Your main concern when making a statistical inference is drawing conclusions about a population, *not* about a sample. For example, a political pollster is interested in the sample results only as a way of estimating the actual proportion of the votes that each candidate will receive from the population of voters. Likewise, as an operations manager for the Oxford Cereal Company, you are only interested in using the sample mean calculated from a sample of cereal boxes for estimating the mean weight contained in a population of boxes.

In practice, you select a single random sample of a predetermined size from the population. The items included in the sample are determined through the use of a random number generator, such as a table of random numbers (see section 7.4 and Table E.1), or by using Microsoft Excel or Minitab (see page 256).

Hypothetically, to use the sample statistic to estimate the population parameter, you should examine *every* possible sample that could occur. A **sampling distribution** is the distribution of the results if you actually selected all possible samples.

7.2 SAMPLING DISTRIBUTION OF THE MEAN

In Chapter 3, several measures of central tendency were discussed. Undoubtedly, the mean is the most widely used measure of central tendency. The sample mean is often used to estimate the population mean. The **sampling distribution of the mean** is the distribution of all possible sample means if you select all possible samples of a certain size.

The Unbiased Property of the Sample Mean

The sample mean is **unbiased** because the mean of all the possible sample means (of a given sample size *n*) is equal to the population mean μ. A simple example concerning a population of four administrative assistants demonstrates this property. Each assistant is asked to type the same page of a manuscript. Table 7.1 presents the number of errors.

TABLE 7.1
Number of Errors Made by Each of Four Administrative Assistants

Administrative Assistant	Number of Errors
Ann	$X_1 = 3$
Bob	$X_2 = 2$
Carla	$X_3 = 1$
Dave	$X_4 = 4$

This population distribution is shown in Figure 7.1.

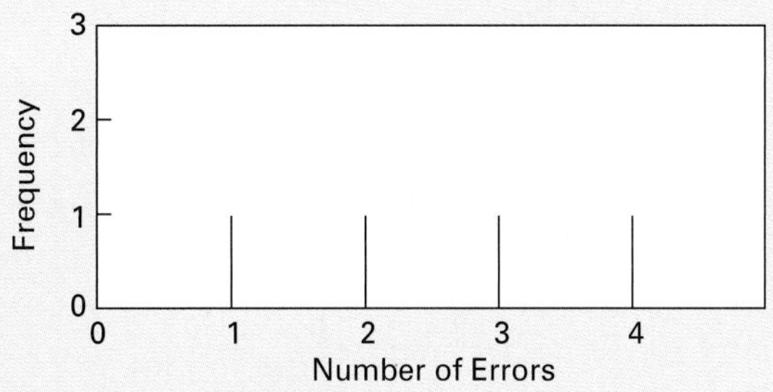

FIGURE 7.1
Number of Errors Made by a Population of Four Administrative Assistants

When you have the data from a population, you compute the mean using Equation (7.1):

POPULATION MEAN

The population mean is the sum of the values in the population divided by the population size *N*.

$$\mu = \frac{\sum_{i=1}^{N} X_i}{N} \quad \textbf{(7.1)}$$

You compute the population standard deviation σ using Equation (7.2):

POPULATION STANDARD DEVIATION

$$\sigma = \sqrt{\frac{\sum_{i=1}^{N} (X_i - \mu)^2}{N}} \quad \textbf{(7.2)}$$

Thus, for the data of Table 7.1,

$$\mu = \frac{3 + 2 + 1 + 4}{4} = 2.5 \text{ errors}$$

and

$$\sigma = \sqrt{\frac{(3 - 2.5)^2 + (2 - 2.5)^2 + (1 - 2.5)^2 + (4 - 2.5)^2}{4}} = 1.12 \text{ errors}$$

If you select samples of two administrative assistants *with* replacement from this population, there are 16 possible samples ($N^n = 4^2 = 16$). Table 7.2 lists the 16 possible sample outcomes. If you average all 16 of these sample means, the mean of these values, $\mu_{\bar{X}}$, is equal to 2.5, which is also the mean of the population μ.

TABLE 7.2

All 16 Samples of $n = 2$ Administrative Assistants from a Population of $N = 4$ Administrative Assistants When Sampling *with* Replacement

Sample	Administrative Assistants	Sample Outcomes	Sample Mean
1	Ann, Ann	3, 3	$\bar{X}_1 = 3$
2	Ann, Bob	3, 2	$\bar{X}_2 = 2.5$
3	Ann, Carla	3, 1	$\bar{X}_3 = 2$
4	Ann, Dave	3, 4	$\bar{X}_4 = 3.5$
5	Bob, Ann	2, 3	$\bar{X}_5 = 2.5$
6	Bob, Bob	2, 2	$\bar{X}_6 = 2$
7	Bob, Carla	2, 1	$\bar{X}_7 = 1.5$
8	Bob, Dave	2, 4	$\bar{X}_8 = 3$
9	Carla, Ann	1, 3	$\bar{X}_9 = 2$
10	Carla, Bob	1, 2	$\bar{X}_{10} = 1.5$
11	Carla, Carla	1, 1	$\bar{X}_{11} = 1$
12	Carla, Dave	1, 4	$\bar{X}_{12} = 2.5$
13	Dave, Ann	4, 3	$\bar{X}_{13} = 3.5$
14	Dave, Bob	4, 2	$\bar{X}_{14} = 3$
15	Dave, Carla	4, 1	$\bar{X}_{15} = 2.5$
16	Dave, Dave	4, 4	$\bar{X}_{16} = 4$
		$\mu_{\bar{X}} = 2.5$	

Since the mean of the 16 sample means is equal to the population mean, the sample mean is an unbiased estimator of the population mean. Therefore, although you do not know how close the sample mean of any particular sample selected comes to the population mean, you are at least assured that the mean of all the possible sample means that could have been selected is equal to the population mean.

Standard Error of the Mean

Figure 7.2 illustrates the variation in the sample mean when selecting all 16 possible samples. In this small example, although the sample mean varies from sample to sample depending on which administrative assistants are selected, the sample mean does not vary as much as the individual values in the population. That the sample means are less variable than the individual values in the population follows directly from the fact that each sample mean averages together all the values in the sample. A population consists of individual outcomes that can take on a wide range of values from extremely small to extremely large. However, if a sample contains an extreme value, although this value will have an effect on the sample mean, the effect is reduced because the value is averaged with all the other values in the sample. As the sample size increases, the effect of a single extreme value becomes smaller because it is averaged with more values.

FIGURE 7.2
Sampling Distribution of the Mean Based on All Possible Samples Containing Two Administrative Assistants

Source: *Data are from Table 7.2.*

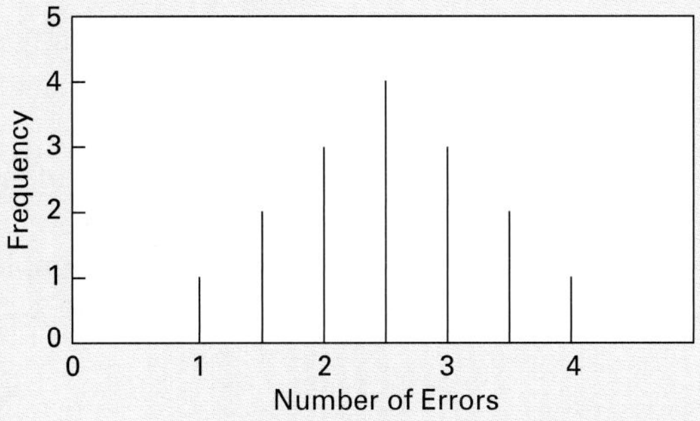

The value of the standard deviation of all possible sample means, called the **standard error of the mean**, expresses how the sample mean varies from sample to sample. Equation (7.3) defines the standard error of the mean when sampling *with* replacement, or *without* replacement (see page 243) from large or infinite populations.

STANDARD ERROR OF THE MEAN

The standard error of the mean $\sigma_{\bar{X}}$ is equal to the standard deviation in the population σ divided by the square root of the sample size n.

$$\sigma_{\bar{X}} = \frac{\sigma}{\sqrt{n}} \qquad \textbf{(7.3)}$$

Therefore, as the sample size increases, the standard error of the mean decreases by a factor equal to the square root of the sample size.

You can also use Equation (7.3) as an approximation to the standard error of the mean when the sample is selected without replacement if the sample contains less than 5% of the entire population. Example 7.1 computes the standard error of the mean for such a situation. (See the section 7.6.pdf file on the CD-ROM that accompanies this text for the case in which more than 5% of the population is contained in a sample selected without replacement.)

EXAMPLE 7.1

COMPUTING THE STANDARD ERROR OF THE MEAN

Return to the cereal-filling process described in the "Using Statistics" scenario on page 228. If you randomly select a sample of 25 boxes without replacement from the thousands of boxes filled during a shift, the sample contains far less than 5% of the population. Given that the standard deviation of the cereal-filling process is 15 grams, compute the standard error of the mean.

SOLUTION Using Equation (7.3) above with $n = 25$ and $\sigma = 15$, the standard error of the mean is

$$\sigma_{\bar{X}} = \frac{\sigma}{\sqrt{n}} = \frac{15}{\sqrt{25}} = \frac{15}{5} = 3$$

The variation in the sample means for samples of $n = 25$ is much less than the variation in the individual boxes of cereal (i.e., $\sigma_{\bar{X}} = 3$ while $\sigma = 15$).

Sampling from Normally Distributed Populations

Now that the concept of a sampling distribution has been introduced and the standard error of the mean has been defined, what distribution will the sample mean $\overline{X}$ follow? If you are sampling from a population that is normally distributed with mean μ and standard deviation σ, regardless of the sample size n, the sampling distribution of the mean is normally distributed with mean $\mu_{\overline{X}} = \mu$ and standard error of the mean $\sigma_{\overline{X}}$.

In the simplest case, if you take samples of size $n = 1$, each possible sample mean is a single value from the population because

$$\overline{X} = \frac{\sum_{i=1}^{n} X_i}{n} = \frac{X_i}{1} = X_i$$

Therefore, if the population is normally distributed with mean μ and standard deviation σ, then the sampling distribution of $\overline{X}$ for samples of $n = 1$ must also follow the normal distribution with mean $\mu_{\overline{X}} = \mu$ and standard error of the mean $\sigma_{\overline{X}} = \sigma/\sqrt{1} = \sigma$. In addition, as the sample size increases, the sampling distribution of the mean still follows a normal distribution with mean $\mu_{\overline{X}} = \mu$, but the standard error of the mean decreases, so that a larger proportion of sample means are closer to the population mean. Figure 7.3 illustrates this reduction in variability in which 500 samples of sizes 1, 2, 4, 8, 16, and 32 were randomly selected from a

FIGURE 7.3

Sampling Distribution of the Mean from 500 Samples of Sizes $n = 1$, 2, 4, 8, 16, and 32 Selected from a Normal Population

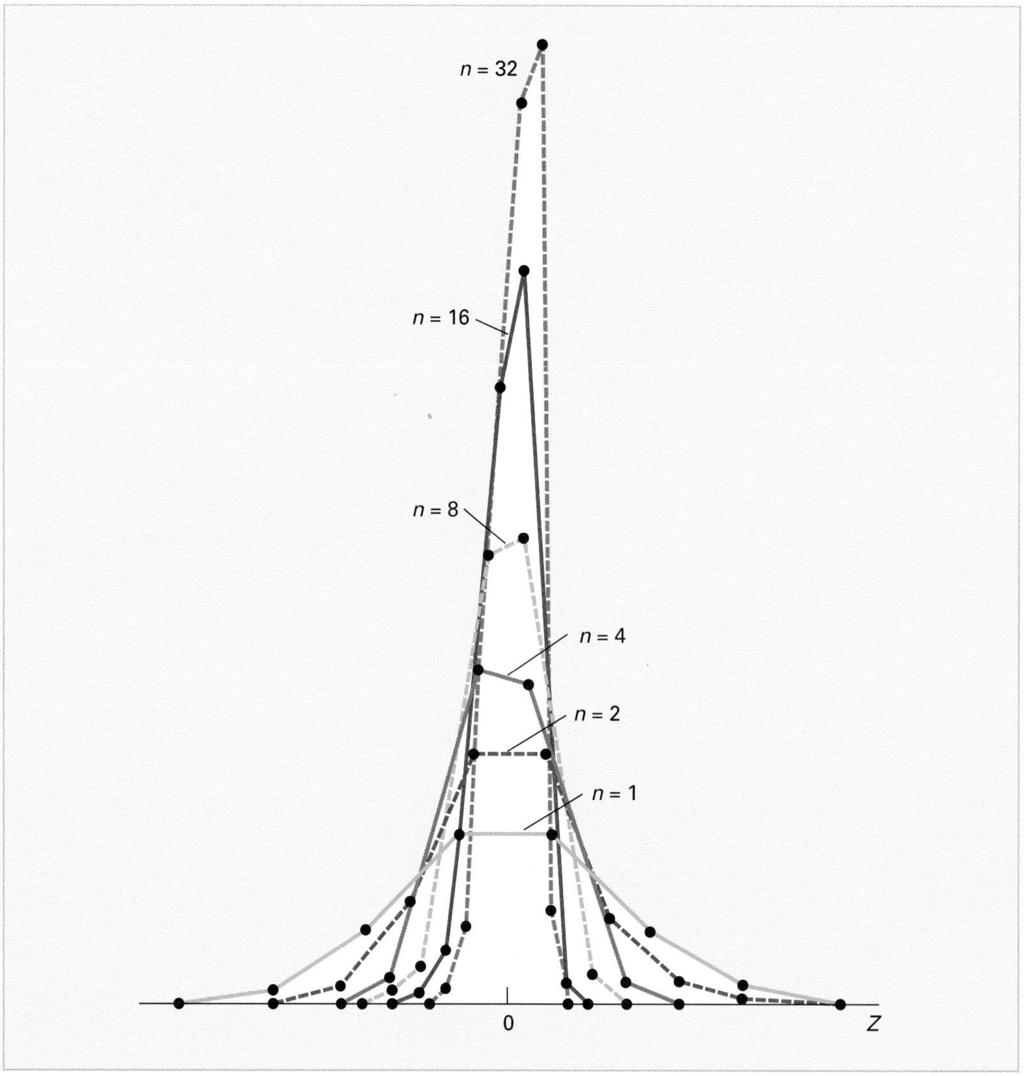

[1]*Remember that "only" 500 samples out of an infinite number of samples have been selected, so that the sampling distributions shown are only approximations of the true distributions.*

normally distributed population. From the polygons in Figure 7.3 you can see that, although the sampling distribution of the mean is approximately[1] normal for each sample size, the sample means are distributed more tightly around the population mean as the sample size is increased.

To examine the concept of the sampling distribution of the mean further, consider the "Using Statistics" scenario described on page 228. The packaging equipment that is filling 368-gram boxes of cereal is set so that the amount of cereal in a box is normally distributed with a mean of 368 grams. From past experience, the population standard deviation for this filling process is 15 grams.

If you randomly select a sample of 25 boxes from the many thousands that are filled in a day and the mean weight is computed for this sample, what type of result could you expect? For example, do you think that the sample mean could be 368 grams? 200 grams? 365 grams?

The sample acts as a miniature representation of the population, so if the values in the population are normally distributed, the values in the sample should be approximately normally distributed. Thus, if the population mean is 368 grams, the sample mean has a good chance of being close to 368 grams.

How can you determine the probability that the sample of 25 boxes will have a mean below 365 grams? From the normal distribution (section 6.2) you know that you can find the area below any value X by converting to standardized Z units.

$$Z = \frac{X - \mu}{\sigma}$$

In the examples in section 6.2, you studied how any single value X differs from the mean. Now, in the cereal-fill example, the value involved is a sample mean $\bar{X}$ and you wish to determine the likelihood that a sample mean is below 365. Thus, by substituting $\bar{X}$ for X, $\mu_{\bar{X}}$ for μ, and $\sigma_{\bar{X}}$ for σ, the appropriate Z value is defined in Equation (7.4):

FINDING Z FOR THE SAMPLING DISTRIBUTION OF THE MEAN

The Z value is equal to the difference between the sample mean $\bar{X}$ and the population mean μ, divided by the standard error of the mean $\sigma_{\bar{X}}$.

$$Z = \frac{\bar{X} - \mu_{\bar{X}}}{\sigma_{\bar{X}}} = \frac{\bar{X} - \mu}{\frac{\sigma}{\sqrt{n}}} \quad \textbf{(7.4)}$$

To find the area below 365 grams, from Equation (7.4),

$$Z = \frac{\bar{X} - \mu_{\bar{X}}}{\sigma_{\bar{X}}} = \frac{365 - 368}{\frac{15}{\sqrt{25}}} = \frac{-3}{3} = -1.00$$

The area corresponding to $Z = -1.00$ in Table E.2 is 0.1587. Therefore, 15.87% of all the possible samples of size 25 have a sample mean below 365 grams.

The above statement is not the same as saying that a certain percentage of *individual* boxes will have less than 365 grams of cereal. You compute that percentage as follows:

$$Z = \frac{X - \mu}{\sigma} = \frac{365 - 368}{15} = \frac{-3}{15} = -0.20$$

The area corresponding to $Z = -0.20$ in Table E.2 is 0.4207. Therefore, 42.07% of the *individual* boxes are expected to contain less than 365 grams. Comparing these results, you see that many more *individual boxes* than *sample means* are below 365 grams. This result is explained by the fact that each sample consists of 25 different values, some small and some large. The averaging process dilutes the importance of any individual value, particularly when the sample size is large. Thus, the chance that the sample mean of 25 boxes is far away from the population mean is less than the chance that a single *box* is far away.

Examples 7.2 and 7.3 show how these results are affected by using a different sample size.

EXAMPLE 7.2

THE EFFECT OF SAMPLE SIZE n ON THE COMPUTATION OF $\sigma_{\overline{X}}$

How is the standard error of the mean affected by increasing the sample size, from 25 to 100 boxes?

SOLUTION If $n = 100$ boxes, then using Equation (7.3) on page 231:

$$\sigma_{\overline{X}} = \frac{\sigma}{\sqrt{n}} = \frac{15}{\sqrt{100}} = \frac{15}{10} = 1.5$$

The fourfold increase in the sample size from 25 to 100 reduces the standard error of the mean by half—from 3 grams to 1.5 grams. This demonstrates that taking a larger sample results in less variability in the sample means from sample to sample.

EXAMPLE 7.3

THE EFFECT OF SAMPLE SIZE n ON THE CLUSTERING OF MEANS IN THE SAMPLING DISTRIBUTION

In the cereal-fill example, if you select a sample of 100 boxes, what is the probability that the sample mean is below 365 grams?

SOLUTION Using Equation (7.4) on page 233,

$$Z = \frac{\overline{X} - \mu_{\overline{X}}}{\sigma_{\overline{X}}} = \frac{365 - 368}{\frac{15}{\sqrt{100}}} = \frac{-3}{1.5} = -2.00$$

From Table E.2, the area less than $Z = -2.00$ is 0.0228. Therefore, 2.28% of the samples of 100 have means below 365 grams, as compared with 15.87% for samples of 25.

Sometimes you need to find the interval that contains a fixed proportion of the sample means. You need to determine a distance below and above the population mean containing a specific area of the normal curve. From Equation (7.4) on page 233,

$$Z = \frac{\overline{X} - \mu}{\frac{\sigma}{\sqrt{n}}}$$

Solving for $\overline{X}$ results in Equation (7.5).

FINDING $\overline{X}$ FOR THE SAMPLING DISTRIBUTION OF THE MEAN

$$\overline{X} = \mu + Z\frac{\sigma}{\sqrt{n}} \quad \textbf{(7.5)}$$

Example 7.4 illustrates the use of Equation (7.5).

EXAMPLE 7.4

DETERMINING THE INTERVAL THAT INCLUDES A FIXED PROPORTION OF THE SAMPLE MEANS

In the cereal-fill example, find an interval around the population mean that will include 95% of the sample means based on samples of 25 boxes.

SOLUTION If 95% of the sample means are in the interval, then 5% are outside the interval. Divide the 5% into two equal parts of 2.5%. The value of Z in Table E.2 corresponding to an area of 0.0250 in the lower tail of the normal curve is −1.96, and the value of Z corresponding to a cumulative area of 0.975 (i.e., 0.025 in the upper tail of the normal curve) is +1.96. The lower value of $\overline{X}$ (called $\overline{X}_L$) and the upper value of $\overline{X}$ (called $\overline{X}_U$) are found by using Equation (7.5):

$$\overline{X}_L = 368 + (-1.96)\frac{15}{\sqrt{25}} = 368 - 5.88 = 362.12$$

$$\overline{X}_U = 368 + (1.96)\frac{15}{\sqrt{25}} = 368 - 5.88 = 373.88$$

Therefore, 95% of all sample means based on samples of 25 boxes are between 362.12 and 373.88 grams.

Sampling from Nonnormally Distributed Populations—The Central Limit Theorem

So far in this section, the sampling distribution of the mean for a normally distributed population has been discussed. However, in many instances, either you know that the population is not normally distributed or it is unrealistic to assume a normal distribution. An important theorem in statistics, the Central Limit Theorem, deals with this situation.

THE CENTRAL LIMIT THEOREM

The **Central Limit Theorem** states that as the sample size (i.e., the number of values in each sample) gets *large enough*, the sampling distribution of the mean is approximately normally distributed. This is true regardless of the shape of the distribution of the individual values in the population.

What sample size is large enough? A great deal of statistical research has gone into this issue. As a general rule, statisticians have found that for many population distributions, when the sample size is at least 30, the sampling distribution of the mean is approximately normal. However, you can apply the Central Limit Theorem for even smaller sample sizes if the population distribution is approximately bell-shaped. In the uncommon case where the distribution is extremely skewed or has more than one mode, you may need sample sizes larger than 30 to ensure normality.

Figure 7.4 illustrates the application of the Central Limit Theorem to different populations. The sampling distributions from three different continuous distributions (normal, uniform, and exponential) for varying sample sizes ($n = 2, 5, 30$) are displayed.

Panel A of Figure 7.4 shows the sampling distribution of the mean selected from a normal population. As mentioned earlier, when the population is normally distributed, the sampling distribution of the mean is normally distributed for any sample size. (You can measure the variability using the standard error of the mean, Equation 7.3 on page 231.) Because of the unbiasedness property, the mean of any sampling distribution is always equal to the mean of the population.

FIGURE 7.4
Sampling Distribution of the Mean for Different Populations for Samples of n = 2, 5, and 30

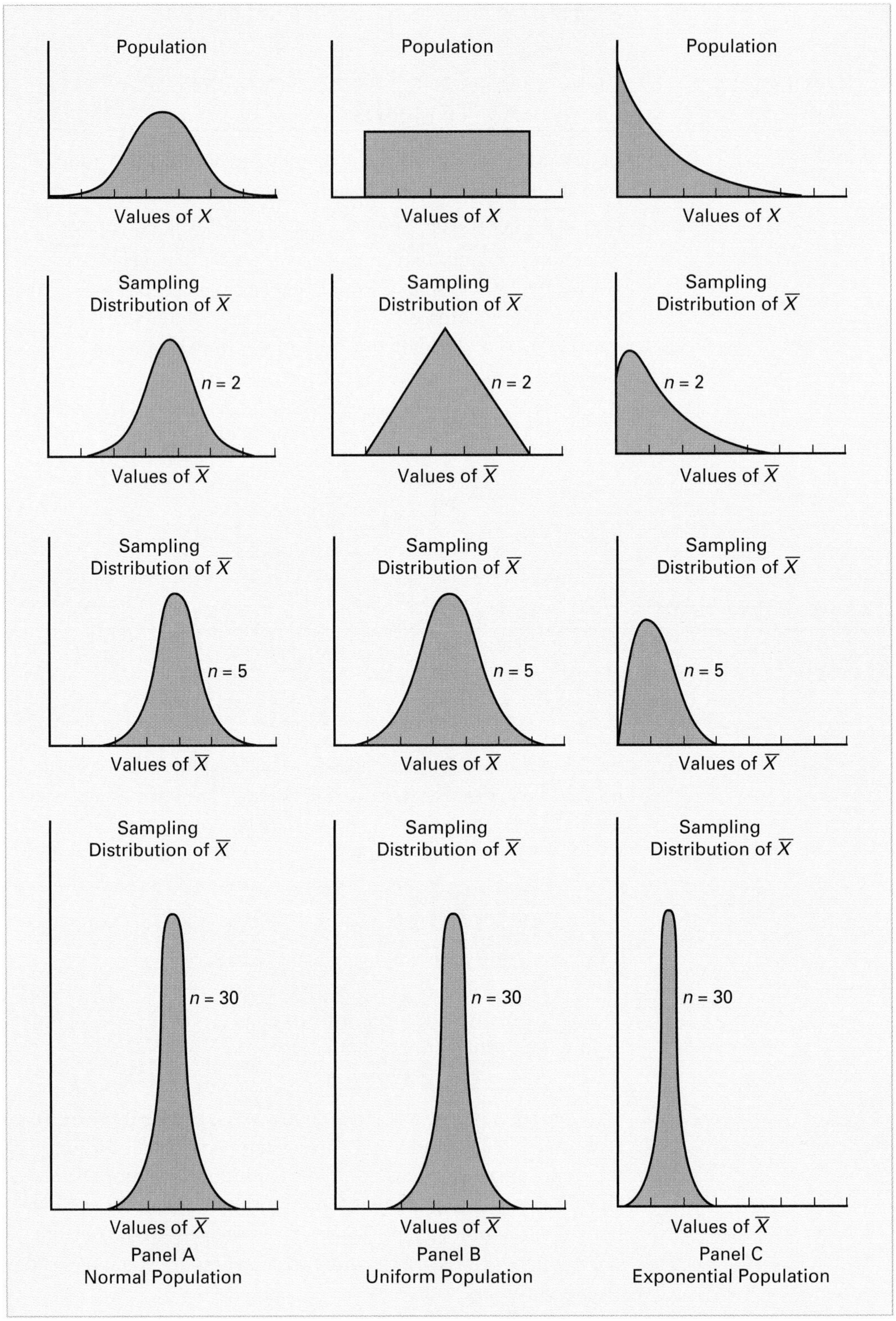

Panel B of Figure 7.4 depicts the sampling distribution from a population with a uniform (or rectangular) distribution (see section 6.4). When samples of size $n = 2$ are selected, there is a peaking or *central limiting* effect already working. For $n = 5$, the sampling distribution is bell-shaped and approximately normal. When $n = 30$, the sampling distribution looks very similar to a normal distribution. In general, the larger the sample size, the more closely the sampling distribution will follow a normal distribution. As with all cases, the mean of each sampling distribution is equal to the mean of the population, and the variability decreases as the sample size increases.

Panel C of Figure 7.4 presents an exponential distribution (see section 6.5). This population is heavily skewed to the right. When $n = 2$, the sampling distribution is still highly skewed to the right but less so than the distribution of the population. For $n = 5$, the sampling distribution is more symmetric with only a slight skew to the right. When $n = 30$, the sampling distribution looks approximately normal. Again, the mean of each sampling distribution is equal to the mean of the population, and the variability decreases as the sample size increases.

Using the results from these well-known statistical distributions (normal, uniform, and exponential), you can make the following conclusions regarding the Central Limit Theorem.

- For most population distributions, regardless of shape, the sampling distribution of the mean is approximately normally distributed if samples of at least 30 are selected.
- If the population distribution is fairly symmetrical, the sampling distribution of the mean is approximately normal for samples as small as 5.
- If the population is normally distributed, the sampling distribution of the mean is normally distributed regardless of the sample size.

The Central Limit Theorem is of crucial importance in using statistical inference to draw conclusions about a population. It allows you to make inferences about the population mean without having to know the specific shape of the population distribution.

VISUAL EXPLORATIONS: Exploring Sampling Distributions

Use the Visual Explorations **Two Dice Probability** procedure to observe the effects of simulated throws on the frequency distribution of the sum of the two dice. Open the Visual Explorations.xla macro workbook (**Visual Explorations.xla**) and select **VisualExplorations → Two Dice Probability** from the Microsoft Excel menu bar. The procedure produces a worksheet that contains an empty frequency distribution table and histogram and a floating control panel (see illustration below).

Click the **Tally** button to tally a set of throws in the frequency distribution table and histogram. Optionally, use the spinner buttons to adjust the number of throws per tally (round).

Click the **Help** button for more information about this simulation. Click **Finish** when you are done with this exploration.

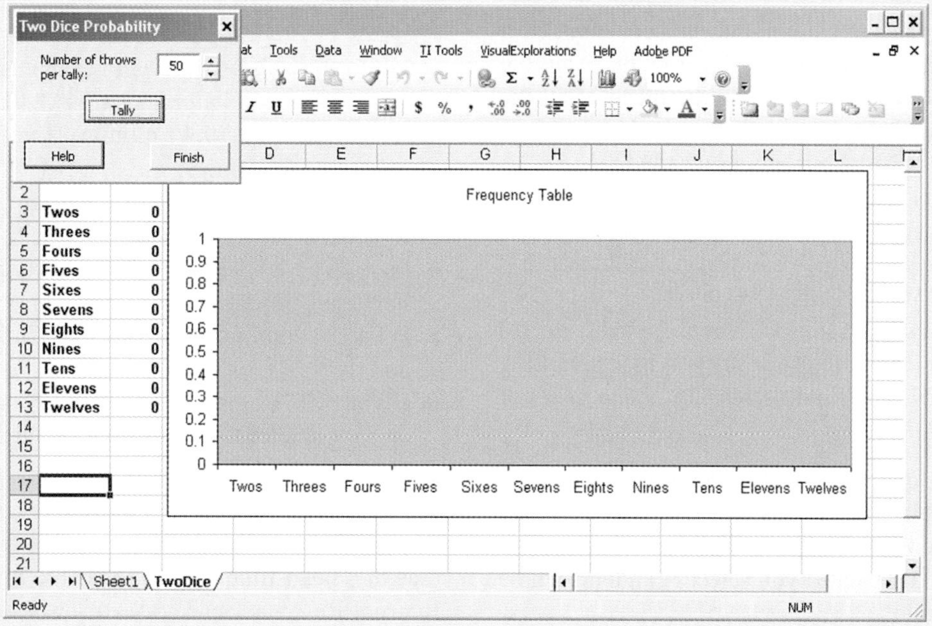

PROBLEMS FOR SECTION 7.2

Learning the Basics

7.1 Given a normal distribution with $\mu = 100$ and $\sigma = 10$, if you select a sample of $n = 25$, what is the probability that $\overline{X}$ is

a. less than 95?
b. between 95 and 97.5?
c. above 102.2?
d. There is a 65% chance that $\overline{X}$ is above what value?

7.2 Given a normal distribution with $\mu = 50$ and $\sigma = 5$, if you select a sample of $n = 100$, what is the probability that $\overline{X}$ is

a. less than 47?
b. between 47 and 49.5?
c. above 51.1?
d. There is a 35% chance that $\overline{X}$ is above what value?

Applying the Concepts

7.3 For each of the following three populations, indicate what the sampling distribution for samples of 25 would consist of.

a. Travel expense vouchers for a university in an academic year
b. Absentee records (days absent per year) in 2004 for employees of a large manufacturing company
c. Yearly sales (in gallons) of unleaded gasoline at service stations located in a particular county

7.4 The following data represent the number of days absent per year in a population of six employees of a small company:

1 3 6 7 9 10

a. Assuming that you sample without replacement, select all possible samples of $n = 2$ and construct the sampling distribution of the mean. Compute the mean of all the sample means and also compute the population mean. Are they equal? What is this property called?
b. Do (a) for all possible samples of $n = 3$.
c. Compare the shape of the sampling distribution of the mean in (a) and (b). Which sampling distribution has less variability? Why?
d. Assuming that you sample with replacement, do (a) through (c) and compare the results. Which sampling distributions have the least variability, those in (a) or (b)? Why?

7.5 The diameter of Ping-Pong balls manufactured at a large factory is approximately normally distributed with a mean of 1.30 inches and a standard deviation of 0.04 inch. If you select a random sample of 16 Ping-Pong balls,

a. what is the sampling distribution of the mean?
b. what is the probability that the sample mean is less than 1.28 inches?
c. what is the probability that the sample mean is between 1.31 and 1.33 inches?
d. The probability is 60% that the sample mean will be between what two values symmetrically distributed around the population mean?

7.6 The U.S. Commerce Department reported that the median price of a new house sold in March 2004 was $201,400, and the mean price $260,000 (Michael Schroeder, "New-Home Sales Increase 8.9%, the Biggest Rise in Nine Months," *The Wall Street Journal*, April 27, 2004, A15). Assume that the standard deviation of the prices is $90,000.

a. If you take samples of $n = 2$, describe the shape of the sampling distribution of $\overline{X}$.
b. If you take samples of $n = 100$, describe the shape of the sampling distribution of $\overline{X}$.
c. If you take a random sample of $n = 100$, what is the probability that the sample mean will be less than $250,000?

7.7 Time spent using e-mail per session is normally distributed with $\mu = 8$ minutes and $\sigma = 2$ minutes. If you select a random sample of 25 sessions,

a. what is the probability that the sample mean is between 7.8 and 8.2 minutes?
b. what is the probability that the sample mean is between 7.5 and 8 minutes?
c. If you select a random sample of 100 sessions, what is the probability that the sample mean is between 7.8 and 8.2 minutes?
d. Explain the difference in the results of (a) and (c).

7.8 The amount of time a bank teller spends with each customer has a population mean $\mu = 3.10$ minutes and standard deviation $\sigma = 0.40$ minute. If you select a random sample of 16 customers,

a. what is the probability that the mean time spent per customer is at least 3 minutes?
b. there is an 85% chance that the sample mean is below how many minutes?
c. What assumption must you make in order to solve (a) and (b)?
d. If you select a random sample of 64 customers, there is an 85% chance that the sample mean is below how many minutes?

7.9 The *New York Times* reported (Laurie J. Flynn, "Tax Surfing," *The New York Times*, March 25, 2002, C10) that the mean time to download the homepage from the Internal Revenue Service Web site **www.irs.gov** was 0.8 second. Suppose that the download time was normally distributed with a standard deviation of 0.2 second. If you select a random sample of 30 download times,

a. what is the probability that the sample mean is less than 0.75 second?

b. what is the probability that the sample mean is between 0.70 and 0.90 second?

c. the probability is 80% that the sample mean is between what two values symmetrically distributed around the population mean?

d. the probability is 90% that the sample mean is less than what value?

7.10 The article discussed in problem 7.9 also reported that the mean download time for the H&R Block Web site **www.hrblock.com** was 2.5 seconds. Suppose that the download time for the H&R Block Web site was normally distributed with a standard deviation of 0.5 second. If you select a random sample of 30 download times,

a. what is the probability that the sample mean is less than 2.75 seconds?

b. what is the probability that the sample mean is between 2.70 and 2.90 seconds?

c. the probability is 80% that the sample mean is between what two values symmetrically distributed around the population mean?

d. the probability is 90% that the sample mean is less than what value?

7.3 SAMPLING DISTRIBUTION OF THE PROPORTION

Consider a categorical variable that has only two categories, such as the customer prefers your brand or the customer prefers the competitor's brand. Of interest is the proportion of items belonging to one of the categories, for example, the proportion of customers that prefers your brand. The population proportion, represented by π, is the proportion of items in the entire population with the characteristic of interest. The sample proportion, represented by p, is the proportion of items in the sample with the characteristic of interest. The sample proportion, a statistic, is used to estimate the population proportion, a parameter. To calculate the sample proportion you assign the two possible outcomes scores of 1 or 0 to represent the presence or absence of the characteristic. You then sum all the 1 and 0 scores and divide by n, the sample size. For example, if, in a sample of five customers, three preferred your brand and two did not, you have three 1's and two 0's. Summing the three 1's and two 0's and dividing by the sample size of 5 gives you a sample proportion of 0.60.

THE SAMPLE PROPORTION

$$p = \frac{X}{n} = \frac{\text{number of items having the characteristic of interest}}{\text{sample size}} \tag{7.6}$$

The sample proportion p takes on values between 0 and 1. If all individuals possess the characteristic, you assign each a score of 1 and p is equal to 1. If half the individuals possess the characteristic, you assign half a score of 1, and assign the other half a score of 0, and p is equal to 0.5. If none of the individuals possesses the characteristic, you assign each a score of 0 and p is equal to 0.

While the sample mean $\overline{X}$ is an unbiased estimator of the population mean μ, the statistic p is an unbiased estimator of the population proportion π. By analogy to the sampling distribution of the mean, the **standard error of the proportion** σ_p is given in Equation (7.7).

STANDARD ERROR OF THE PROPORTION

$$\sigma_p = \sqrt{\frac{\pi(1-\pi)}{n}} \tag{7.7}$$

If you select all possible samples of a certain size, the distribution of all possible sample proportions is referred to as the **sampling distribution of the proportion**. When sampling with replacement from a finite population, the sampling distribution of the proportion follows the binomial distribution as discussed in section 5.3. However, you can use the normal distribution to approximate the binomial distribution when $n\pi$ and $n(1 - \pi)$ are each at least 5 (see section 6.6). In most cases in which inferences are made about the proportion, the sample size is substantial enough to meet the conditions for using the normal approximation (see reference 1). Therefore, in many instances, you can use the normal distribution to estimate the sampling distribution of the proportion. Substituting p for $\bar{X}$, π for μ, and $\sqrt{\frac{\pi(1-\pi)}{n}}$ for $\frac{\sigma}{\sqrt{n}}$ in Equation (7.4) on page 233, results in Equation (7.8).

DIFFERENCE BETWEEN THE SAMPLE PROPORTION AND THE POPULATION PROPORTION IN STANDARDIZED NORMAL UNITS

$$Z = \frac{p - \pi}{\sqrt{\frac{\pi(1-\pi)}{n}}} \quad \textbf{(7.8)}$$

To illustrate the sampling distribution of the proportion, suppose that the manager of the local branch of a savings bank determines that 40% of all depositors have multiple accounts at the bank. If you select a random sample of 200 depositors, the probability that the sample proportion of depositors with multiple accounts is less than 0.30 is calculated as follows.

Because $n\pi = 200(0.40) = 80 \geq 5$ and $n(1 - \pi) = 200(0.60) = 120 \geq 5$, the sample size is large enough to assume that the sampling distribution of the proportion is approximately normally distributed. Using Equation (7.8),

$$\begin{aligned} Z &= \frac{p - \pi}{\sqrt{\frac{\pi(1-\pi)}{n}}} \\ &= \frac{0.30 - 0.40}{\sqrt{\frac{(0.40)(0.60)}{200}}} = \frac{-0.10}{\sqrt{\frac{0.24}{200}}} = \frac{-0.10}{0.0346} \\ &= -2.89 \end{aligned}$$

Using Table E.2, the area under the normal curve less than $Z = -2.89$ is 0.0019. Therefore, the probability that the sample proportion is less than 0.30 is 0.0019—a highly unlikely event. This means that if the true proportion of successes in the population is 0.40, less than one-fifth of 1% of the samples of $n = 200$ are expected to have sample proportions of less than 0.30.

PROBLEMS FOR SECTION 7.3

Learning the Basics

PH Grade ASSIST **7.11** In a random sample of 64 people, 48 are classified as "successful." If the population proportion is 0.70,

a. determine the sample proportion p of "successful" people.
b. determine the standard error of the proportion.

PH Grade ASSIST **7.12** A random sample of 50 households was selected for a telephone survey. The key question asked was, "Do you or any member of your household own a cellular telephone?" Of the 50 respondents, 15 said yes and 35 said no. If the population proportion is 0.40,

a. determine the sample proportion p of households with cellular telephones.
b. determine the standard error of the proportion.

7.13 The following data represent the responses (Y for yes and N for no) from a sample of 40 college students to the question, "Do you currently own shares in any stocks?"

N N Y N N Y N Y N Y N N Y N Y Y N N N Y
N Y N N N N Y N N Y Y N N N Y N N Y N N

If the population proportion is 0.30,

a. determine the sample proportion p of college students who own shares of stock.
b. determine the standard error of the proportion.

Applying the Concepts

PH Grade ASSIST SELF Test **7.14** A political pollster is conducting an analysis of sample results in order to make predictions on election night. Assuming a two-candidate election, if a specific candidate receives at least 55% of the vote in the sample, then that candidate will be forecast as the winner of the election. If you select a random sample of 100 voters, what is the probability that a candidate will be forecast as the winner when

a. the true percentage of her vote is 50.1%?
b. the true percentage of her vote is 60%?
c. the true percentage of her vote is 49% (and she will actually lose the election)?
d. If the sample size is increased to 400, what are your answers to (a), (b), and (c)? Discuss.

PH Grade ASSIST **7.15** You plan to conduct a marketing experiment in which students are to taste one of two different brands of soft drink. Their task is to correctly identify the brand he or she tasted. You select a random sample of 200 students and assume that the students have no ability to distinguish between the two brands. (*Hint:* If an individual has no ability to distinguish between the two soft drinks, then each brand is equally likely to be selected.)

a. What is the probability that the sample will have between 50% and 60% of the identifications correct?
b. The probability is 90% that the sample percentage is contained within what symmetrical limits of the population percentage?
c. What is the probability that the sample percentage of correct identifications is greater than 65%?
d. Which is more likely to occur—more than 60% correct identifications in the sample of 200 or more than 55% correct identifications in a sample of 1,000? Explain.

7.16 A study of women in corporate leadership was conducted by Catalyst, a New York research organization. The study concluded that slightly more than 15% of corporate officers at Fortune 500 companies are women (Carol Hymowitz, "Women Put Noses to the Grindstone, and Miss Opportunities," *The Wall Street Journal*, February 3, 2004, B1). Suppose that you select a random sample of 200 corporate officers and the true proportion held by women is 0.15.

a. What is the probability that in the sample less than 15% of the corporate officers will be women?
b. What is the probability that in the sample between 13% and 17% of the corporate officers will be women?
c. What is the probability that in the sample between 10% and 20% of the corporate officers will be women?
d. If a sample of 100 is taken, how does this change your answers to (a) through (c)?

7.17 The NBC hit comedy *Friends* was TiVo's most popular show during the week of April 18–24, 2004. According to the Nielsen ratings, 29.7% of TiVo owners in the United States either recorded *Friends* or watched it live ("Prime-time Nielsen Ratings," *USA Today*, April 28, 2004, 3D). Suppose you select a random sample of 50 TiVo owners.

a. What is the probability that more than half the people in the sample watched or recorded *Friends*?
b. What is the probability that less than 25% of the people in the sample watched or recorded *Friends*?
c. If a random sample of size 500 is taken, how does this change your answers to (a) and (b)?

7.18 Millions of Americans arrange their travel plans on the Web. According to an article in *USA Today*, 77% of travelers purchase plane tickets on the Web ("Travelers Head Online," *USA Today Snaphots*, July 22, 2003, A1). If you select a random sample of 200 travelers,

a. what is the probability that the sample will have between 75% and 80% purchase plane tickets on the Web?
b. The probability is 90% that the sample percentage will be contained within what symmetrical limits of the population percentage?
c. The probability is 95% that the sample percentage will be contained within what symmetrical limits of the population percentage?

PH Grade ASSIST **7.19** According to the National Restaurant Association, 20% of fine-dining restaurants have instituted policies restricting the use of cell phones ("Business Bulletin," *The Wall Street Journal*, June 1, 2000, A1). If you select a random sample of 100 fine-dining restaurants,

a. what is the probability that the sample has between 15% and 25% that have established policies restricting cell phone use?
b. the probability is 90% that the sample percentage will be contained within what symmetrical limits of the population percentage?
c. the probability is 95% that the sample percentage will be contained within what symmetrical limits of the population percentage?

7.20 An article (P. Kitchen, "Retirement Plan: To Keep Working," *Newsday*, September 24, 2003) discussed the retirement plans of Americans ages 50 to 70 who were employed full-time or part-time. Twenty-nine percent of the respondents said that they did not intend to work for pay at all. If you select a random sample of 400 Americans ages 50 to 70 who were employed full-time or part-time,

a. what is the probability that the sample has between 25% and 30% who do not intend to work for pay at all?
b. If a current sample of 400 Americans ages 50 to 70 who were employed full-time or part-time has 35% who do not intend to work for pay at all, what can you infer about the population estimate of 29%? Explain.

c. If a current sample of 100 Americans ages 50 to 70 who were employed full-time or part-time has 35% who do not intend to work for pay at all, what can you infer about the population estimate of 29%? Explain.
d. Explain the difference in the results in (b) and (c).

7.21 The Internal Revenue Service (IRS) discontinued random audits in 1988. Instead, the IRS conducts audits on returns deemed as questionable by their Discriminant Function System (DFS), a complicated and highly secretive computerized analysis system. In an attempt to reduce the proportion of "no-change" audits (i.e., audits that uncover that no additional taxes are due) the IRS only audits returns the DFS scores as highly questionable. The proportion of "no-change" audits has risen over the years and is currently approximately 0.25 (Tom Herman, "Unhappy Returns: IRS Moves to Bring Back Random Audits," *The Wall Street Journal*, June 20, 2002, A1). Suppose that you select a random sample of 100 audits. What is the probability that the sample will have
a. between 24% and 26% no-change audits?
b. between 20% and 30% no-change audits?
c. more than 30% no-change audits?

7.22 The IRS announced that it planned to resume totally random audits in 2002. Suppose that you select a random sample of 200 totally random audits, and that only 10% of all the returns filed would result in audits requiring additional taxes. What is the probability that the sample has
a. between 89% and 91% no-change audits?
b. between 85% and 95% no-change audits?
c. more than 95% no-change audits?

7.4 TYPES OF SURVEY SAMPLING METHODS

In section 1.1, a sample was defined as the portion of the population that has been selected for analysis. Rather than taking a complete census of the whole population, statistical sampling procedures focus on collecting a small representative group of the larger population. The resulting sample results are used to estimate characteristics of the entire population. The three main reasons for drawing a sample are:

- A sample is less time-consuming than a census.
- A sample is less costly to administer than a census.
- A sample is less cumbersome and more practical to administer than a census.

The sampling process begins by defining the frame. The **frame** is a listing of items that make up the population. Frames are data sources such as population lists, directories, or maps. Samples are drawn from these frames. Inaccurate or biased results can result if the frame excludes certain groups of the population. Using different frames to generate data can lead to opposite conclusions.

Once you select a frame, you draw a sample from the frame. As illustrated in Figure 7.5, there are two kinds of samples: the nonprobability sample and the probability sample.

FIGURE 7.5
Types of Samples

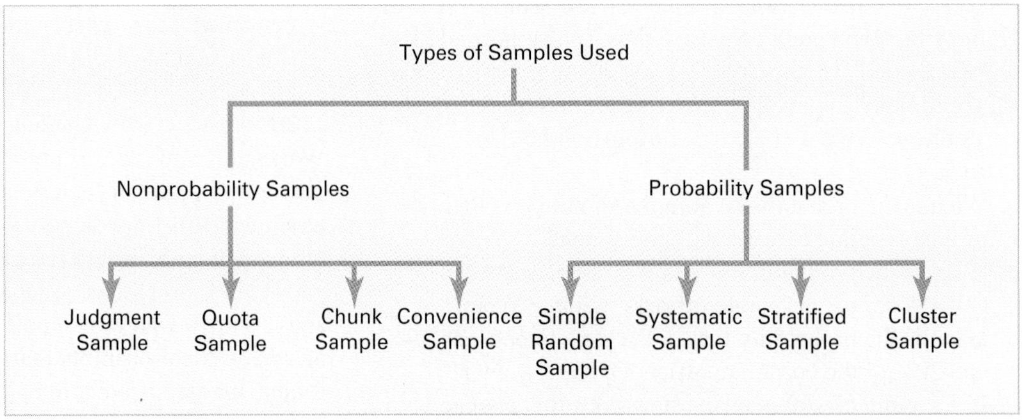

In a **nonprobability sample** you select the items or individuals without knowing their probabilities of selection. Thus, the theory that has been developed for probability sampling cannot be applied to nonprobability samples. A common type of nonprobability sampling is convenience sampling. In **convenience sampling**, items are selected based only on the fact that they are easy, inexpensive, or convenient to sample. In some cases participants are self-selected. For example, many companies conduct surveys by giving visitors to their World Wide Web site the opportunity to complete survey forms and submit them electronically. The response to these surveys can provide large amounts of data quickly, but the sample consists of self-selected Web users. For many studies, only a nonprobability sample such as a judgment sample is available. In a **judgment sample**, you get the opinions of preselected experts in the subject matter. Some other common procedures of nonprobability sampling are quota sampling and chunk sampling. These are discussed in detail in specialized books on sampling methods (see reference 1).

Nonprobability samples can have certain advantages such as convenience, speed, and lower cost. However, their lack of accuracy due to selection bias and lack of generalizability of the results more than offset these advantages. Therefore, you should restrict the use of nonprobability sampling methods to situations in which you want to get rough approximations at low cost in order to satisfy your curiosity about a particular subject or to small-scale studies that precede more rigorous investigations.

In a **probability sample** you select the items based on known probabilities. Whenever possible, you should use probability sampling methods. The samples based on these methods allow you to make unbiased inferences about the population of interest. In practice, it is often difficult or impossible to take a probability sample. However, you should work toward achieving a probability sample and acknowledge any potential biases that might exist. The four types of probability samples most commonly used are simple random, systematic, stratified, and cluster. These sampling methods vary from one another in their cost, accuracy, and complexity.

Simple Random Sample

In a **simple random sample** every item from a frame has the same chance of selection as every other item. In addition, every sample of a fixed size has the same chance of selection as every other sample of that size. Simple random sampling is the most elementary random sampling technique. It forms the basis for the other random sampling techniques.

With simple random sampling, you use n to represent the sample size and N to represent the frame size. You number every item in the frame from 1 to N. The chance that you will select any particular member of the frame on the first draw is $1/N$.

You select samples with replacement or without replacement. **Sampling with replacement** means that after you select an item, you return it to the frame, where it has the same probability of being selected again. Imagine you have a fishbowl with N business cards. On the first selection, you select the card for Judy Craven. You record pertinent information and replace the business card in the bowl. You then shuffle the cards in the bowl and select the second card. On the second selection, Judy Craven has the same probability of being selected again, $1/N$. You repeat this process until you have selected the desired sample size n. However, it is usually more desirable to have a sample of different items than to permit a repetition of measurements on the same item.

Sampling without replacement means that once you select an item it cannot be selected again. The chance you will select any particular item in the frame, say the business card for Judy Craven, on the first draw is $1/N$. The chance you will select any card not previously selected on the second draw is now 1 out of $N - 1$. This process continues until you have selected the desired sample of size n.

Regardless of whether you have sampled with or without replacement, "fishbowl" methods for sample selection have a major drawback—the ability to thoroughly mix the cards and randomly pull the sample. As a result, fishbowl methods are not very useful. You need to use less cumbersome and more scientific methods of selection.

One such method uses a **table of random numbers** (see Table E.1) for selecting the sample. A table of random numbers consists of a series of digits listed in a randomly generated sequence (see reference 8). Because the numeric system uses 10 digits (0, 1, 2, . . . , 9), the chance you will randomly generate any particular digit is equal to the probability of generating any other digit. This probability is 1 out of 10. Hence, if a sequence of 800 digits were generated, you would expect about 80 of them to be the digit 0, 80 to be the digit 1, and so on. In fact, those who use tables of random numbers usually test the generated digits for randomness prior to using them. Table E.1 has met all such criteria for randomness. Because every digit or sequence of digits in the table is random, the table can be read either horizontally or vertically. The margins of the table designate row numbers and column numbers. The digits themselves are grouped into sequences of five in order to make reading the table easier.

To use such a table instead of a fishbowl for selecting the sample, you first need to assign code numbers to the individual members of the frame. Then you get the random sample by reading the table of random numbers and selecting those individuals from the frame whose assigned code numbers match the digits found in the table. You can better understand the process of sample selection by examining Example 7.5.

EXAMPLE 7.5

SELECTING A SIMPLE RANDOM SAMPLE USING A TABLE OF RANDOM NUMBERS

A company wants to select a sample of 32 full-time workers from a population of 800 full-time employees in order to collect information on expenditures concerning a company-sponsored dental plan. How do you select a simple random sample?

SOLUTION The company assumes that not everyone will respond to the survey, so you need to mail more than 32 surveys to get the desired 32 responses. Assuming that 8 out of 10 full-time workers will respond to such a survey (i.e., a response rate of 80%), you decide to mail 40 surveys.

The frame consists of a listing of the names and company mailbox numbers of all $N = 800$ full-time employees taken from the company personnel files. Thus, the frame is an accurate and complete listing of the population. To select the random sample of 40 employees from this frame, you use a table of random numbers. Because the population size (800) is a three-digit number, each assigned code number must also be three digits so that every full-time worker has an equal chance for selection. You give a code of 001 to the first full-time employee in the population listing, a code of 002 to the second full-time employee in the population listing, and so on, until a code of 800 is given to the Nth full-time worker in the listing. Because $N = 800$ is the largest possible coded value, you discard all three-digit code sequences greater than N (i.e., 801 through 999 and 000).

To select the simple random sample, you choose an arbitrary starting point from the table of random numbers. One method you can use is to close your eyes and strike the table of random numbers with a pencil. Suppose you used this procedure and you selected row 06, column 05, of Table 7.3 (which is extracted from Table E.1), as the starting point. Although you can go in any direction, in this example, you will read the table from left to right in sequences of three digits without skipping.

The individual with code number 003 is the first full-time employee in the sample (row 06 and columns 05–07), the second individual has code number 364 (row 06 and columns 08–10), and the third individual has code number 884. Because the highest code for any employee is 800, you discard this number. Individuals with code numbers 720, 433, 463, 363, 109, 592, 470, and 705 are selected third through tenth, respectively.

You continue the selection process until you get the needed sample size of 40 full-time employees. During the selection process, if any three-digit coded sequence repeats, you include the employee corresponding to that coded sequence again as part of the sample if sampling with replacement. You discard the repeating coded sequence if sampling without replacement.

TABLE 7.3
Using a Table of Random Numbers

		Column							
	Row	**00000 12345**	**00001 67890**	**11111 12345**	**11112 67890**	**22222 12345**	**22223 67890**	**33333 12345**	**33334 67890**
	01	49280	88924	35779	00283	81163	07275	89863	02348
	02	61870	41657	07468	08612	98083	97349	20775	45091
	03	43898	65923	25078	86129	78496	97653	91550	08078
	04	62993	93912	30454	84598	56095	20664	12872	64647
	05	33850	58555	51438	85507	71865	79488	76783	31708
Begin	06	97340	03364	88472	04334	63919	36394	11095	92470
selection	07	70543	29776	10087	10072	55980	64688	68239	20461
(row 06,	08	89382	93809	00796	95945	34101	81277	66090	88872
column 5)	09	37818	72142	67140	50785	22380	16703	53362	44940
	10	60430	22834	14130	96593	23298	56203	92671	15925
	11	82975	66158	84731	19436	55790	69229	28661	13675
	12	39087	71938	40355	54324	08401	26299	49420	59208
	13	55700	24586	93247	32596	11865	63397	44251	43189
	14	14756	23997	78643	75912	83832	32768	18928	57070
	15	32166	53251	70654	92827	63491	04233	33825	69662
	16	23236	73751	31888	81718	06546	83246	47651	04877
	17	45794	26926	15130	82455	78305	55058	52551	47182
	18	09893	20505	14225	68514	46427	56788	96297	78822
	19	54382	74598	91499	14523	68479	27686	46162	83554
	20	94750	89923	37089	20048	80336	94598	26940	36858
	21	70297	34135	53140	33340	42050	82341	44104	82949
	22	85157	47954	32979	26575	57600	40881	12250	73742
	23	11100	02340	12860	74697	96644	89439	28707	25815
	24	36871	50775	30592	57143	17381	68856	25853	35041
	25	23913	48357	63308	16090	51690	54607	72407	55538

Source: Partially extracted from The Rand Corporation, A Million Random Digits with 100,000 Normal Deviates *(Glencoe, IL: The Free Press, 1955) and displayed in Table E.1 in Appendix E of this book.*

Systematic Sample

In a **systematic sample**, you partition the N items in the frame into n groups of k items where,

$$k = \frac{N}{n}$$

You round k to the nearest integer. To select a systematic sample, you choose the first item to be selected at random from the first k items in the frame. Then you select the remaining $n - 1$ items by taking every k th item thereafter from the entire frame.

If the frame consists of a listing of prenumbered checks, sales receipts, or invoices, a systematic sample is faster and easier to take than a simple random sample. A systematic sample is also a convenient mechanism for collecting data from telephone books, class rosters, and consecutive items coming off an assembly line.

To take a systematic sample of $n = 40$ from the population of $N = 800$ employees, you partition the frame of 800 into 40 groups, each of which contains 20 employees. You then select a random number from the first 20 individuals, and include every twentieth individual after the first selection in the sample. For example, if the first number you select is 008, your subsequent selections are 028, 048, 068, 088, 108, . . . , 768, and 788.

Although they are simpler to use, simple random sampling and systematic sampling are generally less efficient than other, more sophisticated, probability sampling methods. Even greater possibilities for selection bias and lack of representation of the population characteristics occur from systematic samples than from simple random samples. If there is a pattern in the frame, you could have severe selection biases. To overcome the potential problem of disproportionate representation of specific groups in a sample, you can use either stratified sampling methods or cluster sampling methods.

Stratified Sample

In a **stratified sample**, you first subdivide the N items in the frame into separate subpopulations, or **strata**. A strata is defined by some common characteristic. You select a simple random sample within each of the strata, and combine the results from the separate simple random samples. This method is more efficient than either simple random sampling or systematic sampling because you are ensured of the representation of items across the entire population. The homogeneity of items within each stratum provides greater precision in the estimates of underlying population parameters.

EXAMPLE 7.6 SELECTING A STRATIFIED SAMPLE

A company wants to select a sample of 32 full-time workers from a population of 800 full-time employees in order to estimate expenditures from a company-sponsored dental plan. Of the full-time employees, 25% are managerial and 75% are nonmanagerial workers. How do you select the stratified sample in order for the sample to represent the correct proportion of managerial workers?

SOLUTION If you assume an 80% response rate, you need to distribute 40 surveys to get the desired 32 responses. The frame consists of a listing of the names and company mailbox numbers of all $N = 800$ full-time employees included in the company personnel files. Since 25% of the full-time employees are managerial, you first separate the population frame into two strata: a subpopulation listing of all 200 managerial-level personnel and a separate subpopulation listing of all 600 full-time nonmanagerial workers. Since the first stratum consists of a listing of 200 managers, you assign three-digit code numbers from 001 to 200. Since the second stratum contains a listing of 600 nonmanagerial-level workers, you assign three-digit code numbers from 001 to 600.

To collect a stratified sample proportional to the sizes of the strata, you select 25% of the overall sample from the first stratum and 75% of the overall sample from the second stratum. You take two separate simple random samples, each of which is based on a distinct random starting point from a table of random numbers (Table E.1). In the first sample you select 10 managers from the listing of 200 in the first stratum, and in the second sample you select 30 nonmanagerial workers from the listing of 600 in the second stratum. You then combine the results to reflect the composition of the entire company.

Cluster Sample

In a **cluster sample**, you divide the N items in the frame into several clusters so that each cluster is representative of the entire population. You then take a random sample of clusters, and study all items in each selected cluster. **Clusters** are naturally occurring designations, such as counties, election districts, city blocks, households, or sales territories.

Cluster sampling is often more cost-effective than simple random sampling, particularly if the population is spread over a wide geographic region. However, cluster sampling often requires a larger sample size to produce results as precise as those from simple random sampling or stratified sampling. A detailed discussion of systematic sampling, stratified sampling, and cluster sampling procedures can be found in reference 1.

PROBLEMS FOR SECTION 7.4

Learning the Basics

PH Grade ASSIST **7.23** For a population containing $N = 902$ individuals, what code number would you assign for
a. the first person on the list?
b. the fortieth person on the list?
c. the last person on the list?

7.24 For a population of $N = 902$, verify that by starting in row 05 of the table of random numbers (Table E.1), you need only six rows to select a sample of $n = 60$ *without* replacement.

PH Grade ASSIST **7.25** Given a population of $N = 93$, starting in row 29 of the table of random numbers (Table E.1), and reading across the row, select a sample of $n = 15$
a. *without* replacement.
b. *with* replacement.

Applying the Concepts

7.26 For a study that consists of personal interviews with participants (rather than mail or phone surveys), explain why a simple random sample might be less practical than some other methods.

PH Grade ASSIST **7.27** You want to select a random sample of $n = 1$ from a population of three items (which are called *A*, *B*, and *C*). The rule for selecting the sample is: Flip a coin; if it is heads, pick item *A*; if it is tails, flip the coin again; this time, if it is heads, choose *B*; if it is tails, choose *C*. Explain why this is a random sample but not a simple random sample.

7.28 A population has four members (call them *A*, *B*, *C*, and *D*). You would like to draw a random sample of $n = 2$, which you decide to do in the following way: Flip a coin; if it is heads, the sample will be items *A* and *B*; if it is tails, the sample will be items *C* and *D*. Although this is a random sample, it is not a simple random sample. Explain why. (If you did problem 7.27, compare the procedure described there with the procedure described in this problem.)

SELF Test **7.29** The registrar of a college with a population of $N = 4{,}000$ full-time students is asked by the president to conduct a survey to measure satisfaction with the quality of life on campus. The following table contains a breakdown of the 4,000 registered full-time students by gender and class designation:

	Class Designation				
Gender	**Fr.**	**So.**	**Jr.**	**Sr.**	**Total**
Female	700	520	500	480	2,200
Male	560	460	400	380	1,800
Total	1,260	980	900	860	4,000

The registrar intends to take a probability sample of $n = 200$ students and project the results from the sample to the entire population of full-time students.
a. If the frame available from the registrar's files is an alphabetical listing of the names of all $N = 4{,}000$ registered full-time students, what type of sample could you take? Discuss.
b. What is the advantage of selecting a simple random sample in (a)?
c. What is the advantage of selecting a systematic sample in (a)?
d. If the frame available from the registrar's files is a listing of the names of all $N = 4{,}000$ registered full-time students compiled from eight separate alphabetical lists based on the gender and class designation breakdowns shown in the class designation table, what type of sample should you take? Discuss.
e. Suppose that each of the $N = 4{,}000$ registered full-time students lived in one of the 20 campus dormitories. Each dormitory contains four floors with 50 beds per floor, and therefore accommodates 200 students. It is college policy to fully integrate students by gender and class designation on each floor of each dormitory. If the registrar is able to compile a frame through a listing of all student occupants on each floor within each dormitory, what type of sample should you take? Discuss.

7.30 Prenumbered sales invoices are kept in a sales journal. The invoices are numbered from 0001 to 5,000.
a. Beginning in row 16, column 1, and proceeding horizontally in Table E.1, select a simple random sample of 50 invoice numbers.
b. Select a systematic sample of 50 invoice numbers. Use the random numbers in row 20, columns 5–7, as the starting point for your selection.
c. Are the invoices selected in (a) the same as those selected in (b)? Why or why not?

PH Grade ASSIST **7.31** Suppose that 5,000 sales invoices are separated into four strata. Stratum 1 contains 50 invoices, stratum 2 contains 500 invoices, stratum 3 contains 1,000 invoices, and stratum 4 contains 3,450 invoices. A sample of 500 sales invoices is needed.
a. What type of sampling should you do? Why?
b. Explain how you would carry out the sampling according to the method stated in (a).
c. Why is the sampling in (a) not simple random sampling?

7.5 EVALUATING SURVEY WORTHINESS

Nearly every day, you read or hear about survey or opinion poll results in newspapers, on the Internet, or on radio or television. To identify surveys that lack objectivity or credibility, you must critically evaluate what you read and hear by examining the worthiness of the survey. First, you must evaluate the purpose of the survey, why it was conducted, and for whom it was conducted. An opinion poll or survey conducted to satisfy curiosity is mainly for entertainment. Its result is an end in itself rather than a means to an end. You should be skeptical of such a survey because the result should not be put to further use.

The second step in evaluating the worthiness of a survey is for you to determine whether it was based on a probability or a nonprobability sample (as discussed in section 7.4). You need to remember that the only way to make correct statistical inferences from a sample to a population is through the use of a probability sample. Surveys that use nonprobability sampling methods are subject to serious, perhaps unintentional, biases that may render the results meaningless, as illustrated in the following example from the 1948 U.S. presidential election.

In 1948, major pollsters predicted the outcome of the U.S. presidential election between Harry S. Truman, the incumbent president, and Thomas E. Dewey, then governor of New York, as going to Dewey. The *Chicago Tribune* was so confident of the polls' predictions that it printed its early edition based on the predictions rather than waiting for the ballots to be counted.

An embarrassed newspaper and the pollsters it had relied on had a lot of explaining to do. How had the pollsters been so wrong? Intent on discovering the source of the error, the pollsters found that their use of a nonprobability sampling method was the culprit (see reference 7). As a result, polling organizations adopted probability sampling methods for future elections.

Survey Errors

Even when surveys use random probability sampling methods, they are subject to potential errors. Four types of survey errors are:

- Coverage error
- Nonresponse error
- Sampling error
- Measurement error

Good survey research design attempts to reduce or minimize these various survey errors, often at considerable cost.

Coverage Error The key to proper sample selection is an adequate frame. Remember, a frame is an up-to-date list of all the items from which you will select the sample. **Coverage error** occurs if certain groups of items are excluded from this frame so that they have no chance of being selected in the sample. Coverage error results in a **selection bias**. If the frame is inadequate because certain groups of items in the population were not properly included, any random probability sample selected will provide an estimate of the characteristics of the frame, not the *actual* population.

Nonresponse Error Not everyone is willing to respond to a survey. In fact, research has shown that individuals in the upper and lower economic classes tend to respond less frequently to surveys than do people in the middle class. **Nonresponse error** arises from the failure to collect data on all items in the sample and results in a **nonresponse bias**. Because you cannot generally assume that persons who do not respond to surveys are similar to those who do, you need to follow up on the nonresponses after a specified period of time. You should make several attempts to convince such individuals to complete the survey. The follow-up responses are then compared to the initial responses in order to make valid inferences from the survey (reference 1).

The mode of response you use affects the rate of response. The personal interview and the telephone interview usually produce a higher response rate than does the mail survey—but at a higher cost. The following is a famous example of coverage error and nonresponse bias.

In 1936, the magazine *Literary Digest* predicted that Governor Alf Landon of Kansas would receive 57% of the votes in the U.S. presidential election and overwhelmingly defeat President Franklin D. Roosevelt's bid for a second term. However, Landon was soundly defeated when he received only 38% of the vote. Such an unprecedented error by a magazine with respect to a major poll had never occurred before. As a result, the prediction devastated the magazine's credibility with the public, eventually causing it to go bankrupt. *Literary Digest* thought it had done everything right. It had based its prediction on a huge sample size, 2.4 million respondents, out of a survey sent to 10 million registered voters. What went wrong? There are two answers: selection bias and nonresponse bias.

To understand the role of selection bias, some historical background must be provided. In 1936 the United States was still suffering from the Great Depression. Not accounting for this, the *Literary Digest* compiled its frame from such sources as telephone books, club membership lists, magazine subscriptions, and automobile registrations (reference 7). Inadvertently, it chose a frame primarily composed of the rich and excluded the majority of the voting population who, during the Great Depression, could not afford telephones, club memberships, magazine subscriptions, and automobiles. Thus, the 57% estimate for the Landon vote may have been very close to the frame but certainly not the total U.S. population.

Nonresponse error produced a possible bias when the huge sample of 10 million registered voters produced only 2.4 million responses. A response rate of only 24% is far too low to yield accurate estimates of the population parameters without some mechanism to ensure that the 7.6 million individual nonrespondents have similar opinions. However, the problem of nonresponse bias was secondary to the problem of selection bias. Even if all 10 million registered voters in the sample had responded, this would not have compensated for the fact that the frame differed substantially in composition from the actual voting population.

Sampling Error There are three main reasons for you to select a sample rather than taking a complete census: It is more expedient, less costly, and more efficient. However, chance dictates which individuals or items will or will not be included in the sample. **Sampling error** reflects the heterogeneity, or "chance differences," from sample to sample based on the probability of particular individuals or items being selected in the particular samples.

When you read about the results of surveys or polls in newspapers or magazines, there is often a statement regarding margin of error or precision. For example, "the results of this poll are expected to be within ±4 percentage points of the actual value." This margin of error is the sampling error. You can reduce sampling error by taking larger sample sizes, although this also increases the cost of conducting the survey.

Measurement Error In the practice of good survey research, you design a questionnaire with the intention of gathering meaningful information. But you have a dilemma here—getting meaningful measurements is often easier said than done. Consider the following proverb:

A man with one watch always knows what time it is;
A man with two watches always searches to identify the correct one;
A man with ten watches is always reminded of the difficulty in measuring time.

Unfortunately, the process of getting a measurement is often governed by what is convenient, not what is needed. The measurements are often only a proxy for the ones you really desire. Much attention has been given to measurement error that occurs because of a weakness in question wording (reference 3). A question should be clear, not ambiguous. Furthermore, in order to avoid *leading questions*, you need to present them in a neutral manner.

There are three sources of **measurement error**: ambiguous wording of questions, the halo effect, and respondent error. As an example of ambiguous wording, in November 1993 the Labor Department reported that the unemployment rate in the United States had been underestimated for more than a decade because of poor questionnaire wording in the Current Population Survey. In particular, the wording led to a significant undercount of women in the labor force. Because unemployment rates are tied to benefit programs such as state unemployment compensation systems, it was imperative that government survey researchers rectify the situation by adjusting the questionnaire wording.

The "halo effect" occurs when the respondent feels obligated to please the interviewer. Proper interviewer training can minimize the halo effect.

Respondent error occurs as a result of overzealous or underzealous effort by the respondent. You can minimize this error in two ways: (1) by carefully scrutinizing the data and calling back those individuals whose responses seem unusual and (2) by establishing a program of random callbacks in order to determine the reliability of the responses.

Ethical Issues

Ethical considerations arise with respect to the four types of potential errors that can occur when designing surveys that use probability samples: coverage error, nonresponse error, sampling error, and measurement error. Coverage error can result in selection bias and becomes an ethical issue if particular groups or individuals are *purposely* excluded from the frame so that the survey results are skewed, indicating a position more favorable to the survey's sponsor. Nonresponse error can lead to nonresponse bias and becomes an ethical issue if the sponsor knowingly designs the survey in such a manner that particular groups or individuals are less likely to respond. Sampling error becomes an ethical issue if the findings are purposely presented without reference to sample size and margin of error so that the sponsor can promote a viewpoint that might otherwise be truly insignificant. Measurement error becomes an ethical issue in one of three ways: (1) a survey sponsor chooses leading questions that guide the responses in a particular direction; (2) an interviewer, through mannerisms and tone, purposely creates a halo effect or otherwise guides the responses in a particular direction; (3) a respondent, having a disdain for the survey process, willfully provides false information.

Ethical issues also arise when the results of nonprobability samples are used to form conclusions about the entire population. When you use a nonprobability sampling method, you need to explain the sampling procedures and state that the results cannot be generalized beyond the sample.

PROBLEMS FOR SECTION 7.5

Applying the Concepts

7.32 "A survey indicates that the vast majority of college students own their own personal computers." What information would you want to know before you accepted the results of this survey?

7.33 A simple random sample of $n = 300$ full-time employees is selected from a company list containing the names of all $N = 5,000$ full-time employees in order to evaluate job satisfaction.

a. Give an example of possible coverage error.
b. Give an example of possible nonresponse error.
c. Give an example of possible sampling error.
d. Give an example of possible measurement error.

7.34 According to a survey of 1,000 AOL subscribers (Harry Berkowitz, "Screen Name Loyalty," *Newsday*, December 1, 2002, A42), 92% of the AOL subscribers gave "don't want to change e-mail address" as a reason for sticking with the online service. What information would you want to know before you accepted the results of the survey?

7.35 A survey of online shoppers was conducted by Forrester Research Inc. (Michael Totty, "The Masses Have Arrived," *The Wall Street Journal*, January 27, 2003, R8). For those shoppers who have been buying online for less than one year, 39% have a college degree, 57% are women, and the mean annual income of these shoppers is \$52,300. What information would you want to know before you accepted the results of the survey?

7.36 According to a Maritz poll of 1,004 adult drivers ("Snapshots," *USA Today*, October 23, 2002), 45% admit to often or sometimes eating or drinking while driving and 36% admit to talking on a cellphone. What information would you want to know before you accepted the results of the survey?

7.37 According to a survey conducted by MessageOne, among people who have e-mail at work, almost four in 10 indicate that they cannot live without it (Anne R. Carey and Chad Palmer, "Snapshots," *USA Today*, January 14, 2004, A1). More specifically, 37% said they "Can't live without it," 26% said e-mail was "Important," 19% said "Not essential," 13% said "Don't use," and 5% said "Not important." What information would you want to know before you accepted the results of this survey?

7.38 What do restaurant diners want? In a survey conducted by Caravan for IHOP, 56% responded that they want "Great food." Other responses were "Reasonable prices" 22%, "Atmosphere" 11%, "Quick service" 8%, and "Don't know" 3% "(Darryl Haralson and Jeff Dionise, "Snapshots," *USA Today*, January 16, 2004, A1). What information would you want to know before you accepted the results of this survey?

7.6 (*CD-ROM TOPIC*) SAMPLING FROM FINITE POPULATIONS

In this section, sampling without replacement from finite populations is considered. For further discussion see **section 7.6.pdf** on the CD-ROM that accompanies the text.

SUMMARY

In this chapter you studied the sampling distribution of the sample mean, the Central Limit Theorem, and the sampling distribution of the sample proportion. You learned that the sample mean is an unbiased estimator of the population mean and the sample proportion is an unbiased estimator of the population proportion. By observing the mean weight in a sample of cereal boxes filled by the Oxford Cereal Company, you were able to draw conclusions concerning the mean weight in the population of cereal boxes. You also studied four common survey sampling methods—simple random, systematic, stratified, and cluster. In the next five chapters, techniques commonly used for statistical inference, confidence intervals, and tests of hypotheses are discussed.

KEY FORMULAS

Population Mean

$$\mu = \frac{\sum_{i=1}^{N} X_i}{N} \quad \textbf{(7.1)}$$

Population Standard Deviation

$$\sigma = \sqrt{\frac{\sum_{i=1}^{N} (X_i - \mu)^2}{N}} \quad \textbf{(7.2)}$$

Standard Error of the Mean

$$\sigma_{\overline{X}} = \frac{\sigma}{\sqrt{n}} \quad \textbf{(7.3)}$$

Finding Z for the Sampling Distribution of the Mean

$$Z = \frac{\overline{X} - \mu_{\overline{X}}}{\sigma_{\overline{X}}} = \frac{\overline{X} - \mu}{\frac{\sigma}{\sqrt{n}}} \quad \textbf{(7.4)}$$

Finding $\overline{X}$ for the Sampling Distribution of the Mean

$$\overline{X} = \mu + Z\frac{\sigma}{\sqrt{n}} \quad \textbf{(7.5)}$$

Sample Proportion

$$p = \frac{X}{n} \quad \textbf{(7.6)}$$

Standard Error of the Sample Proportion

$$\sigma_p = \sqrt{\frac{\pi(1-\pi)}{n}} \quad \textbf{(7.7)}$$

Finding Z for the Sampling Distribution of the Proportion

$$Z = \frac{p - \pi}{\sqrt{\frac{\pi(1-\pi)}{n}}} \quad \textbf{(7.8)}$$

KEY TERMS

CHAPTER REVIEW PROBLEMS

Checking Your Understanding

7.39 Why is the sample mean an unbiased estimator of the population mean?

7.40 Why does the standard error of the mean decrease as the sample size *n* increases?

7.41 Why does the sampling distribution of the mean follow a normal distribution for a large enough sample size even though the population may not be normally distributed?

7.42 What is the difference between a probability distribution and a sampling distribution?

7.43 Under what circumstances does the sampling distribution of the proportion approximately follow the normal distribution?

7.44 What is the difference between probability and nonprobability sampling?

7.45 What are some potential problems with using "fishbowl" methods to select a simple random sample?

7.46 What is the difference between sampling *with* replacement versus *without* replacement?

7.47 What is the difference between a simple random sample and a systematic sample?

7.48 What is the difference between a simple random sample and a stratified sample?

7.49 What is the difference between a stratified sample and a cluster sample?

Applying the Concepts

7.50 An industrial sewing machine uses ball bearings that are targeted to have a diameter of 0.75 inch. The lower and upper specification limits under which the ball bearing can operate are 0.74 inch (lower) and 0.76 inch (upper). Past experience has indicated that the actual diameter of the ball bearings is approximately normally distributed with a mean of 0.753 inch and a standard deviation of 0.004 inch. If you select a random sample of 25 ball bearings, what is the probability that the sample mean is

a. between the target and the population mean of 0.753?
b. between the lower specification limit and the target?
c. above the upper specification limit?
d. below the lower specification limit?
e. The probability is 93% that the sample mean diameter will be above what value?

7.51 The fill amount of bottles of soft drink is normally distributed with a mean of 2.0 liters and a standard deviation of 0.05 liter. If you select a random sample of 25 bottles, what is the probability that the sample mean will be

a. between 1.99 and 2.0 liters?
b. below 1.98 liters?
c. above 2.01 liters?
d. The probability is 99% that the sample mean will contain at least how much soft drink?
e. The probability is 99% that the sample mean will contain an amount that is between which two values (symmetrically distributed around the mean)?

7.52 An orange juice producer buys all his oranges from a large orange grove that has one variety of orange. The amount of juice squeezed from each of these oranges is approximately normally distributed with a mean of 4.70 ounces and a standard deviation of 0.40 ounce. Suppose that you select a sample of 25 oranges:

a. What is the probability that the sample mean will be at least 4.60 ounces?
b. The probability is 70% that the sample mean will be contained between what two values symmetrically distributed around the population mean?
c. The probability is 77% that the sample mean will be above what value?

7.53 DiGiorno's frozen pizza has some of the most creative and likeable advertisements on television. *USA Today's* Ad Track claims that 20% of viewers like the ads "a lot" (Theresa Howard, "DiGiorno Campaign Delivers Major Sales," **www.usatoday.com**, April 1, 2002). Suppose that a sample of 400 television viewers is shown the advertisements. What is the probability that the sample will have between

a. 18% and 22% who like the ads "a lot"?
b. 16% and 24% who like the ads "a lot"?
c. 14% and 26% who like the ads "a lot"?
d. 12% and 28% who like the ads "a lot"?

7.54 Mutual funds reported modest earnings in the first quarter of 2004. U.S. diversified equity funds, large baskets of stocks from a wide variety of companies, had a 2.98% return (Michael J. Martinez, "Mutual-fund Returns Minimal in First Quarter," **Cincinnati.com**, April 3, 2004). Assume that the returns for U.S. funds were distributed as a normal random variable with a mean of 2.98 and a standard deviation of 4. If you selected a random sample of 10 funds from this population, what is the probability that the sample would have a mean return

a. less than 0, that is, a loss?
b. between 0 and 6?
c. greater than 10?

7.55 Mutual funds reported modest earnings in the first quarter of 2004. International funds, which are historically slightly more volatile than U.S. funds, had a mean return of 5.13% (Michael J. Martinez, "Mutual-fund Returns Minimal in First Quarter," **Cincinnati.com**, April 3, 2004). Assume that the returns for international funds were distributed as a normal random variable with a mean of 5.13 and a standard deviation of 6. If you select an individual fund from this population, what is the probability that it would have a return

a. less than 0, that is, a loss?
b. between 0 and 6?
c. greater than 10?

If you selected a random sample of 10 funds from this population, what is the probability that the sample would have a mean return

d. less than 0, that is, a loss?
e. between 0 and 6?
f. greater than 10?
g. Compare your results in parts (d) through (f) to (a) through (c).
h. Compare your results in parts (d) through (f) to problem 7.54 (a) through (c).

7.56 Political polling has traditionally used telephone interviews. Researchers at Harris Black International Ltd. argue that Internet polling is less expensive, faster, and offers higher response rates than telephone surveys. Critics are concerned about the scientific reliability of this approach (*The Wall Street Journal*, April 13, 1999). Even amid this strong criticism, Internet polling is becoming more and more common. What concerns, if any, do you have about Internet polling?

7.57 A study by Rajesh Mirani and Albert Lederer ("An Instrument for Accessing the Organizational Benefits of IS Projects," *Decision Sciences*, vol. 29, 1998, 803–838) discusses the organizational benefits of information systems (IS) projects. The researchers mailed 936 questionnaires to randomly selected members of a large nationwide information systems organization. Two hundred valid responses were received, for a response rate of 21%. Of the 200 respondents, 190 answered questions concerning a recently completed IS project. The average budget for these projects was \$3.8 million with a range of \$4,000 to \$100 million. Of these 190 responses, 45% indicated that the CEO was required to give approval before starting the projects.

a. What was the source of the data used in this study?
b. Discuss the sampling method used in this study.
c. What types of survey errors do you think the researchers are most likely to encounter?

7.58 As part of a mediation process overseen by a federal judge to end a lawsuit that accuses Cincinnati, Ohio, of decades of discrimination against African Americans, surveys on how to improve Cincinnati police–community relations were taken. One survey was sent to the 1,020 members of the Cincinnati police force. The survey included a cover letter encouraging participation by the chief of police and president of the Fraternal Order of Police. Respondents could either return a hard copy of the survey or complete the survey online. To the researchers' dismay, only 158 surveys were completed ("Few Cops Fill Out Survey," *The Cincinnati Enquirer*, August 22, 2001, B3).

a. What type of errors or biases should the researchers be especially concerned with?
b. What step(s) should the researchers take to try to overcome the problems noted in (a)?
c. What could have been done differently to improve the survey's worthiness?

7.59 According to a survey conducted by International Communications Research for Capital One Financial, 24% of teens ages 13 to 19 own a cell phone and 10% own a beeper ("USA Snapshots," *USA Today*, August 16, 2001, A1).

a. What other information would you want to know before you accepted the results of this survey?
b. Suppose that you wished to conduct a similar survey for the geographic region you live in. Describe the population for your survey.
c. Explain how you could minimize the chance of a coverage error in this type of survey.

d. Explain how you could minimize the chance of a nonresponse error in this type of survey.
e. Explain how you could minimize the chance of a sampling error in this type of survey.
f. Explain how you could minimize the chance of a measurement error in this type of survey.

7.60 According to Dr. Sarah Beth Estes, sociology professor at the University of Cincinnati, and Dr. Jennifer Glass, sociology professor at the University of Iowa, working women who take advantage of family-friendly schedules can fall behind in wages. More specifically, the sociologists report that in a study of 300 working women who had children and returned to work and opted for flextime, telecommuting, and so on, these women had pay raises that averaged between 16% and 26% less than other workers ("Study: 'Face Time' Can Affect Moms' Raises," *The Cincinnati Enquirer*, August 28, 2001, A1).
a. What other information would you want to know before you accepted the results of this survey?
b. If you were to perform a similar study in the geographic area where you live, define a population, frame, and sampling method you could use.

7.61 **(Class Project)** The table of random numbers is an example of a uniform distribution because each digit is equally likely to occur. Starting in the row corresponding to the day of the month in which you were born, use the table of random numbers (Table E.1) to take one digit at a time.

Select five different samples of $n = 2$, $n = 5$, and $n = 10$. Compute the sample mean of each sample. Develop a frequency distribution of the sample means for the results of the entire class based on samples of sizes $n = 2$, $n = 5$, and $n = 10$.

What can be said about the shape of the sampling distribution for each of these sample sizes?

7.62 **(Class Project)** Toss a coin 10 times and record the number of heads. If each student performs this experiment five times, a frequency distribution of the number of heads can be developed from the results of the entire class. Does this distribution seem to approximate the normal distribution?

7.63 **(Class Project)** The number of cars waiting in line at a car wash is distributed as follows:

Number of Cars	Probability
0	0.25
1	0.40
2	0.20
3	0.10
4	0.04
5	0.01

You can use the table of random numbers (Table E.1) to select samples from this distribution by assigning numbers as follows:
1. Start in the row corresponding to the day of the month in which you were born.
2. Select a two-digit random number.
3. If you select a random number from 00 to 24, record a length of 0; if from 25 to 64, record a length of 1; if from 65 to 84, record a length of 2; if from 85 to 94, record a length of 3; if from 95 to 98, record a length of 4; if it is 99, record a length of 5.

Select samples of $n = 2$, $n = 5$, and $n = 10$. Compute the mean for each sample. For example, if a sample of size 2 results in random numbers 18 and 46, these would correspond to lengths of 0 and 1, respectively, producing a sample mean of 0.5. If each student selects five different samples for each sample size, a frequency distribution of the sample means (for each sample size) can be developed from the results of the entire class. What conclusions can you reach concerning the sampling distribution of the mean as the sample size is increased?

7.64 **(Class Project)** The table of random numbers can simulate the selection of different colored balls from a bowl as follows:
1. Start in the row corresponding to the day of the month in which you were born.
2. Select one-digit numbers.
3. If a random digit between 0 and 6 is selected, consider the ball white; if a random digit is a 7, 8, or 9, consider the ball red.

Select samples of $n = 10$, $n = 25$, and $n = 50$ digits. In each sample, count the number of white balls and compute the proportion of white balls in the sample. If each student in the class selects five different samples for each sample size, a frequency distribution of the proportion of white balls (for each sample size) can be developed from the results of the entire class. What conclusions can you reach about the sampling distribution of the proportion as the sample size is increased?

7.65 **(Class Project)** Suppose that step 3 of problem 7.64 uses the following rule: "If a random digit between 0 and 8 is selected, consider the ball to be white; if a random digit of 9 is selected, consider the ball to be red." Compare and contrast the results in this problem and in problem 7.64.

RUNNING CASE MANAGING THE *SPRINGVILLE HERALD*

Continuing its quality improvement effort first described in the Chapter 6 "Managing the Springville Herald" case, the production department of the newspaper has been monitoring the blackness of the newspaper print. As before, blackness is measured on a standard scale in which the target value is 1.0. Data collected over the past year indicate that the blackness is normally distributed with a mean of 1.005 and a standard deviation of 0.10.

SH7.1 Each day, 25 spots on the first newspaper printed are chosen and the blackness of the spots is measured. Assuming that the distribution has not changed from what it was in the past year, what is the probability that the mean blackness of the spots is:

a. less than 1.0?
b. between 0.95 and 1.0?
c. between 1.0 and 1.05?
d. less than 0.95 or greater than 1.05?
e. Suppose that the mean blackness of today's sample of 25 spots is 0.952. What conclusion can you make about the blackness of the newspaper based on this result? Explain.

WEB CASE

Apply your knowledge about sampling distributions in this Web case that reconsiders the Oxford Cereals "Using Statistics" scenario.

The TriCities Consumers Concerned About Cereal Companies That Cheat organization (TCCACCTC) suspects that cereal companies, including Oxford Cereals, are cheating consumers by packaging cereals at less than labeled weights. Visit the organization's Web site **www.prenhall.com/Springville/CerealCheaters.htm**, examine their claims and supporting data and then answer the following:

1. Are the data collection procedures used by the TCCACCTC to form its conclusions flawed? What procedures could the group follow to make their analysis more rigorous?

2. Assume that the two samples of five cereal boxes (one sample for each of two cereal varieties) listed on the TCCACCTC Web site were collected randomly by organization members. For each sample, do the following:

a. Calculate the sample mean.
b. Assume the standard deviation of the process is 15 grams. Calculate the percentage of all samples for each process that would have a sample mean less than the value you calculated in step (a).
c. Again, assuming the standard deviation is 15 grams, calculate the percentage of individual boxes of cereal that would have a weight less than the value you calculated in step (a).

3. What, if any, conclusions can you form using your calculations about the filling processes of the two different cereals?

4. A representative from Oxford Cereals has asked that the TCCACCTC take down its page discussing shortages in boxes of Oxford Cereals. Is that request reasonable? Why, or why not?

5. Can the techniques of this chapter be used to prove cheating in the manner alleged by the TCCACCTC? Why, or why not?

REFERENCES

1. Cochran, W. G., *Sampling Techniques*, 3rd ed. (New York: Wiley, 1977).
2. Gallup, G. H., *The Sophisticated Poll-Watcher's Guide* (Princeton, NJ: Princeton Opinion Press, 1972).
3. Goleman, D., "Pollsters Enlist Psychologists in Quest for Unbiased Results," *The New York Times*, September 7, 1993, C1 and C11.
4. Levine, D. M., P. Ramsey, and R. Smidt, *Applied Statistics for Engineers and Scientists Using Microsoft Excel and Minitab* (Upper Saddle River, NJ: Prentice Hall, 2001).
5. *Microsoft Excel 2003* (Redmond, WA: Microsoft Corp., 2003).
6. *Minitab for Windows Version 14* (State College, PA: Minitab Inc., 2004).
7. Mosteller, F., et al., *The Pre-Election Polls of 1948* (New York: Social Science Research Council, 1949).
8. Rand Corporation, *A Million Random Digits with 100,000 Normal Deviates* (New York: The Free Press, 1955).

Appendix 7 Using Software for Sampling Distributions

A7.1 MICROSOFT EXCEL

For Sampling Distribution Simulation

Use the Data Analysis ToolPak. Open to the worksheet containing the data for which you want to generate a sampling distribution simulation. Select **Tools → Data Analysis**. From the list that appears in the Data Analysis dialog box, select **Random Number Generation** and click **OK**. In the Random Number Generator dialog box (see Figure A7.1), enter the number of samples as the **Number of Variables** and enter the sample size of each sample as the **Number of Random Numbers**. Select the type of distribution from the **Distribution** drop-down list box and make entries in the **Parameters** area as is necessary. (Parameters vary according to distribution chosen.) Select the **New Worksheet Ply** option button and click **OK**. Figure A7.1 shows the entries for generating 100 samples of $n = 30$ from a uniformly distributed population.

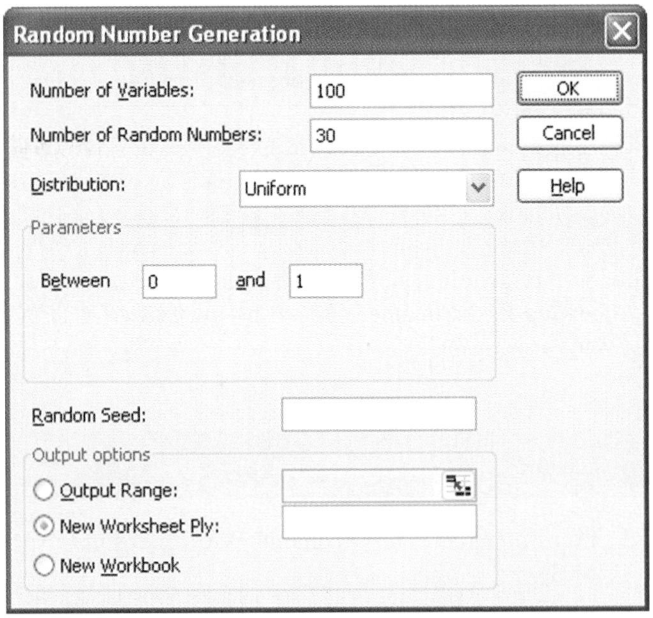

FIGURE A7.1 Data Analysis Random Number Generator Dialog Box

OR See section G.13 (**Sampling Distribution Simulation**) if you want PHStat2 to produce an enhanced version of this worksheet.

A7.2 MINITAB

To develop a simulation of the sampling distribution of the mean from a uniformly distributed population with 100 samples of $n = 30$, select **Calc → Random Data → Uniform**. In the Uniform Distribution dialog box (see Figure A7.2):

1. Enter **100** in the Generate rows of data edit box.
2. Enter **C1–C30** in the Store in column(s): edit box.
3. Enter **0.0** in the Lower endpoint: edit box and **1.0** in the Upper endpoint: edit box. Click the **OK** button.

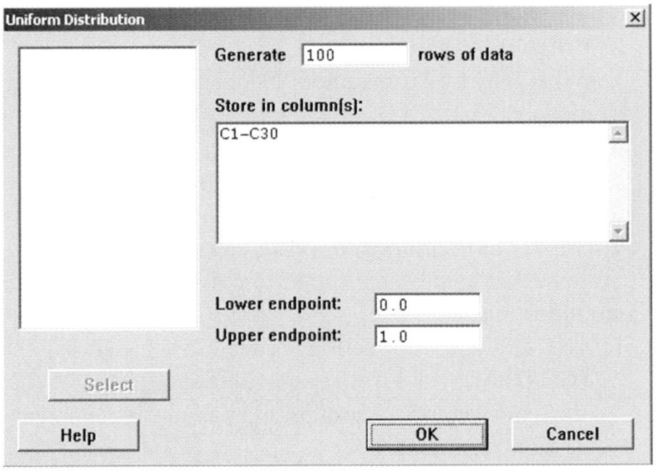

FIGURE A7.2 Minitab Uniform Distribution Dialog Box

One hundred rows of values are now entered in columns C1–C30. To calculate row statistics for each of the 100 samples, select **Calc → Row Statistics**, then in the Row Statistics dialog box (see Figure A7.3):

1. Select the **Mean** option button.
2. Enter **C1–C30** in the Input variables: edit box. Enter **C31** in the Store result in: edit box. Click the **OK** button.

The mean for each of the 100 samples is stored in column C31. To compute statistics for the set of 100 sample means, select **Stat → Basic Statistics → Display Descriptive Statistics**. Enter **C31** in the Variables: edit box. Click the **OK** button.

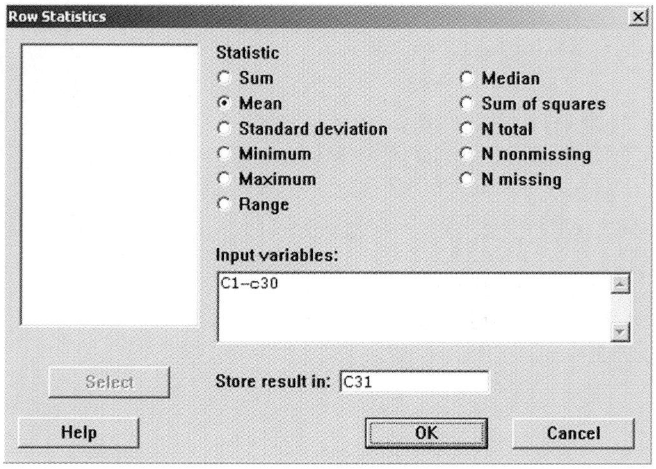

FIGURE A7.3 Minitab Row Statistics Dialog Box

To generate a histogram of the 100 sample means, select **Graph → Histogram** and do the following:

1. In the Histograms dialog box, select **Simple**. Click the **OK** button.
2. In the Histogram–Simple dialog box, enter **C31** in the Graph variables: edit box. Click the **OK** button.

To produce a simulation of the sampling distribution of the mean for a normal population, select **Calc → Random Data → Normal**. Enter a value for μ in the Mean: edit box and for σ in the Standard deviation: edit box. Follow the remainder of the instructions given for the uniform population.

CHAPTER 8

Confidence Interval Estimation

LEARNING OBJECTIVES

In this chapter, you learn:

- To construct and interpret confidence interval estimates for the mean and the proportion
- How to determine the sample size necessary to develop a confidence interval for the mean or proportion
- How to use confidence interval estimates in auditing

USING STATISTICS

Auditing Sales Invoices at the Saxon Home Improvement Company

Saxon Home Improvement Company distributes home improvement supplies in the northeastern United States. As a company accountant, you are responsible for the accuracy of the integrated inventory management and sales information system. You could review the contents of each and every record to check the accuracy of this system, but such a detailed review would be time-consuming and costly. A better approach is to use statistical inference techniques to draw conclusions about the population of all records from a relatively small sample collected during an audit. At the end of each month, you can select a sample of the sales invoices to determine the following:

- The mean dollar amount listed on the sales invoices for the month.
- The total dollar amount listed on the sales invoices for the month.
- Any differences between the dollar amounts on the sales invoice and the amounts entered into the sales information system.
- The frequency of occurrence of various types of errors that violate the internal control policy of the warehouse. These errors include making a shipment when there is no authorized warehouse removal slip, failure to include the correct account number, and shipment of the incorrect home improvement item.

How accurate are the results from the samples and how do you use this information? Are the sample sizes large enough to give you the information you need?

Statistical inference is the process of using sample results to draw conclusions about the characteristics of a population. Inferential statistics enables you to *estimate* unknown population characteristics such as a population mean or a population proportion. There are two types of estimates used to estimate population parameters: point estimates and interval estimates. A **point estimate** is the value of a single sample statistic. A **confidence interval estimate** is a range of numbers, called an interval, constructed around the point estimate. The confidence interval is constructed such that the probability the population parameter is located somewhere within the interval is known.

Suppose that you would like to estimate the mean GPA of all the students in your university. The mean GPA for all the students is an unknown population mean, denoted by μ. You select a sample of students and find that the sample mean is 2.80. The sample mean $\bar{X} = 2.80$ is a point estimate of the population mean μ. How accurate is 2.80? To answer this question you must construct a confidence interval estimate.

In this chapter you will learn how to construct and interpret confidence interval estimates. Recall that the sample mean $\bar{X}$ is a point estimate of the population mean μ. However, the sample mean will vary from sample to sample because it depends on the items selected in the sample. By taking into account the known variability from sample to sample (see section 7.2 on the sampling distribution of the mean), you will learn how to develop the interval estimate for the population mean. The interval constructed will have a specified confidence of correctly estimating the value of the population parameter μ. In other words, there is a specified confidence that μ is somewhere in the range of numbers defined by the interval.

Suppose that after studying this chapter, you find that a 95% confidence interval for the mean GPA at your university is $(2.75 \leq \mu \leq 2.85)$. You can interpret this interval estimate by stating that you are 95% confident that the mean GPA at your university is between 2.75 and 2.85. There is a 5% chance that the mean GPA is below 2.75 or above 2.85.

After learning about the confidence interval for the mean, you will learn how to develop an interval estimate for the population proportion. Then you will learn how large a sample to select when constructing confidence intervals, and how to perform several important estimation procedures accountants use when performing audits.

8.1 CONFIDENCE INTERVAL ESTIMATE FOR THE MEAN (σ KNOWN)

In section 7.2 you used the Central Limit Theorem and knowledge of the population distribution to determine the percentage of sample means that fall within certain distances of the population mean. For instance, in the cereal-filling example used throughout Chapter 7 (see page 235), 95% of all sample means are between 362.12 and 373.88 grams. This statement is based on *deductive reasoning*. However, *inductive reasoning* is what you need here.

You need inductive reasoning because, in statistical inference, you use the results of a single sample to draw conclusions about the population, not vice versa. Suppose that in the cereal-fill example, you wish to estimate the unknown population mean using the information from only a sample. Thus, rather than take $\mu \pm (1.96)(\sigma/\sqrt{n})$ to find the upper and lower limits around μ as in section 7.2, you substitute the sample mean $\bar{X}$ for the unknown μ and use $\bar{X} \pm (1.96)(\sigma/\sqrt{n})$ as an interval to estimate the unknown μ. Although in practice you select a single sample of size n and compute the mean $\bar{X}$, in order to understand the full meaning of the interval estimate, you need to examine a hypothetical set of all possible samples of n values.

Suppose that a sample of $n = 25$ boxes has a mean of 362.3 grams. The interval developed to estimate μ is $362.3 \pm (1.96)(15)/(\sqrt{25})$ or 362.3 ± 5.88. The estimate of μ is

$$356.42 \leq \mu \leq 368.18$$

Because the population mean μ (equal to 368) is included within the interval, this sample has led to a correct statement about μ (see Figure 8.1 below).

FIGURE 8.1
Confidence Interval Estimates for Five Different Samples of $n = 25$ Taken from a Population Where $\mu = 368$ and $\sigma = 15$

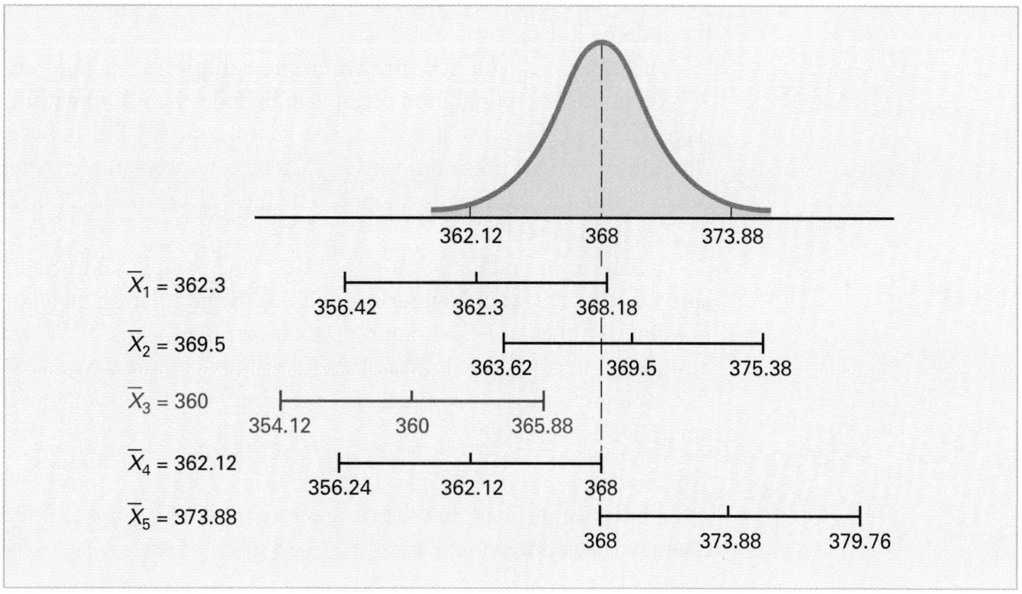

To continue this hypothetical example, suppose that for a different sample of $n = 25$ boxes, the mean is 369.5. The interval developed from this sample is

$$369.5 \pm (1.96)(15)/(\sqrt{25})$$

or 369.5 ± 5.88. The estimate is

$$363.62 \le \mu \le 375.38$$

Because the population mean μ (equal to 368) is also included within this interval, this statement about μ is correct.

Now, before you begin to think that correct statements about μ are always made by developing a confidence interval estimate, suppose a third hypothetical sample of $n = 25$ boxes is selected and the sample mean is equal to 360 grams. The interval developed here is $360 \pm (1.96)(15)/(\sqrt{25})$ or 360 ± 5.88. In this case, the estimate of μ is

$$354.12 \le \mu \le 365.88$$

This estimate is *not* a correct statement, because the population mean μ is not included in the interval developed from this sample (see Figure 8.1 on page 261). Thus, for some samples the interval estimate of μ is correct, but for others it is incorrect. In practice, only one sample is selected, and because the population mean is unknown, you cannot determine whether the interval estimate is correct.

To resolve this dilemma of sometimes having an interval that provides a correct estimate and sometimes having an interval that provides an incorrect estimate, you need to determine the proportion of samples producing intervals that result in correct statements about the population mean μ. To do this, consider two other hypothetical samples: the case in which $\bar{X} = 362.12$ grams and the case in which $\bar{X} = 373.88$ grams. If $\bar{X} = 362.12$, the interval is $362.12 \pm (1.96)(15)/(\sqrt{25})$ or 362.12 ± 5.88. This leads to the following interval

$$356.24 \le \mu \le 368.00$$

Because the population mean of 368 is at the upper limit of the interval, the statement is a correct one (see Figure 8.1).

When $\bar{X} = 373.88$, the interval is $373.88 \pm (1.96)(15)/(\sqrt{25})$ or 373.88 ± 5.88. The interval for the sample mean is

$$368.00 \le \mu \le 379.76$$

In this case, because the population mean of 368 is included at the lower limit of the interval, the statement is correct.

In Figure 8.1 you see that when the sample mean falls anywhere between 362.12 and 373.88 grams, the population mean is included *somewhere* within the interval. In section 7.2 on page 235, you found that 95% of the sample means fall between 362.12 and 373.88 grams. Therefore, 95% of all samples of $n = 25$ boxes have sample means that include the population mean within the interval developed. The interval from 362.12 to 373.88 is referred to as a 95% confidence interval.

Because, in practice, you only select one sample and μ is unknown, you never know for sure whether the specific interval includes the population mean or not. However, if you take all possible samples of n and compute their sample means, 95% of the intervals will include the population mean and only 5% of them will not. In other words, you have 95% confidence that the population mean is somewhere in the interval. Thus, you can interpret the confidence interval above as follows:

"I am 95% confident that the mean amount of cereal in the population of boxes is somewhere between 362.12 and 373.88 grams."

In some situations, you might want a higher degree of confidence (such as 99%) of including the population mean within the interval. In other cases, you might accept less confidence (such as 90%) of correctly estimating the population mean.

In general, the **level of confidence** is symbolized by $(1 - \alpha) \times 100\%$, where α is the proportion in the tails of the distribution that is outside the confidence interval. The proportion in the upper tail of the distribution is $\alpha/2$, and the proportion in the lower tail of the distribution is $\alpha/2$. You use Equation (8.1) to construct a $(1 - \alpha) \times 100\%$ confidence interval estimate of the mean with σ known.

CONFIDENCE INTERVAL FOR A MEAN (σ KNOWN)

$$\bar{X} \pm Z\frac{\sigma}{\sqrt{n}}$$

or

$$\bar{X} - Z\frac{\sigma}{\sqrt{n}} \leq \mu \leq \bar{X} + Z\frac{\sigma}{\sqrt{n}} \quad \textbf{(8.1)}$$

where Z = the value corresponding to a cumulative area of $1 - \alpha/2$ from the standardized normal distribution, that is, an upper-tail probability of $\alpha/2$.

The value of Z needed for constructing a confidence interval is called the **critical value** for the distribution. 95% confidence corresponds to an α value of 0.05. The critical Z value corresponding to a cumulative area of 0.9750 is 1.96 because there is 0.025 in the upper tail of the distribution and the cumulative area less than $Z = 1.96$ is 0.975.

There is a different critical value for each level of confidence $1 - \alpha$. A level of confidence of 95% leads to a Z value of 1.96 (see Figure 8.2). 99% confidence corresponds to an α value of 0.01. The Z value is approximately 2.58 because the upper-tail area is 0.005 and the cumulative area less than $Z = 2.58$ is 0.995 (see Figure 8.3).

FIGURE 8.2
Normal Curve for Determining the Z Value Needed for 95% Confidence

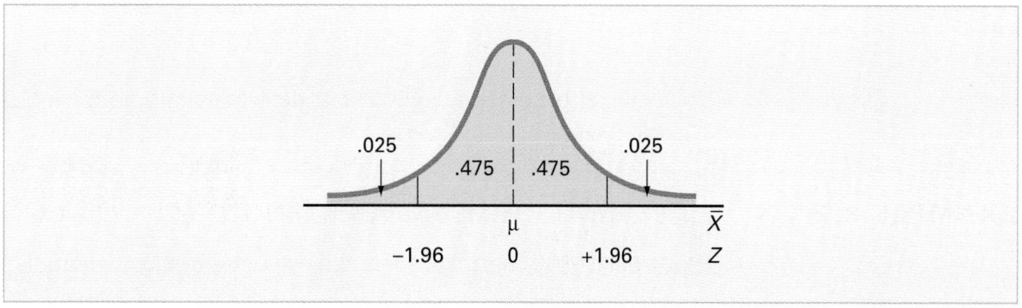

FIGURE 8.3
Normal Curve for Determining the Z Value Needed for 99% Confidence

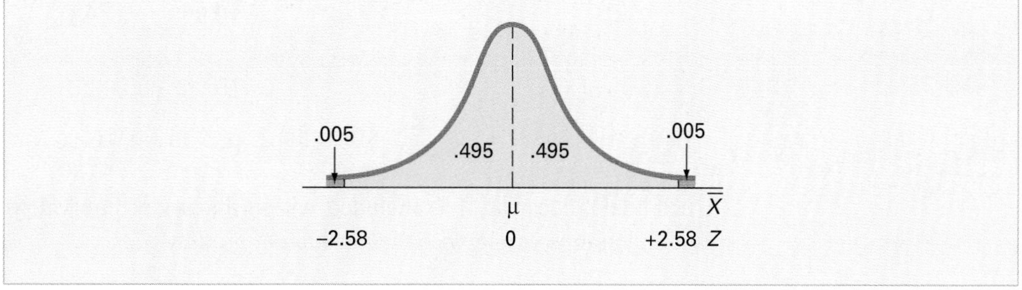

Now that various levels of confidence have been considered, why not make the confidence level as close to 100% as possible? Before doing so, you need to realize that any increase in the level of confidence is achieved only by widening (and making less precise) the confidence interval. There is no "free lunch" here. You would have more confidence that the population mean is within a broader range of values. However, this might make the interpretation of the confidence interval less useful. The trade-off between the width of the confidence interval and the level of confidence is discussed in greater depth in the context of determining the sample size in section 8.4.

Example 8.1 illustrates the application of the confidence interval estimate.

EXAMPLE 8.1

ESTIMATING THE MEAN PAPER LENGTH WITH 95% CONFIDENCE

A manufacturer of computer paper has a production process that operates continuously throughout an entire production shift. The paper is expected to have a mean length of 11 inches and the standard deviation of the length is 0.02 inch. At periodic intervals, a sample is selected to determine whether the mean paper length is still equal to 11 inches or whether something has gone wrong in the production process to change the length of the paper produced. You select a random sample of 100 sheets, and the mean paper length is 10.998 inches. Construct a 95% confidence interval estimate for the population mean paper length.

SOLUTION Using Equation (8.1) on page 263, with $Z = 1.96$ for 95% confidence,

$$\bar{X} \pm Z\frac{\sigma}{\sqrt{n}} = 10.998 \pm (1.96)\frac{0.02}{\sqrt{100}}$$

$$= 10.998 \pm 0.00392$$

$$10.99408 \le \mu \le 11.00192$$

Thus, with 95% confidence, you conclude that the population mean is between 10.99408 and 11.00192 inches. Because the interval includes 11, the value indicating that the production process is working properly, you have no reason to believe that anything is wrong with the production process.

To see the effect of using a 99% confidence interval, examine Example 8.2

EXAMPLE 8.2

ESTIMATING THE MEAN PAPER LENGTH WITH 99% CONFIDENCE

Construct a 99% confidence interval estimate for the population mean paper length.

SOLUTION Using Equation (8.1) on page 263, with $Z = 2.58$ for 99% confidence,

$$\bar{X} \pm Z\frac{\sigma}{\sqrt{n}} = 10.998 \pm (2.58)\frac{0.02}{\sqrt{100}}$$

$$= 10.998 \pm 0.00516$$

$$10.99284 \le \mu \le 11.00316$$

Once again, because 11 is included within this wider interval, you have no reason to believe that anything is wrong with the production process.

PROBLEMS FOR SECTION 8.1

Learning the Basics

PH Grade ASSIST **8.1** If $\bar{X} = 85$, $\sigma = 8$, and $n = 64$, construct a 95% confidence interval estimate of the population mean μ.

PH Grade ASSIST **8.2** If $\bar{X} = 125$, $\sigma = 24$, and $n = 36$, construct a 99% confidence interval estimate of the population mean μ.

8.3 A market researcher states that she has 95% confidence that the mean monthly sales of a product are between $170,000 and $200,000. Explain the meaning of this statement.

8.4 Why is it not possible in Example 8.1 on page 264 to have 100% confidence? Explain.

8.5 From the results of Example 8.1 on page 264 regarding paper production, is it true that 95% of the sample means will fall between 10.99408 and 11.00192 inches? Explain.

8.6 Is it true in Example 8.1 on page 264 that you do not know for sure whether the population mean is between 10.99408 and 11.00192 inches? Explain.

Applying the Concepts

PH Grade ASSIST **8.7** The manager of a paint supply store wants to estimate the actual amount of paint contained in 1-gallon cans purchased from a nationally known manufacturer. It is known from the manufacturer's specifications that the standard deviation of the amount of paint is equal to 0.02 gallon. A random sample of 50 cans is selected, and the sample mean amount of paint per 1-gallon can is 0.995 gallon.

a. Construct a 99% confidence interval estimate of the population mean amount of paint included in a 1-gallon can.
b. On the basis of your results, do you think that the manager has a right to complain to the manufacturer? Why?
c. Must you assume that the population amount of paint per can is normally distributed here? Explain.
d. Construct a 95% confidence interval estimate. How does this change your answer to (b)?

PH Grade ASSIST SELF Test **8.8** The quality control manager at a lightbulb factory needs to estimate the mean life of a large shipment of lightbulbs. The standard deviation is 100 hours. A random sample of 64 lightbulbs indicated a sample mean life of 350 hours.

a. Construct a 95% confidence interval estimate of the population mean life of lightbulbs in this shipment.
b. Do you think that the manufacturer has the right to state that the lightbulbs last an average of 400 hours? Explain.
c. Must you assume that the population of lightbulb life is normally distributed? Explain.
d. Suppose that the standard deviation changed to 80 hours. What are your answers in (a) and (b)?

PH Grade ASSIST **8.9** The inspection division of the Lee County Weights and Measures Department wants to estimate the actual amount of soft drink in 2-liter bottles at the local bottling plant of a large nationally known soft-drink company. The bottling plant has informed the inspection division that the population standard deviation for 2-liter bottles is 0.05 liter. A random sample of 100 2-liter bottles at this bottling plant indicates a sample mean of 1.99 liters.

a. Construct a 95% confidence interval estimate of the population mean amount of soft drink in each bottle.
b. Must you assume that the population of soft-drink fill is normally distributed? Explain.
c. Explain why a value of 2.02 liters for a single bottle is not unusual, even though it is outside the confidence interval you calculated.
d. Suppose that the sample mean had been 1.97 liters. What is your answer to (a)?

8.2 CONFIDENCE INTERVAL ESTIMATION FOR THE MEAN (σ UNKNOWN)

Just as the mean of the population μ is usually unknown, you rarely know the actual standard deviation of the population σ. Therefore, you need to develop a confidence interval estimate of μ using only the sample statistics $\bar{X}$ and S.

Student's *t* Distribution

At the beginning of the twentieth century, a statistician for Guinness Breweries in Ireland (see reference 3) named William S. Gosset wanted to make inferences about the mean when σ was unknown. Because Guinness employees were not permitted to publish research work under their own names, Gosset adopted the pseudonym "Student." The distribution that he developed is known as **Student's *t* distribution**.

If the random variable X is normally distributed, then the following statistic has a t distribution with $n - 1$ **degrees of freedom**:

$$t = \frac{\overline{X} - \mu}{\frac{S}{\sqrt{n}}}$$

This expression has the same form as the Z statistic in Equation (7.4) on page 233, except that S is used to estimate the unknown σ. The concept of *degrees of freedom* is discussed further on page 267.

Properties of the t Distribution

In appearance, the t distribution is very similar to the standardized normal distribution. Both distributions are bell-shaped. However, the t distribution has more area in the tails and less in the center than does the standardized normal distribution (see Figure 8.4). Because the value of σ is unknown and S is used to estimate it, the values of t are more variable than for Z.

FIGURE 8.4
Standardized Normal Distribution and *t* Distribution for 5 Degrees of Freedom

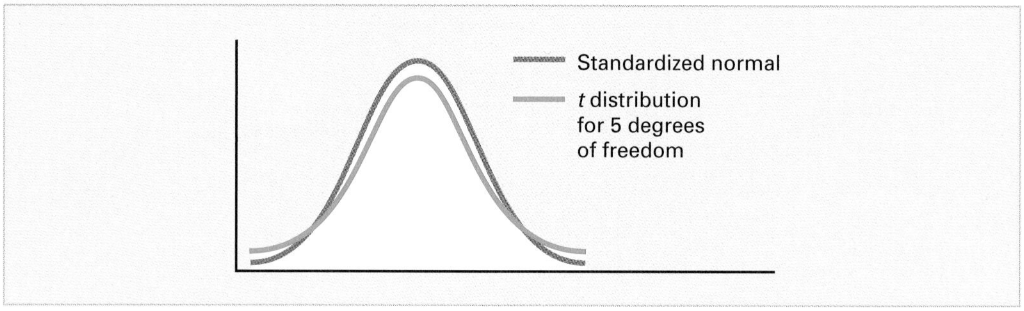

The degrees of freedom $n - 1$ are directly related to the sample size n. As the sample size and degrees of freedom increase, S becomes a better estimate of σ and the t distribution gradually approaches the standardized normal distribution until the two are virtually identical. With a sample size of about 120 or more, S estimates σ precisely enough that there is little difference between the t and Z distributions. For this reason, most statisticians use Z instead of t when the sample size is greater than 120.

As stated earlier, the t distribution assumes that the random variable X is normally distributed. In practice, however, as long as the sample size is large enough and the population is not very skewed, you can use the t distribution to estimate the population mean when σ is unknown. When dealing with a small sample size and a skewed population distribution, the validity of the confidence interval is a concern. To assess the assumption of normality, you can evaluate the shape of the sample data by using a histogram, stem-and-leaf display, box-and-whisker plot, or normal probability plot.

You find the critical values of t for the appropriate degrees of freedom from the table of the t distribution (see Table E.3). The columns of the table represent the area in the upper tail of the t distribution. Each row represents the particular t value for each specific degree of freedom. For example, with 99 degrees of freedom, if you want 95% confidence, you find the appropriate value of t as shown in Table 8.1. The 95% confidence level means that 2.5% of the values (an area of .025) are in each tail of the distribution. Looking in the column for an upper-tail area of .025 and in the row corresponding to 99 degrees of freedom, gives you a critical value for t of 1.9842. Because t is a symmetrical distribution with a mean of 0, if the upper-tail value is +1.9842, the value for the lower-tail area (lower .025) is −1.9842. A t value of −1.9842 means that the probability that t is less than −1.9842 is 0.025, or 2.5% (see Figure 8.5).

TABLE 8.1
Determining the Critical Value from the t Table for an Area of 0.025 in Each Tail with 99 Degrees of Freedom

	Upper-Tail Areas					
Degrees of Freedom	**.25**	**.10**	**.05**	**.025**	**.01**	**.005**
1	1.0000	3.0777	6.3138	12.7062	31.8207	63.6574
2	0.8165	1.8856	2.9200	4.3027	6.9646	9.9248
3	0.7649	1.6377	2.3534	3.1824	4.5407	5.8409
4	0.7407	1.5332	2.1318	2.7764	3.7469	4.6041
5	0.7267	1.4759	2.0150	2.5706	3.3649	4.0322
.	.	.	.	.	.	.
.	.	.	.	.	.	.
.	.	.	.	.	.	.
96	0.6771	1.2904	1.6609	1.9850	2.3658	2.6280
97	0.6770	1.2903	1.6607	1.9847	2.3654	2.6275
98	0.6770	1.2902	1.6606	1.9845	2.3650	2.6269
99	0.6770	1.2902	1.6604	1.9842	2.3646	2.6264
100	0.6770	1.2901	1.6602	1.9840	2.3642	2.6259

Source: Extracted from Table E.3.

FIGURE 8.5
t Distribution with 99 Degrees of Freedom

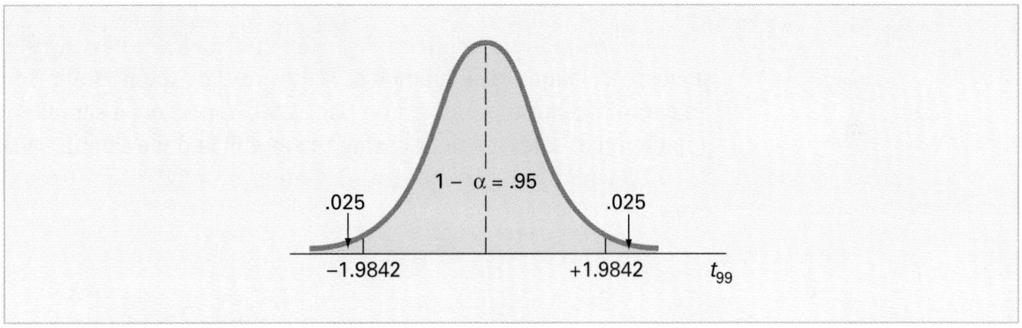

The Concept of Degrees of Freedom

In Chapter 3 you learned that the numerator of the sample variance S^2 (see Equation (3.9) on page 82) requires the computation of

$$\sum_{i=1}^{n}(X_i - \overline{X})^2$$

In order to compute S^2, you first need to know $\overline{X}$. Therefore, only $n - 1$ of the sample values are free to vary. This means that you have $n - 1$ degrees of freedom. For example, suppose a sample of five values has a mean of 20. How many values do you need to know before you can determine the remainder of the values? The fact that $n = 5$ and $\overline{X} = 20$ also tells you that

$$\sum_{i=1}^{n} X_i = 100$$

because

$$\frac{\sum_{i=1}^{n} X_i}{n} = \overline{X}$$

Thus, when you know four of the values, the fifth one will *not* be free to vary because the sum must add to 100. For example, if four of the values are 18, 24, 19, and 16, the fifth value must be 23 so that the sum equals 100.

The Confidence Interval Statement

Equation (8.2) defines the $(1 - \alpha) \times 100\%$ confidence interval estimate for the mean with σ unknown.

CONFIDENCE INTERVAL FOR THE MEAN (σ UNKNOWN)

$$\bar{X} \pm t_{n-1}\frac{S}{\sqrt{n}}$$

or

$$\bar{X} - t_{n-1}\frac{S}{\sqrt{n}} \le \mu \le \bar{X} + t_{n-1}\frac{S}{\sqrt{n}} \qquad \textbf{(8.2)}$$

where t_{n-1} is the critical value of the t distribution with $n - 1$ degrees of freedom for an area of $\alpha/2$ in the upper tail.

To illustrate the application of the confidence interval estimate for the mean when the standard deviation σ is unknown, return to the Saxon Home Improvement Company "Using Statistics" scenario presented on page 260. You select a sample of 100 sales invoices from the population of sales invoices during the month and the sample mean of the 100 sales invoices is \$110.27 with a sample standard deviation of \$28.95. For 95% confidence, the critical value from the t distribution (as shown in Table 8.1) is 1.9842. Using Equation (8.2),

$$\bar{X} \pm t_{n-1}\frac{S}{\sqrt{n}}$$

$$= 110.27 \pm (1.9842)\frac{28.95}{\sqrt{100}}$$

$$= 110.27 \pm 5.74$$

$$\$104.53 \le \mu \le \$116.01$$

A Microsoft Excel worksheet for these data is presented in Figure 8.6.

FIGURE 8.6

Microsoft Excel Worksheet to Compute a Confidence Interval Estimate for the Mean Sales Invoice Amount for the Saxon Home Improvement Company

	A	B	
1	**Estimate for the Mean Sales Invoice Amount**		
2			
3	**Data**		
4	**Sample Standard Deviation**	**28.95**	
5	**Sample Mean**	**110.27**	
6	**Sample Size**	**100**	
7	**Confidence Level**	**95%**	
8			
9	Intermediate Calculations		
10	Standard Error of the Mean	2.8950	**=B4/SQRT(B6)**
11	Degrees of Freedom	99	**=B6 - 1**
12	*t* Value	1.9842	**=TINV(1-B7,B11)**
13	Interval Half Width	5.7443	**=B12 * B10**
14			
15	**Confidence Interval**		
16	**Interval Lower Limit**	**104.53**	**=B5 - B13**
17	**Interval Upper Limit**	**116.01**	**=B5 + B13**

Thus, with 95% confidence, you conclude that the mean amount of all the sales invoices is between \$104.53 and \$116.01. The 95% confidence level indicates that if you selected all possible samples of 100 (something that is never done in practice), 95% of the intervals developed would include the population mean somewhere within the interval. The validity of this confidence interval estimate depends on the assumption of normality for the distribution of the amount of the sales invoices. With a sample of 100, the normality assumption is not overly restrictive and the use of the t distribution is likely appropriate. Example 8.3 further illustrates how you construct the confidence interval for a mean when the population standard deviation is unknown.

EXAMPLE 8.3

ESTIMATING THE MEAN FORCE REQUIRED TO BREAK ELECTRIC INSULATORS

A manufacturing company produces electric insulators. If the insulators break when in use, you are likely to have a short circuit. To test the strength of the insulators, you carry out destructive testing to determine how much *force* is required to break the insulators. You measure force by observing how many pounds are applied to the insulator before it breaks. Table 8.2 lists thirty values from this experiment. FORCE Construct a 95% confidence interval estimate for the population mean force required to break the insulator.

TABLE 8.2
Force (in Pounds) Required to Break the Insulator

1,870	1,728	1,656	1,610	1,634	1,784	1,522	1,696	1,592	1,662
1,866	1,764	1,734	1,662	1,734	1,774	1,550	1,756	1,762	1,866
1,820	1,744	1,788	1,688	1,810	1,752	1,680	1,810	1,652	1,736

SOLUTION

Figure 8.7 shows that the sample mean is $\bar{X} = 1{,}723.4$ pounds and the sample standard deviation is $S = 89.55$ pounds. Using Equation (8.2) on page 268 to construct the confidence interval, you need to determine the critical value from the t table for an area of 0.025 in each tail with 29 degrees of freedom. From Table E.3, you see that $t_{29} = 2.0452$. Thus, using $\bar{X} = 1{,}723.4$, $S = 89.55$, $n = 30$, and $t_{29} = 2.0452$,

$$\bar{X} \pm t_{n-1}\frac{S}{\sqrt{n}}$$

$$= 1{,}723.4 \pm (2.0452)\frac{89.55}{\sqrt{30}}$$

$$= 1{,}723.4 \pm 33.44$$

$$1{,}689.96 \le \mu \le 1{,}756.84$$

FIGURE 8.7
Minitab Confidence Interval Estimate for the Mean Amount of Force Required to Break Electric Insulators

One-Sample T: Force

```
Variable   N     Mean  StDev  SE Mean          95% CI
Force     30  1723.40  89.55    16.35  (1689.96, 1756.84)
```

You conclude with 95% confidence that the mean force required for the population of insulators is between 1,689.96 and 1,756.84 pounds. The validity of this confidence interval estimate depends on the assumption that the force required is normally distributed. Remember, however, that you can slightly relax this assumption for large sample sizes. Thus, with a sample of 30, you can use the t distribution even if the amount of force required is slightly skewed. From the normal probability plot displayed in Figure 8.8 or the box-and-whisker plot displayed in Figure 8.9, the amount of force required appears slightly skewed. Thus, the t distribution is appropriate for these data.

FIGURE 8.8

Minitab Normal Probability Plot for the Amount of Force Required to Break Electric Insulators

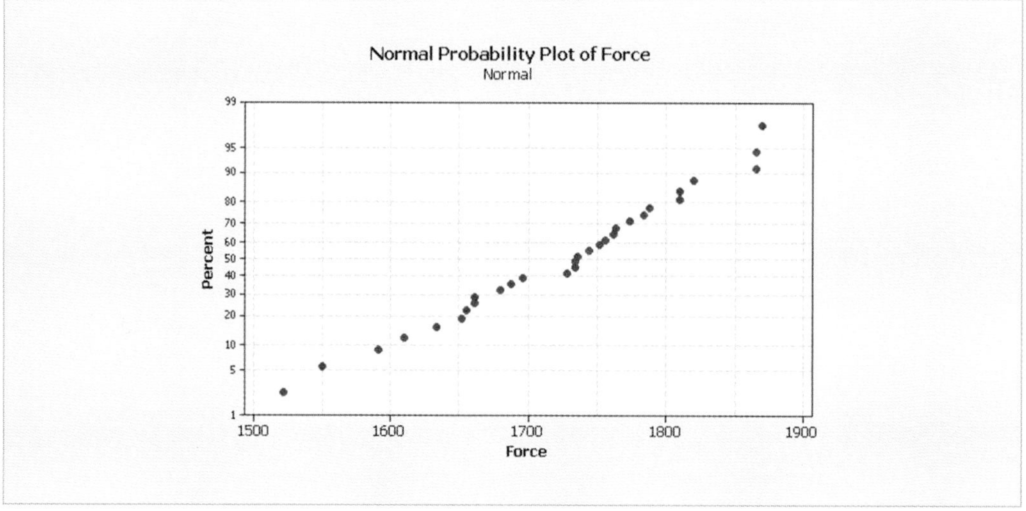

FIGURE 8.9

Minitab Box-and-Whisker Plot for the Amount of Force Required to Break Electric Insulators

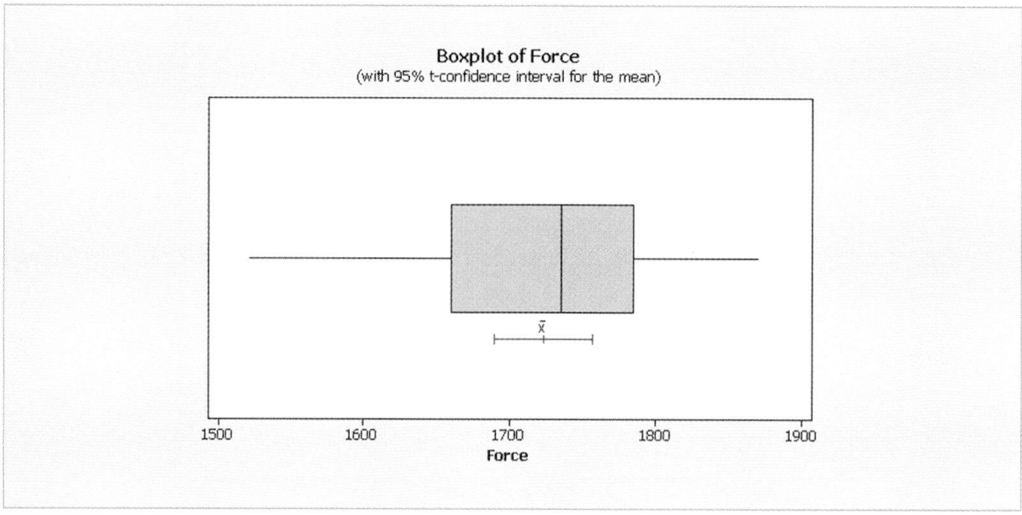

PROBLEMS FOR SECTION 8.2

Learning the Basics

PH Grade ASSIST **8.10** Determine the critical value of t in each of the following circumstances:

a. $1 - \alpha = 0.95$, $n = 10$.
b. $1 - \alpha = 0.99$, $n = 10$.
c. $1 - \alpha = 0.95$, $n = 32$.
d. $1 - \alpha = 0.95$, $n = 65$.
e. $1 - \alpha = 0.90$, $n = 16$.

PH Grade ASSIST **8.11** If $\bar{X} = 75$, $S = 24$, $n = 36$, and assuming that the population is normally distributed, construct a 95% confidence interval estimate of the population mean μ.

PH Grade ASSIST **8.12** If $\bar{X} = 50$, $S = 15$, $n = 16$, and assuming that the population is normally distributed, construct a 99% confidence interval estimate of the population mean μ.

8.13 Construct a 95% confidence interval estimate for the population mean, based on each of the following sets of data, assuming that the population is normally distributed:

Set 1: 1, 1, 1, 1, 8, 8, 8, 8
Set 2: 1, 2, 3, 4, 5, 6, 7, 8

Explain why these data sets have different confidence intervals even though they have the same mean and range.

8.14 Construct a 95% confidence interval for the population mean, based on the numbers 1, 2, 3, 4, 5, 6, and 20. Change the number 20 to 7 and recalculate the confidence interval. Using these results, describe the effect of an outlier (i.e., extreme value) on the confidence interval.

Applying the Concepts

You can solve Problems 8.15–8.22 with or without Microsoft Excel, Minitab, or SPSS. You should use Microsoft Excel, Minitab, or SPSS to solve problem 8. 23.

PH Grade ASSIST SELF Test **8.15** A stationery store wants to estimate the mean retail value of greeting cards that it has in its inventory. A random sample of 20 greeting cards indicates a mean value of \$1.67 and a standard deviation of \$0.32.

a. Assuming a normal distribution, construct a 95% confidence interval estimate of the mean value of all greeting cards in the store's inventory.

b. How are the results in (a) useful in assisting the store owner to estimate the total value of her inventory?

8.16 Southside hospital in Bay Shore, New York, commonly conducts stress tests to study the heart muscle after a person has a heart attack. Members of the diagnostic imaging department conducted a quality improvement project to try to reduce the turnaround time for stress tests. Turnaround time is defined as the time from when the test is ordered to when the radiologist signs off on the test results. Initially the mean turnaround time for a stress test was 68 hours. After incorporating changes into the stress-test process, the quality improvement team collected a sample of 50 turnaround times. In this sample, the mean turnaround time was 32 hours with a standard deviation of 9 hours (Eric Godin, Dennis Raven, Carolyn Sweetapple, and Frank R. Del Guidice, "Faster Test Results," *Quality Progress*, January 2004, 37(1):33–39).

a. Construct a 95% confidence interval for the population mean turnaround time.

b. Interpret the interval constructed in (a)

c. Do you think the quality improvement project was a success? Explain.

PH Grade ASSIST **8.17** The U.S. Department of Transportation requires tire manufacturers to provide tire performance information on the sidewall of the tire to better inform prospective customers when making a purchasing decision. One very important measure of tire performance is the tread wear index, which indicates the tire's resistance to tread wear compared with a tire graded with a base of 100. This means that a tire with a grade of 200 should last twice as long, on average, as a tire graded with a base of 100. A consumer organization wants to estimate the actual tread wear index of a brand name of tires graded 200 that are produced by a certain manufacturer. A random sample of $n = 18$ indicates a sample mean tread wear index of 195.3 and a sample standard deviation of 21.4.

a. Assuming that the population of tread wear indices is normally distributed, construct a 95% confidence interval estimate of the population mean tread wear index for tires produced by this manufacturer under this brand name.

b. Do you think that the consumer organization should accuse the manufacturer of producing tires that do not meet the performance information provided on the sidewall of the tire? Explain.

c. Explain why an observed tread wear index of 210 for a particular tire is not unusual, even though it is outside the confidence interval developed in (a).

8.18 The following data represent the bounced check fee in dollars for a sample of 23 banks for direct-deposit customers who maintain a \$100 balance. BANKCOST1

26	29	20	20	21	22	25	25	18	25	15	20
18	20	25	25	22	30	30	30	15	20	29	

Source: Extracted from "The New Face of Banking," Copyright © 2000 by Consumers Union of U.S. Inc., Yonkers, NY 10703–1057. Adapted with permission from Consumer Reports, *June 2000.*

a. Construct a 95% confidence interval for the population mean bounced check fee.

b. Interpret the interval constructed in (a)

8.19 The following data represent the monthly service fee in dollars if a customer's account falls below the minimum required \$1,500 balance for a sample of 26 banks for direct-deposit customers. BANKCOST2

12	8	5	5	6	6	10	10	9	7	10	7	7
5	0	10	6	9	12	0	5	10	8	5	5	9

Source: Extracted from "The New Face of Banking," Copyright © 2000 by Consumers Union of U.S. Inc., Yonkers, NY 10703–1057. Adapted with permission from Consumer Reports, *June 2000.*

a. Construct a 95% confidence interval for the population mean monthly service fee in dollars if a customer's account falls below the minimum required balance.

b. Interpret the interval constructed in (a).

8.20 One of the major measures of the quality of service provided by any organization is the speed with which it responds to customer complaints. A large family-held department store selling furniture and flooring including carpeting had undergone a major expansion in the past several years. In particular, the flooring department had expanded from 2 installation crews to an installation supervisor, a measurer, and 15 installation crews. Last year there were 50 complaints concerning carpeting installation. The following data FURNITURE represent the number of days between the receipt of the complaint and the resolution of the complaint.

54	5	35	137	31	27	152	2	123	81	74	27
11	19	126	110	110	29	61	35	94	31	26	5
12	4	165	32	29	28	29	26	25	1	14	13
13	10	5	27	4	52	30	22	36	26	20	23
33	68										

a. Construct a 95% confidence interval estimate of the mean number of days between the receipt of the complaint and the resolution of the complaint.
b. What assumption must you make about the population distribution in (a)?
c. Do you think that the assumption made in (b) is seriously violated? Explain.
d. What effect might your conclusion in (c) have on the validity of the results in (a)?

8.21 In New York state, savings banks are permitted to sell a form of life insurance called Savings Bank Life Insurance (SBLI). The approval process consists of underwriting, which includes a review of the application, a medical information bureau check, possible requests for additional medical information and medical exams, and a policy compilation stage where the policy pages are generated and sent to the bank for delivery. The ability to deliver approved policies to customers in a timely manner is critical to the profitability of this service to the bank. During a period of 1 month, a random sample of 27 approved policies was selected INSURANCE and the total processing time in days recorded:

73 19 16 64 28 28 31 90 60 56 31 56 22 18
45 48 17 17 17 91 92 63 50 51 69 16 17

a. Construct a 95% confidence interval estimate of the mean processing time.
b. What assumption must you make about the population distribution in (a)?
c. Do you think that the assumption made in (b) is seriously violated? Explain.

8.22 The following data represent the daily hotel cost and rental car cost for 20 U.S. cities during a week in October 2003. HOTEL-CAR

City	Hotel	Cars
San Francisco	205	47
Los Angeles	179	41
Seattle	185	49
Phoenix	210	38
Denver	128	32
Dallas	145	48
Houston	177	49
Minneapolis	117	41
Chicago	221	56
St. Louis	159	41
New Orleans	205	50
Detroit	128	32
Cleveland	165	34
Atlanta	180	46
Orlando	198	41
Miami	158	40
Pittsburgh	132	39
Boston	283	67
New York	269	69
Washington D.C.	204	40

Source: Extracted from the Wall Street Journal, *October 10, 2003, W4.*

a. Construct a 95% confidence interval for the population mean hotel rate.
b. Construct a 95% confidence interval for the population mean car rental rate.
c. What assumption do you need to make about the populations of interest to construct the intervals in (a) and (b)?
d. Given the data presented, do you think the assumption needed in (a) and (b) is valid? Explain.

8.23 One operation of a mill is to cut pieces of steel into parts that are used later in the frame for front seats in an automobile. The steel is cut with a diamond saw and requires the resulting parts to be within plus or minus 0.005 inch of the length specified by the automobile company. The measurement reported from a sample of 100 steel parts STEEL is the difference in inches between the actual length of the steel part as measured by a laser measurement device, and the specified length of the steel part. For example, the first observation, –0.002, represents a steel part that is 0.002 inch shorter than the specified length.
a. Construct a 95% confidence interval estimate of the mean difference between the actual length of the steel part, and the specified length of the steel part.
b. What assumption must you make about the population distribution in (a)?
c. Do you think that the assumption made in (b) is seriously violated? Explain.
d. Compare the conclusions reached in (a) with those of problem 2.23 on page 42.

8.3 CONFIDENCE INTERVAL ESTIMATION FOR THE PROPORTION

This section extends the concept of the confidence interval to categorical data. Here you are concerned with estimating the proportion of items in a population having a certain characteristic of interest. The unknown population proportion is represented by the Greek letter π. The point estimate for π is the sample proportion, $p = X/n$, where n is the sample size and X is the number of items in the sample having the characteristic of interest. Equation (8.3) defines the confidence interval estimate for the population proportion.

CONFIDENCE INTERVAL ESTIMATE FOR THE PROPORTION

$$p \pm Z\sqrt{\frac{p(1-p)}{n}}$$

or

$$p - Z\sqrt{\frac{p(1-p)}{n}} \le \pi \le p + Z\sqrt{\frac{p(1-p)}{n}} \qquad \textbf{(8.3)}$$

where $p =$ sample proportion $= \dfrac{X}{n} = \dfrac{\text{number of items having the characteristic}}{\text{sample size}}$

$\pi =$ population proportion

$Z =$ critical value from the standardized normal distribution

$n =$ sample size

assuming both X and $n - X$ are greater than 5

You can use the confidence interval estimate of the proportion defined in Equation (8.3) to estimate the proportion of sales invoices that contain errors (see the "Using Statistics" scenario on page 260). Suppose that in a sample of 100 sales invoices, 10 contain errors. Thus, for these data, $p = X/n = 10/100 = 0.10$. Using Equation (8.3) and $Z = 1.96$ for 95% confidence,

$$p \pm Z\sqrt{\frac{p(1-p)}{n}}$$

$$= 0.10 \pm (1.96)\sqrt{\frac{(0.10)(0.90)}{100}}$$

$$= 0.10 \pm (1.96)(0.03)$$

$$= 0.10 \pm 0.0588$$

$$0.0412 \le \pi \le 0.1588$$

Therefore, you have 95% confidence that between 4.12% and 15.88% of all the sales invoices contain errors. Figure 8.10 shows a Microsoft Excel worksheet for these data, while Figure 8.11 illustrates Minitab output.

FIGURE 8.10

Microsoft Excel Worksheet to Form a Confidence Interval Estimate for the Proportion of Sales Invoices that Contain Errors

	A	B	
1	**Proportion of In-Error Sales Invoices**		
2			
3	**Data**		
4	**Sample Size**	**100**	
5	**Number of Successes**	**10**	
6	**Confidence Level**	**95%**	
7			
8	Intermediate Calculations		
9	Sample Proportion	0.1	**=B5/B4**
10	Z Value	-1.9600	**=NORMSINV((1 - B6)/2)**
11	Standard Error of the Proportion	0.03	**=SQRT(B9 * (1 - B9)/B4)**
12	Interval Half Width	0.0588	**=ABS(B10 * B11)**
13			
14	**Confidence Interval**		
15	**Interval Lower Limit**	**0.0412**	**=B9 - B12**
16	**Interval Upper Limit**	**0.1588**	**=B9 + B12**

```
Sample    X    N  Sample p          95% CI
1        10  100  0.100000  (0.041201, 0.158799)
```

FIGURE 8.11 Minitab Confidence Interval Estimate for the Proportion of Sales Invoices that Contain Errors

Example 8.4 illustrates another application of a confidence interval estimate for the proportion.

EXAMPLE 8.4

ESTIMATING THE PROPORTION OF NONCONFORMING NEWSPAPERS PRINTED

A large newspaper wants to estimate the proportion of newspapers printed that have a nonconforming attribute, such as excessive ruboff, improper page setup, missing pages, or duplicate pages. A random sample of 200 newspapers is selected from all the newspapers printed during a single day. For this sample of 200, 35 contain some type of nonconformance. Construct and interpret a 90% confidence interval for the proportion of newspapers printed during the day that have a nonconforming attribute.

SOLUTION Using Equation (8.3):

$$p = \frac{35}{200} = 0.175, \text{ and with a 90\% level of confidence } Z = 1.645$$

$$\begin{aligned}
&p \pm Z\sqrt{\frac{p(1-p)}{n}} \\
&= 0.175 \pm (1.645)\sqrt{\frac{(0.175)(0.825)}{200}} \\
&= 0.175 \pm (1.645)(0.0269) \\
&= 0.175 \pm 0.0442 \\
&\quad 0.1308 \le \pi \le 0.2192
\end{aligned}$$

You conclude with 90% confidence that between 13.08% and 21.92% of the newspapers printed on that day have some type of nonconformance.

Equation (8.3) contains a Z statistic since you can use the normal distribution to approximate the binomial distribution when the sample size is sufficiently large. In Example 8.4, the confidence interval using Z provides an excellent approximation for the population proportion since both X and $n - X$ are greater than 5. However, if you do not have a sufficiently large sample size, then you should use the binomial distribution rather than Equation (8.3) (see references 1, 2, and 7). The exact confidence intervals for various sample sizes and proportions of successes have been tabulated by Fisher and Yates (reference 2) and can be computed using Minitab.

PROBLEMS FOR SECTION 8.3

Learning the Basics

PH Grade ASSIST **8.24** If $n = 200$ and $X = 50$, construct a 95% confidence interval estimate of the population proportion.

PH Grade ASSIST **8.25** If $n = 400$ and $X = 25$, construct a 99% confidence interval estimate of the population proportion.

Applying the Concepts

PH Grade ASSIST SELF Test **8.26** The telephone company wants to estimate the proportion of households that would purchase an additional telephone line if it were made available at a substantially reduced installation cost. A random sample of 500 households is selected. The results indicate that 135 of the households would purchase the additional telephone line at a reduced installation cost.

a. Construct a 99% confidence interval estimate of the population proportion of households that would purchase the additional telephone line.

b. How would the manager in charge of promotional programs concerning residential customers use the results in (a)?

8.27 According to the Center for Work-Life Policy, a survey of 500 highly educated women who left careers for family reasons found 66% wanted to return to work (Anne Marie Chaker and Hilary Stout, "After Years Off, Women Struggle to Revive Careers," *The Wall Street Journal*, May 6, 2004, A1).

a. Construct a 95% confidence interval for the population proportion of highly educated women who left careers for family reasons who want to return to work.

b. Interpret the interval in (a).

8.28 Millions of Americans arrange travel plans on the Web ("Travelers Head Online," *USA Today Snapshots*, July 22, 2003). A recent survey reported that 77% purchase plane tickets on the Web. Suppose the survey was based on 1,000 respondents.

a. Construct a 95% confidence interval estimate for the population proportion of Americans who purchase plane tickets on the Web.

b. Construct a 90% confidence interval estimate for the population proportion of Americans who purchase plane tickets on the Web.

c. Which interval is wider? Explain why this is true.

8.29 The number of older consumers in the United States is growing and they are becoming an even bigger economic force. Many feel overwhelmed when confronted with the task of selecting investments, banking services, health care providers, or phone service providers. A telephone survey of 1,900 older consumers found that 27% said they didn't have enough time to be good money managers ("Seniors Confused by Financial Choices—Study," **msnbc.com**, May 6, 2004).

a. Construct a 95% confidence interval for the population proportion of older consumers that don't think they have enough time to be good money managers.

b. Interpret the interval in (a).

PH Grade ASSIST **8.30** To study the problem of using cellphones while driving ("Drivers Using Cell Phones Have Problems," *USA Today*, May 16, 2001, 1A), a survey of drivers who use cellphones was conducted. In the survey, 46% of the respondents reported having to swerve and 10% knew someone who had a crash while talking on a cellphone. Suppose the survey was based on 500 respondents.

a. Construct a 95% confidence interval for the proportion of all drivers who have had to swerve.

b. Construct a 95% confidence interval for the proportion of all drivers who know someone who had a crash while talking on a cellphone.

8.31 As health-insurance costs and the number of disabled employees increase, more companies are firing these employees. A survey of 723 employers found that 195 dismiss employees as soon as they go on long-term disability (J. Pereira,"To Save on Health-care Costs, Firms Fire Disabled Workers," *The Wall Street Journal*, July, 14, 2003, A1–A7).

a. Construct a 95% confidence interval for the proportion of employers who dismiss employees as soon as they go on long-term disability.

b. Construct a 99% confidence interval for the proportion of employers who dismiss employees as soon as they go on long-term disability.

c. Which interval is wider? Explain why this is true.

8.32 A survey of working women in North America was conducted by the Clinique unit of Estee Lauder Cosmetics. Of the 1,000 women surveyed, 55% believed that companies should hold positions for those on maternity leave for six months or less, and 45% felt that they should hold those positions for more than six months ("Work Week," *The Wall Street Journal*, September 11, 2001, A1).

a. Construct a 95% confidence interval for the proportion of all working women in North America who believe that companies should hold positions for those on maternity leave for more than six months.

b. Interpret the interval constructed in (a).

8.33 A large number of companies are trying to reduce the cost of prescription drug benefits by requiring employees to purchase drugs through a mandatory mail-order program. In a survey of 600 employers, 126 indicated that they either have a mandatory mail-order program in place or are adopting one by the end of 2004 (Barbara Martinez, "Forcing Employees to Buy Drugs Via Mail," *The Wall Street Journal*, February 18, 2004, D1).

a. Construct a 95% confidence interval for the population proportion of employers who have a mandatory mail-order program in place or are adopting one by the end of 2004.

b. Construct a 99% confidence interval for the population proportion of employers who have a mandatory mail-order program in place or are adopting one by the end of 2004.

c. Interpret the intervals in (a) and (b).

d. Discuss the effect on the confidence interval estimate when you change the level of confidence.

8.4 DETERMINING SAMPLE SIZE

In each example of confidence interval estimation, you selected the sample size without regard to the width of the resulting confidence interval. In the business world, determining the proper sample size is a complicated procedure, subject to the constraints of budget, time, and the amount of acceptable sampling error. If, in the Saxon Home Improvement Company example, you want to estimate the mean dollar amount of the sales invoices or the proportion of sales invoices that contain errors, you must determine in advance how large a sampling error to allow in estimating each of the parameters. You must also determine in advance the level of confidence to use in estimating the population parameter.

Sample Size Determination for the Mean

To develop a formula for determining the appropriate sample size needed when constructing a confidence interval estimate of the mean, recall Equation (8.1) on page 263:

$$\bar{X} \pm Z\frac{\sigma}{\sqrt{n}}$$

The amount added to or subtracted from $\bar{X}$ is equal to half the width of the interval. This quantity represents the amount of imprecision in the estimate that results from sampling error. The **sampling error**[1] e is defined as

$$e = Z\frac{\sigma}{\sqrt{n}}$$

[1] In this context, some statisticians refer to e as the "margin of error."

Solving for n gives the sample size needed to construct the appropriate confidence interval estimate for the mean. "Appropriate" means that the resulting interval will have an acceptable amount of sampling error.

SAMPLE SIZE DETERMINATION FOR THE MEAN

The sample size n is equal to the product of the Z value squared and the variance σ^2, divided by the sampling error e squared.

$$n = \frac{Z^2\sigma^2}{e^2} \quad \textbf{(8.4)}$$

To determine the sample size, you must know three factors:

1. The desired confidence level, which determines the value of Z, the critical value from the standardized normal distribution[2]
2. The acceptable sampling error e
3. The standard deviation σ

[2] You use Z instead of t because, to determine the critical value of t, you need to know the sample size, but you do not know it yet. For most studies, the sample size needed is large enough that the standardized normal distribution is a good approximation of the t distribution.

In some business-to-business relationships requiring estimation of important parameters, legal contracts specify acceptable levels of sampling error and the confidence level required. For companies in the food or drug sectors, government regulations often specify sampling errors and confidence levels. In general, however, it is usually not easy to specify the two factors needed to determine the sample size. How can you determine the level of confidence and sampling error? Typically, these questions are answered only by the subject matter expert (i.e., the individual most familiar with the variables under study). Although 95% is the most common confidence level used, if more confidence is desired, then 99% might be more appropriate; if less confidence is deemed acceptable, then 90% might be used. For the sampling error, you should think not of how much sampling error you would like to have (you really do not want any error), but of how much you can tolerate when drawing conclusions from the data.

In addition to specifying the confidence level and the sampling error, you need an estimate of the standard deviation. Unfortunately, you rarely know the population standard deviation σ. In some instances, you can estimate the standard deviation from past data. In other situations, you can make an educated guess by taking into account the range and distribution of the variable. For example, if you assume a normal distribution, the range is approximately equal to 6 σ (i.e., $\pm 3\sigma$ around the mean) so that you estimate σ as the range divided by 6. If you cannot estimate σ in this way, you can conduct a small scale study and estimate the standard deviation from the resulting data.

To explore how to determine the sample size needed for estimating the population mean, consider again the audit at Saxon Home Improvement Company. In section 8.2, you selected a sample of 100 sales invoices and developed a 95% confidence interval estimate of the population mean sales invoice amount. How was this sample size determined? Should you have selected a different sample size?

Suppose that, after consultation with company officials, you determine that a sampling error of no more than ±\$5 is desired along with 95% confidence. Past data indicate that the standard deviation of the sales amount is approximately \$25. Thus, $e = \$5$, $\sigma = \$25$, and $Z = 1.96$ (for 95% confidence). Using Equation (8.4),

$$n = \frac{Z^2\sigma^2}{e^2} = \frac{(1.96)^2(25)^2}{(5)^2}$$

$$= 96.04$$

Because the general rule is to slightly oversatisfy the criteria by rounding the sample size up to the next whole integer, you should select a sample of size 97. Thus, the sample of size 100 used on page 268 is close to what is necessary to satisfy the needs of the company based on the estimated standard deviation, desired confidence level, and sampling error. Because the calculated sample standard deviation is slightly higher than expected, \$28.95 compared to \$25.00, the confidence interval is slightly wider than desired. Figure 8.12 illustrates the Microsoft Excel worksheet to determine the sample size.

FIGURE 8.12
Microsoft Excel Worksheet for Determining Sample Size for Estimating the Mean Sales Invoice Amount for the Saxon Home Improvement Company

	A	B	
1	**For the Mean Sales Invoice Amount**		
2			
3	**Data**		
4	**Population Standard Deviation**	25	
5	**Sampling Error**	5	
6	**Confidence Level**	95%	
7			
8	Intemediate Calculations		
9	Z Value	-1.9600	=NORMSINV((1 - B6)/2)
10	Calculated Sample Size	96.0365	=((B9 * B4)/B5)^2
11			
12	Result		
13	**Sample Size Needed**	97	=ROUNDUP(B10, 0)

Example 8.5 illustrates another application of determining the sample size needed to develop a confidence interval estimate for the mean.

EXAMPLE 8.5

DETERMINING THE SAMPLE SIZE FOR THE MEAN

Returning to Example 8.3 on page 269, suppose you want to estimate the population mean force required to break the insulator to within ±25 pounds with 95% confidence. On the basis of a study taken the previous year, you believe that the standard deviation is 100 pounds. Find the sample size needed.

SOLUTION Using Equation (8.4) on page 276 and $e = 25$, $\sigma = 100$, and $Z = 1.96$ for 95% confidence,

$$n = \frac{Z^2\sigma^2}{e^2} = \frac{(1.96)^2(100)^2}{(25)^2}$$
$$= 61.47$$

Therefore, you should select a sample size of 62 insulators because the general rule for determining sample size is to always round up to the next integer value in order to slightly oversatisfy the criteria desired.

An actual sampling error slightly larger than 25 will result if the sample standard deviation calculated in this sample of 62 is greater than 100, and slightly smaller if the sample standard deviation is less than 100.

Sample Size Determination for the Proportion

So far in this section, you learned how to determine the sample size needed for estimating the population mean. Now suppose that you want to determine the sample size necessary for estimating the proportion of sales invoices at the Saxon Home Improvement Company that contain errors.

To determine the sample size needed to estimate a population proportion (π), you use a method similar to the method for a population mean. Recall that in developing the sample size for a confidence interval for the mean, the sampling error is defined by

$$e = Z\frac{\sigma}{\sqrt{n}}$$

When estimating a proportion, you replace σ by $\sqrt{\pi(1-\pi)}$. Thus, the sampling error is

$$e = Z\sqrt{\frac{\pi(1-\pi)}{n}}$$

Solving for n, you have the sample size necessary to develop a confidence interval estimate for a proportion.

SAMPLE SIZE DETERMINATION FOR THE PROPORTION

The sample size n is equal to the Z value squared times the population proportion π, times 1 minus the population proportion π, divided by the sampling error e squared.

$$n = \frac{Z^2\pi(1-\pi)}{e^2} \tag{8.5}$$

To determine the sample size, you must know three factors:

1. The desired confidence level, which determines the value of Z, the critical value from the standardized normal distribution
2. The acceptable sampling error e
3. The population proportion π

In practice, selecting these quantities requires some planning. Once you determine the desired level of confidence, you can find the appropriate Z value from the standardized normal distribution. The sampling error e indicates the amount of error that you are willing to tolerate in estimating the population proportion. The third quantity π is actually the population parameter that you want to estimate! Thus, how do you state a value for the very thing that you are taking a sample in order to determine?

Here you have two alternatives. In many situations, you may have past information or relevant experiences that provide an educated estimate of π. Or, if you do not have past information or relevant experiences, you can try to provide a value for π that would never *underestimate* the sample size needed. Referring to Equation (8.5), you can see that the quantity $\pi(1 - \pi)$ appears in the numerator. Thus, you need to determine the value of π that will make the quantity $\pi(1 - \pi)$ as large as possible. When $\pi = 0.5$, the product $\pi(1 - \pi)$ achieves its maximum result. To show this, several values of π along with the accompanying products of $\pi(1 - \pi)$ are

When $\pi = 0.9$, then $\pi(1 - \pi) = (0.9)(0.1) = 0.09$
When $\pi = 0.7$, then $\pi(1 - \pi) = (0.7)(0.3) = 0.21$
When $\pi = 0.5$, then $\pi(1 - \pi) = (0.5)(0.5) = 0.25$
When $\pi = 0.3$, then $\pi(1 - \pi) = (0.3)(0.7) = 0.21$
When $\pi = 0.1$, then $\pi(1 - \pi) = (0.1)(0.9) = 0.09$

Therefore, when you have no prior knowledge or estimate of the population proportion π, you should use $\pi = 0.5$ for determining the sample size. This produces the largest possible sample size, and results in the highest possible cost of sampling. Using $\pi = 0.5$ may overestimate the sample size needed because you use the actual sample proportion in developing the confidence interval. You will get a confidence interval narrower than originally intended if the actual sample proportion is different from 0.5. The increased precision comes at the cost of spending more time and money for an increased sample size.

Returning to the Saxon Home Improvement Company "Using Statistics" scenario, suppose that the auditing procedures require you to have 95% confidence in estimating the population proportion of sales invoices with errors to within ± 0.07. The results from past months indicate that the largest proportion has been no more than 0.15. Thus, using Equation (8.5) on page 278 and $e = 0.07$, $\pi = 0.15$, and $Z = 1.96$ for 95% confidence,

$$n = \frac{Z^2\pi(1 - \pi)}{e^2}$$

$$= \frac{(1.96)^2(0.15)(0.85)}{(0.07)^2}$$

$$= 99.96$$

Because the general rule is to round the sample size up to the next whole integer to slightly oversatisfy the criteria, a sample size of 100 is needed. Thus, the sample size needed to satisfy the requirements of the company based on the estimated proportion, desired confidence level, and sampling error is equal to the sample size taken on page 273. The actual confidence interval is narrower than required since the sample proportion is 0.10 while 0.15 was used for π in Equation (8.5). Figure 8.13 below shows a Microsoft Excel worksheet.

FIGURE 8.13

Microsoft Excel Worksheet for Determining Sample Size for Estimating the Proportion of Sales Invoices with Errors for the Saxon Home Improvement Company

	A	B	
1	**For the Proportion of In-Error Sales Invoices**		
2			
3	**Data**		
4	**Estimate of True Proportion**	**0.15**	
5	**Sampling Error**	**0.07**	
6	**Confidence Level**	**95%**	
7			
8	Intermediate Calculations		
9	Z Value	-1.9600	=NORMSINV((1 - B6)/2)
10	Calculated Sample Size	99.9563	=(B9^2 * B4 * (1 - B4))/B5^2
11			
12	**Result**		
13	**Sample Size Needed**	**100**	=ROUNDUP(B10, 0)

Example 8.6 provides a second application of determining the sample size for estimating the population proportion.

EXAMPLE 8.6 DETERMINING THE SAMPLE SIZE FOR THE POPULATION PROPORTION

You want to have 90% confidence of estimating the proportion of office workers who respond to email within an hour to within ±0.05. Because you have not previously undertaken such a study, there is no information available from past data. Determine the sample size needed.

SOLUTION Because no information is available from past data, assume $\pi = 0.50$. Using Equation (8.5) on page 278 and $e = 0.05$, $\pi = 0.50$, and $Z = 1.645$ for 90% confidence,

$$n = \frac{(1.645)^2(0.50)(0.50)}{(0.05)^2}$$

$$= 270.6$$

Therefore, you need a sample of 271 office workers to estimate the population proportion to within ±0.05 with 90% confidence.

PROBLEMS FOR SECTION 8.4

Learning the Basics

PH Grade ASSIST **8.34** If you want to be 95% confident of estimating the population mean to within a sampling error of ±5 and the standard deviation is assumed to be 15, what sample size is required?

PH Grade ASSIST **8.35** If you want to be 99% confident of estimating the population mean to within a sampling error of ±20 and the standard deviation is assumed to be 100, what sample size is required?

PH Grade ASSIST **8.36** If you want to be 99% confident of estimating the population proportion to within an error of ±0.04, what sample size is needed?

PH Grade ASSIST **8.37** If you want to be 95% confident of estimating the population proportion to within an error of ±0.02 and there is historical evidence that the population proportion is approximately 0.40, what sample size is needed?

Applying the Concepts

SELF Test **8.38** A survey is planned to determine the mean annual family medical expenses of employees of a large company. The management of the company wishes to be 95% confident that the sample mean is correct to within ±$50 of the population mean annual family medical expenses. A previous study indicates that the standard deviation is approximately $400.

a. How large a sample size is necessary?

b. If management wants to be correct to within ±$25, what sample size is necessary?

8.39 If the manager of a paint supply store wants to estimate the mean amount of paint in a 1-gallon can to within ±0.004 gallon with 95% confidence and also assumes that the standard deviation is 0.02 gallon, what sample size is needed?

8.40 If a quality control manager wants to estimate the mean life of lightbulbs to within ±20 hours with 95% confidence and also assumes that the process standard deviation is 100 hours, what sample size is needed?

8.41 If the inspection division of a county weights and measures department wants to estimate the mean amount of soft-drink fill in 2-liter bottles to within ±0.01 liter with 95% confidence and also assumes that the standard deviation is 0.05 liter, what sample size is needed?

PH Grade ASSIST **8.42** A consumer group wants to estimate the mean electric bill for the month of July for single-family homes in a large city. Based on studies conducted in other cities, the standard deviation is assumed to be $25. The group wants to estimate the mean bill for July to within ±$5 with 99% confidence.

a. What sample size is needed?

b. If 95% confidence is desired, what sample size is necessary?

8.43 An advertising agency that serves a major radio station wants to estimate the mean amount of time that the station's audience spends listening to the radio daily. From past studies, the standard deviation is estimated as 45 minutes.

a. What sample size is needed if the agency wants to be 90% confident of being correct to within ±5 minutes?

b. If 99% confidence is desired, what sample size is necessary?

8.44 Suppose that a gas utility company wants to estimate its mean waiting time for installation of service to within ±5 days with 95% confidence. The company does not have access to previous data, but suspects that the standard deviation is approximately 20 days. What sample size is needed?

8.45 The U.S. Department of Transportation defines an airline flight as being "on-time" if it landed less than 15 minutes after the scheduled time shown in the carrier's Computerized Reservation System. Cancelled and diverted flights are counted as late. A study of the 10 largest U.S. domestic airlines found Southwest Airlines to have the lowest proportion of late arrivals, 0.1577 (Nikos Tsikriktsis and Janelle Heineke, "The Impact of Process Variation on Customer Dissatisfaction: Evidence from the U.S. Domestic Airline Industry," *Decision Sciences*, Winter 2004, 35(1):129–142). Suppose you were asked to perform a follow-up study for Southwest Airlines in order to update the estimated proportion of late arrivals. What sample size would you use in order to estimate the population proportion to within an error of

a. ±0.06 with 95% confidence?

b. ±0.04 with 95% confidence?

c. ±0.02 with 95% confidence?

8.46 In 2001 it was estimated that 45% of U.S. households purchased groceries at convenience stores and 29% purchased groceries at wholesale clubs ("68th Annual Report of the Grocery Industry," *Progressive Grocer*, April 2002, 29). Consider a follow-up study focusing on the last calendar year.

a. What sample size is needed to estimate the population proportion of U.S. households purchasing groceries at convenience stores to within ±0.02 with 95% confidence?

b. What sample size is needed to estimate the population proportion of U.S. households purchasing groceries at wholesale clubs to within ±0.02 with 95% confidence?

c. Compare the results of (a) and (b). Explain why these results differ.

d. If you were to design the follow-up study, would you use one sample and ask the respondents both questions, or would you select two separate samples? Explain the rationale behind your decision.

8.47 What proportion of people living in the United States use the Internet when planning their vacation? According to a poll conducted by American Express, 35% use the Internet (A. R. Carey, and K. Carter, "Snapshots," *USA Today*, January 14, 2003, 1A).

a. To conduct a follow-up study that would provide 95% confidence that the point estimate is correct to within ±0.04 of the population proportion, how large a sample size is required?

b. To conduct a follow-up study that would provide 99% confidence that the point estimate is correct to within ±0.04 of the population proportion, how large a sample size is required?

c. To conduct a follow-up study that would provide 95% confidence that the point estimate is correct to within ±0.02 of the population proportion, how large a sample size is required?

d. To conduct a follow-up study that would provide 99% confidence that the point estimate is correct to within ±0.02 of the population proportion, how large a sample size is required?

e. Discuss the effects of changing the desired confidence level and the acceptable sampling error on sample size requirements.

8.48 Have you had a business presentation disturbed by a ringing cellphone? In a poll of 326 business men and women, 303 answered this question "yes" and only 23 answered "no" ("You Say," *Presentations: Technology and Techniques for Effective Communication*, January 2003, 18).

a. Construct a 95% confidence interval for the population proportion of business men and women who have their presentations disturbed by cellphones.

b. Interpret the interval constructed in (a).

c. To conduct a follow-up study that would provide 95% confidence that the point estimate is correct to within ±0.04 of the population proportion, how large a sample size is required?

d. To conduct a follow-up study that would provide 99% confidence that the point estimate is correct to within ±0.04 of the population proportion, how large a sample size is required?

8.49 A study conducted by the Federal Reserve reported that 52% of 4,449 families interviewed in 2001 held stocks, either directly or through mutual funds (Barbara Hagenbaugh, "Nation's Wealth Disparity Widens," *USA Today*, January 22, 2003, 1A).

a. Construct a 95% confidence interval for the proportion of families that held stocks in 2001.

b. Interpret the interval constructed in (a).

c. To conduct a follow-up study to estimate the population proportion of families that currently hold stocks to within ±0.01 with 95% confidence, how many families would you interview?

8.5 APPLICATIONS OF CONFIDENCE INTERVAL ESTIMATION IN AUDITING

This chapter has focused on estimating either the population mean or the population proportion. Auditing is one of the areas in business that makes widespread use of statistical sampling for the purposes of estimation.

AUDITING

Auditing is the collection and evaluation of evidence about information relating to an economic entity such as a sole business proprietor, a partnership, a corporation, or a government agency in order to determine and report on how well the information corresponds to established criteria.

Six advantages of statistical sampling in auditing are listed below:

- Results are objective and defensible. Because the sample size is based on demonstrable statistical principles, the audit is defensible before one's superiors and in a court of law.
- Statistical sampling provides an objective way of estimating the sample size in advance.
- Statistical sampling provides an estimate of the sampling error.
- Statistical sampling is often more accurate for drawing conclusions about large populations. Examining large populations is time-consuming and therefore often subject to more nonsampling error than a statistical sample.
- Statistical sampling allows auditors to combine, and then evaluate collectively, samples collected by different individuals.
- Statistical sampling allows auditors to generalize their findings to the population with a known sampling error.

Estimating the Population Total Amount

In auditing applications, you are often more interested in developing estimates of the population **total amount** than the population mean. Equation (8.6) shows how to estimate a population total amount.

ESTIMATING THE POPULATION TOTAL

The point estimate for the population total is equal to the population size N times the sample mean.

$$\text{Total} = N\overline{X} \quad \textbf{(8.6)}$$

Equation (8.7) defines the confidence interval estimate for the population total.

CONFIDENCE INTERVAL ESTIMATE FOR THE TOTAL

$$N\overline{X} \pm N(t_{n-1})\frac{S}{\sqrt{n}}\sqrt{\frac{N-n}{N-1}} \quad \textbf{(8.7)}$$

To demonstrate the application of the confidence interval estimate for the population total amount, return to the Saxon Home Improvement Company "Using Statistics" scenario on page 260. One of the auditing tasks is to estimate the total dollar amount of all sales invoices for the month. If there are 5,000 invoices for that month and $\overline{X} = \$110.27$, then using Equation (8.6),

$$N\overline{X} = (5{,}000)(\$110.27) = \$551{,}350$$

If $n = 100$ and $S = \$28.95$, then using Equation (8.7) with $t_{99} = 1.9842$ for 95% confidence,

$$N\bar{X} \pm N(t_{n-1})\frac{S}{\sqrt{n}}\sqrt{\frac{N-n}{N-1}} = 551{,}350 \pm (5{,}000)(1.9842)\frac{28.95}{\sqrt{100}}\sqrt{\frac{5{,}000-100}{5{,}000-1}}$$

$$= 551{,}350 \pm 28{,}721.295(0.99005)$$

$$= 551{,}350 \pm 28{,}436$$

$$\$522{,}914 \leq \text{population total} \leq \$579{,}786$$

Therefore, with 95% confidence, you estimate that the total amount of sales invoices is between \$522,914 and \$579,786.

To further illustrate the population total, see Example 8.7.

EXAMPLE 8.7

DEVELOPING A CONFIDENCE INTERVAL ESTIMATE FOR THE POPULATION TOTAL

An auditor is faced with a population of 1,000 vouchers and wants to estimate the total value of the population of vouchers. A sample of 50 vouchers is selected with the following results:

$$\text{Mean voucher amount } (\bar{X}) = \$1{,}076.39$$

$$\text{Standard deviation } (S) = \$273.62$$

Construct a 95% confidence interval estimate of the total amount for the population of vouchers.

SOLUTION Using Equation (8.6) on page 282, the point estimate of the population total is

$$N\bar{X} = (1{,}000)(1{,}076.39) = \$1{,}076{,}390$$

From Equation (8.7) on page 282, a 95% confidence interval estimate of the population total amount is:

$$(1{,}000)(1{,}076.39) \pm (1{,}000)(2.0096)\frac{273.62}{\sqrt{50}}\sqrt{\frac{1{,}000-50}{1{,}000-1}}$$

$$= 1{,}076{,}390 \pm 77{,}762.878(0.97517)$$

$$= 1{,}076{,}390 \pm 75{,}832$$

$$\$1{,}000{,}558 \leq \text{population total} \leq \$1{,}152{,}222$$

Therefore, with 95% confidence, you estimate that the total amount of the vouchers is between \$1,000,558 and \$1,152,222.

Difference Estimation

An auditor uses **difference estimation** when he or she believes that errors exist in a set of items and he or she wants to estimate the magnitude of the errors based only on a sample. The following steps are used in difference estimation:

1. Determine the sample size required.
2. Calculate the differences between the values reached during the audit and the original values recorded. The difference in value i, denoted D_i, is equal to 0 if the auditor finds that the original value is correct, is a positive value when the audited value is larger than the original value, and is negative when the audited value is smaller than the original value.
3. Compute the mean difference in the sample ($\bar{D}$) by dividing the total difference by the sample size as shown in Equation (8.8).

MEAN DIFFERENCE

$$\bar{D} = \frac{\sum_{i=1}^{n} D_i}{n} \quad \textbf{(8.8)}$$

where D_i = audited value – original value

4. Compute the standard deviation of the differences (S_D) as shown in Equation (8.9). *Remember that any item that is not in error has a difference value of 0.*

STANDARD DEVIATION OF THE DIFFERENCE

$$S_D = \sqrt{\frac{\sum_{i=1}^{n} (D_i - \bar{D})^2}{n-1}} \quad \textbf{(8.9)}$$

5. Use Equation (8.10) to construct a confidence interval estimate of the total difference in the population.

CONFIDENCE INTERVAL ESTIMATE FOR THE TOTAL DIFFERENCE

$$N\bar{D} \pm N(t_{n-1})\frac{S_D}{\sqrt{n}}\sqrt{\frac{N-n}{N-1}} \quad \textbf{(8.10)}$$

The auditing procedures for Saxon Home Improvement Company require a 95% confidence interval estimate of the difference between the actual dollar amounts on the sales invoice and the amounts entered into the integrated inventory and sales information system. Suppose that in a sample of 100 sales invoices, you have 12 invoices in which the actual amount on the sales invoice and the amount entered into the integrated inventory management and sales information system is different. These 12 differences **PLUMBINV** are:

$9.03 $7.47 $17.32 $8.30 $5.21 $10.80 $6.22 $5.63 $4.97 $7.43 $2.99 $4.63

The other 88 invoices are not in error. Their *differences* are each 0. Thus,

$$\bar{D} = \frac{\sum_{i=1}^{n} D_i}{n} = \frac{90}{100} = 0.90$$

and

$$S_D = \sqrt{\frac{\sum_{i=1}^{n} (D_i - \bar{D})^2}{n-1}}$$

$$= \sqrt{\frac{(9.03-0.9)^2 + (7.47-0.9)^2 + \cdots + (0-0.9)^2}{100-1}}$$

[In the numerator, there are 100 differences. The last 88 are all $(0 - 0.9)^2$]

$$S_D = 2.752$$

Using Equation (8.10), construct the confidence interval estimate for the total difference in the population of 5,000 sales invoices as follows:

$$(5{,}000)(0.90) \pm (5{,}000)(1.9842)\frac{2.752}{\sqrt{100}}\sqrt{\frac{5{,}000 - 100}{5{,}000 - 1}}$$

$$= 4{,}500 \pm 2{,}702.91$$

$$\$1{,}797.09 \le \text{total difference} \le \$7{,}202.91$$

Thus, the auditor estimates with 95% confidence that the total difference between the sales invoices as determined during the audit, and the amount originally entered into the accounting system is between $1,797.09 and $7,202.91.

In the previous example, all 12 differences are positive because the actual amount on the sales invoice is more than the amount entered into the accounting system. In some circumstances you could have negative errors. Example 8.8 illustrates such an occurrence.

EXAMPLE 8.8

DIFFERENCE ESTIMATION

Returning to Example 8.7 on page 283, suppose that 14 vouchers contain errors in the sample of 50 vouchers. The values DIFFTEST of the 14 errors are as follows, in which two differences are negative.

$75.41 $38.97 $108.54 –$37.18 $62.75 $118.32 –$88.84
$127.74 $55.42 $39.03 $29.41 $47.99 $28.73 $84.05

Construct a 95% confidence interval estimate of the total difference in the population of 1,000 vouchers.

SOLUTION For these data,

$$\bar{D} = \frac{\sum_{i=1}^{n} D_i}{n} = \frac{690.34}{50} = 13.8068$$

and

$$S_D = \sqrt{\frac{\sum_{i=1}^{n}(D_i - \bar{D})^2}{n-1}}$$

$$= \sqrt{\frac{(75.41 - 13.8068)^2 + (38.97 - 13.8068)^2 + \cdots + (0 - 13.8068)^2}{50 - 1}}$$

$$= 37.427$$

Using Equation (8.10) on page 284, construct the confidence interval estimate for the total difference in the population,

$$(1{,}000)(13.8068) \pm (1{,}000)(2.0096)\frac{37.427}{\sqrt{50}}\sqrt{\frac{1{,}000 - 50}{1{,}000 - 1}}$$

$$= 13{,}806.8 \pm 10{,}372.4$$

$$\$3{,}434.40 \le \text{total difference} \le \$24{,}179.20$$

Therefore, with 95% confidence you estimate that the total difference in the population of vouchers is between $3,434.40 and $24,179.20.

One-Sided Confidence Interval Estimation of the Rate of Noncompliance with Internal Controls

Organizations use internal control mechanisms to ensure that individuals act in accordance with company guidelines. For example, Saxon Home Improvement Company requires that an authorized warehouse-removal slip is completed before goods are removed from the warehouse. During the monthly audit of the company, the auditing team is charged with the task of estimating the proportion of times goods were removed without proper authorization. This is referred to as the rate of noncompliance with the internal control. To estimate the rate of noncompliance, auditors take a random sample of sales invoices and determine how often merchandise was shipped without an authorized warehouse-removal slip. The auditors then compare their results with a previously established tolerable exception rate, which is the maximum allowable proportion of items in the population not in compliance. When estimating the rate of noncompliance, it is commonplace to use a **one-sided confidence interval**. That is, the auditors estimate an upper bound on the rate of noncompliance. Equation (8.11) defines a one-sided confidence interval for a proportion.

ONE-SIDED CONFIDENCE INTERVAL FOR A PROPORTION

$$\textit{Upper Bound} = p + Z\sqrt{\frac{p(1-p)}{n}}\sqrt{\frac{N-n}{N-1}} \qquad \textbf{(8.11)}$$

where Z = the value corresponding to a cumulative area of $(1 - \alpha)$ from the standardized normal distribution, that is, a right-hand tail probability of α.

If the tolerable exception rate is higher than the upper bound, then the auditor concludes that the company is in compliance with the internal control. If the upper bound is higher than the tolerable exception rate, the auditor concludes that the control noncompliance rate is too high. The auditor may then request a larger sample.

Suppose that in the monthly audit, you select 400 of the sales invoices from a population of 10,000 invoices. In the sample of 400 sales invoices, 20 are in violation of the internal control. If the tolerable exception rate for this internal control is 6%, what should you conclude? Use a 95% level of confidence.

The one-sided confidence interval is computed using $p = 20/400 = 0.05$ and $Z = 1.645$. Using Equation (8.11)

$$\textit{Upper Bound} = p + Z\sqrt{\frac{p(1-p)}{n}}\sqrt{\frac{N-n}{N-1}} = 0.05 + 1.645\sqrt{\frac{0.05(1-0.05)}{400}}\sqrt{\frac{10{,}000-400}{10{,}000-1}}$$

$$= 0.05 + 1.645(0.0109)(0.98) = 0.05 + 0.0176 = 0.0676$$

Thus, you have 95% confidence that the rate of noncompliance is less than 6.76%. Because the tolerable exception rate is 6%, the rate of noncompliance may be too high for this internal control. In other words, it is possible that the noncompliance rate for the population is higher than the rate deemed tolerable. Therefore, you should request a larger sample.

In many cases, the auditor is able to conclude that the rate of noncompliance with the company's internal controls is acceptable. Example 8.9 illustrates such an occurrence.

EXAMPLE 8.9 ESTIMATING THE RATE OF NONCOMPLIANCE

A large electronics firm writes a million checks a year. An internal control policy for the company is that the authorization to sign each check is granted only after an invoice has been initialed by an accounts payable supervisor. The company's tolerable exception rate for this control is 4%. If control deviations are found in 8 of the 400 invoices sampled, what should the auditor do? Use a 95% level of confidence.

SOLUTION The auditor constructs a 95% one-sided confidence interval for the proportion of invoices in noncompliance and compares this to the tolerable exception rate. Using Equation (8.11) on page 286, $p = 8/400 = 0.02$ and $Z = 1.645$ for 95% confidence,

$$\textit{Upper Bound} = p + Z\sqrt{\frac{p(1-p)}{n}}\sqrt{\frac{N-n}{N-1}} = 0.02 + 1.645\sqrt{\frac{0.02(1-0.02)}{400}}\sqrt{\frac{1{,}000{,}000-400}{1{,}000{,}000-1}}$$

$$= 0.02 + 1.645(0.007)(0.9998) = 0.02 + 0.0115 = 0.0315$$

The auditor concludes with 95% confidence that the rate of noncompliance is less than 3.15%. Since this is less than the tolerable exception rate, the auditor concludes that the internal control compliance is adequate. In other words, the auditor is more than 95% confident that the rate of noncompliance is less than 4%.

PROBLEMS FOR SECTION 8.5

Learning the Basics

8.50 A sample of 25 is selected from a population of 500 items. The sample mean is 25.7 and the sample standard deviation is 7.8. Construct a 99% confidence interval estimate of the population total.

8.51 Suppose that a sample of 200 is selected from a population of 10,000 items. Ten items are found to have errors of the following amounts:

13.76 42.87 34.65 11.09 14.54
22.87 25.52 9.81 10.03 15.49

Construct a 95% confidence interval estimate of the total difference in the population. **ITEMERR**

8.52 If $p = 0.04$, $n = 300$, and $N = 5{,}000$, calculate the upper bound for a one-sided confidence interval estimate of the population proportion π using a level of confidence of:

a. 90%.
b. 95%.
c. 99%.

Applying the Concepts

8.53 A stationery store wants to estimate the total retail value of the 300 greeting cards that it has in its inventory. Construct a 95% confidence interval estimate of the population total value of all greeting cards that are in inventory if a random sample of 20 greeting cards indicates an average value of \$1.67 and a standard deviation of \$0.32.

SELF Test **8.54** The personnel department of a large corporation employing 3,000 workers wants to estimate the family dental expenses of its employees to determine the feasibility of providing a dental insurance plan. A random sample of 10 employees reveals the following family dental expenses (in dollars) for the preceding year: **DENTAL**

110 362 246 85 510 208 173 425 316 179

Construct a 90% confidence interval estimate of the total family dental expenses for all employees in the preceding year.

8.55 A branch of a chain of large electronics stores is conducting an end-of-month inventory of the merchandise in stock. There were 1,546 items in inventory at that time. A sample of 50 items was randomly selected and an audit was conducted with the following results:

Value of Merchandise
$\bar{X} = \$252.28$ $S = \$93.67$

Construct a 95% confidence interval estimate of the total value of the merchandise in inventory at the end of the month.

8.56 A customer in the wholesale garment trade is often entitled to a discount for a cash payment for goods. The amount of discount varies by vendor. A sample of 150 items selected from a population of 4,000 invoices at the end of a period of time revealed that in 13 cases the customer failed to take the discount to which he or she was entitled. The amounts (in dollars) of the 13 discounts that were not taken were as follows: **DISCOUNT**

\$6.45 15.32 97.36 230.63 104.18 84.92 132.76
66.12 26.55 129.43 88.32 47.81 89.01

Construct a 99% confidence interval estimate of the population total amount of discounts not taken.

8.57 Econe Dresses is a small company that manufactures women's dresses for sale to specialty stores. There are 1,200 inventory items, and the historical cost is recorded on a first in, first out (FIFO) basis. In the past, approximately 15% of the inventory items were incorrectly priced. However, any misstatements were usually not significant. A sample of 120 items was selected and the historical cost of each item was compared with the audited value. The results indicated that 15 items differed in their historical costs and audited values. These differences were as follows: **FIFO**

Sample Number	Historical Cost ($)	Audited Value ($)	Sample Number	Historical Cost ($)	Audited Value ($)
5	261	240	60	21	210
9	87	105	73	140	152
17	201	276	86	129	112
18	121	110	95	340	216
28	315	298	96	341	402
35	411	356	107	135	97
43	249	211	119	228	220
51	216	305			

Construct a 95% confidence interval estimate of the total population difference in the historical cost and audited value.

8.58 Tom and Brent's Alpine Outfitters conduct an annual audit of their financial records. An internal control policy for the company is that a check can be issued only after the accounts payable manager initials the invoice. The tolerable exception rate for this internal control is 0.04. During an audit, a sample of 300 invoices is examined from a population of 10,000 invoices and 11 invoices are found to violate the internal control.

a. Calculate the upper bound for a 95% one-sided confidence interval estimate for the rate of noncompliance.

b. Based on (a), what should the auditor conclude?

8.59 An internal control policy for Rhonda's Online Fashion Accessories requires a quality assurance check before a shipment is made. The tolerable exception rate for this internal control is 0.05. During an audit, 500 shipping records were sampled from a population of 5,000 shipping records and 12 were found that violated the internal control.

a. Calculate the upper bound for a 95% one-sided confidence interval estimate for the rate of noncompliance.

b. Based on (a), what should the auditor conclude?

8.6 CONFIDENCE INTERVAL ESTIMATION AND ETHICAL ISSUES

Ethical issues relating to the selection of samples and the inferences that accompany them can arise in several ways. The major ethical issue relates to whether or not confidence interval estimates are provided along with the sample statistics. To provide a sample statistic without also including the confidence interval limits (typically set at 95%), the sample size used, and an interpretation of the meaning of the confidence interval in terms that a layperson can understand, raises ethical issues because of their omission. Failure to include a confidence interval estimate might mislead the user of the results into thinking that the point estimate is all that is needed to predict the population characteristic with certainty. Thus, it is important that you indicate the interval estimate in a prominent place in any written communication, along with a simple explanation of the meaning of the confidence interval. In addition, you should highlight the size of the sample.

One of the most common areas where ethical issues concerning estimation occurs is in the publication of the results of political polls. Often, the results of the polls are highlighted on the front page of the newspaper and the sampling error involved along with the methodology used is printed on the page where the article is typically continued, often in the middle of the newspaper. To ensure an ethical presentation of statistical results, the confidence levels, sample size, and confidence limits should be made available for all surveys and other statistical studies.

8.7 (*CD-ROM TOPIC*) ESTIMATION AND SAMPLE SIZE DETERMINATION FOR FINITE POPULATIONS

In this section, confidence intervals are developed and the sample size is determined for situations in which sampling is done without replacement from a finite population. For further discussion, see **section 8.7.pdf** on the CD-ROM that accompanies this text.

SUMMARY

This chapter discusses confidence intervals for estimating the characteristics of a population along with how you can determine the necessary sample size. You learned how an accountant of the Saxon Home Improvement Company can use the sample data from an audit to estimate important population parameters such as the mean dollar amount on invoices and the proportion of shipments that were made without the proper authorization. Table 8.3 provides a list of topics covered in this chapter.

TABLE 8.3
Summary of Topics in Chapter 8

	Type of Data	
Type of Analysis	**Numerical**	**Categorical**
Confidence Interval for a Population Parameter	Confidence Interval Estimate of the Mean (sections 8.1 and 8.2) Confidence Interval Estimate for the Total and the Mean Difference (section 8.5)	Confidence Interval Estimate for the Proportion (section 8.3) One-sided Confidence Interval Estimate for the Proportion (section 8.5)

To determine what equation to use for a particular situation, you need to ask several questions:

- Are you developing a confidence interval or are you determining sample size?
- Do you have a numerical variable or do you have a categorical variable?
- If you have a numerical variable, do you know the population standard deviation? If you do, use the normal distribution. If you do not, use the t distribution.

The next four chapters develop a hypothesis-testing approach that makes decisions about population parameters.

KEY FORMULAS

Confidence Interval for the Mean (σ Known)

$$\bar{X} \pm Z\frac{\sigma}{\sqrt{n}} \quad \textbf{(8.1)}$$

or

$$\bar{X} - Z\frac{\sigma}{\sqrt{n}} \le \mu \le \bar{X} + Z\frac{\sigma}{\sqrt{n}}$$

Confidence Interval for the Mean (σ Unknown)

$$\bar{X} \pm t_{n-1}\frac{S}{\sqrt{n}} \quad \textbf{(8.2)}$$

or

$$\bar{X} - t_{n-1}\frac{S}{\sqrt{n}} \le \mu \le \bar{X} + t_{n-1}\frac{S}{\sqrt{n}}$$

Confidence Interval Estimate for the Proportion

$$p \pm Z\sqrt{\frac{p(1-p)}{n}} \quad \textbf{(8.3)}$$

or

$$p - Z\sqrt{\frac{p(1-p)}{n}} \le \pi \le p + Z\sqrt{\frac{p(1-p)}{n}}$$

Sample Size Determination for the Mean

$$n = \frac{Z^2\sigma^2}{e^2} \quad \textbf{(8.4)}$$

Sample Size Determination for the Proportion

$$n = \frac{Z^2\pi(1-\pi)}{e^2} \quad \textbf{(8.5)}$$

KEY TERMS

auditing 282
confidence interval estimate 260
critical value 263
degrees of freedom 266
difference estimation 283
level of confidence 263
one-sided confidence interval 286
point estimate 260
sampling error 276
Student's t distribution 266
total amount 282

CHAPTER REVIEW PROBLEMS

Checking Your Understanding

8.60 Why is it that you can never really have 100% confidence of correctly estimating the population characteristic of interest?

8.61 When do you use the t distribution to develop the confidence interval estimate for the mean?

8.62 Why is it true that for a given sample size n, an increase in confidence is achieved by widening (and making less precise) the confidence interval?

8.63 Under what circumstances do you use a one-sided confidence interval instead of a two-sided confidence interval?

8.64 When would you want to estimate the population total instead of the population mean?

8.65 How does difference estimation differ from estimating the mean?

Applying the Concepts

You can solve problems 8.66–8.80 with or without Microsoft Excel, Minitab, or SPSS. You should use Microsoft Excel, Minitab, or SPSS to solve problems 8.81–8.87.

8.66 *Redbook* magazine conducted a survey through its Internet Web site (Stacy Kravetz, "Work Week," *The Wall Street Journal*, April 13, 1999, A1). People visiting the Web site were given the opportunity to fill out an on-screen survey form. A total of 665 women responded to a question that asked whether they would prefer a four-day workweek or a 20% pay increase. A four-day workweek was the preference for 412 of the respondents.

a. Define the population from which this sample was drawn.

b. Is this a random sample from this population?

c. Is this a statistically valid study?

d. Describe how you would design a statistically valid study to investigate the proportion of *Redbook* subscribers who would prefer a four-day workweek rather than a 20% pay increase. Use the information above to determine the sample size needed to estimate this population proportion to within ±0.02 with 95% confidence.

8.67 Companies are spending more time screening applicants than in the past. A result of the more in-depth screening process is that companies are finding that many applicants are stretching the truth on their applications. A study conducted by Automatic Data Processing found inaccuracies between the applicant and past employers in 44% of the applications screened. Unfortunately the article did not disclose the sample size used in the study (Stephanie Armour, "Security Checks Worry Workers," *USA Today*, June 19, 2002, B1).

a. Suppose that 500 applications were screened. Construct a 95% confidence interval estimate of the population proportion of applications containing inaccuracies with past employers.

b. Based on (a), is it right to conclude that less than half of all applications contain inaccuracies with past employers?

c. Suppose that the study used a sample size of 200 applications. Construct a 95% confidence interval estimate of the population proportion of applications containing inaccuracies with the past employers.

d. Based on (c), is it right to conclude that less than half of all applications contain inaccuracies with their past employers?

e. Discuss the effect of sample size on your answers to (a) through (d).

8.68 A sales and marketing management magazine conducted a survey on salespeople cheating on their expense reports and other unethical conduct (D. Haralson and Q. Tian, "Cheating Hearts," *USA Today*, February 15, 2001, 1A). In the survey, managers have caught salespeople cheating on an expense report 58% of the time, working a second job on company time 50% of the time, listing a "strip bar" as a restaurant on an expense report 22% of the time, and giving a kickback to a customer 19% of the time. Suppose the survey was based on a sample of 200 managers. Construct a 95% confidence interval estimate of the population proportion of managers who have caught salespeople

a. cheating on an expense report.

b. working a second job on company time.

c. listing a "strip bar" as a restaurant on an expense report.

d. giving a kickback to a customer.

e. Use the information given in this problem to determine the sample size needed to estimate the proportion of managers who have caught salespeople working a second job on company time to within ±0.02 with 95% confidence.

8.69 Starwood Hotels conducted a survey of 401 top executives who play golf (Del Jones, "Many CEOs Bend the Rules (of Golf)," *USA Today*, June 26, 2002). Among the results were the following:

- 329 cheat at golf
- 329 hate others who cheat at golf
- 289 believe business and golf behavior parallel
- 80 would let a client win to get business
- 40 would call in sick to play golf

Construct 95% confidence interval estimates for each of these questions. What conclusions can you reach about CEO's attitudes toward golf from these results?

8.70 A market researcher for a consumer electronics company wants to study the television viewing habits of residents of a particular area. A random sample of 40 respondents is selected, and each respondent is instructed to keep a detailed record of all television viewing in a particular week. The results are as follows:

- Viewing time per week: $\bar{X} = 15.3$ hours, $S = 3.8$ hours.
- 27 respondents watch the evening news on at least 3 weeknights.

a. Construct a 95% confidence interval estimate for the mean amount of television watched per week in this city.
b. Construct a 95% confidence interval estimate for the population proportion who watch the evening news on at least 3 nights per week.

If the market researcher wants to take another survey in a different city, answer these questions:

c. What sample size is required to be 95% confident of estimating the population mean to within ±2 hours and assumes that the population standard deviation is equal to 5 hours?
d. What sample size is needed to be 95% confident of being within ±0.035 of the population proportion who watch the evening news on at least 3 weeknights if no previous estimate is available?
e. Based on (c) and (d), what sample size should the market researcher select if a single survey is being conducted?

8.71 The real estate assessor for a county government wants to study various characteristics of single-family houses in the county. A random sample of 70 houses reveals the following:

- Heated area of the house (in square feet): $\bar{X} = 1{,}759$, $S = 380$.
- 42 houses have central air-conditioning.

a. Construct a 99% confidence interval estimate of the population mean heated area of the house.
b. Construct a 95% confidence interval estimate of the population proportion of houses that have central air-conditioning.

8.72 The personnel director of a large corporation wishes to study absenteeism among clerical workers at the corporation's central office during the year. A random sample of 25 clerical workers reveals the following:

- Absenteeism: $\bar{X} = 9.7$ days, $S = 4.0$ days.
- 12 clerical workers were absent more than 10 days.

a. Construct a 95% confidence interval estimate of the mean number of absences for clerical workers last year.
b. Construct a 95% confidence interval estimate of the population proportion of clerical workers absent more than 10 days last year.

If the personnel director also wishes to take a survey in a branch office, answer these questions:

c. What sample size is needed to have 95% confidence in estimating the population mean to within ±1.5 days if the population standard deviation is 4.5 days?
d. What sample size is needed to have 90% confidence in estimating the population proportion to within ±0.075 if no previous estimate is available?
e. Based on (c) and (d), what sample size is needed if a single survey is being conducted?

8.73 The market research director for Dotty's department store wants to study women's spending on cosmetics. A survey is designed in order to estimate the proportion of women who purchase their cosmetics primarily from Dotty's department store, and the mean yearly amount that women spend on cosmetics. A previous survey found that the standard deviation of the amount women spend on cosmetics in a year is approximately \$18.

a. What sample size is needed to have 99% confidence of estimating the population mean to within ±\$5?
b. What sample size is needed to have 90% confidence of estimating the population proportion to within ±0.045?
c. Based on the results in (a) and (b), how many of the store's credit cardholders should be sampled? Explain.

8.74 The branch manager of a nationwide bookstore chain wants to study characteristics of her store's customers. She decides to focus on two variables: the amount of money spent by customers and whether the customers would consider purchasing educational videotapes relating to graduate preparation exams such as GMAT, GRE, or LSAT. The results from a sample of 70 customers are as follows:

- Amount spent: $\bar{X} = \$28.52$, $S = \$11.39$.
- 28 customers stated that they would consider purchasing the educational videotapes.

a. Construct a 95% confidence interval estimate of the population mean amount spent in the bookstore.
b. Construct a 90% confidence interval estimate of the population proportion of customers who would consider purchasing educational videotapes.

Assume the branch manager of another store in the chain wants to conduct a similar survey in his store.

c. What sample size is needed to have 95% confidence of estimating the population mean amount spent in his store to within ±\$2 if the standard deviation is assumed to be \$10?
d. What sample size is needed to have 90% confidence of estimating the population proportion who would consider purchasing the educational videotapes to within ±0.04?
e. Based on your answers to (c) and (d), how large a sample should the manager take?

8.75 The branch manager of an outlet (store 1) of a nationwide chain of pet supply stores wants to study characteristics of her customers. In particular, she decides to focus on two variables: the amount of money spent by customers and whether the customers own only one dog, only one cat, or more than one dog and/or cat. The results from a sample of 70 customers are shown at the top of page 292:

- Amount of money spent: $\bar{X} = \$21.34$, $S = \$9.22$.
- 37 customers own only a dog.
- 26 customers own only a cat.
- 7 customers own more than one dog and/or cat.

a. Construct a 95% confidence interval estimate of the population mean amount spent in the pet supply store.
b. Construct a 90% confidence interval estimate of the population proportion of customers who own only a cat.

The branch manager of another outlet (store 2) wishes to conduct a similar survey in his store. The manager does not have any access to the information generated by the manager of store 1.

c. What sample size is needed to have 95% confidence of estimating the population mean amount spent in his store to within ±$1.50 if the standard deviation is $10?
d. What sample size is needed to have 90% confidence of estimating the population proportion of customers who own only a cat to within ±0.045?
e. Based on your answers to (c) and (d), how large a sample should the manager take?

8.76 The owner of a restaurant that serves continental food wants to study characteristics of his customers. He decides to focus on two variables: the amount of money spent by customers and whether customers order dessert. The results from a sample of 60 customers are as follows:

- Amount spent: $\bar{X} = \$38.54$, $S = \$7.26$.
- 18 customers purchased dessert.

a. Construct a 95% confidence interval estimate of the population mean amount spent per customer in the restaurant.
b. Construct a 90% confidence interval estimate of the population proportion of customers who purchase dessert.

The owner of a competing restaurant wants to conduct a similar survey in her restaurant. This owner does not have access to the information of the owner of the first restaurant.

c. What sample size is needed to have 95% confidence of estimating the population mean amount spent in her restaurant to within ±$1.50 assuming the standard deviation is $8?
d. What sample size is needed to have 90% confidence of estimating the population proportion of customers who purchase dessert to within ±0.04?
e. Based on your answers to (c) and (d), how large a sample should the owner take?

8.77 The manufacturer of "Ice Melt" claims its product will melt snow and ice at temperatures as low as 0° Fahrenheit. A representative for a large chain of hardware stores is interested in testing this claim. The chain purchases a large shipment of five-pound bags for distribution. The representative wants to know with 95% confidence, within ±0.05, what proportion of bags of Ice Melt perform the job as claimed by the manufacturer.

a. How many bags does the representative need to test? What assumption should be made concerning the population proportion? (This is called *destructive testing*; that is, the product being tested is destroyed by the test and is then unavailable to be sold.) The representative tests 50 bags, and 42 do the job as claimed.
b. Construct a 95% confidence interval estimate for the population proportion that will do the job as claimed.
c. How can the representative use the results of (b) to determine whether to sell the Ice Melt product?

8.78 An auditor needs to estimate the percentage of times a company fails to follow an internal control procedure. A sample of 50 from a population of 1,000 items is selected and in seven instances the internal control procedure was not followed.

a. Construct a 90% one-sided confidence interval estimate of the population proportion of items in which the internal control procedure was not followed.
b. If the tolerable exception rate is 0.15, what should the auditor conclude?

8.79 An auditor for a government agency needs to evaluate payments for doctors' office visits paid by Medicare in a particular zip code during the month of June. A total of 25,056 visits occurred during June in this area. The auditor wants to estimate the total amount paid by Medicare to within ±$5 with 95% confidence. On the basis of past experience, she believes that the standard deviation is approximately $30.

a. What sample size should she select?

Using the sample size selected in (a), an audit is conducted. It is discovered that in 12 of the office visits, an incorrect amount of reimbursement was provided.

Amount of Reimbursement

$\bar{X} = \$93.70$ $S = \$34.55$

For the 12 office visits in which incorrect reimbursement was provided, the differences between the amount reimbursed and the amount that the auditor determined should have been reimbursed were: **MEDICARE**

$17 $25 $14 –$10 $20 $40 $35 $30 $28 $22 $15 $5

b. Construct a 90% confidence interval estimate of the population proportion of reimbursements that contain errors.
c. Construct a 95% confidence interval estimate of the population mean reimbursement per office visit.
d. Construct a 95% confidence interval estimate of the population total amount of reimbursements for this geographic area in June.
e. Construct a 95% confidence interval estimate of the total difference between the amount reimbursed and the amount that the auditor determined should have been reimbursed.

8.80 A large computer store is conducting an end-of-month inventory of the computers in stock. An auditor for the store wants to estimate the mean value of the computers in inventory at that time. She wants to have 99% confidence that her estimate of the mean value is correct to within ±$100. On the basis of past experience, she estimates that the standard deviation of the value of a computer is $200.

a. What sample size should she select?

b. Using the sample size selected in (a), an audit was conducted with the following results:

$$\bar{X} = \$1{,}654.27 \qquad S = \$184.62$$

Construct a 99% confidence interval estimate of the total value of the computers in inventory at the end of the month if there were 258 computers in inventory.

8.81 A quality characteristic of interest for a tea-bag-filling process is the weight of the tea in the individual bags. In this example, the label weight on the package indicates that the mean amount is 5.5 grams of tea in a bag. If the bags are underfilled, two problems arise. First, customers may not be able to brew the tea to be as strong as they wish. Second, the company may be in violation of the truth-in-labeling laws. On the other hand, if the mean amount of tea in a bag exceeds the label weight, the company is giving away product. Getting an exact amount of tea in a bag is problematic because of variation in the temperature and humidity inside the factory, differences in the density of the tea, and the extremely fast filling operation of the machine (approximately 170 bags a minute). The following table provides the weight in grams of a sample of 50 tea bags produced in one hour by a single machine. TEABAGS

Weight of Tea Bags in Grams

5.65	5.44	5.42	5.40	5.53	5.34	5.54	5.45	5.52	5.41
5.57	5.40	5.53	5.54	5.55	5.62	5.56	5.46	5.44	5.51
5.47	5.40	5.47	5.61	5.53	5.32	5.67	5.29	5.49	5.55
5.77	5.57	5.42	5.58	5.58	5.50	5.32	5.50	5.53	5.58
5.61	5.45	5.44	5.25	5.56	5.63	5.50	5.57	5.67	5.36

a. Construct a 99% confidence interval estimate of the population mean weight of the tea bags.

b. Is the company meeting the requirement set forth on the label that the mean amount of tea in a bag is 5.5 grams?

8.82 A manufacturing company produces steel housings for electrical equipment. The main component part of the housing is a steel trough that is made out of a 14-gauge steel coil. It is produced using a 250-ton progressive punch press with a wipe-down operation that puts two 90-degree forms in the flat steel to make the trough. The distance from one side of the form to the other is critical because of weatherproofing in outdoor applications. The data from a sample of 49 troughs follows: TROUGH

Width of Trough (in Inches)

8.312	8.343	8.317	8.383	8.348	8.410	8.351	8.373	8.481	8.422
8.476	8.382	8.484	8.403	8.414	8.419	8.385	8.465	8.498	8.447
8.436	8.413	8.489	8.414	8.481	8.415	8.479	8.429	8.458	8.462
8.460	8.444	8.429	8.460	8.412	8.420	8.410	8.405	8.323	8.420
8.396	8.447	8.405	8.439	8.411	8.427	8.420	8.498	8.409	

a. Construct a 95% confidence interval estimate of the mean width of the troughs.

b. Interpret the interval developed in (a).

8.83 The manufacturer of "Boston" and "Vermont" asphalt shingles know that product weight is a major factor in the customer's perception of quality. The last stage of the assembly line packages the shingles before they are placed on wooden pallets. Once a pallet is full (a pallet for most brands holds 16 squares of shingles), it is weighed and the measurement is recorded. The data file PALLET contains the weight (in pounds) from a sample of 368 pallets of Boston shingles and 330 pallets of Vermont shingles.

a. For the Boston shingles, construct a 95% confidence interval estimate of the mean weight.

b. For the Vermont shingles, construct a 95% confidence interval estimate of the mean weight.

c. Evaluate whether the assumption needed for (a) and (b) has been seriously violated.

d. Based on the results of (a) and (b), what conclusions can you reach concerning the mean weight of the Boston and Vermont shingles?

8.84 The manufacturer of "Boston" and "Vermont" asphalt shingles provide their customers with a 20-year warranty on most of their products. To determine whether a shingle will last as long as the warranty period, accelerated-life testing is conducted at the manufacturing plant. Accelerated-life testing exposes the shingle to the stresses it would be subject to in a lifetime of normal use via a laboratory experiment that takes only a few minutes to conduct. In this test, a shingle is repeatedly scraped with a brush for a short period of time and the amount of shingle granules that are removed by the brushing is weighed (in grams). Shingles that experience low amounts of granule loss are expected to last longer in normal use than shingles that experience high amounts of granule loss. In this situation, a shingle should experience no more than 0.8 grams of granule loss if it is expected to last the length of the warranty period. The data file GRANULE contains a sample of 170 measurements made on the company's Boston shingles, and 140 measurements made on Vermont shingles.

a. For the Boston shingles, construct a 95% confidence interval estimate of the mean granule loss.

b. For the Vermont shingles, construct a 95% confidence interval estimate of the mean granule loss.

c. Evaluate whether the assumption needed for (a) and (b) has been seriously violated.
d. Based on the results of (a) and (b), what conclusions can you reach concerning the mean granule loss of the Boston and Vermont shingles?

8.85 Zagat's publishes restaurant ratings for various locations in the United States. The data file RESTRATE contains the Zagat rating for food, décor, and service, and the price per person for a sample of 50 restaurants located in New York City, and 50 restaurants located on Long Island, a suburb of New York City.

Source: Extracted from Zagat Survey 2002 New York City Restaurants *and* Zagat Survey 2001/2002 Long Island Restaurants.

For New York City and Long Island restaurants separately,
a. Construct 95% confidence interval estimates for the mean food rating, mean décor rating, mean service rating, and mean price per person.
b. What conclusions can you reach about the New York City and Long Island restaurants from the results in (a)?

Report Writing Exercises

8.86 Referring to the results in problem 8.82 on page 293 concerning the width of a steel trough, write a report that summarizes your conclusions.

TEAM PROJECT

8.87 Refer to the team project on page 62. Construct all appropriate confidence interval estimates of population characteristics of low-risk, average-risk, and high-risk mutual funds. Include these estimates in a report to the vice president for research at the financial investment service. MUTUALFUNDS2004

RUNNING CASE MANAGING THE *SPRINGVILLE HERALD*

The marketing department has been considering ways to increase the number of new subscriptions and increase the rate of retention among those customers who agreed to a trial subscription. Following the suggestion of assistant manager Lauren Alfonso, the department staff designed a survey to help determine various characteristics of readers of the newspaper who were not home-delivery subscribers. The survey consists of the following ten questions:

1. Do you or a member of your household ever purchase the *Springville Herald*?
   (1) Yes (2) No
   [If the respondent answers no, the interview is terminated.]
2. Do you receive the *Springville Herald* via home delivery?
   (1) Yes (2) No
   [If no, skip to question 4.]]
3. Do you receive the *Springville Herald*:
   (1) Monday–Saturday (2) Sunday only (3) Every day
   [If every day, skip to question 9.]
4. How often during the Monday–Saturday period do you purchase the *Springville Herald*?
   (1) Every day (2) Most days (3) Occasionally or never
5. How often do you purchase the *Springville Herald* on Sundays?
   (1) Every Sunday
   (2) 2–3 Sundays per month
   (3) No more than once a month
6. Where are you most likely to purchase the *Springville Herald*?
   (1) Convenience store
   (2) Newsstand/candy store
   (3) Vending machine (4) Supermarket (5) Other
7. Would you consider subscribing to the *Springville Herald* for a trial period if a discount was offered?
   (1) Yes (2) No
   [If no, skip to question 9.]
8. The *Springville Herald* currently costs 50 cents on Monday–Saturday and $1.50 on Sunday for a total of $4.50 per week. How much would you be willing to pay per week to get home delivery for a 90-day trial period?
9. Do you read a daily newspaper other than the *Springville Herald*?
   (1) Yes (2) No
10. As an incentive for long-term subscribers, the newspaper is considering the possibility of offering a card that would provide discounts at select restaurants in the Springville area to all subscribers who pay in advance for six months of home delivery. Would you want to get such a card under the terms of this offer?
    (1) Yes (2) No

The group agreed to use a random-digit dialing method to poll 500 local households by telephone. Using this approach, the last four digits of a telephone number are randomly selected to go with an area code and exchange (the first 6 digits of a ten-digit telephone number). Only those pairs of area codes and exchanges that were for the Springville city area were used for this survey.

Of the 500 households selected, 94 households either refused to participate, could not be contacted after repeated attempts, or represented telephone numbers that were not in service. The summary results are as follows:

Households that Purchase the Springville Herald	Frequency
Yes	352
No	54

Households with Home Delivery	Frequency
Yes	136
No	216

Type of Home Delivery Subscription	Frequency
Monday–Saturday	18
Sunday only	25
7 days a week	93

Purchase Behavior of Nonsubscribers for Monday–Saturday Editions	Frequency
Every day	78
Most days	95
Occasionally or never	43

Purchase Behavior of Nonsubscribers for Sunday editions	Frequency
Every Sunday	138
2–3 Sundays a month	54
No more than once a month	24

Nonsubscribers' Purchase Location	Frequency
Convenience store	74
Newsstand/candy store	95
Vending machine	21
Supermarket	13
Other locations	13

Would Consider Trial Subscription if Offered a Discount	Frequency
Yes	46
No	170

Rate Willing to Pay per Week (in Dollars) for a 90-Day Home-Delivery Trial Subscription SH8

$4.15 3.60 4.10 3.60 3.60 3.60 4.40 3.15 4.00 3.75 4.00 3.25 3.75 3.30 3.75 3.65 4.00 4.10 3.90 3.50 3.75 3.00 3.40 4.00 3.80 3.50 4.10 4.25 3.50 3.90 3.95 4.30 4.20 3.50 3.75 3.30 3.85 3.20 4.40 3.80 3.40 3.50 2.85 3.75 3.80 3.90

Read a Daily Newspaper Other than the *Springville Herald*	Frequency
Yes	138
No	214

Would Prepay 6 Months to Receive a Restaurant Discount Card	Frequency
Yes	66
No	286

EXERCISES

SH8.1 Some members of the marketing department are concerned about the random-digit dialing method used to collect survey responses. Prepare a memorandum that examines the following issues:

- The advantages and disadvantages of using the random-digit dialing method.
- Possible alternative approaches for conducting the survey and their advantages and disadvantages.

SH8.2 Analyze the results of the survey of Springville households. Write a report that discusses the marketing implications of the survey results for the *Springville Herald*.

WEB CASE

Apply your knowledge about confidence interval estimation in this Web Case that extends the OnCampus! Web Case from Chapter 6.

Among their other features, the *OnCampus!* Web site allows customers to purchase *OnCampus! LifeStyles* merchandise online. To handle payment processing, the management of *OnCampus!* has contracted with the following firms:

- PayAFriend (PAF): an online payments system in which customers and businesses such as *OnCampus!* register in order to exchange payments in a secure and convenient manner without the need for a credit card.
- Continental Banking Company (Conbanco): a processing services provider that allows *OnCampus!* customers to pay for merchandise using nationally recognized credit cards issued by a financial institution.

To reduce costs, the management is considering eliminating one of these two payment systems. However, Virginia Duffy of the sales department suspects that customers use the two forms of payment in unequal numbers and that customers display different buying behaviors when using the two forms of payment. Therefore, she would like to first determine:

a. The proportion of customers using PAF and the proportion of customers using a credit card to pay for their purchase.

b. The mean purchase amount when using PAF and the mean purchase amount when using a credit card.

Assist Ms. Duffy by preparing an appropriate analysis based on a random sample of 50 transactions that she has prepared and placed in an internal file on the *OnCampus!* Web site **www. prenhall.com/Springville/OnCampus_PymtSample.htm**. Summarize your findings and determine if Ms. Duffy's conjectures of *OnCampus!* customer purchasing behaviors are correct. If you want the sampling error to be no more than \$3, is Ms. Duffy's sample large enough to perform a valid analysis?

REFERENCES

1. Cochran, W. G., *Sampling Techniques*, 3rd ed. (New York: Wiley, 1977).
2. Fisher, R. A., and F. Yates, *Statistical Tables for Biological, Agricultural and Medical Research*, 5th ed. (Edinburgh: Oliver & Boyd, 1957).
3. Kirk, R. E., ed., *Statistical Issues: A Reader for the Behavioral Sciences* (Belmont, CA: Wadsworth, 1972).
4. Larsen, R. L., and M. L. Marx, *An Introduction to Mathematical Statistics and Its Applications*, 2nd ed. (Englewood Cliffs, NJ: Prentice Hall, 1986).
5. *Microsoft Excel 2003* (Redmond, WA: Microsoft Corporation, 2003).
6. *Minitab for Windows Version 14* (State College, PA: Minitab Inc., 2004).
7. Snedecor, G. W., and W. G. Cochran, *Statistical Methods*, 7th ed. (Ames, IA: Iowa State University Press, 1980).
8. *SPSS Base 12.0 Brief Guide* (Upper Saddle River, NJ: Prentice Hall, 2003).

Appendix 8 Using Software for Confidence Intervals and Sample Size Determination

A8.1 MICROSOFT EXCEL

For Confidence Interval Estimate for the Mean (σ known)

Open the **CIE sigma known.xls** file. This worksheet already contains the entries for Example 8.1 on page 264 and uses the NORMSINV and CONFIDENCE functions (see section G.14 for more information). To adapt this worksheet to other problems, change the population standard deviation, sample mean, sample size, and confidence level values in the tinted cells in rows 4 through 7.

OR See section G.14 (**Confidence Interval Estimate for the Mean, sigma known**) if you want PHStat2 to produce a worksheet for you.

For Confidence Interval Estimate for the Mean (σ unknown)

Open the **CIE sigma unknown.xls** file, shown in Figure 8.6 on page 268. The worksheet uses the TINV function to determine the critical value from the *t* distribution (see section G.15 for more information).

To adapt this worksheet to other problems, change the sample statistics and confidence level values in the tinted cells in rows 4 through 7.

OR See section G.15 (**Confidence Interval Estimate for the Mean, sigma unknown**) if you want PHStat2 to produce a worksheet for you.

For Confidence Interval Estimate for the Proportion

Open the **CIE Proportion.xls** file, shown in Figure 8.10 on page 273. The worksheet uses the NORMSINV function to determine the Z value (see section G.16 for more information).

To adapt this worksheet to other problems, change the sample size, number of successes, and confidence level values in the tinted cells in rows 4 through 6.

OR See section G.16 (**Confidence Interval Estimate for the Proportion**) if you want PHStat2 to produce a worksheet for you.

For Sample Size Determination for the Mean

Open the **Sample Size Mean.xls** file, shown in Figure 8.12 on page 277. The worksheet uses the NORMSINV and ROUNDUP functions (see section G.17 for more information).

To adapt this worksheet to other problems, change the population standard deviation, sampling error, and confidence level values in the tinted cells in rows 4 through 6.

OR See section G.17 (**Sample Size Determination for the Mean**) if you want PHStat2 to produce a worksheet for you.

For Sample Size Determination for the Proportion

Open the **Sample Size Proportion.xls** file, shown in Figure 8.13 on page 279. The worksheet uses the NORMSINV and ROUNDUP functions (see section G.18 for more information).

To adapt this worksheet to other problems, change the estimate of true proportion, sampling error, and confidence level values in the tinted cells in rows 4 through 6.

OR See section G.18 (**Sample Size Determination for the Proportion**) if you want PHStat2 to produce a worksheet for you.

For Confidence Interval Estimate for the Population Total

Open the **CIE Total.xls** file. The worksheet uses the TINV function to determine the critical value from the t distribution (see section G.19 for more information).

To adapt this worksheet to other problems, change the population size, sample mean, sample size, sample standard deviation, and confidence level values in the tinted cells in rows 4 through 8.

OR See section G.19 (**Estimate for the Population Total**) if you want PHStat2 to produce a worksheet for you.

For Confidence Interval Estimate for the Total Difference

Open the **CIE Total Difference.xls** file. This two-worksheet file already contains the entries for the Saxon Home Improvement Company example used in section 8.5. To adapt these worksheets to other problems, first change the population size, sample size, and confidence level values in the tinted cells in rows 4 through 6 of the **CIE Total Difference** worksheet. Then select the **Data** worksheet and enter differences data in column A of that worksheet, replacing the data already there for the section 8.5 problem. Finally, adjust the column B formulas copying the formulas down to additional cells, if you have more than 12 differences, or deleting the unneeded formulas if you have fewer than 12 differences.

OR See section G.20 (**Estimate for the Total Difference)** if you want PHStat2 to produce a worksheet for you.

A8.2 MINITAB

Using Minitab for Confidence Interval Estimation for the Mean

You can use Minitab to form a confidence interval estimate for the mean when σ is known by selecting **Stat ➔ Basic Statistics ➔ 1-Sample Z**. Enter the name of the variable in the Samples in columns: edit box. Enter the value for σ in the Standard deviation: edit box. Select the **Options** button and enter the level of confidence in the Confidence level: edit box. Click the **OK** button. Click the **OK** button again.

You can use Minitab to form a confidence interval estimate for the mean when σ is unknown by selecting **Stat ➔ Basic Statistics ➔ 1-Sample t**. To calculate the confidence interval estimate for the population mean force required to break the insulators in Example 8.3 on page 269, open the FORCE.MTW worksheet and do the following:

1. Select **Stat → Basic Statistics → 1-Sample t**.
2. In the 1-Sample t (Test and Confidence Interval) dialog box (see Figure A8.1), select the **Samples in columns**: option button. Enter **C1** or **Force** in the Samples in columns: edit box. (If you have summarized data, select the Summarized data option button and enter the sample size, sample mean, and sample standard deviation in the edit boxes.)
3. Select the **Options** button (see Figure A8.2). In the 1-Sample t-Options dialog box, enter **95.0** in the Confidence level: edit box. Click the **OK** button.
4. Click the **OK** button again.

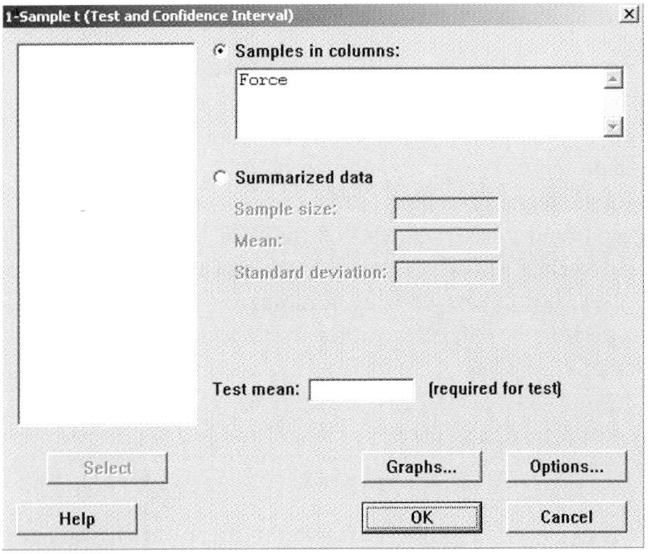

FIGURE A8.1 Minitab 1-Sample t Dialog Box

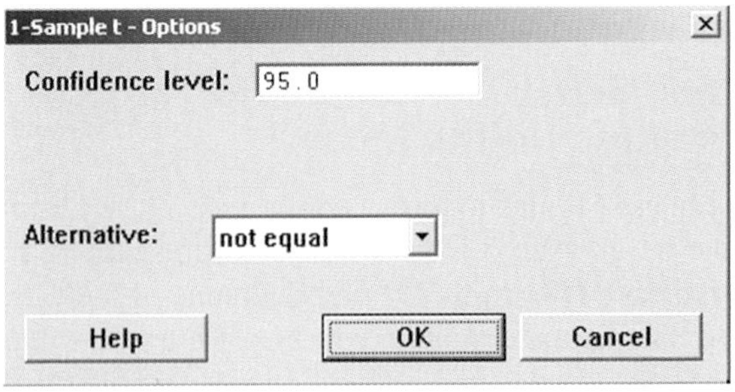

FIGURE A8.2 Minitab 1-Sample t-Options Dialog Box

Using Minitab for Confidence Interval Estimation for the Proportion

You can use Minitab to calculate a confidence interval estimate for the proportion. To calculate the confidence interval estimate for the population proportion for the Saxon Home Improvement Company problem of section 8.3,

1. Select **Stat → Basic Statistics → 1 Proportion**.
2. In the 1 Proportion dialog box (see Figure A8.3), select the **Summarized data** option button. Enter **100** in the Number of trials: edit box. Enter **10** in the Number of events: edit box.
3. Click the **Options** button. In the 1 Proportion-Options dialog box (see Figure A8.4), enter **95** in the Confidence level: edit box. Select the **Use test and interval based on normal distribution** check box. Click the **OK** button to return to the 1 proportion dialog box.
4. Click the **OK** button.

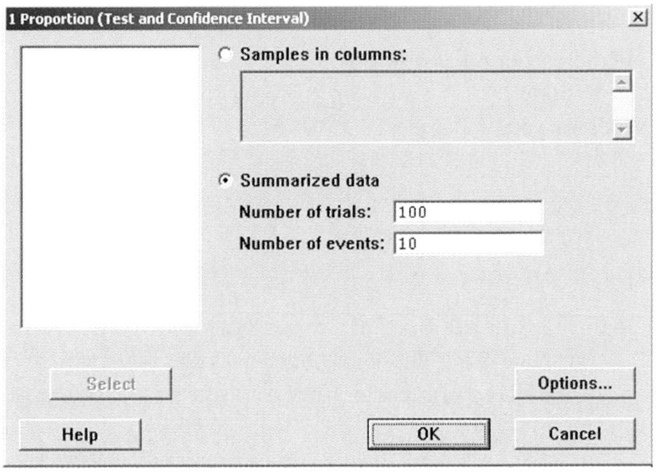

FIGURE A8.3 Minitab 1 Proportion Dialog Box

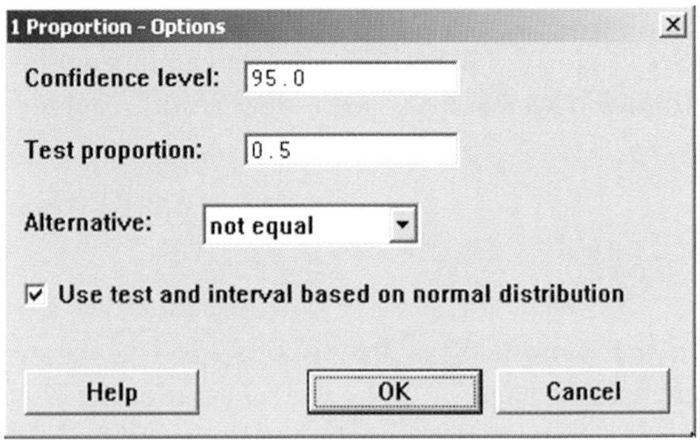

FIGURE A8.4 Minitab 1 Proportion-Options Dialog Box

CHAPTER 9

Fundamentals of Hypothesis Testing: One-Sample Tests

LEARNING OBJECTIVES

In this chapter, you learn:

- The basic principles of hypothesis testing
- How to use hypothesis testing to test a mean or proportion
- The assumptions of each hypothesis-testing procedure, how to evaluate them, and the consequences if they are seriously violated
- How to avoid the pitfalls involved in hypothesis testing
- The ethical issues involved in hypothesis testing

USING STATISTICS

The Oxford Cereal Company Revisited

As in the Chapter 7 "Using Statistics" scenario, imagine that you are an operations manager for the Oxford Cereal Company and are responsible for monitoring the amount in each cereal box filled. You select and weigh a random sample of 25 boxes in order to calculate a sample mean and investigate how close the fill weights are to the company's specifications of a mean of 368 grams. This time you must make a decision and conclude whether (or not) the mean fill weight in the entire process is equal to 368 grams in order to know whether the fill process needs adjustment. How could you rationally make this decision?

Unlike Chapter 7 in which the problem facing the operations manager was to determine if the sample mean was consistent with a known population mean, this "Using Statistics" scenario asks how can the sample mean validate the claim that the population mean is 368 grams? To validate the claim, you need to first state the claim unambiguously. For example, the population mean is 368 grams. In the inferential method known as hypothesis testing you consider the evidence—the sample statistic—to see if the evidence better supports the statement, called the *null hypothesis*, or the mutually exclusive alternative that, in this case, states that the population mean is not 368 grams.

In this chapter the focus is on hypothesis testing, another aspect of statistical inference that, like confidence interval estimation, is based on sample information. A step-by-step methodology is developed that enables you to make inferences about a population parameter by *analyzing differences* between the results observed (the sample statistic) and the results you expect to get if some underlying hypothesis is actually true. For example, is the mean weight of the cereal boxes in the sample taken at Oxford Cereal Company a value consistent with what you would expect if the mean of the entire population of cereal boxes is 368 grams? Or, can you infer that the population mean is not equal to 368 grams because the sample mean is significantly different from 368 grams?

9.1 HYPOTHESIS-TESTING METHODOLOGY

The Null and Alternative Hypotheses

Hypothesis testing typically begins with some theory, claim, or assertion about a particular parameter of a population. For example, your initial hypothesis about the cereal example is that the process is working properly, meaning that the mean fill is 368 grams, and no corrective action is needed.

The hypothesis that the population parameter is equal to the company specification is referred to as the null hypothesis. A **null hypothesis** is always one of status quo, and is identified by the symbol H_0. Here the null hypothesis is that the filling process is working properly, and therefore the mean fill is the 368-gram specification. This is stated as

$$H_0: \mu = 368$$

Even though information is available only from the sample, the null hypothesis is written in terms of the population. Remember, your focus is on the population of all cereal boxes. The sample statistic is used to make inferences about the entire filling process. One inference may be that the results observed from the sample data indicate that the null hypothesis is false. If the null hypothesis is considered false, something else must be true.

Whenever a null hypothesis is specified, an alternative hypothesis is also specified, one that must be true if the null hypothesis is false. The **alternative hypothesis H_1** is the opposite of the null hypothesis H_0. This is stated in the cereal example as

$$H_1: \mu \neq 368$$

The alternative hypothesis represents the conclusion reached by rejecting the null hypothesis. The null hypothesis is rejected when there is sufficient evidence from the sample information that the null hypothesis is false. In the cereal example, if the weights of the sampled boxes are sufficiently above or below the expected 368-gram mean specified by the company, you reject the null hypothesis in favor of the alternative hypothesis that the mean fill is different from 368 grams. You stop production and take whatever action is necessary to correct the problem. If the null hypothesis is not rejected, then you should continue to believe in the status quo that the process is working correctly and therefore no corrective action is necessary. In this second scenario, you have not proven that the process is working correctly. Rather, you have failed to prove that it is working incorrectly and therefore you continue your belief (although unproven) in the null hypothesis.

In the hypothesis-testing methodology, the null hypothesis is rejected when the sample evidence suggests that it is far more likely that the alternative hypothesis is true. However, failure to reject the null hypothesis is not proof that it is true. You can never prove that the null hypothesis is correct because the decision is based only on the sample information, not on the entire population. Therefore, if you fail to reject the null hypothesis, you can only conclude that there is insufficient evidence to warrant its rejection. The following key points summarize the null and alternative hypotheses:

- The null hypothesis H_0 represents the status quo or the current belief in a situation.
- The alternative hypothesis H_1 is the opposite of the null hypothesis and represents a research claim or specific inference you would like to prove.
- If you reject the null hypothesis, you have statistical proof that the alternative hypothesis is correct.
- If you do not reject the null hypothesis, then you have failed to prove the alternative hypothesis. The failure to prove the alternative hypothesis, however, does not mean that you have proven the null hypothesis.
- The null hypothesis H_0 always refers to a specified value of the population parameter (such as μ), not a sample statistic (such as $\bar{X}$).
- The statement of the null hypothesis always contains an equal sign regarding the specified value of the population parameter (e.g., H_0: $\mu = 368$ grams).
- The statement of the alternative hypothesis never contains an equal sign regarding the specified value of the population parameter (e.g., H_1: $\mu \neq 368$ grams).

EXAMPLE 9.1

THE NULL AND ALTERNATIVE HYPOTHESES

You are the manager of a fast-food restaurant. You want to determine whether the waiting time to place an order has changed in the last month from its previous population mean value of 4.5 minutes. State the null and alternative hypotheses.

SOLUTION The null hypothesis is that the population mean has not changed from its previous value of 4.5 minutes. This is stated as

$$H_0: \mu = 4.5$$

The alternative hypothesis is the opposite of the null hypothesis. Since the null hypothesis is that the population mean is 4.5 minutes, the alternative hypothesis is that the population mean is not 4.5 minutes. This is stated as

$$H_1: \mu \neq 4.5$$

The Critical Value of the Test Statistic

The logic behind the hypothesis-testing methodology is to determine how likely the null hypothesis is true by considering the information gathered in a sample. In the Oxford Cereal Company scenario, the null hypothesis is that the mean amount of cereal per box in the entire filling process is 368 grams (i.e., the population parameter specified by the company). You select a sample of boxes from the filling process, weigh each box, and compute the sample mean. This statistic is an estimate of the corresponding parameter (the population mean μ). Even if the null hypothesis is in fact true, the statistic (the sample mean $\bar{X}$) is likely to differ from the value of the parameter (the population mean μ) because of variation due to sampling. However, you expect the sample statistic to be close to the population parameter if the null hypothesis is true. If the sample statistic is close to the population parameter, you have insufficient evidence to reject the null hypothesis. For example, if the sample mean is 367.9, you would conclude that the population mean has not changed (i.e., $\mu = 368$), because a sample mean of 367.9 is very close to the hypothesized value of 368. Intuitively, you think that it is likely that you could get a sample mean of 367.9 from a population whose mean is 368.

On the other hand, if there is a large difference between the value of the statistic and the hypothesized value of the population parameter, you will conclude that the null hypothesis is false. For example, if the sample mean is 320, you would conclude that the population mean is not 368 (i.e., $\mu \neq 368$), because the sample mean is very far from the hypothesized value of 368. In such a case you conclude that it is very unlikely to get a sample mean of 320 if the population mean is really 368. Therefore, it is more logical to conclude that the population mean is not equal to 368. Here you reject the null hypothesis.

Unfortunately, the decision-making process is not always so clear-cut. Determining what is "very close" and what is "very different" is arbitrary without clear definitions. Hypothesis-testing methodology provides clear definitions for evaluating differences. Furthermore, it enables you to quantify the decision-making process by computing the probability of getting a given sample result if the null hypothesis is true. You calculate this probability by determining the sampling distribution for the sample statistic of interest (e.g., the sample mean) and then computing the particular **test statistic** based on the given sample result. Because the sampling distribution for the test statistic often follows a well-known statistical distribution, such as the standardized normal distribution or *t* distribution, you can use these distributions to help determine whether the null hypothesis is true.

Regions of Rejection and Nonrejection

The sampling distribution of the test statistic is divided into two regions, a **region of rejection** (sometimes called the critical region) and a **region of nonrejection** (see Figure 9.1).

FIGURE 9.1
Regions of Rejection and Nonrejection in Hypothesis Testing

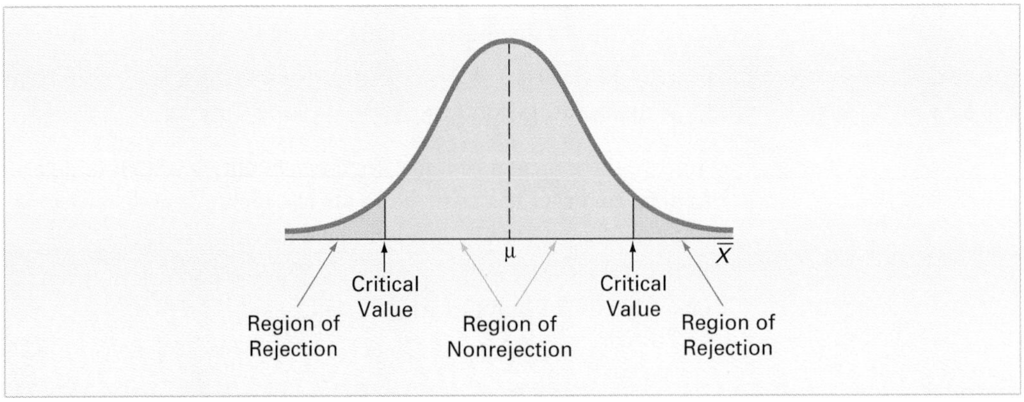

If the test statistic falls into the region of nonrejection, you do not reject the null hypothesis. In the Oxford Cereal Company scenario, you conclude that there is insufficient evidence that the population mean fill is different from 368 grams. If the test statistic falls into the rejection region, you reject the null hypothesis. In this case, you conclude that the population mean is not 368 grams.

The region of rejection consists of the values of the test statistic that are unlikely to occur if the null hypothesis is true. These values are more likely to occur if the null hypothesis is false. Therefore, if a value of the test statistic falls into this *rejection region*, you reject the null hypothesis because that value is unlikely if the null hypothesis is true.

To make a decision concerning the null hypothesis, you first determine the **critical value** of the test statistic. The critical value divides the nonrejection region from the rejection region. Determining this critical value depends on the size of the rejection region. The size of the rejection region is directly related to the risks involved in using only sample evidence to make decisions about a population parameter.

Risks in Decision Making Using Hypothesis-Testing Methodology

When using a sample statistic to make decisions about a population parameter, there is a risk that you will reach an incorrect conclusion. You can make two different types of errors when applying hypothesis-testing methodology, a Type I error and a Type II error.

> A **Type I error** occurs if you reject the null hypothesis H_0 when in fact it is true and should not be rejected. The probability of a Type I error occurring is α.
> A **Type II error** occurs if you do not reject the null hypothesis H_0 when in fact it is false and should be rejected. The probability of a Type II error occurring is β.

In the Oxford Cereal Company scenario, you make a Type I error if you conclude that the population mean fill is *not* 368 when in fact it *is* 368. On the other hand, you make a Type II error if you conclude that the population mean fill *is* 368 when in fact it is *not* 368.

The Level of Significance (α) The probability of committing a Type I error, denoted by α (the lowercase Greek letter *alpha*), is referred to as the **level of significance** of the statistical test. Traditionally, you control the Type I error by deciding the risk level α that you are willing to have in rejecting the null hypothesis when it is true. Because you specify the level of significance before the hypothesis test is performed, the risk of committing a Type I error, α, is directly under your control. Traditionally, you select levels of 0.01, 0.05, or 0.10. The choice of a particular risk level for making a Type I error depends on the cost of making a Type I error. After you specify the value for α, you know the size of the rejection region because α is the probability of rejection under the null hypothesis. From this fact, you can then determine the critical value or values that divide the rejection and nonrejection regions.

The Confidence Coefficient The complement of the probability of a Type I error $(1 - \alpha)$ is called the confidence coefficient. When multiplied by 100%, the confidence coefficient yields the confidence level that was studied when constructing confidence intervals (see section 8.1).

> The **confidence coefficient**, $1 - \alpha$, is the probability that you will not reject the null hypothesis H_0 when it is true and should not be rejected. The **confidence level** of a hypothesis test is $(1 - \alpha) \times 100\%$.

In terms of hypothesis-testing methodology, the confidence coefficient represents the probability of concluding that the value of the parameter as specified in the null hypothesis is plausible when it is true. In the Oxford Cereal Company scenario, the confidence coefficient measures the probability of concluding that the population mean fill is 368 grams when it actually is 368 grams.

The β Risk The probability of committing a Type II error is denoted by β (the lowercase Greek letter *beta*). Unlike the Type I error, which you control by the selection of α, the probability of making a Type II error depends on the difference between the hypothesized and actual value of the population parameter. Because large differences are easier to find than small ones, if the difference between the hypothesized and actual value of the population parameter is large, β is small. For example, if the true population mean is 330 grams, there is a small chance (β) that you will conclude that the mean has not changed from 368. On the other hand, if the difference between the hypothesized and actual value of the parameter is small, the probability you will commit a Type II error is large. Thus, if the population mean is really 367 grams, you have a high probability of making a Type II error by concluding that the population mean fill amount is still 368 grams.

The Power of a Test The complement of the probability of a Type II error (1 − β) is called the **power of a statistical test**.

The **power of a statistical test**, 1 − β, is the probability that you will reject the null hypothesis when in fact it is false and should be rejected.

In the Oxford Cereal Company scenario, the power of the test is the probability you will correctly conclude that the mean fill amount is not 368 grams when it actually is not 368 grams. For a detailed discussion of the power of the test, see section 9.6.

Risks in Decision Making: A Delicate Balance Table 9.1 illustrates the results of the two possible decisions (do not reject H_0 or reject H_0) that you can make in any hypothesis test. Depending on the specific decision, you can make one of two types of errors or you can reach one of two types of correct conclusions.

TABLE 9.1
Hypothesis Testing and Decision Making

	Actual Situation	
Statistical Decision	**H_0 True**	**H_0 False**
Do not reject H_0	Correct decision Confidence = 1 − α	Type II error *P*(Type II error) = β
Reject H_0	Type I error *P*(Type I error) = α	Correct decision Power = 1 − β

One way in which you can reduce the probability of making a Type II error is by increasing the size of the sample. Large samples generally permit you to detect even very small differences between the hypothesized values and the population parameters. For a given level of α, increasing the sample size will decrease β and therefore will increase the power of the test to detect that the null hypothesis H_0 is false. However, there is always a limit to your resources, and this will affect the decision as to how large a sample you can take. Thus, for a given sample size, you must consider the trade-offs between the two possible types of errors. Because you can directly control the risk of Type I error, you can reduce this risk by selecting a smaller value for α. For example, if the negative consequences associated with making a Type I error are substantial, you could select α = 0.01 instead of 0.05. However, when you decrease α, you increase β, so reducing the risk of a Type I error will result in an increased risk of Type II error. If, on the other hand, you wish to reduce β, you could select a larger value for α. Therefore, if it is important to try to avoid a Type II error, you can select α of 0.05 or 0.10 instead of 0.01.

In the Oxford Cereal Company scenario, the risk of a Type I error involves concluding that the mean fill amount has changed from the hypothesized 368 grams when in fact it has not changed. The risk of a Type II error involves concluding that the mean fill amount has not changed from the hypothesized 368 grams when in truth it has changed. The choice of reasonable values for α and β depends on the costs inherent in each type of error. For example, if it is very costly to change the cereal-filling process, then you would want to be very confident that a change is needed before making any changes. In this case, the risk of a Type I error is most important and you would choose a small α. On the other hand, if you want to be very certain of detecting changes from a mean of 368 grams, the risk of a Type II error is most important, and you would choose a higher level of α.

PROBLEMS FOR SECTION 9.1

Learning the Basics

PH Grade ASSIST **9.1** You use the symbol H_0 for which hypothesis?

PH Grade ASSIST **9.2** You use the symbol H_1 for which hypothesis?

PH Grade ASSIST **9.3** What symbol do you use for the level of significance or chance of committing a Type I error?

9.4 What symbol do you use for the chance of committing a Type II error?

PH Grade ASSIST **9.5** What does $1 - \beta$ represent?

9.6 What is the relationship of α to the Type I error?

9.7 What is the relationship of β to the Type II error?

9.8 How is power related to the probability of making a Type II error?

PH Grade ASSIST **9.9** Why is it possible to reject the null hypothesis when in fact it is true?

PH Grade ASSIST **9.10** Why is it possible to not reject the null hypothesis when it is false?

9.11 For a given sample size, if α is reduced from 0.05 to 0.01, what will happen to β?

9.12 For H_0: $\mu = 100$, H_1: $\mu \neq 100$, and for a sample of size n, why is β larger if the actual value of μ is 90 than if the actual value of μ is 75?

Applying the Concepts

PH Grade ASSIST **9.13** In the U.S. legal system, a defendant is presumed innocent until proven guilty. Consider a null hypothesis H_0, that the defendant is innocent, and an alternative hypothesis H_1, that the defendant is guilty. A jury has two possible decisions: Convict the defendant (i.e., reject the null hypothesis) or do not convict the defendant (i.e., do not reject the null hypothesis). Explain the meaning of the risks of committing either a Type I or Type II error in this example.

9.14 Suppose the defendant in problem 9.13 is presumed guilty until proven innocent as in some other judicial systems. How do the null and alternative hypotheses differ from those in problem 9.13? What are the meanings of the risks of committing either a Type I or Type II error here?

9.15 The U.S. Food and Drug Administration (FDA) is responsible for approving new drugs. Many consumer groups feel that the approval process is too easy, and, therefore, too many drugs are approved that are later found to be unsafe. On the other hand, there are a number of industry lobbyists who are pushing for a more lenient approval process so that pharmaceutical companies can get new drugs approved more easily and quickly (Rochelle Sharpe, "FDA Tries to Find Right Balance on Drug Approvals," *The Wall Street Journal*, April 20, 1999, A24). Consider a null hypothesis that a new, unapproved drug is unsafe and an alternative hypothesis that a new, unapproved drug is safe.

a. Explain the risks of committing a Type I or Type II error.
b. Which type of error are the consumer groups trying to avoid? Explain.
c. Which type of error are the industry lobbyists trying to avoid? Explain.
d. How would it be possible to lower the chance of both Type I and Type II errors?

PH Grade ASSIST SELF Test **9.16** As a result of complaints from both students and faculty about lateness, the registrar at a large university wants to adjust the scheduled class times to allow for adequate travel time between classes and is ready to undertake a study. Until now, the registrar has believed that 20 minutes between scheduled classes should be sufficient. State the null hypothesis H_0 and the alternative hypothesis H_1.

9.17 The director of manufacturing at a clothing factory needs to determine whether a new machine is producing a particular type of cloth according to the manufacturer's specifications, which indicate that the cloth should have a mean breaking strength of 70 pounds. A sample of 49 pieces of cloth reveals a sample mean breaking strength of 69.1 pounds. State the null and alternative hypotheses.

9.18 The manager of a paint supply store wants to determine whether the amount of paint contained in 1-gallon cans purchased from a nationally known manufacturer actually averages 1 gallon. It is known from the manufacturer's specifications that the standard deviation of the amount of paint is 0.02 gallon. A random sample of fifty 1-gallon cans is selected, and the sample mean is 0.995 gallon. State the null and alternative hypotheses.

9.19 The quality-control manager at a lightbulb factory needs to determine whether the mean life of a large shipment of lightbulbs is equal to the specified value of 375 hours. It is known that the population standard deviation is 100 hours. A random sample of 64 lightbulbs indicates a sample mean life of 350 hours. State the null and alternative hypotheses.

9.2 *Z* TEST OF HYPOTHESIS FOR THE MEAN (σ KNOWN)

Now that you have been introduced to hypothesis-testing methodology, recall that in the "Using Statistics" scenario on page 300, the Oxford Cereal Company wants to determine whether the cereal-filling process is working properly (that is, whether the mean fill throughout the entire packaging process remains at the specified 368 grams and no corrective action is needed). To evaluate the 368-gram requirement, you take a random sample of 25 boxes, weigh each box, and then evaluate the difference between the sample statistic and the hypothesized population parameter by comparing the mean weight (in grams) from the sample to the expected mean of 368 grams specified by the company. For this filling process, the null and alternative hypotheses are

$$H_0: \mu = 368$$
$$H_1: \mu \neq 368$$

When the standard deviation σ is known, you use the Z test if the population is normally distributed. If the population is not normally distributed, you can still use the Z test if the sample size is large enough for the Central Limit Theorem to take effect (see section 7.2). Equation (9.1) defines the ***Z*-test statistic** for determining the difference between the sample mean $\bar{X}$ and the population mean μ when the standard deviation σ is known.

Z TEST OF HYPOTHESIS FOR THE MEAN (σ KNOWN)

$$Z = \frac{\bar{X} - \mu}{\frac{\sigma}{\sqrt{n}}} \quad \textbf{(9.1)}$$

In Equation (9.1) the numerator measures how far (in an absolute sense) the observed sample mean $\bar{X}$ is from the hypothesized mean μ. The denominator is the standard error of the mean, so Z represents the difference between $\bar{X}$ and μ in standard error units.

The Critical Value Approach to Hypothesis Testing

The observed value of the Z test statistic, Equation (9.1), is compared to critical values. These critical values are expressed as standardized Z values (i.e., in standard-deviation units). For example, if you use a level of significance of 0.05, the size of the rejection region is 0.05. Because the rejection region is divided into the two tails of the distribution (this is called a **two-tail test**), you divide the 0.05 into two equal parts of 0.025 each. A rejection region of 0.025 in each tail of the normal distribution results in a cumulative area of 0.025 below the lower critical value and a cumulative area of 0.975 below the upper critical value. According to the cumulative standardized normal distribution table (Table E.2), the critical values that divide the rejection and nonrejection regions are −1.96 and +1.96. Figure 9.2 illustrates that if the mean is actually 368 grams, as H_0 claims, then the values of the test statistic Z have a standardized normal distribution centered at $Z = 0$ (which corresponds to an $\overline{X}$ value of 368 grams). Values of Z greater than +1.96 or less than −1.96 indicate that $\overline{X}$ is so far from the hypothesized $\mu = 368$ that it is unlikely that such a value would occur if H_0 were true.

FIGURE 9.2
Testing a Hypothesis about the Mean (σ Known) at the 0.05 Level of Significance

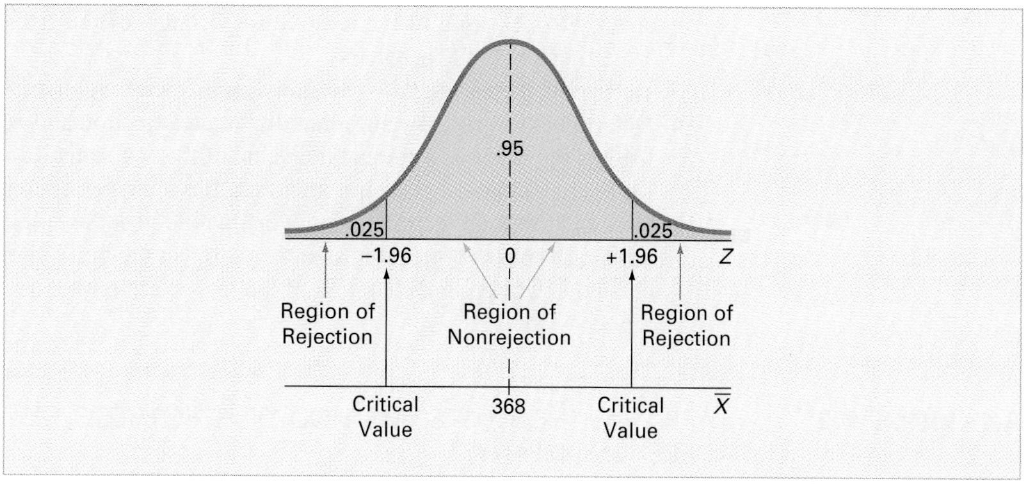

Therefore, the decision rule is

$$\text{Reject } H_0 \text{ if } Z > +1.96$$
$$\text{or if } Z < -1.96;$$
$$\text{otherwise do not reject } H_0.$$

Suppose that the sample of 25 cereal boxes indicates a sample mean $\overline{X} = 372.5$ grams and the population standard deviation σ is assumed to be 15 grams. Using Equation (9.1) on page 306:

$$Z = \frac{\overline{X} - \mu}{\frac{\sigma}{\sqrt{n}}} = \frac{372.5 - 368}{\frac{15}{\sqrt{25}}} = +1.50$$

Because the test statistic $Z = +1.50$ is between −1.96 and +1.96, you do not reject H_0 (see Figure 9.3). You continue to believe that the mean fill amount is 368 grams. To take into account the possibility of a Type II error, you state the conclusion as "there is insufficient evidence that the mean fill is different from 368 grams."

FIGURE 9.3
Testing a Hypothesis about the Mean (σ Known) at 0.05 Level of Significance

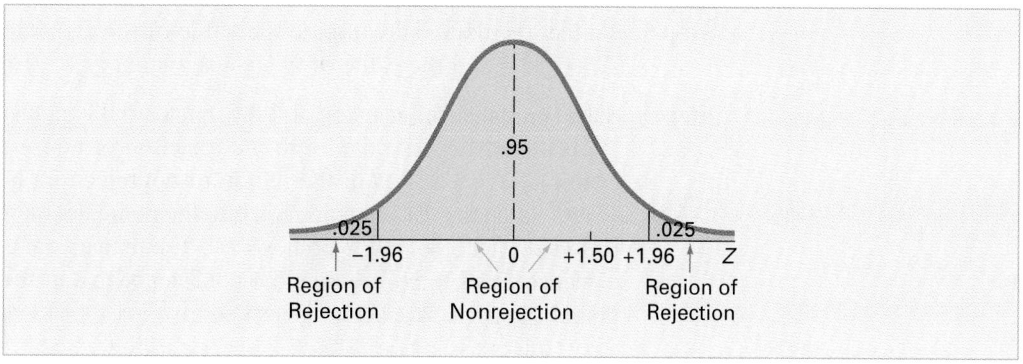

Exhibit 9.1 provides a summary of the critical value approach for hypothesis testing.

EXHIBIT 9.1: THE SIX-STEP METHOD OF HYPOTHESIS TESTING

1. State the null hypothesis, H_0, and the alternative hypothesis, H_1.
2. Choose the level of significance, α, and the sample size, n. The level of significance is specified according to the relative importance of the risks of committing Type I and Type II errors in the problem.
3. Determine the appropriate test statistic and sampling distribution.
4. Determine the critical values that divide the rejection and nonrejection regions.
5. Collect the data and compute the value of the test statistic.
6. Make the statistical decision and state the managerial conclusion. If the test statistic falls into the nonrejection region, you do not reject the null hypothesis H_0. If the test statistic falls into the rejection region, you reject the null hypothesis. The managerial conclusion is written in the context of the real-world problem.

EXAMPLE 9.2

APPLYING THE SIX-STEP METHOD OF HYPOTHESIS TESTING AT THE OXFORD CEREAL COMPANY

State the six-step method of hypothesis testing at the Oxford Cereal Company.

SOLUTION

Step 1: State the null and alternative hypotheses. The null hypothesis H_0 is always stated in statistical terms using population parameters. In testing whether the mean fill is 368 grams, the null hypothesis states that μ equals 368. The alternative hypothesis, H_1, is also stated in statistical terms using population parameters. Therefore, the alternative hypothesis states that μ is not equal to 368 grams.

Step 2: Choose the level of significance and the sample size. You choose the level of significance α according to the relative importance of the risks of committing Type I and Type II errors in the problem. The smaller the value of α, the less risk there is of making a Type I error. In this example, a Type I error is to conclude that the population mean is not 368 grams, when it is 368 grams. Here, $\alpha = 0.05$ is selected. The sample $n = 25$.

Step 3: Select the appropriate test statistic. Because σ is known from information about the filling process, you use the normal distribution and the Z-test statistic.

Step 4: Determine the rejection region. Critical values for the appropriate test statistic are selected so that the rejection region contains a total area of α when H_0 is true, and the nonrejection region contains a total area of $1 - \alpha$ when H_0 is true. Since $\alpha = 0.05$ in the cereal example, the critical values of the Z-test statistic are -1.96 and $+1.96$. The rejection region is therefore $Z < -1.96$ or $Z > +1.96$. The nonrejection region is $-1.96 < Z < +1.96$.

Step 5: Collect the data and compute the value of the test statistic. In the cereal example, $\bar{X} = 372.5$ and the value of the test statistic is $Z = +1.50$.

Step 6: State the statistical decision and the managerial conclusion. First, determine whether the test statistic has fallen into the rejection or nonrejection region. For the cereal example, $Z = +1.50$ is in the region of nonrejection, because $-1.96 < Z = +1.50 < +1.96$. Since the test statistic falls into the nonrejection region, the statistical decision is do not reject the null hypothesis H_0. The managerial conclusion is that insufficient evidence exists to prove with 95% confidence that the mean fill is different from 368 grams. No corrective action is needed.

EXAMPLE 9.3 REJECTING A NULL HYPOTHESIS

You are the manager of a fast-food restaurant. You want to determine whether the waiting time to place an order has changed in the last month from its previous population mean value of 4.5 minutes. From past experience, you can assume that the population standard deviation is 1.2 minutes. You select a sample of 25 orders during a one-hour period. The sample mean is 5.1 minutes. Use the six-step approach of Exhibit 9.1 on page 308 to determine whether there is evidence at the 0.05 level of significance that the mean waiting time to place an order has changed in the last month from its previous population mean value of 4.5 minutes.

SOLUTION

Step 1: The null hypothesis is that the population mean has not changed from its previous value of 4.5 minutes.

$$H_0\text{: } \mu = 4.5$$

The alternative hypothesis is the opposite of the null hypothesis. Since the null hypothesis is that the population mean is 4.5 minutes, the alternative hypothesis is that the population mean is not 4.5 minutes.

$$H_1\text{: } \mu \neq 4.5$$

Step 2: You have selected a sample of $n = 25$. The level of significance is 0.05 (i.e., $\alpha = 0.05$).

Step 3: Because σ is known, you use the normal distribution and the Z-test statistic.

Step 4: Since $\alpha = 0.05$, the critical values of the Z-test statistic are -1.96 and $+1.96$. The rejection region is $Z < -1.96$ or $Z > +1.96$. The nonrejection region is $-1.96 < Z < +1.96$.

Step 5: You collect the data and compute $\bar{X} = 5.1$. Using Equation (9.1) on page 306, you compute the test statistic.

$$Z = \frac{\bar{X} - \mu}{\dfrac{\sigma}{\sqrt{n}}} = \frac{5.1 - 4.5}{\dfrac{1.2}{\sqrt{25}}} = 2.50$$

Step 6: Since $Z = 2.50 > 1.96$, you reject the null hypothesis. You conclude that there is evidence that the waiting time to place an order has changed from its previous population mean value of 4.5 minutes. The mean waiting time for customers is longer now than last month.

The *p*-Value Approach to Hypothesis Testing

Most modern software including Microsoft Excel, Minitab, and SPSS compute the p-value when performing a test of hypothesis.

The ***p*-value** is the probability of getting a test statistic equal to or more extreme than the sample result, given that the null hypothesis H_0 is true.

The p-value, often referred to as the *observed level of significance*, is the smallest level at which H_0 can be rejected.

The decision rules for rejecting H_0 in the p-value approach are.

- If the p-value is greater than or equal to α, you do not reject the null hypothesis.
- If the p-value is less than α, you reject the null hypothesis.

Many people confuse these rules, mistakenly believing that a high p-value is grounds for rejection. You can avoid this confusion by remembering the following mantra.

If the p-value is low, then H_0 must go.

To understand the p-value approach, consider the Oxford Cereal Company scenario. You tested whether or not the mean fill was equal to 368 grams. The test statistic resulted in a Z value of +1.50 and you did not reject the null hypothesis because +1.50 was less than the upper critical value of +1.96 and more than the lower critical value of −1.96.

To use the p-value approach for the *two-tail test*, you find the probability of getting a test statistic Z that is equal to or *more extreme than* 1.50 standard deviation units from the center of a standardized normal distribution. In other words, you need to compute the probability of a Z value greater than +1.50 along with the probability of a Z value less than −1.50. Table E.2 shows that the probability of a Z value below −1.50 is 0.0668. The probability of a value below +1.50 is 0.9332 and the probability of a value above +1.50 is $1 - 0.9332 = 0.0668$. Therefore, the p-value for this two-tail test is $0.0668 + 0.0668 = 0.1336$ (see Figure 9.4). Thus, the probability of a result equal to or more extreme than the one observed is 0.1336. Because 0.1336 is greater than $\alpha = 0.05$, you do not reject the null hypothesis.

FIGURE 9.4

Finding a p-Value for a Two-Tail Test

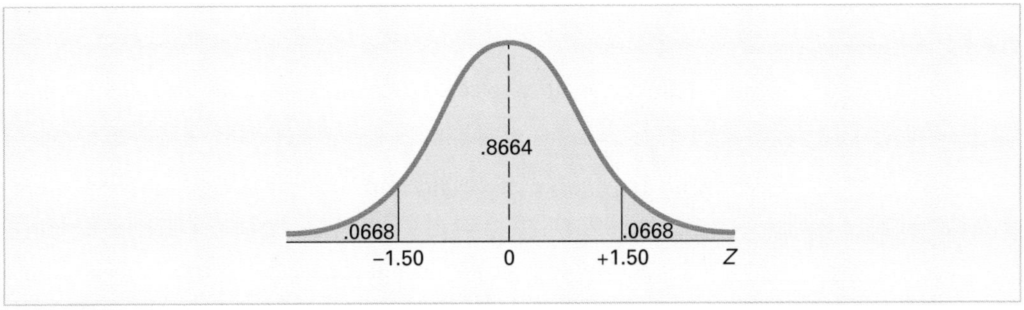

In this example, the observed sample mean is 372.5 grams, 4.5 grams above the hypothesized value and the p-value is 0.1336. Thus, if the population mean is 368 grams, there is a 13.36% chance that the sample mean will be more than 4.5 grams away from 368 (i.e., greater than or equal to 372.5 grams or less than or equal to 363.5 grams). Therefore, while 372.5 is above the hypothesized value of 368, a result as extreme or more extreme than 372.5 is not highly unlikely when the population mean is 368.

Unless you are dealing with a test statistic that follows the normal distribution, the computation of the p-value can be very difficult. However, software such as Microsoft Excel, Minitab or SPSS routinely presents the p-value as part of the output for hypothesis-testing procedures. Figure 9.5 displays a Microsoft Excel worksheet for the cereal-filling example discussed in this section.

FIGURE 9.5

Microsoft Excel Z-Test Worksheet for the Cereal-Filling Example

	A	B	
1	Cereal-Filling Process Hypothesis Test		
2			
3	Data		
4	Null Hypothesis μ=	368	
5	Level of Significance	0.05	
6	Population Standard Deviation	15	
7	Sample Size	25	
8	Sample Mean	372.5	
9			
10	Intermediate Calculations		
11	Standard Error of the Mean	3	=B6/SQRT(B7)
12	Z Test Statistic	1.5	=(B8 - B4)/B11
13			
14	Two-Tail Test		
15	Lower Critical Value	-1.9600	=NORMSINV(B5/2)
16	Upper Critical Value	1.9600	=NORMSINV(1 - B5/2)
17	p-Value	0.1336	=2 * (1 - NORMSDIST(ABS(B12)))
18	Do not reject the null hypothesis		

Exhibit 9.2 provides a summary of the p-value approach for hypothesis testing.

EXHIBIT 9.2: THE FIVE-STEP METHOD OF HYPOTHESIS TESTING USING THE *P*-VALUE APPROACH

1. State the null hypothesis, H_0, and the alternative hypothesis, H_1.
2. Choose the level of significance, α, and the sample size, n. The level of significance is specified according to the relative importance of the risks of committing Type I and Type II errors in the problem.
3. Determine the appropriate test statistic and sampling distribution.
4. Collect the data, compute the value of the test statistic, and compute the p-value.
5. Make the statistical decision and state the managerial conclusion. If the p-value is greater than or equal to α, you do not reject the null hypothesis H_0. If the p-value is less than α, you reject the null hypothesis. Remember the mantra, if the p-value is low, then H_0 must go. The managerial conclusion is written in the context of the real-world problem.

EXAMPLE 9.4

REJECTING A NULL HYPOTHESIS USING THE *P*-VALUE APPROACH

You are the manager of a fast-food restaurant. You want to determine whether the waiting time to place an order has changed in the last month from its previous population mean value of 4.5 minutes. From past experience, you can assume that the population standard deviation is 1.2 minutes. You select a sample of 25 orders during a one-hour period. The sample mean is 5.1 minutes. Use the five-step approach of Exhibit 9.2 above to determine whether there is evidence that the mean waiting time to place an order has changed in the last month from its previous population mean value of 4.5 minutes.

SOLUTION

Step 1: The null hypothesis is that the population mean has not changed from its previous value of 4.5 minutes.

$$H_0: \mu = 4.5$$

The alternative hypothesis is the opposite of the null hypothesis. Since the null hypothesis is that the population mean is 4.5 minutes, the alternative hypothesis is that the population mean is not 4.5 minutes.

$$H_1: \mu \neq 4.5$$

Step 2: You have selected a sample size of $n = 25$. You choose a 0.05 level of significance (i.e., $\alpha = .05$).

Step 3: Select the appropriate test statistic. Because σ is known you use the normal distribution and the Z-test statistic.

Step 4: You collect the data and compute $\bar{X} = 5.1$. Using Equation (9.1) on page 306 you compute the test statistic as follows.

$$Z = \frac{\bar{X} - \mu}{\frac{\sigma}{\sqrt{n}}} = \frac{5.1 - 4.5}{\frac{1.2}{\sqrt{25}}} = 2.50$$

To find the probability of getting a test statistic Z that is equal to or *more extreme than* 2.50 standard deviation units from the center of a standardized normal distribution, you compute the probability of a Z value greater than 2.50 along with the probability of a Z value less than -2.50. From Table E.2, the probability of a Z value below -2.50 is 0.0062. The probability of a value below $+2.50$ is 0.9938. Therefore, the probability of a value above $+2.50$ is $1 - 0.9938 = 0.0062$. Thus, the p-value for this two-tail test is $0.0062 + 0.0062 = 0.0124$.

Step 5: Since the p-value $= 0.0124 < \alpha = 0.05$, you reject the null hypothesis. You conclude that there is evidence that the mean waiting time to place an order has changed from its previous population mean value of 4.5 minutes. The mean waiting time for customers is longer now than last month.

A Connection between Confidence Interval Estimation and Hypothesis Testing

This chapter and Chapter 8 discuss the two major components of statistical inference, confidence interval estimation and hypothesis testing. Although they are based on the same set of concepts, they are used for different purposes. In Chapter 8, you used confidence intervals to estimate parameters. In this chapter, hypothesis testing is used for making decisions about specified values of population parameters. Hypothesis tests are used when trying to prove that a parameter is less than, more than, or not equal to a specified value. Proper interpretation of a confidence interval, however, can also indicate whether a parameter is less than, more than, or not equal to a specified value.

For example, in this section you tested whether the population mean fill amount was different from 368 grams by using Equation (9.1) on page 306:

$$Z = \frac{\bar{X} - \mu}{\frac{\sigma}{\sqrt{n}}}$$

Instead of testing the null hypothesis that $\mu = 368$ grams, you can reach the same conclusion by forming a confidence interval estimate of μ. If the hypothesized value of $\mu = 368$ falls into the interval, you do not reject the null hypothesis because 368 would not be considered an unusual value. On the other hand, if the hypothesized value does not fall into the interval, you reject the null hypothesis, because "368 grams" is then considered an unusual value. Using Equation (8.1) on page 263 and the following data:

$$n = 25,\ \bar{X} = 372.5 \text{ grams},\ \sigma = 15 \text{ grams}$$

For a confidence level of 95% (corresponding to a 0.05 level of significance—i.e., $\alpha = 0.05$),

$$\bar{X} \pm Z\frac{\sigma}{\sqrt{n}}$$

$$372.5 \pm (1.96)\frac{15}{\sqrt{25}}$$

$$372.5 \pm 5.88$$

so that

$$366.62 \le \mu \le 378.38$$

Because the interval includes the hypothesized value of 368 grams, you do not reject the null hypothesis. There is insufficient evidence that the mean fill amount over the entire filling process is not 368 grams. You reached the same decision by using hypothesis testing.

PROBLEMS FOR SECTION 9.2

Learning the Basics

PH Grade ASSIST **9.20** If you use a 0.05 level of significance in a (two-tail) hypothesis test, what will you decide if the computed value of the test statistic Z is +2.21?

PH Grade ASSIST **9.21** If you use a 0.10 level of significance in a (two-tail) hypothesis test, what is your decision rule for rejecting a null hypothesis that the population mean is 500 if you use the Z test?

PH Grade ASSIST **9.22** If you use a 0.01 level of significance in a (two-tail) hypothesis test, what is your decision rule for rejecting H_0: $\mu = 12.5$ if you use the Z test?

PH Grade ASSIST **9.23** What is your decision in problem 9.22 if the computed value of the test statistic Z is -2.61?

PH Grade ASSIST **9.24** Suppose that in a two-tail hypothesis test you compute the value of the test statistic Z as +2.00. What is the p-value?

PH Grade ASSIST **9.25** In problem 9.24, what is your statistical decision if you test the null hypothesis at the 0.10 level of significance?

PH Grade ASSIST **9.26** Suppose that in a two-tail hypothesis test you compute the value of the test statistic Z as -1.38. What is the p-value?

PH Grade ASSIST **9.27** In problem 9.26, what is your statistical decision if you test the null hypothesis at the 0.01 level of significance?

Applying the Concepts

SELF Test **9.28** The director of manufacturing at a clothing factory needs to determine whether a new machine is producing a particular type of cloth according to the manufacturer's specifications, which indicate that the cloth should have a mean breaking strength of 70 pounds and a standard deviation of 3.5 pounds. A sample of 49 pieces of cloth reveals a sample mean breaking strength of 69.1 pounds.

a. Is there evidence that the machine is not meeting the manufacturer's specifications for mean breaking strength? (Use a 0.05 level of significance.)
b. Compute the p-value and interpret its meaning.
c. What is your answer in (a) if the standard deviation is 1.75 pounds?
d. What is your answer in (a) if the sample mean is 69 pounds and the standard deviation is 3.5 pounds?

PH Grade ASSIST **9.29** The manager of a paint supply store wants to determine whether the mean amount of paint contained in 1-gallon cans purchased from a nationally known manufacturer is actually 1 gallon. You know from the manufacturer's specifications that the standard deviation of the amount of paint is 0.02 gallon. You select a random sample of 50 cans, and the mean amount of paint per 1-gallon can is 0.995 gallon.

a. Is there evidence that the mean amount is different from 1.0 gallon (use $\alpha = 0.01$)?
b. Compute the p-value and interpret its meaning.
c. Construct a 99% confidence interval estimate of the population mean amount of paint.
d. Compare the results of (a) and (c). What conclusions do you reach?

9.30 The quality-control manager at a lightbulb factory needs to determine whether the mean life of a large shipment of lightbulbs is equal to 375 hours. The population standard deviation is 100 hours. A random sample of 64 lightbulbs indicates a sample mean life of 350 hours.

a. At the 0.05 level of significance is there evidence that the mean life is different from 375 hours?
b. Compute the p-value and interpret its meaning.
c. Construct a 95% confidence interval estimate of the population mean life of the lightbulbs.
d. Compare the results of (a) and (c). What conclusions do you reach?

9.31 The inspection division of the Lee County Weights and Measures Department is interested in determining whether the proper amount of soft drink has been placed in 2-liter bottles at the local bottling plant of a large nationally known soft-drink company. The bottling plant has informed the inspection division that the standard deviation for 2-liter bottles is 0.05 liter. A random sample of one hundred 2-liter bottles selected from this bottling plant indicates a sample mean of 1.99 liters.

a. At the 0.05 level of significance, is there evidence that the mean amount in the bottles is different from 2.0 liters?
b. Compute the p-value and interpret its meaning.
c. Construct a 95% confidence interval estimate of the population mean amount in the bottles.
d. Compare the results of (a) and (c). What conclusions do you reach?

9.32 A manufacturer of salad dressings uses machines to dispense liquid ingredients into bottles that move along a filling line. The machine that dispenses dressings is working properly when the mean amount dispensed is 8 ounces. The population standard deviation of the amount dispensed is 0.15 ounce. A sample of 50 bottles is selected periodically, and the filling line is stopped if there is evidence that the mean amount dispensed is different from 8 ounces.

Suppose that the mean amount dispensed in a particular sample of 50 bottles is 7.983 ounces.

a. Is there evidence that the population mean amount is different from 8 ounces? (Use a 0.05 level of significance.)

b. Compute the p-value and interpret its meaning.

c. What is your answer in (a) if the standard deviation is 0.05 ounce?

d. What is your answer in (a) if the sample mean is 7.952 ounces and the standard deviation is 0.15 ounce?

9.33 ATMs must be stocked with enough cash to satisfy customers making withdrawals over an entire weekend. But if too much cash is unnecessarily kept in the ATMs, the bank is forgoing the opportunity of investing the money and earning interest. Suppose that at a particular branch the population mean amount of money withdrawn from ATMs per customer transaction over the weekend is \$160 with a population standard deviation of \$30.

a. If a random sample of 36 customer transactions indicates that the sample mean withdrawal amount is \$172, is there evidence to believe that the population mean withdrawal amount is no longer \$160? (Use a 0.05 level of significance.)

b. Compute the p-value and interpret its meaning.

c. What is your answer in (b) if you use a 0.01 level of significance?

d. What is your answer in (b) if the standard deviation is \$24 (use $\alpha = 0.05$)?

9.3 ONE-TAIL TESTS

So far, hypothesis-testing methodology has been used to examine the question of whether or not the population mean amount of cereal filled is 368 grams. The alternative hypothesis (H_1: $\mu \neq 368$) contains two possibilities: Either the mean is less than 368 grams, or the mean is more than 368 grams. For this reason, the rejection region is divided into the two tails of the sampling distribution of the mean.

In many situations, however, the alternative hypothesis focuses on a *particular direction*. One such situation occurs in the following application. A company that makes processed cheese is interested in determining whether some suppliers that provide milk for the processing operation are adding water to their milk to increase the amount supplied to the processing operation. It is known that excess water reduces the freezing point of the milk. The freezing point of natural milk is normally distributed with a mean of −0.545° Celsius (C). The standard deviation of the freezing temperature of natural milk is known to be 0.008°C. Because the cheese company is only interested in determining whether the freezing point of the milk is less than what would be expected from natural milk, the entire rejection region is located in the lower tail of the distribution.

The Critical Value Approach

Suppose you wish to determine whether the mean freezing point of milk is less than −0.545°. To perform this one-tail hypothesis test, you use the six-step method listed in Exhibit 9.1 on page 308.

Step 1: H_0: $\mu \geq -0.545°$
H_1: $\mu < -0.545°$

The alternative hypothesis contains the statement you are trying to prove. If you reject the null hypothesis there is statistical proof that the mean freezing point of the milk is less than the natural freezing point of −0.545°. If the conclusion of the test is "do not reject H_0," then there is insufficient evidence to prove that the mean freezing point is below the natural freezing point of −0.545°.

Step 2: You have selected a sample size of $n = 25$. You decide to use $\alpha = 0.05$.

Step 3: Because σ is known, you use the normal distribution and the Z-test statistic.

Step 4: The rejection region is entirely contained in the lower tail of the sampling distribution of the mean since you want to reject H_0 only when the sample mean is significantly below −0.545°. When the entire rejection region is contained in one tail of the sampling distribution of the test statistic, the test is called a **one-tail** or **directional test**.

When the alternative hypothesis includes the *less than* sign, the critical value of Z must be less than zero. As shown from Table 9.2 and Figure 9.6, because the entire rejection region is in the lower tail of the standardized normal distribution and contains an area of 0.05, the critical value of the Z-test statistic is −1.645, the mean of −1.64 and −1.65.

The decision rule is

$$\text{Reject } H_0 \text{ if } Z < -1.645;$$
$$\text{otherwise do not reject } H_0.$$

TABLE 9.2
Finding the Critical Value of the Z-Test Statistic from the Standardized Normal Distribution for a One-Tail Test with $\alpha = 0.05$

Z	.00	.01	.02	.03	.04	.05	.06	.07	.08	.09
⋮	⋮	⋮	⋮	⋮	⋮	⋮	⋮	⋮	⋮	⋮
−1.8	.0359	.0351	.0344	.0336	.0329	.0322	.0314	.0307	.0301	.0294
−1.7	.0446	.0436	.0427	.0418	.0409	.0401	.0392	.0384	.0375	.0367
−1.6	.0548	.0537	.0526	.0516	.0505	.0495	.0485	.0475	.0465	.0455

Source: Extracted from Table E.2.

FIGURE 9.6
One-Tail Test of Hypothesis for a Mean (σ Known) at the 0.05 Level of Significance

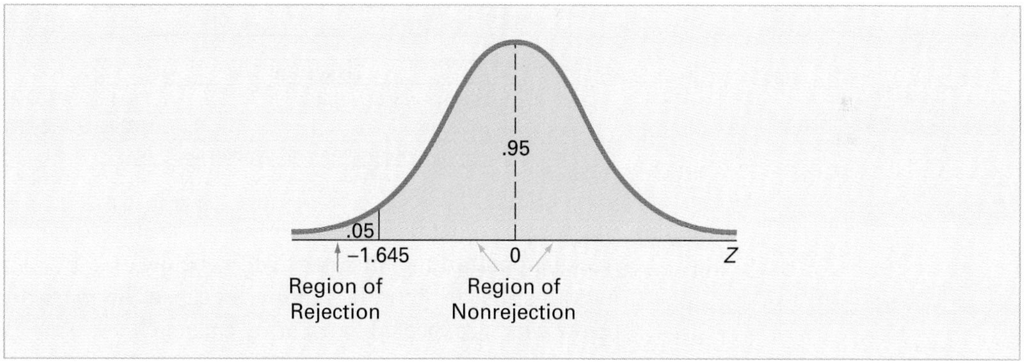

Step 5: You select a sample of 25 containers of milk and compute the sample mean freezing point to be −0.550°. Using $n = 25$, $\bar{X} = -0.550°$, $\sigma = 0.008°$, and Equation (9.1) on page 306,

$$Z = \frac{\bar{X} - \mu}{\frac{\sigma}{\sqrt{n}}} = \frac{-0.550 - (-0.545)}{\frac{0.008}{\sqrt{25}}} = -3.125$$

Step 6: Since $Z = -3.125 < -1.645$, you reject the null hypothesis (see Figure 9.6). You conclude that the mean freezing point of the milk provided is below −0.545°. The company should pursue an investigation of the milk supplier because the mean freezing point is significantly below what is expected to occur by chance.

The *p*-Value Approach

Use the five steps listed in Exhibit 9.2 on page 311 to illustrate the above test using the p-value approach.

Steps 1–3: These steps are the same as in the critical value approach.

Step 4: $Z = -3.125$ (see step 5 of the critical value approach). Since the alternative hypothesis indicates a rejection region entirely in the *lower* tail of the sampling distribution of the Z-test statistic, to compute the p-value you need to find the probability that the Z value will be *below* the test statistic of −3.125. From Table E.2, the probability that the Z value will be below −3.125 is 0.0009 (see Figures 9.7 and 9.8).

FIGURE 9.7
Determining the p-Value for a One-Tail Test

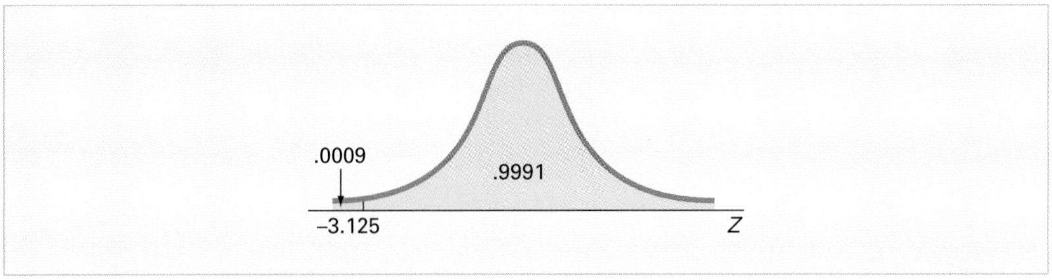

FIGURE 9.8
Microsoft Excel Z-Test Output for the Milk Production Example

	A	B	
1	**Milk Production Hypothesis**		
2			
3	**Data**		
4	**Null Hypothesis** μ=	**-0.545**	
5	**Level of Significance**	**0.05**	
6	**Population Standard Deviation**	**0.008**	
7	**Sample Size**	**25**	
8	**Sample Mean**	**-0.55**	
9			
10	Intermediate Calculations		
11	Standard Error of the Mean	0.0016	=B6/SQRT(B7)
12	*Z* Test Statistic	-3.125	=(B8-B4)/B11
13			
14	**Lower-Tail Test**		
15	**Lower Critical Value**	-1.6449	=NORMSINV(B5)
16	***p*-Value**	0.0009	=NORMSDIST(B12)
17	**Reject the null hypothesis**		

Step 5: The p-value of 0.0009 is less than $\alpha = 0.05$. You reject H_0. You conclude that the mean freezing point of the milk provided is below –0.545°. The company should pursue an investigation of the milk supplier because the mean freezing point is significantly below what is expected to occur by chance.

EXAMPLE 9.5

A ONE-TAIL TEST FOR THE MEAN

A company that manufactures chocolate bars is particularly concerned that the mean weight of the chocolate bar not exceed 6.03 ounces. Past experience allows you to assume that the standard deviation is 0.02 ounces. A sample of 50 chocolate bars is selected and the sample mean is 6.034 ounces. Using the $\alpha = 0.01$ level of significance, is there evidence that the population mean weight of the chocolate bars is greater than 6.03 ounces?

SOLUTION Using the critical value approach,

Step 1: H_0: $\mu \leq 6.03$
H_1: $\mu > 6.03$

Step 2: You have selected a sample size of $n = 50$. You decide to use $\alpha = 0.01$.

Step 3: Because σ is known, you use the normal distribution and the Z-test statistic.

Step 4: The rejection region is entirely contained in the upper tail of the sampling distribution of the mean since you want to reject H_0 only when the sample mean is significantly above 6.03 ounces. Because the entire rejection region is in the upper tail of the standardized normal distribution and contains an area of 0.01, the critical value of the Z-test statistic is 2.33.

The decision rule is

$$\text{Reject } H_0 \text{ if } Z > 2.33;$$
$$\text{otherwise do not reject } H_0.$$

Step 5: You select a sample of 50 chocolate bars and the sample mean weight is 6.034 ounces. Using $n = 50$, $\bar{X} = 6.034$, $\sigma = 0.02$, and Equation (9.1) on page 306:

$$Z = \frac{\bar{X} - \mu}{\frac{\sigma}{\sqrt{n}}} = \frac{6.034 - 6.03}{\frac{0.02}{\sqrt{50}}} = 1.414$$

Step 6: Since $Z = 1.414 < 2.33$, you do not reject the null hypothesis.

There is insufficient evidence to conclude that the population mean weight is above 6.03 ounces.

To perform one-tail tests of hypotheses, you must properly formulate H_0 and H_1. A summary of the null and alternative hypotheses for one-tail tests is as follows.

1. The null hypothesis H_0 represents the status quo or the current belief in a situation.
2. The alternative hypothesis H_1 is the opposite of the null hypothesis and represents a research claim or specific inference you would like to prove.
3. If you reject the null hypothesis, you have statistical proof that the alternative hypothesis is correct.
4. If you do not reject the null hypothesis, then you have failed to prove the alternative hypothesis. The failure to prove the alternative hypothesis, however, does not mean that you have proven the null hypothesis.
5. The null hypothesis (H_0) always refers to a specified value of the *population parameter* (such as μ), not to a *sample statistic* (such as $\bar{X}$).
6. The statement of the null hypothesis *always* contains an equal sign regarding the specified value of the parameter (e.g., H_0: $\mu \geq -0.545$°C).
7. The statement of the alternative hypothesis *never* contains an equal sign regarding the specified value of the parameter (e.g., H_1: $\mu < -0.545$°C).

PROBLEMS FOR SECTION 9.3

Learning the Basics

PH Grade ASSIST **9.34** What is the *upper-tail* critical value of the Z-test statistic at the 0.01 level of significance?

PH Grade ASSIST **9.35** In problem 9.34, what is your statistical decision if the computed value of the Z-test statistic is +2.39?

PH Grade ASSIST **9.36** What is the *lower-tail* critical value of the Z-test statistic at the 0.01 level of significance?

PH Grade ASSIST **9.37** In problem 9.36, what is your statistical decision if the computed value of the Z-test statistic is −1.15?

PH Grade ASSIST **9.38** Suppose that in a one-tail hypothesis test where you reject H_0 only in the *upper* tail, you compute the value of the test statistic Z to be +2.00. What is the p-value?

PH Grade ASSIST **9.39** In problem 9.38, what is your statistical decision if you tested the null hypothesis at the 0.05 level of significance?

PH Grade ASSIST **9.40** Suppose that in a one-tail hypothesis test where you reject H_0 only in the *lower* tail, you compute the value of the test statistic Z as −1.38. What is the p-value?

PH Grade ASSIST **9.41** In problem 9.40, what is your statistical decision if you tested the null hypothesis at the 0.01 level of significance?

9.42 In a one-tail hypothesis test where you reject H_0 only in the *lower* tail, you compute the value of the test statistic Z as +1.38. What is the p-value?

9.43 In problem 9.42, what is the statistical decision if you tested the null hypothesis at the 0.01 level of significance?

Applying the Concepts

PH Grade ASSIST **9.44** The Glen Valley Steel Company manufactures steel bars. If the production process is working properly, it turns out steel bars with mean length of *at least* 2.8 feet with a standard deviation of 0.20 foot (as determined from engineering specifications on the production equipment involved). Longer steel bars can be used or altered, but shorter bars must be scrapped. You select a sample of 25 bars and the mean length is 2.73 feet. Do you need to adjust the production equipment?

a. If you want to test the hypothesis at the 0.05 level of significance, what decision would you make using the critical value approach to hypothesis testing?

b. If you want to test the hypothesis at the 0.05 level of significance, what decision would you make using the *p*-value approach to hypothesis testing?

c. Interpret the meaning of the *p*-value in this problem.

d. Compare your conclusions in (a) and (b).

9.45 You are the manager of a restaurant that delivers pizza to college dormitory rooms. You have just changed your delivery process in an effort to reduce the mean time between the order and completion of delivery from the current 25 minutes. From past experience, you can assume that the population standard deviation is 6 minutes. A sample of 36 orders using the new delivery process yields a sample mean of 22.4 minutes.

a. Using the six-step critical value approach, at the 0.05 level of significance, is there evidence that the mean delivery time has been reduced below the previous population mean value of 25 minutes?

b. At the 0.05 level of significance, use the five-step *p*-value approach.

c. Interpret the meaning of the *p*-value in (b).

d. Compare your conclusions in (a) and (b).

9.46 Children in the United States account directly for \$36 billion in sales annually. When their indirect influence over product decisions from stereos to vacations is considered, the total economic spending impacted by children in the United States is \$290 billion. It is estimated that by age 10, a child makes an average of over five trips a week to a store (M. E. Goldberg, G. J. Gorn, L. A. Peracchio, and G. Bamossy, "Understanding Materialism Among Youth," *Journal of Consumer Psychology*, 2003, 13(3):278–288). Suppose that you want to prove that children in your city average more than five trips a week to a store. Let μ represent the population mean number of times children in your city make trips to a store.

a. State the null and alternative hypothesis.

b. Explain in the context of the above scenario the meaning of the Type I and Type II errors.

c. Suppose that you carry out a study in the city in which you live. Based on past studies, you assume that the standard deviation of the number of trips to the store is 1.6. You take a sample of 100 children and find that the mean number of trips to the store is 5.47. At the 0.01 level of significance, is there evidence that the population mean number of trips to the store is greater than 5 per week?

d. Interpret the meaning of the *p*-value in (c).

9.47 The policy of a particular bank branch is that its ATMs must be stocked with enough cash to satisfy customers making withdrawals over an entire weekend. Customer goodwill depends on such services meeting customer needs. At this branch the population mean amount of money withdrawn from ATMs per customer transaction over the weekend is \$160 with a population standard deviation of \$30. Suppose that a random sample of 36 customer transactions is examined, and you find that the sample mean withdrawal amount is \$172.

a. At the 0.05 level of significance, using the critical value approach to hypothesis testing, is there evidence to believe that the population mean withdrawal amount is greater than \$160?

b. At the 0.05 level of significance, using the *p*-value approach to hypothesis testing, is there evidence to believe that the population mean withdrawal amount is greater than \$160?

c. Interpret the meaning of the *p*-value in this problem.

d. Compare your conclusions in (a) and (b).

9.4 *t* TEST OF HYPOTHESIS FOR THE MEAN (σ UNKNOWN)

In most hypothesis-testing situations dealing with numerical data, you do not know the population standard deviation σ. Instead, you use the sample standard deviation S. If you assume that the population is normally distributed, the sampling distribution of the mean will follow a t distribution with $n - 1$ degrees of freedom. If the population is not normally distributed, you can still use the t test if the sample size is large enough for the Central Limit Theorem to take effect (see section 7.2). Equation (9.2) defines the test statistic t for determining the difference between the sample mean $\bar{X}$ and the population mean μ when the sample standard deviation S is used.

t TEST OF HYPOTHESIS FOR THE MEAN (σ UNKNOWN)

$$t = \frac{\bar{X} - \mu}{\dfrac{S}{\sqrt{n}}} \qquad (9.2)$$

where the test statistic t follows a t distribution having $n - 1$ degrees of freedom.

To illustrate the use of this t test, return to the "Using Statistics" scenario concerning the Saxon Home Improvement Company on page 260. Over the past five years, the mean amount per sales invoice is \$120. As an accountant for the company, you need to inform the finance department if this amount changes. In other words, the hypothesis test is used to try to prove that the mean amount per sales invoice is increasing or decreasing.

The Critical Value Approach

To perform this two-tail hypothesis test, you use the six-step method listed in Exhibit 9.1 on page 308.

Step 1: H_0: $\mu = \$120$
H_1: $\mu \neq \$120$
The alternative hypothesis contains the statement you are trying to prove. If the null hypothesis is rejected, you will have statistical proof that the mean amount per sales invoice is no longer \$120. If the statistical conclusion is "do not reject H_0," then you will conclude that there is insufficient evidence to prove that the mean amount differs from the long-term mean of \$120.

Step 2: You have selected a sample of $n = 12$. You decide to use $\alpha = 0.05$.

Step 3: Because σ is unknown, you use the t distribution and the t-test statistic for this example. You must assume that the population of sales invoices is normally distributed. This assumption is discussed on page 321.

Step 4: For a given sample size n, the test statistic t follows a t distribution with $n - 1$ degrees of freedom. The critical values of the t distribution with $12 - 1 = 11$ degrees of freedom are found in Table E.3, as illustrated in Figure 9.9 and Table 9.3. Because the alternative hypothesis H_1 that $\mu \neq \$120$ is *nondirectional*, the area in the rejection region of the t distribution's left (lower) tail is 0.025, and the area in the rejection region of the t distribution's right (upper) tail is also 0.025.

From the t table as given in Table E.3, a portion of which is shown in Table 9.3, the critical values are ± 2.2010. The decision rule is

$$\text{Reject } H_0 \text{ if } t < -t_{11} = -2.2010$$
$$\text{or if } t > t_{11} = +2.2010;$$
$$\text{otherwise do not reject } H_0.$$

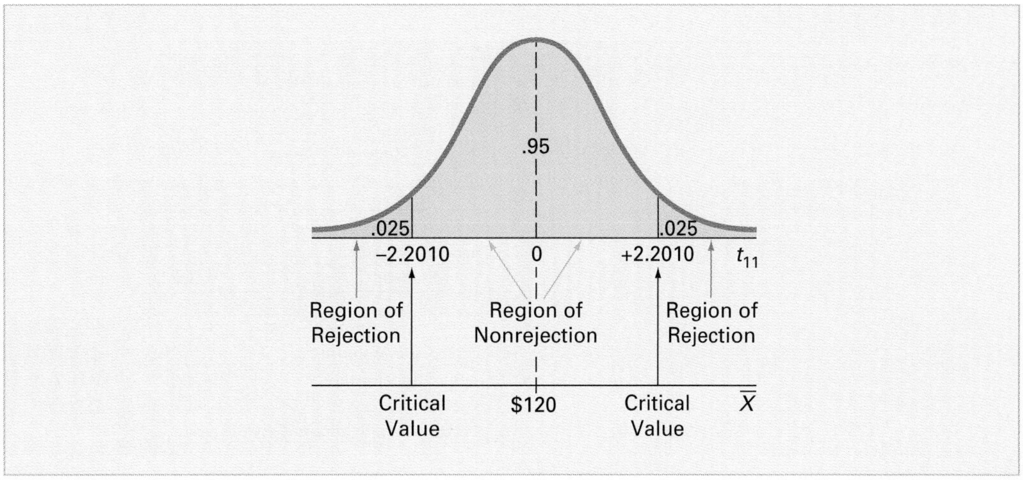

FIGURE 9.9
Testing a Hypothesis about the Mean (σ Unknown) at 0.05 Level of Significance with 11 Degrees of Freedom

TABLE 9.3

Determining the Critical Value from the t Table for an Area of 0.025 in Each Tail with 11 Degrees of Freedom

	Upper-Tail Areas					
Degrees of Freedom	**.25**	**.10**	**.05**	**.025**	**.01**	**.005**
1	1.0000	3.0777	6.3138	12.7062	31.8207	63.6574
2	0.8165	1.8856	2.9200	4.3027	6.9646	9.9248
3	0.7649	1.6377	2.3534	3.1824	4.5407	5.8409
4	0.7407	1.5332	2.1318	2.7764	3.7469	4.6041
5	0.7267	1.4759	2.0150	2.5706	3.3649	4.0322
6	0.7176	1.4398	1.9432	2.4469	3.1427	3.7074
7	0.7111	1.4149	1.8946	2.3646	2.9980	3.4995
8	0.7064	1.3968	1.8595	2.3060	2.8965	3.3554
9	0.7027	1.3830	1.8331	2.2622	2.8214	3.2498
10	0.6998	1.3722	1.8125	2.2281	2.7638	3.1693
11	0.6974	1.3634	1.7959	2.2010	2.7181	3.1058

Source: Extracted from Table E.3.

Step 5: The following data INVOICES are the amounts (in dollars) in a random sample of 12 sales invoices.

108.98	152.22	111.45	110.59	127.46	107.26
93.32	91.97	111.56	75.71	128.58	135.11

Using Equations (3.1) and (3.10) on pages 73 and 82, or the Microsoft Excel output of Figure 9.10 or the Minitab output of Figure 9.11

$$\bar{X} = \frac{\sum_{i=1}^{n} X_i}{n} = \$112.85 \quad \text{and} \quad S = \sqrt{\frac{\sum_{i=1}^{n}(X_i - \bar{X})^2}{n-1}} = \$20.80$$

From Equation (9.2) on page 318,

$$t = \frac{\bar{X} - \mu}{\frac{S}{\sqrt{n}}} = \frac{112.85 - 120}{\frac{20.80}{\sqrt{12}}} = -1.19$$

FIGURE 9.10

Microsoft Excel Worksheet for the One-Sample t Test of Sales Invoices

	A	B	
1	**Mean Amount Per Invoice Hypothesis**		
2			
3	**Data**		
4	**Null Hypothesis** μ=	120	
5	**Level of Significance**	0.05	
6	**Sample Size**	12	
7	**Sample Mean**	112.85	
8	**Sample Standard Deviation**	20.8	
9			
10	Intermediate Calculations		
11	Standard Error of the Mean	6.0044	=B8/SQRT(B6)
12	Degrees of Freedom	11	=B6 - 1
13	***t* Test Statistic**	-1.1908	=(B7 - B4)/B11
14			
15	**Two-Tail Test**		
16	**Lower Critical Value**	-2.2010	=-(TINV(B5, B12))
17	**Upper Critical Value**	2.2010	=TINV(B5, B12)
18	***p*-Value**	0.2588	=TDIST(ABS(B13), B12, 2)
19	**Do not reject the null hypothesis**		

FIGURE 9.11
Minitab Output for the One-Sample *t* Test of Sales Invoices

```
Test of mu = 120 vs not = 120

Variable   N     Mean   StDev  SE Mean        95% CI              T      P
Amount    12  112.851  20.798    6.004  (99.636, 126.065)   -1.19  0.259
```

Step 6: Since $-2.201 < t = -1.19 < 2.201$, you do not reject H_0. You have insufficient evidence to conclude that the mean amount per sales invoice differs from \$120. You should inform the finance department that the audit suggests that the mean amount per invoice has not changed.

The *p*-Value Approach

Steps 1–3: These steps are the same as in the critical value approach.

Step 4: $t = -1.19$ (see step 5 of the critical value approach).

Step 5: The Microsoft Excel worksheet of Figure 9.10 and the Minitab output of Figure 9.11 give the *p*-value for this two-tail test as 0.259. Since the *p*-value of 0.259 is greater than $\alpha = 0.05$, you do not reject H_0. The data provide insufficient evidence to conclude that the mean amount per sales invoice differs from \$120. You should inform the finance department that the audit suggests that the mean amount per invoice has not changed. The *p*-value indicates that if the null hypothesis were true, the probability that a sample of 12 invoices could have a monthly mean that differs by \$7.15 or more from the stated \$120 is 0.259. In other words, if the mean amount per sales invoice is truly \$120, then there is a 25.9% chance of observing a sample mean below \$112.85 or above \$127.15.

In the above example it is incorrect to state that there is a 25.9% chance that the null hypothesis is true. This misinterpretation of the *p*-value is sometimes used by those not properly trained in statistics. Remember that the *p*-value is a conditional probability, calculated by *assuming* that the null hypothesis is true. In general, it is proper to state the following. If the null hypothesis is true, then there is a (*p*-value)*100% chance of observing a sample result at least as contradictory to the null hypothesis as the result observed.

Checking Assumptions

You use the one-sample *t* test when the population standard deviation σ is not known and is estimated using the sample standard deviation[1] *S*. The *t* test is considered a *classical parametric* procedure, one that makes a variety of stringent assumptions that must hold to ensure that the results of the test are valid.

To use the one-sample *t* test, the data are assumed to represent a random sample from a population that is normally distributed. In practice, as long as the sample size is not very small and the population is not very skewed, the *t* distribution provides a good approximation to the sampling distribution of the mean when σ is unknown.

There are several ways to evaluate the normality assumption necessary for using the *t* test. You can observe how closely the sample statistics match the normal distribution's theoretical properties. You can also use a histogram, stem-and-leaf display, box-and-whisker plot, or normal probability plot. For details on evaluating normality, see section 6.3 on page 208.

[1]When a large sample size is available, S estimates σ precisely enough that there is little difference between the t and Z distributions. Therefore, you can use a Z test instead of a t test when the sample size is greater than 120.

Figure 9.12 presents Microsoft Excel output that provides descriptive statistics. Figure 9.13 is a Minitab box-and-whisker plot. Figure 9.14 is a Minitab normal probability plot.

FIGURE 9.12
Microsoft Excel Descriptive Statistics for the Sales Invoice Data

	A	B
1	*Invoice Amount*	
2		
3	Mean	112.8508333
4	Standard Error	6.003863082
5	Median	111.02
6	Mode	#N/A
7	Standard Deviation	20.7979918
8	Sample Variance	432.5564629
9	Kurtosis	0.172707598
10	Skewness	0.13363802
11	Range	76.51
12	Minimum	75.71
13	Maximum	152.22
14	Sum	1354.21
15	Count	12
16	Largest(1)	152.22
17	Smallest(1)	75.71

FIGURE 9.13
Minitab Box-and-Whisker Plot for the Sales Invoice data

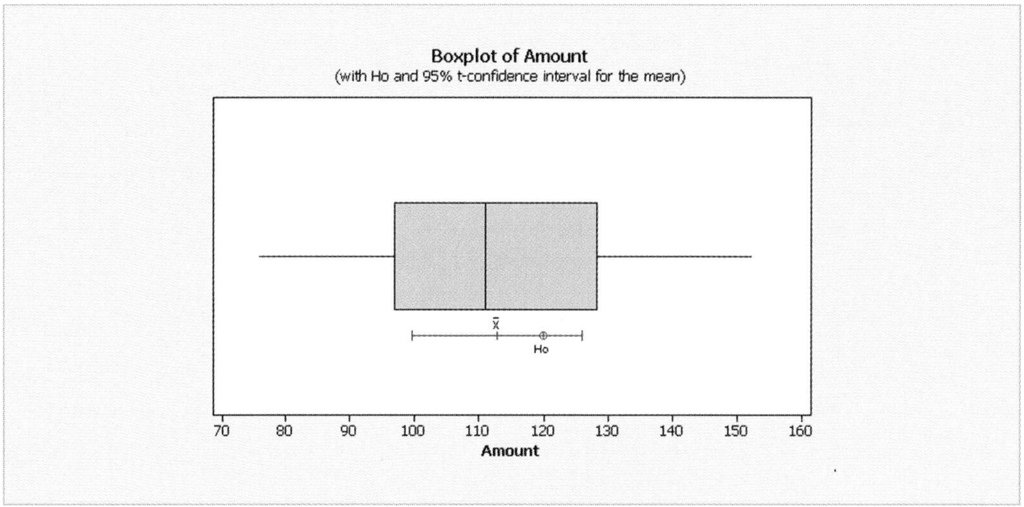

FIGURE 9.14
Minitab Normal Probability Plot for the Sales Invoice Data

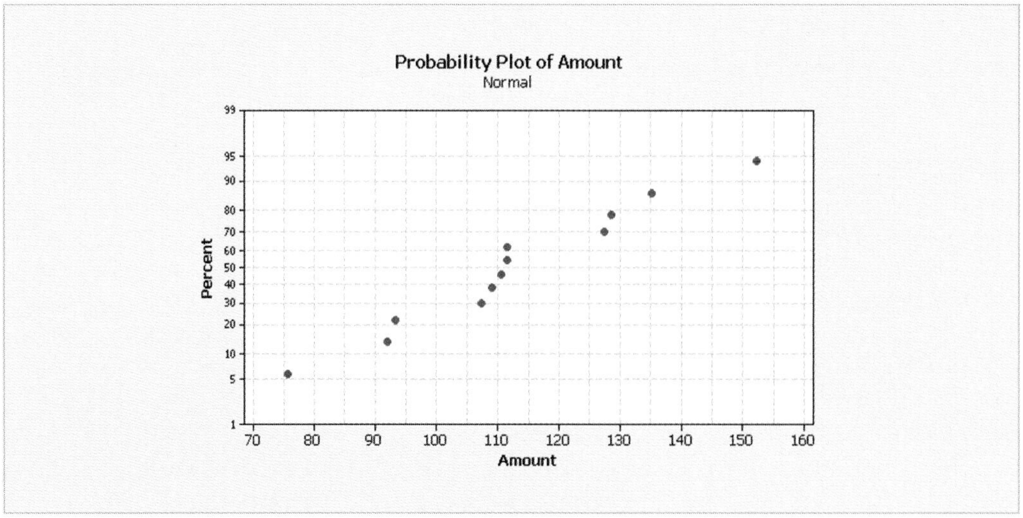

Because the mean is very close to the median, the points on the normal probability plot appear to be increasing approximately in a straight line and the box-and-whisker plot appears approximately symmetrical. You can assume that the population of sales invoices is approximately normally distributed. The normality assumption is valid and therefore the auditor's results are valid.

The t test is a **robust** test. It does not lose power if the shape of the population departs somewhat from a normal distribution, particularly when the sample size is large enough to enable the test statistic t to be influenced by the Central Limit Theorem (see section 7.2). However, you can make erroneous conclusions and can lose statistical power if you use the t test incorrectly. If the sample size n is small (i.e., less than 30) and you cannot easily make the assumption that the underlying population is at least approximately normally distributed, other *nonparametric* testing procedures are more appropriate (see references 1 and 2).

PROBLEMS FOR SECTION 9.4

Learning the Basics

PH Grade ASSIST **9.48** If, in a sample of $n = 16$ selected from a normal population, $\bar{X} = 56$ and $S = 12$, what is the value of the t-test statistic if you are testing the null hypothesis, H_0: $\mu = 50$?

9.49 In problem 9.48, how many degrees of freedom are there in the one-sample t test?

9.50 In problems 9.48 and 9.49, what are the critical values from the t table if the level of significance $\alpha = 0.05$ and the alternative hypothesis H_1 is:

a. $\mu \neq 50$?
b. $\mu > 50$?

9.51 In problems 9.48, 9.49, and 9.50, what is your statistical decision if the alternative hypothesis H_1 is:

a. $\mu \neq 50$?
b. $\mu > 50$?

9.52 If, in a sample of $n = 16$ selected from a left-skewed population, $\bar{X} = 65$ and $S = 21$, would you use the t test to test the null hypothesis, H_0: $\mu = 60$? Discuss.

9.53 If, in a sample of $n = 160$ selected from a left-skewed population, $\bar{X} = 65$ and $S = 21$, would you use the t test to test the null hypothesis, H_0: $\mu = 60$? Discuss.

Applying the Concepts

Problems 9.54–9.56 can be solved manually or by using Microsoft Excel, Minitab, or SPSS. We recommend that you use Microsoft Excel, Minitab, or SPSS to solve problems 9.57–9.65.

SELF Test **9.54** The director of admissions at a large university advises parents of incoming students about the cost of textbooks during a typical semester. He selected a sample of 100 students and recorded their textbook expenses for the semester. He then computed a sample mean cost of \$315.40 and a sample standard deviation of \$43.20.

a. Using the 0.10 level of significance, is there evidence that the population mean is above \$300?
b. What is your answer in (a) if the standard deviation is \$75 and the 0.05 level of significance is used?
c. What is your answer in (a) if the sample mean is \$305.11 and the sample standard deviation is \$43.20?

9.55 In an article (Nanci Hellmich, " 'Supermarket Guru' Has a Simple Mantra," *USA Today*, June 19, 2002, 70) it was claimed that the typical supermarket trip takes a mean of 22 minutes. Suppose that in an effort to test this claim, you select a sample of 50 shoppers at a local supermarket. The mean shopping time for the sample of 50 shoppers was 25.36 minutes with a standard deviation of 7.24 minutes. Using the 0.10 level of significance, is there evidence that the mean shopping time at the local supermarket is different from the claimed value of 22 minutes?

9.56 You are the manager of a restaurant for a fast-food franchise. Last month the mean waiting time at the drive-through window, as measured from the time a customer places an order until the time the customer receives the order, was 3.7 minutes. The franchise helped you institute a new process intended to reduce waiting time. You select a random sample of 64 orders. The sample mean waiting time is 3.57 minutes with a sample standard deviation of 0.8 minute. At the 0.05 level of significance, is there evidence that the population mean waiting time is now less than 3.7 minutes?

9.57 A manufacturer of chocolate candies uses machines to package candies as they move along a filling line. Although the packages are labeled as 8 ounces, the company wants the packages to contain 8.17 ounces so that virtually none of the packages contain less than 8 ounces. A sample of 50 packages is selected periodically, and the packaging process is stopped if there is evidence that the mean amount packaged is different from 8.17 ounces. Suppose that the mean amount dispensed in a particular sample of 50 packages is 8.159 ounces with a sample standard deviation of 0.051 ounce.

a. Is there evidence that the population mean amount is different from 8.17 ounces? (Use a 0.05 level of significance.)

b. Compute the p-value and interpret its meaning.

9.58 A manufacturer of flashlight batteries took a sample of 13 batteries BATTERIES from a day's production and used them continuously until they failed to work. The life of the batteries in hours until failure is:

342 426 317 545 264 451 1,049
631 512 266 492 562 298

a. At the 0.05 level of significance, is there evidence that the mean life of the batteries is more than 400 hours?

b. Determine the p-value in (a) and interpret its meaning.

c. Using the information above, what would you advise if the manufacturer wanted to say in advertisements that these batteries "should last more than 400 hours"?

d. Suppose that the first value was 1,342 instead of 342. Repeat (a) through (c), using this value. Comment on the difference in the results.

9.59 In New York State, savings banks are permitted to sell a form of life insurance called Savings Bank Life Insurance (SBLI). The approval process consists of underwriting, which includes a review of the application, a medical information bureau check, possible requests for additional medical information and medical exams, and a policy compilation stage where the policy pages are generated and sent to the bank for delivery. The ability to deliver approved policies to customers in a timely manner is critical to the profitability of this service. During a period of one month, a random sample of 27 approved policies is selected INSURANCE and the total processing time in days recorded:

73 19 16 64 28 28 31 90 60 56 31 56 22 18
45 48 17 17 17 91 92 63 50 51 69 16 17

a. In the past, the mean processing time averaged 45 days. At the 0.05 level of significance, is there evidence that the mean processing time has changed from 45 days?

b. What assumption about the population distribution is needed in (a)?

c. Do you think that the assumption needed in (b) is seriously violated? Explain.

9.60 The following data represent the amount of soft drink filled in a sample of 50 consecutive 2-liter bottles. DRINK The results, listed horizontally in the order of being filled, were:

2.109 2.086 2.066 2.075 2.065 2.057 2.052 2.044 2.036 2.038
2.031 2.029 2.025 2.029 2.023 2.020 2.015 2.014 2.013 2.014
2.012 2.012 2.012 2.010 2.005 2.003 1.999 1.996 1.997 1.992
1.994 1.986 1.984 1.981 1.973 1.975 1.971 1.969 1.966 1.967
1.963 1.957 1.951 1.951 1.947 1.941 1.941 1.938 1.908 1.894

a. At the 0.05 level of significance, is there evidence that the mean amount of soft drink filled is different from 2.0 liters?

b. Determine the p-value in (a) and interpret its meaning.

c. Evaluate the assumption you made in (a) graphically. Are the results of (a) valid? Why?

d. Examine the values of the 50 bottles in their sequential order as given in the problem. Is there a pattern to the results? If so, what impact might this pattern have on the validity of the results in (a)?

9.61 One of the major measures of the quality of service provided by any organization is the speed with which it responds to customer complaints. A large family-held department store selling furniture and flooring including carpeting had undergone a major expansion in the past several years. In particular, the flooring department had expanded from 2 installation crews to an installation supervisor, a measurer, and 15 installation crews. Last year there were 50 complaints concerning carpeting installation. The following data FURNITURE represent the number of days between the receipt of the complaint and the resolution of the complaint.

54	5	35	137	31	27	152	2	123	81	74	27
11	19	126	110	110	29	61	35	94	31	26	5
12	4	165	32	29	28	29	26	25	1	14	13
13	10	5	27	4	52	30	22	36	26	20	23
33	68										

a. The installation supervisor claims that the mean number of days between the receipt of the complaint and the resolution of the complaint is 20 days or less. At the 0.05 level of significance, is there evidence that the claim is not true (i.e., that the mean number of days is greater than 20)?

b. What assumption about the population distribution must you make in (a)?

c. Do you think that the assumption made in (b) is seriously violated? Explain.

d. What effect might your conclusion in (c) have on the validity of the results in (a)?

9.62 In an article in *Quality Engineering*, the viscosity (resistance to flow) of a chemical product produced in batches was examined. The data for 120 batches are in the data file CHEMICAL.

Source: Holmes and Mergen, "Parabolic Control Limits for the Exponentially Weighted Moving Average Control Charts," Quality Engineering, *1992, 4(4): 487–495.*

a. In the past, the mean viscosity was 15.5. At the 0.10 level of significance, is there evidence that the mean viscosity has changed from 15.5?
b. What assumption about the population distribution do you need to make in (a)?
c. Do you think that the assumption made in (b) has been seriously violated? Explain.

9.63 One operation of a steel mill is to cut pieces of steel into parts that are used in the frame for front seats in an automobile. The steel is cut with a diamond saw and requires the resulting parts to be within ±0.005 inch of the length specified by the automobile company. The data in the file STEEL come from a sample of 100 steel parts. The measurement reported is the difference in inches between the actual length of the steel part, as measured by a laser measurement device, and the specified length of the steel part. For example, a value of −0.002 represents a steel part that is 0.002 inch shorter than the specified length.
a. At the 0.05 level of significance, is there evidence that the mean difference is not equal to 0.0 inches?
b. Determine the p-value in (a) and interpret its meaning.
c. What assumption about the differences between the actual length of the steel part and the specified length of the steel part must you make in (a)?
d. Evaluate the assumption in (c) graphically. Are the results of (a) valid? Why?

9.64 In problem 3.61 on page 115, you were introduced to a tea-bag-filling operation. An important quality characteristic of interest for this process is the weight of the tea in the individual bags. The data in the file TEABAGS is an ordered array of the weight, in grams, of a sample of 50 tea bags produced during an eight-hour shift.
a. Is there evidence that the mean amount of tea per bag is different from 5.5 grams (use $\alpha = 0.01$)?
b. Construct a 99% confidence interval estimate of the population mean amount of tea per bag. Interpret this interval.
c. Compare the conclusions reached in (a) and (b).

9.65 The following table contains a random sample of 30 mutual funds taken from the mutual funds reported in the *The Wall Street Journal* on June 15, 2004. CHANGE2004 For each mutual fund, "Change" is the change (in dollars) in fund value on June 14, 2004.

Mutual Fund	Change
ABN AMRO Growth I	−0.22
Aim Funds HYld	0.00
Amer Advant Int Plan	−0.42
Artisan Funds SmCap	−0.24
Calif Trust S&P500	−0.23
Cohen and Steers Inst Rel	−0.55
Columbia Balance Z	−0.15
Delaware Large Cap Value A	−0.17
Dimension EmgMkt	−0.38
Dodge&Cox Stock	−1.50
Dreyfus It Inc	−0.09
Dreyfus OHMA	−0.02
Eaton Balanced	−0.06
Emerald Gr A	−0.18
Evergreen GLLeadA	−0.26
Federated Cap App	−0.20
Federated Intl Eq	−0.43
Fidelity Banking	−0.58
FirstAmerican LgCapValue	−0.17
Janus Balanced	−0.16
Kinetics Internet	−0.21
Merrill Lynch Equity Inc	−0.12
Nicholas Group Nich	−0.46
Northern Balanced A	−0.09
One Group DivMidCap	−0.24
Prudential Bear	0.03
Putnam Income	−0.02
South Trust Value	−0.14
Third Avenue Real Est	−0.26
Van Kampen Entc	−0.10

Source: Extracted from The Wall Street Journal, *June 15, 2004.*

a. Is there evidence that the population mean fund value changed on June 14, 2004? Use a level of significance of 0.05.
b. What assumptions are made to perform the test in (a)?
c. Determine the p-value and interpret its meaning.

9.5 *Z* TEST OF HYPOTHESIS FOR THE PROPORTION

In some situations, you want to test a hypothesis about the population proportion π of values that are in a particular category rather than testing the population mean. To begin, you select a random sample and compute the **sample proportion**, $p = X/n$. You then compare the value of this statistic to the hypothesized value of the parameter π in order to decide whether to reject the null hypothesis.

If the number of successes (X) and the number of failures ($n - X$) are each at least five, the sampling distribution of a proportion approximately follows a standardized normal distribution. You use the ***Z* test for the proportion** given in Equation (9.3) to perform the hypothesis test for the difference between the sample proportion p and the hypothesized population proportion π.

ONE SAMPLE Z TEST FOR THE PROPORTION

$$Z = \frac{p - \pi}{\sqrt{\dfrac{\pi(1-\pi)}{n}}} \tag{9.3}$$

where $p = \dfrac{X}{n} = \dfrac{\text{number of successes in the sample}}{\text{sample size}}$

= sample proportion of successes

π = hypothesized proportion of successes in the population

The test statistic Z approximately follows a standardized normal distribution.

Alternatively, by multiplying numerator and denominator by n, you can write the Z-test statistic in terms of the number of successes X as shown in Equation (9.4).

Z TEST FOR THE PROPORTION IN TERMS OF THE NUMBER OF SUCCESSES

$$Z = \frac{X - n\pi}{\sqrt{n\pi(1-\pi)}} \tag{9.4}$$

To illustrate the one-sample Z test for a proportion, consider the following study reported in *The Wall Street Journal*. In the study, the question posed was "are an equal number of home-based businesses owned by men and women?" The study of 899 home-based businesses reported that 369 were owned by women (Eleena De Lisser and Dan Morse, "More Men Work at Home than Women, Study Shows," *The Wall Street Journal*, May 18, 1999, B2).

For this study, the null and alternative hypotheses are stated as follows:

H_0: $\pi = 0.50$ (i.e., the proportion of home-based businesses owned by females is 0.50)

H_1: $\pi \neq 0.50$ (i.e., the proportion of home-based businesses owned by females is not 0.50)

The Critical Value Approach

Because you are interested in whether or not the proportion of home-based businesses owned by females is 0.50 (and the proportion owned by males is 0.50), you use a two-tail test. If you select the $\alpha = 0.05$ level of significance, the rejection and nonrejection regions are set up as in Figure 9.15, and the decision rule is

Reject H_0 if $Z < -1.96$ or if $Z > +1.96$;

otherwise do not reject H_0.

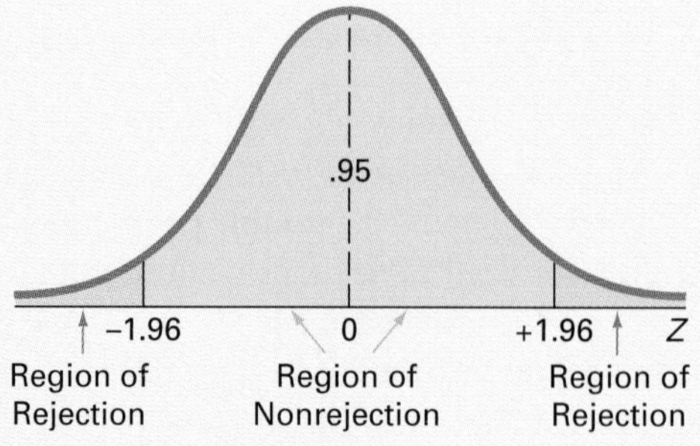

FIGURE 9.15
Two-Tail Test of Hypothesis for the Proportion at the 0.05 Level of Significance

Since 369 of the 899 home-based businesses are owned by females,

$$p = \frac{369}{899} = 0.41046$$

Using Equation (9.3),

$$Z \cong \frac{p - \pi}{\sqrt{\dfrac{\pi(1-\pi)}{n}}} = \frac{0.41046 - 0.50}{\sqrt{\dfrac{0.50(1-0.50)}{899}}} = \frac{-0.08954}{0.0167} = -5.37$$

or, using Equation (9.4),

$$Z \cong \frac{X - n\pi}{\sqrt{n\pi(1-\pi)}} = \frac{369 - (899)(0.50)}{\sqrt{899(0.50)(0.50)}} = \frac{-80.5}{14.99} = -5.37$$

Because $-5.37 < -1.96$, you reject H_0. You can conclude that the proportion of home-based businesses owned by females is not 0.50. Figure 9.16 presents a Microsoft Excel worksheet for these data.

FIGURE 9.16
Microsoft Excel Worksheet for the Study on Home-Based Business Ownership

	A	B	
1	**Ownership Proportion Hypothesis**		
2			
3	**Data**		
4	**Null Hypothesis** *p*=	**0.5**	
5	**Level of Significance**	**0.05**	
6	**Number of Successes**	**369**	
7	**Sample Size**	**899**	
8			
9	Intermediate Calculations		
10	Sample Proportion	0.4105	=B6/B7
11	Standard Error	0.0167	=SQRT(B4 * (1 - B4)/B7)
12	**Z Test Statistic**	-5.3697	=(B10 - B4)/B11
13			
14	**Two-Tail Test**		
15	**Lower Critical Value**	**-1.9600**	=NORMSINV(B5/2)
16	**Upper Critical value**	**1.9600**	=NORMSINV(1 - B5/2)
17	***p*-Value**	**0.0000**	=2 * (1 - NORMSDIST(ABS(B12)))
18	**Reject the null hypothesis**		

The *p*-Value Approach

As an alternative approach toward making a hypothesis-testing decision, you can compute the p-value. For this two-tail test in which the rejection region is located in the lower tail and the upper tail, you need to find the area below a Z value of -5.37 and above a Z value of $+5.37$. Figure 9.16 and Figure 9.17 report a p-value of 0.0000. Because this value is less than the selected level of significance ($\alpha = 0.05$), you reject the null hypothesis. This extremely small p-value indicates that there is virtually no chance that the sample proportion will be as small as 0.41046 if the population proportion is 0.50.

FIGURE 9.17
Minitab Output for the Study of Home-Based Business Ownership

```
Test of p = 0.5 vs p not = 0.5

Sample    X    N  Sample p          95% CI           Z-Value  P-Value
1       369  899  0.410456  (0.378300, 0.442612)     -5.37    0.000
```

EXAMPLE 9.6 TESTING A HYPOTHESIS FOR A PROPORTION

A fast-food chain has just developed a new process to make sure that orders at the drive-through are filled correctly. The previous process filled orders correctly 88% of the time. A sample of 100 orders using the new process is selected and 92 were filled correctly. At the 0.01 level of significance, can you conclude that the new process has increased the proportion of orders filled correctly?

SOLUTION The null and alternative hypotheses are

H_0: $\pi \le 0.88$ (i.e., the proportion of orders filled correctly is less than or equal to 0.88)

H_1: $\pi > 0.88$ (i.e., the proportion of orders filled correctly is greater than 0.88)

Using Equation (9.3) on page 326,

$$p = \frac{X}{n} = \frac{92}{100} = 0.92$$

$$Z = \frac{p - \pi}{\sqrt{\dfrac{\pi(1-\pi)}{n}}} = \frac{0.92 - 0.88}{\sqrt{\dfrac{0.88(1-0.88)}{100}}} = \frac{0.04}{0.0325} = 1.23$$

The p-value for $Z > 1.23$ is 0.1093.

Using the critical value approach, you reject H_0 if $Z > 2.33$. Using the p-value approach, you reject H_0 if the p-value < 0.01. Since $Z = 1.23 < 2.33$ or the p-value $= 0.1093 > 0.01$, you do not reject H_0. You conclude that there is insufficient evidence that the new process has increased the proportion of correct orders above 0.88.

PROBLEMS FOR SECTION 9.5

Learning the Basics

PH Grade ASSIST **9.66** If, in a random sample of 400 items, 88 are defective, what is the sample proportion of defective items?

PH Grade ASSIST **9.67** In problem 9.66, if the null hypothesis is that 20% of the items in the population are defective, what is the value of the Z-test statistic?

PH Grade ASSIST **9.68** In problems 9.66 and 9.67, suppose you are testing the null hypothesis H_0: $\pi = 0.20$ against the two-tail alternative hypothesis H_1: $\pi \ne 0.20$ and you choose the level of significance $\alpha = 0.05$. What is your statistical decision?

Applying the Concepts

PH Grade ASSIST SELF Test **9.69** An article in *The Wall Street Journal* implies that more than half of all Americans would prefer being given $100 rather than a day off from work. This statement is based on a survey conducted by American Express Incentive Services, in which 593 of 1,040 respondents indicated that they would rather have the $100 (Carlos Tejada, "Work Week," *The Wall Street Journal*, July 25, 2000, A1).

a. At the 0.05 level of significance, is there evidence based on the survey data that more than half of all Americans would rather have $100 than a day off from work?

b. Compute the p-value and interpret its meaning.

9.70 Due to a very weak economy, only an estimated 43% of employers in the United States recruited new employees in 2003. But, by the end of the year, the economy showed signs of improvement. A survey by the Society for Human Resource Management indicated that 181 of 362 human-resources professionals planned to recruit new employees in 2004 (Hane J. Kim, "Finally, 2004 May Be the Time to Seek a Raise," *The Wall Street Journal*, January 8, 2004, D4). Conduct a hypothesis test to try to prove that the proportion of employers that planned to hire new employees in 2004 is larger than the 2003 proportion of 0.43. Use the six-step hypothesis-testing method and a 0.05 level of significance.

9.71 A *Wall Street Journal* article suggests that age bias is becoming an even bigger problem in the corporate world (Carol Hymowitz, "Top Executives Chase Youthful Appearance, But Miss Real Issue," *The Wall Street Journal*, February 17, 2004, B1). In 2001, an estimated 78% of executives believed that age bias was a serious problem. In a 2004 study by ExecuNet, 82% of the executives surveyed considered age bias a serious problem. The sample size for the 2004 study was not disclosed. Suppose 50 executives were surveyed.

a. At the 0.05 level of significance, use the six-step hypothesis method to try and prove that the proportion of executives who believe that age bias is a serious problem is higher than the 2001 value of 0.78.

b. Use the five-step p-value approach. Interpret the meaning of the p-value.

c. Suppose that the sample size used was 1,000. Redo (a) and (b).

d. Discuss the effect that sample size had on the outcome of this analysis and, in general, on the effect sample size plays in hypothesis-testing.

9.72 A *Wall Street Journal* poll asked respondents if they trusted energy-efficiency ratings on cars and appliances; 552 responded *yes*, and 531 responded *no* ("What's News Online," *The Wall Street Journal*, March 30, 2004, D7).

a. At the 0.05 level of significance, use the six-step hypothesis testing method to try and prove that the percentage of people who trust energy-efficiency ratings differs from 50%.

b. Use the 5-step p-value approach. Interpret the meaning of the p-value.

9.73 One of the biggest issues facing e-retailers is the ability to turn browsers into buyers (M. Totty, "Making the Sale," *The Wall Street Journal*, September 24, 2001, R6). This is measured by the conversion rate, the percentage of browsers who buy something in their visit to a site. This article reported that the conversion rate for llbean.com was 10.1% and for victoriasecret.com was 8.2%. Suppose that each of these sites were redesigned in an attempt to increase their conversion rates. Samples of 200 browsers at each redesigned site were selected. Suppose that 24 browsers at llbean.com made a purchase and 25 browsers at victoriasecret.com made a purchase.

a. Is there evidence of an increased conversion rate at llbean.com at the 0.05 level of significance?

b. Is there evidence of an increased conversion rate at victoriasecret.com at the 0.05 level of significance?

9.74 More professional women than ever before are foregoing motherhood because of the time constraints of their careers. Yet, many women still manage to find time to climb the corporate ladder *and* set time aside to have children. A survey of 187 attendees at *Fortune Magazine*'s Most Powerful Women in Business summit in March 2002 found that 133 had at least one child (Carol Hymowitz, "Women Plotting Mix of Work and Family Won't Find Perfect Plan," *The Wall Street Journal*, June 11, 2002, B1). Assume that the group of 187 women is a random sample from the population of all successful women executives.

a. What is the sample proportion of successful women executives who have children?

b. At the 0.05 level of significance, can you state that more than half of all successful women executives have children?

c. At the 0.05 level of significance, can you state that more than two-thirds of all successful women executives have children?

d. Do you think the random sample assumption is valid? Explain.

9.6 THE POWER OF A TEST

Section 9.1 defined Type I and Type II errors and their associated risks. Recall that α represents the probability that you reject the null hypothesis when it is true and should not be rejected, and β represents the probability that you do not reject the null hypothesis when it is false and should be rejected. The power of the test, $1 - \beta$, is the probability that you correctly reject a false null hypothesis. This probability depends on how different the actual population mean is from the value being hypothesized (under H_0), the value of α used, and the sample size. If there is a large difference between the population mean and the hypothesized mean, the power of the test will be much greater than if the difference between the population mean and the hypothesized mean is small. Selecting a larger value of α makes it easier to reject H_0 and therefore increases the power of a test. Increasing the sample size increases the precision in the estimates and therefore increases the ability to detect differences in the parameters and increases the power of a test.

In this section, the power of a statistical test is illustrated using the Oxford Cereal Company scenario. The filling process is subject to periodic inspection from a representative of the consumer affairs office. The representative's job is to detect the possible "short weighting" of boxes in which cereal boxes having less than the specified 368 grams are sold. Thus, the representative is interested in determining whether there is evidence that the cereal boxes have a mean weight that is less than 368 grams. The null and alternative hypotheses are as follows:

$$H_0: \mu \geq 368 \text{ (filling process is working properly)}$$

$$H_1: \mu < 368 \text{ (filling process is not working properly)}$$

The representative is willing to accept the company's claim that the standard deviation σ equals 15 grams. Therefore, you can use the Z test. Using Equation (9.1) on page 306, with $\bar{X}_L$ (the lower critical $\bar{X}$ value) substituted for $\bar{X}$, you can find the value of $\bar{X}$ that enables you to reject the null hypothesis:

$$Z = \frac{\bar{X}_L - \mu}{\frac{\sigma}{\sqrt{n}}}$$

$$Z\frac{\sigma}{\sqrt{n}} = \bar{X}_L - \mu$$

$$\bar{X}_L = \mu + Z\frac{\sigma}{\sqrt{n}}$$

Because you have a one-tail test with a level of significance of 0.05, the value of Z is equal to –1.645 (see Figure 9.18). The sample size $n = 25$. Therefore,

$$\bar{X}_L = 368 + (-1.645)\frac{(15)}{\sqrt{25}} = 368 - 4.935 = 363.065$$

The decision rule for this one-tail test is

Reject H_0 if $\bar{X} < 363.065$;

otherwise do not reject H_0.

FIGURE 9.18

Determining the Lower Critical Value for a One-Tail Z Test for a Population Mean at the 0.05 Level of Significance

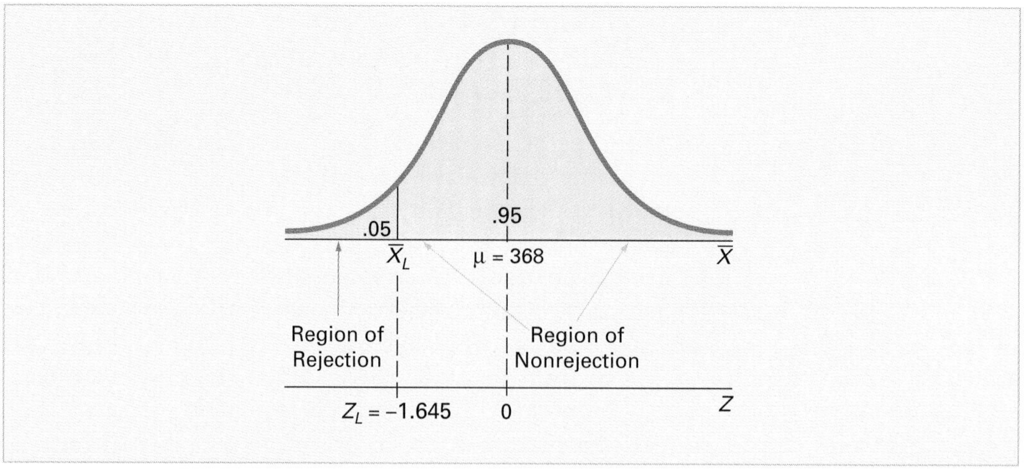

The decision rule states that if in a random sample of 25 boxes, the sample mean is less than 363.065 grams, you reject the null hypothesis, and the representative concludes that the process is not working properly. The power of the test measures the probability of concluding that the process is not working properly for differing values of the true population mean.

What is the power of the test if the actual population mean is 360 grams? To determine the chance of rejecting the null hypothesis when the population mean is 360 grams, you need to determine the area under the normal curve below $\bar{X}_L = 363.065$ grams. Using Equation (9.1), with the population mean $\mu = 360$,

$$Z = \frac{\bar{X} - \mu}{\frac{\sigma}{\sqrt{n}}}$$

$$Z = \frac{363.065 - 360}{\frac{15}{\sqrt{25}}} = 1.02$$

From Table E.2, there is an 84.61% chance that the Z value is less than +1.02. This is the power of the test where μ is the actual population mean (see Figure 9.19). The probability (β) that you will not reject the null hypothesis ($\mu = 368$) is $1 - 0.8461 = 0.1539$. Thus, the probability of committing a Type II error is 15.39%.

FIGURE 9.19
Determining the Power of the Test and the Probability of a Type II Error when μ = 360 Grams

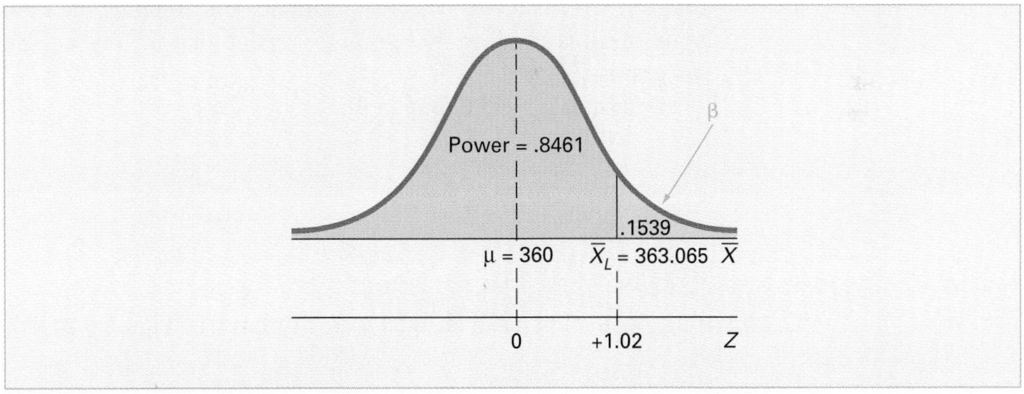

Now that you have determined the power of the test if the population mean were equal to 360, you can calculate the power for any other value of μ. For example, what is the power of the test if the population mean is 352 grams? Assuming the same standard deviation, sample size, and level of significance, the decision rule is

$$\text{Reject } H_0 \text{ if } \bar{X} < 363.065;$$
$$\text{otherwise do not reject } H_0.$$

Once again, because you are testing a hypothesis for a mean, from Equation (9.1)

$$Z = \frac{\bar{X} - \mu}{\frac{\sigma}{\sqrt{n}}}$$

If the population mean shifts down to 352 grams (see Figure 9.20), then

$$Z = \frac{363.065 - 352}{\frac{15}{\sqrt{25}}} = 3.69$$

FIGURE 9.20

Determining the Power of the Test and the Probability of a Type II Error when $\mu = 352$ Grams

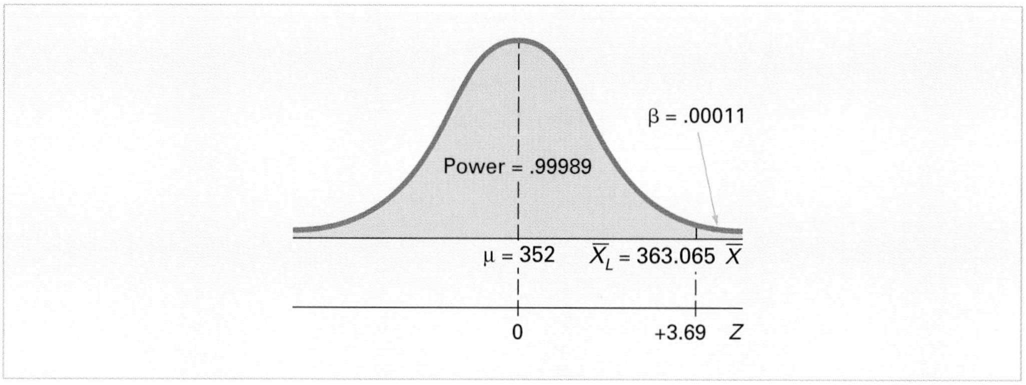

From Table E.2, there is a 99.989% chance that the Z value is less than +3.69. This is the power of the test when the population mean is 352. The probability (β) that you will not reject the null hypothesis ($\mu = 368$) is $1 - 0.99989 = 0.00011$. Thus, the probability of committing a Type II error is only 0.011%.

In the preceding two examples the power of the test is high, and the chance of committing a Type II error is low. In the next example, you compute the power of the test when the population mean is equal to 367 grams—a value that is very close to the hypothesized mean of 368 grams.

Once again, from Equation (9.1),

$$Z = \frac{\bar{X} - \mu}{\frac{\sigma}{\sqrt{n}}}$$

If the population mean is equal to 367 grams (see Figure 9.21), then

$$Z = \frac{363.065 - 367}{\frac{15}{\sqrt{25}}} = -1.31$$

FIGURE 9.21

Determining the Power of the Test and the Probability of a Type II Error when $\mu = 367$ Grams

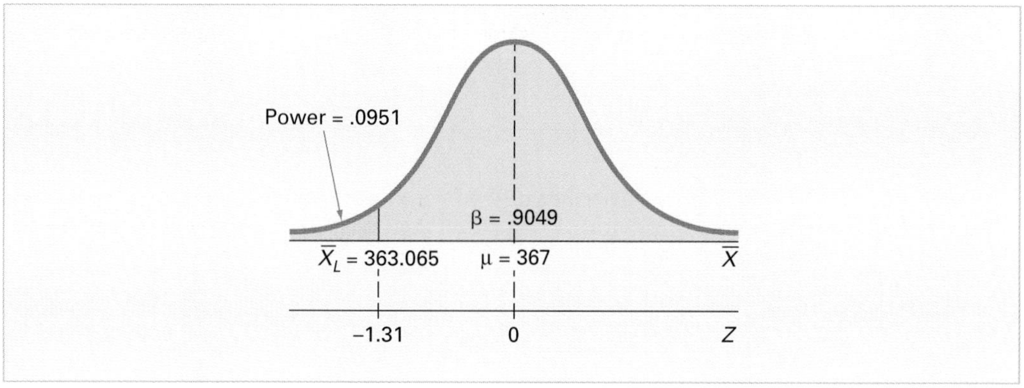

From Table E.2, the probability less than $Z = -1.31$ is 0.0951 (or 9.51%). Because the rejection region is in the lower tail of the distribution, the power of the test is 9.51% and the chance of making a Type II error is 90.49%.

Figure 9.22 illustrates the power of the test for various possible values of μ (including the three values examined). This graph is called a **power curve**.

FIGURE 9.22

Power Curve of the Cereal-Box-Filling Process for H_1: $\mu < 368$ Grams

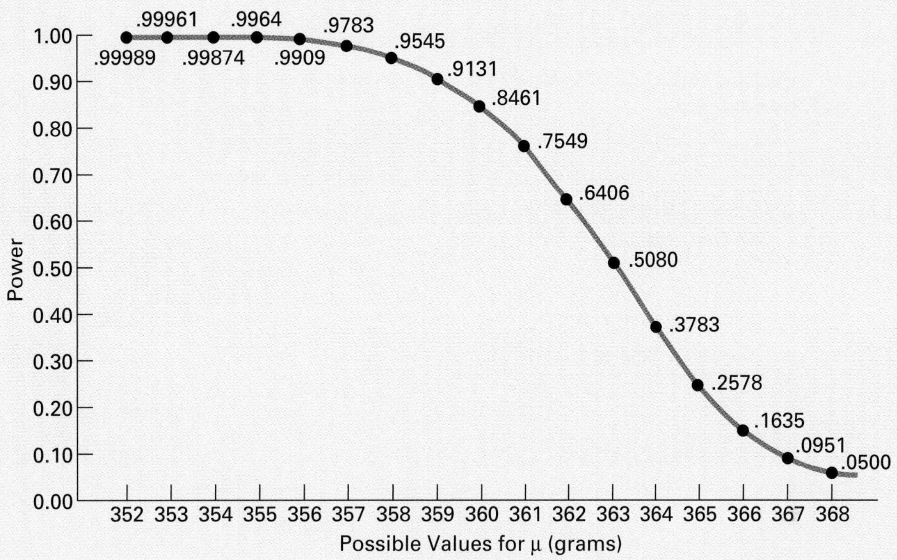

From Figure 9.22, you can see that the power of this one-tail test increases sharply (and approaches 100%) as the population mean takes on values farther below the hypothesized mean of 368 grams. Clearly, for this one-tail test, the smaller the actual mean μ the greater the power to detect this difference.[2] For values of μ close to 368 grams, the power is small because the test cannot effectively detect small differences between the actual population mean and the hypothesized value of 368 grams. When the population mean approaches 368 grams, the power of the test approaches α, the level of significance (which is 0.05 in this example).

Figure 9.23 on page 334 summarizes the computations for the three cases. You can see the drastic changes in the power of the test for differing values of the actual population means by reviewing the different panels of Figure 9.23. From panels *A* and *B* you can see that when the population mean does not greatly differ from 368 grams, the chance of rejecting the null hypothesis, based on the decision rule involved, is not large. However, once the population mean shifts substantially below the hypothesized 368 grams, the power of the test greatly increases, approaching its maximum value of 1 (or 100%).

In the above discussion, a one-tail test with $\alpha = 0.05$ and $n = 25$ was used. The type of statistical test (one-tail versus two-tail), the level of significance, and the sample size all affect the power. Three basic conclusions regarding the power of the test are summarized below:

1. A one-tail test is more powerful than a two-tail test.
2. An increase in the level of significance (α) results in an increase in power. A decrease in α results in a decrease in power.
3. An increase in the sample size *n* results in an increase in power. A decrease in the sample size *n* results in a decrease in power.

[2]For situations involving one-tail tests in which the actual mean μ is greater than the hypothesized mean, the converse is true. The larger the actual mean μ compared with the hypothesized mean, the greater is the power. For two-tail tests, the greater the distance between the actual mean μ and the hypothesized mean, the greater the power of the test.

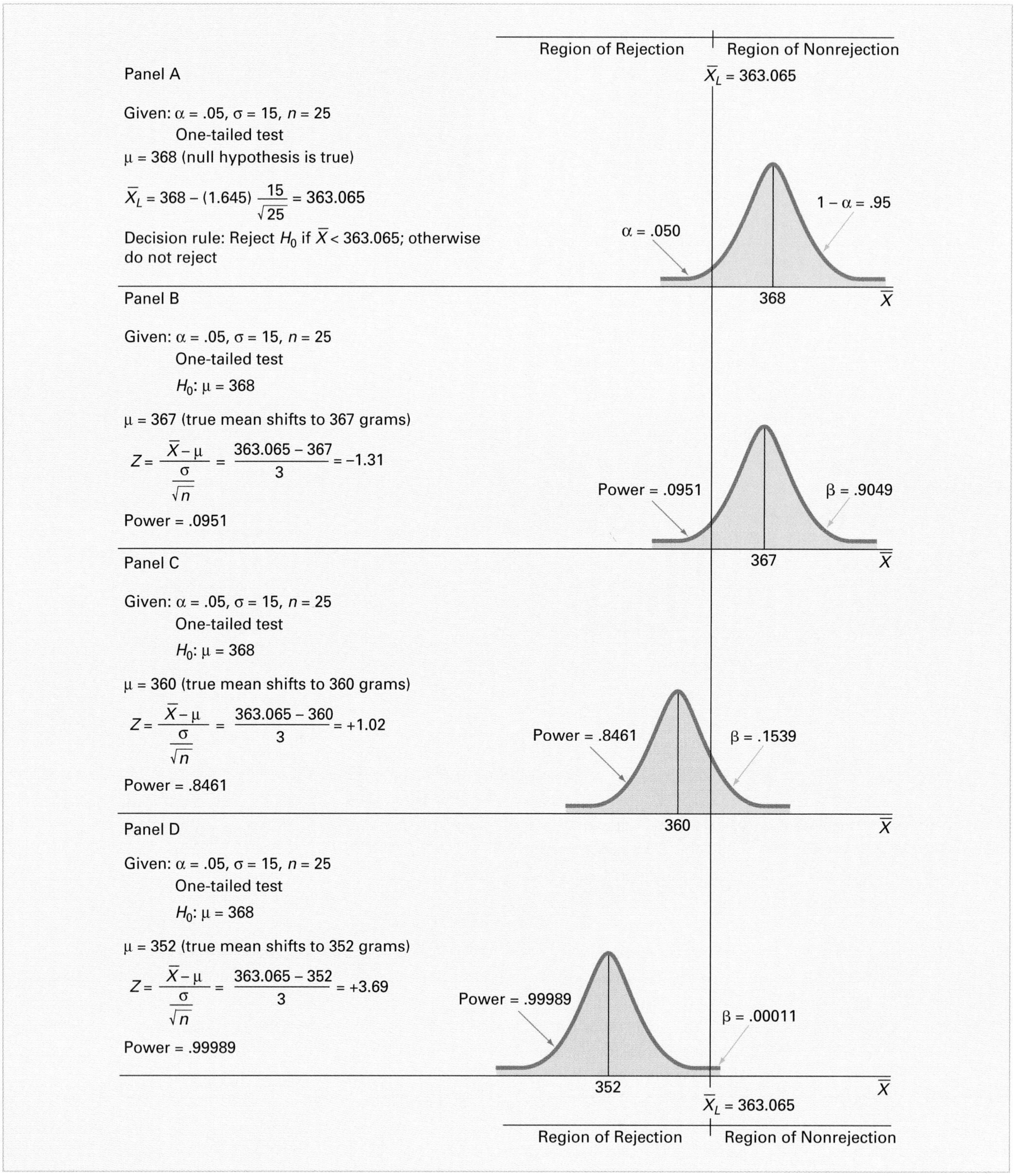

FIGURE 9.23 Determining Statistical Power for Varying Values of the Population Mean

PROBLEMS FOR SECTION 9.6

Applying the Concepts

SELF Test **9.75** A coin-operated soft-drink machine is designed to discharge at least 7 ounces of beverage per cup with a standard deviation of 0.2 ounce. If you select a random sample of 16 cups and you are willing to have an $\alpha = 0.05$ risk of committing a Type I error, compute the power of the test and the probability of a Type II error (β) if the population mean amount dispensed is actually

a. 6.9 ounces per cup.
b. 6.8 ounces per cup.

9.76 Refer to problem 9.75. If you are willing to have an $\alpha = 0.01$ risk of committing a Type I error, compute the power of the test and the probability of a Type II error (β) if the population mean amount dispensed is actually

a. 6.9 ounces per cup.
b. 6.8 ounces per cup.
c. Compare the results in (a) and (b) of this problem and in problem 9.75. What conclusion can you reach?

9.77 Refer to problem 9.75. If you select a random sample of 25 cups and are willing to have an $\alpha = 0.05$ risk of committing a Type I error, compute the power of the test and the probability of a Type II error (β) if the population mean amount dispensed is actually

a. 6.9 ounces per cup.
b. 6.8 ounces per cup.
c. Compare the results in (a) and (b) of this problem and in problem 9.75. What conclusion can you reach?

9.78 A tire manufacturer produces tires that have a mean life of at least 25,000 miles when the production process is working properly. Based on past experience, the standard deviation of the tires is 3,500 miles. The operations manager stops the production process if there is evidence that the mean tire life is below 25,000 miles. If you select a random sample of 100 tires (to be subjected to *destructive testing*) and you are willing to have an $\alpha = 0.05$ risk of committing a Type I error, compute the power of the test and the probability of a Type II error (β) if the population mean life is actually

a. 24,000 miles.
b. 24,900 miles.

9.79 Refer to problem 9.78. If you are willing to have an $\alpha = 0.01$ risk of committing a Type I error, compute the power of the test and the probability of a Type II error (β) if the population mean life is actually

a. 24,000 miles.
b. 24,900 miles.
c. Compare the results in (a) and (b) of this problem and (a) and (b) in problem 9.78. What conclusion can you reach?

9.80 Refer to problem 9.78. If you select a random sample of 25 tires and are willing to have an $\alpha = 0.05$ risk of committing a Type I error, compute the power of the test and the probability of a Type II error (β) if the population mean life is actually

a. 24,000 miles.
b. 24,900 miles.
c. Compare the results in (a) and (b) of this problem and (a) and (b) in problem 9.78. What conclusion can you reach?

9.81 Refer to problem 9.78. If the operations manager stops the process when there is evidence that the mean life is different from 25,000 miles (either less than or greater than) and a random sample of 100 tires is selected along with a level of significance of $\alpha = 0.05$, compute the power of the test and the probability of a Type II error (β) if the population mean life is actually

a. 24,000 miles.
b. 24,900 miles.
c. Compare the results in (a) and (b) of this problem and (a) and (b) in problem 9.78. What conclusion can you reach?

9.7 POTENTIAL HYPOTHESIS-TESTING PITFALLS AND ETHICAL ISSUES

To this point, you have studied the fundamental concepts of hypothesis testing. You used hypothesis testing for analyzing differences between sample estimates (i.e., statistics) and hypothesized population characteristics (i.e., parameters) in order to make decisions about the underlying characteristics. You have also learned how to evaluate the risks involved in making these decisions.

When planning to carry out a test of the hypothesis based on a survey, research study, or designed experiment, you must ask several questions to ensure that proper methodology is used. You need to raise and answer questions like the ones below in the planning stage.

1. What is the goal of the survey, study, or experiment? How can you translate the goal into a null hypothesis and an alternative hypothesis?

2. Is the hypothesis test a two-tail test or one-tail test?
3. Can you select a random sample from the underlying population of interest?
4. What kinds of data will you collect from the sample? Are the variables numerical or categorical?
5. At what significance level, or risk of committing a Type I error, should you conduct the hypothesis test?
6. Is the intended sample size large enough to achieve the desired power of the test for the level of significance chosen?
7. What statistical test procedure should you use and why?
8. What conclusions and interpretations can you make from the results of the hypothesis test?

Therefore, you should consult with a person with substantial statistical training early in the process. All too often such an individual is consulted far too late in the process, after the data have been collected. Typically, all that you can do at such a late stage is to choose the statistical test procedure that is best for the data. You are forced to assume that certain biases built into the study (because of poor planning) are negligible. But this is a large assumption. Good research involves good planning. To avoid biases, adequate controls must be built in from the beginning.

You need to distinguish between poor research methodology and unethical behavior. Ethical considerations arise when the hypothesis-testing process is manipulated. Some of the ethical issues that arise include the data collection method, informed consent from human subjects being "treated," the type of test (one-tail or two-tail), the choice of the level of significance α, data snooping, the cleansing and discarding of data, and the reporting of findings.

Data Collection Method—Randomization To eliminate the possibility of potential biases in the results, you must use proper data collection methods. To draw meaningful conclusions, the data must be the outcome of a random sample from a population or from an experiment in which a **randomization** process was used. Potential respondents should not be permitted to self-select for a study nor should they be purposely selected. Aside from the potential ethical issues that may arise, such a lack of randomization can result in serious coverage errors or selection biases that destroy the integrity of the study.

Informed Consent from Human Respondents Being "Treated" Ethical considerations require that any individual who is to be subjected to some "treatment" in an experiment be made aware of the research endeavor and any potential behavioral or physical side effects. The subject should also provide informed consent with respect to participation.

Type of Test—Two-Tail or One-Tail If prior information is available that leads you to test the null hypothesis against a specifically directed alternative, then a one-tail test is more powerful than a two-tail test. However, if you are interested only in *differences* from the null hypothesis, not in the *direction* of the difference, the two-tail test is the appropriate procedure to use. For example, if previous research and statistical testing have already established the difference in a particular direction, or if an established scientific theory states that it is possible for results to occur in only one direction, then a one-tail test is appropriate. It is never appropriate to change the direction of a test after the data are collected.

Choice of Level of Significance α In a well-designed study, you select the level of significance α before data collection occurs. You cannot alter the level of significance, after the fact, to achieve a specific result. It is also good practice to always report the *p*-value, not just the conclusions of the hypothesis test.

Data Snooping **Data snooping** is never permissible. It is unethical to perform a hypothesis test on a set of data, look at the results, and then decide on the level of significance or decide between a one-tail or two-tail test. You must make these decisions before the data are collected in order for the conclusions to have meaning. In those situations in which you consult a statistician late in the process, with data already available, it is imperative that you establish the null and alternative hypotheses and choose the level of significance prior to carrying out the hypothesis test. In addition, you cannot arbitrarily change or discard extreme or unusual values in order to alter the results of the hypothesis tests.

Cleansing and Discarding of Data Data cleansing is not data snooping. In the data preparation stage of editing, coding, and transcribing, you have an opportunity to review the data for any value whose measurement appears extreme or unusual. After reviewing the unusual observations, you should construct a stem-and-leaf display and/or a box-and-whisker plot in preparation for further data presentation and confirmatory analysis. This exploratory data analysis stage gives you another opportunity to cleanse the data set by flagging possible outliers to double-check against the original data. In addition, the exploratory data analysis enables you to examine the data graphically with respect to the assumptions underlying a particular hypothesis test procedure.

The process of data cleansing raises a major ethical question. Should you ever remove a value from a study? The answer is a qualified yes. If you can determine that a measurement is incomplete or grossly in error because of some equipment problem or unusual behavioral occurrence unrelated to the study, you can discard the value. Sometimes you have no choice—an individual may decide to quit a particular study he or she has been participating in before a final measurement can be made. In a well-designed experiment or study, you should decide, in advance, on all rules regarding the possible discarding of data.

Reporting of Findings In conducting research, you should document both good and bad results. It is inappropriate to report the results of hypothesis tests that show statistical significance, but not those for which there is insufficient evidence in the findings. In those instances where there is insufficient evidence to reject H_0, you must make it clear that this does not prove that the null hypothesis is true. What the result does indicate is that with the sample size used, there is not enough information to *disprove* the null hypothesis.

Statistical Significance versus Practical Significance You need to make the distinction between the existence of a statistically significant result and its practical significance in the context within a field of application. Sometimes, due to a very large sample size, you will get a result that is statistically significant, but has little practical significance. For example, suppose that prior to a national marketing campaign focusing on a series of expensive television commercials, you believe that the proportion of people who recognized your brand was 0.30. At the completion of the campaign, a survey of 20,000 people indicates that 6,168 recognized your brand. A one-tail test trying to prove that the proportion is now greater than 0.30 results in a p-value of 0.0047 and the correct statistical conclusion is that the proportion of consumers recognizing your brand name has now increased. Was the campaign successful? The result of the hypothesis test indicates a statistically significant increase in brand awareness, but is this increase practically important? The population proportion is now estimated at 6,168/20,000 = 0.3084 or 30.84%. This increase is less than 1% more than the hypothesized value of 30%. Did the large expenses associated with the marketing campaign produce a result with a meaningful increase in brand awareness? Because of the minimal real-world impact an increase of less than 1% has on the overall marketing strategy, and the huge expenses associated with the marketing campaign, you should conclude that the campaign was not successful. On the other hand, if the campaign increased brand awareness by 20%, you could conclude that the campaign was successful.

To summarize, in discussing ethical issues concerning hypothesis-testing the key is *intent*. You must distinguish between poor data analysis and unethical practice. Unethical practice occurs when researchers *intentionally* create a selection bias in data collection, manipulate the treatment of human subjects without informed consent, use data snooping to select the type of test (two-tail or one-tail) and/or level of significance, hide the facts by discarding values that do not support a stated hypothesis, or fail to report pertinent findings.

SUMMARY

This chapter presented the foundation of hypothesis-testing. You learned how to perform Z and t tests on the population mean, a Z test on the population proportion, and how an operations manager of a production facility can use hypothesis testing to monitor and improve a cereal-filling process. The three chapters that follow build on the foundation of hypothesis testing discussed here.

Figure 9.24 presents a road map for selecting the correct one-sample hypothesis test to use.

FIGURE 9.24

Road Map for Selecting a One-Sample Test

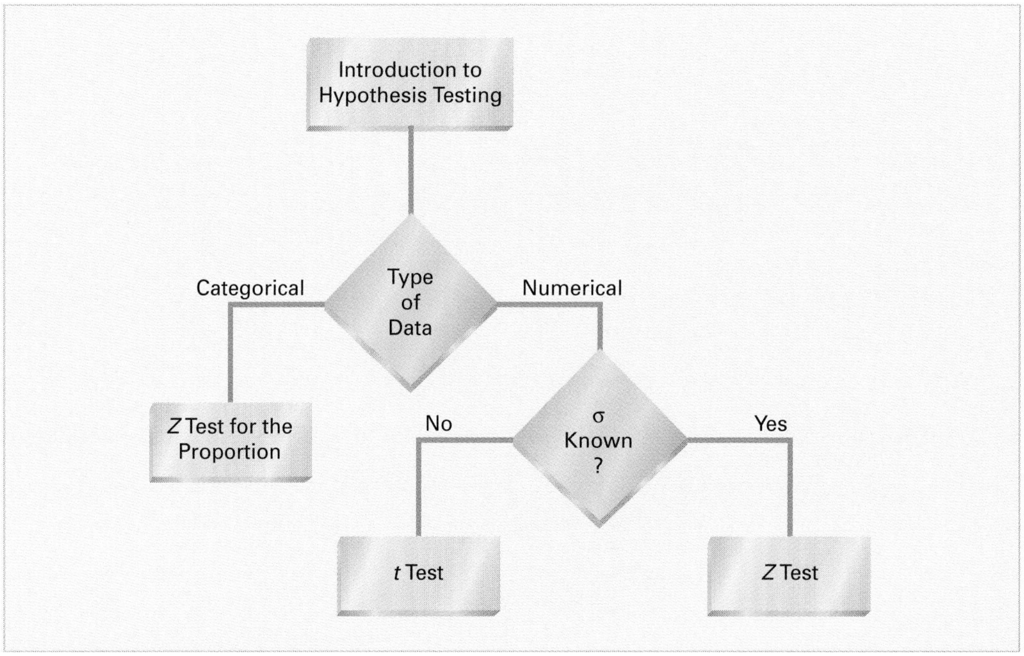

In deciding on which test to use, you should ask the following questions

- Does the test involve a numerical variable or a categorical variable? If the test involves a categorical variable, use the Z test for the proportion.
- If the test involves a numerical variable, do you know the population standard deviation? If you know the population standard deviation, use the Z test for the mean. If you do not know the population standard deviation, use the t test for the mean.

Table 9.4 provides a list of topics covered in this chapter.

TABLE 9.4

Summary of Topics in Chapter 9

	Type of Data	
Type of Analysis	**Numerical**	**Categorical**
Hypothesis Test Concerning a Single Parameter	Z Test of Hypothesis for the Mean (**section 9.2**) t Test of Hypothesis for the Mean (**section 9.4**)	Z Test of Hypothesis for the Proportion (**section 9.5**)

KEY FORMULAS

Z Test of Hypothesis for the Mean (σ Known)

$$Z = \frac{\bar{X} - \mu}{\frac{\sigma}{\sqrt{n}}} \quad \textbf{(9.1)}$$

t Test of Hypothesis for the Mean (σ Unknown)

$$t = \frac{\bar{X} - \mu}{\frac{S}{\sqrt{n}}} \quad \textbf{(9.2)}$$

One Sample Z Test for the Proportion

$$Z \cong \frac{p - \pi}{\sqrt{\frac{\pi(1-\pi)}{n}}} \quad \textbf{(9.3)}$$

Z Test for the Proportion in Terms of the Number of Successes

$$Z \cong \frac{X - n\pi}{\sqrt{n\pi(1-\pi)}} \quad \textbf{(9.4)}$$

KEY TERMS

CHAPTER REVIEW PROBLEMS

Checking Your Understanding

9.82 What is the difference between a null hypothesis H_0 and an alternative hypothesis H_1?

9.83 What is the difference between a Type I and a Type II error?

9.84 What is meant by the power of a test?

9.85 What is the difference between a one-tail and a two-tail test?

9.86 What is meant by a p-value?

9.87 How can a confidence interval estimate for the population mean provide conclusions to the corresponding hypothesis test for the population mean?

9.88 What is the six-step critical value approach to hypothesis testing?

9.89 What are some of the ethical issues to be concerned with in performing a hypothesis test?

9.90 In planning to carry out a test of hypothesis based on a designed experiment or research study under investigation, what are some of the questions that you need to raise in order to ensure that proper methodology will be used?

Applying the Concepts

9.91 An article in *Marketing News* (Thomas T. Semon, "Consider a Statistical Insignificance Test," *Marketing News*, February 1, 1999) argues that the level of significance used when comparing two products is often too low, that is, sometimes you should be using an α value greater than 0.05. Specifically, the article recounts testing the proportion of potential customers with a preference for product 1 over product 2. The null hypothesis is that the population proportion of potential customers preferring product 1 is 0.50, and the alternative hypothesis is that it is not equal to 0.50. The p-value for the test is 0.22. The article suggests that in some cases this should be enough evidence to reject the null hypothesis.

a. State the null and alternative hypotheses in statistical terms.
b. Explain the risks associated with Type I and Type II errors.
c. What are the consequences if you rejected the null hypothesis for a p-value of 0.22?

d. Why do you think the article suggests raising the value of α?
e. What would you do in this situation?
f. What is your answer in (e) if the *p*-value equals 0.12? What if it equals 0.06?

9.92 La Quinta Motor Inns developed a computer model to help predict the profitability of sites that are being considered as locations for new hotels. If the computer model predicts large profits, La Quinta buys the proposed site and builds a new hotel. If the computer model predicts small or moderate profits, La Quinta chooses not to proceed with that site (Sheryl E. Kimes and James A. Fitzsimmons, "Selecting Profitable Hotel Sites at La Quinta Motor Inns," *Interfaces*, Vol. 20, March–April 1990, 12–20). This decision-making procedure can be expressed in the hypothesis-testing framework. The null hypothesis is that the site is not a profitable location. The alternative hypothesis is that the site is a profitable location.
a. Explain the risks associated with committing a Type I error.
b. Explain the risks associated with committing a Type II error.
c. Which type of error do you think the executives at La Quinta Motor Inns are trying hard to avoid? Explain.
d. How do changes in the rejection criterion affect the probabilities of committing Type I and Type II errors?

9.93 A 1999 General Accounting Office (GAO) study found that about a third of the 23.4 million retirees 65 or older supplemented Medicare with some form of employer coverage (Carlos Tejada, "Work Week," *The Wall Street Journal*, June 26, 2002, B5). The article suggests that this proportion is increasing. Suppose that in a current study, a random sample of 500 retirees 65 or older indicated that 185 supplemented Medicare with some form of employer coverage.
a. At the 0.01 level of significance, is there evidence that the proportion of retirees 65 or older that supplement Medicare with some form of employer coverage is now greater than one-third?
b. Compute the *p*-value and interpret its meaning.

9.94 The owner of a gasoline station wants to study gasoline purchasing habits by motorists at his station. You select a random sample of 60 motorists during a certain week with the following results:
- Amount purchased: $\bar{X} = 11.3$ gallons, $S = 3.1$ gallons.
- 11 motorists purchased premium-grade gasoline.

a. At the 0.05 level of significance, is there evidence that the mean purchase is different from 10 gallons?
b. Find the *p*-value in (a).
c. At the 0.05 level of significance, is there evidence that fewer than 20% of all the motorists at his station purchase premium-grade gasoline?
d. What is your answer to (a) if the sample mean equals 10.3 gallons?
e. What is your answer to (c) if seven motorists purchased premium-grade gasoline?

9.95 An auditor for a government agency is assigned the task of evaluating reimbursement for office visits to physicians paid by Medicare. The audit is conducted on a sample of 75 of the reimbursements with the following results:
- In 12 of the office visits, an incorrect amount of reimbursement was provided.
- The amount of reimbursement was: $\bar{X} = \$93.70$, $S = \$34.55$.

a. At the 0.05 level of significance, is there evidence that the mean reimbursement is less than \$100?
b. At the 0.05 level of significance, is there evidence that the proportion of incorrect reimbursements in the population is greater than 0.10?
c. Discuss the underlying assumptions of the test used in (a).
d. What is your answer to (a) if the sample mean equals \$90?
e. What is your answer to (b) if 15 office visits had incorrect reimbursements?

9.96 A bank branch located in a commercial district of a city has developed an improved process for serving customers during the noon to 1:00 P.M. lunch period. The waiting time (defined as the time the customer enters the line until he or she reaches the teller window) of all customers during this hour is recorded over a period of 1 week. A random sample of 15 customers is selected, and the results are as follows: **BANK1**

4.21 5.55 3.02 5.13 4.77 2.34 3.54 3.20
4.50 6.10 0.38 5.12 6.46 6.19 3.79

a. At the 0.05 level of significance, is there evidence that the mean waiting time is less than 5 minutes?
b. What assumption must hold in order to perform the test in (a)?
c. Evaluate this assumption through a graphical approach. Discuss.
d. As a customer walks into the branch office during the lunch hour, she asks the branch manager how long she can expect to wait. The branch manager replies, "Almost certainly not longer than 5 minutes." On the basis of the results of (a), evaluate this statement.

9.97 A manufacturing company produces electrical insulators. If the insulators break when in use, a short circuit is likely to occur. To test the strength of the insulators, destructive testing is carried out to determine how much *force* is required to break the insulators. Force is measured by observing how many pounds is applied to the insulator before it breaks. The following data are from 30 observations from this experiment. **FORCE**

Force (in the Number of Pounds Required to Break the Insulator)

1,870	1,728	1,656	1,610	1,634	1,784	1,522	1,696	1,592	1,662
1,866	1,764	1,734	1,662	1,734	1,774	1,550	1,756	1,762	1,866
1,820	1,744	1,788	1,688	1,810	1,752	1,680	1,810	1,652	1,736

a. At the 0.05 level of significance, is there evidence that the mean force is greater than 1,500 pounds?
b. What assumption must hold in order to perform the test in (a)?
c. Evaluate this assumption through a graphical approach. Discuss.
d. Based on (a), what can you conclude about the strength of the insulators?

9.98 An important quality characteristic used by the manufacturer of "Boston" and "Vermont" asphalt shingles is the amount of moisture the shingles contain when they are packaged. Customers may feel that they have purchased a product lacking in quality if they find moisture and wet shingles inside the packaging. In some cases, excessive moisture can cause the granules attached to the shingle for texture and coloring purposes to fall off the shingle, resulting in appearance problems. To monitor the amount of moisture present, the company conducts moisture tests. A shingle is weighed and then dried. The shingle is then reweighed and, based on the amount of moisture taken out of the product, the pounds of moisture per 100 square feet are calculated. The company would like to show that the mean moisture content is less than 0.35 pounds per 100 square feet. The data file MOISTURE includes 36 measurements (in pounds per 100 square feet) for the "Boston" shingles and 31 for "Vermont" shingles.
a. For the "Boston" shingles, is there evidence at the 0.05 level of significance that the mean moisture content is less than 0.35 pounds per 100 square feet?
b. Interpret the meaning of the p-value in (a).
c. For the "Vermont" shingles, is there evidence at the 0.05 level of significance that the mean moisture content is less than 0.35 pounds per 100 square feet?
d. Interpret the meaning of the p-value in (c).
e. What assumption must hold in order to perform the tests in (a) and (c)?
f. Evaluate the assumption for the "Boston" shingles and the "Vermont" shingles by using a graphical approach. Discuss.

9.99 Studies conducted by a manufacturer of "Boston" and "Vermont" asphalt shingles have shown product weight to be a major factor in the customer's perception of quality. Moreover, the weight represents the amount of raw materials being used and is therefore very important to the company from a cost standpoint. The last stage of the assembly line packages the shingles before the packages are placed on wooden pallets. Once a pallet is full (a pallet for most brands holds 16 squares of shingles), it is weighed and the measurement is recorded. The data file PALLET contains the weight (in pounds) from a sample of 368 pallets of Boston shingles and 330 pallets of Vermont shingles.
a. For the Boston shingles, is there evidence that the mean weight is different from 3,150 pounds?
b. Interpret the meaning of the p-value in (a).
c. For the Vermont shingles, is there evidence that the mean weight is different from 3,700 pounds?
d. Interpret the meaning of the p-value in (c).
e. Is the assumption needed for (a) and (c) seriously violated?

9.100 The manufacturer of "Boston" and "Vermont" asphalt shingles provide their customers with a 20-year warranty on most of their products. To determine whether a shingle will last as long as the warranty period, accelerated-life testing is conducted at the manufacturing plant. Accelerated-life testing exposes the shingle to the stresses it would be subject to in a lifetime of normal use in a laboratory setting via an experiment that takes only a few minutes to conduct. In this test, a shingle is repeatedly scraped with a brush for a short period of time and the amount of shingle granules that are removed by the brushing is weighed (in grams). Shingles that experience low amounts of granule loss are expected to last longer in normal use than shingles that experience high amounts of granule loss. The data file GRANULE contains a sample of 170 measurements made on the company's Boston shingles, and 140 measurements made on Vermont shingles.
a. For the Boston shingles, is there evidence that the mean granule loss is different from 0.50 grams?
b. Interpret the meaning of the p-value in (a).
c. For the Vermont shingles, is there evidence that the mean granule loss is different from 0.50 grams?
d. Interpret the meaning of the p-value in (a).
e. Is the assumption needed for (a) and (c) seriously violated?

Report Writing Exercises

9.101 Referring to the results of problems 9.98–9.100 concerning the "Boston" and "Vermont" shingles, write a report that evaluates the moisture level, weight, and granule loss of the two types of shingles.

RUNNING CASE MANAGING THE *SPRINGVILLE HERALD*

Continuing its monitoring of the blackness of the newspaper print, first described in the Chapter 6 *Managing the Springville Herald* case, the production department of the newspaper wants to ensure that the mean blackness of the print for all newspapers is at least 0.97 on a standard scale in which the target value is 1.0. A random sample of 50 newspapers has been selected and the blackness of each paper has been measured. Calculate the sample statistics and determine if there is evidence that the mean blackness is less than 0.97. Write a memo to management that summarizes your conclusions.

Blackness of 50 Newspapers

0.854	1.023	1.005	1.030	1.219	0.977	1.044	0.778	1.122	1.114
1.091	1.086	1.141	0.931	0.723	0.934	1.060	1.047	0.800	0.889
1.012	0.695	0.869	0.734	1.131	0.993	0.762	0.814	1.108	0.805
1.223	1.024	0.884	0.799	0.870	0.898	0.621	0.818	1.113	1.286
1.052	0.678	1.162	0.808	1.012	0.859	0.951	1.112	1.003	0.972

SH9

WEB CASE

Apply your knowledge about hypothesis testing in this Web Case that continues the cereal-fill-packaging dispute Web Case from Chapter 7.

Oxford Cereals recently conducted a public experiment in which it claims it has successfully debunked the statements of groups such as the TriCities Consumers Concerned About Cereal Companies That Cheat (TCCACCTC) that claimed that Oxford Cereals was cheating consumers by packaging cereals at less than labeled weights. Review the Oxford Cereals' press release and supporting documents that describe the experiment at the company's Web site **www.prenhall.com/Springville/OC_WinTrust.htm** and then answer the following:

1. Are the results of the independent testing valid? Why, or why not? If you were conducting the experiment, is there anything you would change?
2. Do the results support the claim that Oxford Cereals is not cheating its customers?
3. Is the claim of the Oxford CEO that many cereal boxes contain *more* than 368 grams surprising? True?
4. Could there ever be a circumstance in which the results of the public experiment *and* the TCCACCTC's results are both correct? Explain.

REFERENCES

1. Bradley, J. V., *Distribution-Free Statistical Tests* (Englewood Cliffs, NJ: Prentice Hall, 1968).
2. Daniel, W., *Applied Nonparametric Statistics*, 2nd ed. (Boston, MA: Houghton Mifflin, 1990).
3. *Microsoft Excel 2003* (Redmond, WA: Microsoft Corporation, 2003).
4. *Minitab for Windows Version 14* (State College PA: Minitab, Inc., 2004).
5. *SPSS Base 12.0 Brief Guide* (Upper Saddle River, NJ: Prentice Hall, 2003).

Appendix 9 Using Software for One-Sample Tests of Hypothesis

A9.1 MICROSOFT EXCEL

For *Z* Test for the Mean, Sigma Known

Open the **Z Mean.xls** file, shown in Figure 9.5 on page 310. This worksheet already contains the entries for Example 9.2 concerning the Oxford Cereal Company. This worksheet uses the NORMSINV and NORMSDIST functions (see section G.21 for more information). To adapt this worksheet to other problems, change the null hypothesis, level of significance, population standard deviation, sample size, and sample mean values in the tinted cells in rows 4 through 8.

OR See section G.21 (**Z Test for the Mean, sigma known**) if you want PHStat2 to produce a worksheet for you.

For *t* Test for the Mean, Sigma Unknown

Open the **T Mean.xls** file, shown in Figure 9.10 on page 320. This worksheet already contains the entries for the section 9.4 sales invoices example. This worksheet uses the TINV and TDIST functions (see section G.22 for more information). To adapt this worksheet to other problems, change the null hypothesis, level of significance, sample size, sample mean, and sample standard deviation values in the tinted cells in rows 4 through 8.

OR See section G.22 (**t Test for the Mean, sigma unknown**) if you want PHStat2 to produce a worksheet for you.

For *Z* Test for the Proportion

Open the **Z Proportion.xls** file, shown in Figure 9.16 on page 327. This worksheet already contains the entries for the business ownership example of section 9.5. This worksheet uses the NORMSINV and NORMSDIST functions (see section G.23 for more information). To adapt this worksheet to other problems, change the null hypothesis, level of significance, number of successes, and sample size values in the tinted cells in rows 4 through 7.

OR See section G.23 (**Z Test for the Proportion**) if you want PHStat2 to produce a worksheet for you.

A9.2 MINITAB

Using Minitab for the *Z* Test of Hypothesis for the Mean (σ Known)

You can use Minitab for a test of the mean when σ is known. To illustrate the test of hypothesis for the mean when σ is known, return to the Oxford Cereal Company example discussed in section 9.2. Open a blank worksheet. Select **Stat → Basic Statistics → 1-Sample *Z*.**

1. In the 1-Sample *Z* dialog box (see Figure A9.1) select the **Summarized data** option button. Enter **25** in the Sample size: edit box. Enter **372.5** in the Mean: edit box.

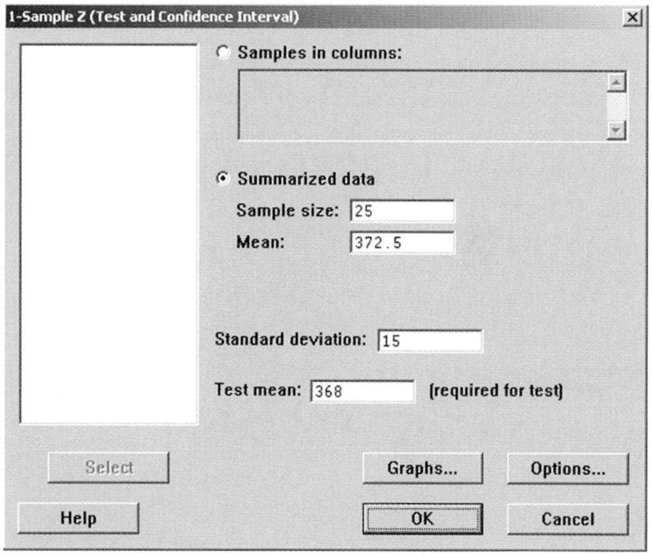

FIGURE A9.1 Minitab 1-Sample *Z* Dialog Box

2. Enter **15** in the Standard deviation: edit box.
3. Enter **368** in the Test mean: edit box.
4. Select the **Options** button. In the 1-Sample Z options dialog box (see Figure A9.2) in the Alternative: drop-down list box, select **not equal** to perform a two-tail test. (Select less than or greater than to perform a one-tail test.) Click the **OK** button to return to the 1-Sample *Z* dialog box. Click the **OK** button.

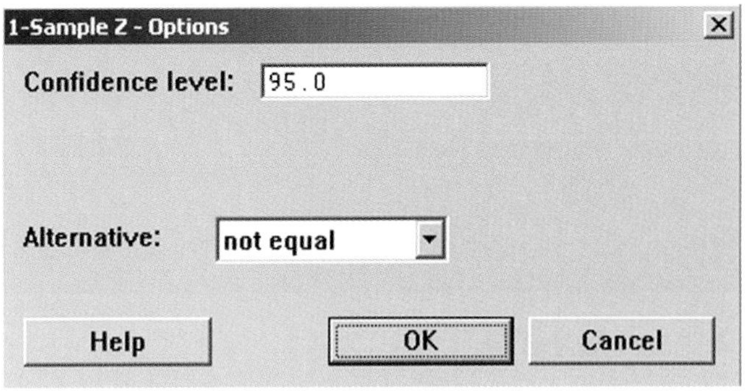

FIGURE A9.2 Minitab 1-Sample Z-Options Dialog Box

Using Minitab for the t Test of Hypothesis for the Mean (σ Unknown)

You can use Minitab for a test of the mean when σ is unknown.

To illustrate the test of hypothesis for the mean when σ is unknown, return to the Saxon Home Improvement Company invoice example discussed in section 9.4. Open the **INVOICES.MTW** worksheet. Select **Stat ➔ Basic Statistics ➔ 1-Sample *t***.

1. In the 1-Sample *t* dialog box (see Figure A9.3) select the Sample in columns: option button. Enter **C1** or **Amount** in the Sample in columns: edit box. In the Test mean: edit box, enter **120**. Select the **Options** button.
2. In the 1-Sample *t* Options dialog box in the Alternative: drop-down list box, select **not equal** to perform a two-tail test. (Select less than or greater than to perform a one-tail test.) Click the **OK** button to return to the 1-Sample *t* dialog box. (To get a box-and-whisker plot, select the **Graphs** option button. In the Minitab 1-sample t - Graphs dialog box, select the **Boxplot of data** option button. Click the **OK** button to return to the 1-Sample *t* dialog box.) Click the **OK** button.

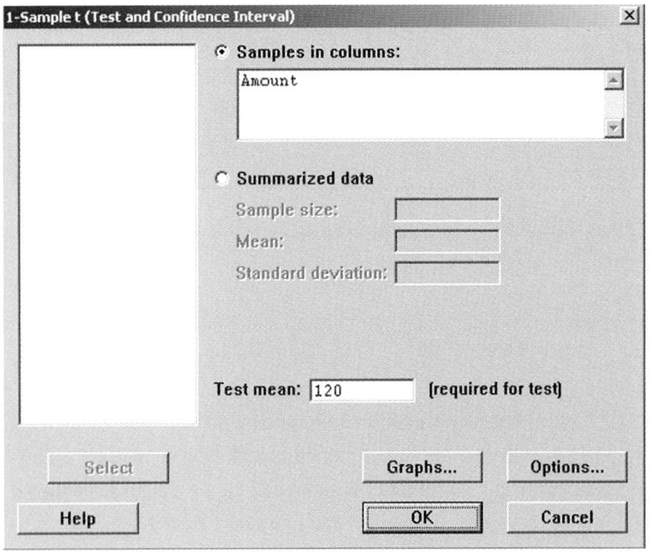

FIGURE A9.3 Minitab 1-Sample *t* Dialog Box

Using Minitab for the *Z* Test for the Proportion

You can use Minitab for the test of hypothesis for the proportion. For example, to perform the *Z* test for the proportion of female-owned business example in section 9.5, select **Stat ➔ Basic Statistics ➔ 1 Proportion**.

1. In the 1 Proportion dialog box (see Figure A9.4), select the **Summarized data** option button. Enter **899** in the Number of trials: edit box. Enter **369** in the Number of events: edit box. Click the **Options** button.

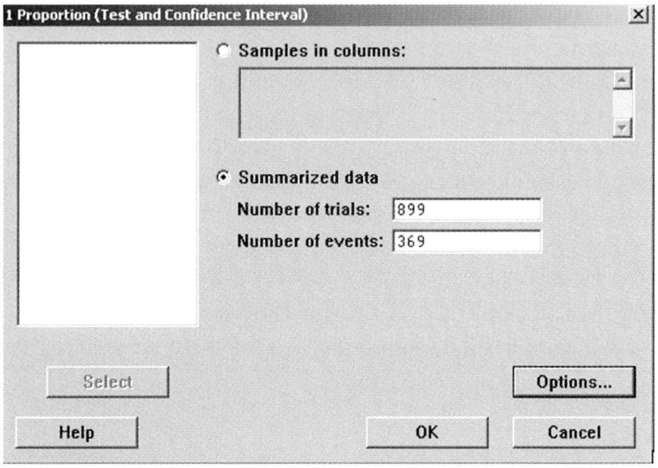

FIGURE A9.4 Minitab 1 Proportion Dialog Box

2. In the 1 Proportion-Options dialog box (see Figure A9.5), enter **0.5** in the Test proportion: edit box. Select the **Use test and interval based on normal distribution** check box. In the Alternative: drop-down list box, select **not equal** to perform a two-tail test. (Select less than or greater than to perform a one-tail test.) Click the **OK** button to return to the 1 Proportion dialog box. Click the **OK** button.

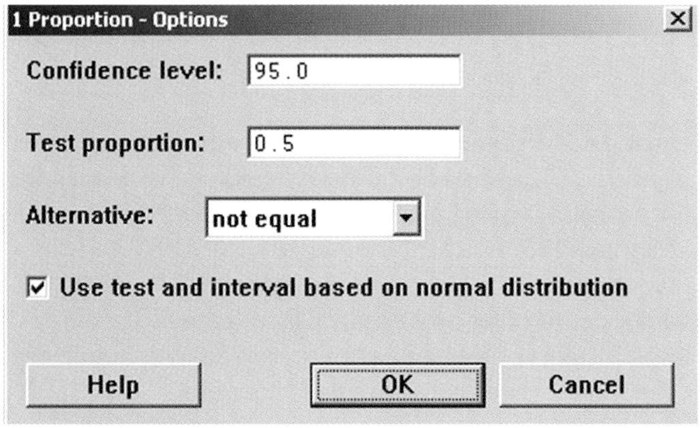

FIGURE A9.5 Minitab 1 Proportion - Options Dialog Box

CHAPTER 10

Two-Sample Tests

USING STATISTICS: Comparing Sales from End-Aisle Displays and Normal Displays

LEARNING OBJECTIVES

In this chapter, you learn how to use hypothesis testing for comparing the difference between:

- The means of two independent populations
- The means of two related populations
- Two proportions
- The variances of two independent populations

USING STATISTICS

Comparing Sales from End-Aisle Displays and Normal Displays

Does the type of display used in a supermarket affect the sales of products? As the regional sales manager for BLK Foods, you are interested in comparing the sales volume of BLK cola when you display the soft drink in the normal shelf location as compared to an end-aisle promotional display. To test the effectiveness of the end-aisle displays, you select 20 stores from the BLK supermarket chain that all experience similar storewide sales volumes. You then randomly assign 10 of the 20 stores to group 1 and 10 to group 2. The managers of the 10 stores in group 1 place the BLK cola in the regular shelf location alongside the other cola products. The 10 stores in group 2 use the special end-aisle promotional display. At the end of one week, the sales of BLK cola are recorded. How can you determine whether sales of BLK cola using the end-aisle displays are the same as those when the cola is placed in the regular shelf location? How can you decide if the variability in BLK cola sales from store to store is the same for the two types of displays? How could you use the answers to these questions to improve sales of BLK colas?

Hypothesis testing provides a *confirmatory* approach to data analysis. In Chapter 9 the focus was on a variety of commonly used hypothesis-testing procedures that relate to a single sample of data selected from a single population. In this chapter, hypothesis testing is extended to procedures that compare statistics from two samples of data drawn from two populations. For example, are the mean weekly sales of BLK cola when using an end-aisle display equal to the mean weekly sales of BLK cola when placed in the normal shelf location?

10.1 COMPARING THE MEANS OF TWO INDEPENDENT POPULATIONS

Z Test for the Difference Between Two Means

Suppose that you take a random sample of n_1 from the first population and a random sample of n_2 from the second population, and the data collected in each sample are from a numerical variable. In the first population, the mean is represented by the symbol μ_1 and the standard deviation is represented by the symbol σ_1. In the second population, the mean is represented by the symbol μ_2 and the standard deviation is represented by the symbol σ_2.

The test statistic used to determine the difference between the population means is based on the difference between the sample means $(\bar{X}_1 - \bar{X}_2)$. If you assume that the samples are randomly and independently selected from populations that are normally distributed, this statistic follows the standardized normal distribution. If the populations are not normally distributed, the Z test is still appropriate if the sample sizes are large enough (typically n_1 and $n_2 \geq 30$; see the Central Limit Theorem in section 7.2). Equation (10.1) defines the ***Z* test for the difference between two means**.

Z TEST FOR THE DIFFERENCE BETWEEN TWO MEANS

$$Z = \frac{(\bar{X}_1 - \bar{X}_2) - (\mu_1 - \mu_2)}{\sqrt{\dfrac{\sigma_1^2}{n_1} + \dfrac{\sigma_2^2}{n_2}}} \qquad \textbf{(10.1)}$$

where $\bar{X}_1$ = mean of the sample taken from population 1

μ_1 = mean of population 1

σ_1^2 = variance of population 1

n_1 = size of the sample taken from population 1

$\bar{X}_2$ = mean of the sample taken from population 2

μ_2 = mean of population 2

σ_2^2 = variance of population 2

n_2 = size of the sample taken from population 2

The test statistic Z follows a standardized normal distribution.

Pooled-Variance *t* Test for the Difference Between Two Means

In most cases the variances of the two populations are not known. The only information you usually have are the sample means and the sample variances. If you assume that the samples are randomly and independently selected from populations that are normally distributed and that the population variances are equal (that is, $\sigma_1^2 = \sigma_2^2$), you can use a **pooled-variance *t* test** to determine whether there is a significant difference between the means of the two populations. If the populations are not normally distributed, the pooled-variance t test is still appropriate if the sample sizes are large enough (typically n_1 and $n_2 \geq 30$; see the Central Limit Theorem in section 7.2).

The pooled-variance t test is so named because the test statistic pools (combines) the two sample variances S_1^2 and S_2^2 to compute S_p^2; the best estimate of the variance common to both populations under the assumption that the two population variances are equal.[1]

To test the null hypothesis of no difference in the means of two independent populations:

$$H_0: \mu_1 = \mu_2 \text{ or } \mu_1 - \mu_2 = 0$$

against the alternative that the means are not the same

$$H_1: \mu_1 \neq \mu_2 \text{ or } \mu_1 - \mu_2 \neq 0$$

you use the pooled-variance t-test statistic.

[1] *When the two sample sizes are equal (that is, $n_1 = n_2$), the formula for the pooled variance can be simplified to* $S_p^2 = \dfrac{S_1^2 + S_2^2}{2}$

POOLED-VARIANCE *t* TEST FOR THE DIFFERENCE BETWEEN TWO MEANS

$$t = \frac{(\bar{X}_1 - \bar{X}_2) - (\mu_1 - \mu_2)}{\sqrt{S_p^2\left(\dfrac{1}{n_1} + \dfrac{1}{n_2}\right)}} \qquad \textbf{(10.2)}$$

where $$S_p^2 = \frac{(n_1 - 1)S_1^2 + (n_2 - 1)S_2^2}{(n_1 - 1) + (n_2 - 1)}$$

and S_p^2 = pooled variance

$\bar{X}_1$ = mean of the sample taken from population 1

S_1^2 = variance of the sample taken from population 1

n_1 = size of the sample taken from population 1

$\bar{X}_2$ = mean of the sample taken from population 2

S_2^2 = variance of the sample taken from population 2

n_2 = size of the sample taken from population 2

The test statistic t follows a t distribution with $n_1 + n_2 - 2$ degrees of freedom.

The pooled-variance t-test statistic follows a t distribution with $n_1 + n_2 - 2$ degrees of freedom. For a given level of significance α, in a two-tail test, you reject the null hypothesis if the computed t-test statistic is greater than the upper-tail critical value from the t distribution or if the computed test statistic is less than the lower-tail critical value from the t distribution. Figure 10.1 displays the regions of rejection. In a one-tail test in which the rejection region is in the lower tail, you reject the null hypothesis if the computed test statistic is less than the lower-tail critical value from the t distribution. In a one-tail test in which the rejection region is in the upper tail, you reject the null hypothesis if the computed test statistic is greater than the upper-tail critical value from the t distribution.

FIGURE 10.1

Regions of Rejection and Nonrejection for the Pooled-Variance t Test for the Difference Between the Means (Two-Tail Test)

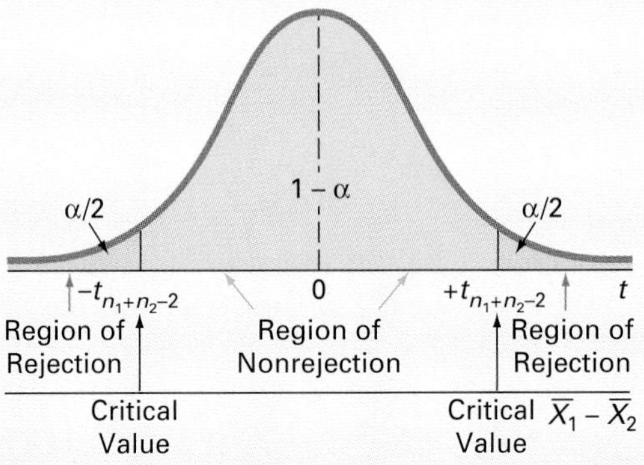

To demonstrate the use of the pooled-variance t test, return to the "Using Statistics" scenario on page 346. The question you want to answer is whether the mean weekly sales of BLK cola are the same when using a normal shelf location and when using an end-aisle display. There are two populations of interest. The first population is the set of all possible weekly sales of BLK cola *if* all the BLK supermarkets used the normal shelf location. The second population is the set of all possible weekly sales of BLK cola *if* all the BLK supermarkets used the end-aisle displays. The first sample contains the weekly sales of BLK cola from the 10 stores selected to use the normal shelf location, while the second sample contains the weekly sales of BLK cola from the 10 stores selected to use the end-aisle display. Table 10.1 contains the cola sales (in number of cases) for the two samples COLA.

TABLE 10.1

Comparing BLK Cola Weekly Sales from Two Different Display Locations (in Number of Cases)

Display Location									
Normal					End-Aisle				
22	34	52	62	30	52	71	76	54	67
40	64	84	56	59	83	66	90	77	84

The null and alternative hypotheses are

$$H_0: \mu_1 = \mu_2 \text{ or } \mu_1 - \mu_2 = 0$$

$$H_1: \mu_1 \neq \mu_2 \text{ or } \mu_1 - \mu_2 \neq 0$$

Assuming that the samples are from underlying normal populations having equal variances, you can use the pooled-variance t test. The t-test statistic follows a t distribution with $10 + 10 - 2 = 18$ degrees of freedom. Using $\alpha = 0.05$ level of significance, you divide the rejection region into the two tails for this two-tail test (i.e., two equal parts of 0.025 each). Table E.3 shows that the critical values for this two-tail test are +2.1009 and −2.1009. As shown in Figure 10.2, the decision rule is:

$$\text{Reject } H_0 \text{ if } t > t_{18} = +2.1009$$

$$\text{or if } t < -t_{18} = -2.1009;$$

$$\text{otherwise do not reject } H_0.$$

FIGURE 10.2
Two-Tail Test of Hypothesis for the Difference Between the Means at the 0.05 Level of Significance with 18 Degrees of Freedom

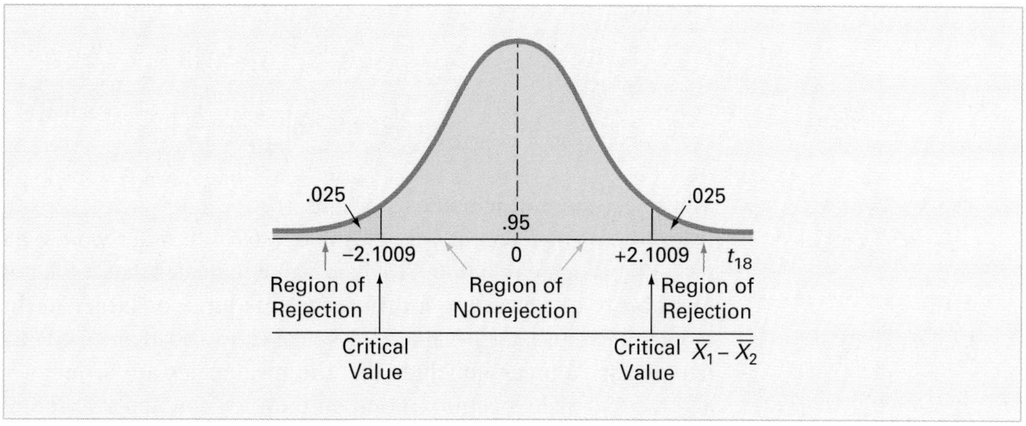

From Figures 10.3 or 10.4, the computed t statistic for this test is −3.04 and the p-value is 0.007.

FIGURE 10.3
Microsoft Excel t-Test Output for the Two Display Locations

	A	B	C
1	t-Test: Two-Sample Assuming Equal Variances		
2			
3		*Normal*	*End-Aisle*
4	Mean	50.3	72
5	Variance	350.6778	157.3333
6	Observations	10	10
7	Pooled Variance	254.0055556	
8	Hypothesized Mean Difference	0	
9	df	18	
10	t Stat	-3.04455	
11	P(T<=t) one-tail	0.00349	
12	t Critical one-tail	1.73406	
13	P(T<=t) two-tail	0.00697	
14	t Critical two-tail	2.10092	

FIGURE 10.4
Minitab t-Test Output for the Two Display Locations

```
Two-sample T for Sales_Normal vs Sales_EndAisle

                 N  Mean  StDev  SE Mean
Sales_Normal    10  50.3   18.7      5.9
Sales_EndAisle  10  72.0   12.5      4.0


Difference = mu (Sales_Normal) - mu (Sales_EndAisle)
Estimate for difference:  -21.7000
95% CI for difference:  (-36.6743, -6.7257)
T-Test of difference = 0 (vs not =): T-Value = -3.04  P-Value = 0.007  DF = 18
Both use Pooled StDev = 15.9376
```

Using Equation (10.2) on page 347 and the descriptive statistics provided in Figure 10.3 or 10.4

$$t = \frac{(\overline{X}_1 - \overline{X}_2) - (\mu_1 - \mu_2)}{\sqrt{S_p^2\left(\frac{1}{n_1} + \frac{1}{n_2}\right)}}$$

where

$$S_p^2 = \frac{(n_1 - 1)S_1^2 + (n_2 - 1)S_2^2}{(n_1 - 1) + (n_2 - 1)}$$

$$= \frac{9(350.6778) + 9(157.3333)}{9 + 9} = 254.0056$$

Therefore,

$$t = \frac{(50.3 - 72.0) - 0.0}{\sqrt{254.0056\left(\frac{1}{10} + \frac{1}{10}\right)}} = \frac{-21.7}{\sqrt{50.801}} = -3.04$$

You reject the null hypothesis because $t = -3.04 < t_{18} = -2.1009$. The p-value (as computed from Microsoft Excel or Minitab) is 0.00697. In other words, the probability that $t > 3.04$ or $t < -3.04$ is equal to 0.00697. This p-value indicates that if the population means are equal, the probability of observing a difference this large or larger in the two sample means is only 0.00697. Because the p-value is less than $\alpha = 0.05$, there is sufficient evidence to reject the null hypothesis. You can conclude that the mean sales are different for the normal shelf location and the end-aisle location. Based on these results, the sales are lower for the normal location (i.e., higher for the end-aisle location).

EXAMPLE 10.1

TESTING FOR THE DIFFERENCE IN THE MEAN DELIVERY TIMES

A local pizza restaurant located close to a college campus advertises that their delivery time to a college dormitory is less than for a local branch of a national pizza chain. In order to determine whether this advertisement is valid, you and some friends have decided to order 10 pizzas from the local pizza restaurant and 10 pizzas from the national chain, all at different times. The delivery times in minutes PIZZATIME are shown in Table 10.2:

TABLE 10.2

Delivery Times for Local Pizza Restaurant and National Pizza Chain

Local	Chain	Local	Chain
16.8	22.0	18.1	19.5
11.7	15.2	14.1	17.0
15.6	18.7	21.8	19.5
16.7	15.6	13.9	16.5
17.5	20.8	20.8	24.0

At the 0.05 level of significance, is there evidence that the mean delivery time is lower for the local pizza restaurant than for the national pizza chain?

SOLUTION Since you want to know whether the mean is *lower* for the local pizza restaurant than for the national pizza chain, you have a one-tail test with the following null and alternative hypotheses:

$H_0: \mu_1 \geq \mu_2$ (The mean time for the local pizza restaurant is equal to or higher than for the national pizza chain)
$H_1: \mu_1 < \mu_2$ (The mean time for the local pizza restaurant is less than for the national pizza chain)

Figure 10.5 displays Microsoft Excel output of the pooled t-test for these data.

FIGURE 10.5

Microsoft Excel Output of the Pooled t-Test for the Pizza Delivery Time Data

	A	B	C
1	t-Test: Two-Sample Assuming Equal Variances		
2			
3		*Local*	*Chain*
4	Mean	16.7	18.88
5	Variance	9.582222	8.215111
6	Observations	10	10
7	Pooled Variance	8.898667	
8	Hypothesized Mean Difference	0	
9	df	18	
10	t Stat	-1.6341	
11	P(T<=t) one-tail	0.059803	
12	t Critical one-tail	1.734063	
13	P(T<=t) two-tail	0.119606	
14	t Critical two-tail	2.100924	

Using Equation (10.2) on page 347,

$$t = \frac{(\bar{X}_1 - \bar{X}_2) - (\mu_1 - \mu_2)}{\sqrt{S_p^2\left(\frac{1}{n_1} + \frac{1}{n_2}\right)}}$$

where

$$S_p^2 = \frac{(n_1 - 1)S_1^2 + (n_2 - 1)S_2^2}{(n_1 - 1) + (n_2 - 1)}$$

$$= \frac{9(9.5822) + 9(8.2151)}{9 + 9} = 8.8987$$

Therefore,

$$t = \frac{(16.7 - 18.88) - 0.0}{\sqrt{8.8987\left(\frac{1}{10} + \frac{1}{10}\right)}} = \frac{-2.18}{\sqrt{1.7797}} = -1.634$$

You do not reject the null hypothesis because $t = -1.634 > t_{18} = -1.734$. The p-value (as computed from Microsoft Excel) is 0.0598. In other words, the probability that $t < -1.634$ is equal to 0.0598. This p-value indicates that if the population means are equal, the probability that the t statistic will be less than -1.634 is 0.0598. Because the p-value is greater than $\alpha = 0.05$, there is insufficient evidence to reject the null hypothesis. Based on these results, there is insufficient evidence for the local pizza restaurant to make the advertising claim that they have a lower delivery time.

In testing for the difference between the means, you assume that the populations are normally distributed with equal variances. For situations in which the two populations have equal variances, the pooled-variance t test is **robust** (i.e., not sensitive) to moderate departures from the assumption of normality, provided that the sample sizes are large. In such situations, you can use the pooled-variance t test without serious effects on its power. However, if you cannot assume that the data in each group are from normally distributed populations, you have two choices. You can use a nonparametric procedure, such as the Wilcoxon rank sum test (covered in section 12.7), that does not depend on the assumption of normality for the two populations, or you can use a normalizing transformation (see reference 6) on each of the outcomes and then use the pooled-variance t test.

To check the assumption of normality in each of the two groups, observe the box-and-whisker plot of the sales for the two aisle locations in Figure 10.6. There appears to be only moderate departure from normality, so the assumption of normality needed for the t test is not seriously violated.

FIGURE 10.6

Minitab Box-and-Whisker Plot of the Sales for Two Aisle Locations

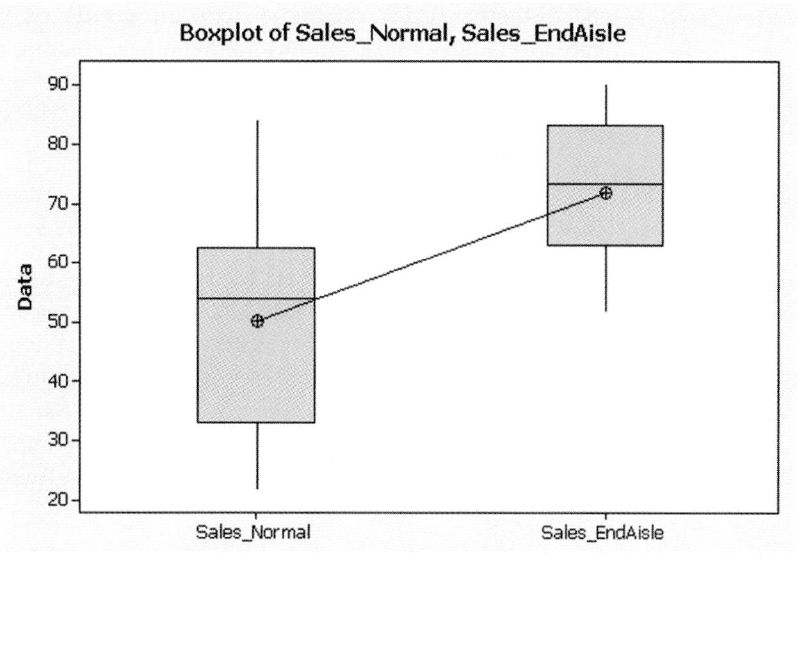

Confidence Interval Estimate for the Difference Between the Means of Two Independent Populations

Instead of, or in addition to, testing for the difference in the means of two independent populations, you can use Equation (10.3) to develop a confidence interval estimate of the difference in the means.

CONFIDENCE INTERVAL ESTIMATE OF THE DIFFERENCE IN THE MEANS OF TWO INDEPENDENT POPULATIONS

$$(\bar{X}_1 - \bar{X}_2) \pm t_{n_1+n_2-2}\sqrt{S_p^2\left(\frac{1}{n_1}+\frac{1}{n_2}\right)} \quad \textbf{(10.3)}$$

or

$$(\bar{X}_1 - \bar{X}_2) - t_{n_1+n_2-2}\sqrt{S_p^2\left(\frac{1}{n_1}+\frac{1}{n_2}\right)} \le \mu_1 - \mu_2 \le (\bar{X}_1 - \bar{X}_2) + t_{n_1+n_2-2}\sqrt{S_p^2\left(\frac{1}{n_1}+\frac{1}{n_2}\right)}$$

where $t_{n_1+n_2-2}$ is the critical value of the t distribution with $n_1 + n_2 - 2$ degrees of freedom for an area of $\alpha/2$ in the upper tail.

Using 95% confidence, the sample statistics reported in Figures 10.3 or 10.4 on page 349, and Equation (10.3),

$$\bar{X}_1 = 50.3, n_1 = 10, \bar{X}_2 = 72, n_2 = 10, S_p^2 = 254.0056, \text{ and } t_{18} = 2.1009:$$

$$(50.3 - 72) \pm (2.1009)\sqrt{254.0056\left(\frac{1}{10}+\frac{1}{10}\right)}$$

$$-21.7 \pm (2.1009)(7.1275)$$

$$-21.7 \pm 14.97$$

$$-36.67 \le \mu_1 - \mu_2 \le -6.73$$

Therefore, you are 95% confident that the difference in mean sales between the normal aisle location and the end-aisle location is between −36.67 cases of cola and −6.73 cases of cola. In other words, the end-aisle location sells, on average, 6.73 to 36.67 cases more than the normal aisle location. From a hypothesis testing perspective, because the interval does not include zero, you reject the null hypothesis of no difference between the means of the two populations.

Separate-Variance t Test for the Difference Between Two Means

In testing for the difference between the means of two independent populations when the population variances are assumed equal, the sample variances are pooled together into a common estimate S_p^2. However, if you cannot make this assumption, then the pooled-variance t test is inappropriate. In this case, it is more appropriate to use the **separate-variance *t* test** developed by Satterthwaite (see reference 5). In the Satterthwaite approximation procedure, you include the two separate sample variances in the computation of the t-test statistic—hence, the name separate-variance t test. Equation (10.4) defines the test statistic for the separate-variance t test.

SEPARATE-VARIANCE *t* TEST FOR THE DIFFERENCE BETWEEN TWO MEANS

$$t = \frac{(\bar{X}_1 - \bar{X}_2) - (\mu_1 - \mu_2)}{\sqrt{\dfrac{S_1^2}{n_1} + \dfrac{S_2^2}{n_2}}} \quad \textbf{(10.4)}$$

where

$\bar{X}_1$ = mean of the sample taken from population 1
S_1^2 = variance of the sample taken from population 1
n_1 = size of the sample taken from population 1
$\bar{X}_2$ = mean of the sample taken from population 2
S_2^2 = variance of the sample taken from population 2
n_2 = size of the sample taken from population 2

The separate-variance *t*-test statistic approximately follows a *t* distribution with degrees of freedom ν equal to the integer portion of the following computation.

COMPUTING DEGREES OF FREEDOM IN THE SEPARATE-VARIANCE *t* TEST

$$\nu = \frac{\left(\dfrac{S_1^2}{n_1} + \dfrac{S_2^2}{n_2}\right)^2}{\dfrac{\left(\dfrac{S_1^2}{n_1}\right)^2}{n_1 - 1} + \dfrac{\left(\dfrac{S_2^2}{n_2}\right)^2}{n_2 - 1}} \quad \textbf{(10.5)}$$

For a given level of significance α, for a two-tail test, you reject the null hypothesis if the computed *t*-test statistic is greater than the upper-tail critical value t_ν from the *t* distribution or if the computed test statistic is less than the lower-tail critical value $-t_\nu$ from the *t* distribution. Thus, the decision rule is

Reject H_0 if $t > t_\nu$
or if $t < -t_\nu$;
otherwise do not reject H_0.

Return to the "Using Statistics" scenario concerning the aisle location of BLK cola. Using Equation (10.5), the separate-variance *t*-test statistic is approximated by a *t* distribution with $\nu = 15$ degrees of freedom, the integer portion of the following computation.

$$\nu = \frac{\left(\dfrac{S_1^2}{n_1} + \dfrac{S_2^2}{n_2}\right)^2}{\dfrac{\left(\dfrac{S_1^2}{n_1}\right)^2}{n_1 - 1} + \dfrac{\left(\dfrac{S_2^2}{n_2}\right)^2}{n_2 - 1}}$$

$$= \frac{\left(\dfrac{350.6778}{10} + \dfrac{157.3333}{10}\right)^2}{\dfrac{\left(\dfrac{350.6778}{10}\right)^2}{9} + \dfrac{\left(\dfrac{157.3333}{10}\right)^2}{9}} = 15.72$$

Using $\alpha = 0.05$, the upper- and lower-critical values for this two-tail test found in Table E.3 are, respectively, +2.1315 and −2.1315. As depicted in Figure 10.7, the decision rule is

$$\text{Reject } H_0 \text{ if } t > t_{15} = +2.1315$$
$$\text{or if } t < -t_{15} = -2.1315;$$
$$\text{otherwise do not reject } H_0.$$

FIGURE 10.7
Two-Tail Test of Hypothesis for the Difference Between the Means at the 0.05 Level of Significance with 15 Degrees of Freedom

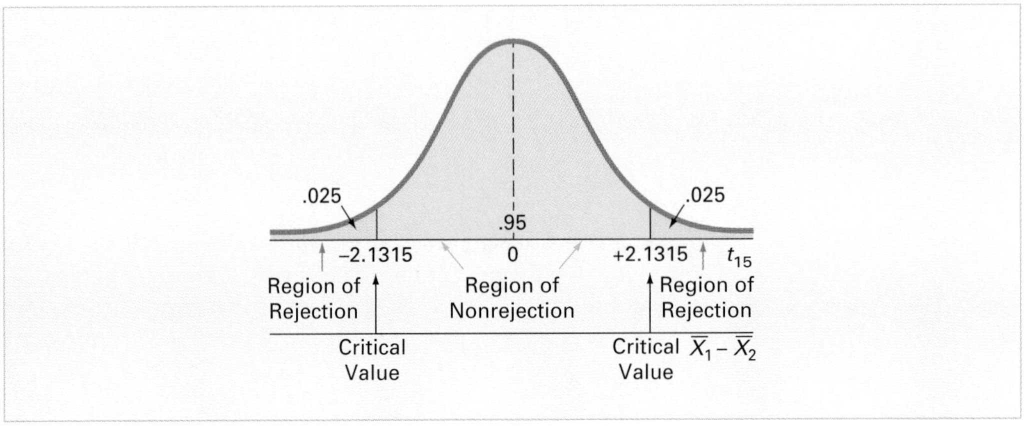

Using Equation (10.4) on page 353 and the descriptive statistics provided in Figures 10.3 and 10.4 on page 349,

$$t = \frac{(\overline{X}_1 - \overline{X}_2) - (\mu_1 - \mu_2)}{\sqrt{\dfrac{S_1^2}{n_1} + \dfrac{S_2^2}{n_2}}}$$

$$= \frac{50.3 - 72}{\sqrt{\left(\dfrac{350.6778}{10} + \dfrac{157.3333}{10}\right)}} = \frac{-21.7}{\sqrt{50.801}} = -3.04$$

Using a 0.05 level of significance, you reject the null hypothesis because $t = -3.04 < -t_{15} = -2.1315$.

Figure 10.8 illustrates Microsoft Excel output for this separate-variance t test, while Figure 10.9 illustrates Minitab output.

FIGURE 10.8
Microsoft Excel Output of the Separate-Variance t Test for the Aisle Location Data

	A	B	C
1	**t-Test: Two-Sample Assuming Unequal Variances**		
2			
3		***Normal***	***End-Aisle***
4	**Mean**	**50.3**	**72**
5	**Variance**	**350.6778**	**157.3333**
6	**Observations**	**10**	**10**
7	**Hypothesized Mean Difference**	**0**	
8	**df**	**16**	
9	**t Stat**	**-3.04455**	
10	**P(T<=t) one-tail**	**0.00386**	
11	**t Critical one-tail**	**1.74588**	
12	**P(T<=t) two-tail**	**0.00773**	
13	**t Critical two-tail**	**2.11990**	

FIGURE 10.9
Minitab Output of the Separate-Variance *t* Test for the Aisle Location Data

```
Two-sample T for Sales_Normal vs Sales_EndAisle

                 N  Mean  StDev  SE Mean
Sales_Normal    10  50.3   18.7      5.9
Sales_EndAisle  10  72.0   12.5      4.0


Difference = mu (Sales_Normal) - mu (Sales_EndAisle)
Estimate for difference:  -21.7000
95% CI for difference:  (-36.8919, -6.5081)
T-Test of difference = 0 (vs not =): T-Value = -3.04  P-Value = 0.008  DF = 15
```

In Figures 10.8 and 10.9 the test statistic is $t = -3.04$ and the p-value is $0.008 < 0.05$. Thus, the results for the separate-variance t test are almost exactly the same as those of the pooled-variance t test. The assumption of equality of population variances had no real effect on the results. Sometimes, however, the results from the pooled- and separate-variance t tests conflict because the assumption of equal variances is violated. Therefore, it is important that you evaluate the assumptions and use those results as a guide in appropriately selecting a test procedure. In section 10.4 the F test to determine whether there is evidence of a difference in the two population variances is discussed. The results of that test can help you determine which of the t tests (pooled-variance or separate-variance) is more appropriate.

PROBLEMS FOR SECTION 10.1

Learning the Basics

PH Grade ASSIST **10.1** Given a sample of $n_1 = 40$ from a population with known standard deviation $\sigma_1 = 20$, and an independent sample of $n_2 = 50$ from another population with known standard deviation $\sigma_2 = 10$, what is the value of the Z-test statistic for testing H_0: $\mu_1 = \mu_2$ if $\bar{X}_1 = 72$ and $\bar{X}_2 = 66$?

PH Grade ASSIST **10.2** What is your decision in problem 10.1 if you are testing H_0: $\mu_1 = \mu_2$ against the two-tail alternative H_1: $\mu_1 \neq \mu_2$ using a level of significance $\alpha = 0.01$?

PH Grade ASSIST **10.3** What is the p-value in problem 10.1 if you are testing H_0: $\mu_1 = \mu_2$ against the two-tail alternative H_1: $\mu_1 \neq \mu_2$?

PH Grade ASSIST **10.4** Assume that you have a sample of $n_1 = 8$ with sample mean $\bar{X}_1 = 42$ and a sample standard deviation of $S_1 = 4$, and you have an independent sample of $n_2 = 15$ from another population with a sample mean of $\bar{X}_2 = 34$ and sample standard deviation $S_2 = 5$,

a. What is the value of the pooled-variance t-test statistic for testing H_0: $\mu_1 = \mu_2$?
b. In finding the critical value of the test statistic t, how many degrees of freedom are there?
c. Using a level of significance $\alpha = 0.01$, what is the critical value for a one-tail test of the hypothesis H_0: $\mu_1 \leq \mu_2$ against the alternative H_1: $\mu_1 > \mu_2$?
d. What is your statistical decision?

10.5 What assumptions about the two populations are necessary in problem 10.4?

10.6 Referring to problem 10.4, construct a 95% confidence interval estimate of the population difference between μ_1 and μ_2.

Applying the Concepts

Problems 10.7–10.17 can be solved manually or by using Microsoft Excel, Minitab, or SPSS. We recommend that you use Microsoft Excel, Minitab, or SPSS to solve problems 10.18–10.21.

PH Grade ASSIST **10.7** The operations manager at a lightbulb factory wants to determine whether there is any difference in the mean life expectancy of bulbs manufactured on two different types of machines. The population standard deviation of machine I is 110 hours and of machine II is 125 hours. A random sample of 25 lightbulbs from machine I indicates a sample mean of 375 hours, and a similar sample of 25 from machine II indicates a sample mean of 362 hours.

a. Using the 0.05 level of significance, is there any evidence of a difference in the mean life of bulbs produced by the two types of machines?
b. Compute the p-value in (a) and interpret its meaning.

PH Grade ASSIST **10.8** The purchasing director for an industrial parts factory is investigating the possibility of purchasing a new type of milling machine. She determines that the new machine will be bought if there is evidence that the parts produced have a higher mean breaking strength than those from the old machine. The population standard deviation of the breaking strength for the old machine is 10 kilograms and for the new machine is 9 kilograms. A sample of 100 parts taken from the old machine indicates a sample mean of 65 kilograms and a similar sample of 100 from the new machine indicates a sample mean of 72 kilograms.

a. Using the 0.01 level of significance, is there evidence that the purchasing director should buy the new machine?

b. Compute the p-value in (a) and interpret its meaning.

10.9 Millions of dollars are spent each year on diet foods. Trends such as the low-fat diet or the low-carb Atkins diet have led to a host of new products. A study by Dr. Linda Stern of the Philadelphia Veterans Administration Hospital compared weight loss between obese patients on a low-fat diet and obese patients on a low-carb diet (R. Bazell, "Study Casts Doubt on Advantages of Atkins Diet," **msnbc.com**, May 17, 2004). Let μ_1 represent the mean number of pounds obese patients on a low-fat diet lose in six months, and μ_2 represent the mean number of pounds obese patients on a low-carb diet lose in six months.

a. State the null and alternative hypotheses if you want to test whether or not the mean weight loss between the two diets are equal.

b. In the context of this study, what is the meaning of a Type I error?

c. In the context of this study, what is the meaning of a Type II error?

d. Suppose that a sample of 100 obese patients on a low-fat diet lost a mean of 7.6 pounds in six months with a standard deviation of 3.2 pounds, while a sample of 100 obese patients on a low-carb diet lost a mean of 6.7 pounds in six months with a standard deviation of 3.9 pounds. At the 0.05 level of significance, is there evidence of a difference in the mean weight loss of obese patients between the low-fat and low-carb diets?

10.10 When do children in the United States develop preferences for brand-name products? In a study reported in the *Journal of Consumer Psychology* (G. W. Achenreiner and D. R. John, "The Meaning of Brand Names to Children: A Developmental Investigation," *Journal of Consumer Psychology*, 2003, 13(3): 205–219), marketers showed children identical pictures of athletic shoes. One picture was labeled Nike and one was labeled K-Mart. The children were asked to evaluate the shoes based on their appearance, quality, price, prestige, favorableness, and preference for owning. A score from 2 (highest product evaluation possible) to –2 (lowest product evaluation possible) was recorded for each child. The following table reports the results of the study.

Age by Brand	Sample Size	Sample Mean	Sample Standard Deviation
Age 8			
Nike	27	0.89	0.98
K-Mart	22	0.86	1.07
Age 12			
Nike	39	0.88	1.01
K-Mart	41	0.09	1.08
Age 16			
Nike	35	0.41	0.81
K-Mart	33	–0.29	0.92

a. Conduct a pooled-variance t test for the difference between two means for each of the three age groups. Use a level of significance of 0.05.

b. Write a brief summary of your findings.

PH Grade ASSIST **10.11** According to a survey conducted in October 2001, consumers were trying to reduce their credit card debt (Margaret Price, "Credit Debts Get Cut Down to Size," *Newsday*, November 25, 2001, F3). Based on a sample of 1,000 consumers in October 2001 and in October 2000, the mean credit card debt was $2,411 in October 2001 as compared to $2,814 in October 2000. Suppose that the standard deviation was $847.43 in October 2001 and $976.93 in October 2000.

a. Assuming that the population variances from both years are equal, is there evidence that the mean credit card debt is lower in October 2001 than in October 2000? (Use the $\alpha = 0.05$ level of significance.)

b. Find the p-value in (a) and interpret its meaning.

c. Assuming that the population variances from both years are equal, construct and interpret a 95% confidence interval estimate of the difference between the population means in October 2001 and October 2000.

SELF Test **10.12** The Computer Anxiety Rating Scale (CARS) measures an individual's level of computer anxiety on a scale from 20 (no anxiety) to 100 (highest level of anxiety). Researchers at Miami University administered CARS to 172 business students. One of the objectives of the study was to determine if there is a difference between the level of computer anxiety experienced by female and male business students.

	Males	Females
$\bar{X}$	40.26	36.85
S	13.35	9.42
n	100	72

Source: Extracted from Travis Broome and Douglas Havelka, "Determinants of Computer Anxiety in Business Students," The Review of Business Information Systems, *Spring 2002, 6(2):9–16.*

a. At the 0.05 level of significance, is there evidence of a difference in the mean computer anxiety experienced by female and male business students?
b. Determine the p-value and interpret its meaning.
c. What assumptions do you have to make about the two populations in order to justify the use of the t test?

10.13 Shipments of meat, meat by-products, and other ingredients are mixed together in several filling lines at a pet food canning factory. After the ingredients are thoroughly mixed, the pet food is placed in eight-ounce cans. Descriptive statistics concerning fill weights from two production lines, from two independent samples are given in the following table.

	Line A	Line B
$\bar{X}$	8.005	7.997
S	0.012	0.005
n	11	16

a. Assuming that the population variances are equal, at the 0.05 level of significance, is there evidence of a difference between the mean weight of cans filled on the two lines?
b. Repeat (a) assuming that the population variances are not equal.
c. Compare the results of (a) and (b).

10.14 A bank with a branch located in a commercial district of a city has developed an improved process for serving customers during the noon to 1 P.M. lunch period. The waiting time (operationally defined as the time elapsed from when the customer enters the line until he or she reaches the teller window) of all customers during this hour is recorded over a period of one week. A random sample of 15 customers **BANK1** is selected, and the results (in minutes) are as follows:

4.21 5.55 3.02 5.13 4.77 2.34 3.54 3.20
4.50 6.10 0.38 5.12 6.46 6.19 3.79

Suppose that another branch located in a residential area is also concerned with the noon to 1 P.M. lunch period. A random sample of 15 customers is selected **BANK2**, and the results are as follows:

9.66 5.90 8.02 5.79 8.73 3.82 8.01 8.35
10.49 6.68 5.64 4.08 6.17 9.91 5.47

a. Assuming that the population variances from both banks are equal, is there evidence of a difference in the mean waiting time between the two branches? (Use $\alpha = 0.05$.)
b. Determine the p-value in (a) and interpret its meaning.
c. What other assumption is necessary in (a)?
d. Assuming that the population variances from both branches are equal, construct and interpret a 95% confidence interval estimate of the difference between the population means in the two branches.

10.15 Repeat problem 10.14(a) assuming that the population variances in the two branches are not equal. Compare the results with those of problem 10.14(a).

10.16 A problem with a telephone line that prevents a customer from receiving or making calls is disconcerting to both the customer and the telephone company. The data **PHONE** represent samples of 20 problems reported to two different offices of a telephone company and the time to clear these problems (in minutes) from the customers' lines:

Central Office I Time to Clear Problems (minutes)

1.48 1.75 0.78 2.85 0.52 1.60 4.15 3.97 1.48 3.10
1.02 0.53 0.93 1.60 0.80 1.05 6.32 3.93 5.45 0.97

Central Office II Time to Clear Problems (minutes)

7.55 3.75 0.10 1.10 0.60 0.52 3.30 2.10 0.58 4.02
3.75 0.65 1.92 0.60 1.53 4.23 0.08 1.48 1.65 0.72

a. Assuming that the population variances from both offices are equal, is there evidence of a difference in the mean waiting time between the two offices? (Use $\alpha = 0.05$.)
b. Find the p-value in (a) and interpret its meaning.
c. What other assumption is necessary in (a)?
d. Assuming that the population variances from both offices are equal, construct and interpret a 95% confidence interval estimate of the difference between the population means in the two offices.

10.17 Repeat problem 10.16(a) assuming that the population variances in the two offices are not equal. Compare the results with those of problem 10.16(a).

10.18 In intaglio printing, a design or figure is carved beneath the surface of hard metal or stone. Suppose that an experiment is designed to compare differences in mean surface hardness of steel plates used in intaglio printing (measured in indentation numbers) based on two different surface conditions—untreated and treated by lightly polishing with emery paper. In the experiment, 40 steel plates are randomly assigned, 20 that are untreated, and 20 that are treated **INTAGLIO**.

Untreated		Treated	
164.368	177.135	158.239	150.226
159.018	163.903	138.216	155.620
153.871	167.802	168.006	151.233
165.096	160.818	149.654	158.653
157.184	167.433	145.456	151.204
154.496	163.538	168.178	150.869
160.920	164.525	154.321	161.657
164.917	171.230	162.763	157.016
169.091	174.964	161.020	156.670
175.276	166.311	167.706	147.920

a. Assuming that the population variances from both conditions are equal, is there evidence of a difference in the mean surface hardness between untreated and treated steel plates? (Use $\alpha = 0.05$.)
b. Find the p-value in (a) and interpret its meaning.
c. What other assumption is necessary in (a)?
d. Assuming that the population variances from untreated and treated steel plates are equal, construct and interpret a 95% confidence interval estimate of the difference between the population means in the two conditions.

10.19 Repeat problem 10.18(a) assuming that the population variances from untreated and treated steel plates are not equal. Compare the results with those of problem 10.18(a).

10.20 The director of training for an electronic equipment manufacturer is interested in determining whether different training methods have an effect on the productivity of assembly-line employees. She randomly assigns 42 recently hired employees into two groups of 21. The first group receives a computer-assisted, individual-based training program and the other receives a team-based training program. Upon completion of the training, the employees are evaluated on the time (in seconds) it takes to assemble a part. The results are in the data file TRAINING.
a. Assuming that the variances in the populations of training methods are equal, is there evidence of a difference between the mean assembly times (in seconds) of employees trained in a computer-assisted, individual-based program and those trained in a team-based program? (Use a 0.05 level of significance.)
b. What other assumption is necessary in (a)?
c. Repeat (a), assuming that the population variances are not equal.
d. Compare the results of (a) and (c).
e. Assuming equal variances, construct and interpret a 95% confidence interval estimate of the difference between the population means of the two training methods.

10.21 Nondestructive evaluation (NDE) is a method that is used to describe the properties of components or materials without causing any permanent physical change to the units. It includes the determination of properties of materials and the classification of flaws by size, shape, type, and location. This method is most effective for detecting surface flaws and characterizing surface properties of electrically conductive materials. Recently, data were collected that classified each component as having a flaw based on manual inspection and operator judgment and also reported the size of the crack in the material. Do the components classified as unflawed have a smaller mean crack size than components classified as flawed? The results in terms of crack size (in inches) are in the data file: CRACK (Source: B. D. Olin and W. Q. Meeker, "Applications of Statistical Methods to Nondestructive Evaluation," *Technometrics*, 38, 1996, 101.)
a. Assuming that the population variances are equal, is there evidence that the mean crack size is smaller for the unflawed specimens than for the flawed specimens? (Use $\alpha = 0.05$.)
b. Repeat (a) assuming that the population variances are not equal.
c. Compare the results of (a) and (b).

10.2 COMPARING THE MEANS OF TWO RELATED POPULATIONS

The hypothesis-testing procedures examined in section 10.1 enable you to make comparisons and examine differences in the means of two *independent* populations. In this section, you will learn about a procedure for analyzing the difference between the means of two populations when you collect sample data from populations that are related, that is, when results of the first population are *not* independent of the results of the second population.

Two approaches that involve related data between populations are possible. In the first approach, items or individuals are **paired** or **matched** according to some characteristic. In the second approach you take **repeated measurements** from the same set of items or individuals. In either case, the variable of interest becomes the *difference between the values* rather than the *values* themselves.

The first approach for analyzing related-samples involves matching or pairing items or individuals according to some characteristic of interest. For example, in test-marketing a product under two different advertising campaigns, a sample of test markets can be *matched* on the basis of the test-market population size and/or demographic variables. By controlling these variables, you are better able to measure the effects of the two different advertising campaigns.

The second approach for analyzing related-samples involves taking repeated measurements on the same items or individuals. Under the theory that the same items or individuals will behave alike if treated alike, your objective is to show that any differences between two measurements of the same items or individuals are due to different treatment conditions. For example, when performing a taste-testing experiment, you can use each person in the sample as his or her own control so that you can have *repeated measurements* on the same individual.

Regardless of whether you have matched (paired) samples or repeated measurements, the objective is to study the difference between two measurements by reducing the effect of the variability that is due to the items or individuals themselves. Table 10.3 shows the differences in the individual values for two related populations. To read this table, let $X_{11}, X_{12}, \ldots X_{1n}$ represent the n values from a sample. And let $X_{21}, X_{22}, \ldots X_{2n}$ represent either the corresponding n matched values from a second sample or the corresponding n repeated measurements from the initial sample. Then, $D_1, D_2, \ldots D_n$ will represent the corresponding set of n difference *scores* such that

$$D_1 = X_{11} - X_{21}, D_2 = X_{12} - X_{22}, \ldots, \text{ and } D_n = X_{1n} - X_{2n}$$

TABLE 10.3

Determining the Difference Between Two Related Populations

	Group		
Value	**1**	**2**	**Difference**
1	X_{11}	X_{21}	$D_1 = X_{11} - X_{21}$
2	X_{12}	X_{22}	$D_2 = X_{12} - X_{22}$
.	.	.	.
.	.	.	.
.	.	.	.
i	X_{1i}	X_{2i}	$D_i = X_{1i} - X_{2i}$
.	.	.	.
.	.	.	.
.	.	.	.
n	X_{1n}	X_{2n}	$D_n = X_{1n} - X_{2n}$

To test for the mean difference between two related populations, you treat the difference scores, the D_i's, as values from a single sample. If you know the population standard deviation of the difference scores, you use the Z test defined in Equation (10.6).[2] This Z test for the mean difference using samples from two related populations is equivalent to the one-sample Z test for the mean of the difference scores [see Equation (9.1) on page 306].

[2] *If the sample size is large, the Central Limit Theorem (see page 235) assures you that the sampling distribution of $\overline{D}$ follows a normal distribution.*

Z TEST FOR THE MEAN DIFFERENCE

$$Z = \frac{\overline{D} - \mu_D}{\frac{\sigma_D}{\sqrt{n}}} \quad \textbf{(10.6)}$$

where $$\overline{D} = \frac{\sum_{i=1}^{n} D_i}{n}$$

μ_D = hypothesized mean difference

σ_D = population standard deviation of the difference scores

n = sample size

The test statistic Z follows a standardized normal distribution.

Paired t Test

In most cases the population standard deviation is unknown. The only information you usually have are the sample mean and the sample standard deviation.

If you assume that the difference scores are randomly and independently selected from a population that is normally distributed, you can use the **paired t test for the mean difference in related populations** to determine whether there is a significant population mean difference. Thus, like the one-sample t test developed in section 9.4 [see Equation (9.2) on page 318], the t-test statistic developed here follows the t distribution with $n - 1$ degrees of freedom. Although you must assume the population is normally distributed, as long as the sample size is not very small and the population is not highly skewed, you can use the paired t test.

To test the null hypothesis that there is no difference in the means of two related populations (i.e., the population mean difference μ_D is 0)

$$H_0: \mu_D = 0 \text{ (where } \mu_D = \mu_1 - \mu_2)$$

against the alternative that the means are not the same (i.e., the population mean difference μ_D is not 0)

$$H_1: \mu_D \neq 0$$

you compute the t-test statistic in Equation (10.7).

PAIRED t TEST FOR THE MEAN DIFFERENCE

$$t = \frac{\bar{D} - \mu_D}{\frac{S_D}{\sqrt{n}}} \tag{10.7}$$

where

$$\bar{D} = \frac{\sum_{i=1}^{n} D_i}{n}$$

and

$$S_D = \sqrt{\frac{\sum_{i=1}^{n} (D_i - \bar{D})^2}{n - 1}}$$

The test statistic t follows a t distribution with $n - 1$ degrees of freedom.

For a two-tail test with a given level of significance α, you reject the null hypothesis if the computed t-test statistic is greater than the upper-tail critical value t_{n-1} from the t distribution or if the computed test statistic is less than the lower-tail critical value $-t_{n-1}$ from the t distribution. That is, the decision rule is

Reject H_0 if $t > t_{n-1}$

or if $t < -t_{n-1}$;

otherwise do not reject H_0.

To illustrate the use of the t test for the mean difference, suppose that a software applications company is developing a new financial applications package. Because computer-processing time is an important decision criterion, the developer wants the new package to have the same features and capabilities as the current market leader while providing results faster than the current leading package. If the new financial package is effective, it will provide the same results as the current market leader but will use less processing time.

What is the best way to design an experiment to compare the processing speed of the new software package to the processing speed of the current package? One approach is to take two independent samples and then use the hypothesis tests discussed in section 10.1. In this approach, you would use one set of financial applications projects to test the new software package. Then you would use a second set of different financial applications projects to test the current package. However, since the first set of financial applications projects used to test the new package might be faster or slower to process than the second set of financial application projects, this is not the best approach. A better approach is to use a repeated measurements experiment. In this experiment, you use one set of financial applications projects. For each of the projects, the new software package is tested and the current package is tested. Measuring the two processing times for each financial applications project serves to reduce the variability in the processing times compared with what would occur if you used two independent sets of financial applications projects. This approach also focuses on the differences between the two processing times for each financial applications project in order to measure the effectiveness of the new software package.

The results displayed in Table 10.4 are for a sample of $n = 10$ financial applications projects used in the experiment COMPTIME.

TABLE 10.4
Repeated Measurements of Time (in Seconds) to Complete Financial Applications Projects on Two Competing Software Packages

	Processing Times (in seconds)		
Applications Project	**By Current Market Leader**	**By New Software Package**	**Difference (D_i)**
1	9.98	9.88	+0.10
2	9.88	9.86	+0.02
3	9.84	9.75	+0.09
4	9.99	9.80	+0.19
5	9.94	9.87	+0.07
6	9.84	9.84	0.00
7	9.86	9.87	−0.01
8	10.12	9.86	+0.26
9	9.90	9.83	+0.07
10	9.91	9.86	+0.05
			+0.84

The question that you must answer is whether the new software package is faster. In other words, is there evidence that the mean processing time is significantly greater when financial applications projects use the current market leader rather than the new software package? Thus, the null and alternative hypotheses are:

H_0: $\mu_D \leq 0$ (mean processing time for the current package is less than or the same as that for the new package)
H_1: $\mu_D > 0$ (mean processing time for the current package is greater than that for the new package)

Choosing a level of significance of $\alpha = 0.05$ and assuming that the differences are normally distributed, you use the paired t test [Equation (10.7) on page 360]. For a sample of $n = 10$ projects, there are $n - 1 = 9$ degrees of freedom. Using Table E.3, the decision rule is:

Reject H_0 if $t > t_9 = 1.8331$;
otherwise do not reject H_0.

For the $n = 10$ differences (see Table 10.4), the sample mean difference is

$$\bar{D} = \frac{\sum_{i=1}^{n} D_i}{n} = \frac{0.84}{10} = 0.084$$

and

$$S_D = \sqrt{\frac{\sum_{i=1}^{n}(D_i - \overline{D})^2}{n-1}} = 0.0844$$

From Equation (10.7) on page 360,

$$t = \frac{\overline{D} - \mu_D}{\frac{S_D}{\sqrt{n}}} = \frac{0.084 - 0}{\frac{0.0844}{\sqrt{10}}} = 3.15$$

Because $t = 3.15$ is greater than 1.8331, you reject the null hypothesis H_0 (see Figure 10.10). There is evidence that the mean processing time is higher for the current market leader than for the new package.

FIGURE 10.10
One-tail Paired *t* Test at the 0.05 Level of Significance with 9 Degrees of Freedom

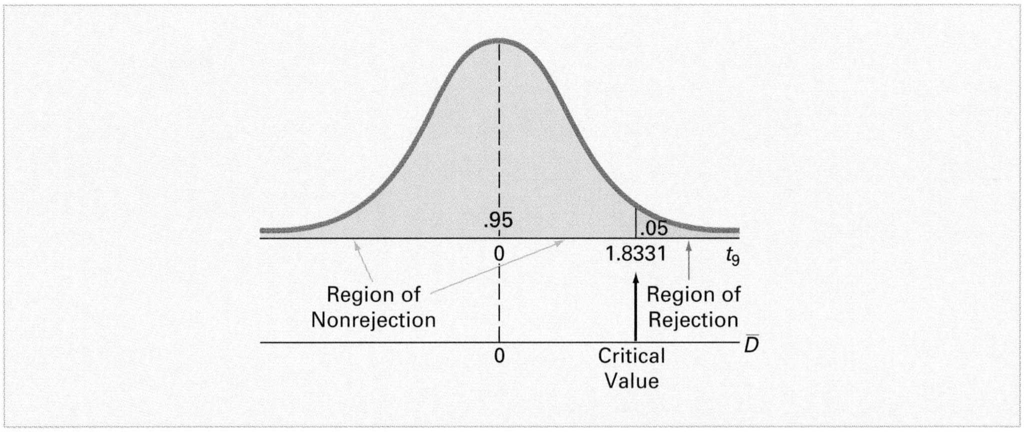

You can compute this test statistic along with the *p*-value using Microsoft Excel or Minitab (see Figures 10.11 and 10.12). Because the *p*-value $= 0.006 < \alpha = 0.05$, you reject H_0. The *p*-value indicates that if the two packages have the same mean processing time, the probability that the new package would outperform the current package by a mean of 0.084 seconds or more is only 0.006. Because this probability is so small, you have a minimal degree of belief in the null hypothesis and you conclude that the alternative hypothesis (i.e., the current package is slower) is true.

FIGURE 10.11
Microsoft Excel Paired *t* Test for the Financial Packages Data

	A	B	C
1	**t-Test: Paired Two Sample for Means**		
2			
3		***Current***	***New***
4	**Mean**	**9.926**	**9.842**
5	**Variance**	**0.0074**	**0.0016**
6	**Observations**	**10**	**10**
7	**Pearson Correlation**	**0.2798**	
8	**Hypothesized Mean Difference**	**0**	
9	**df**	**9**	
10	**t Stat**	**3.14902**	
11	**P(T<=t) one-tail**	**0.00588**	
12	**t Critical one-tail**	**1.83311**	
13	**P(T<=t) two-tail**	**0.01176**	
14	**t Critical two-tail**	**2.26216**	

FIGURE 10.12 PANEL A

Minitab Paired *t* Test for the Financial Packages Data

```
Paired T for Current - New

             N      Mean     StDev   SE Mean
Current     10   9.92600   0.08631   0.02729
New         10   9.84200   0.03994   0.01263
Difference  10  0.084000  0.084354  0.026675


95% lower bound for mean difference: 0.035102
T-Test of mean difference = 0 (vs > 0): T-Value = 3.15  P-Value = 0.006
```

FIGURE 10.12 PANEL B

Minitab Boxplot for the Financial Packages Data

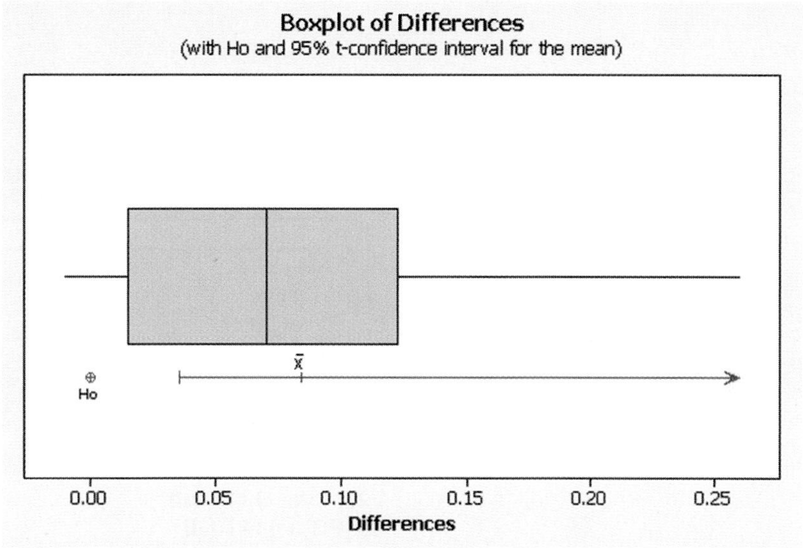

From Figure 10.12, panel B, observe that the box-and-whisker plot shows approximate symmetry between the first and third quartiles, but also has one or more high values. Thus, the data do not greatly contradict the underlying assumption of normality. If an exploratory data analysis reveals that the assumption of underlying normality in the population is severely violated, then the *t* test is inappropriate. If this occurs, you can either use a *nonparametric* procedure that does not make the stringent assumption of underlying normality (see section 12.8) or make a data transformation (see reference 8) and then recheck the assumptions to determine whether you should use the *t* test.

EXAMPLE 10.2 PAIRED *t*-TEST OF PIZZA DELIVERY TIMES

Recall from Example 10.1 on page 350 that a local pizza restaurant located close to a college campus advertises that their delivery time to a college dormitory is less than for a local branch of a national pizza chain. In order to determine whether this advertisement is valid, you and some friends have decided to order 10 pizzas from the local pizza restaurant and 10 pizzas from the national chain. In fact, you and your friends have collected the data at 10 different times. At each time, you ordered a pizza from the local pizza restaurant and also from the national pizza chain. Thus, you have paired data, one measurement for each pizza delivered at each time.

SOLUTION You use the paired *t* test instead of the pooled *t* test to analyze these data PIZZATIME. Table 10.5 summarizes the data.

TABLE 10.5
Delivery Times for Local Pizza Restaurant and National Pizza Chain

Time	Local	Chain	Difference
1	16.8	22.0	−5.2
2	11.7	15.2	−3.5
3	15.6	18.7	−3.1
4	16.7	15.6	1.1
5	17.5	20.8	−3.3
6	18.1	19.5	−1.4
7	14.1	17.0	−2.9
8	21.8	19.5	2.3
9	13.9	16.5	−2.6
10	20.8	24.0	−3.2
			−21.8

FIGURE 10.13
Microsoft Excel Paired t Test Output for the Pizza Delivery Data

	A	B	C
1	t-Test: Paired Two Sample for Means		
2			
3		*Local*	*Chain*
4	Mean	16.7	18.88
5	Variance	9.582222	8.215111
6	Observations	10	10
7	Pearson Correlation	0.714077	
8	Hypothesized Mean Difference	0	
9	df	9	
10	t Stat	-3.04479	
11	P(T<=t) one-tail	0.006955	
12	t Critical one-tail	1.833114	
13	P(T<=t) two-tail	0.01391	
14	t Critical two-tail	2.262159	

Figure 10.13 illustrates Microsoft Excel paired t test output for the pizza delivery data. The null and alternative hypotheses are:

H_0: $\mu_D \geq 0$ (mean delivery time for the local pizza restaurant is higher than or the same as that for the national pizza chain)
H_1: $\mu_D < 0$ (mean delivery time for the local pizza restaurant is lower than for the national pizza chain)

Choosing a level of significance of $\alpha = 0.05$ and assuming that the differences are normally distributed, you use the paired t test [Equation (10.7) on page 360]. For a sample of $n = 10$ delivery times, there are $n - 1 = 9$ degrees of freedom. Using Table E.3, the decision rule is:

$$\text{Reject } H_0 \text{ if } t < t_9 = -1.8331;$$
$$\text{otherwise do not reject } H_0$$

For $n = 10$ differences (see Table 10.5), the sample mean difference is

$$\bar{D} = \frac{\sum_{i=1}^{n} D_i}{n} = \frac{-21.8}{10} = -2.18$$

and the sample standard deviation of the difference is

$$S_D = \sqrt{\frac{\sum_{i=1}^{n}(D_i - \bar{D})^2}{n-1}} = 2.2641$$

From Equation (10.7) on page 360,

$$t = \frac{\overline{D} - \mu_D}{\frac{S_D}{\sqrt{n}}} = \frac{-2.18 - 0}{\frac{2.2641}{\sqrt{10}}} = -3.045$$

Because $t = -3.045$ is less than -1.8331, you reject the null hypothesis H_0 (the p-value is $0.007 < 0.05$). There is evidence that the mean delivery time is lower for the local pizza restaurant than for the national pizza chain.

This conclusion is different from the one you reached when you used the pooled-variance t test for these data. By pairing the delivery times, you were able to focus on the differences between the two pizza delivery services and you had a more powerful statistical procedure that was better able to detect the difference between the two restaurants.

Confidence Interval Estimate for the Mean Difference

Instead of, or in addition to, testing for the difference between the means of two related populations, you can construct a confidence interval estimate of the mean difference as shown in Equation (10.8).

CONFIDENCE INTERVAL ESTIMATE FOR THE MEAN DIFFERENCE

$$\overline{D} \pm t_{n-1}\frac{S_D}{\sqrt{n}} \qquad \textbf{(10.8)}$$

or

$$\overline{D} - t_{n-1}\frac{S_D}{\sqrt{n}} \le \mu_D \le \overline{D} + t_{n-1}\frac{S_D}{\sqrt{n}}$$

Returning to the example concerning the financial applications projects, using Equation (10.8) above, $\overline{D} = 0.084$, $S_D = 0.0844$, $n = 10$, and $t = 2.2622$ (for 95% confidence and $n - 1 = 9$ degrees of freedom),

$$0.084 \pm (2.2622)\frac{0.0844}{\sqrt{10}}$$

$$0.084 \pm 0.0604$$

$$0.0236 \le \mu_D \le 0.1444$$

Thus, with 95% confidence, the mean difference between the two software packages is between 0.0236 and 0.1444 seconds. Since the interval estimate only contains values greater than zero, you can conclude that the mean processing time for the current package is higher than that for the new package.

PROBLEMS FOR SECTION 10.2

Learning the Basics

PH Grade ASSIST **10.22** An experimental design for a paired t test has, as a matched sample, 20 pairs of identical twins. How many degrees of freedom are there in this t test?

PH Grade ASSIST **10.23** An experimental design for a repeated-measures t test requires a measurement before and after the presentation of a stimulus on each of 15 subjects. How many degrees of freedom are there in this t test?

Applying the Concepts

Problems 10.24–10.29 can be solved manually or by using Microsoft Excel, Minitab, or SPSS. We recommend that you use Microsoft Excel, Minitab, or SPSS to solve problem 10.30.

10.24 Travel expenses paid by companies can increase or decrease dramatically when there are changes in the daily rates for hotel rooms. Did these rates stay the same from June 2002 to March 2004? The following data HOTELPRICE2 give typical daily rates for hotels in 18 cities during March 2004 and June 2002.

City	Hotel 2004	Hotel 2002
Atlanta	78.91	173
Boston	112.92	243
Chicago	96.90	257
Dallas	77.43	167
Denver	74.22	139
Detroit	77.71	141
Houston	76.26	180
Los Angeles	95.78	223
Miami	140.61	116
Minneapolis	78.64	167
New Orleans	121.59	142
New York	167.43	273
Orlando	98.57	133
Phoenix	123.19	124
San Francisco	123.51	178
Seattle	95.09	176
St. Louis	74.68	159
Washington	123.27	262

Source: Extracted from "Travel," The Wall Street Journal, *June 7, 2002, W4; and C. Woodyard, "Luxury Costs More These Days,"* USA Today, *April 27, 2004, 5B.*

a. At the 0.05 level of significance, is there evidence of a difference in the mean daily hotel rate in March 2004 and June 2002?
b. What assumption is necessary to perform this test?
c. Determine the p-value in (a) and interpret its meaning.
d. Construct and interpret a 95% confidence interval estimate of the difference in the mean daily hotel rate for March 2004 and June 2002.

10.25 In industrial settings, alternative methods often exist for measuring variables of interest. The data in the file MEASUREMENT (coded to maintain confidentiality) represent measurements in-line (collected from an analyzer during the production process) and from an analytical lab. (M. Leitnaker, "Comparing Measurement Processes: In-line Versus Analytical Measurements," *Quality Engineering*, 13, 2000–2001, 293–298.)

a. At the 0.05 level of significance, is there evidence of a difference in the mean measurements in-line and from an analytical lab?
b. What assumption is necessary to perform this test?
c. Use a graphical method to evaluate the validity of the assumption in (a).
d. Construct and interpret a 95% confidence interval estimate of the difference in the mean measurements in-line and from an analytical lab.

SELF Test **10.26** Can students save money by buying their textbooks at Amazon.com? To investigate this possibility, a random sample of 15 textbooks used during a recent semester at Miami University was selected. The prices for these textbooks at both a local bookstore and through Amazon.com were recorded. The prices for the textbooks, including all relevant taxes and shipping are given below TEXTBOOK:

Textbook	Book Store	Amazon
Access 2000 Guidebook	52.22	57.34
HTML 4.0 CD with Java Script	52.74	44.47
Designing the Physical Education Curriculum	39.04	41.48
Service Management: Operations, Strategy and IT	101.28	73.72
Fundamentals of Real Estate Appraisal	37.45	42.04
Investments	113.41	95.38
Intermediate Financial Management	109.72	119.80
Real Estate Principles	101.28	62.48
The Automobile Age	29.49	32.43
Geographic Information Systems in Ecology	70.07	74.43
Geosystems: An Introduction to Physical Geography	83.87	83.81
Understanding Contemporary Africa	23.21	26.48
Early Childhood Education Today	72.80	73.48
System of Transcendental Idealism (1800)	17.41	20.98
Principles and Labs for Fitness and Wellness	37.72	40.43

a. At the 0.01 level of significance, is there evidence of a difference between the mean price of textbooks at the local bookstore and Amazon.com?
b. What assumption is necessary to perform this test?
c. Construct a 99% confidence interval estimate of the mean difference in price. Interpret the interval.
d. Compare the results of (a) and (c).

10.27 A recent article discussed the new Whole Foods Market in the Time-Warner building in New York City (W. Grimes, "A Pleasure Palace without the Guilt," *The New*

York Times, February 18, 2004, F1, F5). The following data WHOLEFOODS1 compared the price of some kitchen staples at the new Whole Foods Market and at the Fairway supermarket located about 15 blocks from the Time-Warner building.

Item	Whole Foods	Fairway
Half-gallon milk	2.19	1.35
Dozen eggs	2.39	1.69
Tropicana orange juice (64 oz.)	2.00	2.49
Head of Boston lettuce	1.98	1.29
Ground round 1lb.	4.99	3.69
Bumble Bee tuna 6 oz. can	1.79	1.33
Granny Smith apples (1 lb.)	1.69	1.49
Box DeCecco linguini	1.99	1.59
Salmon steak 1 lb.	7.99	5.99
Whole chicken per pound	2.19	1.49

Source: Extracted from W. Grimes, "A Pleasure Palace without the Guilt," The New York Times, *February 18, 2004, F1, F5.*

a. At the 0.01 level of significance, is there evidence that the mean price is higher at Whole Foods Market than at the Fairway Supermarket?
b. Interpret the meaning of the p-value in (a).

10.28 Multiple myeloma or blood plasma cancer is characterized by increased blood vessel formulation (angiogenesis) in the bone marrow that is a prognostic factor in survival. One treatment approach used for multiple myeloma is stem cell transplantation with the patient's own stem cells. The following data MYELOMA represent the bone marrow microvessel density for patients who had a complete response to the stem cell transplant as measured by blood and urine tests. The measurements were taken immediately prior to the stem cell transplant and at the time of the complete response.

Patient	Before	After
1	158	284
2	189	214
3	202	101
4	353	227
5	416	290
6	426	176
7	441	290

Source: S. V. Rajkumar, R. Fonseca, T. E. Witzig, M. A. Gertz, and P. R. Greipp, "Bone Marrow Angiogenesis in Patients Achieving Complete Response After Stem Cell Transplantation for Multiple Myeloma," Leukemia, *1999, 13, 469–472.*

a. At the 0.05 level of significance, is there evidence that the mean bone marrow microvessel density is higher before the stem cell transplant than after the stem cell transplant?
b. Interpret the meaning of the p-value in (a).
c. Construct and interpret a 95% confidence interval estimate of the mean difference in bone marrow microvessel density before and after the stem cell transplant.

10.29 Over the past year the vice president for human resources at a large medical center has run a series of three-month workshops aimed at increasing worker motivation and performance. To check the effectiveness of the workshops, she selected a random sample of 35 employees from the personnel files and recorded their most recent annual performance ratings along with their ratings prior to attending the workshops PERFORM. The Microsoft Excel output in panels A and B below provides both descriptive and inferential information so that you can analyze the results and examine the assumptions of the hypothesis test used.

State your findings and conclusions in a report to the vice president for human resources.

	A	B
1	*Difference*	
2		
3	Mean	-5.25714
4	Standard Error	1.947782
5	Median	-5
6	Mode	-10
7	Standard Deviation	11.52323
8	Sample Variance	132.7849
9	Kurtosis	1.103831
10	Skewness	0.110341
11	Range	61
12	Minimum	-34
13	Maximum	27
14	Sum	-184
15	Count	35
16	Largest(1)	27
17	Smallest(1)	-34

PANEL A

	A	B	C
1	t-Test: Paired Two Sample for Means		
2			
3		*Before*	*After*
4	Mean	74.54286	79.8
5	Variance	80.90252	37.16471
6	Observations	35	35
7	Pearson Correlation	-0.1342	
8	Hypothesized Mean Difference	0	
9	df	34	
10	t Stat	-2.69904	
11	P(T<=t) one-tail	0.005376	
12	t Critical one-tail	1.690923	
13	P(T<=t) two-tail	0.010752	
14	t Critical two-tail	2.032243	

PANEL B

10.30 The data in the file CONCRETE1 represent the compressive strength in thousands of pounds per square inch (psi) of 40 samples of concrete taken two and seven days after pouring.

Source: O. Carrillo-Gamboa and R. F. Gunst, "Measurement-Error-Model Collinearities," Technometrics, *34, 1992, 454–464.*

a. At the 0.01 level of significance, is there evidence that the mean strength is less at two days than at seven days?
b. What assumption is necessary to perform this test?
c. Find the *p*-value in (a) and interpret its meaning.

10.3 COMPARING TWO POPULATION PROPORTIONS

Often you need to make comparisons and analyze differences between two population proportions. You can perform a test for the difference between two proportions selected from independent samples using two different methods. This section presents a procedure whose test statistic Z is approximated by a standardized normal distribution. In section 12.1, a procedure whose test statistic χ^2 is approximated by a chi-square distribution is developed. As you will see, the results from these two tests are equivalent.

Z Test for the Difference Between Two Proportions

In evaluating differences between two population proportions, you can use a ***Z* test for the difference between two proportions**. The test statistic Z is based on the difference between two sample proportions $(p_1 - p_2)$. This test statistic, given in Equation (10.9), approximately follows a standardized normal distribution for large enough sample sizes.

Z TEST FOR THE DIFFERENCE BETWEEN TWO PROPORTIONS

$$Z = \frac{(p_1 - p_2) - (\pi_1 - \pi_2)}{\sqrt{\bar{p}(1 - \bar{p})\left(\frac{1}{n_1} + \frac{1}{n_2}\right)}} \qquad \textbf{(10.9)}$$

with

$$\bar{p} = \frac{X_1 + X_2}{n_1 + n_2} \qquad p_1 = \frac{X_1}{n_1} \qquad p_2 = \frac{X_2}{n_2}$$

where
p_1 = proportion of successes in sample 1
X_1 = number of successes in sample 1
n_1 = sample size of sample 1
π_1 = proportion of successes in population 1
p_2 = proportion of successes in sample 2
X_2 = number of successes in sample 2
n_2 = sample size of sample 2
π_2 = proportion of successes in population 2
$\bar{p}$ = pooled estimate of the population proportion of successes

The test statistic Z approximately follows a standardized normal distribution.

Under the null hypothesis, you assume that the two population proportions are equal ($\pi_1 = \pi_2$). Because the pooled estimate for the population proportion is based on the null hypothesis, you combine or pool the two sample proportions to compute an overall estimate of the common population proportion. This estimate is equal to the number of successes in the two samples combined ($X_1 + X_2$) divided by the total sample size from the two sample groups ($n_1 + n_2$).

As shown in the following table, you can use this Z test for the difference between population proportions to determine whether there is a difference in the proportion of successes in the two groups (two-tail test) or whether one group has a higher proportion of successes than the other group (one-tail test).

Two-Tail Test	**One-Tail Test**	**One-Tail Test**
$H_0: \pi_1 = \pi_2$	$H_0: \pi_1 \geq \pi_2$	$H_0: \pi_1 \leq \pi_2$
$H_1: \pi_1 \neq \pi_2$	$H_1: \pi_1 < \pi_2$	$H_1: \pi_1 > \pi_2$

where π_1 = proportion of successes in population 1

π_2 = proportion of successes in population 2

To test the null hypothesis that there is no difference between the proportions of two independent populations:

$$H_0: \pi_1 = \pi_2$$

against the alternative that the two population proportions are not the same:

$$H_1: \pi_1 \neq \pi_2$$

use the test statistic Z, given by Equation (10.9). For a given level of significance α, reject the null hypothesis if the computed Z-test statistic is greater than the upper-tail critical value from the standardized normal distribution or if the computed test statistic is less than the lower-tail critical value from the standardized normal distribution.

To illustrate the use of the Z test for the equality of two proportions, suppose that you are the manager of T.C. Resort Properties, a collection of five upscale resort hotels located on two resort islands. On one of the islands, T.C. Resort Properties has two hotels, the Beachcomber and the Windsurfer. In tabulating the responses to the single question, "Are you likely to choose this hotel again?" 163 of 227 guests at the Beachcomber responded yes, and 154 of 262 guests at the Windsurfer responded yes. At the 0.05 level of significance, is there evidence of a significant difference in guest satisfaction (as measured by likelihood to return to the hotel) between the two hotels?

The null and alternative hypotheses are

$$H_0: \pi_1 = \pi_2 \text{ or } \pi_1 - \pi_2 = 0$$
$$H_1: \pi_1 \neq \pi_2 \text{ or } \pi_1 - \pi_2 \neq 0$$

Using the 0.05 level of significance, the critical values are −1.96 and +1.96 (see Figure 10.14), and the decision rule is

Reject H_0 if $Z < -1.96$

or if $Z > +1.96$;

otherwise do not reject H_0.

FIGURE 10.14
Regions of Rejection and Nonrejection When Testing a Hypothesis for the Difference Between Two Proportions at the 0.05 Level of Significance

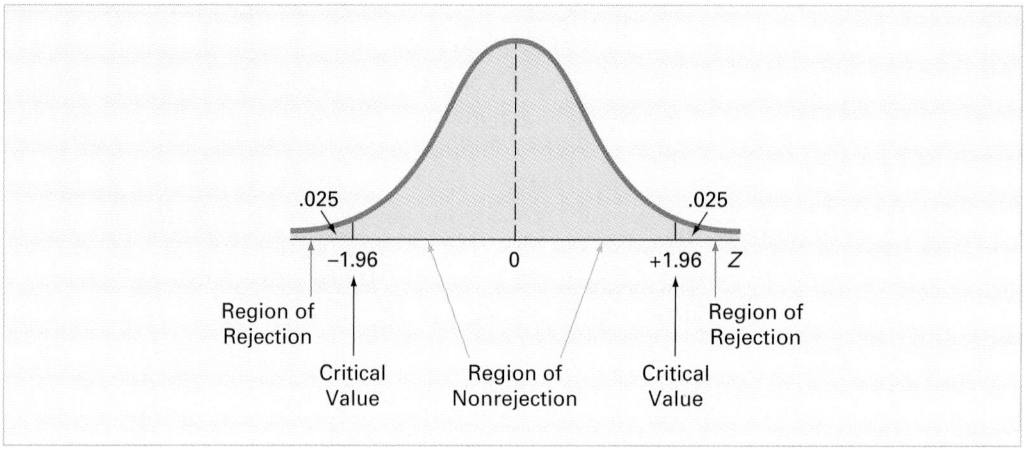

Using Equation (10.9) on page 368,

$$Z = \frac{(p_1 - p_2) - (\pi_1 - \pi_2)}{\sqrt{\bar{p}(1-\bar{p})\left(\frac{1}{n_1} + \frac{1}{n_2}\right)}}$$

where

$$p_1 = \frac{X_1}{n_1} = \frac{163}{227} = 0.718 \qquad p_2 = \frac{X_2}{n_2} = \frac{154}{262} = 0.588$$

and

$$\bar{p} = \frac{X_1 + X_2}{n_1 + n_2} = \frac{163 + 154}{227 + 262} = \frac{317}{489} = 0.648$$

so that

$$Z = \frac{(0.718 - 0.588) - (0)}{\sqrt{0.648(1 - 0.648)\left(\frac{1}{227} + \frac{1}{262}\right)}}$$

$$= \frac{0.13}{\sqrt{(0.228)(0.0082)}}$$

$$= \frac{0.13}{\sqrt{0.00187}}$$

$$= \frac{0.13}{0.0432} = +3.01$$

Using the 0.05 level of significance, reject the null hypothesis because $Z = +3.01 > +1.96$. The p-value is 0.0026 (calculated from Table E.2 or from the Microsoft Excel worksheet of Figure 10.15 or the Minitab output of Figure 10.16). This means that if the null hypothesis is true, the probability that a Z-test statistic is less than -3.01 is 0.0013, and, similarly, the probability that a Z-test statistic is greater than $+3.01$ is 0.0013. Thus, for this two-tail test, the p-value is $0.0013 + 0.0013 = 0.0026$. Because $0.0026 < \alpha = 0.05$, you reject the null hypothesis. There is evidence to conclude that the two hotels are significantly different with respect to guest satisfaction; a greater proportion of guests are willing to return to the Beachcomber than to the Windsurfer.

FIGURE 10.15
Microsoft Excel Worksheet for the Z Test for the Difference Between Two Proportions for the Hotel Guest Satisfaction Problem

	A	B	C	D
1	**Guest Satisfaction Analysis**			
2				
3	**Data**			
4	**Hypothesized Difference**	**0**		
5	**Level of Significance**	**0.05**		
6	**Group 1**			
7	**Number of Successes**	**163**		
8	**Sample Size**	**227**		
9	**Group 2**			
10	**Number of Successes**	**154**		
11	**Sample Size**	**262**		
12				
13	Intermediate Calculations			
14	Group 1 Proportion	0.7181		=B7/B8
15	Group 2 Proportion	0.5878		=B10/B11
16	Difference in Two Proportions	0.1303		=B14 - B15
17	Average Proportion	0.6483		=(B7 + B10)/(B8 + B11)
18	**Z Test Statistic**	**3.0088**		=(B16 - B4)/SQRT(B17 * (1 - B17) * (1/B8 + 1/B11))
19				
20	**Two-Tail Test**			
21	**Lower Critical Value**	**-1.9600**		=NORMSINV(B5/2)
22	**Upper Critical Value**	**1.9600**		=NORMSINV(1 - B5/2)
23	***p*-Value**	**0.0026**		=2 * (1 - NORMSDIST(ABS(B18)))
24	**Reject the null hypothesis**			

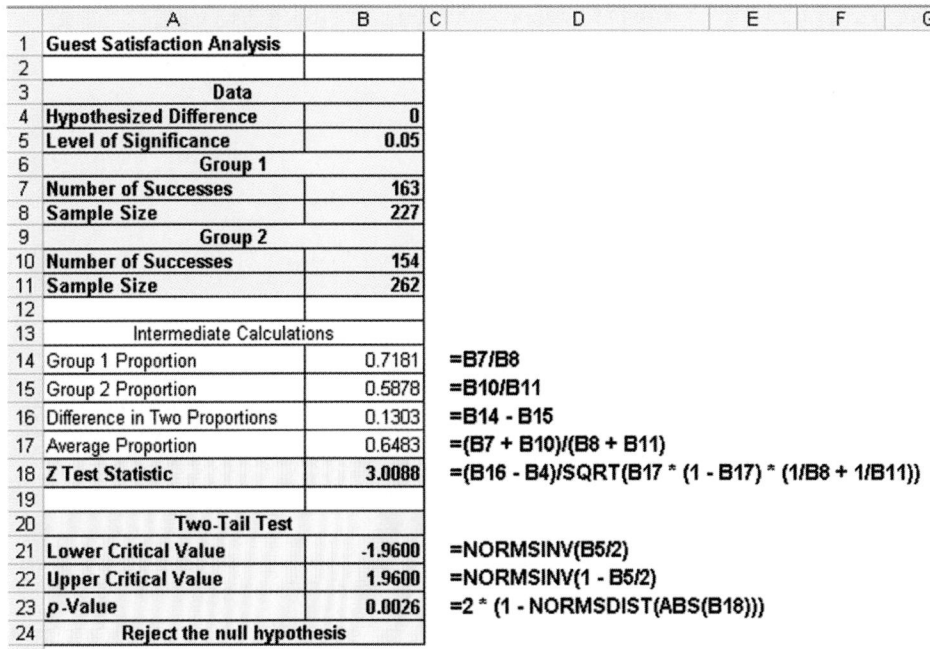

FIGURE 10.16
Minitab Output for the Z Test for the Difference Between Two Proportions for the Hotel Guest Satisfaction Problem

Test and CI for Two Proportions

```
Sample    X    N  Sample p
1       163  227  0.718062
2       154  262  0.587786


Difference = p (1) - p (2)
Estimate for difference:  0.130275
95% CI for difference:  (0.0467379, 0.213813)
Test for difference = 0 (vs not = 0):  Z = 3.01  P-Value = 0.003
```

EXAMPLE 10.3 TESTING FOR THE DIFFERENCE IN TWO PROPORTIONS

Money worries in the United States start at an early age. In a survey, 660 children (330 boys and 330 girls) ages 6 to 14 were asked the question, "Do you worry about having enough money?" Of the boys surveyed, 201 (60.9%) said yes, and 178 (53.9%) of the girls surveyed said yes (Extracted from D. Haralson and K. Simmons, "Snapshots," *USA Today*, May 24, 2004, 1B). At the 0.05 level of significance, is the proportion of boys who worry about having enough money greater than the proportion of girls?

SOLUTION Since you want to know whether there is evidence that the proportion of boys who worry about having enough money is *greater* than the proportion of girls, you have a one-tail test. The null and alternative hypotheses are

H_0: $\pi_1 \leq \pi_2$ (The proportion of boys who worry about having enough money is less than or equal to the proportion of girls)
H_1: $\pi_1 > \pi_2$ (The proportion of boys who worry about having enough money is greater than the proportion of girls)

Using the 0.05 level of significance, for the one-tail test in the upper tail, the critical value is +1.645. The decision rule is

$$\text{Reject } H_0 \text{ if } Z > +1.645;$$
$$\text{otherwise do not reject } H_0.$$

Using Equation (10.9) on page 368,

$$Z = \frac{(p_1 - p_2) - (\pi_1 - \pi_2)}{\sqrt{\bar{p}(1 - \bar{p})\left(\frac{1}{n_1} + \frac{1}{n_2}\right)}}$$

where

$$p_1 = \frac{X_1}{n_1} = \frac{201}{330} = 0.609 \qquad p_2 = \frac{X_2}{n_2} = \frac{178}{330} = 0.539$$

and

$$\bar{p} = \frac{X_1 + X_2}{n_1 + n_2} = \frac{201 + 178}{330 + 330} = \frac{379}{660} = 0.5742$$

so that

$$Z = \frac{(0.609 - 0.539) - (0)}{\sqrt{0.5742(1 - 0.5742)\left(\frac{1}{330} + \frac{1}{330}\right)}}$$

$$= \frac{0.07}{\sqrt{(0.2445)(0.00606)}}$$

$$= \frac{0.07}{\sqrt{0.00148}}$$

$$= \frac{0.07}{0.0385} = +1.818$$

Using the 0.05 level of significance, reject the null hypothesis because $Z = +1.818 > +1.645$. The p-value is 0.0344 (calculated from Table E.2). Therefore, if the null hypothesis is true, the probability that a Z-test statistic is greater than + 1.818 is 0.0344 (which is less than $\alpha = 0.05$). You conclude that there is evidence that the proportion of boys who worry about having enough money is greater than the proportion of girls.

Confidence Interval Estimate for the Difference Between Two Proportions

Instead of or in addition to testing for the difference between the proportions of two independent populations, you can construct a confidence interval estimate of the difference between the two proportions using Equation (10.10).

CONFIDENCE INTERVAL ESTIMATE FOR THE DIFFERENCE BETWEEN TWO PROPORTIONS

$$(p_1 - p_2) \pm Z\sqrt{\frac{p_1(1 - p_1)}{n_1} + \frac{p_2(1 - p_2)}{n_2}} \tag{10.10}$$

or

$$(p_1 - p_2) - Z\sqrt{\frac{p_1(1 - p_1)}{n_1} + \frac{p_2(1 - p_2)}{n_2}} \le (\pi_1 - \pi_2)$$
$$\le (p_1 - p_2) + Z\sqrt{\frac{p_1(1 - p_1)}{n_1} + \frac{p_2(1 - p_2)}{n_2}}$$

To construct a 95% confidence interval estimate of the population difference between the percentages of guests who would return to the Beachcomber and who would return to the Windsurfer, you use the results on page 370 or from Figures 10.15 or 10.16 on page 371.

$$p_1 = \frac{X_1}{n_1} = \frac{163}{227} = 0.718 \quad p_2 = \frac{X_2}{n_2} = \frac{154}{262} = 0.588$$

Using Equation (10.10),

$$(0.718 - 0.588) \pm (1.96)\sqrt{\frac{0.718(1-0.718)}{227} + \frac{0.588(1-0.588)}{262}}$$

$$0.13 \pm (1.96)(0.0426)$$

$$0.13 \pm 0.0835$$

$$0.0465 \le (\pi_1 - \pi_2) \le 0.2135$$

Thus, you have 95% confidence that the difference between the population proportion of guests who would return again to the Beachcomber and the Windsurfer is between 0.0465 and 0.2135. In percentages, the difference is between 4.65% and 21.35%. Guest satisfaction is higher at the Beachcomber than at the Windsurfer.

PROBLEMS FOR SECTION 10.3

Learning the Basics

PH Grade ASSIST **10.31** Assume that $n_1 = 100$, $X_1 = 50$, $n_2 = 100$, and $X_2 = 30$.

a. At the 0.05 level of significance, is there evidence of a significant difference between the two population proportions?

b. Construct a 95% confidence interval estimate of the difference between the two population proportions.

PH Grade ASSIST **10.32** Assume that $n_1 = 100$, $X_1 = 45$, $n_2 = 50$, and $X_2 = 25$.

a. At the 0.01 level of significance, is there evidence of a significant difference between the two population proportions?

b. Construct a 99% confidence interval estimate of the difference between the two population proportions.

Applying the Concepts

PH Grade ASSIST **10.33** A sample of 500 shoppers was selected in a large metropolitan area to determine various information concerning consumer behavior. Among the questions asked was, "Do you enjoy shopping for clothing?" Of 240 males, 136 answered yes. Of 260 females, 224 answered yes.

a. Is there evidence of a significant difference between males and females in the proportion who enjoy shopping for clothing at the 0.01 level of significance?

b. Find the *p*-value in (a) and interpret its meaning.

c. Construct and interpret a 99% confidence interval estimate of the difference between the proportion of males and females who enjoy shopping for clothing.

d. What are your answers to (a) through (c) if 206 males enjoyed shopping for clothing?

PH Grade ASSIST **10.34** *The New York Times* reported a study conducted by the Henry J. Kaiser Family Foundation concerning the role of media in the lives of children (Dylan Loeb McClain, "Where Is Today's Child? Probably Watching TV," *The New York Times*, December 6, 1999, C1). In one question of the study, children were asked if they use a computer each day. From a sample of 1,090 children between the ages of 2 and 7, 283 used a computer each day. From a sample of 2,065 children between the ages of 8 and 18, 1,053 used a computer each day.

a. At the 0.05 level of significance, is there evidence of a significant difference between the two age groups in the proportion of children that use a computer each day?

b. Find the *p*-value in (a) and interpret its meaning.

c. Construct and interpret a 95% confidence interval estimate of the difference between the population proportion of 2- to 7-year-old children and 8- to 18-year-old children who use a computer each day.

PH Grade ASSIST **10.35** The results of a study conducted as part of a yield-improvement effort at a semiconductor manufacturing facility provided defect data for a sample of 450 wafers. The following contingency table presents a summary of the responses to two questions: "Was a particle found on the die that produced the wafer?" and "Is the wafer good or bad?"

	QUALITY OF WAFER		
PARTICLES	**Good**	**Bad**	**Totals**
Yes	14	36	50
No	320	80	400
Totals	334	116	450

Source: Extracted from S.W. Hall, "Analysis of Defectivity of Semiconductor Wafers by Contingency Table," Proceedings Institute of Environmental Sciences, *Vol. 1 (1994), 177–183.*

a. At the 0.05 level of significance, is there evidence of a significant difference between the proportion of good and bad wafers that have particles?
b. Determine the p-value in (a) and interpret its meaning.
c. Construct and interpret a 95% confidence interval estimate of the difference between the population proportion of good and bad wafers that contain particles.
d. What conclusions can you reach from this analysis?

10.36 The percentage of online adults in the United States increased from 63% in 2000 to 69% in December 2003 ("Activities Gaining Popularity on Net," *USA Today Snapshots*, February 3, 2004). In 2000, 25% of online adults used the Internet to gather data about products and services. Suppose that this result was based on a sample of 500 online adults. In December 2003, 299 of 729 online adults who were sampled used the Internet to gather data about products and services.
a. At the 0.05 level of significance, is there evidence that the proportion of online adults who use the Internet to gather data about products and services is higher in December 2003 than in 2000?
b. Find the p-value in (a) and interpret its meaning.

10.37 Is good gas mileage a priority for car shoppers? A survey conducted by Progressive Insurance asked this question to both men and women shopping for new cars. The data were reported as percentages, and no sample sizes were given.

	GENDER	
GAS MILEAGE A PRIORITY?	**Men**	**Women**
Yes	76%	84%
No	24%	16%

Source: Extracted from Snapshots, Usatoday.com, *June 21, 2004.*

a. Assume that 50 men and 50 women were included in the survey. At the 0.05 level of significance, is there evidence of a difference in the population proportion of males and females who make gas mileage a priority?
b. Assume that 500 men and 500 women were included in the survey. At the 0.05 level of significance, is there evidence of a difference between males and females in the proportion who make gas mileage a priority?
c. Discuss the effect of sample size on the Z test for the difference between two proportions.

SELF Test **10.38** Are whites more likely to claim bias? A survey conducted by Barry Goldman ("White Fight: A Researcher Finds Whites are More Likely to Claim Bias," *The Wall Street Journal*, Workweek, April 10, 2001, A1) found that of 56 white workers terminated, 29 claimed bias. Of 407 black workers terminated, 126 claimed bias.
a. At the 0.05 level of significance, is there evidence that white workers are more likely to claim bias than black workers?
b. Find the p-value in (a) and interpret its meaning.

10.39 A study conducted by Ariel Mutual Funds and the Charles Schwab Corporation surveyed 500 African Americans with an annual income above $50,000, and 500 whites with an annual income above $50,000. The results indicated that 74% of the African Americans and 84% of the whites owned stocks (Cheryl Winokur Munk, "Stock-Ownership Race Gap Shrinks, *The Wall Street Journal*, June 13, 2002, B11).
a. Is there evidence of a significant difference between the proportion of African Americans with incomes above $50,000 who invest in stocks and the proportion of whites with incomes above $50,000 who invest in stocks? (Use $\alpha = 0.05$.)
b. Determine the p-value in (a) and interpret its meaning.
c. Construct and interpret a 95% confidence interval estimate of the difference between the population proportion of African Americans with incomes above $50,000 who invest in stocks and the population proportion of whites with incomes above $50,000 who invest in stocks.

10.4 *F* TEST FOR THE DIFFERENCE BETWEEN TWO VARIANCES

Often you need to test whether two independent populations have the same variability. One important reason to test for the difference between the variances of two populations is to determine whether the pooled-variance t test is appropriate.

The test for the difference between the variances of two independent populations is based on the ratio of the two sample variances. If you assume that each population is normally distributed, then the ratio S_1^2/S_2^2 follows the ***F* distribution** (see Table E.5). The critical values of the F distribution in Table E.5 depend on two sets of degrees of freedom. The degrees of freedom in the numerator of the ratio are for the first sample, and the degrees of freedom in the denominator are for the second sample. Equation (10.11) defines the ***F*-test statistic for testing the equality of two variances**.

F STATISTIC FOR TESTING THE EQUALITY OF TWO VARIANCES

The F-test statistic is equal to the variance of sample one divided by the variance of sample two.

$$F = \frac{S_1^2}{S_2^2} \qquad \textbf{(10.11)}$$

where

S_1^2 = variance of sample 1

S_2^2 = variance of sample 2

n_1 = size of sample taken from population 1

n_2 = size of sample taken from population 2

$n_1 - 1$ = degrees of freedom from sample 1 (i.e., the numerator degrees of freedom)

$n_2 - 1$ = degrees of freedom from sample 2 (i.e., the denominator degrees of freedom)

The test statistic F follows an F distribution with $n_1 - 1$ and $n_2 - 1$ degrees of freedom.

For a given level of significance α, to test the null hypothesis of equality of variances

$$H_0: \sigma_1^2 = \sigma_2^2$$

against the alternative hypothesis that the two population variances are not equal

$$H_1: \sigma_1^2 \neq \sigma_2^2$$

you reject the null hypothesis if the computed F-test statistic is greater than the upper-tail critical value F_U from the F distribution with $n_1 - 1$ degrees of freedom in the numerator and $n_2 - 1$ degrees of freedom in the denominator, or if the computed F-test statistic falls below the lower-tail critical value F_L from the F distribution with $n_1 - 1$ and $n_2 - 1$ degrees of freedom in the numerator and denominator, respectively. Thus, the decision rule is

Reject H_0 if $F > F_U$

or if $F < F_L$;

otherwise do not reject H_0.

This decision rule and rejection regions are displayed in Figure 10.17.

FIGURE 10.17
Regions of Rejection and Nonrejection for the Two-Tail *F* Test

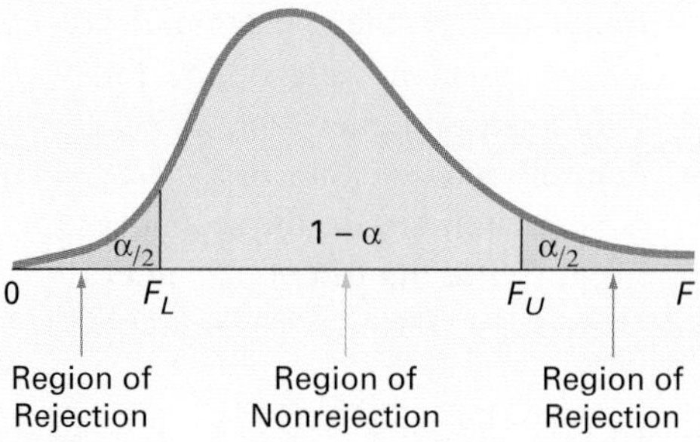

To illustrate how to use the *F* test to determine whether the two variances are equal, return to the "Using Statistics" scenario concerning the sales of BLK cola in two different aisle locations. To determine whether to use the pooled-variance *t* test or the separate-variance *t* test in section 10.1, you can test the equality of the two population variances. The null and alternative hypotheses are:

$$H_0: \sigma_1^2 = \sigma_2^2$$
$$H_1: \sigma_1^2 \neq \sigma_2^2$$

Because this is a two-tail test, the rejection region is split into the lower and upper tails of the *F* distribution. Using a level of significance of $\alpha = 0.05$, each rejection region contains 0.025 of the distribution.

Because there are samples of 10 stores for each of the two display locations, there are $10 - 1 = 9$ degrees of freedom for group 1 and also for group 2. F_U, the upper-tail critical value of the *F* distribution, is found directly from Table E.5, a portion of which is presented in Table 10.6. Because there are 9 degrees of freedom in the numerator and 9 degrees of freedom in the denominator, you find the upper-tail critical value F_U by looking in the column labeled "9" and the row labeled "9." Thus, the upper-tail critical value of this *F* distribution is 4.03.

TABLE 10.6
Finding F_U, Upper-Tail Critical Value of *F* with 9 and 9 Degrees of Freedom for Upper-Tail Area of 0.025

Denominator	Numerator df_1						
df_2	1	2	3	...	7	8	9
1	647.80	799.50	864.20	...	948.20	956.70	963.30
2	38.51	39.00	39.17	...	39.36	39.37	39.39
3	17.44	16.04	15.44	...	14.62	14.54	14.47
.	.	.	.		.	.	.
.	.	.	.		.	.	.
.	.	.	.		.	.	.
7	8.07	6.54	5.89	...	4.99	4.90	4.82
8	7.57	6.06	5.42	...	4.53	4.43	4.36
9	7.21	5.71	5.08	...	4.20	4.10	4.03

Source: Extracted from Table E.5.

Finding Lower-Tail Critical Values

You compute F_L, a lower-tail critical value on the *F* distribution with $n_1 - 1$ degrees of freedom in the numerator and $n_2 - 1$ degrees of freedom in the denominator, by taking the reciprocal of F_{U*}, an upper-tail critical value on the *F* distribution with degrees of freedom "switched" (i.e., $n_2 - 1$ degrees of freedom in the numerator and $n_1 - 1$ degrees of freedom in the denominator). This relationship is shown in Equation (10.12).

FINDING LOWER-TAIL CRITICAL VALUES FROM THE *F* DISTRIBUTION

$$F_L = \frac{1}{F_{U*}} \tag{10.12}$$

where F_{U*} is from an F distribution with $n_2 - 1$ degrees of freedom in the numerator and $n_1 - 1$ degrees of freedom in the denominator.

In the cola sales example, the degrees of freedom are 9 and 9 for the respective numerator sample and denominator sample so there is no "switching" of degrees of freedom; you just take the reciprocal. Therefore, to compute the lower-tail 0.025 critical value, you need to find the upper-tail 0.025 critical value of F with 9 degrees of freedom in the numerator and 9 degrees of freedom in the denominator and take its reciprocal. As shown in Table 10.6 on page 376, this upper-tail value is 4.03. Using Equation (10.12),

$$F_L = \frac{1}{F_{U*}} = \frac{1}{4.03} = 0.248$$

As depicted in Figure 10.18, the decision rule is

Reject H_0 if $F > F_U = 4.03$
or if $F < F_L = 0.248$;
otherwise do not reject H_0

FIGURE 10.18
Regions of Rejection and Nonrejection for Two-Tail *F* Test for Equality of Two Variances at 0.05 Level of Significance with 9 and 9 Degrees of Freedom

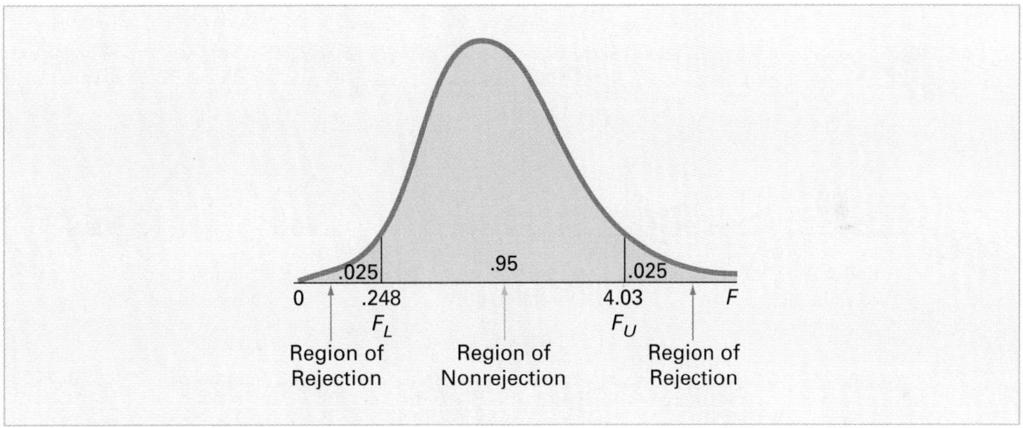

Using Equation (10.11) on page 375 and the cola sales data (see Table 10.1 on page 348), the F-test statistic is

$$F = \frac{S_1^2}{S_2^2}$$

$$= \frac{350.6778}{157.3333} = 2.23$$

Because $F_L = 0.248 < F = 2.23 < F_U = 4.03$, do not reject H_0. The p-value is 0.248 for a two-tail test (twice the p-value for the one-tail test shown in the Microsoft Excel output of Figure 10.19). Figure 10.20 illustrates Minitab output. Since $0.248 > 0.05$, you conclude that there is no significant difference in the variability of the sales of cola for the two display locations.

FIGURE 10.19
Microsoft Excel *F* Test Output for the Cola Sales Data

	A	B	C
1	F-Test Two-Sample for Variances		
2			
3		*Normal*	*End-Aisle*
4	Mean	50.3	72
5	Variance	350.6777778	157.3333333
6	Observations	10	10
7	df	9	9
8	F	2.228884181	
9	P(F<=f) one-tail	0.124104358	
10	F Critical one-tail	3.178893105	

FIGURE 10.20
Minitab *F* Test Output for the Cola Sales Data

```
95% Bonferroni confidence intervals for standard deviations

  Display  N     Lower    StDev    Upper
EndAisle  10    8.2048  12.5433  25.2578
  Normal  10   12.2494  18.7264  37.7085


F-Test (normal distribution)
Test statistic = 0.45, p-value = 0.248
```

In testing for a difference in two variances using the *F* test described in this section, you assume that each of the two populations are normally distributed. The *F* test is very sensitive to the normality assumption. If box-and-whisker plots or normal probability plots suggest even a mild departure from normality for either of the two populations, you should not use the *F* test. In this case, a nonparametric approach is more appropriate (see references 1 and 2).

In testing for the equality of variances, as part of assessing the validity of the pooled-variance *t*-test procedure, the *F* test is a two-tail test. However, when you are interested in the variability itself, the *F* test is often a one-tail test. Thus, in testing the equality of two variances, you can use either a two-tail or one-tail test, depending on whether you are testing whether the two population variances are *different* or whether one variance is *greater than* the other variance. Figure 10.21 illustrates the three possible situations.

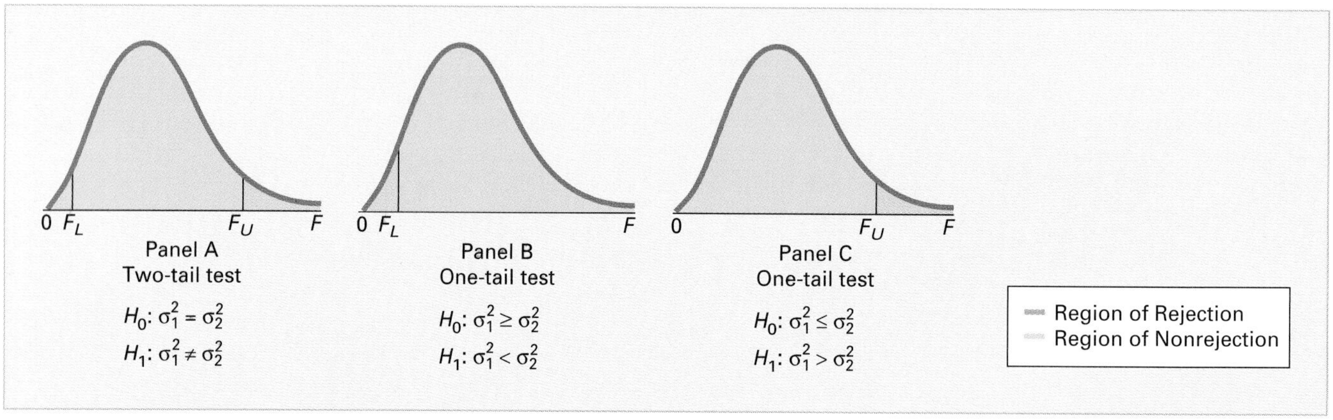

FIGURE 10.21 Determining the Rejection Region for Testing the Equality of Two Population Variances

Often the sample sizes in the two groups will differ. Example 10.4 demonstrates how to find a lower-tail critical value from the *F* distribution in this situation.

EXAMPLE 10.4

FINDING THE LOWER-TAIL CRITICAL VALUE FROM THE *F* DISTRIBUTION IN A TWO-TAIL TEST OF HYPOTHESIS

You select a sample of $n_1 = 8$ from a normally distributed population. The variance for this sample (S_1^2) is 56.0. You select a sample of $n_2 = 10$ from a second normally distributed population (independent of the first population). The variance for this sample (S_2^2) is 24.0. Using a level of significance of $\alpha = 0.05$, test the null hypothesis of no difference in the two population variances against the two-tail alternative that there is evidence of a significant difference in the population variances.

SOLUTION The null and alternative hypotheses are

$$H_0: \sigma_1^2 = \sigma_2^2$$
$$H_1: \sigma_1^2 \neq \sigma_2^2$$

The *F*-test statistic is given by Equation (10.11) on page 375.

$$F = \frac{S_1^2}{S_2^2}$$

You use Table E.5 to find the upper and lower critical values of the *F* distribution. With $n_1 - 1 = 7$ degrees of freedom in the numerator and $n_2 - 1 = 9$ degrees of freedom in the denominator and $\alpha = 0.05$ split equally into the lower- and upper-tail rejection regions of 0.025 each, the upper critical value $F_U = 4.20$ (see Table 10.7).

To find the lower critical value F_L with 7 degrees of freedom in the numerator and 9 degrees of freedom in the denominator, take the reciprocal of F_{U*} with degrees of freedom switched to 9 in the numerator and 7 in the denominator. Thus, from Equation (10.12) on page 377 and Table 10.7,

$$F_L = \frac{1}{F_{U*}} = \frac{1}{4.82} = 0.207$$

TABLE 10.7

Finding F_{U*} and F_L, with 7 and 9 Degrees of Freedom Using a Level of Significance $\alpha = 0.05$

Denominator	**Numerator df_1**						
df_2	**1**	**2**	**3**	**. . .**	**7**	**8**	**9**
1	647.80	799.50	864.20	. . .	948.20	956.70	963.30
2	38.51	39.00	39.17	. . .	39.36	39.37	39.39
3	17.44	16.04	15.44	. . .	14.62	14.54	14.47
.	.	.	.			.	
.	.	.	.			.	
.	.	.	.			.	
7	8.07	6.54	5.89	. . .	4.99	4.90	4.82
8	7.57	6.06	5.42	. . .	4.53	4.43	4.36
9	7.21	5.71	5.08	. . .	4.20	4.10	4.03

Source: Extracted from Table E.5.

The decision rule is

Reject H_0 if $F > F_U = 4.20$

or if $F < F_L = 0.207$;

otherwise do not reject H_0.

From Equation (10.11) on page 375, the *F*-test statistic is

$$F = \frac{S_1^2}{S_2^2} = \frac{56.0}{24.0} = 2.33$$

Because $F_L = 0.207 < F = 2.33 < F_U = 4.20$, you do not reject H_0. Using a 0.05 level of significance, you conclude that there is no evidence of a significant difference in the variances in these two independent populations.

PROBLEMS FOR SECTION 10.4

Learning the Basics

PH Grade ASSIST **10.40** Determine F_U and F_L, the upper- and lower-tail critical values of F in each of the following two-tail tests:

a. $\alpha = 0.10, n_1 = 16, n_2 = 21$
b. $\alpha = 0.05, n_1 = 16, n_2 = 21$
c. $\alpha = 0.02, n_1 = 16, n_2 = 21$
d. $\alpha = 0.01, n_1 = 16, n_2 = 21$

10.41 Determine F_U, the upper-tail critical value of F in each of the following one-tail tests:

a. $\alpha = 0.05, n_1 = 16, n_2 = 21$
b. $\alpha = 0.025, n_1 = 16, n_2 = 21$
c. $\alpha = 0.01, n_1 = 16, n_2 = 21$
d. $\alpha = 0.005, n_1 = 16, n_2 = 21$

10.42 Determine F_L, the lower-tail critical value of F in each of the following one-tail tests:

a. $\alpha = 0.05, n_1 = 16, n_2 = 21$
b. $\alpha = 0.025, n_1 = 16, n_2 = 21$
c. $\alpha = 0.01, n_1 = 16, n_2 = 21$
d. $\alpha = 0.005, n_1 = 16, n_2 = 21$

PH Grade ASSIST **10.43** The following information is available for two samples drawn from independent normally distributed populations:

$$n_1 = 25 \quad S_1^2 = 133.7 \quad n_2 = 25 \quad S_2^2 = 161.9$$

What is the value of the F-test statistic if you are testing the null hypothesis $H_0: \sigma_1^2 = \sigma_2^2$?

PH Grade ASSIST **10.44** In problem 10.43, how many degrees of freedom are there in the numerator and denominator of the F test?

PH Grade ASSIST **10.45** In problems 10.43 and 10.44, what are the critical values for F_U and F_L from Table E.5 if the level of significance α is 0.05 and the alternative hypothesis is $H_1: \sigma_1^2 \neq \sigma_2^2$?

10.46 In problems 10.43 through 10.45, what is your statistical decision?

10.47 The following information is available for two samples selected from independent but very right-skewed populations.

$$n_1 = 16 \quad S_1^2 = 47.3 \quad n_2 = 13 \quad S_2^2 = 36.4$$

Should you use the F test to test the null hypothesis of equality of variances ($H_0: \sigma_1^2 = \sigma_2^2$)? Discuss.

10.48 If the two samples are drawn from independent normally distributed populations in problem 10.47,

a. At the 0.05 level of significance, is there evidence of a difference between σ_1^2 and σ_2^2?
b. Suppose that you want to perform a one-tail test. At the 0.05 level of significance, what is the upper-tail critical value of the F-test statistic to determine whether there is evidence that $\sigma_1^2 > \sigma_2^2$? What is your statistical decision?
c. Suppose that you want to perform a one-tail test. At the 0.05 level of significance, what is the lower-tail critical value of the F-test statistic to determine whether there is evidence that $\sigma_1^2 < \sigma_2^2$? What is your statistical decision?

Applying the Concepts

Problems 10.49–10.54 can be solved manually or by using Microsoft Excel, Minitab, or SPSS.

PH Grade ASSIST **10.49** A professor in the accounting department of a business school claims that there is much more variability in the final exam scores of students taking the introductory accounting course as a requirement than for students taking the course as part of a major in accounting. Random samples of 13 nonaccounting majors (group 1) and 10 accounting majors (group 2) are taken from the professor's class roster in his large lecture, and the following results are computed based on the final exam scores:

$$n_1 = 13 \quad S_1^2 = 210.2 \quad n_2 = 10 \quad S_2^2 = 36.5$$

a. At the 0.05 level of significance, is there evidence to support the professor's claim?
b. Interpret the p-value.
c. What assumptions do you make here about the two populations in order to justify your use of the F test?

SELF Test **10.50** The Computer Anxiety Rating Scale (CARS) measures an individual's level of computer anxiety on a scale from 20 (no anxiety) to 100 (highest level of anxiety). Researchers at Miami University administered CARS to 172 business students. One of the objectives of the study was to determine if there is a difference between the level of computer anxiety experienced by female students and male students.

	Males	Females
$\bar{X}$	40.26	36.85
S	13.35	9.42
n	100	72

Source: Travis Broome and Douglas Havelka, "Determinants of Computer Anxiety in Business Students," The Review of Business Information Systems, *Spring 2002, 6(2):9–16.*

a. At the 0.05 level of significance, is there evidence of a difference in the variability of the computer anxiety experienced by females and males?
b. Interpret the p-value.
c. What assumptions do you need to make about the two populations in order to justify the use of the F test?
d. Based on (a) and (b), which t test defined in section 10.1 should you use to test whether there is a significant difference in mean computer anxiety for female and male students?

10.51 A bank with a branch located in a commercial district of a city **BANK1** has developed an improved process for serving customers during the noon to 1 P.M. lunch period. The waiting time (defined as the time elapsed from when the customer enters the line until he or she reaches the teller window) of all customers during this hour is recorded over a period of one week. A random sample of 15 customers is selected, and the results (in minutes) are as follows:

4.21 5.55 3.02 5.13 4.77 2.34 3.54 3.20
4.50 6.10 0.38 5.12 6.46 6.19 3.79

Suppose that another branch located in a residential area **BANK2** is also concerned with the noon to 1 P.M. lunch period. A random sample of 15 customers is selected, and the results are as follows:

9.66 5.90 8.02 5.79 8.73 3.82 8.01 8.35
10.49 6.68 5.64 4.08 6.17 9.91 5.47

a. Is there evidence of a difference in the variability of the waiting time between the two branches? (Use $\alpha = 0.05$.)
b. Find the p-value in (a) and interpret its meaning.
c. What assumption is necessary in (a)? Is the assumption valid for these data?
d. Based on the results of (a), is it appropriate to use the pooled-variance t-test to compare the means of the two branches?

PH Grade ASSIST **10.52** A problem with a telephone line that prevents a customer from receiving or making calls is disconcerting to both the customer and the telephone company. The following data **PHONE** represent samples of 20 problems reported to two different offices of a telephone company and the time to clear these problems (in minutes) from the customers' lines:

Central Office I Time to Clear Problems (minutes)

1.48 1.75 0.78 2.85 0.52 1.60 4.15 3.97 1.48 3.10
1.02 0.53 0.93 1.60 0.80 1.05 6.32 3.93 5.45 0.97

Central Office II Time to Clear Problems (minutes)

7.55 3.75 0.10 1.10 0.60 0.52 3.30 2.10 0.58 4.02
3.75 0.65 1.92 0.60 1.53 4.23 0.08 1.48 1.65 0.72

a. Is there evidence of a difference in the variability of the waiting time between the two offices? (Use $\alpha = 0.05$.)
b. Find the p-value in (a) and interpret its meaning.
c. What assumption is necessary in (a)? Is the assumption valid for these data?
d. Based on the results of (a), is it appropriate to use the pooled-variance t test to compare the means of the two offices?

10.53 The director of training for a company manufacturing electronic equipment is interested in determining whether different training methods have an effect on the productivity of assembly-line employees. She randomly assigns 42 recently hired employees to two groups of 21. The first group receives a computer-assisted, individual-based training program and the other receives a team-based training program. Upon completion of the training, the employees are evaluated on the time (in seconds) it takes to assemble a part. The results are in the data file: **TRAINING**.
a. Using a 0.05 level of significance, is there evidence of a difference between the variances in assembly times (in seconds) of employees trained in a computer-assisted, individual-based program and those trained in a team-based program?
b. On the basis of the results in (a), is it appropriate to use the pooled-variance t test to compare the means of the two groups? Discuss.

10.54 Shipments of meat, meat by-products, and other ingredients are mixed together in several filling lines at a pet food canning factory. Operations managers suspect that, although the mean amount filled in the can of pet food is usually the same, the variability of the cans filled in line A is greater than that of line B. The following results are from samples of eight-ounce cans:

	Line A	Line B
$\bar{X}$	8.005	7.997
S	0.012	0.005
n	11	16

At the 0.05 level of significance, is there evidence that the variance in line A is greater than the variance in line B? Assume that the population amounts filled are normally distributed.

SUMMARY

In this chapter, you were introduced to statistical test procedures in analyzing possible differences between two independent populations. In addition, you learned a test procedure frequently used when analyzing differences between the means of two related populations. Remember that you need to select the test most appropriate for a given set of conditions, and to critically investigate the validity of the assumptions underlying each of the hypothesis-testing procedures.

The roadmap in Figure 10.22 illustrates the steps needed in determining which two-sample test of hypothesis to use.

1. What type of data do you have? If you are dealing with categorical variables, use the Z test for the difference between two proportions.
2. If you have a numerical variable, determine whether you have independent samples or related samples. If you have related samples, use the paired t test.
3. If you have independent samples, determine whether you can assume that the variances of the two groups are equal. (This assumption can be tested using the F test.)
4. If you can assume that the two groups have equal variances, use the pooled-variance t test. If you cannot assume that the two groups have equal variances, use the separate-variance t test.

Table 10.8 provides a list of topics covered in this chapter.

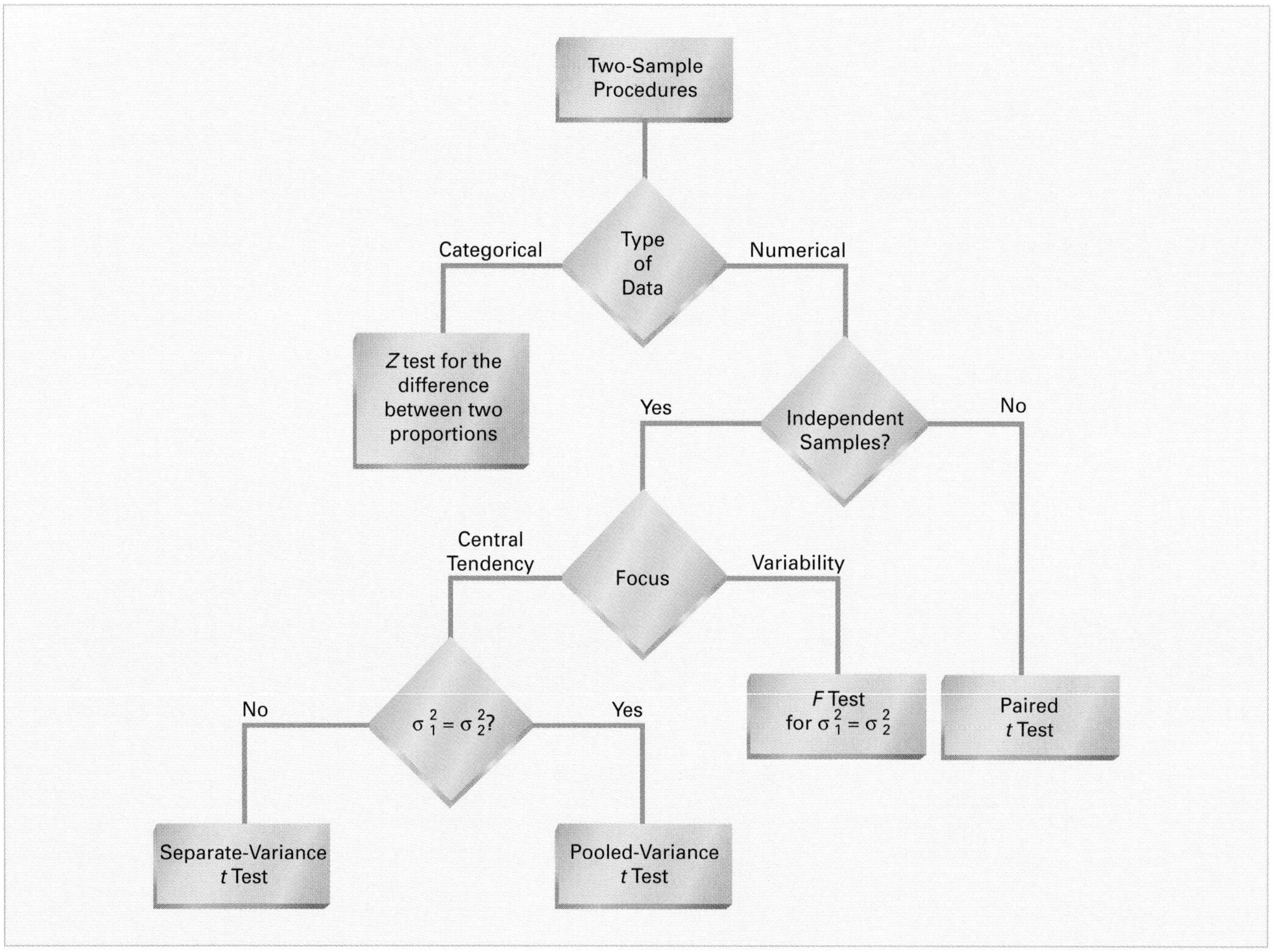

FIGURE 10.22 Roadmap for Selecting a Two-Sample Test of Hypothesis

TABLE 10.8
Summary of Topics in Chapter 10

	Type of Data	
Type of Analysis	**Numerical**	**Categorical**
Comparing Two Populations	Z and t Tests for the Difference in the Means of Two Independent Populations **(section 10.1)** Paired t Test **(section 10.2)** F Test for Differences in Two Variances **(section 10.4)**	Z Test for the Difference Between Two Proportions **(section 10.3)**

KEY FORMULAS

Z Test for the Difference Between Two Means

$$Z = \frac{(\bar{X}_1 - \bar{X}_2) - (\mu_1 - \mu_2)}{\sqrt{\dfrac{\sigma_1^2}{n_1} + \dfrac{\sigma_2^2}{n_2}}} \quad \textbf{(10.1)}$$

Pooled-Variance t Test for the Difference Between Two Means

$$t = \frac{(\bar{X}_1 - \bar{X}_2) - (\mu_1 - \mu_2)}{\sqrt{S_p^2\left(\dfrac{1}{n_1} + \dfrac{1}{n_2}\right)}} \quad \textbf{(10.2)}$$

Confidence Interval Estimate of the Difference in the Means of Two Independent Populations

$$(\bar{X}_1 - \bar{X}_2) \pm t_{n_1+n_2-2}\sqrt{S_p^2\left(\frac{1}{n_1} + \frac{1}{n_2}\right)} \quad \textbf{(10.3)}$$

or

$$(\bar{X}_1 - \bar{X}_2) - t_{n_1+n_2-2}\sqrt{S_p^2\left(\frac{1}{n_1} + \frac{1}{n_2}\right)} \le \mu_1 - \mu_2$$
$$\le (\bar{X}_1 - \bar{X}_2) + t_{n_1+n_2-2}\sqrt{S_p^2\left(\frac{1}{n_1} + \frac{1}{n_2}\right)}$$

Separate-Variance t Test for the Difference Between Two Means

$$t = \frac{(\bar{X}_1 - \bar{X}_2) - (\mu_1 - \mu_2)}{\sqrt{\dfrac{S_1^2}{n_1} + \dfrac{S_2^2}{n_2}}} \quad \textbf{(10.4)}$$

Computing Degrees of Freedom in Separate-Variance t Test

$$\nu = \frac{\left(\dfrac{S_1^2}{n_1} + \dfrac{S_2^2}{n_2}\right)^2}{\dfrac{\left(\dfrac{S_1^2}{n_1}\right)^2}{n_1 - 1} + \dfrac{\left(\dfrac{S_2^2}{n_2}\right)^2}{n_2 - 1}} \quad \textbf{(10.5)}$$

Z Test for the Mean Difference

$$Z = \frac{\bar{D} - \mu_D}{\dfrac{\sigma_D}{\sqrt{n}}} \quad \textbf{(10.6)}$$

Paired t Test for the Mean Difference

$$t = \frac{\bar{D} - \mu_D}{\dfrac{S_D}{\sqrt{n}}} \quad \textbf{(10.7)}$$

Confidence Interval Estimate for the Mean Difference

$$\bar{D} \pm t_{n-1}\frac{S_D}{\sqrt{n}} \quad \textbf{(10.8)}$$

or

$$\bar{D} - t_{n-1}\frac{S_D}{\sqrt{n}} \le \mu_D \le \bar{D} + t_{n-1}\frac{S_D}{\sqrt{n}}$$

Z Test for the Difference Between Two Proportions

$$Z = \frac{(p_1 - p_2) - (\pi_1 - \pi_2)}{\sqrt{\bar{p}(1 - \bar{p})\left(\dfrac{1}{n_1} + \dfrac{1}{n_2}\right)}} \quad \textbf{(10.9)}$$

Confidence Interval Estimate for the Difference Between Two Proportions

$$(p_1 - p_2) \pm Z\sqrt{\frac{p_1(1 - p_1)}{n_1} + \frac{p_2(1 - p_2)}{n_2}} \quad \textbf{(10.10)}$$

or

$$(p_1 - p_2) - Z\sqrt{\frac{p_1(1 - p_1)}{n_1} + \frac{p_2(1 - p_2)}{n_2}} \le (\pi_1 - \pi_2)$$
$$\le (p_1 - p_2) + Z\sqrt{\frac{p_1(1 - p_1)}{n_1} + \frac{p_2(1 - p_2)}{n_2}}$$

F Statistic for Testing the Equality of Two Variances

$$F = \frac{S_1^2}{S_2^2} \quad (10.11)$$

Finding Lower-Tail Critical Values from the F Distribution

$$F_L = \frac{1}{F_{U*}} \quad (10.12)$$

KEY TERMS

F distribution 375
F-test statistic for testing the equality of two variances 375
matched 358
paired 358
paired t test for the mean difference in related populations 360
pooled-variance t test 347
repeated measurements 358
robust 351
separate-variance t test 352
Z test for difference between two means 346
Z test for the difference between two proportions 368

CHAPTER REVIEW PROBLEMS

Checking Your Understanding

10.55 What are some of the criteria used in the selection of a particular hypothesis-testing procedure?

10.56 Under what conditions should you use the pooled-variance t test to examine possible differences in the means of two independent populations?

10.57 Under what conditions should you use the F test to examine possible differences in the variances of two independent populations?

10.58 What is the difference between two independent and two related populations?

10.59 What is the distinction between repeated measurements and matched (or paired) items?

10.60 Under what conditions should the paired t test for the mean difference between two related populations be used?

10.61 Explain the similarities and differences between the test of hypothesis for the difference between the means of two independent populations, and the confidence interval estimate of the difference between the means.

Applying the Concepts

Problems 10.62–10.69 can be solved manually or by using Microsoft Excel, Minitab, or SPSS. We recommend that you use Microsoft Excel, Minitab, or SPSS to solve problems 10.70–10.82.

10.62 A study reported in the *Journal of Business Strategies* compared music compact disk prices for Internet-based retailers and traditional brick-and-mortar retailers (Lee Zoonky, and Sanjay Gosain, "A Longitudinal Price Comparison for Music CDs in Electronic and Brick-and-Mortar Markets: Pricing Strategies in Emergent Electronic Commerce," Spring 2002, 19(1):55–72). Before collecting the data, the researchers carefully defined several research hypotheses including:

1. Price dispersion on the Internet will be lower than the price dispersion in the brick-and-mortar market.
2. Prices in electronic markets will be lower than prices in physical markets.

a. Consider research hypothesis 1. Write the null and alternative hypotheses in terms of population parameters. Carefully define the population parameters used.
b. Define a Type I and Type II error for the hypotheses in (a).
c. What type of statistical test should you use?
d. What assumptions are needed to perform the test you selected?
e. Repeat (a) through (d) for research hypothesis 2.

10.63 The pet-drug market is growing very rapidly. Before introducing new pet drugs into the marketplace, they must be approved by the U.S. Food and Drug Administration (FDA). In 1999, the Norvatis company was trying to get Anafranil, a drug to reduce dog anxiety, approved. According to an article (Elyse Tanouye, "The Ow in Bowwow: With Growing Market in Pet Drugs, Makers Revamp Clinical Trials," *The Wall Street Journal*, April 13, 1999), Norvatis had to find a way to translate a dog's anxiety symptoms into numbers that could be used to prove to the FDA that the drug had a *statistically significant effect* on the condition.

a. What is meant by the phrase statistically significant effect?
b. Consider an experiment where dogs suffering from anxiety are divided into two groups. One group will be given Anafranil, and the other will be given a placebo (a drug without active ingredients). How can you translate a dog's anxiety symptoms into numbers? In other words, define a continuous variable X_1, the measurement of

effectiveness of the drug Anafranil, and X_2, the measurement of the effectiveness of the placebo.

c. Building on your answer to part (b), define the null and alternative hypotheses for this study.

10.64 In response to lawsuits filed against the tobacco industry, many companies such as Philip Morris are running television advertisements that are supposed to educate teenagers about the dangers of smoking. Are these tobacco industry antismoking campaigns successful? Are state-sponsored antismoking commercials more effective? An article (Gordon Fairclough, "Philip Morris's Antismoking Campaign Draws Fire," *The Wall Street Journal*, April 6, 1999, B1) discussed a study in California that compared commercials made by the state of California and commercials produced by Philip Morris. Researchers showed the state ads and the Philip Morris ads to a group of California teenagers and measured the effectiveness of both. The researchers concluded that the state ads were more effective in relaying the dangers of smoking than the Philip Morris ads. The article suggests, however, that the study is not *statistically reliable* because the sample size was too small and because the study specifically selected participants who were considered more likely to start smoking than others.

a. How do you think the researchers measured effectiveness?
b. Define the null and alternative hypotheses for this study.
c. Explain the risks associated with Type I and Type II errors in this study.
d. What type of test is most appropriate in this situation?
e. What do you think is meant by the phrase *statistically reliable*?

10.65 The FedEx St. Jude Classic professional golf tournament is held each year in Memphis, Tennessee. FedEx sponsors this PGA tournament, and part of the proceeds goes to the St. Jude Children's Research Hospital. In 2003, the tournament raised $679,115 for the hospital. This type of corporate sponsorship is known as cause-related marketing. A survey of spectators at the tournament were surveyed and asked to respond to a series of questions on a 5-point scale (1 = Strongly Disagree, 2 = Disagree, 3 = Neutral, 4 = Agree, and 5 = Strongly Agree). Four of the questions asked are listed below:

1. Cause-related marketing creates a positive company image.
2. I would be willing to pay more for a service that supports a cause I care about.
3. Cause marketing should be a standard part of a company's activities.
4. Based on its support of St. Jude, I will be more likely to use FedEx services.

For each question, the researchers tested the null hypothesis that the mean response for males and females is equal. The alternative hypothesis is that the mean response is different for males and females. The following table summarizes the results.

	Sample Mean			
Question	Female (n_1=137)	Male (n_2=305)	t	p-value
1	4.46	4.26	1.907	0.057
2	4.09	3.86	2.105	0.035
3	4.26	3.91	3.258	0.001
4	4.12	4.06	0.567	0.571

Source: Extracted from R. L. Irwin, T. Lachowetz, T. B. Cornwell, and J. S. Cook, 2003, "Cause-Related Sport Sponsorship: An Assessment of Spectator Beliefs, Attitudes, and Behavioral Intentions," Sport Marketing Quarterly, *12(3): 131–139.*

a. Interpret the results of the t test for question 1.
b. Interpret the results of the t test for question 2.
c. Interpret the results of the t test for question 3.
d. Interpret the results of the t test for question 4.
e. Write a short summary about the differences between males and females concerning their views toward cause-related sponsorship.

10.66 A large public utility company wants to compare the consumption of electricity during the summer season for single-family houses in two counties that it services. For each household sampled, the monthly electric bill is recorded with the following results:

	County I	County II
$\bar{X}$	$115	$98
S	$30	$18
n	25	21

Use a level of significance of 0.01.

a. Is there evidence that the mean bill in county II is above $80?
b. Is there evidence of a difference between the variances of bills in county I and in county II?
c. Is there evidence that the mean monthly bill is higher in county I than in county II?
d. Construct and interpret a 99% confidence interval estimate of the difference between the mean monthly bill in county I and county II.

10.67 The manager of computer operations of a large company wants to study computer usage of two departments within the company, the accounting department and the research department. A random sample of five jobs from the accounting department in the past week and six jobs from the research department in the past week are selected, and the processing time (in seconds) for each job is recorded. ACCRES

Department	Processing Time (in Seconds)					
Accounting	9	3	8	7	12	
Research	4	13	10	9	9	6

Use a level of significance of 0.05.

a. Is there evidence that the mean processing time in the research department is greater than 6 seconds?

b. Is there evidence of a difference between the variances in the processing time of the two departments?

c. Is there evidence of a difference between the mean processing time of the accounting department and the research department?

d. Determine the p-values in (a), (b), and (c) and interpret their meanings.

e. Construct and interpret a 95% confidence interval estimate of the difference in the mean processing times between the accounting and research departments.

10.68 A computer information systems professor is interested in studying the amount of time it takes students enrolled in the Introduction to Computers course to write and run a program in Visual Basic. The professor hires you to analyze the following results (in minutes) from a random sample of nine students: VB

10 13 9 15 12 13 11 13 12

a. At the 0.05 level of significance, is there evidence that the population mean amount is greater than 10 minutes? What will you tell the professor?

b. Suppose the computer professor, when checking her results, realizes that the fourth student needed 51 minutes rather than the recorded 15 minutes to write and run the Visual Basic program. At the 0.05 level of significance, reanalyze the question posed in (a) using the revised data. What will you tell the professor now?

c. The professor is perplexed by these paradoxical results and requests an explanation from you regarding the justification for the difference in your findings in (a) and (b). Discuss.

d. A few days later, the professor calls to tell you that the dilemma is completely resolved. The original number 15 (the fourth data value) was correct, and therefore your findings in (a) are being used in the article she is writing for a computer journal. Now she wants to hire you to compare the results from that group of Introduction to Computers students against those from a sample of 11 computer majors in order to determine whether there is evidence that computer majors can write a Visual Basic program in less time than introductory students. The sample mean is 8.5 minutes and the sample standard deviation is 2.0 minutes for the computer majors. At the 0.05 level of significance, completely analyze these data. What will you tell the professor?

e. A few days later the professor calls again to tell you that a reviewer of her article wants her to include the p-value for the "correct" result in (a). In addition, the professor inquires about an unequal variances problem, which the reviewer wants her to discuss in her article. In your own words, discuss the concept of p-value and describe the unequal variances problem. Determine the p-value in (a) and discuss whether or not the unequal variances problem had any meaning in the professor's study.

10.69 Over the last several years, the use of cell phones has increased dramatically. An article in *USA Today* (D. Sharp, "Cellphones Reveal Screaming Lack of Courtesy," *USA Today*, September 2001, 4A) reported that according to a poll, the mean talking time per month for cell phones was 372 minutes for men and 275 minutes for women, while the mean talking time per month for traditional home phones was 334 minutes for men and 510 minutes for women. Suppose that the poll was based on a sample of 100 men and 100 women, and that the standard deviation of the talking time per month for cell phones was 120 minutes for men and 100 minutes for women, while the standard deviation of the talking time per month for traditional home phones was 100 minutes for men and 150 minutes for women.

Use a level of significance of 0.05.

a. Is there evidence of a difference in the mean monthly talking time on cell phones for men and women?

b. Is there evidence of a difference in the mean monthly talking time on traditional home phones for men and women?

c. Construct and interpret a 95% confidence interval estimate of the difference in the mean monthly talking time on cell phones for men and women.

d. Construct and interpret a 95% confidence interval estimate of the difference in the mean monthly talking time on traditional home phones for men and women.

e. Is there evidence of a difference in the variance of the monthly talking time on cell phones for men and women?

f. Is there evidence of a difference in the variance of the monthly talking time on traditional home phones for men and women?

g. Based on the results of (a) through (f), what conclusions can you make concerning cell phone and traditional home phone usage between men and women?

10.70 *Working Woman* magazine conducted a large study to determine typical salaries for men and women in many different types of jobs (Extracted from "Annual Salary Survey," *Working Woman*, July/August 2001, 44–47). A data set containing 114 job titles along with the corresponding typical salaries for men and women is included in the data file SALARIES.

a. At the 0.01 level of significance, is there evidence that the mean salary for men is greater than the mean salary for women?
b. Compute the p-value in (a).

10.71 Use the data in the COLLEGES2002 file to compare public and private universities in terms of first quartile of SAT scores, third quartile of SAT scores, total cost, room and board cost, and total indebtedness at graduation. Use the 0.05 level of significance.

10.72 The lengths of life (in hours) of a sample of forty 100-watt lightbulbs produced by manufacturer A and a sample of forty 100-watt lightbulbs produced by manufacturer B are in the file BULBS. Completely analyze the differences between the life of the bulbs produced by two manufacturers (use $\alpha = 0.05$).

10.73 The data contained in the file PETFOOD2 consist of the cost per serving, cups per can, protein in grams, and fat in grams for 97 varieties of dry and canned dog and cat food. Completely analyze the differences between cat food and dog food for the variables of cost per serving, protein in grams, and fat in grams. Do a similar analysis comparing the differences between dry food and canned food (use $\alpha = 0.05$).

Source: Copyright © 1998 by Consumers Union of U.S., Inc., Yonkers, NY 10703-1057. Adapted with permission from Consumer Reports, *February 1998, 18–19.*

10.74 Use the data in the file AUTO2002 to compare 2002 sports utility vehicles (SUVs) and non-SUVs in terms of miles per gallon, length, width, turning-circle requirement, weight, and luggage capacity. Completely analyze differences between SUVs and non-SUVs (use $\alpha = 0.05$).

Source: "The 2002 cars," Copyright © 2002 by Consumers Union of U.S., Inc., Yonkers, NY 10703-1057. Adapted with permission from Consumer Reports, *April 2002.*

10.75 Zagat's publishes restaurant ratings for various locations in the United States. The data file RESTRATE contains the Zagat rating for food, décor, service, and the price per person for a sample of 50 restaurants located in New York City and 50 restaurants located on Long Island. Completely analyze the differences between New York City and Long Island restaurants for the variables food rating, décor rating, service rating, and price per person, using $\alpha = 0.05$.

Source: Extracted from Zagat Survey 2002 New York City *Restaurants and* Zagat Survey 2001–2002 Long Island Restaurants.

10.76 The data found in the file BEER represent the price of a six-pack of 12-ounce bottles, the calories per 12 fluid ounces, the percentage of alcohol content per 12 fluid ounces, the type of beer (i.e., craft lager, craft ale, imported lager, regular or ice beer, and light or nonalcoholic beer), and the country of origin (U.S. versus imported) for each of the 69 beers that were sampled. Completely analyze the differences between beers brewed in the United States and those that were imported in terms of price, calories, and alcohol content (use $\alpha = 0.05$).

Source: "Beers," Copyright © 1996 by Consumers Union of U.S., Inc., Yonkers, NY 10703-1057. Adapted with permission from Consumer Reports, *June 1996.*

10.77 In manufacturing processes there is a term called *work-in-process* (often abbreviated WIP). In a book manufacturing plant WIP represents the time it takes for sheets from a press to be folded, gathered, sewn, tipped on end sheets, and bound. The following data represent samples of 20 books at each of two production plants and the processing time (operationally defined as the time in days from when the books came off the press to when they were packed in cartons). WIP

Plant A

5.62 5.29 16.25 10.92 11.46 21.62 8.45 8.58 5.41 11.42
11.62 7.29 7.50 7.96 4.42 10.50 7.58 9.29 7.54 8.92

Plant B

9.54 11.46 16.62 12.62 25.75 15.41 14.29 13.13 13.71 10.04
5.75 12.46 9.17 13.21 6.00 2.33 14.25 5.37 6.25 9.71

Completely analyze the differences between the processing times for the two plants using $\alpha = 0.05$, and write a summary of your findings to be presented to the vice president for operations of the company.

10.78 Do marketing promotions, such as bobble-head giveaways, increase attendance at Major League Baseball games? An article reported on the effectiveness of marketing promotions (Extracted from T. C. Boyd and T. C. Krehbiel, "Promotion Timing in Major League Baseball and the Stacking Effects of Factors that Increase Game Attractiveness," *Sport Marketing Quarterly*, March 2003, 12, 173–184). The data file ROYALS includes the following variables for the Kansas City Royals during the 2002 baseball season:

GAME = Home games in the order they were played
ATTENDANCE = Paid attendance for the game
PROMOTION – 1 = if a promotion was held; 0 = if no promotion was held

a. At the 0.05 level of significance, is there evidence of a difference between the variances in the attendance at games with promotions and games without promotions?
b. Based on the result of (a), conduct the appropriate test of hypothesis to determine whether there is a difference in the mean attendance at games with promotions and games without promotions. (Use $\alpha = 0.05$.)
c. Write a brief summary of your results.

10.79 The manufacturer of "Boston" and "Vermont" asphalt shingles know that product weight is a major factor in the customer's perception of quality. Moreover, the weight represents the amount of raw materials being used and is therefore very important to the company from a cost standpoint. The last stage of the assembly-line packages the shingles before they are placed on wooden pallets. Once a pallet is full (a pallet for most brands holds 16 squares of shingles), it is weighed and the measurement is recorded. The data file PALLET contains the weight (in pounds) from a sample of 368 pallets of Boston shingles and 330 pallets of Vermont shingles. Completely analyze the differences in the weight of the Boston and Vermont shingles using $\alpha = 0.05$.

10.80 The manufacturer of "Boston" and "Vermont" asphalt shingles provide their customers with a 20-year warranty on most of their products. To determine whether a shingle will last as long as the warranty period, accelerated-life testing is conducted at the manufacturing plant. Accelerated-life testing exposes the shingle to the stresses it would be subject to in a lifetime of normal use in a laboratory setting via an experiment that takes only a few minutes to conduct. In this test, a shingle is repeatedly scraped with a brush for a short period of time and the amount of shingle granules that are removed by the brushing is weighed (in grams). Shingles that experience low amounts of granule loss are expected to last longer in normal use than shingles that experience high amounts of granule loss. In this situation, a shingle should experience no more than 0.8 grams of granule loss if it is expected to last the length of the warranty period. The data file GRANULE contains a sample of 170 measurements made on the company's Boston shingles, and 140 measurements made on Vermont shingles. Completely analyze the differences in the granule loss of the Boston and Vermont shingles using $\alpha = 0.05$.

Report Writing Exercise

10.81 Referring to the results of problems 10.79 and 10.80 concerning the weight and granule loss of Boston and Vermont shingles, write a report that summarizes your conclusions.

TEAM PROJECT

The data file MUTUALFUNDS2004 contains information regarding 12 variables from a sample of 121 mutual funds. The variables are:

Fund—The name of the mutual fund.
Category—Type of stocks comprising the mutual fund—small cap, mid cap, large cap
Objective—Objective of stocks comprising the mutual fund—growth or value
Assets—In millions of dollars
Fees—Sales charges (no or yes)
Expense ratio—expenses as a percentage of net assets
2003 Return—Twelve-month return in 2003
Three-year return—Annualized return 2001–2003
Five-year return—Annualized return 1999–2003
Risk—Risk-of-loss factor of the mutual fund classified as low, average, or high
Best quarter—Best quarterly performance 1999–2003
Worst quarter—Worst quarterly performance 1999–2003

10.82 Completely analyze the difference between mutual funds without fees and mutual funds with fees in terms of 2003 return, 3-year return, and the 5-year return. Write a report summarizing your findings.

RUNNING CASE MANAGING THE *SPRINGVILLE HERALD*

A marketing department team is charged with improving the telemarketing process in order to increase the number of home-delivery subscriptions sold. After several brainstorming sessions, it was clear that the longer a caller speaks to a respondent the greater is the chance that the caller will sell a home-delivery subscription. Therefore, the team decided to find ways to increase the length of the phone calls.

Initially, the team investigated the impact that time of call might have on the length of the call. Under current arrangements, calls were made in the evening hours between 5:00 P.M. and 9:00 P.M., Monday through Friday. The team wanted to compare length of calls made early in the evening (before 7:00 P.M.) with those made later in the evening (after 7:00 P.M.) to determine whether one of these time periods is more conducive to lengthier calls and, correspondingly, to increased subscription sales. The team selected a sample of 30 female callers who staff the telephone bank on Wednesday evenings and randomly assigned 15 of them to the "early" group and 15 to the "later" group. The callers knew that the team was observing their efforts that evening but didn't know which calls were monitored. The callers had been trained to make their telephone presentations in a structured manner. They were to read from a script and their greeting was personal but informal ("Hi, this is Mary Jones from the *Springville Herald*—may I speak to Bill Richards?").

Measurements were taken on the length of call (defined as the difference, in seconds, between the time the person answers the phone and the time he or she hangs up). The results are presented in Table SH10.1. **SH10**

TABLE SH10.1

Length of Calls in Seconds Based on Time of Call—Early Versus Late in the Evening

Time of Call		Time of Call	
Early	**Late**	**Early**	**Late**
41.3	37.1	40.6	40.7
37.5	38.9	33.3	38.0
39.3	42.2	39.6	43.6
37.4	45.7	35.7	43.8
33.6	42.4	31.3	34.9
38.5	39.0	36.8	35.7
32.6	40.9	36.3	47.4
37.3	40.5		

EXERCISES

SH10.1 Analyze the data in Table SH10.1 and write a report to the marketing department team that indicates your findings. Include an attached appendix in which you discuss the reason you selected a particular statistical test to compare the two independent groups of callers.

SH10.2 Suppose that instead of the research design described here, there were only 15 callers sampled and each caller was to be monitored twice in the evening, once in the early time period and once in the later time period. Suppose that in Table SH10.1 each pair of values represents a particular caller's two measurements. Reanalyze these data and write a report for presentation to the team that indicates your findings.

SH10.3 What other variables should be investigated next? Why?

WEB CASE

Apply your knowledge about hypothesis testing in this Web Case that continues the cereal-fill packaging dispute Web Case from Chapters 7 and 9.

After Oxford Cereals conducted its public experiment about cereal box weights, the TriCities Consumers Concerned About Cereal Companies That Cheat (TCCACCTC) remains unconvinced that Oxford Cereals has not misled the public. The group has created and posted a document in which they claim that cereal boxes produced at Plant Number 2 in Springville always weigh less than the claimed 368 grams. Review the group's document and its data sample at the TCCACCTC Web site **www.prenhall.com/Springville/MoreOnCheaters.htm** and then answer the following:

1. Do the TCCACCTC's results prove that there is a statistical difference in the mean weights of cereal boxes produced at Plant Numbers 1 and 2?
2. Perform the appropriate analysis to test the TCCACCTC's hypothesis. What conclusions can you reach based on their data?

REFERENCES

1. Conover, W. J., *Practical Nonparametric Statistics*, 3rd ed. (New York: Wiley, 2000).
2. Daniel, W., *Applied Nonparametric Statistics*, 2nd ed. (Boston: Houghton Mifflin, 1990).
3. *Microsoft Excel 2003* (Redmond, WA: Microsoft Corp., 2003).
4. *Minitab for Windows Version 14* (State College, PA: Minitab Inc., 2004).
5. Satterthwaite, F. E., "An Approximate Distribution of Estimates of Variance Components," *Biometrics Bulletin*, 2 (1946): 110–114.
6. Snedecor, G. W., and W. G. Cochran, *Statistical Methods*, 7th ed. (Ames, IA: Iowa State University Press, 1980).
7. *SPSS Base 12.0 Brief Guide* (Upper Saddle River, NJ: Prentice Hall, 2003).
8. Winer, B. J., *Statistical Principles in Experimental Design*, 2nd ed. (New York: McGraw-Hill, 1971).

Appendix 10 Using Software for Two-Sample Tests

A10.1 MICROSOFT EXCEL

For Z Test for the Difference Between Two Means

For raw data: Open to the worksheet containing the raw data for the two samples.
Select **Tools → Data Analysis**.
Select **z-Test: Two Sample for Means** from the Data Analysis list and click **OK**. In the procedure's dialog box (see Figure A10.1):

> Enter the cell range of one sample as the **Variable 1 Range**.
> Enter the cell range of the other sample as the **Variable 2 Range**.
> Enter the value of the **Hypothesized Mean Difference**.
> Enter the population variance of the first sample as the **Variable 1 Variance (known)**.
> Enter the population variance of the other sample as the **Variable 2 Variance (known)**.
> Select **Labels** if the variable cell ranges include labels in their first rows.
> Click **OK**.

Results appear on a separate worksheet.

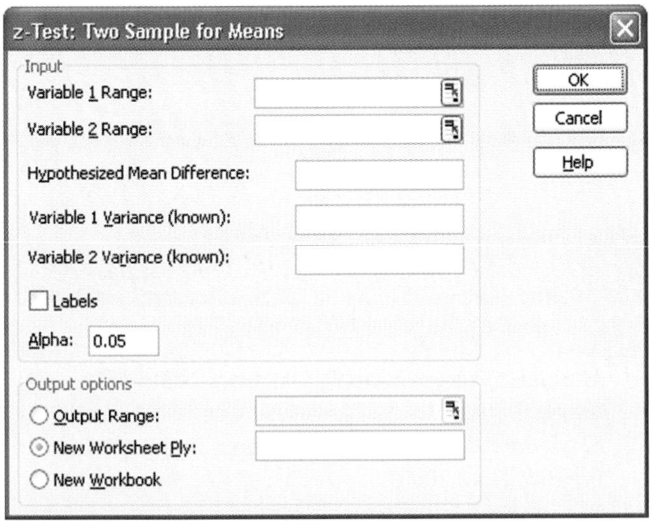

FIGURE A10.1 Microsoft Excel *Z*-Test: Two Sample for Means Dialog Box

For summarized data: Open the **Z Two Means.xls** file. This worksheet already contains the entries for problem 10.1 and uses the NORMSINV and NORMSDIST functions (see section G.24 for more information). To adapt this worksheet to other problems, change the values in the tinted cells in rows 4 through 13.
OR See section G.24 (**Z Test for Differences in Two Means**) if you want PHStat2 to produce a worksheet for you.

For Pooled-Variance *t* Test

For raw data: Open to a worksheet containing the raw data for the two samples such as the Data Worksheet of the **COLA.xls** file.
Select **Tools → Data Analysis**.
Select **t-test: Two-Sample Assuming Equal Variances** from the Data Analysis list and click **OK**. In the procedure's dialog box (see Figure A10.2):

> Enter the cell range of one sample as the **Variable 1 Range**.
> Enter the cell range of the other sample as the **Variable 2 Range**.
> Enter the value for the **Hypothesized Mean Difference**.
> Select **Labels** if the variable cell ranges include labels in their first rows.
> Click **OK**.

Results appear on a separate worksheet.

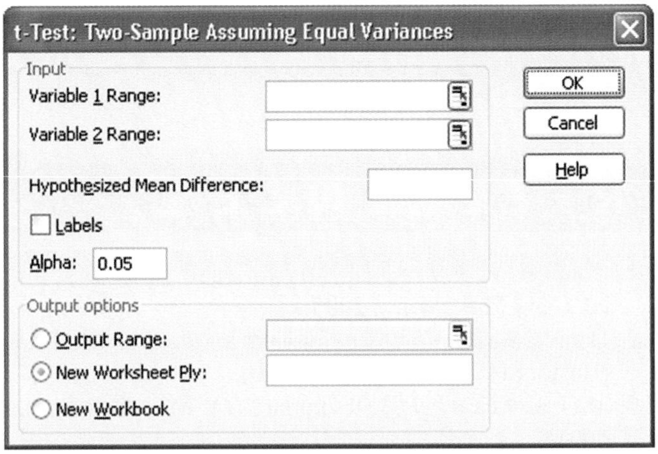

FIGURE A10.2 Microsoft Excel *t*-Test: Two Sample Assuming Equal Variances Dialog Box

For summarized data: Open the **Pooled-Variance T.xls** file. This worksheet already contains the entries for the section 10.1 cola sales example and uses the TINV and TDIST functions (see section G.25 for more information). To adapt this worksheet to other problems, change the values in the tinted cells in rows 4 through 13.

OR See section G.25 (***t* Test for Differences in Two Means**) if you want PHStat2 to produce a worksheet for you.

For Separate-Variance *t* Test for the Difference Between Two Means

Open to the worksheet containing the raw data for the two samples.

Select **Tools → Data Analysis**.

Select **t-test: Two-Sample Assuming Unequal Variances** from the Data Analysis list and click **OK**. In the procedure's dialog box (see Figure A10.3):

Enter the cell range of one sample as the **Variable 1 Range**.
Enter the cell range of the other sample as the **Variable 2 Range**.
Enter the value for the **Hypothesized Mean Difference**.
Select **Labels** if the variable cell ranges include labels in their first rows.
Click **OK**.

Results appear on a separate worksheet.

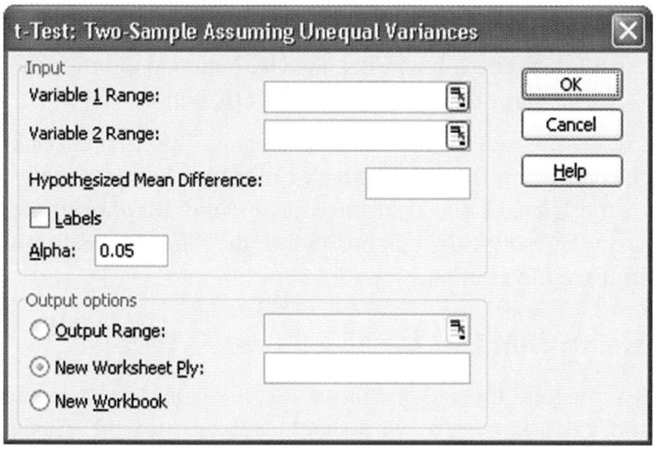

FIGURE A10.3 Microsoft Excel *t*-Test: Two Sample Assuming Unequal Variances Dialog Box

For Paired *t* Test for the Difference Between Two Means

Open to a worksheet containing the raw data for the two samples such as the Data Worksheet of the **COMPTIME.xls** file.

Select **Tools → Data Analysis**.

Select **t-test: Paired Two Sample for Means** from the Data Analysis list and click **OK**. In the procedure's dialog box (see Figure A10.4):

Enter the cell range of one sample as the **Variable 1 Range**.
Enter the cell range of the other sample as the **Variable 2 Range**.
Enter the value of the **Hypothesized Mean Difference**.
Select **Labels** if the variable cell ranges include labels in their first rows.
Click **OK**.

Results appear on a separate worksheet.

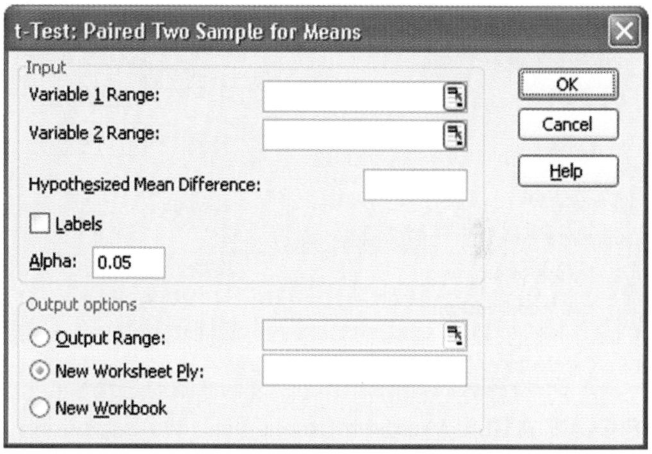

FIGURE A10.4 Microsoft Excel *t*-Test: Paired Two Sample for Means Dialog Box

For *Z* Test for the Difference Between Two Proportions

Open the **Z Two Proportions.xls** file, shown in Figure 10.15 on page 371. This worksheet already contains the entries for the section 10.3 guest satisfaction example. This worksheet uses the NORMSINV and NORMSDIST functions (see section G.26 for more information). To adapt this worksheet to other problems, change the null hypothesis and the level of significance values and the number of successes and sample size for each sample in rows 4 through 11.

OR See section G.26 (**Z Test for the Differences in Two Proportions**) if you want PHStat2 to produce a worksheet for you.

For *F* Test for the Difference Between Two Variances

For raw data: Open to the worksheet containing the raw data for the two samples.

Select **Tools → Data Analysis**.

Select **F-Test Two-Sample for Variances** from the Data Analysis list and click **OK**. In the procedure's dialog box (see Figure A10.5):

Enter the cell range of one sample as the **Variable 1 Range**.
Enter the cell range of the other sample as the **Variable 2 Range**.
Select **Labels** if the variable cell ranges include labels in their first rows.
Click **OK**.

Results appear on a separate worksheet.

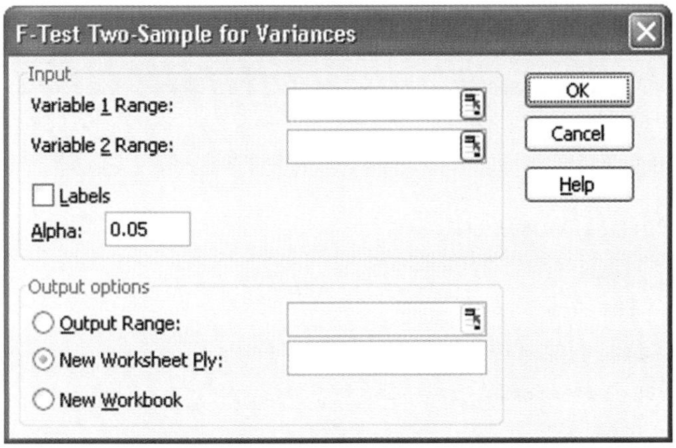

FIGURE A10.5 Microsoft Excel *F*-Test Two-Sample for Variances Dialog Box

For summarized data: Open the **F Two Variances.xls file**. This worksheet already contains the entries for the section 10.4 cola sales example and uses the FINV and FDIST functions (see section G.27 for more information). To adapt this worksheet to other problems, change the values in the tinted cells in rows 4 through 10.
OR See section G.27 (**F Test for Differences in Two Variances**) if you want PHStat2 to produce a worksheet for you.

A10.2 MINITAB

Using Minitab for the t Test for the Difference Between Two Means

To illustrate the use of Minitab for the *t* test of the difference in two means, open the **COLA.MTW** worksheet. Select **Stat ➔ Basic Statistics ➔ 2-Sample t**.

1. If the data are stacked with the values in one column and the categories in a second column, as they are in this worksheet, select the **Samples in one column** option button. (If the samples are in different columns, select the **Samples in different columns** option button and enter the column numbers or names.)
2. In the 2-Sample *t* dialog box (see Figure A10.6), select the **Samples in one column** option button. In the Samples: edit box, enter **C1** or **Sales**. In the Subscripts: edit box, enter **C2** or **Display**. Select the **Assume equal variances** check box for the pooled-variance *t* test. Leave this edit box unchecked for the separate-variance *t* test. Click the **Options** button.

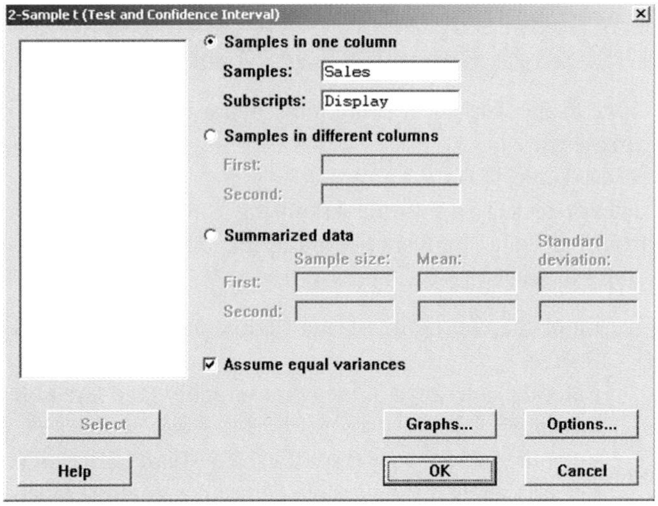

FIGURE A10.6 Minitab 2-Sample *t* Dialog Box

3. In the Alternative: drop-down list box, select **not equal** for the two-tail test performed on page 355. (If performing a one-tail test, select less than or select greater than.) Click the **OK** button.
4. Click the **Graphs** option button, and select the **Boxplots of data** check box. Click the **OK** button to return to the 2-sample *t* dialog box. Click the **OK** button.

(To get the results in Figure 10.4 on page 349, first unstack the data using **Data ➔ Unstack Columns**, then enter **Sales** in the Unstack the data in: edit box and **Display** in the Using Subscripts in: edit box. Select the **After last column in use** option button.)

Using Minitab for the Paired t Test

To illustrate the use of Minitab for the paired *t* test, open the COMPTIME.MTW worksheet. Select **Stat ➔ Basic Statistics ➔ Paired t**.

1. In the Paired *t* dialog box (see Figure A10.7), in the First sample: edit box, enter **C1** or **Current**. In the Second sample: edit box, enter **C2** or **New**. Click the **Options** button.
2. In the Alternative: drop-down list box, select **greater than** for the one-tail test performed on page 363. Click the **OK** button to return to the Paired *t* dialog box. Click the **Graphs** button, and select the **Boxplot of differences** check box. Click the **OK** button to return to the Paired *t* dialog box. Click the **OK button**.

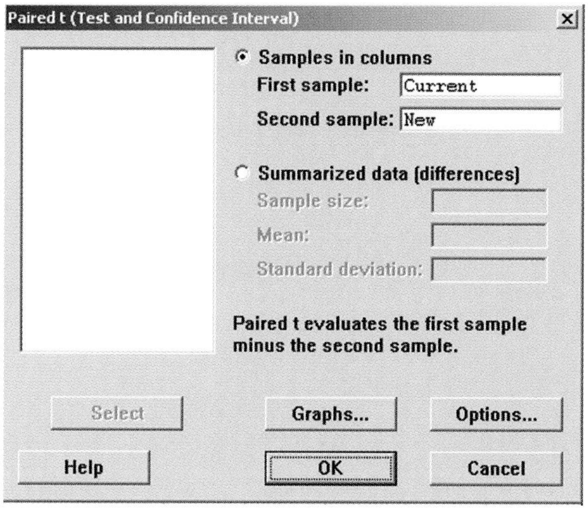

FIGURE A10.7 Minitab Paired *t* Dialog Box

Testing for the Difference Between Two Proportions

To test the hypothesis for the difference in the proportion of guests who would return between the Beachcomber and the Windsurfer illustrated in Figure 10.16 on page 371, select **Stat ➔ Basic Statistics ➔ 2 Proportions**.

1. In the 2 Proportions (Test and Confidence Interval) dialog box (see Figure A10.8), select the **Summarized data** option button.
2. In the First: row, enter **227** in the Trials: edit box and **163** in the Events: edit box. In the Second: row, enter **262** in the Trials: edit box and **154** in the Events: edit box.

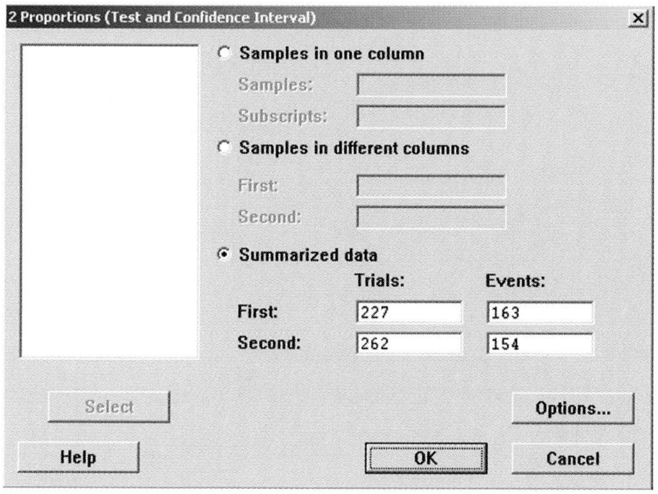

FIGURE A10.8 2 Proportions (Test and Confidence Interval) Dialog Box

3. Click the **Options** button. In the 2 Proportions - Options dialog box (see Figure A10.9) enter **95** in the Confidence level: edit box. Enter **0.0** in the Test difference: edit box. Select the **Use pooled estimate of p for test** option button. In the Alternative: drop down list box, select **not equal** for a two-tail test. Click the **OK** button to return to 2-Proportions (Test and Confidence Interval) dialog box. Click the **OK** button.

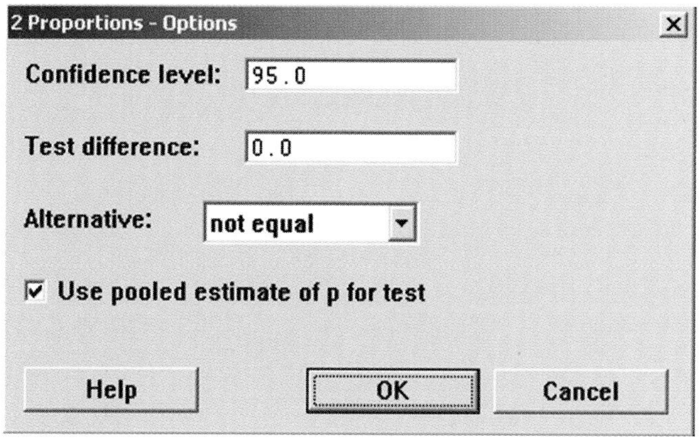

FIGURE A10.9 2 Proportions—Options Dialog Box

Using Minitab for the *F* Test for the Difference Between Two Variances

To illustrate the use of Minitab for the *F* test of the difference in two variances, open the COLA.MTW worksheet. Select **Stat ➔ Basic Statistics ➔ 2 Variances**.

1. In the 2 Variances dialog box (see Figure A10.10), select the **Samples in one column** option button. In the Samples: edit box, enter **C1** or **Sales**. In the Subscripts: edit box, enter **C2** or **Display**. (If you have samples in different columns, select the **Samples in different columns** option button and enter the names of the two columns in the edit boxes. If you have summarized data instead of the actual data, select the **Summarized data** option button and enter values for the sample size and variance.)
2. Click the **OK** button.

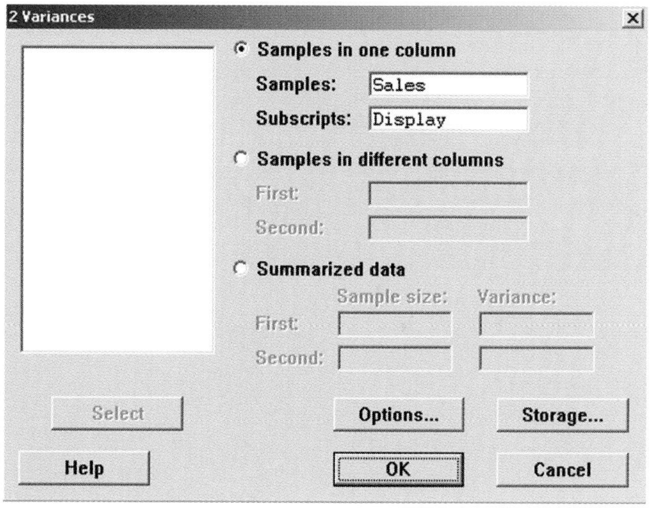

FIGURE A10.10 Minitab 2 Variances Dialog Box

CHAPTER 11

Analysis of Variance

USING STATISTICS: The Perfect Parachute Company

LEARNING OBJECTIVES

In this chapter, you learn:

- The basic concepts of experimental design
- How to use the one-way analysis of variance to test for differences among the means of several groups
- When to use a randomized block design
- How to use the two-way analysis of variance and interpret the interaction

USING STATISTICS

The Perfect Parachute Company

You are the production manager of the Perfect Parachute Company. Parachutes are woven in your factory using a synthetic fiber purchased from one of four different suppliers. For obvious reasons, one of the most important quality characteristics of a parachute is its strength. You need to decide whether the synthetic fibers from your four suppliers result in parachutes of equal strength. Furthermore, there are two types of looms in the factory: the *Jetta* and the *Turk*. Are the parachutes woven on the Jetta looms and those woven on the Turk looms equally strong? Moreover, are any differences in the strength of the parachute that can be attributed to the four suppliers dependent on the type of loom used? To help answer these questions, you have decided to design an experiment to test the strength of parachutes woven from the synthetic fibers from the four suppliers and the two different types of looms. You will need to incorporate the information from the analysis of the experimental data to determine which supplier and type of loom to use in order to manufacture the strongest parachutes.

In Chapter 10 you used hypothesis-testing to draw conclusions about possible differences between two populations. Frequently, however, you need to evaluate differences among several populations (referred to as groups in this chapter).

This chapter begins by examining the *completely randomized design*, which has one *factor* (such as type of tire, marketing strategy, brand of drug, or different suppliers as in the "Using Statistics" scenario), with several groups (such as the four suppliers in the "Using Statistics" scenario). The *randomized block design*, which is also used to study one factor, is introduced next. The completely randomized design is then extended to the *factorial design*, where more than one factor at a time is simultaneously studied in a single experiment. Throughout the chapter, emphasis is placed on the assumptions behind the use of the various testing procedures.

11.1 THE COMPLETELY RANDOMIZED DESIGN: ONE-WAY ANALYSIS OF VARIANCE

Many applications involve experiments in which you consider more than two **groups** pertaining to one **factor** of interest. Groups are defined by assigning different **levels** of the factor. For example, a factor such as baking temperature may have several *numerical levels* (e.g., 300°, 350°, 400°, 450°) or a factor such as preferred supplier for a parachute manufacturer may have several *categorical levels* (e.g., Supplier 1, Supplier 2, Supplier 3, Supplier 4). One-factor experiments are also called **completely randomized designs**.

F Test for Differences Among More Than Two Means

When you are analyzing a numeral variable and certain assumptions are met, you use the **analysis of variance (ANOVA)** to compare the means of the groups. The ANOVA procedure used for the completely randomized design is referred to as a **one-way ANOVA** and is an extension of the *t* test for the difference between two means discussed in section 10.1. Although ANOVA is an acronym for **AN**alysis **O**f **VA**riance, the term is misleading because the objective is to analyze differences among the group means *not* the variances. However, by analyzing the variation among and within the groups, you can make conclusions about possible differences in group means. In ANOVA, the total variation is subdivided into variation that is due to differences *among* the groups and variation that is due to differences *within* the groups (see Figure 11.1). **Within-group variation** is considered **random error**. **Among-group variation** is due to differences from group to group, also known as the **treatment effect**. The symbol c is used to indicate the number of groups.

FIGURE 11.1

Partitioning the Total Variation in a Completely Randomized Design

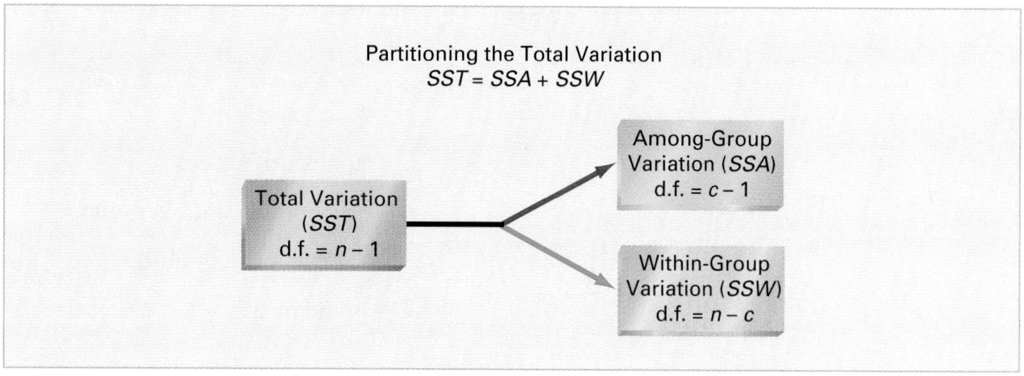

Assuming that the c groups represent populations whose values are randomly and independently selected, follow a normal distribution, and have equal variances, the null hypothesis of no differences in the population means

$$H_0: \mu_1 = \mu_2 = \cdots = \mu_c$$

is tested against the alternative that not all the c population means are equal.

$$H_1: \text{Not all } \mu_j \text{ are equal (where } j = 1, 2, \ldots, c)$$

To perform an ANOVA test of equality of population means, you subdivide the total variation in the values into two parts, that which is due to variation among the groups and that which is due to variation within the groups. The **total variation** is represented by the **sum of squares total (*SST*)**. Because the population means of the c groups are presumed to be equal under the null hypothesis, you compute the total variation among all the values by summing the squared differences between each individual value and the **grand mean** $\bar{\bar{X}}$. The grand mean is the mean of all the values in all the groups combined. Equation (11.1) shows the computation of the total variation.

TOTAL VARIATION IN ONE-WAY ANOVA

$$SST = \sum_{j=1}^{c}\sum_{i=1}^{n_j}(X_{ij} - \bar{\bar{X}})^2 \qquad \textbf{(11.1)}$$

where $\bar{\bar{X}} = \dfrac{\sum_{j=1}^{c}\sum_{i=1}^{n_j} X_{ij}}{n}$ = grand mean

X_{ij} = ith value in group j

n_j = number of values in group j

n = total number of values in all groups combined (that is, $n = n_1 + n_2 + \ldots + n_c$)

c = number of groups

You compute the among-group variation, usually called the **sum of squares among groups (*SSA*)**, by summing the squared differences between the sample mean of each group $\bar{X}_j$ and the grand mean $\bar{\bar{X}}$, weighted by the sample size n_j in each group. Equation (11.2) shows the computation of the among-group variation.

AMONG-GROUP VARIATION IN ONE-WAY ANOVA

$$SSA = \sum_{j=1}^{c} n_j(\bar{X}_j - \bar{\bar{X}})^2 \qquad \textbf{(11.2)}$$

where c = number of groups

n_j = number of values in group j

$\bar{X}_j$ = sample mean of group j

$\bar{\bar{X}}$ = grand mean

The within-group variation, usually called the **sum of squares within groups (*SSW*)**, measures the difference between each value and the mean of its own group and sums the squares of these differences over all groups. Equation (11.3) shows the computation of the within-group variation.

WITHIN-GROUP VARIATION IN ONE-WAY ANOVA

$$SSW = \sum_{j=1}^{c}\sum_{i=1}^{n_j}(X_{ij} - \bar{X}_j)^2 \qquad \textbf{(11.3)}$$

where X_{ij} = ith value in group j

$\bar{X}_j$ = sample mean of group j

Because you are comparing c groups, there are $c - 1$ degrees of freedom associated with the sum of squares among groups. Since each of the c groups contributes $n_j - 1$ degrees of freedom, there are $n - c$ degrees of freedom associated with the sum of squares within groups. In addition, there are $n - 1$ degrees of freedom associated with the sum of squares total because you are comparing each value X_{ij} to the grand mean $\bar{\bar{X}}$ based on all n values.

If you divide each of these sums of squares by its associated degrees of freedom, you have three variances or **mean square** terms—***MSA*** (Mean Square Among), ***MSW*** (Mean Square Within), and ***MST*** (Mean Square Total).

COMPUTING THE MEAN SQUARES IN ONE-WAY ANOVA

$$MSA = \frac{SSA}{c - 1} \qquad \textbf{(11.4a)}$$

$$MSW = \frac{SSW}{n - c} \qquad \textbf{(11.4b)}$$

$$MST = \frac{SST}{n - 1} \qquad \textbf{(11.4c)}$$

Although you want to compare the means of the c groups to determine whether a difference exists among them, the ANOVA procedure derives its name from the fact that you are comparing variances. If the null hypothesis is true and there are no real differences in the c group means, all three mean square terms—*MSA, MSW*, and *MST*—which themselves are *variances*, provide estimates of the overall variance in the data. Thus, to test the null hypothesis

$$H_0: \mu_1 = \mu_2 = \cdots = \mu_c$$

against the alternative

$$H_1: \text{Not all } \mu_j \text{ are equal (where } j = 1, 2, \ldots, c)$$

you compute the **one-way ANOVA *F*-test statistic** as the ratio of *MSA* to *MSW* as in Equation (11.5).

ONE-WAY ANOVA *F*-TEST STATISTIC

$$F = \frac{MSA}{MSW} \quad \textbf{(11.5)}$$

The F-test statistic follows an ***F* distribution** with $c - 1$ degrees of freedom corresponding to *MSA* in the numerator and $n - c$ degrees of freedom corresponding to *MSW* in the denominator. For a given level of significance α, you reject the null hypothesis if the F-test statistic computed in Equation (11.5) is greater than the upper-tail critical value F_U from the F distribution having $c - 1$ degrees of freedom in the numerator and $n - c$ in the denominator (see Table E.5). Thus, as shown in Figure 11.2, the decision rule is

Reject H_0 if $F > F_U$;

otherwise do not reject H_0

FIGURE 11.2

Regions of Rejection and Nonrejection When Using ANOVA

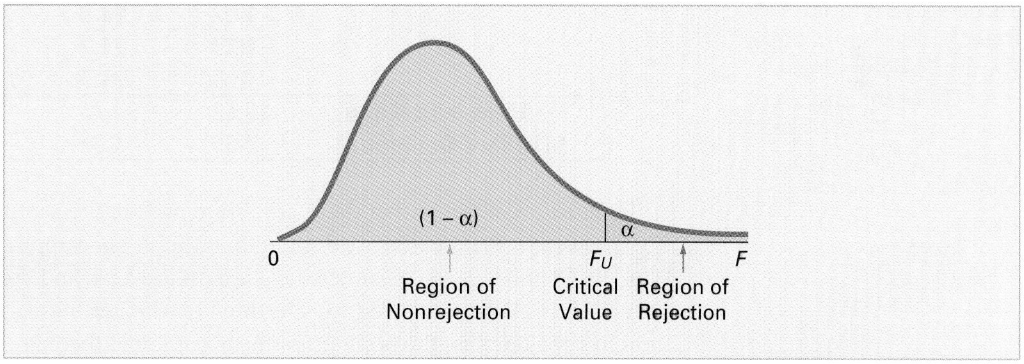

If the null hypothesis is true, the computed F statistic is expected to be approximately equal to 1, because both the numerator and denominator mean square terms are estimating the overall variance in the data. If H_0 is false (and there are real differences in the means), the computed F statistic is expected to be substantially larger than 1 because the numerator, *MSA*, is estimating the treatment effect or differences among groups in addition to the overall variability in the values, while the denominator, *MSW*, is measuring only the overall variability. Thus, the ANOVA procedure provides an F test in which you reject the null hypothesis at a selected α level of significance only if the computed F statistic is greater than F_U, the upper-tail critical value of the F distribution having $c - 1$ and $n - c$ degrees of freedom, as illustrated in Figure 11.2.

The results of an analysis of variance are usually displayed in an **ANOVA summary table**, as shown in Table 11.1. The entries in this table include the sources of variation (i.e., among-group, within-group, and total), the degrees of freedom, the sums of squares, the mean squares (i.e., the variances), and the computed F statistic. In addition, the p-value (i.e., the probability of having an F statistic as large as or larger than the one computed, given that the null hypothesis is true) is included in the ANOVA summary table. The p-value allows you to make direct conclusions about the null hypothesis without referring to a table of critical values of the F distribution. If the p-value is less than the chosen level of significance α, you reject the null hypothesis.

TABLE 11.1
Analysis-of-Variance Summary Table

Source	Degrees of Freedom	Sum of Squares	Mean Square (Variance)	F
Among groups	$c - 1$	SSA	$MSA = \dfrac{SSA}{c-1}$	$F = \dfrac{MSA}{MSW}$
Within groups	$n - c$	SSW	$MSW = \dfrac{SSW}{n-c}$	
Total	$n - 1$	SST		

To illustrate the one-way ANOVA F test, return to the "Using Statistics" scenario concerning the Perfect Parachute Company (see page 396). An experiment was conducted to determine if any significant differences exist in the strength of parachutes woven from synthetic fibers from the different suppliers. Five parachutes were woven for each group—Supplier 1, Supplier 2, Supplier 3, and Supplier 4. The strength of the parachutes is measured by placing them in a testing device that pulls on both ends of a parachute until it tears apart. The amount of force required to tear the parachute is measured on a tensile-strength scale where the larger the value the stronger the parachute. The results of this experiment (in terms of tensile strength) are contained in the data file PARACHUTE and are displayed in Figure 11.3 along with some summary computations of key descriptive statistics.

FIGURE 11.3
Microsoft Excel Worksheet of the Tensile Strength for Parachutes Woven with Synthetic Fibers from Four Different Suppliers Along with the Sample Mean and Sample Standard Deviation

	Supplier 1	Supplier 2	Supplier 3	Supplier 4
	18.5	26.3	20.6	25.4
	24.0	25.3	25.2	19.9
	17.2	24.0	20.8	22.6
	19.9	21.2	24.7	17.5
	18.0	24.5	22.9	20.4
Arithmetic Mean	19.52	24.26	22.84	21.16
Standard Deviation	2.69	1.92	2.13	2.98

In Figure 11.3, observe that there are differences in the sample means for the four suppliers. For Supplier 1, the mean tensile strength is 19.52. For Supplier 2, the mean tensile strength is 24.26. For Supplier 3, the mean tensile strength is 22.84, and, for Supplier 4, the mean tensile strength is 21.16. What you need to determine is whether these sample results are sufficiently different to conclude that the *population* means are not all equal.

In the scatter plot shown in Figure 11.4, you can visually inspect the data and see how the measurements of tensile strength distribute. You can also observe differences among the groups as well as within groups. If the sample sizes in each group were larger, you could develop stem-and-leaf displays, box-and-whisker plots, and normal probability plots.

The null hypothesis states that there is no difference in mean tensile strength among the four suppliers.

$$H_0: \mu_1 = \mu_2 = \mu_3 = \mu_4$$

The alternative hypothesis states that there is a treatment effect; that is, at least one of the suppliers differs with respect to the mean tensile strength.

$$H_1: \text{Not all the means are equal}$$

FIGURE 11.4
Microsoft Excel Scatter Plot of Tensile Strengths for Four Different Suppliers

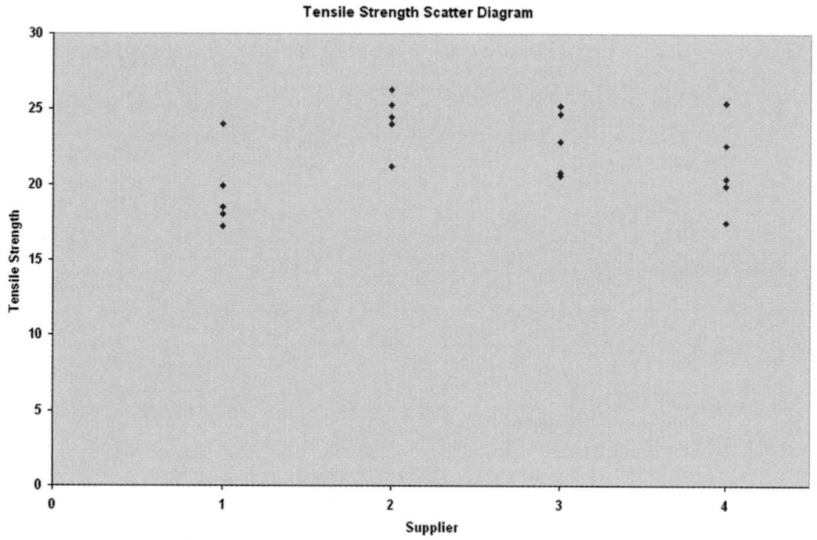

To construct the ANOVA summary table, you first compute the sample means in each group (see Figure 11.3 on page 400). Then you compute the grand mean by summing all 20 values and dividing by the total number of values.

$$\bar{\bar{X}} = \frac{\sum_{j=1}^{c}\sum_{i=1}^{n_j} X_{ij}}{n} = \frac{438.9}{20} = 21.945$$

Then, using Equations (11.1) through (11.3) on pages 397–398 you compute the sum of squares:

$$\begin{aligned} SSA &= \sum_{j=1}^{c} n_j(\bar{X}_j - \bar{\bar{X}})^2 \\ &= (5)(19.52 - 21.945)^2 + (5)(24.26 - 21.945)^2 + (5)(22.84 - 21.945)^2 \\ &\quad + (5)(21.16 - 21.945)^2 \\ &= 63.2855 \end{aligned}$$

$$\begin{aligned} SSW &= \sum_{j=1}^{c}\sum_{i=1}^{n_j} (X_{ij} - \bar{X}_j)^2 \\ &= (18.5 - 19.52)^2 + \cdots + (18 - 19.52)^2 + (26.3 - 24.26)^2 + \cdots + (24.5 - 24.26)^2 \\ &\quad + (20.6 - 22.84)^2 + \cdots + (22.9 - 22.84)^2 + (25.4 - 21.16)^2 + \cdots + (20.4 - 21.16)^2 \\ &= 97.504 \end{aligned}$$

$$\begin{aligned} SST &= \sum_{j=1}^{c}\sum_{i=1}^{n_j} (X_{ij} - \bar{\bar{X}})^2 \\ &= (18.5 - 21.945)^2 + (24 - 21.945)^2 + \cdots + (20.4 - 21.945)^2 \\ &= 160.7895 \end{aligned}$$

You compute the mean square terms by dividing the sum of squares by the corresponding degrees of freedom [see Equation (11.4) on page 398]. Because $c = 4$ and $n = 20$,

$$MSA = \frac{SSA}{c-1} = \frac{63.2855}{4-1} = 21.095$$

$$MSW = \frac{SSW}{n-c} = \frac{97.504}{20-4} = 6.094$$

so that using Equation (11.5) on page 399

$$F = \frac{MSA}{MSW} = \frac{21.095}{6.094} = 3.46$$

For a selected level of significance α, you find the upper-tail critical value F_U from the F distribution using Table E.5. A portion of Table E.5 is presented in Table 11.2. In the parachute supplier example, there are 3 degrees of freedom in the numerator of the F ratio and 16 degrees of freedom in the denominator. F_U, the upper-tail critical value at the 0.05 level of significance, is 3.24. Because the computed test statistic $F = 3.46$ is greater than $F_U = 3.24$, you reject the null hypothesis (see Figure 11.5). You conclude that there is a significant difference in the mean tensile strength among the four suppliers.

TABLE 11.2

Finding the Critical Value of F with 3 and 16 Degrees of Freedom at the 0.05 Level of Significance

	Numerator, df_1								
Denominator, df_2	**1**	**2**	**3**	**4**	**5**	**6**	**7**	**8**	**9**
	.	.	.	.	.	.	.	.	.
	.	.	.	.	.	.	.	.	.
	.	.	.	.	.	.	.	.	.
11	4.84	3.98	3.59	3.36	3.20	3.09	3.01	2.95	2.90
12	4.75	3.89	3.49	3.26	3.11	3.00	2.91	2.85	2.80
13	4.67	3.81	3.41	3.18	3.03	2.92	2.83	2.77	2.71
14	4.60	3.74	3.34	3.11	2.96	2.85	2.76	2.70	2.65
15	4.54	3.68	3.29	3.06	2.90	2.79	2.71	2.64	2.59
16	4.49	3.63	3.24	3.01	2.85	2.74	2.66	2.59	2.54

Source: Extracted from Table E.5.

FIGURE 11.5

Regions of Rejection and Nonrejection for the Analysis of Variance at the 0.05 Level of Significance with 3 and 16 Degrees of Freedom

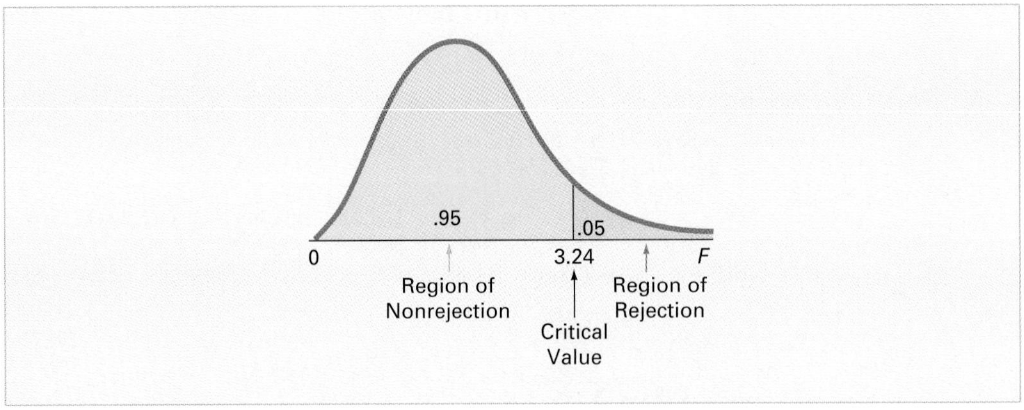

Figure 11.6 shows the Microsoft Excel ANOVA summary table and p-value. Figure 11.7 shows Minitab output. The p-value, or probability of getting an F statistic of 3.46 or larger when the null hypothesis is true, is 0.041. Because this p-value is less than the specified α of

0.05, you reject the null hypothesis. The p-value of 0.041 indicates that there is a 4.1% chance of observing differences this large or larger if the population means for the four suppliers are all equal.

FIGURE 11.6
Microsoft Excel Analysis of Variance for the Parachute Example

	A	B	C	D	E	F	G
1	Anova: Single Factor						
2							
3	SUMMARY						
4	*Groups*	*Count*	*Sum*	*Average*	*Variance*		
5	Supplier 1	5	97.6	19.52	7.237		
6	Supplier 2	5	121.3	24.26	3.683		
7	Supplier 3	5	114.2	22.84	4.553		
8	Supplier 4	5	105.8	21.16	8.903		
9							
10							
11	ANOVA						
12	*Source of Variation*	*SS*	*df*	*MS*	*F*	*P-value*	*F crit*
13	Between Groups	63.2855	3	21.09516667	3.461628925	0.0413656	3.238866952
14	Within Groups	97.504	16	6.094			
15							
16	Total	160.7895	19				

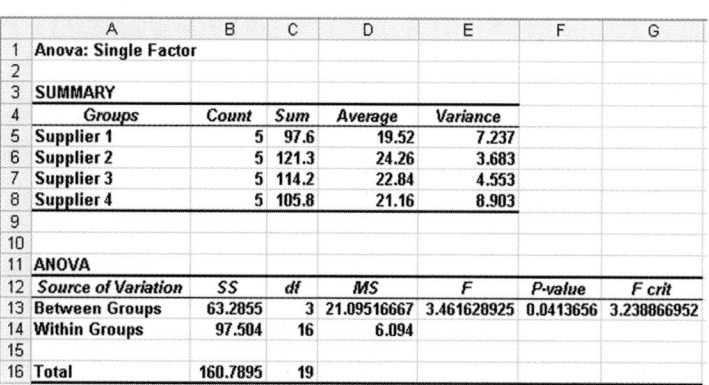

FIGURE 11.7
Minitab Analysis of Variance for the Parachute Example

```
One-way ANOVA: 1, 2, 3, 4

Source  DF      SS     MS     F      P
Factor   3   63.29  21.10  3.46  0.041
Error   16   97.50   6.09
Total   19  160.79

S = 2.469   R-Sq = 39.36%   R-Sq(adj) = 27.99%


                              Individual 95% CIs For Mean Based on
                              Pooled StDev
Level  N    Mean  StDev   -+---------+---------+---------+--------
1      5  19.520  2.690   (--------*--------)
2      5  24.260  1.919                  (--------*--------)
3      5  22.840  2.134             (--------*---------)
4      5  21.160  2.984         (---------*--------)
                          -+---------+---------+---------+--------
                         17.5      20.0      22.5      25.0

Pooled StDev = 2.469
```

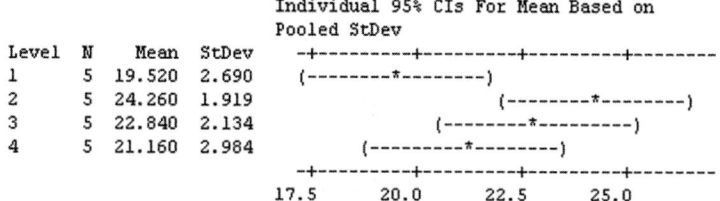

After performing the one-way ANOVA and finding a significant difference among the suppliers, what you do not yet know is *which* suppliers differ. All that you know is that there is sufficient evidence to state that the population means are not all the same. In other words, at least one or more population means are significantly different. To determine which suppliers differ, you can use a multiple comparison procedure such as the Tukey-Kramer procedure.

Multiple Comparisons: The Tukey-Kramer Procedure

In the "Using Statistics" scenario, you used the one-way ANOVA F test to determine that there was a difference among the suppliers. The next step is to use **multiple comparisons** to determine which groups are different.

Although many procedures are available (see references 5, 10, and 11), this text uses the **Tukey-Kramer multiple comparison procedure** to determine which of the c means are significantly different. This procedure was developed by John Tukey (and later modified, independently, by Tukey and by C. Y. Kramer, for situations in which the sample sizes differ—see references 6, 10, and 13). The Tukey-Kramer method is an example of a **post-hoc** comparison procedure, because the hypotheses of interest are formulated *after* the data have been inspected.

The Tukey-Kramer procedure enables you to simultaneously make comparisons between all pairs of groups. First, you compute the differences, $\bar{X}_j - \bar{X}_{j'}$ (where $j \neq j'$), among all $c(c-1)/2$ pairs of means. Then you compute the **critical range** for the Tukey-Kramer procedure using Equation (11.6).

THE CRITICAL RANGE FOR THE TUKEY-KRAMER PROCEDURE

$$\text{Critical range} = Q_U\sqrt{\frac{MSW}{2}\left(\frac{1}{n_j}+\frac{1}{n_{j'}}\right)} \quad \textbf{(11.6)}$$

where Q_U is the upper-tail critical value from a **Studentized range distribution** having c degrees of freedom in the numerator and $n - c$ degrees of freedom in the denominator. Values for the Studentized range distribution are found in Table E.10.

If the sample sizes differ, you compute a critical range for each pairwise comparison of sample means. Finally, you compare each of the $c\ (c - 1)/2$ pairs of means against its corresponding critical range. You declare a specific pair significantly different if the absolute difference in the sample means $|\bar{X}_j - \bar{X}_{j'}|$ is greater than the critical range.

In the parachute example there are four suppliers. Thus, there are $4(4 - 1)/2 = 6$ pairwise comparisons. To apply the Tukey-Kramer procedure, you first compute the absolute mean differences for all six pairwise comparisons. Using Figure 11.3 on page 400,

1. $|\bar{X}_1 - \bar{X}_2| = |19.52 - 24.26| = 4.74$
2. $|\bar{X}_1 - \bar{X}_3| = |19.52 - 22.84| = 3.32$
3. $|\bar{X}_1 - \bar{X}_4| = |19.52 - 21.16| = 1.64$
4. $|\bar{X}_2 - \bar{X}_3| = |24.26 - 22.84| = 1.42$
5. $|\bar{X}_2 - \bar{X}_4| = |24.26 - 21.16| = 3.10$
6. $|\bar{X}_3 - \bar{X}_4| = |22.84 - 21.16| = 1.68$

You need to compute only one critical range because the sample sizes in the four groups are equal. From the ANOVA summary table (Figure 11.6 or 11.7 on page 403) $MSW = 6.094$ and $n_j = n_{j'} = 5$. From Table E.10, for $\alpha = 0.05$, $c = 4$ and $n - c = 20 - 4 = 16$, Q_U, the upper-tail critical value of the test statistic is 4.05 (see Table 11.3). From Equation (11.6),

$$\text{Critical range} = 4.05\sqrt{\left(\frac{6.094}{2}\right)\left(\frac{1}{5}+\frac{1}{5}\right)} = 4.471$$

Because $4.74 > 4.471$, you conclude that there is a significant difference between the means of Suppliers 1 and 2. All other pairwise differences are small enough that they may be due to chance. You can conclude that parachutes woven using fiber from Supplier 1 have a lower mean tensile strength than those from Supplier 2.

TABLE 11.3

Finding the Studentized Range Q_U Statistic for $\alpha = 0.05$ with 4 and 16 Degrees of Freedom

	Numerator Degrees of Freedom							
Denominator Degrees of Freedom	**2**	**3**	**4**	**5**	**6**	**7**	**8**	**9**
.	.	.	.	.	.	.	.	.
.	.	.	.	.	.	.	.	.
.	.	.	.	.	.	.	.	.
11	3.11	3.82	4.26	4.57	4.82	5.03	5.20	5.35
12	3.08	3.77	4.20	4.51	4.75	4.95	5.12	5.27
13	3.06	3.73	4.15	4.45	4.69	4.88	5.05	5.19
14	3.03	3.70	4.11	4.41	4.64	4.83	4.99	5.13
15	3.01	3.67	4.08	4.37	4.60	4.78	4.94	5.08
16	3.00	3.65	4.05	4.33	4.56	4.74	4.90	5.03

Source: Extracted from Table E.10.

These results are summarized in the Microsoft Excel output presented in Figure 11.8 and the Minitab output of Figure 11.9. Minitab gives confidence intervals for the differences between two groups instead of presenting the critical range. For example, the confidence interval for the population mean of supplier 2 minus the population mean for supplier 1,

$$(\bar{X}_2 - \bar{X}_1) \pm Q_U \sqrt{\frac{MSW}{2}\left(\frac{1}{n_j} + \frac{1}{n_{j'}}\right)} = (0.269, 9.211)$$

Since both endpoints of this interval are positive, you can conclude that the population mean of supplier 2 is greater than the population mean of supplier 1.

FIGURE 11.8
Microsoft Excel Output of the Tukey-Kramer Procedure for the Parachute Example

	A	B	C	D	E	F	G	H	I
1	**Parachute Tensile-Strength Analysis**								
2									
3		**Sample**	**Sample**			Absolute	Std. Error	Critical	
4	**Group**	**Mean**	**Size**		**Comparison**	Difference	of Difference	Range	**Results**
5	1	19.52	5		**Group 1 to Group 2**	4.74	1.10399275	4.4712	**Means are different**
6	2	24.26	5		**Group 1 to Group 3**	3.32	1.10399275	4.4712	**Means are not different**
7	3	22.84	5		**Group 1 to Group 4**	1.64	1.10399275	4.4712	**Means are not different**
8	4	21.16	5		**Group 2 to Group 3**	1.42	1.10399275	4.4712	**Means are not different**
9					**Group 2 to Group 4**	3.1	1.10399275	4.4712	**Means are not different**
10	Other Data				**Group 3 to Group 4**	1.68	1.10399275	4.4712	**Means are not different**
11	Level of significance	0.05							
12	Numerator d.f.	4							
13	Denominator d.f.	16							
14	MSW	6.094							
15	Q Statistic	4.05							

FIGURE 11.9
Minitab Output of the Tukey-Kramer Procedure for the Parachute Example

```
All Pairwise Comparisons

Individual confidence level = 98.87%

1 subtracted from:

    Lower  Center  Upper  -----+---------+---------+---------+----
2   0.269   4.740  9.211                 (--------*--------)
3  -1.151   3.320  7.791              (--------*--------)
4  -2.831   1.640  6.111          (--------*--------)
                          -----+---------+---------+---------+----
                            -5.0       0.0       5.0      10.0

2 subtracted from:

    Lower  Center  Upper  -----+---------+---------+---------+----
3  -5.891  -1.420  3.051     (--------*--------)
4  -7.571  -3.100  1.371  (--------*--------)
                          -----+---------+---------+---------+----
                            -5.0       0.0       5.0      10.0

3 subtracted from:

    Lower  Center  Upper  -----+---------+---------+---------+----
4  -6.151  -1.680  2.791     (--------*--------)
```

ANOVA Assumptions

In Chapters 9 and 10, you learned the assumptions made in the application of each hypothesis-testing procedure and the consequences of departures from these assumptions. To use the one-way ANOVA F test, you must also make certain assumptions about the data. These three assumptions are

- Randomness and independence
- Normality
- Homogeneity of variance

The first assumption, **randomness and independence**, is critically important. The validity of any experiment depends on random sampling and/or the randomization process. To avoid biases in the outcomes, you need to either select random samples from the c populations or randomly assign the items or individuals to the c levels of the factor. Selecting a random sample, or randomly assigning the levels, will ensure that a value from one group is independent of any other value in the experiment. Departures from this assumption can seriously affect inferences from the analysis of variance. These problems are discussed more thoroughly in references 5 and 10.

The second assumption, **normality**, states that the sample values in each group are from a normally distributed population. Just as in the case of the t test, the one-way ANOVA F test is fairly robust against departures from the normal distribution. As long as the distributions are not extremely different from a normal distribution, the level of significance of the ANOVA F test is usually not greatly affected, particularly for large samples. You can assess the normality of each of the c samples by constructing a normal probability plot or a box-and-whisker plot.

The third assumption, **homogeneity of variance**, states that the population variances of the c groups are equal (i.e., $\sigma_1^2 = \sigma_2^2 = \cdots = \sigma_c^2$). If you have equal sample sizes in each group, inferences based on the F distribution are not seriously affected by unequal variances. However, if you have unequal sample sizes, then unequal variances can have a serious effect on inferences developed from the ANOVA procedure. Thus, when possible, you should have equal sample sizes in all groups. The Levene test for homogeneity of variance presented below is one method to test whether the variances of the c populations are equal.

When only the normality assumption is violated, the Kruskal-Wallis rank test, a nonparametric procedure discussed in section 12.9, is appropriate. When only the homogeneity-of-variance assumption is violated, procedures similar to those used in the separate-variance t test of section 10.1 are available (see references 1 and 2). When both the normality and homogeneity-of-variance assumptions have been violated, then you need to use an appropriate data transformation that will both normalize the data and reduce the differences in variances (see reference 11) or use a more general nonparametric procedure (see references 2 and 3).

Levene's Test for Homogeneity of Variance

Although the one-way ANOVA F test is relatively robust with respect to the assumption of equal group variances, large differences in the group variances can seriously affect the level of significance and the power of the F test. Many procedures are available to test the assumption of homogeneity of variance. The modified **Levene test** (see references 1, 4, 9, and 12) is a procedure with high statistical power. To test for the equality of the c population variances, you use the following hypotheses:

$$H_0: \sigma_1^2 = \sigma_2^2 = \cdots = \sigma_c^2$$

against the alternative

$$H_1: \text{Not all } \sigma_j^2 \text{ are equal } (j = 1, 2, \ldots, c)$$

To test the null hypothesis of equal variances, you first compute the absolute value of the difference between each value and the median of the group. Then you perform a one-way analysis of variance on these *absolute differences*. To illustrate the modified Levene test, return to the "Using Statistics" scenario concerning the tensile strength of parachutes first presented on page 396. Table 11.4 summarizes the absolute differences from the median of each supplier.

TABLE 11.4
Absolute Differences from the Median Tensile Strength for Four Suppliers

Supplier 1 (Median = 18.5)	Supplier 2 (Median = 24.5)	Supplier 3 (Median = 22.9)	Supplier 4 (Median = 20.4)
|18.5 – 18.5|= 0.0	|26.3 – 24.5|= 1.8	|20.6 – 22.9|= 2.3	|25.4 – 20.4|= 5.0
|24.0 – 18.5|= 5.5	|25.3 – 24.5|= 0.8	|25.2 – 22.9|= 2.3	|19.9 – 20.4|= 0.5
|17.2 – 18.5|= 1.3	|24.0 – 24.5|= 0.5	|20.8 – 22.9|= 2.1	|22.6 – 20.4|= 2.2
|19.9 – 18.5|= 1.4	|21.2 – 24.5|= 3.3	|24.7 – 22.9|= 1.8	|17.5 – 20.4|= 2.9
|18.0 – 18.5|= 0.5	|24.5 – 24.5|= 0.0	|22.9 – 22.9|= 0.0	|20.4 – 20.4|= 0.0

Using the absolute differences given in Table 11.4, you perform a one-way analysis of variance. Figure 11.10 presents Microsoft Excel output and Figure 11.11 shows Minitab output.

FIGURE 11.10
Microsoft Excel Output of the Analysis of Variance of the Absolute Differences for the Parachute Data

	A	B	C	D	E	F	G
1	**Parachute Tensile-Strength Analysis**						
2							
3	**SUMMARY**						
4	*Groups*	*Count*	*Sum*	*Average*	*Variance*		
5	**Supplier 1**	5	8.7	1.74	4.753		
6	**Supplier 2**	5	6.4	1.28	1.707		
7	**Supplier 3**	5	8.5	1.7	0.945		
8	**Supplier 4**	5	10.6	2.12	4.007		
9							
10							
11	**ANOVA**						
12	*Source of Variation*	*SS*	*df*	*MS*	*F*	*P-value*	*F crit*
13	**Between Groups**	1.77	3	0.59	0.2068	0.890189	3.238867
14	**Within Groups**	45.648	16	2.853			
15							
16	**Total**	47.418	19				

FIGURE 11.11
Minitab Output of the Levene Test for the Parachute Data

```
Levene's Test (any continuous distribution)
Test statistic = 0.21, p-value = 0.890
```

From Figures 11.10 and 11.11, observe that $F = 0.21 < 3.238867$ (or the p-value $= 0.89 > 0.05$). Thus, do not reject H_0. There is no evidence of a significant difference among the four variances. In other words, it is reasonable to assume that the materials from the four suppliers produce parachutes with an equal amount of variability. Therefore, the homogeneity of variance assumption for the ANOVA procedure is justified.

EXAMPLE 11.1

ANALYSIS OF VARIANCE OF THE SPEED OF DRIVE-THROUGH SERVICE AT FAST-FOOD CHAINS

For fast-food restaurants, the drive-through window is an increasing source of revenue. The chain that offers the fastest service is likely to attract additional customers. In a study of drive-through times (from menu board to departure) at fast-food chains, the mean time was 150 seconds for Wendy's, 167 seconds for McDonald's, 169 seconds for Checkers, 171 seconds for Burger King, and 172 seconds for Long John Silver's (J. Ordonez, "An Efficiency Drive: Fast-Food Lanes are Getting Even Faster," *The Wall Street Journal*, May 18, 2000, A1, A10). Suppose the study was based on 20 customers for each fast-food chain and the ANOVA table given in Table 11.5 was developed.

TABLE 11.5
Analysis of Variance Table of the Speed of Drive-Through Service at Fast-Food Chains

Source	Degrees of Freedom	Sum of Squares	Mean Squares	F	p-value
Among chains	4	6,536	1,634.0	12.51	0.0000
Within chains	95	12,407	130.6		

At the 0.05 level of significance, is there evidence of a difference in the mean drive-through times of the five chains?

SOLUTION

H_0: $\mu_1 = \mu_2 = \mu_3 = \mu_4 = \mu_5$ where 1 = Wendy's, 2 = McDonald's, 3 = Checkers, 4 = Burger King, 5 = Long John Silver's

H_1: Not all μ_j are equal where $j = 1, 2, 3, 4, 5$

Decision Rule: If p-value < 0.05, reject H_0. Since the p-value is virtually 0, reject H_0. You have sufficient evidence to conclude that the mean drive-through times of the five chains are not all equal.

To determine which of the means are significantly different from one another, use the Tukey-Kramer procedure [Equation (11.6) on page 404] to establish the critical range:

$$Q_{U(c,n-c)} = Q_{U(5,95)} = 3.92$$

$$\text{Critical range} = Q_{U(c,n-c)}\sqrt{\frac{MSW}{2}\cdot\left(\frac{1}{n_j}+\frac{1}{n_{j'}}\right)} = (3.92)\sqrt{\left(\frac{130.6}{2}\right)\left(\frac{1}{20}+\frac{1}{20}\right)}$$

$$= 10.017$$

The drive-through times are different between Wendy's (mean of 150 seconds) and each of the other four chains. With 95% confidence, you can conclude that the drive-through time for Wendy's is faster than McDonald's, Burger King, Checkers, and Long John Silver's, but the drive-through times for McDonald's, Burger King, Checkers, and Long John Silver's are not statistically different.

PROBLEMS FOR SECTION 11.1

Learning the Basics

PH Grade ASSIST **11.1** You are working with an experiment that has a single factor with five groups, and seven values in each group.

a. How many degrees of freedom are there in determining the among-group variation?
b. How many degrees of freedom are there in determining the within-group variation?
c. How many degrees of freedom are there in determining the total variation?

PH Grade ASSIST **11.2** You are working with the same experiment as in problem 11.1:

a. If $SSA = 60$ and $SST = 210$, what is SSW?
b. What is MSA?
c. What is MSW?
d. What is the value of the test statistic F?

PH Grade ASSIST **11.3** You are working with the same experiment as in problems 11.1 and 11.2:

a. Form the ANOVA summary table and fill in all values in the body of the table.
b. At the 0.05 level of significance, what is the upper-tail critical value from the F distribution?
c. State the decision rule for testing the null hypothesis that all five groups have equal population means.
d. What is your statistical decision?

11.4 You are working with an experiment that has one factor containing three groups with seven values in each:

a. How many degrees of freedom are there in determining the among-group variation?
b. How many degrees of freedom are there in determining the within-group variation?
c. How many degrees of freedom are there in determining the total variation?

PH Grade ASSIST **11.5** You are conducting an experiment with one factor containing four groups, with eight values in each group. For the ANOVA summary table below, fill in all the missing results.

Source	Degrees of Freedom	Sum of Squares	Mean Square (Variance)	F
Among groups	$c - 1 = ?$	$SSA = ?$	$MSA = 80$	$F = ?$
Within groups	$n - c = ?$	$SSW = 560$	$MSW = ?$	
Total	$n - 1 = ?$	$SST = ?$		

11.6 You are working with the same experiment as in problem 11.5:

a. At the 0.05 level of significance, state the decision rule for testing the null hypothesis that all four groups have equal population means.
b. What is your statistical decision?
c. At the 0.05 level of significance, what is the upper-tail critical value from the Studentized range distribution?
d. To perform the Tukey-Kramer procedure, what is the critical range?

Applying the Concepts

Problems 11.7–11.14 can be solved manually or by using Microsoft Excel, Minitab, or SPSS.

11.7 The Computer Anxiety Rating Scale (CARS) measures an individual's level of computer anxiety on a scale from 20 (no anxiety) to 100 (highest level of anxiety). Researchers at Miami University administered CARS to 172 business students. One of the objectives of the study was to determine if there are differences in the amount of computer anxiety experienced by students with different majors.

Source	Degrees of Freedom	Sum of Squares	Mean Squares	F
Among Majors	5	3,172		
Within Majors	166	21,246		
Total	171	24,418		

Major	N	Mean
Marketing	19	44.37
Management	11	43.18
Other	14	42.21
Finance	45	41.80
Accountancy	36	37.56
MIS	47	42.21

Source: Travis Broome and Douglas Havelka, "Determinants of Computer Anxiety in Business Students," The Review of Business Information Systems, *Spring 2002, 6(2):9–16.*

a. Complete the ANOVA summary table.
b. At the 0.05 level of significance, is there evidence of a difference in the mean computer anxiety experienced by different majors?
c. If the results in (b) indicate that it is appropriate, use the Tukey-Kramer procedure to determine which majors differ in mean computer anxiety. Discuss your findings.

PH Grade ASSIST **11.8** Periodically, *The Wall Street Journal* conducted a stock-picking contest. The last one was conducted in March 2001. In this experiment, three different methods were used to select stocks that were expected to perform well during the next five months. Four Wall Street professionals, considered experts on picking stocks, selected four stocks. Four randomly chosen readers of the *Wall Street Journal* also selected four stocks. Finally, four stocks were selected by flinging darts at a table containing a list of stocks. The returns of the selected stocks for March 20, 2001 to August 31, 2001 (in percentage return) are given in the following table. Note that during this period the Dow Jones Industrial Average gained 2.4% (Georgette Jasen, "In Picking Stocks, Dartboard Beats the Pros," *The Wall Street Journal*, September 27, 2001, C1, C10). **CONTEST2001**

Experts	Readers	Darts
+39.5	−31.0	+39.0
−1.1	−20.7	+31.9
−4.5	−45.0	+14.1
−8.0	−73.3	+5.4

Source: Extracted from The Wall Street Journal.

a. Is there evidence of a significant difference in the mean return for the three categories? (Use $\alpha = 0.05$.)
b. If appropriate, determine which categories differ in mean return.
c. Comment on the validity of the inference implied by the title of the article, which suggests that the dartboard was better than the professionals.
d. Is there evidence of a significant difference in the variation in the return for the three categories? (Use $\alpha = 0.05$.)

11.9 The following data represent the price of regular gasoline at self-service stations in four counties in New York City and two suburban counties during the week of May 17, 2004. **GASPRICE**

Counties					
Manhattan	Bronx	Queens	Brooklyn	Nassau	Suffolk
2.339	2.199	2.239	2.159	2.099	2.179
2.299	2.139	2.239	2.199	2.199	2.159
2.239	2.239	2.179	2.359	2.259	2.119
2.199	2.159	2.299	2.159	2.239	2.159
2.199	2.179	2.279	1.999	2.239	2.219

a. At the 0.05 level of significance, is there evidence of a difference in the mean price of gasoline in the six counties?
b. If appropriate, determine which counties differ in mean gasoline price.
c. At the 0.05 level of significance, is there evidence of a difference in the variation in gasoline price among the six counties?

SELF Test **11.10** Students in a business statistics course performed a completely randomized design to test the strength of four brands of trash bags. One-pound weights were placed into a bag one at a time until the bag broke. A total of 40 bags, 10 for each brand, were used. The data in the file **TRASHBAGS** give the weight (in pounds) required to break the trash bags.
a. At the 0.05 level of significance, is there evidence of a difference in the mean strength of the four brands of trash bags?
b. If appropriate, determine which brands differ in mean strength.
c. At the 0.05 level of significance, is there evidence of a difference in the variation in strength among the four brands of trash bags?
d. Which brand(s) should you buy and which brand(s) should you avoid? Explain.

11.11 The data on the next page represent the lifetime of four different alloys. **ALLOY**

Alloy			
1	2	3	4
999	1,022	1,026	974
1,010	973	1,008	1,015
995	1,023	1,005	1,009
998	1,023	1,007	1,011
1,001	996	981	995

Source: P. Wludyka, P. Nelson, and P. Silva, "Power Curves for the Analysis of Means for Variances," Journal of Quality Technology, *33, 2001, 60–65.*

a. At the 0.05 level of significance, is there evidence of a difference in the mean lifetime of the four alloys?
b. If appropriate, determine which alloys differ in mean lifetime.
c. At the 0.05 level of significance, is there evidence of a difference in the variation in lifetime among the four alloys?
d. What effect does your result in (c) have on the validity of the results in (a) and (b)?

11.12 An advertising agency has been hired by a manufacturer of pens to develop an advertising campaign for the upcoming holiday season. To prepare for this project, the research director decides to initiate a study of the effect of advertising on product perception. An experiment is designed to compare five different advertisements. Advertisement *A* greatly undersells the pen's characteristics. Advertisement *B* slightly undersells the pen's characteristics. Advertisement *C* slightly oversells the pen's characteristics. Advertisement *D* greatly oversells the pen's characteristics. Advertisement *E* attempts to correctly state the pen's characteristics. A sample of 30 adult respondents, taken from a larger focus group, is randomly assigned to the five advertisements (so that there are six respondents to each). After reading the advertisement and developing a sense of "product expectation," all respondents unknowingly receive the same pen to evaluate. The respondents are permitted to test their pen and the plausibility of the advertising copy. The respondents are then asked to rate the pen from 1 to 7 on the product characteristic scales of appearance, durability, and writing performance. The *combined* scores of three ratings (appearance, durability, and writing performance) for the 30 respondents **PEN** are as follows.

A	B	C	D	E
15	16	8	5	12
18	17	7	6	19
17	21	10	13	18
19	16	15	11	12
19	19	14	9	17
20	17	14	10	14

a. At the 0.05 level of significance, is there evidence of a difference in the mean rating of the five advertisements?
b. If appropriate, determine which advertisements differ in mean rating.
c. At the 0.05 level of significance, is there evidence of a difference in the variation in rating among the five advertisements?
d. Which advertisement(s) should you use and which advertisement(s) should you avoid? Explain.

11.13 The retailing manager of a supermarket chain wants to determine whether product location has any effect on the sale of pet toys. Three different aisle locations are considered: front, middle, and rear. A random sample of 18 stores is selected with 6 stores randomly assigned to each aisle location. The size of the display area and price of the product are constant for all stores. At the end of a 1-month trial period, the sales volumes (in thousands of dollars) of the product in each store were as follows: **LOCATE**

Aisle Location		
Front	Middle	Rear
8.6	3.2	4.6
7.2	2.4	6.0
5.4	2.0	4.0
6.2	1.4	2.8
5.0	1.8	2.2
4.0	1.6	2.8

a. At the 0.05 level of significance, is there evidence of a significant difference in mean sales among the various aisle locations?
b. If appropriate, which aisle locations appear to differ significantly in mean sales?
c. At the 0.05 level of significance, is there evidence of a significant difference in the variation in sales among the various aisle locations?
d. What should the retailing manager conclude? Fully describe the retailing manager's options with respect to aisle locations.

11.14 A sporting goods manufacturing company wanted to compare the distance traveled by golf balls produced by each of four different designs. Ten balls were manufactured with each design and were brought to the local golf course for the club professional to test. The order in which the balls were hit with a driver from the first tee was randomized so that the pro did not know which type of ball was being hit. All 40 balls were hit in a short period of time during which the environmental conditions were essentially the same. The results (distance traveled in yards) for the four designs were as follows: **GOLFBALL**

Designs			
1	2	3	4
206.32	203.81	217.08	213.90
226.77	223.85	230.55	231.10
207.94	206.75	221.43	221.28
224.79	223.97	227.95	221.53
206.19	205.68	218.04	229.43
229.75	234.30	231.84	235.45
204.45	204.49	224.13	213.54
228.51	219.50	224.87	228.35
209.65	210.86	211.82	214.51
221.44	233.00	229.49	225.09

a. At the 0.05 level of significance, is there evidence of a difference in the mean distance traveled by the golf balls with different designs?

b. If the results in (a) indicate that it is appropriate, use the Tukey-Kramer procedure to determine which designs differ in mean distance.

c. What assumptions are necessary in (a)?

d. At the 0.05 level of significance, is there evidence of a difference in the variation of the distance traveled by the golf balls differing in design?

e. What golf ball design should the manufacturing manager choose? Explain.

11.2 THE RANDOMIZED BLOCK DESIGN

In section 11.1, you used the one-way ANOVA F test to evaluate differences among the means of more than two independent groups. In section 10.2 you used the paired t test when you had repeated measurements or matched samples in order to evaluate the difference between the means of two groups. In this section, a method to analyze more than two groups using repeated measures or matched samples is developed. The heterogeneous sets of items or individuals that have been matched (or on whom repeated measurements have been taken) are called **blocks**. Experimental situations that use blocks are called **randomized block designs**.

Although groups and blocks are both used in a randomized block design, the focus of the analysis is on the differences among the different groups. As is the case in completely randomized designs, groups are often different levels pertaining to a factor of interest. For example, if the factor of interest is advertising medium, three groups could be subject to the following different levels: television, radio, and newspaper. In this experiment, different cities could be used as blocks. The purpose of blocking is to remove as much variability as possible from the random error so that the differences among the groups are more evident. A randomized block design is often more efficient statistically than a completely randomized design and therefore produces more precise results (see references 1, 5, 10, and 11).

To compare a completely randomized design with a randomized block design, return to the "Using Statistics" scenario concerning the Perfect Parachute Company. Suppose that a completely randomized design is used with 12 parachutes woven during a 24 hour period. Any variability among the shifts becomes part of the random error, and therefore differences among the four suppliers might be hard to detect. To reduce the random error, a randomized block experiment is designed where three shifts are used and four parachutes are woven during each shift (one parachute using fibers from Supplier 1, one parachute using fibers from Supplier 2, and so forth). The three shifts are considered blocks, while the treatment factor is still the four suppliers. The advantage of the randomized block design is that the variability among the three shifts is removed from the random error. Therefore, this design should provide more precise results concerning differences among the four suppliers.

Tests for the Treatment and Block Effects

Recall from Figure 11.1 on page 397 that, in the completely randomized design, the total variation (SST) is subdivided into variation due to differences *among* the c groups (SSA) and variation due to variation *within* the c groups (SSW). Within-group variation is considered experimental error, and among-group variation is due to treatment effects.

To remove the effects of the blocking from the random error in the randomized block design, the within-group variation (SSW) is subdivided into variation due to differences among the blocks ($SSBL$) and variation due to random error (SSE). Therefore, as presented in Figure 11.12, in a randomized block design, the total variation is the sum of three components—among-group variation (*SSA*), among-block variation (*SSBL*), and random error (*SSE*).

FIGURE 11.12

Partitioning the Total Variation in a Randomized Block Design Model

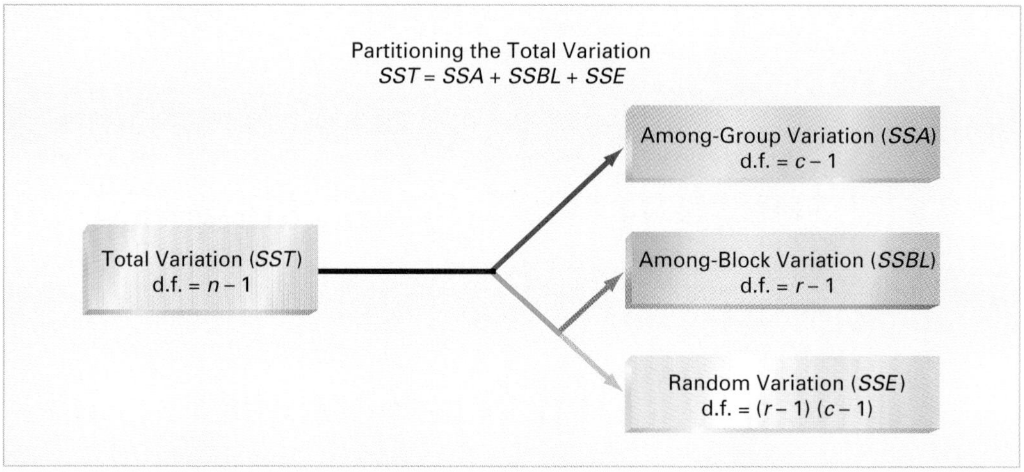

The following definitions are needed to develop the ANOVA procedure for the randomized block design:

r = the number of blocks

c = the number of groups

n = the total number of values (where $n = rc$)

X_{ij} = the value in the ith block for the jth group

$\bar{X}_{i.}$ = the mean of all the values in block i

$\bar{X}_{.j}$ = the mean of all the values for group j

$$\sum_{j=1}^{c}\sum_{i=1}^{r} X_{ij} = \text{the grand total}$$

The total variation, also called sum of squares total (*SST*), is a measure of the variation among all the values. You compute *SST* by summing the squared differences between each individual value and the grand mean $\bar{\bar{X}}$ that is based on all n values. Equation (11.7) shows the computation for total variation.

TOTAL VARIATION IN RANDOMIZED BLOCK DESIGN

$$SST = \sum_{j=1}^{c}\sum_{i=1}^{r}(X_{ij} - \bar{\bar{X}})^2 \qquad \textbf{(11.7)}$$

where

$$\bar{\bar{X}} = \frac{\sum_{j=1}^{c}\sum_{i=1}^{r} X_{ij}}{rc} \quad \text{(i.e., the grand mean)}$$

You compute the among-group variation, also called the sum of squares among groups (*SSA*), by summing the squared differences between the sample mean of each group $\bar{X}_{.j}$ and the grand mean $\bar{\bar{X}}$, weighted by the number of blocks r. Equation (11.8) shows the computation for the among-group variation.

AMONG-GROUP VARIATION IN RANDOMIZED BLOCK DESIGN

$$SSA = r\sum_{j=1}^{c}(\bar{X}_{.j} - \bar{\bar{X}})^2 \tag{11.8}$$

where
$$\bar{X}_{.j} = \frac{\sum_{i=1}^{r} X_{ij}}{r}$$

You compute the **among-block variation**, also called the **sum of squares among blocks (*SSBL*)**, by summing the squared differences between the mean of each block $\bar{X}_{i.}$ and the grand mean $\bar{\bar{X}}$, weighted by the number of groups c. Equation (11.9) shows the computation for the among-block variation.

AMONG-BLOCK VARIATION IN RANDOMIZED BLOCK DESIGN

$$SSBL = c\sum_{i=1}^{r}(\bar{X}_{i.} - \bar{\bar{X}})^2 \tag{11.9}$$

where
$$\bar{X}_{i.} = \frac{\sum_{j=1}^{c} X_{ij}}{c}$$

You compute the random variation, also called the **sum of squares error (*SSE*)**, by summing the squared differences among all the values after the effect of the particular treatments and blocks have been accounted for. Equation (11.10) shows the computation for random error.

RANDOM ERROR IN RANDOMIZED BLOCK DESIGN

$$SSE = \sum_{j=1}^{c}\sum_{i=1}^{r}(X_{ij} - \bar{X}_{.j} - \bar{X}_{i.} + \bar{\bar{X}})^2 \tag{11.10}$$

Since you are comparing c groups, there are $c - 1$ degrees of freedom associated with the sum of squares among groups (*SSA*). Similarly, since there are r blocks, there are $r - 1$ degrees of freedom associated with the sum of squares among blocks (*SSBL*). Moreover, there are $n - 1$ degrees of freedom associated with the sum of squares total (*SST*) because you are comparing each value X_{ij} to the grand mean $\bar{\bar{X}}$ based on all n values. Therefore, since the degrees of freedom for each of the sources of variation must add to the degrees of freedom for the total variation, you compute the degrees of freedom for the sum of squares error (*SSE*) component by subtraction and algebraic manipulation. Thus, the degrees of freedom associated with the sum of squares error is $(r - 1)(c - 1)$.

If you divide each of the component sum of squares by its associated degrees of freedom, you have the three *variances* or mean square terms (***MSA***, ***MSBL***, and ***MSE***). Equations (11.11a–c) give the mean square terms needed for the ANOVA table.

THE MEAN SQUARES IN RANDOMIZED BLOCK DESIGN

$$MSA = \frac{SSA}{c-1} \tag{11.11a}$$

$$MSBL = \frac{SSBL}{r-1} \tag{11.11b}$$

$$MSE = \frac{SSE}{(r-1)(c-1)} \tag{11.11c}$$

If the assumptions of the analysis of variance are valid, the null hypothesis of no differences in the c population means (i.e., no treatment effects):

$$H_0: \mu_{.1} = \mu_{.2} = \cdots = \mu_{.c}$$

is tested against the alternative that not all the c population means are equal (i.e., there are treatment effects):

$$H_1: \text{Not all } \mu_{.j} \text{ are equal (where } j = 1, 2, \ldots, c)$$

by computing the test statistic F given in Equation (11.12).

RANDOMIZED BLOCK *F*-TEST STATISTIC

$$F = \frac{MSA}{MSE} \tag{11.12}$$

The F-test statistic follows an F distribution with $c - 1$ degrees of freedom for the MSA term and $(r - 1)(c - 1)$ degrees of freedom for the MSE term. For a given level of significance α, you reject the null hypothesis if the computed F-test statistic is greater than the upper-tail critical value F_U from the F distribution with $c - 1$ and $(r - 1)(c - 1)$ degrees of freedom (see Table E.5). The decision rule is:

Reject H_0 if $F > F_U$;

otherwise do not reject H_0.

To examine whether the randomized block design was advantageous to use, some statisticians suggest that you perform the ***F* test for block effects**. The null hypothesis of no block effects:

$$H_0: \mu_{1.} = \mu_{2.} = \cdots = \mu_{r.}$$

is tested against the alternative:

$$H_1: \text{Not all } \mu_{i.} \text{ are equal (where } i = 1, 2, \ldots, r)$$

Equation (11.13) gives the F statistic for the block effects.

F-TEST STATISTIC FOR BLOCK EFFECTS

$$F = \frac{MSBL}{MSE} \tag{11.13}$$

You reject the null hypothesis at the α level of significance if the F-test statistic is greater than the upper-tail critical value F_U from the F distribution with $r - 1$ and $(r - 1)(c - 1)$ degrees of freedom (see Table E.5). That is, the decision rule is:

$$\text{Reject } H_0 \text{ if } F > F_U$$
$$\text{otherwise do not reject } H_0.$$

The results of the analysis-of-variance procedure are usually displayed in an ANOVA summary table as shown in Table 11.6.

TABLE 11.6
Analysis-of-Variance Table for the Randomized Block Design

Source	Degrees of Freedom	Sum of Squares	Mean Square (Variance)	F
Among treatments	$c - 1$	SSA	$MSA = \dfrac{SSA}{c - 1}$	$F = \dfrac{MSA}{MSE}$
Among blocks	$r - 1$	$SSBL$	$MSBL = \dfrac{SSBL}{r - 1}$	$F = \dfrac{MSBL}{MSE}$
Error	$(r - 1)(c - 1)$	SSE	$MSE = \dfrac{SSE}{(r - 1)(c - 1)}$	
Total	$rc - 1$	SST		

To illustrate the randomized block design, suppose that a fast-food chain wants to evaluate the service at four restaurants. The customer service director for the chain hires six investigators with varied experiences in food-service evaluations to act as raters. To reduce the effect of the variability from rater-to-rater, you use a randomized block design with raters serving as the blocks. The four restaurants are the groups of interest.

The six raters evaluate the service at each of the four restaurants in a random order. A rating scale from 0 (low) to 100 (high) is used. Table 11.7 summarizes the results FFCHAIN, along with the group totals, group means, block totals, block means, grand total, and grand mean.

TABLE 11.7
Ratings at Four Restaurants of a Fast-Food Chain

	Restaurants					
Raters	A	B	C	D	Totals	Means
1	70	61	82	74	287	71.75
2	77	75	88	76	316	79.00
3	76	67	90	80	313	78.25
4	80	63	96	76	315	78.75
5	84	66	92	84	326	81.50
6	78	68	98	86	330	82.50
Totals	465	400	546	476	1,887	
Means	77.50	66.67	91.00	79.33	78.625	

In addition, from Table 11.7,

$$r = 6 \qquad c = 4 \qquad n = rc = 24$$

and,

$$\bar{\bar{X}} = \frac{\sum_{j=1}^{c}\sum_{i=1}^{r} X_{ij}}{rc} = \frac{1,887}{24} = 78.625$$

Figure 11.13 illustrates output from Microsoft Excel and Figure 11.14 presents Minitab output for this randomized block design.

FIGURE 11.13

Microsoft Excel Output for the Fast-Food-Chain Study

	A	B	C	D	E	F	G
1	Anova: Two-Factor Without Replication						
2							
3	*SUMMARY*	*Count*	*Sum*	*Average*	*Variance*		
4	Rater 1	4	287	71.75	76.25		
5	Rater 2	4	316	79	36.67		
6	Rater 3	4	313	78.25	90.92		
7	Rater 4	4	315	78.75	184.92		
8	Rater 5	4	326	81.5	121		
9	Rater 6	4	330	82.5	161		
10							
11	Restaurant A	6	465	77.5	21.5		
12	Restaurant B	6	400	66.67	23.47		
13	Restaurant C	6	546	91	33.2		
14	Restaurant D	6	476	79.33	23.47		
15							
16							
17	ANOVA						
18	*Source of Variation*	*SS*	*df*	*MS*	*F*	*P-value*	*F crit*
19	Rows	283.375	5	56.675	3.782	0.020	2.901
20	Columns	1787.458	3	595.819	39.758	0.00	3.287
21	Error	224.792	15	14.986			
22							
23	Total	2295.625	23				

FIGURE 11.14

Minitab Output for the Fast-Food-Chain Study

```
Source   DF       SS       MS      F      P
Raters    5   283.38   56.675   3.78  0.020
Restrat   3  1787.46  595.819  39.76  0.000
Error    15   224.79   14.986
Total    23  2295.63

S = 3.871   R-Sq = 90.21%   R-Sq(adj) = 84.99%

   Individual 95% CIs For Mean Based on Pooled StDev
Raters    Mean  -----+---------+---------+---------+----
1        71.75  (--------*-------)
2        79.00                (-------*-------)
3        78.25              (--------*-------)
4        78.75               (--------*-------)
5        81.50                     (-------*-------)
6        82.50                       (-------*-------)
                -----+---------+---------+---------+----
                   70.0      75.0      80.0      85.0

Individual 95% CIs For Mean Based on Pooled StDev
Restrat      Mean   -+---------+---------+---------+--------
A         77.5000                  (---*---)
B         66.6667   (---*----)
C         91.0000                                  (---*---)
D         79.3333                    (---*---)
```

Using the 0.05 level of significance to test for differences among the restaurants, you reject the null hypothesis (H_0: $\mu_{.1} = \mu_{.2} = \mu_{.3} = \mu_{.4}$) if the computed F value is greater than 3.29, the upper-tail critical value from the F distribution with 3 and 15 degrees of freedom in the numerator and denominator, respectively (see Figure 11.15). Since $F = 39.758 > F_U = 3.29$, or since the p-value $= 0.000 < 0.05$, you reject H_0 and conclude that there is evidence of a difference in the mean rating among the different restaurants. The extremely small p-value indicates that if the means from the four restaurants are equal, there is virtually no chance that you will get differences as large or larger among the sample means as observed in this study. Thus, there is little degree of belief in the null hypothesis. You conclude that the alternative hypothesis is correct, the mean ratings among the four restaurants are different.

As a check on the effectiveness of blocking, you can test for a difference among the raters. The decision rule, using the 0.05 level of significance, is to reject the null hypothesis (H_0: $\mu_{1.} = \mu_{2.} = \ldots = \mu_{6.}$) if the computed F value is greater than 2.90, the upper-tail critical value from the F distribution with 5 and 15 degrees of freedom (see Figure 11.16). Since $F = 3.782 > F_U = 2.90$, or the p-value $= 0.02 < 0.05$, you reject H_0 and conclude that there is evidence of a difference among the raters. Thus, you conclude that the blocking has been advantageous in reducing the random error.

FIGURE 11.15
Regions of Rejection and Nonrejection for the Fast-Food-Chain Study at the 0.05 Level of Significance with 3 and 15 Degrees of Freedom

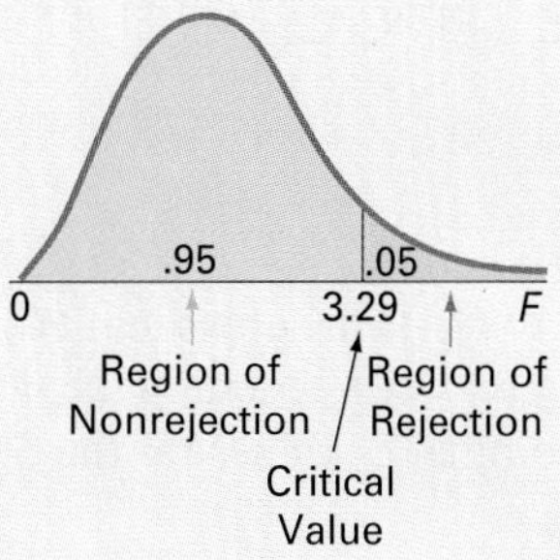

FIGURE 11.16
Regions of Rejection and Nonrejection for the Fast-Food-Chain Study at the 0.05 Level of Significance with 5 and 15 Degrees of Freedom

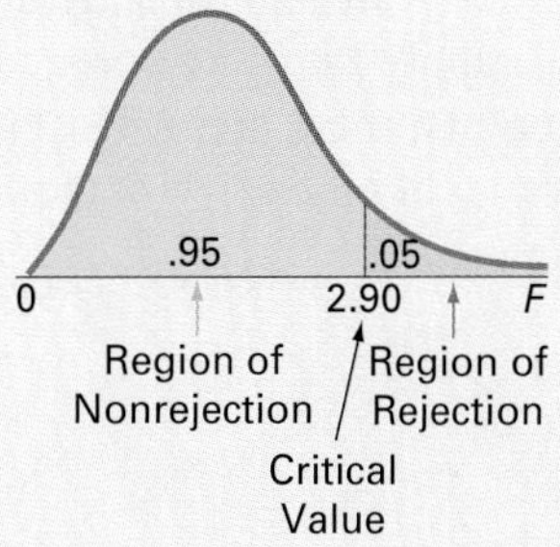

The assumptions of the one-way analysis of variance discussed on page 405 (randomness and independence, normality, and homegeneity of variance) also apply to the randomized block design. If the normality assumption is violated, you can use the Friedman Rank Test discussed in section 12.10. In addition, you need to assume that there is no *interacting effect* between the treatments and the blocks. In other words, you need to assume that any differences between the treatments (the restaurants) are consistent across the entire set of blocks (the raters). The concept of *interaction* is discussed further in section 11.3.

Did the blocking result in an increase in precision in comparing the different treatment groups? To answer this question, use Equation (11.14) to calculate the **estimated relative efficiency (*RE*)** of the randomized block design as compared with the completely randomized design.

ESTIMATED RELATIVE EFFICIENCY

$$RE = \frac{(r-1)MSBL + r(c-1)MSE}{(rc-1)MSE} \qquad \textbf{(11.14)}$$

Using Figure 11.13 or Figure 11.14,

$$RE = \frac{(5)(56.675) + (6)(3)(14.986)}{(23)(14.986)} = 1.60$$

This value for relative efficiency means that it would take 1.6 times as many observations in a one-way ANOVA design as compared to the randomized block design in order to have the same precision in comparing the restaurants.

Multiple Comparisons: The Tukey Procedure

As in the case of the completely randomized design, once you reject the null hypothesis of no differences between the groups, you need to determine *which* groups are significantly different from the others. For the randomized block design, you can use a procedure developed by John Tukey (see references 10, 11, and 13). Equation (11.15) gives the critical range for the **Tukey procedure**.

THE CRITICAL RANGE FOR THE RANDOMIZED BLOCK DESIGN

$$\text{Critical range} = Q_U\sqrt{\frac{MSE}{r}} \quad \textbf{(11.15)}$$

where Q_U is the upper-tail critical value from a Studentized range distribution having c degrees of freedom in the numerator and $(r - 1)(c - 1)$ degrees of freedom in the denominator. Values for the Studentized range distribution are found in Table E.10.

Compare each of the $c(c - 1)/2$ pairs against the critical range. If the absolute difference in a specific pair of sample means, say $|\bar{X}_{.j} - \bar{X}_{.j'}|$, is greater than the critical range, then group j and group j' are significantly different.

To apply the Tukey procedure, return to the fast-food-chain study. Since there are four restaurants, there are $4(4 - 1)/2 = 6$ possible pairwise comparisons. From Figure 11.13 or Figure 11.14 on page 416, the absolute mean differences are

1. $|\bar{X}_{.1} - \bar{X}_{.2}| = |77.50 - 66.67| = 10.83$
2. $|\bar{X}_{.1} - \bar{X}_{.3}| = |77.50 - 91.00| = 13.50$
3. $|\bar{X}_{.1} - \bar{X}_{.4}| = |77.50 - 79.33| = 1.83$
4. $|\bar{X}_{.2} - \bar{X}_{.3}| = |66.67 - 91.00| = 24.33$
5. $|\bar{X}_{.2} - \bar{X}_{.4}| = |66.67 - 79.33| = 12.66$
6. $|\bar{X}_{.3} - \bar{X}_{.4}| = |91.00 - 79.33| = 11.67$

Locate $MSE = 14.986$ and $r = 6$ in Figure 11.13 or Figure 11.14 to determine the critical range. From Table E.10 [for $\alpha = .05$, $c = 4$, and $(r - 1)(c - 1) = 15$], Q_U, the upper-tail critical value of the test statistic with 4 and 15 degrees of freedom, is 4.08. Using Equation (11.15),

$$\text{Critical range} = 4.08\sqrt{\frac{14.986}{6}} = 6.448$$

All pairwise comparisons except $|\bar{X}_{.1} - \bar{X}_{.4}|$ are greater than the critical range. Therefore, you conclude that there is evidence of a significant difference in the mean rating between all pairs of restaurant branches except for branches *A* and *D*. In addition, branch *C* has the highest ratings (i.e., is most preferred) and branch *B* has the lowest (i.e., is least preferred).

PROBLEMS FOR SECTION 11.2

Learning the Basics

PH Grade ASSIST **11.15** Given a randomized block experiment with five treatment levels and seven blocks, answer the following.

a. How many degrees of freedom are there in determining the among-group variation?
b. How many degrees of freedom are there in determining the among-block variation?
c. How many degrees of freedom are there in determining the inherent random variation or error?
d. How many degrees of freedom are there in determining the total variation?

PH Grade ASSIST **11.16** From problem 11.15:

a. If $SSA = 60$, $SSBL = 75$, and $SST = 210$, what is SSE?
b. What are MSA, $MSBL$, and MSE?
c. What is the value of the test statistic F for the difference in the five means?
d. What is the value of the test statistic F for the block effects?

PH Grade ASSIST **11.17** From problems 11.15 and 11.16:

a. Construct the ANOVA summary table and fill in all values in the body of the table.

b. At the 0.05 level of significance, is there evidence of a difference in the population means?

c. At the 0.05 level of significance, is there evidence of a difference due to blocks?

PH Grade ASSIST **11.18** From problems 11.15, 11.16, and 11.17:

a. To perform the Tukey procedure, how many degrees of freedom are there in the numerator and how many degrees of freedom are there in the denominator of the Studentized range distribution?

b. At the 0.05 level of significance, what is the upper-tail critical value from the Studentized range distribution?

c. To perform the Tukey procedure, what is the critical range?

11.19 Given a randomized block experiment with three treatment levels and seven blocks:

a. How many degrees of freedom are there in determining the among-group variation?

b. How many degrees of freedom are there in determining the among-block variation?

c. How many degrees of freedom are there in determining the random variation or error?

d. How many degrees of freedom are there in determining the total variation?

11.20 From problem 11.19, if $SSA = 36$ and the randomized block F-test statistic is 6.0:

a. What is MSE and SSE?

b. What is $SSBL$ if the F-test statistic for block effects is 4.0?

c. What is SST?

d. At the 0.01 level of significance, is there evidence of a treatment effect and is there evidence of a block effect?

11.21 Given a randomized block experiment with four treatment levels and eight blocks, from the ANOVA summary table below, fill in all the missing results.

Source	Degrees of Freedom	Sum of Squares	Mean Square (Variance)	F
Among treatments	$c - 1 = ?$	$SSA = ?$	$MSA = 80$	$F = ?$
Among blocks	$r - 1 = ?$	$SSBL = 540$	$MSBL = ?$	$F = 5.0$
Error	$(r - 1)(c - 1) = ?$	$SSE = ?$	$MSE = ?$	
Total	$rc - 1 = ?$	$SST = ?$		

11.22 From problem 11.21:

a. At the 0.05 level of significance, is there evidence of a difference among the four treatment level means?

b. At the 0.05 level of significance, is there evidence of an effect due to blocks?

Applying the Concepts

Problems 11.23–11.28 can be solved manually or by using Microsoft Excel, Minitab, or SPSS.

PH Grade ASSIST **11.23** Nine experts rated four brands of Colombian coffee in a taste-testing experiment. A rating on a 7-point scale (1 = extremely unpleasing, 7 = extremely pleasing) is given for each of four characteristics: taste, aroma, richness, and acidity. The following table displays the summated ratings—accumulated over all four characteristics. COFFEE

	BRAND			
EXPERT	**A**	**B**	**C**	**D**
C.C.	24	26	25	22
S.E.	27	27	26	24
E.G.	19	22	20	16
B.L.	24	27	25	23
C.M.	22	25	22	21
C.N.	26	27	24	24
G.N.	27	26	22	23
R.M.	25	27	24	21
P.V.	22	23	20	19

At the 0.05 level of significance, completely analyze the data to determine whether there is evidence of a difference in the summated ratings of the four brands of Colombian coffee and, if so, which of the brands are rated highest (i.e., best). What can you conclude?

PH Grade ASSIST SELF Test **11.24** A student team in a business statistics course designed an experiment to investigate whether the brand of bubblegum used affected the size of bubbles they could blow. The students believed that Kyle was an expert at blowing bubbles, and his expertise might negatively affect the results of a completely randomized design. Thus, to reduce the person-to-person variability, they decided to use a randomized block design using themselves as blocks. Four brands of bubblegum were tested. A student chewed two pieces of a brand of gum and then blew two bubbles, attempting to make them as big as possible. Another student measured the diameters of the bubbles at their biggest point. The following table gives the combined diameters of the bubbles (in inches) for the 16 observations. BUBBLEGUM

	BRAND OF BUBBLEGUM			
STUDENT	**Bazooka**	**Bubbletape**	**Bubbleyum**	**Bubblicious**
Kyle	8.75	9.50	8.50	11.50
Sarah	9.50	4.00	8.50	11.00
Leigh	9.25	5.50	7.50	7.50
Isaac	9.50	8.50	7.50	7.50

a. Using a 0.05 level of significance, is there evidence of a difference in the mean diameter of the bubbles produced by the different brands?
b. If appropriate, use the Tukey procedure to determine which brands of bubblegum differ. (Use $\alpha = 0.05$.)
c. Do you think that there was a significant block effect in this experiment? Explain.
d. Do you think Kyle is the best at blowing big bubbles?

11.25 An article discussed the new Whole Foods Market in the Time-Warner building in New York City (W. Grimes, "A Pleasure Palace without the Guilt," *The New York Times*, February 18, 2004, F1, F5). The following data WHOLEFOODS2 compared the price of some kitchen staples at the new Whole Foods Market, at the Gristede's supermarket, at the Fairway supermarket (both located about 15 blocks from the Time-Warner building), and at a Stop & Shop supermarket located in an outer borough of New York City.

Item	Whole Foods	Gristede's	Fair-way	Stop & Shop
Half-gallon milk	2.19	1.59	1.35	1.89
Dozen eggs	2.39	1.59	1.69	2.29
Tropicana orange juice (64 oz.)	2.00	2.99	2.49	2.00
Head of Boston lettuce	1.98	1.49	1.29	1.29
Ground round 1lb.	4.99	3.99	3.69	3.59
Bumble Bee tuna 6 oz. can	1.79	0.79	1.33	1.50
Granny Smith apples (1 lb.)	1.69	1.99	1.49	0.99
Box DeCecco linguini	1.99	1.50	1.59	1.79
Salmon steak 1 lb.	7.99	4.99	5.99	5.99
Whole chicken per pound	2.19	1.19	1.49	1.49

Source: Extracted from W. Grimes, "A Pleasure Palace without the Guilt," The New York Times, *February 18, 2004, F1, F5.*

a. At the 0.05 level of significance, use the randomized block design to determine if there is evidence of a difference in the mean prices for these kitchen staples at the four supermarkets.
b. What assumptions are necessary to perform this test?
c. If appropriate, use the Tukey procedure to determine which supermarkets differ. (Use $\alpha = 0.05$.)
d. Do you think that there was a significant block effect in this experiment? Explain.

11.26 An article (J. Graham, "Prices Going Up, But It's Not Gas; It's Online Music," *USA Today*, May 13, 2004, 1B) stated that the cost of a legitimate music download is increasing. The following data MUSICONLINE represent the prices of five albums at five digital music services.

Album/Artist	Itunes	Wal-Mart	Music Now	Music-match	Napster
D12 World—D12	11.99	9.44	13.99	20.79	9.95
Damita Jo—Janet Jackson	13.99	9.44	13.99	12.49	13.95
30 #1 Hits—Elvis Presley	9.99	27.28	13.99	9.99	9.95
Feels Like Home—Norah Jones	12.87	9.44	9.99	11.99	13.95
Up—Shania Twain	9.99	17.44	13.99	9.99	18.81

Source: Extracted from J. Graham, "Prices Going Up, But It's Not Gas; It's Online Music," USA Today, *May 13, 2004, p. 1B.*

a. At the 0.05 level of significance, use the randomized block design to determine if there is evidence of a difference in the mean prices for albums at the five digital music services.
b. What assumptions are necessary to perform this test?
c. If appropriate, use the Tukey procedure to determine which digital music services differ. (Use $\alpha = 0.05$.)
d. Do you think that there was a significant block effect in this experiment? Explain.

11.27 Philips Semiconductors is a leading European manufacturer of integrated circuits. Integrated circuits are produced on silicon wafers, which are ground to target thickness early in the production process. The wafers are positioned in different locations on a grinder and kept in place through vacuum decompression. One of the goals of process improvement is to reduce the variability in the thickness of the wafers in different positions and in different batches. Data were collected from a sample of 30 batches. In each batch the thickness of the wafers on positions 1 and 2 (outer circle), 18 and 19 (middle circle), and 28 (inner circle) was measured. The results are given in the CIRCUITS file. At the 0.01 level of significance, completely analyze the data to determine whether there is evidence of a difference in the mean thickness of the wafers for the five positions and, if so, which of the positions are different. What can you conclude?

Source: K. C. B. Roes, and R. J. M. M. Does, "Shewhart-type charts in nonstandard situations," Technometrics, *37, 1995, 15–24.*

11.28 The data in the CONCRETE2 file represent the compressive strength in thousands of pounds per square inch of 40 samples of concrete taken 2, 7, and 28 days after pouring.

Source: O. Carrillo-Gamboa and R. F. Gunst, "Measurement-Error-Model Collinearities," Technometrics, *34, 1992, 454–464.*

a. At the 0.05 level of significance, is there evidence of a difference in the mean compressive strength after 2, 7, and 28 days?

b. If appropriate, use the Tukey procedure to determine the days that differ in mean compressive strength. (Use $\alpha = 0.05$.)

c. Determine the relative efficiency of the randomized block design as compared with the one-way ANOVA (completely randomized) design.

d. Construct box-and-whisker plots of the compressive strength for the different time periods.

e. Based on the results of (a), (b), and (d), is there a pattern in the compressive strength over the three time periods?

11.3 THE FACTORIAL DESIGN: TWO-WAY ANALYSIS OF VARIANCE

In section 11.1 you learned about the one-way analysis of variance, and in section 11.2 you studied the randomized block design. In this section, the analysis of variance is extended to the **two-factor factorial design**, in which two factors are simultaneously evaluated. Each factor is evaluated at two or more treatment levels. For example, the Perfect Parachute Company is interested in simultaneously evaluating four suppliers and two types of looms. While discussion here is limited to two factors, you can extend factorial designs to three or more factors (see references 4, 5, and 10).

Data from a two-factor factorial design are analyzed using **two-way ANOVA**. Because of the complexity of the calculations involved, you should use software such as Microsoft Excel or Minitab when conducting this analysis. Nevertheless, for purposes of illustration and a better conceptual understanding of two-way ANOVA, the decomposition of the total variation is presented below. The following definitions are needed to develop the two-way ANOVA procedure:

r = the number of levels of factor A

c = the number of levels of factor B

n' = the number of values (replications) for each cell (combination of a particular level of factor A and a particular level of factor B)

n = total number of values in the experiment (where $n = rcn'$)

X_{ijk} = value of the kth observation for level i of factor A and level j of factor B

$$\bar{\bar{X}} = \frac{\sum_{i=1}^{r}\sum_{j=1}^{c}\sum_{k=1}^{n'} X_{ijk}}{rcn'} = \text{grand mean}$$

$$\bar{X}_{i..} = \frac{\sum_{j=1}^{c}\sum_{k=1}^{n'} X_{ijk}}{cn'} = \text{mean of the } i\text{th level of factor } A \text{ (where } i = 1, 2, \ldots, r)$$

$$\bar{X}_{.j.} = \frac{\sum_{i=1}^{r}\sum_{k=1}^{n'} X_{ijk}}{rn'} = \text{mean of the } j\text{th level of factor } B \text{ (where } j = 1, 2, \ldots, c)$$

$$\bar{X}_{ij.} = \sum_{k=1}^{n'} \frac{X_{ijk}}{n'} = \text{mean of the cell } ij\text{, the combination of the } i\text{th level of factor } A \text{ and the } j\text{th level of factor } B$$

This text deals only with situations in which there are an equal number of **replicates** (i.e., sample sizes n') for each combination of the levels of factor A with those of factor B. (See references 1 and 11 for a discussion of two-factor factorial designs with unequal sample sizes.)

Testing for Factor and Interaction Effects

There is an **interaction** between factors A and B if the effect of factor A is dependent on the level of factor B. Thus, the decomposition of total variation needs to account for a possible interaction effect, as well as factor A, factor B, and random error. Therefore, the total variation (SST) is subdivided into sum of squares due to factor A (or SSA), sum of squares due to factor B (or SSB), sum of squares due to the interaction effect of A and B (or $SSAB$), and sum of squares due to random error (or SSE). This decomposition of the total variation (SST) is displayed in Figure 11.17.

FIGURE 11.17
Partitioning the Total Variation in a Two-Factor Factorial Design

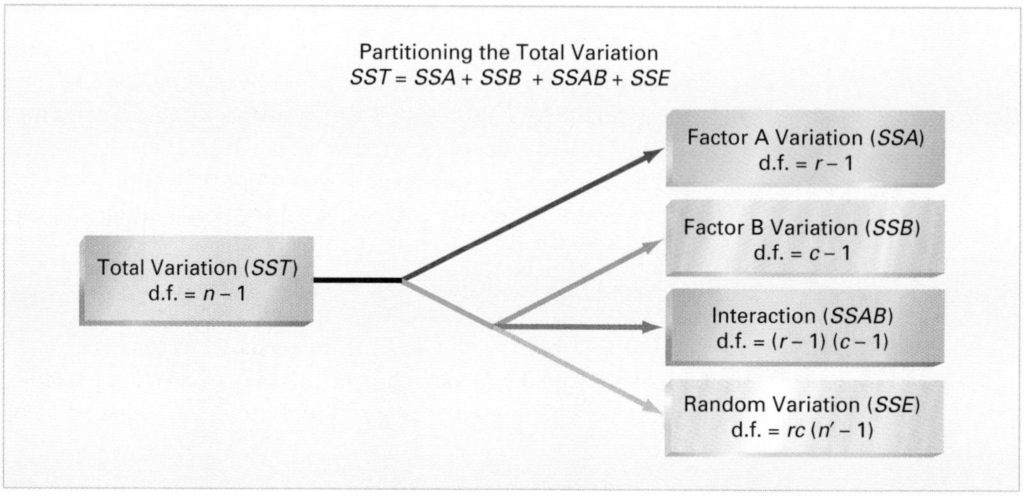

The sum of squares total (SST) represents the total variation among all the values around the grand mean. Equation (11.16) shows the computation for total variation.

TOTAL VARIATION IN TWO-WAY ANOVA

$$SST = \sum_{i=1}^{r}\sum_{j=1}^{c}\sum_{k=1}^{n'}(X_{ijk} - \bar{\bar{X}})^2 \quad \textbf{(11.16)}$$

The **sum of squares due to factor A (SSA)** represents the differences among the various levels of factor A and the grand mean. Equation (11.17) shows the computation for factor A variation.

FACTOR A VARIATION

$$SSA = cn'\sum_{i=1}^{r}(\bar{X}_{i..} - \bar{\bar{X}})^2 \quad \textbf{(11.17)}$$

The **sum of squares due to factor B (SSB)** represents the differences among the various levels of factor B and the grand mean. Equation (11.18) shows the computation for factor B variation.

FACTOR B VARIATION

$$SSB = rn'\sum_{j=1}^{c}(\bar{X}_{.j.} - \bar{\bar{X}})^2 \quad \textbf{(11.18)}$$

The **sum of squares due to interaction (*SSAB*)** represents the interacting effect of specific combinations of factor A and factor B. Equation (11.19) shows the computation for interaction variation.

INTERACTION VARIATION

$$SSAB = n'\sum_{i=1}^{r}\sum_{j=1}^{c}(\bar{X}_{ij.} - \bar{X}_{i..} - \bar{X}_{.j.} + \bar{\bar{X}})^2 \qquad \textbf{(11.19)}$$

The sum of squares error (SSE) represents the differences among the values within each cell and the corresponding cell mean. Equation (11.20) shows the computation for random error.

RANDOM ERROR IN TWO-WAY ANOVA

$$SSE = \sum_{i=1}^{r}\sum_{j=1}^{c}\sum_{k=1}^{n'}(X_{ijk} - \bar{X}_{ij.})^2 \qquad \textbf{(11.20)}$$

Because there are r treatment levels of factor A, there are $r - 1$ degrees of freedom associated with SSA. Similarly, because there are c treatment levels of factor B, there are $c - 1$ degrees of freedom associated with SSB. Because there are n' replications in each of the rc cells, there are $rc(n' - 1)$ degrees of freedom associated with the SSE term. Carrying this further, there are $n - 1$ degrees of freedom associated with the sum of squares total (SST) because you are comparing each value X_{ijk} to the grand mean $\bar{\bar{X}}$ based on all n values. Therefore, because the degrees of freedom for each of the sources of variation must add to the degrees of freedom for the total variation (SST), you calculate the degrees of freedom for the interaction component ($SSAB$) by subtraction. The degrees of freedom for interaction are $(r - 1)(c - 1)$.

If you divide each of the sums of squares by its associated degrees of freedom, you have the four variances or mean square terms (***MSA, MSB, MSAB***, and ***MSE***). Equations (11.21a–d) give the mean square terms needed for the two-way ANOVA table.

THE MEAN SQUARES IN TWO-WAY ANOVA

$$MSA = \frac{SSA}{r - 1} \qquad \textbf{(11.21a)}$$

$$MSB = \frac{SSB}{c - 1} \qquad \textbf{(11.21b)}$$

$$MSAB = \frac{SSAB}{(r - 1)(c - 1)} \qquad \textbf{(11.21c)}$$

$$MSE = \frac{SSE}{rc(n' - 1)} \qquad \textbf{(11.21d)}$$

There are three distinct tests to perform in a two-way ANOVA.

1. To test the hypothesis of no difference due to factor A

$$H_0: \mu_{1..} = \mu_{2..} = \cdots = \mu_{r..}$$

against the alternative

$$H_1: \text{Not all } \mu_{i..} \text{ are equal}$$

you use the F statistic in Equation (11.22).

F TEST FOR FACTOR A EFFECT

$$F = \frac{MSA}{MSE} \tag{11.22}$$

You reject the null hypothesis at the α level of significance if

$$F = \frac{MSA}{MSE} > F_U$$

the upper-tail critical value from an F distribution with $r - 1$ and $rc(n' - 1)$ degrees of freedom.

2. To test the hypothesis of no difference due to factor B

$$H_0: \mu_{.1.} = \mu_{.2.} = \cdots = \mu_{.c.}$$

against the alternative

$$H_1 \text{ Not all } \mu_{.j.} \text{ are equal}$$

you use the F statistic in Equation (11.23).

F TEST FOR FACTOR B EFFECT

$$F = \frac{MSB}{MSE} \tag{11.23}$$

You reject the null hypothesis at the α level of significance if

$$F = \frac{MSB}{MSE} > F_U$$

the upper-tail critical value from an F distribution with $c - 1$ and $rc(n' - 1)$ degrees of freedom.

3. To test the hypothesis of no interaction of factors A and B

H_0: The interaction of A and B is equal to zero

against the alternative

H_1: The interaction of A and B is not equal to zero

you use the F statistic in Equation (11.24).

F TEST FOR INTERACTION EFFECT

$$F = \frac{MSAB}{MSE} \tag{11.24}$$

You reject the null hypothesis at the α level of significance if

$$F = \frac{MSAB}{MSE} > F_U$$

the upper-tail critical value from an F distribution with $(r-1)(c-1)$ and $rc(n'-1)$ degrees of freedom.

Table 11.8 summarizes the entire set of steps.

TABLE 11.8
Analysis-of-Variance Table for the Two-Factor Factorial Design

Source	Degrees of Freedom	Sum of Squares	Mean Square (Variance)	F
A	$r-1$	SSA	$MSA = \frac{SSA}{r-1}$	$F = \frac{MSA}{MSE}$
B	$c-1$	SSB	$MSB = \frac{SSB}{c-1}$	$F = \frac{MSB}{MSE}$
AB	$(r-1)(c-1)$	$SSAB$	$MSAB = \frac{SSAB}{(r-1)(c-1)}$	$F = \frac{MSAB}{MSE}$
Error	$rc(n'-1)$	SSE	$MSE = \frac{SSE}{rc(n'-1)}$	
Total	$n-1$	SST		

To examine a two-way ANOVA, return to the "Using Statistics" scenario on page 396. As operations manager, you have decided not only to evaluate the different suppliers but also to determine whether parachutes woven on the Jetta looms are as strong as those woven on the Turk looms. In addition, you need to determine whether any differences among the four suppliers in the strength of the parachutes are dependent on the type of loom being used. Thus, you have decided to perform an experiment in which five different parachutes from each supplier are manufactured on each of the two different looms. The results PARACHUTE2 are given in Table 11.9.

TABLE 11.9
Tensile Strengths of Parachutes Woven by Two Types of Looms of Synthetic Fibers from Four Suppliers

	SUPPLIER			
LOOM	**1**	**2**	**3**	**4**
Jetta	20.6	22.6	27.7	21.5
	18.0	24.6	18.6	20.0
	19.0	19.6	20.8	21.1
	21.3	23.8	25.1	23.9
	13.2	27.1	17.7	16.0
Turk	18.5	26.3	20.6	25.4
	24.0	25.3	25.2	19.9
	17.2	24.0	20.8	22.6
	19.9	21.2	24.7	17.5
	18.0	24.5	22.9	20.4

Figure 11.18 presents Microsoft Excel output and Figure 11.19 presents Minitab output. In Figure 11.18, the summary tables provide the sample size, sum, mean, and variance for each combination of supplier and loom. The total column in the first two tables provides these statistics for each loom, and the third table provides them for each supplier. In addition, in the ANOVA table *df* represents degrees of freedom, *SS* refers to sum of squares, *MS* stands for mean squares, and F is the computed F-test statistic.

FIGURE 11.18

Microsoft Excel Two-Way ANOVA Output for the Parachute Manufacturing Example

	A	B	C	D	E	F	G
1	**Anova: Two-Factor With Replication**						
2							
3	**SUMMARY**	**Supplier 1**	**Supplier 2**	**Supplier 3**	**Supplier 4**	**Total**	
4	*Jetta*						
5	Count	5	5	5	5	20	
6	Sum	92.1	117.7	109.9	102.5	422.2	
7	Average	18.42	23.54	21.98	20.5	21.11	(jetta loom average)
8	Variance	10.202	7.568	18.397	8.355	13.1283	
9							
10	*Turk*						
11	Count	5	5	5	5	20	
12	Sum	97.6	121.3	114.2	105.8	438.9	
13	Average	19.52	24.26	22.84	21.16	21.945	(turk loom average)
14	Variance	7.237	3.683	4.553	8.903	8.4626	
15							
16	*Total*						
17	Count	10	10	10	10		
18	Sum	189.7	239	224.1	208.3		
19	Average (supplier)	18.97	23.9	22.41	20.83		
20	Variance	8.0868	5.1444	10.4054	7.7912		
21							
22							
23	**ANOVA**						
24	*Source of Variation*	*SS*	*df*	*MS*	*F*	*P-value*	*F crit*
25	Sample (Factor A: looms)	6.97225	1	6.9723	0.8096	0.3750	4.1491
26	Columns (Factor B: suppliers)	134.34875	3	44.7829	5.1999	0.0049	2.9011
27	Interaction	0.28675	3	0.0956	0.0111	0.9984	2.9011
28	Within	275.592	32	8.6123			
29							
30	Total	417.19975	39				

FIGURE 11.19

Minitab Two-Way ANOVA Output for the Parachute Manufacturing Example

Two-way ANOVA: Strength versus Loom, Supplier

```
Source       DF       SS       MS     F      P
Loom          1    6.972   6.9722  0.81  0.375
Supplier      3  134.349  44.7829  5.20  0.005
Interaction   3    0.287   0.0956  0.01  0.998
Error        32  275.592   8.6123
Total        39  417.200

S = 2.935   R-Sq = 33.94%   R-Sq(adj) = 19.49%

                 Individual 95% CIs For Mean Based on
                 Pooled StDev
Loom      Mean   --+---------+---------+---------+-------
jetta   21.110   (------------*------------)
turk    21.945           (------------*-------------)
                 --+---------+---------+---------+-------
                 20.0      21.0      22.0      23.0
                 Individual 95% CIs For Mean Based on
                   Pooled StDev
Supplier   Mean   --+---------+---------+---------+-------
1         18.97   (-------*------)
2         23.90                      (-------*------)
3         22.41                (-------*------)
4         20.83          (------*-------)
```

[1] Table E.5 does not provide the upper-tail critical values from the F distribution with 32 degrees of freedom in the denominator. Table E.5 gives critical values either for F distributions with 30 degrees of freedom in the denominator or for F distributions with 40 degrees of freedom in the denominator. When the desired degrees of freedom are not provided in the table, you can round to the closest value that is given or use the p-value approach.

To interpret the results, you start by testing whether there is an interaction effect between factor A (loom) and factor B (supplier). If the interaction effect is significant, further analysis will refer only to this interaction. If the interaction effect is not significant, you can focus on the **main effects**—potential differences in looms (factor A) and potential differences in suppliers (factor B).

Using the 0.05 level of significance, to determine whether there is evidence of an interaction effect, you reject the null hypothesis of no interaction between loom and supplier if the computed F value is greater than 2.92, the approximate upper-tail critical value from the F distribution with 3 and 32 degrees of freedom (see Figure 11.20).[1]

FIGURE 11.20

Regions of Rejection and Nonrejection at the 0.05 Level of Significance with 3 and 32 Degrees of Freedom

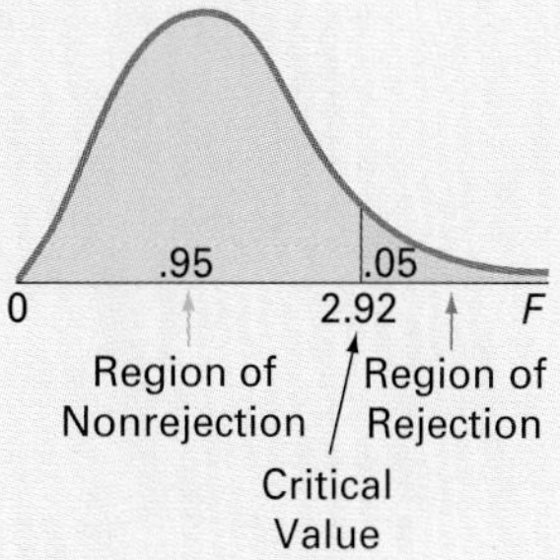

Because $F = 0.011 < F_U = 2.92$ or the p-value $= 0.998 > 0.05$, do not reject H_0. You conclude that there is insufficient evidence of an interaction effect between loom and supplier. The focus is now on the main effects.

Using the 0.05 level of significance and testing for a difference between the two looms (factor A), you reject the null hypothesis if the calculated F value is greater than 4.17, the (approximate) upper-tail critical value from the F distribution with 1 and 32 degrees of freedom (see Figure 11.21). Because $F = 0.81 < F_U = 4.17$ or the p-value $= 0.375 > 0.05$, you do not reject H_0. You conclude that there is insufficient evidence of a difference between the two looms in terms of the mean tensile strengths of the parachutes manufactured.

FIGURE 11.21

Regions of Rejection and Nonrejection at the 0.05 Level of Significance with 1 and 32 Degrees of Freedom

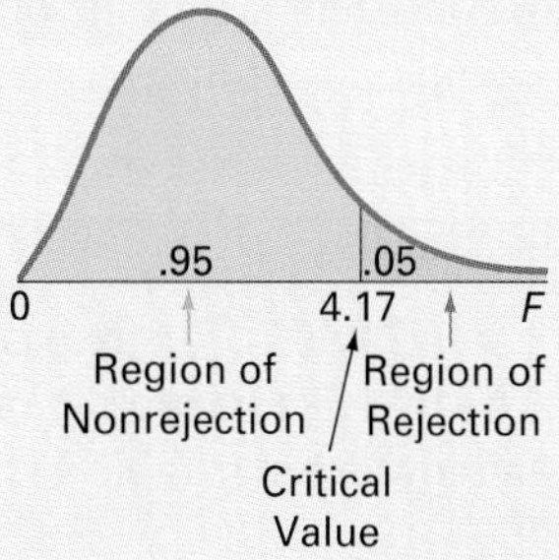

Using the 0.05 level of significance and testing for a difference among the suppliers (factor B), you reject the null hypothesis of no difference if the calculated F value exceeds 2.92, the approximate upper-tail critical value from the F distribution with 3 degrees of freedom in the numerator and 32 degrees of freedom in the denominator (see Figure 11.20). Since $F = 5.20 > F_U = 2.92$ or the p-value $= 0.005 < 0.05$, you reject H_0. You conclude that there is evidence of a difference among the suppliers in terms of the mean tensile strength of the parachutes.

Interpreting Interaction Effects

You can get a better understanding of the interpretation of the interaction by plotting the cell means (i.e., the means of specific treatment level combinations) as shown in Figure 11.22 on page 428. Figure 11.18 on page 426 provides the cell means for the loom–supplier combinations. From the plot of the mean tensile strength for each combination of loom and supplier, observe that the two lines (representing the two looms) are roughly parallel. This indicates that the *difference* between mean tensile strength of the two looms is virtually the same for the four suppliers. In other words, there is no *interaction* between these two factors, as was clearly substantiated from the F test.

FIGURE 11.22
Microsoft Excel Cell Means Plot of Tensile Strength Based on Loom and Supplier

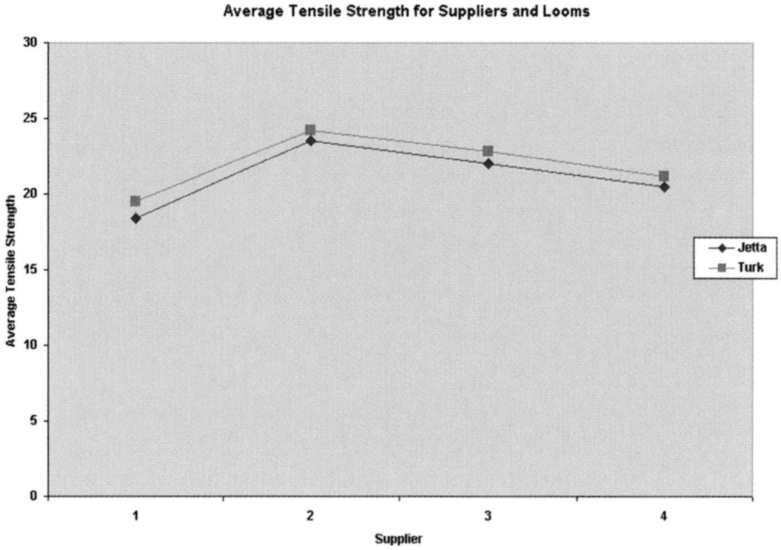

What is the interpretation if there is an interaction? In such a situation, some levels of factor A would respond better with certain levels of factor B. For example, with respect to tensile strength, suppose that some suppliers were better for the Jetta loom and other suppliers were better for the Turk loom. If this were true, the lines of Figure 11.22 would not be nearly as parallel and the interaction effect might be statistically significant. In such a situation, the difference between the looms is no longer the same for all suppliers. Such an outcome would also complicate the interpretation of the *main effects* because differences in one factor (i.e., loom) are not consistent across the other factor (i.e., supplier).

Example 11.2 illustrates a situation with a significant interaction effect.

EXAMPLE 11.2

INTERPRETING SIGNIFICANT INTERACTION EFFECTS

The data in Table 11.10 are based on an experiment concerning the manufacture of roller bearings. The two factors are the outer ring osculation and the heat treatment, each examined at a high and low setting. Table 11.10 shows the length of the bearings' lives **BEARING**.

TABLE 11.10
Lives of Bearings for Different Ring Osculations and Heat Treatments

	HEAT TREATMENT	
RING OSCULATION	**Low**	**High**
Low	12	26
	24	16
High	18	101
	28	113

What are the effects of ring osculation and heat treatment on the life of the roller bearings?

SOLUTION

Figure 11.23 presents Microsoft Excel output for these data.

In addition to the ANOVA summary table, Microsoft Excel has computed the mean value for each combination of ring osculation and heat treatment, as well as the mean for each level of the ring osculation and heat treatment factors. To interpret the results, begin by testing whether there is evidence of an interacting effect between factor A (i.e., ring osculation) and factor B (i.e., heat treatment). Using the level of significance $\alpha = 0.05$, you reject the null hypothesis of no interacting effect because the p-value (0.0018) is less than $\alpha = 0.05$, the selected level of significance. (The computed F-test statistic 53.779 is greater than 7.71, the upper-tail critical value from the F distribution with 1 and 4 degrees of freedom.)

FIGURE 11.23
Microsoft Excel Two-Way ANOVA Output for the Roller Bearing Example

	A	B	C	D	E	F	G
1	Anova: Two-Factor With Replication						
2							
3	SUMMARY	Heat-Low	Heat-High	Total			
4	*Ring-Low*						
5	Count	2	2	4			
6	Sum	36	42	78			
7	Average	18	21	19.5			
8	Variance	72	50	43.66667			
9							
10	*Ring-High*						
11	Count	2	2	4			
12	Sum	46	214	260			
13	Average	23	107	65			
14	Variance	50	72	2392.667			
15							
16	*Total*						
17	Count	4	4				
18	Sum	82	256				
19	Average	20.5	64				
20	Variance	49	2506				
21							
22							
23	ANOVA						
24	*Source of Variation*	*SS*	*df*	*MS*	*F*	*P-value*	*F crit*
25	Sample (*Factor A*)	4140.5	1	4140.5	67.877	0.0012	7.709
26	Columns (*Factor B*)	3784.5	1	3784.5	62.041	0.0014	7.709
27	Interaction	3280.5	1	3280.5	53.779	0.0018	7.709
28	Within	244	4	61			
29							
30	Total	11449.5	7				

You can see this significant interaction effect between ring osculation and heat treatment in Figure 11.24, which plots the mean life for each type of ring osculation according to type of heat treatment (low versus high). Because the lines that represent mean life for the different heat treatments are not parallel for the two ring osculations, the differences between mean life in low and high ring osculation are not constant for the two types of heat treatments. There is a large difference in mean life between low and high heat treatments for high ring osculation but very little difference when there is low ring osculation.

FIGURE 11.24
Microsoft Excel Cell Means Plot of Bearing Life Based on Ring Osculation and Heat Treatment

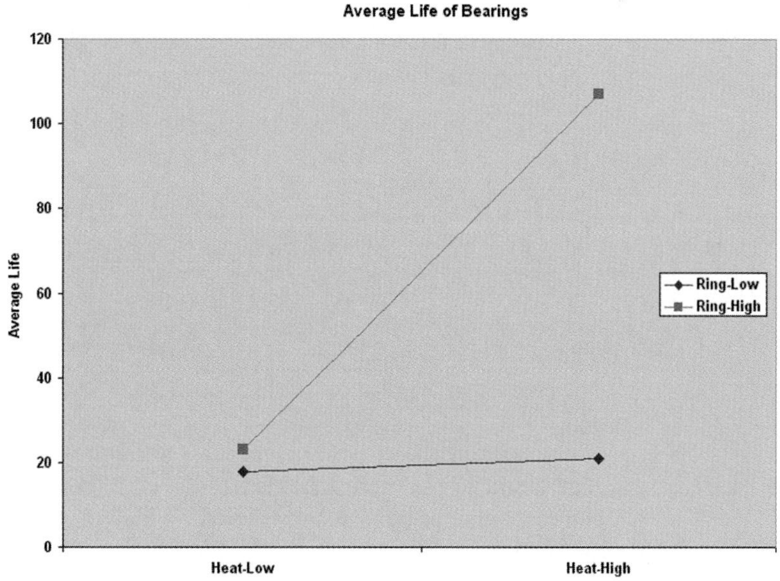

The existence of this significant interaction effect complicates the interpretation of the main effects. You cannot directly conclude that there is a significant difference between mean life of low and high ring osculation because the difference is not the same over all heat treatments. Likewise, you cannot directly conclude that there is a significant difference between the mean lives of the heat treatments because this difference is not the same for both types of ring osculations.

Multiple Comparisons: The Tukey Procedure

Unless there is a significant interaction effect, you can determine the particular levels that are significantly different by using a procedure developed by John Tukey (see references 10 and 13). Equation (11.25) gives the critical range for factor *A*.

THE CRITICAL RANGE FOR FACTOR *A*

$$\text{Critical range} = Q_U\sqrt{\frac{MSE}{cn'}} \tag{11.25}$$

where Q_U is the upper-tail critical value from a Studentized range distribution having r and $rc(n' - 1)$ degrees of freedom. Values for the Studentized range distribution are found in Table E.10.

Equation (11.26) gives the critical range for factor *B*.

THE CRITICAL RANGE FOR FACTOR *B*

$$\text{Critical range} = Q_U\sqrt{\frac{MSE}{rn'}} \tag{11.26}$$

where Q_U is the upper-tail critical value from a Studentized range distribution having c and $rc(n' - 1)$ degrees of freedom. Values for the Studentized range distribution are found in Table E.10.

To use the Tukey procedure, return to the parachute manufacturing data of Table 11.9 on page 425. In Figures 11.18 or 11.19 on page 426 (the ANOVA summary table provided by Microsoft Excel and Minitab), only one of the main effects is significant. Using an α level of 0.05, there is no evidence of a significant difference between the two looms (Jetta and Turk) that comprise factor *A*, but there is evidence of a significant difference among the four suppliers that comprise factor *B*. Thus, you can use the multiple comparison procedure to further analyze differences between the suppliers of factor *B*.

Because there are four levels of factor *B*, there are $4(4 - 1)/2 = 6$ pairwise comparisons. Using the calculations presented in Figures 11.18 or 11.19, the absolute mean differences are as follows.

1. $|\bar{X}_{.1.} - \bar{X}_{.2.}| = |18.97 - 23.90| = 4.93$
2. $|\bar{X}_{.1.} - \bar{X}_{.3.}| = |18.97 - 22.41| = 3.44$
3. $|\bar{X}_{.1.} - \bar{X}_{.4.}| = |18.97 - 20.83| = 1.86$
4. $|\bar{X}_{.2.} - \bar{X}_{.3.}| = |23.90 - 22.41| = 1.49$
5. $|\bar{X}_{.2.} - \bar{X}_{.4.}| = |23.90 - 20.83| = 3.07$
6. $|\bar{X}_{.3.} - \bar{X}_{.4.}| = |22.41 - 20.83| = 1.58$

To determine the critical range, refer to Figures 11.18 or 11.19 to find $MSE = 8.612$, $r = 2$, $c = 4$, and $n' = 5$. From Table E.10 [for $\alpha = 0.05$, $c = 4$, and $rc(n' - 1) = 32$], Q_U, the upper-tail critical value of the test statistic with 4 and 32 degrees of freedom, is approximated as 3.84. Using Equation (11.26):

$$\text{Critical range} = 3.84\sqrt{\frac{8.612}{10}} = 3.56$$

Because 4.93 > 3.56, only the means of Suppliers 1 and 2 are different. You can conclude that the mean tensile strength is lower for Supplier 1 than for Supplier 2.

PROBLEMS FOR SECTION 11.3

Learning the Basics

PH Grade ASSIST **11.29** Given a two-factor factorial design having three levels in factor A, three levels in factor B, and four replicates in each of the nine treatment combinations of factors A and B,

a. How many degrees of freedom are there in determining the factor A variation and the factor B variation?

b. How many degrees of freedom are there in determining the interaction variation?

c. How many degrees of freedom are there in determining the random error variation?

d. How many degrees of freedom are there in determining the total variation?

PH Grade ASSIST **11.30** Assume that you are working with the results from problem 11.29:

a. If $SSA = 120$, $SSB = 110$, $SSE = 270$, and $SST = 540$, what is $SSAB$?

b. What is MSA and MSB?

c. What is $MSAB$?

d. What is MSE?

11.31 Assume that you are working with the results of problem 11.30.

a. What is the value of the test statistic F for the interaction effect?

b. What is the value of the test statistic F for the factor A effect?

c. What is the value of the test statistic F for the factor B effect?

d. Form the ANOVA summary table and fill in all values in the body of the table.

PH Grade ASSIST **11.32** Given the results from problems 11.29 through 11.31,

a. At the 0.05 level of significance, is there an effect due to factor A?

b. At the 0.05 level of significance, is there an effect due to factor B?

c. At the 0.05 level of significance, is there an interaction effect?

11.33 Given a two-way ANOVA with two treatment levels for factor A and five treatment levels for factor B and four replicates in each of the 10 treatment combinations of factors A and B, with $SSA = 18$, $SSB = 64$, $SSE = 60$, and $SST = 150$,

a. Form the ANOVA summary table and fill in all values in the body of the table.

b. At the 0.05 level of significance, is there an effect due to factor A?

c. At the 0.05 level of significance, is there an effect due to factor B?

d. At the 0.05 level of significance, is there an interaction effect?

PH Grade ASSIST **11.34** Given a two-factor factorial experiment and the ANOVA summary table that follows, fill in all the missing results.

Source	Degrees of Freedom	Sum of Squares	Mean Square (Variance)	F
Factor A	$r - 1 = 2$	$SSA = ?$	$MSA = 80$	$F = ?$
Factor B	$c - 1 = ?$	$SSB = 220$	$MSB = ?$	$F = 11.0$
AB interaction	$(r - 1)(c - 1) = 8$	$SSAB = ?$	$MSAB = 10$	$F = ?$
Error	$rc(n' - 1) = 30$	$SSE = ?$	$MSE = ?$	
Total	$n - 1 = ?$	$SST = ?$		

PH Grade ASSIST **11.35** From the results of problem 11.34,

a. At the 0.05 level of significance, is there an effect due to factor A?

b. At the 0.05 level of significance, is there an effect due to factor B?

c. At the 0.05 level of significance, is there an interaction effect?

Applying the Concepts

Problems 11.36–11.40 can be solved manually or by using Microsoft Excel, Minitab, or SPSS.

11.36 The effects of developer strength (factor A) and development time (factor B) on the density of photographic plate film were being studied. Two strengths and two development times were used, and four replicates for each treatment combination were run. The results (with larger being best) are as follows: PHOTO

	DEVELOPMENT TIME (MINUTES)			DEVELOPMENT TIME (MINUTES)	
DEVELOPER STRENGTH	10	14	DEVELOPER STRENGTH	10	14
1	0	1	2	4	6
	5	4		7	7
	2	3		6	8
	4	2		5	7

At the 0.05 level of significance,

a. Is there an interaction between developer strength and development time?

b. Is there an effect due to developer strength?

c. Is there an effect due to development time?

d. Plot the mean density of each developer strength for each development time.
e. What can you conclude about the effect of developer strength and development time on density?

PH Grade ASSIST **11.37** A chef in a restaurant that specializes in pasta dishes was experiencing difficulty in getting brands of pasta to be *al dente*, that is, cooked enough so as not to feel starchy or hard but still feel firm when bitten into. She decided to conduct an experiment in which two brands of pasta, one American and one Italian, were cooked for either 4 or 8 minutes. The response variable measured was weight of the pasta, since the cooking of the pasta enables it to absorb water. A pasta with a faster rate of water absorption may provide a shorter interval in which the pasta is *al dente*, thereby increasing the chance that it might be overcooked. The experiment was conducted by using 150 grams of uncooked pasta. Each trial began by bringing a pot containing 6 quarts of cold unsalted water to a moderate boil. The 150 grams of uncooked pasta was added and then weighed after a given period of time by lifting the pasta from the pot via a built-in strainer. The results (in terms of weight in grams) for two replications of each type of pasta and cooking time are as follows: PASTA

	COOKING TIME (MINUTES)			COOKING TIME (MINUTES)	
TYPE OF PASTA	**4**	**8**	**TYPE OF PASTA**	**4**	**8**
American	265	310	Italian	250	300
	270	320		245	305

At the 0.05 level of significance,
a. Is there an interaction between type of pasta and cooking time?
b. Is there an effect due to type of pasta?
c. Is there an effect due to cooking time?
d. Plot the mean weight for each type of pasta for each cooking time.
e. What conclusions can you reach concerning the importance of each of these two factors on the weight of the pasta?

PH Grade ASSIST SELF Test **11.38** A student team in a business statistics course performed a factorial experiment to investigate the time required for pain-relief tablets to dissolve in a glass of water. The two factors of interest were brand name (Equate, Kroger, Alka-Seltzer) and temperature of the water (hot or cold). The experiment consisted of four observations for each of the six factor combinations. The following table PAIN-RELIEF gives the time a tablet took to dissolve (in seconds) for the 24 tablets used in the experiment.

	BRAND OF PAIN-RELIEF TABLET		
WATER	**Equate**	**Kroger**	**Alka-Seltzer**
Cold	85.87	75.98	100.11
	78.69	87.66	99.65
	76.42	85.71	100.83
	74.43	86.31	94.16
Hot	21.53	24.10	23.80
	26.26	25.83	21.29
	24.95	26.32	20.82
	21.52	22.91	23.21

At the 0.05 level of significance:
a. Is there an interaction between brand of pain-reliever and temperature of water?
b. Is there an effect that is due to brand?
c. Is there an effect that is due to the temperature of the water?
d. Plot the mean dissolving time for each brand for the two temperatures.
e. Discuss the results of (a) through (d).

11.39 Integrated circuits are manufactured on silicon wafers through a process that involves a series of steps. An experiment was carried out to study the effect of the cleansing and etching steps on the yield (coded to maintain confidentiality). The results YIELD are as follows.

	ETCHING STEP	
CLEANSING STEP	**New**	**Standard**
New One	38	34
	34	19
	38	28
New Two	29	20
	35	35
	34	37
Standard	31	29
	23	32
	38	30

Source: J. Ramirez and W. Taam, "An Autologistic Model for Integrated Circuit Manufacturing," Journal of Quality Technology, *2000, 32, 254–262.*

At the 0.05 level of significance:
a. Is there an interaction between the cleansing step and the etching step?
b. Is there an effect that is due to the cleansing step?

c. Is there an effect that is due to the etching step?
d. Plot the mean yield for each cleansing step for the two etching steps.
e. Discuss the results of (a) through (d).

11.40 An experiment was conducted to study the distortion of drive gears in automobiles. Two factors were studied, the tooth size of the gear and the part positioning. The results GEAR are as follows.

	PART POSITIONING	
TOOTH SIZE	**Low**	**High**
Low	18.0	13.5
	16.5	8.5
	26.0	11.5
	22.5	16.0
	21.5	–4.5
	21.0	4.0
	30.0	1.0
	24.5	9.0
High	27.5	17.5
	19.5	11.5
	31.0	10.0
	27.0	1.0
	17.0	14.5
	14.0	3.5
	18.0	7.5
	17.5	6.5

Source: D. R. Bingham and R. R. Sitter, " Design Issues in Fractional Factorial Split-Plot Experiments," Journal of Quality Technology, *33, 2001, 2–15.*

At the 0.05 level of significance:
a. Is there an interaction between the tooth size and the part positioning?
b. Is there an effect that is due to tooth size?
c. Is there an effect that is due to the part positioning?
d. Plot the mean yield for each tooth size for the two part positions.
e. Discuss the results of (a) through (d).

SUMMARY

In this chapter, various statistical procedures were used to analyze the effect of one or two factors of interest. You learned how the production manager of the Perfect Parachute Company could use these procedures to identify the best suppliers and loom types in order to increase the quality (i.e., strength) of their parachutes. The assumptions required for using these procedures were discussed in detail. Remember that you need to critically investigate the validity of the assumptions underlying the hypothesis test procedures. Table 11.11 summarizes the topics covered in this chapter.

TABLE 11.11
Summary of Topics in Chapter 11

Type of Analysis	**Type of Data** **Numerical**
Comparing More than Two Groups	One-Way Analysis of Variance **(section 11.1)** Randomized Block Design **(section 11.2)** Two-Way Analysis of Variance **(section 11.3)**

KEY FORMULAS

Total Variation in One-Way ANOVA

$$SST = \sum_{j=1}^{c} \sum_{i=1}^{n_j} (X_{ij} - \bar{\bar{X}})^2 \quad \textbf{(11.1)}$$

Among-Group Variation in One-Way ANOVA

$$SSA = \sum_{j=1}^{c} n_j(\bar{X}_j - \bar{\bar{X}})^2 \quad \textbf{(11.2)}$$

Within-Group Variation in One-Way ANOVA

$$SSW = \sum_{j=1}^{c}\sum_{i=1}^{n_j}(X_{ij} - \overline{X}_j)^2 \quad \textbf{(11.3)}$$

The Mean Squares in One-Way ANOVA

$$MSA = \frac{SSA}{c-1} \quad \textbf{(11.4a)}$$

$$MSW = \frac{SSW}{n-c} \quad \textbf{(11.4b)}$$

$$MST = \frac{SST}{n-1} \quad \textbf{(11.4c)}$$

One-Way ANOVA F-Test Statistic

$$F = \frac{MSA}{MSW} \quad \textbf{(11.5)}$$

The Critical Range for the Tukey-Kramer Procedure

$$\text{Critical range} = Q_U\sqrt{\frac{MSW}{2}\left(\frac{1}{n_j} + \frac{1}{n_{j'}}\right)} \quad \textbf{(11.6)}$$

Total Variation in Randomized Block Design

$$SST = \sum_{j=1}^{c}\sum_{i=1}^{r}(X_{ij} - \overline{\overline{X}})^2 \quad \textbf{(11.7)}$$

Among-Group Variation in Randomized Block Design

$$SSA = r\sum_{j=1}^{c}(\overline{X}_{.j} - \overline{\overline{X}})^2 \quad \textbf{(11.8)}$$

Among-Block Variation in Randomized Block Design

$$SSBL = c\sum_{i=1}^{r}(\overline{X}_{i.} - \overline{\overline{X}})^2 \quad \textbf{(11.9)}$$

Random Error in Randomized Block Design

$$SSE = \sum_{j=1}^{c}\sum_{i=1}^{r}(X_{ij} - \overline{X}_{.j} - \overline{X}_{i.} + \overline{\overline{X}})^2 \quad \textbf{(11.10)}$$

The Mean Squares in Randomized Block Design

$$MSA = \frac{SSA}{c-1} \quad \textbf{(11.11a)}$$

$$MSBL = \frac{SSBL}{r-1} \quad \textbf{(11.11b)}$$

$$MSE = \frac{SSE}{(r-1)(c-1)} \quad \textbf{(11.11c)}$$

Randomized Block F-Test Statistic

$$F = \frac{MSA}{MSE} \quad \textbf{(11.12)}$$

F-Test Statistic for Block Effects

$$F = \frac{MSBL}{MSE} \quad \textbf{(11.13)}$$

Estimated Relative Efficiency

$$RE = \frac{(r-1)MSBL + r(c-1)MSE}{(rc-1)MSE} \quad \textbf{(11.14)}$$

The Critical Range for the Randomized Block Design

$$\text{Critical range} = Q_U\sqrt{\frac{MSE}{r}} \quad \textbf{(11.15)}$$

Total Variation in Two-Way ANOVA

$$SST = \sum_{i=1}^{r}\sum_{j=1}^{c}\sum_{k=1}^{n'}(X_{ijk} - \overline{\overline{X}})^2 \quad \textbf{(11.16)}$$

Factor A Variation

$$SSA = cn'\sum_{i=1}^{r}(\overline{X}_{i..} - \overline{\overline{X}})^2 \quad \textbf{(11.17)}$$

Factor B Variation

$$SSB = rn'\sum_{j=1}^{c}(\overline{X}_{.j.} - \overline{\overline{X}})^2 \quad \textbf{(11.18)}$$

Interaction Variation

$$SSAB = n'\sum_{i=1}^{r}\sum_{j=1}^{c}(\overline{X}_{ij.} - \overline{X}_{i..} - \overline{X}_{.j.} + \overline{\overline{X}})^2 \quad \textbf{(11.19)}$$

Random Error in Two-Way ANOVA

$$SSE = \sum_{i=1}^{r}\sum_{j=1}^{c}\sum_{k=1}^{n'}(X_{ijk} - \overline{X}_{ij.})^2 \quad \textbf{(11.20)}$$

The Mean Squares in Two-Way ANOVA

$$MSA = \frac{SSA}{r-1} \quad \textbf{(11.21a)}$$

$$MSB = \frac{SSB}{c-1} \quad \textbf{(11.21b)}$$

$$MSAB = \frac{SSAB}{(r-1)(c-1)} \quad \textbf{(11.21c)}$$

$$MSE = \frac{SSE}{rc(n'-1)} \quad \textbf{(11.21d)}$$

F Test for Factor A Effect

$$F = \frac{MSA}{MSE} \quad \textbf{(11.22)}$$

F Test for Factor B Effect

$$F = \frac{MSB}{MSE} \quad \textbf{(11.23)}$$

F Test for Interaction Effect

$$F = \frac{MSAB}{MSE} \quad \textbf{(11.24)}$$

The Critical Range for Factor *A*

$$\text{Critical range} = Q_U \sqrt{\frac{MSE}{cn'}} \quad \textbf{(11.25)}$$

The Critical Range for Factor *B*

$$\text{Critical range} = Q_U \sqrt{\frac{MSE}{rn'}} \quad \textbf{(11.26)}$$

KEY TERMS

CHAPTER REVIEW PROBLEMS

Checking Your Understanding

11.41 In a one-way ANOVA, what is the difference between the among-groups variance *MSA* and the within-groups variance *MSW*?

11.42 What is the difference between the randomized block design and the completely randomized design?

11.43 What are the distinguishing features of the completely randomized design, randomized block design, and two-factor factorial designs?

11.44 What are the assumptions of ANOVA?

11.45 Under what conditions should you select the one-way ANOVA *F* test to examine possible differences among the means of *c* independent populations?

11.46 When and how should you use multiple comparison procedures for evaluating pairwise combinations of the group means?

11.47 What is the difference between the randomized block design and the two-factor factorial design?

11.48 What is the difference between the one-way ANOVA *F* test and the Levene test?

11.49 Under what conditions should you use the two-way ANOVA *F* test to examine possible differences among the means of each factor in a factorial design?

11.50 What is meant by the concept of interaction in a two-factor factorial design?

11.51 How can you determine whether there is an interaction in the two-factor factorial design?

Applying the Concepts

Problems 11.52–11.63 can be solved manually or by using Microsoft Excel, Minitab, or SPSS.

11.52 The operations manager for an appliance manufacturer wants to determine the optimal length of time for the washing cycle of a household clothes washer. An experiment is designed to measure the effect of detergent brand and washing cycle time on the amount of dirt removed from standard household laundry loads. Four brands of detergent (*A, B, C, D*) and four levels of washing cycle (18, 20, 22, and 24 minutes) are specifically selected for analysis. In order to run the experiment, 32 standard household laundry loads (having equal weight and dirt) are randomly assigned, 2 each, to the 16 detergent-washing cycle time combinations. The results (in pounds of dirt removed) are as follows. LAUNDRY

	WASHING CYCLE TIME (IN MINUTES)			
DETERGENT BRAND	**18**	**20**	**22**	**24**
A	0.11	0.13	0.17	0.17
	0.09	0.13	0.19	0.18
B	0.12	0.14	0.17	0.19
	0.10	0.15	0.18	0.17
C	0.08	0.16	0.18	0.20
	0.09	0.13	0.17	0.16
D	0.11	0.12	0.16	0.15
	0.13	0.13	0.17	0.17

At the 0.05 level of significance:

a. Is there an interaction between detergent and washing cycle time?
b. Is there an effect that is due to detergent?
c. Is there an effect that is due to washing cycle time?
d. Plot the mean amount of dirt removed (in pounds) for each detergent for each washing cycle time.
e. If appropriate, use the Tukey procedure to determine differences between detergent brands and between washing cycle times.
f. What washing cycle should be used for this type of household clothes washer?
g. Repeat the analysis using washing cycle time as the only factor. Compare your results to those of (c), (e), and (f).

11.53 The quality control director for a clothing manufacturer wants to study the effect of operators and machines on the breaking strength (in pounds) of wool serge material. A batch of the material is cut into square-yard pieces and these are randomly assigned, 3 each, to all 12 combinations of 4 operators and 3 machines chosen specifically for the experiment. The results are shown at the top of the next column: BREAKSTW

	MACHINE		
OPERATOR	**I**	**II**	**III**
A	115	111	109
	115	108	110
	119	114	107
B	117	105	110
	114	102	113
	114	106	114
C	109	100	103
	110	103	102
	106	101	105
D	112	105	108
	115	107	111
	111	107	110

At the 0.05 level of significance:

a. Is there an interaction between operator and machine?
b. Is there an effect that is due to operator?
c. Is there an effect that is due to machine?
d. Plot the mean breaking strength for each operator for each machine.
e. If appropriate, use the Tukey procedure to examine differences among operators and among machines.
f. What can you conclude about the effects of operators and machines on breaking strength? Explain.
g. Repeat the analysis using machines as the only factor. Compare your results to those of (c), (e), and (f).

11.54 An operations manager wants to examine the effect of air-jet pressure (in psi) on the breaking strength of the yarn. Three different levels of air-jet pressure are to be considered: 30 psi, 40 psi, and 50 psi. A random sample of 18 homogeneous filling yarns are selected from the same batch and the yarns are randomly assigned, 6 each, to the 3 levels of air-jet pressure. The breaking strength scores are in the data file: YARN

a. Is there evidence of a significant difference in the variances of the breaking strengths for the three air-jet pressures? (Use $\alpha = 0.05$).
b. At the 0.05 level of significance, is there evidence of a difference among mean breaking strengths for the three air-jet pressures?
c. If appropriate, use the Tukey-Kramer procedure to determine which air-jet pressures significantly differ with respect to mean breaking strength. (Use $\alpha = 0.05$.)
d. What should the operations manager conclude?

11.55 Suppose that, when setting up his experiment in problem 11.54, the operations manager had access to only six samples of yarn from the batch but was able to divide each yarn sample into three parts and randomly assign them, one each, to the three air-jet pressure levels. Thus, instead of the one-factor completely randomized design model in problem 11.54, he used a randomized block design with the six yarn samples being the blocks and one yarn part each assigned to the three pressure treatments. The breaking-strength scores are in the data file: YARN

a. At the 0.05 level of significance, is there evidence of a difference in the mean breaking strengths for the three air-jet pressures?

b. If appropriate, use the Tukey procedure to determine the air-jet pressures that differ in mean breaking strength. (Use $\alpha = 0.05$.)

c. At the 0.05 level of significance, is there evidence of a blocking effect?

d. Determine the relative efficiency of the randomized block design as compared with the completely randomized design.

e. Compare your results in problem 11.54 (b) versus problem 11.55(a). Given your results in problem 11.55 (c) and (d), what do you think happened here? What can you conclude about the "impact of blocking" when the blocks themselves are not really different from each other?

11.56 Suppose that, when setting up his experiment in problem 11.54, the operations manager is able to study the effect of side-to-side aspect in addition to air-jet pressure. Thus, instead of the one-factor completely randomized design in problem 11.54, he used a two-factor factorial design, the first factor, side-to-side aspects, having two levels (nozzle and opposite) and the second factor, air-jet pressure, having three levels (30 psi, 40 psi, and 50 psi). A sample of 18 yarns is randomly assigned, 3 to each of the 6 side-to-side aspect and pressure level combinations. The breaking-strength scores are as follows: YARN

	AIR-JET PRESSURE		
SIDE-TO-SIDE ASPECT	**30 psi**	**40 psi**	**50 psi**
Nozzle	25.5	24.8	23.2
	24.9	23.7	23.7
	26.1	24.4	22.7
Opposite	24.7	23.6	22.6
	24.2	23.3	22.8
	23.6	21.4	24.9

At the 0.05 level of significance, answer the following:

a. Is there an interaction between side-to-side aspect and air-jet pressure?

b. Is there an effect that is due to side-to-side aspect?

c. Is there an effect that is due to air-jet pressure?

d. Plot the mean yarn breaking strength for the two levels of side-to-side aspect for each level of air-jet pressure.

e. If appropriate, use the Tukey procedure to study differences among the air-jet pressures.

f. On the basis of the results of (a) through (e), what conclusions can you reach concerning yarn breaking strength? Discuss.

g. Compare your results in (a) through (e) with those from the one-way completely randomized design in problem 11.54 (a) through (e). Discuss fully.

11.57 Modern software applications require rapid data access capabilities. An experiment was conducted to test the effect of data file size on the ability to access the files (as measured by read time in milliseconds). Three different levels of data file size were considered: small—50,000 characters; medium—75,000 characters; or large—100,000 characters. A sample of eight files of each size was evaluated. The access read times in milliseconds are in the data file: ACCESS

a. Is there evidence of a significant difference in the variance of the access read times for the three file sizes? (Use $\alpha = 0.05$).

b. At the 0.05 level of significance, is there evidence of a difference among mean access read times for the three file sizes?

c. If appropriate, use the Tukey-Kramer procedure to determine which file sizes significantly differ with respect to mean access read time. (Use $\alpha = 0.05$.)

d. What conclusions can you reach?

11.58 Suppose that in problem 11.57, when setting up the experiment, the effects of the size of the input/output buffer was studied in addition to the effects of the data file size. Thus, instead of the one-factor completely randomized design given in problem 11.57, the experiment used a two-factor factorial design, the first factor, buffer size, having two levels (20 kilobytes and 40 kilobytes) and the second factor, data file size, having three levels (small, medium, and large). That is, there are two factors under consideration: buffer size and data file size. A sample of four programs (replications) were evaluated for each buffer size and data file size combination. The access read times in milliseconds are in the data file: ACCESS

At the 0.05 level of significance, answer the following:

a. Is there an interaction between buffer size and data file size?

b. Is there an effect that is due to buffer size?

c. Is there an effect that is due to data file size?

d. Plot the mean access read times (in milliseconds) for the two buffer size levels for each of the three data file size levels. Describe the interaction and discuss why you can or cannot interpret the main effects in (b) and (c).

e. On the basis of the results of (a) through (d), what conclusions can you reach concerning access read time? Discuss.
f. Compare and contrast your results in (a) through (e) with those from the completely randomized design in problem 11.57(a) through(d). Discuss fully.

11.59 A student team in a business statistics course conducted an experiment to test the download times of the three different types of computers (MAC, iMAC, and Dell) available at the university library. The students randomly selected one computer of each type. The students went to the Microsoft game Web site and clicked on the download link for the NBA game. The time (in seconds) between clicking on the link and the completion of the download was recorded. After each download, the file was deleted and the trash folder emptied. A total of 30 downloads were completed in a random order. COMPUTERS Completely analyze the data given in the following table.

COMPUTER		
MAC	iMAC	Dell
156	160	236
166	165	238
148	184	257
160	192	242
139	197	282
151	172	253
158	189	270
167	179	256
142	200	267
219	193	259

11.60 In a second experiment performed by the team in problem 11.59, the browser (Netscape Communicator or Internet Explorer) was added as a second factor. In this experiment only two of the types of computers were used (MAC and Dell). Eight downloads from each of the four factor combinations were completed. COMPUTERS2 Completely analyze the data in the following table.

	COMPUTER	
BROWSER	MAC	Dell
Netscape Communicator	142	284
	132	304
	125	273
	136	340
	127	326
	138	301
	147	291
	143	285
Internet Explorer	198	285
	210	292
	199	305
	202	325
	196	297
	213	301
	207	285
	201	290

11.61 A recent wine tasting was held by the J. S. Wine Club in which eight wines were rated by club members. Information concerning the country of origin and the price were not known to the club members until after the tasting took place. The wines rated (and the prices paid for them) were

1. French white $10.59
2. Italian white $8.50
3. Italian red $8.50
4. French burgundy (red) $10.69
5. French burgundy (red) $11.75
6. California Beaujolais (red) $10.50
7. French white $9.75
8. California white $13.59

The summated ratings over several characteristics for the 12 club members are in the data file: WINE

a. At the 0.01 level of significance, is there evidence of a difference in the mean rating scores among the wines?
b. What assumptions are necessary in order to do (a) of this problem? Comment on the validity of these assumptions.
c. If appropriate, use the Tukey procedure to determine the wines that differ in mean rating. (Use $\alpha = 0.01$.)
d. Based upon your results in (c), do you think that country of origin, the type of wine, or the price has had an effect on the ratings?
e. Determine the relative efficiency of the randomized block design as compared with the completely randomized design.

11.62 Ignore the blocking variable in problem 11.61.

a. "Erroneously" reanalyze the data as a one-factor completely randomized design where the one factor (brands of wines) has eight levels and each level contains a sample of 12 independent observations.
b. Compare the *SSBL* and *SSE* terms in problem 11.61(a) to the *SSW* term in (a). Discuss.
c. Using the results in problem 11.61(a) and this problem, describe the issues that can arise when analyzing data if the wrong procedures are applied.

Report Writing Exercises

11.63 The data in the file BEER represent the price of a six-pack of 12-ounce bottles, the calories per 12 fluid ounces, the percentage of alcohol content per 12 fluid ounces, the type of beer (craft lager, craft ale, imported lager, regular and ice beer, and light and nonalcoholic beer), and the country of origin (U.S. versus imported) for each of 69 beers.

Your task is to write a report based on a complete evaluation comparing and contrasting price, calories, and alcohol content based on type of beer—craft lager, craft ale, imported lager, regular and ice beer, and light and nonalcoholic beer.

Source: "Beers." Copyright © 1996 by Consumers Union of U.S., Inc., Yonkers, NY 10703-1057. Adapted with permission from Consumer Reports, *June 1996.*

TEAM PROJECT

11.64 The data file MUTUALFUNDS2004 contains information regarding 12 variables from a sample of 121 mutual funds. The variables are:

Fund—The name of the mutual fund.
Category—Type of stocks comprising the mutual fund—small cap, mid cap, large cap
Objective—Objective of stocks comprising the mutual fund—growth or value
Assets—In millions of dollars
Fees—Sales charges (no or yes)
Expense ratio—Expenses as a percentage of net assets
2003 Return—Twelve-month return in 2003
Three-year return—Annualized return 2001–2003
Five-year return—Annualized return 1999–2003
Risk—Risk-of-loss factor of the mutual fund classified as low, average, or high
Best quarter—Best quarterly performance 1999–2003
Worst quarter—Worst quarterly performance 1999–2003

How did the different types of mutual funds as categorized by their type (small cap, mid cap, large cap) perform during 2003, the three-year period from 2001–2003, and the five-year period from 1999–2003? Completely analyze these data at the 0.05 level of significance and write a report summarizing your conclusions.

RUNNING CASE MANAGING THE *SPRINGVILLE HERALD*

Phase 1

In studying the home delivery solicitation process the marketing department team determined that the so-called "later" calls made between 7:00 P.M. and 9:00 P.M. were significantly more conducive to lengthier calls than those made earlier in the evening (between 5:00 P.M. to 7:00 P.M.).

Knowing that the 7:00 P.M. to 9:00 P.M. time period is superior, the team sought to investigate the effect of the type of presentation on the length of the call. A group of 24 female callers were randomly assigned, 8 each, to one of three presentation plans—structured, semistructured, and unstructured—and were trained to make the telephone presentation. All calls were made between 7:00 P.M. and 9:00 P.M., the later time period, and the callers were to provide an introductory greeting that was personal but informal ("Hi, this is Mary Jones from the *Springville Herald*—may I speak to Bill Richards?"). The callers knew that the team was observing their efforts that evening but didn't know which particular calls were monitored. Measurements were taken on the length of call (defined as the difference, in seconds, between the time the person answers the phone and the time he or she hangs up). Table SH11.1 presents the results. SH11-1

TABLE SH11.1
Length of Calls (in Seconds) Based on Presentation Plan

Presentation Plan		
Structured	**Semi-structured**	**Unstructured**
38.8	41.8	32.9
42.1	36.4	36.1
45.2	39.1	39.2
34.8	28.7	29.3
48.3	36.4	41.9
37.8	36.1	31.7
41.1	35.8	35.2
43.6	33.7	38.1

EXERCISE

SH11.1 Analyze these data and write a report to the team that indicates your findings. Be sure to include your recommendations based on your findings. Also, include an appendix in which you discuss the reason you selected a particular statistical test to compare the three independent groups of callers.

DO NOT CONTINUE UNTIL THE PHASE 1 EXERCISE HAS BEEN COMPLETED.

Phase 2

Analyzing the data of Table SH11.1, the marketing department team observed that the structured presentation plan resulted in a significantly longer call than either the semistructured or unstructured plans. The team decided to tentatively recommend that all solicitations should be completely structured calls made later in the evening, from 7:00 P.M. to 9:00 P.M. The team also decided to study the effect of two additional factors on the length of call:

- Gender of the caller: male versus female.
- Type of greeting: personal but formal (i.e., "Hello, my name is Mary Jones from the *Springville Herald*—may I speak to Mr. Richards?"), personal but informal (i.e., "Hi, this is Mary Jones from the *Springville Herald*—may I speak to Bill Richards?"), or impersonal (i.e., "I represent the *Springville Herald* . . . ").

The team acknowledged that in its previous studies it had controlled for these variables. Only female callers were selected to participate in the studies, and they were trained to use a personal but informal greeting style. However, the team wondered if this choice of gender and greeting type was in fact best.

The team designed a study in which a total of 30 callers, 15 males and 15 females, were chosen to participate. The callers were randomly assigned to one of the three greeting-style training groups so that there were five callers in each of the six combinations of the two factors, gender and greeting style (personal but formal—PF; personal but informal—PI; and impersonal—Imp). The callers knew that the team was observing their efforts that evening but didn't know which particular calls were monitored.

Measurements were taken on the length of call (defined as the difference, in seconds, between the time the person answers and the time he or she hangs up the phone). Table SH11.2 summarizes the results. **SH11-2**

TABLE SH11.2
Length of Calls (in Seconds) Based on Gender and Type of Greeting

	GREETING		
GENDER	PF	PI	Imp.
M	45.6	41.7	35.3
	49.0	42.8	37.7
	41.8	40.0	41.0
	35.6	39.6	28.7
	43.4	36.0	31.8
F	44.1	37.9	43.3
	40.8	41.1	40.0
	46.9	35.8	43.1
	51.8	45.3	39.6
	48.5	40.2	33.2

EXERCISES

SH11.2 Completely analyze these data and write a report to the team that indicates the importance of each of the two factors and/or the interaction between them on the length of the call. Include recommendations for future experiments to perform.

SH11.3 Do you believe the length of the telephone call is the most appropriate outcome to study? What other variables should be investigated next? Discuss.

WEB CASE

Apply your knowledge about analysis of variance in this Web Case that continues the cereal-fill packaging dispute Web Case from Chapters 7, 9, and 10.

After the TCCACCTC's latest posting, Oxford Cereals has complained that the group is guilty of using selective data. Review the company's response on its Web site **www.prenhall.com/Springville/OC_SelectiveData.htm** and then answer the following:

1. Does Oxford Cereals have a legitimate argument? Why, or why not?
2. Assuming that the samples the company has posted were randomly selected, perform the appropriate analysis to resolve the ongoing weight dispute.
3. What conclusions can you draw from your results? If you were called as an expert witness, would you support the claims of the TCCACCTC or the claims of Oxford Cereals? Explain.

REFERENCES

1. Berenson, M. L., D. M. Levine, and M. Goldstein, *Intermediate Statistical Methods and Applications: A Computer Package Approach* (Englewood Cliffs, NJ: Prentice Hall, 1983).
2. Conover, W. J., *Practical Nonparametric Statistics*, 3rd ed. (New York: Wiley, 2000).
3. Daniel, W. W., *Applied Nonparametric Statistics*, 2nd ed. (Boston: PWS Kent, 1990).
4. Gitlow, H. S., and D. M. Levine, *Six Sigma for Green Belts and Champions: Foundations, DMAIC, Tools, Cases, and Certification* (Upper Saddle River, NJ: Financial Times—Prentice-Hall, 2005).
5. Hicks, C. R., and K. V. Turner, *Fundamental Concepts in the Design of Experiments*, 5th ed. (New York: Oxford University Press, 1999).
6. Kramer, C. Y., "Extension of Multiple Range Tests to Group Means with Unequal Numbers of Replications," *Biometrics* 12 (1956): 307–310.
7. *Microsoft Excel 2003* (Redmond, WA: Microsoft Corporation, 2003).
8. Miller, R. G., *Simultaneous Statistical Inference*, 2nd ed. (New York: Springer-Verlag, 1980).
9. *Minitab for Windows Version 14* (State College, PA: Minitab, Inc., 2004).
10. Montgomery, D. M., *Design and Analysis of Experiments*, 5th ed. (New York: John Wiley, 2001).
11. Neter, J., M. H. Kutner, C. Nachtsheim, and W. Wasserman, *Applied Linear Statistical Models*, 4th ed. (Homewood, IL: Richard D. Irwin, 1996).
12. *SPSS Version 12.0 Brief Guide* (Upper Saddle River, NJ: Prentice Hall, 2003).
13. Tukey, J. W., "Comparing Individual Means in the Analysis of Variance," *Biometrics* 5 (1949): 99–114.

Appendix 11 Using Software for ANOVA

A11.1 MICROSOFT EXCEL

For Completely Randomized Design: One-Way ANOVA

Open to a worksheet in which the data to be analyzed have been arranged column-wise such as the Data Worksheet of the **Parachute.xls** file.
Select **Tools → Data Analysis.**
Select **Anova: Single Factor** from the Data Analysis list and click **OK**. In the procedure's dialog box (see Figure A11.1 below):

> Enter the cell range of the data as the **Input Range**.
> Select **Columns**.
> Select **Labels in First Row** if the variable cell ranges include labels in their first rows.
> Click **OK**.

Results appear on a separate worksheet.

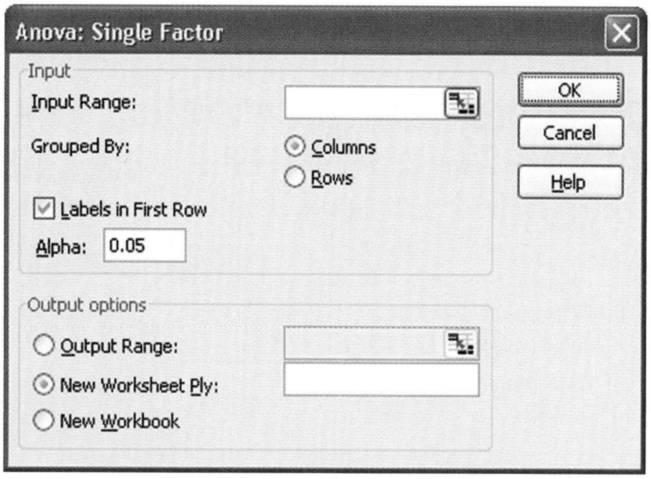

FIGURE A11.1 Microsoft Excel Anova: Single Factor Dialog Box

For Completely Randomized Design: Tukey-Kramer Procedure

Begin by following the "For Completely Randomized Design: One-Way ANOVA" instructions to produce an ANOVA worksheet containing results to be used later. Next, look up the Studentized Range Q statistic from Table E.10 for the level of significance of 0.05. Finally, enter the sample statistics and other data into the worksheet of the **Tukey-Kramer.xls** file for the proper number of groups.

OR See section G.28 (**Tukey-Kramer Procedure**) if you want PHStat2 to run the ANOVA and produce a multiple-comparisons worksheet for you.

For Levene's Test for Homogeneity of Variance

Using Table 11.4 on page 406 as your model, first prepare a worksheet that contains the absolute differences from the median value for each group. Then follow the "For Completely Randomized Design: ANOVA" instructions to produce a worksheet that contains the results of the test.

OR See section G.29 (**Levene's Test**) if you want PHStat2 to calculate the absolute differences and run the ANOVA for you.

For Randomized Block Design

Open to the worksheet in which the data to be analyzed have been arranged column-wise (and rows represent the blocks). The first row of these columns must contain a column heading (label). (See the **FFCHAIN.xls** workbook as an example.)
Select **Tools ➔ Data Analysis**.
Select **Anova: Two-Factor Without Replication** from the Data Analysis list and click **OK**. In the procedure's dialog box (see Figure A11.2 below):

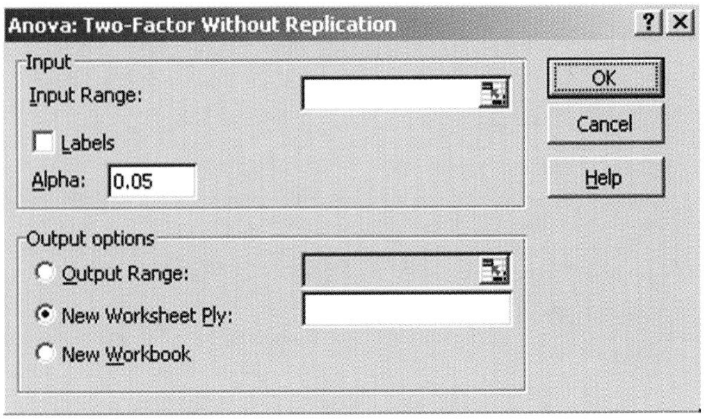

FIGURE A11.2 Microsoft Excel Anova: Two-Factor Without Replication Dialog Box

> Enter the cell range of the data as the **Input Range**.
> Select **Labels** if the variable cell ranges include labels in their first rows.
> Click **OK**.

Results appear on a separate worksheet.

For Two-Way ANOVA

Open to a worksheet in which the Factor *A* data appear as rows and the Factor *B* data appear as columns and the first row contains column headings and the first column contains identifiers for Factor *A*. Such a worksheet is the Data Worksheet of the **Parachute2.xls** file.
Select **Tools ➔ Data Analysis**.
Select **Anova: Two-Factor With Replication** from the Data Analysis list and click **OK**.
In the procedure's dialog box (see Figure A11.3):

> Enter the cell range of the data as the **Input Range**.
> Enter the **Rows per sample**.
> Click **OK**.

Results appear on a separate worksheet.

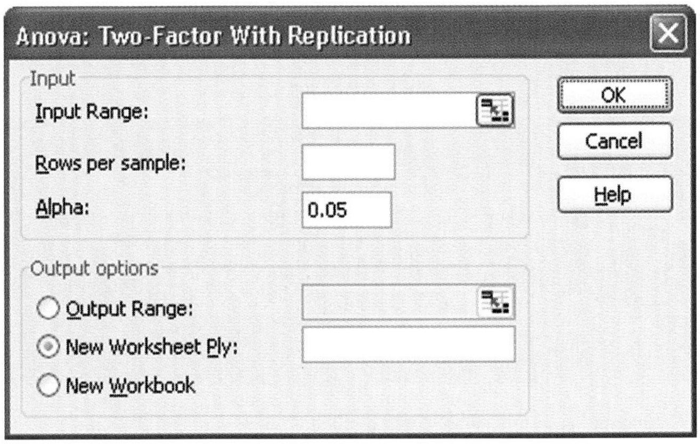

FIGURE A11.3 Microsoft Excel Anova: Two-factor With Replication Dialog Box

A11.2 MINITAB

Using Minitab for the One-Way Anova with Multiple Comparisons

To illustrate the use of Minitab for the one-factor ANOVA, open the **PARACHUTE.MTW** worksheet. The data in this worksheet have been stored in an unstacked format with each level in a separate column. Select **Stat ➔ ANOVA ➔ One-Way (Unstacked)**.

1. In the One-Way Analysis of Variance dialog box (see Figure A11.4), in the Responses (in separate columns): edit box, enter **C1 C2 C3 C4**.

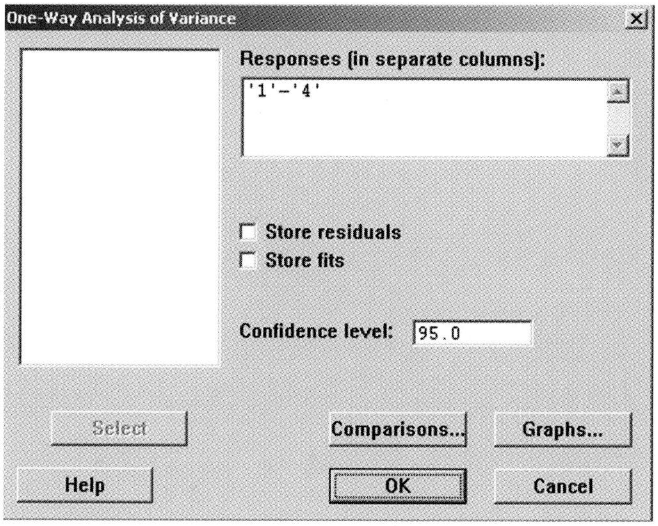

FIGURE A11.4 Minitab One-Way Analysis of Variance Dialog Box

2. Select the **Graphs** button and select the **Boxplots of data** check box. Click the **OK** button to return to the One-Way Analysis of Variance dialog box.
3. Select the **Comparisons** button. In the One-Way Multiple Comparisons dialog box (see Figure A11.5), select the **Tukey's, family error rate:** check box. Enter **5** in the edit box for 95% simultaneous confidence intervals. Click the **OK** button to return to the One-Way Analysis of Variance dialog box.
4. Click the **OK** button.

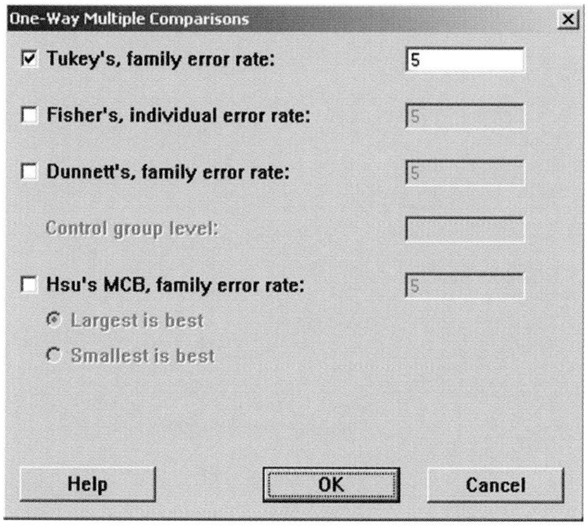

FIGURE A11.5 Minitab One-Way Multiple Comparisons Dialog Box

If the data are stored in a stacked format, select **Stat ➔ ANOVA ➔ One-Way**. Enter the column in which the response variable is stored in the Response: edit box and the column in which the factor is stored in the Factor: edit box.

Using Minitab for the Levene Test

To illustrate the use of Minitab for the Levene test, open the **PARACHUTE.MTW** worksheet. The data have been stored in an unstacked format with each level in a separate column. In order to perform the Levene test, you need to stack the data. Select **Data ➔ Stack ➔ Columns**.

1. In the Stack Columns dialog box (see Figure A11.6), enter **C1–C4** in the Stack the following columns: edit box.
2. Select the **Column of current worksheet** option button. Enter **C5** in the edit box. Enter **C6** in the Store subscripts in: edit box. Click the **OK** button.

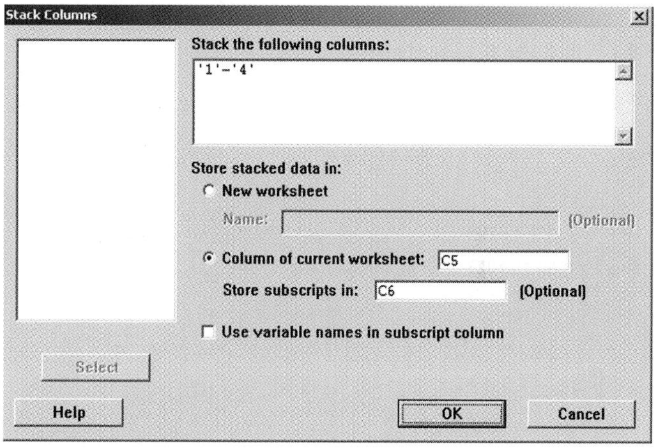

FIGURE A11.6 Minitab Stack Columns Dialog Box

3. Enter the column labels **Strength** in C5 and **Supplier** in C6.
4. Select **Stat ➔ ANOVA ➔ Test for Equal Variances**. In the Test for Equal Variances dialog box (see Figure A11.7), enter **C5** or **Strength** in the Response: edit box and **C6** or **Supplier** in the Factors: edit box. Enter **95.0** in the Confidence level: edit box. Click the **OK** button.

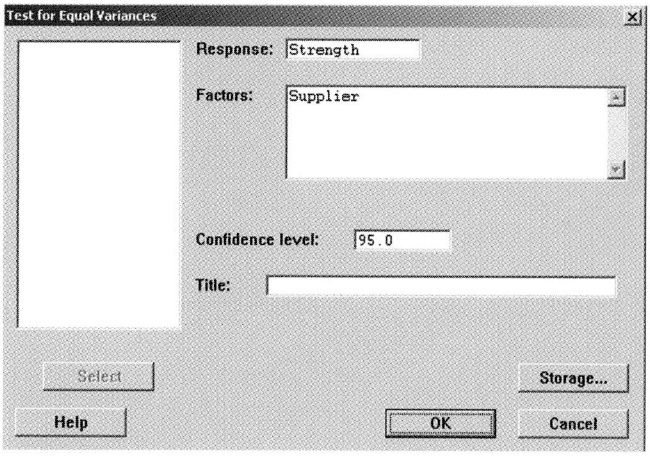

FIGURE A11.7 Minitab Test for Equal Variances Dialog Box

Using Minitab for the Randomized Block and Two-Factor Designs

The procedure for using Minitab is similar for the randomized block design and the two-factor design. To illustrate the use of Minitab for the two-factor design, open the **PARACHUTE2.MTW** worksheet. Note that Loom is stored in C1, Supplier in C2, and Strength in C3. Select **Stat → ANOVA → Two-Way**.

1. In the Two-Way Analysis of Variance dialog box (see Figure A11.8), enter **C3** or **Strength** in the Response: edit box, **C1** or **Loom** in the Row factor: edit box, and **C2** or **Supplier** in the Column factor: edit box. Select the **Display Means** check boxes for the row and column factors.
2. Click the **OK** button.

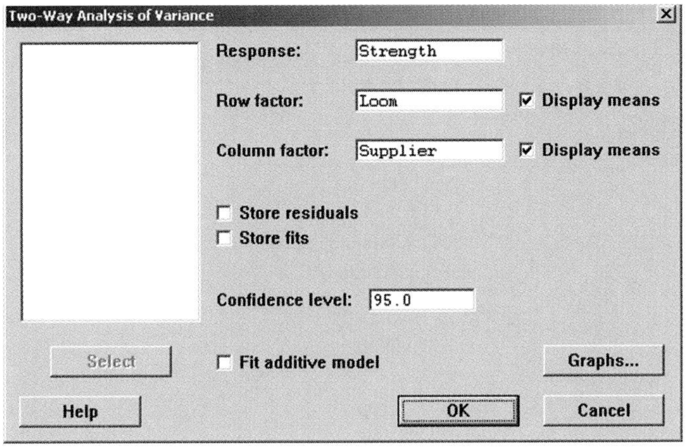

FIGURE A11.8 Minitab Two-Way Analysis of Variance Dialog Box

CHAPTER 12

Chi-Square Tests and Nonparametric Tests

LEARNING OBJECTIVES

In this chapter, you learn:

- How and when to use the chi-square test for contingency tables
- How to use the Marascuilo procedure for determining pairwise differences when evaluating more than two proportions
- How to use the chi-square test to evaluate the goodness of fit of a set of data to a specific probability distribution
- How and when to use nonparametric tests

USING STATISTICS

Guest Satisfaction at T.C. Resort Properties

You are the manager of T.C. Resort Properties, a collection of five upscale hotels located on two resort islands. Guests who are satisfied with the quality of services during their stay are more likely to return on a future vacation and to recommend the hotel to friends and relatives. To assess the quality of services being provided by your hotels, guests are encouraged to complete a satisfaction survey when they check out. You need to analyze the data from these surveys to determine the overall satisfaction with the services provided, the likelihood that the guests will return to the hotel, and the reasons some guests indicate that they will not return. For example, on one island, T.C. Resort Properties operates the Beachcomber and Windsurfer hotels. Is the perceived quality at the Beachcomber Hotel the same as the Windsurfer Hotel? If a difference is present, how can you use this information to improve the overall quality of service at T.C. Resort Properties? Furthermore, if guests indicate that they are not planning to return, what are the most common reasons given for this decision? Are the reasons given unique to a certain hotel or common to all hotels operated by T.C. Resort Properties?

In the preceding three chapters you used hypothesis-testing procedures to analyze both numerical and categorical data. Chapter 9 presented a variety of one-sample tests. Chapter 10 developed several two-sample tests. Chapter 11 used the analysis of variance (ANOVA) to study one or two factors of interest. This chapter extends hypothesis-testing to analyze differences between population proportions based on two or more samples, as well as the hypothesis of *independence* in the joint responses to two categorical variables. In addition, the chi-square test is used to test a population variance and to determine whether a set of data fits a specific probability distribution. The chapter concludes with nonparametric tests as alternatives to several hypothesis tests considered in Chapters 10 and 11.

12.1 CHI-SQUARE TEST FOR THE DIFFERENCE BETWEEN TWO PROPORTIONS (INDEPENDENT SAMPLES)

In section 10.3, you studied the Z test for the difference between two proportions. In this section, the data are examined from a different perspective. The hypothesis-testing procedure uses a test statistic that is approximated by a chi-square (χ^2) distribution. The results of this χ^2 test are equivalent to those of the Z test described in section 10.3.

If you are interested in comparing the counts of categorical responses between two independent groups, you can develop a two-way **cross-classification table** (see section 2.4) to display the frequency of occurrence of successes and failures for each group. This table is called a **contingency table**, and was used in Chapter 4 to define and study probability.

To illustrate the contingency table, return to the "Using Statistics" scenario concerning T.C. Resort Properties above. On one of the islands, T.C. Resort Properties has two hotels (the Beachcomber and the Windsurfer). In tabulating the responses to the single question, "Are you likely to choose this hotel again?" 163 of 227 guests at the Beachcomber responded yes, and 154 of 262 guests at the Windsurfer responded yes. At the 0.05 level of significance, is there evidence of a significant difference in guest satisfaction (as measured by likelihood to return to the hotel) between the two hotels?

The contingency table displayed in Table 12.1 has two rows and two columns and is called a **2 × 2 table**. The cells in the table indicate the frequency for each row and column combination.

TABLE 12.1
Layout of a 2 × 2 Contingency Table

	COLUMN VARIABLE (GROUP)		
ROW VARIABLE	**1**	**2**	**Totals**
Successes	X_1	X_2	X
Failures	$n_1 - X_1$	$n_2 - X_2$	$n - X$
Totals	n_1	n_2	n

where

X_1 = number of successes in group 1

X_2 = number of successes in group 2

$n_1 - X_1$ = number of failures in group 1

$n_2 - X_2$ = number of failures in group 2

$X = X_1 + X_2$ is the total number of successes

$n - X = (n_1 - X_1) + (n_2 - X_2)$ is the total number of failures

n_1 = the sample size in group 1

n_2 = the sample size in group 2

$n = n_1 + n_2$ = the total sample size

Table 12.2 contains the contingency table for the hotel guest satisfaction study. The contingency table has two rows, indicating whether the guests would return to the hotel (i.e., success) or would not return to the hotel (i.e., failure), and two columns, one for each hotel. The cells in the table indicate the frequency of each row and column combination. The row totals indicate the number of guests who would return to the hotel and those who would not return to the hotel. The column totals are the sample sizes for each hotel location.

TABLE 12.2
2 × 2 Contingency Table for the Guest Satisfaction Survey

	HOTEL		
CHOOSE HOTEL AGAIN?	**Beachcomber**	**Windsurfer**	**Total**
Yes	163	154	317
No	64	108	172
Total	227	262	489

To test whether the population proportion of guests who would return to the Beachcomber π_1 is equal to the population proportion of guests who would return to the Windsurfer π_2, you can use the χ^2 test for equality of proportions. To test the null hypothesis that there is no difference between the two population proportions:

$$H_0: \pi_1 = \pi_2$$

against the alternative that the two population proportions are not the same:

$$H_1: \pi_1 \neq \pi_2$$

you use the χ^2-test statistic, shown in Equation (12.1).

χ^2 TEST FOR THE DIFFERENCE BETWEEN TWO PROPORTIONS

The χ^2-test statistic is equal to the squared difference between the observed and expected frequencies, divided by the expected frequency in each cell of the table, summed over all cells of the table.

$$\chi^2 = \sum_{\text{all cells}} \frac{(f_0 - f_e)^2}{f_e} \tag{12.1}$$

where f_0 = **observed frequency** in a particular cell of a contingency table
f_e = **expected frequency** in a particular cell if the null hypothesis is true

The test statistic χ^2 approximately follows a chi-square distribution with 1 degree of freedom.

To compute the expected frequency, f_e, in any cell, you need to understand that if the null hypothesis is true, the proportion of successes in the two populations will be equal. Then the sample proportions you compute from each of the two groups would differ from each other only by chance and each would provide an estimate of the common population parameter π. A statistic that combines these two separate estimates together into one overall estimate of the population parameter π provides more information than either one of the two separate estimates could provide by itself. This statistic, given by the symbol $\bar{p}$, represents the estimated overall proportion of successes for the two groups combined (i.e., the total number of successes divided by the total sample size). The complement of $\bar{p}$, $1 - \bar{p}$, represents the overall proportion of failures in the two groups. Using the notation presented in Table 12.1 on page 447, Equation (12.2) defines $\bar{p}$.

COMPUTING THE ESTIMATED OVERALL PROPORTION

$$\bar{p} = \frac{X_1 + X_2}{n_1 + n_2} = \frac{X}{n} \tag{12.2}$$

To compute the expected frequency, f_e, for each cell pertaining to success (i.e., the cells in the first row in the contingency table), multiply the sample size (or column total) for a group by $\bar{p}$. To compute the expected frequency, f_e, for each cell pertaining to failure (i.e., the cells in the second row in the contingency table), multiply the sample size (or column total) for a group by $(1 - \bar{p})$.

The test statistic shown in Equation (12.1) approximately follows a **chi-square distribution** (see Table E.4) with one degree of freedom. Using a level of significance α, you reject the null hypothesis if the computed χ^2 test statistic is greater than χ^2_U, the upper-tail critical value from the χ^2 distribution having one degree of freedom. Thus, the decision rule is

$$\text{reject } H_0 \text{ if } \chi^2 > \chi^2_U;$$
$$\text{otherwise do not reject } H_0.$$

Figure 12.1 illustrates the decision rule.

FIGURE 12.1
Regions of Rejection and Nonrejection When Using the Chi-Square Test for the Difference Between Two Proportions with Level of Significance α

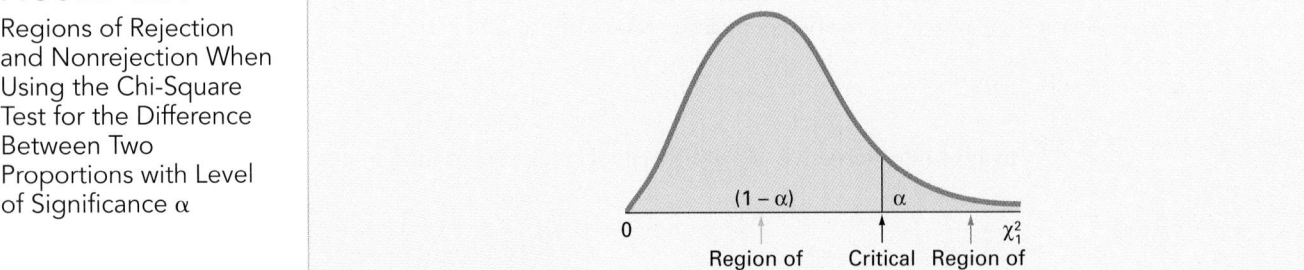

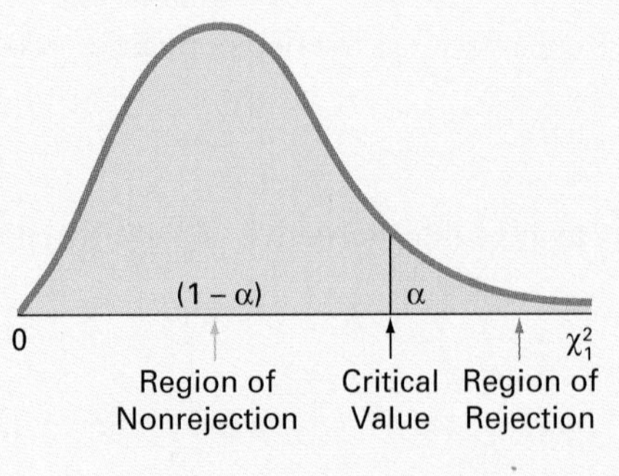

If the null hypothesis is true, the computed χ^2 statistic should be close to zero because the squared difference between what is actually observed in each cell, f_0, and what is theoretically expected, f_e, should be very small. If H_0 is false, and there are real differences in the population proportions, the computed χ^2 statistic is expected to be large. However, what constitutes a large difference in a cell is relative. The same actual difference between f_0 and f_e from a cell with a small number of expected frequencies contributes more to the χ^2 test statistic than a cell with a large number of expected frequencies.

To illustrate the use of the chi-square test for the difference between two proportions, return to the "Using Statistics" scenario, concerning T.C. Resort Properties and the corresponding contingency table displayed in Table 12.2 on page 447. The null hypothesis (H_0: $\pi_1 = \pi_2$) states that there is no difference between the proportion of guests who are likely to choose either of these hotels again. From Equation (12.2), you use $\bar{p}$ to estimate the common parameter π, the population proportion of guests who are likely to choose the Beachcomber Hotel again and the proportion of guests who are likely to choose the Windsurfer Hotel again. To begin, calculate $\bar{p}$ using Equation (12.2):

$$\bar{p} = \frac{X_1 + X_2}{n_1 + n_2} = \frac{163 + 154}{227 + 262} = \frac{317}{489} = 0.6483$$

$\bar{p}$ is the estimate of the common parameter π, the population proportion of guests who are likely to choose either of these hotels again if the null hypothesis is true. The estimated proportion of guests who are *not* likely to choose these hotels again is the complement of $\bar{p}$, $1 - 0.6483 = 0.3517$. Multiplying these two proportions by the sample size for the Beachcomber Hotel gives the number of guests expected to choose the Beachcomber again and the number not expected to choose this hotel again. In a similar manner, multiplying the two respective proportions by the Windsurfer Hotel's sample size yields the corresponding expected frequencies for that group.

EXAMPLE 12.1

COMPUTING THE EXPECTED FREQUENCIES

Compute the expected frequencies for each of the four cells of Table 12.2 on page 447.

SOLUTION

Yes—Beachcomber: $\bar{p} = 0.6483$ and $n_1 = 227$, so $f_e = 147.16$
Yes—Windsurfer: $\bar{p} = 0.6483$ and $n_2 = 262$, so $f_e = 169.84$
No—Beachcomber: $1 - \bar{p} = 0.3517$ and $n_1 = 227$, so $f_e = 79.84$
No—Windsurfer: $1 - \bar{p} = 0.3517$ and $n_2 = 262$, so $f_e = 92.16$

Table 12.3 presents these expected frequencies next to the corresponding observed frequencies.

TABLE 12.3
2 × 2 Contingency Table for Comparing the Observed (f_0) and Expected (f_e) Frequencies

HOTEL	BEACHCOMBER		WINDSURFER		
CHOOSE HOTEL AGAIN?	**Observed**	**Expected**	**Observed**	**Expected**	**Total**
Yes	163	147.16	154	169.84	317
No	64	79.84	108	92.16	172
Total	227	227.00	262	262.00	489

To test the null hypothesis that the population proportions are equal

$$H_0: \pi_1 = \pi_2$$

against the alternative that the population proportions are not equal

$$H_1: \pi_1 \neq \pi_2$$

you use the observed and expected frequencies from Table 12.3 to compute the χ^2-test statistic given by Equation (12.1) on page 448. Table 12.4 presents the calculations.

TABLE 12.4
Computation of χ^2-Test Statistic for the Guest Satisfaction Survey

f_0	f_e	$(f_0 - f_e)$	$(f_0 - f_e)^2$	$(f_0 - f_e)^2/f_e$
163	147.16	15.84	250.9056	1.705
154	169.84	−15.84	250.9056	1.477
64	79.84	−15.84	250.9056	3.143
108	92.16	15.84	250.9056	2.723
				9.048

The chi-square distribution is a right-skewed distribution whose shape depends solely on the number of degrees of freedom. As the number of degrees of freedom increases, the chi-square distribution becomes more symmetrical. You find the critical value of the χ^2-test statistic from Table E.4, a portion of which is presented as Table 12.5.

The values in Table 12.5 refer to selected upper-tail areas of the χ^2 distribution. A 2×2 contingency table has $(2 - 1)(2 - 1) = 1$ degree of freedom. Using $\alpha = 0.05$, with one degree of freedom, the critical value of χ^2 from Table 12.5 is 3.841. You reject H_0 if the computed χ^2 statistic is greater than 3.841 (see Figure 12.2). Since $9.048 > 3.841$, you reject H_0. You conclude that there is a difference in the proportion of guests who would return to the Beachcomber and the Windsurfer.

TABLE 12.5
Finding the χ^2 Critical Value from the Chi-Square Distribution with 1 Degree of Freedom Using the 0.05 Level of Significance

	Upper-Tail Area						
Degrees of Freedom	.995	.99	. . .	.05	.025	.01	.005
1			. . .	3.841	5.024	6.635	7.879
2	0.010	0.020	. . .	5.991	7.378	9.210	10.597
3	0.072	0.115	. . .	7.815	9.348	11.345	12.838
4	0.207	0.297	. . .	9.488	11.143	13.277	14.860
5	0.412	0.554	. . .	11.071	12.833	15.086	16.750

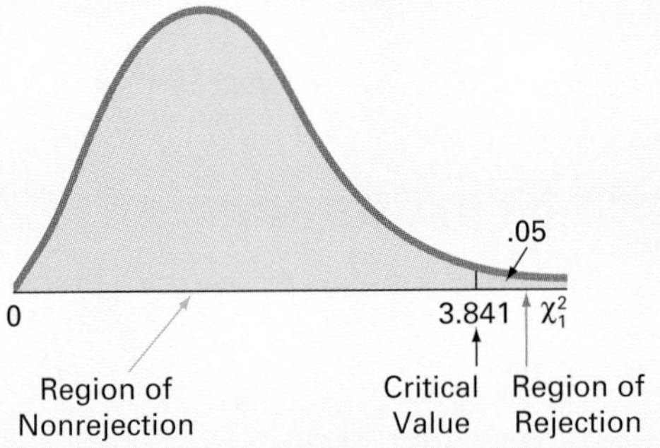

FIGURE 12.2
Regions of Rejection and Nonrejection When Finding the χ^2 Critical Value with 1 Degree of Freedom at the 0.05 Level of Significance

Figure 12.3 represents a Microsoft Excel worksheet for the guest satisfaction contingency table (Table 12.2 on page 447), while Figure 12.4 illustrates Minitab output. These outputs include the expected frequencies, χ^2-test statistic, degrees of freedom, and p-value. The χ^2-test statistic is 9.05, which is greater than the critical value of 3.841 (or the p-value $= 0.0026 < 0.05$), so you reject the null hypothesis that there is no difference in guest satisfaction between the two hotels. The p-value of 0.0026 is the probability of observing sample proportions as different as or more different than the actual difference ($0.718 - 0.588 = 0.13$ observed in the sample data), if the population proportions for the Beachcomber and Windsurfer hotels are equal. Thus, there is strong evidence to conclude that the two hotels are significantly different with respect to guest satisfaction as measured by whether the guest is likely to return to the hotel again. An examination of Table 12.3 on page 449 indicates that a greater proportion of guests at the Beachcomber are likely to return than at the Windsurfer.

FIGURE 12.3
Microsoft Excel Worksheet for the Guest Satisfaction Data

	A	B	C	D	E	F	G
1	**Guest Satisfaction Analysis**						
2							
3	**Observed Frequencies**						
4		**Hotel**				Calculations	
5	**Choose Again?**	**Beachcomber**	**Windsurfer**	**Total**		fo-fe	
6	**Yes**	**163**	**154**	**317**		15.84458	-15.8446
7	**No**	**64**	**108**	**172**		-15.8446	15.84458
8	**Total**	**227**	**262**	**489**			
9							
10	Expected Frequencies						
11		Hotel					
12	Choose Again?	Beachcomber	Windsurfer	Total		(fo-fe)^2/fe	
13	Yes	147.1554	169.8446	317		1.706024	1.47812
14	No	79.8446	92.1554	172		3.144243	2.72421
15	Total	227	262	489			
16							
17	**Data**						
18	**Level of Significance**	**0.05**					
19	Number of Rows	2					
20	Number of Columns	2					
21	Degrees of Freedom	1					
22							
23	**Results**						
24	**Critical Value**	**3.8415**					
25	**Chi-Square Test Statistic**	**9.0526**					
26	***p*-Value**	**0.0026**					
27	**Reject the null hypothesis**						
28							
29	***Expected frequency assumption***						
30	***is met.***						

FIGURE 12.4
Minitab Output for the Guest Satisfaction Data

Chi-Square Test: C1, C2

```
Expected counts are printed below observed counts
Chi-Square contributions are printed below expected counts

            C1       C2  Total
    1      163      154    317
        147.16   169.84
         1.706    1.478

    2       64      108    172
         79.84    92.16
         3.144    2.724

Total      227      262    489

Chi-Sq = 9.053, DF = 1, P-Value = 0.003
```

For the χ^2 test to give accurate results for a 2×2 table, you must assume that each expected frequency is at least 5. If this assumption is not satisfied, you can use alternative procedures such as Fisher's exact test (see references 1 and 2).

In the hotel guest satisfaction survey, both the Z test based on the standardized normal distribution (see section 10.3) and the χ^2 test based on the chi-square distribution have led to the same conclusion. You can explain this result by the interrelationship between the standardized normal distribution and a chi-square distribution with 1 degree of freedom. For such situations, the χ^2-test statistic is the square of the Z-test statistic. For instance, in the guest satisfaction study, the computed Z-test statistic is +3.01 and the computed χ^2-test statistic is 9.05. Except for rounding error, this latter value is the square of +3.01 [i.e., $(+3.01)^2 = 9.05$]. Also, if you compare the critical values of the test statistics from the two distributions, at the 0.05 level of significance, the χ^2_1 value of 3.841 is the square of the Z value of ± 1.96 (i.e., $\chi^2_1 = Z^2$). Furthermore, the p-values for both tests are equal. Therefore, when testing the null hypothesis of equality of proportions:

$$H_0: \pi_1 = \pi_2$$

against the alternative that the population proportions are not equal:

$$H_1: \pi_1 \neq \pi_2$$

the Z test and the χ^2 test are equivalent methods. However, if you are interested in determining whether there is evidence of a *directional* difference, such as $\pi_1 > \pi_2$, then you must use the Z test with the entire rejection region located in one tail of the standardized normal distribution. In section 12.2, the χ^2 test is extended to make comparisons and evaluate differences between the proportions among more than two groups. However, you cannot use the Z test if there are more than two groups.

PROBLEMS FOR SECTION 12.1

Learning the Basics

12.1 Determine the critical value of χ^2 in each of the following circumstances:

a. $\alpha = 0.01, n = 16$
b. $\alpha = 0.025, n = 11$
c. $\alpha = 0.05, n = 8$

12.2 Determine the critical value of χ^2 in each of the following circumstances:

a. $\alpha = 0.95, n = 28$
b. $\alpha = 0.975, n = 21$
c. $\alpha = 0.99, n = 5$

PH Grade ASSIST **12.3** In this problem, use the following contingency table:

	A	B	Total
1	20	30	50
2	30	45	75
Total	50	75	125

a. Find the expected frequency for each cell.
b. Compare the observed and expected frequencies for each cell.
c. Compute the χ^2 statistic. Is it significant at $\alpha = 0.05$?

PH Grade ASSIST **12.4** For this problem, use the following contingency table:

	A	B	Total
1	20	30	50
2	30	20	50
Total	50	50	100

a. Find the expected frequency for each cell.
b. Find the χ^2 statistic for this contingency table. Is it significant at $\alpha = 0.05$?

Applying the Concepts

You can solve problems 12.5–12.10 manually or by using Microsoft Excel or Minitab.

PH Grade ASSIST **12.5** A sample of 500 shoppers was selected in a large metropolitan area to determine various information concerning consumer behavior. Among the questions asked was, "Do you enjoy shopping for clothing?" The results are summarized in the following contingency table:

	GENDER		
ENJOY SHOPPING FOR CLOTHING	**Male**	**Female**	**Total**
Yes	136	224	360
No	104	36	140
Total	240	260	500

a. Is there evidence of a significant difference between the proportion of males and females who enjoy shopping for clothing at the 0.01 level of significance?
b. Find the p-value in (a) and interpret its meaning.
c. What is your answer to (a) and (b) if 206 males enjoyed shopping for clothing?
d. Compare the results of (a) through (c) to those of problem 10.33 (a) through (c) on page 373.

12.6 Is good gas mileage a priority for car shoppers? A survey conducted by Progressive Insurance asked this question of both men and women shopping for new cars. The data were reported as percentages, and no sample sizes were given.

	GENDER	
GAS MILEAGE A PRIORITY?	**Men**	**Women**
Yes	76%	84%
No	24%	16%

Source: Extracted from Snapshots, *USAtoday.com, June 21, 2004.*

a. Assume that 50 men and 50 women were included in the survey. At the 0.05 level of significance, is there a difference between males and females in the proportion who make gas mileage a priority?
b. Assume that 500 men and 500 women were included in the survey. At the 0.05 level of significance, is there a difference between males and females in the proportion who make gas mileage a priority?
c. Discuss the effect of sample size on the chi-square test.

12.7 The results of a yield improvement study at a semiconductor manufacturing facility provided defect data for a sample of 450 wafers. The following contingency table presents a summary of the responses to two questions: "Was a particle found on the die that produced the wafer?" and "Is the wafer good or bad?"

	QUALITY OF WAFER		
PARTICLES	**Good**	**Bad**	**Totals**
Yes	14	36	50
No	320	80	400
Totals	334	116	450

Source: S. W. Hall, "Analysis of Defectivity of Semiconductor Wafers by Contingency Table," Proceedings Institute of Environmental Sciences, *Vol. 1 (1994), 177–183.*

a. At the 0.05 level of significance, is there a difference between the proportion of good and bad wafers that have particles?
b. Determine the p-value in (a) and interpret its meaning.
c. What conclusions can you draw from this analysis?
d. Compare the results of (a) and (b) to those of problem 10.35 on page 374.

SELF Test **12.8** A study conducted by Ariel Mutual Funds and the Charles Schwab Corporation surveyed 500 African Americans with an annual income above $50,000, and 500 whites with an annual income above $50,000. The results indicated that 74% of the African Americans and 84% of the whites owned stocks (Cheryl Winokur Munk, "Stock-Ownership Race Gap Shrinks," *The Wall Street Journal*, June 13, 2002, B11).
a. Is there a difference between the proportion of African Americans with incomes above $50,000 who invest in stocks and the proportion of whites with incomes above $50,000 who invest in stocks? (Use $\alpha = 0.05$.)
b. Determine the p-value in (a) and interpret its meaning.
c. Compare the results of (a) and (b) to those of problem 10.39 on page 374.

12.9 Nonresponses are a problem for most mail surveys. Researchers at John Carroll University conducted a study to see if prenotification sent via postcards one week prior to sending a survey would decrease the nonresponse rate (Paul R. Murphy and James M. Daley, "Postcard Prenotification in Industrial Surveys: Further evidence," *Mid-American Journal of Business*, Spring 2002, 17(1):51–57). A total of 345 U.S.-based international freight forwarders were divided into two groups. One group was prenotified via a postcard that a mail survey concerning contemporary issues facing international freight forwarders would be arriving in one week. The second group received no prenotification of the mail survey. Results from the study are given in the following table.

	TREATMENT GROUP		
RESULT	**Prenotified**	**Not Prenotified**	**Total**
Responded	39	41	80
Nonresponse	142	123	265
Total	181	164	345

Source: Extracted from Paul R. Murphy and James M. Daley, "Postcard Prenotification in Industrial Surveys: Further Evidence," Mid-American Journal of Business, *Spring 2002, 17(1):51–57.*

a. At the 0.05 level of significance, is there a difference between the two treatment groups in the proportion of survey recipients who responded to the survey?
b. Find the p-value in (a) and interpret its meaning.

PH Grade ASSIST **12.10** A survey suggests that brand names are less important to consumers when purchasing clothing than they used to be. Of 7,500 apparel customers, 57% said that logos, labels, and trademarks of clothing have less personal importance today than a few years ago, while only 10% said they have more (Shelly Branch, "What's in a Name? Not Much According to Clothes Shoppers," *The Wall Street Journal*, July 16, 2002, B4). The study also investigated whether this change in the importance of brand name differed between female shoppers and male shoppers. The following table indicates the gender of respondents and whether they said that logos, labels, and trademarks of clothing have more personal importance today than a few years ago:

	GENDER		
IMPORTANCE OF BRAND NAME	**Male**	**Female**	**Total**
More	450	300	750
Equal or less	3,300	3,450	6,750
Total	3,750	3,750	7,500

Source: Extracted from Shelly Branch, "What's in a Name? Not Much According to Clothes Shoppers," The Wall Street Journal, *July 16, 2002, B4.*

a. Is there a difference between the proportion of males and females who place more importance on brand names today than a few years ago? (Use $\alpha = 0.05$.)
b. Find the p-value in (a) and interpret its meaning.

12.2 CHI-SQUARE TEST FOR DIFFERENCES AMONG MORE THAN TWO PROPORTIONS

In this section, the χ^2 test is extended to compare more than two independent populations. The letter c is used to represent the number of independent populations under consideration. Thus, the contingency table now has two rows and c columns. To test the null hypothesis that there are no differences among the c proportions:

$$H_0 : \pi_1 = \pi_2 = \cdots = \pi_c$$

against the alternative that not all the c population proportions are equal:

$$H_1\text{: Not all } \pi_j \text{ are equal (where } j = 1, 2, \ldots, c)$$

use Equation (12.1) on page 448

$$\chi^2 = \sum_{all\ \text{cells}} \frac{(f_0 - f_e)^2}{f_e}$$

where f_0 = observed frequency in a particular cell of a $2 \times c$ contingency table

f_e = expected frequency in a particular cell if the null hypothesis is true

If the null hypothesis is true and the proportions are equal across all c populations, then the c sample proportions should differ only by chance. In such a situation, a statistic that combines these c separate estimates into one overall estimate of the population proportion π provides more information than any one of the c separate estimates alone. To expand on Equation (12.2) on page 448, the statistic $\bar{p}$ in Equation (12.3) represents the estimated overall proportion for all c groups combined.

COMPUTING THE ESTIMATED OVERALL PROPORTION FOR c GROUPS

$$\bar{p} = \frac{X_1 + X_2 + \cdots + X_c}{n_1 + n_2 + \cdots + n_c} = \frac{X}{n} \qquad \textbf{(12.3)}$$

To compute the expected frequency f_e for each cell in the first row in the contingency table, multiply each sample size (or column total) by $\bar{p}$. To compute the expected frequency f_e for each cell in the second row in the contingency table, multiply each sample size (or column total) by $(1 - \bar{p})$. The test statistic shown in Equation (12.1) on page 448 approximately follows a chi-square distribution with degrees of freedom equal to the number of rows in the contingency table minus 1, times the number of columns in the table minus 1. For a **$2 \times c$ contingency table**, there are $c - 1$ degrees of freedom:

$$\text{Degrees of freedom} = (2 - 1)(c - 1) = c - 1$$

Using a level of significance α, you reject the null hypothesis if the computed χ^2-test statistic is greater than χ^2_U, the upper-tail critical value from a chi-square distribution having $c - 1$ degrees of freedom. Therefore, the decision rule is

$$\text{Reject } H_0 \text{ if } \chi^2 > \chi^2_U;$$

otherwise do not reject H_0.

Figure 12.5 illustrates the decision rule.

FIGURE 12.5
Regions of Rejection and Nonrejection When Testing for Differences Among *c* Proportions Using the χ^2 Test

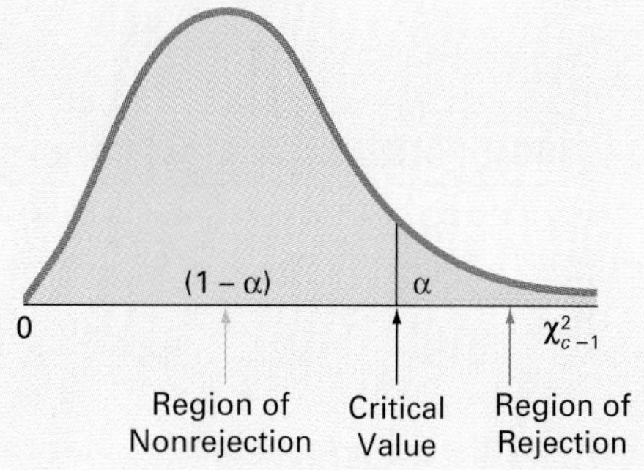

To illustrate the χ^2 test for equality of proportions when there are more than two groups, return to the "Using Statistics" scenario concerning T.C. Resort Properties. A similar survey was recently conducted on a different island in which T.C. Resort Properties has three different hotels. Table 12.6 presents the responses to a question concerning whether guests would be likely to choose their hotel again.

TABLE 12.6
2 × 3 Contingency Table for Guest Satisfaction Survey

	HOTEL			
CHOOSE HOTEL AGAIN	**Golden Palm**	**Palm Royale**	**Palm Princess**	**Total**
Yes	128	199	186	513
No	88	33	66	187
Total	216	232	252	700

Since the null hypothesis states that there are no differences among the three hotels with respect to the proportion of guests who would likely return again, use Equation (12.3) to calculate an estimate of π, the population proportion of guests who would likely return again.

$$\bar{p} = \frac{X_1 + X_2 + \cdots + X_c}{n_1 + n_2 + \cdots + n_c} = \frac{X}{n}$$

$$= \frac{(128 + 199 + 186)}{(216 + 232 + 252)} = \frac{513}{700}$$

$$= 0.733$$

The estimated overall proportion of guests who would *not* be likely to return again is the complement $(1 - \bar{p})$, or 0.267. Multiplying these two proportions by the sample size taken at each hotel yields the expected number of guests who would and would not likely return.

EXAMPLE 12.2

COMPUTING THE EXPECTED FREQUENCIES

Compute the expected frequencies for each of the six cells in Table 12.6.

SOLUTION

Yes—Golden Palm: $\bar{p} = 0.733$ and $n_1 = 216$, so $f_e = 158.30$
Yes—Palm Royale: $\bar{p} = 0.733$ and $n_2 = 232$, so $f_e = 170.02$
Yes—Palm Princess: $\bar{p} = 0.733$ and $n_3 = 252$, so $f_e = 184.68$
No—Golden Palm: $1 - \bar{p} = 0.267$ and $n_1 = 216$, so $f_e = 57.70$
No—Palm Royale: $1 - \bar{p} = 0.267$ and $n_2 = 232$, so $f_e = 61.98$
No—Palm Princess: $1 - \bar{p} = 0.267$ and $n_3 = 252$, so $f_e = 67.32$

Table 12.7 presents these expected frequencies.

TABLE 12.7
Cross-Classification of Expected Frequencies from a Guest Satisfaction Survey of Three Hotels

	HOTEL			
CHOOSE HOTEL AGAIN?	**Golden Palm**	**Palm Royale**	**Palm Princess**	**Total**
Yes	158.30	170.02	184.68	513
No	57.70	61.98	67.32	187
Total	216.00	232.00	252.00	700

To test the null hypothesis that the proportions are equal:

$$H_0: \pi_1 = \pi_2 = \pi_3$$

against the alternative that not all three proportions are equal:

$$H_1: \text{Not all } \pi_j \text{ are equal (where } j = 1, 2, 3)$$

use the observed and expected frequencies from Table 12.6 on page 455 and Table 12.7 above to compute the χ^2-test statistic given by Equation (12.1) on page 448. Table 12.8 presents the computations.

TABLE 12.8
Computation of χ^2-Test Statistic for the Guest Satisfaction Survey of Three Hotels

f_0	f_e	$(f_0 - f_e)$	$(f_0 - f_e)^2$	$(f_0 - f_e)^2/f_e$
128	158.30	−30.30	918.0900	5.800
199	170.02	28.98	839.8404	4.940
186	184.68	1.32	1.7424	0.009
88	57.70	30.30	918.0900	15.911
33	61.98	−28.98	839.8404	13.550
66	67.32	−1.32	1.7424	0.026
				40.236

You find the critical value of the χ^2-test statistic from Table E.4. In the guest satisfaction survey, because three hotels are evaluated, there are $(2 - 1)(3 - 1) = 2$ degrees of freedom. Using $\alpha = 0.05$, the χ^2 critical value with 2 degrees of freedom is 5.991. Because the computed test statistic ($\chi^2 = 40.236$) is greater than this critical value, you reject the null hypothesis (see Figure 12.6). Microsoft Excel (see Figure 12.7) and Minitab (see Figure 12.8) also report the p-value. Since the p-value is approximately 0.0000, which is less than $\alpha = 0.05$, you reject the null hypothesis. Further, this p-value indicates that there is virtually no chance to see differences this large or larger among the three sample proportions, if the population proportions for the three hotels are equal. Thus, there is sufficient evidence to conclude that the hotel properties are different with respect to the proportion of guests who are likely to return.

FIGURE 12.6
Regions of Rejection and Nonrejection When Testing for Differences in Three Proportions at the 0.05 Level of Significance with 2 Degrees of Freedom

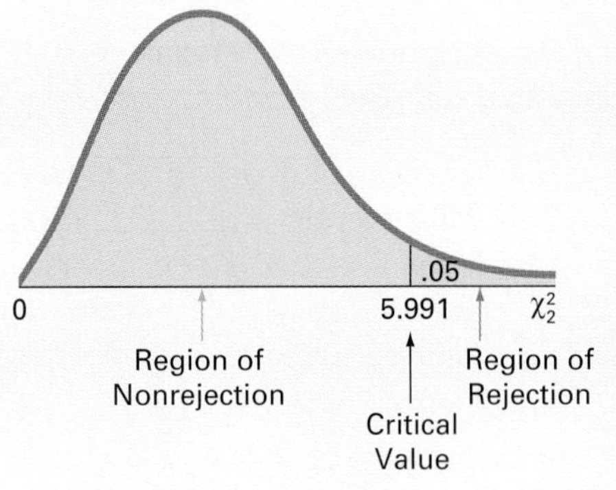

FIGURE 12.7
Microsoft Excel Worksheet for the Guest Satisfaction Data of Table 12.6

	A	B	C	D	E	F	G	H	I
1	**Guest Satisfaction (3-Hotels) Analysis**								
2									
3		**Observed Frequencies**							
4		**Hotel**					Calculations		
5	**Choose Again?**	**Golden Palm**	**Palm Royale**	**Palm Princess**	**Total**			fo-fe	
6	**Yes**	128	199	186	513		-30.2971	28.97714	1.32
7	**No**	88	33	66	187		30.29714	-28.9771	-1.32
8	**Total**	216	232	252	700				
9									
10		Expected Frequencies							
11		Hotel							
12	Choose Again?	Golden Palm	Palm Royale	Palm Princess	Total			(fo-fe)^2/fe	
13	Yes	158.2971	170.0229	184.68	513		5.798695	4.9386	0.009435
14	No	57.7029	61.9771	67.32	187		15.90765	13.54814	0.025882
15	Total	216	232	252	700				
16									
17	**Data**								
18	**Level of Significance**	0.05							
19	Number of Rows	2							
20	Number of Columns	3							
21	Degrees of Freedom	2							
22									
23	**Results**								
24	**Critical Value**	5.9915							
25	**Chi-Square Test Statistic**	40.2284							
26	***p*-Value**	0.0000							
27	**Reject the null hypothesis**								
28									
29	***Expected frequency assumption***								
30	***is met.***								

FIGURE 12.8
Minitab Output for the Guest Satisfaction Data of Table 12.6

Chi-Square Test: C1, C2, C3

```
Expected counts are printed below observed counts
Chi-Square contributions are printed below expected counts

            C1      C2      C3  Total
    1      128     199     186    513
        158.30  170.02  184.68
         5.799   4.939   0.009

    2       88      33      66    187
         57.70   61.98   67.32
        15.908  13.548   0.026

Total      216     232     252    700

Chi-Sq = 40.228, DF = 2, P-Value = 0.000
```

For the χ^2 test to give accurate results when dealing with $2 \times c$ contingency tables, all expected frequencies must be large. For such situations there is much debate among statisticians about the definition of *large*. Some statisticians (see reference 5) have found that the test gives accurate results as long as all expected frequencies equal or exceed 0.5. Other statisticians, more conservative in their approach, require that no more than 20% of the cells contain expected frequencies less than 5 and no cells have expected frequencies less than 1 (see reference 3). A reasonable compromise between these points of view is to make sure that each expected frequency is at least 1. To accomplish this, you may need to collapse two or more low-expected frequency categories into one category in the contingency table before performing the test. Such merging of categories usually results in expected frequencies sufficiently large to conduct the χ^2 test accurately. If the combining of categories is undesirable, alternative procedures are available (see references 2 and 7).

The Marascuilo Procedure

Rejecting the null hypothesis in a χ^2 test of equality of proportions in a $2 \times c$ table only allows you to reach the conclusion that not all c population proportions are equal. But *which* of the proportions differ? Because the result of the χ^2 test for equality of proportions does not specifically answer this question, a *post-hoc* multiple comparison procedure is needed. One such approach that follows rejection of the null hypothesis of equal proportions is the Marascuilo procedure.

The **Marascuilo procedure** enables you to make comparisons between all pairs of groups. First, you need to compute the observed differences $p_j - p_{j'}$ (where $j \neq j'$) among all $c(c-1)/2$ pairs. Then you use Equation (12.4) to compute the corresponding critical ranges for the Marascuilo procedure.

CRITICAL RANGE FOR THE MARASCUILO PROCEDURE

$$\text{Critical range} = \sqrt{\chi^2_U}\sqrt{\frac{p_j(1-p_j)}{n_j} + \frac{p_{j'}(1-p_{j'})}{n_{j'}}} \quad \textbf{(12.4)}$$

You need to compute a different critical range for each pairwise comparison of sample proportions. In the final step you compare each of the $c(c-1)/2$ pairs of sample proportions against its corresponding critical range. You declare a specific pair significantly different if the absolute difference in the sample proportions $|p_j - p_{j'}|$ is greater than its critical range.

To apply the Marascuilo procedure, return to the guest satisfaction survey. Using the χ^2 test, you concluded that there was evidence of a significant difference among the population proportions. Because there are three hotels, there are $(3)(3-1)/2 = 3$ pairwise comparisons. From Table 12.6 on page 455 the three sample proportions are:

$$p_1 = \frac{X_1}{n_1} = \frac{128}{216} = 0.593$$

$$p_2 = \frac{X_2}{n_2} = \frac{199}{232} = 0.858$$

$$p_3 = \frac{X_3}{n_3} = \frac{186}{252} = 0.738$$

Using Table E.4 and an overall level of significance of 0.05, the upper-tail critical value of the χ^2-test statistic for a chi-square distribution having $(c-1) = 2$ degrees of freedom is 5.991. Thus,

$$\sqrt{\chi^2_U} = \sqrt{5.991} = 2.448$$

Next, you compute the three pairs of absolute differences in sample proportions and their corresponding critical ranges. If the absolute difference is greater than its critical range, the proportions are significantly different.

Absolute Difference in Proportions	**Critical Range**
$\lvert p_j - p_{j'}\rvert$	$2.448\sqrt{\frac{p_j(1-p_j)}{n_j} + \frac{p_{j'}(1-p_{j'})}{n_{j'}}}$
$\lvert p_1 - p_2\rvert = \lvert 0.593 - 0.858\rvert = 0.265$	$2.448\sqrt{\frac{(0.593)(0.407)}{216} + \frac{(0.858)(0.142)}{232}} = 0.099$
$\lvert p_1 - p_3\rvert = \lvert 0.593 - 0.738\rvert = 0.145$	$2.448\sqrt{\frac{(0.593)(0.407)}{216} + \frac{(0.738)(0.262)}{252}} = 0.106$
$\lvert p_2 - p_3\rvert = \lvert 0.858 - 0.738\rvert = 0.120$	$2.448\sqrt{\frac{(0.858)(0.142)}{232} + \frac{(0.738)(0.262)}{252}} = 0.088$

Figure 12.9 illustrates a Microsoft Excel worksheet for the Marascuilo procedure.

FIGURE 12.9
Microsoft Excel Worksheet for the Marascuilo Procedure

	A	B	C	D
1	**Marascuilo Procedure**			
2	**Guest Satisfaction (3-Hotels) Analysis**			
3	Level of Significance	0.05		
4	Square Root of Critical Value	2.4477		
5				
6	Sample Proportions			
7	Group 1	0.5926		
8	Group 2	0.8578		
9	Group 3	0.7381		
10				
11	MARASCUILO TABLE			
12	**Proportions**	**Absolute Differences**	**Critical Range**	
13	**| Group 1 - Group 2 |**	**0.2652**	**0.0992**	**Significant**
14	**| Group 1 - Group 3 |**	**0.1455**	**0.1063**	**Significant**
15				
16	**| Group 2 - Group 3 |**	**0.1197**	**0.0880**	**Significant**

You can conclude, using a 0.05 overall level of significance, that guest satisfaction is higher at the Palm Royale ($p_2 = 0.858$) than at either the Golden Palm ($p_1 = 0.593$) or the Palm Princess ($p_3 = 0.738$). Guest satisfaction is also higher at the Palm Princess than at the Golden Palm. These results clearly suggest that management should study the reasons for these differences and, in particular, should try to determine why satisfaction is significantly lower at the Golden Palm than at the other two hotels.

PROBLEMS FOR SECTION 12.2

Learning the Basics

PH Grade ASSIST **12.11** Consider a contingency table with two rows and five columns.

a. Find the degrees of freedom.
b. Find the critical value for $\alpha = 0.05$.
c. Find the critical value for $\alpha = 0.01$.

PH Grade ASSIST **12.12** For this problem, use the following contingency table:

	A	*B*	*C*	**Total**
1	10	30	50	90
2	40	45	50	135
Total	50	75	100	225

a. Compute the expected frequencies for each cell.
b. Compute the χ^2 statistic for this contingency table. Is it significant at $\alpha = 0.05$?
c. If appropriate, use the Marascuilo procedure and $\alpha = 0.05$ to determine which groups are different.

12.13 For this problem, use the following contingency table:

	A	*B*	*C*	**Total**
1	20	30	25	75
2	30	20	25	75
Total	50	50	50	150

a. Compute the expected frequencies for each cell.
b. Compute the χ^2 statistic for this contingency table. Is it significant at $\alpha = 0.05$?

Applying the Concepts

You can solve problems 12.14–12.23 manually or by using Microsoft Excel or Minitab.

12.14 A survey was conducted in five countries. The percentages of respondents who said that they eat out once a week or more are as follows:

Germany	10%
France	12%
United Kingdom	28%
Greece	39%
United States	57%

Source: Adapted from M. Kissel, "Americans Are Keen on Cocooning," The Wall Street Journal, *July 22, 2003, D3.*

Suppose that the survey was based on 1,000 respondents in each country.

a. At the 0.05 level of significance, determine whether there is a significant difference in the proportion of people who eat out at least once a week in the various countries.
b. Find the *p*-value in (a) and interpret its meaning.
c. If appropriate, use the Marascuilo procedure and $\alpha = 0.05$ to determine which countries are different. Discuss your results.

12.15 Is the degree to which students withdraw from introductory business statistics courses the same for online courses and traditional courses taught in a classroom? Professor Constance McLaren at Indiana State University collected data for five semesters to investigate this question. The following table cross-classifies introductory business statistics students by the type of course (classroom, online) and student persistence (active, dropped, vanished).

	STUDENT PERSISTENCE		
TYPE OF COURSE	Active	Dropped	Vanished
Classroom	127	8	4
Online	81	51	20

Source: Constance McLaren, "A Comparison of Student Persistence and Performance in Online and Classroom Business Statistics Experiences," Decision Sciences Journal of Innovative Education, *Spring 2004, 2(1):1–10. Published by the Decision Sciences Institute, headquartered at Georgia State University, Atlanta, GA.*

a. Is there evidence of a difference in student persistence (active, dropped, vanished) based on type of course? (Use $\alpha = 0.01$.)
b. Compute the p-value and interpret its meaning.

SELF Test **12.16** More shoppers do their majority of grocery shopping on Saturday than any other day of the week. However, is the day of the week a person does the majority of grocery shopping dependent on age? A study cross-classified grocery shoppers by age and major shopping day ("Major Shopping by Day," *Progressive Grocer Annual Report*, April 30, 2002). The data were reported as percentages and no sample sizes were given.

	AGE		
MAJOR SHOPPING DAY	Under 35	35–54	Over 54
Saturday	24%	28%	12%
A day other than Saturday	76%	72%	88%

Source: Extracted from "Major Shopping by Day," Progressive Grocer Annual Report, *April 30, 2002.*

Assume that 200 shoppers for each age category were surveyed.
a. Is there evidence of a significant difference among the age groups with respect to major grocery shopping day? (Use $\alpha = 0.05$.)
b. Determine the p-value in (a) and interpret its meaning.
c. If appropriate, use the Marascuilo procedure and $\alpha = 0.05$ to determine which age groups are different. Discuss your results.
d. Discuss the managerial implications of (a) and (c). How can grocery stores use this information to improve marketing and sales? Be specific.

12.17 Repeat (a) through (b) of problem 12.16 assuming that only 50 shoppers for each age category were surveyed. Discuss the implications of sample size on the χ^2 test for differences among more than two populations.

12.18 The health care industry and consumer advocacy groups are at odds over the sharing of a patient's medical records without the patient's consent. The health care industry believes that no consent should be necessary to openly share data among doctors, hospitals, pharmacies, and insurance companies. A phone survey by the Gallup Organization asked respondents if they objected to their medical records being shared without consent to a variety of different types of companies or institutions (Laura Landro, "Medical-Privacy Rules Leave Consumers' Data Vulnerable," *The Wall Street Journal*, June 6, 2002, D3). The following table contains a partial listing of the results:

	ORGANIZATION		
OBJECT TO SHARING?	Insurance Companies	Pharmacies	Medical Researchers
Yes	820	590	670
No	180	410	330

Source: Extracted from Laura Landro, "Medical-Privacy Rules Leave Consumers' Data Vulnerable," The Wall Street Journal, *June 6, 2002, D3.*

a. Determine whether there is a difference in the proportion of people who object to their medical records being shared with the three organizations listed in the table. (Use $\alpha = 0.05$.)
b. If appropriate, use the Marascuilo procedure and $\alpha = 0.05$ to determine which organizations are different. Discuss your results.

12.19 J. C. Schaefer travels the world to rate the services of luxury hotels. Some of his results appear to vary with the city where the hotel is located. Three of the items he rates the hotels on and results from his findings are as follows. The data were reported as percentages, and no sample sizes were given.

(1) Front Desk Uses the Guest's Name During Check-In

	Hong Kong	New York	Paris
Yes	26%	39%	28%
No	74%	61%	72%

(2) Minibar Charges Are Correctly Posted at Checkout

	Hong Kong	New York	Paris
Yes	86%	76%	78%
No	14%	24%	22%

(3) Bathroom Tub and Shower Are Spotlessly Clean

	Hong Kong	New York	Paris
Yes	81%	76%	79%
No	19%	24%	21%

Source: Extracted from N. Templin, "Undercover with a Hotel Spy," The Wall Street Journal, *May 12, 1999.*

Assume that 100 luxury hotels in each city were rated.

a. At the 0.05 level of significance, is there evidence of a difference in the proportion of hotels that use the guest's name among the three cities?

b. Find the p-value in (a) and interpret its meaning.

12.20 Referring to problem 12.19,

a. At the 0.05 level of significance, is there evidence of a difference in the proportion of hotels that correctly post minibar charges among the three cities?

b. Find the p-value in (a) and interpret its meaning.

12.21 Referring to problem 12.19,

a. At the 0.05 level of significance, is there evidence of a difference in the proportion of hotels with spotless bathroom tub and shower among the three cities?

b. Find the p-value in (a) and interpret its meaning.

12.22 Referring to problems 12.19 through 12.21, if appropriate, use the Marascuilo procedure and $\alpha = 0.05$ to determine which hotels are different in the proportion that use the guest's name, the proportion of hotels that correctly post minibar charges, and the proportion of hotels with spotless bathroom tub and shower.

12.23 Referring to problems 12.19 through 12.22, assume that the sample size in each city was 200 instead of 100.

a. Repeat problems 12.19 through 12.22 with this sample size.

b. Discuss the effect that sample size has on tests for differences among the three proportions.

12.3 CHI-SQUARE TEST OF INDEPENDENCE

In sections 12.1 and 12.2 you used the χ^2 test to evaluate potential differences among population proportions. For a contingency table that has r rows and c columns, you can generalize the χ^2 test as a *test of independence* for two categorical variables.

As a test of independence, the null and alternative hypotheses follow:

H_0: The two categorical variables are independent (i.e., there is no relationship between them).
H_1: The two categorical variables are dependent (i.e., there is a relationship between them).

Once again you use Equation (12.1) on page 448 to compute the test statistic:

$$\chi^2 = \sum_{all\ \text{cells}} \frac{(f_0 - f_e)^2}{f_e}$$

You reject the null hypothesis at the α level of significance if the computed value of the χ^2-test statistic is greater than χ_U^2, the upper-tail critical value from a chi-square distribution with $(r-1)(c-1)$ degrees of freedom (see Figure 12.10). Thus, the decision rule is

$$\text{Reject } H_0 \text{ if } \chi^2 > \chi_U^2;$$

otherwise do not reject H_0.

FIGURE 12.10
Regions of Rejection and Nonrejection When Testing for Independence in an $r \times c$ Contingency Table Using the χ^2 Test

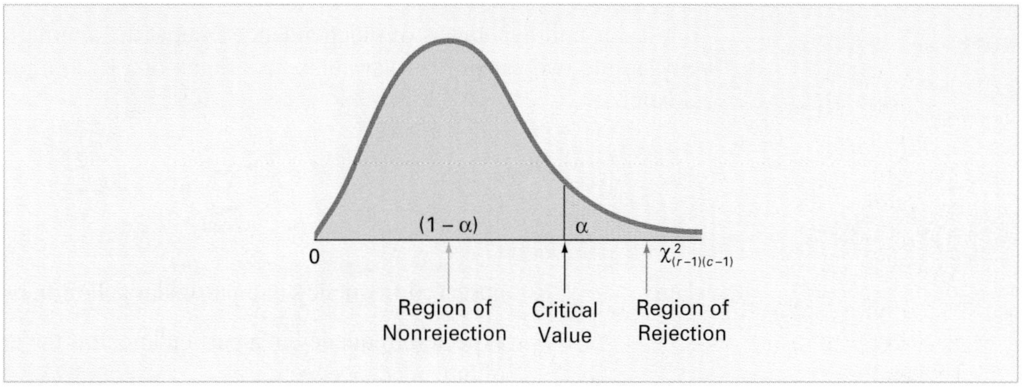

The χ^2 **test of independence** is similar to the χ^2 test for equality of proportions. The test statistics and the decision rules are the same, but the stated hypotheses and the conclusion to be drawn are different. For example, in the guest satisfaction survey of sections 12.1 and 12.2, there is evidence of a significant difference between the hotels with respect to the proportion of guests who would return. From a different viewpoint, you could conclude that there is a significant relationship between the hotels and the likelihood that the guest would return. Nevertheless, there is a fundamental difference between the two types of tests. The major difference is in the sampling scheme used.

In a test for equality of proportions, there is one factor of interest with two or more levels. These levels represent samples drawn from independent populations. The categorical responses in each sample group or level are classified into two categories—*success* and *failure*. The objective is to make comparisons and evaluate differences between the proportions of success among the various levels.

However, in a test for independence, there are two factors of interest, each of which has two or more levels. You select one sample, and tally the joint responses to the two categorical variables into the cells of a contingency table.

To illustrate the χ^2 test for independence, suppose that in the survey on hotel guest satisfaction, a second question was asked of all respondents who indicated that they were not likely to return. These guests were asked to indicate the primary reason for their response. Table 12.9 presents the resulting 4×3 contingency table.

TABLE 12.9

Observed Frequency of Responses Cross-Classifying Primary Reason for Not Returning and Hotel

	HOTEL			
PRIMARY REASON FOR NOT RETURNING	**Golden Palm**	**Palm Royale**	**Palm Princess**	**Total**
Price	23	7	37	67
Location	39	13	8	60
Room accommodation	13	5	13	31
Other	13	8	8	29
Total	88	33	66	187

In Table 12.9 observe that of the primary reasons for not planning to return to the hotel, 67 were due to price, 60 were due to location, 31 were due to room accommodation, and 29 were due to other reasons. As in Table 12.6 on page 455, there were 88 guests in the Golden Palm, 33 guests in the Palm Royale, and 66 guests in the Palm Princess who were not planning to return. The observed frequencies in the cells of the 4×3 contingency table represent the joint tallies of the sampled guests with respect to primary reason for not returning and the hotel.

The null and alternative hypotheses are:

H_0: There is no relationship between the primary reason for not returning and the hotel.
H_1: There is a relationship between the primary reason for not returning and the hotel.

To test this null hypothesis of independence against the alternative that there is a relationship between the two categorical variables, use Equation (12.1) on page 448 to compute the test statistic:

$$\chi^2 = \sum_{\textit{all}\text{ cells}} \frac{(f_0 - f_e)^2}{f_e}$$

where f_0 = observed frequency in a particular cell of the $r \times c$ contingency table

f_e = expected frequency in a particular cell if the null hypothesis of independence were true

To compute the expected frequency f_e in any cell, use the multiplication rule for independent events discussed on page 140 [see Equation (4.7)]. For example, under the null hypothesis of independence, the probability of responses expected in the upper-left-corner cell representing primary reason of price for the Golden Palm is the product of the two separate probabilities:

$$P(\text{price } \textit{and} \text{ Golden Palm}) = P(\text{price}) \times P(\text{Golden Palm})$$

Here, the proportion of reasons that are due to price, P(price), is $67/187 = 0.3583$, and the proportion of all responses from the Golden Palm, P(Golden Palm), is $88/187 = 0.4706$. If the null hypothesis is true, then the primary reason for not returning and the hotel are independent, and the probability P(price *and* Golden Palm) is the product of the separate probabilities, $0.3583 \times 0.4706 = 0.1686$. The expected frequency is the product of the overall sample size n and this probability, $187 \times 0.1686 = 31.53$. The f_e values for the remaining cells are calculated in a similar manner (see Table 12.10).

Equation (12.5) presents a simpler way to compute expected frequencies.

COMPUTING THE EXPECTED FREQUENCIES

The expected frequency in a cell is the product of its row total and column total divided by the overall sample size.

$$f_e = \frac{\text{row total} \times \text{column total}}{n} \quad \textbf{(12.5)}$$

where row total = sum of all the frequencies in the row

column total = sum of all the frequencies in the column

n = overall sample size

For example, using Equation (12.5) for the upper-left-corner cell (price for the Golden Palm),

$$f_e = \frac{\text{row total} \times \text{column total}}{n} = \frac{(67)(88)}{187} = 31.53$$

and for the lower-right-corner cell (other reason for the Palm Princess),

$$f_e = \frac{\text{row total} \times \text{column total}}{n} = \frac{(29)(66)}{187} = 10.24$$

Table 12.10 lists the entire set of f_e values.

TABLE 12.10

Expected Frequency of Responses to the Survey Cross-Classifying Primary Reason for Not Returning with Hotel

	HOTEL			
PRIMARY REASON FOR NOT RETURNING	**Golden Palm**	**Palm Royale**	**Palm Princess**	**Total**
Price	31.53	11.82	23.65	67
Location	28.24	10.59	21.18	60
Room accommodation	14.59	5.47	10.94	31
Other	13.65	5.12	10.24	29
Total	88.00	33.00	66.00	187

To perform the test of independence, you use the χ^2-test statistic shown in Equation (12.1) on page 448. Here the test statistic approximately follows a chi-square distribution with degrees of freedom equal to the number of rows in the contingency table minus 1, times the number of columns in the table minus 1. Thus, for an $r \times c$ contingency table:

$$\text{Degrees of freedom} = (r - 1)(c - 1)$$

Table 12.11 illustrates the computations for the χ^2-test statistic.

TABLE 12.11
Computation of χ^2-Test Statistic for the Test of Independence

Cell	f_0	f_e	$(f_0 - f_e)$	$(f_0 - f_e)^2$	$(f_0 - f_e)^2/f_e$
Price/Golden Palm	23	31.53	−8.53	72.7609	2.308
Price/Palm Royale	7	11.82	−4.82	23.2324	1.966
Price/Palm Princess	37	23.65	13.35	178.2225	7.536
Location/Golden Palm	39	28.24	10.76	115.7776	4.100
Location/Palm Royale	13	10.59	2.41	5.8081	0.548
Location/Palm Princess	8	21.18	−13.18	173.7124	8.202
Room/Golden Palm	13	14.59	−1.59	2.5281	0.173
Room/Palm Royale	5	5.47	−0.47	0.2209	0.040
Room/Palm Princess	13	10.94	2.06	4.2436	0.388
Other/Golden Palm	13	13.65	−0.65	0.4225	0.031
Other/Palm Royale	8	5.12	2.88	8.2944	1.620
Other/Palm Princess	8	10.24	−2.24	5.0176	0.490
					27.402

Using a level of significance of $\alpha = 0.05$, the upper-tail critical value from the chi-square distribution with $(4 - 1)(3 - 1) = 6$ degrees of freedom is 12.592 (see Table E.4). Since the computed test statistic $\chi^2 = 27.402 > 12.592$, you reject the null hypothesis of independence (see Figure 12.11). Similarly, you can use the Microsoft Excel output in Figure 12.12 or the Minitab output in Figure 12.13 to use the p-value approach. Since the p-value $= 0.00012 < 0.05$, you reject the null hypothesis of independence. This p-value indicates that there is virtually no chance of having a relationship this large or larger between hotels and primary reasons for not returning in a sample, if the primary reasons for not returning are independent of the specific hotels in the entire population. Thus, there is strong evidence of a relationship between primary reason for not returning and the hotel. Examination of the observed and expected frequencies (see Table 12.11) reveals that price is underrepresented as a reason for not returning to the Golden Palm (i.e., $f_0 = 23$ and $f_e = 31.53$) but is overrepresented at the Palm Princess.

FIGURE 12.11
Regions of Rejection and Nonrejection When Testing for Independence in the Hotel Guest Satisfaction Survey Example at the 0.05 Level of Significance with 6 Degrees of Freedom

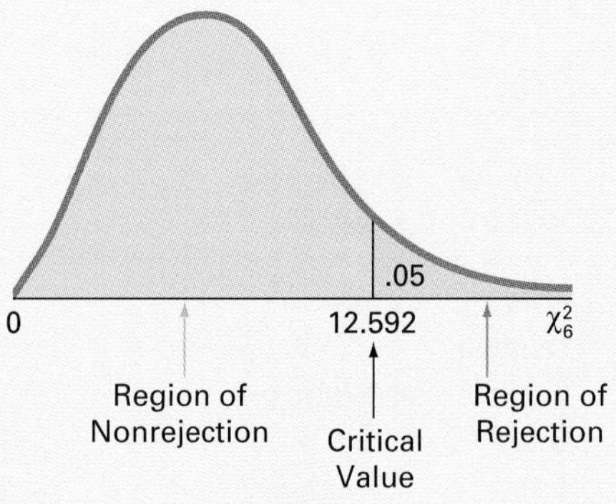

Guests are more satisfied with the price at the Golden Palm compared to the Palm Princess. Location is overrepresented as a reason for not returning to the Golden Palm but greatly underrepresented at the Palm Princess. Thus, guests are much more satisfied with the location of the Palm Princess than the Golden Palm.

FIGURE 12.12
Microsoft Excel Worksheet for the 4 × 3 Contingency Table for Primary Reason for Not Returning and Hotel

	A	B	C	D	E	F	G	H	I
1	**Cross-Classification Hotel Analysis**								
2									
3		**Observed Frequencies**							
4			**Hotel**				Calculations		
5	**Reason for Not Returning**	**Golden Palm**	**Palm Royale**	**Palm Princess**	**Total**			fo-fe	
6	**Price**	**23**	**7**	**37**	**67**		-8.52941	-4.82353	13.35294
7	**Location**	**39**	**13**	**8**	**60**		10.76471	2.411765	-13.1765
8	**Room accommodation**	**13**	**5**	**13**	**31**		-1.58824	-0.47059	2.058824
9	**Other**	**13**	**8**	**8**	**29**		-0.64706	2.882353	-2.23529
10	**Total**	**88**	**33**	**66**	**187**				
11									
12		Expected Frequencies							
13			Hotel						
14	Reason for Not Returning	Golden Palm	Palm Royale	Palm Princess	Total			(fo-fe)^2/fe	
15	Price	31.5294	11.8235	23.6471	67		2.307397	1.967808	7.540094
16	Location	28.2353	10.5882	21.1765	60		4.104044	0.549346	8.198693
17	Room accommodation	14.5882	5.4706	10.9412	31		0.172913	0.040481	0.387413
18	Other	13.6471	5.1176	10.2353	29		0.03068	1.623394	0.488168
19	Total	88	33	66	187				
20									
21	**Data**								
22	**Level of Significance**	**0.05**							
23	Number of Rows	4							
24	Number of Columns	3							
25	Degrees of Freedom	6							
26									
27	**Results**								
28	**Critical Value**	**12.5916**							
29	**Chi-Square Test Statistic**	**27.4104**							
30	***p*-Value**	**0.00012**							
31	**Reject the null hypothesis**								
32									
33	***Expected frequency assumption***								
34	***is met.***								

FIGURE 12.13
Minitab Output for the 4 × 3 Contingency Table for Primary Reason for Not Returning and Hotel

```
Chi-Square contributions are printed below expected counts

          C1      C2      C3   Total
  1       23       7      37      67
       31.53   11.82   23.65
       2.307   1.968   7.540

  2       39      13       8      60
       28.24   10.59   21.18
       4.104   0.549   8.199

  3       13       5      13      31
       14.59    5.47   10.94
       0.173   0.040   0.387

  4       13       8       8      29
       13.65    5.12   10.24
       0.031   1.623   0.488

Total     88      33      66     187

Chi-Sq = 27.410, DF = 6, P-Value = 0.000
```

To ensure accurate results, all expected frequencies need to be large in order to use the χ^2 test when dealing with $r \times c$ contingency tables. As in the case of $2 \times c$ contingency tables on page 457, all expected frequencies should be at least 1. For cases in which one or more expected frequencies are less than 1, you can use the test after collapsing two or more low-frequency rows into one row (or collapsing two or more low-frequency columns into one column). Merging of rows or columns usually results in expected frequencies sufficiently large to conduct the χ^2 test accurately.

PROBLEMS FOR SECTION 12.3

Learning the Basics

PH Grade ASSIST **12.24** If a contingency table has three rows and four columns, how many degrees of freedom are there for the χ^2 test for independence?

PH Grade ASSIST **12.25** When performing a χ^2 test for independence in a contingency table with r rows and c columns, determine the upper-tail critical value of the χ^2-test statistic in each of the following circumstances:

a. $\alpha = 0.05$, $r = 4$ rows, $c = 5$ columns
b. $\alpha = 0.01$, $r = 4$ rows, $c = 5$ columns
c. $\alpha = 0.01$, $r = 4$ rows, $c = 6$ columns
d. $\alpha = 0.01$, $r = 3$ rows, $c = 6$ columns
e. $\alpha = 0.01$, $r = 6$ rows, $c = 3$ columns

Applying the Concepts

You can solve problems 12.26–12.31 manually or by using Microsoft Excel or Minitab.

12.26 During the Vietnam War a lottery system was instituted to choose males to be drafted into the military. Numbers representing days of the year were "randomly" selected; men born on days of the year with low numbers were drafted first; those with high numbers were not drafted. The following shows how many low (1–122), medium (123–244), and high (245–366) numbers were drawn for birth dates in each quarter of the year:

NUMBER SET	QUARTER OF YEAR				
	Jan.–Mar.	Apr.–Jun.	Jul.–Sep.	Oct.–Dec.	Total
Low	21	28	35	38	122
Medium	34	22	29	37	122
High	36	41	28	17	122
Total	91	91	92	92	366

a. Is there evidence that the numbers selected were significantly related to the time of year? (Use $\alpha = 0.05$.)
b. Would you conclude that the lottery drawing appears to have been random?
c. What are your answers to (a) and (b) if the frequencies are

23	30	32	37
27	30	34	31
41	31	26	24

12.27 *USA Today* reported on preferred types of office communication by different age groups ("Talking Face to Face vs. Group Meetings," *USA Today*, October 13, 2003, A1). Suppose the results were based on a survey of 500 respondents in each age group. The results are cross-classified in the following table.

	TYPE OF COMMUNICATION PREFERRED				
AGE GROUP	Group Meetings	Face-to-face Meetings with Individuals	E-Mails	Other	Total
Generation Y	180	260	50	10	500
Generation X	210	190	65	35	500
Boomer	205	195	65	35	500
Mature	200	195	50	55	500
Total	795	840	230	135	2,000

Source: Extracted from "Talking Face to Face vs. Group Meetings," USA Today, *October 13, 2003, A1.*

At the 0.05 level of significance, is there evidence of a relationship between age group and type of communication preferred?

SELF Test PH Grade ASSIST **12.28** A large corporation is interested in determining whether a relationship exists between the commuting time of its employees and the level of stress-related problems observed on the job. A study of 116 assembly-line workers reveals the following:

	STRESS LEVEL			
COMMUTING TIME	High	Moderate	Low	Total
Under 15 min	9	5	18	32
15–45 min	17	8	28	53
Over 45 min	18	6	7	31
Total	44	19	53	116

a. At the 0.01 level of significance, is there evidence of a significant relationship between commuting time and stress level?
b. What is your answer to (a) if you used the 0.05 level of significance?

12.29 With business moving at lightning speed, marketing managers often struggle with demands that they reduce the time it takes to craft and launch (cycle time) a marketing campaign. A survey of 175 U.S. and U.K. marketing managers revealed that a marketing campaign has an average cycle time of 2.5 months and slightly more than 16% have cycle times less than one month (Dana James, "Picking Up the Pace," *Marketing News*, April 1, 2002, 3). The study results suggest that longer is not necessarily better. The marketing managers indicated that a long development time might miss the mark because the data become outdated. On the other hand, they indicated that development time of less than one month can also impair the campaign's effectiveness. Suppose that a cross-classification of the most recent marketing campaign by cycle time and effectiveness resulted in the cross-classification table shown at the top of page 467.

	CYCLE TIME				
EFFECTIVENESS	< 1 month	1–2 months	2–4 months	> 4 months	Total
Highly Effective	15	28	24	6	73
Effective	9	26	33	19	87
Ineffective	5	2	3	5	15
Total	29	56	60	30	175

Source: Extracted from Dana James, "Picking Up the Pace," Marketing News, *April 1, 2002, 3.*

At the 0.05 level of significance, is there evidence of a significant relationship between the length of cycle time and the effectiveness of a marketing campaign? If so, explain the relationship.

PH Grade ASSIST **12.30** *USA Today* reported on when the decision of what to have for dinner is made ("What's for dinner," *USA Today*, January 10, 2000). Suppose the results were based on a survey of 1,000 respondents and considered whether the household included any children under 18 years old. The results were cross-classified in the following table:

	TYPE OF HOUSEHOLD		
WHEN DECISION MADE	One Adult/ No Children	Adult/ Children	Two or More Adults/ No Children
Just before eating	162	54	154
In the afternoon	73	38	69
In the morning	59	58	53
A few days before	21	64	45
The night before	15	50	45
Always eat the same thing on this night	2	16	2
Not sure	7	6	7

Source: Extracted from "What's for Dinner," USA Today, *January 10, 2000.*

At the 0.05 level of significance, is there evidence of a significant relationship between when the decision is made of what to have for dinner and the type of household?

PH Grade ASSIST **12.31** An article in *USA Today* reported on what drivers most want in a car to make driving time as enjoyable as possible ("Drivers Just Want to Have Fun," *USA Today*, May 25, 2000). The survey also considered whether they primarily drove a sedan, a sporty car, or a sports utility vehicle. Suppose the results were based on a survey of 1,000 respondents and the results are cross-classified into the following table:

	TYPE OF CAR DRIVEN		
TECHNOLOGY DESIRED	Sedan	Sporty Car	Sports Utility Vehicle
CD player/change	178	58	54
Quality stereo system	80	54	46
Cell phone	100	8	22
Global positioning system	70	4	16
Movie/game player	68	6	6
Internet access	24	6	10
Radar detector	16	34	20
Don't know	80	26	14

Source: Extracted from "Drivers Just Want to Have Fun," USA Today, *May 25, 2000.*

At the 0.05 level of significance, is there evidence of a significant relationship between the type of technology desired and the type of car driven?

12.4 McNEMAR TEST FOR THE DIFFERENCE BETWEEN TWO PROPORTIONS (RELATED SAMPLES)

In section 10.3 you used the Z test and in section 12.1 you used the chi-square test to test for the difference between two proportions. These tests require that the samples are independent from one another. However, sometimes when you are testing differences between two proportions, the data are from repeated measurements or matched samples, and therefore the samples are related. Such situations arise often in marketing when you want to determine whether there has been a change in attitude, perception, or behavior from one time period to another.

The **McNemar test** is used to determine whether there is evidence of a difference between the proportions of two related samples. Although you could use a test statistic that follows a chi-square distribution, the McNemar test uses a test statistic that approximately follows the normal distribution enabling you to carry out either a one-tail or two-tail test.

Consider the 2×2 table presented in Table 12.12.

TABLE 12.12
2 × 2 Contingency Table for the McNemar Test

	CONDITION (GROUP 2)		
CONDITION (GROUP 1)	**Yes**	**No**	**Totals**
Yes	A	B	$A + B$
No	C	D	$C + D$
Totals	$A + C$	$B + D$	n

where A = number of respondents that answer yes to condition 1 and yes to condition 2
B = number of respondents that answer yes to condition 1 and no to condition 2
C = number of respondents that answer no to condition 1 and yes to condition 2
D = number of respondents that answer no to condition 1 and no to condition 2
n = number of respondents in the sample

The sample proportions of interest are

$$p_1 = \frac{A + B}{n} = \text{proportion of respondents in the sample who answer yes to condition 1}$$

$$p_2 = \frac{A + C}{n} = \text{proportion of respondents in the sample who answer yes to condition 2}$$

The population proportions of interest are

$$\pi_1 = \text{proportion in the population who would answer yes to condition 1}$$

$$\pi_2 = \text{proportion in the population who would answer yes to condition 2}$$

Equation (12.6) presents the McNemar test statistic used to test $H_0 : \pi_1 = \pi_2$.

MCNEMAR TEST

$$Z = \frac{B - C}{\sqrt{B + C}} \qquad \textbf{(12.6)}$$

where the test statistic Z is approximately normally distributed

To illustrate the McNemar test, return to the "Using Statistics" scenario from Chapter 4 in which you studied respondents who planned to purchase a big-screen television and actually purchased a big-screen television. Table 12.13 presents the results.

TABLE 12.13
Purchase Behavior for Big-Screen Television Sets

	ACTUALLY PURCHASED		
PLANNED TO PURCHASE	**Yes**	**No**	**Total**
Yes	200	50	250
No	100	650	750
Total	300	700	1,000

You use the McNemar test for these data because you have repeated measurements from the same set of respondents. Each respondent gave a response about whether he or she planned to purchase a big-screen television set and then whether he or she actually purchased a big-screen television.

You want to determine whether there is a difference between the population proportion who planned to purchase a big-screen television π_1 and the population proportion who actually purchased a big-screen television π_2. The null and alternative hypotheses are

$$H_0: \pi_1 = \pi_2$$
$$H_1: \pi_1 \neq \pi_2$$

Using a 0.05 level of significance, the critical values are -1.96 and $+1.96$ (see Figure 12.14) and the decision rule is

Reject H_0 if $Z < -1.96$ or if $Z > +1.96$;

otherwise, reject H_0.

FIGURE 12.14
Two-Tail McNemar Test at the 0.05 Level of Significance

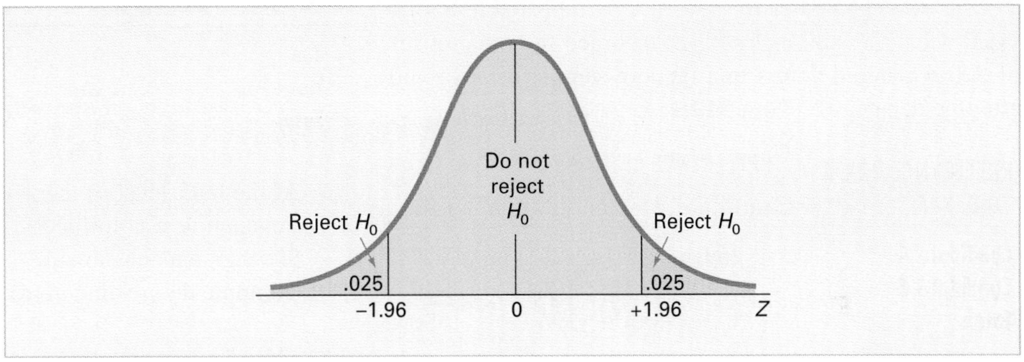

For the data of Table 12.13,

$$A = 200 \qquad B = 50 \qquad C = 100 \qquad D = 650$$

so that

$$p_1 = \frac{A + B}{n} = \frac{200 + 50}{1{,}000} = \frac{250}{1{,}000} = 0.25 \text{ and } p_2 = \frac{A + C}{n} = \frac{200 + 100}{1{,}000} = \frac{300}{1{,}000} = 0.30$$

Using Equation (12.6)

$$Z = \frac{B - C}{\sqrt{B + C}} = \frac{50 - 100}{\sqrt{50 + 100}} = \frac{-50}{\sqrt{150}} = \frac{-50}{12.247} = -4.08$$

Since $Z = -4.08 < -1.96$, you reject H_0. Using the p-value approach, the p-value is less than 0.0001 [from the cumulative standardized normal table (Table E.2)]. Since $0.0001 < 0.05$, you reject H_0. You can conclude that the proportion who intended to purchase the big-screen television is different from the proportion who actually purchased the big-screen television. In fact, from Table 12.13, observe that more respondents actually purchased a big-screen television than intended to purchase a big-screen television set.

PROBLEMS FOR SECTION 12.4

Applying the Concepts

SELF Test **12.32** A market researcher wanted to determine whether the proportion of coffee drinkers who preferred Brand A increased as the result of an advertising campaign. A random sample of 200 coffee drinkers was selected. The results indicating preference for Brand A or Brand B prior to the beginning of the advertising campaign and after its completion are shown at the top of page 470:

PREFERENCE PRIOR TO ADVERTISING CAMPAIGN	PREFERENCE AFTER COMPLETION OF ADVERTISING CAMPAIGN		
	Brand A	Brand B	Total
Brand A	101	9	110
Brand B	22	68	90
Total	123	77	200

a. At the 0.05 level of significance, is there evidence that the proportion of coffee drinkers who prefer Brand A is lower at the beginning of the advertising campaign than at the end of the advertising campaign?

b. Compute the p-value in (a) and interpret its meaning.

12.33 Two candidates for governor participated in a televised debate. A political pollster recorded the preferences of 500 registered voters in a random sample prior to and after the debate.

PREFERENCE PRIOR TO DEBATE	PREFERENCE AFTER DEBATE		
	Candidate A	Candidate B	Total
Candidate A	269	21	290
Candidate B	36	174	210
Total	305	195	500

a. At the 0.01 level of significance, is there evidence of a difference in the proportion of voters who favor Candidate A prior to and after the debate?

b. Compute the p-value in (a) and interpret its meaning.

12.34 A taste-testing experiment compared two brands of Chilean merlot wines. After the initial comparison, 60 preferred Brand A and 40 preferred Brand B. The 100 respondents were then exposed to a very professional and powerful advertisement promoting Brand A. The 100 respondents were then asked to taste the two wines again and declare which brand they preferred.

PREFERENCE PRIOR TO ADVERTISING	PREFERENCE AFTER COMPLETION OF ADVERTISING		
	Brand A	Brand B	Total
Brand A	55	5	60
Brand B	15	25	40
Total	70	30	100

a. At the 0.05 level of significance, is there evidence that the proportion who prefer Brand A is lower before the advertising than after the advertising?

b. Compute the p-value in (a) and interpret its meaning.

12.35 The CEO of a large metropolitan health care facility would like to assess the effects of recent implementation of Six Sigma management on customer satisfaction. A random sample of 100 patients is selected from a list of thousands of patients who had been to the facility the past week and also a year ago.

SATISFIED LAST YEAR	SATISFIED NOW		
	Yes	No	Total
Yes	67	5	72
No	20	8	28
Total	87	13	100

a. At the 0.05 level of significance, is there evidence that satisfaction was lower last year prior to introduction of Six Sigma management?

b. Compute the p-value in (a) and interpret its meaning.

12.36 The personnel director of a large department store wants to reduce absenteeism among sales associates. She decided to institute an incentive plan that provides financial rewards for sales associates who are absent less than five days in a given calendar year. A sample of 100 sales associates selected at the end of the trial year revealed the following:

YEAR 1	YEAR 2		
	<5 days Absent	≥ 5 Days Absent	Total
<5 days Absent	32	4	36
≥ 5 Days Absent	25	39	64
Total	57	43	100

a. At the 0.05 level of significance, is there evidence that the proportion of employees absent less than 5 days was lower in year 1 than in year 2?

b. Compute the p-value in (a) and interpret its meaning.

12.5 CHI-SQUARE TEST FOR A VARIANCE OR STANDARD DEVIATION

When analyzing numerical data, sometimes you need to draw conclusions about the population variance or standard deviation. For example, recall that in the cereal-filling process described in section 9.2, you assumed the population standard deviation σ was equal to 15 grams. To see if the variability of the process has changed, you need to test whether the standard deviation has changed from the previously specified level of 15 grams.

Assuming that the data are normally distributed, Equation (12.7) gives the χ^2-test statistic used to test whether or not the population variance or standard deviation is equal to a specified value.

χ^2 TEST FOR THE VARIANCE OR STANDARD DEVIATION

$$\chi^2 = \frac{(n-1)S^2}{\sigma^2} \tag{12.7}$$

where n = sample size

S^2 = sample variance

σ^2 = hypothesized population variance

The test statistic χ^2 follows a chi-square distribution with $n - 1$ degrees of freedom.

To apply the test of hypothesis, return to the cereal-filling example. You are interested in determining whether the standard deviation has changed from the previously specified level of 15 grams. Thus, you use a two-tail test with the following null and alternative hypotheses:

$$H_0: \sigma = 15 \text{ grams (or } \sigma^2 = 225 \text{ grams squared)}$$

$$H_1: \sigma \neq 15 \text{ grams (or } \sigma^2 \neq 225 \text{ grams squared)}$$

If you select a sample of 25 cereal boxes, you reject the null hypothesis if the χ^2 test statistic falls into either the lower or upper tail of a chi-square distribution with $25 - 1 = 24$ degrees of freedom as shown in Figure 12.15. From Equation (12.7), observe that the χ^2-test statistic falls into the lower tail of the chi-square distribution if the sample standard deviation (S) is sufficiently smaller than the hypothesized σ of 15 grams, and it falls into the upper tail if S is sufficiently larger than 15 grams. From Table 12.14 on page 472 (extracted from Table E.4), if you select a level of significance of 0.05, the lower (χ_L^2) and upper (χ_U^2) critical values are 12.401 and 39.364, respectively. Therefore, the decision rule is:

$$\text{Reject } H_0 \text{ if } \chi^2 < \chi_L^2 = 12.401 \text{ or if } \chi^2 > \chi_U^2 = 39.364;$$

otherwise, do not reject H_0.

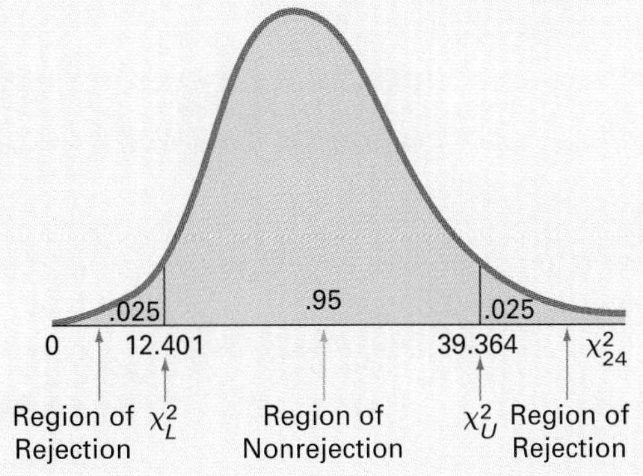

FIGURE 12.15 Determining the Lower and Upper Critical Values of a Chi-Square Distribution with 24 Degrees of Freedom Corresponding to a 0.05 Level of Significance for a Two-Tail Test of Hypothesis about a Population Variance or Standard Deviation

TABLE 12.14
Finding the Critical Values Corresponding to a 0.05 Level of Significance for a Two-Tail Test from the Chi-Square Distribution with 24 Degrees of Freedom

	Upper-Tail Areas							
Degrees of Freedom	**.995**	**.99**	**.975**	**.95**	**.90**	**.10**	**.05**	**.025**
1	. . .	. . .	0.001	0.004	0.016	2.706	3.841	5.024
2	0.010	0.020	0.051	0.103	0.211	4.605	5.991	7.378
3	0.072	0.115	0.216	0.352	0.584	6.251	7.815	9.348
.	.	.	.	.	.	.	.	.
.	.	.	.	.	.	.	.	.
.	.	.	.	.	.	.	.	.
23	9.260	10.196	11.689	13.091	14.848	32.007	35.172	38.076
24	9.886	10.856	12.401	13.848	15.659	33.196	36.415	39.364
25	10.520	11.524	13.120	14.611	16.473	34.382	37.652	40.646

Source: Extracted from Table E.4.

Suppose that in the sample of 25 cereal boxes, the standard deviation S is 17.7 grams. Using Equation (12.7) on page 471

$$\chi^2 = \frac{(n-1)S^2}{\sigma^2} = \frac{(25-1)(17.7)^2}{(15)^2} = 33.42$$

Since $\chi_L^2 = 12.401 < \chi^2 = 33.42 < \chi_U^2 = 39.364$, or since the p-value $= 0.0956 > 0.05$ (see Figure 12.16), you do not reject H_0. You conclude that there is insufficient evidence that the population standard deviation is different from 15 grams.

FIGURE 12.16
Microsoft Excel Worksheet for Testing the Variance in the Cereal-Filling Process

	A		B	
1	**Cereal-Filling Analysis**			
2				
3	**Data**			
4	**Null Hypothesis**	σ^2=	**225**	
5	**Level of Significance**		**0.05**	
6	**Sample Size**		**25**	
7	**Sample Standard Deviation**		**17.7**	
8				
9	Intermediate Calculations			
10	Degrees of Freedom		24	=B6 - 1
11	Half Area		0.025	=B5/2
12	Chi-Square Statistic		33.4176	=B10 * B7^2/B4
13				
14	**Two-Tail Test**			
15	**Lower Critical Value**		**12.4012**	=CHIINV(1 - B11, B10)
16	**Upper Critical Value**		**39.3641**	=CHIINV(B11, B10)
17	***p*-Value**		**0.0956**	=IF(B12 - B15 < 0, 1 - CHIDIST(B12, B10), CHIDIST(B12, B10))
18	**Do not reject the null hypothesis**			

In testing a hypothesis about a population variance or standard deviation, you assume that the data in the population are normally distributed. Unfortunately, the χ^2-test statistic is quite sensitive to departures from this assumption (i.e., it is not a robust test). Thus, if the population is not normally distributed, particularly for small sample sizes, the accuracy of the test can be seriously affected.

PROBLEMS FOR SECTION 12.5

Learning the Basics

12.37 Determine the lower- and upper-tail critical values of χ^2 for each of the following two-tail tests:

a. $\alpha = 0.01$, $n = 26$
b. $\alpha = 0.05$, $n = 17$
c. $\alpha = 0.10$, $n = 14$

12.38 Determine the lower- and upper-tail critical values of χ^2 for each of the following two-tail tests:

a. $\alpha = 0.01$, $n = 24$
b. $\alpha = 0.05$, $n = 20$
c. $\alpha = 0.10$, $n = 16$

12.39 In a sample of $n = 25$ selected from an underlying normal population, $S = 150$. What is the value of the χ^2-test statistic if you are testing the null hypothesis H_0: $\sigma = 100$?

12.40 In a sample of $n = 16$ selected from an underlying normal population, $S = 10$. What is the value of the χ^2-test statistic if you are testing the null hypothesis H_0 that $\sigma = 12$?

12.41 In problem 12.40, how many degrees of freedom are there in the one-sample χ^2 test?

12.42 In problems 12.40 and 12.41, what are the critical values from Table E.4 if the level of significance $\alpha = 0.05$ and H_1 is as follows:

a. $\sigma \neq 12$?
b. $\sigma < 12$?

12.43 In problems 12.40, 12.41, and 12.42, what is your statistical decision if H_1 is

a. $\sigma \neq 12$?
b. $\sigma < 12$?

12.44 If, in a sample of size $n = 16$ selected from a very left-skewed population, the sample standard deviation is $S = 24$, would you use the one-sample χ^2 test in order to test H_0: $\sigma = 20$? Discuss.

Applying the Concepts

12.45 A manufacturer of candy must monitor the temperature at which the candies are baked. Too much variation will cause inconsistency in the taste of the candy. Past records show that the standard deviation of the temperature has been 1.2°F. A random sample of 30 batches of candy is selected and the sample standard deviation of the temperature is 2.1°F.

a. At the 0.05 level of significance, is there evidence that the population standard deviation has increased above 1.2°F?
b. What assumption do you need to make in order to perform this test?
c. Compute the p-value in (a) and interpret its meaning.

SELF Test **12.46** A market researcher for an automobile dealer intends to conduct a nationwide survey concerning car repairs. Among the questions included in the survey is the following: "What was the cost of all repairs performed on your car last year?" In order to determine the sample size necessary, he needs to provide an estimate of the standard deviation. Using his past experience and judgment, he estimates that the standard deviation of the amount of repairs is \$200. Suppose that a small scale study of 25 auto owners selected at random indicates a sample standard deviation of \$237.52.

a. At the 0.05 level of significance, is there evidence that the population standard deviation is different from \$200?
b. What assumption do you need to make in order to perform this test?
c. Compute the p-value in part (a) and interpret its meaning.

12.47 The marketing manager of a branch office of a local telephone operating company wants to study characteristics of residential customers served by her office. In particular, she wants to estimate the mean monthly cost of calls within the local calling region. In order to determine the sample size necessary, she needs an estimate of the standard deviation. On the basis of her past experience and judgment, she estimates that the standard deviation is equal to \$12. Suppose that a small scale study of 15 residential customers indicates a sample standard deviation of \$9.25.

a. At the 0.10 level of significance, is there evidence that the population standard deviation is different from \$12?
b. What assumption do you need to make in order to perform this test?
c. Compute the p-value in part (a) and interpret its meaning.

12.48 A manufacturer of doorknobs has a production process that is designed to provide a doorknob with a target diameter of 2.5 inches. In the past, the standard deviation of the diameter has been 0.035 inch. In an effort to reduce the variation in the process, various studies have resulted in a redesigned process. A sample of 25 doorknobs produced under the new process indicates a sample standard deviation of 0.025 inch.

a. At the 0.05 level of significance, is there evidence that the population standard deviation is less than 0.035 inch in the new process?
b. What assumption do you need to make in order to perform this test?
c. Compute the p-value in part (a) and interpret its meaning.

12.49 The standard deviation of the weight of raisins packaged per box is 0.25 ounce. The operations manager selects a sample of 30 raisin packages filled during the last 24 hours. Their weights are in the file **RAISINS**.

a. At the 0.05 level of significance, is there evidence that the population standard deviation differs from 0.25 ounce?
b. What assumption do you need to make in order to perform this test?
c. Compute the p-value in part (a) and interpret its meaning.

12.50 A manufacturer claims that the standard deviation in capacity of a certain type of battery the company produces is 2.5 ampere-hours. An independent consumer protection agency wishes to test the credibility of the manufacturer's claim and measures the capacity of a random sample of 20 batteries from a recently produced batch. The results are in the file **AMPHRS**.

a. At the 0.05 level of significance is there evidence that the population standard deviation in battery capacity is greater than 2.5 ampere-hours?
b. What assumption do you need to make in order to perform this test?
c. Compute the p-value in part (a) and interpret its meaning.

12.6 CHI-SQUARE GOODNESS-OF-FIT TESTS

In this section, χ^2 **goodness-of-fit tests** are used to determine whether a set of data matches a specific probability distribution. Goodness-of-fit tests compare the observed frequencies in a category to the frequencies that are theoretically expected if the data follows a specific probability distribution.

To perform a χ^2 goodness-of-fit test, you follow several steps. First, you determine the specific probability distribution to compare to the data. Second, you estimate (from a sample) or hypothesize the value of each parameter (such as the mean) of the selected probability distribution. Next, you determine the theoretical probability in each category using the selected probability distribution. Finally, you use a χ^2-test statistic to test whether the selected distribution is a good fit to the data.

Chi-Square Goodness-of-Fit Test for a Poisson Distribution

Recall from section 5.4 that you used the Poisson distribution to find the probability of a specific number of arrivals per minute at a bank located in the central business district of a city. Suppose that you recorded the actual arrivals per minute in 200 one-minute periods over the course of a week. Table 12.15 summarizes the results.

TABLE 12.15
Frequency Distribution of Arrivals per Minute

Arrivals	Frequency
0	14
1	31
2	47
3	41
4	29
5	21
6	10
7	5
8	2
	200

To determine whether the number of arrivals per minute follows a Poisson distribution, the null and alternative hypotheses are:

H_0: The number of arrivals per minute follows a Poisson distribution.
H_1: The number of arrivals per minute does not follow a Poisson distribution.

The Poisson distribution has one parameter, its mean λ, and you need to specify its value in the null and alternative hypotheses. You can use either a λ value based on past knowledge, or use a λ value estimated from sample data. In this example, you estimate λ using the data in Table 12.15. Using Equation (3.16) on page 99 and the computations in Table 12.16.

$$\bar{X} = \frac{\sum_{j=1}^{c} m_j f_j}{n}$$

$$= \frac{580}{200} = 2.90$$

TABLE 12.16
Computation of the Sample Mean Number of Arrivals from the Frequency Distribution of Arrivals per Minute

Arrivals	Frequency f_j	$m_j f_j$
0	14	0
1	31	31
2	47	94
3	41	123
4	29	116
5	21	105
6	10	60
7	5	35
8	2	16
	200	580

To find the probabilities from the tables of the Poisson distribution (Table E.7), you use 2.90 as an estimate of λ. You then compute the expected frequency for each number of arrivals by multiplying the appropriate Poisson probability by the sample size $n = 200$. Table 12.17 summarizes these results.

TABLE 12.17
Observed and Expected Frequencies of the Arrivals per Minute

Arrivals	Observed Frequency f_0	Probability, $P(X)$, for Poisson Distribution with $\lambda = 2.9$	Expected Frequency $f_e = n \cdot P(X)$
0	14	0.0550	11.00
1	31	0.1596	31.92
2	47	0.2314	46.28
3	41	0.2237	44.74
4	29	0.1622	32.44
5	21	0.0940	18.80
6	10	0.0455	9.10
7	5	0.0188	3.76
8	2	0.0068	1.36
9 or more	0	0.0030	0.60

Observe in Table 12.17 that the expected frequency of 9 or more arrivals is less than 1.0. In order to have all categories contain a frequency of 1.0 or greater, you need to combine the category 9 or more with the category of 8 arrivals.

Equation (12.8) defines the chi-square test for determining whether the data follow a specific probability distribution.

CHI-SQUARE GOODNESS-OF-FIT TEST

$$\chi^2_{k-p-1} = \sum_k \frac{(f_0 - f_e)^2}{f_e} \quad \textbf{(12.8)}$$

where f_0 = observed frequency

f_e = expected frequency

k = number of categories or classes remaining after combining classes

p = number of parameters estimated

The test statistic χ^2 follows a chi-square distribution with $k - p - 1$ degrees of freedom.

Returning to the example concerning the arrivals at the bank, nine categories remain (0, 1, 2, 3, 4, 5, 6, 7, 8 or more). Since you estimated the mean of the Poisson distribution from the data, the number of degrees of freedom are $k - p - 1 = 9 - 1 - 1 = 7$.

Using the 0.05 level of significance, from Table E.4, the critical value of χ^2 with 7 degrees of freedom is 14.067. The decision rule is

$$\text{Reject } H_0 \text{ if } \chi^2 > 14.067; \text{ otherwise do not reject } H_0.$$

Table 12.18 shows the computation of the χ^2 statistic. Since $\chi^2 = 2.28954 < 14.067$, you do not reject H_0. There is insufficient evidence to conclude that the arrivals per minute do not fit a Poisson distribution.

TABLE 12.18

Computation of the χ^2-Test Statistic for the Arrivals per Minute

Arrivals	f_o	f_e	$(f_o - f_e)$	$(f_o - f_e)^2$	$(f_o - f_e)^2/f_e$
0	14	11.00	3.00	9.0000	0.81818
1	31	31.92	–0.92	0.8464	0.02652
2	47	46.28	0.72	0.5184	0.01120
3	41	44.74	–3.74	13.9876	0.31264
4	29	32.44	–3.44	11.8336	0.36478
5	21	18.80	2.20	4.8400	0.25745
6	10	9.10	0.90	0.8100	0.08901
7	5	3.76	1.24	1.5376	0.40894
8 or more	2	1.96	0.04	0.0016	0.00082
					2.28954

Chi-Square Goodness-of-Fit Test for a Normal Distribution

In Chapters 9 through 11, when testing hypotheses about numerical variables, you assumed that the underlying population was normally distributed. While the box-and-whisker plot and the normal probability plot are useful for evaluating the validity of this assumption, when you have a large sample size, you can also use the χ^2 goodness-of-fit test.

As an example of how you can use the χ^2 goodness-of-fit test for a normal distribution, refer to the 2003 returns achieved by the 121 funds summarized in Table 2.8 on page 34. To test whether these returns follow a normal distribution, the null and alternative hypotheses are:

H_0: The 2003 returns follow a normal distribution.
H_1: The 2003 returns do not follow a normal distribution.

The normal distribution has two parameters, the mean μ and the standard deviation σ. You must specify the values of these parameters in the null and alternative hypotheses. You can use values based on past knowledge or use values estimated from sample data. In this example, from the sample data: $\bar{X} = 42.621$ and $S = 13.037$. Table 2.8 uses class interval widths of 10 with class boundaries beginning at 10.0. Since the normal distribution is continuous, you need to determine the area in each class interval. In addition, since a normally distributed variable theoretically ranges from $-\infty$ to $+\infty$, you must account for the area beyond the first and last class interval. Thus, the area below 10 is the area below the Z value:

$$Z = \frac{X - \bar{X}}{S}$$

$$= \frac{10.0 - 42.621}{13.037} = -2.50$$

From Table E.2, the area below $Z = -2.50$ is approximately 0.0062.

To find the area between 10.0 and 20.0, you compute the area below 20.0 as follows:

$$Z = \frac{X - \bar{X}}{S}$$

$$= \frac{20.0 - 42.621}{13.037} = -1.74$$

From Table E.2, the area below $Z = -1.74$ is approximately 0.0414. Thus, the area between 10.0 and 20.0 is the difference in the area below 20.0 and the area below 10.0, which is 0.0414 – 0.0062 = 0.0352.

Continuing, to find the area between 20.0 and 30.0, you compute the area below 30.0 as follows:

$$Z = \frac{X - \bar{X}}{S}$$

$$= \frac{30.0 - 42.621}{13.037} = -0.97$$

From Table E.2, the area below $Z = -0.97$ is approximately 0.1665. Thus, the area between 20.0 and 30.0 is the difference in the area below 30.0 and the area below 20.0, which is 0.1665 – 0.0414 = 0.1251. In a similar manner, you can compute the area in each class interval. Table 12.19 summarizes the complete set of computations needed to find the area and expected frequency in each class.

TABLE 12.19

Computation of the Area and Expected Frequencies in Each Class Interval for the 2003 Returns

Classes	X	$X - \bar{X}$	Z	Area Below	$P(X)$ Area in Class	$f_e = n \times P(X)$
Below 10	10	−32.621	−2.50	0.0062	0.0062	0.74675
10 but < 20	20	−22.621	−1.74	0.0414	0.0352	4.25758
20 but < 30	30	−12.621	−0.97	0.1665	0.1251	15.14211
30 but < 40	40	−2.621	−0.20	0.4203	0.2538	30.71378
40 but < 50	50	7.379	0.57	0.7143	0.2940	35.57064
50 but < 60	60	17.379	1.33	0.9087	0.1944	23.52699
60 but < 70	70	27.379	2.10	0.9821	0.0734	8.88109
70 but < 80	80	37.379	2.87	0.9979	0.0158	1.91046
80 or more	–		∞	1.0000	0.0021	0.25059

Observe in Table 12.19, that the expected frequency in the category below 10.0 and in the category 80.0 or more are each less than 1.0. Since all categories need to have a frequency of 1.0 or greater, you combine the category below 10.0 with the category 10.0 to 20.0, and the category 80 or more with the category 70.0 to 80.0.

Using Equation (12.8) on page 475, you now compute the chi-square test statistic. In this example, after combining classes, 7 classes remain. Since you used the sample mean and sample standard deviation to estimate the population mean and population standard deviation, the number of degrees of freedom is equal to $k - p - 1 = 7 - 2 - 1 = 4$. Using a level of significance of 0.05, the critical value of chi-square with 4 degrees of freedom is 9.488. Table 12.20 summarizes the computations for the chi-square test. Since $\chi^2 = 6.97321 < 9.488$, do not reject H_0. There is insufficient evidence to conclude that the 2003 returns do not fit a normal distribution.

TABLE 12.20
Computation of the Chi-Square Test Statistic for the 2003 Returns

Classes	f_O	f_e	$(f_O - f_e)$	$(f_O - f_e)^2$	$(f_O - f_e)^2/f_e$
Below 20	4	5.0043	−1.0043	1.00868	0.20156
20 but < 30	18	15.1421	2.8579	8.16754	0.53939
30 but < 40	33	30.7138	2.2862	5.22678	0.17018
40 but < 50	35	35.5706	−0.5706	0.32563	0.00915
50 but < 60	15	23.5270	−8.5270	72.70953	3.09047
60 but < 70	14	8.8811	5.1189	26.20323	2.95045
70 or more	2	2.1611	−0.1611	0.02594	0.01200
					6.97321

PROBLEMS FOR SECTION 12.6

Applying the Concepts

12.51 The manager of a computer network has collected data on the number of times that service has been interrupted on each day over the past 500 days. The results are as follows:

Interruptions per Day	Number of Days
0	160
1	175
2	86
3	41
4	18
5	12
6	8
	500

Does the distribution of service interruptions follow a Poisson distribution? (Use the 0.01 level of significance.)

12.52 Referring to the data in problem 12.51, at the 0.01 level of significance, does the distribution of service interruptions follow a Poisson distribution with a population mean of 1.5 interruptions per day?

12.53 The manager of a commercial mortgage department of a large bank has collected data during the past two years concerning the number of commercial mortgages approved per week. The results from these two years (104 weeks) indicate the following:

Number of Commercial Mortgages Approved	Frequency
0	13
1	25
2	32
3	17
4	9
5	6
6	1
7	1
	104

Does the distribution of commercial mortgages approved per week follow a Poisson distribution? (Use the 0.01 level of significance.)

SELF Test **12.54** A random sample of 500 car batteries revealed the following distribution of battery life (in years).

Life (in Years)	Frequency
0—under 1	12
1—under 2	94
2—under 3	170
3—under 4	188
4—under 5	28
5—under 6	8
	500

For these data, $\bar{X} = 2.80$ and $S = 0.97$. At the 0.05 level of significance, does battery life follow a normal distribution?

12.55 A random sample of 500 long distance telephone calls revealed the following distribution of call length (in minutes).

Length (in Minutes)	Frequency
0—under 5	48
5—under 10	84
10—under 15	164
15—under 20	126
20—under 25	50
25—under 30	28
	500

a. Compute the mean and standard deviation of this frequency distribution.
b. At the 0.05 level of significance, does call length follow a normal distribution?

12.7 WILCOXON RANK SUM TEST: NONPARAMETRIC ANALYSIS FOR TWO INDEPENDENT POPULATIONS

"A nonparametric procedure is a statistical procedure that has (certain) desirable properties that hold under relatively mild assumptions regarding the underlying population(s) from which the data are obtained."

Myles Hollander and Douglas A. Wolfe (Reference 4, page 1)

In section 10.1 you used the t test for the difference between the means of two independent populations. If sample sizes are small and you cannot or do not want to make the assumption that the data in each sample are taken from normally distributed populations, two choices exist:

- Use the Wilcoxon rank sum test that does not depend on the assumption of normality for the two populations
- Use the pooled-variance t test following some *normalizing transformation* on the data (see reference 11)

This section introduces the **Wilcoxon rank sum test** for testing whether there is a difference between two medians. The Wilcoxon rank sum test is almost as powerful as the pooled-variance and separate-variance t tests under conditions appropriate to these tests and is likely to be more powerful when the assumptions of those t tests are not met. In addition, you can use the Wilcoxon rank sum test when you have only ordinal data, as often happens when dealing with studies in consumer behavior and marketing research.

To perform the Wilcoxon rank sum test, replace the values in the two samples of size n_1 and n_2 with their combined ranks (unless the data contained the ranks initially). Define $n = n_1 + n_2$ as the total sample size. Next, assign the ranks so that rank 1 is given to the smallest of the n combined values, rank 2 is given to the second-smallest, and so on, until rank n is given to the largest. If several values are tied, you assign each the average of the ranks that otherwise would have been assigned had there been no ties.

For convenience, whenever the two sample sizes are unequal, let n_1 represent the smaller sample and n_2 the larger sample. The Wilcoxon rank sum test statistic T_1 is defined as the sum of the ranks assigned to the n_1 values in the smaller sample. (For equal samples, either sample may be selected for determining T_1.) For any integer value n, the sum of the first n consecutive integers is $n(n + 1)/2$. Therefore, the test statistic T_1 plus T_2, the sum of the ranks assigned to the n_2 items in the second sample, must equal $n(n + 1)/2$. Use Equation (12.9) to check the accuracy of your rankings.

CHECKING THE RANKINGS

$$T_1 + T_2 = \frac{n(n+1)}{2} \qquad \textbf{(12.9)}$$

The Wilcoxon rank sum test can be either a two-tail test or a one-tail test, depending on whether you are testing whether the two population medians are *different* or whether one median is *greater than* the other median.

Two-Tail Test	One-Tail Test	One-Tail Test
$H_0: M_1 = M_2$	$H_0: M_1 \ge M_2$	$H_0: M_1 \le M_2$
$H_1: M_1 \ne M_2$	$H_1: M_1 < M_2$	$H_1: M_1 > M_2$

where M_1 = median of population 1

M_2 = median of population 2

When both samples n_1 and n_2 are ≤ 10, use Table E.8 to find the critical values of the test statistic T_1. For a two-tail test, you reject the null hypothesis (see panel A of Figure 12.17) if the computed value of T_1 equals or is greater than the upper critical value or is less than or equal to the lower critical value. For one-tail tests having the alternative H_1: $M_1 < M_2$, you reject the null hypothesis if the observed value of T_1 is less than or equal to the lower critical value (see panel B of Figure 12.17). For one-tail tests having the alternative H_1: $M_1 > M_2$, you reject the null hypothesis if the observed value of T_1 equals or is greater than the upper critical value (see panel C of Figure 12.17).

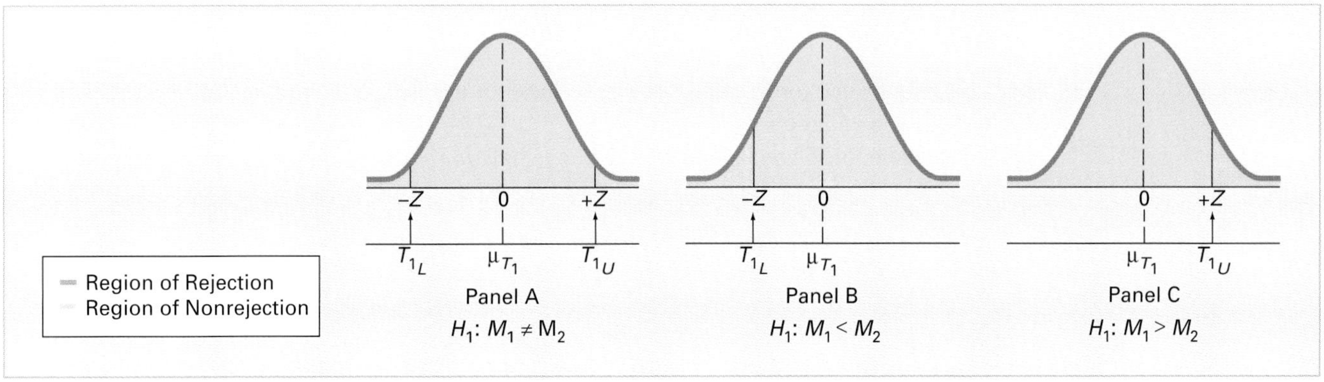

FIGURE 12.17 Regions of Rejection and Nonrejection Using the Wilcoxon Rank Sum Test

For large sample sizes, the test statistic T_1 is approximately normally distributed with mean μ_{T_1} equal to

$$\mu_{T_1} = \frac{n_1(n+1)}{2}$$

and standard deviation σ_{T_1}, equal to

$$\sigma_{T_1} = \sqrt{\frac{n_1\, n_2(n+1)}{12}}$$

Therefore, Equation (12.10) defines the standardized Z-test statistic.

LARGE SAMPLE WILCOXON RANK SUM TEST

$$Z = \frac{T_1 - \frac{n_1(n+1)}{2}}{\sqrt{\frac{n_1\, n_2(n+1)}{12}}} \tag{12.10}$$

where the test statistic Z approximately follows a standardized normal distribution.

You use Equation (12.10) for testing the null hypothesis when the sample sizes are outside the range of Table E.8. Based on α, the level of significance selected, you reject the null hypothesis if the computed Z value falls in the region of rejection.

Return to the "Using Statistics" scenario of Chapter 10 concerning sales of BLK cola for the two locations: normal display and end-aisle location (see page 346). If you do not think that the populations are normally distributed, you can use the Wilcoxon rank sum test for evaluating possible differences in the median sales for the two aisle locations.[1] The data are shown in Table 12.21 below COLA.

[1] To test for differences in the median sales between the two locations, you must assume that the distributions of sales in both populations are identical except for differences in location (i.e., the medians).

Because you have not specified in advance which aisle location is likely to have a higher median, use a two-tail test with the following null and alternative hypotheses:

$$H_0: M_1 = M_2 \text{ (i.e., the median sales are equal)}$$

$$H_1: M_1 \neq M_2 \text{ (i.e., the median sales are different)}$$

To perform the Wilcoxon rank sum test, compute the rankings for the sales from the $n_1 = 10$ stores with a normal display and the $n_2 = 10$ stores with an end-aisle display. Table 12.21 provides the combined rankings.

TABLE 12.21
Forming the Combined Rankings

Sales			
Normal Display ($n_1 = 10$)	**Combined Rank**	**End-Aisle Display ($n_2 = 10$)**	**Combined Rank**
22	1.0	52	5.5
34	3.0	71	14.0
52	5.5	76	15.0
62	10.0	54	7.0
30	2.0	67	13.0
40	4.0	83	17.0
64	11.0	66	12.0
84	18.5	90	20.0
56	8.0	77	16.0
59	9.0	84	18.5

Source: Data are taken from Table 10.1 on page 348.

The next step is to compute T_1, the sum of the ranks assigned to the *smaller* sample. When the sample sizes are equal, as in this example, you can identify either sample as the group from which to compute T_1. Choosing the normal display as the first sample,

$$T_1 = 1 + 3 + 5.5 + 10 + 2 + 4 + 11 + 18.5 + 8 + 9 = 72$$

As a check on the ranking procedure, you compute T_2 from

$$T_2 = 5.5 + 14 + 15 + 7 + 13 + 17 + 12 + 20 + 16 + 18.5 = 138$$

and then use Equation (12.9) on page 479 to show that the sum of the first $n = 20$ integers in the combined ranking is equal to $T_1 + T_2$:

$$T_1 + T_2 = \frac{n(n+1)}{2}$$

$$72 + 138 = \frac{20(21)}{2} = 210$$

$$210 = 210$$

To test the null hypothesis that there is no difference between the median sales of the two populations, use Table E.8 to determine the lower- and upper-tail critical values for the test statistic T_1. From Table 12.22, a portion of Table E.8, observe that for a level of significance of 0.05, the critical values are 78 and 132. The decision rule is

$$\text{Reject } H_0 \text{ if } T_1 \leq 78 \text{ or if } T_1 \geq 132;$$

$$\text{otherwise do not reject } H_0.$$

TABLE 12.22

Finding the Lower- and Upper-Tail Critical Values for the Wilcoxon Rank Sum Test Statistic T_1 Where $n_1 = 10$, $n_2 = 10$, and $\alpha = 0.05$

	α		n_1						
n_2	**One-Tail**	**Two-Tail**	**4**	**5**	**6**	**7** (Lower, Upper)	**8**	**9**	**10**
9	.05	.10	16,40	24,51	33,63	43,76	54,90	66,105	
	.025	.05	14,42	22,53	31,65	40,79	51,93	62,109	
	.01	.02	13,43	20,55	28,68	37,82	47,97	59,112	
	.005	.01	11,45	18,57	26,70	35,84	45,99	56,115	
10	.05	.10	17,43	26,54	35,67	45,81	56,96	69,111	82,128
	.025	.05	15,45	23,57	32,70	42,84	53,99	65,115	78,132
	.01	.02	13,47	21,59	29,73	39,87	49,103	61,119	74,136
	.005	.01	12,48	19,61	27,75	37,89	47,105	58,122	71,139

Source: Extracted from Table E.8.

Since the test statistic $T_1 = 72 < 78$, reject H_0. There is evidence of a significant difference in the median sales for the two displays. Since the sum of the ranks is higher for the end-aisle display, you conclude that median sales are higher for the end-aisle display. From the Microsoft Excel worksheet in Figure 12.18, observe that the p-value is approximately 0.0126, which is less than $\alpha = 0.05$. The p-value indicates that if the medians of the two populations are equal, the chance of finding a difference at least this large in the samples is only 0.0126.

FIGURE 12.18

Microsoft Excel Wilcoxon Rank Sum Test Worksheet for the Sales for Two Displays

	A	B	
1	**Display Location Analysis**		
2			
3	**Data**		
4	**Level of Significance**	**0.05**	
5			
6	**Population 1 Sample**		
7	**Sample Size**	**10**	
8	**Sum of Ranks**	**72**	
9	**Population 2 Sample**		
10	**Sample Size**	**10**	
11	**Sum of Ranks**	**138**	
12			
13	Intermediate Calculations		
14	Total Sample Size n	20	=B7 + B10
15	T1 Test Statistic	72	=IF(B7<= B10, B8, B11)
16	T1 Mean	105	=IF(B7<= B10, B7 * (B14 + 1)/2, B10 * (B14 + 1)/2)
17	Standard Error of T1	13.2288	=SQRT(B7 * B10 * (B14 + 1)/12)
18	**Z Test Statistic**	**-2.4946**	=(B15 - B16)/B17
19			
20	**Two-Tail Test**		
21	**Lower Critical Value**	**-1.9600**	=NORMSINV(B4/2)
22	**Upper Critical Value**	**1.9600**	=NORMSINV(1 - B4/2)
23	***p*-Value**	**0.0126**	=2 * (1 - NORMSDIST(ABS(B18)))
24	**Reject the null hypothesis**		

FIGURE 12.19
Minitab Wilcoxon Rank Sum Test Output for the Sales for Two Displays

```
Mann-Whitney Test and CI: Normal, EndAisle

            N  Median
Normal     10   73.50
EndAisle   10   54.00


Point estimate for ETA1-ETA2 is 21.50
95.5 Percent CI for ETA1-ETA2 is (6.00,37.01)
W = 138.0
Test of ETA1 = ETA2 vs ETA1 not = ETA2 is significant at 0.0140
The test is significant at 0.0139 (adjusted for ties)
```

In Figure 12.19, Minitab provides output for the *Mann-Whitney test*, which is numerically equivalent to the Wilcoxon rank sum test (see references 1, 2, and 9). From Figure 12.19, ETA1 and ETA2 refer to the hypothesized population medians M_1 and M_2. Also, the test statistic $W = 138$ is the sum of the ranks for the end-aisle display. The actual p-value for this two-tail hypothesis test is 0.0139, adjusted for ties in the rankings. Although you should use these adjusted p-values in making statistical decisions, a discussion of such adjustments is beyond the scope of this text (see references 1 and 2).

Table E.8 shows the lower and upper critical values of the Wilcoxon rank sum test statistic T_1, but it provides critical values only for situations involving small samples, where both n_1 and n_2 are less than or equal to 10. If either one or both of the sample sizes are greater than 10, you *must* use the large-sample Z approximation formula [Equation (12.10) on page 480]. However, you can also use this approximation formula for small sample sizes. To demonstrate the large-sample Z approximation formula, consider the BLK cola sales data. Using Equation (12.10),

$$Z = \frac{T_1 - \dfrac{n_1(n+1)}{2}}{\sqrt{\dfrac{n_1 n_2(n+1)}{12}}}$$

$$= \frac{72 - \dfrac{(10)(21)}{2}}{\sqrt{\dfrac{(10)(10)(21)}{12}}}$$

$$= \frac{72 - 105}{13.23} = -2.49$$

Since $Z = -2.49 < -1.96$, the critical value of Z at the 0.05 level of significance, you reject H_0.

PROBLEMS FOR SECTION 12.7

Learning the Basics

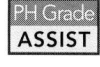

PH Grade ASSIST **12.56** Using Table E.8, determine the lower- and upper-tail critical values for the Wilcoxon rank sum test statistic T_1 in each of the following two-tail tests:

a. $\alpha = 0.10$, $n_1 = 6$, $n_2 = 8$
b. $\alpha = 0.05$, $n_1 = 6$, $n_2 = 8$
c. $\alpha = 0.01$, $n_1 = 6$, $n_2 = 8$
d. Given your results in (a) through (c), what do you conclude regarding the width of the region of nonrejection as the selected level of significance α gets smaller?

12.57 Using Table E.8, determine the upper-tail critical value for the Wilcoxon rank sum test statistic T_1 in each of the following one-tail tests:

a. $\alpha = 0.05, n_1 = 6, n_2 = 8$
b. $\alpha = 0.025, n_1 = 6, n_2 = 8$
c. $\alpha = 0.01, n_1 = 6, n_2 = 8$
d. $\alpha = 0.005, n_1 = 6, n_2 = 8$

12.58 Using Table E.8, determine the lower-tail critical value for the Wilcoxon rank sum test statistic T_1 in each of the following one-tail tests:
a. $\alpha = 0.05, n_1 = 6, n_2 = 8$
b. $\alpha = 0.025, n_1 = 6, n_2 = 8$
c. $\alpha = 0.01, n_1 = 6, n_2 = 8$
d. $\alpha = 0.005, n_1 = 6, n_2 = 8$

12.59 The following information is available for two samples selected from independent populations:

Sample 1: $n_1 = 7$ Assigned ranks: 4 1 8 2 5 10 11
Sample 2: $n_2 = 9$ Assigned ranks: 7 16 12 9 3 14 13 6 15

What is the value of T_1 if you are testing the null hypothesis H_0: $M_1 = M_2$?

PH Grade ASSIST **12.60** In problem 12.59, what are the lower- and upper-tail critical values for the test statistic T_1 from Table E.8 if you use a 0.05 level of significance and the alternative hypothesis is H_1: $M_1 \neq M_2$?

12.61 In problems 12.59 and 12.60, what is your statistical decision?

12.62 The following information is available for two samples selected from independent and similarly shaped right-skewed populations.

Sample 1: $n_1 = 5$ 1.1 2.3 2.9 3.6 14.7

Sample 2: $n_2 = 6$ 2.8 4.4 4.4 5.2 6.0 18.5

a. Replace the observed values with the corresponding ranks (where 1 = smallest value; $n = n_1 + n_2 = 11$ = largest value) in the combined samples.
b. What is the value of the test statistic T_1?
c. Compute the value of T_2, the sum of the ranks in the larger sample.
d. To check the accuracy of your rankings, use Equation (12.9) on page 479 to demonstrate that

$$T_1 + T_2 = \frac{n(n+1)}{2}$$

PH Grade ASSIST **12.63** From problem 12.62, at a level of significance of 0.05, determine the lower-tail critical value for the Wilcoxon rank sum test statistic T_1 if you want to test the null hypothesis H_0: $M_1 \geq M_2$ against the one-tail alternative H_1: $M_1 < M_2$.

12.64 In problems 12.62 and 12.63, what is your statistical decision?

Applying the Concepts

Problems 12.65–12.66 can be solved manually or by using Microsoft Excel, Minitab, or SPSS. We recommend that you use Microsoft Excel, Minitab, or SPSS to solve problems 12.67–12.70.

12.65 A vice president for marketing recruited 20 college graduates for management training. The 20 individuals are randomly assigned, 10 each, to one of two groups. A "traditional" method of training (T) is used in one group and an "experimental" method (E) is used in the other. After six months on the job, the vice president ranks the individuals on the basis of their performance from 1 (worst) to 20 (best) with the following results: TESTRANK

T 1 2 3 5 9 10 12 13 14 15
E 4 6 7 8 11 16 17 18 19 20

Is there evidence of a difference in the median performance between the two methods? (Use $\alpha = 0.05$.)

SELF Test **12.66** A New York television station decides to produce a story comparing two commuter railroads in the area—the Long Island Rail Road (LIRR) and New Jersey Transit (NJT). Data analysts at the television station sample the performance of several scheduled train runs of each railroad, 10 for the LIRR and 12 for NJT. The data on how many minutes each train is early (negative numbers) or late (positive numbers) are presented below: TRAIN2

LIRR: 5 −1 39 9 12 21 15 52 18 23
NJT: 8 4 10 4 12 5 4 9 15 33 14 7

a. Is there evidence that the railroads differ in their median tendencies to be late? (Use $\alpha = 0.01$.)
b. What assumptions must you make in (a)?
c. What conclusions can you make about the lateness of the two railroads?

12.67 The director of training for an electronic equipment manufacturer wants to determine whether different training methods have an effect on the productivity of assembly-line employees. She randomly assigns 42 recently hired employees to two groups of 21. The first group received a computer-assisted, individual-based training program and the other received a team-based training program. Upon completion of the training, the employees are evaluated on the time (in seconds) it takes to assemble a part. The results are in the data file TRAINING.
a. Using a 0.05 level of significance, is there evidence of a difference in the median assembly times (in seconds) between employees trained in a computer-assisted, individual-based program and those trained in a team-based program?

b. What assumptions must you make in order to do (a) of this problem?
c. Compare the results of (a) of problem 10.20 on page 358 with the results of (a) in this problem. Discuss.

12.68 Nondestructive evaluation (NDE) is a method that is used to describe the properties of components or materials without causing any permanent physical change to the units. It includes the determination of properties of materials and the classification of flaws by size, shape, type, and location. This method is most effective for detecting surface flaws and characterizing surface properties of electrically conductive materials. Recently, data were collected that classified each component as having a flaw based on manual inspection and operator judgment and also reported the size of the crack in the material. Do the components classified as unflawed have a smaller median crack size than components classified as flawed? The results in terms of crack size (in inches) are as follows. CRACK

Unflawed

0.003 0.004 0.012 0.014 0.021 0.023 0.024 0.030 0.034
0.041 0.041 0.042 0.043 0.045 0.057 0.063 0.074 0.076

Flawed

0.022 0.026 0.026 0.030 0.031 0.034 0.042 0.043 0.044
0.046 0.046 0.052 0.055 0.058 0.060 0.060 0.070 0.071
0.073 0.073 0.078 0.079 0.079 0.083 0.090 0.095 0.095
0.096 0.100 0.102 0.103 0.105 0.114 0.119 0.120 0.130
0.160 0.306 0.328 0.440

Source: B. D. Olin, and W. Q. Meeker, "Applications of Statistical Methods to Nondestructive Evaluation," Technometrics, *38, 1996, 101.*

a. Using a 0.05 level of significance, is there evidence that the median crack size is less for unflawed components than for the flawed components?
b. What assumptions must you make in (a)?
c. Compare the results of problem 10.21(a) on page 358 with the results of (a) in this problem. Discuss.

12.69 A bank with a branch located in a commercial district of a city BANK1 has developed an improved process for serving customers during the noon to 1 P.M. lunch period. The waiting time (defined as the time elapsed from when the customer enters the line until he or she reaches the teller window) of all customers during this hour is recorded over a period of 1 week. A random sample of 15 customers is selected, and the results (in minutes) are as follows:

4.21 5.55 3.02 5.13 4.77 2.34 3.54 3.20
4.50 6.10 0.38 5.12 6.46 6.19 3.79

Another branch located in a residential area BANK2 is also concerned with the noon to 1 P.M. lunch period. A random sample of 15 customers is selected, and the results are as follows:

9.66 5.90 8.02 5.79 8.73 3.82 8.01 8.35
10.49 6.68 5.64 4.08 6.17 9.91 5.47

a. Is there evidence of a difference in the median waiting time between the two branches? (Use $\alpha = 0.05$.)
b. What assumptions must you make in (a)?
c. Compare the results of problem 10.14(a) on page 357 with the results of (a) in this problem. Discuss.

12.70 A problem with a telephone line that prevents a customer from receiving or making calls is upsetting to both the customer and the telephone company. The following data PHONE represent samples of 20 problems reported to two different offices of a telephone company and the time to clear these problems (in minutes) from the customers' lines:

Central Office I Time to Clear Problems (minutes)

1.48 1.75 0.78 2.85 0.52 1.60 4.15 3.97 1.48 3.10
1.02 0.53 0.93 1.60 0.80 1.05 6.32 3.93 5.45 0.97

Central Office II Time to Clear Problems (minutes)

7.55 3.75 0.10 1.10 0.60 0.52 3.30 2.10 0.58 4.02
3.75 0.65 1.92 0.60 1.53 4.23 0.08 1.48 1.65 0.72

a. Is there evidence of a difference in the median time to clear these problems between the two offices? (Use $\alpha = 0.05$.)
b. What assumptions must you make in (a)?
c. Compare the results of problem 10.16(a) on page 357 with the results of (a) in this problem. Discuss.

12.8 WILCOXON SIGNED RANKS TEST: NONPARAMETRIC ANALYSIS FOR TWO RELATED POPULATIONS

In section 10.2, you used the paired *t* test to compare the means of two populations when you had repeated measures or matched samples. The paired *t* test assumes that the data are measured on an interval or a ratio scale and are normally distributed. If you cannot make these assumptions, you can use the nonparametric **Wilcoxon signed ranks test** to test for the median difference. The Wilcoxon signed ranks test requires that the differences are approximately symmetric and that the data are measured on an ordinal, interval, or ratio scale. When the assumptions for the Wilcoxon signed ranks test are met, but the assumptions of the t test are violated, the Wilcoxon signed ranks test is usually more powerful in detecting a difference in the two populations. Even under conditions appropriate to the paired t test, the Wilcoxon signed ranks test is almost as powerful.

The Wilcoxon signed ranks test uses the test statistic W. Exhibit 12.1 lists the steps required to compute W.

EXHIBIT 12.1: STEPS IN COMPUTING THE WILCOXON SIGNED RANKS TEST STATISTIC *W*

1. For each item in a sample of n items, compute a difference score D_i between the two paired values.
2. Neglect the "+" and "–" signs and compute a set of n absolute differences $|D_i|$.
3. Omit from further analysis any absolute difference score of zero, thereby yielding a set of n' nonzero absolute difference scores, where $n' \leq n$. After you remove values with absolute difference scores of zero, n' becomes the actual sample size.
4. Assign ranks R_i from 1 to n' to each of the $|D_i|$ such that the smallest absolute difference score gets rank 1 and the largest gets rank n'. If two or more $|D_i|$ are equal, assign each of them the mean rank of the ranks they would have been assigned individually had ties in the data not occurred.
5. Reassign the symbol "+" or "–" to each of the n' ranks R_i, depending on whether D_i was originally positive or negative.
6. Compute the Wilcoxon test statistic W as the sum of the positive ranks [see Equation (12.11)].

WILCOXON SIGNED RANKS TEST STATISTIC *W*

The Wilcoxon test statistic W is computed as the sum of the positive ranks.

$$W = \sum_{i=1}^{n'} R_i^{(+)} \qquad \textbf{(12.11)}$$

The null and alternative hypotheses for the Wilcoxon signed rank test are:

Two-Tail Test	One-Tail Test	One-Tail Test
H_0: $M_D = 0$	H_0: $M_D \geq 0$	H_0: $M_D \leq 0$
H_1: $M_D \neq 0$	H_1: $M_D < 0$	H_1: $M_D > 0$

Because the sum of the first n' integers $(1, 2, \ldots, n')$ equals $n'(n' + 1)/2$, the Wilcoxon test statistic W ranges from a minimum of 0 (where all the difference scores are negative) to a maximum of $n'(n' + 1)/2$ (where all the difference scores are positive). If the null hypothesis is true, the test statistic W is expected to be close to its mean $\mu_W = n'(n' + 1)/4$. If the null hypothesis is false, the value of the test statistic is expected to be close to one of the extremes.

You use Table E.9 to find the critical values of the test statistic W for both one- and two-tail tests for samples of $n' \leq 20$. For a two-tail test, you reject the null hypothesis (panel A of Figure 12.20) if W equals or is greater than the upper critical value or is equal to or less than the lower critical value. For a one-tail test in the negative direction, you reject the null hypothesis if W is less than or equal to the lower critical value (panel B of Figure 12.20). For a one-tail test in the positive direction, the decision rule is to reject the null hypothesis if W equals or is greater than the upper critical value (panel C of Figure 12.20).

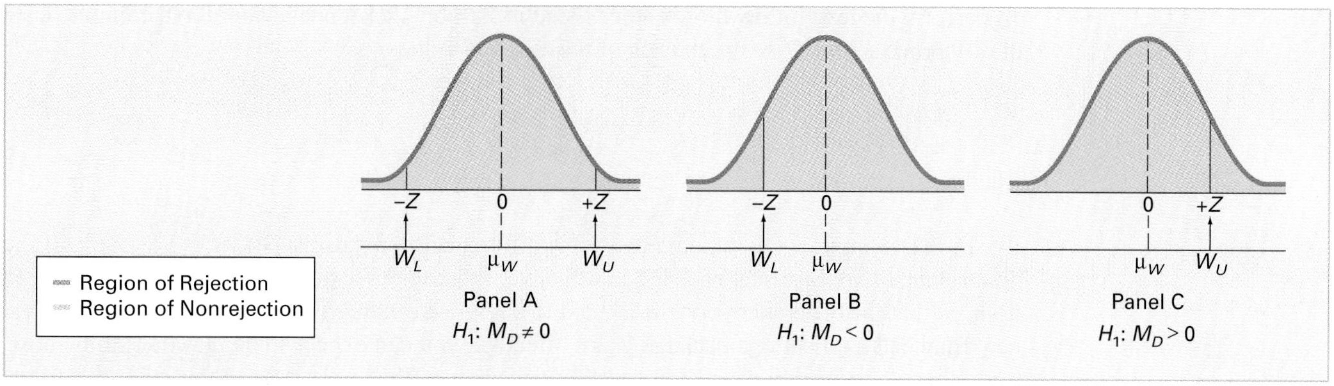

FIGURE 12.20 Regions of Rejection and Nonrejection Using the Wilcoxon Signed Ranks Test

For samples of $n' > 20$, the test statistic W is approximately normally distributed with mean μ_W and standard deviation σ_W. The mean of the test statistic W is

$$\mu_W = \frac{n'(n'+1)}{4}$$

and the standard deviation of the test statistic W is

$$\sigma_W = \sqrt{\frac{n'(n'+1)(2n'+1)}{24}}$$

Therefore, Equation (12.12) defines the standardized Z-test statistic.

LARGE-SAMPLE WILCOXON SIGNED RANKS TEST

$$Z = \frac{W - \dfrac{n'(n'+1)}{4}}{\sqrt{\dfrac{n'(n'+1)(2n'+1)}{24}}} \qquad \textbf{(12.12)}$$

You use this large-sample approximation formula when sample sizes are outside the range of Table E.9. You reject the null hypothesis if the computed Z value falls in the region of rejection. The region of rejection used depends on the level of significance and whether the test is one-tail or two-tail (see Figure 12.20).

To demonstrate how to use the Wilcoxon signed ranks test, return to the example concerning the financial applications packages discussed in section 10.2. If you cannot assume that the differences are taken from normally distributed populations, you can use the Wilcoxon signed ranks test to determine whether the current market leader has a higher median processing time than the new package.

The null and alternative hypotheses for this one-tail test are:

$$H_0: M_D \leq 0$$
$$H_1: M_D > 0$$

To perform the test, follow the six steps listed in Exhibit 12.1 on page 486. First, compute a set of difference scores D_i between each of the n paired values:

$$D_i = X_{1i} - X_{2i}$$

where $$i = 1, 2, \ldots, n$$

In this example, you compute a set of n difference scores using $D_i = X_{current} - X_{new}$. If the new financial applications package is effective, the computer processing time is expected to drop, so that the difference scores will tend to be *positive* values (and you will reject H_0). If the new financial applications package is not effective, you can expect some D_i values to be positive, others to be negative, and some to show no change (that is, $D_i = 0$). In this case, the difference scores will cluster near zero (i.e., $\overline{D} \cong 0$) and you will not reject H_0.

The remaining steps of the six-step procedure are developed in Table 12.23 COMPTIME.

TABLE 12.23

Setting Up the Wilcoxon Signed Ranks Test for the Median Difference

Processing Time (in Seconds)

Applications Project	Current Leader X_{1i}	New Package X_{2i}	$D_i = X_{1i} - X_{2i}$	$\lvert D_i \rvert$	R_i	Sign of D_i
1	9.98	9.88	+0.10	0.10	7.0	+
2	9.88	9.86	+0.02	0.02	2.0	+
3	9.84	9.75	+0.09	0.09	6.0	+
4	9.99	9.80	+0.19	0.19	8.0	+
5	9.94	9.87	+0.07	0.07	4.5	+
6	9.84	9.84	0.00	0.00	—	Discard
7	9.86	9.87	–0.01	0.01	1.0	–
8	10.12	9.86	+0.26	0.26	9.0	+
9	9.90	9.83	+0.07	0.07	4.5	+
10	9.91	9.86	+0.05	0.05	3.0	+

Discard application project 6 from the study because its difference score is zero. Eight of the remaining $n' = 9$ difference scores have a positive sign. You now compute the test statistic W as the sum of the positive ranks:

$$W = \sum_{i=1}^{n'} R_i^{(+)} = 7 + 2 + 6 + 8 + 4.5 + 9 + 4.5 + 3 = 44$$

Because $n' = 9 \leq 20$, use Table E.9 to determine the upper-tail critical value. Using $\alpha = 0.05$ for this one-tail test, the upper-tail critical value is 37 (see Table 12.24, which is a portion of Table E.9).

TABLE 12.24

Finding the Upper-Tail Critical Value for the Wilcoxon Signed Ranks Test Statistic W Where $n' = 9$ and $\alpha = 0.05$

One-Tail:	$\alpha = 0.05$	$\alpha = 0.025$	$\alpha = 0.01$	$\alpha = 0.005$
Two-Tail:	$\alpha = 0.10$	$\alpha = 0.05$	$\alpha = 0.02$	$\alpha = 0.010$
n'		(Lower, Upper)		
5	0,15	—	—	—
6	2,19	0,21	—	—
7	3,25	2,26	0,28	—
8	5,31	3,33	1,35	0,36
9	8,37	5,40	3,42	1,44
10	10,45	8,47	5,50	3,52

Source: Extracted from Table E.9.

Because $W = 44 > 37$, reject the null hypothesis. There is evidence that the median processing time using the new financial applications package is significantly faster than that using the current market leader. From the Minitab output in Figure 12.21, the p-value is 0.006, which is less than 0.05.

FIGURE 12.21

Minitab Wilcoxon Signed Ranks Test for the Financial Applications Example

Wilcoxon Signed Rank Test: Difference

Test of median = 0.000000 versus median > 0.000000

	N	N for Test	Wilcoxon Statistic	P	Estimated Median
Difference	10	9	44.0	0.006	0.07000

Table E.9 provides critical values only for situations involving small samples (where $n' \le 20$). If the sample size n' is greater than 20, you must use the large-sample Z approximation formula [Equation (12.12) on page 487]. For small samples, you can use either Table E.9 or the Z approximation approach.

To demonstrate the large-sample Z approximation approach, even for samples as small as nine, consider again the financial applications package data. Using Equation (12.12):

$$Z = \frac{W - \dfrac{n'(n'+1)}{4}}{\sqrt{\dfrac{n'(n'+1)(2n'+1)}{24}}}$$

$$= \frac{44 - \dfrac{(9)(10)}{4}}{\sqrt{\dfrac{(9)(10)(19)}{24}}}$$

$$= \frac{44 - 22.5}{8.44} = +2.55$$

The decision rule is

$$\text{Reject } H_0 \text{ if } Z > 1.645;$$
$$\text{otherwise do not reject } H_0.$$

Because $Z = +2.55 > 1.645$, reject H_0. The p-value is 0.0054. Because the p-value is less than $\alpha = 0.05$, you reject the null hypothesis. Thus, without having to assume that the original population of difference scores are normally distributed, you can conclude that the new financial applications program is faster than the current market leader.

The Wilcoxon signed ranks test makes fewer and less stringent assumptions than does the paired t test. The assumptions are:

- The data are a random sample of n independent difference scores. The difference scores are the result of repeated measures or matched pairs.
- The underlying variable is continuous.
- The data are measured on an ordinal, interval, or ratio scale.
- The distribution of the population of difference scores is approximately symmetric.

PROBLEMS FOR SECTION 12.8

Learning the Basics

12.71 Using Table E.9, determine the lower- and upper-tail critical values for the Wilcoxon signed ranks test statistic W in each of the following two-tail tests:

a. $\alpha = 0.10, n' = 11$
b. $\alpha = 0.05, n' = 11$
c. $\alpha = 0.02, n' = 11$
d. $\alpha = 0.01, n' = 11$

PH Grade ASSIST **12.72** Using Table E.9, determine the upper-tail critical value for the Wilcoxon signed ranks test statistic W in each of the following one-tail tests:

a. $\alpha = 0.05, n' = 11$
b. $\alpha = 0.025, n' = 11$
c. $\alpha = 0.01, n' = 11$
d. $\alpha = 0.005, n' = 11$

PH Grade ASSIST **12.73** Using Table E.9, determine the lower-tail critical value for the Wilcoxon signed ranks test statistic W in each of the following one-tail tests:

a. $\alpha = 0.05, n' = 11$
b. $\alpha = 0.025, n' = 11$
c. $\alpha = 0.01, n' = 11$
d. $\alpha = 0.005, n' = 11$

12.74 Consider the following $n = 12$ difference scores from two related samples:

Difference scores (D_i): +3.2, +1.7, +4.5, 0.0, +11.1, −0.8, +2.3, −2.0, 0.0, +14.8, +5.6, +1.7

What is the value of the test statistic W if you are testing H_0: $M_D = 0$?

12.75 In problem 12.74, what are the lower- and upper-tail critical values for the test statistic W from Table E.9 if the level of significance α is 0.05 and the alternative hypothesis is H_1: $M_D \neq 0$?

12.76 In problems 12.74 and 12.75, what is your statistical decision?

12.77 Consider the following $n' = 12$ signed ranks (R_i) computed from the difference scores (D_i) from two related samples:

Signed ranks (R_i): +5, +6.5, +4, +11, −8, +2.5, −2.5, +1, +12, +6.5, +10, +9

What is the value of the test statistic W if you are testing H_0: $M_D \leq 0$?

12.78 For problem 12.77, at a level of significance of 0.05, determine the upper-tail critical value for the Wilcoxon signed ranks test statistic W if you want to test H_0: $M_D \leq 0$ against H_1: $M_D > 0$.

12.79 For problems 12.77 and 12.78, what is your statistical decision?

Applying the Concepts

Problems 12.80–12.82 can be solved manually or by using Minitab or SPSS. We recommend that you use Minitab or SPSS to solve problems 12.83–12.85.

12.80 Travel expenses paid by companies can increase or decrease dramatically when there are changes in the daily rates for hotel rooms. Did these rates stay the same from June 2002 to March 2004? The following data HOTELPRICE2 give typical daily rates for hotels in 18 cities during March 2004 and June 2002.

City	March 2004	June 2002
Atlanta	78.91	173
Boston	112.92	243
Chicago	96.90	257
Dallas	77.43	167
Denver	74.22	139
Detroit	77.71	141
Houston	76.26	180
Los Angeles	95.78	223
Miami	140.61	116
Minneapolis	78.64	167
New Orleans	121.59	142
New York	167.43	273
Orlando	98.57	133
Phoenix	123.19	124
San Francisco	123.51	178
Seattle	95.09	176
St. Louis	74.68	159
Washington	123.27	262

Source: Extracted from "Travel," The Wall Street Journal, *June 7, 2002, W4; and C. Woodyard, "Luxury Costs More These Days,"* USA Today, *April 27, 2004, 5B.*

a. At the 0.05 level of significance, is there evidence of a difference in the median daily hotel rate in March 2004 and June 2002?
b. Compare the results of (a) with those of problem 10.24 on page 366.

12.81 An article discussed the new Whole Foods Market in the Time-Warner building in New York City (W. Grimes, "A Pleasure Palace Without the Guilt," *The New York Times*, February 18, 2004, F1, F5). The following data WHOLEFOODS1 compared the price of some kitchen staples at the new Whole Foods Market and at the Fairway supermarket located about 15 blocks from the Time-Warner building.

Item	Whole Foods	Fairway
Half-gallon milk	2.19	1.35
Dozen eggs	2.39	1.69
Tropicana orange juice (64 oz.)	2.00	2.49
Head of Boston lettuce	1.98	1.29
Ground round 1lb.	4.99	3.69
Bumble Bee tuna 6 oz. can	1.79	1.33
Granny Smith apples (1 lb.)	1.69	1.49
Box DeCecco linguini	1.99	1.59
Salmon steak 1 lb.	7.99	5.99
Whole chicken per pound	2.19	1.49

Source: Extracted from W. Grimes, "A Pleasure Palace Without the Guilt," The New York Times, *February 18, 2004, F1, F5.*

a. At the 0.01 level of significance, is there evidence that the median price is higher at Whole Foods Market than at the Fairway supermarket?

b. Compare the results of (a) with those of problem 10.27 on page 366.

12.82 In industrial settings, alternative methods often exist for measuring variables of interest. The data in the file MEASUREMENT (coded to maintain confidentiality) represent measurements in-line (collected from an analyzer during the production process) and from an analytical lab. (*Source:* M. Leitnaker, "Comparing Measurement Processes: In-Line versus Analytical Measurements," *Quality Engineering*, 13, 2000–2001, 293–298.)

a. At the 0.05 level of significance, is there evidence of a difference in the median measurements in-line and from an analytical lab?

b. Compare the results of (a) with those of problem 10.25 on page 366.

SELF Test **12.83** Can students save money by buying their textbooks at Amazon.com? To investigate this possibility, a random sample of 15 textbooks used during a recent semester at Miami University was selected. The prices for these textbooks at both a local bookstore and through Amazon.com were recorded. The prices for the textbooks, including all relevant taxes and shipping, are in the data file TEXTBOOK.

a. At the 0.01 level of significance, is there evidence of a difference in the median price of textbooks between the local bookstore and Amazon.com?

b. Compare the results of (a) with those of problem 10.26 on page 366.

12.84 Over the past year the vice president for human resources at a large medical center ran a series of three-month workshops aimed at increasing worker motivation and performance. As a check on the effectiveness of the workshops, she selected a random sample of 35 employees from the personnel files and recorded their most recent annual performance ratings along with the ratings attained prior to attending the workshops. The data are stored in the PERFORM file.

a. At the 0.05 level of significance, is there evidence of a difference in the median performance ratings?

b. Compare the results of (a) with those of problem 10.29 on page 367.

12.85 The data in the file CONCRETE1 represent the compressive strength in thousands of pounds per square inch (psi) of 40 samples of concrete taken two and seven days after pouring.

Source: O. Carrillo-Gamboa and R. F. Gunst, "Measurement-Error-Model Collinearities," Technometrics, *34, 1992, 454–464.*

a. At the 0.01 level of significance, is there evidence that the median strength is less at two days than at seven days?

b. Compare the results of (a) with those of problem 10.30 on page 368.

12.9 KRUSKAL-WALLIS RANK TEST: NONPARAMETRIC ANALYSIS FOR THE ONE-WAY ANOVA

If the assumptions of the one-way ANOVA F test are not met, you can use the Kruskal-Wallis rank test. The Kruskal-Wallis rank test for differences among c medians (where $c > 2$) is an extension of the Wilcoxon rank sum test for two independent populations, discussed in section 12.7. Thus, the Kruskal-Wallis test has the same power relative to the one-way ANOVA F test that the Wilcoxon rank sum test has relative to the pooled-variance t test for two independent populations.

You use the **Kruskal-Wallis rank test** to test whether c independent groups have equal medians. The null hypothesis is

$$H_0: M_1 = M_2 = \cdots = M_c$$

and the alternative hypothesis is

$$H_1: \text{Not all } M_j \text{ are equal (where } j = 1, 2, \ldots, c)$$

To use the Kruskal-Wallis test, you need to assume only that the c groups have the same variability and shape.

To use the Kruskal-Wallis rank test, you first replace the values in the c samples with their combined ranks (if necessary). Rank 1 is given to the smallest of the combined values and rank n to the largest of the combined values (where $n = n_1 + n_2 + \ldots + n_c$). If any values are tied, you assign them the mean of the ranks they would have otherwise been assigned if ties had not been present in the data.

The Kruskal-Wallis test is an alternative to the one-way ANOVA F test. Thus, the Kruskal-Wallis test statistic H is conceptually similar to SSA, the *among-group variation* term [see Equation (11.2) on page 398], which is part of the test statistic F [see Equation (11.5) on page 399]. Instead of comparing each of the c group means $\bar{X}_j$ against the grand mean $\bar{\bar{X}}$ the Kruskal-Wallis test compares the mean rank in each of the c groups against the overall mean rank based on all n combined values. If there is a significant difference among the c groups, the mean rank will differ considerably from group to group. In the process of squaring these differences, the test statistic H becomes large. If there is no difference present, the test statistic H will theoretically have a value of 0. In practice, because of random variation, the H-test statistic will be small because the mean of the ranks assigned in each group should be very similar from group to group.

Equation (12.13) defines the Kruskal-Wallis test statistic H.

KRUSKAL-WALLIS RANK TEST FOR DIFFERENCES IN c MEDIANS

$$H = \left[\frac{12}{n(n+1)} \sum_{j=1}^{c} \frac{T_j^2}{n_j}\right] - 3(n+1) \tag{12.13}$$

where n = the total number of values over the combined samples

n_j = number of values in the jth sample ($j = 1, 2, \ldots, c$)

T_j = sum of the ranks assigned to the jth sample

T_j^2 = square of the sum of the ranks assigned to the jth sample

c = number of groups

As the sample sizes in each group get large (i.e., greater than 5), you can approximate the test statistic H by the chi-square distribution with $c - 1$ degrees of freedom. Thus, you reject the null hypothesis if the computed value of H is greater than the χ_U^2 upper-tail critical value (see Figure 12.22). That is,

$$\text{Reject } H_0 \text{ if } H > \chi_U^2;$$
$$\text{otherwise do not reject } H_0.$$

The critical values from the chi-square distribution are given in Table E.4.

FIGURE 12.22
Determining the Rejection Region for the Kruskal-Wallis Test

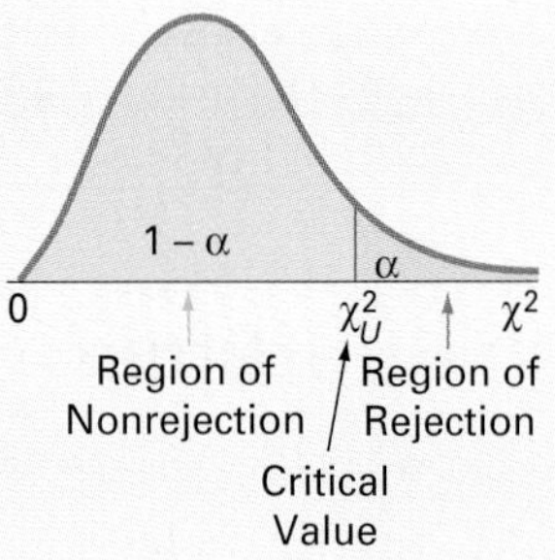

To illustrate the Kruskal-Wallis rank test for differences among c medians, return to the "Using Statistics" scenario of section 11.1 concerning the strength of parachutes. If you cannot assume that the tensile strength is normally distributed in all c groups, you can use the Kruskal-Wallis rank test.

The null hypothesis is that the median tensile strength of parachutes for the four suppliers are equal. The alternative hypothesis is that at least one of the suppliers differs from the others.

$$H_0: M_1 = M_2 = M_3 = M_4$$

$$H_1: \text{Not all } M_j \text{ are equal (where } j = 1, 2, 3, 4)$$

Table 12.25 displays the results in terms of tensile strength along with the corresponding ranks **PARACHUTE**.

TABLE 12.25

Tensile Strength and Ranks of Parachutes Woven from Synthetic Fibers from Four Suppliers

Supplier 1		2		3		4	
Amount	**Rank**	**Amount**	**Rank**	**Amount**	**Rank**	**Amount**	**Rank**
18.5	4	26.3	20	20.6	8	25.4	19
24.0	13.5	25.3	18	25.2	17	19.9	5.5
17.2	1	24.0	13.5	20.8	9	22.6	11
19.9	5.5	21.2	10	24.7	16	17.5	2
18.0	3	24.5	15	22.9	12	20.4	7

In converting the 20 tensile strengths to ranks, observe in Table 12.25 that the third parachute for Supplier 1 has the lowest tensile strength, 17.2. It is given a rank of 1. The fourth value for Supplier 1 and the second value for Supplier 4 each have a value of 19.9. Because they are tied for ranks 5 and 6, they are assigned the rank 5.5. Finally, the first value for Supplier 2 is the largest value, 26.3, and is assigned a rank of 20.

After all the ranks are assigned, compute the sum of the ranks for each group:

Rank sums: $T_1 = 27$ $T_2 = 76.5$ $T_3 = 62$ $T_4 = 44.5$

As a check on the rankings,

$$T_1 + T_2 + T_3 + T_4 = \frac{n(n+1)}{2}$$

$$27 + 76.5 + 62 + 44.5 = \frac{(20)(21)}{2}$$

$$210 = 210$$

Using Equation (12.13) to test the null hypothesis of equal population medians,

$$H = \left[\frac{12}{n(n+1)} \sum_{j=1}^{c} \frac{T_j^2}{n_j}\right] - 3(n+1)$$

$$= \left\{\frac{12}{(20)(21)}\left[\frac{(27)^2}{5} + \frac{(76.5)^2}{5} + \frac{(62)^2}{5} + \frac{(44.5)^2}{5}\right]\right\} - 3(21)$$

$$= \left(\frac{12}{420}\right)(2{,}481.1) - 63 = 7.889$$

The statistic H approximately follows a chi-square distribution with $c - 1$ degrees of freedom. Using a 0.05 level of significance, χ^2_U, the upper-tail critical value of the chi-square distribution with $c - 1 = 3$ degrees of freedom is 7.815 (see Table 12.26). Because the computed value of the test statistic $H = 7.889$ is greater than the critical value, reject the null hypothesis and conclude that not all the suppliers are the same with respect to median tensile strength. The same conclusion is reached using the p-value approach. In Figure 12.23 and Figure 12.24, observe that the p-value $= 0.048 < 0.05$.

TABLE 12.26
Finding χ^2_U, the Upper-Tail Critical Value for the Kruskal-Wallis Test at the 0.05 Level of Significance with 3 Degrees of Freedom

	Upper-Tail Area									
Degrees of Freedom	**.995**	**.99**	**.975**	**.95**	**.90**	**.75**	**.25**	**.10**	**.05**	**.025**
1	—	—	0.001	0.004	0.016	0.102	1.323	2.706	3.841	5.024
2	0.010	0.020	0.051	0.103	0.211	0.575	2.773	4.605	5.991	7.378
3	0.072	0.115	0.216	0.352	0.584	1.213	4.108	6.251	7.815	9.348
4	0.207	0.297	0.484	0.711	1.064	1.923	5.385	7.779	9.488	11.143
5	0.412	0.554	0.831	1.145	1.610	2.675	6.626	9.236	11.071	12.833

Source: Extracted from Table E.4.

FIGURE 12.23
Microsoft Excel Worksheet for the Kruskal-Wallis Rank Test for Differences Among the Four Medians in the Parachute Example

	A	B	C	D	
1	**Tensile-Strength Analysis**				
2					
3	**Data**				
4	**Level of Significance**	**0.05**			
5					
6	**Group**	**Sample Size**	**Sum of Ranks**	**Mean Ranks**	
7	1	5	27	5.4	
8	2	5	76.5	15.3	
9	3	5	62	12.4	
10	4	5	44.5	8.9	
11					
12	Intermediate Calculations				
13	Sum of Squared Ranks/Sample Size	2481.1	=(D7 * C7) + (D8 * C8) + (D9 * C9) + (D10 * C10)		
14	Sum of Sample Sizes	20	=SUM(B7:B10)		
15	Number of Groups	4			
16					
17	**Test Result**				
18	**H Test Statistic**	**7.8886**	=(12/(B14 * (B14 + 1))) * B13 - (3 * (B14 + 1))		
19	**Critical Value**	**7.8147**	=CHIINV(B4, B15 - 1)		
20	**p-Value**	**0.0484**	=CHIDIST(B18, B15 - 1)		
21	**Reject the null hypothesis**				

FIGURE 12.24
Minitab Output of the Kruskal-Wallis Rank Test for Differences Among the Four Medians in the Parachute Example

Kruskal-Wallis Test: Strength versus Supplier

```
Kruskal-Wallis Test on Strength

Supplier   N  Median  Ave Rank      Z
1          5   18.50       5.4  -2.23
2          5   24.50      15.3   2.09
3          5   22.90      12.4   0.83
4          5   20.40       8.9  -0.70
Overall   20              10.5

H = 7.89  DF = 3  P = 0.048
H = 7.90  DF = 3  P = 0.048  (adjusted for ties)
```

You rejected the null hypothesis and concluded that there was evidence of a significant difference among the suppliers with respect to the median tensile strength. At this point, you could simultaneously compare all pairs of suppliers to determine which ones differ (see reference 2).

The following assumptions are needed to use the Kruskal-Wallis rank test.

- The c samples are randomly and independently selected from their respective populations.
- The underlying variable is continuous.
- The data provide at least a set of ranks, both within and among the c samples.
- The c populations have the same variability.
- The c populations have the same shape.

The Kruskal-Wallis procedure makes less stringent assumptions than does the F test. If you ignore the last two assumptions (variability and shape), you can still use the Kruskal-Wallis rank test to test whether at least one of the populations differs from the other populations in some characteristic—such as central tendency, variation, or shape. On the other hand, to use the F test, you must assume that the c samples are from underlying normal populations that have equal variances.

When the more stringent assumptions of the F test hold, you should use the F test instead of the Kruskal-Wallis test because it is slightly more powerful in its ability to detect significant differences among groups. However, if the assumptions of the F test do not hold, you should use the Kruskal-Wallis test.

PROBLEMS FOR SECTION 12.9

Learning the Basics

12.86 What is the upper-tail critical value from the chi-square distribution if you use the Kruskal-Wallis rank test for comparing the medians in six populations at the 0.01 level of significance?

12.87 Using the results of problem 12.86:

a. State the decision rule for testing the null hypothesis that all six groups have equal population medians.

b. What is your statistical decision if the computed value of the test statistic H is 13.77?

Applying the Concepts

Problems 12.88–12.91 can be solved manually or by using Microsoft Excel, Minitab, or SPSS. We recommend that you use Microsoft Excel, Minitab, or SPSS to solve problem 12.92.

12.88 An industrial psychologist wants to test whether the reaction times of assembly-line workers are equivalent under three different learning methods. From a group of 25 new employees, 9 are randomly assigned to method A, 8 to method B, and 8 to method C. After the learning period, the workers are given a task to complete and their reaction times are measured. The rankings of the reaction times from 1 (fastest) to 25 (slowest) are in the data file: **INDPSYCH**

Is there evidence of a significant difference between the median reaction times for these learning methods? (Use $\alpha = 0.01$.)

12.89 An operations manager in a company that manufactures electronic audio equipment is inspecting a new type of battery. A batch of 20 batteries are randomly assigned to four groups (so that there are 5 batteries per group). Each group of batteries is then subjected to a particular pressure level—low, normal, high, or very high. The batteries are simultaneously tested under these pressure levels and the following times to failure (in hours) are recorded. **BATFAIL**

Pressure			
Low	**Normal**	**High**	**Very High**
8.0	7.6	6.0	5.1
8.1	8.2	6.3	5.6
9.2	9.8	7.1	5.9
9.4	10.9	7.7	6.7
11.7	12.3	8.9	7.8

The operations manager, by experience, knows that such data come from populations that are not normally distributed.

a. At the 0.05 level of significance, analyze the data to determine whether there is evidence of a significant difference in the median battery life between the four pressure levels.

b. Recommend a warranty policy with respect to battery life.

SELF Test **12.90** The retailing manager of a supermarket chain wants to determine whether product location has an effect on the sale of pet toys. Three different aisle locations are considered: front, middle, and rear. A random sample of 18 stores is selected with 6 stores randomly assigned to each aisle location. The size of the display area and the price of the product are constant for all stores. At the end of a 1-month trial period, the sales volume (in thousands of dollars) of the product in each store is as follows: LOCATE

Aisle Location		
Front	**Middle**	**Rear**
8.6	3.2	4.6
7.2	2.4	6.0
5.4	2.0	4.0
6.2	1.4	2.8
5.0	1.8	2.2
4.0	1.6	2.8

a. At the $\alpha = 0.05$ level of significance, is there evidence of a significant difference between the median sales of the various aisle locations?

b. Compare the results of (a) to those of problem 11.13 on page 410. Discuss.

12.91 The following data represent the lifetime of four different alloys. ALLOY

Alloy			
1	**2**	**3**	**4**
999	1,022	1,026	974
1,010	973	1,008	1,015
995	1,023	1,005	1,009
998	1,023	1,007	1,011
1,001	996	981	995

Source: P. Wludyka, P. Nelson, and P. Silva, " Power Curves for the Analysis of Means for Variances," Journal of Quality Technology, *33, 2001, 60–65.*

a. At the 0.05 level of significance, is there evidence of a difference in the median lifetime of the four alloys?

b. Compare the results in (a) to those in problem 11.11 on page 409.

12.92 Students in a business statistics course performed an experiment to test the strength of four brands of trash bags. One-pound weights were placed into a bag, one at a time, until the bag broke. A total of 40 bags were used (10 for each brand). The data file TRASHBAGS gives the weight (in pounds) required to break the trash bags.

a. At the 0.05 level of significance, is there evidence of a difference in the median strength of the four brands of trash bags?

b. Compare the results in (a) to those in problem 11.10 on page 409.

12.10 FRIEDMAN RANK TEST: NONPARAMETRIC ANALYSIS FOR THE RANDOMIZED BLOCK DESIGN

When analyzing a randomized block design, sometimes you cannot assume normality or the data consists of only the ranks within each block. In these situations, you can use the **Friedman rank test**.

You use the Friedman rank test to determine whether c groups (i.e., the treatment levels) have been selected from populations having equal medians. That is, you test

$$H_0: M_{.1} = M_{.2} = \cdots = M_{.c}$$

against the alternative

$$H_1: \text{Not all } M_{.j} \text{ are equal (where } j = 1, 2, \ldots, c)$$

To conduct the test, you first replace the data by their ranks *in each block*. In each of the r independent blocks, the c values are replaced by their corresponding ranks such that you assign rank 1 to the smallest value in the block and rank c to the largest. If any values in a block are tied, you assign them the mean of the ranks that they would otherwise have been given. Thus, R_{ij} is the rank (from 1 to c) associated with the jth group in the ith block.

Equation (12.14) defines the test statistic for the Friedman rank test.

FRIEDMAN RANK TEST FOR DIFFERENCES AMONG c MEDIANS

$$F_R = \frac{12}{rc(c+1)} \sum_{j=1}^{c} R_{.j}^2 - 3r(c+1) \quad \textbf{(12.14)}$$

where $R_{.j}^2$ = the square of the total of the ranks for group j ($j = 1, 2, \ldots, c$)

r = the number of blocks

c = the number of groups

As the number of blocks in the experiment gets large (greater than 5), you can approximate the test statistic F_R by the chi-square distribution with $c - 1$ degrees of freedom. Thus, for any selected level of significance α, you reject the null hypothesis if the computed value of F_R is greater than χ_U^2, the upper-tail critical value for the chi-square distribution having $c - 1$ degrees of freedom as shown in Figure 12.25. That is,

Reject H_0 if $F_R > \chi_U^2$;

otherwise do not reject H_0.

The critical values from the chi-square distribution are given in Table E.4.

FIGURE 12.25
Determining the Rejection Region for the Friedman Test

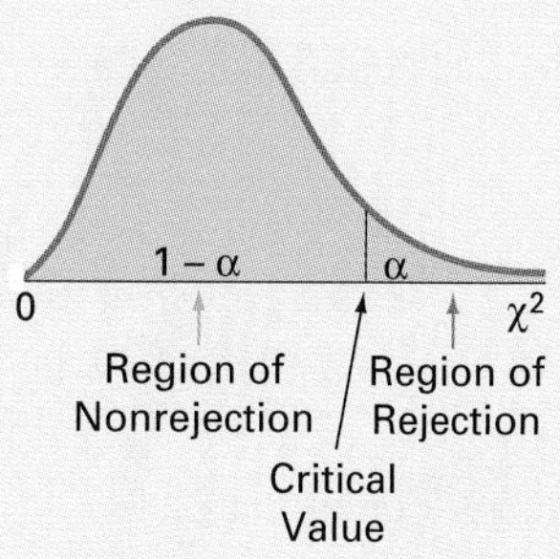

To illustrate the Friedman rank test, return to the fast-food-chain study from section 11.2 in which six raters (blocks) evaluated four restaurants (treatments). The results of the experiment are displayed in Table 12.27 along with some summary computations. If you cannot make the assumption that the service ratings are normally distributed for each restaurant, the Friedman rank test is more appropriate than the F test.

The null hypothesis is that the median service ratings for the four restaurants are equal. The alternative hypothesis is that at least one of the restaurants differs from at least one of the others.

$$H_0: M_{.1} = M_{.2} = M_{.3} = M_{.4}$$

$$H_1: \text{Not all the medians are equal}$$

Table 12.27 provides the 24 service ratings from Table 11.6 along with the ranks assigned within each block **FFCHAIN**.

TABLE 12.27
Converting Data to Ranks Within Blocks

	Restaurants							
	A		B		C		D	
Blocks of Raters	**Rating**	**Rank**	**Rating**	**Rank**	**Rating**	**Rank**	**Rating**	**Rank**
1	70	**2.0**	61	**1.0**	82	**4.0**	74	**3.0**
2	77	**3.0**	75	**1.0**	88	**4.0**	76	**2.0**
3	76	**2.0**	67	**1.0**	90	**4.0**	80	**3.0**
4	80	**3.0**	63	**1.0**	96	**4.0**	76	**2.0**
5	84	**2.5**	66	**1.0**	92	**4.0**	84	**2.5**
6	78	**2.0**	68	**1.0**	98	**4.0**	86	**3.0**
Rank total		**14.5**		**6.0**		**24.0**		**15.5**

From Table 12.27, you compute the following rank totals for each group:

Rank Totals: $R_{.1} = 14.5$ $R_{.2} = 6.0$ $R_{.3} = 24.0$ $R_{.4} = 15.5$

Equation (12.15) provides a check on the rankings.

CHECKING THE RANKINGS IN THE FRIEDMAN TEST

$$R_{.1} + R_{.2} + R_{.3} + R_{.4} = \frac{rc(c+1)}{2} \quad \textbf{(12.15)}$$

For the fast-food-chain data,

$$14.5 + 6 + 24 + 15.5 = \frac{(6)(4)(5)}{2}$$

$$60 = 60$$

Using Equation (12.14),

$$\begin{aligned} F_R &= \frac{12}{rc(c+1)} \sum_{j=1}^{c} R_{.j}^2 - 3r(c+1) \\ &= \left\{ \frac{12}{(6)(4)(5)} [14.5^2 + 6.0^2 + 24.0^2 + 15.5^2] \right\} - (3)(6)(5) \\ &= \left(\frac{12}{120} \right)(1{,}062.5) - 90 = 16.25 \end{aligned}$$

Since $F_R = 16.25 > 7.815$, the upper-tail critical value χ_U^2 of the chi-square distribution with $c - 1 = 3$ degrees of freedom (see Table E.4), reject the null hypothesis at the $\alpha = 0.05$ level. You conclude that there are significant differences (as perceived by the raters) in the median service rating at the four restaurants. Using the Minitab output of Figure 12.26, since the p-value $= 0.001 < 0.05$, you reject the null hypothesis. As a follow-up to the Friedman rank test, you can use a *post hoc* multiple comparison procedure developed by Nemenyi (see reference 7).

FIGURE 12.26
Minitab Output of Friedman Rank Test for Differences Among the Four Medians in the Fast-Food-Chain Study

Friedman Test: Rating versus Restrat blocked by Raters

```
S = 16.25  DF = 3  P = 0.001
S = 16.53  DF = 3  P = 0.001 (adjusted for ties)

                           Sum
                    Est     of
Restrat  N  Median   Ranks
A        6   76.88    14.5
B        6   66.50     6.0
C        6   90.38    24.0
D        6   79.25    15.5

Grand median = 78.25
```

The Minitab output includes the sample size, the estimated median rating, and the rank total (Sum of Ranks) for each restaurant along with the Friedman test statistic S (which is equivalent to the statistic F_R), the degrees of freedom (df), and the p-value. If there are ties in the rankings, as is the case here, Minitab provides an adjustment to the test statistic S along with an adjusted p-value. This adjustment has a minimal impact on these results.

The following assumptions are needed to use the Friedman rank test:

- The r blocks are independent so that the values in one block have no influence on the values in any other block.
- The underlying variable is continuous.
- The data constitute at least an ordinal scale of measurement within each of the r blocks.
- There is no interaction between the r blocks and the c treatment levels.
- The c populations have the same variability.
- The c populations have the same shape.

The Friedman procedure makes less stringent assumptions than does the randomized block F test. If you ignore the last two assumptions (variability and shape), you could still use the Friedman rank test to test whether at least one of the populations differs from the other populations in some characteristic—either central tendency, variation, or shape.

On the other hand, the F test requires that the level of measurement is an interval or ratio scale and that the c samples are from underlying normal populations having equal variances. Both the F test and the Friedman test assume that there is no interaction between the treatments and the blocks.

When the more stringent assumptions of the F test hold, you should select it over the Friedman test because it has more power to detect significant treatment effects. However, if the assumptions of the F test are inappropriate, you should use the Friedman rank test.

PROBLEMS FOR SECTION 12.10

Learning the Basics

12.93 What is the upper-tail critical value χ^2_U when testing for the equality of the medians in six populations using $\alpha = 0.10$?

12.94 For problem 12.93:

a. State the decision rule for testing the null hypothesis that all six groups have equal population medians.

b. What is your statistical decision if $F_R = 11.56$?

Applying the Concepts

Problems 12.95–12.97 can be solved manually or by using Minitab or SPSS. We recommend that you use Minitab or SPSS to solve problems 12.98–12.99.

12.95 Nine experts rated four brands of Colombian coffee in a taste-testing experiment. A rating on a 7-point scale (1 = extremely unpleasing, 7 = extremely pleasing) is given for each of the following four characteristics: taste, aroma, richness, and acidity. The following table displays the summated ratings—accumulated over all four characteristics. **COFFEE**

	BRAND			
EXPERT	A	B	C	D
C.C.	24	26	25	22
S.E.	27	27	26	24
E.G.	19	22	20	16
B.L.	24	27	25	23
C.M.	22	25	22	21
C.N.	26	27	24	24
G.N.	27	26	22	23
R.M.	25	27	24	21
P.V.	22	23	20	19

a. At the 0.05 level of significance, is there evidence of a difference in the median summated ratings of the four brands of Colombian coffee?

b. Are there any differences in the results of (a) from those of problem 11.23 on page 419? Discuss.

SELF Test **12.96** A student team in a business statistics course designed an experiment to investigate whether the brand of bubblegum used affected the size of bubbles they could blow. The students believed that Kyle was an expert at blowing bubbles, and his expertise might negatively affect the results of a completely randomized design. Thus, to reduce the person-to-person variability, they decided to use a randomized block design using themselves as blocks. Four brands of bubblegum were tested. A student chewed two pieces of a brand of gum and then blew two bubbles, attempting to make them as big as possible. Another student measured the diameter of the bubbles at their biggest point. The following table gives the combined diameters of the bubbles (in inches) for the 16 observations. **BUBBLEGUM**

	BRAND OF BUBBLEGUM			
STUDENT	**Bazooka**	**Bubbletape**	**Bubbleyum**	**Bubblicious**
Kyle	8.75	9.50	8.50	11.50
Sarah	9.50	4.00	8.50	11.00
Leigh	9.25	5.50	7.50	7.50
Isaac	9.50	8.50	7.50	7.50

a. Using a 0.05 level of significance, is there evidence of a difference in the median diameter of the bubbles produced by the different brands?

b. Are there any differences in the results of (a) from those of problem 11.24 on page 419? Discuss.

12.97 An article discussed the new Whole Foods Market in the Time-Warner building in New York City. The following data **WHOLEFOODS2** compared the price of some kitchen staples at the new Whole Foods Market, at the Gristede's supermarket, at the Fairway supermarket (both located about 15 blocks from the Time-Warner building), and at a Stop & Shop supermarket located in one of the outer boroughs of New York City.

Item	**Whole Foods**	**Gristede's**	**Fair-way**	**Stop & Shop**
Half-gallon milk	2.19	1.59	1.35	1.89
Dozen eggs	2.39	1.59	1.69	2.29
Tropicana orange juice (64 oz.)	2.00	2.99	2.49	2.00
Head of Boston lettuce	1.98	1.49	1.29	1.29
Ground round 1lb.	4.99	3.99	3.69	3.59
Bumble Bee tuna 6 oz. can	1.79	0.79	1.33	1.50
Granny Smith apples (1 lb.)	1.69	1.99	1.49	0.99
Box DeCecco linguini	1.99	1.50	1.59	1.79
Salmon steak 1 lb.	7.99	4.99	5.99	5.99
Whole chicken per pound	2.19	1.19	1.49	1.49

Source: Extracted from W. Grimes, "A Pleasure Palace Without the Guilt," The New York Times, *February 18, 2004, F1, F5.*

a. At the 0.05 level of significance, use the Friedman rank test to determine if there is evidence of a difference in the median prices for these kitchen staples at the four stores?

b. Are there any differences in the results of (a) from those of problem 11.25 on page 420? Discuss.

12.98 Philips Semiconductors is a leading European manufacturer of integrated circuits. Integrated circuits are produced on silicon wafers, which are ground to target thickness early in the production process. The wafers are positioned in various locations on a grinder and kept in place through vacuum decompression. One of the goals of process improvement is to reduce the variability in the thickness of the wafers in different positions and in different batches. Data were collected from a sample of 30 batches. In each batch, the thickness of the wafers on positions 1 and 2 (outer circle), 18 and 19 (middle circle), and 28 (inner circle) was measured. The results are given in the **CIRCUITS** file.

Source: K. C. B. Roes and R. J. M. M. Does, "Shewhart-Type Charts in Nonstandard Situations," Technometrics, *37, 1995, 15–24.*

a. At the 0.05 level of significance, is there evidence of a difference in the median thickness of the wafers for the five positions?

b. Are there any differences in the results of (a) from those of problem 11.27 on page 420? Discuss.

12.99 The data in the CONCRETE2 file represent the compressive strength in thousands of pounds per square inch of 40 samples of concrete taken 2, 7, and 28 days after pouring.

Source: O. Carrillo-Gamboa and R. F. Gunst, "Measurement-Error-Model Collinearities," Technometrics, *34, 1992, 454–464.*

a. At the 0.05 level of significance, is there evidence of a difference in the median compressive strength after 2, 7, and 28 days?

b. Are there any differences in the results of (a) from those of problem 11.28 on page 420? Discuss.

SUMMARY

Figure 12.27 presents a road map for this chapter. First, you used hypothesis testing for analyzing categorical response data from two samples (independent and related) and from more than two independent samples. In addition, the rules of probability from section 4.2 were extended to the hypothesis of independence in the joint responses to two categorical variables. You applied these methods to the surveys conducted by T.C. Resort Properties. You concluded that a greater proportion of guests are willing to return to the Beachcomber Hotel than to the Windsurfer; that the Golden Palm, Palm Royale, and Palm Princess hotels are different with respect to the proportion of guests who are likely to return; and that the reasons given for not returning to a hotel are dependent on the hotel the guests visited. These inferences will allow T.C. Resort Properties to improve the quality of service it provides. In addition to

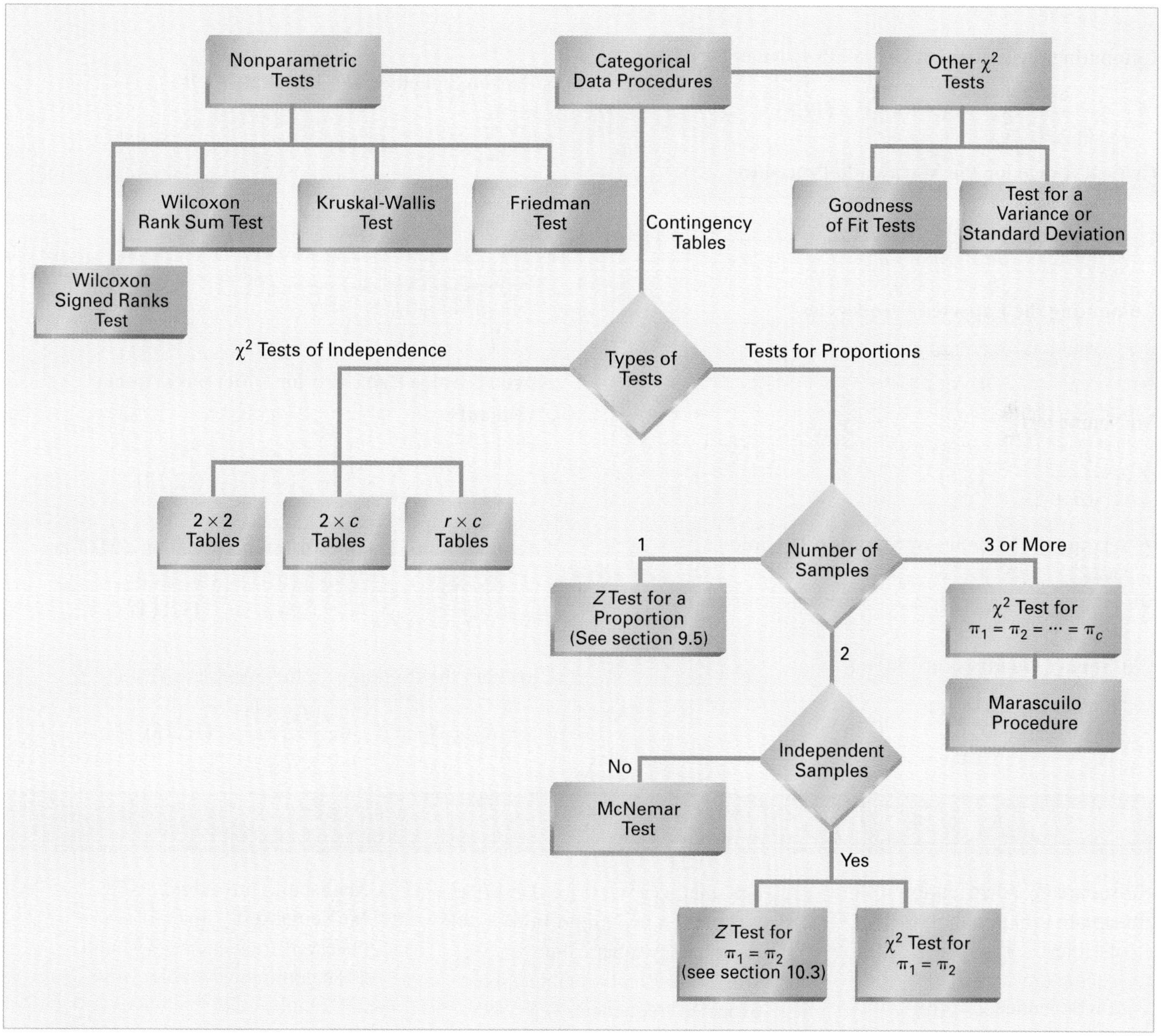

FIGURE 12.27 Road Map of Chapter 12

these tests involving contingency tables, you also used the chi-square distribution to test a variance and to test the goodness of fit of a set of data to a specific distribution. As in all statistical tests, the assumptions behind these methods must be met in order for the conclusions to be valid.

In addition to the chi-square tests, you also studied four nonparametric tests. You used the Wilcoxon rank sum test when the assumptions of the t test for two independent samples were violated; the Wilcoxon signed ranks test when the assumptions of the paired t test were violated; the Kruskal-Wallis test when the assumptions of the one-way ANOVA were violated; and the Friedman test when the assumptions of the randomized block model were violated.

KEY FORMULAS

χ^2 Test for the Difference Between Two Proportions

$$\chi^2 = \sum_{all\text{ cells}} \frac{(f_0 - f_e)^2}{f_e} \quad \textbf{(12.1)}$$

Computing the Estimated Overall Proportion

$$\bar{p} = \frac{X_1 + X_2}{n_1 + n_2} = \frac{X}{n} \quad \textbf{(12.2)}$$

Computing the Estimated Overall Proportion for c Groups

$$\bar{p} = \frac{X_1 + X_2 + \cdots + X_c}{n_1 + n_2 + \cdots + n_c} = \frac{X}{n} \quad \textbf{(12.3)}$$

Critical Range for the Marascuilo Procedure

$$\text{Critical range} = \sqrt{\chi^2_U}\sqrt{\frac{p_j(1-p_j)}{n_j} + \frac{p_{j'}(1-p_{j'})}{n_{j'}}} \quad \textbf{(12.4)}$$

Computing the Expected Frequencies

$$f_e = \frac{\text{row total} \times \text{column total}}{n} \quad \textbf{(12.5)}$$

McNemar test

$$Z = \frac{B - C}{\sqrt{B + C}} \quad \textbf{(12.6)}$$

χ^2 Test for the Variance or Standard Deviation

$$\chi^2 = \frac{(n-1)S^2}{\sigma^2} \quad \textbf{(12.7)}$$

Chi-Square Goodness of Fit Test

$$\chi^2_{k-p-1} = \sum_k \frac{(f_0 - f_e)^2}{f_e} \quad \textbf{(12.8)}$$

Checking the Rankings

$$T_1 + T_2 = \frac{n(n+1)}{2} \quad \textbf{(12.9)}$$

Large Sample Wilcoxon Rank Sum Test

$$Z = \frac{T_1 - \frac{n_1(n+1)}{2}}{\sqrt{\frac{n_1 n_2(n+1)}{12}}} \quad \textbf{(12.10)}$$

Wilcoxon Signed Ranks Test Statistic W

$$W = \sum_{i=1}^{n'} R_i^{(+)} \quad \textbf{(12.11)}$$

Large-Sample Wilcoxon Signed Ranks Test

$$Z = \frac{W - \frac{n'(n'+1)}{4}}{\sqrt{\frac{n'(n'+1)(2n'+1)}{24}}} \quad \textbf{(12.12)}$$

Kruskal-Wallis Rank Test for Differences Among c Medians

$$H = \left[\frac{12}{n(n+1)}\sum_{j=1}^{c}\frac{T_j^2}{n_j}\right] - 3(n+1) \quad \textbf{(12.13)}$$

Friedman Rank Test for Differences Among c Medians

$$F_R = \frac{12}{rc(c+1)}\sum_{j=1}^{c} R_{.j}^2 - 3r(c+1) \quad \textbf{(12.14)}$$

Checking the Rankings in the Friedman Test

$$R_{.1} + R_{.2} + R_{.3} + R_{.4} = \frac{rc(c+1)}{2} \quad \textbf{(12.15)}$$

KEY TERMS

CHAPTER REVIEW PROBLEMS

Checking Your Understanding

12.100 Under what conditions should you use the χ^2 test to determine whether there is a difference between the proportions of two independent populations?

12.101 Under what conditions should you use the χ^2 test to determine whether there is a difference between the proportions of more than two independent populations?

12.102 Under what conditions should you use the χ^2 test of independence?

12.103 Under what conditions should you use the McNemar test?

12.104 Under what conditions should you use the Wilcoxon rank sum test?

12.105 Under what conditions should you use the Wilcoxon signed ranks test?

12.106 Under what conditions should you use the Kruskal-Wallis rank test?

12.107 Under what conditions should you use the Friedman rank test?

Applying the Concepts

12.108 Undergraduate students at Miami University in Oxford, Ohio, were surveyed in order to evaluate the effect of gender and price on purchasing a pizza from Pizza Hut. Students were told to suppose that they were planning on having a large two-topping pizza delivered to their residence that evening. The students had to decide between ordering from Pizza Hut at a reduced price of $8.49 (the regular price for a large two-topping pizza from the Oxford Pizza Hut at this time was $11.49) and ordering a pizza from a different pizzeria. The results from this question are summarized in the following contingency table.

	PIZZERIA		
GENDER	**Pizza Hut**	**Other**	**Total**
Female	4	13	17
Male	6	12	18
Total	10	25	35

The survey also evaluated purchase decisions at other prices. These results are summarized in the following contingency table.

	PRICE			
PIZZERIA	**8.49**	**11.49**	**14.49**	**Total**
Pizza Hut	10	5	2	17
Other	25	23	27	75
Total	35	28	29	92

a. Using a level of significance of 0.05 and using the data in the first contingency table, is there evidence of a significant relationship between a student's gender and his or her pizzeria selection?

b. What is your answer to (a) if 9 of the male students had selected Pizza Hut and 9 selected other?

c. Using a level of significance of 0.05 and using the data in the second contingency table, is there evidence of a difference in pizzeria selection based on price?

d. Determine the p-value in (c) and interpret its meaning.

e. If appropriate, use the Marascuilo procedure and $\alpha = 0.05$ to determine which prices are different in terms of pizzeria preference.

12.109 A 2004 study by the American Society for Quality investigated executives' views toward quality. Top executives were asked whether they view quality as a profession in the way law, medicine, engineering, and accounting are viewed, or whether they see practicing quality more as the ability to understand and use a variety of tools and techniques to produce a result. Table (1) provides the responses to this question, cross-classified by the type of industry with which the executive is involved. A second question asked whether their companies actually measure the impact of process improvement initiatives designed to raise the quality of their products or services. Table (2) provides the results to this question.

(1) Do you believe that quality is a profession?

	Manufacturing	**Service**	**Healthcare**
Yes	108	88	49
No	72	132	50

(2) Does your company measure the impact of process improvement initiatives?

	Manufacturing	**Service**	**Healthcare**
Yes	132	129	54
No	48	91	46

Source: Adapted from Greg Weiler, "What Do CEOs Think About Quality?" Quality Progress, *May 2004, 37(5):52–56.*

a. Is there a significant difference among the three industries with respect to the proportion of top executives who believe quality is a profession? (Use $\alpha = 0.05$.)

b. If appropriate, apply the Marascuilo procedure to (a) using $\alpha = 0.05$.

c. Is there a significant difference among the different industries with respect to the proportion of companies that measure the impact of process improvement initiatives? (Use $\alpha = 0.05$.)

d. If appropriate, apply the Marascuilo procedure to question (c) using $\alpha = 0.05$.

12.110 Money worries in the United States start at an early age. In a survey 660 children (330 boys and 330 girls) ages 6 to 14 were asked the question, "Do you worry about having enough money?" Of the boys surveyed, 61% said yes, and 54% of the girls surveyed said yes (D. Haralson and K. Simmons, "Snapshots," *USA Today*, May 24, 2004, 1B).

a. At the 0.05 level of significance, is there a significant difference between the proportion of boys and girls who worry about having enough money?

b. Find the p-value in (a) and interpret its meaning.

12.111 A company that produces and markets videotaped continuing education programs for the financial industry has traditionally mailed sample tapes that contain previews of the programs to prospective customers. Customers then agree to purchase the program tapes or return the sample tapes. A group of sales representatives studied how to increase sales and found that many prospective customers believed it was difficult to tell from the sample tape alone whether the educational programs would meet their needs. The sales representatives performed an experiment to test whether sending the complete program tapes for review by customers would increase sales. They selected 80 customers from the mailing list and randomly assigned 40 to receive the sample tapes and 40 to receive the full-program tapes for review. They then determined the number of tapes that were purchased and returned in each group. The results of the experiment are in the following table.

	TYPE OF VIDEOTAPE RECEIVED		
ACTION	**Sample**	**Full**	**Total**
Purchased	6	14	20
Returned	34	26	60
Total	40	40	80

a. At the 0.05 level of significance, is there evidence of a difference in the proportion of tapes purchased on the basis of the type of tape sent to the customer?

b. On the basis of the results of (a), which tape do you think a representative should send in the future? Explain the rationale for your decision.

The sales representatives also wanted to determine which of three initial sales approaches result in the most sales: (1) a video sales-information tape mailed to prospective customers, (2) a personal sales call to prospective customers, and (3) a telephone call to prospective customers. A random sample of 300 prospective customers was selected, and 100 were randomly assigned to each of the three sales approaches. The results in terms of purchases of the full-program tapes are as follows:

	SALES APPROACH			
ACTION	**Videotape**	**Personal Sales Call**	**Telephone**	**Total**
Purchase	19	27	14	60
Don't purchase	81	73	86	240
Total	100	100	100	300

c. At the 0.05 level of significance, is there evidence of a difference in the proportion of tapes purchased on the basis of the sales strategy used?

d. If appropriate, use the Marascuilo procedure and $\alpha = 0.05$ to determine which sales approaches are different.

e. On the basis of the results of (c) and (d), which sales approach do you think a representative should use in the future? Explain the rationale for your decision.

12.112 In October 2000, the Markle Foundation sponsored a telephone survey concerning important issues facing the Internet. One part of the survey separated the respondents into either "general public" or "Internet experts." They were then read statements concerning common Internet practices and asked if they thought the statement was a serious concern or not. The responses to two of these statements, cross-classified by type of user, are given in the following tables:

Statement 1: "Most Web sites place a small file in your computer, called a cookie, that makes it possible for Internet businesses to keep track of all the Web sites you have visited."

	SERIOUSNESS OF CONCERN OVER STATEMENT 1		
TYPE OF USER	**Serious**	**Not Serious**	**Total**
General Public	67	28	95
Internet Experts	46	54	100
Total	113	82	195

Source: Extracted from ***www.markle.com***.

Statement 2: "Three quarters of all large companies repeatedly monitor the e-mail and Internet use of their employees."

SERIOUSNESS OF CONCERN OVER STATEMENT 2			
TYPE OF USER	**Serious**	**Not Serious**	**Total**
General Public	54	42	96
Internet Experts	37	63	100
Total	91	105	196

Source: Extracted from ***www.markle.com***.

a. At the 0.05 level of significance, is there evidence of a significant relationship between type of user and the seriousness of concern over the first statement?

b. Determine the p-value in (a) and interpret its meaning.

c. At the 0.05 level of significance, is there evidence of a significant relationship between type of user and the seriousness of concern over the second statement?

d. Determine the p-value in (c) and interpret its meaning.

12.113 The National Coffee Association conducts an annual winter survey of 3,300 people 10 years of age or older. An article discussing this survey (Nikhil Drogun, "Joe Wakes Up, Smells the Soda," *The Wall Street Journal*, June 8, 1999, B1, B16) indicates that soft drinks have become the nation's beverage of choice and that coffee is primarily a breakfast beverage.

a. The survey indicated that 49% of Americans drank coffee the previous day as compared to 75% in 1959. Suppose that the 1959 survey was based on 2,000 respondents. At the 0.01 level of significance, is there evidence of a significant difference between the percentage of Americans who drank coffee in 1999 and the percentage that drank coffee in 1959?

b. Find the p-value in (a) and interpret its meaning.

c. The survey indicated that about 23% of 18- to 24-year-olds in the survey drank coffee the previous day as compared to 74% of those over 60 years of age. Suppose there were 300 respondents 18 to 24 years old in the survey and 500 respondents over 60 years of age. At the 0.01 level of significance, is there evidence of a significant difference in the percentage of 18- to 24-year-olds and those over 60 years of age who drank coffee the previous day?

d. Find the p-value in (c) and interpret its meaning.

e. The survey indicated that of meals at home in 1998, 35% of breakfast meals, 4% of lunch meals, and 3% of supper meals were served with coffee. This compared to 40% of breakfast meals, 9% of lunch meals, and 7% of supper meals in 1988. Suppose that the 1988 survey also was based on 3,300 respondents. At the 0.01 level of significance, is there evidence of a significant difference in the proportion of meals served with coffee between 1988 and 1998? (*Hint*: Do a separate analysis for breakfast, lunch, and supper.)

f. Find the p-values in (e) and interpret its meaning.

12.114 A company is considering an organizational change by adopting the use of self-managed work teams. To assess the attitudes of employees of the company toward this change, a sample of 400 employees is selected and asked whether they favor the institution of self-managed work teams in the organization. Three responses were permitted: favor, neutral, or oppose. The results of the survey, cross-classified by type of job and attitude toward self-managed work teams, are summarized as follows:

ATTITUDE TOWARD SELF-MANAGED WORK TEAMS				
TYPE OF JOB	**Favor**	**Neutral**	**Oppose**	**Total**
Hourly worker	108	46	71	225
Supervisor	18	12	30	60
Middle management	35	14	26	75
Upper management	24	7	9	40
Total	185	79	136	400

a. At the 0.05 level of significance, is there evidence of a relationship between attitude toward self-managed work teams and type of job?

The survey also asked respondents about their attitudes toward instituting a policy whereby an employee could take one additional vacation day per month without pay. The results, cross-classified by type of job, are as follows:

ATTITUDE TOWARD VACATION TIME WITHOUT PAY				
TYPE OF JOB	**Favor**	**Neutral**	**Oppose**	**Total**
Hourly worker	135	23	67	225
Supervisor	39	7	14	60
Middle management	47	6	22	75
Upper management	26	6	8	40
Total	247	42	111	400

b. At the 0.05 level of significance, is there evidence of a relationship between attitude toward vacation time without pay and type of job?

12.115 Researchers studied the goals and outcomes of 349 work teams from various manufacturing companies in Ohio. In the first table, teams are categorized as to whether or not they had specified environmental improvements as a goal and also according to one of four types of manufacturing processes that best described their workplace. The following three tables indicate different outcomes the teams accomplished based on whether or not the team had specified cost cutting as one of the team goals.

ENVIRONMENTAL GOAL			
TYPE OF MANUFACTURING PROCESS	**Yes**	**No**	**Total**
Job shop or batch	2	42	44
Repetitive batch	4	57	61
Discrete process	15	147	162
Continuous process	17	65	82
Total	38	311	349

	COST-CUTTING GOAL		
OUTCOME	Yes	No	Total
Improved environmental performance	77	52	129
Environmental performance not improved	91	129	220
Total	168	181	349

	COST-CUTTING GOAL		
OUTCOME	Yes	No	Total
Improved profitability	70	68	138
Profitability not improved	98	113	211
Total	168	181	349

	COST-CUTTING GOAL		
OUTCOME	Yes	No	Total
Improved morale	67	55	122
Morale not improved	101	126	227
Total	168	181	349

Source: Extracted from M. Hanna, W. Newman and P. Johnson, "Linking Operational and Environmental Improvement Thru Employee Involvement," International Journal of Operations and Production Management *20, (2000), 148–165.*

a. At the 0.05 level of significance, determine whether there is evidence of a significant relationship between the presence of environmental goals and the type of manufacturing process.
b. Calculate the p-value in (a) and interpret its meaning.
c. At the 0.05 level of significance, is there evidence of a difference in improved environmental performance for teams with a specified goal of cutting costs?
d. Calculate the p-value in (c) and interpret its meaning.
e. At the 0.05 level of significance, is there evidence of a difference in improved profitability for teams with a specified goal of cutting costs?
f. Calculate the p-value in (e) and interpret its meaning.
g. At the 0.05 level of significance, is there evidence of a difference in improved morale for teams with a specified goal of cutting costs?
h. Calculate the p-value in (g) and interpret its meaning.

12.116 In summer 2000, a growing number of warranty claims on Firestone tires sold on Ford SUVs prompted Firestone and Ford to issue a major recall. The 2,030 warranty claims for the 23575R15 tires can be categorized into ATX models and Wilderness models. The type of incident leading to a warranty claim, by model type, is summarized in the following table.

Incident	ATX Model Warranty Claims	Wilderness Warranty Claims
Tread Separation	1,365	59
Blow out	77	41
Other/Unknown	422	66
Total	1,864	166

Source: Extracted from Robert L. Simison, "Ford Steps Up Recall Without Firestone," The Wall Street Journal, *August 14, 2000, A3.*

At the 0.05 level of significance, is there evidence of a significant relationship between type of incident and type of model?

12.117 A market researcher was interested in studying the effect of advertisements on brand preference of new car buyers. Prospective purchasers of new cars were first asked whether they preferred Toyota or GM and then watched video advertisements of comparable models of the two manufacturers. After viewing the ads, the prospective customers again indicated their preference. The results are summarized in the following table.

	PREFERENCE AFTER ADS		
PREFERENCE BEFORE ADS	Toyota	GM	Total
Toyota	97	3	100
GM	11	89	100
Total	108	92	200

a. Is there evidence of a difference in the proportion of respondents who prefer Toyota before and after viewing the ads? (Use $\alpha = 0.05$.)
b. Compute the p-value and interpret its meaning.

The following table was derived from the table above.

	MANUFACTURER		
PREFERENCE	Toyota	GM	Total
Before ad	100	100	200
After ad	108	92	200
Total	208	192	400

c. Show how this table is derived from the table above.
d. Using the second table, is there evidence of a difference in preference for Toyota before and after viewing the ads? (Use $\alpha = 0.05$.)
e. Compute the p-value and interpret its meaning.
f. Explain the difference in the results of (a) and (d). Which method of analyzing the data should you use? Why?

12.118 A market researcher investigated consumer preferences for Coca-Cola and Pepsi-Cola before a taste test and after a taste test. The following table summarizes the results from a sample of 200 respondents.

PREFERENCE BEFORE TASTE TEST	PREFERENCE AFTER TASTE TEST		
	Coca-Cola	Pepsi-Cola	Total
Coca-Cola	104	6	110
Pepsi-Cola	14	76	90
Total	118	82	200

a. Is there evidence of a difference in the proportion of respondents who prefer Coca-Cola before and after the taste tests? (Use $\alpha = 0.10$.)

b. Compute the p-value and interpret its meaning.

The following table was derived from the table above.

PREFERENCE	SOFT DRINK		
	Coca-Cola	Pepsi-Cola	Total
Before taste test	110	90	200
After taste test	118	82	200
Total	228	172	400

c. Show how this table is derived from the table above.

d. Using the second table, is there evidence of a difference in preference for Coca-Cola before and after the taste test? (Use $\alpha = 0.05$.)

e. Compute the p-value and interpret its meaning.

f. Explain the difference in the results of (a) and (d). Which method of analyzing the data should you use? Why?

TEAM PROJECT

The data file MUTUALFUNDS2004 contains information regarding 12 variables from a sample of 121 mutual funds. The variables are:

Fund—The name of the mutual fund
Category—Type of stocks comprising the mutual fund—small cap, mid cap, large cap
Objective—Objective of stocks comprising the mutual fund—growth or value
Assets—In millions of dollars
Fees—Sales charges (no or yes)
Expense ratio—ratio of expenses to net assets in percentage
2003 Return—Twelve-month return in 2003
Three-year return—Annualized return 2001–2003
Five-year return—Annualized return 1999–2003
Risk—Risk-of-loss factor of the mutual fund classified as low, average, or high
Best quarter—Best quarterly performance 1999–2003
Worst quarter—Worst quarterly performance 1999–2003

12.119 a. Construct a 2×2 contingency table using fees as the row variable and objective as the column variable.

b. At the 0.05 level of significance, is there evidence of a significant relationship between the objective of a mutual fund and whether or not there is a fee?

12.120 a. Construct a 2×3 contingency table using fees as the row variable and risk as the column variable.

b. At the 0.05 level of significance, is there evidence of a significant relationship between the perceived risk of a mutual fund and whether there is a fee?

12.121 a. Construct a 3×2 contingency table using risk as the row variable and objective as the column variable.

b. At the 0.05 level of significance, is there evidence of a significant relationship between the objective of a mutual fund and its perceived risk?

12.122 Does the three-year return fit a normal distribution ($\alpha = 0.05$)?

12.123 Does the five-year return fit a normal distribution ($\alpha = 0.05$)?

RUNNING CASE MANAGING THE *SPRINGVILLE HERALD*

Phase 1

Reviewing the results of their research, the marketing department concluded that a segment of Springville households might be interested in a discounted trial home subscription to the *Herald*. The team decided to test various discounts before determining the type of discount to offer during the trial period. They decided to conduct an experiment using three types of discounts plus a plan that offered no discount during the trial period. These plans were:

1. No discount for the newspaper. Subscribers would pay \$4.50 per week for the newspaper during the 90-day trial period.
2. Moderate discount for the newspaper. Subscribers would pay \$4.00 per week for the newspaper during the 90-day trial period.
3. Substantial discount for the newspaper. Subscribers would pay \$3.00 per week for the newspaper during the 90-day trial period.

4. Discount restaurant card. Subscribers would be given a card providing a discount of 15% at selected restaurants in Springville during the trial period.

Each participant in the experiment was randomly assigned to a discount plan. A random sample of 100 subscribers to each plan during the trial period was tracked to determine how many would continue to subscribe to the *Herald* after the trial period. Table SH12.1 summarizes the results.

TABLE SH12.1

Number of Subscribers Who Continue Subscriptions after Four Trial Discount Plans

CONTINUE SUBSCRIPTIONS AFTER TRIAL PERIOD	DISCOUNT PLANS				
	No Discount	Moderate Discount	Substantial Discount	Restaurant Card	Total
Yes	34	37	38	61	170
No	66	63	62	39	230
Total	100	100	100	100	400

EXERCISE

SH12.1 Analyze the results of the experiment. Write a report to the team that includes your recommendation for which discount plan to use. Be prepared to discuss the limitations and assumptions of the experiment.

DO NOT CONTINUE UNTIL THE PHASE 1 EXERCISE HAS BEEN COMPLETED.

Phase 2

The marketing department team discussed the results of the survey presented in Chapter 8 on pages 294–295. The team realized that the evaluation of individual questions were providing only limited information. In order to further understand the market for home-delivery subscriptions, the data were organized in the following cross-classification tables.

HOME DELIVERY	READ OTHER NEWSPAPER		
	Yes	No	Total
Yes	61	75	136
No	77	139	216
Total	138	214	352

HOME DELIVERY	RESTAURANT CARD		
	Yes	No	Total
Yes	26	110	136
No	40	176	216
Total	66	286	352

INTEREST IN TRIAL SUBSCRIPTION	MONDAY–SATURDAY PURCHASE BEHAVIOR			
	Every Day	Most Days	Occasionally or Never	Total
Yes	29	14	3	46
No	49	81	40	170
Total	78	95	43	216

INTEREST IN TRIAL SUBSCRIPTION	SUNDAY PURCHASE BEHAVIOR			
	Every Sunday	2–3/ Month	No More Than Once/ Month	Total
Yes	35	10	1	46
No	103	44	23	170
Total	138	54	24	216

WHERE PURCHASED	INTEREST IN TRIAL SUBSCRIPTION		
	Yes	No	Total
Convenience store	12	62	74
Newstand/candy store	15	80	95
Vending machine	10	11	21
Supermarket	5	8	13
Other locations	4	9	13
Total	46	170	216

SUNDAY PURCHASE BEHAVIOR	MONDAY–SATURDAY PURCHASE BEHAVIOR			
	Every Day	Most Days	Occasionally or Never	Total
Every Sunday	55	65	18	138
2–3/month	19	23	12	54
Once/month	4	7	13	24
Total	78	95	43	216

EXERCISE

SH12.2 Analyze the results of the cross-classification tables. Write a report for the marketing department team, and discuss the marketing implications of the results for the *Springville Herald*.

WEB CASE

Apply your knowledge of testing for the difference between two proportions in this Web Case that extends the T.C. Resort Properties "Using Statistics" scenario of this chapter.

As it tries to improve its customer service, T.C. Resort Properties faces new competition from SunLow Resorts. SunLow has recently opened resort hotels on the islands where T.C. Resorts Properties has its five hotels. SunLow is currently advertising that a random survey of 300 customers revealed that about 60% percent of the customers preferred its "Concierge Class" travel reward program over T.C. Resorts' "TCPass Plus" program.

Visit the SunLow Web site **www.prenhall.com/Springville/SunLowHome.htm**, and examine the survey data. Then answer the following:

1. Are the claims made by SunLow valid?

2. What analyses of the survey data would lead to a more favorable impression about T.C. Resort Properties?

3. Perform one of the analyses identified in your answer to step 2.

4. Given the data about T.C. Resorts Properties' customers discussed in this chapter, are there any other factors that you might include in a future survey of travel reward programs? Explain.

REFERENCES

1. Conover, W. J., *Practical Nonparametric Statistics*, 3rd ed. (New York: John Wiley, 2000).
2. Daniel, W. W., *Applied Nonparametric Statistics*, 2nd ed. (Boston: PWS Kent, 1990).
3. Dixon, W. J., and F. J. Massey, Jr., *Introduction to Statistical Analysis*, 4th ed. (New York: McGraw-Hill, 1983).
4. Hollander, M., and D. A. Wolfe, *Nonparametric Statistical Methods* (New York: John Wiley and Sons, 1973).
5. Lewontin, R. C., and J. Felsenstein, "Robustness of Homogeneity Tests in 2 × *n* Tables," *Biometrics* 21 (March 1965): 19–33.
6. Marascuilo, L. A., "Large-Sample Multiple Comparisons," *Psychological Bulletin* 65 (1966): 280–290.
7. Marascuilo, L. A., and M. McSweeney, *Nonparametric and Distribution-Free Methods for the Social Sciences* (Monterey, CA: Brooks/Cole, 1977).
8. *Microsoft Excel 2003* (Redmond, WA: Microsoft Corp., 2003).
9. *Minitab for Windows Version 14* (State College, PA: Minitab, Inc., 2004).
10. *SPSS 12.0 for Students Brief Guide* (Upper Saddle River, NJ: Prentice Hall, 2003).
11. Winer, B. J., *Statistical Principles in Experimental Design*, 2nd ed. (New York: McGraw-Hill, 1971).

Appendix 12 Using Software for Chi-Square Tests and Nonparametric Tests

A12.1 MICROSOFT EXCEL

For χ^2 Test for the Difference Between Two Proportions

Open the **Chi-Square.xls** file. This worksheet already contains the entries for the section 12.1 guest satisfaction example. This worksheet uses the CHIINV and CHIDIST functions (see section G.30 for more information). To adapt this worksheet to other problems, change the tinted labels and values in the observed frequencies table in rows 4 through 7 and the level of significance value in tinted cell B18.

OR See section G.30 (**Chi-Square Test for the Differences in Two Proportions**) if you want PHStat2 to produce a worksheet for you.

For χ^2 Test for the Differences in More Than Two Proportions and for the χ^2 Test of Independence

Open the **Chi-Square Worksheets.xls** file to the worksheet that contains the appropriate number of rows and columns for your problem. For the section 12.2 guest satisfaction example that requires two rows and three columns, open to the **ChiSquare2×3** worksheet. For the section 12.3 guest satisfaction survey example that requires four rows and three columns, open to the **ChiSquare4×3** worksheet. Enter the observed frequency table data and, optionally, the level of significance value in the tinted cell in column B. (#DIV/0! messages in individual cells will disappear after you enter the observed frequencies.) These worksheets use the CHIINV and CHIDIST functions (see section G.31 for more information).

OR See section G.31 (**Chi-Square Test**) if you want PHStat2 to produce a worksheet for you.

For the χ^2 Test for the Variance or Standard Deviation

Open the **Chi-Square Variance.xls** file, shown in Figure 12.16 on page 472. This worksheet uses the CHIINV and CHIDIST functions (see section G.32 for more information). To adapt this worksheet to other problems, change the null hypothesis, level of significance, sample size, and sample standard deviation values in the tinted cells in rows 4 through 7.

OR See section G.32 (**Chi-Square Test for a Variance**) if you want PHStat2 to produce a worksheet for you.

For Wilcoxon Rank Sum Test

Open the **Wilcoxon.xls** file, shown in Figure 12.18 on page 482. This worksheet uses the NORMSINV and NORMSDIST functions (see section G.33 for more information). To adapt this worksheet to other problems, change the level of significance, sample sizes, and rank sum values in the tinted cells in rows 4 through 11.

OR See section G.33 (**Wilcoxon Rank Sum Test**) if you want PHStat2 to determine (and sum) the ranks of your sample data and produce a worksheet for you.

For Kruskal-Wallis Rank Test

Open the **Kruskal-Wallis Worksheets.xls** file to the worksheet that contains the appropriate number of groups (populations). For example, for the section 12.9 tensile strength example contains four different populations, open to the **Kruskal-Wallis4** worksheet. Enter the sample size, sum of ranks, and mean ranks in the tinted cells in columns E through G and the level of significance value in Cell B4. (#DIV/0! messages in several cells will disappear after you enter your data.) These worksheets use the CHIINV and CHIDIST functions (see section G.34 for more information). A slight variant of these worksheets is shown in Figure 12.23 on page 494.

OR See section G.34 (**Kruskal-Wallis Rank Test**) if you want PHStat2 to produce a worksheet for you.

A12.2 MINITAB

Using Minitab for Chi-Square Tests

To construct a two-way contingency table from raw data, open the file of interest. Select **Stat ➔ Tables ➔ Cross Tabulation and Chi-Square**.

1. In the Cross Tabulation and Chi-Square dialog box (see Figure A12.1), enter the **row variable** in the For rows: edit box and the **column variable** in the For columns: edit box. Select the **Counts, Row percents, Column percents,** and **Total percents** check boxes. Select the **Chi-Square** button.

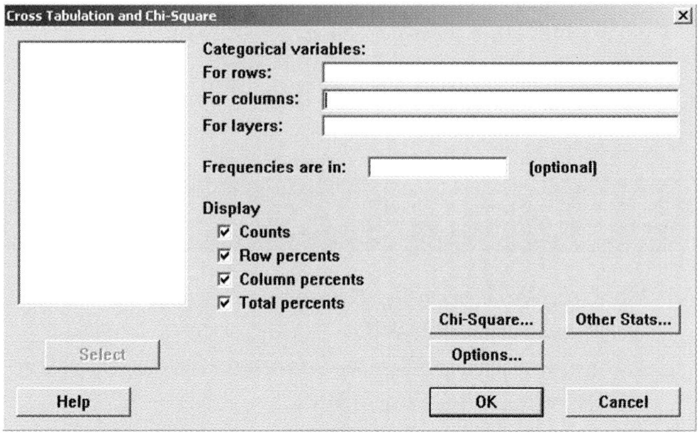

FIGURE A12.1 Minitab Cross Tabulation and Chi-Square Dialog Box

2. In the Cross Tabulation—Chi-Square dialog box (see Figure A12.2), select the **Chi-Square analysis** and **Expected cell counts** check boxes. Click the **OK** button to return to the Cross Tabulation and Chi-Square dialog box. Click the **OK** button again.

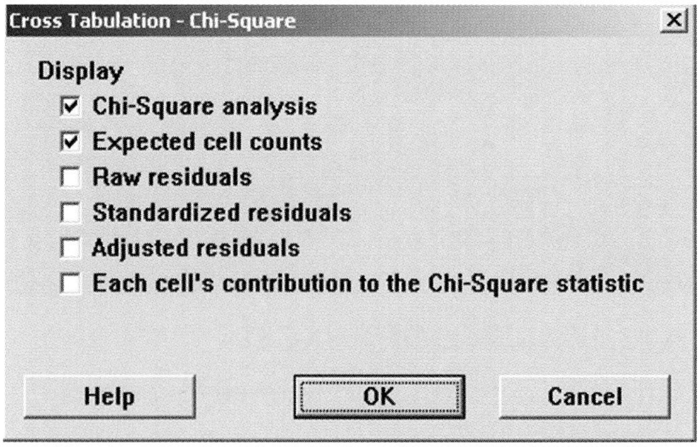

FIGURE A12.2 Minitab Cross Tabulation—Chi-Square Dialog Box

If you have only the cell frequencies, as in the hotel satisfaction survey example, enter them in columns in a Minitab worksheet. For the hotel satisfaction 2 × 2 table, enter **163** and **64** in column C1 and **154** and **108** in column C2. Select **Stat ➔ Tables ➔ Chi-Square Test (Table in Worksheet)**. In the Chi-Square Test (Table in Worksheet) dialog box (see Figure A12.3), in the Columns containing the table: edit box, enter **C1** and **C2**. Click the **OK** button.

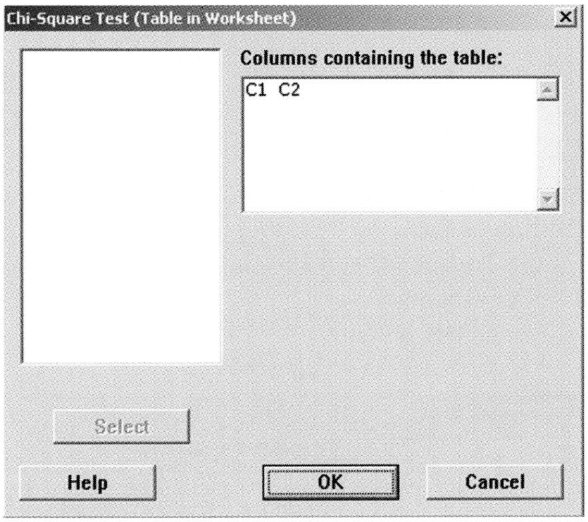

FIGURE A12.3 Minitab Chi-Square Test (Tables in Worksheet) Dialog Box

Using Minitab for the Wilcoxon Rank Sum Test

To illustrate the use of Minitab for the Wilcoxon rank sum test, open the **COLA.MTW** worksheet. If the data are stacked with the values in one column and the categories in a second column as they are in this worksheet, you must first unstack the data. Select **Data ➔ Unstack Columns**.

1. In the Unstack Columns dialog box (see Figure A12.4), enter **Sales** or **C1** in the Unstack the data in: edit box. Enter **Display** or **C2** in the Using subscripts in: edit box.
2. Select the **After last column in use** option button. Click the **OK** button. Deselect the **Name the columns containing the unstacked data** check box.

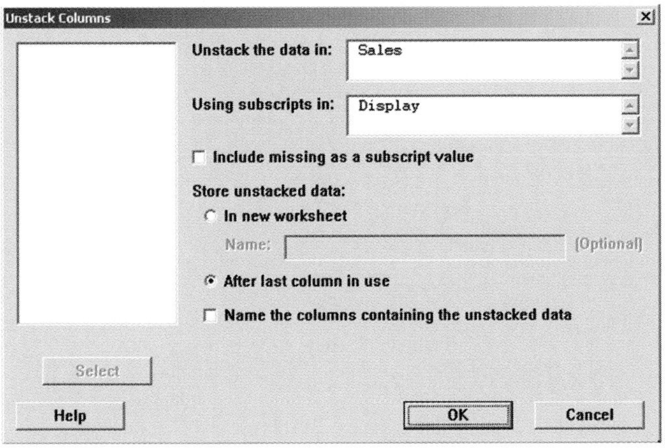

FIGURE A12.4 Minitab Unstack Columns Dialog Box

3. Enter the column labels **End Aisle** in C3 and **Normal** in C4.
4. Select **Stat ➔ Nonparametrics ➔ Mann-Whitney**. In the Mann-Whitney dialog box (see Figure A12.5) in the First Sample: edit box, enter **Normal** or **C4**. In the Second Sample: edit box, enter **EndAisle** or **C3**. Click the **OK** button.

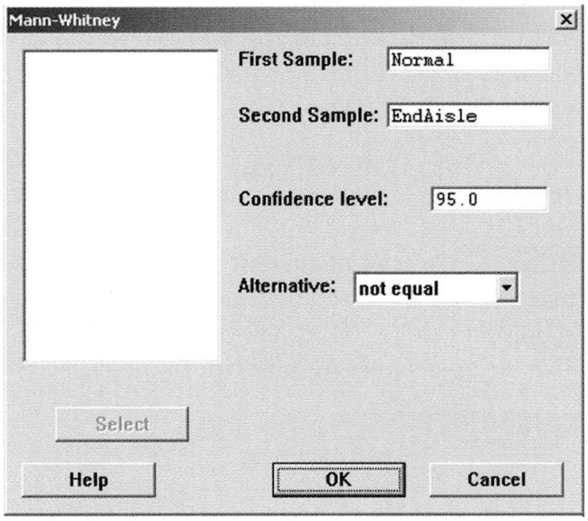

FIGURE A12.5

Using Minitab for the Wilcoxon Signed Ranks Test

To illustrate the use of Minitab for the Wilcoxon signed ranks test, open the **COMPTIME.MTW** worksheet.

1. To compute the differences between the current and new packages, select **Calc ➔ Calculator**. In the Store result in variable: edit box, enter **C3**. In the Expression: edit box, enter **C1–C2** or **Current–New**. Click the **OK** button. Enter the column label **Difference** for **C3**.
2. Select **Stat ➔ Nonparametrics ➔ 1-Sample Wilcoxon**. In the 1-Sample Wilcoxon dialog box (see Figure A12.6), in the Variables: edit box, enter **Difference** or **C3**. Select the **Test median**: option button, and enter **0** in the edit box. In the Alternative: drop-down list box, select **greater than** to perform the one-tail test. Click the **OK** button.

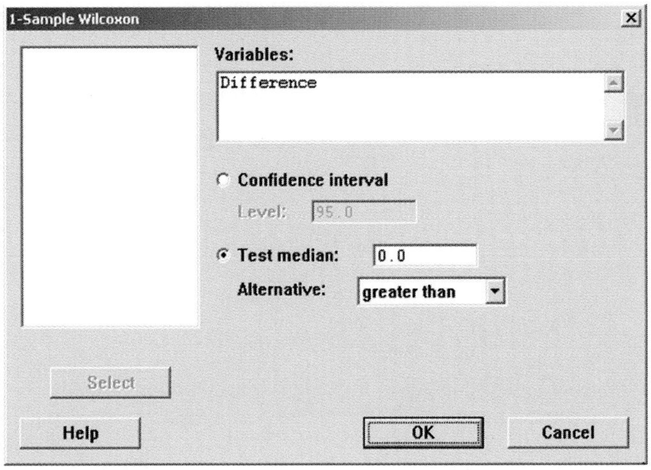

FIGURE A12.6 Minitab 1-Sample Wilcoxon Dialog Box

Using Minitab for the Kruskal-Wallis Test

To illustrate the use of Minitab for the Kruskal-Wallis test, open the **PARACHUTE.MTW** worksheet. Note that the data are stored in an unstacked format with each level in a separate column. In order to perform the Kruskal-Wallis test, you need to stack the data. Select **Data → Stack → Columns**.

1. In the Stack Columns dialog box (see Figure A12.7), enter **C1–C4** in the Stack the following columns: edit box.
2. Select the **Column of current worksheet**: option button. Enter **C5** in the edit box. Enter **C6** in the Store subscripts in: edit box. Click the **OK** button. Deselect the **Use variable names in subscript column** check box.

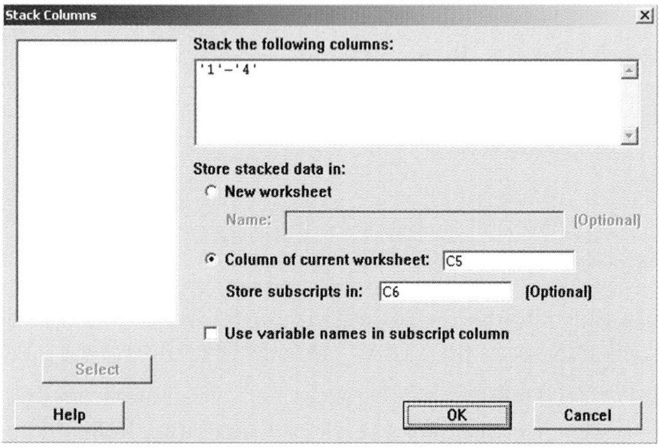

FIGURE A12.7 Minitab Stack Columns Dialog Box

3. Enter the column labels **Strength** in **C5** and **Supplier** in **C6**.
4. Select **Stat → Nonparametrics → Kruskal-Wallis**. In the Kruskal-Wallis dialog box (see Figure A12.8), enter **Strength** or **C5** in the Response: edit box and **Supplier** or **C6** in the Factor: edit box. Click the **OK** button.

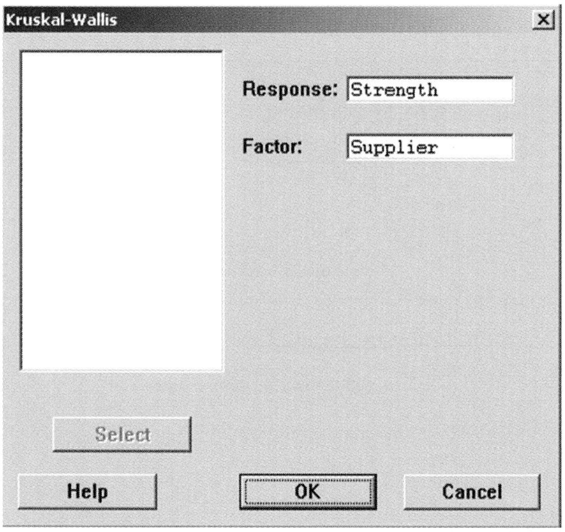

FIGURE A12.8 Minitab Kruskal-Wallis Dialog Box

Using Minitab for the Friedman Test

To illustrate the use of Minitab for the Friedman test, open the **FFCHAIN.MTW** worksheet. Select **Stat → Nonparametrics → Friedman**.

1. In the Friedman dialog box (see Figure A12.9), enter **Rating** or **C3** in the Response: edit box, **Restrat** or **C2** in the Treatment: edit box, and **Raters** or **C1** in the Blocks: edit box.
2. Click the **OK** button.

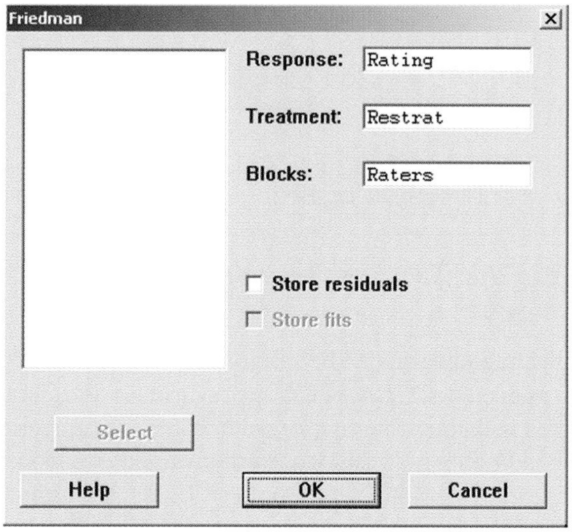

FIGURE A12.9 Minitab Friedman Dialog Box

CHAPTER 13

Simple Linear Regression

LEARNING OBJECTIVES

In this chapter, you learn:

- How to use regression analysis to predict the value of a dependent variable based on an independent variable
- The meaning of the regression coefficients b_0 and b_1
- How to evaluate the assumptions of regression analysis and know what to do if the assumptions are violated
- To make inferences about the slope and correlation coefficient
- To estimate mean values and predict individual values

USING STATISTICS

Forecasting Sales for a Clothing Store

The sales for Sunflowers, a chain of apparel stores for women, have increased during the past 12 years as the chain expanded the number of stores open. Until now, Sunflowers senior managers selected sites based on subjective factors such as the availability of a good lease or the perception that a location seemed ideal for an apparel store. As the new director of planning, you need to develop a systematic approach to selecting new sites that will allow Sunflowers to make better-informed decisions for opening additional stores. This plan must be able to forecast annual sales for all potential stores under consideration. You believe that the size of the store significantly contributes to the success of a store and you want to use this relationship in the decision-making process. How can you use statistics so that you can forecast the annual sales of a proposed store based on the size of that store?

In this and the following two chapters, you will learn how **regression analysis** allows you to develop a model to predict the values of a numerical variable based on the values of one or more other variables. For example, in the "Using Statistics" scenario above, you may wish to predict sales for a Sunflowers store based on the size of the store. Other examples include predicting your college GPA based on your S.A.T. score, and predicting a professor's salary based on his or her years of experience.

In regression analysis, the variable you wish to predict is called the **dependent variable**. The variables used to make the prediction are called **independent variables**. In addition to predicting values of the dependent variable, regression analysis also allows you to identify the type of mathematical relationship that exists between a dependent and independent variable, to quantify the effect that changes in the independent variable have on the dependent variable, and to identify unusual observations. This chapter discusses **simple linear regression** in which a *single* numerical independent variable X is used to predict the numerical dependent variable Y, such as using the size of a store to predict the annual sales of the store. Chapters 14 and 15 discuss *multiple regression models* that use several independent variables to predict a numerical dependent variable Y.

13.1 TYPES OF REGRESSION MODELS

In section 2.5, you used the **scatter diagram** to plot the relationship between an X variable on the horizontal axis and a Y variable on the vertical axis. The nature of the relationship between two variables can take many forms, ranging from simple to extremely complicated mathematical functions. The simplest relationship consists of a straight-line or **linear relationship**. An example of this relationship is shown in Figure 13.1.

FIGURE 13.1
A Positive Straight-Line Relationship

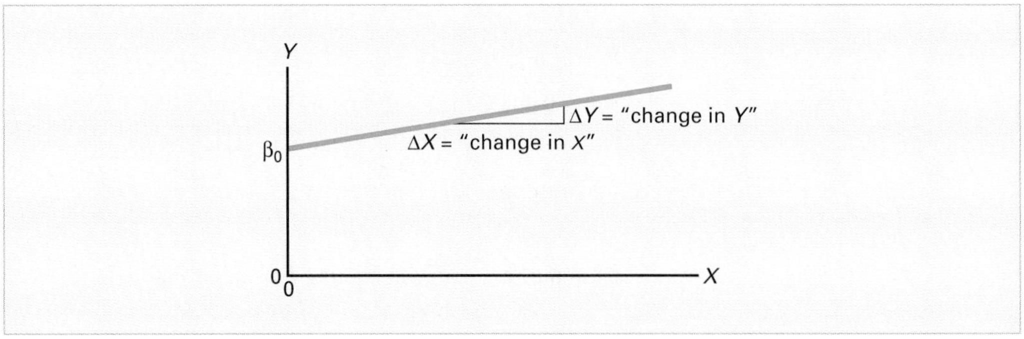

Equation (13.1) represents the straight-line (linear) model.

SIMPLE LINEAR REGRESSION MODEL

$$Y_i = \beta_0 + \beta_1 X_i + \varepsilon_i \quad \textbf{(13.1)}$$

where β_0 = Y intercept for the population

β_1 = slope for the population

ε_i = random error in Y for observation i

Y_i = dependent variable (sometimes referred to as the **response variable**)

X_i = independent variable (sometimes referred to as the **explanatory variable**)

$Y_i = \beta_0 + \beta_1 X_i$ is the equation of a straight line. The **slope** of the line β_1 represents the expected change in Y per unit change in X. It represents the mean amount that Y changes (either positively or negatively) for a one-unit change in X. The ***Y* intercept** β_0 represents the mean value of Y when X equals 0. The last component of the model, ε_i, represents the random error in Y for each observation i that occurs. In other words, ε_i is the vertical distance Y_i is above or below the line.

The selection of the proper mathematical model depends on the distribution of the X and Y values on the scatter diagram. In panel A of Figure 13.2, the values of Y are generally increasing linearly as X increases. This panel is similar to Figure 13.3 on page 517, which illustrates the positive relationship between the square footage (i.e., store size available) and the annual sales at branches of the Sunflowers women's clothing store chain.

FIGURE 13.2

Examples of Types of Relationships Found in Scatter Diagrams

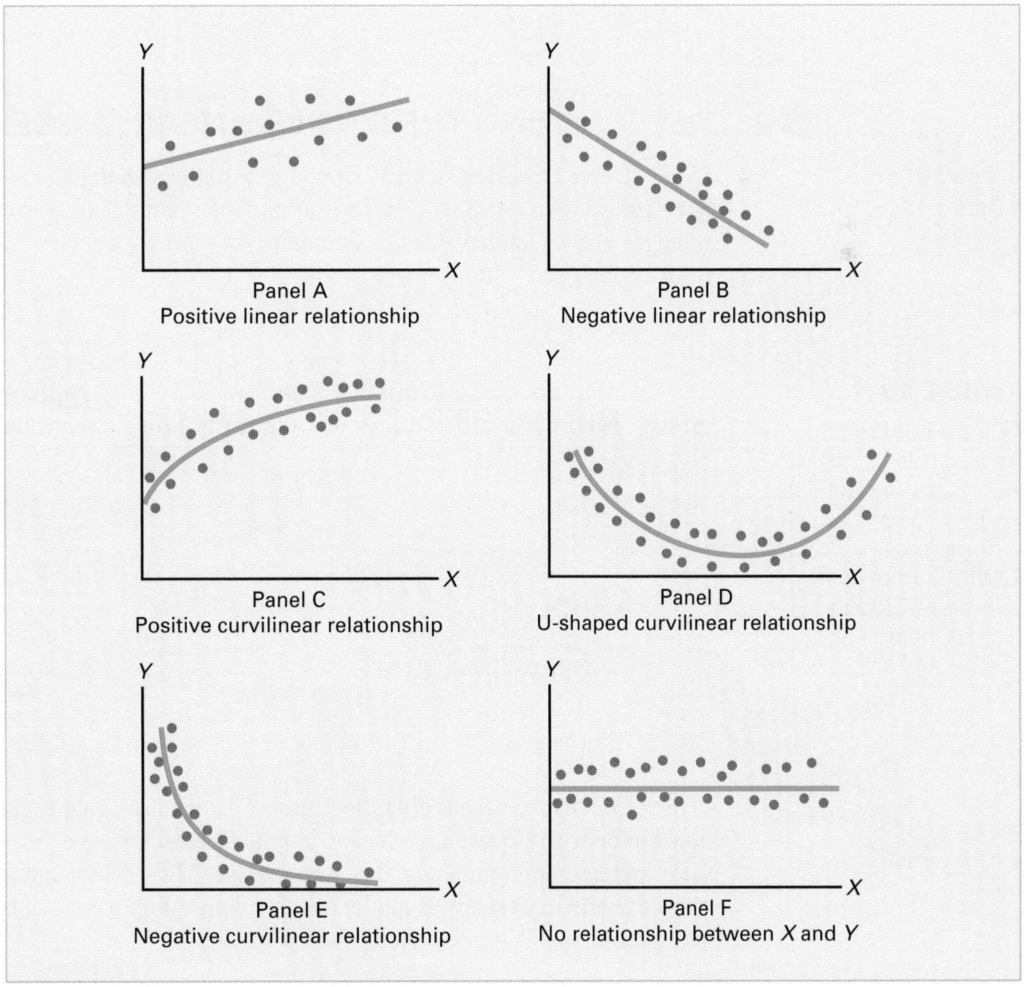

Panel *B* is an example of a negative linear relationship. As *X* increases, the values of *Y* are generally decreasing. An example of this type of relationship might be the price of a particular product and the amount of sales.

The data in panel *C* show a positive curvilinear relationship between *X* and *Y*. The values of *Y* increase as *X* increases, but this increase tapers off beyond certain values of *X*. An example of this positive curvilinear relationship might be the age and maintenance cost of a machine. As a machine gets older, the maintenance cost may rise rapidly at first but then level off beyond a certain number of years.

Panel *D* shows a U-shaped relationship between *X* and *Y*. As *X* increases, at first *Y* generally decreases; but as *X* continues to increase, *Y* not only stops decreasing but actually increases above its minimum value. An example of this type of relationship might be the number of errors per hour at a task and the number of hours worked. The number of errors per hour decreases as the individual becomes more proficient at the task but then it increases beyond a certain point because of factors such as fatigue and boredom.

Panel *E* indicates an exponential relationship between *X* and *Y*. In this case, *Y* decreases very rapidly as *X* first increases, but then it decreases much less rapidly as *X* increases further. An example of this exponential relationship could be the resale value of an automobile and its age. In the first year the resale value drops drastically from its original price; however, the resale value then decreases much less rapidly in subsequent years.

Finally, Panel *F* shows a set of data in which there is very little or no relationship between *X* and *Y*. High and low values of *Y* appear at each value of *X*.

In this section, a variety of different models that represent the relationship between two variables were briefly examined. Although scatter diagrams are useful in visually showing the mathematical form of a relationship, more sophisticated statistical procedures are available to determine the most appropriate model for a set of variables. The rest of this chapter discusses the model used when there is a *linear* relationship between variables.

13.2 DETERMINING THE SIMPLE LINEAR REGRESSION EQUATION

In the "Using Statistics" scenario on page 514, the stated goal is to forecast annual sales for all new stores based on store size. To examine the relationship between the store size (i.e., square footage) and its annual sales, a sample of 14 stores was selected. Table 13.1 summarizes the results for these 14 stores SITE.

TABLE 13.1

Square Footage (in Thousands of Square Feet) and Annual Sales (in Millions of Dollars) for a Sample of 14 Branches of the Sunflowers Women's Clothing Store Chain

Store	Square Feet (000)	Annual Sales (in Millions of Dollars)	Store	Square Feet (000)	Annual Sales (in Millions of Dollars)
1	1.7	3.7	8	1.1	2.7
2	1.6	3.9	9	3.2	5.5
3	2.8	6.7	10	1.5	2.9
4	5.6	9.5	11	5.2	10.7
5	1.3	3.4	12	4.6	7.6
6	2.2	5.6	13	5.8	11.8
7	1.3	3.7	14	3.0	4.1

Figure 13.3 displays the scatter diagram for the data in Table 13.1. Observe the increasing relationship between square feet (*X*) and annual sales (*Y*). As the size of the store increases, annual sales increase approximately as a straight line. Thus, you can assume that a straight line provides a useful mathematical model of this relationship. Now you need to determine the specific straight line that is the *best* fit to these data.

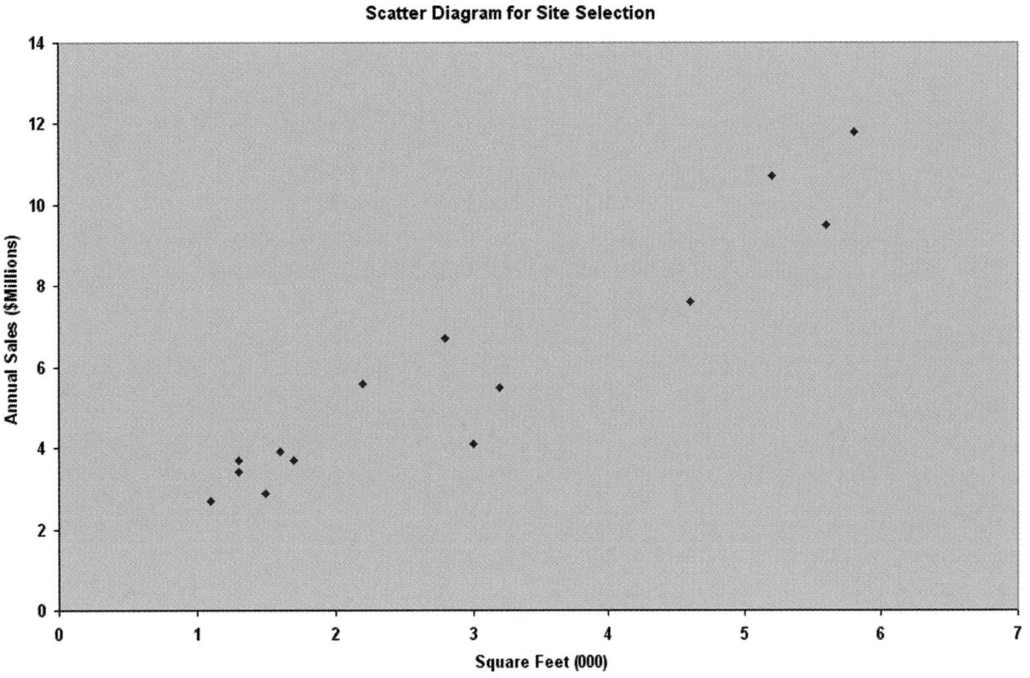

FIGURE 13.3
Microsoft Excel Scatter Diagram for the Site Selection Data

The Least-Squares Method

In the preceding section, a statistical model is hypothesized to represent the relationship between two variables, square footage and sales, in the entire population of Sunflowers women's clothing stores. However, as shown in Table 13.1, the data are from only a random sample of stores. If certain assumptions are valid (see section 13.4), you can use the sample Y intercept b_0 and the sample slope b_1 as estimates of the respective population parameters β_0 and β_1. Equation (13.2) uses these estimates to form the **simple linear regression equation**. This straight line is often referred to as the **prediction line**.

SIMPLE LINEAR REGRESSION EQUATION: THE PREDICTION LINE

The predicted value of Y equals the Y intercept plus the slope times the value of X.

$$\hat{Y}_i = b_0 + b_1 X_i \qquad \textbf{(13.2)}$$

where

$\hat{Y}_i$ = predicted value of Y for observation i

X_i = value of X for observation i

b_0 = sample Y intercept

b_1 = sample slope

Equation (13.2) requires the determination of two **regression coefficients**—b_0 (the sample Y intercept) and b_1 (the sample slope). The most common approach to find b_0 and b_1 is the method of least squares. This method minimizes the sum of the squared differences between the actual values (Y_i) and the predicted values ($\hat{Y}_i$) using the simple linear regression equation [i.e., the prediction line; see Equation (13.2)]. This sum of squared differences is equal to:

$$\sum_{i=1}^{n}(Y_i - \hat{Y}_i)^2$$

Since $\hat{Y}_i = b_0 + b_1 X_i$

$$\sum_{i=1}^{n}(Y_i - \hat{Y}_i)^2 = \sum_{i=1}^{n}[Y_i - (b_0 + b_1 X_i)]^2$$

Since this equation has two unknowns, b_0 and b_1, the sum of squared differences is a function of the sample Y-intercept b_0 and the sample slope b_1. The **least-squares method** determines what values of b_0 and b_1 minimize the sum of squared differences. Any values for b_0 and b_1 other than those determined by the least-squares method result in a greater sum of squared differences between the actual value of Y and the predicted value of Y.

In this text, Microsoft Excel spreadsheet software and Minitab statistical software are used to perform the computations involved in the least-squares method. For the data of Table 13.1, Figure 13.4 represents Microsoft Excel output and Figure 13.5 illustrates Minitab output. However, to understand how the results are computed, many of the computations involved are illustrated in Examples 13.3 and 13.4 on pages 522–523 and 528–530.

FIGURE 13.4

Microsoft Excel Output for the Site Selection Problem

	A	B	C	D	E	F	G
1	**Site Selection Analysis**						
2							
3	***Regression Statistics***						
4	**Multiple R**	**0.95088**					
5	**R Square**	**0.90418**					
6	**Adjusted R Square**	**0.89619**					
7	**Standard Error**	**0.96638** — S_{YX}					
8	**Observations**	**14** — n					
9						*p*-value	
10	**ANOVA**			*SSR*			
11		***df***	***SS***	***MS***	***F***	***Significance F***	
12	**Regression**	*SSE* **1**	**105.74761**	**105.74761**	**113.23351**	**1.82269E-07**	
13	**Residual**	**12**	**11.20668**	**0.93389**			
14	**Total**	**13**	**116.95429**		*SST*		
15							
16		***Coefficients***	***Standard Error***	***t Stat***	***P-value***	***Lower 95%***	***Upper 95%***
17	**Intercept** b_0	**0.96447**	**0.52619**	**1.83293**	**0.09173**	**-0.18200**	**2.11095**
18	**Square Feet** b_1	**1.66986**	**0.15693**	**10.64112**	**0.00000**	**1.32795**	**2.01177**

FIGURE 13.5

Minitab Output for the Site Selection Problem

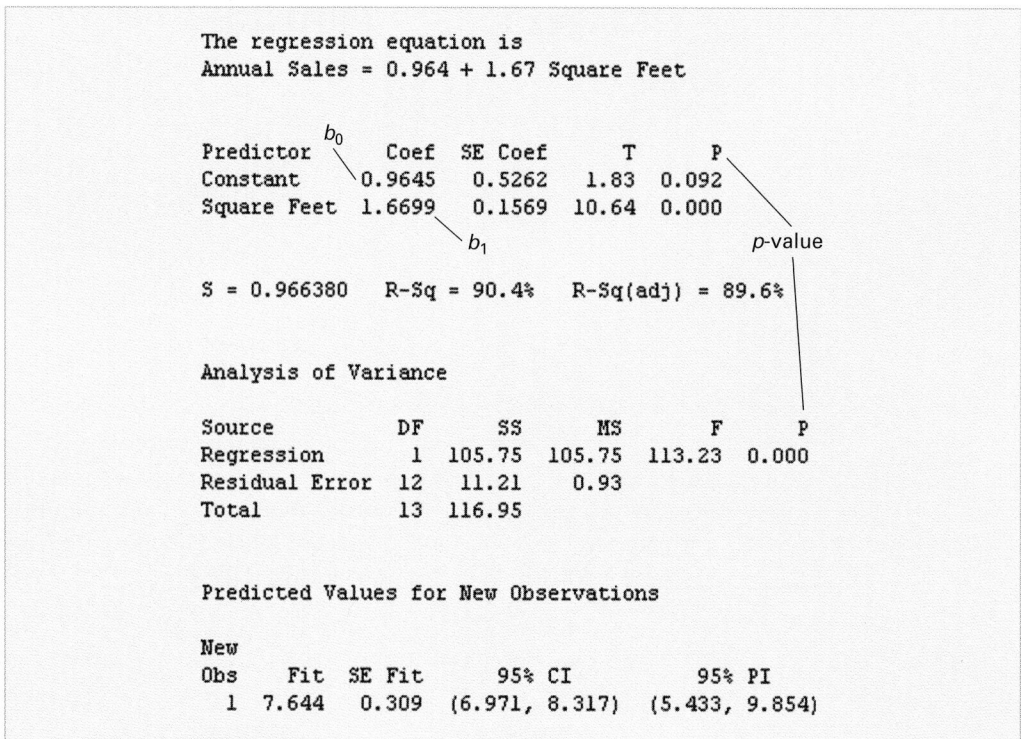

```
The regression equation is
Annual Sales = 0.964 + 1.67 Square Feet

                  b0
Predictor        Coef   SE Coef       T      P
Constant       0.9645    0.5262    1.83  0.092
Square Feet    1.6699    0.1569   10.64  0.000
                     b1                          p-value

S = 0.966380   R-Sq = 90.4%   R-Sq(adj) = 89.6%


Analysis of Variance

Source           DF      SS      MS       F      P
Regression        1  105.75  105.75  113.23  0.000
Residual Error   12   11.21    0.93
Total            13  116.95


Predicted Values for New Observations

New
Obs    Fit  SE Fit       95% CI          95% PI
  1  7.644   0.309  (6.971, 8.317)  (5.433, 9.854)
```

In Figure 13.4 or 13.5, observe that $b_0 = 0.964$ and $b_1 = 1.670$. Thus, the prediction line [see Equation (13.2) on page 517] for these data is

$$\hat{Y}_i = 0.964 + 1.670X_i$$

The slope b_1 is +1.670. This means that for each increase of 1 unit in X, the mean value of Y is estimated to increase by 1.670 units. In other words, for each increase of 1.0 thousand square feet in the size of the store, the mean annual sales are estimated to increase by 1.670 millions of dollars. Thus, the slope represents the portion of the annual sales that are estimated to vary according to the size of the store.

The Y intercept b_0 is +0.964. The Y intercept represents the mean value of Y when X equals 0. Because the square footage of the store cannot be 0, this Y intercept has no practical interpretation. Also, the Y intercept for this example is outside the range of the observed values of the X variable, and therefore interpretations of the value of b_0 should be made cautiously. Figure 13.6 displays the actual observations and the prediction line. To illustrate a situation where there is a direct interpretation for the Y intercept b_0, see Example 13.1.

FIGURE 13.6

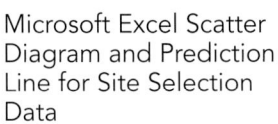

Microsoft Excel Scatter Diagram and Prediction Line for Site Selection Data

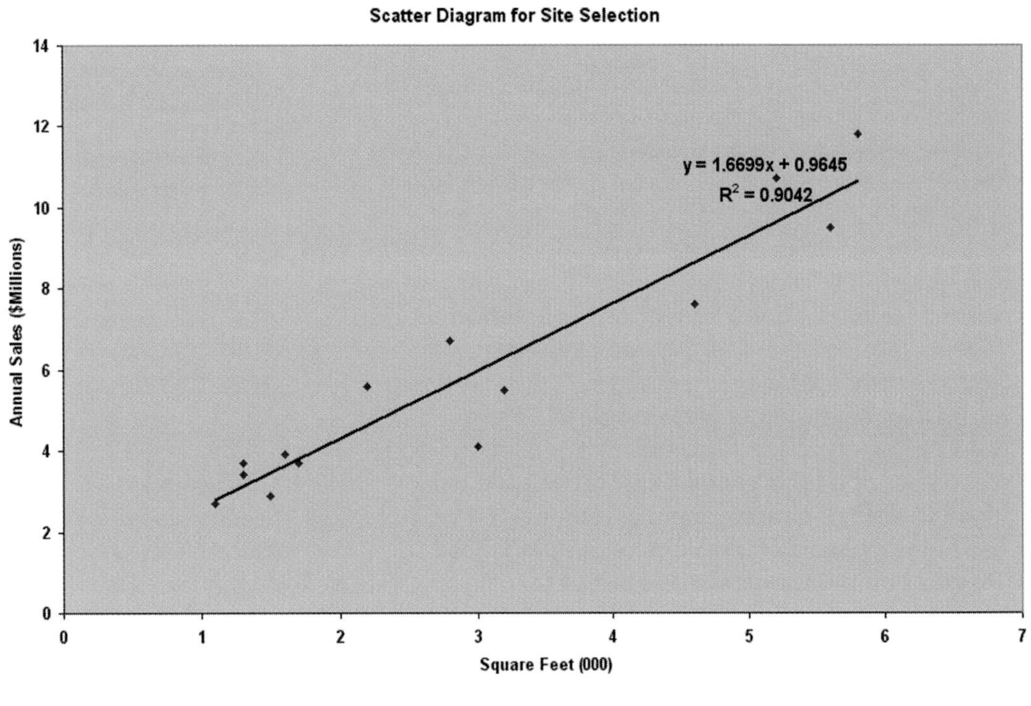

EXAMPLE 13.1

INTERPRETING THE Y INTERCEPT b_0 AND THE SLOPE b_1

A statistics professor wants to use the number of hours a student studies for a statistics final exam (X) to predict the final exam score (Y). A regression model was fit based upon data collected for a class during the previous semester with the following results:

$$\hat{Y}_i = 35.0 + 3X_i$$

What is the interpretation of the Y intercept b_0 and the slope b_1?

SOLUTION The Y intercept $b_0 = 35.0$ indicates that when the student does not study for the final exam, the mean final exam score is 35.0. The slope $b_1 = 3$ indicates that for each increase of one hour in studying time, the mean change in the final exam score is predicted to be +3.0. In other words, the final exam score is predicted to increase by 3 points for each one hour increase in studying time.

VISUAL EXPLORATIONS: Exploring Simple Linear Regression Coefficients

Use the Visual Explorations Simple Linear Regression procedure to produce a prediction line that is as close as possible to the prediction line defined by the least-squares solution. Open the **Visual Explorations.xla** macro workbook and:

Select **VisualExplorations → Simple Linear Regression** from the Microsoft Excel menu bar.

When a scatter diagram of the site selection data of Table 13.1 on page 516 with an initial prediction line appears (shown below), click the spinner buttons to change the values for b_1, the slope of the prediction line, and b_0, the Y intercept of the prediction line.

Try to produce a prediction line that is as close as possible to the prediction line defined by the least-squares estimates, using the chart display and the Difference from Target SSE value as feedback (see page 527 for an explanation of SSE). Click **Finish** when you are done with this exploration.

At anytime, click **Reset** to reset the b_1 and b_0 values, **Help** for more information, or **Solution** to reveal the prediction line defined by the least-squares estimates.

Using Your own regression data. To use Visual Explorations to find a prediction line for your own data, open the **Visual Explorations.xla** workbook (if it is not already open) and then:

Select **VisualExplorations → Simple Linear Regression with your worksheet data**.

In the Simple Linear Regression (your data) dialog box (shown below).

Enter the Y variable cell range in the **Y Variable Cell Range** edit box.

Enter the X variable cell range in the **X Variable Cell Range** edit box.

Select the **First cells in both ranges contain a label** check box, if appropriate.

Enter a title in the **Title** edit box.

Click the **OK** button.

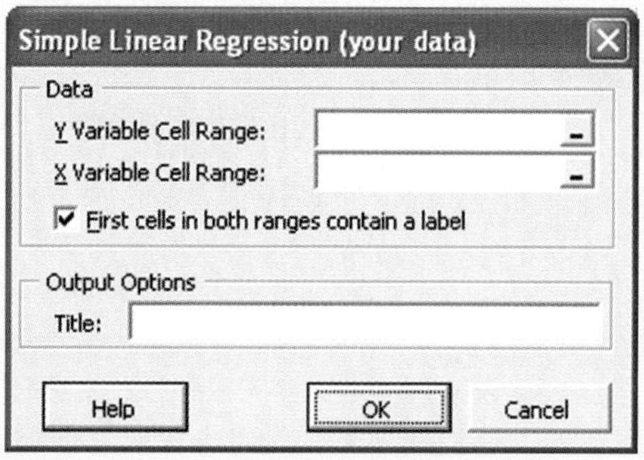

When the scatter diagram with an initial prediction line appears, use the instructions in the first part of this section to try to produce the prediction line defined by the least-squares estimate.

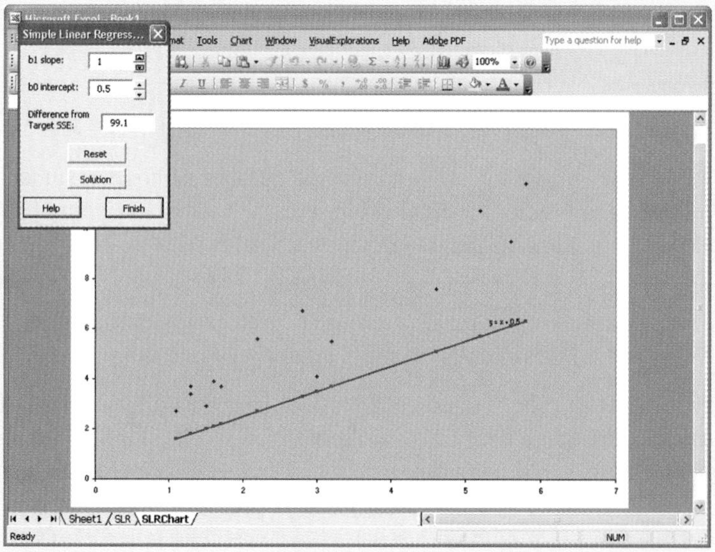

Return to the "Using Statistics" scenario concerning the Sunflowers clothing stores. Example 13.2 illustrates how you use the prediction equation to predict the mean annual sales.

EXAMPLE 13.2

PREDICTING MEAN ANNUAL SALES BASED ON SQUARE FOOTAGE

Use the prediction line to predict the mean annual sales for a store with 4,000 square feet.

SOLUTION You can determine the predicted value by substituting $X = 4$ (thousands of square feet) into the simple linear regression equation,

$$\hat{Y}_i = 0.964 + 1.670X_i$$

$$\hat{Y}_i = 0.964 + 1.670(4) = 7.644 \text{ or } \$7{,}644{,}000$$

Thus, the predicted mean annual sales of a store with 4,000 square feet is $7,644,000.

Predictions in Regression Analysis: Interpolation versus Extrapolation

When using a regression model for prediction purposes, you need to consider only the **relevant range** of the independent variable in making predictions. This relevant range includes all values from the smallest to the largest X used in developing the regression model. Hence, when predicting Y for a given value of X, you can interpolate within this relevant range of the X values, but you should not extrapolate beyond the range of X values. When you use the square footage to predict annual sales, the square footage (in thousands of square feet) varies from 1.1 to 5.8 (see Table 13.1 on page 516). Therefore, you should predict annual sales *only* for stores whose size is between 1.1 and 5.8 thousands of square feet. Any prediction of annual sales for stores outside this range assumes that the observed relationship between sales and store size for store sizes from 1.1 to 5.8 thousand square feet is the same as for stores outside this range. For example, you cannot extrapolate the linear relationship beyond 5,800 square feet in Example 13.2. It would be improper to use the prediction line to forecast the sales for a new store containing 8,000 square feet. It is quite possible that store size has a point of diminishing returns. If that were true, as square footage increases beyond 5,800 square feet, the effect on sales might become smaller and smaller.

Computing the Y Intercept b_0 and the Slope b_1

For small data sets it is possible to perform the least squres method using a hand calculator. Equations (13.3) and (13.4) give the values of b_0 and b_1, which minimize

$$\sum_{i=1}^{n}(Y_i - \hat{Y}_i)^2 = \sum_{i=1}^{n}[Y_i - (b_0 + b_1X_i)]^2$$

COMPUTATIONAL FORMULA FOR THE SLOPE b_1

$$b_1 = \frac{SSXY}{SSX} \qquad \textbf{(13.3)}$$

where

$$SSXY = \sum_{i=1}^{n}(X_i - \bar{X})(Y_i - \bar{Y}) = \sum_{i=1}^{n}X_iY_i - \frac{\left(\sum_{i=1}^{n}X_i\right)\left(\sum_{i=1}^{n}Y_i\right)}{n}$$

$$SSX = \sum_{i=1}^{n}(X_i - \bar{X})^2 = \sum_{i=1}^{n}X_i^2 - \frac{\left(\sum_{i=1}^{n}X_i\right)^2}{n}$$

COMPUTATIONAL FORMULA FOR THE Y INTERCEPT b_0

$$b_0 = \bar{Y} - b_1\bar{X} \quad (13.4)$$

where

$$\bar{Y} = \frac{\sum_{i=1}^{n} Y_i}{n} \text{ and } \bar{X} = \frac{\sum_{i=1}^{n} X_i}{n}$$

EXAMPLE 13.3

COMPUTING THE Y INTERCEPT b_0 AND THE SLOPE b_1

Compute the Y intercept b_0 and the slope b_1 for the site selection problem.

SOLUTION Examining Equations (13.3) and (13.4), you see that five quantities must be calculated to determine b_1 and b_0. These are n, the sample size; $\sum_{i=1}^{n} X_i$, the sum of the X values; $\sum_{i=1}^{n} Y_i$, the sum of the Y values; $\sum_{i=1}^{n} X_i^2$, the sum of the squared X values; and $\sum_{i=1}^{n} X_iY_i$, the sum of the product of X and Y. For the site selection data, the number of square feet is used to predict the annual sales in a store. Table 13.2 presents the computations of the various sums needed (including $\sum_{i=1}^{n} Y_i^2$, the sum of the squared Y values that will be used to compute SST in section 13.3).

TABLE 13.2
Computations for the Site Selection Problem

Store	Square Feet(X)	Annual Sales(Y)	X^2	Y^2	XY
1	1.7	3.7	2.89	13.69	6.29
2	1.6	3.9	2.56	15.21	6.24
3	2.8	6.7	7.84	44.89	18.76
4	5.6	9.5	31.36	90.25	53.20
5	1.3	3.4	1.69	11.56	4.42
6	2.2	5.6	4.84	31.36	12.32
7	1.3	3.7	1.69	13.69	4.81
8	1.1	2.7	1.21	7.29	2.97
9	3.2	5.5	10.24	30.25	17.60
10	1.5	2.9	2.25	8.41	4.35
11	5.2	10.7	27.04	114.49	55.64
12	4.6	7.6	21.16	57.76	34.96
13	5.8	11.8	33.64	139.24	68.44
14	3.0	4.1	9.00	16.81	12.30
Totals	40.9	81.8	157.41	594.90	302.30

Using Equations (13.3) and (13.4), you can compute the values of b_0 and b_1:

$$b_1 = \frac{SSXY}{SSX}$$

$$SSXY = \sum_{i=1}^{n}(X_i - \bar{X})(Y_i - \bar{Y}) = \sum_{i=1}^{n} X_iY_i - \frac{\left(\sum_{i=1}^{n} X_i\right)\left(\sum_{i=1}^{n} Y_i\right)}{n}$$

$$SSXY = 302.3 - \frac{(40.9)(81.8)}{14}$$
$$= 302.3 - 238.97285$$
$$= 63.32715$$

$$SSX = \sum_{i=1}^{n}(X_i - \bar{X})^2 = \sum_{i=1}^{n} X_i^2 - \frac{\left(\sum_{i=1}^{n} X_i\right)^2}{n}$$
$$= 157.41 - \frac{(40.9)^2}{14}$$
$$= 157.41 - 119.48642$$
$$= 37.92358$$

so that

$$b_1 = \frac{63.32715}{37.92358}$$
$$= 1.66986$$

and

$$b_0 = \bar{Y} - b_1\bar{X}$$
$$\bar{Y} = \frac{\sum_{i=1}^{n} Y_i}{n} = \frac{81.8}{14} = 5.842857$$
$$\bar{X} = \frac{\sum_{i=1}^{n} X_i}{n} = \frac{40.9}{14} = 2.92143$$
$$b_0 = 5.842857 - (1.66986)(2.92143)$$
$$= 0.964478$$

PROBLEMS FOR SECTION 13.2

Learning the Basics

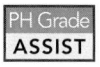

PH Grade ASSIST **13.1** Fitting a straight line to a set of data yields the following prediction line

$$\hat{Y}_i = 2 + 5X_i$$

a. Interpret the meaning of the Y intercept b_0.
b. Interpret the meaning of the slope b_1.
c. Predict the mean value of Y for $X = 3$.

PH Grade ASSIST **13.2** If the values of X in problem 13.1 range from 2 to 25, should you use this model to predict the mean value of Y when X equals

a. 3?
b. −3?
c. 0?
d. 24?

13.3 Fitting a straight line to a set of data yields the following prediction line

$$\hat{Y}_i = 16 - 0.5X_i$$

a. Interpret the meaning of the Y intercept b_0.
b. Interpret the meaning of the slope b_1.
c. Predict the mean value of Y for $X = 6$.

Applying the Concepts

Problems 13.4–13.10 can be solved manually or by using Microsoft Excel, Minitab, or SPSS.

PH Grade ASSIST SELF Test **13.4** The marketing manager of a large supermarket chain would like to use shelf space to predict the sales of pet food. A random sample of 12 equal-sized stores **PETFOOD** is selected, with the following results:

Store	Shelf Space (X) (Feet)	Weekly Sales (Y) (Hundreds of Dollars)
1	5	1.6
2	5	2.2
3	5	1.4
4	10	1.9
5	10	2.4
6	10	2.6
7	15	2.3
8	15	2.7
9	15	2.8
10	20	2.6
11	20	2.9
12	20	3.1

a. Construct a scatter diagram.
For these data $b_0 = 1.45$ and $b_1 = 0.074$.
b. Interpret the meaning of the slope b_1 in this problem.
c. Predict the mean weekly sales (in hundreds of dollars) of pet food for stores with 8 feet of shelf space for pet food.

13.5 Circulation is the lifeblood of the publishing business. The larger the sales of a magazine, the more it can charge advertisers. Recently, a circulation gap has appeared between the publishers' reports of magazines newsstand sales and subsequent audits by the Audit Bureau of Circulations. The following data **CIRCULATION** represent the reported and audited newsstand sales (in thousands) in 2001 for the following ten magazines.

Magazine	Reported (X)	Audited (Y)
YM	621.0	299.6
CosmoGirl	359.7	207.7
Rosie	530.0	325.0
Playboy	492.1	336.3
Esquire	70.5	48.6
TeenPeople	567.0	400.3
More	125.5	91.2
Spin	50.6	39.1
Vogue	353.3	268.6
Elle	263.6	214.3

Source: Extracted from M. Rose, "In Fight for Ads, Publishers Often Overstate Their Sales," The Wall Street Journal, *August 6, 2003, A1, A10.*

a. Construct a scatter diagram.
For these data $b_0 = 26.724$ and $b_1 = 0.5719$.
b. Interpret the meaning of the slope b_1 in this problem.
c. Predict the mean audited newsstand sales for a magazine that reports newsstand sales of 400,000.

13.6 The owner of an intracity moving company typically has his most experienced manager predict the total number of labor hours that will be required to complete an upcoming move. This approach has proved useful in the past, but he would like to be able to develop a more accurate method of predicting the labor hours by using the amount of cubic feet moved. In a preliminary effort to provide a more accurate method, he has collected data for 36 moves in which the origin and destination were within the borough of Manhattan in New York City, and the travel time was an insignificant portion of the hours worked. **MOVING**

a. Construct a scatter diagram.
b. Assuming a linear relationship, use the least-squares method to find the regression coefficients b_0 and b_1.
c. Interpret the meaning of the slope b_1 in this problem.
d. Predict the mean labor hours for moving 500 cubic feet.

PH Grade ASSIST **13.7** A large mail-order house believes that there is a linear relationship between the weight of the mail it receives and the number of orders to be filled. It would like to investigate the relationship in order to predict the number of orders based on the weight of the mail. From an operational perspective, knowledge of the number of orders will help in the planning of the order-fulfillment process. A sample of 25 mail shipments is selected within a range of 200 to 700 pounds. The results are as follows. **MAIL**

Weight of Mail (Pounds)	Orders (In Thousands)	Weight of Mail (Pounds)	Orders (In Thousands)
216	6.1	432	13.6
283	9.1	409	12.8
237	7.2	553	16.5
203	7.5	572	17.1
259	6.9	506	15.0
374	11.5	528	16.2
342	10.3	501	15.8
301	9.5	628	19.0
365	9.2	677	19.4
384	10.6	602	19.1
404	12.5	630	18.0
426	12.9	652	20.2
482	14.5		

a. Construct a scatter diagram.
b. Assuming a linear relationship, use the least-squares method to find the regression coefficients b_0 and b_1.
c. Interpret the meaning of the slope b_1 in this problem.

d. Predict the mean number of orders when the weight of the mail is 500 pounds.

13.8 The value of a sports franchise is directly related to the amount of revenue that a franchise can generate. The data in the file **BBREVENUE** represent the estimated value in 2004 (in millions of dollars) and the estimated annual revenue (in millions of dollars) for the 30 baseball franchises. Suppose you want to develop a simple linear regression model to predict franchise value based on annual revenue generated.

a. Construct a scatter diagram.
b. Use the least-squares method to find the regression coefficients b_0 and b_1.
c. Interpret the meaning of b_0 and b_1 in this problem.
d. Predict the mean value of a baseball franchise that generates $150 million of annual revenue.

13.9 An agent for a residential real estate company in a large city would like to be able to predict the monthly rental cost for apartments based on the size of the apartment as defined by square footage. A sample of 25 apartments **RENT** in a particular residential neighborhood was selected, and the information gathered revealed the following:

Apartment	Monthly Rent ($)	Size (Square Feet)	Apartment	Monthly Rent ($)	Size (Square Feet)
1	950	850	9	875	700
2	1,600	1,450	10	1,150	956
3	1,200	1,085	11	1,400	1,100
4	1,500	1,232	12	1,650	1,285
5	950	718	13	2,300	1,985
6	1,700	1,485	14	1,800	1,369
7	1,650	1,136	15	1,400	1,175
8	935	726	16	1,450	1,225

Apartment	Monthly Rent ($)	Size (Square Feet)	Apartment	Monthly Rent ($)	Size (Square Feet)
17	1,100	1,245	22	1,650	1,040
18	1,700	1,259	23	1,200	755
19	1,200	1,150	24	800	1,000
20	1,150	896	25	1,750	1,200
21	1,600	1,361			

a. Construct a scatter diagram.
b. Use the least-squares method to find the regression coefficients b_0 and b_1.
c. Interpret the meaning of b_0 and b_1 in this problem.
d. Predict the mean monthly rent for an apartment that has 1,000 square feet.
e. Why would it not be appropriate to use the model to predict the monthly rent for apartments that have 500 square feet?
f. Your friends Jim and Jennifer are considering signing a lease for an apartment in this residential neighborhood. They are trying to decide between two apartments, one with 1,000 square feet for a monthly rent of $1,275 and the other with 1,200 square feet for a monthly rent of $1,425. What would you recommend to them? Why?

13.10 The data in the file **HARDNESS** provide measurements on the hardness and tensile strength for 35 specimens of die-cast aluminum. It is believed that hardness (measured in Rockwell E units) can be used to predict tensile strength (measured in thousands of pounds per square inch).

a. Construct a scatter diagram.
b. Assuming a linear relationship, use the least-squares method to find the regression coefficients b_0 and b_1.
c. Interpret the meaning of the slope b_1 in this problem.
d. Predict the mean tensile strength for die-cast aluminum that has a hardness of 30 Rockwell E units.

13.3 MEASURES OF VARIATION

Computing the Sum of Squares

When using the least squares method to find the regression coefficients for a set of data, there are three measures of variation that you need to compute. The first measure, the **total sum of squares (*SST*)**, is a measure of variation of the Y_i values around their mean $\bar{Y}$. In a regression analysis, the **total variation** or total sum of squares is subdivided into **explained variation** or **regression sum of squares (*SSR*)**, that which is due to the relationship between X and Y, and **unexplained variation** or **error sum of squares (*SSE*)**, that which is due to factors other than the relationship between X and Y. Figure 13.7 shows these different measures of variation.

FIGURE 13.7
Measures of Variation

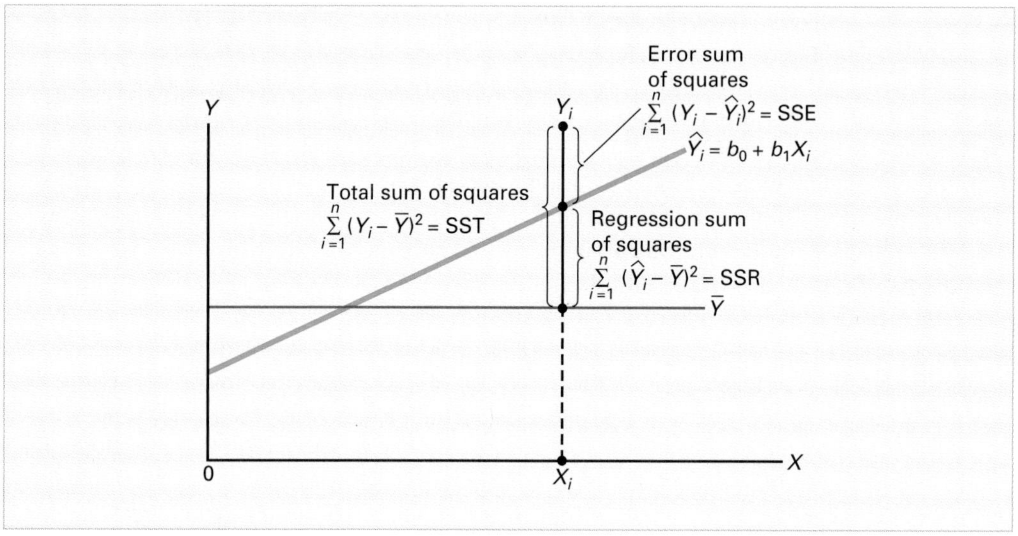

The regression sum of squares (*SSR*) is equal to the difference between $\hat{Y}_i$ (the value of Y that is predicted from the prediction line) and $\bar{Y}$ (the mean value of Y). The error sum of squares (*SSE*) represents the part of the variation in Y that is not explained by the regression. It is based on the difference between Y_i and $\hat{Y}_i$. Equations (13.5), (13.6), (13.7), and (13.8) define these measures of variation.

MEASURES OF VARIATION IN REGRESSION

Total sum of squares = regression sum of squares + error sum of squares.

$$SST = SSR + SSE \tag{13.5}$$

TOTAL SUM OF SQUARES (*SST*)

The total sum of squares (*SST*) is equal to the sum of the squared differences between each observed Y value and $\bar{Y}$, the mean value of Y.

$$SST = \text{total sum of squares} = \sum_{i=1}^{n}(Y_i - \bar{Y})^2 \tag{13.6}$$

REGRESSION SUM OF SQUARES (*SSR*)

The regression sum of squares (*SSR*) is equal to the sum of the squared differences between the predicted value of Y and $\bar{Y}$, the mean value of Y.

$$SSR = \text{explained variation or regression sum of squares} \tag{13.7}$$

$$= \sum_{i=1}^{n}(\hat{Y}_i - \bar{Y})^2$$

ERROR SUM OF SQUARES (*SSE*)

The error sum of squares (*SSE*) is equal to the sum of the squared differences between the observed value of *Y* and the predicted value of *Y*.

$$SSE = \text{unexplained variation or error sum of squares} \quad \textbf{(13.8)}$$

$$= \sum_{i=1}^{n}(Y_i - \hat{Y}_i)^2$$

Figure 13.8 represents the sum of squares portion of the Microsoft Excel output for the site selection problem, while Figure 13.9 illustrates Minitab output.

FIGURE 13.8
Microsoft Excel Sum of Squares for the Site Selection Problem

10	**ANOVA**					
11		*df*	*SS*	*MS*	*F*	*Significance F*
12	**Regression**	1	105.74761	105.74761	113.23351	1.82269E-07
13	**Residual**	12	11.20668	0.93389		
14	**Total**	13	116.95429			

FIGURE 13.9
Minitab Sum of Squares for the Site Selection Problem

```
Analysis of Variance

Source           DF      SS      MS       F      P
Regression        1  105.75  105.75  113.23  0.000
Residual Error   12   11.21    0.93
Total            13  116.95
```

From Figure 13.8 or 13.9, you see that

$$SSR = 105.7476,\ SSE = 11.2067,\ \text{and}\ SST = 116.9543$$

From Equation (13.5) on page 526,

$$SST = SSR + SSE$$

$$116.9543 = 105.7476 + 11.2067$$

The *SST* is equal to 116.9543. This amount is subdivided into the sum of squares that is explained by the regression (*SSR*), equal to 105.7476, and the sum of squares that is unexplained by the regression (*SSE*), equal to 11.2067.

In data sets with a large number of significant digits, the output may be displayed using a numerical format known as *scientific notation*. This type of format is used to display very small or very large values. The number after the letter E represents the number of digits that the decimal point needs to be moved to the left (for a negative number) or to the right (for a positive number). For example, the number 3.7431E+02 means that the decimal point should be moved two places to the right, producing the number 374.31. The number 3.7431E–02 means that the decimal point should be moved two places to the left, producing the number 0.037431. When scientific notation is used, fewer significant digits are usually displayed, and the numbers may appear to be rounded.

The Coefficient of Determination

By themselves, *SSR, SSE*, and *SST* provide little information. However, the ratio of the regression sum of squares (*SSR*) to the total sum of squares (*SST*) measures the proportion of variation in *Y* that is explained by the independent variable *X* in the regression model. This ratio is called the **coefficient of determination** r^2 and is defined in Equation (13.9).

COEFFICIENT OF DETERMINATION

The coefficient of determination is equal to the regression sum of squares (i.e., explained variation) divided by the total sum of squares (i.e., total variation).

$$r^2 = \frac{\text{regression sum of squares}}{\text{total sum of squares}} = \frac{SSR}{SST} \tag{13.9}$$

The coefficient of determination measures the proportion of variation in *Y* that is explained by the independent variable *X* in the regression model. For the site selection example, with $SSR = 105.7476$; $SSE = 11.2067$; and $SST = 116.9543$,

$$r^2 = \frac{105.7476}{116.9543} = 0.904$$

Therefore, 90.4% of the variation in annual sales is explained by the variability in the size of the store as measured by the square footage. This large r^2 indicates a strong positive linear relationship between two variables because the use of a regression model has reduced the variability in predicting annual sales by 90.4%. Only 9.6% of the sample variability in annual sales is due to factors other than what is accounted for by the linear regression model that uses square footage.

Figure 13.10 represents the coefficient of determination portion of the Microsoft Excel output for the site selection problem and Figure 13.11 illustrates Minitab output.

FIGURE 13.10

Partial Microsoft Excel Regression Output for the Site Selection Problem

3	*Regression Statistics*	
4	Multiple R	0.95088
5	R Square	0.90418
6	Adjusted R Square	0.89619
7	Standard Error	S_{YX} — 0.96638
8	Observations	14.00000

FIGURE 13.11

Partial Minitab Regression Output for the Site Selection Problem

```
Predictor      Coef  SE Coef      T      P
Constant     0.9645   0.5262   1.83  0.092
Square Feet  1.6699   0.1569  10.64  0.000


S = 0.966380   R-Sq = 90.4%   R-Sq(adj) = 89.6%
```

EXAMPLE 13.4 COMPUTING THE COEFFICIENT OF DETERMINATION

Compute the coefficient of determination r^2 for the site selection problem.

SOLUTION You can compute *SST, SSR*, and *SSE* that were defined in Equations (13.6), (13.7), and (13.8) on pages 526 and 527 by using Equations (13.10), (13.11), and (13.12).

COMPUTATIONAL FORMULA FOR *SST*

$$SST = \sum_{i=1}^{n}(Y_i - \bar{Y})^2 = \sum_{i=1}^{n}Y_i^2 - \frac{\left(\sum_{i=1}^{n}Y_i\right)^2}{n} \tag{13.10}$$

COMPUTATIONAL FORMULA FOR *SSR*

$$SSR = \sum_{i=1}^{n}(\hat{Y}_i - \bar{Y})^2 \tag{13.11}$$

$$= b_0\sum_{i=1}^{n}Y_i + b_1\sum_{i=1}^{n}X_iY_i - \frac{\left(\sum_{i=1}^{n}Y_i\right)^2}{n}$$

COMPUTATIONAL FORMULA FOR *SSE*

$$SSE = \sum_{i=1}^{n}(Y_i - \hat{Y})^2 \tag{13.12}$$

$$= \sum_{i=1}^{n}Y_i^2 - b_0\sum_{i=1}^{n}Y_i - b_1\sum_{i=1}^{n}X_iY_i$$

Using the summary results from Table 13.2 on page 522,

$$SST = \sum_{i=1}^{n}(Y_i - \bar{Y})^2 = \sum_{i=1}^{n}Y_i^2 - \frac{\left(\sum_{i=1}^{n}Y_i\right)^2}{n}$$

$$= 594.9 - \frac{(81.8)^2}{14}$$

$$= 594.9 - 477.94571$$

$$= 116.95429$$

$$SSR = \sum_{i=1}^{n}(\hat{Y}_i - \bar{Y})^2$$

$$= b_0\sum_{i=1}^{n}Y_i + b_1\sum_{i=1}^{n}X_iY_i - \frac{\left(\sum_{i=1}^{n}Y_i\right)^2}{n}$$

$$= (0.964478)(81.8) + (1.66986)(302.3) - \frac{(81.8)^2}{14}$$

$$= 105.74726$$

$$SSE = \sum_{i=1}^{n}(Y_i - \hat{Y}_i)^2$$

$$= \sum_{i=1}^{n} Y_i^2 - b_0 \sum_{i=1}^{n} Y_i - b_1 \sum_{i=1}^{n} X_i Y_i$$

$$= 594.9 - (0.964478)(81.8) - (1.66986)(302.3)$$

$$= 11.207$$

Therefore,

$$r^2 = \frac{105.74726}{116.9543} = 0.904$$

Standard Error of the Estimate

Although the least-squares method results in the line that fits the data with the minimum amount of variation, unless all the observed data points fall on a straight line, the prediction line is not a perfect predictor. Just as all data values cannot be expected to be exactly equal to their mean, neither can they be expected to fall exactly on the prediction line. Therefore, a statistic that measures the variability of the actual Y values from the predicted Y values needs to be developed, in the same way that the standard deviation was developed in Chapter 3 as a measure of the variability of each value around the mean. This standard deviation around the prediction line is called the **standard error of the estimate**.

Figure 13.6 on page 519 illustrates the variability around the prediction line for the site selection data. Observe that, although many of the actual values of Y fall near the prediction line, no values are exactly on the line.

The standard error of the estimate, represented by the symbol S_{YX}, is defined in Equation (13.13).

STANDARD ERROR OF THE ESTIMATE

$$S_{YX} = \sqrt{\frac{SSE}{n-2}} = \sqrt{\frac{\sum_{i=1}^{n}(Y_i - \hat{Y}_i)^2}{n-2}} \qquad \textbf{(13.13)}$$

where Y_i = actual value of Y for a given X_i

$\hat{Y}_i$ = predicted value of Y for a given X_i

SSE = error sum of squares

From Equation (13.8), with $SSE = 11.2067$,

$$S_{YX} = \sqrt{\frac{11.2067}{14-2}} = 0.9664$$

This standard error of the estimate, equal to 0.9664 millions of dollars (i.e., $966,400), is labeled Standard Error on the Microsoft Excel output of Figure 13.10 and as S in the Minitab output of Figure 13.11 (on page 528). The standard error of the estimate represents a measure of the variation around the prediction line. It is measured in the same units as the dependent variable Y. The interpretation of the standard error of the estimate is similar to that of the standard deviation. Just as the standard deviation measures variability around the mean, the standard error of the estimate measures variability around the prediction line. As you will see in sections 13.7 and 13.8, the standard error of the estimate is used to determine whether a statistically significant relationship exists between the two variables and also to make inferences about future values of Y.

PROBLEMS FOR SECTION 13.3

Learning the Basics

PH Grade ASSIST **13.11** How do you interpret a coefficient of determination r^2 equal to 0.80?

PH Grade ASSIST **13.12** If $SSR = 36$ and $SSE = 4$, find SST, and then compute the coefficient of determination r^2 and interpret its meaning.

PH Grade ASSIST **13.13** If $SSR = 66$ and $SST = 88$, compute the coefficient of determination r^2 and interpret its meaning.

PH Grade ASSIST **13.14** If $SSE = 10$ and $SSR = 30$, compute the coefficient of determination r^2 and interpret its meaning.

13.15 If $SSR = 120$, why is it impossible for SST to equal 110?

Applying the Concepts

Problems 13.16–13.22 can be solved manually or by using Microsoft Excel, Minitab, or SPSS.

PH Grade ASSIST SELF Test **13.16** In problem 13.4 on page 523, the marketing manager used shelf space for pet food to predict weekly sales. PETFOOD For that data, $SSR = 2.0535$ and $SST = 3.0025$.

a. Determine the coefficient of determination r^2 and interpret its meaning.
b. Determine the standard error of the estimate.
c. How useful do you think this regression model is for predicting sales?

13.17 In problem 13.5 on page 524, you used reported magazine newsstand sales to predict audited sales. CIRCULATION For that data, $SSR = 130{,}301.41$ and $SST = 144{,}538.64$.

a. Determine the coefficient of determination r^2 and interpret its meaning.
b. Determine the standard error of the estimate.
c. How useful do you think this regression model is for predicting audited sales?

13.18 In problem 13.6 on page 524, an owner of a moving company wanted to predict labor hours based on the cubic feet moved. MOVING Using the results of that problem,

a. Determine the coefficient of determination r^2 and interpret its meaning.
b. Determine the standard error of the estimate.
c. How useful do you think this regression is for predicting labor hours?

PH Grade ASSIST **13.19** In problem 13.7 on page 524, you used the weight of mail to predict the number of orders received. MAIL Using the results of that problem,

a. Determine the coefficient of determination r^2 and interpret its meaning.
b. Find the standard error of the estimate.
c. How useful do you think this regression model is for predicting the number of orders?

13.20 In problem 13.8 on page 525, you used annual revenues to predict the value of a baseball franchise. BBREVENUE Using the results of that problem,

a. Determine the coefficient of determination r^2 and interpret its meaning.
b. Determine the standard error of the estimate.
c. How useful do you think this regression model is for predicting the value of a baseball franchise?

13.21 In problem 13.9 on page 525, an agent for a real estate company wanted to predict the monthly rent for apartments based on the size of the apartment. RENT Using the results of that problem,

a. Determine the coefficient of determination r^2 and interpret its meaning.
b. Determine the standard error of the estimate.
c. How useful do you think this regression model is for predicting the monthly rent?

13.22 In problem 13.10 on page 525, you used hardness to predict the tensile strength of die-cast aluminum. HARDNESS Using the results of that problem,

a. Determine the coefficient of determination r^2 and interpret its meaning.
b. Find the standard error of the estimate.
c. How useful do you think this regression model is for predicting the tensile strength of die-cast aluminum?

13.4 ASSUMPTIONS

The discussion of hypothesis testing and the analysis of variance emphasized the importance of the assumptions to the validity of any conclusions reached. The assumptions necessary for regression are similar to those of the analysis of variance because both topics fall under the general heading of *linear models* (reference 5).

The four **assumptions of regression** (known by the acronym LINE) are as follows.

- **L**inearity
- **I**ndependence of errors
- **N**ormality of error
- **E**qual variance (also called homoscedasticity)

The first assumption, **linearity**, states that the relationship between variables is linear. Relationships between variables that are not linear are discussed in Chapter 15.

The second assumption, **independence of errors**, requires that the errors (ε_i's) are independent from one another. This assumption is particularly important when data are collected over a period of time. In such situations, the errors for a specific time period are often correlated with those of the previous time period.

The third assumption, **normality**, requires that the errors (ε_i's) are normally distributed at each value of X. Like the t test and the ANOVA F test, regression analysis is fairly robust against departures from the normality assumption. As long as the distribution of the errors at each level of X is not extremely different from a normal distribution, inferences about β_0 and β_1 are not seriously affected.

The fourth assumption, **equal variance** or **homoscedasticity**, requires that the variance of the errors (ε_i 's) are constant for all values of X. In other words, the variability of Y values will be the same when X is a low value as when X is a high value. The equal variance assumption is important when making inferences about β_0 and β_1. If there are serious departures from this assumption, you can use either data transformations or weighted least-squares methods (see reference 5).

13.5 RESIDUAL ANALYSIS

In section 13.1 regression analysis was introduced. In sections 13.2 and 13.3, a model was developed and estimated using the least-squares approach for the site selection data. Is this the correct model for these data? Are the assumptions introduced in section 13.4 valid? In this section, a graphical approach called **residual analysis** is used to evaluate the assumptions and thus determine whether the regression model selected is an appropriate model.

The **residual** or estimated error value e_i is the difference between the observed (Y_i) and predicted ($\hat{Y}_i$) values of the dependent variable for a given value of X_i. Graphically, a residual appears on a scatter diagram as the vertical distance between an observed value of Y and the prediction line. Equation (13.14) defines the residual.

THE RESIDUAL

The residual is equal to the difference between the observed value of Y and the predicted value of Y.

$$e_i = Y_i - \hat{Y}_i \qquad \textbf{(13.14)}$$

Evaluating the Assumptions

Recall from section 13.4 that the four assumptions of regression (known by the acronym LINE) are linearity, independence, normality, and equal variance.

Linearity To evaluate linearity, plot the residuals on the vertical axis against the corresponding X_i values of the independent variable on the horizontal axis. If the linear model is appropriate for the data, there will be no apparent pattern in this plot. However, if the linear model is not appropriate, there will be a relationship between the X_i values and the residuals e_i. You can see such a pattern in Figure 13.12. Panel A shows a situation in which, although there is an increasing trend in Y as X increases, the relationship seems curvilinear because the upward trend decreases for increasing values of X. This quadratic effect is highlighted in panel B where there is a clear relationship between X_i and e_i. By plotting the residuals, the linear trend of X with Y has been removed, thereby exposing the lack of fit in the simple linear model. Thus, a quadratic model is a better fit and should be used in place of the simple linear model (see section 15.1 for further discussions of fitting quadratic models).

FIGURE 13.12
Studying the Appropriateness of the Simple Linear Regression Model

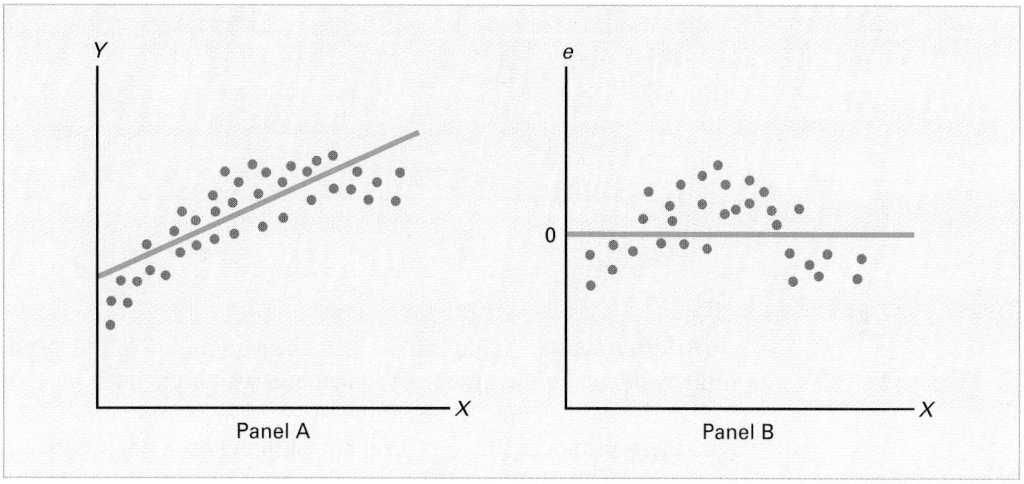

[1]Other more computationally complex residuals are available in Minitab and SPSS (see references 7 and 9)

To determine whether the simple linear regression model is appropriate, return to the evaluation of the site selection data. Figure 13.13 provides the predicted and residual values of the response variable (annual sales) computed by Microsoft Excel.[1]

FIGURE 13.13
Microsoft Excel Residual Statistics for the Site Selection Problem

22	RESIDUAL OUTPUT		
23			
24	*Observation*	*Predicted Annual Sales*	*Residuals*
25	1	3.803239598	-0.103239598
26	2	3.636253367	0.263746633
27	3	5.640088147	1.059911853
28	4	10.31570263	-0.815702635
29	5	3.135294672	0.264705328
30	6	4.638170757	0.961829243
31	7	3.135294672	0.564705328
32	8	2.801322208	-0.101322208
33	9	6.308033074	-0.808033074
34	10	3.469267135	-0.569267135
35	11	9.647757708	1.052242292
36	12	8.645840318	-1.045840318
37	13	10.6496751	1.150324902
38	14	5.974060611	-1.874060611

To assess linearity, the residuals are plotted against the independent variable (store size in thousands of square feet) in Figure 13.14. Although there is widespread scatter in the residual plot, there is no apparent pattern or relationship between the residuals and X_i. The residuals appear to be evenly spread above and below 0 for the differing values of X. You can conclude that the linear model is appropriate for the site selection data.

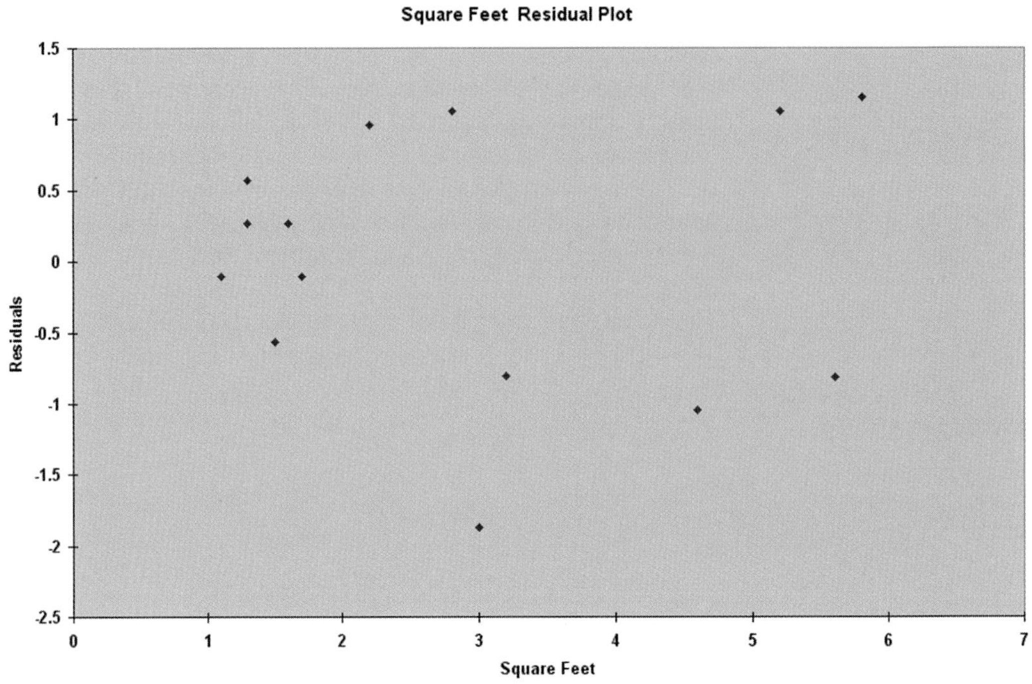

FIGURE 13.14
Microsoft Excel Plot of Residuals Against the Square Footage of a Store for the Site Selection Problem

Independence You can evaluate the assumption of independence of the errors by plotting the residuals in the order or sequence in which the data were collected. Data collected over periods of time sometimes exhibit an *autocorrelation* effect among successive observations. In these instances, there is a relationship between consecutive residuals. If this relationship exists (which violates the assumption of independence), it will be apparent in the plot of the residuals versus the time in which the data were collected. You can also test for autocorrelation using the Durbin-Watson statistic, which is the subject of section 13.6. For the site selection data considered thus far in this chapter, the data were collected during the same time period. Therefore, you do not need to evaluate the independence assumption for these data.

Normality You can evaluate the assumption of normality in the errors by tallying the residuals into a frequency distribution and displaying the results in a histogram (see section 2.3). For the site selection data, the residuals have been tallied into a frequency distribution as shown in Table 13.3. (There are an insufficient number of values, however, to construct a histogram.) You can also evaluate the normality assumption by comparing the actual versus theoretical values of the residuals, or by constructing a normal probability plot of the residuals (see section 6.3). Figure 13.15 is a normal probability plot of the residuals for the site selection data.

TABLE 13.3
Frequency Distribution of 14 Residual Values for the Site Selection Data

Residuals	Frequency
−2.25 but less than −1.75	1
−1.75 but less than −1.25	0
−1.25 but less than −0.75	3
−0.75 but less than −0.25	1
−0.25 but less than +0.25	2
+0.25 but less than +0.75	3
+0.75 but less than +1.25	4
Total	14

FIGURE 13.15
Minitab Normal Probability Plot of Residuals for the Site Selection Data

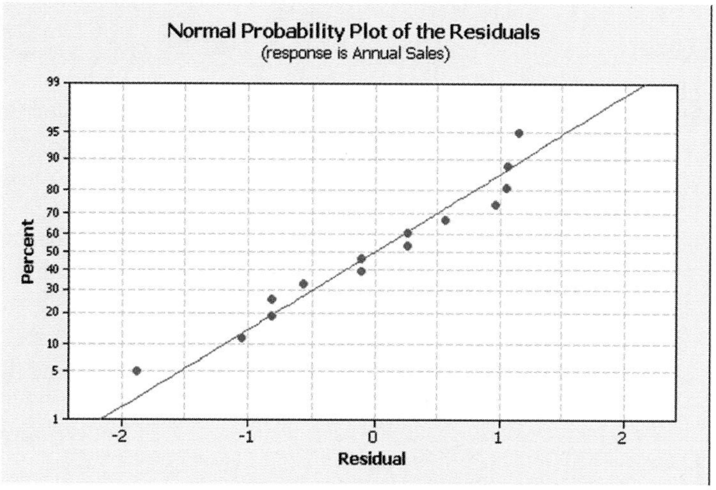

It is difficult to evaluate the normality assumption for a sample of only 14 values regardless of whether you use a histogram, stem-and-leaf display, box-and-whisker plot, or normal probability plot. You can see from Figure 13.15 that the data do not appear to depart substantially from a normal distribution. The robustness of regression analysis to modest departures from normality enables you to conclude that you should not be overly concerned about departures from this normality assumption in the site selection data.

Equal Variance You can evaluate the assumption of equal variance from a plot of the residuals with X_i. For the site selection data of Figure 13.14 on page 534 there do not appear to be major differences in the variability of the residuals for different X_i values. Thus, you can conclude that there is no apparent violation in the assumption of equal variance at each level of X.

To examine a case in which the equal variance assumption is violated, observe Figure 13.16, which is a plot of the residuals with X_i for a hypothetical set of data. In this plot, the variability of the residuals increases dramatically as X increases, demonstrating the lack of homogeneity in the variances of Y_i at each level of X. For these data, the equal variance assumption is invalid.

FIGURE 13.16
Violation of Equal Variance

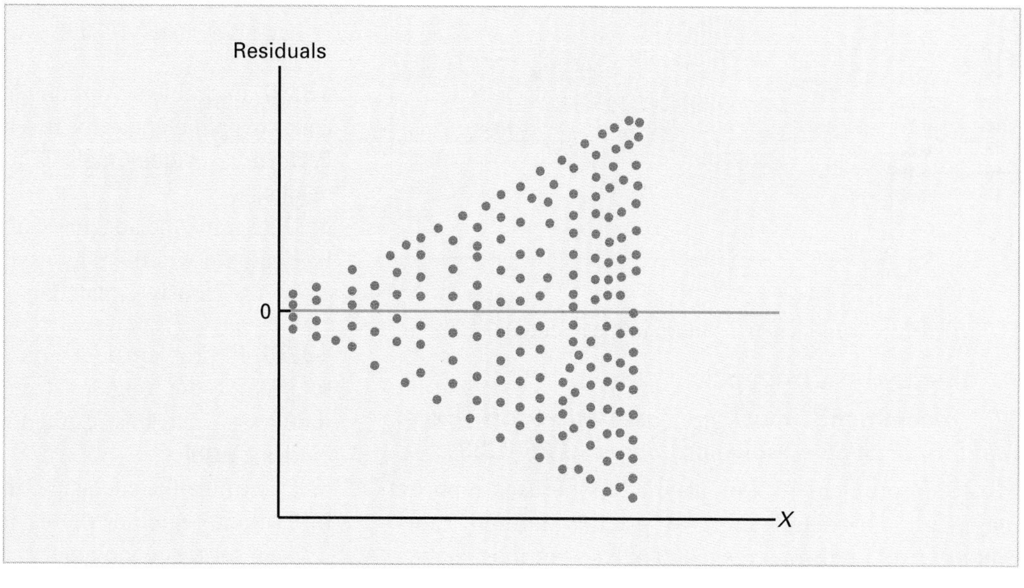

PROBLEMS FOR SECTION 13.5

Learning the Basics

13.23 The following computer output contains the X values, residuals, and a residual plot from a regression analysis.

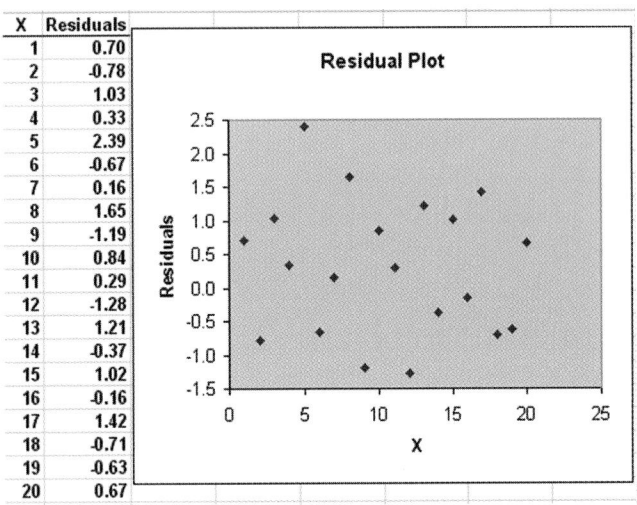

X	Residuals
1	0.70
2	-0.78
3	1.03
4	0.33
5	2.39
6	-0.67
7	0.16
8	1.65
9	-1.19
10	0.84
11	0.29
12	-1.28
13	1.21
14	-0.37
15	1.02
16	-0.16
17	1.42
18	-0.71
19	-0.63
20	0.67

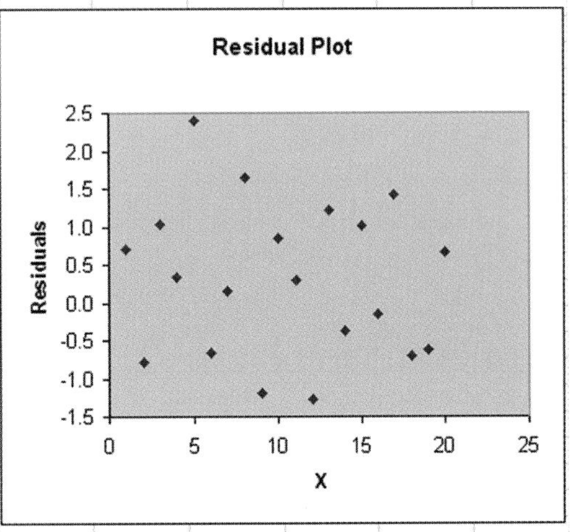

Is there any evidence of a pattern in the residuals? Explain.

13.24 The following computer output contains the X values, residuals, and a residual plot from a regression analysis.

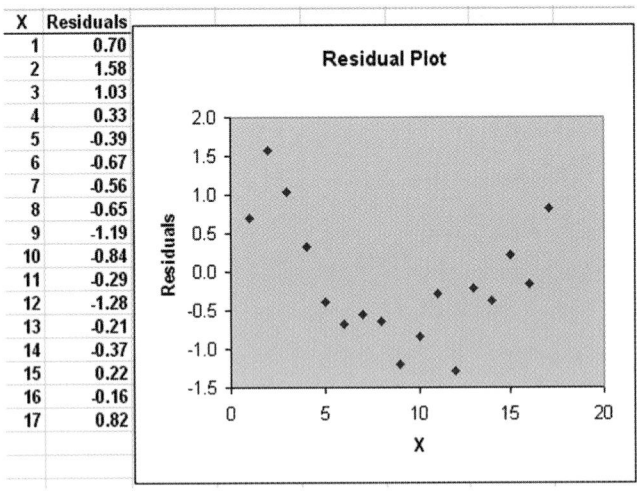

X	Residuals
1	0.70
2	1.58
3	1.03
4	0.33
5	-0.39
6	-0.67
7	-0.56
8	-0.65
9	-1.19
10	-0.84
11	-0.29
12	-1.28
13	-0.21
14	-0.37
15	0.22
16	-0.16
17	0.82

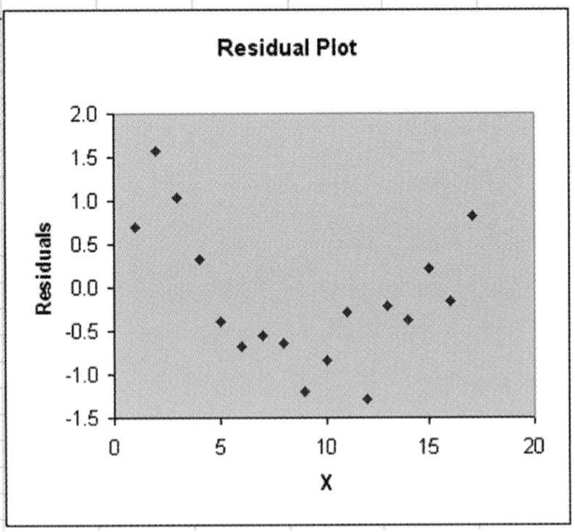

Is there any evidence of a pattern in the residuals? Explain.

Applying the Concepts

We recommend that you use Microsoft Excel, Minitab, or SPSS to solve problems 13.25–13.31.

13.25 In problem 13.5 on page 524, you used reported magazine newsstand sales to predict audited sales. CIRCULATION Perform a residual analysis for these data.

a. Determine the adequacy of the fit of the model.
b. Evaluate whether the assumptions of regression have been seriously violated.

SELF Test **13.26** In problem 13.4 on page 523, the marketing manager used shelf space for pet food to predict weekly sales. PETFOOD Perform a residual analysis for these data.

a. Determine the adequacy of the fit of the model.
b. Evaluate whether the assumptions of regression have been seriously violated.

13.27 In problem 13.7 on page 524, you used the weight of mail to predict the number of orders received. Perform a residual analysis for these data. MAIL Based on these results,

a. Determine the adequacy of the fit of the model.
b. Evaluate whether the assumptions of regression have been seriously violated.

13.28 In problem 13.6 on page 524, the owner of a moving company wanted to predict labor hours based on the cubic feet moved. Perform a residual analysis for these data. MOVING Based on these results,

a. Determine the adequacy of the fit of the model.
b. Evaluate whether the assumptions of regression have been seriously violated.

13.29 In problem 13.9 on page 525, an agent for a real estate company wanted to predict the monthly rent for apartments based on the size of the apartment. Perform a residual analysis for these data. RENT Based on these results,

a. Determine the adequacy of the fit of the model.
b. Evaluate whether the assumptions of regression have been seriously violated.

13.30 In problem 13.8 on page 525, you used annual revenues to predict the value of a baseball franchise. BBREVENUE Perform a residual analysis for these data. Based on these results,

a. Determine the adequacy of the fit of the model.
b. Evaluate whether the assumptions of regression have been seriously violated.

13.31 In problem 13.10 on page 525, you used hardness to predict the tensile strength of die-cast aluminum. HARDNESS Perform a residual analysis for these data. Based on these results,

a. Determine the adequacy of the fit of the model.
b. Evaluate whether the assumptions of regression have been seriously violated.

13.6 MEASURING AUTOCORRELATION: THE DURBIN-WATSON STATISTIC

One of the basic assumptions of the regression model is the independence of the errors. This assumption is sometimes violated when data are collected over sequential periods of time because a residual at any one point in time may tend to be similar to residuals at adjacent points in time. This pattern in the residuals is called **autocorrelation**. When a set of data has substantial autocorrelation, the validity of a regression model can be in serious doubt.

Residual Plots to Detect Autocorrelation

As mentioned in section 13.5, one way to detect autocorrelation is to plot the residuals in time order. If a positive autocorrelation effect is present, there will be clusters of residuals with the same sign and you will readily detect an apparent pattern. If negative autocorrelation exists, residuals will tend to jump back and forth from positive to negative to positive, and so on. This type of pattern is very rarely seen in regression analysis. Thus, the focus of this section is on positive autocorrelation. To illustrate positive autocorrelation, consider the following example.

The manager of a package delivery store wants to predict weekly sales based on the number of customers making purchases for a period of 15 weeks. In this situation, because data are collected over a period of 15 consecutive weeks at the same store, you need to determine whether autocorrelation is present. Table 13.4 summarizes the data for this store CUSTSALE. Figure 13.17 illustrates Excel output and Figure 13.18 illustrates Minitab output.

TABLE 13.4
Customers and Sales for Period of 15 Consecutive Weeks

Week	Customers	Sales (in thousands of dollars)	Week	Customers	Sales (in thousands of dollars)
1	794	9.33	9	880	12.07
2	799	8.26	10	905	12.55
3	837	7.48	11	886	11.92
4	855	9.08	12	843	10.27
5	845	9.83	13	904	11.80
6	844	10.09	14	950	12.15
7	863	11.01	15	841	9.64
8	875	11.49			

FIGURE 13.17
Microsoft Excel Output for the Package Delivery Store Data of Table 13.4

	A	B	C	D	E	F	G
1	Package Delivery Store Sales Analysis						
2							
3	*Regression Statistics*						
4	Multiple R	0.81083					
5	R Square	0.65745					
6	Adjusted R Square	0.63109					
7	Standard Error	0.93604					
8	Observations	15					
9							
10	ANOVA						
11		*df*	*SS*	*MS*	*F*	*Significance F*	
12	Regression	1	21.86043	21.86043	24.95014	0.00025	
13	Residual	13	11.39014	0.87616			
14	Total	14	33.25057333				
15							
16		*Coefficients*	*Standard Error*	*t Stat*	*P-value*	*Lower 95%*	*Upper 95%*
17	Intercept	-16.03219	5.31017	-3.01915	0.00987	-27.50411	-4.56028
18	Customers	0.03076	0.00616	4.99501	0.00025	0.01746	0.04406

FIGURE 13.18
Minitab Output for the Package Delivery Store Data of Table 13.4

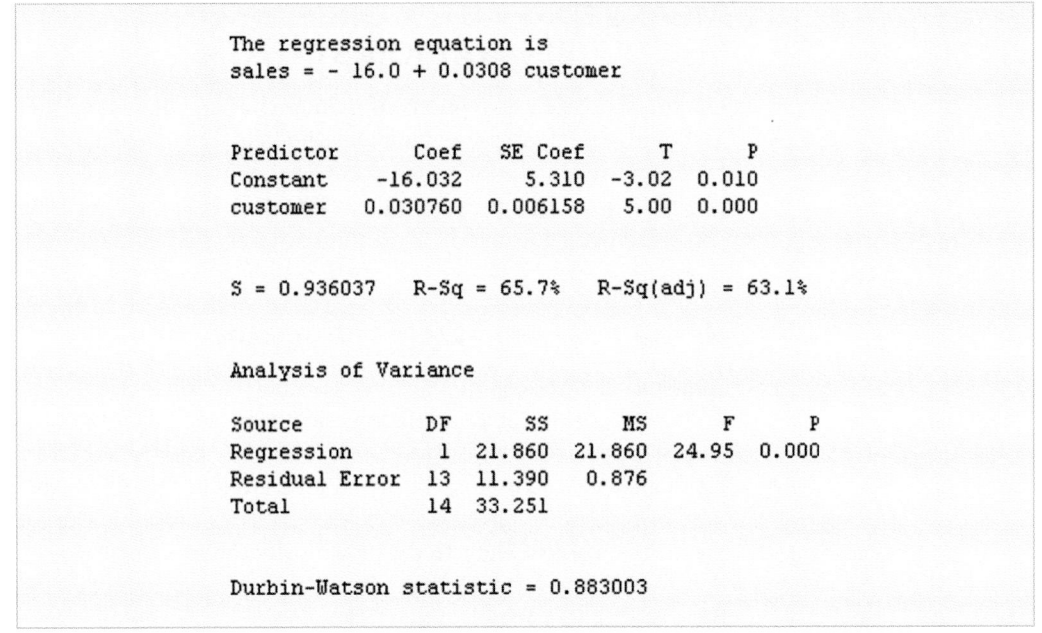

```
The regression equation is
sales = - 16.0 + 0.0308 customer


Predictor        Coef    SE Coef      T      P
Constant      -16.032      5.310  -3.02  0.010
customer     0.030760   0.006158   5.00  0.000


S = 0.936037   R-Sq = 65.7%   R-Sq(adj) = 63.1%


Analysis of Variance

Source          DF      SS      MS      F      P
Regression       1  21.860  21.860  24.95  0.000
Residual Error  13  11.390   0.876
Total           14  33.251


Durbin-Watson statistic = 0.883003
```

From Figure 13.17 or 13.18 observe that r^2 is 0.657, indicating that 65.7% of the variation in sales is explained by variation in the number of customers. In addition, the Y intercept b_0 is −16.032, and the slope b_1 is 0.03076. However, before using this model for prediction, you must undertake proper analyses of the residuals. Because the data have been collected over a consecutive period of 15 weeks, in addition to checking the linearity, normality, and equal variance assumptions, you must investigate the independence of errors assumption. Figure 13.19 plots the residuals versus time to see whether a pattern exists. In Figure 13.19, the residuals tend to fluctuate up and down in a cyclical pattern. This cyclical pattern provides strong cause for concern about the autocorrelation of the residuals and, hence, a violation of the independence of errors assumption.

FIGURE 13.19
Microsoft Excel Residual Plot for the Package Delivery Store Data of Table 13.4

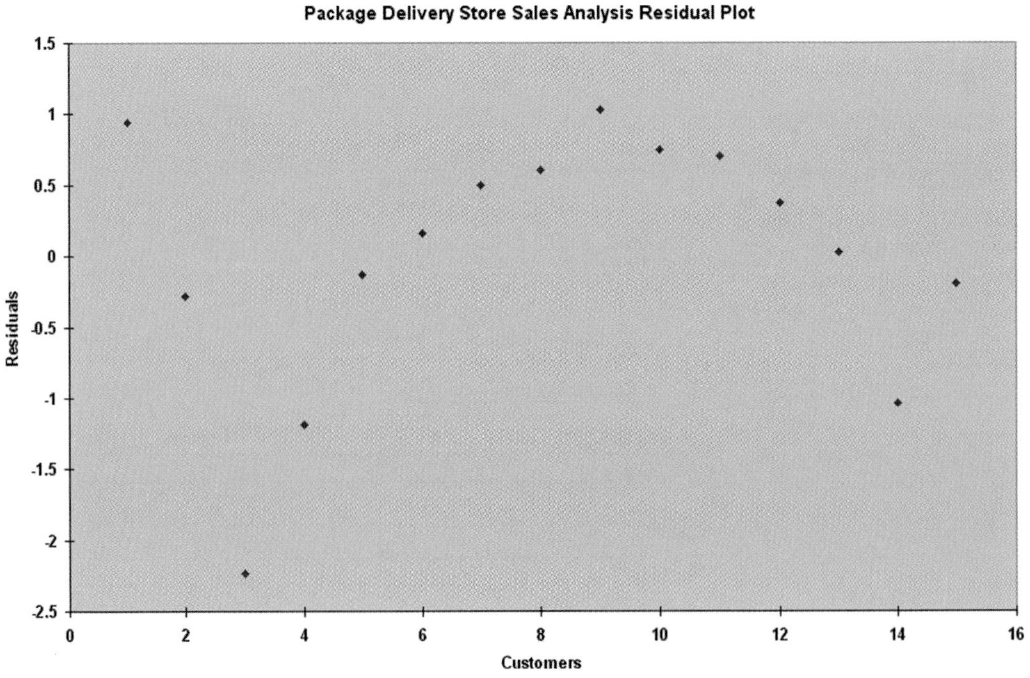

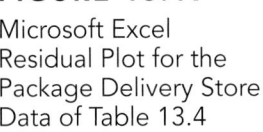

The Durbin-Watson Statistic

The **Durbin-Watson statistic** is used to detect autocorrelation. This statistic measures the correlation between each residual and the residual for the time period immediately preceding the one of interest. Equation (13.15) defines the Durbin-Watson statistic.

DURBIN-WATSON STATISTIC

$$D = \frac{\sum_{i=2}^{n} (e_i - e_{i-1})^2}{\sum_{i=1}^{n} e_i^2} \quad \textbf{(13.15)}$$

where e_i = residual at the time period i

To better understand the Durbin-Watson statistic D, examine the composition of the statistic presented in Equation (13.15). The numerator $\sum_{i=2}^{n} (e_i - e_{i-1})^2$ represents the squared difference between two successive residuals, summed from the second value to the nth value. The denominator $\sum_{i=1}^{n} e_i^2$ represents the sum of the squared residuals. When successive residuals are positively autocorrelated, the value of D will approach 0. If the residuals are not correlated, the value of D will be close to 2. (If there is negative autocorrelation, D will be greater than 2 and could even approach its maximum value of 4.) For the package delivery store data, the Durbin-Watson statistic, $D = 0.883$, is computed by Microsoft Excel as illustrated in Figure 13.20 and in the Minitab output of Figure 13.18 on page 538.

FIGURE 13.20
Microsoft Excel Output of the Durbin-Watson Statistic for the Sales Data

	A	B
1	**Durbin-Watson Calculations**	
2		
3	Sum of Squared Difference of Residuals	10.0575
4	Sum of Squared Residuals	11.3901
5		
6	**Durbin-Watson Statistic**	**0.8830**

You need to determine when the autocorrelation is large enough to make the Durbin-Watson statistic D fall sufficiently below 2 to conclude that there is significant positive autocorrelation. Whether D is significant is dependent on n, the sample size, and k, the number of independent variables in the model (in simple linear regression, $k = 1$). Table 13.5 has been extracted from Table E.11, the table of the Durbin-Watson statistic.

TABLE 13.5
Finding Critical Values of the Durbin-Watson Statistic

$\alpha = .05$

	$k = 1$		$k = 2$		$k = 3$		$k = 4$		$k = 5$	
n	d_L	d_U	d_L	d_U	d_L	d_U	d_L	d_U	d_L	d_U
15	1.08	1.36	.95	1.54	.82	1.75	.69	1.97	.56	2.21
16	1.10	1.37	.98	1.54	.86	1.73	.74	1.93	.62	2.15
17	1.13	1.38	1.02	1.54	.90	1.71	.78	1.90	.67	2.10
18	1.16	1.39	1.05	1.53	.93	1.69	.82	1.87	.71	2.06

In Table 13.5 two values are shown for each combination of α (level of significance), n (sample size), and k (number of independent variables in the model). The first value, d_L, represents the lower critical value. If D is below d_L, you conclude that there is evidence of positive autocorrelation among the residuals. Under such a circumstance, the least-squares method used in this chapter is inappropriate, and you should use alternative methods (reference 5). The second value, d_U, represents the upper critical value of D, above which you would conclude that there is no evidence of positive autocorrelation among the residuals. If D is between d_L and d_U, you are unable to arrive at a definite conclusion.

For the data concerning the package delivery stores, with one independent variable ($k = 1$) and 15 values ($n = 15$), $d_L = 1.08$ and $d_U = 1.36$. Because $D = 0.883 < 1.08$, you conclude that there is positive autocorrelation among the residuals. The least-squares regression analysis of the data is inappropriate because of the presence of significant positive autocorrelation among the residuals. In other words, the independence of errors assumption is invalid. You need to use alternative approaches discussed in reference 5.

PROBLEMS FOR SECTION 13.6

Learning the Basics

PH Grade ASSIST **13.32** The residuals for 10 consecutive time periods are as follows:

Time Period	Residual	Time Period	Residual
1	−5	6	+1
2	−4	7	+2
3	−3	8	+3
4	−2	9	+4
5	−1	10	+5

a. Plot the residuals over time. What conclusion can you reach about the pattern of the residuals over time?
b. Based on (a), what conclusion can you reach about the autocorrelation of the residuals?

PH Grade ASSIST **13.33** The residuals for 15 consecutive time periods are as follows:

Time Period	Residual	Time Period	Residual
1	+4	9	+6
2	−6	10	−3
3	−1	11	+1
4	−5	12	+3
5	+2	13	0
6	+5	14	−4
7	−2	15	−7
8	+7		

a. Plot the residuals over time. What conclusion can you reach about the pattern of the residuals over time?
b. Compute the Durbin-Watson statistic. At the 0.05 level of significance, is there evidence of positive autocorrelation among the residuals?
c. Based on (a) and (b), what conclusion can you reach about the autocorrelation of the residuals?

Applying the Concepts

We recommend that you use Microsoft Excel, Minitab, or SPSS to solve problems 13.35–13.39.

PH Grade ASSIST **13.34** In problem 13.4 (pet food sales) on page 523, the marketing manager used shelf space for pet food to predict weekly sales.
a. Is it necessary to compute the Durbin-Watson statistic? Explain.
b. Under what circumstances is it necessary to compute the Durbin-Watson statistic before proceeding with the least-squares method of regression analysis?

13.35 The owner of a single-family home in a suburban county in the northeastern United States would like to develop a model to predict electricity consumption in his all-electric house (lights, fans, heat, appliances, and so on) based on average atmospheric temperature (in degrees Fahrenheit). Monthly kilowatt usage and temperature information are available for a period of 24 consecutive months in the file ELECUSE.
a. Assuming a linear relationship, use the least-squares method to find the regression coefficients b_0 and b_1.
b. Predict the mean kilowatt usage when the average atmospheric temperature is 50 degrees Fahrenheit.
c. Plot the residuals versus the time period.
d. Compute the Durbin-Watson statistic. At the 0.05 level of significance, is there evidence of positive autocorrelation among the residuals?
e. Based on the results of (c) and (d), is there reason to question the validity of the model?

SELF Test **13.36** A mail-order catalog business that sells personal computer supplies, software, and hardware maintains a centralized warehouse for the distribution of products ordered. Management is currently examining the process of distribution from the warehouse and is interested in studying the factors that affect warehouse distribution costs. Currently, a small handling fee is added to the order, regardless of the amount of the order. Data have been collected over the past 24 months indicating the warehouse distribution costs and the number of orders received. WARECOST The results are as follows:

Months	Distribution Cost (Thousands of Dollars)	Number of Orders
1	52.95	4,015
2	71.66	3,806
3	85.58	5,309
4	63.69	4,262
5	72.81	4,296
6	68.44	4,097
7	52.46	3,213
8	70.77	4,809
9	82.03	5,237
10	74.39	4,732
11	70.84	4,413
12	54.08	2,921
13	62.98	3,977
14	72.30	4,428
15	58.99	3,964
16	79.38	4,582
17	94.44	5,582
18	59.74	3,450
19	90.50	5,079
20	93.24	5,735
21	69.33	4,269
22	53.71	3,708
23	89.18	5,387
24	66.80	4,161

a. Assuming a linear relationship, use the least-squares method to find the regression coefficients b_0 and b_1.
b. Predict the monthly warehouse distribution costs when the number of orders is 4,500.
c. Plot the residuals versus the time period.
d. Compute the Durbin-Watson statistic. At the 0.05 level of significance, is there evidence of positive autocorrelation among the residuals?
e. Based on the results of (c) and (d), is there reason to question the validity of the model?

13.37 A freshly brewed shot of espresso has three distinct components, the heart, body, and crema. The separation of these three components typically lasts only 10 to 20 seconds. To use the espresso shot in making a latte, cappuccino, or other drinks, the shot must be poured into the beverage during the separation of the heart, body, and cream. If the shot is used after the separation occurs, the drink becomes excessively bitter and acidic, ruining the final drink. Thus, a longer separation time allows the drink-maker more time to pour the shot and ensure that the beverage will meet expectations. An employee at a coffee shop hypothesized that the harder the espresso grounds were tamped down into the portafilter before brewing, the longer the separation time would be. An experiment using 24 observations was conducted to test this relationship. ESPRESSO The independent variable Tamp measures the distance in inches between the espresso grounds and the top of the portafilter (i.e., the harder the tamp the larger the distance). The dependent variable Time is the number of seconds the heart, body, and crema are separated (i.e., the amount of time after the shot is poured before it must be used for the customer's beverage).

Shot	Tamp	Time	Shot	Tamp	Time
1	0.20	14	13	0.50	18
2	0.50	14	14	0.50	13
3	0.50	18	15	0.35	19
4	0.20	16	16	0.35	19
5	0.20	16	17	0.20	17
6	0.50	13	18	0.20	18
7	0.20	12	19	0.20	15
8	0.35	15	20	0.20	16
9	0.50	9	21	0.35	18
10	0.35	15	22	0.35	16
11	0.50	11	23	0.35	14
12	0.50	16	24	0.35	16

a. Determine the prediction line using Time as the dependent variable and Tamp as the independent variable.
b. Predict the mean separation time for a Tamp distance of 0.50 inch.
c. Plot the residuals versus the time order of experimentation. Are there any noticeable patterns?
d. Compute the Durbin-Watson statistic. At the 0.05 level of significance, is there evidence of positive autocorrelation among the residuals?
e. Based on the results of (c) and (d), is there reason to question the validity of the model?

13.38 The owner of a chain of ice cream stores would like to study the effect of atmospheric temperature on sales during the summer season. A sample of 21 consecutive days is selected, with the results stored in the data file ICECREAM.

(*Hint:* Determine which are the independent and dependent variables.)
a. Assuming a linear relationship, use the least-squares method to find the regression coefficients b_0 and b_1.
b. Predict the sales per store for a day in which the temperature is 83°F.
c. Plot the residuals versus the time period.
d. Compute the Durbin-Watson statistic. At the 0.05 level of significance, is there evidence of positive autocorrelation among the residuals?
e. Based on the results of (c) and (d), is there reason to question the validity of the model?

13.39 Gasoline prices in the United States hit record highs in May of 2004. Experts blamed the high demand for gasoline worldwide, a lack of production capacity in U.S. refineries, geopolitical unrest, and most importantly, the high price of crude oil on the world market. The data file OIL-GAS contains the U.S. retail price for regular gasoline (cents/gallon) and the world crude oil price (dollars/barrel) for 100 weeks ending May 17, 2004 (U.S. Department of Energy, Energy Information Administration, **www.eia.doe.gov**, May 25, 2004.)

a. Construct a scatter diagram with crude oil price on the X axis and gasoline price on the Y axis. Discuss any patterns present in the data.

b. Determine the prediction line using crude oil price as the independent variable and gasoline price as the dependent variable.

c. Plot the residuals versus the time period and interpret the plot.

d. Compute the Durbin-Watson statistic. At the 0.05 level of significance, is there evidence of positive autocorrelation among the residuals?

e. Based on the result of (c) and (d), is there reason to question the validity of the model?

13.7 INFERENCES ABOUT THE SLOPE AND CORRELATION COEFFICIENT

In sections 13.1 through 13.3, you used regression solely for the purpose of description. You learned how the least-squares method determines the regression coefficients, and how to predict Y for a given value of X. In addition, you learned how to compute and interpret the standard error of the estimate and the coefficient of determination.

When residual analysis, as discussed in section 13.5, indicates that the assumptions of a least-squares regression model are not seriously violated and that the straight-line model is appropriate, you can make inferences about the linear relationship between the variables in the population.

t Test for the Slope

To determine the existence of a significant linear relationship between the X and Y variables, you can test whether β_1 (the population slope) is equal to 0. The null and alternative hypotheses are as follows:

$$H_0: \beta_1 = 0 \text{ (There is no linear relationship.)}$$

$$H_1: \beta_1 \neq 0 \text{ (There is a linear relationship.)}$$

If you reject the null hypothesis, you conclude that there is evidence of a linear relationship. Equation (13.16) defines the test statistic.

TESTING A HYPOTHESIS FOR A POPULATION SLOPE β_1 USING THE t TEST

The t statistic equals the difference between the sample slope and hypothesized value of the population slope divided by the standard error of the slope.

$$t = \frac{b_1 - \beta_1}{S_{b_1}} \qquad \textbf{(13.16)}$$

where

$$S_{b_1} = \frac{S_{YX}}{\sqrt{SSX}}$$

$$SSX = \sum_{i=1}^{n}(X_i - \bar{X})^2$$

The test statistic t follows a t distribution with $n - 2$ degrees of freedom.

Return to the "Using Statistics" scenario concerning the site selection data. To test whether there is a significant relationship between the size of the store and the annual sales at the 0.05 level of significance, refer to the Microsoft Excel output for the t test presented in Figure 13.21 or the Minitab output presented in Figure 13.22. From Figure 13.21 or 13.22,

$$b_1 = +1.670 \quad n = 14 \quad S_{b_1} = 0.157$$

and

$$t = \frac{b_1 - \beta_1}{S_{b_1}}$$

$$= \frac{1.670 - 0}{0.157} = 10.64$$

Microsoft Excel labels this t statistic t Stat (see Figure 13.21) and Minitab labels it T (see Figure 13.22). Using the 0.05 level of significance, the critical value of t with $n - 2 = 12$ degrees of freedom is 2.1788. Because $t = 10.64 > 2.1788$, reject H_0 (see Figure 13.23). Using the p-value, you reject H_0 because the p-value is approximately 0. Hence, you can conclude that there is a significant linear relationship between mean annual sales and the size of the store.

FIGURE 13.21
Microsoft Excel t Test for the Slope for the Site Selection Data

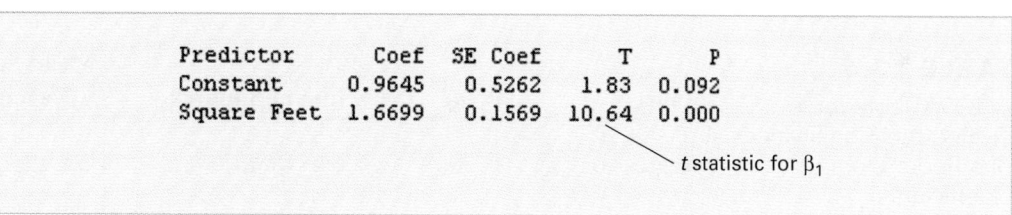

16		Coefficients	Standard Error	t Stat	P-value	Lower 95%	Upper 95%
17	Intercept	0.96447	0.52619	1.83293	0.09173	-0.18200	2.11095
18	Square Feet	1.66986	0.15693	10.64112	0.00000	1.32795	2.01177

FIGURE 13.22
Minitab t Test for the Slope for the Site Selection Data

Predictor	Coef	SE Coef	T	P
Constant	0.9645	0.5262	1.83	0.092
Square Feet	1.6699	0.1569	10.64	0.000

t statistic for β_1

FIGURE 13.23
Testing a Hypothesis About the Population Slope at the 0.05 Level of Significance with 12 Degrees of Freedom

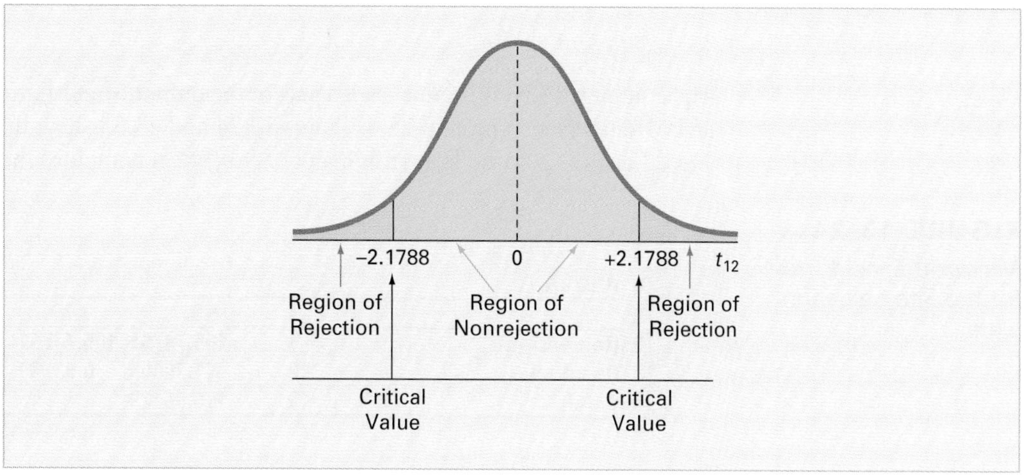

F Test for the Slope

You can also use an F test to determine whether the slope in simple linear regression is statistically significant. Recall from section 10.4 that you use the F distribution to test the ratio of two variances. In testing for the significance of the slope, the F test, defined in Equation (13.17), is the ratio of the variance that is due to the regression (MSR) divided by the error variance $(MSE = S_{YX}^2)$.

TESTING A HYPOTHESIS FOR A POPULATION SLOPE β_1 USING THE F TEST

The F statistic is equal to the regression mean square (MSR) divided by the error mean square (MSE).

$$F = \frac{MSR}{MSE} \tag{13.17}$$

where

$$MSR = \frac{SSR}{k}$$

$$MSE = \frac{SSE}{n - k - 1}$$

k = number of independent variables in the regression model

The test statistic F follows an F distribution with k and $n - k - 1$ degrees of freedom.

Using a level of significance α, the decision rule is

Reject H_0 if $F > F_U$;

otherwise do not reject H_0.

Table 13.6 organizes the complete set of results into an ANOVA table.

TABLE 13.6
ANOVA Table for Testing the Significance of a Regression Coefficient

Source	df	Sum of Squares	Mean Square (Variance)	F
Regression	k	SSR	$MSR = \frac{SSR}{k}$	$F = \frac{MSR}{MSE}$
Error	$n - k - 1$	SSE	$MSE = \frac{SSE}{n - k - 1}$	
Total	$n - 1$	SST		

The completed ANOVA table is also part of the output from Microsoft Excel (see Figure 13.24) and Minitab (see Figure 13.25). Figures 13.24 and 13.25 show that the computed F statistic is 113.23 and the p-value is approximately zero (Excel computes the p-value as 0.000000182).

FIGURE 13.24
Microsoft Excel F Test for the Site Selection Data

10	ANOVA					
11		df	SS	MS	F	Significance F
12	Regression	1	105.74761	105.74761	113.23351	1.82269E-07
13	Residual	12	11.20668	0.93389		
14	Total	13	116.95429			

MSR (callout to 105.74761 in MS column); MSE (callout to 0.93389 in MS column)

FIGURE 13.25
Minitab F Test for the Site Selection Data

Analysis of Variance

Source	DF	SS	MS	F	P
Regression	1	105.75	105.75	113.23	0.000
Residual Error	12	11.21	0.93		
Total	13	116.95			

MSR (105.75); MSE (0.93)

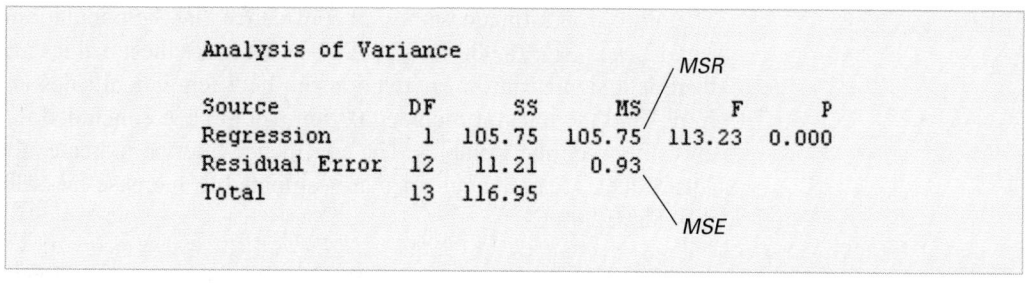

Using a level of significance of 0.05, from Table E.5 the critical value of the F distribution with 1 and 12 degrees of freedom is 4.75 (see Figure 13.26). Because $F = 113.23 > 4.75$ or because the p-value $= 0.000000182 < 0.05$, you reject H_0 and conclude that the size of the store is significantly related to annual sales.

FIGURE 13.26
Regions of Rejection and Nonrejection When Testing for Significance of Slope at the 0.05 Level of Significance with 1 and 12 Degrees of Freedom

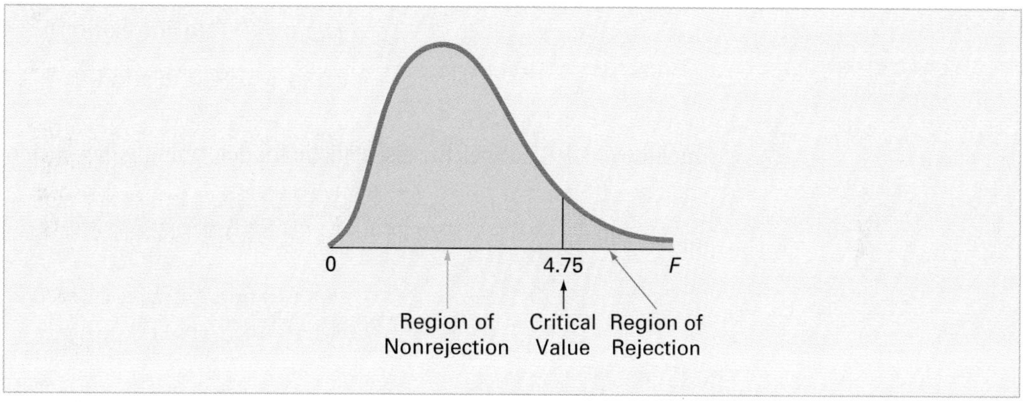

Confidence Interval Estimate of the Slope (β_1)

As an alternative to testing the existence of a linear relationship between the variables, you can set up a confidence interval estimate of β_1 and determine whether the hypothesized value ($\beta_1 = 0$) is included in the interval. Equation (13.18) defines the confidence interval estimate of β_1.

CONFIDENCE INTERVAL ESTIMATE OF THE SLOPE β_1

The confidence interval estimate for the slope can be formed by taking the sample slope b_1 and adding and subtracting the critical t value multiplied by the standard error of the slope.

$$b_1 \pm t_{n-2} S_{b_1} \qquad \textbf{(13.18)}$$

From the Microsoft Excel output of Figure 13.21 or the Minitab output of Figure 13.22, on page 543,

$$b_1 = +1.670 \quad n = 14 \quad S_{b_1} = 0.157$$

To construct a 95% confidence interval estimate, $\alpha/2 = 0.025$, and from Table E.3, $t_{12} = 2.1788$. Thus,

$$\begin{aligned} b_1 \pm t_{n-2} S_{b_1} &= +1.670 \pm (2.1788)(0.157) \\ &= +1.670 \pm 0.342 \\ &+1.328 \le \beta_1 \le +2.012 \end{aligned}$$

Therefore, you estimate with 95% confidence that the population slope is between +1.328 and +2.012 (i.e., \$1,328,000 to \$2,012,000). Because these values are above 0, you conclude that there is a significant linear relationship between annual sales and square footage size of the store. Had the interval included 0, you would have concluded that no significant relationship exists between the variables. The confidence interval indicates that for each increase of 1,000 square feet, mean annual sales are estimated to increase by at least \$1,328,000 but no more than \$2,012,000.

t Test for a Correlation Coefficient

In section 3.5 on page 110 the strength of the relationship between two numerical variables was measured using the **correlation coefficient** r. You can use the correlation coefficient to determine whether there is a statistically significant linear relationship between X and Y. You hypothesize that the population correlation coefficient ρ is 0. Thus, the null and alternative hypotheses are

$$H_0: \rho = 0 \text{ (No correlation)}$$

$$H_1: \rho \neq 0 \text{ (Correlation)}$$

Equation (13.19) defines the test statistic for determining the existence of a significant correlation.

TESTING FOR THE EXISTENCE OF CORRELATION

$$t = \frac{r - \rho}{\sqrt{\dfrac{1 - r^2}{n - 2}}} \qquad \textbf{(13.19)}$$

where

$$r = +\sqrt{r^2} \text{ if } b_1 > 0$$

or

$$r = -\sqrt{r^2} \text{ if } b_1 < 0$$

The test statistic t follows a t distribution with $n - 2$ degrees of freedom.

In the site selection problem, $r^2 = 0.904$ and $b_1 = +1.670$ (see Figure 13.4 or 13.5 on page 518). Since $b_1 > 0$, the correlation coefficient for annual sales and store size is the positive square root of r^2, that is, $r^2 = +\sqrt{0.904} = +0.951$. Testing the null hypothesis that there is no correlation between these two variables results in the following observed t statistic:

$$t = \frac{r - 0}{\sqrt{\dfrac{1 - r^2}{n - 2}}}$$

$$= \frac{0.951 - 0}{\sqrt{\dfrac{1 - (0.951)^2}{14 - 2}}} = 10.64$$

Using the 0.05 level of significance, since $t = 10.64 > 2.1788$, you reject the null hypothesis. You conclude that there is evidence of an association between annual sales and store size. This t statistic is equivalent to the t statistic found when testing whether the population slope β_1 is equal to zero (Figures 13.21 and 13.22 on page 543).

When inferences concerning the population slope were discussed, confidence intervals and tests of hypothesis were used interchangeably. However, developing a confidence interval for the correlation coefficient is more complicated because the shape of the sampling distribution of the statistic r varies for different values of the population correlation coefficient. Methods for developing a confidence interval estimate for the correlation coefficient are presented in reference 5.

PROBLEMS FOR SECTION 13.7

Learning the Basics

PH Grade ASSIST **13.40** You are testing the null hypothesis that there is no relationship between two variables X and Y. From your sample of $n = 18$, you determine that $b_1 = +4.5$ and $S_{b_1} = 1.5$.

a. What is the value of the t-test statistic?
b. At the $\alpha = 0.05$ level of significance, what are the critical values?
c. Based on your answers to (a) and (b), what statistical decision should you make?
d. Construct a 95% confidence interval estimate of the population slope β_1.

PH Grade ASSIST **13.41** You are testing the null hypothesis that there is no relationship between two variables X and Y. From your sample of $n = 20$, you determine that $SSR = 60$ and $SSE = 40$.

a. What is the value of the F-test statistic?
b. At the $\alpha = 0.05$ level of significance, what is the critical value?
c. Based on your answers to (a) and (b), what statistical decision should you make?
d. Calculate the correlation coefficient by first calculating r^2 and assuming b_1 is negative.
e. At the 0.05 level of significance, is there a significant correlation between X and Y?

Applying the Concepts

Problems 13.42–13.48 and 13.51–13.54 can be solved manually or by using Microsoft Excel, Minitab, or SPSS.

SELF Test **13.42** In problem 13.4 on page 523, the marketing manager used shelf space for pet food to predict weekly sales. PETFOOD From the results of that problem, $b_1 = 0.074$ and $S_{b_1} = 0.0159$.

a. At the 0.05 level of significance, is there evidence of a linear relationship between shelf space and sales?
b. Construct a 95% confidence interval estimate of the population slope β_1.

13.43 In problem 13.5 on page 524, you used reported magazine newsstand sales to predict audited sales. CIRCULATION Using the results of that problem, $b_1 = 0.5719$ and $S_{b_1} = 0.0668$.

a. At the 0.05 level of significance, is there evidence of a linear relationship between reported sales and audited sales?
b. Construct a 95% confidence interval estimate of the population slope β_1.

13.44 In problem 13.6 on page 524, the owner of a moving company wanted to predict labor hours based on the number of cubic feet moved. MOVING Using the results of that problem,

a. At the 0.05 level of significance, is there evidence of a linear relationship between the number of cubic feet moved and labor hours?
b. Construct a 95% confidence interval estimate of the population slope β_1.

PH Grade ASSIST **13.45** In problem 13.7 on page 524, you used the weight of mail to predict the number of orders received. MAIL Using the results of that problem,

a. At the 0.05 level of significance, is there evidence of a linear relationship between the weight of mail and the number of orders received?
b. Construct a 95% confidence interval estimate of the population slope β_1.

13.46 In problem 13.8 on page 525, you used annual revenues to predict the value of a baseball franchise. BBREVENUE Using the results of that problem,

a. At the 0.05 level of significance, is there evidence of a linear relationship between annual revenue and franchise value?
b. Construct a 95% confidence interval estimate of the population slope β_1.

13.47 In problem 13.9 on page 525, an agent for a real estate company wanted to predict the monthly rent for apartments based on the size of the apartment. RENT Using the results of that problem,

a. At the 0.05 level of significance, is there evidence of a linear relationship between the size of the apartment and the monthly rent?
b. Construct a 95% confidence interval estimate of the population slope β_1.

13.48 In problem 13.10 on page 525, you used hardness to predict the tensile strength of die-cast aluminum. HARDNESS Using the results of that problem,

a. At the 0.05 level of significance, is there evidence of a linear relationship between hardness and tensile strength?

b. Construct a 95% confidence interval estimate of the population slope β_1.

13.49 The volatility of a stock is often measured by its beta value. You can estimate the beta value of a stock by developing a simple linear regression model using the percentage weekly change in the stock as the dependent variable and the percentage weekly change in a market index as the independent variable. The Standard & Poor's (S&P) 500 Index is a common index to use. For example, if you wanted to estimate the beta for IBM, you could use the following model, which is sometimes referred to as a *market model.*

$$(\text{\% weekly change in IBM}) = \beta_0 + \beta_1 (\text{\% weekly change in S \& P 500 index}) + \varepsilon$$

The least-squares regression estimate of the slope b_1 is the estimate of the beta for IBM. A stock with a beta of 1.0 tends to move the same as the overall market. A stock with a beta value of 1.5 tends to move 50% more than the overall market, and a stock with a beta of 0.6 tends to move only 60% as much as the overall market. Stocks with negative betas tend to move in a direction opposite to that of the overall market. The following table gives some beta values for some widely held stocks.

Company	Beta
Sears, Roebuck and Company	0.603
Disney Company	1.109
Ford Motor Company	1.340
IBM	1.449
LSI Logic	2.421

Source: Extracted from ***finance.yahoo.com****, July 27, 2004.*

a. For each of the five companies, interpret the value of the beta.

b. How can investors use the beta value as a guide for investing?

13.50 Index funds are mutual funds that try to mimic the movement of leading indexes such as the S&P 500 index, the NASDAQ index, or the Russell 2000 Index. The beta values for these funds (as described in problem 13.49) are therefore approximately 1.0. The estimated market models for these funds are approximately:

$$(\text{\% weekly change in index fund}) = 0.0 + 1.0 (\text{\% weekly change in the index})$$

Leveraged index funds are designed to magnify the movement of major indexes. An article in *Mutual Funds* (Lynn O'Shaughnessy, "Reach for Higher Returns," *Mutual Funds*, July 1999, 44–49) described some of the risks and rewards associated with these funds and gave details on some of the most popular leveraged funds including those in the following table.

Name (Ticker Symbol)	Fund Description
Potomac Small Cap Plus (POSCX)	125% of Russell 2000 Index
Rydex "Inv" Nova (RYNVX)	150% of the S&P 500 Index
ProFund UltraOTC "Inv" (UOPIX)	Double (200%) the NASDAQ 100 Index

Thus, estimated market models for these funds are approximately:

$$(\text{\% weekly change in POSCX}) = 0.0 + 1.25 (\text{\% weekly change in the Russell 2000 index})$$

$$(\text{\% weekly change in RYNVX}) = 0.0 + 1.50 (\text{\% weekly change in the S\&P 500 index})$$

$$(\text{\% weekly change in UOPIX fund}) = 0.0 + 2.0 (\text{\% weekly change in the NASDAQ 100 index})$$

Thus, if the Russell 2000 Index gains 10% over a period of time, the leveraged mutual fund POSCX gains approximately 12.5%. On the downside, if the same index loses 20%, POSCX loses approximately 25%.

a. Consider the leveraged mutual fund ProFund UltraBull "Inv" (ULPIX), whose description is 200% of the performance of the S&P 500 Index. What is its approximate market model?

b. If the S&P gains 30% in a year, what return do you expect ULPIX to have?

c. If the S&P loses 35% in a year, what return do you expect ULPIX to have?

d. What type of investors should be attracted to leveraged funds? Which type of investors should stay away from these funds?

13.51 The data in the file REFRIGERATOR represent the approximate retail price and the energy cost per year of 10 medium-size top-freezer refrigerators.

Source: "For the Chill of It," Copyright © 2002 by Consumers Union of U.S., Inc., Yonkers, NY 10703–1057. Adapted with permission from Consumer Reports, *August 2002, 26.*

a. Compute the coefficient of correlation r.

b. At the 0.05 level of significance, is there a significant linear relationship between the retail price (in \$) and the energy cost per year (in \$) of medium-size top-freezer refrigerators?

13.52 The data in the file **SECURITY** represent the turnover rate of pre-boarding screeners at airports in 1998–1999 and the security violations detected per million passengers.

Source: Extracted from *Alan B. Krueger, "A Small Dose of Common Sense Would Help Congress Break the Gridlock Over Airport Security,"* The New York Times, *November 15, 2001, C2.*

a. Compute the coefficient of correlation r.
b. At the 0.05 level of significance, is there a significant linear relationship between the turnover rate of pre-boarding screeners and the security violations detected?
c. What conclusions can you reach about the relationship between the turnover rate of pre-boarding screeners and the security violations detected?

13.53 The data in the file **CELLPHONE** represents the digital-mode talk time in hours and the battery capacity in milliampere-hours of cellphones.

Talk Time	Battery Capacity	Talk Time	Battery Capacity
4.50	800	1.50	450
4.00	1,500	2.25	900
3.00	1,300	2.25	900
2.00	1,550	3.25	900
2.75	900	2.25	700
1.75	875	2.25	800
1.75	750	2.50	800
2.25	1,100	2.25	900
1.75	850	2.00	900

Source: "Service Shortcomings," Copyright 2002 by Consumers Union of U.S., Inc., Yonkers, NY 10703-1057. Adapted with permission from Consumer Reports, *February 2002, 25.*

a. Compute the coefficient of correlation r.
b. At the 0.05 level of significance, is there a significant linear relationship between the battery capacity and the digital-mode talk time?
c. What conclusions can you reach about the relationship between the battery capacity and the digital-mode talk time?
d. You would expect the cellphones with higher battery capacity to have a higher talk time. Do the data support your expectations?

13.54 The data in the file **BATTERIES2** represent the price and cold-cranking amps (which denotes the starting current the battery can deliver) of automobile batteries.

Source: "Leading the Charge," Copyright 2001 by Consumers Union of U.S., Inc., Yonkers, NY 10703-1057. Adapted with permission from Consumer Reports, *October 2001, 25.*

a. Compute the coefficient of correlation r.
b. At the 0.05 level of significance, is there a significant linear relationship between the cold-cranking amps and the price?
c. What conclusions can you reach about the relationship between the cold-cranking amps and the price?
d. You would expect the batteries with higher cold-cranking amps to have a higher price. Do the data support your expectations?

13.8 ESTIMATION OF MEAN VALUES AND PREDICTION OF INDIVIDUAL VALUES

This section presents methods of making inferences about the mean of Y and the prediction of individual values of Y.

The Confidence Interval Estimate

In Example 13.2 on page 521, you used the prediction line to make predictions about the value of Y for a given X. The mean yearly sales for stores with 4,000 square feet was predicted to be 7.644 millions of dollars ($7,644,000). This estimate, however, is a *point estimate* of the population mean value. In Chapter 8, you studied the concept of the confidence interval as an estimate of the population mean. In a similar fashion, Equation (13.20) defines the **confidence interval estimate for the mean response** for a given X.

CONFIDENCE INTERVAL ESTIMATE FOR THE MEAN OF Y

$$\hat{Y}_i \pm t_{n-2} S_{YX} \sqrt{h_i} \qquad \textbf{(13.20)}$$

$$\hat{Y}_i - t_{n-2} S_{YX} \sqrt{h_i} \le \mu_{Y|X=X_i} \le \hat{Y}_i + t_{n-2} S_{YX} \sqrt{h_i}$$

where
$$h_i = \frac{1}{n} + \frac{(X_i - \bar{X})^2}{SSX}$$

$\hat{Y}_i$ = predicted value of Y; $\hat{Y}_i = b_0 + b_1 X_i$

S_{YX} = standard error of the estimate

n = sample size

X_i = given value of X

$\mu_{Y|X=X_i}$ = mean value of Y when $X = X_i$

$$SSX = \sum_{i=1}^{n} (X_i - \bar{X})^2$$

The width of the confidence interval in Equation (13.20) depends on several factors. For a given level of confidence, increased variation around the prediction line, as measured by the standard error of the estimate, results in a wider interval. However, as you would expect, increased sample size reduces the width of the interval. In addition, the width of the interval also varies at different values of X. When you predict Y for values of X close to $\bar{X}$, the interval is narrower than for predictions for X values more distant from $\bar{X}$.

In the site selection example, suppose you want a 95% confidence interval estimate of the mean annual sales for the entire population of stores that contain 4,000 square feet ($X = 4$). Using the simple linear regression equation:

$$\hat{Y}_i = 0.964 + 1.670 X_i$$
$$= 0.964 + 1.670(4) = 7.644 \text{(millions of dollars)}$$

Also, given the following

$$\bar{X} = 2.9214 \qquad S_{YX} = 0.9664$$

$$SSX = \sum_{i=1}^{n} (X_i - \bar{X})^2 = 37.9236$$

From Table E.3, $t_{12} = 2.1788$. Thus,

$$\hat{Y}_i \pm t_{n-2} S_{YX} \sqrt{h_i}$$

where

$$h_i = \frac{1}{n} + \frac{(X_i - \bar{X})^2}{SSX}$$

so that

$$\hat{Y}_i \pm t_{n-2} S_{YX} \sqrt{\frac{1}{n} + \frac{(X_i - \bar{X})^2}{SSX}}$$

$$= 7.644 \pm (2.1788)(0.9664)\sqrt{\frac{1}{14} + \frac{(4 - 2.9214)^2}{37.9236}}$$

$$= 7.644 \pm 0.673$$

so

$$6.971 \leq \mu_{Y|X=4} \leq 8.317$$

Therefore, the 95% confidence interval estimate is that the mean annual sales are between 6.971 and 8.317 (millions of dollars) for the population of stores with 4,000 square feet.

The Prediction Interval

In addition to the need for a confidence interval estimate for the mean value, you often want to predict the response for an individual value. Although the form of the prediction interval is similar to the confidence interval estimate of Equation (13.20), the prediction interval is predicting an individual value, not estimating a parameter. Equation (13.21) defines the **prediction interval for an individual response *Y***, at a particular value X_i, denoted by $Y_{X=X_i}$.

PREDICTION INTERVAL FOR AN INDIVIDUAL RESPONSE *Y*

$$\hat{Y}_i \pm t_{n-2} S_{YX} \sqrt{1 + h_i} \qquad \textbf{(13.21)}$$

$$\hat{Y}_i - t_{n-2} S_{YX} \sqrt{1 + h_i} \leq Y_{X=X_i} \leq \hat{Y}_i + t_{n-2} S_{YX} \sqrt{1 + h_i}$$

where h_i, $\hat{Y}_i$, S_{YX}, n, and X_i are defined as in Equation (13.20) on page 550 and $Y_{X=X_i}$ = a future value of Y when $X = X_i$.

To construct a 95% prediction interval of the annual sales for an individual store that contains 4,000 square feet ($X = 4$), you first compute $\hat{Y}_i$. Using the prediction line:

$$\hat{Y}_i = 0.964 + 1.670X_i$$

$$= 0.964 + 1.670(4)$$

$$= 7.644 \text{ (millions of dollars)}$$

Also, given the following:

$$\bar{X} = 2.9214 \qquad S_{YX} = 0.9664$$

$$SSX = \sum_{i=1}^{n} (X_i - \bar{X})^2 = 37.9236$$

From Table E.3, $t_{12} = 2.1788$. Thus,

$$\hat{Y}_i \pm t_{n-2} S_{YX} \sqrt{1 + h_i}$$

where

$$h_i = \frac{1}{n} + \frac{(X_i - \bar{X})^2}{\sum_{i=1}^{n} (X_i - \bar{X})^2}$$

so that

$$\hat{Y}_i \pm t_{n-2} S_{YX} \sqrt{1 + \frac{1}{n} + \frac{(X_i - \bar{X})^2}{SSX}}$$

$$= 7.644 \pm (2.1788)(0.9664)\sqrt{1 + \frac{1}{14} + \frac{(4 - 2.9214)^2}{37.9236}}$$

$$= 7.644 \pm 2.210$$

so

$$5.433 \le Y_{X=4} \le 9.854$$

Therefore, with 95% confidence you predict that the annual sales for an individual store with 4,000 square feet is between \$5,433,000 and \$9,854,000.

Figure 13.27 is a Microsoft Excel worksheet that illustrates the confidence interval estimate and the prediction interval for the site selection problem, and Figure 13.28 shows Minitab output. If you compare the results of the confidence interval estimate and the prediction interval, you see that the width of the prediction interval for an individual store is much wider than the confidence interval estimate for the mean store. Remember that there is much more variation in predicting an individual value than in estimating a mean value.

FIGURE 13.27

Microsoft Excel Confidence Interval Estimate and Prediction Interval for the Site Selection Problem

	A	B	
1	**Site Selection Analysis**		
2			
3	**Data**		
4	**X Value**	**4**	
5	**Confidence Level**	**95%**	
6			
7	Intermediate Calculations		
8	Sample Size	14	**=DataCopy!F2**
9	Degrees of Freedom	12	**=B8 - 2**
10	t Value	2.1788	**=TINV(1 - B5, B9)**
11	Sample Mean	2.9214	**=DataCopy!F3**
12	Sum of Squared Difference	37.9236	**=DataCopy!F4**
13	Standard Error of the Estimate	0.9664	*(regression worksheet cell B7 value)*
14	h Statistic	0.1021	**=1/B8 + (B4 - B11)^2/B12**
15	Predicted Y (YHat)	7.6439	**=DataCopy!F5**
16			
17	**For Average Y**		
18	Interval Half Width	0.6728	**=B10 * B13 * SQRT(B14)**
19	**Confidence Interval Lower Limit**	**6.9711**	**=B15 - B18**
20	**Confidence Interval Upper Limit**	**8.3167**	**=B15 + B18**
21			
22	**For Individual Response Y**		
23	Interval Half Width	2.2104	**=B10 * B13 * SQRT(1 + B14)**
24	**Prediction Interval Lower Limit**	**5.4335**	**=B15 - B23**
25	**Prediction Interval Upper Limit**	**9.8544**	**=B15 + B23**

FIGURE 13.28
Minitab Confidence Interval Estimate and Prediction Interval for the Site Selection Problem

```
Predicted Values for New Observations

New
Obs     Fit  SE Fit       95% CI            95% PI
  1   7.644   0.309  (6.971, 8.317)  (5.433, 9.854)


Values of Predictors for New Observations

New  Square
Obs    Feet
  1    4.00
```

PROBLEMS FOR SECTION 13.8

Learning the Basics

PH Grade ASSIST **13.55** Based on a sample of $n = 20$, the least-squares method was used to develop the following prediction line: $\hat{Y}_i = 5 + 3X_i$. In addition,

$$S_{YX} = 1.0,\ \bar{X} = 2,\ \text{and}\ \sum_{i=1}^{n}(X_i - \bar{X})^2 = 20$$

a. Construct a 95% confidence interval estimate of the mean response for $X = 2$.
b. Construct a 95% prediction interval of an individual response for $X = 2$.

PH Grade ASSIST **13.56** Based on a sample of $n = 20$, the least-squares method was used to develop the following prediction line: $\hat{Y}_i = 5 + 3X_i$. In addition,

$$S_{YX} = 1.0,\ \bar{X} = 2,\ \text{and}\ \sum_{i=1}^{n}(X_i - \bar{X})^2 = 20$$

a. Construct a 95% confidence interval estimate of the population mean response for $X = 4$.
b. Construct a 95% prediction interval of an individual response for $X = 4$.
c. Compare the results of (a) and (b) with those of problem 13.55 (a) and (b). Which interval is wider? Why?

Applying the Concepts

Problems 13.57–13.63 can be solved manually or by using Microsoft Excel, Minitab, or SPSS.

13.57 In problem 13.5 on page 524, you used reported sales to predict audited sales of magazines. CIRCULATION For these data $S_{YX} = 42.186$ and $h_i = 0.108$.
a. Construct a 95% confidence interval estimate of the mean audited sales for magazines that report newsstand sales of 400,000.
b. Construct a 95% prediction interval of the audited sales for an individual magazine that reports newsstand sales of 400,000.
c. Explain the difference in the results in (a) and (b).

PH Grade ASSIST SELF Test **13.58** In problem 13.4 on page 523, the marketing manager used shelf space for pet food to predict weekly sales. PETFOOD For these data $S_{YX} = 0.3081$ and $h_i = 0.1373$.
a. Construct a 95% confidence interval estimate of the mean weekly sales for all stores that have 8 feet of shelf space for pet food.
b. Construct a 95% prediction interval of the weekly sales of an individual store that has 8 feet of shelf space for pet food.
c. Explain the difference in the results in (a) and (b).

13.59 In problem 13.7 on page 524, you used the weight of mail to predict the number of orders received. MAIL
a. Construct a 95% confidence interval estimate of the mean number of orders received for all packages with a weight of 500 pounds.
b. Construct a 95% prediction interval of the number of orders received for an individual package with a weight of 500 pounds.
c. Explain the difference in the results in (a) and (b).

13.60 In problem 13.6 on page 524, the owner of a moving company wanted to predict labor hours based on the number of cubic feet moved. MOVING
a. Construct a 95% confidence interval estimate of the mean labor hours for all moves of 500 cubic feet.
b. Construct a 95% prediction interval of the labor hours of an individual move that has 500 cubic feet.
c. Explain the difference in the results in (a) and (b).

13.61 In problem 13.9 on page 525, an agent for a real estate company wanted to predict the monthly rent for apartments based on the size of the apartment. RENT
a. Construct a 95% confidence interval estimate of the mean monthly rental for all apartments that are 1,000 square feet in size.
b. Construct a 95% prediction interval of the monthly rental of an individual apartment that is 1,000 square feet in size.
c. Explain the difference in the results in (a) and (b).

13.62 In problem 13.8 on page 525, you predicted the value of a baseball franchise based on current revenue. BBREVENUE

a. Construct a 95% confidence interval estimate of the mean value of all baseball franchises that generate $150 million of annual revenue.

b. Construct a 95% prediction interval of the value of an individual baseball franchise that generates $150 million of annual revenue.

c. Explain the difference in the results in (a) and (b).

13.63 In problem 13.10 on page 525, you used hardness to predict the tensile strength of die-cast aluminum. HARDNESS

a. Construct a 95% confidence interval estimate of the mean tensile strength for all specimens with a hardness of 30 Rockwell E units.

b. Construct a 95% prediction interval of the tensile strength for an individual specimen that has a hardness of 30 Rockwell E units.

c. Explain the difference in the results in (a) and (b).

13.9 PITFALLS IN REGRESSION AND ETHICAL ISSUES

Some of the pitfalls involved in using regression analysis are as follows.

- Lacking an awareness of the assumptions of least-squares regression
- Not knowing how to evaluate the assumptions of least-squares regression
- Not knowing what the alternatives to least-squares regression are if a particular assumption is violated
- Using a regression model without knowledge of the subject matter
- Extrapolating outside the relevant range
- Concluding that a significant relationship identified in an observational study is due to a cause-and-effect relationship

The widespread availability of spreadsheet and statistical software has removed the computational hurdle that prevented many users from applying regression analysis. However, for many users this enhanced availability of software has not been accompanied by an understanding of how to use regression analysis properly. A user who is not familiar with either the assumptions of regression or how to evaluate the assumptions cannot be expected to know what the alternatives to least-squares regression are if a particular assumption is violated.

The data in Table 13.7 illustrates the necessity of using scatter plots and residual analysis to go beyond the basic number crunching of computing the Y intercept, the slope, and r^2.

TABLE 13.7
Four Sets of Artificial Data ANSCOMBE

Data Set A		Data Set B		Data Set C		Data Set D	
X_i	Y_i	X_i	Y_i	X_i	Y_i	X_i	Y_i
10	8.04	10	9.14	10	7.46	8	6.58
14	9.96	14	8.10	14	8.84	8	5.76
5	5.68	5	4.74	5	5.73	8	7.71
8	6.95	8	8.14	8	6.77	8	8.84
9	8.81	9	8.77	9	7.11	8	8.47
12	10.84	12	9.13	12	8.15	8	7.04
4	4.26	4	3.10	4	5.39	8	5.25
7	4.82	7	7.26	7	6.42	19	12.50
11	8.33	11	9.26	11	7.81	8	5.56
13	7.58	13	8.74	13	12.74	8	7.91
6	7.24	6	6.13	6	6.08	8	6.89

Source: F. J. Anscombe, "Graphs in Statistical Analysis," American Statistician, *Vol. 27 (1973), 17–21.*

Anscombe (reference 1) showed that all four data sets given in Table 13.7 have the following identical results:

$$\hat{Y}_i = 3.0 + 0.5X_i$$

$$S_{YX} = 1.237$$

$$S_{b_1} = 0.118$$

$$r^2 = 0.667$$

$$SSR = \text{explained variation} = \sum_{i=1}^{n}(\hat{Y}_i - \overline{Y})^2 = 27.51$$

$$SSE = \text{unexplained variation} = \sum_{i=1}^{n}(Y_i - \hat{Y}_i)^2 = 13.76$$

$$SST = \text{total variation} = \sum_{i=1}^{n}(Y_i - \overline{Y})^2 = 41.27$$

Thus, with respect to these statistics associated with a simple linear regression, the four data sets are identical. Were you to stop the analysis at this point, you would lose valuable information in the data. By examining Figure 13.29, which represents scatter diagrams for the four data sets, and Figure 13.30, which represents residual plots for the four data sets, you can clearly see that each of the four data sets has a different relationship between X and Y.

From the scatter diagrams of Figure 13.29 and the residual plots of Figure 13.30, you see how different the data sets are. The only data set that seems to follow an approximate straight line is data set A. The residual plot for data set A does not show any obvious patterns or outlying residuals. This is certainly not true for data sets B, C, and D. The scatter plot for data set B shows that a quadratic regression model (see section 15.1) is more appropriate. This conclusion is reinforced by the clear parabolic form of the residual plot for B. The scatter diagram and the residual plot for data set C clearly show an outlying observation. If this is the case, you may want to remove the outlier and reestimate the regression model. (Section 15.3 introduces influence analysis, a formal methodology to help determine whether the removal of observations is appropriate.) Reestimating the model will uncover a much different relationship. Similarly, the scatter diagram for data set D also represents the situation in which the model is heavily dependent on the outcome of a single response ($X_8 = 19$ and $Y_8 = 12.50$). You would have to cautiously evaluate any regression model since its regression coefficients are heavily dependent on a single observation.

FIGURE 13.29

Scatter Diagrams for Four Data Sets

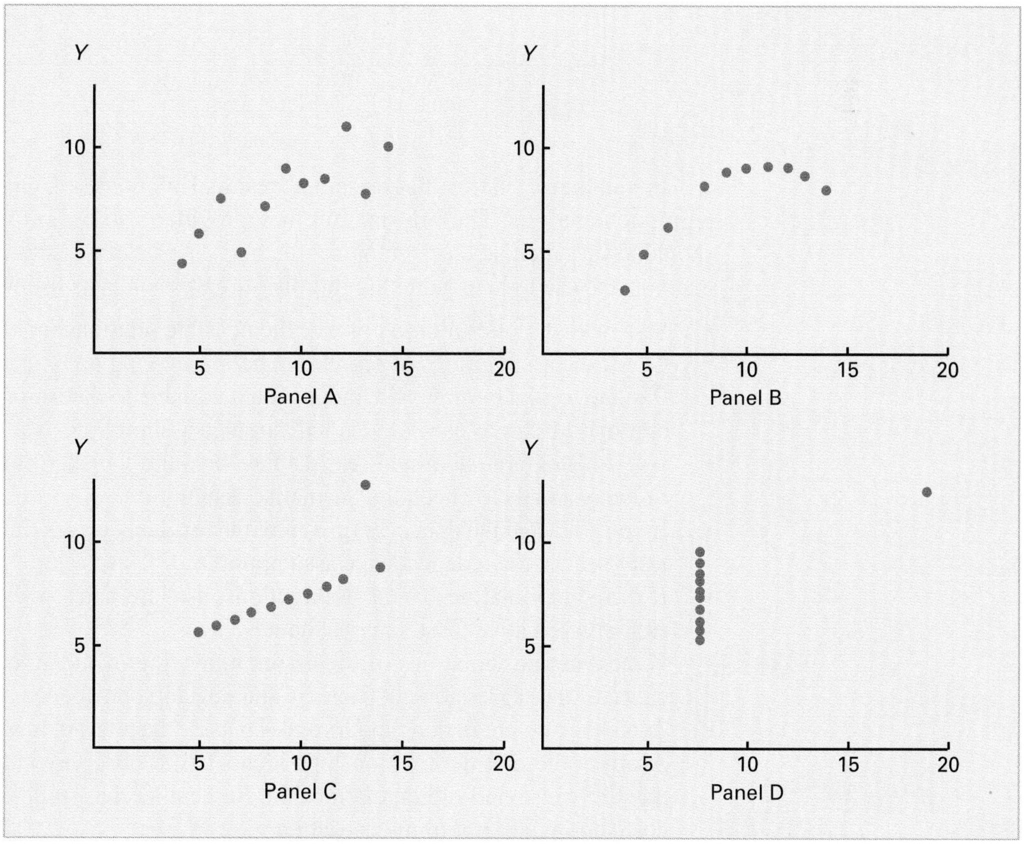

FIGURE 13.30
Residual Plots for Four Data Sets

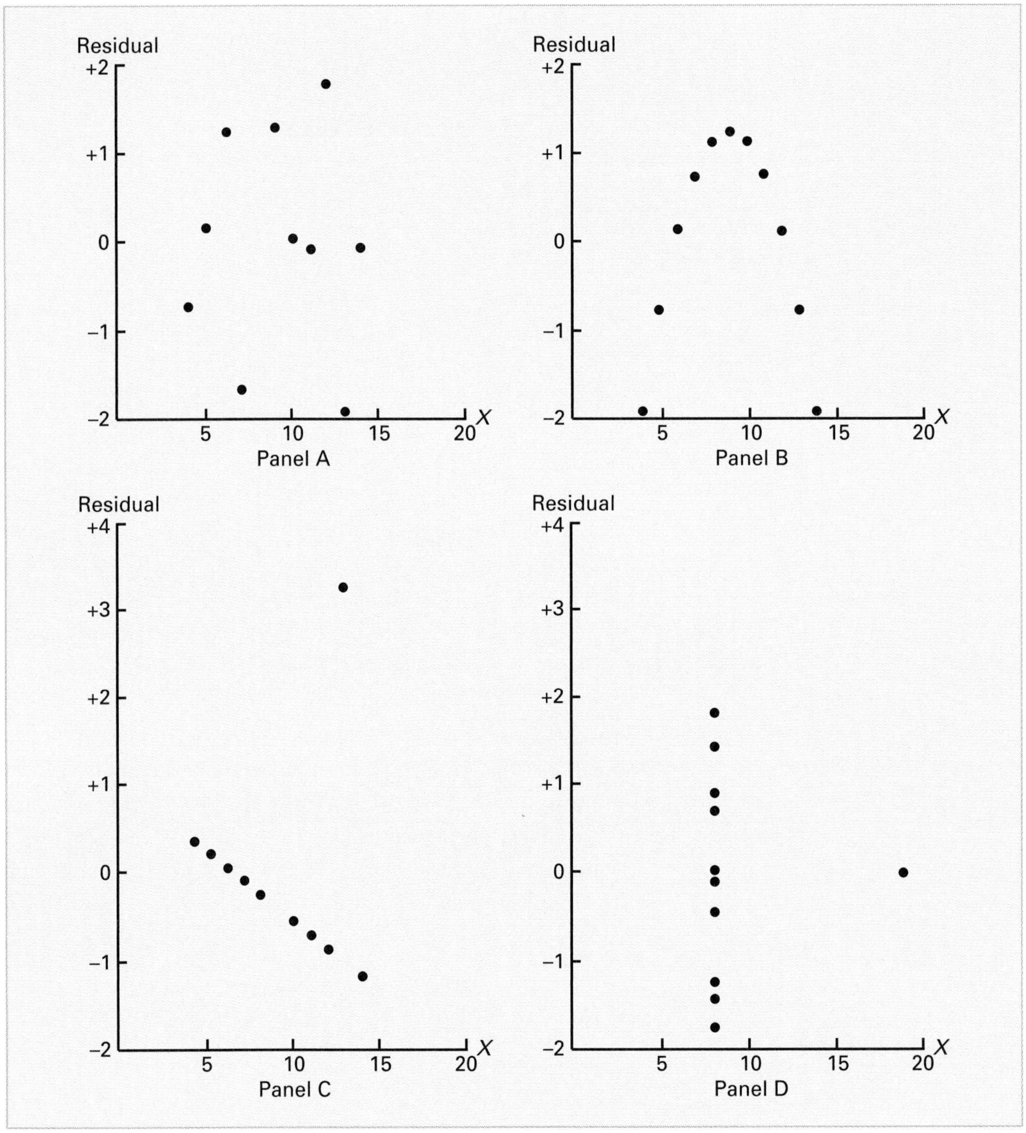

In summary, scatter diagrams and residual plots are of vital importance to a complete regression analysis. The information they provide is so basic to a credible analysis that you should always include these graphical methods as part of a regression analysis. Thus, a strategy that you can use to help avoid the pitfalls of regression is as follows.

- Start with a scatter plot to observe the possible relationship between X and Y.
- Check the assumptions of regression before moving on to using the results of the model.
- Plot the residuals versus the independent variable to determine whether the linear model is appropriate and to check the equal variance assumption.
- Use a histogram, stem-and-leaf display, box-and-whisker plot, or normal probability plot of the residuals to check the normality assumption.
- If you collected the data over time, plot the residuals versus time and use the Durbin-Watson test to check the independence assumption.
- If there are violations of the assumptions, use alternative methods to least-squares regression or alternative least-squares models.
- If there are no violations of the assumptions, then you can carry out tests for the significance of the regression coefficients and develop confidence and prediction intervals.
- Avoid making predictions and forecasts outside the relevant range of the independent variable.
- Always note that the relationships identified in observational studies may or may not be due to a cause-and-effect relationship. Remember that while causation implies correlation, correlation does not imply causation.

SUMMARY

As you can see from the chapter road map in Figure 13.31, this chapter develops the simple linear regression model and discusses the assumptions and how to evaluate them. Once you are assured that the model is appropriate, you can predict values using the prediction line and test for the significance of the slope.

You have learned how the director of planning for a clothing store can use regression analysis to investigate the relationship between the size of a store and its annual sales. You have used this analysis to make better decisions when selecting new sites for stores, as well as forecasting sales for existing stores. In Chapter 14, regression analysis is extended to situations where more than one independent variable is used to predict the value of a dependent variable.

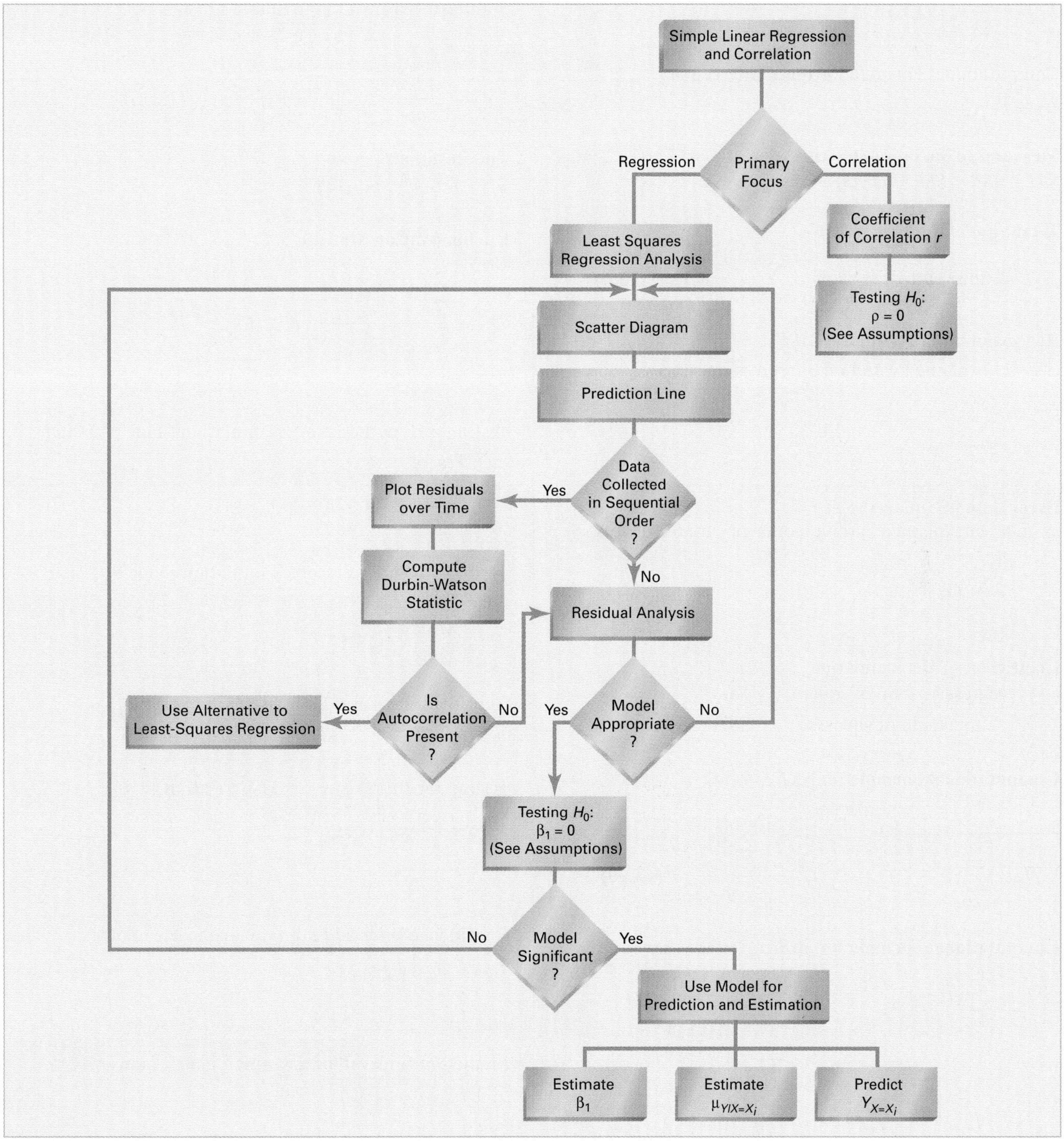

FIGURE 13.31 Road Map for Simple Linear Regression

KEY FORMULAS

Simple Linear Regression Model

$Y_i = \beta_0 + \beta_1 X_i + \varepsilon_i$ **(13.1)**

Simple Linear Regression Equation: The Prediction Line

$\hat{Y}_i = b_0 + b_1 X_i$ **(13.2)**

Computational Formula for the Slope b_1

$$b_1 = \frac{SSXY}{SSX} \quad \textbf{(13.3)}$$

Computational Formula for the Y Intercept b_0

$b_0 = \bar{Y} - b_1 \bar{X}$ **(13.4)**

Measures of Variation in Regression

$SST = SSR + SSE$ **(13.5)**

Total Sum of Squares (SST)

$$SST = \text{total sum of squares} = \sum_{i=1}^{n} (Y_i - \bar{Y})^2 \quad \textbf{(13.6)}$$

Regression Sum of Squares (SSR)

$SSR =$ explained variation or regression sum of squares

$$= \sum_{i=1}^{n} (\hat{Y}_i - \bar{Y})^2 \quad \textbf{(13.7)}$$

Error Sum of Squares (SSE)

$SSE =$ unexplained variation or error sum of squares

$$= \sum_{i=1}^{n} (Y_i - \hat{Y}_i)^2 \quad \textbf{(13.8)}$$

Coefficient of Determination

$$r^2 = \frac{\text{regression sum of squares}}{\text{total sum of squares}} = \frac{SSR}{SST} \quad \textbf{(13.9)}$$

Computational Formula for SST

$$SST = \sum_{i=1}^{n} (Y_i - \bar{Y})^2 = \sum_{i=1}^{n} Y_i^2 - \frac{\left(\sum_{i=1}^{n} Y_i\right)^2}{n} \quad \textbf{(13.10)}$$

Computational Formula for SSR

$$SSR = \sum_{i=1}^{n} (\hat{Y}_i - \bar{Y})^2 \quad \textbf{(13.11)}$$

$$= b_0 \sum_{i=1}^{n} Y_i + b_1 \sum_{i=1}^{n} X_i Y_i - \frac{\left(\sum_{i=1}^{n} Y_i\right)^2}{n}$$

Computational Formula for SSE

$$SSE = \sum_{i=1}^{n} (Y_i - \hat{Y})^2 \quad \textbf{(13.12)}$$

$$= \sum_{i=1}^{n} Y_i^2 - b_0 \sum_{i=1}^{n} Y_i - b_1 \sum_{i=1}^{n} X_i Y_i$$

Standard Error of the Estimate

$$S_{YX} = \sqrt{\frac{SSE}{n-2}} = \sqrt{\frac{\sum_{i=1}^{n} (Y_i - \hat{Y}_i)^2}{n-2}} \quad \textbf{(13.13)}$$

The Residual

$e_i = Y_i - \hat{Y}_i$ **(13.14)**

Durbin-Watson Statistic

$$D = \frac{\sum_{i=2}^{n} (e_i - e_{i-1})^2}{\sum_{i=1}^{n} e_i^2} \quad \textbf{(13.15)}$$

Testing a Hypothesis for a Population Slope β_1 Using the t Test

$$t = \frac{b_1 - \beta_1}{S_{b_1}} \quad \textbf{(13.16)}$$

Testing a Hypothesis for a Population Slope β_1 Using the F Test

$$F = \frac{MSR}{MSE} \quad \textbf{(13.17)}$$

Confidence Interval Estimate of the Slope β_1

$b_1 \pm t_{n-2} S_{b_1}$ **(13.18)**

Testing for the Existence of Correlation

$$t = \frac{r - \rho}{\sqrt{\dfrac{1 - r^2}{n-2}}} \quad \textbf{(13.19)}$$

Confidence Interval Estimate for the Mean of Y

$$\hat{Y}_i \pm t_{n-2} S_{YX} \sqrt{h_i} \quad \textbf{(13.20)}$$

$$\hat{Y}_i - t_{n-2} S_{YX} \sqrt{h_i} \le \mu_{Y|X=X_i} \le \hat{Y}_i + t_{n-2} S_{YX} \sqrt{h_i}$$

Prediction Interval for an Individual Response Y

$$\hat{Y}_i \pm t_{n-2} S_{YX} \sqrt{1 + h_i} \quad \textbf{(13.21)}$$

$$\hat{Y}_i - t_{n-2} S_{YX} \sqrt{1 + h_i} \le Y_{X=X_i} \le \hat{Y}_i + t_{n-2} S_{YX} \sqrt{1 + h_i}$$

KEY TERMS

CHAPTER REVIEW PROBLEMS

Checking Your Understanding

13.64 What is the interpretation of the *Y* intercept and the slope in the simple linear regression equation?

13.65 What is the interpretation of the coefficient of determination?

13.66 When will the unexplained variation (i.e., error sum of squares) be equal to 0?

13.67 When will the explained variation (i.e., sum of squares due to regression) be equal to 0?

13.68 Why should you always carry out a residual analysis as part of a regression model?

13.69 What are the assumptions of regression analysis?

13.70 How do you evaluate the assumptions of regression analysis?

13.71 When and how do you use the Durbin-Watson statistic?

13.72 What is the difference between a confidence interval estimate of the mean response $\mu_{Y|X=X_i}$ and a prediction interval of $Y_{X=X_i}$?

Applying the Concepts

Problems 13.74–13.92 can be solved manually or by using Microsoft Excel, Minitab, or SPSS.

13.73 Researchers from the Lubin School of Business at Pace University in New York City conducted a study on Internet-supported courses. In one part of the study, four numerical variables were collected on 108 students in an introductory management course that met once a week for an entire semester. One variable collected was *hit consistency*. To measure hit consistency, the researchers did the following:

If a student did not visit the Internet site between classes, the student was given a zero for that time period. If a student visited the Internet site one or more times between classes, the student was given a one for that time period. Since there were 13 time periods, a student's score on hit consistency could range from 0 to 13.

The other three variables included the student's course average, the student's cumulative GPA, and the total number of hits the student had on the Internet site supporting the course. The following table gives the correlation coefficient for all pairs of variables. Note that correlations marked with an * are statistically significant using $\alpha = 0.001$.

Variables	Correlation
Course Average, Cumulative GPA	0.72*
Course Average, Total Hits	0.08
Course Average, Hit Consistency	0.37*
Cumulative GPA, Total Hits	0.12
Cumulative GPA, Hit Consistency	0.32*
Total Hits, Hit Consistency	0.64*

Source: Adapted from Baugher, D., Varanelli, A. and E. Weisbord, "Student Hits in an Internet-Supported Course: How Can Instructors Use Them and What Do They Mean?" Decision Sciences Journal of Innovative Education, *Fall 2003, 1(2): 159–179.*

a. What conclusions can you reach from this correlation analysis?

b. Are you surprised by the results or are they consistent with your own observations and experiences?

13.74 Management of a soft-drink bottling company wants to develop a method for allocating delivery costs to customers. Although one cost clearly relates to travel time within a particular route, another variable cost reflects the time required to unload the cases of soft drink at the delivery point. A sample of 20 deliveries within a territory was selected. The delivery time and the number of cases delivered were recorded: DELIVERY

Customer	Number of Cases	Delivery Time (Minutes)	Customer	Number of Cases	Delivery Time (Minutes)
1	52	32.1	11	161	43.0
2	64	34.8	12	184	49.4
3	73	36.2	13	202	57.2
4	85	37.8	14	218	56.8
5	95	37.8	15	243	60.6
6	103	39.7	16	254	61.2
7	116	38.5	17	267	58.2
8	121	41.9	18	275	63.1
9	143	44.2	19	287	65.6
10	157	47.1	20	298	67.3

Develop a regression model to predict delivery time based on the number of cases delivered.

a. Use the least-squares method to compute the regression coefficients b_0 and b_1.
b. Interpret the meaning of b_0 and b_1 in this problem.
c. Predict the delivery time for 150 cases of soft drink.
d. Would it be appropriate to use the model to predict the delivery time for a customer who is receiving 500 cases of soft drink? Why?
e. Determine the coefficient of determination r^2 and explain its meaning in this problem.
f. Perform a residual analysis. Is there any evidence of a pattern in the residuals? Explain.
g. At the 0.05 level of significance, is there evidence of a linear relationship between delivery time and the number of cases delivered?
h. Construct a 95% confidence interval estimate of the mean delivery time for 150 cases of soft drink.
i. Construct a 95% prediction interval of the delivery time for a single delivery of 150 cases of soft drink.
j. Construct a 95% confidence interval estimate of the population slope.
k. Explain how the results in (a) through (j) can help allocate delivery costs to customers.

13.75 A brokerage house wants to predict the number of trade executions per day using the number of incoming phone calls as a predictor variable. Data were collected over a period of 35 days. TRADES

a. Use the least-squares method to compute the regression coefficients b_0 and b_1.
b. Interpret the meaning of b_0 and b_1 in this problem.
c. Predict the number of trades executed for a day in which the number of incoming calls is 2,000.
d. Is it appropriate to use the model to predict the number of trades executed for a day in which the number of incoming calls is 5,000? Why?
e. Determine the coefficient of determination r^2 and explain its meaning in this problem.
f. Plot the residuals against the number of incoming calls and also against the days. Is there any evidence of a pattern in the residuals with either of these variables? Explain.
g. Determine the Durbin-Watson statistic for these data.
h. Based on the results of (f) and (g), is there reason to question the validity of the model? Explain.
i. At the 0.05 level of significance, is there evidence of a linear relationship between the volume of trade executions and the number of incoming calls?
j. Construct a 95% confidence interval estimate of the mean number of trades executed for days in which the number of incoming calls is 2,000.
k. Construct a 95% prediction interval of the number of trades executed for a particular day in which the number of incoming calls is 2,000.
l. Construct a 95% confidence interval estimate of the population slope.
m. Based on the results of (a) through (l), do you think the brokerage house should focus on a strategy of increasing the total number of incoming calls or on a strategy that relies on trading by a small number of heavy traders? Explain.

13.76 You want to develop a model to predict the selling price of homes based on assessed value. A sample of 30 recently sold single-family houses in a small city is selected to study the relationship between selling price (in thousands of dollars) and assessed value (in thousands of dollars). The houses in the city had been reassessed at full value one year prior to the study. The results are in the file HOUSE1. (*Hint:* First, determine which are the independent and dependent variables.)

a. Plot a scatter diagram and, assuming a linear relationship, use the least-squares method to compute the regression coefficients b_0 and b_1.
b. Interpret the meaning of the Y intercept b_0 and the slope b_1 in this problem.
c. Use the prediction line developed in (a) to predict the selling price for a house whose assessed value is $70,000.
d. Determine the coefficient of determination r^2 and interpret its meaning in this problem.
e. Perform a residual analysis on your results and determine the adequacy of the fit of the model.
f. At the 0.05 level of significance, is there evidence of a linear relationship between selling price and assessed value?
g. Construct a 95% confidence interval estimate of the mean selling price for houses with an assessed value of $70,000.
h. Construct a 95% prediction interval of the selling price of an individual house with an assessed value of $70,000.
i. Construct a 95% confidence interval estimate of the population slope.

13.77 You want to develop a model to predict the assessed value of houses based on heating area. A sample of 15 single-family houses is selected in a city. The assessed value (in thousands of dollars) and the heating area of the houses (in thousands of square feet) are recorded with the following results: HOUSE2

House	Assessed Value ($000)	Heating Area of Dwelling (Thousands of Square Feet)
1	84.4	2.00
2	77.4	1.71
3	75.7	1.45
4	85.9	1.76
5	79.1	1.93
6	70.4	1.20
7	75.8	1.55
8	85.9	1.93
9	78.5	1.59
10	79.2	1.50
11	86.7	1.90
12	79.3	1.39
13	74.5	1.54
14	83.8	1.89
15	76.8	1.59

(*Hint:* First, determine which are the independent and dependent variables.)

a. Plot a scatter diagram and, assuming a linear relationship, use the least-squares method to compute the regression coefficients b_0 and b_1.

b. Interpret the meaning of the Y intercept b_0 and the slope b_1 in this problem.

c. Use the prediction line developed in (a) to predict the assessed value for a house whose heating area is 1,750 square feet.

d. Determine the coefficient of determination r^2 and interpret its meaning in this problem.

e. Perform a residual analysis on your results and determine the adequacy of the fit of the model.

f. At the 0.05 level of significance, is there evidence of a linear relationship between assessed value and heating area?

g. Construct a 95% confidence interval estimate of the mean assessed value for houses with a heating area of 1,750 square feet.

h. Construct a 95% prediction interval of the assessed value of an individual house with a heating area of 1,750 square feet.

i. Construct a 95% confidence interval estimate of the population slope.

13.78 The director of graduate studies at a large college of business would like to predict the grade point index (GPI) of students in an MBA program based on the Graduate Management Aptitude Test (GMAT) score. A sample of 20 students who had completed 2 years in the program is selected. The results are as follows: GPIGMAT

Observation	GMAT Score	GPI	Observation	GMAT Score	GPI
1	688	3.72	11	567	3.07
2	647	3.44	12	542	2.86
3	652	3.21	13	551	2.91
4	608	3.29	14	573	2.79
5	680	3.91	15	536	3.00
6	617	3.28	16	639	3.55
7	557	3.02	17	619	3.47
8	599	3.13	18	694	3.60
9	616	3.45	19	718	3.88
10	594	3.33	20	759	3.76

(*Hint:* First, determine which are the independent and dependent variables.)

a. Plot a scatter diagram and, assuming a linear relationship, use the least-squares method to compute the regression coefficients b_0 and b_1.

b. Interpret the meaning of the Y intercept b_0 and the slope b_1 in this problem.

c. Use the prediction line developed in (a) to predict the GPI for a student with a GMAT score of 600.

d. Determine the coefficient of determination r^2 and interpret its meaning in this problem.

e. Perform a residual analysis on your results and determine the adequacy of the fit of the model.

f. At the 0.05 level of significance, is there evidence of a linear relationship between GMAT score and GPI?

g. Construct a 95% confidence interval estimate for the mean GPI of students with a GMAT score of 600.

h. Construct a 95% prediction interval for a particular student with a GMAT score of 600.

i. Construct a 95% confidence interval estimate of the population slope.

13.79 The manager of the purchasing department of a large banking organization would like to develop a model to predict the amount of time it takes to process invoices. Data are collected from a sample of 30 days, and the number of invoices processed and completion time in hours is recorded. INVOICE (*Hint:* First, determine which are the independent and dependent variables.)

a. Assuming a linear relationship, use the least-squares method to compute the regression coefficients b_0 and b_1.

b. Interpret the meaning of the Y intercept b_0 and the slope b_1 in this problem.

c. Use the prediction line developed in (a) to predict the amount of time it would take to process 150 invoices.

d. Determine the coefficient of determination r^2 and interpret its meaning.

e. Plot the residuals against the number of invoices processed and also against time.

f. Based on the plots in (e), does the model seem appropriate?

g. Compute the Durbin-Watson statistic and, at the 0.05 level of significance, determine whether there is any autocorrelation in the residuals.

h. Based on the results of (e) through (g), what conclusions can you reach concerning the validity of the model?
i. At the 0.05 level of significance, is there evidence of a linear relationship between the amount of time and the number of invoices processed?
j. Construct a 95% confidence interval estimate of the mean amount of time it would take to process 150 invoices.
k. Construct a 95% prediction interval of the amount of time it would take to process 150 invoices on a particular day.

13.80 On January 28, 1986, the space shuttle *Challenger* exploded and seven astronauts were killed. Prior to the launch, the predicted atmospheric temperature was for freezing weather at the launch site. Engineers for Morton Thiokol (the manufacturer of the rocket motor) prepared charts to make the case that the launch should not take place due to the cold weather. These arguments were rejected and the launch tragically took place. Upon investigation after the tragedy, experts agreed that the disaster occurred because of leaky rubber O-rings that did not seal properly due to the cold temperature. Data indicating the atmospheric temperature at the time of 23 previous launches and the O-ring damage index are as follows: **O-RING**

Flight Number	Temperature (°F)	O-Ring Damage Index
1	66	0
2	70	4
3	69	0
5	68	0
6	67	0
7	72	0
8	73	0
9	70	0
41-B	57	4
41-C	63	2
41-D	70	4
41-G	78	0
51-A	67	0
51-B	75	0
51-C	53	11
51-D	67	0
51-F	81	0
51-G	70	0
51-I	67	0
51-J	79	0
61-A	75	4
61-B	76	0
61-C	58	4

Note: Data from flight 4 is omitted due to unknown O-ring condition.
Source: Extracted from Report of the Presidential Commission on the Space Shuttle Challenger Accident, *Washington, DC, 1986, Vol. II (H1–H3) and Vol. IV (664),* Post Challenger Evaluation of Space Shuttle Risk Assessment and Management, *Washington, DC, 1988, 135–136.*

a. Construct a scatter diagram for the seven flights in which there was O-ring damage (O-ring damage index $\neq 0$). What conclusions, if any, can you draw about the relationship between atmospheric temperature and O-ring damage?
b. Construct a scatter diagram for all 23 flights.
c. Explain any differences in the interpretation of the relationship between atmospheric temperature and O-ring damage in (a) and (b).
d. Based on the scatter diagram in (b), provide reasons for why a prediction should not be made for an atmospheric temperature of 31°F, the temperature on the morning of the launch of the *Challenger*.
e. Although the assumption of a linear relationship may not be valid, fit a simple linear regression model to predict O-ring damage based on atmospheric temperature.
f. Plot the straight line found in (e) on the scatter diagram developed in (b).
g. Based on the results of (f), do you think a straight line is an appropriate model for these data? Explain.
h. Perform a residual analysis. What conclusions do you reach?

13.81 Crazy Dave, a well-known baseball analyst, would like to study various team statistics for the 2003 baseball season to determine which variables might be useful in predicting the number of wins achieved by teams during the season. He has decided to begin by using the team earned run average (ERA), a measure of pitching performance, to predict the number of wins. The data for the 30 major league teams are in the file **BB2003**.

(*Hint:* First, determine which are the independent and dependent variables.)
a. Assuming a linear relationship, use the least-squares method to compute the regression coefficients b_0 and b_1.
b. Interpret the meaning of the Y intercept b_0 and the slope b_1 in this problem.
c. Use the prediction line developed in (a) to predict the number of wins for a team with an ERA of 4.50.
d. Compute the coefficient of determination r^2 and interpret its meaning.
e. Perform a residual analysis on your results and determine the adequacy of the fit of the model.
f. At the 0.05 level of significance, is there evidence of a linear relationship between the number of wins and the ERA?
g. Construct a 95% confidence interval estimate of the mean number of wins expected for teams with an ERA of 4.50.
h. Construct a 95% prediction interval of the number of wins for an individual team that has an ERA of 4.50.
i. Construct a 95% confidence interval estimate of the slope.
j. The 30 teams constitute a population. In order to use statistical inference [as in (f)–(i)], the data must be assumed to represent a random sample. What "population" would this sample be drawing conclusions about?

k. What other independent variables might you consider for inclusion in the model?

13.82 During the fall harvest season in the United States, pumpkins are sold in large quantities at farm stands. Often, instead of weighing the pumpkins prior to sale, the farm stand operator will just place the pumpkin in the appropriate circular cutout on the counter. When asked why this was done, one farmer replied, "I can tell the weight of the pumpkin from its circumference." To determine whether this was really true, a sample of 23 pumpkins were measured for circumference and weighed with the following results. **PUMPKIN**

Circumference (cm)	Weight (Grams)	Circumference (cm)	Weight (Grams)
50	1,200	57	2,000
55	2,000	66	2,500
54	1,500	82	4,600
52	1,700	83	4,600
37	500	70	3,100
52	1,000	34	600
53	1,500	51	1,500
47	1,400	50	1,500
51	1,500	49	1,600
63	2,500	60	2,300
33	500	59	2,100
43	1,000		

a. Assuming a linear relationship, use the least-squares method to compute the regression coefficients b_0 and b_1.
b. Interpret the meaning of the slope b_1 in this problem.
c. Predict the mean weight for a pumpkin that is 60 centimeters in circumference.
d. Do you think it is a good idea for the farmer to sell pumpkins by circumference instead of weight? Explain.
e. Determine the coefficient of determination r^2 and interpret its meaning.
f. Perform a residual analysis for these data and determine the adequacy of the fit of the model.
g. At the 0.05 level of significance, is there evidence of a linear relationship between the circumference and the weight of a pumpkin?
h. Construct a 95% confidence interval estimate of the population slope β_1.
i. Construct a 95% confidence interval estimate of the population mean weight for pumpkins that have a circumference of 60 centimeters.
j. Construct a 95% prediction interval of the weight for an individual pumpkin that has a circumference of 60 centimeters.

13.83 Can demographic information be helpful in predicting the sales for sporting goods stores? The data stored in **SPORTING** are the monthly sales totals from a random sample of 38 stores in a large chain of nationwide sporting goods stores. All stores in the franchise, and thus within the sample, are approximately the same size and carry the same merchandise. The county, or in some cases counties, in which the store draws the majority of its customers is referred to here as the customer base. For each of the 38 stores, demographic information about the customer base is provided. The data are real but the name of the franchise is not used at the request of the company. The variables in the data set are:

Sales: Latest one-month sales total (dollars)
Age: Median age of customer base (years)
HS: Percentage of customer base with a high school diploma
College: Percentage of customer base with a college diploma
Growth: Annual population growth rate of customer base over the past 10 years
Income: Median family income of customer base (dollars)

a. Construct a scatter diagram using sales as the dependent variable and median family income as the independent variable. Discuss the scatter diagram.
b. Assuming a linear relationship, use the least-squares method to compute the regression coefficients b_0 and b_1.
c. Interpret the meaning of the Y intercept b_0 and the slope b_1 in this problem.
d. Compute the coefficient of determination r^2 and interpret its meaning.
e. Perform a residual analysis on your results and determine the adequacy of the fit of the model.
f. At the 0.05 level of significance, is there evidence of a linear relationship between the independent variable and the dependent variable?
g. Construct a 95% confidence interval estimate of the slope and interpret its meaning.

13.84 For the data of problem 13.83, repeat (a) through (g) using age as the independent variable.

13.85 For the data of problem 13.83, repeat (a) through (g) using high school graduation rate as the independent variable.

13.86 For the data of problem 13.83, repeat (a) through (g) using college graduation rate as the independent variable.

13.87 For the data of problem 13.83, repeat (a) through (g) using growth as the independent variable.

13.88 Zagat's publishes restaurant ratings for various locations in the United States. The data file **RESTRATE** contains the Zagat rating for food, décor, service, and the price per person for a sample of 50 restaurants located in New York City and 50 restaurants located on Long Island. Develop a regression model to predict the price per person based on a variable that represents the sum of the ratings for food, décor, and service.

Source: Extracted from Zagat Survey 2002 New York City Restaurants *and* Zagat Survey 2001–2002, Long Island Restaurants.

a. Assuming a linear relationship, use the least-squares method to compute the regression coefficients b_0 and b_1.
b. Interpret the meaning of the Y intercept b_0 and the slope b_1 in this problem.
c. Use the prediction line developed in (a) to predict the price per person for a restaurant with a summated rating of 50.
d. Compute the coefficient of determination r^2 and interpret its meaning.
e. Perform a residual analysis on your results and determine the adequacy of the fit of the model.
f. At the 0.05 level of significance, is there evidence of a linear relationship between the price per person and the summated rating?
g. Construct a 95% confidence interval estimate of the mean price per person for all restaurants with a summated rating of 50.
h. Construct a 95% prediction interval of the price per person for a restaurant with a summated rating of 50.
i. Construct a 95% confidence interval estimate of the slope.
j. How useful do you think the summated rating is as a predictor of price? Explain.

13.89 Refer to the discussion of beta values and market models in problem 13.49 on page 548. The 2003 weekly data for the S&P 500 and three individual stocks are in the data file SP500. Note that the *weekly percentage change* for both the S&P 500 and the individual stocks is measured as the percentage change from the previous week's closing value to the current week's closing value. The variables included are:

Week: Current week
SP500: Weekly percentage change in the S&P 500 Index
SEARS: Weekly percentage change in stock price of Sears, Roebuck, and Company
TARGET: Weekly percentage change in stock price of the Target Corporation
SARALEE: Weekly percentage change in stock price of the Sara Lee Corporation

Source: Extracted from ***finance.yahoo.com****, January 20, 2004.*

a. Estimate the market model for Sears, Roebuck, and Company (*Hint:* Use the percentage change in the S&P 500 Index as the independent variable and the percentage change in Sears' stock price as the dependent variable.)
b. Interpret the beta value for Sears, Roebuck, and Company.
c. Repeat (a) and (b) for Target, Inc.
d. Repeat (a) and (b) for Sara Lee, Inc.
e. Write a brief summary of your findings.

13.90 The data file RETURNS contains the stock price of four companies collected weekly for 53 consecutive weeks ending July 26, 2004. The variables are:

Week: Closing date for stock prices
MSFT: Stock price for Microsoft, Inc.
Ford: Stock price of Ford Motor Company
GM: Stock price of General Motors, Inc.
IAL: Stock price of International Aluminum, Inc.

Source: Extracted from ***finance.yahoo.com****, July 30, 2004.*

a. Calculate the correlation coefficient r for each pair of stocks. (There are six of them.)
b. Interpret the meaning of r for each pair.
c. Is it a good idea to have all the stocks in an individual's portfolio be strongly positively correlated among each other? Explain.

13.91 Is the daily performance of stocks and bonds correlated? The data file STOCKS&BONDS contains information concerning the closing value of the Dow Jones Industrial Average and the Vanguard Long-Term Bond Index Fund for 60 consecutive business days ending July 29, 2004. The variables included are:

Date: Current day
Bonds: Closing price of Vanguard Long-Term Bond Index Fund DJIA
Stocks: Closing price of the Dow Jones Industrial Average

Source: Extracted from ***finance.yahoo.com****, July 29, 2004.*

a. Compute and interpret the correlation coefficient r for the variables stocks and bonds.
b. At the 0.05 level of significance, is there a relationship between these two variables? Explain.

Report Writing Exercises

13.92 In problems 13.83–13.87 on page 563, you developed regression models to predict monthly sales at a sporting goods store. Now, write a report based on the models you developed. Append to your report all appropriate charts and statistical information.

RUNNING CASE MANAGING THE *SPRINGVILLE HERALD*

To ensure that as many trial subscriptions as possible are converted to regular subscriptions, the *Herald* marketing department works closely with the distribution department to accomplish a smooth initial delivery process for the trial subscription customers. To assist in this effort, the marketing department needs to accurately forecast the number of new regular subscriptions for the coming months.

A team consisting of managers from the marketing and distribution departments was convened to develop a better method of forecasting new subscriptions. Previously, after examining new subscription data for the prior three months, a group of three managers would develop a subjective forecast of the number of new subscriptions. Lauren Hall, who was recently hired by the company to provide special skills in quantitative forecasting methods, suggested that the department look for factors that might help in predicting new subscriptions.

Members of the team found that the forecasts in the last year had been particularly inaccurate because in some months much more time was spent on telemarketing than in other months. In particular, in the last month, only 1,055 hours were completed since callers were busy during the first week of the month attending training sessions on the personal but formal greeting style and a new standard presentation guide (see "Managing the Springville Herald" for Chapter 11). Lauren collected the data for the number of new subscriptions and hours spent on telemarketing for each month for the past 2 years. **SH13**

EXERCISES

SH13.1 What criticism can you make concerning the method of forecasting that involved taking the new subscriptions for the past 3 months as the basis for future projections?

SH13.2 What factors other than number of telemarketing hours spent might be useful in predicting the number of new subscriptions? Explain.

SH13.3 **a.** Analyze the data and develop a statistical model to predict the mean number of new subscriptions for a month based on the number of hours spent on telemarketing for new subscriptions.

b. If you expect to spend 1,200 hours on telemarketing per month, estimate the mean number of new subscriptions for the month. Indicate the assumptions upon which this prediction is based. Do you think these assumptions are valid? Explain.

c. What would be the danger of predicting the number of new subscriptions for a month in which 2,000 hours were spent on telemarketing?

WEB CASE

Apply your knowledge of simple linear regression in this Web Case that extends the Sunflowers Clothing Stores "Using Statistics" scenario from this chapter.

Leasing agents from the Triangle Mall Management Corporation have suggested that Sunflowers consider several locations in some of Triangle's newly renovated "lifestyle" malls located in areas of higher than mean disposable income. Although the locations are smaller than the typical Sunflowers location, the leasing agents argue that higher than mean disposable income in the surrounding community is a better predictor of higher sales than store size. The leasing agents maintain that sample data from 14 Sunflowers stores prove that this is true.

Review the leasing agents' proposal and supporting documents that describe the data at the company's Web site **www.prenhall.com/Springville/Triangle_Sunflower.htm** and then answer the following:

1. Should mean disposable income be used to predict sales based on the sample of 14 Sunflowers stores?
2. Should the management of Sunflowers accept the claims of Triangle's leasing agents? Why or why not?
3. Is it possible that the mean disposable income of the surrounding area is not an important factor in leasing new locations? Explain.
4. Are there any other factors not mentioned by the leasing agents that might be relevant to the store leasing decision?

REFERENCES

1. Anscombe, F. J., "Graphs in Statistical Analysis," *The American Statistician* 27 (1973): 17–21.
2. Hoaglin, D. C., and R. Welsch, "The Hat Matrix in Regression and ANOVA," *The American Statistician* 32 (1978): 17–22.
3. Hocking, R. R., "Developments in Linear Regression Methodology: 1959–1982," *Technometrics* 25 (1983): 219–250.
4. Hosmer, D. W., and S. Lemeshow, *Applied Logistic Regression*, 2nd ed. (New York: Wiley, 2001).
5. Kutner, M. H., C. J. Nachtsheim, J. Neter, and W. Li, *Applied Linear Statistical Models*, 5th ed. (New York: McGraw-Hill/Irwin, 2005).
6. *Microsoft Excel 2003* (Redmond, WA: Microsoft Corporation, 2002).
7. *Minitab for Windows Version 14* (State College, PA: Minitab Inc., 2004).
8. Ramsey, P. P., and P. H. Ramsey, "Simple Tests of Normality in Small Samples," *Journal of Quality Technology* 22 (1990): 299–309.
9. *SPSS Base 12 Brief Guide* (Upper Saddle River, NJ: Prentice Hall, 2003).

Appendix 13 Using Software for Simple Linear Regression

A13.1 MICROSOFT EXCEL

For Simple Linear Regression

Open to a worksheet containing the data for the regression analysis (such as the Data worksheet of the **SITE.xls** file).
Select **Tools → Data Analysis**.
Select **Regression** from the Data Analysis list and click **OK**. In the Regression dialog box (see Figure A13.1):

Enter the cell range of the *Y* variable data as the **Input Y Range**.
Enter the cell range of the *X* variable data as the **Input X Range**.
Select **Labels** if the variable cell ranges include labels in their first rows.
Select **Confidence Level**.
Select **Residuals** and **Residual Plots** if you want to perform a residual analysis.
Click **OK**.

Results appear on a separate worksheet.

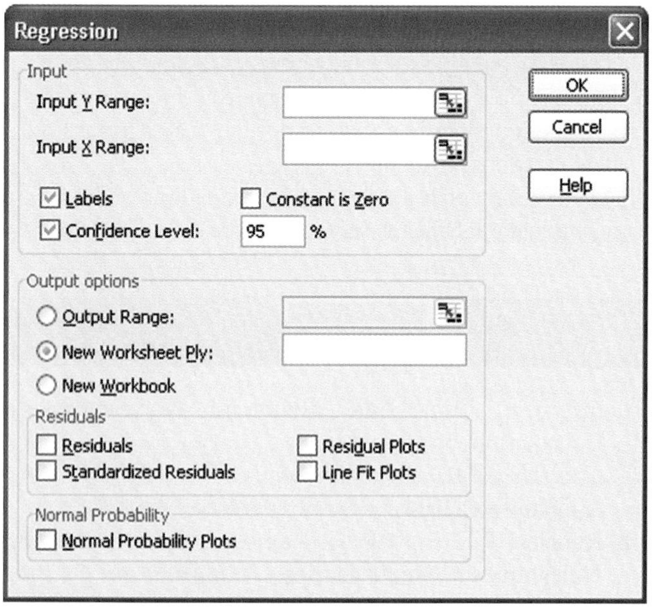

FIGURE A13.1 Microsoft Excel Regression Dialog Box

OR See section G.35 (**Simple Linear Regression**) if you want PHStat2 to produce a worksheet for you.

For Scatter Diagram and Regression Analysis

Use the Microsoft Excel Chart Wizard to generate a scatter diagram for a regression analysis. Open to the worksheet containing the data for the regression analysis. (This worksheet should place the data for each variable in separate columns and should be arranged so that the data for the *X* variable appears before the data for the *Y* variable.) Select **Insert → Chart**, and make these entries in the Chart Wizard dialog boxes:

Step 1: Click **XY (Scatter)** from the **Standard Types Chart type** box and leave the first **Chart sub-type** labeled as **Scatter. Compares pairs of values** selected. Click **Next**.

Step 2: Select the **Data Range** tab and enter the cell range of the data to be plotted as the **Data range**. Select **Columns** (for variable data that is column-wise) and click **Next**. (Reminder: the first column of the **Data range** must be the data for the *X* variable in order to create a correct scatter diagram.)

Step 3: Select the formatting and labeling chart options for the chart. (See "Using the Chart Wizard" in section A1.2 on page 18 for suggestions.) Click **Next**.

Step 4: Select **As new sheet** and click **Finish**.

If there are *Y* values that are less than zero, the Chart Wizard will place the *X* axis (and its labels) at $Y = 0$ and not at the bottom of the chart. To relocate the *X*-axis, right-click the *Y*-axis and select **Format Axis** from the shortcut menu. In the dialog box that appears, change the value at which the *X* axis crosses the *Y* axis in the **Scale** tab. Click **Ok**.

To add the prediction line to the scatter diagram that the wizard produces, open to the scatter diagram and select **Chart → Add Trendline** and in the Add Trendline dialog box (see Figure A13.2):

Type tab: Select **Linear**.
Options tab: Select **Automatic, Display equation on chart**, and **Display R-squared value on chart**.

(If your data range contains a label, that label and not the label **Series1** will appear in the Add Trendline dialog box.)

OR See section G.35 (**Simple Linear Regression**) if you want PHStat2 to produce a scatter diagram that includes a prediction line.

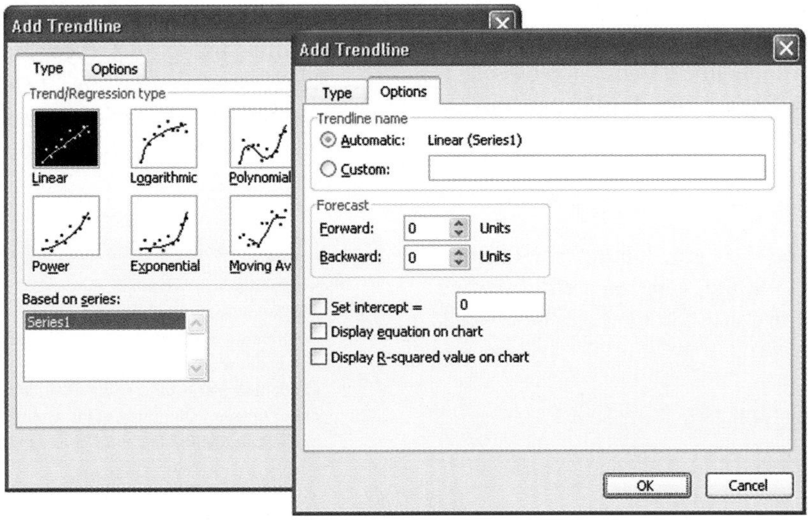

FIGURE A13.2 Microsoft Excel Add Trendline Dialog Box

For Durbin-Watson Statistic

Open the **Durbin-Watson.xls** file to the **DurbinWatson** worksheet, shown in Figure 13.20 on page 539. This worksheet contains results based on the package store sales analysis regression model. This worksheet uses the SUMXMY2 function, the template of which is **SUMXMY2 (*cell range 1, cell range 2*)**, to calculate the sum of squared difference of residuals and the SUMSQ function, the template of which is **SUMSQ (*residuals cell range*)** to calculate the sum of squared residuals. To adapt this file to other problems, do the following:

Perform a simple linear regression analysis, specifying residual output.
Copy the regression results worksheet to the **Durbin-Watson.xls** file.
Open to the **DurbinWatson** worksheet of the **Durbin-Watson.xls** file.
Edit ***cell range 1*** (in cell B3 formula) to be the regression results worksheet cells that contain the second through the last residual.
Edit ***cell range 2*** (in cell B3 formula) to be the regression results worksheet that contain the first through the second-to-last residual.
Edit ***residuals cell range*** (in cell B4 formula) to be the column C cells that contain the residuals in the residual output table.

OR See section G.35 (**Simple Linear Regression**) if you want PHStat2 to produce a worksheet for you.

For Predicting *Y* Values

Open the **SLR CIEPI.xls** file to the **CIEPI** worksheet, shown in Figure 13.27 on page 552. This worksheet uses the TINV function to determine the t value and contains several formulas that reference the **DataCopy** worksheet column F cells shown in Figure A13.3.

DataCopy worksheet cell F5 uses the TREND function, the template of which is **TREND(*Y variable cell range, X variable cell range, X value*)** to calculate the predicted Y value. To adapt this file to other problems, do the following:

Perform a simple linear regression analysis.
Transfer the standard error value found in the regression results worksheet cell B7 to **CIEPI** worksheet cell B13.
Change the X value in **CIEPI** worksheet cell B4.
Open to the **DataCopy** worksheet (in the **SLR CIEPI.xls** file).
Enter your X values in **column A** and Y values in **column B**, first following the instructions found in that worksheet if your sample size is not 14 as in the site selection example.

OR See section G.35 (**Simple Linear Regression**) if you want PHStat2 to produce the two worksheets for you.

	E	F	
1			
2	Sample Size	14	=COUNT(B:B)
3	Sample Mean	2.92143	=AVERAGE(A:A)
4	Sum of Squared Difference	37.92357	=SUM(C:C)
5	Predicted Y (YHat)	7.64392	=TREND(B2:B15, A2:A15, CIEPI!B4)

FIGURE A13.3 Microsoft Excel Worksheet for Predicting Y

A13.2 MINITAB

You can use Minitab for simple linear regression by selecting **Stat → Regression → Regression**. To illustrate the use of Minitab for simple linear regression with the site selection example of this chapter, open the **SITE.MTW** worksheet and select **Stat → Regression → Regression**.

1. In the Regression dialog box (see Figure A13.4), enter **'Annual Sales'** or **C3** in the Response: edit box and **'Square Feet'** or **C2** in the Predictors: edit box. Click the **Graphs** button.

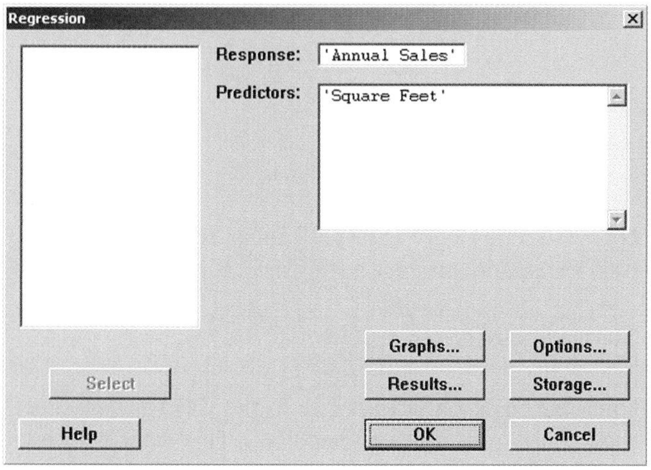

FIGURE A13.4 Minitab Regression Dialog Box

2. In the Regression-Graphs dialog box (see Figure A13.5 below), select the **Regular** option button under Residuals for Plots. Under Residual Plots, select the **Histogram of residuals, Normal plot of residuals, Residuals versus fits**, and **Residuals versus order** check boxes. In the Residuals versus the variables: edit box, enter **'Square Feet'** or **C2**. Click the **OK** button to return to the Regression dialog box.

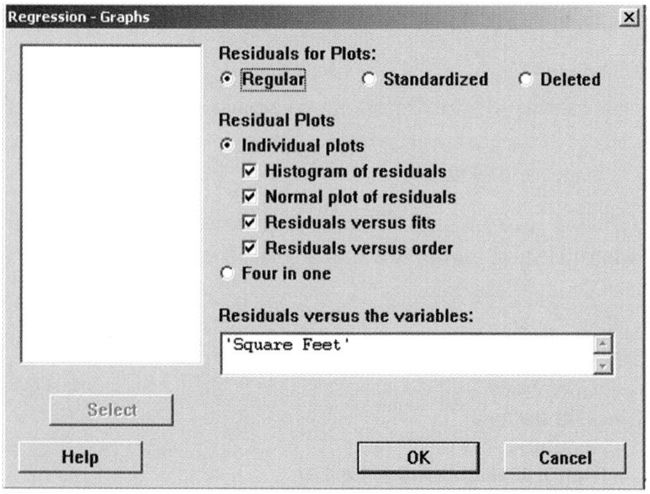

FIGURE A13.5 Minitab Regression-Graphs Dialog Box

3. Click the **Results** option button. In the Regression-Results dialog box (see Figure A13.6), select the **Regression equation, table of coefficients, s, R-squared, and basic analysis of variance** option button. Click the **OK** button to return to the Regression dialog box.

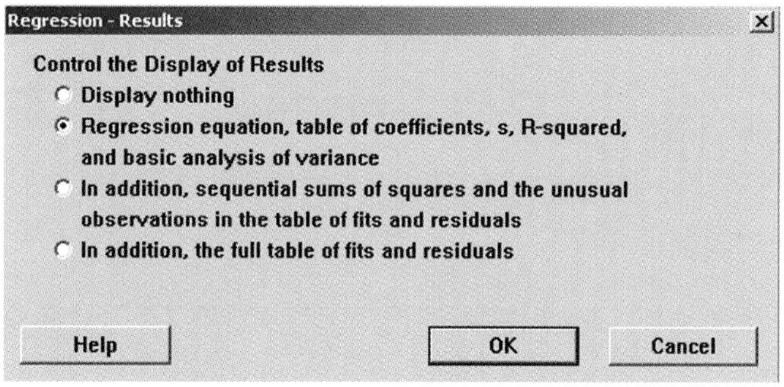

FIGURE A13.6 Minitab Regression-Results Dialog Box

4. Click the **Options** button. In the Regression-Options dialog box (see Figure A13.7), in the Prediction intervals for new observations: edit box, enter **4**. In the Confidence level: edit box, enter **95**. (If the data had been collected over time, you would want to select the Durbin-Watson statistic check box under Display.) Click the **OK** button to return to the Regression dialog box. Click the **OK** button.

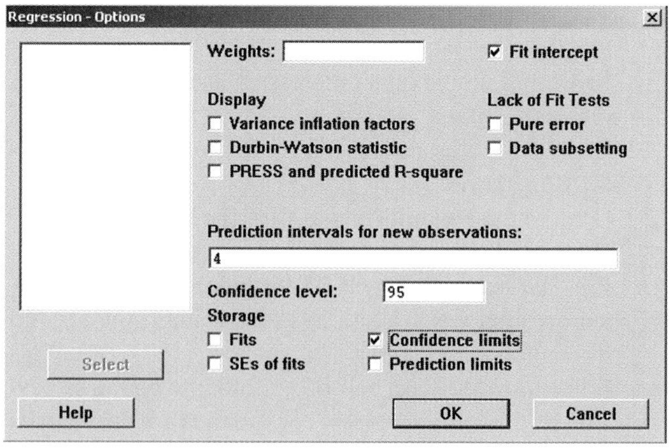

FIGURE A13.7 Minitab Regression-Options Dialog Box

CHAPTER 14

Introduction to Multiple Regression

LEARNING OBJECTIVES

In this chapter, you learn:

- How to develop a multiple regression model
- How to interpret the regression coefficients
- How to determine which independent variables to include in the regression model
- How to determine which independent variables are more important in predicting a dependent variable
- How to use categorical variables in a regression model
- How to predict a categorical dependent variable using logistic regression

USING STATISTICS

Predicting OmniPower Sales

You are the marketing manager for Omni Foods, a large food products company. Omni Foods is preparing a nationwide introduction for a new product, a high-energy bar called OmniPower. Although originally marketed to runners, mountain climbers, and other athletes, high-energy bars are now popular with college students, young business professionals, and other individuals who never seem to have enough time for a traditional breakfast or lunch.

Sales of competitors' high-energy bars have skyrocketed. Before introducing the bar nationwide, you need to determine the effect that price and in-store promotions will have on the sales of OmniPower. You plan to use a sample of 34 stores in a supermarket chain for a test-market study of OmniPower sales. How can you extend the linear regression methods discussed in Chapter 13 to incorporate the effects of price *and* promotion into the same model? How can you use this model to improve the success of the nationwide introduction of OmniPower?

Chapter 13 focused on simple linear regression models that use *one* numerical independent variable X to predict the value of a numerical dependent variable Y. Often you can make better predictions by using *more than one* independent variable. This chapter introduces you to **multiple regression models** that use two or more independent variables to predict the value of a dependent variable.

14.1 DEVELOPING THE MULTIPLE REGRESSION MODEL

A sample of 34 stores in a supermarket chain is selected for a test-market study of OmniPower. All the stores selected have approximately the same monthly sales volume. Two independent variables are considered here—the price of an OmniPower bar as measured in cents (X_1) and the monthly budget for in-store promotional expenditures measured in dollars (X_2). In-store promotional expenditures typically include signs and displays, in-store coupons, and free samples. The dependent variable Y is the number of OmniPower bars sold in a month. Table 14.1 presents the results OMNI of the test-market study.

TABLE 14.1
Monthly OmniPower Sales, Price, and Promotional Expenditures

Store	Sales	Price	Promotion	Store	Sales	Price	Promotion
1	4,141	59	200	18	2,730	79	400
2	3,842	59	200	19	2,618	79	400
3	3,056	59	200	20	4,421	79	400
4	3,519	59	200	21	4,113	79	600
5	4,226	59	400	22	3,746	79	600
6	4,630	59	400	23	3,532	79	600
7	3,507	59	400	24	3,825	79	600
8	3,754	59	400	25	1,096	99	200
9	5,000	59	600	26	761	99	200
10	5,120	59	600	27	2,088	99	200
11	4,011	59	600	28	820	99	200
12	5,015	59	600	29	2,114	99	400
13	1,916	79	200	30	1,882	99	400
14	675	79	200	31	2,159	99	400
15	3,636	79	200	32	1,602	99	400
16	3,224	79	200	33	3,354	99	600
17	2,295	79	400	34	2,927	99	600

With two independent variables and a dependent variable, the data are in three dimensions. Figure 14.1 illustrates a three-dimensional graph constructed by Minitab.

FIGURE 14.1

Minitab Three-Dimensional Graph of Monthly Omnipower Sales, Price, and Promotional Expenditures

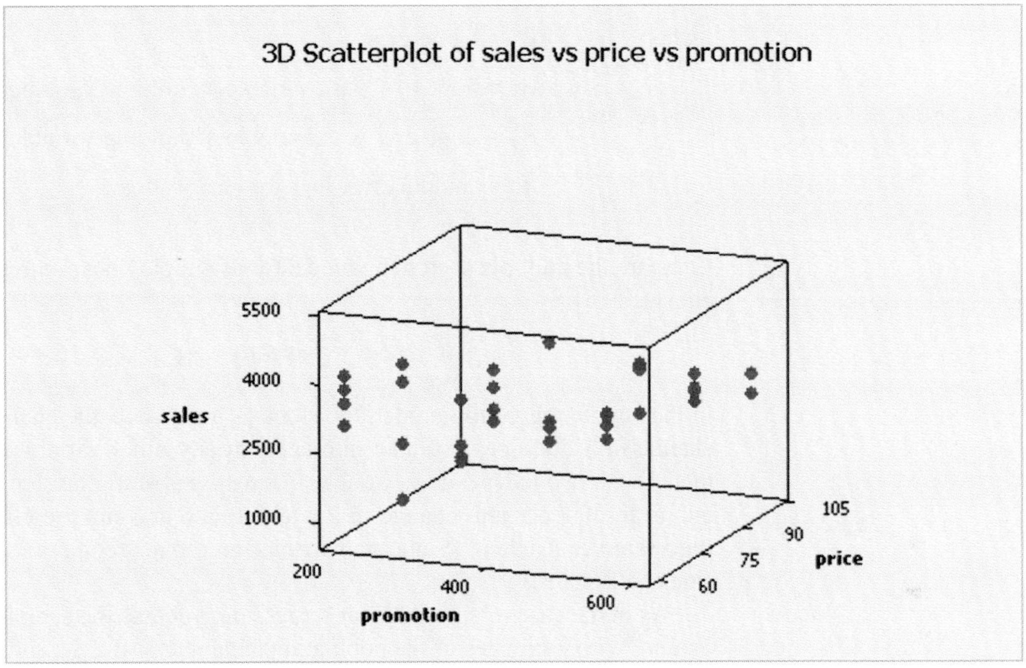

Interpreting the Regression Coefficients

When there are several independent variables, you can extend the simple linear regression model of Equation (13.1) on page 515 by assuming a linear relationship between each independent variable and the dependent variable. For example, with k independent variables, the multiple regression model is expressed in Equation (14.1).

MULTIPLE REGRESSION MODEL WITH k INDEPENDENT VARIABLES

$$Y_i = \beta_0 + \beta_1 X_{1i} + \beta_2 X_{2i} + \beta_3 X_{3i} + \cdots + \beta_k X_{ki} + \varepsilon_i \qquad \textbf{(14.1)}$$

where $\beta_0 = Y$ intercept

β_1 = slope of Y with variable X_1 holding variables $X_2, X_3, \ldots, X_k$ constant

β_2 = slope of Y with variable X_2 holding variables $X_1, X_3, \ldots, X_k$ constant

β_3 = slope of Y with variable X_3 holding variables $X_1, X_2, X_4, \ldots, X_k$ constant

.

.

.

β_k = slope of Y with variable X_k holding variables $X_1, X_2, X_3, \ldots, X_{k-1}$ constant

ε_i = random error in Y for observation i

Equation (14.2) defines the multiple regression model with two independent variables.

MULTIPLE REGRESSION MODEL WITH TWO INDEPENDENT VARIABLES

$$Y_i = \beta_0 + \beta_1 X_{1i} + \beta_2 X_{2i} + \varepsilon_i \qquad \textbf{(14.2)}$$

where $\beta_0 = Y$ intercept

β_1 = slope of Y with variable X_1 holding variable X_2 constant

β_2 = slope of Y with variable X_2 holding variable X_1 constant

ε_i = random error in Y for observation i

Compare the multiple regression model to the simple linear regression model [Equation (13.1) on page 515],

$$Y_i = \beta_0 + \beta_1 X_i + \varepsilon_i$$

In the simple regression model, the slope β_1 represents the change in the mean of Y per unit change in X and does not take into account any other variables. In the multiple regression model with two independent variables [Equation (14.2)], the slope β_1 represents the change in the mean of Y per unit change in X_1, taking into account the effect of X_2. β_1 is called a **net regression coefficient**. (Some statisticians refer to net regression coefficients as partial regression coefficients.)

As in the case of simple linear regression, you use the sample regression coefficients (b_0, b_1, and b_2) as estimates of the population parameters (β_0, β_1, and β_2). Equation (14.3) defines the regression equation for a multiple regression model with two independent variables.

MULTIPLE REGRESSION EQUATION WITH TWO INDEPENDENT VARIABLES

$$\hat{Y}_i = b_0 + b_1 X_{1i} + b_2 X_{2i} \qquad \textbf{(14.3)}$$

You can use Microsoft Excel or Minitab to compute the values of the three regression coefficients using the least-squares method. Figure 14.2 presents Microsoft Excel output for the OmniPower sales data and Figure 14.3 illustrates Minitab output.

From Figure 14.2 or 14.3, the computed values of the regression coefficients are

$$b_0 = 5{,}837.52 \qquad b_1 = -53.2173 \qquad b_2 = 3.6131$$

FIGURE 14.2

Partial Microsoft Excel Output for OmniPower Sales Data

	A	B	C	D	E	F	G
1	**OmniPower Sales Analysis**						
2							
3	*Regression Statistics*						
4	Multiple R	0.870475					
5	R Square	0.757726					
6	Adjusted R Square	0.742095					
7	Standard Error	638.06529					
8	Observations	34					
9							
10	ANOVA		SSR				
11		*df*	*SS*	*MS*	*F*	*Significance F*	
12	Regression	SSE 2	39472730.77	19736365.387	48.47713433	2.86258E-10	
13	Residual	31	12620946.67	407127.312			
14	Total	33	52093677.44				
15				SST			
16		*Coefficients*	*Standard Error*	*t Stat*	*P-value*	*Lower 95%*	*Upper 95%*
17	Intercept b_0	5837.5208	628.150	9.29319	1.79101E-10	4556.39921	7118.64230
18	Price b_1	-53.21734	6.85222	-7.76644	9.20016E-09	-67.19254	-39.24213
19	Promotion b_2	3.61306	0.68522	5.27283	9.82196E-06	2.21554	5.01058

FIGURE 14.3
Partial Minitab Output for OmniPower Sales Data

```
The regression equation is
sales = 5838 - 53.2 price + 3.61 promotion

             b0
Predictor      Coef  SE Coef      T      P
Constant     5837.5    628.2   9.29  0.000
price   b1 -53.217     6.852  -7.77  0.000
promotion    3.6131   0.6852   5.27  0.000
                   b2

S = 638.065   R-Sq = 75.8%   R-Sq(adj) = 74.2%

Analysis of Variance
                          SSR
Source            DF        SS        MS      F      P
Regression         2  39472731  19736365  48.48  0.000
Residual Error    31  12620947    407127
Total             33  52093677  SSE
                  SST
Predicted Values for New Observations

New  Ŷi
Obs   Fit  SE Fit      95% CI          95% PI
  1  3079     110  (2854, 3303)  (1758, 4399)
```

Therefore, the multiple regression equation is

$$\hat{Y}_i = 5{,}837.52 - 53.2173X_{1i} + 3.6131X_{2i}$$

where $\hat{Y}_i$ = predicted monthly sales of OmniPower bars for store i

X_{1i} = price of OmniPower bar (in cents) for store i

X_{2i} = monthly in-store promotional expenditures (in dollars) for store i

The sample Y intercept ($b_0 = 5{,}837.52$) estimates the number of OmniPower bars sold in a month if the price is \$0.00 and the total amount spent on promotional expenditures is also \$0.00. Because these values of price and promotion are outside the range of price and promotion used in the test-market study, and are nonsensical, the value of b_0 has no practical interpretation.

The slope of price with OmniPower sales ($b_1 = -53.2173$) indicates that, for a given amount of monthly promotional expenditures, the mean sales of OmniPower are estimated to decrease by 53.2173 bars per month for each 1-cent increase in the price. The slope of monthly promotional expenditures with OmniPower sales ($b_2 = 3.6131$) indicates that, for a given price, the mean sales of OmniPower are estimated to increase by 3.6131 bars for each additional \$1 spent on promotions. These estimates allow you to better understand the likely effect that price and promotion decisions will have in the marketplace. For example, a 10-cent decrease in price is estimated to increase mean sales by 532.173 bars, with a fixed amount of monthly promotional expenditures. A \$100 increase in promotional expenditures is estimated to increase mean sales by 361.31 bars, for a given price.

Regression coefficients in multiple regression are called net regression coefficients and measure the mean change in Y per unit change in a particular X, *holding constant the effect of the other X variables*. For example, in the study of OmniPower bar sales, for a store with a given amount of promotional expenditures, the mean sales are estimated to decrease by 53.22 bars per month for each 1-cent increase in the price of the OmniPower bar. Another way to interpret this "net effect" is to think of two stores with an equal amount of promotional expenditures. If the first store charges 1-cent more than the other store, the "net effect" of this difference is that the first store is predicted to sell 53.22 less bars per month than the second store. To interpret the net effect of promotional expenditures, you can consider two stores that are charging the same price. If the first store spends \$1 more on promotional expenditures, the net effect of this difference is that the first store is predicted to sell 3.61 more bars per month than the second store.

Predicting the Dependent Variable *Y*

You can use the multiple regression equation computed by Microsoft Excel or Minitab to predict values of the dependent variable. For example, what is the predicted sales for a store charging 79 cents during a month in which promotional expenditures are \$400? Using the multiple regression equation

$$\hat{Y}_i = 5{,}837.52 - 53.2173X_{1i} + 3.6131X_{2i}$$

with $X_{1i} = 79$ and $X_{2i} = 400$,

$$\hat{Y}_i = 5{,}837.52 - 53.2173(79) + 3.6131(400)$$
$$= 3{,}078.57$$

Thus, your sales prediction for stores charging 79 cents and spending \$400 in promotional expenditures is 3,078.57 OmniPower bars per month.

After you have predicted *Y* and done a residual analysis (see section 14.3), the next step often involves a confidence interval estimate of the mean response and a prediction interval for an individual response. The computation of these intervals is too complex to perform by hand, and you should use Minitab or Microsoft Excel to perform the calculations. Figure 14.3 on page 573 presents confidence and prediction intervals computed using Minitab for predicting the sales of OmniPower bars. Figure 14.4 illustrates Microsoft Excel output.

FIGURE 14.4
Microsoft Excel Confidence Interval Estimate and Prediction Interval for the OmniPower Example

	A	B	C	D
1	**Confidence Interval Estimate and Prediction Interval**			
2				
3	**Data**			
4	**Confidence Level**	**95%**		
5		1		
6	**Price given value**	**79**		
7	**Promotion given value**	**400**		
8				
9	X'X	34	2646	13200
10		2646	214674	1018800
11		13200	1018800	6000000
12				
13	Inverse of X'X	0.969163	-0.00941	-0.00053
14		-0.009408	0.000115	1.12E-06
15		-0.000535	1.12E-06	1.15E-06
16				
17	X'G times Inverse of X'X	0.012054	0.000149	1.49E-05
18				
19	[X'G times Inverse of X'X] times XG	0.029762		
20	t Statistic	2.0395		
21	**Predicted Y (YHat)**	**3078.57**		
22				
23	**For Average Predicted Y (YHat)**			
24	**Interval Half Width**	**224.50**		
25	**Confidence Interval Lower Limit**	**2854.07**		
26	**Confidence Interval Upper Limit**	**3303.08**		
27				
28	**For Individual Response Y**			
29	**Interval Half Width**	**1320.57**		
30	**Prediction Interval Lower Limit**	**1758.01**		
31	**Prediction Interval Upper Limit**	**4399.14**		

The 95% confidence interval estimate of the mean OmniPower sales for all stores charging 79 cents and spending \$400 in promotional expenditures is 2,854.07 to 3,303.08 bars. The prediction interval for an individual store is 1,758.01 to 4,399.14 bars.

PROBLEMS FOR SECTION 14.1

Learning the Basics

PH Grade ASSIST **14.1** For this problem, use the following multiple regression equation:

$$\hat{Y}_i = 10 + 5X_{1i} + 3X_{2i}$$

a. Interpret the meaning of the slopes.
b. Interpret the meaning of the Y intercept.

PH Grade ASSIST **14.2** For this problem, use the following multiple regression equation:

$$\hat{Y}_i = 50 + 2X_{1i} + 7X_{2i}$$

a. Interpret the meaning of the slopes.
b. Interpret the meaning of the Y intercept.

Applying the Concepts

You need to use Microsoft Excel, Minitab, or SPSS to solve problems 14.4–14.8.

PH Grade ASSIST **14.3** A marketing analyst for a shoe manufacturer is considering the development of a new brand of running shoes. The marketing analyst wants to determine which variables to use in predicting durability (i.e., the effect of long-term impact). Two independent variables under consideration are X_1 (FOREIMP), a measurement of the forefoot shock-absorbing capability, and X_2 (MIDSOLE), a measurement of the change in impact properties over time. The dependent variable Y is LTIMP, a measure of the shoe's durability after a repeated impact test. A random sample of 15 types of currently manufactured running shoes was selected for testing with the following results.

ANOVA	*df*	SS	*MS*	*F*	Significance *F*
Regression	2	12.61020	6.30510	97.69	0.0001
Error	12	0.77453	0.06454		
Total	14	13.38473			

Variable	Coefficients	Standard Error	*t* Stat	*p*-Value
Intercept	−0.02686	0.06905	−0.39	0.7034
Foreimp	0.79116	0.06295	12.57	0.0000
Midsole	0.60484	0.07174	8.43	0.0000

a. State the multiple regression equation.
b. Interpret the meaning of the slopes in this problem.

SELF Test **14.4** A mail-order catalog business selling personal computer supplies, software, and hardware maintains a centralized warehouse. Management is currently examining the process of distribution from the warehouse and wants to study the factors that affect warehouse distribution costs. Currently, a small handling fee is added to each order, regardless of the amount of the order. Data collected over the past 24 months indicate the warehouse distribution costs (in thousands of dollars), the sales (in thousands of dollars), and the number of orders received. WARECOST

a. State the multiple regression equation.
b. Interpret the meaning of the slopes b_1 and b_2 in this problem.
c. Explain why the regression coefficient b_0 has no practical meaning in the context of this problem.
d. Predict the mean monthly warehouse distribution cost when sales are \$400,000 and the number of orders is 4,500.
e. Construct a 95% confidence interval estimate for the mean monthly warehouse distribution cost when sales are \$400,000 and the number of orders is 4,500.
f. Construct a 95% prediction interval for the monthly warehouse distribution cost for a particular month when sales are \$400,000 and the number of orders is 4,500.

PH Grade ASSIST **14.5** A consumer organization wants to develop a regression model to predict gasoline mileage (as measured by miles per gallon) based on the horsepower of the car's engine and the weight of the car in pounds. A sample of 50 recent car models was selected and the results recorded. AUTO

a. State the multiple regression equation.
b. Interpret the meaning of the slopes b_1 and b_2 in this problem.
c. Explain why the regression coefficient b_0 has no practical meaning in the context of this problem.
d. Predict the mean miles per gallon for cars that have 60 horsepower and weigh 2,000 pounds.
e. Construct a 95% confidence interval estimate for the mean miles per gallon for cars that have 60 horsepower and weigh 2,000 pounds.
f. Construct a 95% prediction interval for the miles per gallon for an individual car that has 60 horsepower and weighs 2,000 pounds.

PH Grade ASSIST **14.6** A consumer products company wants to measure the effectiveness of different types of advertising media in the promotion of its products. Specifically, the company is interested in the effectiveness of radio advertising and newspaper advertising (including the cost of discount coupons). A sample of 22 cities with approximately equal populations is selected for study during a test period of one month. Each city is allocated a specific expenditure level both for radio advertising and for newspaper advertising. The sales of the product (in thousands of dollars) and also the levels of media expenditure (in thousands of dollars) during the test month are recorded with the following results: ADVERTISE

City	Sales ($000)	Radio Advertising ($000)	Newspaper Advertising ($000)
1	973	0	40
2	1,119	0	40
3	875	25	25
4	625	25	25
5	910	30	30
6	971	30	30
7	931	35	35
8	1,177	35	35
9	882	40	25
10	982	40	25
11	1,628	45	45
12	1,577	45	45
13	1,044	50	0
14	914	50	0
15	1,329	55	25
16	1,330	55	25
17	1,405	60	30
18	1,436	60	30
19	1,521	65	35
20	1,741	65	35
21	1,866	70	40
22	1,717	70	40

a. State the multiple regression equation.
b. Interpret the meaning of the slopes b_1 and b_2 in this problem.
c. Interpret the meaning of the regression coefficient b_0.
d. Predict the sales for a city in which radio advertising is $20,000 and newspaper advertising is $20,000.
e. Construct a 95% confidence interval estimate for the mean sales for cities in which radio advertising is $20,000 and newspaper advertising is $20,000.
f. Construct a 95% prediction interval for the sales for an individual city in which radio advertising is $20,000 and newspaper advertising is $20,000.

14.7 The director of broadcasting operations for a television station wants to study the issue of standby hours, i.e., hours in which unionized graphic artists at the station are paid but are not actually involved in any activity. The variables in the study include:

Standby hours (Y)—the total number of standby hours in a week
Total staff present (X_1)—the weekly total of people-days in a week.
Remote hours (X_2)—the total number of hours worked by employees at locations away from the central plant

The results for a period of 26 weeks are in the data file STANDBY.

a. State the multiple regression equation.
b. Interpret the meaning of the slopes b_1 and b_2 in this problem.
c. Explain why the regression coefficient b_0 has no practical meaning in the context of this problem.
d. Predict the standby hours for a week in which the total staff present is 310 people-days and the remote hours are 400.
e. Construct a 95% confidence interval estimate for the mean standby hours for weeks in which the total staff present is 310 people-days and the remote hours are 400.
f. Construct a 95% prediction interval for the standby hours for a single week in which the total staff present is 310 people-days and the remote hours are 400.

14.8 Nassau County is located approximately 25 miles east of New York City. Until all residential property was reassessed in 2002, property taxes were assessed based on actual value in 1938 or when the property was built if it was constructed after 1938. Data in the file GLENCOVE include the appraised value (in 2002), land area of the property (acres), and age in years for a sample of 30 single-family homes located in Glen Cove, a small city in Nassau County. Develop a multiple linear regression model to predict appraised value based on land area of the property and age in years.

a. State the multiple regression equation.
b. Interpret the meaning of the slopes b_1 and b_2 in this problem.
c. Explain why the regression coefficient b_0 has no practical meaning in the context of this problem.
d. Predict the appraised value for a house that has a land area of 0.25 acres and is 45 years old.
e. Construct a 95% confidence interval estimate for the mean appraised value for houses that have a land area of 0.25 acres and are 45 years old.
f. Construct a 95% prediction interval estimate for the appraised value for an individual house that has a land area of 0.25 acres and is 45 years old.

14.2 r^2, ADJUSTED r^2, AND THE OVERALL F TEST

Coefficient of Multiple Determination

Recall from section 13.3 that the coefficient of determination r^2 measures the variation in Y that is explained by the independent variable X in the simple linear regression model. In multiple regression, the **coefficient of multiple determination** represents the proportion of the variation in Y that is explained by the set of independent variables. Equation (14.4) defines the coefficient of multiple determination for a multiple regression model with two or more independent variables.

THE COEFFICIENT OF MULTIPLE DETERMINATION

The coefficient of multiple determination is equal to the regression sum of squares (SSR) divided by the total sum of squares (SST).

$$r^2 = \frac{SSR}{SST} \tag{14.4}$$

where SSR = regression sum of squares

SST = total sum of squares

In the OmniPower example, from Figure 14.2 or 14.3 on pages 572–573, $SSR = 39{,}472{,}730.77$, and $SST = 52{,}093{,}677.44$. Thus,

$$r^2 = \frac{SSR}{SST} = \frac{39{,}472{,}730.77}{52{,}093{,}677.44} = 0.7577$$

The coefficient of multiple determination ($r^2 = 0.7577$) indicates that 75.77% of the variation in sales is explained by the variation in the price and the variation in the promotional expenditures.

However, when dealing with multiple regression models, some statisticians suggest that you should use the **adjusted r^2** to reflect both the number of independent variables in the model and the sample size. Reporting the adjusted r^2 is extremely important when you are comparing two or more regression models that predict the same dependent variable but have a different number of independent variables. Equation (14.5) defines the adjusted r^2.

ADJUSTED r^2

$$r^2_{adj} = 1 - \left[(1 - r^2)\frac{n - 1}{n - k - 1}\right] \tag{14.5}$$

where k is the number of independent variables in the regression equation.

Thus, for the OmniPower data, because $r^2 = 0.7577$, $n = 34$, and $k = 2$,

$$\begin{aligned} r^2_{adj} &= 1 - \left[(1 - r^2)\frac{(34 - 1)}{(34 - 2 - 1)}\right] \\ &= 1 - \left[(1 - 0.7577)\frac{33}{31}\right] \\ &= 1 - 0.2579 \\ &= 0.7421 \end{aligned}$$

Hence, 74.21% of the variation in sales is explained by the multiple regression model—adjusted for number of independent variables and sample size.

Test for the Significance of the Overall Multiple Regression Model

You use the **overall F test** to test for the significance of the overall multiple regression model. This test determines whether there is a significant relationship between the dependent variable and the entire set of independent variables. Because there is more than one independent variable, you use the following null and alternative hypotheses:

$H_0: \beta_1 = \cdots = \beta_k = 0$ (No linear relationship between the dependent variable and the independent variables.)

H_1: At least one $\beta_j \neq 0$ (Linear relationship between the dependent variable and at least one of the independent variables.)

Equation (14.6) defines the statistic for the overall F test. Table 14.2 presents the associated ANOVA summary table.

OVERALL F-TEST STATISTIC

The F statistic is equal to the regression mean square (MSR) divided by the error mean square (MSE).

$$F = \frac{MSR}{MSE} \quad \textbf{(14.6)}$$

where F = test statistic from an F distribution with k and $n - k - 1$ degrees of freedom
k = number of independent variables in the regression model

TABLE 14.2
ANOVA Summary Table for the Overall F Test

Source	Degrees of Freedom	Sum of Squares	Mean Square (Variance)	F
Regression	k	SSR	$MSR = \frac{SSR}{k}$	$F = \frac{MSR}{MSE}$
Error	$n - k - 1$	SSE	$MSE = \frac{SSE}{n - k - 1}$	
Total	$n - 1$	SST		

The decision rule is

Reject H_0 at the α level of significance if $F > F_{U(k,n-k-1)}$;
otherwise, do not reject H_0.

Using a 0.05 level of significance, the critical value of the F distribution with 2 and 31 degrees of freedom found from Table E.5 is approximately 3.32 (see Figure 14.5). From Figure 14.2 or 14.3 on pages 572–573, the F statistic given in the ANOVA summary table is 48.48. Because 48.48 > 3.32, or because the p-value = 0.000 < 0.05, you reject H_0 and conclude that at least one of the independent variables (price and/or promotional expenditures) is related to sales.

FIGURE 14.5
Testing for Significance of a Set of Regression Coefficients at the 0.05 Level of Significance with 2 and 31 Degrees of Freedom

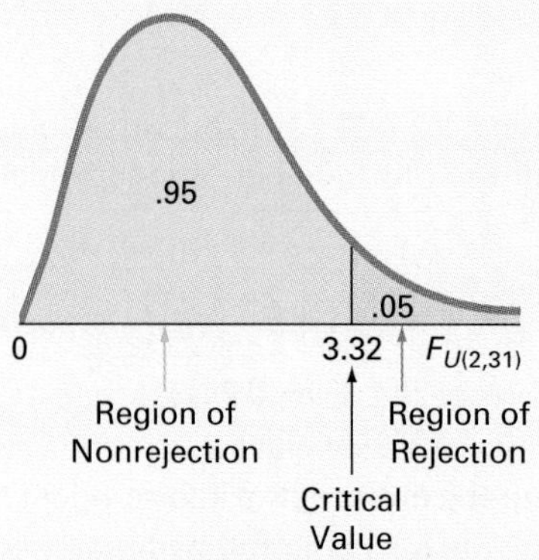

PROBLEMS FOR SECTION 14.2

Learning the Basics

PH Grade ASSIST **14.9** The following ANOVA summary table is for a multiple regression model with two independent variables.

Source	Degrees of Freedom	Sum of Squares	Mean Squares	F
Regression	2	60		
Error	18	120		
Total	20	180		

a. Determine the mean square that is due to regression and the mean square that is due to error.
b. Compute the F statistic.
c. Determine whether there is a significant relationship between Y and the two independent variables at the 0.05 level of significance.
d. Compute the coefficient of multiple determination r^2 and interpret its meaning.
e. Compute the adjusted r^2.

PH Grade ASSIST **14.10** The following ANOVA summary table is for a multiple regression model with two independent variables.

Source	Degrees of Freedom	Sum of Squares	Mean Squares	F
Regression	2	30		
Error	10	120		
Total	12	150		

a. Determine the mean square that is due to regression and the mean square that is due to error.
b. Compute the F statistic.
c. Determine whether there is a significant relationship between Y and the two independent variables at the 0.05 level of significance.
d. Compute the coefficient of multiple determination r^2 and interpret its meaning.
e. Compute the adjusted r^2.

Applying the Concepts

You need to use Microsoft Excel, Minitab, or SPSS to solve problems 14.13–14.17.

14.11 Eileen M. Van Aken and Brian M. Kleiner, professors at Virginia Polytechnic Institute and State University, investigated the factors that contribute to the effectiveness of teams ("Determinants of Effectiveness for Cross-Functional Organizational Design Teams," *Quality Management Journal*, 1997, 4, 51–79). The researchers studied 34 independent variables such as team skills, diversity, meeting frequency, and clarity in expectations. For each of the teams studied, each of the variables was given a value of 1 through 100 based on the results of interviews and survey data, where 100 represents the highest rating. The dependent variable, team performance, was also given a value of 1 through 100, with 100 representing the highest rating. Many different regression models were explored, including the following:

Model 1

$$\text{Team Performance} = \beta_0 + \beta_1 \text{(Team Skills)} + \varepsilon,$$
$$r^2_{\text{adj}} = 0.68$$

Model 2

$$\text{Team Performance} = \beta_0 + \beta_1 \text{(Clarity in Expectations)} + \varepsilon,$$
$$r^2_{\text{adj}} = 0.78$$

Model 3

$$\text{Team Performance} = \beta_0 + \beta_1 \text{(Team Skills)} + \beta_2 \text{(Clarity in Expectations)} + \varepsilon,$$
$$r^2_{\text{adj}} = 0.97$$

a. Interpret the adjusted r^2 for each of the three models.
b. Which of these three models do you think is the best predictor of team performance?

PH Grade ASSIST **14.12** In problem 14.3 on page 575, you predicted the durability of a brand of running shoe based on the forefoot shock-absorbing capability and the change in impact properties over time. The regression analysis resulted in the following ANOVA summary table:

ANOVA	Degrees of Freedom	SS	MS	F	Significance
Regression	2	12.61020	6.30510	97.69	0.0001
Error	12	10.77453	0.06454		
Total	14	13.38473			

a. Determine whether there is a significant relationship between durability and the two independent variables at the 0.05 level of significance.
b. Interpret the meaning of the p-value.
c. Compute the coefficient of multiple determination r^2 and interpret its meaning.
d. Compute the adjusted r^2.

PH Grade ASSIST **14.13** In problem 14.5 on page 575, you used horsepower and weight to predict gasoline mileage. Using the computer output from that problem, AUTO

a. Determine whether there is a significant relationship between gasoline mileage and the two independent variables (horsepower and weight) at the 0.05 level of significance.
b. Interpret the meaning of the p-value.
c. Compute the coefficient of multiple determination r^2 and interpret its meaning.
d. Compute the adjusted r^2.

PH Grade ASSIST **14.14** In problem 14.4 on page 575, you used sales and number of orders to predict distribution costs at a mail-order catalog business. Using the computer output from that problem, WARECOST
a. Determine whether there is a significant relationship between distribution costs and the two independent variables (sales and number of orders) at the 0.05 level of significance.
b. Interpret the meaning of the p-value.
c. Compute the coefficient of multiple determination r^2 and interpret its meaning.
d. Compute the adjusted r^2.

14.15 In problem 14.7 on page 576, you used the total staff present and remote hours to predict standby hours. Using the computer output from that problem, STANDBY
a. Determine whether there is a significant relationship between standby hours and the two independent variables (total staff present and remote hours) at the 0.05 level of significance.
b. Interpret the meaning of the p-value.
c. Compute the coefficient of multiple determination r^2 and interpret its meaning.
d. Compute the adjusted r^2.

14.16 In problem 14.6 on page 575, you used radio advertising and newspaper advertising to predict sales. Using the computer output from that problem, ADVERTISE
a. Determine whether there is a significant relationship between sales and the two independent variables (radio advertising and newspaper advertising) at the 0.05 level of significance.
b. Interpret the meaning of the p-value.
c. Compute the coefficient of multiple determination r^2 and interpret its meaning.
d. Compute the adjusted r^2.

14.17 In problem 14.8 on page 576, you used the land area of a property and the age of a house to predict appraised value. Using the computer output from that problem, GLENCOVE
a. Determine whether there is a significant relationship between appraised value and the two independent variables (land area of a property and the age of a house) at the 0.05 level of significance.
b. Interpret the meaning of the p-value.
c. Compute the coefficient of multiple determination r^2 and interpret its meaning.
d. Compute the adjusted r^2.

14.3 RESIDUAL ANALYSIS FOR THE MULTIPLE REGRESSION MODEL

In section 13.5, you used residual analysis to evaluate the appropriateness of using the simple linear regression model for a set of data. For the multiple regression model with two independent variables, you need to construct and analyze the following residual plots.

1. Residuals versus $\hat{Y}_i$
2. Residuals versus X_{1i}
3. Residuals versus X_{2i}
4. Residuals versus time

The first residual plot examines the pattern of residuals versus the predicted values of Y. If the residuals show a pattern for different predicted values of Y, there is evidence of a possible quadratic effect in at least one independent variable, a possible violation to the assumption of equal variance (see Figure 13.16 on page 535), and/or the need to transform the Y variable.

The second and third residual plots involve the independent variables. Patterns in the plot of the residuals versus an independent variable may indicate the existence of a quadratic effect and, therefore, indicate the need to add a quadratic independent variable to the multiple regression model. The fourth plot is used to investigate patterns in the residuals in order to validate the independence assumption when the data are collected in time order. Associated with this residual plot, as in section 13.6, you can compute the Durbin-Watson statistic to determine the existence of positive autocorrelation among the residuals.

You can use statistical and spreadsheet software to plot the residuals. Figure 14.6 illustrates the Microsoft Excel residual plots for the OmniPower sales example. In Figure 14.6, there is very little or no pattern in the relationship between the residuals and the predicted value of Y, the value of X_1 (price), or the value of X_2 (promotional expenditures). Thus, you can conclude that the multiple regression model is appropriate for predicting sales.

FIGURE 14.6
Microsoft Excel Residual Plots for the OmniPower Example: Panel *A*, Residuals versus Predicted *Y*; Panel *B*, Residuals versus Price; Panel *C*, Residuals versus Promotional Expenditures

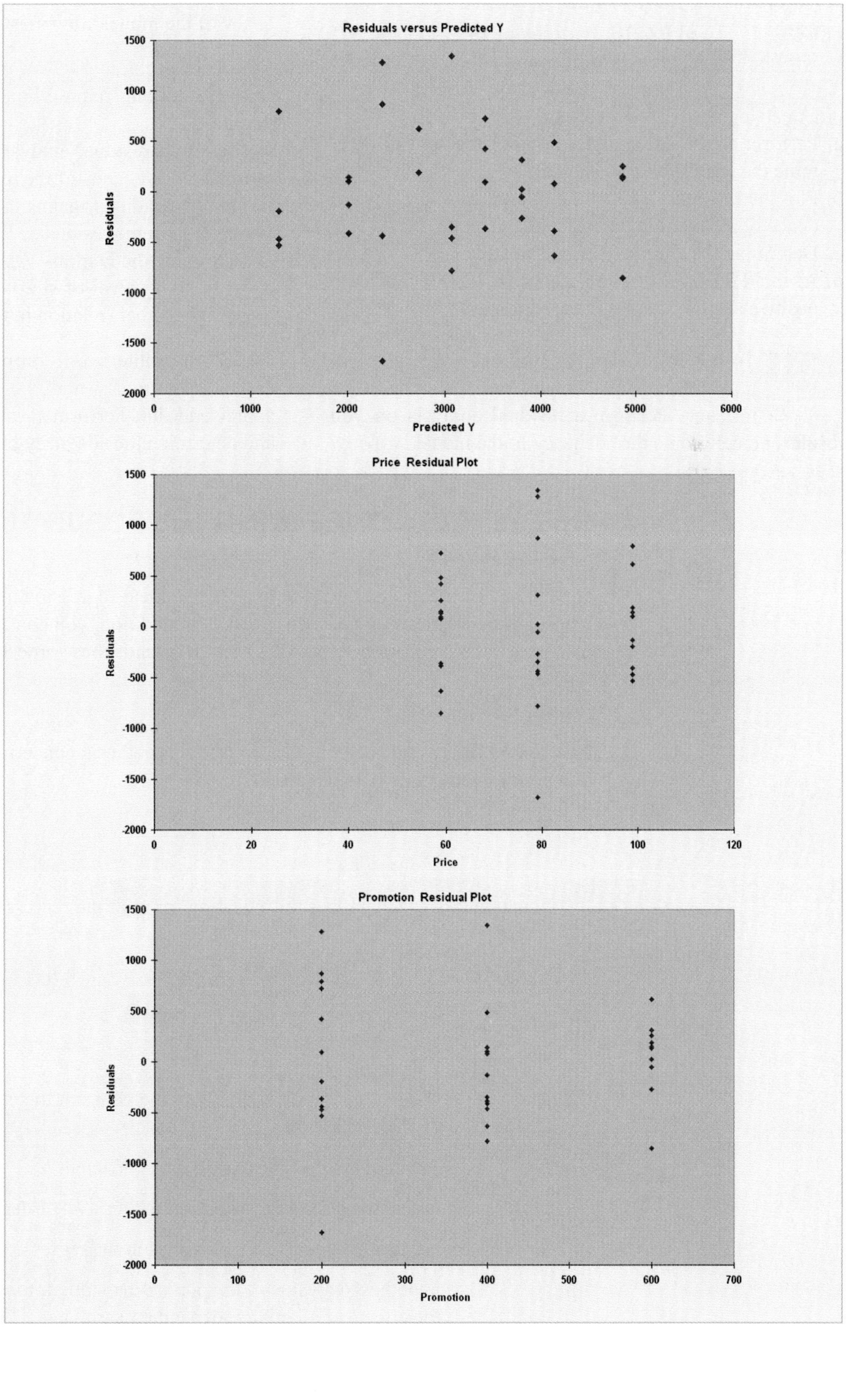

PROBLEMS FOR SECTION 14.3

Applying the Concepts

You need to use Microsoft Excel, Minitab, or SPSS to solve problems 14.18–14.22

PH Grade ASSIST SELF Test **14.18** In problem 14.4 on page 575, you used sales and number of orders to predict distribution costs at a mail-order catalog business. WARECOST

a. Perform a residual analysis on your results and determine the adequacy of the model.
b. Plot the residuals against the months. Is there any evidence of a pattern in the residuals? Explain.
c. Determine the Durbin-Watson statistic.
d. At the 0.05 level of significance, is there evidence of positive autocorrelation in the residuals?

PH Grade ASSIST **14.19** In problem 14.5 on page 575, you used horsepower and weight to predict gasoline mileage. Perform a residual analysis on your results and determine the adequacy of the model. AUTO

14.20 In problem 14.6 on page 575, you used radio advertising and newspaper advertising to predict sales. Perform a residual analysis on your results and determine the adequacy of the model. ADVERTISE

14.21 In problem 14.7 on page 576, you used the total staff present and remote hours to predict standby hours. STANDBY

a. Perform a residual analysis on your results and determine the adequacy of the model.
b. Plot the residuals against the weeks. Is there evidence of a pattern in the residuals? Explain.
c. Determine the Durbin-Watson statistic.
d. At the 0.05 level of significance, is there evidence of positive autocorrelation in the residuals?

14.22 In problem 14.8 on page 576, you used the land area of a property and the age of a house to predict appraised value. Perform a residual analysis on your results and determine the adequacy of the model. GLENCOVE

14.4 INFERENCES CONCERNING THE POPULATION REGRESSION COEFFICIENTS

In section 13.7 you tested the slope in a simple linear regression model to determine the significance of the relationship between X and Y. In addition, you constructed a confidence interval estimate of the population slope. This section extends these procedures to multiple regression.

Tests of Hypothesis

In a simple linear regression model, to test a hypothesis concerning the population slope β_1, you used Equation (13.16) on page 542:

$$t = \frac{b_1 - \beta_1}{S_{b_1}}$$

Equation (14.7) generalizes this equation for multiple regression.

TESTING FOR THE SLOPE IN MULTIPLE REGRESSION

$$t = \frac{b_j - \beta_j}{S_{b_j}} \tag{14.7}$$

where b_j = slope of variable j with Y, holding constant the effects of all other independent variables

s_{b_j} = standard error of the regression coefficient b_j

t = test statistic for a t distribution with $n - k - 1$ degrees of freedom

k = number of independent variables in the regression equation

β_j = hypothesized value of the population slope for variable j, holding constant the effects of all other independent variables

Microsoft Excel and Minitab output provide the results of the t test for each of the independent variables included in the regression model (see Figure 14.2 or 14.3 on pages 572–573).

To determine whether variable X_2 (amount of promotional expenditures) has a significant effect on sales, taking into account the price of OmniPower bars, the null and alternative hypotheses are

$$H_0: \beta_2 = 0$$
$$H_1: \beta_2 \neq 0$$

From Equation (14.7), and Figure 14.2 or 14.3,

$$t = \frac{b_2 - \beta_2}{S_{b_2}}$$
$$= \frac{3.6131 - 0}{0.6852} = 5.27$$

If you select a level of significance of 0.05, the critical values of t for 31 degrees of freedom from Table E.3 are −2.0395 and +2.0395 (see Figure 14.7).

FIGURE 14.7

Testing for Significance of a Regression Coefficient at the 0.05 Level of Significance with 31 Degrees of Freedom

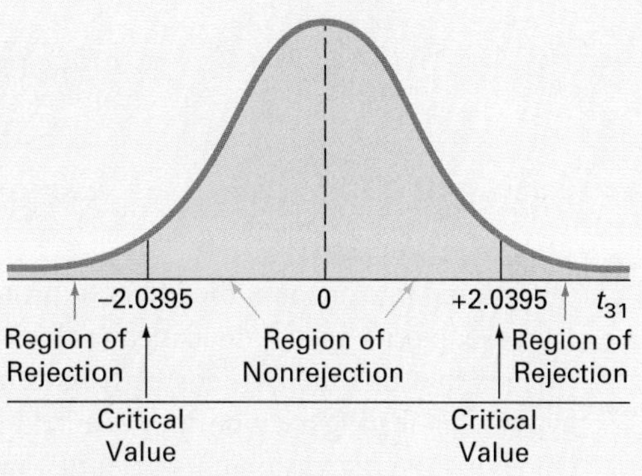

From Figure 14.2 or 14.3, the p-value is 0.00000982 (or 9.82E-06 in scientific notation). Because $t = 5.27 > 2.0395$ or the p-value of $0.00000982 < 0.05$, you reject H_0 and conclude that there is a significant relationship between variable X_2 (promotional expenditures) and sales, taking into account the price X_1. This extremely small p-value allows you to strongly reject the null hypothesis that there is no linear relationship between sales and promotional expenditures.

Example 14.1 presents the test for the significance of β_1, the slope of sales with price.

EXAMPLE 14.1 TESTING FOR THE SIGNIFICANCE OF THE SLOPE OF SALES WITH PRICE

At the 0.05 level of significance, is there evidence that the slope of sales with price is different from zero?

SOLUTION From Figure 14.2 or 14.3 on pages 572–573, $t = -7.766 < -2.0395$ (the critical value for $\alpha = 0.05$) or the p-value $= 0.0000000092 < 0.05$. Thus, there is a significant relationship between price X_1 and sales, taking into account the promotional expenditures X_2.

As seen with each of these two X variables, the test of significance for a particular regression coefficient is actually a test for the significance of adding a particular variable into a regression model given that the other variable is included. Therefore, the t test for the regression coefficient is equivalent to testing for the contribution of each independent variable.

Confidence Interval Estimation

Instead of testing the significance of a population slope, you may want to estimate the value of a population slope. Equation (14.8) defines the confidence interval estimate for a population slope in multiple regression.

CONFIDENCE INTERVAL ESTIMATE FOR THE SLOPE

$$b_j \pm t_{n-k-1} S_{b_j} \qquad \textbf{(14.8)}$$

For example, if you want to construct a 95% confidence interval estimate of the population slope β_1 (the effect of price X_1 on sales Y, holding constant the effect of promotional expenditures X_2), from Equation (14.8) and Figure 14.2 or 14.3 on pages 572–573,

$$b_1 \pm t_{n-k-1} S_{b_1}$$

Because the critical value of t at the 95% confidence level with 31 degrees of freedom is 2.0395 (see Table E.3),

$$-53.2173 \pm (2.0395)(6.8522)$$

$$-53.2173 \pm 13.9752$$

$$-67.1925 \leq \beta_1 \leq -39.2421$$

Taking into account the effect of promotional expenditures, the estimated effect of a 1-cent increase in price is to reduce mean sales by approximately 39.2 to 67.2 bars. You have 95% confidence that this interval correctly estimates the relationship between these variables. From a hypothesis-testing viewpoint, because this confidence interval does not include 0, you conclude that the regression coefficient β_1 has a significant effect.

Example 14.2 constructs and interprets a confidence interval estimate for the slope of sales with promotional expenditures.

EXAMPLE 14.2

CONSTRUCTING A CONFIDENCE INTERNAL ESTIMATE FOR THE SLOPE OF SALES WITH PROMOTIONAL EXPENDITURES

Construct a 95% confidence interval estimate of the population slope of sales with promotional expenditures.

SOLUTION The critical value of t at the 95% confidence level with 31 degrees of freedom is 2.0395 (see Table E.3). Using Equation (14.8)

$$3.6131 \pm (2.0395)(0.6852)$$

$$3.6131 \pm 1.3975$$

$$2.2156 \leq \beta_2 \leq 5.0106$$

Thus, taking into account the effect of price, the estimated effect of each additional dollar of promotional expenditures is to increase mean sales by approximately 2.2 to 5.0 bars. You have 95% confidence that this interval correctly estimates the relationship between these variables. From a hypothesis-testing viewpoint, because this confidence interval does not include 0, you can conclude that the regression coefficient β_2 has a significant effect.

PROBLEMS FOR SECTION 14.4

Learning the Basics

PH Grade ASSIST **14.23** Given the following information from a multiple regression analysis:

$$n = 25, \quad b_1 = 5, \quad b_2 = 10, \quad S_{b_1} = 2, \quad S_{b_2} = 8$$

a. Which variable has the largest slope in units of a *t* statistic?
b. Construct a 95% confidence interval estimate of the population slope β_1.
c. At the 0.05 level of significance, determine whether each independent variable makes a significant contribution to the regression model. On the basis of these results, indicate the independent variables to include in this model.

14.24 Given the following information from a multiple regression analysis:

$$n = 20, \quad b_1 = 4, \quad b_2 = 3, \quad S_{b_1} = 1.2, \quad S_{b_2} = 0.8$$

a. Which variable has the largest slope in units of a *t* statistic?
b. Construct a 95% confidence interval estimate of the population slope β_1.
c. At the 0.05 level of significance, determine whether each independent variable makes a significant contribution to the regression model. On the basis of these results, indicate the independent variables to include in this model.

Applying the Concepts

You need to use Microsoft Excel, Minitab, or SPSS to solve problems 14.26–14.30.

PH Grade ASSIST **14.25** In problem 14.3 on page 575, you predicted the durability of a brand of running shoe based on the forefoot shock-absorbing capability (Foreimp) and the change in impact properties over time (Midsole) for a sample of 15 pairs of shoes. Use the following results:

Variable	Coefficients	Standard Error	*t* stat	*p*-value
Intercept	−0.02686	0.06905	−0.39	0.7034
Foreimp	0.79116	0.06295	12.57	0.0000
Midsole	0.60484	0.07174	8.43	0.0000

a. Construct a 95% confidence interval estimate of the population slope between durability and forefoot shock-absorbing capability.
b. At the 0.05 level of significance, determine whether each independent variable makes a significant contribution to the regression model. On the basis of these results, indicate the independent variables to include in this model.

PH Grade ASSIST SELF Test **14.26** In problem 14.4 on page 575, you used sales and number of orders to predict distribution costs at a mail-order catalog business. Using the computer output from that problem, WARECOST
a. Construct a 95% confidence interval estimate of the population slope between distribution cost and sales.
b. At the 0.05 level of significance, determine whether each independent variable makes a significant contribution to the regression model. On the basis of these results, indicate the independent variables to include in this model.

14.27 In problem 14.5 on page 575, you used horsepower and weight to predict gasoline mileage. Using the computer output from that problem, AUTO
a. Construct a 95% confidence interval estimate of the population slope between gasoline mileage and horsepower.
b. At the 0.05 level of significance, determine whether each independent variable makes a significant contribution to the regression model. On the basis of these results, indicate the independent variables to include in this model.

14.28 In problem 14.6 on page 575, you used radio advertising and newspaper advertising to predict sales. Using the computer output from that problem, ADVERTISE
a. Construct a 95% confidence interval estimate of the population slope between sales and radio advertising.
b. At the 0.05 level of significance, determine whether each independent variable makes a significant contribution to the regression model. On the basis of these results, indicate the independent variables to include in this model.

14.29 In problem 14.7 on page 576, you used the total number of staff present and remote hours to predict standby hours. Using the computer output from that problem, STANDBY
a. Construct a 95% confidence interval estimate of the population slope between standby hours and total number of staff present.
b. At the 0.05 level of significance, determine whether each independent variable makes a significant contribution to the regression model. On the basis of these results, indicate the independent variables to include in this model.

14.30 In problem 14.8 on page 576, you used the land area of a property and the age of a house to predict appraised value. Using the computer output from that problem, GLENCOVE
a. Construct a 95% confidence interval estimate of the population slope between appraised value and land area of a property.
b. At the 0.05 level of significance, determine whether each independent variable makes a significant contribution to the regression model. On the basis of these results, indicate the independent variables to include in this model.

14.5 TESTING PORTIONS OF THE MULTIPLE REGRESSION MODEL

In developing a multiple regression model, you want to use only those independent variables that sufficiently reduce the error in predicting the value of a dependent variable. If an independent variable does not improve the prediction, you can delete it from the multiple regression model and use a model with fewer independent variables.

The **partial *F*-test** is an alternative method to the *t* test discussed in section 14.4 for determining the contribution of an independent variable. It involves determining the contribution to the regression sum of squares made by each independent variable after all the other independent variables have been included in the model. The new independent variable is included only if it significantly improves the model.

To conduct partial *F* tests in the OmniPower sales example, you first need to evaluate the contribution of promotional expenditures (X_2) after price (X_1) has been included in the model, and the contribution of price (X_1) after promotional expenditures (X_2) have been included in the model.

In general, if there are several independent variables, you determine the contribution of each independent variable by taking into account the regression sum of squares of a model that includes all independent variables except the one of interest, *SSR* (all variables except *j*). Equation (14.9) determines the contribution of variable *j*, assuming that all other variables are already included.

DETERMINING THE CONTRIBUTION OF AN INDEPENDENT VARIABLE TO THE REGRESSION MODEL

$$SSR(X_j \mid \text{all variables } except\ j) = SSR\text{ (all variables } including\ j) - SSR\text{ (all variables } except\ j) \quad \textbf{(14.9)}$$

If there are two independent variables, use Equations (14.10a) and (14.10b) to determine the contribution of each.

CONTRIBUTION OF VARIABLE X_1 GIVEN THAT X_2 HAS BEEN INCLUDED:

$$SSR(X_1 \mid X_2) = SSR(X_1 \text{ and } X_2) - SSR(X_2) \quad \textbf{(14.10a)}$$

CONTRIBUTION OF VARIABLE X_2 GIVEN THAT X_1 HAS BEEN INCLUDED:

$$SSR(X_2 \mid X_1) = SSR(X_1 \text{ and } X_2) - SSR(X_1) \quad \textbf{(14.10b)}$$

The term $SSR(X_2)$ represents the sum of squares that is due to regression for a model that includes only the independent variable X_2 (promotional expenditures). Similarly, $SSR(X_1)$ represents the sum of squares that is due to regression for a model that includes only the independent variable X_1 (price). Figures 14.8 and 14.9 present Microsoft Excel output for these two models. Figures 14.10 and 14.11 illustrate Minitab output.

FIGURE 14.8

Microsoft Excel Output of Simple Linear Regression Model for Sales and Promotional Expenditures $SSR(X_2)$

	A	B	C	D	E	F	G
1	Sales & Promotional Expenses Analysis						
2							
3	*Regression Statistics*						
4	Multiple R	0.535095					
5	R Square	0.286327					
6	Adjusted R Square	0.264024					
7	Standard Error	1077.872084					
8	Observations	34					
9							
10	ANOVA						
11		*df*	*SS*	*MS*	*F*	*Significance F*	
12	Regression	1	14915814.102	14915814.102	12.83845	0.0011115	
13	Residual	32	37177863.339	1161808.229			
14	Total	33	52093677.441				
15							
16		*Coefficients*	*Standard Error*	*t Stat*	*P-value*	*Lower 95%*	*Upper 95%*
17	Intercept	1496.01613	483.978853	3.09107747	0.004111	510.1843006	2481.847958
18	Promotion	4.12806	1.152100	3.58307795	0.001111	1.7813154	6.474814

FIGURE 14.9
Microsoft Excel Output of Simple Linear Regression Model for Sales and Price $SSR(X_1)$

	A	B	C	D	E	F	G
1	Sales & Price Analysis						
2							
3	*Regression Statistics*						
4	Multiple R	0.735146					
5	R Square	0.540440					
6	Adjusted R Square	0.526078					
7	Standard Error	864.945650					
8	Observations	34					
9							
10	ANOVA						
11		*df*	*SS*	*MS*	*F*	*Significance F*	
12	Regression	1	28153486.15	28153486.15	37.63176099	7.35855E-07	
13	Residual	32	23940191.29	748130.98			
14	Total	33	52093677.44				
15							
16		*Coefficients*	*Standard Error*	*t Stat*	*P-value*	*Lower 95%*	*Upper 95%*
17	Intercept	7512.34798	734.6188701	10.22618434	1.30793E-11	6015.97958	9008.716388
18	Price	-56.71384	9.2451043	-6.13447316	7.35855E-07	-75.54549	-37.882199

FIGURE 14.10
Minitab Output of Simple Linear Regression Model for Sales and Promotional Expenditures $SSR(X_2)$

```
The regression equation is
sales = 1496 + 4.13 promotion


Predictor     Coef  SE Coef     T      P
Constant    1496.0    484.0  3.09  0.004
promotion    4.128    1.152  3.58  0.001


S = 1077.87   R-Sq = 28.6%   R-Sq(adj) = 26.4%


Analysis of Variance

Source           DF        SS        MS      F      P
Regression        1  14915814  14915814  12.84  0.001
Residual Error   32  37177863   1161808
Total            33  52093677
```

FIGURE 14.11
Minitab Output of Simple Linear Regression Model for Sales and Price $SSR(X_1)$

```
The regression equation is
sales = 7512 - 56.7 price


Predictor     Coef  SE Coef      T      P
Constant    7512.3    734.6  10.23  0.000
price      -56.714    9.245  -6.13  0.000


S = 864.946   R-Sq = 54.0%   R-Sq(adj) = 52.6%


Analysis of Variance

Source           DF        SS        MS      F      P
Regression        1  28153486  28153486  37.63  0.000
Residual Error   32  23940191    748131
Total            33  52093677
```

From Figure 14.8 or 14.10, $SSR\ (X_2) = 14{,}915{,}814.10$ and from Figure 14.2 or 14.3 on pages 572 and 573, $SSR\ (X_1 \text{ and } X_2) = 39{,}472{,}730.77$. Then, using Equation (14.10a),

$$\begin{aligned} SSR(X_1|X_2) &= SSR(X_1 \text{ and } X_2) - SSR(X_2) \\ &= 39{,}472{,}730.77 - 14{,}915{,}814.10 \\ &= 24{,}556{,}916.67 \end{aligned}$$

To determine whether X_1 significantly improves the model after X_2 has been included, you divide the regression sum of squares into two component parts as shown in Table 14.3.

TABLE 14.3
ANOVA Table Dividing the Regression Sum of Squares Into Components to Determine the Contribution of Variable X_1

Source	Degrees of Freedom	Sum of Squares	Mean Square (Variance)	F
Regression	2	39,472,730.77	19,736,365.39	
X_2	1	14,915,814.10		
$X_1 \mid X_2$	1	24,556,916.67	24,556,916.67	60.32
Error	31	12,620,946.67	407,127.31	
Total	33	52,093,677.44		

The null and alternative hypotheses to test for the contribution of X_1 to the model are:

H_0: Variable X_1 does not significantly improve the model after variable X_2 has been included.
H_1: Variable X_1 significantly improves the model after variable X_2 has been included.

Equation (14.11) defines the partial F-test statistic for testing the contribution of an independent variable.

PARTIAL F-TEST STATISTIC

$$F = \frac{SSR(X_j \mid \text{all variables } except\ j)}{MSE} \tag{14.11}$$

The partial F-test statistic follows an F distribution with 1 and $n - k - 1$ degrees of freedom.

From Table 14.3,

$$F = \frac{24{,}556{,}916.67}{407{,}127.31} = 60.32$$

The partial F-test statistic has 1 and $n - k - 1 = 34 - 2 - 1 = 31$ degrees of freedom. Using a level of significance of 0.05, the critical value from Table E.5 is approximately 4.17 (see Figure 14.12).

FIGURE 14.12
Testing for Contribution of a Regression Coefficient to Multiple Regression Model at the 0.05 Level of Significance with 1 and 31 Degrees of Freedom

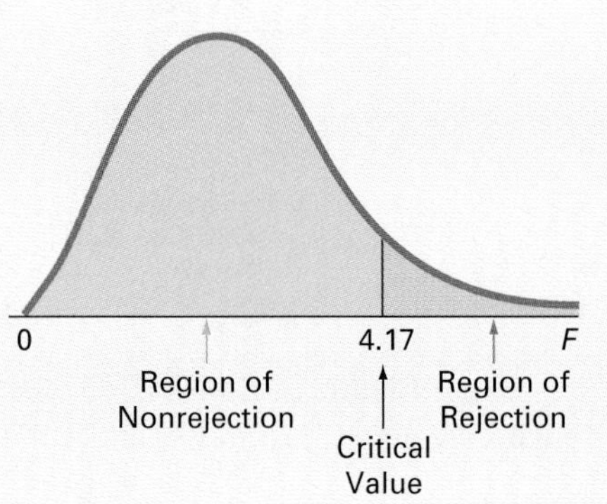

Because the partial F-test statistic is greater than this critical F value (60.32 > 4.17), you reject H_0 and conclude that the addition of variable X_1 (price) significantly improves a regression model that already contains variable X_2 (promotional expenditures).

To evaluate the contribution of variable X_2 (promotional expenditures) to a model in which variable X_1 (price) has been included, you need to use Equation (14.10b). First, from Figure 14.9 or 14.11 on page 587, observe that $SSR(X_1) = 28{,}153{,}486.15$. Second, from Table 14.3 observe that $SSR(X_1 \text{ and } X_2) = 39{,}472{,}730.77$. Then, using Equation (14.10b) on page 586,

$$SSR(X_2 \mid X_1) = 39{,}472{,}730.77 - 28{,}153{,}486.15 = 11{,}319{,}244.62$$

To determine whether X_2 significantly improves a model after X_1 has been included, you can divide the regression sum of squares into two component parts as shown in Table 14.4.

TABLE 14.4

ANOVA Table Dividing the Regression Sum of Squares into Components to Determine the Contribution of Variable X_2

Source	Degrees of Freedom	Sum of Squares	Mean Square (Variance)	F
Regression	2	39,472,730.77	19,736,365.39	
$\left\{\begin{matrix} X_1 \\ X_2 \mid X_1 \end{matrix}\right\}$	$\left\{\begin{matrix} 1 \\ 1 \end{matrix}\right\}$	$\left\{\begin{matrix} 28{,}153{,}486.15 \\ 11{,}319{,}244.62 \end{matrix}\right\}$	11,319,244.62	27.80
Error	31	12,620,946.67	407,127.31	
Total	33	52,093,677.44		

The null and alternative hypotheses to test for the contribution of X_2 to the model are:

H_0: Variable X_2 does not significantly improve the model after variable X_1 has been included.
H_1: Variable X_2 significantly improves the model after variable X_1 has been included.

Using Equation (14.11) and Table 14.4,

$$F = \frac{11{,}319{,}244.62}{407{,}127.31} = 27.80$$

In Figure 14.12 on page 588, you can see that using a 0.05 level of significance, the critical value of F with 1 and 31 degrees of freedom is approximately 4.17. Since the partial F-test statistic is greater than this critical value (27.80 > 4.17), you reject H_0 and conclude that the addition of variable X_2 (promotional expenditures) significantly improves the multiple regression model already containing X_1 (price).

Thus, by testing for the contribution of each independent variable after the other has been included in the model, you determine that each of the two independent variables significantly improves the model. Therefore, the multiple regression model should include both price X_1 and promotional expenditures X_2.

The partial F-test statistic developed in this section and the t-test statistic of Equation (14.7) on page 582 are both used to determine the contribution of an independent variable to a multiple regression model. In fact, the hypothesis tests associated with these two statistics always result in the same decision (i.e., the p-values are identical). The t values for the OmniPower regression model are −7.77 and +5.27, and the corresponding F values are 60.32 and 27.80. Equation (14.12) illustrates the relationship between t and F.

THE RELATIONSHIP BETWEEN A t STATISTIC AND AN F STATISTIC

$$t_a^2 = F_{1,a} \quad \textbf{(14.12)}$$

where a = degrees of freedom

Coefficients of Partial Determination

Recall from section 14.1 that the coefficient of multiple determination r^2 measures the proportion of the variation in Y that is explained by variation in the two independent variables. Now, the contribution of each independent variable to the multiple regression model, while holding constant the other variable, is examined. The **coefficients of partial determination** ($r^2_{Y1.2}$ and $r^2_{Y2.1}$) measure the proportion of the variation in the dependent variable that is explained by each independent variable while controlling for, or holding constant, the other independent variable. Equation (14.13) defines the coefficients of partial determination for a multiple regression model with two independent variables.

COEFFICIENTS OF PARTIAL DETERMINATION FOR A MULTIPLE REGRESSION MODEL CONTAINING TWO INDEPENDENT VARIABLES

$$r^2_{Y1.2} = \frac{SSR(X_1|X_2)}{SST - SSR(X_1 \text{ and } X_2) + SSR(X_1|X_2)} \quad \textbf{(14.13a)}$$

and

$$r^2_{Y2.1} = \frac{SSR(X_2|X_1)}{SST - SSR(X_1 \text{ and } X_2) + SSR(X_2|X_1)} \quad \textbf{(14.13b)}$$

where $SSR(X_1 \mid X_2)$ = sum of squares of the contribution of variable X_1 to the regression model given that variable X_2 has been included in the model

SST = total sum of squares for Y

$SSR(X_1 \text{ and } X_2)$ = regression sum of squares when variable X_1 and X_2 are both included in the multiple regression model

$SSR(X_2 \mid X_1)$ = sum of squares of the contribution of variable X_2 to the regression model given that variable X_1 has been included in the model

Equation (14.14) defines the coefficient of partial determination for the jth variable in a multiple regression model containing several (k) independent variables.

COEFFICIENT OF PARTIAL DETERMINATION FOR A MULTIPLE REGRESSION MODEL CONTAINING k INDEPENDENT VARIABLES

$$r^2_{Y_j.(\text{all variables } except\ j)} = \frac{SSR(X_j|\text{all variables } except\ j)}{SST - SSR(\text{all variables } including\ j) + SSR(X_j|\text{all variables } except\ j)} \quad \textbf{(14.14)}$$

For the OmniPower sales example,

$$r^2_{Y1.2} = \frac{24{,}556{,}916.67}{52{,}093{,}677.44 - 39{,}472{,}730.77 + 24{,}556{,}916.67}$$
$$= 0.6605$$

and

$$r^2_{Y2.1} = \frac{11{,}319{,}244.62}{52{,}093{,}677.44 - 39{,}472{,}730.77 + 11{,}319{,}244.62}$$
$$= 0.4728$$

The coefficient of partial determination of variable Y with X_1 while holding X_2 constant ($r^2_{Y1.2}$) is 0.6605. Thus, for a given (constant) amount of promotional expenditures, 66.05% of the variation in OmniPower sales is explained by the variation in the price. The coefficient of partial determination of variable Y with X_2 while holding X_1 constant ($r^2_{Y2.1}$) is 0.4728. Thus, for a given (constant) price, 47.28% of the variation in sales of OmniPower bars is explained by variation in the amount of promotional expenditures.

PROBLEMS FOR SECTION 14.5

Learning the Basics

PH Grade ASSIST **14.31** The following is the ANOVA summary table for a multiple regression model with two independent variables:

Source	Degrees of Freedom	Sum of Squares	Mean Squares	F
Regression	2	60		
Error	18	120		
Total	20	180		

If $SSR(X_1) = 45$ and $SSR(X_2) = 25$:

a. Determine whether there is a significant relationship between Y and each of the independent variables at the 0.05 level of significance.

b. Compute the coefficients of partial determination $r^2_{Y1.2}$ and $r^2_{Y2.1}$ and interpret their meaning.

PH Grade ASSIST **14.32** The following is the ANOVA summary table for a multiple regression model with two independent variables:

Source	Degrees of Freedom	Sum of Squares	Mean Squares	F
Regression	2	30		
Error	10	120		
Total	12	150		

If $SSR(X_1) = 20$ and $SSR(X_2) = 15$:

a. Determine whether there is a significant relationship between Y and each of the independent variables at the 0.05 level of significance.

b. Compute the coefficients of partial determination $r^2_{Y1.2}$ and $r^2_{Y2.1}$ and interpret their meaning.

Applying the Concepts

You need to use Microsoft Excel, Minitab, or SPSS to solve problems 14.33–14.37.

14.33 In problem 14.5 on page 575, you used horsepower and weight to predict gasoline mileage. Using the computer output from that problem, AUTO

a. At the 0.05 level of significance, determine whether each independent variable makes a significant contribution to the regression model. On the basis of these results, indicate the most appropriate regression model for this set of data.

b. Compute the coefficients of partial determination $r^2_{Y1.2}$ and $r^2_{Y2.1}$ and interpret their meaning.

SELF Test **14.34** In problem 14.4 on page 575, you used sales and number of orders to predict distribution costs at a mail-order catalog business. Using the computer output from that problem, WARECOST

a. At the 0.05 level of significance, determine whether each independent variable makes a significant contribution to the regression model. On the basis of these results, indicate the most appropriate regression model for this set of data.

b. Compute the coefficients of partial determination $r^2_{Y1.2}$ and $r^2_{Y2.1}$ and interpret their meaning.

14.35 In problem 14.7 on page 576, you used the total staff present and remote hours to predict standby hours. Using the computer output from that problem, STANDBY

a. At the 0.05 level of significance, determine whether each independent variable makes a significant contribution to the regression model. On the basis of these results, indicate the most appropriate regression model for this set of data.

b. Compute the coefficients of partial determination $r^2_{Y1.2}$ and $r^2_{Y2.1}$ and interpret their meaning.

14.36 In problem 14.6 on page 575, you used radio advertising and newspaper advertising to predict sales. Using the computer output from that problem, ADVERTISE

a. At the 0.05 level of significance, determine whether each independent variable makes a significant contribution to the regression model. On the basis of these results, indicate the most appropriate regression model for this set of data.

b. Compute the coefficients of partial determination $r^2_{Y1.2}$ and $r^2_{Y2.1}$ and interpret their meaning.

14.37 In problem 14.8 on page 576, you used the land area of a property and the age of a house to predict appraised value. Using the computer output from that problem, GLENCOVE

a. At the 0.05 level of significance, determine whether each independent variable makes a significant contribution to the regression model. On the basis of these results, indicate the most appropriate regression model for this set of data.

b. Compute the coefficients of partial determination $r^2_{Y1.2}$ and $r^2_{Y2.1}$ and interpret their meaning.

14.6 USING DUMMY VARIABLES AND INTERACTION TERMS IN REGRESSION MODELS

The multiple regression models discussed in sections 14.1 through 14.5 assumed that each independent variable is numerical. However, in some situations you might want to include categorical variables as independent variables in the regression model. For example, in section 14.1, you used price and promotional expenditures to predict the monthly sales of OmniPower high-energy bars. In addition to these numerical independent variables, you may want to include the effect of the shelf location in the store (e.g., end-aisle display or no end-aisle display) when developing a model to predict OmniPower sales.

The use of **dummy variables** allows you to include categorical independent variables as part of the regression model. If a given categorical independent variable has two categories, then you need only one dummy variable to represent the two categories. A dummy variable X_d is defined as;

$$X_d = 0 \text{ if the observation is in category 1}$$

$$X_d = 1 \text{ if the observation is in category 2}$$

To illustrate the application of dummy variables in regression, consider a model for predicting the assessed value from a sample of 15 houses based on the size of the house (in thousands of square feet) and whether or not the house has a fireplace. To include the categorical variable concerning the presence of a fireplace, the dummy variable X_2 is defined as:

$$X_2 = 0 \text{ if the house does not have a fireplace}$$

$$X_2 = 1 \text{ if the house has a fireplace}$$

In the last column of Table 14.5, you can see how the categorical data are converted to numerical values. **HOUSE3**

TABLE 14.5

Predicting Assessed Value Based on Size of the House and Presence of a Fireplace

House	Y = Assessed Value (\$000)	X_1 = Size of Dwelling (Thousands of Square Feet)	Fireplace	X_2 = Fireplace
1	84.4	2.00	Yes	1
2	77.4	1.71	No	0
3	75.7	1.45	No	0
4	85.9	1.76	Yes	1
5	79.1	1.93	No	0
6	70.4	1.20	Yes	1
7	75.8	1.55	Yes	1
8	85.9	1.93	Yes	1
9	78.5	1.59	Yes	1
10	79.2	1.50	Yes	1
11	86.7	1.90	Yes	1
12	79.3	1.39	Yes	1
13	74.5	1.54	No	0
14	83.8	1.89	Yes	1
15	76.8	1.59	No	0

Assuming that the slope of assessed value with the size of the house is the same for houses that have and do not have a fireplace, the multiple regression model is

$$Y_i = \beta_0 + \beta_1 X_{1i} + \beta_2 X_{2i} + \varepsilon_i$$

where Y_i = assessed value in thousands of dollars for house i

β_0 = Y intercept

X_{1i} = size of the house in thousands of square feet for house i

β_1 = slope of assessed value with size of the house, holding constant the effect of the presence of a fireplace

X_{2i} = dummy variable representing the presence or absence of a fireplace for house i

β_2 = incremental effect of the presence of a fireplace, holding constant the effect of the size of the house

ε_i = random error in Y for house i

Figure 14.13 illustrates the Microsoft Excel output for this model. Figure 14.14 shows Minitab output.

FIGURE 14.13
Microsoft Excel Output for the Regression Model that Includes Size of the House and Presence of Fireplace

	A	B	C	D	E	F	G
1	**Assessed Value Analysis**						
2							
3	*Regression Statistics*						
4	Multiple R	0.90059					
5	R Square	0.81106					
6	Adjusted R Square	0.77957					
7	Standard Error	2.26260					
8	Observations	15					
9							
10	ANOVA						
11		*df*	*SS*	*MS*	*F*	*Significance F*	
12	Regression	2	263.70391	131.85196	25.75565	4.54968E-05	
13	Residual	12	61.43209	5.11934			
14	Total	14	325.136				
15							
16		*Coefficients*	*Standard Error*	*t Stat*	*P-value*	*Lower 95%*	*Upper 95%*
17	Intercept	50.09049	4.351658	11.510668	7.67943E-08	40.60904	59.57194
18	Size	16.18583	2.574442	6.287124	4.02437E-05	10.57661	21.79506
19	Fireplace	3.85298	1.241223	3.104183	0.00912	1.14859	6.55737

FIGURE 14.14
Minitab Output for the Regression Model that Includes Size of the House and Presence of Fireplace

```
The regression equation is
Value = 50.1 + 16.2 Size + 3.85 Fireplace


Predictor     Coef  SE Coef      T      P
Constant    50.090    4.352  11.51  0.000
Size        16.186    2.574   6.29  0.000
Fireplace    3.853    1.241   3.10  0.009


S = 2.26260   R-Sq = 81.1%   R-Sq(adj) = 78.0%


Analysis of Variance

Source          DF      SS      MS      F      P
Regression       2  263.70  131.85  25.76  0.000
Residual Error  12   61.43    5.12
Total           14  325.14
```

From Figure 14.13 or 14.14, the regression equation is

$$\hat{Y}_i = 50.09 + 16.186X_{1i} + 3.853X_{2i}$$

For houses without a fireplace, you substitute $X_2 = 0$ into the regression equation

$$\begin{aligned}\hat{Y}_i &= 50.09 + 16.186X_{1i} + 3.853X_{2i} \\ &= 50.09 + 16.186X_{1i} + 3.853(0) \\ &= 50.09 + 16.186X_{1i}\end{aligned}$$

For houses with a fireplace, you substitute $X_2 = 1$ into the regression equation

$$\begin{aligned}\hat{Y}_i &= 50.09 + 16.186X_{1i} + 3.853X_{2i} \\ &= 50.09 + 16.186X_{1i} + 3.853(1) \\ &= 53.943 + 16.186X_{1i}\end{aligned}$$

In this model, the regression coefficients are interpreted as follows:

1. Holding constant whether or not a house has a fireplace, for each increase of 1.0 thousand square feet in the size of the house, the mean assessed value is estimated to increase by 16.186 thousand dollars (or $16,186).
2. Holding constant the size of the house, the presence of a fireplace is estimated to increase the mean value of the house by 3.853 thousand dollars (or $3,853).

In Figure 14.13 or 14.14, the t statistic for the slope of the size of the house with assessed value is 6.29 and the p-value is approximately 0.000; the t statistic for presence of a fireplace is 3.10 and the p-value is 0.009. Thus, each of the two variables makes a significant contribution to the model at a level of significance of 0.01. In addition, the coefficient of multiple determination indicates that 81.1% of the variation in assessed value is explained by variation in the size of the house and whether the house has a fireplace.

EXAMPLE 14.3

MODELING A THREE-LEVEL CATEGORICAL VARIABLE

Define a multiple regression model using sales as the dependent variable and package-design and price as independent variables. Package-design is a three-level categorical variable with designs A, B, or C.

SOLUTION To model a three-level categorical variable, two dummy variables are needed:

$$X_{1i} = 1 \text{ if package design } A \text{ is used in observation } i,\ 0 \text{ otherwise}$$
$$X_{2i} = 1 \text{ if package design } B \text{ is used in observation } i,\ 0 \text{ otherwise}$$

If observation i is for package-design A, then $X_{1i} = 1$ and $X_{2i} = 0$; for package-design B, then $X_{1i} = 0$ and $X_{2i} = 1$; and for package-design C, then $X_{1i} = X_{2i} = 0$. A third independent variable is used for price:

$$X_{3i} = \text{price for observation } i$$

Thus, the regression model for this example is:

$$Y_i = \beta_0 + \beta_1 X_{1i} + \beta_2 X_{2i} + \beta_3 X_{3i} + \varepsilon_i$$

where Y_i = sales for observation i

β_0 = Y intercept

β_1 = difference between the mean of design A and the mean of design C, holding the price constant

β_2 = difference between the mean of design B and the mean of design C, holding the price constant

β_3 = slope of sales with price, holding the package design constant

ε_i = random error in Y for observation i

Interactions

In all the regression models discussed so far, the *effect* an independent variable has on the dependent variable was assumed to be statistically independent of the other independent variables in the model. An **interaction** occurs if the *effect* of an independent variable on the response variable is dependent on the *value* of a second independent variable. For example, it is possible for advertising to have a large effect on the sales of a product when the price of a product is low. However, if the price of the product is too high, increases in advertising will not dramatically change sales. In this case, sales and advertising are said to interact. In other words, you cannot make general statements about the effect of advertising on sales. The effect that advertising has on sales is *dependent* on the price. You use an **interaction term** (sometimes referred to as a **cross-product term**) to model an interaction effect in a regression model.

To illustrate the concept of interaction and use of an interaction term, return to the example concerning the assessed values of homes discussed on pages 592–594. In the regression model, you assumed that the effect the size of the home has on the assessed value is independent of whether or not the house has a fireplace. In other words, you assumed that the slope of assessed value with size is the same for houses with fireplaces as it is for houses without fireplaces. If these two slopes are different, an interaction between size of the home and fireplace exists.

To evaluate a hypothesis of equal slopes of a Y variable with X, you first define an interaction term that consists of the product of the independent variable X_1 and the dummy variable X_2. You then test whether this interaction variable makes a significant contribution to a regression model that contains the other X variables. If the interaction is significant, you cannot use the original model for prediction. For the data of Table 14.5 on page 592, let

$$X_3 = X_1 * X_2$$

Figure 14.15 illustrates Microsoft Excel output for this regression model, which includes the size of the house X_1, the presence of a fireplace X_2, and the interaction of X_1 and X_2 (which is defined as X_3). Figure 14.16 displays Minitab output.

To test for the existence of an interaction, you use the null hypothesis H_0: $\beta_3 = 0$ versus the alternative hypothesis H_1: $\beta_3 \neq 0$. In Figure 14.15 or 14.16, the t statistic for the interaction of size and fireplace is 1.48. Because the p-value $= 0.166 > 0.05$, you do not reject the null hypothesis. Therefore, the interaction does not make a significant contribution to the model given that size and presence of a fireplace are already included.

Regression models can have several numerical independent variables. Example 14.4 illustrates a regression model in which there are two numerical independent variables as well as a categorical independent variable.

FIGURE 14.15

Microsoft Excel Output for a Regression Model that Includes Size, Presence of Fireplace, and Interaction of Size and Fireplace

	A	B	C	D	E	F	G
1	**Assessed Value Analysis**						
2							
3	*Regression Statistics*						
4	Multiple R	0.91791					
5	R Square	0.84255					
6	Adjusted R Square	0.79961					
7	Standard Error	2.15727					
8	Observations	15					
9							
10	ANOVA						
11		*df*	*SS*	*MS*	*F*	*Significance F*	
12	Regression	3	273.94410	91.31470	19.62150	0.00010	
13	Residual	11	51.19190	4.65381			
14	Total	14	325.136				
15							
16		*Coefficients*	*Standard Error*	*t Stat*	*P-value*	*Lower 95%*	*Upper 95%*
17	Intercept	62.95218	9.61218	6.54921	4.13993E-05	41.79591	84.10845
18	Size	8.36242	5.81730	1.43751	0.17841	-4.44137	21.16621
19	Fireplace	-11.84036	10.64550	-1.11224	0.28975	-35.27097	11.59024
20	Size * Fireplace	9.51800	6.41647	1.48337	0.16605	-4.60456	23.64056

FIGURE 14.16

Minitab Output for a Regression Model that Includes Size, Presence of Fireplace, and Interaction of Size and Fireplace

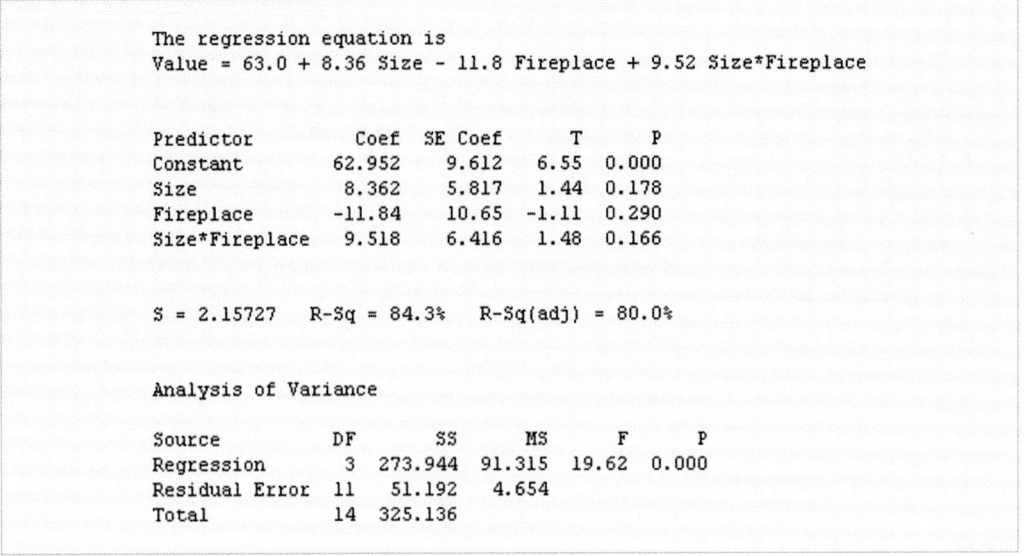

```
The regression equation is
Value = 63.0 + 8.36 Size - 11.8 Fireplace + 9.52 Size*Fireplace

Predictor          Coef  SE Coef      T      P
Constant         62.952    9.612   6.55  0.000
Size              8.362    5.817   1.44  0.178
Fireplace        -11.84    10.65  -1.11  0.290
Size*Fireplace    9.518    6.416   1.48  0.166

S = 2.15727   R-Sq = 84.3%   R-Sq(adj) = 80.0%

Analysis of Variance

Source          DF       SS      MS      F      P
Regression       3  273.944  91.315  19.62  0.000
Residual Error  11   51.192   4.654
Total           14  325.136
```

EXAMPLE 14.4 STUDYING A REGRESSION MODEL THAT CONTAINS A DUMMY VARIABLE

A real estate developer wants to predict heating oil consumption based on atmospheric temperature X_1 and the amount of attic insulation X_2. Suppose that, of 15 houses selected, houses 1, 4, 6, 7, 8, 10, and 12 are ranch-style houses. Develop and analyze an appropriate regression model using these three independent variables, X_1, X_2, and X_3 (the dummy variable for ranch-style houses). **HTNGOIL**

SOLUTION Define X_3, a dummy variable for ranch-style house as follows

$$X_3 = 0 \text{ if the style is not ranch}$$
$$X_3 = 1 \text{ if the style is ranch}$$

Assuming that the slope between home heating oil consumption and atmospheric temperature X_1 and between home heating oil consumption and the amount of attic insulation X_2 is the same for both styles of houses, the regression model is:

$$Y_i = \beta_0 + \beta_1 X_{1i} + \beta_2 X_{2i} + \beta_3 X_{3i} + \varepsilon_i$$

where Y_i = monthly heating oil consumption in gallons for house i

β_0 = Y intercept

β_1 = slope of heating oil consumption with atmospheric temperature, holding constant the effect of attic insulation and the style of the house

β_2 = slope of heating oil consumption with attic insulation, holding constant the effect of atmospheric temperature and the style of the house

β_3 = incremental effect of the presence of a ranch-style house, holding constant the effect of atmospheric temperature and attic insulation

ε_i = random error in Y for house i

Figure 14.17 displays Microsoft Excel output. Figure 14.18 illustrates Minitab output.

FIGURE 14.17

Microsoft Excel Output for a Regression Model that Includes Temperature, Insulation, and Style for the Heating Oil Data

	A	B	C	D	E	F	G
1	Heating Oil Consumption Analysis						
2							
3	*Regression Statistics*						
4	Multiple R	0.99421					
5	R Square	0.98845					
6	Adjusted R Square	0.98529					
7	Standard Error	15.74894					
8	Observations	15					
9							
10	ANOVA						
11		*df*	*SS*	*MS*	*F*	*Significance F*	
12	Regression	3	233406.90935	77802.30312	313.68217	6.21548E-11	
13	Residual	11	2728.31998	248.02909			
14	Total	14	236135.22933				
15							
16		*Coefficients*	*Standard Error*	*t Stat*	*P-value*	*Lower 95%*	*Upper 95%*
17	Intercept	592.54012	14.33698	41.32948	2.02317E-13	560.98461	624.09562
18	Temperature	-5.52510	0.20443	-27.02670	2.07188E-11	-5.97505	-5.07515
19	Insulation	-21.37613	1.44802	-14.76232	1.34816E-08	-24.56320	-18.18906
20	Ranch-style	-38.97267	8.35844	-4.66267	0.00069	-57.36947	-20.57586

FIGURE 14.18

Minitab Output for a Regression Model that Includes Temperature, Insulation, and Style for the Heating Oil Data

```
The regression equation is
Gallons = 593 - 5.53 Temp(F) - 21.4 Insulation - 39.0 Style


Predictor       Coef  SE Coef       T      P
Constant      592.54    14.34   41.33  0.000
Temp(F)      -5.5251   0.2044  -27.03  0.000
Insulation   -21.376    1.448  -14.76  0.000
Style        -38.973    8.358   -4.66  0.001


S = 15.7489   R-Sq = 98.8%   R-Sq(adj) = 98.5%


Analysis of Variance

Source          DF      SS     MS       F      P
Regression       3  233407  77802  313.68  0.000
Residual Error  11    2728    248
Total           14  236135
```

From the output in Figure 14.17 or 14.18, the regression equation is:

$$\hat{Y}_i = 592.5401 - 5.5251X_{1i} - 21.3761X_{2i} - 38.9727X_{3i}$$

For houses that are not ranch-style, since $X_3 = 0$, this reduces to:

$$\hat{Y}_i = 592.5401 - 5.5251X_{1i} - 21.3761X_{2i}$$

For houses that are ranch-style, since $X_3 = 1$, this reduces to:

$$\hat{Y}_i = 553.5674 - 5.5251X_{1i} - 21.3761X_{2i}$$

The regression coefficients are interpreted as follows:

1. Holding constant the effect of attic insulation and the house style, for each additional 1°F increase in atmospheric temperature, you estimate that the mean oil consumption decreases by 5.5251 gallons.
2. Holding constant the effect of atmospheric temperature and the house style, for each additional 1-inch increase in attic insulation, you estimate that the mean oil consumption decreases by 21.376 gallons.
3. b_3 measures the effect on oil consumption of having a ranch-style house ($X_3 = 1$) compared with having a house that is not ranch-style ($X_3 = 0$). Thus, with atmospheric temperature and attic insulation held constant, you estimate that the mean oil consumption is 38.973 gallons less for a ranch-style house than for a house that is not ranch-style.

The three t statistics representing the slopes for temperature, insulation, and ranch-style are –27.03, –14.76, and –4.66. Each of the corresponding p-values is extremely small, all being less than 0.001. Thus, each of the three variables makes a significant contribution to the model. In addition, the coefficient of multiple determination indicates that 98.8% of the variation in oil usage is explained by variation in the temperature, insulation, and whether the house is ranch-style.

Before you can use the model in Example 14.4, you need to determine whether the independent variables interact with each other. In Example 14.5, three interaction terms are added to the model.

EXAMPLE 14.5

EVALUATING A REGRESSION MODEL WITH SEVERAL INTERACTIONS

For the data of Example 14.4, determine whether adding the interaction terms make a significant contribution to the regression model.

SOLUTION To evaluate possible interactions between the independent variables, three interaction terms are constructed. Let $X_4 = X_1 * X_2$, $X_5 = X_1 * X_3$, and $X_6 = X_2 * X_3$. The regression model is now

$$Y_i = \beta_0 + \beta_1 X_{1i} + \beta_2 X_{2i} + \beta_3 X_{3i} + \beta_4 X_{4i} + \beta_5 X_{5i} + \beta_6 X_{6i} + \varepsilon_i$$

where X_1 is temperature, X_2 is insulation, X_3 is the dummy variable ranch-style, X_4 is the interaction between temperature and insulation, X_5 is the interaction between temperature and ranch-style, and X_6 is the interaction between insulation and ranch-style.

To test whether the three interactions significantly improve the regression model, you use the partial F test. The null and alternative hypotheses are:

H_0: $\beta_4 = \beta_5 = \beta_6 = 0$ (There are no interactions among X_1, X_2, and X_3.)
H_1: $\beta_4 \neq 0$ and/or $\beta_5 \neq 0$ and/or $\beta_6 \neq 0$ (X_1 interacts with X_2, and/or X_1 interacts with X_3, and/or X_2 interacts with X_3.)

From Figure 14.19 or 14.20

$$SSR(X_1, X_2, X_3, X_4, X_5, X_6) = 234{,}510.58 \text{ with 6 degrees of freedom}$$

and from Figure 14.17 or 14.18 on page 597, $SSR(X_1, X_2, X_3) = 233{,}406.91$ with 3 degrees of freedom.

Thus, $SSR(X_1, X_2, X_3, X_4, X_5, X_6) - SSR(X_1, X_2, X_3) = 234{,}510.58 - 233{,}406.91 = 1{,}103.67$. The difference in degrees of freedom is $6 - 3 = 3$.

FIGURE 14.19

Microsoft Excel Output for a Regression Model that Includes Temperature X_1, Insulation X_2, the Dummy Variable Ranch-Style X_3, the Interaction of Temperature and Insulation X_4, the Interaction of Temperature and Ranch-Style X_5, and the Interaction of Insulation and Ranch-Style X_6

	A	B	C	D	E	F	G
1	**Heating Oil Consumption Analysis**						
2							
3	*Regression Statistics*						
4	Multiple R	0.99655					
5	R Square	0.99312					
6	Adjusted R Square	0.98796					
7	Standard Error	14.25065					
8	Observations	15					
9							
10	ANOVA						
11		*df*	*SS*	*MS*	*F*	*Significance F*	
12	Regression	6	234510.58185	39085.09697	192.46069	3.32423E-08	
13	Residual	8	1624.64749	203.08094			
14	Total	14	236135.22933				
15							
16		*Coefficients*	*Standard Error*	*t Stat*	*P-value*	*Lower 95%*	*Upper 95%*
17	Intercept	642.88670	26.70590	24.07283	9.45284E-09	581.30273	704.47066
18	Temperature	-6.92627	0.75311	-9.19687	1.58014E-05	-8.66295	-5.18959
19	Insulation	-27.88251	3.58011	-7.78818	5.29456E-05	-36.13826	-19.62677
20	Style	-84.60882	29.99556	-2.82071	0.02247	-153.77875	-15.43889
21	Temperature * Insulation	0.17021	0.08863	1.92039	0.09106	-0.03418	0.37460
22	Temperature * Ranch-style	0.65957	0.46168	1.42862	0.19097	-0.40507	1.72420
23	Insulation * Ranch-style	4.98698	3.51368	1.41930	0.19358	-3.11559	13.08955

FIGURE 14.20

Minitab Output for a Regression Model that Includes Temperature X_1, Insulation X_2, the Dummy Variable Ranch-Style X_3, the Interaction of Temperature and Insulation X_4, the Interaction of Temperature and Ranch-Style X_5, and the Interaction of Insulation and Ranch-Style X_6

```
The regression equation is
Gallons = 643 - 6.93 Temp(F) - 27.9 Insulation - 84.6 Style + 0.170 Temp*Insu
          + 0.660 Temp*Style + 4.99 Insu*Style


Predictor        Coef   SE Coef       T      P
Constant       642.89     26.71   24.07  0.000
Temp(F)       -6.9263    0.7531   -9.20  0.000
Insulation    -27.883     3.580   -7.79  0.000
Style          -84.61     30.00   -2.82  0.022
Temp*Insu     0.17021   0.08863    1.92  0.091
Temp*Style     0.6596    0.4617    1.43  0.191
Insu*Style      4.987     3.514    1.42  0.194


S = 14.2506   R-Sq = 99.3%   R-Sq(adj) = 98.8%


Analysis of Variance

Source          DF      SS     MS       F      P
Regression       6  234511  39085  192.46  0.000
Residual Error   8    1625    203
Total           14  236135
```

To use the partial F test for the simultaneous contribution of three variables to a model you use an extension of Equation (14.11) on page 588.[1] The partial F test statistic is

$$F = \frac{[SSR(X_1, X_2, X_3, X_4, X_5, X_6) - SSR(X_1, X_2, X_3)]/3}{MSE(X_1, X_2, X_3, X_4, X_5, X_6)} = \frac{1{,}103.67/3}{203.08} = 1.81$$

You compare the F statistic of 1.81 to the critical F value for 3 and 8 degrees of freedom. Using a level of significance of 0.05, the critical F value from Table E.5 is 4.07. Because $1.81 < 4.07$, you conclude that the interactions do not make a significant contribution to the model, given that the model already includes temperature X_1, insulation X_2, and whether the house is ranch-style X_3. Therefore, the multiple regression model using X_1, X_2, and X_3 but no interaction terms is the better model. Had you rejected this null hypothesis, you would then test the contribution of each interaction separately in order to determine which interaction terms to include in the model.

[1] *In general, if the model has several independent variables and you want to test whether an additional set of independent variables contribute to the model, the numerator of the F test is SSR (for all independent variables) – SSR (for the initial set of variables) divided by the number of independent variables whose contribution is being tested.*

PROBLEMS FOR SECTION 14.6

Learning the Basics

PH Grade ASSIST **14.38** Suppose X_1 is a numerical variable and X_2 is a dummy variable and the following regression equation for a sample of $n = 20$ is:

$$\hat{Y}_i = 6 + 4X_{1i} + 2X_{2i}$$

a. Interpret the meaning of the slope for variable X_1.
b. Interpret the meaning of the slope for variable X_2.
c. Suppose that the t statistic for testing the contribution of variable X_2 is 3.27. At the 0.05 level of significance, is there evidence that variable X_2 makes a significant contribution to the model?

Applying the Concepts

You need to use Microsoft Excel, Minitab, or SPSS to solve problems 14.40–14.49.

14.39 The chair of the accounting department wants to develop a regression model to predict the grade-point average in accounting for graduating accounting majors based on the student's SAT score and whether or not the student received a grade of B or higher in the introductory statistics course (0 = no and 1 = yes).
a. Explain the steps involved in developing a regression model for these data. Be sure to indicate the particular models you need to evaluate and compare.
b. Suppose the regression coefficient for the variable of whether or not the student received a grade of B or higher in the introductory statistics course is +0.30. How do you interpret this result?

14.40 A real estate association in a suburban community would like to study the relationship between the size of a single-family house (as measured by the number of rooms) and the selling price of the house (in thousands of dollars). Two different neighborhoods are included in the study, one on the east side of the community (=0) and the other on the west side (=1). A random sample of 20 houses was selected with the following results given in the file: NEIGHBOR
a. State the multiple regression equation.
b. Interpret the meaning of the slopes in this problem.
c. Predict the selling price for a house with nine rooms that is located in an east-side neighborhood, and construct a 95% confidence interval estimate and a 95% prediction interval.
d. Perform a residual analysis on the results, and determine the adequacy of the model.
e. Is there a significant relationship between selling price and the two independent variables (rooms and neighborhood) at the 0.05 level of significance?
f. At the 0.05 level of significance, determine whether each independent variable makes a contribution to the regression model. Indicate the most appropriate regression model for this set of data.
g. Construct 95% confidence interval estimates of the population slope for the relationship between selling price and number of rooms, and between selling price and neighborhood.
h. Interpret the meaning of the coefficient of multiple determination.
i. Compute the adjusted r^2.
j. Compute the coefficients of partial determination and interpret their meaning.
k. What assumption do you need to make about the slope of selling price with number of rooms?
l. Add an interaction term to the model, and, at the 0.05 level of significance, determine whether it makes a significant contribution to the model.
m. On the basis of the results of (f) and (l), which model is most appropriate? Explain.

14.41 The marketing manager of a large supermarket chain would like to determine the effect of shelf space and whether the product was placed at the front or back of the aisle on the sales of pet food. A random sample of 12 equal-sized stores is selected with the following results: PETFOOD

Store	Shelf Space (Feet)	Location	Weekly Sales (Hundreds of Dollars)
1	5	Back	1.6
2	5	Front	2.2
3	5	Back	1.4
4	10	Back	1.9
5	10	Back	2.4
6	10	Front	2.6
7	15	Back	2.3
8	15	Back	2.7
9	15	Front	2.8
10	20	Back	2.6
11	20	Back	2.9
12	20	Front	3.1

a. State the multiple regression equation.
b. Interpret the meaning of the slopes in this problem.
c. Predict the weekly sales of pet food for a store with 8 feet of shelf space situated at the back of the aisle, and construct a 95% confidence interval estimate and a 95% prediction interval.

d. Perform a residual analysis on the results and determine the adequacy of the model.
e. Is there is a significant relationship between sales and the two independent variables (shelf space and aisle position) at the 0.05 level of significance?
f. At the 0.05 level of significance, determine whether each independent variable makes a contribution to the regression model. Indicate the most appropriate regression model for this set of data.
g. Construct 95% confidence interval estimates of the population slope for the relationship between sales and shelf space, and between sales and aisle location.
h. Compare the slope in (b) with the slope for the simple linear regression model of problem 13.4 on page 523. Explain the difference in the results.
i. Interpret the meaning of the coefficient of multiple determination r^2.
j. Compute the adjusted r^2.
k. Compare r^2 with the r^2 value computed in problem 13.16(a) on page 531.
l. Compute the coefficients of partial determination and interpret their meaning.
m. What assumption about the slope of shelf space with sales do you need to make in this problem?
n. Add an interaction term to the model, and at the 0.05 level of significance, determine whether it makes a significant contribution to the model.
o. On the basis of the results of (f) and (n), which model is most appropriate? Explain.

14.42 In mining engineering, holes are often drilled through rock using drill bits. As the drill hole gets deeper, additional rods are added to the drill bit to enable additional drilling to take place. It is expected that drilling time increases with depth. This increased drilling time could be caused by several factors, including the mass of the drill rods that are strung together. A key question relates to whether drilling is faster using dry drilling holes or wet drilling holes. Dry drilling holes involve forcing compressed air down the drill rods to flush the cuttings and drive the hammer. Using wet drilling holes involves forcing water rather than air down the hole. The data file DRILL contains measurements of the time to drill each additional five feet (in minutes), the depth (in feet), and whether the hole was a dry drilling hole or a wet drilling hole for a sample of 50 drill holes. Develop a model to predict additional drilling time based on depth and type of drilling hole (dry or wet).

Source: R. Penner, and D. G. Watts, "Mining Information," The American Statistician, *45, 1991, 4–9.*

a. State the multiple regression equation.
b. Interpret the meaning of the slopes in this problem.
c. Predict the additional drilling time for a dry drilling hole at a depth of 100 feet, and construct a 95% confidence interval estimate and a 95% prediction interval.
d. Perform a residual analysis on the results and determine the adequacy of the model.
e. Is there a significant relationship between additional drilling time and the two independent variables (depth and type of drilling hole) at the 0.05 level of significance?
f. At the 0.05 level of significance, determine whether each independent variable makes a contribution to the regression model. Indicate the most appropriate regression model for this set of data.
g. Construct 95% confidence interval estimates of the population slope for the relationship between additional drilling time and depth, and between additional drilling time and type of drilling hole.
h. Interpret the meaning of the coefficient of multiple determination.
i. Compute the adjusted r^2.
j. Compute the coefficients of partial determination and interpret their meaning.
k. What assumption about the slope of additional drilling time with depth do you need to make?
l. Add an interaction term to the model and, at the 0.05 level of significance, determine whether it makes a significant contribution to the model.
m. On the basis of the results of (f) and (l), which model is most appropriate? Explain.

14.43 The file COLLEGES2002 contains data on 80 colleges and universities. Among the variables included are the annual total cost (in thousands of dollars), the first quartile score on the Scholastic Aptitude Test (SAT), and whether the school is public or private (0 = public; 1 = private). Develop a model to predict the annual total cost based on first quartile SAT score and whether the school is public or private.

a. State the multiple regression equation.
b. Interpret the meaning of the slopes in this problem.
c. Predict the annual total cost for a school with a first quartile SAT score of 1,000 that is a public institution, and construct a 95% confidence interval estimate and a 95% prediction interval.
d. Perform a residual analysis on the results and determine the adequacy of the model.
e. Is there a significant relationship between annual total cost and the two independent variables (first quartile SAT score and whether the school is public or private) at the 0.05 level of significance?
f. At the 0.05 level of significance, determine whether each independent variable makes a contribution to the regression model. Indicate the most appropriate regression model for this set of data.
g. Construct 95% confidence interval estimates of the population slope for the relationship between annual total cost and first quartile SAT score, and between annual total cost and whether the school is public or private.

h. Interpret the meaning of the coefficient of multiple determination.
i. Compute the adjusted r^2.
j. Compute the coefficients of partial determination and interpret their meaning.
k. What assumption about the slope of annual total cost with first quartile SAT score do you need to make?
l. Add an interaction term to the model and, at the 0.05 level of significance, determine whether it makes a significant contribution to the model.
m. On the basis of the results of (f) and (l), which model is most appropriate? Explain.

PH Grade ASSIST SELF Test **14.44** In problem 14.4 on page 575, you used sales and orders to predict distribution cost. Develop a regression model to predict distribution cost that includes the sales, orders, and the interaction of sales and orders. WARECOST
a. At the 0.05 level of significance, is there evidence that the interaction term makes a significant contribution to the model?
b. Which regression model is more appropriate, the one used in this problem or the one used in problem 14.4? Explain.

14.45 Zagat's publishes restaurant ratings for various locations in the United States. The data file RESTRATE contains the Zagat rating for food, décor, service, and the price per person for a sample of 50 restaurants located in New York City ($X_d = 0$) and 50 restaurants located on Long Island ($X_d = 1$). Develop a regression model to predict the price per person based on a variable that represents the sum of the ratings for food, décor, and service and a dummy variable concerning location (New York City or Long Island).

Source: Extracted from Zagat Survey 2002 New York City Restaurants *and* Zagat Survey 2001–2002 Long Island Restaurants.

a. State the multiple regression equation.
b. Interpret the meaning of the slopes in this problem.
c. Predict the price for a restaurant with a summated rating of 60 that is located in New York City and construct a 95% confidence interval estimate and 95% prediction interval.
d. Perform a residual analysis on the results and determine the adequacy of the model.
e. Is there a significant relationship between price and the two independent variables (summated rating and location) at the 0.05 level of significance?
f. At the 0.05 level of significance, determine whether each independent variable makes a contribution to the regression model. Indicate the most appropriate regression model for this set of data.
g. Construct 95% confidence interval estimates of the population slope for the relationship between price and summated rating, and between price and location.
h. Compare the slope in (b) with the slope for the simple linear regression model of problem 13.88 on page 563. Explain the difference in the results.
i. Interpret the meaning of the coefficient of multiple determination.
j. Compute the adjusted r^2.
k. Compare r^2 with the r^2 value computed in problem 13.88(d) on page 564.
l. Compute the coefficients of partial determination and interpret their meaning.
m. What assumption about the slope of price with summated rating do you need to make in this problem?
n. Add an interaction term to the model and, at the 0.05 level of significance, determine whether it makes a significant contribution to the model.
o. On the basis of the results of (f) and (n), which model is most appropriate? Explain.

14.46 In problem 14.6 on page 575, you used radio advertising and newspaper advertising to predict sales. Develop a regression model to predict sales that includes radio advertising, newspaper advertising, and the interaction of radio advertising and newspaper advertising. ADVERTISE
a. At the 0.05 level of significance, is there evidence that the interaction term makes a significant contribution to the model?
b. Which regression model is more appropriate, the one used in this problem or the one used in problem 14.6? Explain.

PH Grade ASSIST **14.47** In problem 14.5 on page 575, horsepower and weight were used to predict miles per gallon. Develop a regression model that includes horsepower, weight, and the interaction of horsepower and weight to predict miles per gallon. AUTO
a. At the 0.05 level of significance, is there evidence that the interaction term makes a significant contribution to the model?
b. Which regression model is more appropriate, the one used in this problem or the one used in problem 14.5? Explain.

14.48 In problem 14.7 on page 576, you used total staff present and remote hours to predict standby hours. Develop a regression model to predict standby hours that includes total staff present, remote hours, and the interaction of total staff present and remote hours. STANDBY
a. At the 0.05 level of significance, is there evidence that the interaction term makes a significant contribution to the model?
b. Which regression model is more appropriate, the one used in this problem or the one used in problem 14.7? Explain.

14.49 The director of a training program for a large insurance company is evaluating three different methods of training underwriters. The three methods are traditional, CD-ROM-based, and Web-based. She divides 30 trainees into three randomly assigned groups of 10. Before the start of the training, each trainee is given a proficiency exam that measures mathematics and computer skills. At the end of the training, all students take the same end-of-training exam. The results UNDERWRITING are as follows.

Proficiency Exam	End-of-Training Exam	Method
94	14	Traditional
96	19	Traditional
98	17	Traditional
100	38	Traditional
102	40	Traditional
105	26	Traditional
109	41	Traditional
110	28	Traditional
111	36	Traditional
130	66	Traditional
80	38	CD-ROM-based
84	34	CD-ROM-based
90	43	CD-ROM-based
97	43	CD-ROM-based
97	61	CD-ROM-based
112	63	CD-ROM-based
115	93	CD-ROM-based
118	74	CD-ROM-based
120	76	CD-ROM-based
120	79	CD-ROM-based
92	55	Web-based
96	53	Web-based
99	55	Web-based
101	52	Web-based
102	35	Web-based
104	46	Web-based
107	57	Web-based
110	55	Web-based
111	42	Web-based
118	81	Web-based

Develop a multiple regression model to predict the score on the end-of-training exam based on the score on the proficiency exam and the method of training used.

a. State the multiple regression equation.

b. Interpret the meaning of the slopes in this problem.

c. Predict the end-of-training exam score for a student with a proficiency exam score of 100 who had Web-based training.

d. Perform a residual analysis on your results and determine the adequacy of the model.

e. Is there a significant relationship between the end-of-training exam score and the independent variables (proficiency score and training method) at the 0.05 level of significance.

f. At the 0.05 level of significance, determine whether each independent variable makes a contribution to the regression model. Indicate the most appropriate regression model for this set of data.

g. Construct 95% confidence interval estimates of the population slope for the relationship between end-of-training exam score and each independent variable.

h. Interpret the meaning of the coefficient of multiple determination.

i. Compute the adjusted r^2.

j. Compute the coefficients of partial determination, and interpret their meaning.

k. What assumption about the slope of proficiency score with end-of-training exam score do you need to make in this problem?

l. Add interaction terms to the model, and at the 0.05 level of significance determine whether any interaction terms make a significant contribution to the model.

m. On the basis of the results of (f) and (l), which model is most appropriate? Explain.

14.7 LOGISTIC REGRESSION

The discussion of the simple linear regression model in Chapter 13 and the multiple regression models in sections 14.1 through 14.6 only considered numerical response variables. However, in many instances, the response variable is a categorical variable that takes on one of only two possible values. For example, a customer prefers Brand *A* or a customer prefers Brand *B*. The use of simple or multiple least-squares regression for a categorical response variable violates the normality assumption and can predict *Y* values that are impossible.

An alternative approach, **logistic regression**, originally applied to survival data in the health sciences (see reference 2), enables you to use regression models to predict the probability of a particular categorical response for a given set of explanatory variables. This logistic regression model is based on the **odds ratio**, which represents the probability of a success compared with the probability of failure. Equation (14.15) defines the odds ratio.

ODDS RATIO

$$\text{Odds ratio} = \frac{\text{probability of success}}{1 - \text{probability of success}} \qquad \textbf{(14.15)}$$

Using Equation (14.15), if the probability of success for an event is 0.50, the odds ratio is

$$\text{Odds ratio} = \frac{0.50}{1 - 0.50} = 1.0, \text{ or 1 to 1}$$

If the probability of success for an event is 0.75, the odds ratio is

$$\text{Odds ratio} = \frac{0.75}{1 - 0.75} = 3.0, \text{ or 3 to 1}$$

The logistic regression model is based on the natural logarithm (ln) of this odds ratio.[2] Equation (14.16) defines the logistic regression model for k independent variables.

[2] *For more information on logarithms see Appendix A.3.*

LOGISTIC REGRESSION MODEL

$$\ln(\text{odds ratio}) = \beta_0 + \beta_1 X_{1i} + \beta_2 X_{2i} + \cdots + \beta_k X_{ki} + \varepsilon_i \quad \textbf{(14.16)}$$

where k = number of independent variables in the model
ε_i = random error in observation i

A mathematical method called *maximum likelihood estimation* is usually used to develop a regression equation to predict the natural logarithm of this odds ratio. Equation (14.17) defines the logistic regression equation.

LOGISTIC REGRESSION EQUATION

$$\ln(\text{estimated odds ratio}) = b_0 + b_1 X_{1i} + b_2 X_{2i} + \cdots + b_k X_{ki} \quad \textbf{(14.17)}$$

Once you have determined the logistic regression equation, use Equation (14.18) to compute the estimated odds ratio.

ESTIMATED ODDS RATIO

$$\text{Estimated odds ratio} = e^{\ln(\text{estimated odds ratio})} \quad \textbf{(14.18)}$$

Once you have computed the estimated odds ratio, use Equation (14.19) to find the estimated probability of success.

ESTIMATED PROBABILITY OF SUCCESS

$$\text{Estimated probability of success} = \frac{\text{estimated odds ratio}}{1 + \text{estimated odds ratio}} \quad \textbf{(14.19)}$$

To illustrate the logistic regression model, the marketing department for a credit card company is about to embark on a campaign to convince existing holders of the company's standard credit card to upgrade to the company's premium card for a nominal annual fee. A major question facing the marketing department is "Which of the existing standard credit cardholders should be the target for the campaign?" Data available from a sample of 30 cardholders who were contacted during last year's campaign indicate whether the cardholder upgraded to a premium card (0 = no, 1 = yes). The marketing department wants to predict the categorical variable purchase behavior (i.e., did the customer upgrade to a premium card?) using two explanatory variables: total amount of credit card purchases (in thousands of dollars) in the prior year (X_1), and whether the cardholder possesses additional credit cards (which involves an extra cost) for other members of the household (X_2: 0 = no, 1 = yes). Table 14.6 presents the data. **LOGPURCH**

TABLE 14.6
Purchase Behavior, Annual Credit Card Spending, and Possession of Additional Credit Cards

Observation	Purchase Behavior	Annual Spending ($000)	Possession of Additional Credit Cards	Observation	Purchase Behavior	Annual Spending ($000)	Possession of Additional Credit Cards
1	0	32.1007	0	16	0	23.7609	0
2	1	34.3706	1	17	0	35.0388	1
3	0	4.8749	0	18	1	49.7388	1
4	0	8.1263	0	19	0	24.7372	0
5	0	12.9783	0	20	1	26.1315	1
6	0	16.0471	0	21	0	31.3220	1
7	0	20.6648	0	22	1	40.1967	1
8	1	42.0483	1	23	0	35.3899	0
9	0	42.2264	1	24	0	30.2280	0
10	1	37.9900	1	25	1	50.3778	0
11	1	53.6063	1	26	0	52.7713	0
12	0	38.7936	0	27	0	27.3728	0
13	0	27.9999	0	28	1	59.2146	1
14	1	42.1694	0	29	1	50.0686	1
15	1	56.1997	1	30	1	35.4234	1

Figure 14.21 represents Minitab output for the logistic regression model. There are two independent variables, X_1 (Spending) and X_2 (Extra).

FIGURE 14.21
Minitab Logistic Regression Output for the Data of Table 14.6

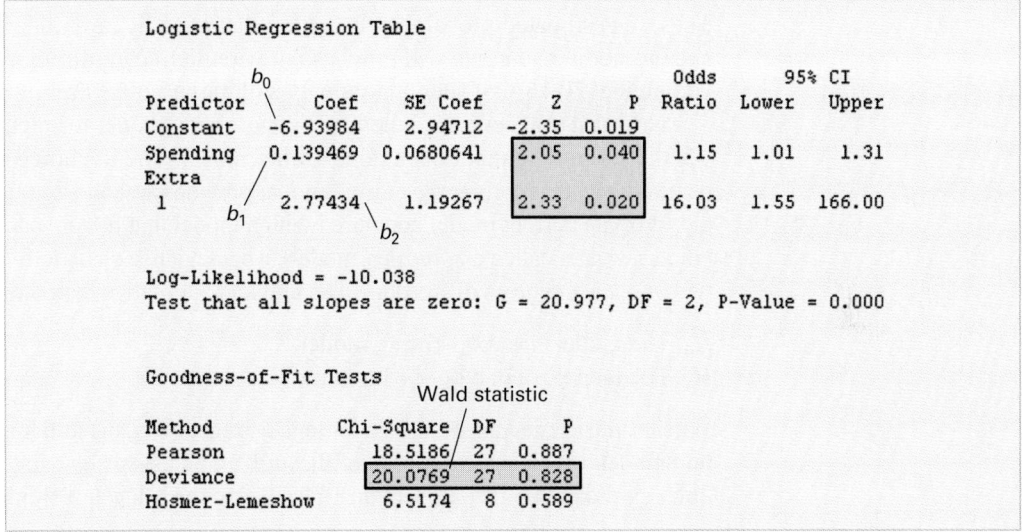

```
Logistic Regression Table

              b0                                              Odds      95% CI
Predictor      \    Coef    SE Coef       Z       P   Ratio  Lower   Upper
Constant   -6.93984     2.94712   -2.35   0.019
Spending    0.139469  0.0680641    2.05   0.040    1.15   1.01    1.31
Extra
  1        b1   2.77434    1.19267    2.33   0.020   16.03   1.55  166.00
                       b2

Log-Likelihood = -10.038
Test that all slopes are zero: G = 20.977, DF = 2, P-Value = 0.000


Goodness-of-Fit Tests
                               Wald statistic
Method              Chi-Square  DF      P
Pearson                18.5186  27  0.887
Deviance               20.0769  27  0.828
Hosmer-Lemeshow         6.5174   8  0.589
```

The regression coefficients b_0, b_1, and b_2 are interpreted as follows:

1. The regression constant b_0 is –6.940. Thus, for a credit cardholder who did not charge any purchases last year and who does not have additional cards, the estimated natural logarithm of the odds ratio of purchasing the premium card is –6.940.
2. The regression coefficient b_1 is 0.13947. Therefore, holding constant the effect of whether the credit cardholder has additional cards for members of the household, for each increase of $1,000 in annual credit card spending using the company's card, the estimated natural logarithm of the odds ratio of purchasing the premium card increases by 0.13947. Therefore, cardholders who charged more in the previous year are more likely to upgrade to a premium card.
3. The regression coefficient b_2 is 2.774. Therefore, holding constant the annual credit card spending, the estimated natural logarithm of the odds ratio of purchasing the premium card increases by 2.774 for a credit cardholder who has additional cards for members of the household compared with one who does not have additional cards. Therefore, cardholders possessing additional cards for other members of the household are much more likely to upgrade to a premium card.

Based on 2 and 3 above, the credit card company should develop a marketing campaign that targets cardholders who tend to charge large amounts to their cards, and to households that possess more than one card.

As was the case with least-squares regression models, a main purpose of performing logistic regression analysis is to provide predictions of a response variable. For example, consider a cardholder who charged $36,000 last year and possesses additional cards for members of the household. What is the probability the cardholder will upgrade to the premium card during the marketing campaign? Using $X_1 = 36$, $X_2 = 1$, Equation (14.17) on page 604, and the results displayed in Figure 14.21 on page 605,

$$\begin{aligned}\ln(\text{estimated odds of purchasing versus not purchasing}) &= -6.94 + (0.13947)(36) + (2.774)(1)\\ &= 0.85492\end{aligned}$$

Using Equation (14.18) on page 604,

$$\text{estimated odds ratio} = e^{0.85492} = 2.3512$$

Therefore, the odds are 2.3512 to 1 that a credit cardholder who spent $36,000 last year and has additional cards will purchase the premium card during the campaign. Using Equation (14.19) on page 604, you can convert this odds ratio to a probability.

$$\begin{aligned}\text{estimated probability of purchasing premium card} &= \frac{2.3512}{1 + 2.3512}\\ &= 0.7016\end{aligned}$$

Thus, the estimated probability is 0.7016 that a credit cardholder who spent $36,000 last year and has additional cards will purchase the premium card during the campaign. In other words, you predict 70.16% of such individuals will purchase the premium card.

Now that you have used the logistic regression model for prediction, you need to determine whether or not the model is a good-fitting model. The **deviance statistic** is frequently used to determine whether or not the current model provides a good fit to the data. This statistic measures the fit of the current model compared with a model that has as many parameters as there are data points (what is called a *saturated* model). The deviance statistic follows a chi-square distribution with $n - k - 1$ degrees of freedom. The null and alternative hypotheses are:

H_0: The model is a good-fitting model.
H_1: The model is not a good-fitting model.

When using the deviance statistic for logistic regression, the null hypothesis represents a good-fitting model, which is the opposite of the null hypothesis when using the overall F test for the multiple regression model (see section 14.2). Using the α level of significance, the decision rule is:

$$\text{Reject } H_0 \text{ if deviance} > \chi^2 \text{ with } n - k - 1 \text{ degrees of freedom;}$$

$$\text{otherwise, do not reject } H_0.$$

The critical value for a χ^2 statistic with $n - k - 1 = 30 - 2 - 1 = 27$ degrees of freedom is 40.113 (see Table E.4). From Figure 14.21 on page 605, the deviance = $20.08 < \chi^2 = 40.113$, or p-value = $0.828 > 0.05$. Thus, you do not reject H_0, and you conclude that the model is a good-fitting one.

Now that you have concluded that the model is a good-fitting one, you need to evaluate whether each of the independent variables makes a significant contribution to the model in the presence of the others. As was the case with linear regression in sections 13.7 and 14.4, the test statistic is based on the ratio of the regression coefficient to the standard error of the regression coefficient. In logistic regression, this ratio is defined by the **Wald statistic**, which approximately follows the normal distribution. From Figure 14.21, the Wald statistic is 2.05 for X_1 and 2.33 for X_2. Each of these is greater than the critical value of +1.96 for the normal distribution at the 0.05 level of significance (the p-values are 0.04 and 0.02). You can conclude that each of the two independent variables makes a contribution to the model in the presence of the other. Therefore, you should include both these independent variables in the model.

PROBLEMS FOR SECTION 14.7

Learning the Basics

14.50 Interpret the meaning of a logistic regression slope coefficient equal to 2.2.

14.51 Given an estimated odds ratio of 2.5, find the estimated probability of success.

14.52 Given an estimated odds ratio of 0.75, find the estimated probability of success.

14.53 Consider the following logistic regression equation:

$$\ln(\text{estimated odds ratio}) = 0.1 + 0.5X_{1i} + 0.2X_{2i}$$

a. Interpret the meaning of the logistic regression coefficients.
b. If $X_1 = 2$ and $X_2 = 1.5$, find the estimated odds ratio and interpret its meaning.
c. On the basis of the results of (b), compute the estimated probability of success.

Applying the Concepts

You need to use Minitab to solve problems 14.56–14.58.

SELF Test **14.54** Refer to Figure 14.21 on page 605.
a. Predict the probability that a cardholder who charged \$36,000 last year and does not have any additional credit cards for members of the household will purchase the premium card during the marketing campaign.
b. Compare the results in (a) with those for a person with additional credit cards.
c. Predict the probability that a cardholder who charged \$18,000 and does not have any additional credit cards for members of the household will purchase the premium card during the marketing campaign.
d. Compare the results of (a) and (c) and indicate what implications these results might have for the strategy for the marketing campaign.

14.55 Researchers at John Carroll University investigated reasons companies adopt electronic data interchange (EDI) with their supply chain partners (Charles A. Watts, Patrick T. Hogan, and Mark D. Treleven, "Issues Influencing Use of Electronic Data Interchange Technology: An Empirical Study," *Mid-American Journal of Business*, Fall 1998, 13(2): 7–13). The researchers used a logistic regression model to predict whether a firm adopts EDI. The dependent variable was equal to 1 if a company uses EDI and 0 otherwise. Independent variables in their logistic regression model are:

X_1 = The company's resistance to change (the larger the value of X_1, the more resistance to change present in the organization)

X_2 = The importance a company places on technology infrastructure (the larger the value of X_2, the more importance placed on technology within the organization)

X_3 = The financial hurdles involved in implementing EDI (the larger the value of X_3, the more expensive it is for a company to implement EDI)

X_4 = The amount of opinion leadership (the larger the value of X_4, the more contact the company has with sources that have previous experience with EDI such as customers and user groups)

A partial listing of the logistic regression analysis is given in the following table.

$n = 67$ Variable	Coefficient Estimates	p-Value
X_1	–0.95	< 0.01
X_2	0.06	> 0.05
X_3	0.73	< 0.05
X_4	–0.53	> 0.05

Explain the meaning of the regression coefficient estimates. Be sure to write your answers in a business context.

14.56 The director of graduate studies at a college of business wants to predict the success of students in an MBA program using two explanatory variables, undergraduate grade point average and GMAT score. A random sample of 30 students indicates that 20 successfully completed the program (coded as 1) and 10 did not (coded as 0). **MBA**

Success in MBA Program	Undergraduate Grade Point Average	GMAT Score	Success in MBA Program	Undergraduate Grade Point Average	GMAT Score
0	2.93	617	1	3.17	639
0	3.05	557	1	3.24	632
0	3.11	599	1	3.41	639
0	3.24	616	1	3.37	619
0	3.36	594	1	3.46	665
0	3.41	567	1	3.57	694
0	3.45	542	1	3.62	641
0	3.60	551	1	3.66	594
0	3.64	573	1	3.69	678
0	3.57	536	1	3.70	624
1	2.75	688	1	3.78	654
1	2.81	647	1	3.84	718
1	3.03	652	1	3.77	692
1	3.10	608	1	3.79	632
1	3.06	680	1	3.97	784

a. Use Minitab to estimate a logistic regression model to predict the probability of successful completion of the MBA program based on undergraduate grade point average and GMAT score.

b. Explain the meaning of the regression coefficients for the model fit in (a).

c. Predict the probability of successful completion of the program for a student with an undergraduate grade point average of 3.25 and a GMAT score of 600.

d. At the 0.05 level of significance, is there evidence that a logistic regression model that uses undergraduate grade point average and GMAT score to predict probability of success in the MBA program is a good-fitting model?

e. At the 0.05 level of significance, is there evidence that undergraduate grade point average and GMAT score each make a significant contribution to the logistic regression model?

f. Use Minitab to estimate a logistic regression model that includes only undergraduate grade point average to predict probability of success in the MBA program.

g. Use Minitab to estimate a logistic regression model that includes only GMAT score to predict probability of success in the MBA program.

h. Compare the models in (a), (f) and (g). Evaluate the differences among the models.

14.57 The marketing manager for a nationally franchised lawn service company would like to study the characteristics that differentiate home owners who do and do not have a lawn service. A random sample of 30 home owners located in a suburban area near a large city was selected; 15 did not have a lawn service (code 0) and 15 had a lawn service (code 1). Additional information available concerning these 30 home owners includes family income (in thousands of dollars), lawn size (in thousands of square feet), attitude toward outdoor recreational activities (0 = unfavorable, 1 = favorable), number of teenagers in the household, and age of the head of the household. LAWN

a. Use Minitab to estimate a logistic regression model to predict the probability of using a lawn service based on family income (in thousands of dollars), lawn size (in thousands of square feet), attitude toward outdoor recreational activities (0 = unfavorable, 1 = favorable), number of teenagers in the household, and age of the head of the household.

b. Explain the meaning of the regression coefficients for the model fit in (a).

c. Predict the probability of purchasing a lawn service for a 48-year-old home owner with a family income of \$100,000, a lawn size of 5,000 square feet, a negative attitude toward outdoor recreation, and one teenager in the household.

d. At the 0.05 level of significance, is there evidence that a logistic regression model that uses family income, lawn size, attitude toward outdoor recreation, number of teenagers in the household, and age of the head of the household is a good-fitting model?

e. At the 0.05 level of significance, is there evidence that each of the five explanatory variables (family income, lawn size, attitude toward outdoor recreation, number of teenagers in the household, and age of the head of the household) makes a significant contribution to the logistic regression model?

14.58 Undergraduate students at Miami University in Oxford, Ohio, were surveyed in order to evaluate the effect of price on the purchase of a pizza from Pizza Hut. Two hundred twenty students were asked to suppose that they were going to have a large 2-topping pizza delivered to their residence. They were asked to select from either Pizza Hut or another pizzeria of their choice. The price they would have to pay to get a Pizza Hut pizza differed from survey to survey. For example, some surveys used the price currently charged by the Oxford Pizza Hut, \$11.49. Other prices investigated were \$8.49, \$9.49, \$10.49, \$12.49, \$13.49, and \$14.49. The dependent variable for this study is whether or not a student will select Pizza Hut. Possible independent variables are the price of Pizza Hut and the gender of the student. The data set PIZZAHUT has 220 observations and 3 variables:

Gender (1 = male, 0 = female)
Price (8.49, 9.49, 10.49, 11.49, 12.49, 13.49, or 14.49)
Purchase (1 = the student selected Pizza Hut, 0 = the student selected another pizzeria)

a. Use Minitab to estimate a logistic regression model to predict the probability of a student selecting Pizza Hut based on the price of the pizza. Is price an important indicator of purchase selection?

b. Use Minitab to estimate a logistic regression model to predict the probability of a student selecting Pizza Hut based on the price of the pizza and the gender of the student. Is price an important indicator of purchase selection? Is gender an important indicator of purchase selection?

c. Compare the results from (a) and (b). Which model would you choose? Discuss.

d. Using the model selected in (c), predict the probability that a student will select Pizza Hut if the price is \$8.99.

e. Using the model selected in (c), predict the probability that a student will select Pizza Hut if the price is \$11.49.

f. Using the model selected in (c), predict the probability that a student will select Pizza Hut if the price is \$13.99.

SUMMARY

In this chapter, you learned how a marketing manager can use multiple regression analysis to understand the effects of price and promotional expenses on the sales of a new product. You also learned how to include categorical independent variables and interaction terms in regression models. In addition, you used the logistic regression model to predict a categorical response variable. Figure 14.22 represents a road map of the chapter.

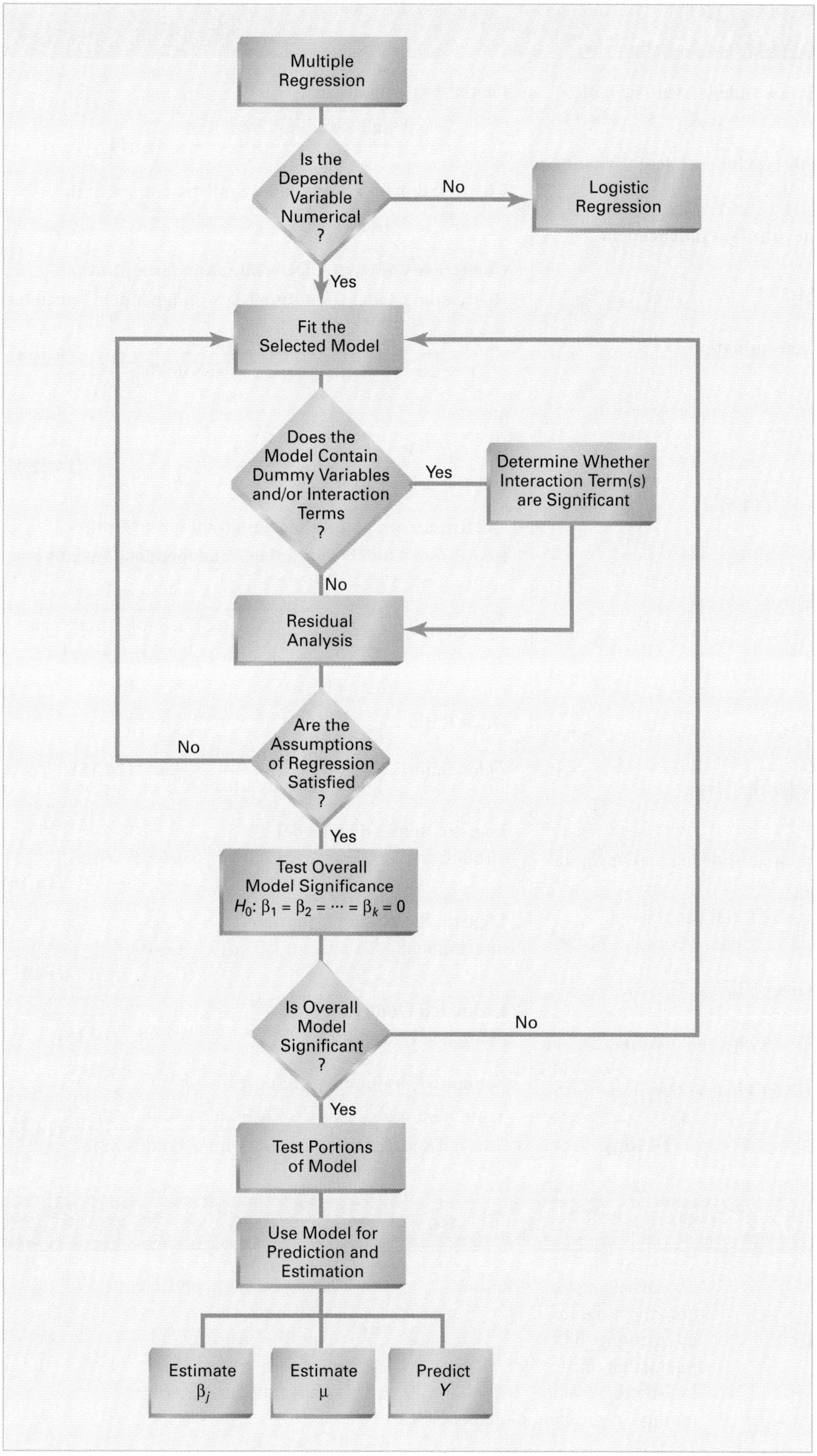

FIGURE 14.22
Road Map for Multiple Regression

KEY FORMULAS

Multiple Regression Model with k Independent Variables

$$Y_i = \beta_0 + \beta_1 X_{1i} + \beta_2 X_{2i} + \beta_3 X_{3i} + \cdots + \beta_k X_{ki} + \varepsilon_i \quad \textbf{(14.1)}$$

Multiple Regression Model with Two Independent Variables

$$Y_i = \beta_0 + \beta_1 X_{1i} + \beta_2 X_{2i} + \varepsilon_i \quad \textbf{(14.2)}$$

Multiple Regression Equation with Two Independent Variables

$$\hat{Y}_i = b_0 + b_1 X_{1i} + b_2 X_{2i} \quad \textbf{(14.3)}$$

The Coefficient of Multiple Determination

$$r^2 = \frac{SSR}{SST} \quad \textbf{(14.4)}$$

Adjusted r^2

$$r^2_{\text{adj}} = 1 - \left[(1 - r^2)\frac{n-1}{n-k-1}\right] \quad \textbf{(14.5)}$$

Overall F-Test Statistic

$$F = \frac{MSR}{MSE} \quad \textbf{(14.6)}$$

Testing for the Slope in Multiple Regression

$$t = \frac{b_j - \beta_j}{S_{b_j}} \quad \textbf{(14.7)}$$

Confidence Interval Estimate for the Slope

$$b_j \pm t_{n-k-1} S_{b_j} \quad \textbf{(14.8)}$$

Determining the Contribution of an Independent Variable to the Regression Model

$$SSR(X_j \mid \text{all variables } except\ j) = SSR\ (\text{all variables } including\ j) - SSR\ (\text{all variables } except\ j) \quad \textbf{(14.9)}$$

Contribution of Variable X_1 Given That X_2 Has Been Included:

$$SSR(X_1 \mid X_2) = SSR(X_1 \text{ and } X_2) - SSR(X_2) \quad \textbf{(14.10a)}$$

Contribution of Variable X_2 Given That X_1 Has Been Included:

$$SSR(X_2 \mid X_1) = SSR(X_1 \text{ and } X_2) - SSR(X_1) \quad \textbf{(14.10b)}$$

Partial F-Test Statistic

$$F = \frac{SSR(X_j \mid \text{all variables } except\ j)}{MSE} \quad \textbf{(14.11)}$$

The Relationship Between a t Statistic and an F Statistic

$$t_a^2 = F_{1,a} \quad \textbf{(14.12)}$$

Coefficients of Partial Determination for a Multiple Regression Model Containing Two Independent Variables

$$r^2_{Y1.2} = \frac{SSR(X_1 \mid X_2)}{SST - SSR(X_1\ and\ X_2) + SSR(X_1 \mid X_2)} \quad \textbf{(14.13a)}$$

and

$$r^2_{Y2.1} = \frac{SSR(X_2 \mid X_1)}{SST - SSR(X_1\ and\ X_2) + SSR(X_2 \mid X_1)} \quad \textbf{(14.13b)}$$

Coefficient of Partial Determination for a Multiple Regression Model Containing k Independent Variables

$$r^2_{Yj.(\text{all variables } except\ j)} = \frac{SSR(X_j \mid \text{all variables } except\ j)}{SST - SSR(\text{all variables } including\ j) + SSR(X_j \mid \text{all variables } except\ j)} \quad \textbf{(14.14)}$$

Odds Ratio

$$\text{Odds ratio} = \frac{\text{probability of success}}{1 - \text{probability of success}} \quad \textbf{(14.15)}$$

Logistic Regression Model

$$\ln(\text{odds ratio}) = \beta_0 + \beta_1 X_{1i} + \beta_2 X_{2i} + \cdots + \beta_k X_{ki} + \varepsilon_i \quad \textbf{(14.16)}$$

Logistic Regression Equation

$$\ln(\text{estimated odds ratio}) = b_0 + b_1 X_{1i} + b_2 X_{2i} + \cdots + b_k X_{ki} \quad \textbf{(14.17)}$$

Estimated Odds Ratio

$$\text{Estimated odds ratio} = e^{\ln(\text{estimated odds ratio})} \quad \textbf{(14.18)}$$

Estimated Probability of Success

$$\text{Estimated probability of success} = \frac{\text{estimated odds ratio}}{1 + \text{estimated odds ratio}} \quad \textbf{(14.19)}$$

KEY TERMS

CHAPTER REVIEW PROBLEMS

Checking Your Understanding

14.59 How does the interpretation of the regression coefficients differ in multiple regression and simple regression?

14.60 How does testing the significance of the entire regression model differ from testing the contribution of each independent variable in the multiple regression model?

14.61 How do the coefficients of partial determination differ from the coefficient of multiple determination?

14.62 What is the difference between least-squares regression and logistic regression?

14.63 Why and how do you use dummy variables?

14.64 How can you evaluate whether the slope of the response variable with an independent variable is the same for each level of the dummy variable?

14.65 Under what circumstances do you include an interaction in a regression model?

14.66 When a dummy variable is included in a regression model, what assumption do you need to make concerning the slope between the response variable Y and the numerical explanatory variable X?

Applying the Concepts

You need to use Microsoft Excel, Minitab, or SPSS to solve problems 14.67–14.75.

14.67 Professional basketball has truly become a sport that generates interest among fans around the world. More and more players come from outside the United States to play in the National Basketball Association (NBA). In 2004, nine of the 29 players selected in the first round of the NBA draft were players from outside the United States. You want to develop a regression model to predict the number of wins achieved by each NBA team based on field goal (shots made) percentage for the team and for the opponent. NBA2004

a. State the multiple regression equation.
b. Interpret the meaning of the slopes in this equation.
c. Predict the number of wins for a team that has a field goal percentage of 45% and an opponent field goal percentage of 44%.
d. Perform a residual analysis on your results, and determine the adequacy of the fit of the model.
e. Is there a significant relationship between number of wins and the two independent variables (field goal percentage for the team and for the opponent) at the 0.05 level of significance?
f. Determine the p-value in (e) and interpret its meaning.
g. Interpret the meaning of the coefficient of multiple determination in this problem.
h. Determine the adjusted r^2.
i. At the 0.05 level of significance, determine whether each independent variable makes a significant contribution to the regression model. Indicate the most appropriate regression model for this set of data.
j. Determine the p-values in (i) and interpret their meaning.
k. Compute and interpret the coefficients of partial determination.

14.68 A sample of 30 recently sold single-family houses in a small city is selected. Develop a model to predict the selling price (in thousands of dollars) using the assessed value (in thousands of dollars) as well as time period (in months since reassessment). The houses in the city had been reassessed at full value one year prior to the study. The results are contained in the file HOUSE1.

a. State the multiple regression equation.
b. Interpret the meaning of the slopes in this equation.
c. Predict the selling price for a house that has an assessed value of $70,000 and was sold in time period 12.
d. Perform a residual analysis on your results, and determine the adequacy of the model.
e. Determine whether there is a significant relationship between selling price and the two independent variables (assessed value and time period) at the 0.05 level of significance.
f. Determine the p-value in (e) and interpret its meaning.
g. Interpret the meaning of the coefficient of multiple determination in this problem.
h. Determine the adjusted r^2.
i. At the 0.05 level of significance, determine whether each independent variable makes a significant contribution to the regression model. Indicate the most appropriate regression model for this set of data.
j. Determine the p-values in (i) and interpret their meaning.
k. Construct a 95% confidence interval estimate of the population slope between selling price and assessed value. How does the interpretation of the slope here differ from that in problem 13.76 on page 560?
l. Compute and interpret the coefficients of partial determination.

14.69 Measuring the height of a California redwood tree is a very difficult undertaking because these trees grow to heights of over 300 feet. People familiar with these trees understand that the height of a California redwood tree is related to other characteristics of the tree, including the diameter of the tree at the breast height of a person and the thickness of the bark of the tree. The data in the file REDWOOD represent the height, diameter at breast height of a person, and bark thickness for a sample of 21 California redwood trees.

a. State the multiple regression equation.

b. Interpret the meaning of the slopes in this equation.
c. Predict the height for a tree that has a breast diameter of 25 inches and a bark thickness of 2 inches.
d. Interpret the meaning of the coefficient of multiple determination in this problem.
e. Perform a residual analysis on the results and determine the adequacy of the model.
f. Determine whether there is a significant relationship between height of redwood trees and the two independent variables (breast diameter and the bark thickness) at the 0.05 level of significance?
g. Construct a 95% confidence interval estimate of the population slope between the height of the redwood trees and breast diameter, and between the height of redwood trees and the bark thickness.
h. At the 0.05 level of significance, determine whether each independent variable makes a significant contribution to the regression model. Indicate the independent variables to include in this model.
i. Construct a 95% confidence interval estimate of the mean height for trees that have a breast diameter of 25 inches and a bark thickness of 2 inches along with a prediction interval for an individual tree.
j. Compute and interpret the coefficients of partial determination.

14.70 Develop a model to predict the assessed value (in thousands of dollars) using the size of the houses (in thousands of square feet) and the age of the houses (in years) from the following table: HOUSE2

House	Assessed Value ($000)	Size of Dwelling (Thousands of Square Feet)	Age (Years)
1	84.4	2.00	3.42
2	77.4	1.71	11.50
3	75.7	1.45	8.33
4	85.9	1.76	0.00
5	79.1	1.93	7.42
6	70.4	1.20	32.00
7	75.8	1.55	16.00
8	85.9	1.93	2.00
9	78.5	1.59	1.75
10	79.2	1.50	2.75
11	86.7	1.90	0.00
12	79.3	1.39	0.00
13	74.5	1.54	12.58
14	83.8	1.89	2.75
15	76.8	1.59	7.17

a. State the multiple regression equation.
b. Interpret the meaning of the slopes in this equation.
c. Predict the assessed value for a house that has a size of 1,750 square feet and is 10 years old.
d. Perform a residual analysis on the results and determine the adequacy of the model.
e. Determine whether there is a significant relationship between assessed value and the two independent variables (size and age) at the 0.05 level of significance.
f. Determine the p-value in (e) and interpret its meaning.
g. Interpret the meaning of the coefficient of multiple determination in this problem.
h. Determine the adjusted r^2.
i. At the 0.05 level of significance, determine whether each independent variable makes a significant contribution to the regression model. Indicate the most appropriate regression model for this set of data.
j. Determine the p-values in (i) and interpret their meaning.
k. Construct a 95% confidence interval estimate of the population slope between assessed value and size. How does the interpretation of the slope here differ from that of problem 13.77 on page 561?
l. Compute and interpret the coefficients of partial determination.
m. The real estate assessor's office has been publicly quoted as saying that the age of a house has no bearing on its assessed value. Based on your answers to (a) through (l), do you agree with this statement? Explain.

14.71 The file COLLEGES2002 contains data on 80 colleges and universities. Among the variables included are the annual total cost (in thousands of dollars), the first-quartile (Q_1) and third-quartile (Q_3) score on the Scholastic Aptitude Test (SAT), and the room-and-board expenses (in thousands of dollars). Develop a model to predict the annual total cost based on first quartile SAT score and room-and-board expenses.
a. State the multiple regression equation.
b. Interpret the meaning of the slopes in this equation.
c. Predict the annual total cost for a school that has a first-quartile SAT score of 1,100 and a room-and-board expense of $5,000.
d. Perform a residual analysis on the results and determine the adequacy of the model.
e. Is there is a significant relationship between annual total cost and the two independent variables (first-quartile SAT score and room-and-board expenses) at the 0.05 level of significance?
f. Determine the p-value in (e) and interpret its meaning.
g. Interpret the meaning of the coefficient of multiple determination in this problem.
h. Determine the adjusted r^2.
i. At the 0.05 level of significance, determine whether each independent variable makes a significant contribution to the regression model. Indicate the most appropriate regression model for this set of data.
j. Determine the p-values in (i) and interpret their meaning.

k. Construct a 95% confidence interval estimate of the population slope between annual total cost and first-quartile SAT score.
l. Compute and interpret the coefficients of partial determination.
m. Explain why the slope for total annual cost with room-and-board expenses seems substantially different from 1.0.
n. What other factors that are not included in the model might account for the strong positive relationship between total cost and first quartile SAT score?

14.72 The file **AUTO2002** contains data on 121 automobile models from the year 2002. Among the variables included are the gasoline mileage (in miles per gallon), the length (in inches), and the weight (in pounds) of each automobile. Develop a model to predict the gasoline mileage based on the length and weight of each automobile.
a. State the multiple regression equation.
b. Interpret the meaning of the slopes in this equation.
c. Predict the gasoline mileage for an automobile that has a length of 195 inches and a weight of 3,000 pounds.
d. Perform a residual analysis on the results and determine the adequacy of the model.
e. Is there is a significant relationship between gasoline mileage and the two independent variables (length and weight) at the 0.05 level of significance?
f. Determine the p-value in (e) and interpret its meaning.
g. Interpret the meaning of the coefficient of multiple determination in this problem.
h. Determine the adjusted r^2.
i. At the 0.05 level of significance, determine whether each independent variable makes a significant contribution to the regression model. Indicate the most appropriate regression model for this set of data.
j. Determine the p-values in (i) and interpret their meaning.
k. Construct a 95% confidence interval estimate of the population slope between gasoline mileage and weight.
l. Compute and interpret the coefficients of partial determination.

14.73 Crazy Dave, the well-known baseball analyst, wants to determine which variables are important in predicting a team's wins in a given season. He has collected data related to wins, earned-run average (ERA), and runs scored for the 2003 season (stored in the file **BB2003**). Develop a model to predict the number of wins based on ERA and runs scored.
a. State the multiple regression equation.
b. Interpret the meaning of the slopes in this equation.
c. Predict the number of wins for a team that has an ERA of 4.50 and has scored 750 runs.
d. Perform a residual analysis on the results and determine the adequacy of the model.
e. Is there a significant relationship between number of wins and the two independent variables (ERA and runs scored) at the 0.05 level of significance?
f. Determine the p-value in (e) and interpret its meaning.
g. Interpret the meaning of the coefficient of multiple determination in this problem.
h. Determine the adjusted r^2.
i. At the 0.05 level of significance, determine whether each independent variable makes a significant contribution to the regression model. Indicate the most appropriate regression model for this set of data.
j. Determine the p-values in (i) and interpret their meaning.
k. Construct a 95% confidence interval estimate of the population slope between wins and ERA.
l. Compute and interpret the coefficients of partial determination.
m. Which is more important in predicting wins, pitching as measured by ERA or offense as measured by runs scored? Explain.

14.74 Referring to problem 14.73, suppose that in addition to using earned run average (ERA) to predict the number of wins, Crazy Dave wants to include the league (American versus National) as an independent variable. Develop a model to predict wins based on ERA and league: **BB2003**
a. State the multiple regression equation.
b. Interpret the meaning of the slopes in this problem.
c. Predict the number of wins for a team with an ERA of 4.50 in the American League. Construct a 95% confidence interval estimate for all teams and a 95% prediction interval for an individual team.
d. Perform a residual analysis on the results, and determine the adequacy of the model.
e. Is there a significant relationship between wins and the two independent variables (ERA and league) at the 0.05 level of significance?
f. At the 0.05 level of significance, determine whether each independent variable makes a contribution to the regression model. Indicate the most appropriate regression model for this set of data.
g. Construct 95% confidence interval estimates of the population slope for the relationship between wins and ERA, and between wins and league.
h. Interpret the meaning of the coefficient of multiple determination.
i. Determine the adjusted r^2.
j. Compute and interpret the coefficients of partial determination.
k. What assumption do you have to make about the slope of wins with ERA?
l. Add an interaction term to the model, and, at the 0.05 level of significance, determine whether it makes a significant contribution to the model.
m. On the basis of the results of (f) and (l), which model is most appropriate? Explain.

14.75 You are a real estate broker who wants to compare property values in Glen Cove and Roslyn (which are located approximately 8 miles apart). In order to do so, you will analyze a data set that includes samples for Glen Cove and Roslyn. GCROSLYN Making sure to include the dummy variable for location (Glen Cove or Roslyn) in the regression model, develop a regression model to predict appraised value based on the land area of a property, the age of a house, and location. Be sure to determine whether any interaction terms need to be included in the model.

RUNNING CASE: MANAGING THE *SPRINGVILLE HERALD*

In its continuing study of the home-delivery subscription solicitation process, a marketing department team wants to test the effects of two types of structured sales presentations (personal formal and personal informal) and the number of hours spent on telemarketing on the number of new subscriptions. The staff has recorded these data SH14 for the past 24 weeks. You can find this data set at **http://www.prenhall.com/HeraldCase/EffectsData.htm** and in the **EffectsData.htm** file in the HeraldCase folder on the CD that accompanies this text.

Analyze these data and develop a statistical model to predict the number of new subscriptions for a week based on the number of hours spent on telemarketing and the sales presentation type. Write a report giving detailed findings concerning the regression model used.

WEB CASE

Apply your knowledge of multiple regression models in this Web Case that extends the "Using Statistics" *OmniPower scenario from this chapter.*

To ensure a successful test-marketing of its OmniPower energy bars, the marketing department of OmniFoods has arranged for the In-Store Placements Group (ISPG), a merchandising consultancy, to work with the grocery store chain conducting the test-market study. Using the same 34-store sample used in the test-market study, ISPG claims that the choice of shelf location and the presence of in-store OmniPower coupon dispensers both increase sale of the energy bars.

Review the ISPG claims and supporting data at OmniFoods internal Web site **www.prenhall.com/Springville/Omni_ISPGMemo.htm** and then answer the following:

1. Are the supporting data consistent to ISPG's claims? Perform an appropriate statistical analysis to confirm (or discredit) the stated relationship between sales and the two independent variables of product shelf location and the presence of in-store OmniPower coupon dispensers.
2. If you were advising OmniFoods, would you recommend using a specific shelf location and in-store coupon dispensers to sell OmniPower bars?
3. What additional data would you advise collecting in order to determine the effectiveness of the sales promotion techniques used by ISPG?

REFERENCES

1. Hocking, R. R., "Developments in Linear Regression Methodology: 1959–1982," *Technometrics* 25 (1983): 219–250.
2. Hosmer, D. W., and S. Lemeshow, *Applied Logistic Regression*, 2nd ed. (New York: Wiley, 2001).
3. Kutner, M., C. Nachtsheim, J. Neter, and W. Li, *Applied Linear Statistical Models*, 5th ed. (New York: McGraw-Hill/Irwin, 2005).
4. *Microsoft Excel 2003* (Redmond, WA: Microsoft Corp., 2002).
5. *Minitab for Windows Version 14* (State College, PA: Minitab, Inc., 2004).
6. *SPSS Base 12.0 Brief Guide* (Upper Saddle River, NJ: Prentice Hall, 2003).

Appendix 14 Using Software for Multiple Regression

A14.1 MICROSOFT EXCEL

For Multiple Regression

Open to a worksheet that contains the regression analysis data arranged in columns such as the Data worksheet of the **OMNI.XLS** workbook.
Select **Tools → Data Analysis**.
Select **Regression** from the Data Analysis list and click **OK**.
In the procedure's dialog box (see Figure A14.1 below):

Enter the cell range of the *Y* variable data as the **Input Y Range**.
Enter the cell range of the *X* variable data as the **Input X Range**.
Select **Labels** if the variable cell ranges include labels in their first rows.
Select **Confidence Level**.
Select **Residuals** and **Residual Plots** if you want to perform a residual analysis.
Click **OK**.

Results appear on a separate worksheet.

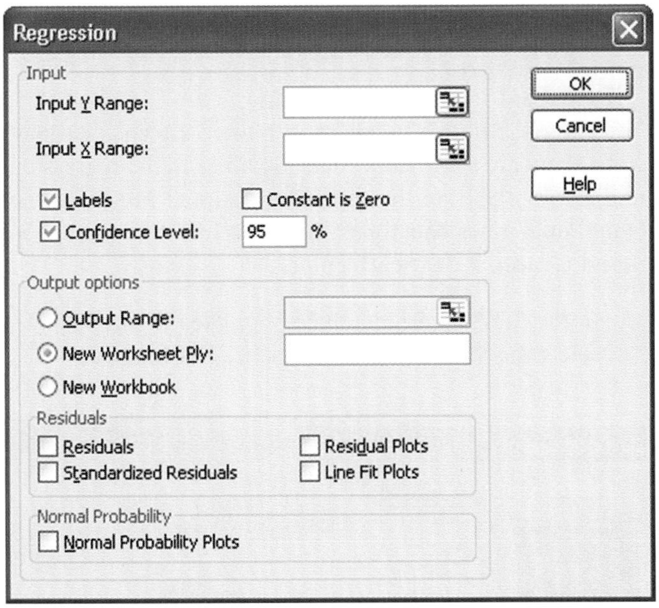

FIGURE A14.1 Microsoft Excel Regression Dialog Box

OR See section G.36 (**Multiple Regression**) if you want PHStat2 to produce a worksheet for you.

For Confidence Interval Estimate and Prediction Interval

See section G.36 (**Multiple Regression**) if you want PHStat2 to produce a confidence interval estimate and prediction interval worksheet or open the **MR CIEPI.xls** file to inspect such a worksheet for the OmniPower multiple regression example of section 14.1. (This worksheet uses matrix and cross-worksheet formulas that are beyond the scope of this book to discuss or modify.)

For Coefficients of Partial Determination

Open the **CPD.xls** file. This worksheet uses simple formulas to calculate the coefficients of a multiple regression that contains two explanatory variables. Cells B5 through B10 already contain the entries for the OmniPower example of section 14.5 (see Figure A14.2). To adapt this file to other problems, first produce the necessary regression worksheets and then follow the instructions in column D (not shown in Figure A14.2) to transfer data from those worksheets.

	A	B	
1	**OmniPower Sales Analysis**		
2	**Coefficients of Partial Determination**		
3			
4	Intermediate Calculations		
5	SSR(X1, X2)	39472730.7730	
6	SST	52093677.4412	
7	SSR(X2)	14915814.1025	
8	SSR(X1)	28153486.1482	
9	SSR(X1 | X2)	24556916.6706	**=B5 - B7**
10	SSR(X2 | X1)	11319244.6249	**=B5 - B8**
11			
12	**Coefficients**		
13	**r2 Y1.2**	**0.6605**	**=B9/(B6 - B5 + B9)**
14	**r2 Y2.1**	**0.4728**	**=B10/(B6 - B5 + B10)**

FIGURE A14.2 Microsoft Excel Coefficient of Partial Determination Worksheet

OR See section G.36 (**Multiple Regression**) if you want PHStat2 to produce a worksheet for you.

For Creating Dummy Variables

To create a dummy variable, do the following:

Open to the worksheet that contains your regression data.
Copy the entire contents of a column that contains a categorical variable to the first empty column.
Select the newly created column (very important!).
Select **Edit → Replace** to display the Replace dialog box (see Figure A14.3).

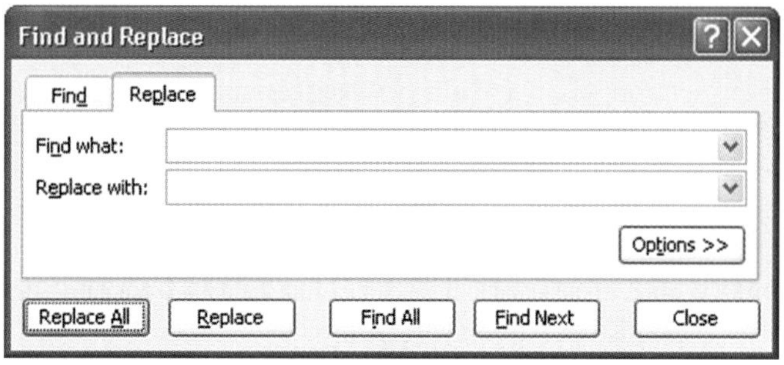

FIGURE A14.3 Microsoft Excel Find and Replace Dialog Box

For each unique categorical value in the column:

- Enter the categorical value as the **Find what** value.
- Enter 0 or 1 to represent the categorical value as the **Replace with** value.
- Click **Replace All**.
- If a message box to confirm the replacement appears, click **OK** to continue.

Click **Close** when you are finished making all replacements.

Double-check your work for consistency by comparing the original values with the coded values in the new column. After you double-check your work, delete the original categorical variable column. Use your new column as part of the **Input X Range** (for the Data Analysis Regression procedure) or **X Variable Cell Range** (for the PHStat2 Multiple Regression command).

For Creating Interaction Terms

To create interaction terms, enter formulas in the form *=cell1 * cell2* in a new column. The table below shows column D formulas for the size and fireplace interaction for the **Data** worksheet of the **House3.xls** file. The original categorical Fireplace variable was converted into a dummy variable with the values 1 for Yes and 0 for No.

	A	B	C	D
1	Value	Size	Fireplace	Size * Fireplace
2	84.4	2.00	1	=B2 * C2
3	77.4	1.71	0	=B3 * C3
15	83.8	1.89	1	=B15 * C15
16	76.8	1.59	0	=B16 * C16

A14.2 MINITAB

Generating a Multiple Regression Equation

In Appendix A13.2, instructions are provided for using Minitab for simple linear regression. The same set of instructions is valid in using Minitab for multiple regression. To carry out a regression analysis for the OmniPower sales data, open the **OMNI.MTW** worksheet and select **Stat → Regression → Regression**.

1. Enter **C1** or **Sales** in the Response: edit box and **C2** or **Price** and **C3** or **Promotion** in the Predictors: edit box. Click the **Graphs** button.
2. In the Residuals for Plots: edit box, select the **Regular** option button. To check for normality, select the **Histogram of residuals** check box. In the Residuals versus the variables: edit box, select **C2** or **Price** and **C3** or **Promotion**. Click the **OK** button to return to the Regression dialog box.
3. Click the **Results** option button. In the Regression-Results dialog box, click the **In addition, the full table of fits and residuals** option button. Click the **OK** button to return to the Regression dialog box.
4. Click the **Options** button. (If the data have been collected over time, under Display, select the Durbin-Watson statistic check box.) In the Prediction intervals for new observations: edit box, enter **79 400**. Enter **95** in the Confidence level: edit box. Click the **OK** button to return to the Regression dialog box. Click the **OK** button.

Using Minitab for a Three-Dimensional Plot

You can use Minitab to construct a three-dimensional plot when there are two independent variables in the regression model. To illustrate the three-dimensional plot with the OmniPower sales data, open the **OMNI.MTW** worksheet. Select **Graph → 3D Scatterplot**.

1. In the 3D Scatterplots dialog box (see Figure A14.4), select the **Simple** button. Click the **OK** button.

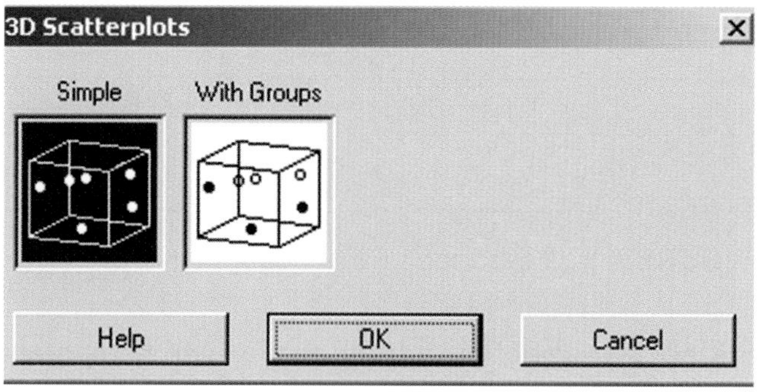

FIGURE A14.4 Minitab 3D Scatterplots Dialog Box

2. In the 3D Scatterplot - Simple dialog box (see Figure A14.5), enter **C1** or **Sales** in the Z variable: edit box, **C2** or **Price** in the Y variable: edit box, and **C3** or **Promotion** in the X variable: edit box. Click the **OK** button.

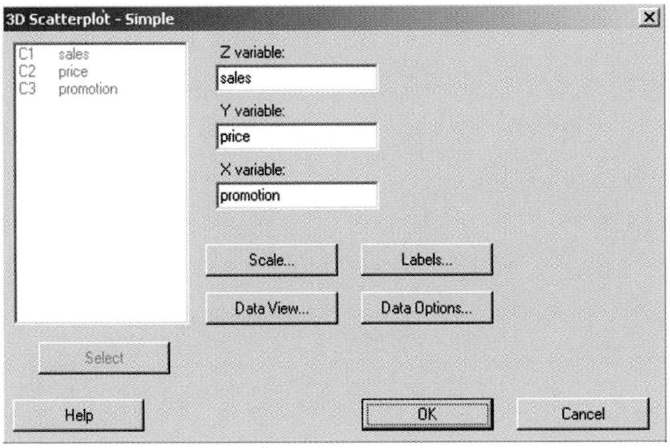

FIGURE A14.5 Minitab 3D Scatterplot - Simple Dialog Box

Using Minitab for Logistic Regression

To illustrate the use of Minitab for logistic regression with the credit card upgrade example on page 605, open the **LOGPURCH.MTW** worksheet. To perform a logistic regression, select **Stat ➔ Regression ➔ Binary Logistic Regression**.

In the Binary Logistic Regression dialog box (see Figure A14.6), in the Response: edit box, enter **C2** or **Purchase**. In the Model: edit box, enter **C1** or **Spending** and **C3** or **Extra**. In the Factors: edit box, enter **C3** or **Extra** because it is a categorical variable. Click the **OK** button.

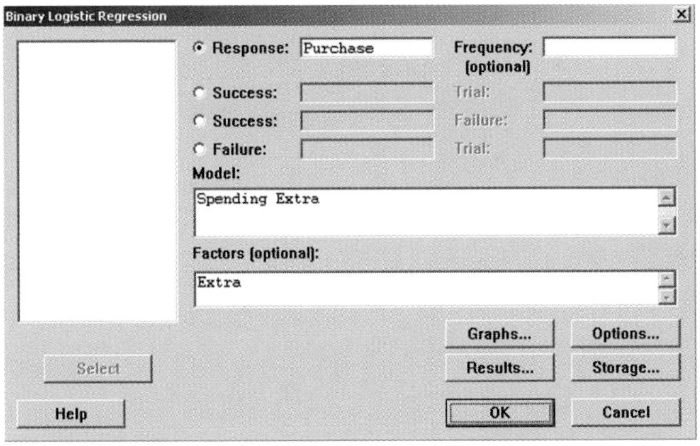

FIGURE A14.6 Minitab Binary Logistic Regression Dialog Box

Using Minitab for Dummy Variables and Interactions

In order to perform regression analysis with dummy variables, the categories of the dummy variable are coded as 0 and 1. If the dummy variable has not already been recoded as a 0–1 variable, Minitab can recode the variable. As an illustration with the data of Table 14.5 on page 592, open the **HOUSE3.MTW** worksheet. In this worksheet, the fireplace variable in column C3 has been entered as Yes and No. To recode this variable using Minitab, select **Calc ➔ Make Indicator Variables**. In the Indicator variables for: edit box, enter **Firepl** or **C3**. In the Store results in: edit box, enter **C4 C5**, because you need to specify a column for each possible definition of the dummy variable even though only one column (C5) will be used in the regression analysis. Click the **OK** button. Observe that No is coded as 1 in C4, and Yes is coded as 1 in C5. Enter the label **Fireplace** for C5.

To define an interaction term that is the product of heating area and the dummy variable fireplace, select **Calc ➔ Calculator**. In the Store result in variable: edit box enter **C6**. In the Expression: edit box, enter **Size** * **Fireplace** or **C2** * **C5**. Click the **OK** button. C6 now contains a new *X* variable that is the product of C2 and C5. Enter a label for C6.

CHAPTER 15

Multiple Regression Model Building

LEARNING OBJECTIVES

In this chapter, you learn:

- To use quadratic terms in a regression model
- To use transformed variables in a regression model
- To examine the effect of each observation on the regression model
- To measure the correlation among the independent variables
- To build a regression model using either the stepwise or best-subsets approach
- To avoid the pitfalls involved in developing a multiple regression model

USING STATISTICS

Predicting Standby Hours for Unionized Graphic Artists

As part of your job as the director of operations for a television station, you look for ways to reduce labor expenses. Currently, the unionized graphic artists at the station receive hourly pay for a significant number of hours in which they are idle. These hours are called standby hours. You have collected data concerning standby hours and four factors that you suspect are related to the excessive number of standby hours the station is currently experiencing: the total number of staff present, remote hours, Dubner hours, and total labor hours.

You plan to build a multiple regression model to help determine which factors most heavily affect standby hours. You believe that an appropriate model will help you to predict the number of future standby hours, identify the root causes for excessive amounts of standby hours, and allow you to reduce the total number of future standby hours. How do you build the model with the most appropriate mix of independent variables? Are there statistical techniques to help you identify a "best" model without having to consider all possible models? How do you begin?

In Chapter 14 you studied multiple regression models that contained two independent variables. In this chapter, regression analysis is extended to models containing more than two independent variables. In order to help you learn to develop the best model when confronted with a large set of data (such as the one described in the "Using Statistics" scenario), this chapter introduces you to various topics related to model building. These topics include quadratic independent variables, transformations of either the dependent or independent variables, identification of influential observations, and stepwise regression.

15.1 THE QUADRATIC REGRESSION MODEL

The simple regression model discussed in Chapter 13 and the multiple regression model discussed in Chapter 14 assumed that the relationship between Y and each independent variable is linear. However, in section 13.1, several different types of nonlinear relationships between variables were introduced. One of the most common nonlinear relationships is a quadratic relationship between two variables in which Y increases (or decreases) at a changing rate for various values of X (see Figure 13.2, panels C–E, on page 515). You can use the quadratic regression model defined in Equation (15.1) to analyze this type of a relationship between X and Y.

THE QUADRATIC REGRESSION MODEL

$$Y_i = \beta_0 + \beta_1 X_{1i} + \beta_2 X_{1i}^2 + \varepsilon_i \qquad \textbf{(15.1)}$$

where

β_0 = Y intercept

β_1 = coefficient of the linear effect on Y

β_2 = coefficient of the quadratic effect on Y

ε_i = random error in Y for observation i

This **quadratic regression model** is similar to the multiple regression model with two independent variables [see Equation (14.2) on page 572] except that the second independent variable is the square of the first independent variable. Once again, you use the sample regression coefficients (b_0, b_1, and b_2) as estimates of the population parameters (β_0, β_1, and β_2). Equation (15.2) defines the regression equation for the quadratic model with one independent variable (X_1) and a dependent variable (Y).

QUADRATIC REGRESSION EQUATION

$$\hat{Y}_i = b_0 + b_1X_{1i} + b_2X_{1i}^2 \tag{15.2}$$

In Equation (15.2), the first regression coefficient, b_0, represents the Y intercept; the second regression coefficient, b_1, represents the linear effect; and the third regression coefficient, b_2, represents the quadratic effect.

Finding the Regression Coefficients and Predicting *Y*

To illustrate the quadratic regression model, consider the following experiment conducted to study the effect of various amounts of an ingredient called fly ash on the strength of concrete. A sample of 18 specimens of 28-day-old concrete FLYASH was collected in which the percentage of fly ash in the concrete varies from 0 to 60%. Table 15.1 summarizes the data.

TABLE 15.1
Fly Ash % and Strength of 18 Samples of 28-day-old Concrete

Fly Ash %	Strength (psi)
0	4,779
0	4,706
0	4,350
20	5,189
20	5,140
20	4,976
30	5,110
30	5,685
30	5,618
40	5,995
40	5,628
40	5,897
50	5,746
50	5,719
50	5,782
60	4,895
60	5,030
60	4,648

To help you select the proper model for expressing the relationship between fly ash percentage and strength, you plot a scatter diagram as shown in Figure 15.1. Figure 15.1 indicates an initial increase in the strength of the concrete as the percentage of fly ash increases. The strength appears to level off and then drop after achieving maximum strength at about 40% fly ash. Strength for 50% fly ash is slightly below strength at 40%, but strength at 60% is substantially below strength at 50%. Therefore, a quadratic model is a more appropriate choice than a linear model to estimate strength based on fly ash percentage.

FIGURE 15.1

Microsoft Excel Scatter Diagram of Fly Ash % (X) and Strength (Y)

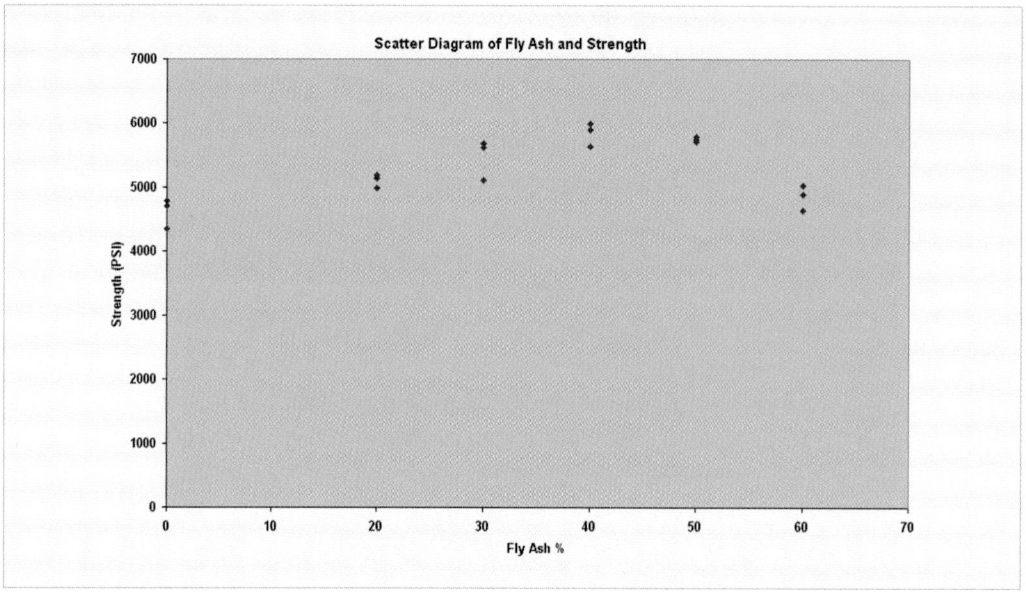

You need to use a spreadsheet program or a statistical package to perform the least-squares method needed to compute the three sample regression coefficients (b_0, b_1, and b_2) for this example. Figure 15.2 presents a Microsoft Excel worksheet and Figure 15.3 illustrates Minitab output. From Figure 15.2 or 15.3,

$$b_0 = 4{,}486.361 \quad b_1 = 63.005 \quad b_2 = -0.876$$

FIGURE 15.2

Microsoft Excel Output for the Concrete Strength Data

	A	B	C	D	E	F	G
1	Concrete Strength Analysis						
2							
3	*Regression Statistics*						
4	Multiple R	0.80527					
5	R Square	0.64847					
6	Adjusted R Square	0.60160					
7	Standard Error	312.11291					
8	Observations	18					
9							
10	ANOVA						
11		*df*	*SS*	*MS*	*F*	*Significance F*	
12	Regression	2	2695473.49	1347736.745	13.83508	0.00039	
13	Residual	15	1461217.01	97414.46735			
14	Total	17	4156690.5				
15							
16		*Coefficients*	*Standard Error*	*t Stat*	*P-value*	*Lower 95%*	*Upper 95%*
17	Intercept	4486.36111	174.75315	25.67256	8.24736E-14	4113.88337	4858.83886
18	Fly Ash%	63.00524	12.37255	5.09234	0.00013	36.63377	89.37671
19	Fly Ash% ^2	-0.87647	0.19661	-4.45784	0.00046	-1.29554	-0.45740

FIGURE 15.3

Minitab Output for the Concrete Strength Data

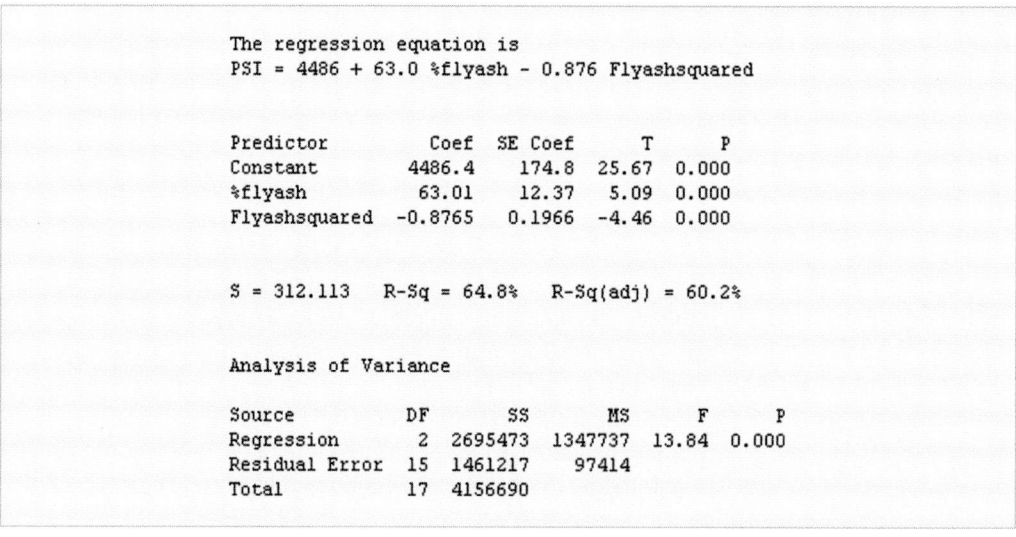

```
The regression equation is
PSI = 4486 + 63.0 %flyash - 0.876 Flyashsquared


Predictor          Coef  SE Coef       T      P
Constant         4486.4    174.8   25.67  0.000
%flyash           63.01    12.37    5.09  0.000
Flyashsquared   -0.8765   0.1966   -4.46  0.000


S = 312.113   R-Sq = 64.8%   R-Sq(adj) = 60.2%


Analysis of Variance

Source          DF       SS       MS      F      P
Regression       2  2695473  1347737  13.84  0.000
Residual Error  15  1461217    97414
Total           17  4156690
```

Therefore, the quadratic regression equation is

$$\hat{Y}_i = 4{,}486.361 + 63.005X_{1i} - 0.876X_{1i}^2$$

where $\hat{Y}_i$ = predicted strength for sample i

X_{1i} = percentage of fly ash for sample i

Figure 15.4 plots this quadratic regression equation on the scatter diagram to show the fit of the quadratic regression model to the original data.

FIGURE 15.4

Microsoft Excel Scatter Diagram Expressing the Quadratic Relationship Between Fly Ash Percentage and Strength for the Concrete Data

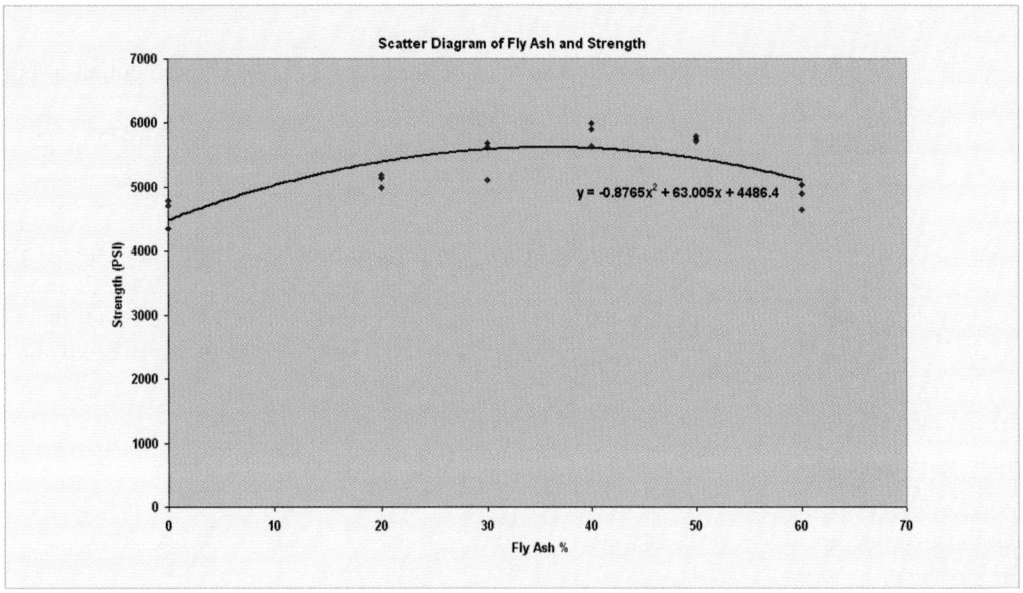

From the quadratic regression equation and Figure 15.4, the Y intercept ($b_0 = 4{,}486.361$) is the predicted strength when the percentage of fly ash is 0. To interpret the coefficients b_1 and b_2, observe that after an initial increase, strength decreases as fly ash percentage increases. This nonlinear relationship is further demonstrated by predicting the strength for fly ash percentages of 20, 40, and 60. Using the quadratic regression equation,

$$\hat{Y}_i = 4{,}486.361 + 63.005X_{1i} - 0.876X_{1i}^2$$

for $X_{1i} = 20$,

$$\hat{Y}_i = 4{,}486.361 + 63.005(20) - 0.876(20)^2 = 5{,}395.9$$

for $X_{1i} = 40$,

$$\hat{Y}_i = 4{,}486.361 + 63.005(40) - 0.876(40)^2 = 5{,}604.2$$

and for $X_{1i} = 60$,

$$\hat{Y}_i = 4{,}486.361 + 63.005(60) - 0.876(60)^2 = 5{,}111.4$$

Thus, the predicted concrete strength for 40% fly ash is 208.3 psi above the predicted strength for 20% fly ash, but the predicted strength for 60% fly ash is 492.8 psi below the predicted strength for 40% fly ash.

Testing for the Significance of the Quadratic Model

After you calculate the quadratic regression equation, you can test whether there is a significant overall relationship between strength, Y, and fly ash percentage, X. The null and alternative hypotheses are as follows:

H_0: $\beta_1 = \beta_2 = 0$ (No overall relationship between X_1 and Y)
H_1: β_1 and/or $\beta_2 \neq 0$ (Overall relationship between X_1 and Y)

Equation (14.6) on page 578 defines the overall F statistic used for this test:

$$F = \frac{MSR}{MSE}$$

From the Excel output in Figure 15.2 or the Minitab output in Figure 15.3 on page 622,

$$F = \frac{MSR}{MSE} = \frac{1{,}347{,}736.75}{97{,}414.47} = 13.84$$

If you choose a level of significance of 0.05, then from Table E.5, the critical value of the F distribution with 2 and 15 degrees of freedom, is 3.68 (see Figure 15.5). Since $F = 13.84 > 3.68$, or since the p-value $= 0.00039 < 0.05$, you reject the null hypothesis (H_0) and conclude that there is a significant quadratic relationship between strength and fly ash percentage.

FIGURE 15.5
Testing for the Existence of the Overall Relationship at the 0.05 Level of Significance with 2 and 15 Degrees of Freedom

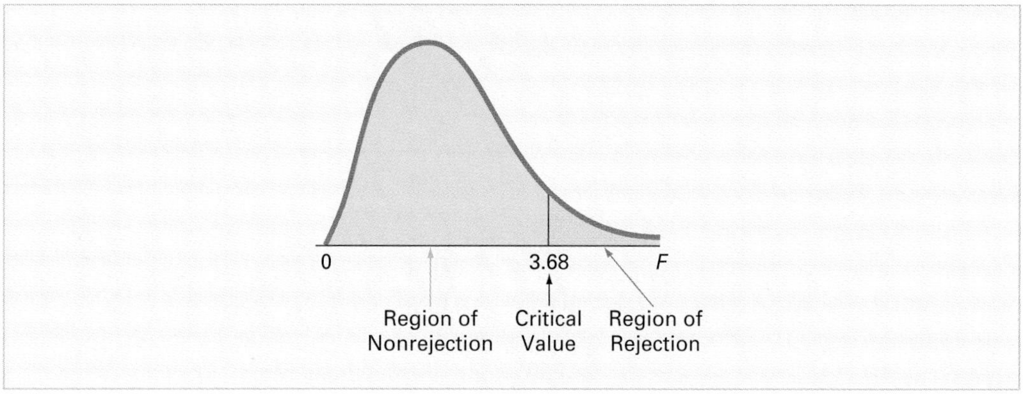

Testing the Quadratic Effect

In using a regression model to examine a relationship between two variables, you want to find not only the most accurate model, but also the simplest model expressing that relationship. Therefore, you need to examine whether there is a significant difference between the quadratic model

$$Y_i = \beta_0 + \beta_1 X_{1i} + \beta_2 X_{1i}^2 + \varepsilon_i$$

and the linear model

$$Y_i = \beta_0 + \beta_1 X_{1i} + \varepsilon_i$$

In section 14.4, you used the t test to determine whether each particular variable makes a significant contribution to the regression model. To test the significance of the contribution of the quadratic effect, you use the following null and alternative hypotheses:

H_0: Including the quadratic effect does not significantly improve the model ($\beta_2 = 0$).
H_1: Including the quadratic effect significantly improves the model ($\beta_2 \neq 0$).

The standard error of each regression coefficient and its corresponding t statistic are part of the Excel or Minitab output (see Figure 15.2 or 15.3 on page 622). Equation (14.7) defines the t-test statistic on page 582,

$$t = \frac{b_2 - \beta_2}{S_{b_2}}$$

$$= \frac{-0.8765 - 0}{0.1966} = -4.46$$

If you select a level of significance of 0.05, then from Table E.3, the critical values for the t distribution with 15 degrees of freedom are -2.1315 and $+2.1315$ (see Figure 15.6).

FIGURE 15.6
Testing for the Contribution of the Quadratic Effect to a Regression Model at the 0.05 Level of Significance with 15 Degrees of Freedom

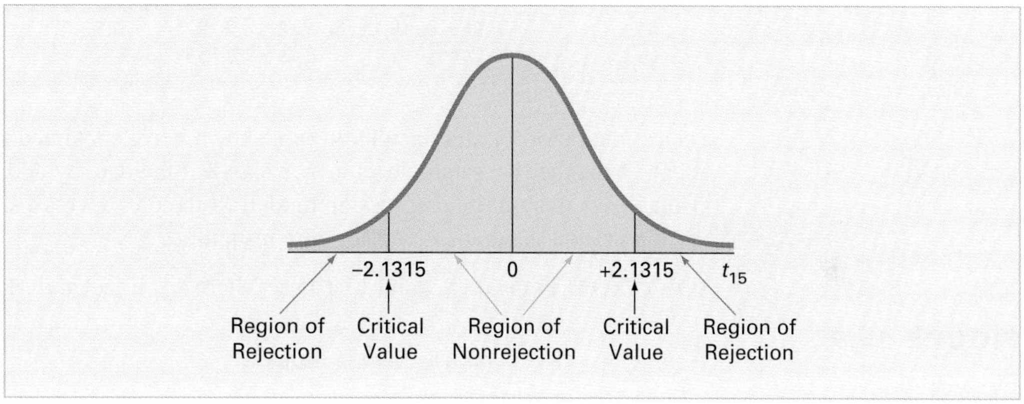

Since $t = -4.46 < -2.1315$, or since the p-value $= 0.00046 < 0.05$, you reject H_0 and conclude that the quadratic model is significantly better than the linear model for representing the relationship between strength and fly ash percentage.

Example 15.1 provides an additional illustration of a possible quadratic effect.

EXAMPLE 15.1

STUDYING THE QUADRATIC EFFECT IN A MULTIPLE REGRESSION MODEL

A real estate developer studying the effect of atmospheric temperature and the amount of attic insulation on the amount of heating oil consumed selects a random sample of 15 single-family houses. The data, including the consumption of heating oil, are stored in the file **HTNGOIL**. Figure 15.7 contains the partial results of a multiple regression analysis using the two independent variables computed by Microsoft Excel. Figure 15.8 presents Minitab output.

FIGURE 15.7
Microsoft Excel Output for the Monthly Consumption of Heating Oil

	A	B	C	D	E	F	G
1	**Heating Oil Consumption Analysis**						
2							
3	*Regression Statistics*						
4	Multiple R	0.98265					
5	R Square	0.96561					
6	Adjusted R Square	0.95988					
7	Standard Error	26.01378					
8	Observations	15					
9							
10	ANOVA						
11		*df*	*SS*	*MS*	*F*	*Significance F*	
12	Regression	2	228014.62632	114007.31316	168.47120	1.65411E-09	
13	Residual	12	8120.60302	676.71692			
14	Total	14	236135.22933				
15							
16		*Coefficients*	*Standard Error*	*t Stat*	*P-value*	*Lower 95%*	*Upper 95%*
17	Intercept	562.15101	21.09310	26.65094	4.77868E-12	516.19308	608.10893
18	Temperature	-5.43658	0.33622	-16.16990	1.64178E-09	-6.16913	-4.70403
19	Insulation	-20.01232	2.34251	-8.54313	1.90731E-06	-25.11620	-14.90844

FIGURE 15.8

Minitab Output for the Monthly Consumption of Heating Oil

```
The regression equation is
Gallons = 562 - 5.44 Temp(F) - 20.0 Insulation

Predictor       Coef  SE Coef       T      P
Constant      562.15    21.09   26.65  0.000
Temp(F)      -5.4366   0.3362  -16.17  0.000
Insulation   -20.012    2.343   -8.54  0.000

S = 26.0138   R-Sq = 96.6%   R-Sq(adj) = 96.0%

Analysis of Variance

Source          DF      SS      MS       F      P
Regression       2  228015  114007  168.47  0.000
Residual Error  12    8121     677
Total           14  236135
```

The residual plot for attic insulation (not shown here) contained some evidence of a quadratic effect. Thus, the real estate developer reanalyzed the data by adding a quadratic term for attic insulation to the multiple regression model. At the 0.05 level of significance, is there evidence of a significant quadratic effect for attic insulation?

SOLUTION Figures 15.9 and 15.10 use Microsoft Excel and Minitab to illustrate the results.

FIGURE 15.9

Microsoft Excel Output for the Multiple Regression Model with a Quadratic Term for Attic Insulation

	A	B	C	D	E	F	G
1	**Quadratic Effect for Insulation Variable?**						
2							
3	*Regression Statistics*						
4	Multiple R	0.98616					
5	R Square	0.97251					
6	Adjusted R Square	0.96501					
7	Standard Error	24.29378					
8	Observations	15					
9							
10	ANOVA						
11		*df*	*SS*	*MS*	*F*	*Significance F*	
12	Regression	3	229643.2	76547.72149	129.70063	7.26403E-09	
13	Residual	11	6492.1	590.1877159			
14	Total	14	236135.2				
15							
16		*Coefficients*	*Standard Error*	*t Stat*	*P-value*	*Lower 95%*	*Upper 95%*
17	Intercept	624.58642	42.43515952	14.71860664	1.39E-08	531.18722	717.98562
18	Temperature	-5.36260	0.31713	-16.90988	3.21E-09	-6.06060	-4.66461
19	Insulation	-44.58679	14.95469	-2.98146	0.01249	-77.50185	-11.67172
20	Insulation ^2	1.86670	1.12376	1.66113	0.12489	-0.60667	4.34007

FIGURE 15.10

Minitab Output for the Multiple Regression Model with a Quadratic Term for Attic Insulation

```
The regression equation is
Gallons = 625 - 5.36 Temp(F) - 44.6 Insulation + 1.87 Insulationsquared

Predictor             Coef  SE Coef       T      P
Constant            624.59    42.44   14.72  0.000
Temp(F)            -5.3626   0.3171  -16.91  0.000
Insulation          -44.59    14.95   -2.98  0.012
Insulationsquared    1.867    1.124    1.66  0.125

S = 24.2938   R-Sq = 97.3%   R-Sq(adj) = 96.5%

Analysis of Variance

Source          DF      SS     MS       F      P
Regression       3  229643  76548  129.70  0.000
Residual Error  11    6492    590
Total           14  236135
```

The multiple regression equation is

$$\hat{Y}_i = 624.5864 - 5.3626X_{1i} - 44.5868X_{2i} + 1.8667X_{2i}^2$$

To test for the significance of the quadratic effect,

H_0: Including the quadratic effect does not significantly improve the model ($\beta_3 = 0$).
H_1: Including the quadratic effect significantly improves the model ($\beta_3 \neq 0$).

From Figure 15.9 or 15.10, $t = 1.661 < 2.201$ or the p-value $= 0.1249 > 0.05$. Therefore, you do not reject the null hypothesis. You conclude that there is insufficient evidence that the quadratic effect for attic insulation is different from zero. In the interest of keeping the model as simple as possible, you should use the multiple regression equation computed in Figure 15.7 and 15.8 on pages 625 and 626,

$$\hat{Y}_i = 562.151 - 5.43658X_{1i} - 20.0123X_{2i}.$$

The Coefficient of Multiple Determination

In the multiple regression model, the coefficient of multiple determination r^2 (see section 14.2) represents the proportion of variation in Y that is explained by variation in the independent variables. Consider the quadratic regression model you used to predict the strength of concrete using fly ash and fly ash squared. You compute r^2 using Equation (14.4) on page 577:

$$r^2 = \frac{SSR}{SST}$$

From Figure 15.2 or 15.3 on page 622,

$$SSR = 2{,}695{,}473.5 \text{ and } SST = 4{,}156{,}690.5$$

Thus,

$$r^2 = \frac{SSR}{SST} = \frac{2{,}695{,}473.5}{4{,}156{,}690.5} = 0.6485$$

This coefficient of multiple determination indicates that 64.85% of the variation in strength is explained by the quadratic relationship between strength and the percentage of fly ash. You should also compute the r_{adj}^2 to account for the number of independent variables and the sample size. In the quadratic regression model, $k = 2$ since there are two independent variables, X_1 and X_1^2. Thus, using Equation (14.5) on page 577

$$\begin{aligned} r_{adj}^2 &= 1 - \left[\left(1 - r^2\right)\frac{(n-1)}{(n-k-1)}\right] \\ &= 1 - \left[(1 - 0.6485)\frac{17}{15}\right] \\ &= 1 - 0.3984 \\ &= 0.6016 \end{aligned}$$

PROBLEMS FOR SECTION 15.1

Learning the Basics

15.1 The following quadratic regression equation is for a sample of $n = 25$:

$$\hat{Y}_i = 5 + 3X_{1i} + 1.5X_{1i}^2$$

a. Predict the Y for $X_1 = 2$.
b. Suppose the t statistic for the quadratic regression coefficient is 2.35. At the 0.05 level of significance, is there evidence that the quadratic model is better than the linear model?
c. Suppose the t statistic for the quadratic regression coefficient is 1.17. At the 0.05 level of significance, is there evidence that the quadratic model is better than the linear model?
d. Suppose the regression coefficient for the linear effect is -3.0. Predict Y for $X_1 = 2$.

Applying the Concepts

You need to use Microsoft Excel, Minitab, or SPSS to solve problems 15.2–15.5.

SELF Test **15.2** A research analyst for an oil company wants to develop a model to predict miles per gallon based on highway speed. An experiment is designed in which a test car is driven at speeds ranging from 10 miles per hour to 75 miles per hour. The results are in the data file SPEED.

Assume a quadratic relationship between speed and mileage.
a. Construct a scatter diagram for speed and miles per gallon.
b. State the quadratic regression equation.
c. Predict the miles per gallon when the car is driven at 55 miles per hour.
d. Perform a residual analysis on the results, and determine the adequacy of the model.
e. At the 0.05 level of significance, is there a significant quadratic relationship between miles per gallon and speed?
f. At the 0.05 level of significance, determine whether the quadratic model is a better fit than the linear regression model.
g. Interpret the meaning of the coefficient of multiple determination.
h. Compute the adjusted r^2.

15.3 The marketing department of a large supermarket chain wants to study the price elasticity for packages of disposable razors. A sample of 15 stores with equivalent store traffic and product placement (i.e., at the checkout counter) is selected. Five stores are randomly assigned to each of three price levels (79, 99, and 119 cents) for the package of razors. The number of packages sold over a full week and the price at each store are in the following table. DISPRAZ

Sales	Price (cents)	Sales	Price (cents)
142	79	115	99
151	79	126	99
163	79	77	119
168	79	86	119
176	79	95	119
91	99	100	119
100	99	106	119
107	99		

Assume a quadratic relationship between price and sales.
a. Construct a scatter diagram for price and sales.
b. State the quadratic regression equation.
c. Predict the weekly sales for a price of 79 cents.
d. Perform a residual analysis on the results, and determine the adequacy of the model.
e. At the 0.05 level of significance, is there a significant quadratic relationship between weekly sales and price?
f. At the 0.05 level of significance, determine whether the quadratic model is a better fit than the linear model.
g. Interpret the meaning of the coefficient of multiple determination.
h. Compute the adjusted r^2.

15.4 An agronomist designed a study in which tomatoes were grown using six different amounts of fertilizer: 0, 20, 40, 60, 80, and 100 pounds per 1,000 square feet. These fertilizer application rates were then randomly assigned to plots of land with the following yields of tomatoes (in pounds). TOMYLD2

Plot	Fertilizer Application Rate	Yield	Plot	Fertilizer Application Rate	Yield
1	0	6	7	60	46
2	0	9	8	60	50
3	20	19	9	80	48
4	20	24	10	80	54
5	40	32	11	100	52
6	40	38	12	100	58

Assume a quadratic relationship between the fertilizer application rate and yield.
a. Construct a scatter diagram for fertilizer application rate and yield.
b. State the quadratic regression equation.
c. Predict the yield for a plot fertilized with 70 pounds per 1,000 square feet.
d. Perform a residual analysis on the results, and determine the adequacy of the model.

e. At the 0.05 level of significance, is there a significant overall relationship between the fertilizer application rate and tomato yield?
f. What is the p-value in (e)? Interpret its meaning.
g. At the 0.05 level of significance, determine whether there is a significant quadratic effect.
h. What is the p-value in (g)? Interpret its meaning.
i. Interpret the meaning of the coefficient of multiple determination.
j. Compute the adjusted r^2.

15.5 An auditor for a county government would like to develop a model to predict the county taxes based on the age of single-family houses. She selects a random sample of 19 single-family houses, with the results in the data file TAXES.

Assume a quadratic relationship between the age and county taxes.
a. Construct a scatter diagram between age and county taxes.
b. State the quadratic regression equation.
c. Predict the county taxes for a house that is 20 years old.
d. Perform a residual analysis on the results, and determine the adequacy of the model.
e. At the 0.05 level of significance, is there is a significant overall relationship between age and county taxes?
f. What is the p-value in (e)? Interpret its meaning.
g. At the 0.05 level of significance, determine whether the quadratic model is superior to the linear model.
h. What is the p-value in (g)? Interpret its meaning.
i. Interpret the meaning of the coefficient of multiple determination.
j. Compute the adjusted r^2.

15.2 USING TRANSFORMATIONS IN REGRESSION MODELS

This section introduces regression models in which the independent variable, the dependent variable, or both are transformed in order to either overcome violations of the assumptions of regression or to make a model linear in form. Among the many transformations available (see reference 7) are the square-root transformation and transformations involving the common logarithm (base 10) and the natural logarithm (base e).[1]

[1] *For more information on logarithms, see Appendix A.3.*

The Square-Root Transformation

The **square-root transformation** is often used to overcome violations of the equal variance assumption, as well as to transform a model that is not linear in form into one that is linear. Equation (15.3) shows a regression model that uses a square-root transformation of the independent variable.

REGRESSION MODEL WITH A SQUARE-ROOT TRANSFORMATION

$$Y_i = \beta_0 + \beta_1\sqrt{X_{1i}} + \varepsilon_i \qquad \textbf{(15.3)}$$

Example 15.2 illustrates the use of a square-root transformation.

EXAMPLE 15.2 USING THE SQUARE-ROOT TRANSFORMATION

Given the following values for Y and X, use a square-root transformation for the X variable. Construct a scatter diagram for X and Y and for the square root of X and Y.

Y	X	Y	X
42.7	1	100.4	3
50.4	1	104.7	4
69.1	2	112.3	4
79.8	2	113.6	5
90.0	3	123.9	5

SOLUTION Figure 15.11, panel A, displays the scatter diagram of X and Y, and panel B shows the square root of X versus Y.

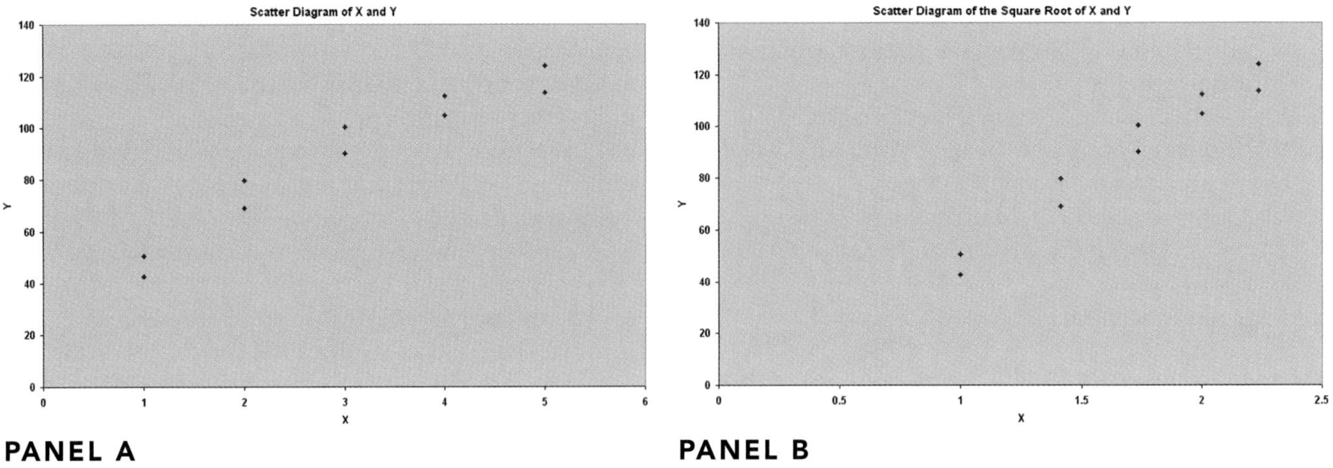

FIGURE 15.11 Panel *A*: Scatter Diagram of *X* and *Y*; Panel *B*: Scatter Diagram of the Square Root of *X* and *Y*

You can see that the square-root transformation has transformed a nonlinear relationship into a linear relationship.

The Log Transformation

The **logarithmic transformation** is often used to overcome violations to the equal variance assumption. You can also use the logarithmic transformation to change a nonlinear model into a linear model. Equation (15.4) shows a multiplicative model.

ORIGINAL MULTIPLICATIVE MODEL

$$Y_i = \beta_0 X_{1i}^{\beta_1} X_{2i}^{\beta_2} \varepsilon_i \qquad (15.4)$$

By taking base 10 logarithms of both the dependent and independent variables, you can transform Equation (15.4) to the model shown in Equation (15.5).

TRANSFORMED MULTIPLICATIVE MODEL

$$\begin{aligned} \log Y_i &= \log(\beta_0 X_{1i}^{\beta_1} X_{2i}^{\beta_2} \varepsilon_i) \qquad (15.5) \\ &= \log \beta_0 + \log(X_{1i}^{\beta_1}) + \log(X_{2i}^{\beta_2}) + \log \varepsilon_i \\ &= \log \beta_0 + \beta_1 \log X_{1i} + \beta_2 \log X_{2i} + \log \varepsilon_i \end{aligned}$$

Hence, Equation (15.5) is linear in the logarithms. In a similar fashion, you can transform the exponential model shown in Equation (15.6) to linear form by taking the natural logarithm of both sides of the equation.

ORIGINAL EXPONENTIAL MODEL

$$Y_i = e^{\beta_0 + \beta_1 X_{1i} + \beta_2 X_{2i}} \varepsilon_i \qquad (15.6)$$

TRANSFORMED EXPONENTIAL MODEL

$$\begin{aligned}\ln Y_i &= \ln(e^{\beta_0+\beta_1 X_{1i}+\beta_2 X_{2i}}\varepsilon_i) \\ &= \ln(e^{\beta_0+\beta_1 X_{1i}+\beta_2 X_{2i}}) + \ln \varepsilon_i \\ &= \beta_0 + \beta_1 X_{1i} + \beta_2 X_{2i} + \ln \varepsilon_i\end{aligned} \quad \textbf{(15.7)}$$

EXAMPLE 15.3 USING THE NATURAL LOG TRANSFORMATION

Given the following values for Y and X, use a natural logarithm transformation for the Y variable. Construct a scatter diagram for X and Y, and for X and the natural logarithm of Y.

Y	X	Y	X
0.7	1	4.8	3
0.5	1	12.9	4
1.6	2	11.5	4
1.8	2	32.1	5
4.2	3	33.9	5

SOLUTION As shown in Figure 15.12, the natural logarithm transformation has transformed a nonlinear relationship into a linear relationship.

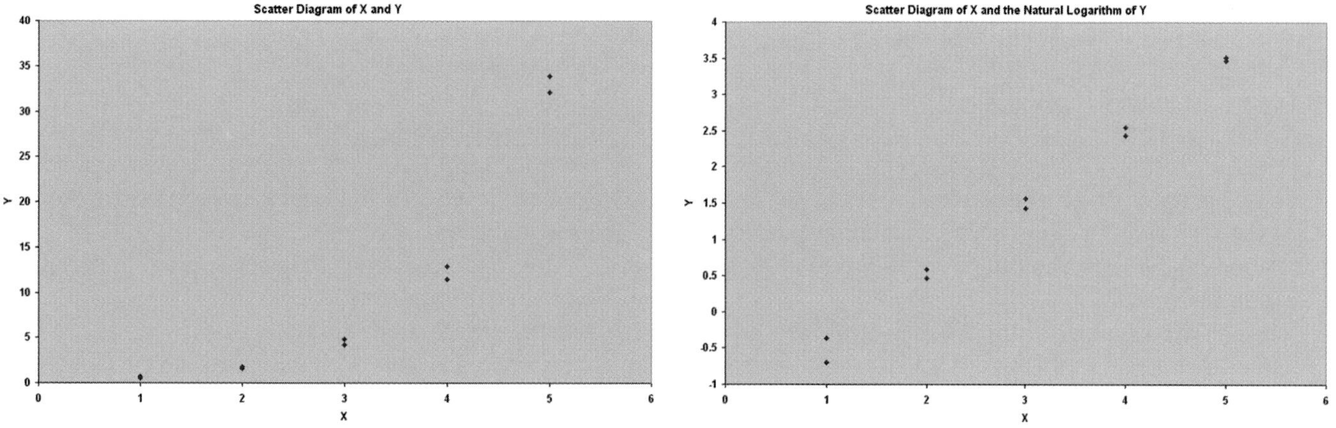

PANEL A **PANEL B**

FIGURE 15.12 Panel *A*: Scatter Diagram of *X* and *Y*; Panel *B*: Scatter Diagram of *X* and the Natural Logarithm of *Y*

PROBLEMS FOR SECTION 15.2

Learning the Basics

15.6 Consider the following regression equation:

$$\log \hat{Y}_i = \log 3.07 + 0.9 \log X_{1i} + 1.41 \log X_{2i}$$

a. Predict the value of Y when $X_1 = 8.5$ and $X_2 = 5.2$.
b. Interpret the meaning of the slopes.

15.7 Consider the following regression equation:

$$\ln \hat{Y}_i = 4.62 + 0.5X_{1i} + 0.7X_{2i}$$

a. Predict the value of Y when $X_1 = 8.5$ and $X_2 = 5.2$.
b. Interpret the meaning of the slopes.

Applying the Concepts

You need to use Microsoft Excel, Minitab, or SPSS to solve problems 15.8–15.11.

15.8 Referring to the data of problem 15.2 on page 628, and using the file SPEED, perform a square-root transformation of the independent variable (speed) and reanalyze the data using this model.

a. State the regression equation.
b. Predict the miles per gallon when the car is driven at 55 miles per hour.
c. Perform a residual analysis of the results, and determine the adequacy of the model.
d. At the 0.05 level of significance, is there a significant relationship between miles per gallon and the square root of speed?
e. Interpret the meaning of the coefficient of determination r^2 in this problem.
f. Compute the adjusted r^2.
g. Compare your results with those in problem 15.2. Which model is better? Why?

15.9 Referring to the data of problem 15.2 on page 628, using the file SPEED, perform a natural logarithmic transformation of the dependent variable (miles per gallon) and reanalyze the data using this model.

a. State the regression equation.
b. Predict the miles per gallon when the car is driven at 55 miles per hour.
c. Perform a residual analysis of the results and determine the adequacy of the fit of the model.
d. At the 0.05 level of significance, is there a significant relationship between the natural logarithm of miles per gallon and speed?
e. Interpret the meaning of the coefficient of determination r^2 in this problem.
f. Compute the adjusted r^2.
g. Compare your results with those in problems 15.2 and 15.8. Which model is best? Why?

15.10 Referring to the data of problem 15.4 on page 628, and using the file TOMYLD2, perform a natural logarithm transformation of the dependent variable (yield) and reanalyze the data using this model.

a. State the regression equation.
b. Predict the yield when 55 pounds of fertilizer is applied per 1,000 square feet.
c. Perform a residual analysis of the results and determine the adequacy of the model.
d. At the 0.05 level of significance, is there a significant relationship between the natural logarithm of yield and the fertilizer application rate?
e. Interpret the meaning of the coefficient of determination r^2 in this problem.
f. Compute the adjusted r^2.
g. Compare your results with those in problem 15.4. Which model is better? Why?

15.11 Referring to the data of problem 15.4 on page 628, and using the file TOMYLD2, perform a square-root transformation of the independent variable (fertilizer application rate) and reanalyze the data using this model.

a. State the regression equation.
b. Predict the yield when 55 pounds of fertilizer is applied per 1,000 square feet.
c. Perform a residual analysis of the results and determine the adequacy of the model.
d. At the 0.05 level of significance, is there a significant relationship between yield and the square root of the fertilizer application rate?
e. Interpret the meaning of the coefficient of determination r^2 in this problem.
f. Compute the adjusted r^2.
g. Compare your results with those of problems 15.4 and 15.10. Which model is best? Why?

15.3 INFLUENCE ANALYSIS

In sections 13.5 and 14.3, you used residual analysis to evaluate the regression assumptions. This section introduces several methods that measure the influence of individual observations:

- The hat matrix elements h_i
- The Studentized deleted residuals t_i
- Cook's distance statistic D_i

Figure 15.13 presents the values of these statistics computed by Minitab for the OmniPower sales data. In Figure 15.13, certain influence statistics are highlighted for further analysis.

↓	C1	C2	C3	C4	C5	C6
	sales	price	promotion	TRES1	HI1	COOK1
1	4141	59	200	1.21248	0.119048	0.065232
2	3842	59	200	0.69829	0.119048	0.022333
3	3056	59	200	-0.60203	0.119048	0.016669
4	3519	59	200	0.16218	0.119048	0.001223
5	4226	59	400	0.13285	0.069940	0.000457
6	4630	59	400	0.78667	0.069940	0.015706
7	3507	59	400	-1.03461	0.069940	0.026771
8	3754	59	400	-0.62580	0.069940	0.010013
9	5000	59	600	0.22031	0.113095	0.002128
10	5120	59	600	0.41780	0.113095	0.007623
11	4011	59	600	-1.44695	0.113095	0.085960
12	5015	59	600	0.24494	0.113095	0.002630
13	1916	79	200	-0.70923	0.069940	0.012814
14	675	79	200	-3.08402	0.069940	0.187056
15	3636	79	200	2.20612	0.069940	0.108468
16	3224	79	200	1.43451	0.069940	0.049881
17	2295	79	400	-1.25842	0.029762	0.015893
18	2730	79	400	-0.54832	0.029762	0.003145
19	2618	79	400	-0.72723	0.029762	0.005491
20	4421	79	400	2.27527	0.029762	0.046648
21	4113	79	600	0.50383	0.081845	0.007729
22	3746	79	600	-0.08881	0.081845	0.000242
23	3532	79	600	-0.43448	0.081845	0.005760
24	3825	79	600	0.03832	0.081845	0.000045
25	1096	99	200	-0.32079	0.113095	0.004505
26	761	99	200	-0.87981	0.113095	0.033144
27	2088	99	200	1.34235	0.113095	0.074659
28	820	99	200	-0.77987	0.113095	0.026183
29	2114	99	400	0.16060	0.081845	0.000791
30	1882	99	400	-0.21292	0.081845	0.001390
31	2159	99	400	0.23315	0.081845	0.001666
32	1602	99	400	-0.66819	0.081845	0.013508
33	3354	99	600	1.04633	0.142857	0.060637
34	2927	99	600	0.31720	0.142857	0.005757

FIGURE 15.13 Minitab Influence Statistics for the OmniPower Sales Data

The Hat Matrix Elements h_i

In section 13.8, h_i was defined for the simple linear regression model when constructing the confidence interval estimate of the mean response. For multiple regression models, the formula for calculating the **hat matrix diagonal elements h_i** requires the use of matrix algebra and is beyond the scope of this text (see references 4, 5, and 7).

The hat matrix diagonal element for observation i, denoted h_i, reflects the possible influence of X_i on the regression equation. If potentially influential observations are present, you may need to delete them from the model. In a regression model containing k independent variables, Hoaglin and Welsch (see reference 5) suggest the following decision rule:

$$\text{If } h_i > 2(k+1)/n$$

then X_i is an influential observation and is a candidate for removal from the model.

For the OmniPower sales data, because $n = 34$ and $k = 2$, you flag any h_i value greater than $2(2 + 1)/34 = 0.1765$. Referring to Figure 15.13, you see that none of the h_i values are greater than 0.1429. Therefore, none of the observations are candidates for removal from the analysis.

The Studentized Deleted Residuals t_i

Recall from section 13.5 that a residual is the difference between the observed value of Y and the predicted value of Y [see Equation (13.14) on page 532]. Studentized residuals are the residuals divided by the standard error of the estimate S_{YX} and adjusted for the distance from $\bar{X}$. The **Studentized deleted residual**, expressed as a t statistic in Equation (15.8), measures the difference of each Y_i from the value predicted by a model that includes all observations *except* observation i.

STUDENTIZED DELETED RESIDUAL

$$t_i = e_i \sqrt{\frac{n-k-1}{SSE(1-h_i)-e_i^2}} \qquad (15.8)$$

where e_i = the residual for observation i

k = number of independent variables

SSE = error sum of squares of the regression model fitted

h_i = hat matrix diagonal element for observation i

Hoaglin and Welsch (see Reference 5) suggest that if $t_i > t_{n-k-2}$ or $t_i < -t_{n-k-2}$ (using a two-tail test with a level of significance of 0.10), then the observed and predicted values are so different that observation i is highly influential on the regression equation and is a candidate for removal.

For the OmniPower sales data, $n = 34$ and $k = 2$. Thus, you flag any t_i whose absolute value is greater than 1.6973 (see Table E.3). In Figure 15.13 on page 633, $t_{14} = -3.08402$, $t_{15} = 2.20612$, and $t_{20} = 2.27527$ are highlighted. Thus, the 14th, 15th, and 20th observations may each have an adverse effect on the model. These observations were not previously flagged according to the h_i criterion. Since h_i and t_i measure different aspects of influence, neither criterion is sufficient by itself. When h_i is small, t_i may be large. When h_i is large, t_i may be moderate or small because the observed Y_i is consistent with the rest of the data. The next section introduces a third method for identifying influential observations.

Cook's Distance Statistic D_i

Cook's D_i statistic, based on both h_i and the Studentized residual, is a third criterion for identifying influential observations. To decide whether an observation flagged by either the h_i or t_i criterion is unduly affecting the model, Cook and Weisberg (see reference 4) developed the Cook's D_i statistic.

COOK'S D_i STATISTIC

$$D_i = \frac{e_i^2}{k\,MSE}\left[\frac{h_i}{(1-h_i)^2}\right] \tag{15.9}$$

where e_i = the residual for observation i

k = number of independent variables

MSE = mean square error of the regression model fitted

h_i = hat matrix diagonal element for observation i

Cook and Weisberg suggest that if $D_i > F_{k+1,n-k-1}$, the critical value of the F distribution having $k + 1$ degrees of freedom in the numerator and $n - k - 1$ degrees of freedom in the denominator at a 0.50 level of significance, then the observation is highly influential on the regression equation and is a candidate for removal.

For the OmniPower sales data, $n = 34$ and $k = 2$. Thus, any $D_i > F_{3,31} = 0.807$ is flagged. Referring to Figure 15.13, you see that none of the D_i values exceed 0.187, and therefore no observations are identified as influential using Cook's D_i statistic.

Overview

This section discussed three criteria for evaluating the influence of each observation on the multiple regression model. The various statistics did not lead to a consistent set of conclusions. According to both the h_i and the D_i criteria, none of the observations is a candidate for removal. Under such circumstances, most statisticians would conclude that there is insufficient evidence for the removal of any observation from the analysis.

In addition to the three criteria presented here, there are other measures of influence (see references 1 and 6). Although different statisticians seem to prefer particular measures, currently there is no consensus as to the "best" measure.

PROBLEMS FOR SECTION 15.3

Applying the Concepts

You need to use Microsoft Excel, Minitab, or SPSS to solve problems 15.12–15.16.

SELF Test **15.12** In problem 14.4 on page 575, you used sales and number of orders to predict distribution costs at a mail-order catalog business WARECOST. Perform an influence analysis on your results and determine whether any observations should be deleted from the analysis. If necessary, reanalyze the regression model after deleting these observations and compare your results.

15.13 In problem 14.5 on page 575, you used horsepower and weight to predict gasoline mileage AUTO. Perform an influence analysis on your results and determine whether any observations should be deleted from the analysis. If necessary, reanalyze the regression model after deleting these observations and compare your results.

15.14 In problem 14.6 on page 575, you used the amount of radio advertising and newspaper advertising to predict sales ADVERTISE. Perform an influence analysis on your results and determine whether any observations should be deleted from the analysis. If necessary, reanalyze the regression model after deleting these observations and compare your results.

15.15 In problem 14.7 on page 576, you used the total staff present and remote hours to predict standby hours STANDBY. Perform an influence analysis on your results and determine whether any observations should be deleted from the analysis. If necessary, reanalyze the regression model after deleting these observations and compare your results.

15.16 In problem 14.8 on page 576, you used the land area of the property and age in years to predict appraised value GLENCOVE. Perform an influence analysis on your results and determine whether any observations should be deleted from the analysis. If necessary, reanalyze the regression model after deleting these observations and compare your results.

15.4 COLLINEARITY

One important problem in the application of multiple regression analysis involves the possible **collinearity** of the independent variables. This condition refers to situations in which one or more of the independent variables are highly correlated with each other. In such situations, collinear variables do not provide unique information, and it becomes difficult to separate the effect of such variables on the dependent variable. When collinearity exists, the values of the regression coefficients for the correlated variables may fluctuate drastically, depending on which independent variables are included in the model.

One method of measuring collinearity is the **variance inflationary factor (*VIF*)** for each independent variable. Equation (15.10) defines VIF_j, the variance inflationary factor for variable j.

VARIANCE INFLATIONARY FACTOR

$$VIF_j = \frac{1}{1 - R_j^2} \quad \textbf{(15.10)}$$

where R_j^2 is the coefficient of multiple determination of independent variable X_j with all other X variables

If there are only two independent variables, R_1^2 is the coefficient of determination between X_1 and X_2. It is identical to R_2^2, which is the coefficient of determination between X_2 and X_1. If, for example, there are three independent variables, then R_1^2 is the coefficient of multiple determination of X_1 with X_2 and X_3; R_2^2 is the coefficient of multiple determination of X_2 with X_1 and X_3; and R_3^2 is the coefficient of multiple determination of X_3 with X_1 and X_2.

If a set of independent variables is uncorrelated, each VIF_j is equal to 1. If the set is highly correlated, then a VIF_j might even exceed 10. Marquardt (see reference 8) suggests that if VIF_j is greater than 10, there is too much correlation between the variable X_j and the other independent variables. However, other statisticians suggest a more conservative criterion.

Snee (see reference 11) recommends using alternatives to least-squares regression if the maximum VIF_j exceeds 5.

You need to proceed with extreme caution when using a multiple regression model that has one or more large *VIF* values. You can use the model to predict values of the dependent variable *only* in the case where the values of the independent variables used in the prediction are in the relevant range of the values in the data set. However, you cannot extrapolate to values of the independent variables not observed in the sample data. And since the independent variables contain overlapping information, you should always avoid interpreting the regression coefficient estimates separately (i.e., there is no way to accurately estimate the individual effects of the independent variables). One solution to the problem is to delete the variable with the largest *VIF* value. The reduced model (i.e., the model with the independent variable with the largest *VIF* value deleted) is often free of collinearity problems. If you determine that all the independent variables are needed in the model, you can use methods discussed in references 7 and 8.

In the OmniPower sales data, the correlation between the two independent variables, price and promotional expenditure, is –0.0968. Because there are only two independent variables in the model, from Equation (15.10) on page 635:

$$VIF_1 = VIF_2 = \frac{1}{1-(-0.0968)^2}$$

$$= 1.009$$

Thus, you can conclude that there is no problem with collinearity for the OmniPower sales data.

In models containing quadratic and interaction terms, collinearity is usually present. The linear and quadratic terms of an independent variable are usually highly correlated with each other, and an interaction term is often correlated with one or both of the independent variables making up the interaction. Thus, you cannot interpret individual parameter estimates separately. You need to interpret the linear and quadratic parameter estimates together in order to understand the nonlinear relationship. Likewise, you need to interpret an interaction parameter estimate in conjunction with the two parameter estimates associated with the variables comprising the interaction. In summary, large *VIF*s in quadratic or interaction models do not necessarily mean that the model is a poor one. It does, however, require you to carefully interpret the parameter estimates.

PROBLEMS FOR SECTION 15.4

Learning the Basics

15.17 If the coefficient of determination between two independent variables is 0.80, what is the *VIF*?

15.18 If the coefficient of determination between two independent variables is 0.20, what is the *VIF*?

15.19 If the coefficient of determination between two independent variables is 0.50, what is the *VIF*?

Applying the Concepts

You need to use Microsoft Excel, Minitab, or SPSS to solve problems 15.20–15.24.

SELF Test **15.20** Refer to problem 14.4 on page 575. Perform a multiple regression analysis using the file WARECOST and determine the *VIF* for each independent variable in the model. Is there reason to suspect the existence of collinearity?

15.21 Refer to problem 14.5 on page 575. Perform a multiple regression analysis using the file AUTO and determine the *VIF* for each independent variable in the model. Is there reason to suspect the existence of collinearity?

15.22 Refer to problem 14.6 on page 575. Perform a multiple regression analysis using the file ADVERTISE and determine the *VIF* for each independent variable in the model. Is there reason to suspect the existence of collinearity?

15.23 Refer to problem 14.7 on page 576. Perform a multiple regression analysis using the file STANDBY and determine the *VIF* for each independent variable in the model. Is there reason to suspect the existence of collinearity?

15.24 Refer to problem 14.8 on page 576. Perform a multiple regression analysis using the file GLENCOVE and determine the *VIF* for each independent variable in the model. Is there reason to suspect the existence of collinearity?

15.5 MODEL BUILDING

This chapter and Chapter 14 have introduced you to many different topics in regression analysis, including quadratic terms, dummy variables, interaction terms, and influential observations. In this section, you will learn a structured approach to building the most appropriate regression model. As you will see, successful model building incorporates many of the topics you have studied so far.

To begin, refer to the "Using Statistics" scenario introduced at the beginning of the chapter in which four independent variables (total staff present, remote hours, Dubner hours, and total labor hours) are considered in developing a regression model to predict standby hours of unionized graphic artists. Table 15.2 presents the data STANDBY.

TABLE 15.2
Predicting Standby Hours Based on Total Staff Present, Remote Hours, Dubner Hours, and Total Labor Hours

Week	Standby Hours	Total Staff Present	Remote Hours	Dubner Hours	Total Labor Hours
1	245	338	414	323	2,001
2	177	333	598	340	2,030
3	271	358	656	340	2,226
4	211	372	631	352	2,154
5	196	339	528	380	2,078
6	135	289	409	339	2,080
7	195	334	382	331	2,073
8	118	293	399	311	1,758
9	116	325	343	328	1,624
10	147	311	338	353	1,889
11	154	304	353	518	1,988
12	146	312	289	440	2,049
13	115	283	388	276	1,796
14	161	307	402	207	1,720
15	274	322	151	287	2,056
16	245	335	228	290	1,890
17	201	350	271	355	2,187
18	183	339	440	300	2,032
19	237	327	475	284	1,856
20	175	328	347	337	2,068
21	152	319	449	279	1,813
22	188	325	336	244	1,808
23	188	322	267	253	1,834
24	197	317	235	272	1,973
25	261	315	164	223	1,839
26	232	331	270	272	1,935

Before you develop a model to predict standby hours, you need to consider the principle of parsimony. **Parsimony** means that you want to develop a regression model that includes the fewest number of independent variables that permit an adequate interpretation of the dependent variable. Regression models with fewer independent variables are easier to interpret, particularly because they are less likely to be affected by collinearity problems (described in section 15.4).

The selection of an appropriate model when many independent variables are under consideration involves complexities that are not present with a model with only two independent variables. The evaluation of all possible regression models is more computationally complex. Although you can quantitatively evaluate competing models, there may not be a *uniquely* best model, but rather several *equally appropriate* models.

To begin analyzing the standby-hours data, you compute the variance inflationary factors [see Equation (15.10) on page 635] to measure the amount of collinearity among the independent variables. Figure 15.14 shows Microsoft Excel output of the *VIF* values along with the regression equation. Figure 15.15 illustrates Minitab output. Observe that all the *VIF* values are relatively small, ranging from a high of 2.0 for the total labor hours to a low of 1.2 for remote hours. Thus, on the basis of the criteria developed by Snee that all *VIF* values should be less than 5.0 (see reference 11), there is little evidence of collinearity among the set of independent variables.

Standby Hours Analysis

	Regression Statistics			
	Total Staff and all other X	Remote and all other X	Dubner and all other X	Total Labor and all other X
Multiple R	0.64368	0.43490	0.56099	0.70698
R Square	0.41433	0.18914	0.31471	0.49982
Adjusted R Square	0.33446	0.07856	0.22126	0.43161
Standard Error	16.47151	124.93921	57.55254	114.41183
Observations	26	26	26	26
VIF	1.70743	1.23325	1.45924	1.99928

PANEL A

	A	B	C	D	E	F	G
1	Standby Hours Analysis						
2							
3	*Regression Statistics*						
4	Multiple R	0.78935					
5	R Square	0.62308					
6	Adjusted R Square	0.55128					
7	Standard Error	31.83501					
8	Observations	26					
9							
10	ANOVA						
11		*df*	*SS*	*MS*	*F*	*Significance F*	
12	Regression	4	35181.79373	8795.44843	8.67857	0.00027	
13	Residual	21	21282.82166	1013.46770			
14	Total	25	56464.61538				
15							
16		*Coefficients*	*Standard Error*	*t Stat*	*P-value*	*Lower 95%*	*Upper 95%*
17	Intercept	-330.83184	110.89536	-2.98328	0.00709	-561.451405	-100.212285
18	Total Staff	1.24563	0.41206	3.02293	0.00647	0.388704	2.102554
19	Remote	-0.11842	0.05432	-2.17983	0.04080	-0.231392	-0.005444
20	Dubner	-0.29706	0.11793	-2.51891	0.01995	-0.542310	-0.051807
21	Total Labor	0.13053	0.05932	2.20041	0.03911	0.007166	0.253904

PANEL B

	A	B
1	Durbin-Watson Calculations	
2		
3	Sum of Squared Difference of Residuals	47241.61261
4	Sum of Squared Residuals	21282.82166
5		
6	Durbin-Watson Statistic	2.21971

PANEL C

FIGURE 15.14 Microsoft Excel Regression Model to Predict Standby Hours Based on Four Independent Variables

FIGURE 15.15 Minitab Regression Model to Predict Standby Hours Based on Four Independent Variables

```
The regression equation is
Standby = - 331 + 1.25 Staff - 0.118 Remote - 0.297 Dubner + 0.131 Labor


Predictor       Coef   SE Coef       T      P  VIF
Constant      -330.8     110.9   -2.98  0.007
Staff         1.2456    0.4121    3.02  0.006  1.7
Remote      -0.11842   0.05432   -2.18  0.041  1.2
Dubner       -0.2971    0.1179   -2.52  0.020  1.5
Labor        0.13053   0.05932    2.20  0.039  2.0


S = 31.8350   R-Sq = 62.3%   R-Sq(adj) = 55.1%


Analysis of Variance

Source          DF     SS     MS     F      P
Regression       4  35182   8795  8.68  0.000
Residual Error  21  21283   1013
Total           25  56465


Durbin-Watson statistic = 2.21971
```

The Stepwise Regression Approach to Model Building

You continue your analysis of the standby hours data by attempting to determine if a subset of all independent variables yields an adequate and appropriate model. The first approach described here is **stepwise regression**, which attempts to find the "best" regression model without examining all possible models.

The first step of stepwise regression is to find the best model that uses one independent variable. The next step is to find the best of the remaining independent variables to add to the model selected in the first step. An important feature of the stepwise approach is that an independent variable that has entered into the model at an early stage may subsequently be removed after other independent variables are considered. Thus, in stepwise regression, variables are either added to or deleted from the regression model at each step of the model-building process. The partial F-test statistic (see section 14.5) is used to determine if variables are added or deleted. The stepwise procedure terminates with the selection of a best-fitting model when no additional variables can be added to or deleted from the last model evaluated.

Figure 15.16 represents Microsoft Excel stepwise regression output for the standby-hours data while Figure 15.17 illustrates Minitab output. For this example, a significance level of 0.05 is used to enter a variable into the model or to delete a variable from the model. The first variable entered into the model is total staff, the variable that correlates most highly with the dependent variable standby hours. Because the p-value of 0.001 is less than 0.05, total staff is included in the regression model.

FIGURE 15.16
Microsoft Excel Stepwise Regression Output for the Standby Hours Data

	A	B	C	D	E	F	G	H
1	Stepwise Analysis for Standby Hours							
2	Table of Results for General Stepwise							
3								
4	Total Staff entered.							
5								
6			*df*	*SS*	*MS*	*F*	*Significance F*	
7		Regression	1	20667.39798	20667.39798	13.85631586	0.00106	
8		Residual	24	35797.21741	1491.550725			
9		Total	25	56464.61538				
10								
11			*Coefficients*	*Standard Error*	*t Stat*	*P-value*	*Lower 95%*	*Upper 95%*
12		Intercept	-272.38165	124.24020	-2.19238	0.03829	-528.80077	-15.96253
13		Total Staff	1.42405	0.38256	3.72241	0.00106	0.63448	2.21362
14								
15								
16	Remote entered.							
17								
18			*df*	*SS*	*MS*	*F*	*Significance F*	
19		Regression	2	27662.54287	13831.27143	11.04501	0.00043	
20		Residual	23	28802.07251	1252.26402			
21		Total	25	56464.61538				
22								
23			*Coefficients*	*Standard Error*	*t Stat*	*P-value*	*Lower 95%*	*Upper 95%*
24		Intercept	-330.67483	116.48022	-2.83889	0.00930	-571.63220	-89.71747
25		Total Staff	1.76486	0.37904	4.65619	0.00011	0.98077	2.54896
26		Remote	-0.13897	0.05880	-2.36347	0.02693	-0.26060	-0.01733
27								
28								
29	No other variables could be entered into the model. Stepwise ends.							

FIGURE 15.17
Minitab Stepwise Regression Output for the Standby Hours Data

```
Alpha-to-Enter: 0.05  Alpha-to-Remove: 0.05


Response is Standby on 4 predictors, with N = 26


Step              1       2
Constant     -272.4  -330.7

Staff          1.42    1.76
T-Value        3.72    4.66
P-Value       0.001   0.000

Remote               -0.139
T-Value               -2.36
P-Value               0.027

S              38.6    35.4
R-Sq          36.60   48.99
R-Sq(adj)     33.96   44.56
Mallows C-p    13.3     8.4
```

The next step involves selecting a second independent variable for the model. The second variable chosen is one that makes the largest contribution to the model, given that the first variable has been selected. For this model, the second variable is remote hours. Because the p-value of 0.027 for remote hours is less than 0.05, remote hours is included in the regression model.

After remote hours is entered into the model, the stepwise procedure determines whether total staff is still an important contributing variable or whether it can be eliminated from the model. Because the p-value of 0.0001 for total staff is less than 0.05, total staff remains in the regression model.

The next step involves selecting a third independent variable for the model. Because none of the other variables meets the 0.05 criterion for entry into the model, the stepwise procedure terminates with a model that includes total staff present and the number of remote hours.

This stepwise regression approach to model building was originally developed more than three decades ago, in an era in which regression analysis on mainframe computers involved the costly use of large amounts of processing time. Under such conditions, stepwise regression became widely used, although it provides a limited evaluation of alternative models. With today's extremely fast personal computers, the evaluation of many different regression models is completed quickly at a very small cost. Thus, a more general way of evaluating alternative regression models, in this era of fast computers, is the best subsets approach discussed below. Stepwise regression is not obsolete, however. Today, many businesses use stepwise regression as part of a new research technique called **data mining**, where huge data sets are explored to discover significant statistical relationships among a large number of variables. These data sets are so large that the best-subsets approach is impractical.

The Best-Subsets Approach to Model Building

The **best-subsets approach** evaluates all possible regression models for a given set of independent variables. Figure 15.18 represents Microsoft Excel output of all possible regression models for the standby hours data. Figure 15.19 illustrates Minitab output.

A criterion often used in model building is the adjusted r^2, which adjusts the r^2 of each model to account for the number of independent variables in the model as well as for the sample size (see section 14.1). Because model building requires you to compare models with different numbers of independent variables, the adjusted r^2 is more appropriate than r^2.

FIGURE 15.18

Microsoft Excel Best-Subsets Regression Output for the Standby Hours Data (Note: For an extremely small r^2, it is possible to get a negative adjusted r^2.)

	A	B	C	D	E	F
1	Best Subsets Analysis for Standby Hours					
2						
3	Intermediate Calculations					
4	R2T	0.62308				
5	1 - R2T	0.37692				
6	n	26				
7	T	5				
8	n - T	21				
9						
10	Model	Cp	k+1	R Square	Adj. R Square	Std. Error
11	X1	13.32152	2	0.36602	0.33961	38.6206
12	X1X2	8.41933	3	0.48991	0.44555	35.38734
13	X1X2X3	7.84181	4	0.53617	0.47292	34.50286
14	X1X2X3X4	5.00000	5	0.62308	0.55128	31.83501
15	X1X2X4	9.34492	4	0.50919	0.44227	35.49212
16	X1X3	10.64856	3	0.44990	0.40206	36.74905
17	X1X3X4	7.75166	4	0.53779	0.47476	34.44263
18	X1X4	14.79818	3	0.37542	0.32111	39.15789
19	X2	33.20781	2	0.00909	-0.03220	48.28359
20	X2X3	32.30673	3	0.06116	-0.02048	48.00868
21	X2X3X4	12.13813	4	0.45906	0.38529	37.26076
22	X2X4	23.24809	3	0.22375	0.15625	43.65405
23	X3	30.38835	2	0.05970	0.02052	47.03452
24	X3X4	11.82309	3	0.42882	0.37915	37.44658
25	X4	24.18460	2	0.17105	0.13651	44.16192

FIGURE 15.19
Minitab Best-Subsets Regression Output for the Standby Hours Data

Best Subsets Regression: Standby versus Staff, Remote, Dubner, Labor

Response is Standby

Vars	R-Sq	R-Sq(adj)	Mallows C-p	S	Staff	Remote	Dubner	Labor
1	36.6	34.0	13.3	38.621	X			
1	17.1	13.7	24.2	44.162				X
1	6.0	2.1	30.4	47.035			X	
2	49.0	44.6	8.4	35.387	X	X		
2	45.0	40.2	10.6	36.749	X		X	
2	42.9	37.9	11.8	37.447			X	X
3	53.8	47.5	7.8	34.443	X		X	X
3	53.6	47.3	7.8	34.503	X	X	X	
3	50.9	44.2	9.3	35.492	X	X		X
4	62.3	55.1	5.0	31.835	X	X	X	X

Referring to Figure 15.18 or Figure 15.19, you see that the adjusted r^2 reaches a maximum value of 0.551 when all four independent variables plus the intercept term (for a total of five estimated parameters) are included in the model.

A second criterion often used in the evaluation of competing models is the C_p statistic developed by Mallows (see reference 7). The C_p statistic, defined in Equation (15.11), measures the differences between a fitted regression model and a *true* model, along with random error. Equation (15.11) defines the **C_p statistic**.

THE C_p STATISTIC

$$C_p = \frac{(1 - R_k^2)(n - T)}{1 - R_T^2} - [n - 2(k + 1)] \qquad \textbf{(15.11)}$$

where k = number of independent variables included in a regression model

T = total number of parameters (including the intercept) to be estimated in the full regression model

R_k^2 = coefficient of multiple determination for a regression model that has k independent variables

R_T^2 = coefficient of multiple determination for a full regression model that contains all T estimated parameters

Using Equation (15.11) to compute C_p for the model containing total staff present and remote hours,

$$n = 26 \quad k = 2 \quad T = 4 + 1 = 5 \quad R_k^2 = 0.490 \quad R_T^2 = 0.623$$

so that

$$C_p = \frac{(1 - 0.49)(26 - 5)}{1 - 0.623} - [26 - 2(2 + 1)]$$

$$= 8.42$$

When a regression model with k independent variables contains only random differences from a *true* model, the mean value of C_p is $k + 1$, the number of parameters. Thus, in evaluating many alternative regression models, the goal is to find models whose C_p is close to or less than $k + 1$.

In Figure 15.18 on page 640 or Figure 15.19 on page 641, you see that only the model with all four independent variables considered contains a C_p value close to or below $k + 1$. Therefore, you should choose this model. Although it was not the case here, the C_p statistic often provides several alternative models for you to evaluate in greater depth using other criteria such as parsimony, interpretability, and departure from model assumptions (as evaluated by residual analysis). The model selected using stepwise regression has a C_p value of 8.4, which is substantially above the suggested criterion of $k + 1 = 3$ for that model.

Since the data were collected in time order, you need to compute the Durbin-Watson statistic to determine if there is autocorrelation in the residuals (see section 13.6). From Figure 15.14 or 15.15 on page 638, you see that the Durbin-Watson D statistic is 2.22. Since D is greater than 2.0, there is no indication of positive correlation in the residuals.

When you have finished selecting the independent variables to include in the model, you should perform a residual analysis to evaluate the regression assumptions. Figure 15.20 presents Microsoft Excel residual analysis output.

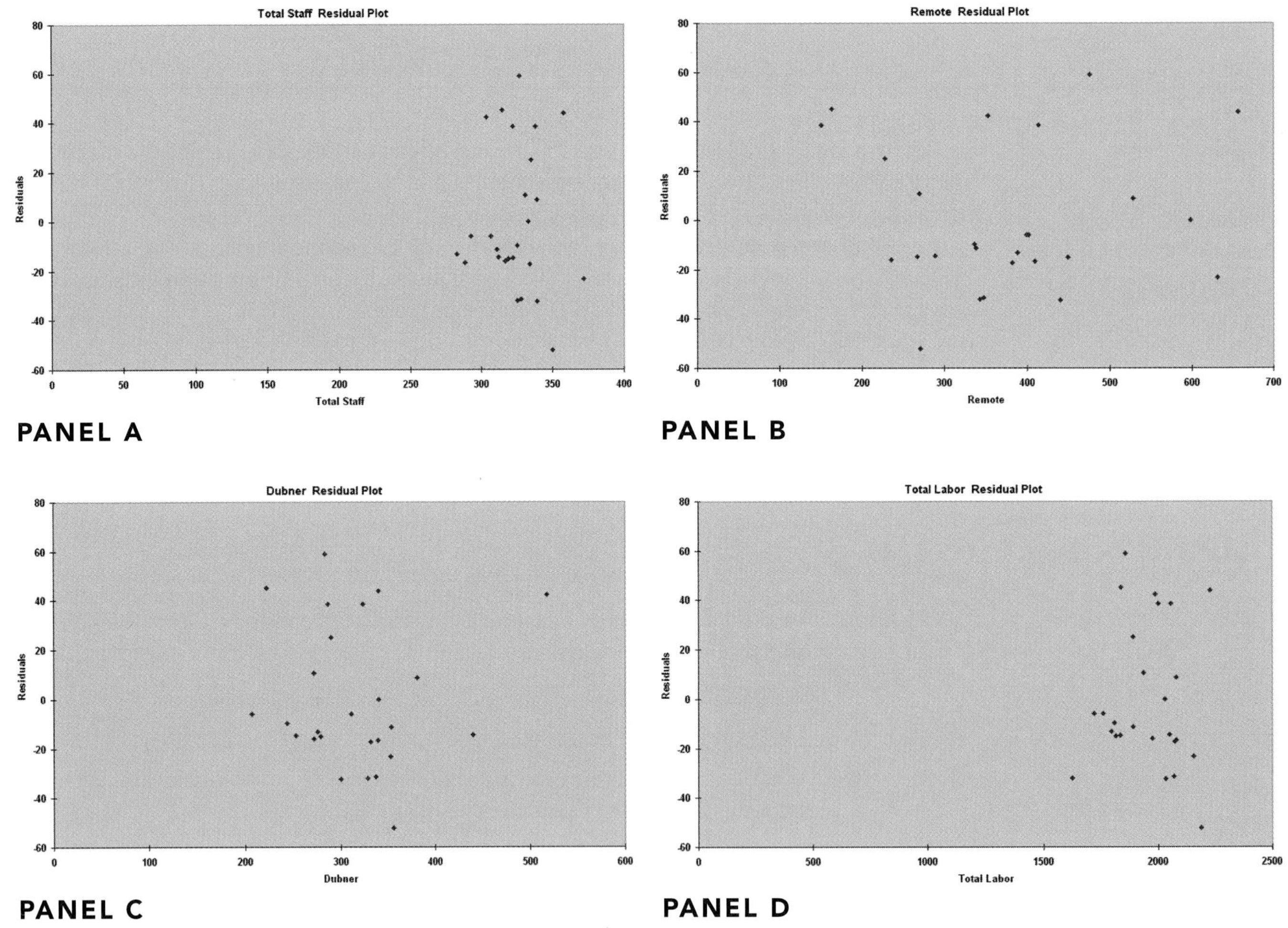

FIGURE 15.20 Microsoft Excel Residual Plots for the Standby Hours Data

None of the plots of the residuals versus the total staff, the remote hours, the Dubner hours, and the total labor hours reveal an apparent pattern. In addition, a histogram of the residuals (not shown here) indicates only moderate departure from normality. Because the residual analysis appears to confirm the aptness of the model, you can now use various influence measures to determine whether any of the values unduly influence the regression equation. Figure 15.21 presents the values of the h_i, t, and Cook's D_i statistics for the fitted model, with certain observations highlighted.

FIGURE 15.21

Minitab Influence Statistics for the Standby Hours Data

↓	C1	C2	C3	C4	C5	C6	C7	C8
	Standby	Staff	Remote	Dubner	Labor	TRES1	HI1	COOK1
1	245	338	414	323	2001	1.26648	0.057851	0.019147
2	177	333	598	340	2030	-0.00450	0.159009	0.000001
3	271	358	656	340	2226	1.74109	0.308626	0.246769
4	211	372	631	352	2154	-0.88635	0.317663	0.073903
5	196	339	528	380	2078	0.28517	0.117963	0.002275
6	135	289	409	339	2080	-0.66415	0.404904	0.061665
7	195	334	382	331	2073	-0.55135	0.066692	0.004493
8	118	293	399	311	1758	-0.20251	0.177819	0.001859
9	116	325	343	328	1624	-1.38665	0.453697	0.305928
10	147	311	338	353	1889	-0.36082	0.080160	0.002367
11	154	304	353	518	1988	2.06732	0.521708	0.806606
12	146	312	289	440	2049	-0.50740	0.239542	0.016814
13	115	283	388	276	1796	-0.47510	0.267786	0.017142
14	161	307	402	207	1720	-0.20841	0.219149	0.002554
15	274	322	151	287	2056	1.42189	0.241543	0.122799
16	245	335	228	290	1890	0.84801	0.155157	0.026771
17	201	350	271	355	2187	-2.00716	0.240144	0.222548
18	183	339	440	300	2032	-1.06250	0.073308	0.017752
19	237	327	475	284	1856	2.10290	0.101265	0.085690
20	175	328	347	337	2068	-1.02770	0.071640	0.016257
21	152	319	449	279	1813	-0.49356	0.105621	0.005969
22	188	325	336	244	1808	-0.31659	0.107193	0.002515
23	188	322	267	253	1834	-0.48404	0.100514	0.005434
24	197	317	235	272	1973	-0.52597	0.123908	0.008105
25	261	315	164	223	1839	1.63790	0.193025	0.118818
26	232	331	270	272	1935	0.34618	0.094110	0.002599

Using the decision rule suggested by Hoaglin and Welsch (see section 15.3), you flag any h_i value greater than $2(k + 1)/n = 2(4 + 1)/26 = 0.3846$. Referring to Figure 15.21, $h_6 = 0.4049$, $h_9 = 0.4537$, and $h_{11} = 0.5217$ have h_i values that are greater than 0.3846 and are therefore candidates for deletion from the analysis.

Now consider the Studentized deleted residual measure t_i. Using the decision rule suggested by Hoaglin and Welsch (see section 15.3), you flag any t_i value greater than 1.7247 or less than −1.7247 (see Table E.3 to find the critical t value with $n - k - 2 = 26 - 4 - 2 = 20$ degrees of freedom). Referring to Figure 15.21, you see that $t_3 = 1.7411$, $t_{11} = 2.0673$, $t_{17} = -2.0072$, and $t_{19} = 2.1029$. Thus, these observations may have an adverse effect on the model. Observation 11 was also flagged according to the h_i criterion.

Now you should consider a third criterion, Cook's D_i statistic, which is based on h_i and the standardized residual. For the model in which $k = 4$ and $n = 26$, using the decision rule suggested by Cook and Weisberg (see section 15.3), you flag any $D_i > F = 0.899$, the critical value for the F statistic having 5 and 21 degrees of freedom at the 0.50 level of significance (see Table E.12). Referring to Figure 15.21, none of the D_i values exceed 0.899 although D_i for observation 11 is 0.807. So according to Cook's D_i statistic, no values are candidates for deletion. Hence, you have no clear basis for removing any observations from the analysis. Thus, from Figure 15.14 or Figure 15.15 on page 638, the regression equation is

$$\hat{Y}_i = -330.83 + 1.2456X_{1i} - 0.1184X_{2i} - 0.2971X_{3i} + 0.1305X_{4i}$$

Example 15.4 presents a situation where there are several alternative models in which the C_p statistic is close to or less than $k + 1$.

EXAMPLE 15.4

CHOOSING AMONG ALTERNATIVE REGRESSION MODELS

Given the output in Table 15.3 from a best-subsets regression analysis of a regression model with seven independent variables, determine which regression model you would choose as the *best* model.

TABLE 15.3

Partial Output from Best-Subsets Regression

Number of Variables	r^2(%)	Adjusted r^2(%)	C_p	Variables Included
1	12.1	11.9	113.9	X_4
1	9.3	9.0	130.4	X_1
1	8.3	8.0	136.2	X_3
2	21.4	21.0	62.1	$X_3 X_4$
2	19.1	18.6	75.6	$X_1 X_3$
2	18.1	17.7	81.0	$X_1 X_4$
3	28.5	28.0	22.6	$X_1 X_3 X_4$
3	26.8	26.3	32.4	$X_3 X_4 X_5$
3	24.0	23.4	49.0	$X_2 X_3 X_4$
4	30.8	30.1	11.3	$X_1 X_2 X_3 X_4$
4	30.4	29.7	14.0	$X_1 X_3 X_4 X_6$
4	29.6	28.9	18.3	$X_1 X_3 X_4 X_5$
5	31.7	30.8	8.2	$X_1 X_2 X_3 X_4 X_5$
5	31.5	30.6	9.6	$X_1 X_2 X_3 X_4 X_6$
5	31.3	30.4	10.7	$X_1 X_3 X_4 X_5 X_6$
6	32.3	31.3	6.8	$X_1 X_2 X_3 X_4 X_5 X_6$
6	31.9	30.9	9.0	$X_1 X_2 X_3 X_4 X_5 X_7$
6	31.7	30.6	10.4	$X_1 X_2 X_3 X_4 X_6 X_7$
7	32.4	31.2	8.0	$X_1 X_2 X_3 X_4 X_5 X_6 X_7$

SOLUTION From Table 15.3, you need to determine which models have C_p values that are less than or close to $k + 1$. Two models meet this criterion. The model with six independent variables ($X_1, X_2, X_3, X_4, X_5, X_6$) has a C_p value of 6.8, which is less than $k + 1 = 6 + 1 = 7$, and the full model with seven independent variables ($X_1, X_2, X_3, X_4, X_5, X_6, X_7$) has a C_p value of 8.0. One way you can choose among models that meet these criteria is to determine whether the models contain a subset of variables that are common. Then you test whether the contribution of the additional variables is significant. In this case, because the models differ only by the inclusion of variable X_7 in the full model, you test whether variable X_7 makes a significant contribution to the regression model given that the variables X_1, X_2, X_3, X_4, X_5, and X_6 are already included in the model. If the contribution is statistically significant, then you should include variable X_7 in the regression model. If variable X_7 does not make a statistically significant contribution, you should not include it in the model.

Exhibit 15.1 summarizes the steps involved in model building.

EXHIBIT 15.1: STEPS INVOLVED IN MODEL BUILDING

1. Compile a listing of all independent variables under consideration.
2. Fit a regression model that includes all the independent variables under consideration and determine the variance inflationary factor (*VIF*) for each independent variable.
3. Determine whether any independent variables have a $VIF > 5$. Three possible results can occur:
   a. None of the independent variables have a $VIF > 5$; proceed to step 4.
   b. One of the independent variables has a $VIF > 5$; eliminate that independent variable and proceed to step 4.
   c. More than one of the independent variables has a $VIF > 5$; eliminate the independent variable that has the highest *VIF*, and go back to step 2.
4. Perform a best-subsets regression with the remaining independent variables and determine the C_p statistic and/or the adjusted r^2 for each model.
5. List all models that have C_p close to or less than $(k + 1)$ and/or a high adjusted r^2.

6. From those models listed in step 5, choose a best model.
7. Perform a complete analysis of the model chosen, including a residual analysis and influence analysis.
8. Depending on the results of the residual analysis and influence analysis, add quadratic terms, transform variables, delete individual observations if necessary, and reanalyze the data.
9. Use the selected model for prediction and inference.

Figure 15.22 represents a road map for the steps involved in model building.

FIGURE 15.22
Road Map for Model Building

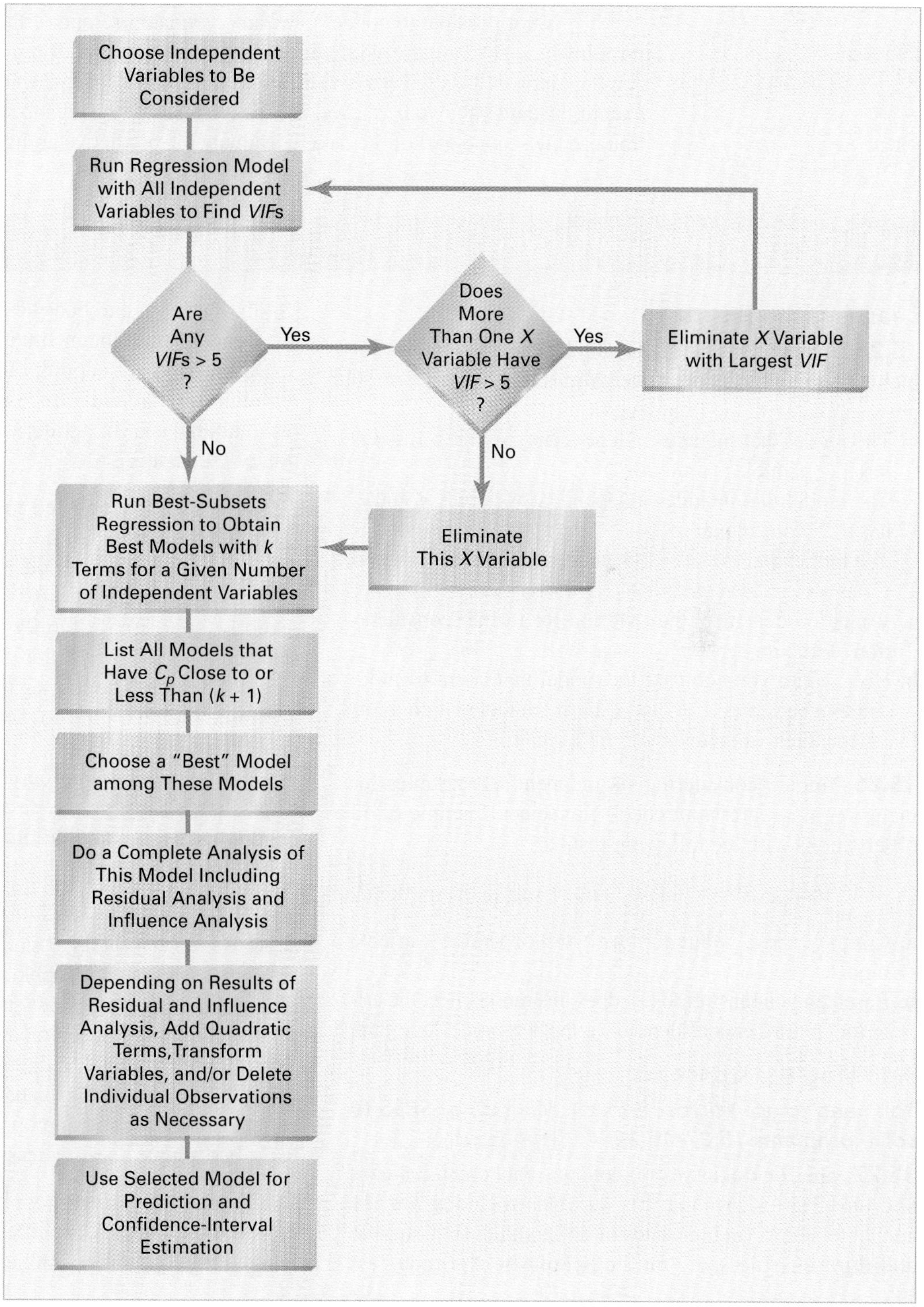

Model Validation

The final step in the model-building process is to validate the selected regression model. This step involves checking the model against data that was not part of the sample analyzed. Several ways of validating a regression model are:

- Collect new data and compare the results.
- Compare the results of the regression model to previous results.
- If the data set is large, split the data into two parts and cross-validate the results.

Perhaps the best way of validating a regression model is by collecting new data. If the results with new data are consistent to the selected regression model, you have strong reason to believe that the fitted regression model is applicable in a wide set of circumstances.

If it is not possible to collect new data, you can use one of the two other approaches. One possibility is to compare your regression coefficients and predictions to previous results. Another approach, called **cross-validation**, that you can use when the data set is large enough, is to split the data into two parts. You use the first part of the data to develop the regression model. You use the second part of the data to evaluate the predictive ability of the regression model.

PROBLEMS FOR SECTION 15.5

Learning the Basics

15.25 You are considering four independent variables for inclusion in a regression model. You select a sample of 30 observations with the following results:

The model that includes independent variables *A* and *B* has a C_p value equal to 4.6.

The model that includes independent variables *A* and *C* has a C_p value equal to 2.4.

The model that includes independent variables *A, B*, and *C* has a C_p value equal to 2.7.

a. Which models meet the criterion for further consideration? Explain.

b. How would you compare the model that contains independent variables *A, B*, and *C* to the model that contains independent variables *A* and *B*? Explain.

15.26 You are considering six independent variables for inclusion in a regression model. You select a sample of 40 observations with the following results:

$$n = 40 \quad k = 2 \quad T = 6 + 1 = 7 \quad R_k^2 = 0.274 \quad R_T^2 = 0.653$$

a. Compute the C_p value for this two-independent-variable model.

b. Based on your answer to (a), does this model meet the criterion for further consideration as the best model? Explain.

Applying the Concepts

You need to use Microsoft Excel, Minitab, or SPSS to solve problems 15.27–15.32.

15.27 The file COLLEGES2002 contains data on 80 colleges and universities. Among the variables included are the annual total cost (in thousands of dollars), the first quartile and third quartile score on the Scholastic Aptitude Test (SAT), the room and board expenses (in thousands of dollars), and whether the institution is public or private.

Develop the most appropriate multiple regression model to predict annual total cost. Be sure to perform a thorough residual analysis. In addition, provide a detailed explanation of the results.

SELF Test **15.28** You need to develop a model to predict the selling price of houses based on assessed value, time period in which a house is sold, and whether the house is new (0 = no; 1 = yes). A sample of 30 recently sold single-family houses in a small city is selected to study the relationship between selling price and assessed value. (The houses in the city were reassessed at full value one year prior to the study.) The results are in the data file HOUSE1.

Develop the most appropriate multiple regression model to predict selling price. Be sure to perform a thorough residual analysis. In addition, provide a detailed explanation of the results.

15.29 In problems 13.83–13.87 on page 563, you constructed simple linear regression models to investigate the relationship between demographic information and monthly sales for a chain of sporting goods stores. Develop the most appropriate multiple regression model to predict a store's monthly sales. Be sure to include a thorough residual analysis. In addition, provide a detailed explanation of the results, including a comparison of the most appropriate multiple regression model to the best simple linear regression model. SPORTING

15.30 The file AUTO2002 contains data on 121 automobile models from 2002. Among the variables included are gasoline mileage, weight, width, length of each automobile, and whether or not the car is a sports utility vehicle (SUV). Develop the most appropriate multiple regression model to predict gasoline mileage. Be sure to perform a thorough residual analysis. In addition, provide a detailed explanation of your results.

15.31 The Human Resources (HR) director for a large company that produces highly technical industrial instrumentation devices is interested in using regression modeling to help in recruiting decisions concerning their sales managers. The company has 45 sales regions, each headed by a sales manager. Many of the sales managers have degrees in electrical engineering, and due to the technical nature of the product line, several company officials believe that only applicants with degrees in electrical engineering should be considered. At the time of their application, candidates are asked to take the Strong-Campbell Interest Inventory Test and the Wonderlic Personnel Test. Due to the time and money involved with the testing, some discussion has taken place about dropping one or both of the tests. To start, the HR director gathered information on each of the 45 current sales managers, including years of selling experience, electrical engineering background, and the scores from both the Wonderlic and Strong-Campbell tests. The dependent variable was "sales-index" score, which is the ratio of the regions' actual sales divided by the target sales. The target values are constructed each year by upper management in consultation with the sales managers, and are based on past performance and market potential within each region. The data file MANAGERS contains information on the 45 current sales managers. The variables included are:

Sales: Ratio of yearly sales divided by the target sales value for that region. The target values were mutually agreed upon "realistic expectations."

Wonder: Score from the Wonderlic Personnel Test. The higher the score, the higher the applicant's perceived ability to manage.

SC: Score on the Strong-Campbell Interest Inventory Test. The higher the score, the higher the applicant's perceived interest in sales.

Experience: Number of years of selling experience prior to becoming a sales manager.

Engineer: Dummy variable that equals 1 if the sales manager has a degree in electrical engineering and 0 otherwise.

a. Develop the most appropriate model to predict sales.

b. Do you think that the company should continue administering the Wonderlic and Strong-Campbell tests? Explain.

c. Do the data support the argument that electrical engineers outperform the other sales managers? Would you support the idea to only hire electrical engineers? Explain.

d. How important is prior selling experience? Explain.

e. Discuss in detail how the HR director should incorporate the regression model you developed into the recruiting process.

15.32 The data in the file PRINTERS contains the price, text speed, text cost, color photo time, and color photo cost of 15 printers. Develop the most appropriate multiple regression model to predict the cost of printers.

15.6 PITFALLS IN MULTIPLE REGRESSION AND ETHICAL ISSUES

Pitfalls in Multiple Regression

Model building is an art as well as a science. Different individuals may not always agree on the best multiple regression model. Nevertheless, you should use the process described in Exhibit 15.1 on page 644. In doing so, you must avoid certain pitfalls that can interfere with the development of a useful model. Section 13.9 discussed pitfalls in simple linear regression and strategies for avoiding them. Now that you have studied a variety of multiple regression models, you need to take some additional precautions. To avoid pitfalls in multiple regression, you need to:

- Interpret the regression coefficient for a particular independent variable from a perspective in which the values of all other independent variables are held constant.
- Evaluate residual plots for each independent variable.
- Evaluate interaction terms.
- Compute the *VIF* for each independent variable before determining which independent variables to include in the model.
- Examine several alternative models using best-subsets regression.
- Use influence analysis to determine whether to remove any observations from the analysis.
- Use logistic regression instead of least-squares regression when the response variable is categorical.

Ethical Considerations

Ethical considerations arise when a user who wants to make predictions manipulates the development process of the multiple regression model. The key here is intent. In addition to the situations discussed in section 13.9, unethical behavior occurs when someone uses multiple regression analysis and *willfully fails* to remove from consideration variables that exhibit a high collinearity with other independent variables or *willfully fails* to use methods other than least-squares regression when the assumptions necessary for least-squares regression are seriously violated.

SUMMARY

In this chapter, various multiple regression topics were considered (see the Figure 15.23) including quadratic regression models, interactions, transformations, collinearity, and model building. You have learned how a director of broadcasting operations can build a multiple regression model as an aid in reducing operational expenses.

KEY FORMULAS

The Quadratic Regression Model

$$Y_i = \beta_0 + \beta_1 X_{1i} + \beta_2 X_{1i}^2 + \varepsilon_i \quad \textbf{(15.1)}$$

Quadratic Regression Equation

$$\hat{Y}_i = b_0 + b_1 X_{1i} + b_2 X_{1i}^2 \quad \textbf{(15.2)}$$

Regression Model with a Square-Root Transformation

$$Y_i = \beta_0 + \beta_1 \sqrt{X_{1i}} + \varepsilon_i \quad \textbf{(15.3)}$$

Original Multiplicative Model

$$Y_i = \beta_0 X_{1i}^{\beta_1} X_{2i}^{\beta_2} \varepsilon_i \quad \textbf{(15.4)}$$

Transformed Multiplicative Model

$$\begin{aligned} \log Y_i &= \log(\beta_0 X_{1i}^{\beta_1} X_{2i}^{\beta_2} \varepsilon_i) \\ &= \log \beta_0 + \log(X_{1i}^{\beta_1}) + \log(X_{2i}^{\beta_2}) + \log \varepsilon_i \\ &= \log \beta_0 + \beta_1 \log X_{1i} + \beta_2 \log X_{2i} + \log \varepsilon_i \end{aligned} \quad \textbf{(15.5)}$$

Original Exponential Model

$$Y_i = e^{\beta_0 + \beta_1 X_{1i} + \beta_2 X_{2i}} \varepsilon_i \quad \textbf{(15.6)}$$

Transformed Exponential Model

$$\begin{aligned} \ln Y_i &= \ln(e^{\beta_0 + \beta_1 X_{1i} + \beta_2 X_{2i}} \varepsilon_i) \\ &= \ln(e^{\beta_0 + \beta_1 X_{1i} + \beta_2 X_{2i}}) + \ln \varepsilon_i \\ &= \beta_0 + \beta_1 X_{1i} + \beta_2 X_{2i} + \ln \varepsilon_i \end{aligned} \quad \textbf{(15.7)}$$

Studentized Deleted Residual

$$t_i = e_i \sqrt{\frac{n - k - 1}{SSE(1 - h_i) - e_i^2}} \quad \textbf{(15.8)}$$

Cook's D_i Statistic

$$D_i = \frac{e_i^2}{k\, MSE}\left[\frac{h_i}{(1 - h_i)^2}\right] \quad \textbf{(15.9)}$$

Variance Inflationary Factor

$$VIF_j = \frac{1}{1 - R_j^2} \quad \textbf{(15.10)}$$

The C_p Statistic

$$C_p = \frac{(1 - R_k^2)(n - T)}{1 - R_T^2} - [n - 2(k + 1)] \quad \textbf{(15.11)}$$

KEY TERMS

best-subsets approach 640
Cook's D_i statistic 634
C_p statistic 641
collinearity 635
cross-validation 646
data mining 640
hat matrix diagonal elements h_i 633
logarithmic transformation 630
parsimony 637
quadratic regression model 621
square-root transformation 629
stepwise regression 638
Studentized deleted residual 633
variance inflationary factor (VIF) 635

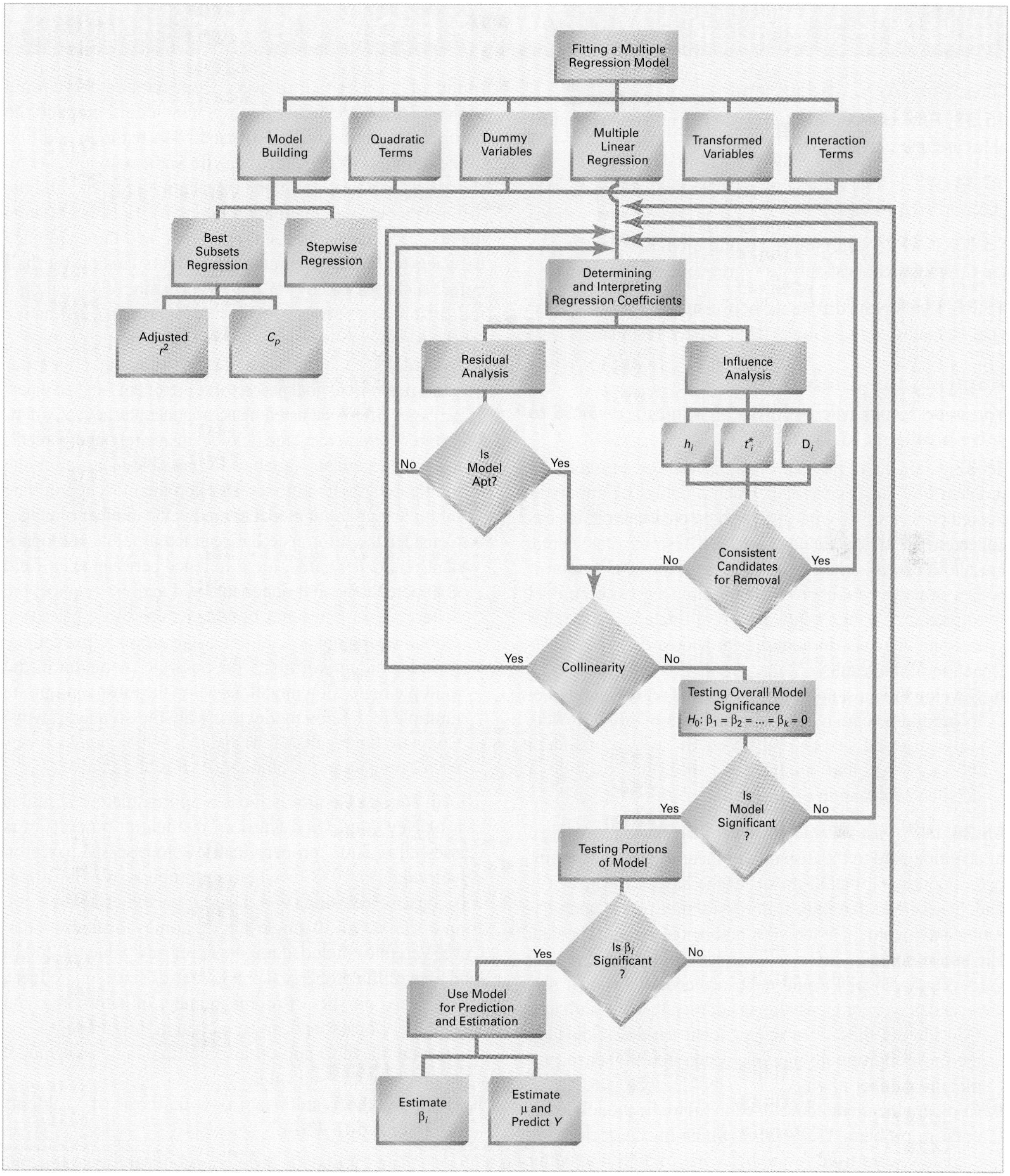

FIGURE 15.23 Road Map for Multiple Regression

CHAPTER REVIEW PROBLEMS

Checking Your Understanding

15.33 How can you evaluate whether independent variables are intercorrelated?

15.34 What is the difference between stepwise regression and best-subsets regression?

15.35 How do you choose among models according to the C_p statistic in best-subsets regression?

15.36 Explain the difference between the hat matrix diagonal elements and the Studentized deleted residuals.

Applying the Concepts

You need to use Microsoft Excel, Minitab, or SPSS to solve problems 15.37–15.57.

15.37 Crazy Dave has expanded his analysis, presented in problem 14.73 on page 613, of which variables are important in predicting a team's wins in a given baseball season. He has collected data BB2003 related to wins, ERA, saves, runs scored, hits allowed, walks allowed, and errors for the 2003 season.

a. Develop the most appropriate multiple regression model to predict a team's wins. Be sure to include a thorough residual analysis. In addition, provide a detailed explanation of the results.

b. Develop the most appropriate multiple regression model to predict a team's ERA on the basis of hits allowed, walks allowed, errors, and saves. Be sure to include a thorough residual analysis. In addition, provide a detailed explanation of the results.

15.38 In the last several years there has been a great deal of attention paid to the disparity in income and player compensation among the 30 major league baseball teams. It is widely assumed that only teams with high player compensation and income win the most number of games. The data file BB2001 contains information relating to regular season gate receipts, local TV and radio revenue, other local revenue, and player compensation in addition to team statistics.

a. Considering these four independent variables, develop the most appropriate multiple regression model to predict the number of wins.

b. Compare the model developed in (a) to the model developed in problem 15.37 (a) in which on-the-field team statistics were used to predict wins. Which is a better predictor of wins—income and player compensation variables or on-the-field team statistics? Explain.

15.39 Professional basketball has truly become a sport that generates interest among fans around the world. More and more players come from outside the United States to play in the National Basketball Association (NBA). In 2004, nine of the 29 players chosen in the first round of the NBA draft were from outside the United States. There are many factors that could impact the numbers of wins achieved by each NBA team. In addition to the number of wins, the data file NBA2004 contains team statistics for points per game (for team, opponent, and the difference between team and opponent), field goal (shots made) percentage (for team, opponent, and the difference between team and opponent), turnovers (losing the ball before a shot is taken) per game (for team, opponent, and the difference between team and opponent), offensive rebound percentage, and defensive rebound percentage.

a. Consider team points per game, opponent points per game, team field goal percentage, opponent field goal percentage, difference in team and opponent turnovers, offensive rebound percentage, and defensive rebound percentage as independent variables for possible inclusion in the multiple regression model. Develop the most appropriate multiple regression model to predict the number of wins.

b. Consider the difference between team points and opponent points per game, the difference between team field goal percentage and opponent field goal percentage, the difference in team and opponent turnovers, offensive rebound percentage, and defensive rebound percentage as independent variables for possible inclusion in the multiple regression model. Develop the most appropriate multiple regression model to predict the number of wins.

c. Compare the results of (a) and (b). Which model is better for predicting the number of wins? Explain.

15.40 Nassau County is located approximately 25 miles east of New York City. Until all residential property was reassessed in 2002, property taxes were assessed based on actual value in 1938 or when the property was built if it was constructed after 1938. Data in the file GLENCOVE are from a sample of 30 single-family homes located in Glen Cove. Variables included are the appraised value (in 2002), land area of the property (acres), interior size (square feet), age in years, number of rooms, number of bathrooms, and the number of cars that can be parked in the garage.

a. Develop the most appropriate multiple regression model to predict appraised value.

b. Compare the results in (a) with those of problems 15.41(a) and 15.42(a).

15.41 Data similar to problem 15.40 are available for homes located in Roslyn. ROSLYN

a. Perform an analysis similar to that of problem 15.40.

b. Compare the results in (a) with those of problems 15.40(a) and 15.42 (a).

15.42 Data similar to problem 15.40 are available for homes located in Freeport. FREEPORT

a. Perform an analysis similar to that of problem 15.40.

b. Compare the results in (a) with those of problems 15.40(a) and 15.41(a).

15.43 You are a real estate broker who wants to compare property values in Glen Cove and Roslyn (which are located approximately 8 miles apart). Use the data in the file. GCROSLYN Make sure to include the dummy variable for location (Glen Cove or Roslyn) in the regression model.

a. Develop the most appropriate multiple regression model to predict appraised value.
b. What conclusions can you reach concerning the differences in appraised value between Glen Cove and Roslyn?

15.44 You are a real estate broker who wants to compare property values in Glen Cove, Freeport, and Roslyn. Use the data in the file. GCFREEROSLYN

a. Develop the most appropriate multiple regression model to predict appraised value.
b. What conclusions can you reach concerning the differences in appraised value between Glen Cove, Freeport, and Roslyn.

15.45 The file COLLEGES2002 contains data on 80 colleges and universities. Among the variables included are the annual total cost (in thousands of dollars), first quartile (Q_1) and third quartile (Q_3) scores on the Scholastic Aptitude Test (SAT), room and board expenses (in thousands of dollars), whether the institution is public or private, and average indebtedness at graduation.

Develop the most appropriate multiple regression model to predict average indebtedness at graduation. Be sure to perform a thorough residual analysis. In addition, provide a detailed explanation of the results.

15.46 The file AUTO2002 contains data on automobile models from 2002. Among the variables included are gasoline mileage, horsepower, weight, width, length, cargo volume, turning circle, and whether the car is a sports utility vehicle.

Develop the most appropriate multiple regression model to predict gasoline mileage. Be sure to perform a thorough residual analysis. In addition, provide a detailed explanation of your results.

15.47 Over the past 30 years, public awareness and concern about air pollution have escalated dramatically. Venturi scrubbers are used for the removal of submicron particulate matter found in dust, fogs, fumes, odors, and smoke from gas streams. An experiment was conducted to determine the effect of air flow rate, water flow rate (liters/minute), the recirculating water flow rate (liters/minute), and the orifice size (mm) in the air side of the pneumatic nozzle on the performance of the scrubber as measured by the number of transfer units. The results are provided in the file SCRUBBER.

Develop the most appropriate multiple regression model to predict the number of transfer units. Be sure to perform a thorough residual analysis. In addition, provide a detailed explanation of your results.

Source: D. A. Marshall, R. J. Sumner, and C. A. Shook, "Removal of SiO_2 Particles with an Ejector Venturi Scrubber," Environmental Progress, *14, 1995, 28–32.*

15.48 A flux chamber is a Plexiglas dome about two feet in diameter that is placed over contaminated soil to sample soil gases. A study was carried out at a suspected radon hot spot to predict radon concentration (pCi/L) based on solar radiation (Ly/Day), soil temperature (°F), vapor pressure (mBar), wind speed (mph), relative humidity (%), dew point (°F), and ambient air temperature (°F). The data are contained in the RADON file.

Develop the most appropriate multiple regression model to predict radon concentration. Be sure to perform a thorough residual analysis. In addition, provide a detailed explanation of your results.

15.49 Oxford, Ohio, which is located 45 miles northwest of Cincinnati, is the home of Miami University. In addition to the 16,000 university students, the city has approximately 20,000 permanent residents. The data file HOMES contains information on all the single-family houses sold in the city limits for one year. The variables included are:

Price: Selling price of home in dollars
Location: Rating of the location from 1 to 5, with 1 the worst and 5 the best
Condition: Rating of the condition of the home from 1 to 5, with 1 the worst and 5 the best
Bedrooms: Number of bedrooms in the home
Bathrooms: Number of bathrooms in the home
Other Rooms: Number of rooms in the home other than bedrooms and bathrooms

Perform a multiple regression analysis using selling price as the dependent variable and the five remaining variables as independent variables.

a. State the multiple regression equation.
b. Interpret the meaning of the regression coefficients in this equation.
c. At the 0.05 level of significance, determine whether each explanatory variable makes a significant contribution to the regression model.
d. Determine the p-values in (c) and interpret their meaning.
e. Predict the price of a home with 3 bedrooms, 2.5 bathrooms, 4 other rooms, a location rating of 4, and a condition rating of 4. Construct a 95% confidence interval estimate and a 95% prediction interval.
f. Detemine and interpret the coefficient of multiple determination.
g. Determine the adjusted r^2.
h. Do a residual analysis on the results and determine the adequacy of the model.

i. Delete any independent variables that are not making a significant contribution to the regression model based on the best-subsets approach. Repeat (a) through (h) using this more parsimonious model. Which model do you think is better? Explain.

j. Perform an influence analysis on your results and determine whether to delete any observations from the analysis. If necessary, reanalyze the regression model after deleting these observations and compare your results.

15.50 Many factors determine the attendance at Major League Baseball games. These factors can include when the game is played, the weather, the opponent, whether or not the team is having a good season, and whether or not a marketing promotion is held. Popular promotions during the 2002 season were the traditional hat days and poster days, and the new craze, bobble-heads of star players (T. C. Boyd and T. C. Krehbiel, "Promotion Timing in Major League Baseball and the Stacking Effects of Factors That Increase Game Attractiveness," *Sport Marketing Quarterly*, 12, 2003, 173–184). The data file BASEBALL includes the following variables for the 2002 major league baseball season:

TEAM = (Kansas City Royals, Philadelphia Phillies, Chicago Cubs, or Cincinnati Reds)
ATTENDANCE = Paid attendance for the game
TEMP = High temperature for the day
WIN% = Team's winning percentage at the time of the game
OPWIN% = Opponent team's winning percentage at the time of the game
WEEKEND: 1 if game played on Friday, Saturday or Sunday; 0 otherwise
PROMOTION: 1 = if a promotion was held; 0 = if no promotion was held

a. Construct a multiple regression model for the Kansas City Royals, using attendance as the dependent variable and the remaining five variables as the independent variables.
b. State the multiple regression equation.
c. Interpret the meaning of the regression coefficients.
d. At the 0.05 level of significance, determine whether each independent variable makes a significant contribution to the regression model.
e. Determine and interpret the adjusted r^2.
f. Do a residual analysis on the results and determine the adequacy of the model.
g. Based on the best-subsets approach, delete any independent variables that are not making a significant contribution to the regression model. Repeat (b) through (f) using this more parsimonious model. Which model do you think is better? Explain.

15.51 Repeat problem 15.50 for the Philadelphia Phillies.

15.52 Repeat problem 15.50 for the Chicago Cubs.

15.53 Repeat problem 15.50 for the Cincinnati Reds.

15.54 Referring to problems 15.50–15.53, in terms of increasing attendance, which team ran the most effective promotions in 2002?

15.55 A headline on page 1 of *The New York Times* of March 4, 1990, read: "Wine equation puts some noses out of joint." The article explained that Professor Orley Ashenfelter, a Princeton University economist, had developed a multiple regression model to predict the quality of French Bordeaux based on the amount of winter rain, the average temperature during the growing season, and the harvest rain. The sample multiple regression equation is

$$Q = -12.145 + .00117WR + .6164TMP - .00386HR$$

where Q = logarithmic index of quality
WR = winter rain (October through March) in millimeters
TMP = average temperature during the growing season (April through September) in degrees Celsius
HR = harvest rain (August to September) in millimeters

You are at a cocktail party, sipping a glass of wine, when one of your friends mentions to you that she has read the article. She asks you to explain the meaning of the coefficients in the equation and also asks you about analyses that might have been done and were not included in the article. What is your reply?

Report Writing Exercises

15.56 In problem 15.29 on page 646, you developed a multiple regression model to predict monthly sales at sporting goods stores SPORTING. Your task is to write a report based on the model you developed. Append all appropriate charts and statistical information to your report.

TEAM PROJECT

15.57 The data file MUTUALFUNDS2004 contains information regarding 12 variables from a sample of 121 mutual funds. The variables are:

Fund—The name of the mutual fund.
Categories—Type of stocks comprising the mutual fund—small cap, mid cap, large cap
Objective—Objective of stocks comprising the mutual fund—growth or value
Assets—In millions of dollars
Fees—Sales charges (no or yes)
Expense ratio—Ratio of expenses to net assets (in percentage)
2003 Return—Twelve-month return in 2003
Three-year return—Annualized return 2001–2003
Five-year return—Annualized return 1999–2003

Risk—Risk-of-loss factor of the mutual fund classified as low, average, or high
Best quarter—Best quarterly performance 1999–2003
Worst quarter—Worst quarterly performance 1999–2003

Develop regression models to predict the 2003 return, the 3-year return, and the 5-year return based on fees, expense ratio, objective, and risk (for the average and low funds; do not include the high-risk funds in the analysis). Be sure to perform a thorough residual analysis. In addition, provide a detailed explanation of your results. Append all appropriate charts and statistical information to your report.

CASE STUDY
THE MOUNTAIN STATES POTATO COMPANY

Mountain States Potato Company sells a by-product of its potato-processing operation, called a filter cake, to area feedlots as cattle feed. Recently, the feedlot owners have noticed that the cattle are not gaining weight as quickly. They believe that the root cause of the problem is that the percentage of solids in the filter cake is too low.

Historically the percentage of solids in the filter cakes runs slightly above 12%. Lately, however, the solids are running in the 11% range. What is actually affecting the solids is a mystery, but something has to be done quickly. Individuals involved in the process were asked to identify variables that might affect the percentage of solids. This review turned up the six variables listed below. Data collected by monitoring the process several times daily for 20 days are stored in the POTATO file.

Variable	Comments
SOLIDS	Percentage solids in the filter cake.
PH	Acidity. This measure of acidity indicates bacterial action in the clarifier and is controlled by the amount of downtime in the system. As bacterial action progresses, organic acids are produced that can be measured using pH.
LOWER	Pressure of the vacuum line below the fluid line on the rotating drum.
UPPER	Pressure of the vacuum line above the fluid line on the rotating drum.
THICK	Filter cake thickness measured on the drum.
VARIDRIV	Setting used to control the drum speed. May differ from DRUMSPD because of mechanical inefficiencies.
DRUMSPD	Speed at which the drum is rotating when collecting filter cake. Measured with a stop-watch.

1. Thoroughly analyze the data and develop a regression model to predict the percentage of solids.
2. Write an executive summary concerning your findings to the president of the Mountain States Potato Company. Include specific recommendations on how to get the percentage of solids back above 12%.

WEB CASE

Apply your knowledge of multiple regression model building in this Web Case that extends the "Using Statistics" scenario concerning OmniPower energy bars from Chapter 14.

Still concerned about ensuring a successful test-marketing of its OmniPower energy bars, the marketing department of OmniFoods has also consulted with Connect2Coupons (C2C), another merchandising consultancy. C2C suggests that earlier analysis done by In-Store Placements Group (ISPG) was faulty because it did not use the correct type of data. C2C claims that its Internet-based viral marketing will have an even greater effect on OmniPower energy bar sales, as new data from the same 34-store sample will show. In response, ISPG says its earlier claims are valid and has reported to the OmniFood marketing department that it can discern no simple relation between C2C's viral marketing and increased OmniPower sales.

Review all these claims on the message board at the OmniFoods' internal Web site **www.prenhall.com/Springville/Omni_OmniPowerMB.htm** and then answer the following:

1. Which of the claims are true? False? True, but misleading? Support your answer by performing an appropriate statistical analysis.
2. If the grocery store chain allowed OmniFoods to use an unlimited number of sales techniques, which techniques should it use? Explain.
3. If the grocery store chain allowed OmniFoods to use only one sales technique, which technique should it use? Explain.

REFERENCES

1. Andrews, D. F., and D. Pregibon, "Finding the Outliers that Matter," *Journal of the Royal Statistical Society* 40 (Ser. B., 1978): 85–93.
2. Atkinson, A. C., "Robust and Diagnostic Regression Analysis," *Communications in Statistics* 11 (1982): 2559–2572.
3. Belsley, D. A., E. Kuh, and R. Welsch, *Regression Diagnostics: Identifying Influential Data and Sources of Collinearity* (New York: Wiley, 1980).
4. Cook, R. D., and S. Weisberg, *Residuals and Influence in Regression* (New York: Chapman and Hall, 1982).
5. Hoaglin, D. C., and R. Welsch, "The Hat Matrix in Regression and ANOVA," *The American Statistician*, 32 (1978): 17–22.
6. Hocking, R. R., "Developments in Linear Regression Methodology: 1959–1982," *Technometrics* 25 (1983): 219–250.
7. Kutner, M., C. Nachtsheim, J. Neter, and W. Li, *Applied Linear Statistical Models*, 5th ed. (New York: McGraw-Hill/Irwin, 2005).
8. Marquardt, D. W., "You Should Standardize the Predictor Variables in Your Regression Models," discussion of "A Critique of Some Ridge Regression Methods," by G. Smith and F. Campbell, *Journal of the American Statistical Association* 75 (1980): 87–91.
9. *Microsoft Excel 2003* (Redmond, WA: Microsoft Corp., 2002).
10. *Minitab for Windows Version 14* (State College, PA: Minitab, Inc., 2004).
11. Snee, R. D., "Some Aspects of Nonorthogonal Data Analysis, Part I. Developing Prediction Equations," *Journal of Quality Technology* 5 (1973): 67–79.
12. *SPSS Base 12.0 Brief Guide* (Upper Saddle River, NJ: Prentice Hall, 2003).

Appendix 15 Software for Multiple Regression Model Building

A15.1 MICROSOFT EXCEL

For Creating Quadratic Terms

To create a quadratic term, do the following:

Open the worksheet that contains your regression data in column wise order.
Locate the column to the right of the column that contains the data for the first explanatory variable of the quadratic regression model.
If this column is not blank, select **Insert → Columns**.
Enter a label in the row 1 cell of the blank column.
Starting in row 2, enter formulas in the form ***=previouscolumncell^2*** down the column through the row containing the data for the last observation.
Run a multiple regression (see Chapter 14) that includes this new column of formulas as the second explanatory variable.

For example, if your first explanatory variable was in column C, enter the formula **=C2^2** in cell D2 and copy the formula down through the row containing the data for the last observation.

For Creating Transformations

To transform a column of data, enter formulas in the form ***=FUNCTION(previouscolumncell)*** in a blank column to the right of the column containing the data. Use either the SQRT (square root), LOG (common logarithm), or LN (natural logarithm) functions. Follow the first four instructions in the Creating Quadratic Terms section to insert a blank column, if such a column does not already exist.

For example, to create a square-root transformation for an independent variable (that is in column B) in column C, enter the formula **=SQRT(B2)** in cell C2 and copy the formula down through the row containing the data for the last observation.

For Variance Inflationary Factors

Run a regression for every combination of *X* variables. On the regression results worksheets, enter the label **VIF** in cell **A9** and enter the formula **=1/(1 – B5)** into cell **B9**.

OR See section G.36 (**Multiple Regression**) if you want PHStat2 to do this for you.

For Stepwise Regression

See section G.37 (**Stepwise Regression**) if you want PHStat2 to produce a stepwise regression analysis. (There are no Microsoft Excel commands that directly produce a stepwise analysis.)

For Best-Subsets Regression

See section G.38 (**Best Subsets**) if you want PHStat2 to produce a best-subsets regression analysis. (There are no Microsoft Excel commands that directly produce a best-subsets analysis.)

A15.2 MINITAB

In Appendices A13.2 and A14.2, instructions are provided for using Minitab for simple linear regression and multiple regression. You can use those instructions for this chapter.

Using Minitab for the Quadratic Regression

To create a new X variable that is the square of another X variable, select **Calc ➔ Calculator**. In the Store result in variable: edit box, enter the column number or name for the new variable. In the Expression: edit box, enter {name or column number of the X variable you wish to square} ** 2. Click the **OK** button. A new X variable that is the square of the original X variable is now in the specified column. Continue with the regression analysis as discussed previously.

Using Minitab for Transforming Variables

To transform a variable, select **Calc ➔ Calculator**. In the Store result in variable: edit box, enter the column number or name for the new variable. Select the function to use for the transformation such as Log 10, Natural log, or Square root. After you have selected the function, enter the name of the X variable to transform in the parentheses of the function now displayed in the Expression: edit box. Click the **OK** button. Continue with the regression analysis as discussed previously.

Using Minitab for Influence Analysis

To compute the results of Figure 15.13 on page 633, open the **OMNI.MTW** worksheet. In addition to the instructions given in Appendix A14.2, click the **Storage** button. Select the **Deleted t residuals, Hi (leverages)**, and **Cook's distance** check boxes. Click the **OK** button to return to the Regression dialog box. Click the **OK** button.

Using Minitab to Compute the Variance Inflationary Factors (*VIF*)

Open the **STANDBY.MTW** worksheet. Select **Stat ➔ Regression ➔ Regression**. In addition to the other selections in the Regression dialog box, click the **Options** button. In the Options dialog box, select the **Variance inflation factors** check box. Click **OK** to return to the Regression dialog box.

Using Minitab for Stepwise Regression and Best-Subsets Regression

You can use Minitab for model building with either stepwise regression or best-subsets regression. To illustrate model building with the standby hours data, open the **STANDBY.MTW** worksheet. To perform stepwise regression, select **Stat ➔ Regression ➔ Stepwise**.

1. In the Stepwise Regression dialog box (see Figure A15.1) in the Response: edit box, enter **STANDBY** or **C1**.
2. In the Predictors: edit box, enter **Staff** or **C2, Remote** or **C3, Dubner** or **C4**, and **Labor** or **C5**. Click the **Methods** button.
3. In the Stepwise-Methods dialog box (see Figure A15.2), select the **Use alpha values and Stepwise** option buttons. Enter **0.05** in the Alpha to enter: edit box and **0.05** in the Alpha to remove edit box. Click the **OK** button to return to the Stepwise Regression dialog box. Click the **OK** button.

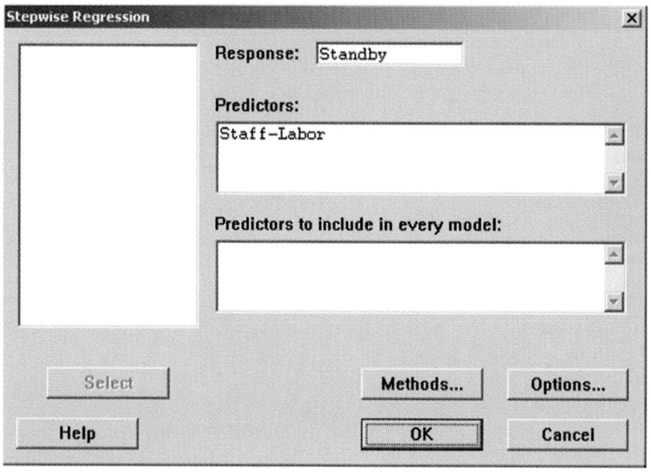

FIGURE A15.1 Minitab Stepwise Regression Dialog Box

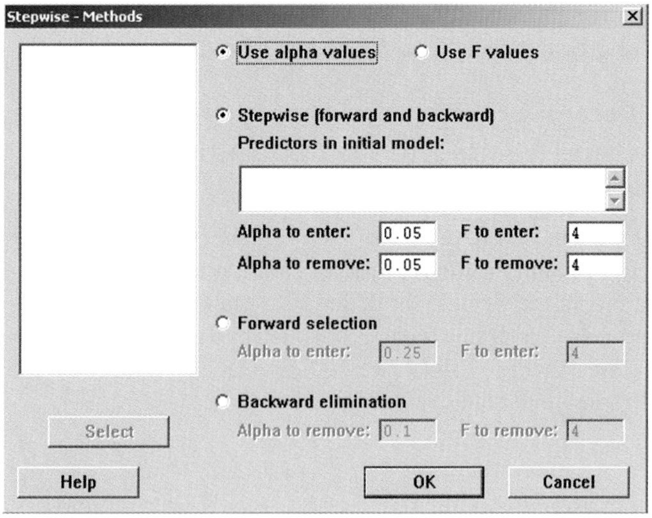

FIGURE A15.2 Minitab Stepwise-Methods Dialog Box

To perform a best-subsets regression, select **Stat ➔ Regression ➔ Best Subsets**. In the Best Subsets Regression dialog box (see Figure A15.3), in the Response: edit box, enter **Standby** or **C1**. In the Free Predictors: edit box, enter **Staff** or **C2, Remote** or **C3, Dubner** or **C4**, and **Labor** or **C5**. Click the **Options** button. Enter **3** in the Models of each size to print: edit box. Click the **OK** button to return to the Best Subsets Regression dialog box. Click the **OK** button.

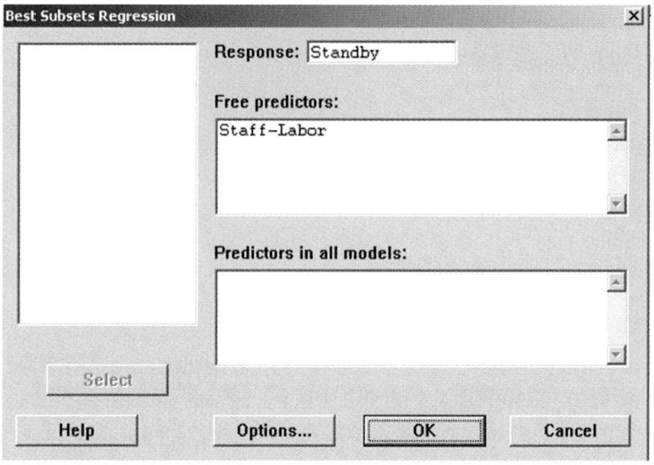

FIGURE A15.3 Minitab Best Subsets Regression Dialog Box

CHAPTER 16

Time-Series Forecasting and Index Numbers

LEARNING OBJECTIVES

In this chapter, you learn:

- About eight different forecasting models: moving averages, exponential smoothing, linear trend, quadratic trend, exponential trend, Holt-Winters, autoregressive, and least-squares model for seasonal data.
- To choose the most appropriate time-series forecasting model
- About price indexes and the difference between aggregated and unaggregated indexes

USING STATISTICS

Forecasting Revenues for Three Companies

You are a financial analyst for a large financial services company. You need to forecast revenues for three companies in order to better evaluate investment opportunities for your clients. To assist in the forecasting, you have collected time-series data on three companies: Wm. Wrigley Jr. Company, Cabot Corporation, and Wal-Mart. Each time series has unique characteristics due to the different types of business activities and growth patterns experienced by the three companies. You understand that you can use several different types of forecasting models. How do you decide which type of forecasting model is best for each company? How do you use the information gained from the forecasting models to evaluate investment opportunities for your clients?

In Chapters 13 through 15, you used regression analysis as a tool for model building and prediction. In this chapter, regression analysis and other statistical methodologies are applied to time-series data. A **time series** is a set of numerical data collected over time. Due to differences in the features of data for various companies such as the three companies described in the "Using Statistics" scenario, you need to consider several different approaches to forecasting time-series data.

Discussion of time series begins with annual time-series data. Two techniques for smoothing a series are illustrated—moving averages and exponential smoothing (see section 16.3). The analysis of annual time series continues with the use of least-squares trend fitting and forecasting (see section 16.4) and other, more sophisticated forecasting methods (see sections 16.5 and 16.6). These trend-fitting and forecasting models are then extended to a monthly or quarterly time series (see section 16.8).

16.1 THE IMPORTANCE OF BUSINESS FORECASTING

Forecasting is needed to monitor the changes that occur over time. Forecasting is commonly used in both the for-profit and nonprofit sectors of the economy. For example, officials in government forecast unemployment, inflation, industrial production, and revenues from income taxes in order to formulate policies. Marketing executives of a retailing corporation forecast product demand, sales revenues, consumer preferences, inventory, and so on, in order to make timely decisions regarding promotions and strategic planning. And the administrators of a college or university forecast student enrollments in order to plan for the construction of dormitories and other academic facilities, plan for student and faculty recruitment, and make assessments of other needs.

There are two common approaches to forecasting: *qualitative* and *quantitative*. **Qualitative forecasting methods** are especially important when historical data are unavailable. Qualitative forecasting methods are considered to be highly subjective and judgmental.

Quantitative forecasting methods make use of historical data. The goal of these methods is to use past data to predict future values. Quantitative forecasting methods are subdivided into two types: *time series* and *causal*. **Time-series forecasting methods** involve the forecast of future values of a variable based entirely on the past and present values of that variable. For example, the daily closing prices of a particular stock on the New York Stock Exchange constitute a time series.

Other examples of economic or business time series are the monthly publication of the Consumer Price Index, the quarterly gross domestic product (GDP), and the annual sales revenues of a particular company.

Causal forecasting methods involve the determination of factors that relate to the variable you are trying to forecast. These include multiple regression analysis with lagged variables, econometric modeling, leading indicator analysis, diffusion indexes, and other economic barometers that are beyond the scope of this text (see references 1–4). The primary emphasis in this chapter is on time-series forecasting methods.

16.2 COMPONENT FACTORS OF THE CLASSICAL MULTIPLICATIVE TIME-SERIES MODEL

The basic assumption of time-series forecasting is that the factors that have influenced activities in the past and present will continue to do so in more or less the same way in the future. Thus, the major goals of time-series forecasting are to identify and isolate these influencing factors in order to make predictions.

To achieve these goals, many mathematical models are available for exploring the fluctuations among the component factors of a time series. Perhaps the most basic is the **classical multiplicative model** for annual, quarterly, or monthly data. To demonstrate the classical multiplicative time-series model, Figure 16.1 plots the actual gross revenues for Wm. Wrigley Jr. Company from 1984 through 2003.

FIGURE 16.1
Microsoft Excel Plot of Actual Gross Revenues (in Millions of Current Dollars) for the Wm. Wrigley Jr. Company (1984–2003)

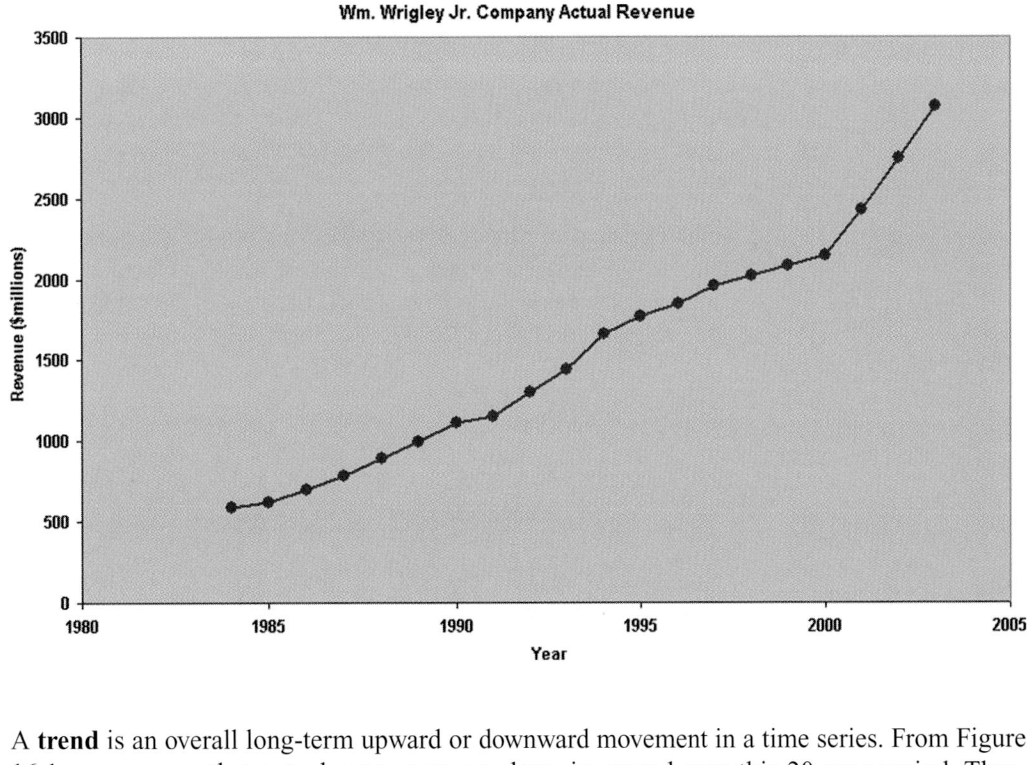

A **trend** is an overall long-term upward or downward movement in a time series. From Figure 16.1 you can see that actual gross revenues have increased over this 20-year period. Thus, actual gross revenues for the Wrigley Company exhibit an upward trend.

Trend is not the only component factor that influences data in a time series. Two other factors, the cyclical component and the irregular component, are also present in the data. The **cyclical component** depicts the up-and-down swings or movements through the series. Cyclical movements vary in length, usually lasting from 2 to 10 years. They differ in intensity and are often correlated with a business cycle. In some years, the values are higher than would be predicted by a trend line (i.e., they are at or near the peak of a cycle); in other years the values are lower than would be predicted by a trend line (i.e., they are at or near the bottom of a cycle). Any data that do not follow the trend modified by the cyclical component are considered part of the **irregular** or **random component**. When you have monthly or quarterly data, an additional component, the **seasonal component**, is considered along with the trend, cyclical, and irregular components.

Table 16.1 summarizes the four component factors that can influence an economic or business time series.

TABLE 16.1 Factors Influencing Time-Series Data

Component	Classification of Component	Definition	Reason for Influence	Duration
Trend	Systematic	Overall or persistent, long-term upward or downward pattern of movement	Changes in technology, population, wealth, value	Several years
Seasonal	Systematic	Fairly regular periodic fluctuations that occur within each 12-month period year after year	Weather conditions, social customs, religious customs, school schedules	Within 12 months (or monthly or quarterly data)
Cyclical	Systematic	Repeating up-and-down swings or movements through four phases: from peak (prosperity) to contraction (recession) to trough (depression) to expansion (recovery or growth)	Interactions of numerous combinations of factors that influence the economy	Usually 2–10 years, with differing intensity for a complete cycle
Irregular	Unsystematic	The erratic, or "residual" fluctuations in a series that exist after taking into account the systematic effects	Random variations in data due to unforeseen events such as strikes, natural disasters, and wars	Short duration and nonrepeating

The **classical multiplicative time-series model** states that any value in a time series is the product of these components. When forecasting an annual time series, you do not include the seasonal component. Equation (16.1) defines Y_i, the value of an annual time series recorded in year i, as the product of the trend, cyclical, and irregular components.

CLASSICAL MULTIPLICATIVE TIME-SERIES MODEL FOR ANNUAL DATA

$$Y_i = T_i \times C_i \times I_i \tag{16.1}$$

where in year i

T_i = value of the trend component

C_i = value of the cyclical component

I_i = value of the irregular component

When forecasting quarterly or monthly data, you include the seasonal component in the model. Equation (16.2) defines Y_i, a value recorded in time period i, as the product of all four components.

CLASSICAL MULTIPLICATIVE TIME-SERIES MODEL FOR DATA WITH A SEASONAL COMPONENT

$$Y_i = T_i \times S_i \times C_i \times I_i \tag{16.2}$$

where T_i, C_i, I_i = value of the trend, cyclical, and irregular components in time period i

S_i = value of the seasonal component in time period i

Your first step in a time-series analysis is to plot the data and observe any patterns that occur over time. You must determine whether there is a long-term upward or downward movement in the series (i.e., a trend). If there is no obvious long-term upward or downward trend, then you can use the method of moving averages or the method of exponential smoothing to smooth the series and provide an overall long-term impression (see section 16.3). If a trend is present, you can consider several time-series forecasting methods (see sections 16.4–16.6 for forecasting annual data, and section 16.8 for forecasting monthly or quarterly time series).

16.3 SMOOTHING THE ANNUAL TIME SERIES

One of the companies of interest in the "Using Statistics" scenario is the Cabot Corporation. Headquartered in Boston, Massachusetts, the Cabot Corporation is a global company with businesses specializing in the manufacture and distribution of chemicals, performance materials, specialty fluids, microelectronic materials, and liquefied natural gas. The company operates 36 manufacturing facilities in 21 countries. The Cabot Corporation **w1.cabot-corp.com** is traded on the New York Stock Exchange with ticker symbol CBT. Revenues in 2003 were approximately $1.8 billion. Table 16.2 gives the total revenues in millions of dollars for 1982 to 2003 CABOT. Figure 16.2 presents the time-series plot.

TABLE 16.2
Revenues (in Millions of Dollars) for the Cabot Corporation from 1982–2003

Year	Revenue	Year	Revenue	Year	Revenue
1982	1,588	1990	1,685	1998	1,653
1983	1,558	1991	1,488	1999	1,699
1984	1,753	1992	1,562	2000	1,698
1985	1,408	1993	1,619	2001	1,523
1986	1,310	1994	1,687	2002	1,557
1987	1,424	1995	1,841	2003	1,795
1988	1,677	1996	1,865		
1989	1,937	1997	1,637		

Source: Extracted from Moody's Handbook of Common Stocks, *1992 and* Mergent's Handbook of Common Stocks, *2004.*

FIGURE 16.2
Microsoft Excel Plot of Revenues (in Millions of Dollars) for the Cabot Corporation from 1982–2003

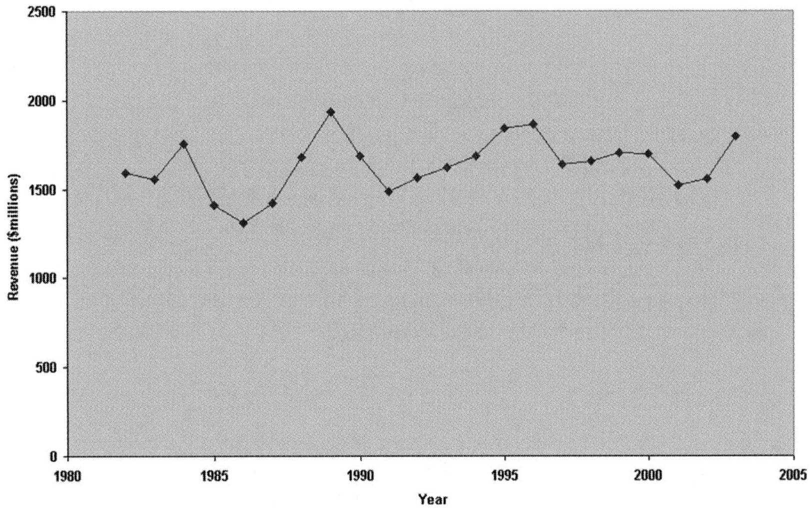

When you examine annual data, your visual impression of the long-term trend in the series is sometimes obscured by the amount of variation from year to year. Often, you cannot judge whether any long-term upward or downward trend exists in the series. To get a better overall impression of the pattern of movement in the data over time, you can use the methods of moving averages or exponential smoothing.

Moving Averages

Moving averages for a chosen period of length L consist of a series of means computed over time such that each mean is calculated for a sequence of L observed values. Moving averages are represented by the symbol $MA(L)$.

The method of moving averages for smoothing a time series is highly subjective and dependent on L, the length of the period selected for constructing the averages. If cyclical fluctuations are present in the data, you should choose an integer value of L that corresponds to (or is a multiple of) the estimated length of a cycle in the series.

To illustrate, suppose you want to compute 5-year moving averages from a series that has $n = 11$ years. Since $L = 5$, the 5-year moving averages consist of a series of means computed by averaging consecutive sequences of five values. You compute the first 5-year moving average by summing the values for the first 5 years in the series and dividing by 5:

$$MA(5) = \frac{Y_1 + Y_2 + Y_3 + Y_4 + Y_5}{5}$$

You compute the second 5-year moving average by summing the values of years 2 through 6 in the series and then dividing by 5:

$$MA(5) = \frac{Y_2 + Y_3 + Y_4 + Y_5 + Y_6}{5}$$

You continue this process until you have computed the last of these 5-year moving averages by summing the values of the last 5 years in the series (i.e., years 7 through 11) and then dividing by 5:

$$MA(5) = \frac{Y_7 + Y_8 + Y_9 + Y_{10} + Y_{11}}{5}$$

When you are dealing with annual time-series data, L, the length of the period chosen for constructing the moving averages, should be an *odd* number of years. By following this rule, you are not able to compute any moving averages for the first $(L - 1)/2$ years or the last $(L - 1)/2$ years of the series. Thus, for a 5-year moving average, you cannot make computations for the first 2 years or the last 2 years of the series.

When graphing moving averages, you plot each of the computed values against the middle year of the sequence of years used to compute it. If $n = 11$ and $L = 5$, the first moving average is centered on the third year, the second moving average is centered on the fourth year, and the last moving average is centered on the ninth year. Example 16.1 illustrates the computation of the 5-year moving averages.

EXAMPLE 16.1 COMPUTING A 5-YEAR MOVING AVERAGE

The following data represent total revenues (in millions of *constant* 1995 dollars) for a car rental agency over the 11-year period 1994 to 2004:

4.0	5.0	7.0	6.0	8.0	9.0	5.0	2.0	3.5	5.5	6.5

Compute the 5-year moving averages for this annual time series.

SOLUTION To compute a 5-year moving average, you first compute the 5-year moving total and then divide this total by 5. The first of the 5-year moving averages is:

$$MA(5) = \frac{Y_1 + Y_2 + Y_3 + Y_4 + Y_5}{5} = \frac{4.0 + 5.0 + 7.0 + 6.0 + 8.0}{5} = \frac{30.0}{5} = 6.0$$

The moving average is then centered on the middle value—the third year of this time series.

To compute the second of the 5-year moving averages, you compute the moving total of the second through sixth years and divide this value by 5:

$$MA(5) = \frac{Y_2 + Y_3 + Y_4 + Y_5 + Y_6}{5} = \frac{5.0 + 7.0 + 6.0 + 8.0 + 9.0}{5} = \frac{35.0}{5} = 7.0$$

This moving average is centered on the new middle value—the fourth year of the time series. The remaining moving averages are:

$$MA(5) = \frac{Y_3 + Y_4 + Y_5 + Y_6 + Y_7}{5} = \frac{7.0 + 6.0 + 8.0 + 9.0 + 5.0}{5} = \frac{35.0}{5} = 7.0$$

$$MA(5) = \frac{Y_4 + Y_5 + Y_6 + Y_7 + Y_8}{5} = \frac{6.0 + 8.0 + 9.0 + 5.0 + 2.0}{5} = \frac{30.0}{5} = 6.0$$

$$MA(5) = \frac{Y_5 + Y_6 + Y_7 + Y_8 + Y_9}{5} = \frac{8.0 + 9.0 + 5.0 + 2.0 + 3.5}{5} = \frac{27.5}{5} = 5.5$$

$$MA(5) = \frac{Y_6 + Y_7 + Y_8 + Y_9 + Y_{10}}{5} = \frac{9.0 + 5.0 + 2.0 + 3.5 + 5.5}{5} = \frac{25.0}{5} = 5.0$$

$$MA(5) = \frac{Y_7 + Y_8 + Y_9 + Y_{10} + Y_{11}}{5} = \frac{5.0 + 2.0 + 3.5 + 5.5 + 6.5}{5} = \frac{22.5}{5} = 4.5$$

These moving averages are then centered on their respective middle values, the fifth, sixth, seventh, eighth, and ninth years in the time series. By using the 5-year moving averages, you are unable to compute a moving average for the first two or last two values in the time series.

In practice, you should use computer software such as Microsoft Excel or Minitab when computing moving averages to avoid the tedious computations. Figure 16.3 presents the annual Cabot Corporation revenue data for the 22-year period from 1982 through 2003, the computations for 3- and 7-year moving averages, and a plot of the original data and the moving averages.

FIGURE 16.3

Microsoft Excel 3-Year and 7-Year Moving Averages for Cabot Corporation Revenues

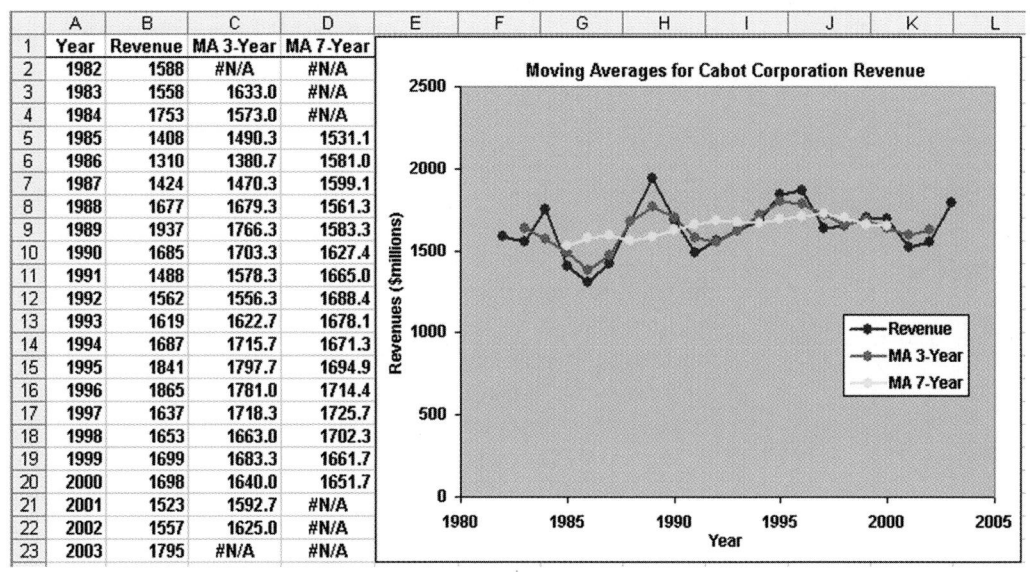

	A	B	C	D
1	Year	Revenue	MA 3-Year	MA 7-Year
2	1982	1588	#N/A	#N/A
3	1983	1558	1633.0	#N/A
4	1984	1753	1573.0	#N/A
5	1985	1408	1490.3	1531.1
6	1986	1310	1380.7	1581.0
7	1987	1424	1470.3	1599.1
8	1988	1677	1679.3	1561.3
9	1989	1937	1766.3	1583.3
10	1990	1685	1703.3	1627.4
11	1991	1488	1578.3	1665.0
12	1992	1562	1556.3	1688.4
13	1993	1619	1622.7	1678.1
14	1994	1687	1715.7	1671.3
15	1995	1841	1797.7	1694.9
16	1996	1865	1781.0	1714.4
17	1997	1637	1718.3	1725.7
18	1998	1653	1663.0	1702.3
19	1999	1699	1683.3	1661.7
20	2000	1698	1640.0	1651.7
21	2001	1523	1592.7	#N/A
22	2002	1557	1625.0	#N/A
23	2003	1795	#N/A	#N/A

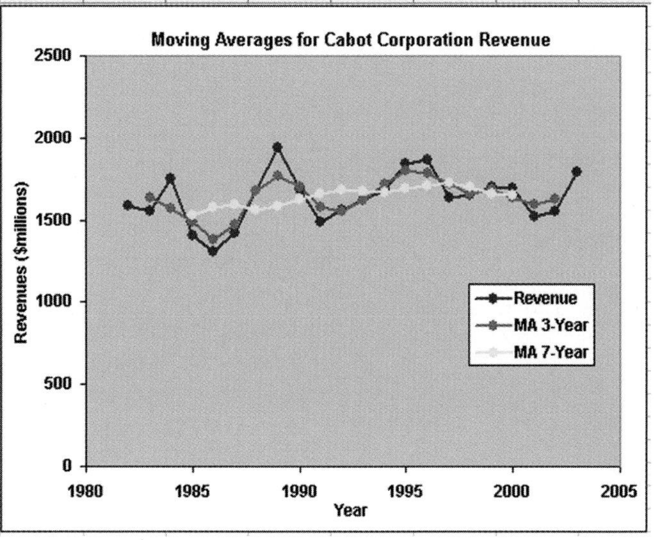

In Figure 16.3, there is no 3-year moving average for the first and the last year and no 7-year moving average for the first three years and last three years. You can see that the 7-year moving averages smooth the series a great deal more than the 3-year moving averages, because the period is longer. Unfortunately, the longer the period, the fewer the number of moving averages that you can compute. Therefore, selecting moving averages that are longer than 7 years is usually undesirable because too many moving average values are missing at the beginning and end of the series. This makes it more difficult to get an overall impression of the entire series.

Exponential Smoothing

Exponential smoothing is also used to smooth a time series. In addition, you can use exponential smoothing to compute short-term (one period into the future) forecasts when it is questionable what type of long-term trend effect, if any, is present in a time series. In this respect, exponential smoothing has a distinct advantage over the method of moving averages.

The name exponential smoothing comes from the fact that it consists of a series of *exponentially weighted* moving averages. The weights assigned to the values decrease, so that the most recent value receives the highest weight, the previous value receives the second highest weight, and so on, with the first value receiving the lowest weight. Throughout the series each exponentially smoothed value depends on all previous values, which is another advantage of exponential smoothing over the method of moving averages. Although the computations involved in exponential smoothing seem formidable, you can use Microsoft Excel or Minitab for the computations.

The equation developed for exponentially smoothing a series in any time period i is based on only three terms—the current value in the time series Y_i, the previously computed exponentially smoothed value, E_{i-1}, and an assigned weight or smoothing coefficient W. You use Equation (16.3) to exponentially smoothe a time series.

COMPUTING AN EXPONENTIALLY SMOOTHED VALUE IN TIME PERIOD *i*

$$E_1 = Y_1 \qquad \textbf{(16.3)}$$

$$E_i = WY_i + (1 - W)E_{i-1} \qquad i = 2, 3, 4, \ldots$$

where E_i = value of the exponentially smoothed series being computed in time period i

E_{i-1} = value of the exponentially smoothed series already computed in time period $i - 1$

Y_i = observed value of the time series in period i

W = subjectively assigned weight or smoothing coefficient (where $0 < W < 1$)

Choosing the smoothing coefficient (i.e., weight) that you assign to the time series is critical. Unfortunately, this selection is somewhat subjective. If your goal is only to smooth a series by eliminating unwanted cyclical and irregular variations, you should select a small value for W (close to 0). If your goal is forecasting, you should choose a large value for W (close to 1). In the former case, the overall long-term tendencies of the series will be more apparent; in the latter case, future short-term directions may be more adequately predicted.

Figure 16.4 presents the Microsoft Excel exponentially smoothed values (with smoothing coefficients of $W = 0.50$ and $W = 0.25$) for annual revenue of the Cabot Corporation over the 22-year period 1982 to 2003, along with a plot of the original data and the two exponentially smoothed time series.

FIGURE 16.4
Microsoft Excel Plot of Exponentially Smoothed Series (W = 0.50 and W = 0.25) of the Cabot Corporation

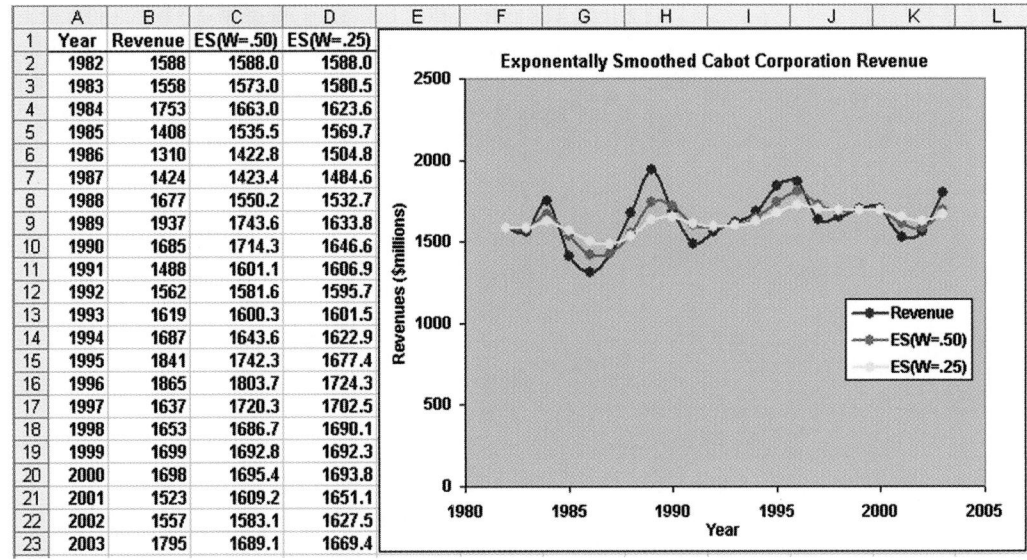

	A	B	C	D
1	Year	Revenue	ES(W=.50)	ES(W=.25)
2	1982	1588	1588.0	1588.0
3	1983	1558	1573.0	1580.5
4	1984	1753	1663.0	1623.6
5	1985	1408	1535.5	1569.7
6	1986	1310	1422.8	1504.8
7	1987	1424	1423.4	1484.6
8	1988	1677	1550.2	1532.7
9	1989	1937	1743.6	1633.8
10	1990	1685	1714.3	1646.6
11	1991	1488	1601.1	1606.9
12	1992	1562	1581.6	1595.7
13	1993	1619	1600.3	1601.5
14	1994	1687	1643.6	1622.9
15	1995	1841	1742.3	1677.4
16	1996	1865	1803.7	1724.3
17	1997	1637	1720.3	1702.5
18	1998	1653	1686.7	1690.1
19	1999	1699	1692.8	1692.3
20	2000	1698	1695.4	1693.8
21	2001	1523	1609.2	1651.1
22	2002	1557	1583.1	1627.5
23	2003	1795	1689.1	1669.4

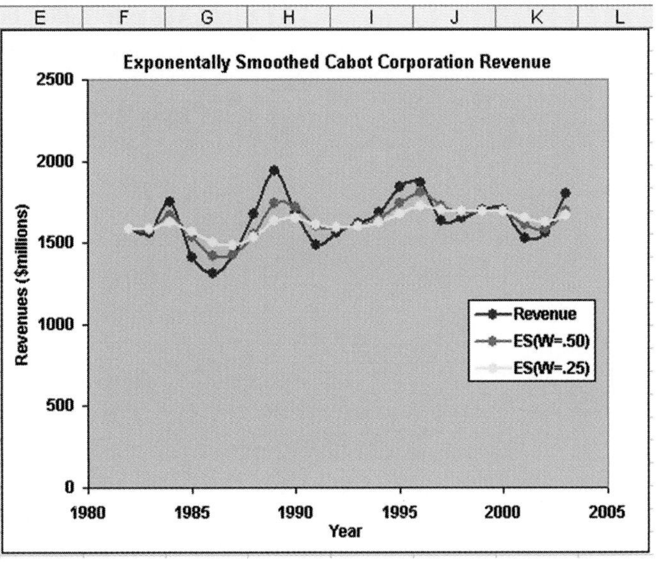

To illustrate these exponential smoothing computations for a smoothing coefficient of $W = 0.25$, begin with the initial value $Y_{1982} = 1{,}588$ as the first smoothed value ($E_{1982} = 1{,}588$). Then, using the value of the time series for 1983 ($Y_{1983} = 1{,}558$), you smooth the series for 1983 by computing

$$\begin{aligned} E_{1983} &= WY_{1983} + (1 - W)E_{1982} \\ &= (0.25)(1{,}558) + (0.75)(1{,}588) = 1{,}580.5 \end{aligned}$$

To smooth the series for 1984

$$\begin{aligned} E_{1984} &= WY_{1984} + (1 - W)E_{1983} \\ &= (0.25)(1{,}753) + (0.75)(1{,}580.5) = 1{,}623.6 \end{aligned}$$

To smooth the series for 1985

$$\begin{aligned} E_{1985} &= WY_{1985} + (1 - W)E_{1984} \\ &= (0.25)(1{,}408) + (0.75)(1{,}623.6) = 1{,}569.7 \end{aligned}$$

This process continues until you have computed all the exponentially smoothed values for all 22 years in the series, as shown in Figure 16.4.

To use exponential smoothing for forecasting, you use the smoothed value in the current time period as the forecast of the value in the following period.

FORECASTING TIME PERIOD i + 1

$$\hat{Y}_{i+1} = E_i \qquad \textbf{(16.4)}$$

To forecast the revenue of Cabot Corporation during 2004 using a smoothing coefficient of $W = 0.25$, you use the smoothed value for 2003 as its estimate. Figure 16.4 above shows that this projection is \$1,669.4 million. (How close is this forecast? Look up Cabot Corporation's revenue in a recent *Mergent's Handbook of Common Stocks* or search the World Wide Web to find out.)

When the value for 2004 becomes available, you can use Equation (16.3) to make a forecast for 2005 by computing the smoothed value for 2004 as follows.

$$\text{Current smoothed value} = (W)(\text{current value}) + (1 - W)(\text{previous smoothed value})$$

$$E_{2004} = WY_{2004} + (1 - W)E_{2003}$$

Or in terms of forecasting, you compute the following.

$$\text{New forecast} = (W)(\text{current value}) + (1 - W)(\text{current forecast})$$

$$\hat{Y}_{2005} = WY_{2004} + (1 - W)\hat{Y}_{2004}$$

PROBLEMS FOR SECTION 16.3

Learning the Basics

16.1 If you are using exponential smoothing for forecasting an annual time series of revenues, what is your forecast for next year if the smoothed value for this year is 32.4 million dollars?

PH Grade ASSIST **16.2** Consider a 9-year moving average used to smooth a time series that was first recorded in 1955.

a. Which year serves as the first centered value in the smoothed series?
b. How many years of values in the series are lost when computing all the 9-year moving averages?

PH Grade ASSIST **16.3** You are using exponential smoothing on an annual time series concerning total revenues (in millions of *constant* 1995 dollars). If you use a smoothing coefficient of $W = 0.20$ and the exponentially smoothed value for 2004 is $E_{2004} = (0.20)(12.1) + (0.80)(9.4)$.

a. What is the smoothed value of this series in 2004?
b. What is the smoothed value of this series in 2005 if the value of the series in that year is 11.5 millions of *constant* 1995 dollars?

Applying the Concepts

You can solve problems 16.4–16.8 manually or by using Microsoft Excel, Minitab, or SPSS.

SELF Test **16.4** The following data represent the price of a gallon of milk (in 2004 dollars) from 1996–2004 **MILK**.

Year	Price	Year	Price	Year	Price
1996	3.03	1999	3.05	2002	2.90
1997	3.05	2000	3.01	2003	2.72
1998	3.07	2001	3.02	2004	3.02

Source: Bureau of Labor Statistics, 2004, ***www.bls.gov****.*

a. Plot the time series.
b. Fit a 3-year moving average to the data and plot the results.
c. Using a smoothing coefficient of $W = 0.50$, exponentially smooth the series and plot the results.
d. Repeat (c) using $W = 0.25$.
e. Compare the results of (c) and (d).

16.5 The data in the following table represent the mean number of television viewers (excluding local broadcasts) per game (in millions) for the National Football League (NFL). **SPORTSTV**

Year	NFL	Year	NFL
1995	19.6	1999	18.3
1996	18.5	2000	17.0
1997	17.4	2001	16.9
1998	18.1	2002	18.6

Source: Extracted from S. Fatsis, "Salaries, Promos, and Flying Solo," The Wall Street Journal, *February 9, 2004, R.4,*

a. Plot the time series.
b. Fit a 3-year moving average to the data and plot the results.
c. Using a smoothing coefficient of $W = 0.50$, exponentially smooth the series and plot the results.
d. Repeat (c) using $W = 0.25$.
e. Compare the results of (c) and (d).

16.6 The NASDAQ stock market includes small- and medium-sized companies, many of which are in high-tech industries. Because of the nature of these companies, the NASDAQ tends to be more volatile than the Dow Jones Industrial Average or the S&P 500. The weekly values for the NASDAQ during the first 25 weeks of 2004 are listed in the data file **NASDAQWEEKLY**.

Week Ending	NASDAQ	Week Ending	NASDAQ
2-Jan-04	2,007	2-Apr-04	2,057
9-Jan-04	2,087	8-Apr-04	2,053
16-Jan-04	2,140	16-Apr-04	1,996
23-Jan-04	2,184	23-Apr-04	2,050
30-Jan-04	2,066	30-Apr-04	1,920
6-Feb-04	2,064	7-May-04	1,918
13-Feb-04	2,054	14-May-04	1,904
20-Feb-04	2,038	21-May-04	1,912
27-Feb-04	2,030	28-May-04	1,987
5-Mar-04	2,048	4-Jun-04	1,979
12-Mar-04	1,985	10-Jun-04	2,000
19-Mar-04	1,940	18-Jun-04	1,998
26-Mar-04	1,960		

*Source: Extracted from **finance.Yahoo.com**.*

a. Plot the time series.
b. Fit a 3-period moving average to the data and plot the results.
c. Using a smoothing coefficient of $W = 0.50$, exponentially smooth the series and plot the results.
d. Repeat (c) using $W = 0.25$.
e. What conclusions can you reach concerning the presence or absence of trends during the first 25 weeks of 2004?

16.7 The following data represent the three-month treasury bill rates in the United States from 1991 to 2003. TREASURY

Year	Rate	Year	Rate
1991	5.38	1998	4.78
1992	3.43	1999	4.64
1993	3.00	2000	5.82
1994	4.25	2001	3.40
1995	5.49	2002	1.61
1996	5.01	2003	1.01
1997	5.06		

*Source: Board of Governors of the Federal Reserve System **federalreserve.gov**.*

a. Plot the data.
b. Fit a 3-year moving average to the data and plot the results.
c. Using a smoothing coefficient of $W = 0.50$, exponentially smooth the series and plot the results.
d. What is your exponentially smoothed forecast for 2004?
e. Repeat (c), using a smoothing coefficient of $W = 0.25$.
f. Compare the results of (d) and (e).

16.8 The following data represent the average residential prices of electricity in cost per kilowatt hour in the October–March winter months from 1994–95 to 2003–04. ELECTRICITY

Year	Cost
1994–1995	8.16
1995–1996	8.10
1996–1997	8.17
1997–1998	8.12
1998–1999	7.94
1999–2000	7.98
2000–2001	8.11
2001–2002	8.37
2002–2003	8.20
2003–2004	8.49

*Source: Energy Information Administration, Department of Energy, **www.eia.doe.gov**.*

a. Plot the data.
b. Fit a 3-year moving average to the data and plot the results.
c. Using a smoothing coefficient of $W = 0.50$, exponentially smooth the series and plot the results.
d. What is your exponentially smoothed forecast for 2004–2005?
e. Repeat (c), using a smoothing coefficient of $W = 0.25$.
f. Compare the results of (d) and (e).

16.4 LEAST-SQUARES TREND-FITTING AND FORECASTING

The component factor of a time series most often studied is trend. You study trend in order to make intermediate and long-range forecasts. To get a visual impression of the overall long-term movements in a time series, you construct a time-series plot (see Figure 16.1 on page 659). If a straight-line trend adequately fits the data, the two most widely used methods of trend-fitting are the methods of least squares (see section 13.2) and *double exponential smoothing* (reference 1). If the time-series data indicate some long-run downward or upward quadratic movement, the two most widely used trend-fitting methods are the method of least squares (see section 15.1) and the method of *triple exponential smoothing* (reference 1). The focus of this section is forecasting linear, quadratic, and exponential trends using the method of least squares.

The Linear Trend Model

The **linear trend model**

$$Y_i = \beta_0 + \beta_1 X_i + \varepsilon_i$$

is the simplest forecasting model. Equation (16.5) defines the linear trend forecasting equation.

LINEAR TREND FORECASTING EQUATION

$$\hat{Y}_i = b_0 + b_1 X_i \quad \textbf{(16.5)}$$

Recall that in linear regression analysis, you used the method of least squares to compute the sample slope b_1 and the sample Y intercept b_0. You then substitute the values for X into Equation (16.5) to predict Y.

When using the least-squares method for fitting trends in a time series, you can simplify the interpretation of the coefficients by coding the X values so that the first value is assigned a code value of $X = 0$. You then assign all successive values consecutively increasing integer codes: 1, 2, 3, . . . , so that the nth and last value in the series has code $n - 1$. For example, for time-series data recorded annually over 20 years, you assign the first year a coded value of 0, the second year a coded value of 1, the third year a coded value of 2, and so on, with the final (20th) year assigned a coded value of 19.

In the "Using Statistics" scenario on page 658, one of the companies of interest is the Wm. Wrigley Jr. Company. Wrigley's is the world's largest manufacturer and marketer of chewing gum. Headquartered in Chicago, Illinois, Wrigley's operates factories in 12 countries and markets its products in over 150 countries. Revenues in 2003 topped $3 billion dollars (Wm. Wrigley Jr. Company, **www.wrigley.com**, June 3, 2004). Table 16.3 lists the actual gross revenues (in millions of current dollars) from 1984–2003 WRIGLEY. This time series is plotted in Figure 16.1 on page 659. To adjust for inflation, you then use the Bureau of Labor Statistics Consumer Price Index (CPI) to convert (and deflate) actual gross revenue dollars into real gross revenue dollars. This adjustment is achieved by multiplying each actual gross revenue value by the corresponding quantity $\left(\frac{100}{\text{CPI}}\right)$. The revised values are real gross revenues data in millions of constant 1982 to 1984 dollars. Figure 16.5 plots the real gross revenues along with the *actual* gross revenues in millions of current dollars.

TABLE 16.3

Actual Gross Revenues (in Millions of Dollars) for the Wm. Wrigley Jr. Company (1984–2003)

Year	Actual Revenue	Year	Actual Revenue
1984	591	1994	1,661
1985	620	1995	1,770
1986	699	1996	1,851
1987	781	1997	1,954
1988	891	1998	2,023
1989	993	1999	2,079
1990	1,111	2000	2,146
1991	1,149	2001	2,430
1992	1,301	2002	2,746
1993	1,440	2003	3,069

Source: Extracted from Moody's Handbook of Common Stocks, *1992, and* Mergent's Handbook of Common Stocks, *2004. Reprinted by permission of Financial Information Services, a division of Financial Communications Company, Inc.*

FIGURE 16.5
Microsoft Excel Time-Series Plots of Actual and Real Gross Revenues at Wm. Wrigley Jr. Company (in Millions of Dollars), 1984–2003

	A	B	C	D	E
1	Year	Coded Year	Actual Revenue	CPI-U	Real Revenue
2	1984	0	591	103.9	569
3	1985	1	620	107.6	576
4	1986	2	699	109.6	638
5	1987	3	781	113.6	688
6	1988	4	891	118.3	753
7	1989	5	993	124.0	801
8	1990	6	1111	130.7	850
9	1991	7	1149	136.2	844
10	1992	8	1301	140.3	927
11	1993	9	1440	144.5	997
12	1994	10	1661	148.2	1121
13	1995	11	1770	152.4	1161
14	1996	12	1851	156.9	1180
15	1997	13	1954	160.5	1217
16	1998	14	2023	163.0	1241
17	1999	15	2079	166.6	1248
18	2000	16	2146	172.2	1246
19	2001	17	2430	177.1	1372
20	2002	18	2746	179.9	1526
21	2003	19	3069	184.0	1668

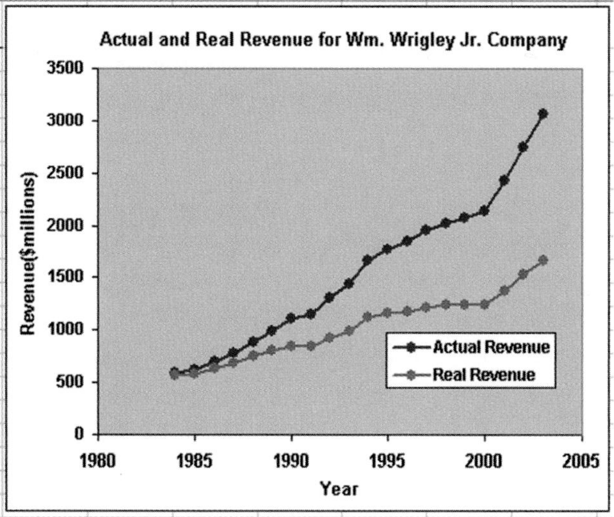

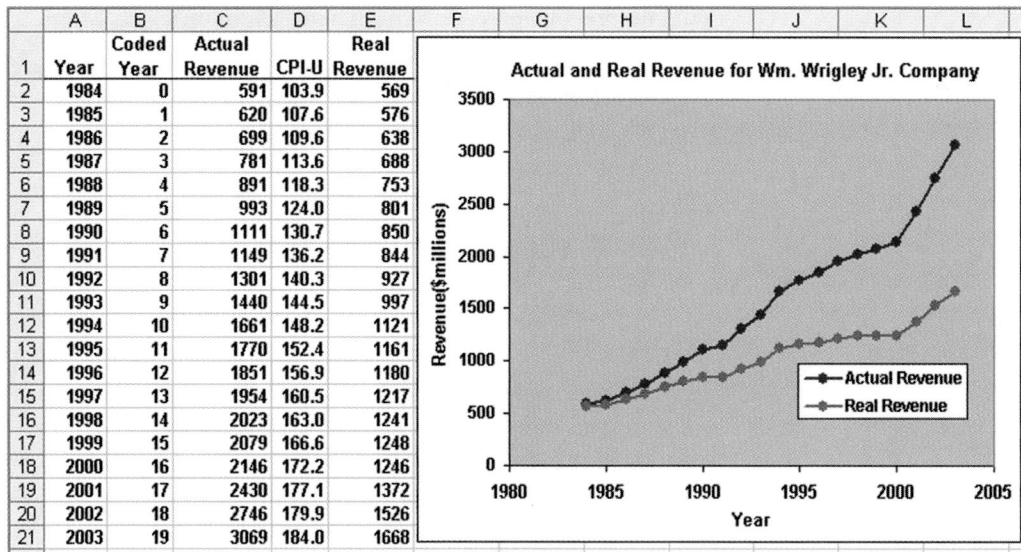

Coding the consecutive X values from 0 through 19 and then using Microsoft Excel or Minitab to perform a simple linear regression analysis on the adjusted time series (see Figure 16.6 or 16.7) results in the following linear trend forecasting equation.

$$\hat{Y}_i = 532.578 + 52.48X_i$$

where year zero is 1984.

FIGURE 16.6
Microsoft Excel Output for a Linear Regression Model to Forecast Real Gross Revenues (in Millions of Constant 1982–1984 Dollars) at Wm. Wrigley Jr. Company

	A	B	C	D	E	F	G
1	Linear Trend Model for Real Annual Gross Revenue Model						
2							
3	*Regression Statistics*						
4	Multiple R	0.9844					
5	R Square	0.9691					
6	Adjusted R Square	0.9674					
7	Standard Error	56.9842					
8	Observations	20					
9							
10	ANOVA						
11		*df*	*SS*	*MS*	*F*	*Significance F*	
12	Regression	1	1831527.9928	1831527.9928	564.0340	0.0000	
13	Residual	18	58449.5019	3247.1945			
14	Total	19	1889977.4947				
15							
16		*Coefficients*	*Standard Error*	*t Stat*	*P-value*	*Lower 95%*	*Upper 95%*
17	Intercept	532.5783	24.5571	21.6874	0.0000	480.9858	584.1708
18	Coded Year	52.4803	2.2098	23.7494	0.0000	47.8377	57.1228

FIGURE 16.7
Minitab Output for a Linear Regression Model to Forecast Real Gross Revenues (in Millions of Constant 1982–1984 Dollars) at Wm. Wrigley Jr. Company

```
The regression equation is
Real Revenue = 533 + 52.5 Coded Year


Predictor      Coef  SE Coef      T      P
Constant     532.58    24.56  21.69  0.000
Coded Year   52.480    2.210  23.75  0.000


S = 56.9842   R-Sq = 96.9%   R-Sq(adj) = 96.7%


Analysis of Variance

Source           DF       SS       MS       F      P
Regression        1  1831528  1831528  564.03  0.000
Residual Error   18    58450     3247
Total            19  1889977
```

You interpret the regression coefficients as follows:

- The Y intercept $b_0 = 532.578$ is the predicted real gross revenues (in millions of constant 1982–1984 dollars) at Wm. Wrigley Jr. during the origin or base year, 1984.
- The slope $b_1 = 52.48$ indicates that real gross revenues are predicted to increase by 52.48 million dollars per year.

To project the trend in the real gross revenues at Wm. Wrigley Jr. to 2004, substitute $X_2 = 20$, the code for 2004, into the linear trend forecasting equation

$$\hat{Y}_i = 532.578 + 52.48(20) = 1{,}582.178 \text{ millions of constant 1982–1984 dollars}$$

The trend line is plotted in Figure 16.8 along with the observed values of the time series. There is a strong upward linear trend and the adjusted r^2 is 0.967, indicating that virtually all of the variation in real gross revenues is explained by the linear trend over the time series. To investigate whether a different trend model might provide an even better fit, a *quadratic* trend model and an *exponential* trend model are presented next.

FIGURE 16.8

Microsoft Excel Least-Squares Trend Line for Wm. Wrigley Jr. Company Real Gross Revenues Data

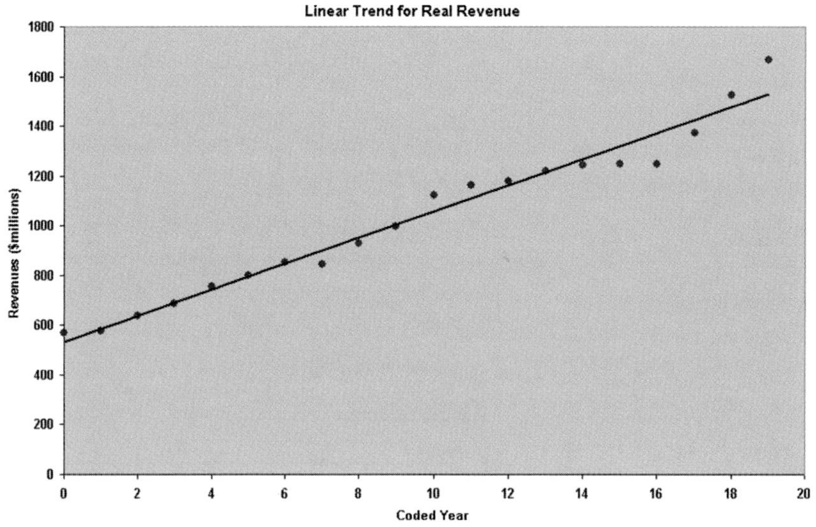

The Quadratic Trend Model

A **quadratic trend model**

$$Y_i = \beta_0 + \beta_1 X_i + \beta_2 X_i^2 + \varepsilon_i$$

is the simplest nonlinear model. Using the least-squares method described in section 15.1, you can compute a quadratic trend forecasting equation presented in Equation (16.6).

QUADRATIC TREND FORECASTING EQUATION

$$\hat{Y}_i = b_0 + b_1 X_i + b_2 X_i^2 \qquad \textbf{(16.6)}$$

where b_0 = estimated Y intercept

b_1 = estimated *linear* effect on Y

b_2 = estimated *quadratic* effect on Y

Once again, you use Microsoft Excel or Minitab to compute the quadratic trend forecasting equation. Figure 16.9 provides Excel output for the quadratic trend model used to forecast real gross revenues at the Wm. Wrigley Jr. Company. Figure 16.10 illustrates Minitab output.

FIGURE 16.9
Microsoft Excel Output for a Quadratic Regression Model to Forecast Real Gross Revenues at Wm. Wrigley Jr. Company

	A	B	C	D	E	F	G
1	Quadratic Trend Model for Real Annual Gross Revenue Model						
2							
3	*Regression Statistics*						
4	Multiple R	0.9854					
5	R Square	0.9710					
6	Adjusted R Square	0.9675					
7	Standard Error	56.8204					
8	Observations	20					
9							
10	ANOVA						
11		*df*	*SS*	*MS*	*F*	*Significance F*	
12	Regression	2	1835091.9785	917545.9892	284.1967	0.0000	
13	Residual	17	54885.5162	3228.5598			
14	Total	19	1889977.4947				
15							
16		*Coefficients*	*Standard Error*	*t Stat*	*P-value*	*Lower 95%*	*Upper 95%*
17	Intercept	558.2604	34.5989	16.1352	0.0000	485.2631	631.2576
18	Coded Year	43.9196	8.4406	5.2034	0.0001	26.1115	61.7276
19	Coded Year ^2	0.4506	0.4288	1.0507	0.3081	-0.4542	1.3553

FIGURE 16.10
Minitab Output for a Quadratic Regression Model to Forecast Real Gross Revenues at Wm. Wrigley Jr. Company

```
The regression equation is
Real Revenue = 558 + 43.9 Coded Year + 0.451 CodedYearSquare

Predictor           Coef   SE Coef      T      P
Constant          558.26     34.60  16.14  0.000
Coded Year        43.920     8.441   5.20  0.000
CodedYearSquare   0.4506    0.4288   1.05  0.308

S = 56.8204   R-Sq = 97.1%   R-Sq(adj) = 96.8%

Analysis of Variance

Source          DF       SS      MS       F      P
Regression       2  1835092  917546  284.20  0.000
Residual Error  17    54886    3229
Total           19  1889977
```

In Figure 16.9 or 16.10,

$$\hat{Y}_i = 558.26 + 43.92X_i + 0.45X_i^2$$

where year zero is 1984.

To use the quadratic trend equation for forecasting, you substitute the appropriate coded X values into this equation. For example, to predict the trend in real gross revenues for 2004 (i.e., $X_{20} = 20$),

$$\hat{Y}_i = 558.26 + 43.92(20) + 0.45(20)^2 = 1{,}616.66 \text{ millions of dollars}$$

Figure 16.11 plots the quadratic trend forecasting equation along with the time series for the actual data. This quadratic trend model does not provide a better fit (adjusted $r^2 = 0.968$) to the time series than does the linear trend model. The t statistic for the contribution of the quadratic term to the model is 1.05 (p-value = 0.308).

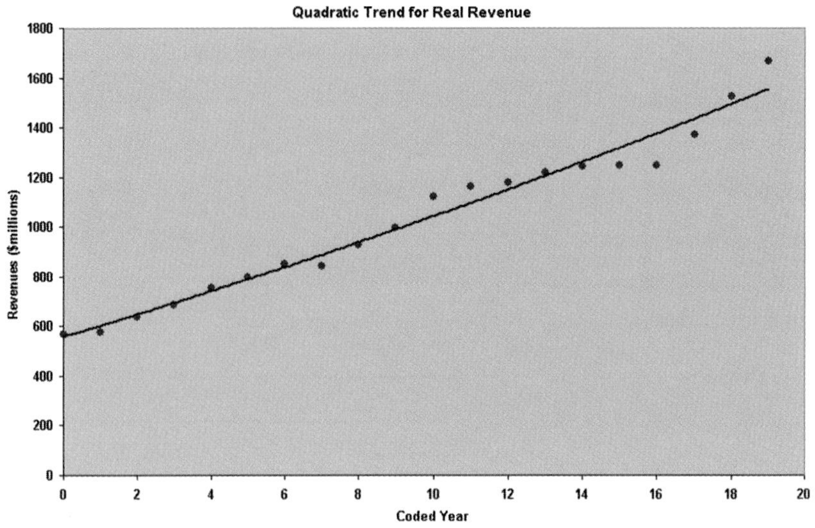

FIGURE 16.11
Microsoft Excel Fitted Quadratic Trend Forecasting Equation for the Wm. Wrigley Jr. Company

The Exponential Trend Model

When a time series increases at a rate such that the percentage difference from value to value is constant, an exponential trend is present. Equation (16.7) defines the **exponential trend model**.

EXPONENTIAL TREND MODEL

$$Y_i = \beta_0 \beta_1^{X_i} \varepsilon_i \qquad \textbf{(16.7)}$$

where $\beta_0 = Y$ intercept

$(\beta_1 - 1) \times 100\%$ is the annual compound growth rate (in %)

The model in Equation (16.7) is not in the form of a linear regression model. To transform this nonlinear model to a linear model, you use a base-10 logarithm transformation.[1] Taking the logarithm of each side of Equation (16.7) yields Equation (16.8).

[1]*Alternatively, you can use base e logarithms. For more information on logarithms see Appendix A.*

TRANSFORMED EXPONENTIAL TREND MODEL

$$\begin{aligned} \log(Y_i) &= \log(\beta_0 \beta_1^{X_i} \varepsilon_i) \qquad \textbf{(16.8)} \\ &= \log(\beta_0) + \log(\beta_1^{X_i}) + \log(\varepsilon_i) \\ &= \log(\beta_0) + X_i \log(\beta_1) + \log(\varepsilon_i) \end{aligned}$$

Equation (16.8) is a linear model you can estimate using the least-squares method with $\log(Y_i)$ as the dependent variable and X_i as the independent variable. This results in Equation (16.9).

EXPONENTIAL TREND FORECASTING EQUATION

$$\log(\hat{Y}_i) = b_0 + b_1 X_i \qquad \textbf{(16.9a)}$$

where b_0 = estimate of $\log(\beta_0)$ and thus $10^{b_0} = \hat{\beta}_0$

b_1 = estimate of $\log(\beta_1)$ and thus $10^{b_1} = \hat{\beta}_1$

therefore

$$\hat{Y}_i = \hat{\beta}_0 \hat{\beta}_1^{X_i} \qquad \textbf{(16.9b)}$$

where $(\hat{\beta}_1 - 1) \times 100\%$ is the estimated annual compound growth rate (in %)

Figure 16.12 represents Excel output for an exponential trend model of real gross revenues at the Wm. Wrigley Jr Company. Figure 16.13 illustrates Minitab output.

FIGURE 16.12
Microsoft Excel Output for an Exponential Regression Model to Forecast Real Gross Revenues at Wm. Wrigley Jr. Company

	A	B	C	D	E	F	G
1	**Exponential Trend Model for Real Annual Gross Revenue Model**						
2							
3	*Regression Statistics*						
4	Multiple R	0.9851					
5	R Square	0.9705					
6	Adjusted R Square	0.9688					
7	Standard Error	0.0245					
8	Observations	20					
9							
10	ANOVA						
11		*df*	*SS*	*MS*	*F*	*Significance F*	
12	Regression	1	0.3545	0.3545	591.9493	0.0000	
13	Residual	18	0.0108	0.0006			
14	Total	19	0.3653				
15							
16		*Coefficients*	*Standard Error*	*t Stat*	*P-value*	*Lower 95%*	*Upper 95%*
17	Intercept	2.7736	0.0105	263.0045	0.0000	2.7515	2.7958
18	Coded Year	0.0231	0.0009	24.3300	0.0000	0.0211	0.0251

FIGURE 16.13
Minitab Output for an Exponential Regression Model to Forecast Real Gross Revenues at Wm. Wrigley Jr. Company

```
The regression equation is
Log(RealRevenue) = 2.77 + 0.0231 Coded Year


Predictor        Coef     SE Coef       T      P
Constant      2.77365     0.01055  263.00  0.000
Coded Year  0.0230886   0.0009490   24.33  0.000


S = 0.0244718   R-Sq = 97.0%   R-Sq(adj) = 96.9%


Analysis of Variance

Source          DF       SS        MS       F      P
Regression       1  0.35450   0.35450  591.95  0.000
Residual Error  18  0.01078   0.00060
Total           19  0.36528
```

Using Equation (16.9a) and the results from Figure 16.12 or 16.13,

$$\log(\hat{Y}_i) = 2.7736 + 0.0231X_i$$

where year zero is 1984.

You compute the values for $\hat{\beta}_0$ and $\hat{\beta}_1$ by taking the antilog of the regression coefficients (b_0 and b_1):

$$\hat{\beta}_0 = \text{antilog } b_0 = \text{antilog}(2.7736) = 10^{2.7736} = 593.745$$
$$\hat{\beta}_1 = \text{antilog } b_1 = \text{antilog}(0.0231) = 10^{(0.0231)} = 1.0546$$

Thus, using Equation (16.9b), the exponential trend forecasting equation is

$$\hat{Y}_i = (593.745)(1.0546)^{X_i}$$

where year zero is 1984.

The Y intercept $\hat{\beta}_0 = 593.745$ millions of dollars is the real gross revenues forecast for the base year 1984. The value $(\hat{\beta}_1 - 1) \times 100\% = 5.46\%$ is the annual compound growth rate in real gross revenues at Wm. Wrigley Jr.

For forecasting purposes, you can substitute the appropriate coded X values into either Equation (16.9a) or Equation (16.9b). For example, to forecast real gross revenues for 2004 (i.e., $X_i = 20$) using Equation (16.9a),

$$\log(\hat{Y}_i) = 2.7736 + 0.0231(20) = 3.2356$$
$$\hat{Y}_i = \text{antilog}(3.2356) = 10^{3.2356} = 1{,}720.283 \text{ millions of dollars}$$

Figure 16.14 plots the exponential trend forecasting equation along with the time series for the real revenue data. The adjusted r^2 for the exponential trend model (0.9688) is approximately equal to the adjusted r^2 for the linear trend model (0.9674).

FIGURE 16.14

Fitting an Exponential Trend Equation Using Microsoft Excel for the Wm. Wrigley Jr. Company Real Gross Revenues Data

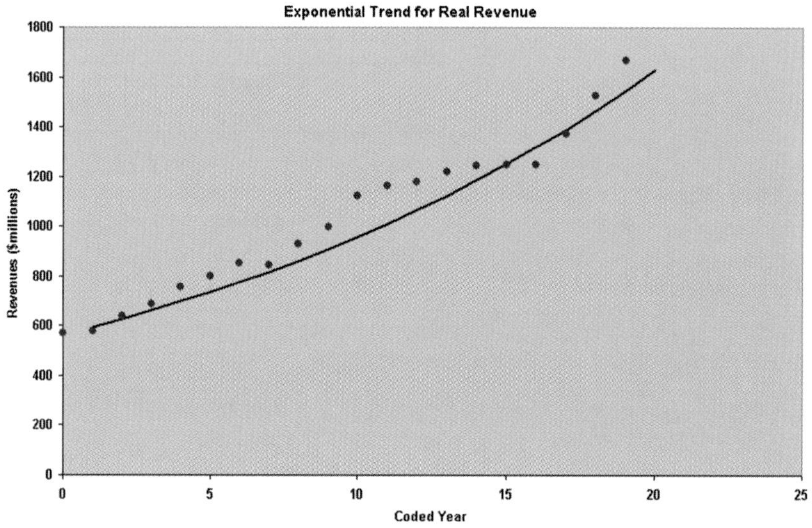

Model Selection Using First, Second, and Percentage Differences

You have used the linear, quadratic, and exponential models to forecast real gross revenues data for the Wm. Wrigley Jr. Company. How can you determine which of these models is the most appropriate model? In addition to visually inspecting scatter plots and comparing adjusted r^2 values, you can compute and examine first, second, and percentage differences. The identifying features of linear, quadratic, and exponential trend models are as follows:

- If a linear trend model provides a perfect fit to a time series, then the first differences are constant. Thus, the differences between consecutive values in the series are the same throughout:

$$(Y_2 - Y_1) = (Y_3 - Y_2) = \cdots = (Y_n - Y_{n-1})$$

- If a quadratic trend model provides a perfect fit to a time series, then the second differences are constant. Thus,

$$[(Y_3 - Y_2) - (Y_2 - Y_1)] = [(Y_4 - Y_3) - (Y_3 - Y_2)] = \cdots = [(Y_n - Y_{n-1}) - (Y_{n-1} - Y_{n-2})]$$

- If an exponential trend model provides a perfect fit to a time series, then the percentage differences between consecutive values are constant. Thus,

$$\frac{Y_2 - Y_1}{Y_1} \times 100\% = \frac{Y_3 - Y_2}{Y_2} \times 100\% = \cdots = \frac{Y_n - Y_{n-1}}{Y_{n-1}} \times 100\%$$

Although you should not expect a perfectly fitting model for any particular set of time-series data, you can consider the first differences, second differences, and percentage differences for a given series as guides in choosing an appropriate model. Examples 16.2, 16.3, and 16.4 illustrate applications of linear, quadratic, and exponential trend models having perfect fits to their respective data sets.

EXAMPLE 16.2

A LINEAR TREND MODEL WITH A PERFECT FIT

Given the following annual time series of the number of passengers (in millions) on a particular airline:

	Year									
	1995	**1996**	**1997**	**1998**	**1999**	**2000**	**2001**	**2002**	**2003**	**2004**
Passengers	30.0	33.0	36.0	39.0	42.0	45.0	48.0	51.0	54.0	57.0

Using first differences, show that the linear trend model provides a perfect fit to these data.

SOLUTION The following table shows the solution:

	Year									
	1995	**1996**	**1997**	**1998**	**1999**	**2000**	**2001**	**2002**	**2003**	**2004**
Passengers	30.0	33.0	36.0	39.0	42.0	45.0	48.0	51.0	54.0	57.0
First differences		3.0	3.0	3.0	3.0	3.0	3.0	3.0	3.0	3.0

The differences between consecutive values in the series are the same throughout.

EXAMPLE 16.3

A QUADRATIC TREND MODEL WITH A PERFECT FIT

Given the following annual time series of the number of passengers (in millions) on a particular airline:

	Year									
	1995	**1996**	**1997**	**1998**	**1999**	**2000**	**2001**	**2002**	**2003**	**2004**
Passengers	30.0	31.0	33.5	37.5	43.0	50.0	58.5	68.5	80.0	93.0

Using second differences, show that the quadratic trend model provides a perfect fit to these data.

SOLUTION The following table shows the solution:

	Year									
	1995	**1996**	**1997**	**1998**	**1999**	**2000**	**2001**	**2002**	**2003**	**2004**
Passengers	30.0	31.0	33.5	37.5	43.0	50.0	58.5	68.5	80.0	93.0
First differences		1.0	2.5	4.0	5.5	7.0	8.5	10.0	11.5	13.0
Second differences			1.5	1.5	1.5	1.5	1.5	1.5	1.5	1.5

The second differences between consecutive pairs of values in the series are the same throughout.

EXAMPLE 16.4

AN EXPONENTIAL TREND MODEL WITH A PERFECT FIT

Given the following annual time series of the number of passengers (in millions) on a particular airline:

	Year									
	1995	**1996**	**1997**	**1998**	**1999**	**2000**	**2001**	**2002**	**2003**	**2004**
Passengers	30.0	31.5	33.1	34.8	36.5	38.3	40.2	42.2	44.3	46.5

Using percentage differences, show that the exponential trend model provides a perfect fit to these data.

SOLUTION The following table shows the solution:

	Year									
	1995	1996	1997	1998	1999	2000	2001	2002	2003	2004
Passengers	30.0	31.5	33.1	34.8	36.5	38.3	40.2	42.2	44.3	46.5
First differences		1.5	1.6	1.7	1.7	1.8	1.9	2.0	2.1	2.2
Percentage differences		5.0	5.1	5.1	4.9	4.9	5.0	5.0	5.0	5.0

The percentage differences between consecutive values in the series are approximately the same throughout.

Figure 16.15 presents the first, second, and percentage differences for the real gross revenues data at Wm. Wrigley Jr. Company. Neither the first differences, second differences, or percentage differences are constant across the series. Even though the r^2 for each of the three models considered in this section is approximately 0.97, other models (including those considered in sections 16.5 and 16.6) may be more appropriate.

FIGURE 16.15

Comparing First, Second, and Percentage Differences in Real Annual Gross Revenues (in Millions of Constant 1982–1984 Dollars) for Wm. Wrigley Jr. Company (1984–2003)

Year	Real Revenue	First Difference	Second Difference	Percentage Difference
1984	569			
1985	576	7		1.23%
1986	638	62	55	10.76%
1987	688	50	-12	7.84%
1988	753	65	15	9.45%
1989	801	48	-17	6.37%
1990	850	49	1	6.12%
1991	844	-6	-55	-0.71%
1992	927	83	89	9.83%
1993	997	70	-13	7.55%
1994	1121	124	54	12.44%
1995	1161	40	-84	3.57%
1996	1180	19	-21	1.64%
1997	1217	37	18	3.14%
1998	1241	24	-13	1.97%
1999	1248	7	-17	0.56%
2000	1246	-2	-9	-0.16%
2001	1372	126	128	10.11%
2002	1526	154	28	11.22%
2003	1668	142	-12	9.31%

PROBLEMS FOR SECTION 16.4

Learning the Basics

16.9 If using the method of least squares for fitting trends in an annual time series containing 25 consecutive yearly values:

a. What coded value do you assign to X for the first year in the series?

b. What coded value do you assign to X for the fifth year in the series?

c. What coded value do you assign to X for the most recent recorded year in the series?

d. What coded value do you assign to X if you want to project the trend and make a forecast 5 years beyond the last observed value?

PH Grade ASSIST **16.10** The linear trend forecasting equation for an annual time series containing 20 values (from 1985 to 2004) on real total revenues (in millions of constant 1995 dollars) is

$$\hat{Y}_i = 4.0 + 1.5X_i$$

a. Interpret the Y intercept b_0.

b. Interpret the slope b_1.

c. What is the fitted trend value for the fifth year?

d. What is the fitted trend value for the most recent year?

e. What is the projected trend forecast 3 years after the last value?

16.11 The linear trend forecasting equation for an annual time series containing 40 values (from 1965 to 2004) on real net sales (in billions of constant 1995 dollars) is

$$\hat{Y}_i = 1.2 + 0.5X_i$$

a. Interpret the Y intercept b_0.
b. Interpret the slope b_1.
c. What is the fitted trend value for the tenth year?
d. What is the fitted trend value for the most recent year?
e. What is the projected trend forecast 2 years after the last value?

Applying the Concepts

You should use Microsoft Excel, Minitab, or SPSS to solve problems 16.12–16.22.

16.12 The following data reflect the annual values of the Consumer Price Index (CPI) in the United States over the 39-year period 1965 through 2003 using 1982 through 1984 as the base period. This index measures the average change in prices over time in a fixed "market basket" of goods and services purchased by all urban consumers including urban wage earners (i.e., clerical, professional, managerial, and technical workers; self-employed individuals; and short-term workers), unemployed individuals, and retirees. CPI-U

Year	CPI	Year	CPI	Year	CPI
1965	31.5	1978	65.2	1991	136.2
1966	32.4	1979	72.6	1992	140.3
1967	33.4	1980	82.4	1993	144.5
1968	34.8	1981	90.9	1994	148.2
1969	36.7	1982	96.5	1995	152.4
1970	38.8	1983	99.6	1996	156.9
1971	40.5	1984	103.9	1997	160.5
1972	41.8	1985	107.6	1998	163.0
1973	44.4	1986	109.6	1999	166.6
1974	49.3	1987	113.6	2000	172.2
1975	53.8	1988	118.3	2001	177.1
1976	56.9	1989	124.0	2002	179.9
1977	60.6	1990	130.7	2003	184.0

Source: Bureau of Labor Statistics, U.S. Department of Labor ***www.bls.gov***

a. Plot the data.
b. Describe the movement in this time series over the 39-year period.

16.13 Gross domestic product (GDP) is a major indicator of the nation's overall economic activity. It consists of personal consumption expenditures, gross domestic investment, net exports of goods and services, and government consumption expenditures. The gross domestic product (in billions of current dollars) for the United States from 1980 to 2003 is in the data file GDP.

Source: Bureau of Economic Analysis, U. S. Department of Commerce ***www.bea.gov***

a. Plot the data.
b. Compute a linear trend forecasting equation and plot the trend line.
c. What are your forecasts for 2004 and 2005?
d. What conclusions can you reach concerning the trend in gross domestic product?

16.14 The data in the file FEDRECPT represent federal receipts from 1978 through 2002 in billions of current dollars from individual and corporate income tax, social insurance, excise tax, estate and gift tax, customs duties, and federal reserve deposits.

Source: Tax Policy Center ***taxpolicycenter.org***.

a. Plot the series of data.
b. Compute a linear trend forecasting equation and plot the trend line.
c. What are your forecasts of the federal receipts for 2003 and 2004?
d. What conclusions can you reach concerning the trend in federal receipts between 1978 and 2002?

16.15 The data in the file STRATEGIC represent the amount of oil in millions of barrels held in the U.S. strategic oil reserve from 1981 through 2003.

Source: Energy Information Administration, U. S. Department of Energy ***eia.doe.gov***.

a. Plot the data.
b. Compute a linear trend forecasting equation and plot the trend line.
c. Compute a quadratic trend forecasting equation and plot the results.
d. Compute an exponential trend forecasting equation and plot the results.
e. Which model is the most appropriate?
f. Using the most appropriate model, forecast the number of barrels in millions for 2004. Check how accurate your forecast is by locating the true value on the Internet or in your library.

16.16 The data given in the file COCACOLA represent the annual net operating revenues (in billions of current dollars) at Coca-Cola Company from 1975 through 2003.

Source: Extracted from Moody's Handbook of Common Stocks, *1980, 1989, 1999 and* Mergent's Handbook of Common Stocks, *2003, and the Coca-Cola Company* ***www.cocacola.com***.

a. Plot the data.
b. Compute a linear trend forecasting equation and plot the trend line.
c. Compute a quadratic trend forecasting equation and plot the results.
d. Compute an exponential trend forecasting equation and plot the results.
e. Using the models in (b), (c), and (d), what are your annual trend forecasts of net operating revenues for 2004 and 2005?

16.17 The data in the following table represent the closing value of the Dow Jones Industrial Average (DJIA) from 1979 through 2003. DJIA

Year	DJIA	Year	DJIA	Year	DJIA
1979	838.7	1987	1,938.8	1995	5,117.1
1980	964.0	1988	2,168.6	1996	6,448.3
1981	875.0	1989	2,753.2	1997	7,908.3
1982	1,046.5	1990	2,633.7	1998	9,181.4
1983	1,258.6	1991	3,168.8	1999	11,497.1
1984	1,211.6	1992	3,301.1	2000	10,788.0
1985	1,546.7	1993	3,754.1	2001	10,021.5
1986	1,896.0	1994	3,834.4	2002	8,341.6
				2003	10,453.9

Source: Extracted from ***finance.Yahoo.com***.

a. Plot the data.
b. Compute a linear trend forecasting equation and plot the trend line.
c. Compute a quadratic trend forecasting equation and plot the results.
d. Compute an exponential trend forecasting equation and plot the results.
e. Which model is the most appropriate?
f. Using the most appropriate model, forecast the closing value for the DJIA in 2004. Discuss the accuracy of your forecast and try to explain the difference between the forecast and the actual value.

16.18 Procter & Gamble (P&G) is a worldwide producer and marketer of a broad range of consumer products with a market capitalization of over 112.9 billion and annual revenues of over $40 billion (*Yahoo.com*, September 2, 2002). The data in the file P&G represent the stock price as of January 1 for Proctor & Gamble Company for the 34-year period from January 1, 1970, to January 1, 2003. (These values have been adjusted for stock splits and dividends.)

Source: Extracted from ***finance.Yahoo.com***.

a. Plot the data.
b. Compute a linear trend forecasting equation and plot the trend line.
c. Compute a quadratic trend forecasting equation and plot the results.
d. Compute an exponential trend forecasting equation and plot the results.
e. Which model is the most appropriate?
f. Using the most appropriate model, forecast the stock price for January 1, 2004.

16.19 Although you should not expect a perfectly fitting model for any time-series data, you can consider the first differences, second differences, and percentage differences for a given series as guides in choosing an appropriate model. For this problem, use each of the time-series presented in the following table: TSMODEL1

	Year				
	1995	**1996**	**1997**	**1998**	**1999**
Time series I	10.0	15.1	24.0	36.7	53.8
Time series II	30.0	33.1	36.4	39.9	43.9
Time series III	60.0	67.9	76.1	84.0	92.2
	2000	**2001**	**2002**	**2003**	**2004**
Time series I	74.8	100.0	129.2	162.4	199.0
Time series II	48.2	53.2	58.2	64.5	70.7
Time series III	100.0	108.0	115.8	124.1	132.0

a. Determine the most appropriate model.
b. Compute the forecasting equation.
c. Forecast the value for 2005.

16.20 A time-series plot often helps you to determine the appropriate model to use. For this problem, use each of the time-series presented in the following table: TSMODEL2

	Year				
	1995	**1996**	**1997**	**1998**	**1999**
Time series I	100.0	115.2	130.1	144.9	160.0
Time series II	100.0	115.2	131.7	150.8	174.1
	2000	**2001**	**2002**	**2003**	**2004**
Time series I	175.0	189.8	204.9	219.8	235.0
Time series II	200.0	230.8	266.1	305.5	351.8

a. Plot the observed data (Y) over time (X), and plot the logarithm of the observed data (log Y) over time (X) to determine whether a linear trend model or an exponential trend model is more appropriate. (*Hint*: Recall that if the plot of log Y versus X appears to be linear, an exponential trend model provides an appropriate fit.)
b. Compute the appropriate forecasting equation.
c. Forecast the value for 2005.

16.21 The data given in the file GROSSREV represent the revenues (in millions of constant 1995 dollars) of a utility company for 1991 through 2004:
a. Compare the first differences, second differences, and percentage differences to determine the most appropriate model.
b. Compute the appropriate forecasting equation.
c. What has been the annual growth in revenues over the 14 years?
d. Forecast the revenues for 2005, 2006, and 2007.

SELF Test **16.22** An article (B. Horovitz, "What's Next? Fast-Food Giants Hunt for New Products to Tempt Consumers," *USA Today*, July 3, 2002, 1A–2A) discussed the need of fast-food chains to constantly develop new products in order to increase sales. The following table provides U.S. sales (in $ billions) from 1992 to 2001. **FASTFOODSALES**

Year	Sales	Year	Sales
1992	70.6	1997	88.8
1993	74.9	1998	92.5
1994	78.5	1999	97.5
1995	82.5	2000	101.4
1996	85.9	2001	105.5

a. Plot the data.

b. Form a new table of adjusted sales by multiplying each of the sales by 100/CPI (see problem 16.12 on page 677 for CPI values).

c. Plot the adjusted sales data.

d. Compute a linear trend forecasting equation and plot the results.

e. Compute a quadratic trend forecasting equation and plot the results.

f. Compute an exponential trend forecasting equation and plot the results.

g. Using the forecasting equations in (d), (e), and (f), what are your annual forecasts of adjusted sales for 2002 and 2003?

16.5 THE HOLT-WINTERS METHOD FOR TREND-FITTING AND FORECASTING

The **Holt-Winters method** extends the exponential smoothing approach described in section 16.3 by including the future trend. To use the Holt-Winters method at any time period i, you must continuously estimate the level of the series (i.e., the smoothed value E_i) and the trend value (T_i). Equations (16.10a) and (16.10b) estimate the level and trend values in the Holt-Winters method.

THE HOLT-WINTERS METHOD

$$\text{Level: } E_i = U(E_{i-1} + T_{i-1}) + (1 - U)Y_i \quad \textbf{(16.10a)}$$

$$\text{Trend: } T_i = VT_{i-1} + (1 - V)(E_i - E_{i-1}) \quad \textbf{(16.10b)}$$

where E_i = level of the smoothed series being computed in time period i

E_{i-1} = level of the smoothed series already computed in time period $i - 1$

T_i = value of the trend component being computed in time period i

T_{i-1} = value of the trend component already computed in time period $i - 1$

Y_i = observed value of the time series in period i

U = subjectively assigned smoothing constant (where $0 < U < 1$)

V = subjectively assigned smoothing constant (where $0 < V < 1$)

To begin computations, define $E_2 = Y_2$ and $T_2 = Y_2 - Y_1$ and choose smoothing constants for U and V. Then compute E_i and T_i for all i years, $i = 3, 4, \ldots, n$.

The choices for the smoothing constants U and V affect the results. Smaller values of U give more weight to the more recent levels of the time series and less weight to earlier levels in the series. Smaller values of V give more weight to the current trend in the time series and less weight to past trends in the series. Larger values of U and V have the opposite effect on the Holt-Winters method.

To illustrate the Holt-Winters method, return to the time series for the Wm. Wrigley Jr. Company discussed in section 16.4 (see Figure 16.5 on page 669). To compute the level and the trend for the third and fourth years (1986 and 1987) using the selected smoothing constants of $U = 0.3$ and $V = 0.3$, begin by setting

$$E_2 = Y_2 = 576$$

and

$$T_2 = Y_2 - Y_1 = 576 - 569 = 7.00$$

Choosing smoothing constants $U = 0.3$ and $V = 0.3$, Equations (16.10a) and (16.10b) become

$$E_i = (0.3)(E_{i-1} + T_{i-1}) + (0.7)(Y_i)$$

and

$$T_i = (0.3)(T_{i-1}) + (0.7)(E_i - E_{i-1})$$

For 1986, the third year, $i = 3$ and

$$E_3 = (0.3)(576 + 7) + (0.7)(638) = 621.50$$

and

$$T_3 = (0.3)(7) + (0.7)(621.50 - 576.00) = 33.95$$

For 1987, the fourth year, $i = 4$ and

$$E_4 = (0.3)(621.50 + 33.95) + (0.7)(688) = 678.24$$

and

$$T_4 = (0.3)(33.95) + (0.7)(678.24 - 621.50) = 49.90$$

Fortunately, you can use Microsoft Excel or Minitab to compute these values. Figure 16.16 displays the calculated values for level and trend for the entire series.

FIGURE 16.16

Using the Holt-Winters Method on Real Annual Gross Revenues (in Millions of Constant 1982–1984 Dollars) for Wm. Wrigley Jr. Company (1984–2003)

	A	B	C	D	E	F
1	Year	Real Revenue	HW Prediction	T		
2	1984	569	#N/A	#N/A	U	0.3
3	1985	576	576.00	7.00	V	0.3
4	1986	638	621.50	33.95		
5	1987	688	678.24	49.90		
6	1988	753	745.54	62.08		
7	1989	801	802.99	58.84		
8	1990	850	853.55	53.04		
9	1991	844	862.78	22.37		
10	1992	927	914.45	42.88		
11	1993	997	985.10	62.32		
12	1994	1121	1098.93	98.38		
13	1995	1161	1171.89	80.59		
14	1996	1180	1201.74	45.07		
15	1997	1217	1225.95	30.46		
16	1998	1241	1245.62	22.91		
17	1999	1248	1254.16	12.85		
18	2000	1246	1252.30	2.56		
19	2001	1372	1336.86	59.95		
20	2002	1526	1487.24	123.26		
21	2003	1668	1650.75	151.43		

To use the Holt-Winters method for forecasting, you assume that all future trend movements will continue from the most recent smoothed level E_n. Thus, you can use Equation (16.11) to forecast j years into the future.

USING THE HOLT-WINTERS METHOD FOR FORECASTING

$$\hat{Y}_{n+j} = E_n + j(T_n) \quad \textbf{(16.11)}$$

where $\hat{Y}_{n+j}$ = forecast value j years into the future

E_n = level of the smoothed series computed in the most recent time period n

T_n = value of the trend component computed in the most recent time period n

j = number of years into the future

To illustrate the process of forecasting with the Holt-Winters method, you can forecast real gross revenues for the Wm. Wrigley Jr. Company in 2004 and 2005. Using the values of level and trend based on smoothing constants of $U = 0.3$ and $V = 0.3$ in Figure 16.16, from Equation (16.11), the forecasts of real gross revenues for 2004 and 2005 are

$$\hat{Y}_{n+j} = E_n + j(T_n)$$

2004: 1 year ahead $\hat{Y}_{21} = E_{20} + (1)(T_{20}) = 1{,}650.75 + (1)(151.43) = 1{,}802.18$ millions of dollars
2005: 2 years ahead $\hat{Y}_{22} = E_{20} + (2)(T_{20}) = 1{,}650.75 + (2)(151.43) = 1{,}953.61$ millions of dollars

Figure 16.17 plots the real gross revenues and the forecasted values.

FIGURE 16.17
Plot of Real Gross Revenues and Forecasted Real Gross Revenues Using the Holt-Winters Method on the Wm. Wrigley Jr. Company Data

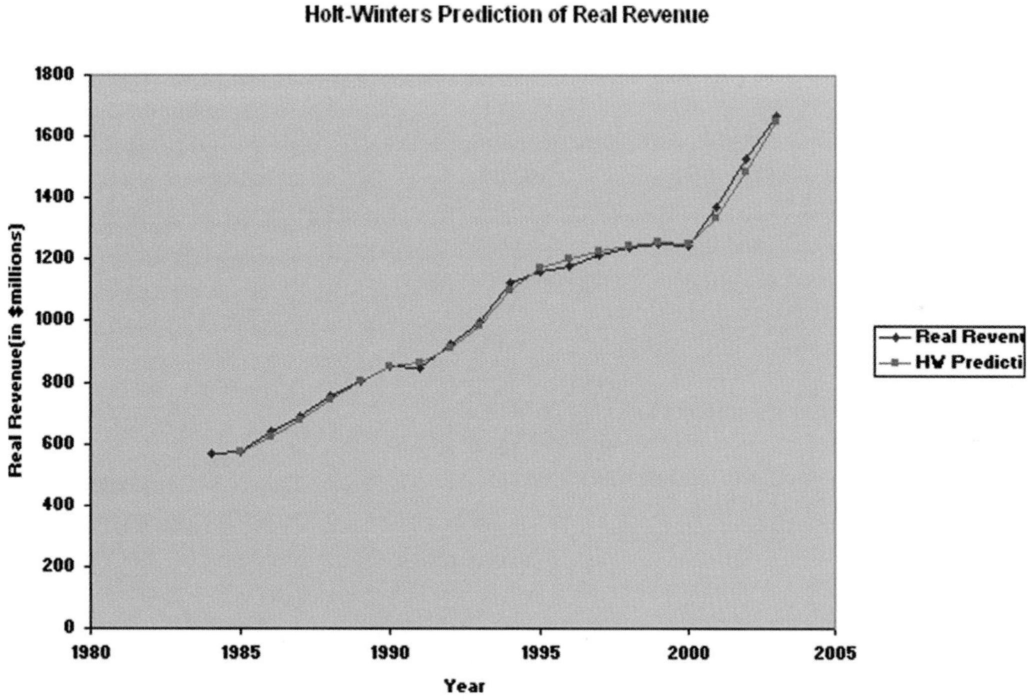

PROBLEMS FOR SECTION 16.5

Learning the Basics

PH Grade ASSIST **16.23** Consider an annual time series with 20 consecutive values. If the smoothed level for the most recent value is 34.2 and the corresponding trend level is 5.6,

a. What is your forecast for the coming year?
b. What is your forecast five years from now?

16.24 Given the following series from $n = 15$ consecutive time periods:

3 5 6 8 10 10 12 15 16 13 16 17 22 19 24

Use the Holt-Winters method (with $U = 0.30$ and $V = 0.30$) to forecast the 16th through 20th periods.

16.25 Given the following series from $n = 10$ consecutive time periods:

137 125 116 110 103 96 86 79 72 66

Use the Holt-Winters method (with $U = 0.20$ and $V = 0.20$) to forecast the 11th through 14th periods.

Applying the Concepts

You should use Microsoft Excel, Minitab, or SPSS to solve problems 16.26–16.30.

SELF Test **16.26** Refer to the data of problem 16.13 on page 677 concerning gross domestic product. GDP

a. Forecast gross domestic product for 2004 and 2005 using the Holt-Winters method with $U = 0.30$ and $V = 0.30$.
b. Do (a) with $U = 0.70$ and $V = 0.70$.
c. Do (a) with $U = 0.30$ and $V = 0.70$.
d. Which of these sets of forecasts would you select, given the historical movement of the time series? Discuss.
e. Compare the results of (a) through (c) with those of problem 16.13(c).
f. Go to the library or the Internet and find the actual gross domestic product for 2004 and 2005. Which method provided the best forecasts?

16.27 Refer to the data of problem 16.15 on page 677 concerning the number of barrels of oil in the U.S. strategic reserve. STRATEGIC

a. Forecast the number of barrels for 2004 using the Holt-Winters method with $U = 0.30$ and $V = 0.30$.
b. Do (a) with $U = 0.70$ and $V = 0.70$.
c. Do (a) with $U = 0.30$ and $V = 0.70$.
d. Which of these sets of forecasts would you select, given the historical movement of the time series? Discuss.
e. Compare the results of (a) through (c) with those of problem 16.15(f).
f. Go to the library or the Internet and find the actual number of barrels of oil in the U.S. strategic reserve for 2004. Which method provided the best forecast?

16.28 Refer to the data of problem 16.16 on page 677 concerning real operating revenue of Coca-Cola. COCACOLA

a. Forecast real operating revenues (in billions of *constant* 1982–1984 dollars) for 2004 and 2005 using the Holt-Winters method with $U = 0.30$ and $V = 0.30$.
b. Do (a) with $U = 0.70$ and $V = 0.70$.
c. Do (a) with $U = 0.30$ and $V = 0.70$.
d. Which of these sets of forecasts would you select, given the historical movement of the time series? Discuss.
e. Compare the results of (a) through (c) with those of problem 16.16(e).
f. Go to the library or the Internet and find the real operating revenue (in billions of *constant* 1982–1984 dollars) of Coca-Cola for 2004 and 2005. Which method provided the best forecasts?

16.29 Refer to the data of problem 16.17 on page 678 concerning the closing value of the Dow Jones Industrial Average (DJIA). DJIA

a. Forecast the closing value of the Dow Jones Industrial Average (DJIA) for 2004 using the Holt-Winters method with $U = 0.30$ and $V = 0.30$.
b. Do (a) with $U = 0.70$ and $V = 0.70$.
c. Do (a) with $U = 0.30$ and $V = 0.70$.
d. Which of these sets of forecasts would you select, given the historical movement of the time series? Discuss.
e. Compare the results of (a) through (c) with those of problem 16.17(f).
f. Go to the library or the Internet and find the actual Dow Jones Industrial Average (DJIA) for 2004. Which method provided the best forecast?

16.30 Refer to the data of problem 16.18 on page 678 concerning the stock price of Proctor & Gamble. P&G

a. Forecast the stock price of Proctor & Gamble for January 1, 2004 using the Holt-Winters method with $U = 0.30$ and $V = 0.30$.
b. Do (a) with $U = 0.70$ and $V = 0.70$.
c. Do (a) with $U = 0.30$ and $V = 0.70$.
d. Which of these sets of forecasts would you select, given the historical movement of the time series? Discuss.
e. Compare the results of (a) through (c) with those of problem 16.18(f).

16.6 AUTOREGRESSIVE MODELING FOR TREND-FITTING AND FORECASTING

Frequently, the values of a time-series are highly correlated with the values that precede and succeed them. This type of correlation is called autocorrelation. **Autoregressive modeling**[2] is a technique used to forecast time series with autocorrelation. A **first-order autocorrelation** refers to the association between consecutive values in a time series. A **second-order autocorrelation** refers to the relationship between values that are two periods apart. A ***p*th-order autocorrelation** refers to the correlation between values in a time series that are p periods apart. You can take advantage of the autocorrelation in data by using autoregressive modeling methods.

Equations (16.12), (16.13), and (16.14) define first-order, second-order, and p-th order autoregressive models.

[2]The exponential smoothing model described in section 16.3 and the autoregressive models described in this section are special cases of autoregressive integrated moving average (ARIMA) models developed by Box and Jenkins (reference 2).

FIRST-ORDER AUTOREGRESSIVE MODEL

$$Y_i = A_0 + A_1 Y_{i-1} + \delta_i \tag{16.12}$$

SECOND-ORDER AUTOREGRESSIVE MODEL

$$Y_i = A_0 + A_1 Y_{i-1} + A_2 Y_{i-2} + \delta_i \tag{16.13}$$

*p*TH-ORDER AUTOREGRESSIVE MODELS

$$Y_i = A_0 + A_1 Y_{i-1} + A_2 Y_{i-2} + \cdots + A_p Y_{i-p} + \delta_i \tag{16.14}$$

where Y_i = the observed value of the series at time i

Y_{i-1} = the observed value of the series at time $i - 1$

Y_{i-2} = the observed value of the series at time $i - 2$

Y_{i-p} = the observed value of the series at time $i - p$

$A_0, A_1, A_2, \ldots, A_p$ = autoregression parameters to be estimated from least-squares regression analysis

δ_i = a nonautocorrelated random error component (with mean = 0 and constant variance)

The **first-order autoregressive model** [Equation (16.12)] is similar in form to the simple linear regression model [Equation (13.1) on page 515]. The **second-order autoregressive model** [Equation (16.13)] is similar to the multiple regression model with two independent variables [Equation (14.2) on page 572]. The ***p*th-order autoregressive model** [Equation (16.14)] is similar to the multiple regression model [Equation (14.1) on page 571]. In the regression models, the regression parameters are given by the symbols $\beta_0, \beta_1, \ldots, \beta_k$, with corresponding estimates denoted by $b_0, b_1, \ldots, b_k$. In the autoregressive models, the parameters are given by the symbols $A_0, A_1, \ldots, A_p$, with corresponding estimates denoted by $a_0, a_1, \ldots, a_p$.

Selecting an appropriate autoregressive model is not easy. You must weigh the advantages that are due to simplicity against the concern of failing to take into account important autocorrelation in the data. You must be equally concerned with selecting a higher-order model requiring the estimation of numerous, unnecessary parameters—especially if n, the number of values in the series, is small. The reason for this concern is that p out of n data values are lost in computing an estimate of A_p when comparing each data value with another data value, which are p periods apart.

Examples 16.5 and 16.6 illustrate this loss of data values.

EXAMPLE 16.5

COMPARISON SCHEMA FOR A FIRST-ORDER AUTOREGRESSIVE MODEL

Consider the following series of $n = 7$ consecutive annual values:

	Year						
	1	**2**	**3**	**4**	**5**	**6**	**7**
Series	31	34	37	35	36	43	40

Demonstrate the comparisons needed for a first-order autoregressive model.

SOLUTION

Year	First-Order Autoregressive Model
i	(Y_i versus Y_{i-1})
1	31 ↔ . . .
2	34 ↔ 31
3	37 ↔ 34
4	35 ↔ 37
5	36 ↔ 35
6	43 ↔ 36
7	40 ↔ 43

Because there is no value recorded prior to Y_1, this value is lost for regression analysis. Therefore, the first-order autoregressive model is based on six pairs of values.

EXAMPLE 16.6 COMPARISON SCHEMA FOR A SECOND-ORDER AUTOREGRESSIVE MODEL

Consider the following series of $n = 7$ consecutive annual values:

	Year						
	1	**2**	**3**	**4**	**5**	**6**	**7**
Series	31	34	37	35	36	43	40

Demonstrate the comparisons needed for a second-order autoregressive model.

SOLUTION

Year	Second-Order Autoregressive Model
i	(Y_i versus Y_{i-1} and Y_i versus Y_{i-2})
1	31 ↔ . . . and 31 ↔ . . .
2	34 ↔ 31 and 34 ↔ . . .
3	37 ↔ 34 and 37 ↔ 31
4	35 ↔ 37 and 35 ↔ 34
5	36 ↔ 35 and 36 ↔ 37
6	43 ↔ 36 and 43 ↔ 35
7	40 ↔ 43 and 40 ↔ 36

Because there is no value recorded prior to Y_1, two values are lost for regression analysis. Therefore, the second-order autoregressive model is based on five pairs of values.

After selecting a model and using the least-squares method to compute estimates of the parameters, you need to determine the appropriateness of the model. Either you can select a particular pth-order autoregressive model based on previous experiences with similar data, or, as a starting point, you can choose a model with several parameters and then eliminate the parameters that do not significantly contribute to the model. In this latter approach, you use a t test for the significance of A_p, the highest-order autoregressive parameter in the current model under consideration. The null and alternative hypotheses are:

$$H_0: A_p = 0$$
$$H_1: A_p \neq 0$$

Equation (16.15) defines the test statistic.

t TEST FOR SIGNIFICANCE OF THE HIGHEST-ORDER AUTOREGRESSIVE PARAMETER A_p

$$t = \frac{a_p - A_p}{S_{a_p}} \tag{16.15}$$

where A_p = hypothesized value of the highest-order parameter A_p in the regression model

a_p = the estimate of the highest-order parameter A_p in the autoregressive model

S_{a_p} = the standard deviation of a_p

and the test statistic t follows a t distribution having $n - 2p - 1$ degrees of freedom.[3]

[3] *In addition to the degrees of freedom lost for each of the p population parameters you are estimating, p additional degrees of freedom are lost because there are p fewer comparisons to be made out of the original n values in the time series.*

For a given level of significance α, you reject the null hypothesis if the computed t-test statistic is greater than the upper-tail critical value from the t distribution, or if the computed test statistic is less than the lower-tail critical value from the t distribution. Thus, the decision rule is

$$\text{Reject } H_0 \text{ if } t > t_{n-2p-1} \text{ or if } t < -t_{n-2p-1};$$
$$\text{otherwise do not reject } H_0.$$

Figure 16.18 illustrates the decision rule and regions of rejection and nonrejection.

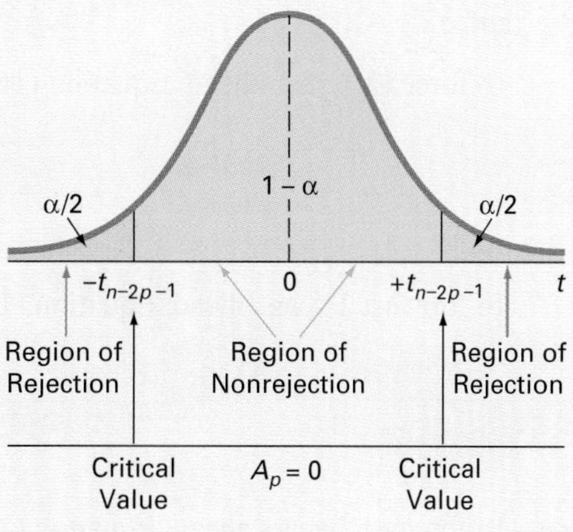

FIGURE 16.18
Rejection Regions for a Two-Tail Test for the Significance of the Highest-Order Autoregressive Parameter A_p

If you do not reject the null hypothesis that $A_p = 0$, you conclude that the selected model contains too many estimated parameters. You then discard the highest-order term, and estimate an autoregressive model of order $p - 1$ using the least-squares method. You then repeat the test of the hypothesis that the new highest-order term is 0. This testing and modeling continues until you reject H_0. When this occurs, you know that the remaining highest-order parameter is significant, and you can use that model for forecasting purposes.

Equation (16.16) defines the fitted pth-order autoregressive equation.

FITTED pTH-ORDER AUTOREGRESSIVE EQUATION

$$\hat{Y}_i = a_0 + a_1Y_{i-1} + a_2Y_{i-2} + \cdots + a_pY_{i-p} \quad \textbf{(16.16)}$$

where

$\hat{Y}_i$ = fitted values of the series at time i

Y_{i-1} = observed value of the series at time $i - 1$

Y_{i-2} = observed value of the series at time $i - 2$

Y_{i-p} = observed value of the series at time $i - p$

$a_0, a_1, a_2, \ldots, a_p$ = regression estimates of the parameters $A_0, A_1, A_2, \ldots, A_p$

You use Equation (16.17) to forecast j years into the future from the current nth time period.

pTH-ORDER AUTOREGRESSIVE FORECASTING EQUATION

$$\hat{Y}_{n+j} = a_0 + a_1\hat{Y}_{n+j-1} + a_2\hat{Y}_{n+j-2} + \cdots + a_p\hat{Y}_{n+j-p} \quad \textbf{(16.17)}$$

where $a_0, a_1, a_2, \ldots, a_p$ = regression estimates of the parameters $A_0, A_1, A_2, \ldots, A_p$

j = number of years into the future

$\hat{Y}_{n+j-p}$ = forecast of Y_{n+j-p} from the current time period for $j - p > 0$ or

$\hat{Y}_{n+j-p}$ = observed value for Y_{n+j-p} for $j - p \leq 0$

Thus, to make forecasts j years into the future using a third-order autoregressive model, you need only the most recent $p = 3$ values (Y_n, Y_{n-1}, and Y_{n-2}) and the regression estimates a_0, a_1, a_2, and a_3.

To forecast 1 year ahead, Equation (16.17) becomes:

$$\hat{Y}_{n+1} = a_0 + a_1Y_n + a_2Y_{n-1} + a_3Y_{n-2}$$

To forecast 2 years ahead, Equation (16.17) becomes:

$$\hat{Y}_{n+2} = a_0 + a_1\hat{Y}_{n+1} + a_2Y_n + a_3Y_{n-1}$$

To forecast 3 years ahead, Equation (16.17) becomes:

$$\hat{Y}_{n+3} = a_0 + a_1\hat{Y}_{n+2} + a_2\hat{Y}_{n+1} + a_3Y_n$$

and so on.

Autoregressive modeling is a powerful forecasting technique for time series that have autocorrelation. Although slightly more complicated than other methods, the following step-by-step approach should guide you through the analysis.

1. Choose a value for p, the highest-order parameter in the autoregressive model to be evaluated, realizing that the t test for significance is based on $n - 2p - 1$ degrees of freedom.
2. Form a series of p "lagged predictor" variables such that the first variable lags by 1 time period, the second variable lags by 2 time periods, and so on and the last predictor variable lags by p time periods (see Figure 16.19 below).
3. Use Microsoft Excel or Minitab to perform a least-squares analysis of the multiple regression model containing all p lagged predictor variables.
4. Test for the significance of A_p, the highest-order autoregressive parameter in the model.
   a. If you do not reject the null hypothesis, discard the pth variable, and repeat steps 3 and 4. The test for the significance of the new highest-order parameter is based on a t distribution whose degrees of freedom are revised to correspond with the new number of predictors.
   b. If you reject the null hypothesis, select the autoregressive model with all p predictors for fitting [see Equation (16.16)] and forecasting [see Equation (16.17)].

To demonstrate the autoregressive modeling approach, return to the time series concerning the real gross revenues (in millions of constant 1982 to 1984 dollars) for the Wm. Wrigley Jr. Company over the 20-year period 1984 through 2003. Figure 16.19 displays the real revenues and the setup for the first-order, second-order, and third-order autoregressive models. All the columns in this table are needed for fitting the third-order autoregressive model. The last column is omitted when fitting second-order autoregressive models, and the last two columns are omitted when fitting first-order autoregressive models. Thus, $p = 1$, 2, or 3 values out of $n = 20$ are lost in the comparisons needed for developing the first-order, second-order, and third-order autoregressive models.

FIGURE 16.19

Developing First-Order, Second-Order, and Third-Order Autoregressive Models on Real Revenues for Wm. Wrigley Jr. Company (1984–2003)

	A	B	C	D	E
1	Year	Real Revenue	Lag1	Lag2	Lag3
2	1984	569	#N/A	#N/A	#N/A
3	1985	576	569	#N/A	#N/A
4	1986	638	576	569	#N/A
5	1987	688	638	576	569
6	1988	753	688	638	576
7	1989	801	753	688	638
8	1990	850	801	753	688
9	1991	844	850	801	753
10	1992	927	844	850	801
11	1993	997	927	844	850
12	1994	1121	997	927	844
13	1995	1161	1121	997	927
14	1996	1180	1161	1121	997
15	1997	1217	1180	1161	1121
16	1998	1241	1217	1180	1161
17	1999	1248	1241	1217	1180
18	2000	1246	1248	1241	1217
19	2001	1372	1246	1248	1241
20	2002	1526	1372	1246	1248
21	2003	1668	1526	1372	1246

Selecting an autoregressive model that best fits the annual time series begins with the third-order autoregressive model depicted in Figure 16.20 using Microsoft Excel and in Figure 16.21 using Minitab. From Figure 16.20 or 16.21, the fitted third-order autoregressive equation is

$$\hat{Y}_i = 3.2005 + 1.5132Y_{i-1} - 0.6258Y_{i-2} + 0.1486Y_{i-3}$$

where year zero is 1987.

FIGURE 16.20
Microsoft Excel Output for the Third-Order Autoregressive Model for the Wm. Wrigley Jr. Real Gross Revenues Data

	A	B	C	D	E	F	G
1	**Third-Order Autoregressive Model**						
2							
3	*Regression Statistics*						
4	Multiple R	0.9874					
5	R Square	0.9750					
6	Adjusted R Square	0.9693					
7	Standard Error	48.2712					
8	Observations	17					
9							
10	ANOVA						
11		*df*	*SS*	*MS*	*F*	*Significance F*	
12	Regression	3	1183197.5526	394399.1842	169.2619	0.0000	
13	Residual	13	30291.4486	2330.1114			
14	Total	16	1213489.0012				
15							
16		*Coefficients*	*Standard Error*	*t Stat*	*P-value*	*Lower 95%*	*Upper 95%*
17	Intercept	3.2005	51.1498	0.0626	0.9511	-107.3021	113.7030
18	X Variable 1	1.5132	0.2758	5.4873	0.0001	0.9174	2.1089
19	X Variable 2	-0.6258	0.4766	-1.3131	0.2119	-1.6553	0.4038
20	X Variable 3	0.1486	0.3209	0.4632	0.6509	-0.5447	0.8420

FIGURE 16.21
Minitab Output for the Third-Order Autoregressive Model for the Wm. Wrigley Jr. Real Gross Revenues Data

```
The regression equation is
Real Revenue = 3.2 + 1.51 Lag1 - 0.626 Lag2 + 0.149 Lag3


17 cases used, 3 cases contain missing values


Predictor      Coef  SE Coef      T      P
Constant       3.20    51.15   0.06  0.951
Lag1         1.5132   0.2758   5.49  0.000
Lag2        -0.6258   0.4766  -1.31  0.212
Lag3         0.1486   0.3209   0.46  0.651


S = 48.2712   R-Sq = 97.5%   R-Sq(adj) = 96.9%


Analysis of Variance

Source          DF       SS      MS       F      P
Regression       3  1183198  394399  169.26  0.000
Residual Error  13    30291    2330
Total           16  1213489
```

Next, you test for the significance of A_3, the highest-order parameter. The highest-order parameter estimate a_3 for the fitted third-order autoregressive model is 0.1486, with a standard error of 0.3209.

The null and alternative hypotheses are

$$H_0: A_3 = 0$$

against

$$H_1: A_3 \neq 0$$

From Equation (16.15) on page 685 and the Microsoft Excel or Minitab output given in Figures 16.20 and 16.21,

$$t = \frac{a_3 - A_3}{S_{a_3}} = \frac{0.1486 - 0}{0.3209} = 0.463$$

Using a 0.05 level of significance, the two-tail t test with 13 degrees of freedom has critical values of ± 2.1604. Because $-2.1604 < t = 0.463 < +2.1604$ or because the p-value $= 0.6509 > \alpha = 0.05$, you do not reject H_0. You conclude that the third-order parameter of the autoregressive model is not significant and can be deleted.

Using Microsoft Excel or Minitab once again (see Figures 16.22 and 16.23), you fit a second-order autoregressive model.

FIGURE 16.22

Microsoft Excel Output for the Second-Order Autoregressive Model for the Wm. Wrigley Jr. Company Real Gross Revenues Data

	A	B	C	D	E	F	G
1	**Second-Order Autoregressive Model**						
2							
3	*Regression Statistics*						
4	Multiple R	0.9887					
5	R Square	0.9775					
6	Adjusted R Square	0.9745					
7	Standard Error	46.1543					
8	Observations	18					
9							
10	ANOVA						
11		*df*	*SS*	*MS*	*F*	*Significance F*	
12	Regression	2	1390574.2713	695287.1357	326.3917	0.0000	
13	Residual	15	31953.3444	2130.2230			
14	Total	17	1422527.6157				
15							
16		*Coefficients*	*Standard Error*	*t Stat*	*P-value*	*Lower 95%*	*Upper 95%*
17	Intercept	15.6941	43.2427	0.3629	0.7217	-76.4755	107.8637
18	X Variable 1	1.4406	0.2488	5.7893	0.0000	0.9102	1.9710
19	X Variable 2	-0.4184	0.2594	-1.6130	0.1276	-0.9712	0.1345

FIGURE 16.23

Minitab Output for the Second-Order Autoregressive Model for the Wm. Wrigley Jr. Company Real Gross Revenues Data

```
The regression equation is
Real Revenue = 15.7 + 1.44 Lag1 - 0.418 Lag2


18 cases used, 2 cases contain missing values


Predictor      Coef  SE Coef       T      P
Constant      15.69    43.24    0.36  0.722
Lag1         1.4406   0.2488    5.79  0.000
Lag2        -0.4184   0.2594   -1.61  0.128


S = 46.1543   R-Sq = 97.8%   R-Sq(adj) = 97.5%


Analysis of Variance

Source          DF       SS      MS       F      P
Regression       2  1390574  695287  326.39  0.000
Residual Error  15    31953    2130
Total           17  1422528
```

The fitted second-order autoregressive equation is

$$\hat{Y}_i = 15.6941 + 1.4406Y_{i-1} - 0.4184Y_{i-2}$$

where year zero is 1986.

From the Microsoft Excel or Minitab output, the highest-order parameter estimate is $a_2 = -0.4184$ with a standard error $= 0.2594$.

To test

$$H_0: A_2 = 0$$

against

$$H_1: A_2 \neq 0$$

from Equation (16.15) on page 685

$$t = \frac{a_2 - A_2}{S_{a_2}} = \frac{-0.4184 - 0}{0.2594} = -1.613$$

To test at the 0.05 level of significance, the two-tail t test with 15 degrees of freedom has critical values t_{15} of $\pm$ 2.1315. Because $-2.1315 < t = -1.613 < +2.1315$ or because the p-value = $0.1276 > \alpha = 0.05$, you do not reject H_0. You conclude that the second-order parameter of the autoregressive model is not significant and should be deleted from the model.

Using Microsoft Excel or Minitab once again (see Figures 16.24 and 16.25), you fit a first-order autoregressive model.

FIGURE 16.24

Microsoft Excel Output for the First-Order Autoregressive Model for the Wm. Wrigley Jr. Company Real Gross Revenues Data

	A	B	C	D	E	F	G
1	First-Order Autoregressive Model						
2							
3	*Regression Statistics*						
4	Multiple R	0.9884					
5	R Square	0.9769					
6	Adjusted R Square	0.9756					
7	Standard Error	47.5219					
8	Observations	19					
9							
10	ANOVA						
11		*df*	*SS*	*MS*	*F*	*Significance F*	
12	Regression	1	1626592.2166	1626592.2166	720.2636	0.0000	
13	Residual	17	38391.5959	2258.3292			
14	Total	18	1664983.8125				
15							
16		*Coefficients*	*Standard Error*	*t Stat*	*P-value*	*Lower 95%*	*Upper 95%*
17	Intercept	3.5951	40.6821	0.0884	0.9306	-82.2365	89.4268
18	X Variable 1	1.0544	0.0393	26.8377	0.0000	0.9715	1.1373

FIGURE 16.25

Minitab Output for the First-Order Autoregressive Model for the Wm. Wrigley Jr. Company Real Gross Revenues Data

```
The regression equation is
Real Revenue = 3.6 + 1.05 Lag1


19 cases used, 1 cases contain missing values


Predictor      Coef  SE Coef      T      P
Constant       3.60    40.68   0.09  0.931
Lag1        1.05438  0.03929  26.84  0.000


S = 47.5219   R-Sq = 97.7%   R-Sq(adj) = 97.6%


Analysis of Variance

Source          DF       SS       MS       F      P
Regression       1  1626592  1626592  720.26  0.000
Residual Error  17    38392     2258
Total           18  1664984
```

The first-order autoregressive equation is

$$\hat{Y}_i = 3.5951 + 1.0544Y_{i-1}$$

From the Microsoft Excel or Minitab output, the highest-order parameter estimate is $a_1 = 1.0544$ and $S_{a_1} = 0.0393$.

To test

$$H_0 : A_1 = 0$$

against

$$H_1 : A_1 \neq 0$$

from Equation (16.15) on page 685

$$t = \frac{a_1 - A_1}{S_{a_1}} = \frac{1.0544 - 0}{0.0393} = 26.84$$

To test at the 0.05 level of significance, the two-tail t test with 17 degrees of freedom has critical values t_{17} of $\pm$ 2.1098. Because $t = 26.84 > 2.1098$, or because the p-value $= 0.0000 < \alpha = 0.05$, you reject H_0. You conclude that the first-order parameter of the autoregressive model is significant and should remain in the model. The model-building approach has led to the selection of the first-order autoregressive model as the most appropriate for the given data. Using the estimates $a_0 = 3.5951$, and $a_1 = 1.0544$, as well as the most recent data value $Y_{20} = 1{,}668$, the forecasts of real gross revenues at the Wm. Wrigley Jr. Company for 2004 and 2005 from Equation (16.17) on page 686 are:

$$\hat{Y}_{n+j} = 3.5951 + 1.0544\hat{Y}_{n+j-1}$$

2004: 1 year ahead $\hat{Y}_{21} = 3.5951 + 1.0544(1{,}668) = 1{,}762.33$ millions of dollars
2005: 2 years ahead $\hat{Y}_{22} = 3.5951 + 1.0544(1{,}762.33) = 1{,}861.80$ millions of dollars

Figure 16.26 displays the predicted Y values from the first-order autoregressive model.

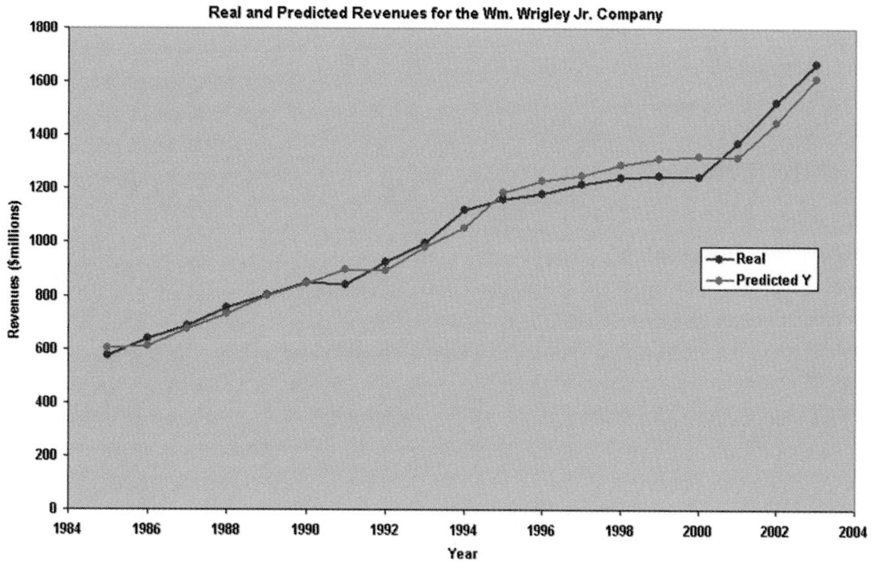

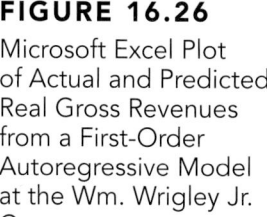

FIGURE 16.26
Microsoft Excel Plot of Actual and Predicted Real Gross Revenues from a First-Order Autoregressive Model at the Wm. Wrigley Jr. Company

PROBLEMS FOR SECTION 16.6

Learning the Basics

16.31 You are given an annual time series with 40 consecutive values and asked to fit a fifth-order autoregressive model.

a. How many comparisons are lost in the development of the autoregressive model?
b. How many parameters do you need to estimate?
c. Which of the original 40 values do you need for forecasting?
d. State the model.
e. Write an equation to indicate how you would forecast j years into the future.

PH Grade ASSIST **16.32** A third-order autoregressive model is fitted to an annual time series with 17 values and has the following estimated parameters and standard deviations:

$$a_0 = 4.50 \quad a_1 = 1.80 \quad a_2 = 0.80 \quad a_3 = 0.24$$
$$S_{a_1} = 0.50 \quad S_{a_2} = 0.30 \quad S_{a_3} = 0.10$$

At the 0.05 level of significance, test the appropriateness of the fitted model.

16.33 Refer to problem 16.32. The three most recent values are

$$Y_{15} = 23 \quad Y_{16} = 28 \quad Y_{17} = 34$$

Forecast the values for the next year and the following year.

16.34 Refer to problem 16.32. Suppose, when testing for the appropriateness of the fitted model, the standard deviations are

$$S_{a_1} = 0.45 \quad S_{a_2} = 0.35 \quad S_{a_3} = 0.15$$

a. What conclusions can you make?
b. Discuss how to proceed if forecasting is still your main objective.

Applying the Concepts

You should use Microsoft Excel, Minitab, or SPSS to solve problems 16.35–16.39.

16.35 Refer to the data given in problem 16.15 on page 677 that represent the amount of oil (in millions of barrels) held in the U.S. strategic reserve from 1981 through 2003. STRATEGIC

a. Fit a third-order autoregressive model to the amount of oil and test for the significance of the third-order autoregressive parameter. (Use $\alpha = 0.05$.)
b. If necessary, fit a second-order autoregressive model to the amount of oil and test for the significance of the second-order autoregressive parameter. (Use $\alpha = 0.05$.)
c. If necessary, fit a first-order autoregressive model to the amount of oil and test for the significance of the first-order autoregressive parameter. (Use $\alpha = 0.05$.)
d. If appropriate, forecast the barrels held in 2004 and 2005.

16.36 Refer to the data introduced in problem 16.16 on page 677 concerning annual net operating revenues (in billions of current dollars) at Coca-Cola Company over the 29-year period 1975 through 2003. COCACOLA

a. Fit a third-order autoregressive model to the annual net operating revenues and test for the significance of the third-order autoregressive parameter. (Use $\alpha = 0.05$.)
b. If necessary, fit a second-order autoregressive model to the annual net operating revenues and test for the significance of the second-order autoregressive parameter. (Use $\alpha = 0.05$.)
c. If necessary, fit a first-order autoregressive model to the annual net operating revenues and test for the significance of the first-order autoregressive parameter. (Use $\alpha = 0.05$.)
d. If appropriate, forecast net operating revenues for 2004 and 2005.

16.37 Refer to the data given in problem 16.17 on page 678 that represent the closing value of the Dow Jones Industrial Average (DJIA) from 1979 to 2003. DJIA

a. Fit a third-order autoregressive model to the DJIA and test for the significance of the third-order autoregressive parameter. (Use $\alpha = 0.05$.)
b. If necessary, fit a second-order autoregressive model to the DJIA and test for the significance of the second-order autoregressive parameter. (Use $\alpha = 0.05$.)
c. If necessary, fit a first-order autoregressive model to the DJIA and test for the significance of the first-order autoregressive parameter. (Use $\alpha = 0.05$.)
d. If appropriate, forecast the DJIA for 2004 and 2005.

16.38 Refer to the data given in problem 16.18 on page 678 that represent the stock price on January 1 for P&G from 1970 through 2003. P&G

a. Fit a third-order autoregressive model to the stock price and test for the significance of the third-order autoregressive parameter. (Use $\alpha = 0.05$.)
b. If necessary, fit a second-order autoregressive model to the stock price and test for the significance of the second-order autoregressive parameter. (Use $\alpha = 0.05$.)
c. If necessary, fit a first-order autoregressive model to the stock price and test for the significance of the first-order autoregressive parameter. (Use $\alpha = 0.05$.)
d. If appropriate, forecast the stock price for January 1, 2004.

SELF Test **16.39** Refer to the data given in problem 16.22 on page 679 that represent the U.S. fast-food sales over the 10-year period 1992 through 2001. First, deflate the series by multiplying the fast-food sales by (100/CPI). FASTFOODSALES

a. Fit a third-order autoregressive model to the adjusted sales and test for the significance of the third-order autoregressive parameter. (Use $\alpha = 0.05$.)

b. If necessary, fit a second-order autoregressive model to the adjusted sales and test for the significance of the second-order autoregressive parameter. (Use $\alpha = 0.05$.)

c. If necessary, fit a first-order autoregressive model to the adjusted sales and test for the significance of the first-order autoregressive parameter. (Use $\alpha = 0.05$.)

d. If appropriate, forecast the adjusted sales from 2002 through 2004.

16.7 CHOOSING AN APPROPRIATE FORECASTING MODEL

In sections 16.4 to 16.6, you studied seven time-series forecasting methods: the linear trend model, the quadratic trend model, and the exponential trend model in section 16.4; the Holt-Winters method in section 16.5; and the first-order, second-order, and *p*th-order autoregressive models in section 16.6. Is there a *best* model? Among these models which one should you select for forecasting? The following guidelines are provided for determining the adequacy of a particular forecasting model. These guidelines are based on a judgment of how well the model fits the past data of a given time series, and assumes that future movements in the time series can be projected by a study of the past data.

- Perform a residual analysis.
- Measure the magnitude of the residual error through squared differences.
- Measure the magnitude of the residual error through absolute differences.
- Use the principle of parsimony.

A discussion of these guidelines follows.

Performing a Residual Analysis

Recall from sections 13.5 and 14.2 that residuals are the differences between the observed and predicted values. After fitting a particular model to a time series, you plot the residuals over the *n* time periods. As shown in panel A of Figure 16.27, if the particular model fits adequately, the residuals represent the irregular component of the time series. Therefore, they should be randomly distributed throughout the series. However, as illustrated in the three remaining panels of Figure 16.27, if the particular model does not fit adequately, the residuals may show a systematic pattern such as a failure to account for trend (panel B), a failure to account for cyclical variation (panel C), or, with monthly or quarterly data, a failure to account for seasonal variation (panel D).

FIGURE 16.27

Residual Analysis for Studying Error Patterns

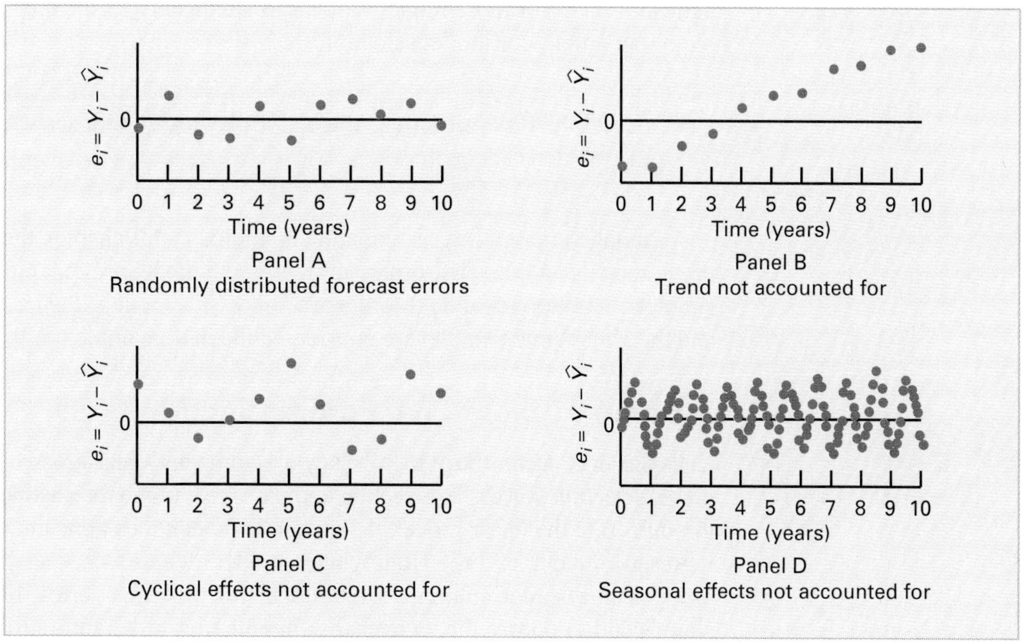

Measuring the Magnitude of the Residual Error through Squared or Absolute Differences

If, after performing a residual analysis, you still believe that two or more models appear to fit the data adequately, you can use additional methods for model selection. Numerous measures based on the residual error are available (see references 1 and 4). However, there is no consensus among statisticians as to which particular measure is best for determining the most appropriate forecasting model.

Based on the principle of least-squares, one measure that you have already used in regression analysis (see section 13.3) is the standard error of the estimate (S_{YX}). For a particular model, this measure is based on the sum of squared differences between the actual and predicted values in a time series. If a model fits the time-series data perfectly, then the standard error of the estimate is zero. If a model fits the time-series data poorly, then S_{YX} is large. Thus, when comparing the adequacy of two or more forecasting models, you can select the model with the minimum S_{YX} as most appropriate.

However, a major drawback to using S_{YX} when comparing forecasting models is that it penalizes a model too much for a large individual forecasting error. Thus, whenever there is a large difference between even a single Y_i and $\hat{Y}_i$, the value of S_{YX} (see page 530) becomes magnified through the squaring process. For this reason, many statisticians prefer the **mean absolute deviation (*MAD*)**. Equation (16.18) defines the *MAD* as the mean of the absolute differences between the actual and predicted values in a time series.

MEAN ABSOLUTE DEVIATION

$$MAD = \frac{\sum_{i=1}^{n} |Y_i - \hat{Y}_i|}{n} \tag{16.18}$$

If a model fits the time-series data perfectly, the *MAD* is zero. If a model fits the time-series data poorly, the *MAD* is large. When comparing two or more forecasting models, you can select the one with the minimum *MAD* as most appropriate.

Principle of Parsimony

If, after performing a residual analysis and comparing the S_{YX} and *MAD* measures, you still believe that two or more models appear to adequately fit the data, then you can use the principle of parsimony for model selection.

The **principle of parsimony** is the belief that you should select the simplest model that gets the job done adequately.

Among the seven forecasting models studied in this chapter, the least-squares linear and quadratic models and the first-order autoregressive model are regarded by most statisticians as the simplest. The second- and *p*th-order autoregressive models, the least-squares exponential model, and the Holt-Winters model are considered the more complex of the techniques presented.

A Comparison of Five Forecasting Methods

Consider once again the Wm. Wrigley Jr. Company's real gross revenues data. To illustrate the model selection process, this section compares five of the forecasting models used in sections 16.4 through 16.6: the linear model, the quadratic model, the exponential model, the first-order autoregressive model, and the Holt-Winters model. (There is no need to further study the second-order or third-order autoregressive model for this time series, because these models did not significantly improve the fit over the simpler first-order autoregressive model.)

Figure 16.28 displays the residual plots for the five models. In drawing conclusions from these residual plots, you must use caution because there are only 20 values.

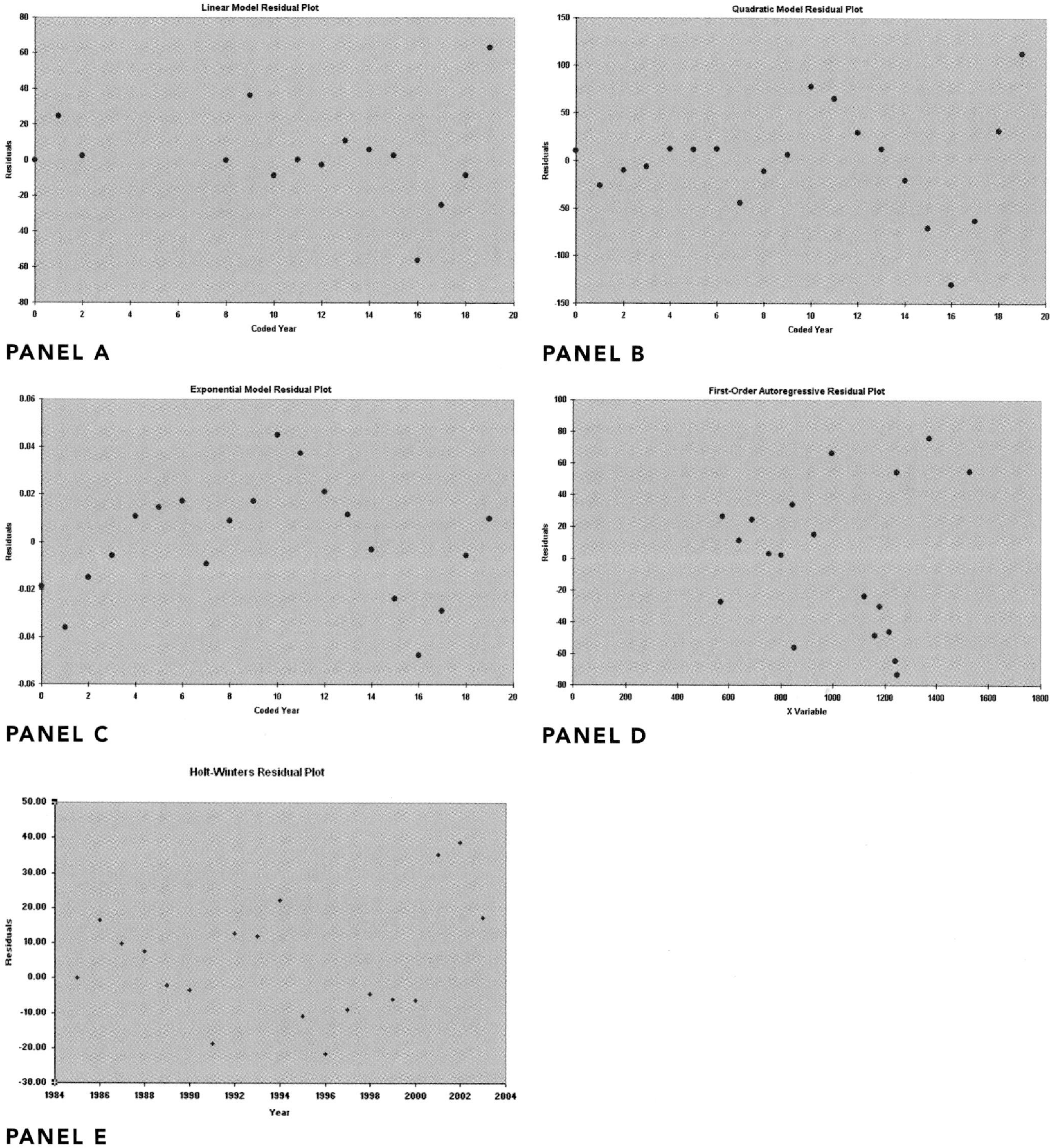

FIGURE 16.28 Microsoft Excel Residual Plots for Five Forecasting Methods

In Figure 16.28, observe the systematic structure of the residuals in the quadratic model (panel B) and the exponential model (panel C). For the linear model (panel A), the first-order autoregressive model (panel D), and the Holt-Winters model (panel E), the residuals appear more random.

To summarize, on the basis of the residual analysis of all five forecasting models, it appears that the linear model, the Holt-Winters model, and the first-order autoregressive model are the most appropriate and the quadratic and exponential models are the least appropriate. For further verification, you can compare the five models with respect to the magnitude of their residuals. Figure 16.29 provides the actual values (Y_i) along with the predicted values ($\hat{Y}_i$), the residuals (e_i), the error sum of squares (*SSE*), the standard error of the estimate (S_{YX}), and the mean absolute deviation (*MAD*) for each of the five models.

FIGURE 16.29

Comparison of Five Forecasting Methods Using *SSE*, S_{YX}, and *MAD*

		Linear		Quadratic		Exponential		Autoreg-First order		Holt-Winters	
Year	Real	Predicted	Residual	Predicted	Residual	Predicted	Residual	Predicted	Residual	Predicted	Residual
1984	569	532.578	36.238	558.260	10.556	593.810	-24.993	#N/A	#N/A	#N/A	#N/A
1985	576	585.059	-8.850	602.630	-26.422	626.233	-50.025	603.345	-27.137	576.00	0.00
1986	638	637.539	0.235	647.902	-10.128	660.426	-22.653	611.139	26.635	621.50	16.50
1987	688	690.019	-2.519	694.074	-6.574	696.487	-8.987	676.052	11.448	678.24	9.76
1988	753	742.499	10.671	741.148	12.022	734.516	18.654	728.483	24.687	745.54	7.46
1989	801	794.980	5.827	789.122	11.684	774.622	26.184	797.724	3.082	802.99	-1.99
1990	850	847.460	2.578	837.998	12.040	816.918	33.120	847.951	2.087	853.55	-3.55
1991	844	899.940	-56.328	887.775	-44.163	861.523	-17.911	899.860	-56.248	862.78	-18.78
1992	927	952.420	-25.122	938.453	-11.154	908.564	18.734	893.085	34.214	914.45	12.55
1993	997	1004.901	-8.361	990.032	6.508	958.174	38.366	981.322	15.217	985.10	11.90
1994	1121	1057.381	63.402	1042.512	78.270	1010.492	110.291	1054.329	66.454	1098.93	22.07
1995	1161	1109.861	51.556	1095.894	65.524	1065.667	95.751	1185.329	-23.911	1171.89	-10.89
1996	1180	1162.341	17.391	1150.176	29.556	1123.854	55.878	1228.173	-48.441	1201.74	-21.74
1997	1217	1214.822	2.624	1205.360	12.086	1185.219	32.227	1247.484	-30.039	1225.95	-8.95
1998	1241	1267.302	-26.198	1261.445	-20.340	1249.934	-8.830	1287.248	-46.144	1245.62	-4.62
1999	1248	1319.782	-71.883	1318.430	-70.531	1318.183	-70.283	1312.194	-64.294	1254.16	-6.16
2000	1246	1372.262	-126.037	1376.317	-130.092	1390.158	-143.933	1319.358	-73.133	1252.30	-6.30
2001	1372	1424.743	-52.637	1435.106	-62.999	1466.063	-93.957	1317.593	54.513	1336.86	35.14
2002	1526	1477.223	49.181	1494.795	31.609	1546.113	-19.710	1450.320	76.084	1487.24	38.76
2003	1668	1529.703	138.232	1555.385	112.550	1630.534	37.401	1613.008	54.927	1650.75	17.25
		SSE:	58449.502	SSE:	54885.516	SSE:	69838.069	SSE:	38391.596	SSE:	5384.01
		SYX:	56.984	SYX:	56.820	SYX:	62.289	SYX:	47.522	SYX:	17.80
		MAD:	37.793	MAD:	38.240	MAD:	46.394	MAD:	38.879	MAD:	13.39

For this time series, the *SSE*, S_{YX}, and *MAD* provide similar results. A comparison of the *SSE*, S_{YX}, and *MAD* clearly indicates that the exponential model provides the poorest fit. The Holt-Winters model provides the best fit. Considering the results of the residual analysis and *SSE*, S_{YX}, and *MAD*, the choice for the best model is the Holt-Winters model, the first-order autoregressive model, or the linear trend model. Although the Holt-Winters model is clearly superior in terms of the historical fit of the data, since the choice of the smoothing constants is subjective, you might want to proceed cautiously with this model and instead consider either the linear model or the first-order autoregressive model.

Once you have selected a particular forecasting model, you need to continually monitor your forecasts. If large errors occur between forecasted and actual values, the underlying structure of the time series may have changed. Remember that the forecasting methods presented in this chapter assume that the patterns inherent in the past will continue into the future. Large forecast errors are an indication that this assumption is no longer true.

PROBLEMS FOR SECTION 16.7

Learning the Basics

PH Grade ASSIST **16.40** The following residuals are from a linear trend model used to forecast sales.

2.0 −0.5 1.5 1.0 0.0 1.0 −3.0 1.5 −4.5 2.0 0.0 −1.0

a. Compute S_{YX} and interpret your findings.

b. Compute the *MAD* and interpret your findings.

PH Grade ASSIST **16.41** Refer to problem 16.40. Suppose the first residual is 12.0 (instead of 2.0) and the last value is −11.0 (instead of −1.0).

a. Compute S_{YX} and interpret your findings

b. Compute the *MAD* and interpret your findings.

Applying the Concepts

You should use Microsoft Excel, Minitab, or SPSS to solve problems 16.42–16.47.

16.42 Refer to the results in problem 16.13 on page 677. GDP

a. Perform a residual analysis.
b. Compute the standard error of the estimate (S_{YX}).
c. Compute the *MAD*.
d. On the basis of (a), (b), and (c), are you satisfied with your linear trend forecasts in problem 16.13? Discuss.

16.43 Refer to the results in problem 16.15 on page 677, problem 16.27 on page 682, and problem 16.35 on page 692 concerning the number of barrels of oil in the U.S. strategic oil reserve. STRATEGIC

a. Perform a residual analysis for each model.
b. Compute the standard error of the estimate (S_{YX}) for each model.
c. Compute the *MAD* for each model.
d. On the basis of (a), (b), (c), and parsimony, which forecasting model would you select? Discuss.

16.44 Refer to the results in problem 16.16 on page 677, problem 16.28 on page 682, and problem 16.36 on page 692 concerning annual net operating revenues at Coca-Cola. COCACOLA

a. Perform a residual analysis for each model.
b. Compute the standard error of the estimate (S_{YX}) for each model.
c. Compute the *MAD* for each model.
d. On the basis of (a), (b), (c), and parsimony, which forecasting model would you select? Discuss.

16.45 Refer to the results in problem 16.17 on page 678, problem 16.29 on page 682, and problem 16.37 on page 692 concerning the Dow Jones Industrial Average. DJIA

a. Perform a residual analysis for each model.
b. Compute the standard error of the estimate (S_{YX}) for each model.
c. Compute the *MAD* for each model.
d. On the basis of (a), (b), (c), and parsimony, which forecasting model would you select? Discuss.

16.46 Refer to the results in problem 16.18 on page 678, problem 16.30 on page 682, and problem 16.38 on page 692 concerning the price per share for Procter & Gamble stock. P&G

a. Perform a residual analysis for each model.
b. Compute the standard error of the estimate (S_{YX}) for each model.
c. Compute the *MAD* for each model.
d. On the basis of (a), (b), (c), and parsimony, which forecasting model would you select? Discuss.

SELF Test **16.47** Refer to the results in problem 16.22 on page 679 and problem 16.39 on page 693 concerning the sales at fast-food chains in the United States. FASTFOODSALES

a. Perform a residual analysis for each model.
b. Compute the standard error of the estimate (S_{YX}) for each model.
c. Compute the *MAD* for each model.
d. On the basis of (a), (b), (c), and the principle of parsimony, which forecasting model would you select? Discuss.

16.8 TIME-SERIES FORECASTING OF SEASONAL DATA

So far this chapter has focused on time-series forecasting with annual data. However, numerous time series are collected quarterly or monthly, and others are collected weekly, daily, and even hourly. When a time series is collected quarterly or monthly, you must consider the impact of seasonal effects (see Table 16.1 on page 660). In this section, regression model building is used to forecast monthly or quarterly data.

One of the companies of interest in the "Using Statistics" scenario is Wal-Mart Stores, Inc. In 2004, Wal-Mart operated over 5,000 Wal-Marts, Supercenters, Sam's Clubs, or Neighborhood Markets. Revenues in 2004 exceeded $256 billion (Wal-Mart Stores, Inc., **investor.walmartstores.com**, April 8, 2004). Sales for Wal-Mart are very seasonal, and therefore you need to analyze quarterly revenue. The fiscal year for the company ends on January 31. Thus, the fourth quarter of 2004 includes November and December of 2003 and January of 2004. Table 16.4 provides the quarterly revenues in billions of dollars from 1996 to 2004. WALMART Figure 16.30 displays the time series.

For quarterly time series such as these, the *classical multiplicative time-series model* includes the seasonal component in addition to the trend, cyclical, and irregular components. It is expressed by Equation (16.2) on page 660 as

$$Y_i = T_i \times S_i \times C_i \times I_i$$

TABLE 16.4
Quarterly Revenues for Wal-Mart Stores, Inc., in Billions of Dollars (1996–2004)

Quarter	1996	1997	1998	1999	2000	2001	2002	2003	2004
1	20.4	22.8	25.4	29.8	34.7	43.0	48.6	55.0	56.7
2	22.7	25.6	28.4	33.5	38.2	46.1	53.3	59.7	62.5
3	22.9	25.6	28.8	33.5	40.4	45.7	51.8	58.8	62.5
4	27.6	30.9	35.4	40.8	51.4	56.6	64.2	71.1	74.5

Source: Extracted from Standard & Poor's Stock Reports, *November 1995, November 1999. New York: McGraw-Hill, Inc., and Wal-Mart Stores, Inc.,* ***investor.walmartstores.com****, April 8, 2004.*

FIGURE 16.30
Microsoft Excel Plot of Quarterly Revenues for Wal-Mart Stores, Inc., in Billions of Dollars (1996–2004)

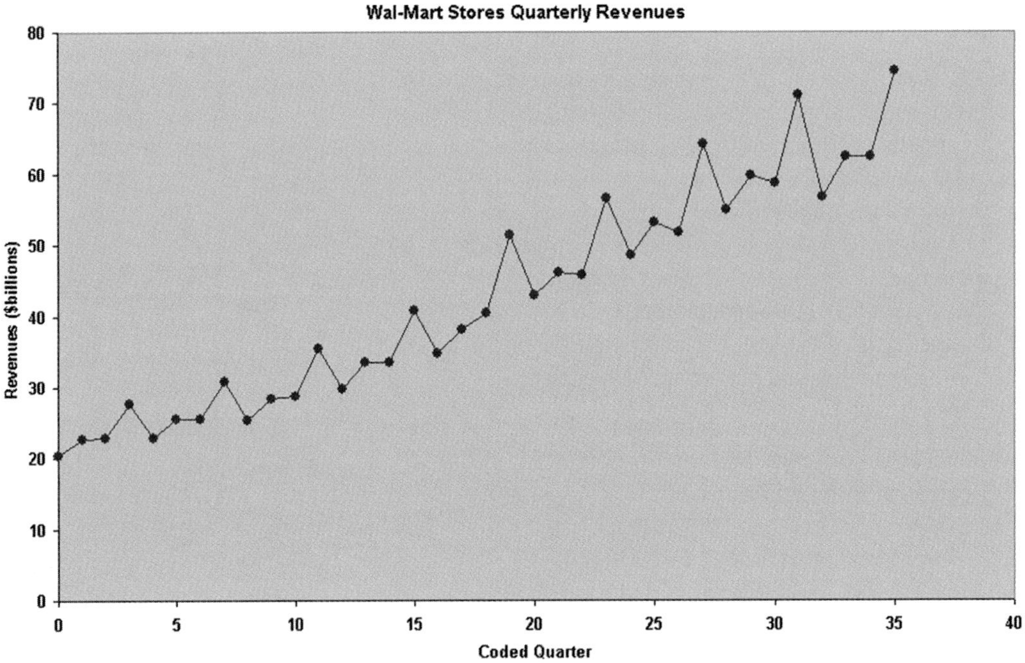

Least-Squares Forecasting with Monthly or Quarterly Data

To develop a least-squares regression model that includes trend, seasonal, cyclical, and irregular components, the approach to least-squares trend-fitting in section 16.4 is combined with the approach to model building using categorical independent variables (see section 14.6) to model the seasonal component.

Equation (16.19) defines the exponential trend model for quarterly data.

EXPONENTIAL MODEL WITH QUARTERLY DATA

$$Y_i = \beta_0 \beta_1^{X_i} \beta_2^{Q_1} \beta_3^{Q_2} \beta_4^{Q_3} \varepsilon_i \qquad \textbf{(16.19)}$$

where X_i = coded quarterly value, $i = 0, 1, 2, \ldots$

Q_1 = 1 if first quarter, 0 if not first quarter

Q_2 = 1 if second quarter, 0 if not second quarter

Q_3 = 1 if third quarter, 0 if not third quarter

β_0 = Y intercept

$(\beta_1 - 1) \times 100\%$ is the quarterly compound growth rate (in %)

β_2 = multiplier for first quarter relative to fourth quarter

β_3 = multiplier for second quarter relative to fourth quarter

β_4 = multiplier for third quarter relative to fourth quarter

ε_i = value of the irregular component for time period i

The model in Equation (16.19) is not in the form of a linear regression model. To transform this nonlinear model to a linear model, you use a base 10 logarithmic transformation.[4] Taking the logarithm of each side of Equation (16.19) yields Equation (16.20).

[4]*Alternatively, you can use base e logarithms. For more information on logarithms, see Appendix A.*

TRANSFORMED EXPONENTIAL MODEL WITH QUARTERLY DATA

$$\begin{aligned}\log(Y_i) &= \log(\beta_0\beta_1^{X_i}\beta_2^{Q_1}\beta_3^{Q_2}\beta_4^{Q_3}\varepsilon_i) \qquad \textbf{(16.20)}\\ &= \log(\beta_0) + \log(\beta_1^{X_i}) + \log(\beta_2^{Q_1}) + \log(\beta_3^{Q_2}) + \log(\beta_4^{Q_3}) + \log(\varepsilon_i)\\ &= \log(\beta_0) + X_i\log(\beta_1) + Q_1\log(\beta_2) + Q_2\log(\beta_3) + Q_3\log(\beta_4) + \log(\varepsilon_i)\end{aligned}$$

Equation (16.20) is a linear model that you can estimate using least-squares regression. Performing the regression using $\log(Y_i)$ as the dependent variable and X_i, Q_1, Q_2, and Q_3 as the independent variables results in Equation (16.21).

EXPONENTIAL GROWTH WITH QUARTERLY DATA FORECASTING EQUATION

$$\log(\hat{Y}_i) = b_0 + b_1X_i + b_2Q_1 + b_3Q_2 + b_4Q_3 \qquad \textbf{(16.21)}$$

where

b_0 = estimate of $\log(\beta_0)$ and thus $10^{b_0} = \hat{\beta}_0$

b_1 = estimate of $\log(\beta_1)$ and thus $10^{b_1} = \hat{\beta}_1$

b_2 = estimate of $\log(\beta_2)$ and thus $10^{b_2} = \hat{\beta}_2$

b_3 = estimate of $\log(\beta_3)$ and thus $10^{b_3} = \hat{\beta}_3$

b_4 = estimate of $\log(\beta_4)$ and thus $10^{b_4} = \hat{\beta}_4$

Equation (16.22) is used for monthly data.

EXPONENTIAL MODEL WITH MONTHLY DATA

$$Y_i = \beta_0\beta_1^{X_i}\beta_2^{M_1}\beta_3^{M_2}\beta_4^{M_3}\beta_5^{M_4}\beta_6^{M_5}\beta_7^{M_6}\beta_8^{M_7}\beta_9^{M_8}\beta_{10}^{M_9}\beta_{11}^{M_{10}}\beta_{12}^{M_{11}}\varepsilon_i \qquad \textbf{(16.22)}$$

where

X_i = coded monthly value, $i = 0, 1, 2, \ldots$

M_1 = 1 if January, 0 if not January

M_2 = 1 if February, 0 if not February

M_3 = 1 if March, 0 if not March

.
.
.

M_{11} = 1 if November, 0 if not November

β_0 = Y intercept

$(\beta_1 - 1) \times 100\%$ is the monthly compound growth rate (in %)

β_2 = multiplier for January relative to December

β_3 = multiplier for February relative to December

β_4 = multiplier for March relative to December

.
.
.

β_{12} = multiplier for November relative to December

ε_i = value of the irregular component for time period i

The model in Equation (16.22) is not in the form of a linear regression model. To transform this nonlinear model to a linear model, you can use a base 10 logarithm transformation. Taking the logarithm of each side of Equation (16.22) yields Equation (16.23).

TRANSFORMED EXPONENTIAL MODEL WITH MONTHLY DATA

$$\log(Y_i) = \log(\beta_0 \beta_1^{X_i} \beta_2^{M_1} \beta_3^{M_2} \beta_4^{M_3} \beta_5^{M_4} \beta_6^{M_5} \beta_7^{M_6} \beta_8^{M_7} \beta_9^{M_8} \beta_{10}^{M_9} \beta_{11}^{M_{10}} \beta_{12}^{M_{11}} \varepsilon_i) \quad \textbf{(16.23)}$$

$$\begin{aligned} &= \log(\beta_0) + X_i \log(\beta_1) + M_1 \log(\beta_2) + M_2 \log(\beta_3) \\ &\quad + M_3 \log(\beta_4) + M_4 \log(\beta_5) + M_5 \log(\beta_6) + M_6 \log(\beta_7) \\ &\quad + M_7 \log(\beta_8) + M_8 \log(\beta_9) + M_9 \log(\beta_{10}) + M_{10} \log(\beta_{11}) \\ &\quad + M_{11} \log(\beta_{12}) + \log(\varepsilon_i) \end{aligned}$$

Equation (16.23) is a linear model that you can estimate using the least-squares method. Performing the regression using $\log(Y_i)$ as the dependent variable and X_i, M_1, M_2, . . . , and M_{11} as the independent variables results in Equation (16.24).

EXPONENTIAL GROWTH WITH MONTHLY DATA FORECASTING EQUATION

$$\begin{aligned} \log(\hat{Y}_i) = b_0 + b_1 X_i + b_2 M_1 + b_3 M_2 + b_4 M_3 + b_5 M_4 + b_6 M_5 + b_7 M_6 \\ + b_8 M_7 + b_9 M_8 + b_{10} M_9 + b_{11} M_{10} + b_{12} M_{11} \end{aligned} \quad \textbf{(16.24)}$$

where

b_0 = estimate of $\log(\beta_0)$ and thus $10^{b_0} = \hat{\beta}_0$

b_1 = estimate of $\log(\beta_1)$ and thus $10^{b_1} = \hat{\beta}_1$

b_2 = estimate of $\log(\beta_2)$ and thus $10^{b_2} = \hat{\beta}_2$

b_3 = estimate of $\log(\beta_3)$ and thus $10^{b_3} = \hat{\beta}_3$

.
.
.

b_{12} = estimate of $\log(\beta_{12})$ and thus $10^{b_{12}} = \hat{\beta}_{12}$

Q_1, Q_2, and Q_3 are the three dummy variables needed to represent the four quarter periods in a quarterly time series. M_1, M_2, M_3, . . . , M_{11} are the 11 dummy variables needed to represent the 12 months in a monthly time series. In building the model, you use $\log(Y_i)$ instead of Y_i values and then find the actual regression coefficients by taking the antilog of the regression coefficients developed from Equations (16.21) and (16.24).

Although at first glance these regression models look imposing, when fitting or forecasting in any one time period, the values of all or all but one of the dummy variables in the model are set equal to zero, and the equations simplify dramatically. In establishing the dummy variables for quarterly time-series data, the fourth quarter is the base period and has a coded value of zero for each dummy variable. With a quarterly time series, Equation (16.21) reduces as follows.

For any first quarter: $\log(\hat{Y}_i) = b_0 + b_1 X_i + b_2$

For any second quarter: $\log(\hat{Y}_i) = b_0 + b_1 X_i + b_3$

For any third quarter: $\log(\hat{Y}_i) = b_0 + b_1 X_i + b_4$

For any fourth quarter: $\log(\hat{Y}_i) = b_0 + b_1 X_i$

When establishing the dummy variables for each month, December serves as the base period and has a coded value of 0 for each dummy variable. For example, with a monthly time series, Equation (16.24) reduces as follows.

For any January: $\log(\hat{Y}_i) = b_0 + b_1X_i + b_2$

For any December: $\log(\hat{Y}_i) = b_0 + b_1X_i$

To demonstrate the process of model building and least-squares forecasting with a quarterly time series, return to the Wal-Mart revenue data (in billions of dollars) originally displayed in Table 16.4 on page 698. The data are from each quarter from the first quarter of 1996 through the last quarter of 2004. Microsoft Excel output for the quarterly exponential trend model is displayed in Figure 16.31. Figure 16.32 illustrates Minitab output.

FIGURE 16.31
Microsoft Excel Output for Fitting and Forecasting with the Quarterly Wal-Mart Revenue Data

	A	B	C	D	E	F	G
1	**Quarterly Revenues Regression Analysis for Wal-Mart Stores Inc 1996-2004**						
2							
3	*Regression Statistics*						
4	**Multiple R**	0.9944					
5	**R Square**	0.9888					
6	**Adjusted R Square**	0.9874					
7	**Standard Error**	0.0183					
8	**Observations**	36					
9							
10	**ANOVA**						
11		*df*	*SS*	*MS*	*F*	*Significance F*	
12	**Regression**	4	0.9151	0.2288	686.7545	0.0000	
13	**Residual**	31	0.0103	0.0003			
14	**Total**	35	0.9255				
15							
16		*Coefficients*	*Standard Error*	*t Stat*	*P-value*	*Lower 95%*	*Upper 95%*
17	**Intercept**	1.3964	0.0083	168.9281	0.0000	1.3796	1.4133
18	**Coded Q**	0.0147	0.0003	49.9943	0.0000	0.0141	0.0153
19	**Q1**	-0.0875	0.0086	-10.1137	0.0000	-0.1051	-0.0698
20	**Q2**	-0.0593	0.0086	-6.8778	0.0000	-0.0769	-0.0417
21	**Q3**	-0.0728	0.0086	-8.4529	0.0000	-0.0903	-0.0552

FIGURE 16.32
Minitab Output for Fitting and Forecasting with the Quarterly Wal-Mart Revenue Data

```
The regression equation is
Log(Revenues) = 1.40 + 0.0147 Coded Quarter - 0.0875 Q1 - 0.0593 Q2 - 0.0728 Q3


Predictor            Coef     SE Coef       T      P
Constant          1.39643     0.00827  168.93  0.000
Coded Quarter   0.0147253   0.0002945   49.99  0.000
Q1              -0.087477    0.008649  -10.11  0.000
Q2              -0.059316    0.008624   -6.88  0.000
Q3              -0.072772    0.008609   -8.45  0.000


S = 0.0182520   R-Sq = 98.9%   R-Sq(adj) = 98.7%


Analysis of Variance

Source           DF        SS        MS       F      P
Regression        4   0.91513   0.22878  686.75  0.000
Residual Error   31   0.01033   0.00033
Total            35   0.92546
```

From Figure 16.31 or 16.32, the model fits the data extremely well. The coefficient of determination $r^2 = 98.9\%$ and the adjusted $r^2 = 98.7\%$, and the overall F test results in an F statistic of 686.75 (p-value = 0.000). Looking further, at the 0.05 level of significance, each regression coefficient is statistically significant and contributes to the classical multiplicative time-series model. Taking the antilogs of all the regression coefficients, you have the following summary:

Regression Coefficients	$b_i = \log \hat{\beta}_i$	$\hat{\beta}_i = \text{antilog}(b_i) = 10^{b_i}$
b_0: Y intercept	1.3964	24.9115
b_1: coded quarter	0.0147	1.0344
b_2: first quarter	−0.0875	0.8175
b_3: second quarter	−0.0593	0.8724
b_4: third quarter	−0.0728	0.8457

The interpretations for $\hat{\beta}_0$, $\hat{\beta}_1$, $\hat{\beta}_2$, $\hat{\beta}_3$, and $\hat{\beta}_4$ are as follows:

- The Y intercept $\hat{\beta}_0 = 24.9115$ (in billions of dollars) is the *unadjusted* forecast for quarterly revenues in the first quarter of 1996, the initial quarter in the time series. *Unadjusted* means that the seasonal component is not incorporated in the forecast.
- The value $(\hat{\beta}_1 - 1) \times 100\% = 0.0344$ or 3.44% is the estimated *quarterly compound growth rate* in revenues, after adjusting for the seasonal component.
- $\hat{\beta}_2 = 0.8175$ is the seasonal multiplier for the first quarter relative to the fourth quarter; it indicates that there is 18.25% less revenue for the first quarter as compared with the fourth quarter.
- $\hat{\beta}_3 = 0.8724$ is the seasonal multiplier for the second quarter relative to the fourth quarter; it indicates that there is 12.76% less revenue for the second quarter as compared with the fourth quarter.
- $\hat{\beta}_4 = 0.8457$ is the seasonal multiplier for the third quarter relative to the fourth quarter; it indicates that there is 15.43% less revenue for the third quarter than the fourth quarter. Thus, the fourth quarter, which includes the holiday shopping season, has the strongest sales.

Using the regression coefficients b_0, b_1, b_2, b_3, b_4, and Equation (16.21) on page 699, you can make forecasts for selected quarters. As an example, to predict revenues for the fourth quarter of 2004 ($X_i = 35$):

$$\begin{aligned}\log(\hat{Y}_i) &= b_0 + b_1 X_i \\ &= 1.3964 + (0.0147)(35) \\ &= 1.9109\end{aligned}$$

Thus,

$$\hat{Y}_i = 10^{1.9109} = 81.4517$$

The predicted revenue for the fourth quarter of 2004 is 81.4517 billion dollars. To make a forecast for a future time period, such as the first quarter of 2005 ($X_i = 36$, $Q_1 = 1$):

$$\begin{aligned}\log(\hat{Y}_i) &= b_0 + b_1 X_i + b_2 Q_1 \\ &= 1.3964 + (0.0147)(36) + (-0.0875)(1) \\ &= 1.8381\end{aligned}$$

Thus,

$$\hat{Y}_i = 10^{1.8381} = 68.8811$$

The predicted revenue for the first quarter of 2005 is 68.8811 billion dollars.

PROBLEMS FOR SECTION 16.8

Learning the Basics

PH Grade ASSIST **16.48** In forecasting a monthly time series over a 5-year period from January 2000 to December 2004, the exponential trend forecasting equation for January is

$$\log \hat{Y}_i = 2.0 + 0.01X_i + 0.10 \text{ January}$$

Take the antilog of the appropriate coefficient from the above equation and interpret

a. the Y intercept $\hat{\beta}_0$

b. the monthly compound growth rate.

c. the January multiplier.

16.49 If forecasting weekly time-series data, how many dummy variables are needed to account for the seasonal categorical variable week?

PH Grade ASSIST **16.50** In forecasting a quarterly time series over the 5-year period from the first-quarter 2000 through the fourth-quarter 2004, the exponential trend forecasting equation is given by

$$\log \hat{Y}_i = 3.0 + 0.10X_i - 0.25Q_1 + 0.20Q_2 + 0.15Q_3$$

where quarter zero is first-quarter 2000. Take the antilog of the appropriate coefficient from the above equation and interpret the

a. Y intercept $\hat{\beta}_0$.

b. quarterly compound growth rate.

c. second-quarter multiplier.

16.51 Refer to the exponential model given in problem 16.50.

a. What is the fitted value of the series in the fourth quarter of 2002?

b. What is the fitted value of the series in the first quarter of 2003?

c. What is the forecast in the fourth quarter of 2005?

d. What is the forecast in the first quarter of 2006?

Applying the Concepts

You should use Microsoft Excel, Minitab, or SPSS to solve problems 16.52–16.57.

16.52 The data given in the following table represent Standard & Poor's Composite Stock Price Index recorded at the end of each quarter from 1994 through the second quarter of 2004. S&PSTKIN

	Year					
Quarter	**1994**	**1995**	**1996**	**1997**	**1998**	**1999**
1	445.77	500.71	645.50	757.12	1,101.75	1,286.37
2	444.27	544.75	670.63	885.14	1,133.84	1,372.71
3	462.69	584.41	687.31	947.28	1,017.01	1,282.71
4	459.27	615.93	740.74	970.43	1,229.23	1,469.25

	Year				
Quarter	**2000**	**2001**	**2002**	**2003**	**2004**
1	1,498.58	1,160.33	1,147.38	848.18	1,126.21
2	1,454.60	1,224.38	989.81	974.51	1,140.81
3	1,436.51	1,040.94	815.28	995.97	
4	1,320.28	1,148.08	879.28	1,111.92	

Source: Extracted from ***www.yahoo.com***

a. Plot the data.

b. Develop an exponential trend forecasting equation with quarterly components.

c. What is the fitted value in the first quarter of 2004?

d. What is the fitted value in the second quarter of 2004?

e. What are the forecasts for the last two quarters of 2004 and all four quarters of 2005?

f. Interpret the quarterly compound growth rate.

g. Interpret the second-quarter multiplier.

16.53 Are gasoline prices higher during the height of the summer vacation season? The following table contains the mean monthly price (in dollars per gallon) for unleaded gasoline in the United States from 2000 to 2003. UNLEADED

	Year			
Month	**2000**	**2001**	**2002**	**2003**
January	1.301	1.472	1.139	1.473
February	1.369	1.484	1.130	1.641
March	1.541	1.447	1.241	1.748
April	1.506	1.564	1.407	1.659
May	1.498	1.729	1.421	1.542
June	1.617	1.640	1.404	1.514
July	1.593	1.482	1.412	1.524
August	1.510	1.427	1.423	1.628
September	1.582	1.531	1.422	1.728
October	1.559	1.362	1.449	1.603
November	1.555	1.263	1.448	1.535
December	1.489	1.131	1.394	1.494

Source: Bureau of Labor Statistics, U.S. Department of Labor ***bls.gov***.

a. Construct a time-series plot.

b. Develop an exponential trend forecasting equation for monthly data.

c. Interpret the monthly compound growth rate.

d. Interpret the monthly multipliers.

e. Write a short summary of your findings.

16.54 The U.S. Bureau of Labor Statistics compiles data on a wide variety of workforce issues. The data in the file UERATE gives the monthly seasonally adjusted civilian unemployment rate for the United States from 1996 through 2003.

Source: Bureau of Labor Statistics, U.S. Department of Labor ***bls.gov***.

a. Plot the time-series data.
b. Develop an exponential trend forecasting equation with monthly components.
c. What is the fitted value in December 2003?
d. What are the forecasts for all 12 months of 2004?
e. Interpret the monthly compound growth rate.
f. Interpret the July multiplier.
g. Go to your library or the Internet and locate the actual unemployment rate in 2004. Discuss.

SELF Test **16.55** The following data are monthly credit card charges (in millions of dollars) for a popular credit card issued by a large bank. (The name of which is not disclosed at its request.) CREDIT

	Year		
Month	**2001**	**2002**	**2003**
January	31.9	39.4	45.0
February	27.0	36.2	39.6
March	31.3	40.5	
April	31.0	44.6	
May	39.4	46.8	
June	40.7	44.7	
July	42.3	52.2	
August	49.5	54.0	
September	45.0	48.8	
October	50.0	55.8	
November	50.9	58.7	
December	58.5	63.4	

a. Construct the time-series plot.
b. Describe the monthly pattern that is evident in the data.
c. In general, would you say that the overall dollar amounts charged on the bank's credit cards is increasing or decreasing? Explain.
d. Note that December 2002 charges were over $63 million but February 2003 was under $40 million. Was February's total close to what you should have expected?
e. Develop an exponential trend forecasting equation with monthly components.
f. Interpret the monthly compound growth rate.
g. Interpret the January multiplier.
h. What is the predicted value for March 2003?
i. What is the predicted value for April 2003?
j. How can this type of time-series forecasting benefit the bank?

16.56 The data in the file TOYS-REV are quarterly revenues (in millions of dollars) for Toys Я Us from 1992 through 2003. The fiscal year for the company ends on February 1st. Thus, the fourth quarter of 2003 includes November and December of 2003 and January of 2004.

Source: Extracted from Standard & Poor's Stock Reports, *November 1995, November 1998, April 2002. New York: McGraw-Hill, Inc., and Toys Я Us, Inc.* ***www4.toysrus.com***.

a. Do you think that the revenues for Toys Я Us are subject to seasonal variation? Explain.
b. Plot the data. Does this chart support your answer to (a)?
c. Develop an exponential trend forecasting equation with quarterly components.
d. Interpret the quarterly compound growth rate.
e. Interpret the quarter multipliers.
f. What are the forecasts for all four quarters of 2004?

16.57 The data in the file FORD-REV are quarterly revenues (in millions of dollars) for the Ford Motor Company, from 1992 through 2003.

Source: Standard & Poor's Stock Reports, *November 1995, November 2000, April 2002. New York: McGraw-Hill, Inc., the Ford Motor Company* ***ford.com***.

a. Do you think that the revenues for the Ford Motor Company are subject to seasonal variation? Explain.
b. Plot the data. Does this chart support your answer to (a)?
c. Develop an exponential trend forecasting equation with quarterly components.
d. Interpret the quarterly compound growth rate.
e. Interpret the quarter multipliers.
f. What are the forecasts for all four quarters of 2004?

16.9 INDEX NUMBERS

This chapter has presented various methods to study time series. In this section, index numbers are used to compare a value of a time series relative to another value of a time series. **Index numbers** measure the value of an item (or group of items) at a particular point in time as a percentage of an item's (or group of items) value at another point in time. They are commonly used in business and economics as indicators of changing business or economic activity. There are many kinds of index numbers including price indexes, quantity indexes, value indexes, and sociological indexes. In this section, only the price index is considered. In addition to allowing comparison of prices at different points in time, price indexes are also used to deflate the effect of inflation on a time series in order to compare values in real dollars instead of actual dollars.

The Price Index

A **price index** compares the price of a commodity in a given period of time to the price paid for that commodity at a particular point of time in the past. A **simple price index** tracks the price of a single commodity. An **aggregate price index** tracks the prices for a group of commodities (called a market basket) at a given period of time to the price paid for that group of commodities at a particular point of time in the past. The **base period** is the point of time in the past against which all comparisons are made. In selecting the base period for a particular index, if possible, you select a period of economic stability, rather than one at or near the peak of an expanding economy or the bottom of a recession or declining economy. In addition, the base period should be recent, so that comparisons are not greatly affected by changing technology and consumer attitudes and habits. Equation (16.25) defines the simple price index.

SIMPLE PRICE INDEX

$$I_i = \frac{P_i}{P_{base}} \times 100 \qquad \textbf{(16.25)}$$

where

I_i = price index for year i

P_i = price for year i

P_{base} = price for the base year

As an example of the simple price index, consider the price per gallon of unleaded gasoline in the United States from 1980 to 2003. GASOLINE Table 16.5 presents the price data plus two sets of index numbers. To illustrate the computation of the simple price index for 2003, using 1980 as the base year, from Equation (16.25) and Table 16.5,

$$I_{2003} = \frac{P_{2003}}{P_{1980}} \times 100 = \frac{1.59}{1.25} \times 100 = 127.2$$

TABLE 16.5

Price per Gallon of Unleaded Gasoline in the United States and Simple Price Index with 1980 and 1995 as the Base Years (1980–2003)

Year	Gasoline Price	Price Index—1980	Price Index—1995
1980	1.25	100.0	108.7
1981	1.38	110.4	120.0
1982	1.30	104.0	113.0
1983	1.24	99.2	107.8
1984	1.21	96.8	105.2
1985	1.20	96.0	104.3
1986	0.93	74.4	80.9
1987	0.95	76.0	82.6
1988	0.95	76.0	82.6
1989	1.02	81.6	88.7
1990	1.16	92.8	100.9
1991	1.14	91.2	99.1
1992	1.14	91.2	99.1
1993	1.11	88.8	96.5
1994	1.11	88.8	96.5
1995	1.15	92.0	100.0
1996	1.23	98.4	107.0
1997	1.23	98.4	107.0
1998	1.06	84.8	92.2
1999	1.17	93.6	101.7
2000	1.51	120.8	131.3
2001	1.46	116.8	127.0
2002	1.36	108.8	118.3
2003	1.59	127.2	138.3

Source: Bureau of Labor Statistics, U.S. Department of Labor ***www.bls.gov***.

Therefore, the price per gallon of unleaded gasoline in the United States in 2003 was 27.2% higher than in 1980. An examination of the price indexes for 1980–2003 in Table 16.5 indicates that the price of unleaded gasoline increased in 1981 and 1982 over the base year of 1980, but then was below the 1980 price every year until 2000. Since the base period for the index numbers in Table 16.5 is the initial year 1980, you should use a base year closer to the present. The price remained fairly constant from 1990–1995, thus you should use 1995 as a base year. Equation (16.26) is used to develop index numbers with a new base.

SHIFTING THE BASE FOR A SIMPLE PRICE INDEX

$$I_{new} = \frac{I_{old}}{I_{new\,base}} \times 100 \tag{16.26}$$

where I_{new} = new price index

I_{old} = old price index

$I_{new\,base}$ = value of the old price index for the new base year

To change the base year to 1995, $I_{new\,base} = 92.0$. Using Equation (16.26) to find the new price index for 2003,

$$I_{new} = \frac{I_{old}}{I_{new\,base}} \times 100 = \frac{127.2}{92.0} \times 100 = 138.3$$

Thus, the 2003 price for unleaded gasoline in the United States was 38.3% higher than it was in 1995. See Table 16.5 for the complete set of price indexes.

Aggregate Price Indexes

What is more important than a price index for any individual commodity is an index that consists of a group of commodities taken together. There are two basic types of aggregate price indexes: unweighted aggregate price indexes and weighted aggregate price indexes. An **unweighted aggregate price index**, defined in Equation (16.27), places equal weight on all the items in the market basket.

UNWEIGHTED AGGREGATE PRICE INDEX

$$I_U^{(t)} = \frac{\sum_{i=1}^{n} P_i^{(t)}}{\sum_{i=1}^{n} P_i^{(0)}} \times 100 \tag{16.27}$$

where t = time period (0, 1, 2, . . .)

i = item (1, 2, . . . , n)

n = total number of items under consideration

$\sum_{i=1}^{n} P_i^{(t)}$ = sum of the prices paid for each of the n commodities at time period t

$\sum_{i=1}^{n} P_i^{(0)}$ = sum of the prices paid for each of the n commodities at time period 0

$I_U^{(t)}$ = value of the unweighted price index at time period t

Table 16.6 presents the mean prices for three fruit items for selected periods from 1980 to 2000. **FRUIT**

TABLE 16.6
Prices (in Dollars per Pound) for Three Fruit Items

	Year				
Fruit	**1980** $P_i^{(0)}$	**1985** $P_i^{(1)}$	**1990** $P_i^{(2)}$	**1995** $P_i^{(3)}$	**2000** $P_i^{(4)}$
Apples	0.692	0.684	0.719	0.835	0.927
Bananas	0.342	0.367	0.463	0.490	0.509
Oranges	0.365	0.533	0.570	0.625	0.638

Source: Bureau of Labor Statistics, U.S. Department of Labor ***www.bls.gov***.

To calculate the unweighted aggregate price index for the various years, using Equation (16.27) and 1980 as the base period:

$$1980: I_U^{(0)} = \frac{\sum_{i=1}^{3} P_i^{(0)}}{\sum_{i=1}^{3} P_i^{(0)}} \times 100 = \frac{0.692 + 0.342 + 0.365}{0.692 + 0.342 + 0.365} \times 100 = \frac{1.399}{1.399} \times 100 = 100.0$$

$$1985: I_U^{(1)} = \frac{\sum_{i=1}^{3} P_i^{(1)}}{\sum_{i=1}^{3} P_i^{(0)}} \times 100 = \frac{0.684 + 0.367 + 0.533}{0.692 + 0.342 + 0.365} \times 100 = \frac{1.584}{1.399} \times 100 = 113.2$$

$$1990: I_U^{(2)} = \frac{\sum_{i=1}^{3} P_i^{(2)}}{\sum_{i=1}^{3} P_i^{(0)}} \times 100 = \frac{0.719 + 0.463 + 0.570}{0.692 + 0.342 + 0.365} \times 100 = \frac{1.752}{1.399} \times 100 = 125.2$$

$$1995: I_U^{(3)} = \frac{\sum_{i=1}^{3} P_i^{(3)}}{\sum_{i=1}^{3} P_i^{(0)}} \times 100 = \frac{0.835 + 0.490 + 0.625}{0.692 + 0.342 + 0.365} \times 100 = \frac{1.950}{1.399} \times 100 = 139.4$$

$$2000: I_U^{(4)} = \frac{\sum_{i=1}^{3} P_i^{(4)}}{\sum_{i=1}^{3} P_i^{(0)}} \times 100 = \frac{0.927 + 0.509 + 0.638}{0.692 + 0.342 + 0.365} \times 100 = \frac{2.074}{1.399} \times 100 = 148.2$$

Thus, in 2000, the combined price of a pound of apples, a pound of bananas, and a pound of oranges was 48.2% more than it was in 1980.

An unweighted aggregate price index represents the changes in prices, over time, for an entire group of commodities. However, an unweighted aggregate price index has two shortcomings. First, the index considers each commodity in the group as equally important. Thus, the most expensive commodities per unit are overly influential. Second, not all the commodities are consumed at the same rate. In the unweighted index, changes in the price of the least consumed commodities are overly influential.

Weighted Aggregate Price Indexes

Due to the shortcomings of the unweighted aggregate price index, weighted aggregate price indexes are generally preferable. **Weighted aggregate price indexes** account for differences in the magnitude of prices per unit and differences in the consumption levels of the items in the market basket. There are two types of weighted aggregate price indexes commonly used in business and economics: the Laspeyres price index and the Paasche price index. Equation (16.28) defines the **Laspeyres price index** that uses the consumption quantities associated with the base year in the calculation of all price indexes in the series.

LASPEYRES PRICE INDEX

$$I_L^{(t)} = \frac{\sum_{i=1}^{n} P_i^{(t)} Q_i^{(0)}}{\sum_{i=1}^{n} P_i^{(0)} Q_i^{(0)}} \times 100 \qquad \textbf{(16.28)}$$

where t = time period (0, 1, 2, . . .)

i = item (1, 2, . . . , n)

n = total number of items under consideration

$Q_i^{(0)}$ = quantity of item i at time period 0

$I_L^{(t)}$ = value of the Laspeyres price index at time t

Table 16.7 gives the price and per capita consumption in pounds for the three fruit items comprising the market basket of interest. **FRUIT**

TABLE 16.7
Prices (in Dollars per Pound) and Quantities (Annual per Capita Consumption in Pounds) for Three Fruit Items

	Year				
Fruit	**1980** $P_i^{(0)}, Q_i^{(0)}$	**1985** $P_i^{(1)}, Q_i^{(1)}$	**1990** $P_i^{(2)}, Q_i^{(2)}$	**1995** $P_i^{(3)}, Q_i^{(3)}$	**2000** $P_i^{(4)}, Q_i^{(4)}$
Apples	0.692, 19.2	0.684, 17.3	0.719, 19.6	0.835, 18.9	0.927, 17.5
Bananas	0.342, 20.2	0.367, 23.5	0.463, 24.4	0.490, 27.4	0.509, 28.5
Oranges	0.365, 14.3	0.533, 11.6	0.570, 12.4	0.625, 12.0	0.638, 11.7

Source: Bureau of Labor Statistics, U.S. Department of Labor ***www.bls.gov****, and Statistical Abstract of the United States, U.S. Census Bureau* ***www.census.gov****.*

Using 1980 as the base year, you calculate the Laspeyres price index for 2000 (t = 4) using Equation (16.28):

$$I_L^{(4)} = \frac{\sum_{i=1}^{3} P_i^{(4)} Q_i^{(0)}}{\sum_{i=1}^{3} P_i^{(0)} Q_i^{(0)}} \times 100 = \frac{(0.927 \times 19.2) + (0.509 \times 20.2) + (0.638 \times 14.3)}{(0.692 \times 19.2) + (0.342 \times 20.2) + (0.365 \times 14.3)} \times 100$$

$$= \frac{37.2036}{25.4143} \times 100 = 146.4$$

Thus, the Laspeyres price index is 146.4, indicating that the cost of purchasing these three items in 2000 was 46.4% more than in 1980. This index is less than the unweighted index, 148.2, since the least purchased item, oranges, increased in price over the time span more than apples and bananas. In other words, since the more heavily consumed fruit experienced less of a price increase than oranges, the overall impact on purchasing the market basket is less using the Laspeyres method than the unweighted method.

The **Paasche price index** uses the consumption quantities in the year of interest instead of using the initial quantities. Thus, the Paasche index is a more accurate reflection of total consumption costs at that point in time. However, there are two major drawbacks of the Paasche index. First, accurate consumption values for current purchases are often hard to obtain. Thus, many important indexes such as the Consumer Price Index use the Laspeyres method. Second, if a particular product increases greatly in price compared to the other items in the market basket, consumers will avoid the high-priced item out of necessity, not because of changes in what they might prefer to purchase. Equation (16.29) defines the Paasche price index.

PAASCHE PRICE INDEX

$$I_P^{(t)} = \frac{\sum_{i=1}^{n} P_i^{(t)} Q_i^{(t)}}{\sum_{i=1}^{n} P_i^{(0)} Q_i^{(t)}} \times 100 \qquad \textbf{(16.29)}$$

where

t = time period (0, 1, 2, . . .)

i = item (1, 2, . . . , n)

n = total number of items under consideration

$Q_i^{(t)}$ = quantity of item i at time period t

$I_P^{(t)}$ = value of the Paasche price index at time period t

To calculate the Paasche price index in 2000 using 1980 as a base year, use $t = 4$ in Equation (16.29):

$$I_P^{(4)} = \frac{\sum_{i=1}^{3} P_i^{(4)} Q_i^{(4)}}{\sum_{i=1}^{3} P_i^{(0)} Q_i^{(4)}} \times 100 = \frac{(0.927 \times 17.5) + (0.509 \times 28.5) + (0.638 \times 11.7)}{(0.692 \times 17.5) + (0.342 \times 28.5) + (0.365 \times 11.7)} \times 100$$

$$= \frac{38.1936}{26.1275} \times 100 = 146.2$$

The Paasche price index for this market basket is 146.2. Thus, the cost of these three fruit items in 2000 was 46.2% higher in 2000 than in 1980 when using 2000 quantities.

Some Common Price Indexes

Various price indexes are commonly used in business and economics. The Consumer Price Index is the most familiar index in the United States. This index is officially referred to as the CPI-U to reflect that it measures the prices "urban" residents are subject to, but is commonly referred to as the CPI. The CPI, published monthly by the U.S. Bureau of Labor Statistics, is the primary measure of changes in the cost of living in the United States. The CPI is a weighted aggregate price index, using the Laspeyres method, for 400 commonly purchased food, clothing, transportation, medical, and housing items. Currently computed using 1982–1984 averages as a base year, the CPI was 184 in 2003. (See data file CPI-U for a listing of the CPI-U for 1965–2003.)

An important use of the CPI is as a price deflator. The CPI is used to convert (and deflate) actual dollars into real dollars by multiplying each dollar value in a time series by the quantity (100/CPI). For example, the gross revenues of the Wm. Wrigley Jr. Company were transformed from actual revenues to real gross revenues in Figure 16.5 on page 669. This transformation allowed you to see that the increase in revenues for Wrigley's were actually *real* increases, not simply increases that could be explained by an increase in the cost of living.

Another important price index published by the U.S. Bureau of Labor Statistics is the Producer Price Index (PPI). The PPI is a weighted aggregate price index, also using the Laspeyres method, for prices of commodities sold by wholesalers. The PPI is considered a leading indicator of the CPI. In other words, increases in the PPI tend to precede increases in the CPI, and similarly, decreases in the PPI tend to precede decreases in the CPI. Financial indexes such as the Dow Jones Industrial Average, the S&P 500, and the NASDAQ index are price indexes for different sets of stocks in the United States. Many indexes measure the performance of international stock markets including the Nikkei Index for Japan, the Dax 30 for Germany, and the SSE Composite for China.

PROBLEMS FOR SECTION 16.9

Learning the Basics

16.58 The simple price index for a commodity in 2002, using 1995 as the base year, is 175. Interpret this index number.

16.59 The following are prices for a commodity from 2002–2004:

2002	\$5
2003	\$8
2004	\$7

a. Calculate the simple price indexes for 2002–2004, using 2002 as the base year.
b. Calculate the simple price indexes for 2002–2004, using 2003 as the base year.

PH Grade ASSIST **16.60** The following are prices and consumption quantities for three commodities in 1995 and 2004:

	Year	
Commodity	1995 Price, Quantity	2004 Price, Quantity
A	\$2, 20	\$3, 21
B	\$18, 3	\$36, 2
C	\$3, 18	\$4, 23

a. Calculate the unweighted aggregate price index for 2004, using 1995 as the base year.
b. Calculate the Laspeyres aggregate price index for 2004, using 1995 as the base year.
c. Calculate the Paasche aggregate price index for 2004, using 1995 as the base year.

Applying the Concepts

16.61 The data in the file CPI-U reflect the annual values of the Consumer Price Index (CPI) in the United States constructed over the 39-year period from 1965 through 2003 using the years 1982 through 1984 as the base period.

Source: Bureau of Labor Statistics, U.S. Department of Labor ***bls.gov***.

a. Calculate the price index for the Consumer Price Index in the United States with 1965 as the base year.
b. Shift the base of the Consumer Price Index in the United States to 1990 and recalculate the price index.
c. Compare the results of (a) and (b). Which price index do you think is more useful in understanding the changes in the Consumer Price Index in the United States? Explain.

16.62 The data in the following table represent the closing value of the Dow Jones Industrial Average (DJIA) from 1979 through 2003. DJIA

Year	DJIA	Year	DJIA	Year	DJIA
1979	838.7	1987	1,938.8	1995	5,117.1
1980	964.0	1988	2,168.6	1996	6,448.3
1981	875.0	1989	2,753.2	1997	7,908.3
1982	1,046.5	1990	2,633.7	1998	9,181.4
1983	1,258.6	1991	3,168.8	1999	11,497.1
1984	1,211.6	1992	3,301.1	2000	10,788.0
1985	1,546.7	1993	3,754.1	2001	10,021.5
1986	1,896.0	1994	3,834.4	2002	8,341.6
				2003	10,453.9

Source: Extracted from ***finance.Yahoo.com****.*

a. Calculate the price index for the Dow Jones Industrial Average (DJIA) with 1979 as the base year.
b. Shift the base of the Dow Jones Industrial Average (DJIA) to 1990 and recalculate the price index.
c. Compare the results of (a) and (b). Which price index do you think is more useful in understanding the changes in the Dow Jones Industrial Average (DJIA)? Explain.

16.63 The consumer price index for Japan from 1992-2003 is given in the following table. JAPANCPI

Year	CPI	Year	CPI
1992	96.7	1998	101.0
1993	98.0	1999	100.7
1994	98.6	2000	100.0
1995	98.5	2001	99.3
1996	98.6	2002	98.4
1997	100.4	2003	98.1

Source: Statistics Bureau, Ministry of Public Management ***stat.go.jp****.*

a. Calculate the price index for the consumer price index in Japan with 1992 as the base year.
b. Discuss the changes in consumer prices during this 12-year period.

SELF Test **16.64** The following data UKCPI represent the consumer price index for the United Kingdom from 1990 to 2003.

Year	CPI	Year	CPI
1990	129.9	1997	160.0
1991	135.7	1998	164.4
1992	139.2	1999	167.3
1993	141.9	2000	172.2
1994	146.0	2001	173.4
1995	150.7	2002	176.3
1996	154.4	2003	181.4

Source: National Statistics Online ***statistics.gov.uk****.*

a. Calculate the price index for the Consumer Price Index in the United Kingdom with 1990 as the base year.
b. Shift the base of the Consumer Price Index in the United Kingdom to 2003 and recalculate the price index.
c. Compare the results of (a) and (b). Which price index do you think is more useful in understanding the changes in the Consumer Price Index in the United Kingdom? Explain.
d. Compare the results for the Consumer Price Index in the United Kingdom in (a) through (c) to those of Japan in problem 16.63.

16.65 The data in the file COFFEEPRICE represent the mean price per pound of coffee in the United States from 1980 to 2004.

Source: Bureau of Labor Statistics, U.S. Department of Labor ***www.bls.gov****.*

a. Calculate the simple price indexes for 1980 to 2004, using 1980 as the base year.
b. Interpret the simple price index for 2004, using 1980 as the base year.
c. Recalculate the simple price indexes found in (a) using Equation (16.26) on page 706 with 1990 as the base year.
d. Interpret the simple price index for 2004, using 1990 as the base year.
e. Would it be a good idea to use 1995 as the base year? Explain.
f. Describe the trends in coffee costs from 1980 to 2004.

16.66 The following data represent the mean price per pound of fresh tomatoes in the United States from 1980 to 2004. TOMATOES

Year	Price	Year	Price	Year	Price
1980	0.703	1988	0.871	1996	1.103
1981	0.792	1989	0.797	1997	1.213
1982	0.763	1990	1.735	1998	1.452
1983	0.726	1991	0.912	1999	1.904
1984	0.854	1992	0.936	2000	1.443
1985	0.697	1993	1.141	2001	1.414
1986	1.104	1994	1.604	2002	1.451
1987	0.943	1995	1.323	2003	1.711
				2004	1.472

Source: Bureau of Labor Statistics, U.S. Department of Labor ***www.bls.gov****.*

a. Calculate the simple price indexes for 1980 to 2004, using 1980 as the base year.
b. Interpret the simple price index for 2004, using 1980 as the base year.
c. Recalculate the simple price indexes found in (a) using Equation (16.26) on page 706 with 1990 as the base year.
d. Interpret the simple price index for 2004, using 1990 as the base year.

e. Describe the trends in the cost of fresh tomatoes from 1980 to 2004.

16.67 The data in the file ENERGY2 represent the mean price for three types of energy products in the United States from 1992 to 2004. Included are electricity (dollars per 500 KWH), natural gas (dollars per 40 therms), and fuel oil (dollars per gallon).

Source: Bureau of Labor Statistics, U.S. Department of Labor ***www.bls.gov***.

a. Calculate the 1992–2004 simple price indexes for electricity, natural gas, and fuel oil using 1992 as the base year.

b. Recalculate the price indexes in (a) using 1996 as the base year.

c. Calculate the 1992–2002 unweighted aggregate price indexes for the group of three energy items.

d. Calculate the 2004 Laspeyres price index for the group of three energy items for a family that consumed 5,000 KWH of electricity (10 units), 960 therms of natural gas (24 units), and 400 gallons of fuel oil (400 units) in 1992.

e. Calculate the 2004 Laspeyres price index for the group of three energy items for a family that consumed 6,500 KWH of electricity, 1,040 therms of natural gas, and 235 gallons of fuel oil in 1992.

16.10 PITFALLS CONCERNING TIME-SERIES FORECASTING

The value of time-series forecasting methodology, which uses past and present information as guides to the future, was recognized and most eloquently expressed more than two centuries ago by the U.S. statesman Patrick Henry, who said:

> I have but one lamp by which my feet are guided, and that is the lamp of experience. I know no way of judging the future but by the past. [Speech at Virginia Convention (Richmond), March 23, 1775]

However, critics of time-series forecasting argue that these techniques are overly naïve and mechanical; that is, a mathematical model based on the past should not be used to mechanically extrapolate trends into the future without considering personal judgments, business experiences, or changing technologies, habits, and needs (see problem 16.81 on page 715). Thus, in recent years econometricians have developed highly sophisticated computerized models of economic activity incorporating such factors for forecasting purposes. Such forecasting methods, however, are beyond the scope of this text (references 1, 2, and 3).

Nevertheless, as you have seen from the preceding sections of this chapter, time-series methods provide useful guides for projecting future trends (on long- and short-term bases). If used properly and, in conjunction with other forecasting methods as well as with business judgment and experience, time-series methods will continue to be excellent tools for forecasting.

SUMMARY

In this chapter, you used time-series methods to develop forecasts for the Wm. Wrigley Jr. Company, Cabot Corporation, and Wal-Mart. Figure 16.33 provides a summary chart for the time-series and index number methods discussed in this chapter.

If you are using time-series forecasting, you need to ask the following question:

- Is there a trend in the data? If there is no trend, then you should use moving averages or exponential smoothing. If there is a trend, then you can use the linear, quadratic, and exponential trend models, the Holt-Winters method, and the autoregressive model.

If you are developing index numbers, you need to ask the following questions:

- Are you developing an aggregate index of more than one commodity? If the answer is no, then you can use a simple price index.
- If the answer is yes, then you need to determine whether you will develop a weighted price index. If not, you can develop an unweighted price index. If you are developing a weighted price index, you can use a Laspeyres price index or a Paasche price index.

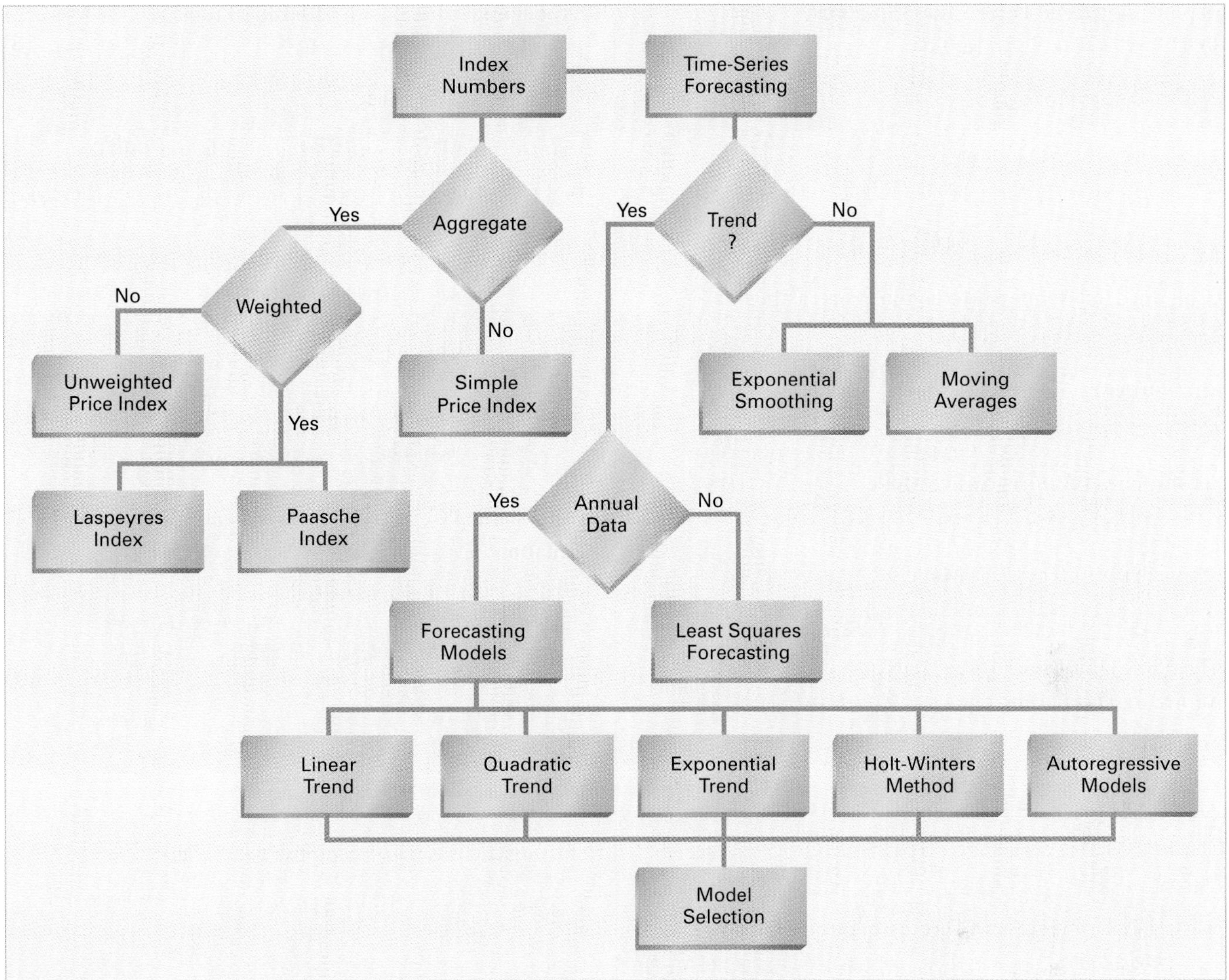

FIGURE 16.33 Summary Chart of Time Series and Index Number Methods

KEY FORMULAS

Classical Multiplicative Time-Series Model for Annual Data

$Y_i = T_i \times C_i \times I_i$ **(16.1)**

Classical Multiplicative Time-Series Model for Data with a Seasonal Component

$Y_i = T_i \times S_i \times C_i \times I_i$ **(16.2)**

Computing an Exponentially Smoothed Value in Time Period *i*

$$E_1 = Y_1 \qquad \textbf{(16.3)}$$

$$E_i = WY_i + (1 - W)E_{i-1} \qquad i = 2, 3, 4, \ldots$$

Forecasting Time Period *i* + 1

$\hat{Y}_{i+1} = E_i$ **(16.4)**

Linear Trend Forecasting Equation

$\hat{Y}_i = b_0 + b_1 X_i$ **(16.5)**

Quadratic Trend Forecasting Equation

$\hat{Y}_i = b_0 + b_1 X_i + b_2 X_i^2$ **(16.6)**

Exponential Trend Model

$Y_i = \beta_0 \beta_1^{X_i} \varepsilon_i$ **(16.7)**

Transformed Exponential Trend Model

$$\begin{aligned} \log(Y_i) &= \log(\beta_0 \beta_1^{X_i} \varepsilon_i) \qquad \textbf{(16.8)} \\ &= \log(\beta_0) + \log(\beta_1^{X_i}) + \log(\varepsilon_i) \\ &= \log(\beta_0) + X_i \log(\beta_1) + \log(\varepsilon_i) \end{aligned}$$

Exponential Trend Forecasting Equation

$$\log(\hat{Y}_i) = b_0 + b_1 X_i \quad \textbf{(16.9a)}$$

$$\hat{Y}_i = \hat{\beta}_0 \hat{\beta}_1^{X_i} \quad \textbf{(16.9b)}$$

The Holt-Winters Method

$$\text{Level}: E_i = U(E_{i-1} + T_{i-1}) + (1 - U)Y_i \quad \textbf{(16.10a)}$$

$$\text{Trend}: T_i = VT_{i-1} + (1 - V)(E_i - E_{i-1}) \quad \textbf{(16.10b)}$$

Using the Holt-Winters Method for Forecasting

$$\hat{Y}_{n+j} = E_n + j(T_n) \quad \textbf{(16.11)}$$

First-Order Autoregressive Model

$$Y_i = A_0 + A_1 Y_{i-1} + \delta_i \quad \textbf{(16.12)}$$

Second-Order Autoregressive Model

$$Y_i = A_0 + A_1 Y_{i-1} + A_2 Y_{i-2} + \delta_i \quad \textbf{(16.13)}$$

***p*th-Order Autoregressive Models**

$$Y_i = A_0 + A_1 Y_{i-1} + A_2 Y_{i-2} + \cdots + A_p Y_{i-p} + \delta_i \quad \textbf{(16.14)}$$

***t* Test for Significance of the Highest-Order Autoregressive Parameter A_p**

$$t = \frac{a_p - A_p}{S_{a_p}} \quad \textbf{(16.15)}$$

Fitted *p*th-Order Autoregressive Equation

$$\hat{Y}_i = a_0 + a_1 Y_{i-1} + a_2 Y_{i-2} + \cdots + a_p Y_{i-p} \quad \textbf{(16.16)}$$

***p*th-Order Autoregressive Forecasting Equation**

$$\hat{Y}_{n+j} = a_0 + a_1 \hat{Y}_{n+j-1} + a_2 \hat{Y}_{n+j-2} + \cdots + a_p \hat{Y}_{n+j-p} \quad \textbf{(16.17)}$$

Mean Absolute Deviation

$$MAD = \frac{\sum_{i=1}^{n} \left| Y_i - \hat{Y}_i \right|}{n} \quad \textbf{(16.18)}$$

Exponential Model with Quarterly Data

$$Y_i = \beta_0 \beta_1^{X_i} \beta_2^{Q_1} \beta_3^{Q_2} \beta_4^{Q_3} \varepsilon_i \quad \textbf{(16.19)}$$

Transformed Exponential Model with Quarterly Data

$$\begin{aligned} \log(Y_i) &= \log(\beta_0 \beta_1^{X_i} \beta_2^{Q_1} \beta_3^{Q_2} \beta_4^{Q_3} \varepsilon_i) \quad \textbf{(16.20)} \\ &= \log(\beta_0) + \log(\beta_1^{X_i}) + \log(\beta_2^{Q_1}) \\ &\quad + \log(\beta_3^{Q_2}) + \log(\beta_4^{Q_3}) + \log(\varepsilon_i) \\ &= \log(\beta_0) + X_i \log(\beta_1) + Q_1 \log(\beta_2) \\ &\quad + Q_2 \log(\beta_3) + Q_3 \log(\beta_4) + \log(\varepsilon_i) \end{aligned}$$

Exponential Growth with Quarterly Data Forecasting Equation

$$\log(\hat{Y}_i) = b_0 + b_1 X_i + b_2 Q_1 + b_3 Q_2 + b_4 Q_3 \quad \textbf{(16.21)}$$

Exponential Model with Monthly Data

$$Y_i = \beta_0 \beta_1^{X_i} \beta_2^{M_1} \beta_3^{M_2} \beta_4^{M_3} \beta_5^{M_4} \beta_6^{M_5} \beta_7^{M_6} \beta_8^{M_7} \beta_9^{M_8} \beta_{10}^{M_9} \beta_{11}^{M_{10}} \beta_{12}^{M_{11}} \varepsilon_i \quad \textbf{(16.22)}$$

Transformed Exponential Model with Monthly Data

$$\begin{aligned} \log(Y_i) &= \log(\beta_0 \beta_1^{X_i} \beta_2^{M_1} \beta_3^{M_2} \beta_4^{M_3} \beta_5^{M_4} \beta_6^{M_5} \beta_7^{M_6} \beta_8^{M_7} \beta_9^{M_8} \beta_{10}^{M_9} \beta_{11}^{M_{10}} \beta_{12}^{M_{11}} \varepsilon_i) \quad \textbf{(16.23)} \\ &= \log(\beta_0) + X_i \log(\beta_1) + M_1 \log(\beta_2) \\ &\quad + M_2 \log(\beta_3) + M_3 \log(\beta_4) \\ &\quad + M_4 \log(\beta_5) + M_5 \log(\beta_6) \\ &\quad + M_6 \log(\beta_7) + M_7 \log(\beta_8) \\ &\quad + M_8 \log(\beta_9) + M_9 \log(\beta_{10}) \\ &\quad + M_{10} \log(\beta_{11}) + M_{11} \log(\beta_{12}) \\ &\quad + \log(\varepsilon_i) \end{aligned}$$

Exponential Growth with Monthly Data Forecasting Equation

$$\begin{aligned} \log(\hat{Y}_i) &= b_0 + b_1 X_i + b_2 M_1 + b_3 M_2 + b_4 M_3 \quad \textbf{(16.24)} \\ &\quad + b_5 M_4 + b_6 M_5 + b_7 M_6 + b_8 M_7 \\ &\quad + b_9 M_8 + b_{10} M_9 + b_{11} M_{10} + b_{12} M_{11} \end{aligned}$$

Simple Price Index

$$I_i = \frac{P_i}{P_{base}} \times 100 \quad \textbf{(16.25)}$$

Shifting the Base for a Simple Price Index

$$I_{new} = \frac{I_{old}}{I_{new\,base}} \times 100 \quad \textbf{(16.26)}$$

Unweighted Aggregate Price Index

$$I_U^{(t)} = \frac{\sum_{i=1}^{n} P_i^{(t)}}{\sum_{i=1}^{n} P_i^{(0)}} \times 100 \quad \textbf{(16.27)}$$

Laspeyres Price Index

$$I_L^{(t)} = \frac{\sum_{i=1}^{n} P_i^{(t)} Q_i^{(0)}}{\sum_{i=1}^{n} P_i^{(0)} Q_i^{(0)}} \times 100 \quad \textbf{(16.28)}$$

Paasche Price Index

$$I_P^{(t)} = \frac{\sum_{i=1}^{n} P_i^{(t)} Q_i^{(t)}}{\sum_{i=1}^{n} P_i^{(0)} Q_i^{(t)}} \times 100 \quad \textbf{(16.29)}$$

KEY TERMS

CHAPTER REVIEW PROBLEMS

Checking Your Understanding

16.68 Why is forecasting so important?

16.69 What is a time series?

16.70 What are the distinguishing features among the various components of the classical multiplicative time-series model?

16.71 What is the difference between moving averages and exponential smoothing?

16.72 Under what circumstances is the exponential trend model most appropriate?

16.73 How does the least-squares linear trend forecasting model developed in this chapter differ from the least-squares linear regression model considered in Chapter 13?

16.74 How does autoregressive modeling differ from the other approaches to forecasting?

16.75 What are the different approaches to choosing an appropriate forecasting model?

16.76 What is the major difference between using S_{YX} and *MAD* for evaluating how well a particular model fits the data?

16.77 How does forecasting for monthly or quarterly data differ from forecasting for annual data?

16.78 What is an index number?

16.79 What is the difference between a simple price index and an aggregate price index?

16.80 What is the difference between a Paasche price index and a Laspeyres price index?

Applying the Concepts

You need to use Microsoft Excel, Minitab, or SPSS to solve problems 16.81–16.89.

16.81 The following table represents the annual incidence rates (per 100,000 persons) of reported acute poliomyelitis recorded over 5-year periods from 1915 to 1955. POLIO

Year	1915	1920	1925	1930	1935	1940	1945	1950	1955
Rate	3.1	2.2	5.3	7.5	8.5	7.4	10.3	22.1	17.6

Source: Data are taken from B. Wattenberg, ed., The Statistical History of the United States: From Colonial Times to the Present, *ser. B303 (New York: Basic Books, 1976).*

a. Plot the data.
b. Compute the linear trend forecasting equation and plot the trend line.
c. What are your forecasts for 1960, 1965, and 1970?
d. Go to a library or the Internet and find the actually reported incidence rates of acute poliomyelitis for 1960, 1965, and 1970. Record your results.
e. Why are the forecasts you made in (c) not useful? Discuss.

16.82 The U.S. Department of Labor gathers and publishes statistics concerning the labor market. The data file WORKFORCE contains the U.S. civilian noninstitutional population of people 16 years and over (in thousands) and the U.S. civilian noninstitutional workforce of people 16 years and over (in thousands) for 1984–2003. The workforce variable reports the number of people in the population who have a job or are actively looking for a job.

Year	Population	Workforce	Year	Population	Workforce
1984	176,383	113,544	1994	196,814	131,056
1985	178,206	115,461	1995	198,584	132,304
1986	180,587	117,834	1996	200,591	133,943
1987	182,753	119,865	1997	203,133	136,297
1988	184,613	121,669	1998	205,220	137,673
1989	186,393	123,869	1999	207,753	139,368
1990	189,164	125,840	2000	212,577	142,583
1991	190,925	126,346	2001	215,092	143,734
1992	192,805	128,105	2002	217,570	144,863
1993	194,838	129,200	2003	221,168	146,510

Source: Bureau of Labor Statistics, U.S. Department of Labor ***www.bls.gov****.*

a. Plot the time series for the U.S. civilian noninstitutional population of people 16 years and older.
b. Compute the linear trend forecasting equation.
c. Forecast the U.S. civilian noninstitutional population of people 16 years and older for 2004 and 2005.
d. Repeat (a) through (c) for the U.S. civilian noninstitutional workforce of people 16 years and older.

16.83 The quarterly price for natural gas (dollars per 40 therms) in the United States from 1994 through 2003 is given in the data file **NATURALGAS**.

Source: Bureau of Labor Statistics, U.S. Department of Labor ***www.bls.gov****.*

a. Do you think the price for natural gas has a seasonal component?
b. Plot the time series. Does this chart support your answer in (a)?
c. Compute an exponential trend forecasting equation for quarterly data.
d. Interpret the quarterly compound growth rate.
e. Interpret the quarter multipliers. Do the multipliers support your answers to (a) and (b)?

16.84 The data in the following table represent the gross revenues (in billions of current dollars) of McDonald's Corporation over the 29-year period from 1975 through 2003. **MCDONALD**

Year	Revenues	Year	Revenues	Year	Revenues
1975	1.0	1985	3.8	1995	9.8
1976	1.2	1986	4.2	1996	10.7
1977	1.4	1987	4.9	1997	11.4
1978	1.7	1988	5.6	1998	12.4
1979	1.9	1989	6.1	1999	13.3
1980	2.2	1990	6.8	2000	14.2
1981	2.5	1991	6.7	2001	14.9
1982	2.8	1992	7.1	2002	15.4
1983	3.1	1993	7.4	2003	17.1
1984	3.4	1994	8.3		

Source: Extracted from Moody's Handbook of Common Stocks, *1980, 1989, 1999,* Mergent's Handbook of Common Stocks, *Spring 2002 and Spring 2004.*

a. Plot the data.
b. Compute the linear trend forecasting equation.
c. Compute the quadratic trend forecasting equation.
d. Compute the exponential trend forecasting equation.
e. Find the best-fitting autoregressive model using $\alpha = 0.05$.
f. Perform a residual analysis for each of the models in (b) through (e).
g. Compute the standard error of the estimate (S_{YX}) and the *MAD* for each corresponding model in (f).
h. On the basis of your results in (f) and (g), along with a consideration of parsimony, which model would you select for purposes of forecasting? Discuss.
i. Using the selected model in (h), forecast gross revenues for 2004.

16.85 The data in the file **SEARS** represent the gross revenues (in billions of current dollars) of Sears, Roebuck & Company over the 29-year period from 1975 through 2003.

Source: Extracted from Moody's Handbook of Common Stocks, *1980, 1989, 1999,* Mergent's Handbook of Common Stocks, *Spring 2002 and Spring 2004.*

a. Plot the data.
b. Compute the linear trend forecasting equation.
c. Compute the quadratic trend forecasting equation.
d. Compute the exponential trend forecasting equation.
e. Find the best-fitting autoregressive model using $\alpha = 0.05$.
f. Use the Holt-Winters method with $U = 0.3$ and $V = 0.3$ to compute forecasts.
g. Perform a residual analysis for each of the models in (b) through (f).
h. Compute the standard error of the estimate (S_{YX}) and the *MAD* for each corresponding model in (g).
i. On the basis of your results in (g) and (h), along with a consideration of parsimony, which model would you select for purposes of forecasting? Discuss.
j. Using the selected model in (i), forecast gross revenues for 2004.

16.86 Teachers Retirement System of New York City offers several types of investments for its members. Among the choices are investments with fixed and variable rates of return. There are currently two categories of variable return investments. Variable *A* consists of investments that are primarily made in stocks, while variable *B* consists of investments in corporate bonds and other types of lower risk instruments. The following data **TRSNYC** represent the value of a unit of each type of variable return investment at the beginning of each year from 1984 to 2004.

Year	A	B	Year	A	B
1984	13.111	10.342	1995	30.830	17.351
1985	13.176	11.073	1996	39.644	17.682
1986	16.526	11.925	1997	45.389	18.004
1987	18.652	12.694	1998	54.882	18.341
1988	15.564	13.352	1999	64.790	18.678
1989	20.827	13.919	2000	74.220	18.962
1990	24.738	14.557	2001	67.534	19.320
1991	22.678	15.213	2002	57.709	19.673
1992	28.549	15.883	2003	44.843	19.735
1993	29.829	16.510	2004	55.993	19.609
1994	32.199	16.970			

Source: ***www.trs.nyc.ny.us***.

For each of the two time series:

a. Plot the data.

b. Compute the linear trend forecasting equation.

c. Compute the quadratic trend forecasting equation.

d. Compute the exponential trend forecasting equation.

e. Find the best-fitting autoregressive model using $\alpha = 0.05$.

f. Perform a residual analysis for each of the models in (b) through (e).

g. Compute the standard error of the estimate (S_{YX}) and the *MAD* for each corresponding model in (f).

h. On the basis of your results in (f) and (g), along with a consideration of parsimony, which model would you select for purposes of forecasting? Discuss.

i. Using the selected model in (h), forecast the unit values for 2005.

j. Based on the results of (a) through (i), what investment strategy would you recommend for a member of Teachers Retirement System of New York City? Explain.

16.87 The data file BASKET contains the prices of a basket of food items from 1992 to 2004. Included are the prices (in dollars) for a one-pound loaf of white bread, a pound of beef (ground chuck), a dozen grade-A large eggs, and one pound of iceberg lettuce.

Year	Bread	Beef	Eggs	Lettuce
1992	0.726	1.926	0.933	0.573
1993	0.748	1.970	0.898	0.625
1994	0.768	1.892	0.917	0.506
1995	0.767	1.847	0.882	0.821
1996	0.860	1.799	1.155	0.769
1997	0.862	1.850	1.148	0.651
1998	0.855	1.818	1.120	1.072
1999	0.872	1.834	1.053	0.649
2000	0.907	1.903	0.975	0.748
2001	0.982	2.037	1.011	0.736
2002	1.001	2.151	0.973	1.003
2003	1.042	2.131	1.175	0.734
2004	0.946	2.585	1.573	0.876

Source: Bureau of Labor Statistics, U.S. Department of Labor ***www.bls.gov***.

a. Compute the 1992–2004 simple price indexes for bread, beef, eggs, and lettuce using 1992 as the base year.

b. Recalculate the price indexes in (a) using 1996 as the base year.

c. Compute the 1992–2004 unweighted aggregate price indexes for the basket of these four food items.

d. Compute the 2004 Laspeyres price index for the basket of these four food items for a family that consumed 50 loafs of bread, 22 pounds of beef, 24 dozen eggs, and 18 pounds of lettuce in 1992.

e. Compute the 2004 Paasche price index for the basket of these four food items for a family that consumed 55 loafs of bread, 17 pounds of beef, 20 dozen eggs, and 28 pounds of lettuce in 2004.

Report Writing Exercises

16.88 Labor negotiations between Major League Baseball players and owners of Major League Baseball teams have been contentious over the years with periodic strikes and work stoppages. The data in the file BBSALARY represent the mean salaries (in thousands of dollars) from 1979 to 2003, the median salary (in thousands of dollars) from 1983 to 2002, and the minimum salaries (in thousands of dollars) from 1979 to 2003.

a. You have been hired by the owners to prepare a report that shows that the salaries have skyrocketed over the years. Develop a forecasting model that provides evidence for that conclusion.

b. You have been hired by the players union to prepare a report that shows that after adjusting for the Consumer Price Index, salaries have not increased a substantial amount over the years. Develop a forecasting model that provides evidence for that conclusion.

c. You have been hired by a sports television network to study salaries. Develop a report that does not presuppose any conclusions.

16.89 As a consultant to an investment company trading in various currencies, you have been assigned the task of studying the long-term trends in the exchange rates of the Canadian dollar, the Japanese yen, and the English pound. Data have been collected for the 37-year period from 1967 to 2003 and are contained in the file CURRENCY. The Canadian dollar and the Japanese yen are expressed in units per U.S. dollar. The English pound is expressed in cents per pound.

Develop a forecasting model for the exchange rate of each of these three currencies and provide forecasts for 2004 and 2005 for each currency. Write an executive summary for a presentation to be given to the investment company. Append to this executive summary a discussion regarding possible limitations that may exist in these models.

RUNNING CASE MANAGING THE *SPRINGVILLE HERALD*

As part of the continuing strategic initiative to increase home-delivery subscriptions, the circulation department is closely monitoring the number of such subscriptions. The circulation department wants to forecast future home-delivery subscriptions. To accomplish this task, the circulation department compiled the number of home-delivery subscriptions for the most recent 24-month period. SH16

EXERCISE

SH16.1 **a.** Analyze these data and develop a model to forecast home-delivery subscriptions. Present your findings in a report that includes the assumptions of the model and its limitations. Forecast home-delivery subscriptions for the next four months.

b. Would you be willing to use the model developed to forecast home-delivery subscriptions one year into the future? Explain.

c. Compare the trend in home-delivery subscriptions to the number of new subscriptions per month provided in the data file SH13. What explanation can you provide for any differences?

WEB CASE

Apply your knowledge about time-series forecasting in this Web case that extends the running Managing the Springville Herald *case of this text.*

For a number of years, the *Springville Herald* has been competing with the newer *Oxford Glen Journal* (*OGJ*) for readership in the Tri-Cities area. Recently, the circulation staff at the *OGJ* claimed that their newspaper's circulation and subscription base is growing faster than that of the *Herald* and that local advertisers would do better transferring their advertisements from the *Herald* to the *OGJ*. The circulation department of the *Herald* has complained to the Springville Chamber of Commerce about *OGJ*'s claims and has asked the chamber to investigate, a request that was welcomed by *OGJ*'s circulation staff.

Review the information collected by the Springville Chamber of Commerce about the circulation dispute at the chamber's Web site **www.prenhall.com/Springville/SCC_CirculationDispute.htm** and then answer the following:

1. Which newspaper would you say has the right to claim the fastest-growing circulation and subscription base? Support your answer by performing and summarizing an appropriate statistical analysis.
2. What is the single most positive fact about the *Herald*'s circulation and subscription base? What is the single most positive fact about the *OGJ*'s circulation and subscription base? Explain your answers.
3. What additional data would be helpful in investigating the circulation claims made by the staffs of each newspaper.

REFERENCES

1. Bowerman, B. L., R. T. O'Connell, and A. Koehler, *Forecasting, Time Series, and Regression*, 4th ed. (Belmont, CA: Duxbury Press, 2005).
2. Box, G. E. P., G. M. Jenkins, and G. C. Reinsel, *Time Series Analysis: Forecasting and Control*, 3rd ed. (Englewood Cliffs, NJ: Prentice Hall, 1994).
3. Frees, E. W., *Data Analysis Using Regression Models: The Business Perspective* (Upper Saddle River, NJ: Prentice Hall, 1996).
4. Hanke, J. E., D. W. Wichern, and A. G. Reitsch, *Business Forecasting*, 7th ed. (Upper Saddle River, NJ: Prentice Hall, 2001).
5. Ittig, P., "A Seasonal Index for Business," *Decision Sciences* 28 (1997): 335–355.
6. Mahmoud, E., "Accuracy in Forecasting: A Survey," *Journal of Forecasting* 3 (1984): 139–159.
7. *Microsoft Excel 2003* (Redmond, WA: Microsoft Corp., 2003).
8. *Minitab for Windows Version 14* (State College, PA: Minitab, Inc. 2004).
9. *SPSS Base 12.0 Brief Guide* (Upper Saddle River, NJ: Prentice Hall, 2003).

Appendix 16 Using Software for Time-Series Forecasting and Index Numbers

A16.1 MICROSOFT EXCEL

For Creating Moving Averages

To create a moving average, do the following:

Open the worksheet that contains your time-series data in row order (each row is one time period).
In a blank column, enter formulas that use the AVERAGE function to average values in a cell range that matches your chosen period of length *L*. Enter ranges such that each formula is located in the middle row of the cell range that the formula averages.
Enter the special Excel value **#N/A** (not available) in cells at the beginning and end of the column for which no moving average can be calculated.

For example, to create three-year and seven-year moving averages for the Cabot revenue data, open the **Cabot.xls** workbook and add formulas to columns C and D. The entries for the first four and last four data rows of these columns are shown below. Note the use of **#N/A** in rows 2 and 23, and in cell ranges D3:D4 and D21:D22.

	A	B	C	D
1	Year	Revenue	MA 3-Year	MA 7-Year
2	1982	1588	#N/A	#N/A
3	1983	1558	=AVERAGE (B2:B4)	#N/A
4	1984	1753	=AVERAGE (B3:B5)	#N/A
5	1985	1408	=AVERAGE (B4:B6)	=AVERAGE (B2:B8)
20	2000	1698	=AVERAGE (B19:B21)	=AVERAGE (B17:B23)
21	2001	1523	=AVERAGE (B20:B22)	#N/A
22	2002	1557	=AVERAGE (B21:B23)	#N/A
23	2003	1795	#N/A	#N/A

For Creating Time-Series Plots

To create a time-series plot, open to your time-series data and select **Insert → Chart**. Make these entries in the Chart Wizard dialog boxes:

Step 1: Click **XY (Scatter)** from the **Standard Types Chart type** box. Select the first choice of the third row from the **Chart sub-types**, described as **Scatter with data points connected by lines**. Click **Next**.

Step 2: Select the **Data Range** tab. Enter the cell range of the original data and the data created by the formulas as the **Data range**. The first column of this range should contain the time periods. Select the **Columns** option and click **Next**.

Step 3: Select the **Legend** tab and select **Show legend**. For selections in other tabs, see "Using the Chart Wizard" on page 17 for general suggestions.

Step 4: Select **As new sheet** and click **Finish**.

If field buttons appear on the chart, right-click any field button and select **Hide PivotChart Field Buttons** from the shortcut menu.

For Creating Exponentially Smoothed Values

Use the Data Analysis ToolPak. Open to the worksheet containing your time-series data in row-wise order. Select **Tools → Data Analysis**. From the list that appears in the Data Analysis dialog box, select **Exponential Smoothing** and click **OK**. In the Exponential Smoothing dialog box (see Figure A16.1 on page 720),

Enter the cell range of the time-series variable to be smoothed as the **Input Range**. Enter either **.5** (for $W = .50$) or **.75** (for $W = 0.25$, as $1 - 0.25$ equals 0.75) as the **damping factor**.
Select **Labels** if the variable cell ranges include labels in their first rows.
Enter the cell range in which you will place the smoothed values as the **Output Range**. If you selected **Labels**, this cell range should begin in row 2.
If you selected **Labels**, enter a column heading in row 1 cell above the **Output Range**.
Click **OK**.

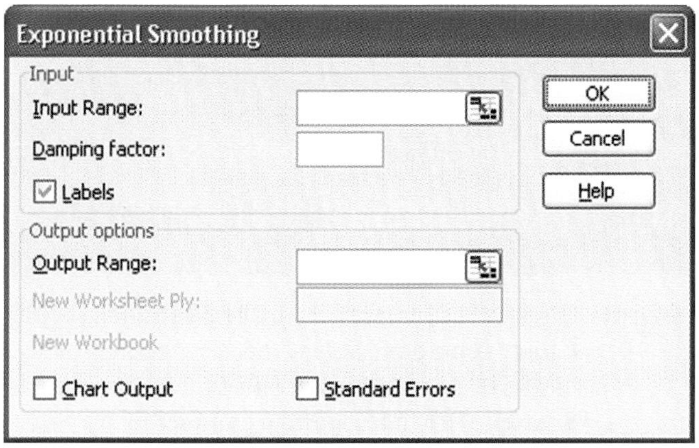

FIGURE A16.1 **Data Analysis Exponential Smoothing Dialog Box**

The column of formulas created shows exponentially smoothed values for a particular year in the row below the year. To adjust the column so that each exponentially smoothed value appears in the same row with its associated year and observed values, select the **row 2 cell** for that column and then select **Edit → Delete**. In the Delete dialog box, select **Shift cells up** and click **OK**. Then copy the formula in the second-to-last cell of the column down to the last cell.

For Creating Coded *X* Variables

Open to the worksheet containing your time-series data in row-wise order. To create a coded *X* variable, add a new column that contains an integer series starting with zero. In a blank column, enter a column label in the row 1 cell and the first integer in the series in the row 2 cell (0 if the series begins with 0). Automate the entry of the other integers by reselecting the row 2 cell and selecting **Edit → Fill → Series**. In the Series dialog box (Figure A16.2), select the **Columns** and **Linear** options, enter the appropriate **Step value** and **Stop value** and click **OK**.

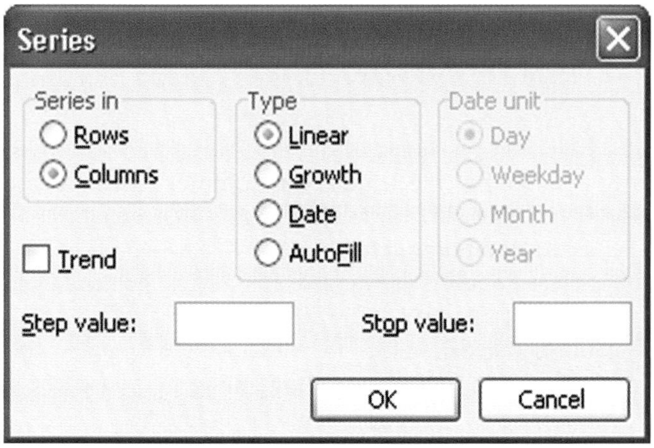

FIGURE A16.2 **Series Dialog Box**

For example, enter **1** as the **Step value** and **21** as the **Stop value** to complete the twenty-two integer series 0 through 21.

For Creating Quadratic and Exponential Terms

See Appendix A15.1 to review the methods for adding quadratic and exponential terms.

For Least-Squares Linear Trend Fitting

See the "For Simple Linear Regression" section in Appendix A13.1, using the cell range of the coded variable as the cell range for the *X* variable.

OR See section G.35 (**Simple Linear Regression**) if you want PHStat2 to produce the regression results worksheet for you.

For Least-Squares Quadratic Trend Fitting

See the "For Multiple Regression" section in Appendix A14.1, using the cell range of the coded variable and the squared coded variable as the cell range for the *X* variables.

OR See section G.36 (**Multiple Regression**) if you want PHStat2 to produce the regression results worksheet for you.

For Least-Squares Exponential Trend Fitting

See the "For Simple Linear Regression" section in Appendix A13.1, using the cell range of the log *Y* values as the cell range for the Y variable and use the cell range of the coded variable for the *X* variable cell range.

OR See section G.35 (**Simple Linear Regression**) if you want PHStat2 to produce the regression results worksheet for you.

For Holt-Winters Method (defining E_2 as Y_2 and T_2 as $Y_2 - Y_1$)

Open to the worksheet containing your time-series data in row-wise order. Add two new columns, one for *E*, the level of the smoothed series and the other for *T*, the value of the trend component.

For the column containing E:

Enter a column label in the row 1 cell and the special value #N/A in the row 2 cell.
In the row 3 cell, enter the formula that refers to the original data for that period. For example, if the original data was in column B, enter **=B3**.
In the row 4 cell, enter a formula in the form ***=UValue * (PreviousRowEValue + PreviousRowTValue) + (1 – UValue) * this row's original data*** and copy this formula down through the last row of time-series data.

For the column containing T:

Enter a column label in the row 1 cell and the special value #N/A in the row 2 cell.
In the row 3 cell, enter the formula that subtracts the original data for the previous period (in row 2) from the original data in row 3. For example, if the original data was in column B, enter **=B3 – B2**.
In the row 4 cell, enter a formula in the form ***=VValue * PreviousRowTValue + (1 – V Value) * (This Row'sEValue – PreviousRowEValue)*** and copy this formula down through the last row of time-series data.

For example, if column C holds the *E* values and column D holds the *T* values, then if $U = 0.3$ and $V = 0.4$, you would enter **=0.3 * (C3 + D3) + (1 – 0.3) * B4** in cell C4 and **=0.4 * D3 + (1 – 0.4) * (C4 – C3)** in cell D4.

For Creating Lagged Predictor Variables

Open to the worksheet containing your time-series data in row-wise order. For each lagged predictor variable you want to add, enter formulas that refer to a previous row's (previous time period) *Y* value.

For example, the table below shows lagged variables for the first-order, second-order, and third-order autoregressive models in columns C, D, and E that are based on *Y* values found in column B. The time-series extends from row 2 through row 21. Note the use of the special value **#N/A** (not available) in cells C2, D2:D3, and E2:E4. These entries will trigger an error message should their cells be inadvertently included in a regression cell range.

	C	D	E
1	Lag1	Lag2	Lag3
2	#N/A	#N/A	#N/A
3	=B2	#N/A	#N/A
4	=B3	=B2	#N/A
5	=B4	=B3	=B2
21	=B20	=B19	=B18

For First-Order Autoregressive Models

Use the "For Simple Linear Regression" section instructions in Appendix A13.1 with the modification below.

OR See section G.35 (**Simple Linear Regression**) if you want PHStat2 to produce the regression results worksheet for you using the modification below.

Modification: Use the cell range of the first-order lagged variable as the cell range of the *X* variable. Do not select the **Labels** (Data Analysis) or **First cells in both ranges contain label** (PHStat2) check boxes.

For Second-Order or Third-Order Autoregressive Models

Use the "For Multiple Regression" section instructions in Appendix A14.1 with the modification below.

OR See section G.36 (**Multiple Regression**) if you want PHStat2 to produce the regression results worksheet for you using the modification below.

Modification: Use the cell range of the lagged variables as the cell range of the *X* variables. Use the cell range of the first-order and second-order lagged variables for a second-order model and use the cell range of the first-order, second-order, and third-order lagged variables for the third-order model. For all models, do *not* select the **Labels** (Data Analysis) or **First cells in both ranges contain label** (PHStat2) check boxes.

For Mean Absolute Deviation (MAD)

For a linear, quadratic, or autoregressive model.

First, perform the appropriate regression analysis. From the regression statistics worksheet, copy the predicted and residuals values found in columns B and C of the Residual Output section to blank columns on the source data worksheet for the regression model. Add a column of formulas in the form ***=ABS(residual)*** to calculate the absolute value of the residuals. Then add a single formula in the form ***=AVERAGE(cell range of residual absolute values)*** to calculate the MAD.

For an exponential model. Perform the regression analysis for the exponential model. The results of this model cannot be immediately used because the dependent variable and the predicted *Y* value are in logarithms. The logarithm of the predicted *Y* values will need to be converted to the original units of the *Y* values in order to calculate the *MAD*.

To the Residual output section of the regression statistics worksheet, add a column of formulas that use the

POWER function, the template of which is **POWER(10, *LogValue*)**, to convert the predicted *Y* logarithms to the predicted *Y* values. Next, copy the original *Y* values to the next blank column. With the predicted and original *Y* values in the residual output section table, the residuals can now be calculated. To calculate these residuals, add a column of formulas in the form ***=ABS(YValueCell – PredictedYValueCell)*** to calculate the absolute value of the residuals. Then add a single formula in the form ***=AVERAGE(cell range of the absolute values of the residuals)*** to calculate the MAD.

For Creating Dummy Variables for Monthly or Quarterly Data

Open to the worksheet containing your time-series data in row-wise order. Add columns that use the IF function, the template of which is **IF(*comparison, value to use if comparison is true, value to use if comparison is false*)**.

For example, the first table below shows three quarterly dummy variables Q1, Q2, and Q3 in columns F, G, and H that are based on coded quarterly values found in column B. The time-series extends from row 2 through row 37. For example, the IF function in cell F2, =IF(B2 = 1, 1, 0) means that if the value in cell B2 = 1, then the value in cell F2 is 1, otherwise it is 0.

The second table shows two monthly variables M1 and M11 in columns J and K that are based on monthly values found in column C.

	F	G	H
1	Q1	Q2	Q3
2	=IF(B2 = 1, 1, 0)	=IF(B2 = 2, 1, 0)	=IF(B2 = 3, 1, 0)
37	=IF(B37 = 1, 1, 0)	=IF(B37 = 2, 1, 0)	=IF(B37 = 3, 1, 0)

	J	K
1	M1	M11
2	=IF(C2 = "January", 1, 0)	=IF(C2 = "November", 1, 0)
37	=IF(C37 = "January", 1, 0)	=IF(C37 = "November", 1, 0)

For Index Numbers

Use the SUM function or the arithmetic operators in formulas to calculate index numbers. Open the **Index Numbers.xls** file and examine its worksheets for specific ways in which the price index calculations discussed in section 16.9 can be calculated and presented.

A16.2 MINITAB

Using Minitab for Moving Averages

To illustrate the computation of moving averages, open the **CABOT.MTW** worksheet. Select **Stat → Time Series → Moving Average**.

1. In the Moving Average dialog box (see Figure A16.3), enter **C2** or **Revenue** in the Variable: edit box. Enter **3** in the MA length: edit box for a 3-year moving average. Select the **Center the moving averages** check box. Click the **Results** button.

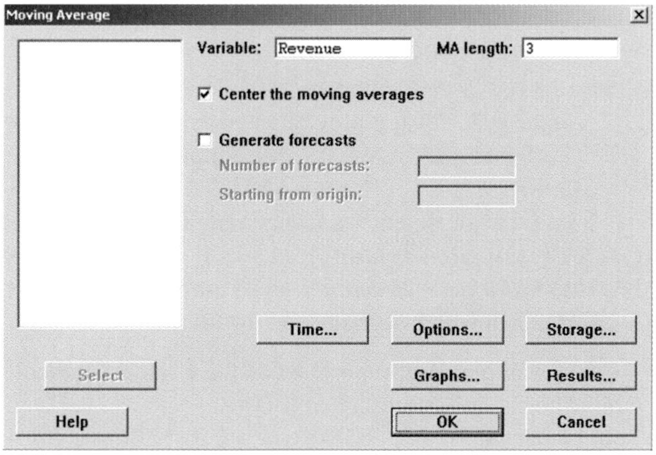

FIGURE A16.3 Minitab Moving Average Dialog Box

2. Select the **Summary table and results table** option button. Click the **OK** button to return to the Moving Average dialog box. Click the **OK** button again.

Using Minitab for Exponential Smoothing

To illustrate the computation of exponentially smoothed values, open the **CABOT.MTW** worksheet. Select **Stat → Time Series → Single Exp Smoothing**.

1. In the Single Exponential Smoothing dialog box (see Figure A16.4), enter **C2** or **Revenue** in the Variable: edit box. In the Weight to Use in Smoothing section, select the **Use**: option button and enter **0.25** for a *W* value of 0.25. Click the **Results** button.
2. In the Results dialog box, select the **Summary table and results table** option button. Click the **OK** button to return to the Single Exponential Smoothing dialog box. Click the **Options** button.
3. In the Single Exponential Smoothing—Options dialog box, enter **1** in the Use average of first: edit box. Click the **OK** button to return to the Single Exponential Smoothing dialog box. Click the **OK** button again.

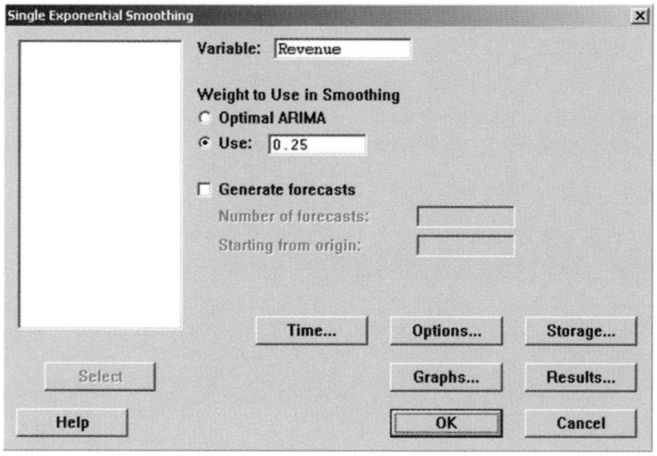

FIGURE A16.4 Minitab Single Exponential Smoothing Dialog Box

Using Minitab for Least-Squares Trend-Fitting

In Chapters 13 through 15, you used Minitab for the simple linear regression model and for a variety of multiple regression models. In this chapter, time-series models were developed assuming either a linear, quadratic, or exponential trend. For the linear trend model, see Appendix A13.2. For the quadratic model and the exponential model, see Appendices A14.2 and A15.2.

Using Minitab for Autoregressive Modeling

In order to use Minitab for autoregressive models, you need to create lagged variables. Open the **WRIGLEY.MTW** worksheet. Select **Stat → Time Series → Lag**.

1. In the Lag dialog box, enter **C5** or '**Real Revenue**' in the Series: edit box. Enter **C6** in the Store lags in: edit box. For a 1-period lag, enter **1** in the Lag: edit box. Click the **OK** button.

2. Repeat this procedure for 2- and 3-period lags by changing the lags to 2 and then 3, respectively, and then store the lags in C7 and C8, respectively. Once the lagged variables are formed, use **Stat → Regression → Regression** as in Appendices A13.2 and A14.2.

CHAPTER 17

Decision Making

USING STATISTICS: Selecting a Stock

LEARNING OBJECTIVES

In this chapter, you learn:

- To use payoff tables and decision trees to evaluate alternative courses of action
- To use several criteria to select an alternative course of action
- To use Bayes' theorem to revise probabilities in light of sample information
- About the concept of utility

USING STATISTICS

Selecting a Stock

As the manager of a stock mutual fund, you are responsible for purchasing and selling stocks for the fund. The investors in your fund expect a large return on their investment and, at the same time, they want to minimize their risk. At the present time, you need to decide between two stocks to purchase. An economist for your company has evaluated the potential one-year returns for both stocks under four economic conditions: recession, stability, moderate growth, and boom. She has also estimated the probability of each economic condition occurring. How can you use the information provided by the economist to determine which stock to choose in order to maximize return and minimize risk?

In Chapter 4, you studied various rules of probability and used Bayes' theorem to revise probabilities. In Chapter 5, you learned about discrete probability distributions and how to compute the expected value. In this chapter, these probability rules and probability distributions are applied to a decision-making process for evaluating alternative courses of action. In this context, you can consider the four basic features of a decision-making situation:

- **Alternative courses of action.** The decision-maker must have two or more possible choices to evaluate prior to selecting one course of action. For example, as a manager of a mutual fund in the "Using Statistics" scenario, you must decide whether to purchase stock *A* or stock *B*.
- **Events or states of the world.** The decision-maker must list the events that can occur and consider each event's probability of occurring. To aid in selecting which stock to purchase in the "Using Statistics" scenario, an economist for your company has listed four possible economic conditions and the probability of each occurring in the next year.
- **Payoffs.** In order to evaluate each course of action, the decision-maker must associate a value or payoff with the result of each event. In business applications, this payoff is usually expressed in terms of profits or costs, although other payoffs such as units of satisfaction or utility are sometimes considered. In the "Using Statistics" scenario, the payoff is the return on investment.
- **Decision criteria.** The decision-maker must determine how to select the best course of action. Section 17.2 discusses three criteria for decision making; expected monetary value, expected opportunity loss, and return-to-risk ratio.

17.1 PAYOFF TABLES AND DECISION TREES

In order to evaluate the various alternative courses of action for the complete set of events, you need to develop a payoff table or construct a decision tree. A **payoff table** contains each possible event that can occur for each alternative course of action. You must associate a value or payoff for each combination of an event and course of action. Example 17.1 shows a payoff table for a marketing manager trying to decide whether or not to introduce a new model of a television set.

EXAMPLE 17.1

A PAYOFF TABLE FOR DECIDING WHETHER TO MARKET A TELEVISION SET

As the marketing manager of a consumer electronics company, you are considering whether to introduce a new large screen television set into the market. You are aware of the risks in deciding whether to market this television set. For example, you could decide to market the television set and then, for any of a number of reasons, the introduction turns out unsuccessful. Second, you could decide not to market the television set when, in reality, it would have been successful. There is a fixed cost of \$3 (millions of dollars) incurred prior to making a final decision to market the television set. Based on past experience, if the television set is successful, you expect to profit \$45 million dollars. If the television set is not successful, you expect to lose \$36 million. Construct a payoff table for these two alternative courses of action.

SOLUTION Table 17.1 is a payoff table for the television set marketing example.

TABLE 17.1
Payoff Table for the Television Set Marketing Example (in Millions of Dollars)

	Alternative Courses of Action	
Event E_i	**Market, A_1**	**Do Not Market, A_2**
Successful television set, E_1	+\$45	–\$3
Unsuccessful television set, E_2	–\$36	–\$3

A **decision tree** is another way of representing the events for each alternative course of action. The decision tree pictorially represents the events and courses of action through a set of branches and nodes. Example 17.2 illustrates a decision tree.

EXAMPLE 17.2

THE DECISION TREE FOR THE TELEVISION SET MARKETING DECISION

Given the payoff table for the television marketing example, construct a decision tree.

SOLUTION Figure 17.1 is the decision tree for the payoff table shown in Table 17.1.

FIGURE 17.1
Decision Tree for the Television Set Marketing Example

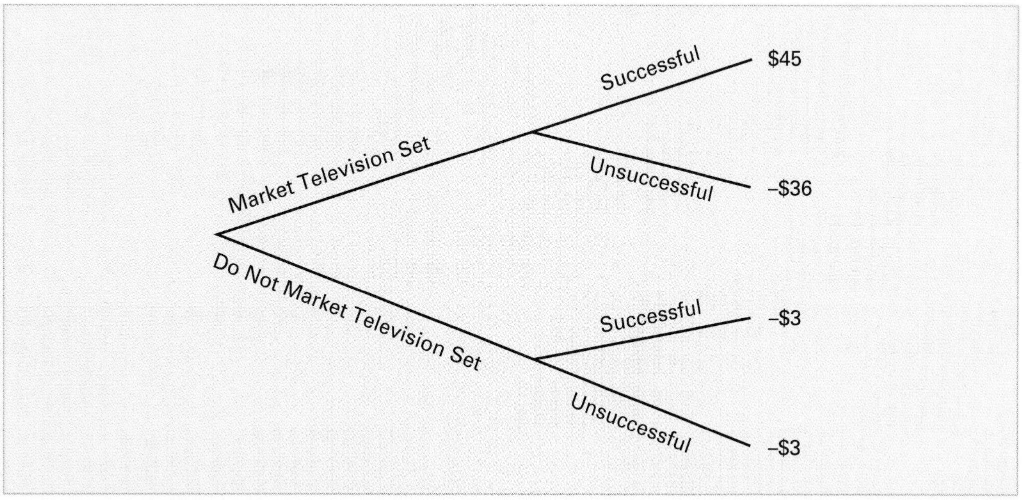

In Figure 17.1, the first set of branches relates to the two alternative courses of action, market the television set or do not market the television set. The second set of branches represents the possible events of successful television set and unsuccessful television set. These events occur for each of the alternative courses of action on the decision tree.

The decision structure for the television set marketing example contains only two possible alternative courses of action and two possible events. In general, there can be several alternative courses of action and events. As a manager of a mutual fund in the "Using Statistics" scenario, you need to decide between two stocks to purchase for a short-term investment of one year. An economist in the company has predicted returns for the two stocks under the four economic conditions—recession, stability, moderate growth, and boom. Table 17.2 presents the predicted 1-year return of a $1,000 investment in each stock under each economic condition. Figure 17.2 shows the decision tree for this payoff table. The decision (which stock to purchase) is the first branch of the tree and the second set of branches represents the four events (the economic conditions).

TABLE 17.2

Predicted 1-Year Return on $1,000 Investment in Each of Two Stocks under Four Economic Conditions

	Stocks	
Economic Conditions	***A***	***B***
Recession	$30	–$50
Stable economy	70	30
Moderate growth	100	250
Boom	150	400

FIGURE 17.2

Decision Tree for the Stock Selection Payoff Table

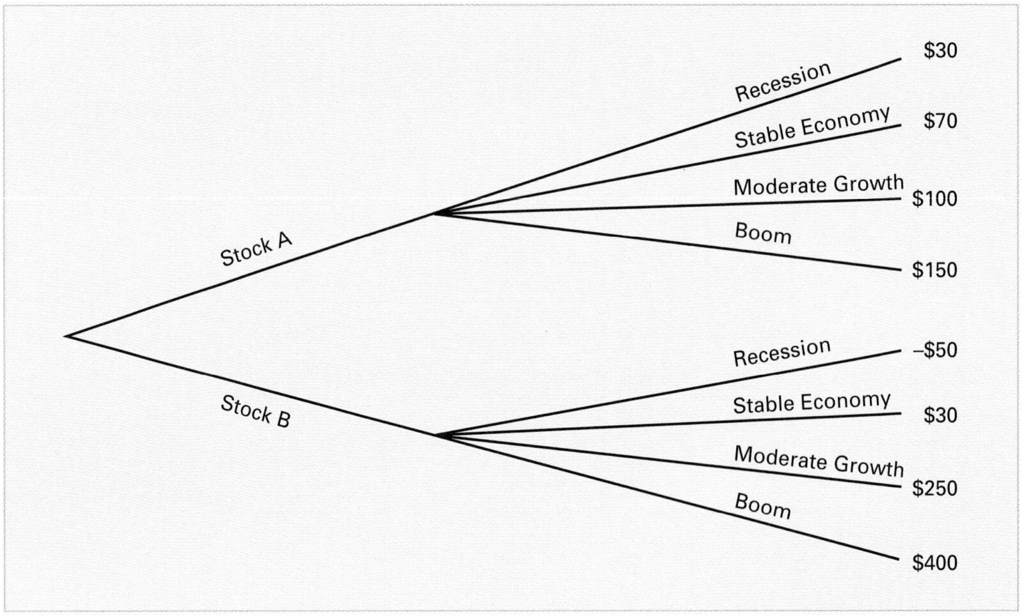

You use payoff tables and decision trees as decision-making tools to help determine the best course of action. For example, when deciding whether to market a television set, you would market it if you knew that the television set was going to be successful. Certainly, you would not market it if you knew that it was not going to be successful. For each event, you can determine the amount of profit that will be lost if the best alternative course of action is not taken. This is called opportunity loss.

The **opportunity loss** is the difference between the highest possible profit for an event and the actual profit for an action taken.

Example 17.3 illustrates the computation of opportunity loss.

EXAMPLE 17.3

FINDING OPPORTUNITY LOSS IN THE TELEVISION SET MARKETING EXAMPLE

Using the payoff table from Example 17.1 on page 727, construct an opportunity loss table.

SOLUTION For the event "successful television set," the maximum profit occurs when the product is marketed (+$45 million). The opportunity that is lost by not marketing the television set is the difference between $45 million and –$3 million, which is $48 million. If the television is unsuccessful, the best action is not to market the television set (–$3 million profit). The opportunity that is lost by making the incorrect decision of marketing the television set is –$3 – (–$36) = $33 million. The opportunity loss is always a nonnegative number, because it represents the difference between the profit under the best action and any other course of action that is taken for the particular event. Table 17.3 shows the complete opportunity loss table for the television set marketing example.

TABLE 17.3
Opportunity Loss Table for the Television Set Marketing Example (in Millions of Dollars)

			Alternative Courses of Action	
Event E_i	Optimum Action	Profit of Optimum Action	Market	Do Not Market
Successful	Market	$45	$45 – $45 = $ 0	$45 – (–$3) = $48
Unsuccessful	Do not market	–$3	–$3 – (–$36) = $33	–$3 – (–$3) = $ 0

Figure 17.3 represents the Microsoft Excel opportunity loss table.

FIGURE 17.3
Microsoft Excel Opportunity Loss Table

	A	B	C	D	E
1	Television Marketing Opportunity Loss				
2					
3	Payoff Table:				
4		Market	Do not market		
5	Successful	45	-3		
6	Unsuccessful	-36	-3		
7					
8					
9	Opportunity Loss Table:				
10		Optimum	Optimum	Alternatives	
11		Action	Profit	Market	Do not market
12	Successful	Market	45	0	48
13	Unsuccessful	Do not market	-3	33	0

You can also develop an opportunity loss table for the stock selection problem in the "Using Statistics" scenario. Here, there are four possible events or economic conditions that will affect the 1-year return for each of the two stocks. In a recession, stock *A* is best, providing a return of $30 as compared to a loss of $50 from stock *B*. In a stable economy, stock *A* again is better than stock *B* because it provides a return of $70 compared to $30 for stock *B*. However, under conditions of moderate growth or boom, stock *B* is superior to stock *A*. In a moderate growth period, stock *B* provides a return of $250 as compared to $100 from stock *A*, while in boom conditions the difference between stocks is even greater with stock *B* providing a return of $400 as compared to $150 for stock *A*. Table 17.4 summarizes the complete set of opportunity losses.

TABLE 17.4
Opportunity Loss Table for Two Stocks under Four Economic Conditions

			Alternative Courses of Action	
Event E_i	Optimum Action	Profit of Optimum Action	*A*	*B*
Recession	*A*	$30	30 – 30 = 0	30 – (–50) = 80
Stable economy	*A*	70	70 – 70 = 0	70 – 30 = 40
Moderate growth	*B*	250	250 – 100 = 150	250 – 250 = 0
Boom	*B*	400	400 – 150 = 250	400 – 400 = 0

PROBLEMS FOR SECTION 17.1

Learning the Basics

PH Grade ASSIST **17.1** For this problem, use the following payoff table:

	Action	
Event	***A*** **($)**	***B*** **($)**
1	50	100
2	200	125

a. Construct an opportunity loss table.
b. Construct a decision tree.

PH Grade ASSIST **17.2** For this problem, use the following payoff table:

	Action	
Event	***A*** **($)**	***B*** **($)**
1	50	10
2	300	100
3	500	200

a. Construct an opportunity loss table.
b. Construct a decision tree.

Applying the Concepts

PH Grade ASSIST **17.3** A manufacturer of designer jeans must decide whether to build a large factory or a small factory in a particular location. The profit per pair of jeans manufactured is estimated as \$10. A small factory will incur an annual cost of \$200,000, with a production capacity of 50,000 jeans per year. A large factory will incur an annual cost of \$400,000 with a production capacity of 100,000 jeans per year. Four levels of manufacturing demand are considered likely: 10,000, 20,000, 50,000, and 100,000 pairs of jeans per year.

a. Determine the possible levels of production for a small factory and the payoffs for each possible level of production.
b. Determine the possible levels of production for a large factory and the payoffs for each possible level of production.
c. Based on the results of (a) and (b), construct a payoff table indicating the events and alternative courses of action.
d. Construct a decision tree.
e. Construct an opportunity loss table.

PH Grade ASSIST SELF Test **17.4** An author is trying to choose between two publishing companies that are competing for the marketing rights to her new novel. Company *A* has offered the author \$10,000 plus \$2 per book sold. Company *B* has offered the author \$2,000 plus \$4 per book sold. The author believes that five levels of demand for the book are possible: 1,000, 2,000, 5,000, 10,000, and 50,000 books sold.

a. Compute the payoffs for each level of demand for company *A* and company *B*.
b. Construct a payoff table indicating the events and alternative courses of action.
c. Construct a decision tree.
d. Construct an opportunity loss table.

17.5 The LeFleur Garden Center purchases and sells Christmas trees during the holiday season. It purchases the trees for \$10 each and sells them for \$20 each. Any trees not sold by Christmas Day are sold for \$2 each to a company that makes wood chips. The garden center estimates that four levels of demand are possible: 100, 200, 500, and 1,000 trees.

a. Compute the payoffs for purchasing 100, 200, 500, or 1,000 trees for each of the four levels of demand.
b. Construct a payoff table indicating the events and alternative courses of action.
c. Construct a decision tree.
d. Construct an opportunity loss table.

17.2 CRITERIA FOR DECISION MAKING

After you compute the profit and opportunity loss for each event under each alternative course of action, you need to determine the criteria for selecting the most desirable course of action. To determine which alternative to choose, you first assign a probability to each event. The probability assigned is based on information available from past data, from the opinions of the decision-maker, or from knowledge about the probability distribution that the event may follow. Using these probabilities, along with the payoffs or opportunity losses of each event–action combination, you select the best course of action according to a particular criterion. In this section, three decision criteria are presented—expected monetary value, expected opportunity loss, and the return-to-risk ratio.

Expected Monetary Value

In section 5.1, Equation (5.1) on page 159 shows how to compute the expected value of a probability distribution. Now you use this formula to compute the expected monetary value for each alternative course of action. The **expected monetary value (*EMV*)** for a course of action j is the payoff (x_{ij}) for each combination of event i and action j times P_i, the probability of occurrence of the event i, summed over all events [see Equation (17.1)].

EXPECTED MONETARY VALUE

$$EMV(j) = \sum_{i=1}^{N} x_{ij}P_i \quad \textbf{(17.1)}$$

where $EMV(j)$ = expected monetary value of action j

x_{ij} = payoff that occurs when course of action j is selected and event i occurs

P_i = probability of occurrence of event i

N = number of events

Criterion: Select the course of action with the largest *EMV*.

Example 17.4 illustrates the application of expected monetary value to the television set marketing example.

EXAMPLE 17.4

COMPUTING THE EXPECTED MONETARY VALUE (*EMV*) IN THE TELEVISION SET MARKETING EXAMPLE

Returning to the payoff table for deciding whether to market a television set (Example 17.1 on page 727), suppose that the probability is 0.40 that the television will be successful (so that the probability is 0.60 that the television will not be successful). Compute the expected monetary value for each alternative course of action and determine whether or not to market the television set.

SOLUTION You use Equation (17.1) to determine the expected monetary value for each alternative course of action. Table 17.5 summarizes these computations.

TABLE 17.5

Expected Monetary Value (in Millions of Dollars) for Each Alternative for the Television Set Marketing Example

		Alternative Courses of Action			
Event E_i	**P_i**	**Market, A_1**	**$x_{ij}P_i$**	**Do Not Market, A_2**	**$x_{ij}P_i$**
Successful E_1	0.40	+\$45	\$45(0.4) = \$18	−\$3	−\$3(0.4) = −\$1.2
Unsuccessful E_2	0.60	−\$36	−\$36(0.6) = −\$21.6	−\$3	−\$3(0.6) = −\$1.8
			$EMV(A_1)$ = −\$3.6		$EMV(A_2)$ = −\$3

The expected monetary value for marketing the television is −\$3.6 million, and the expected monetary value for not marketing the television set is −\$3 million. Thus, if your objective is to choose the action that maximizes the expected monetary value, you would choose the action of not marketing the television set because its profit is highest (or in this case its loss is lowest). Note, however, if the probability the television is successful, P_1, is slightly greater than the assumed value of 0.40, you would make a different decision. Specifically, if $P_1 = 0.41$, then $EMV(A_1) = -\$2.79$ million, $EMV(A_2) = -\$3$ million, and the best decision is to market the television. This change in the optimal decision, when such a small change in the assumed probability of success occurs, illustrates the importance of accuracy when determining the probabilities. You need to consider the *closeness* of the decision-making criterion before making a final decision.

As a second application of expected monetary value, return to the "Using Statistics" scenario and the payoff table presented in Table 17.2 on page 728. Suppose the company economist assigns the following probabilities to the different economic conditions:

$$P(\text{Recession}) = 0.10$$
$$P(\text{Stable economy}) = 0.40$$
$$P(\text{Moderate growth}) = 0.30$$
$$P(\text{Boom}) = 0.20$$

Table 17.6 shows the computations of the expected monetary value for each of the two stocks.

TABLE 17.6
Expected Monetary Value for Each of Two Stocks under Four Economic Conditions

		Alternative Courses of Action			
Event E_i	P_i	A	$x_{ij}P_i$	B	$x_{ij}P_i$
Recession	0.10	30	30(0.1) = 3	−50	−50(0.1) = −5
Stable economy	0.40	70	70(0.4) = 28	30	30(0.4) = 12
Moderate growth	0.30	100	100(0.3) = 30	250	250(0.3) = 75
Boom	0.20	150	150(0.2) = 30	400	400(0.2) = 80
			$EMV(A) = 91$		$EMV(B) = 162$

Thus, the expected monetary value or profit for stock A is \$91, and the expected monetary value or profit for stock B is \$162. Using these results, you should choose stock B because the expected monetary value on stock B is almost twice that of stock A. In terms of expected rate of return on the \$1,000 investment, stock B is 16.2% compared to 9.1% for stock A.

Expected Opportunity Loss

Earlier, you learned how to use the expected monetary value criterion when making a decision. An equivalent criterion, based on opportunity losses, is introduced next. Payoffs and opportunity losses can be viewed as two sides of the same coin. It all depends on whether you wish to view the problem in terms of *maximizing* expected monetary value or *minimizing* expected opportunity loss. The **expected opportunity loss (*EOL*)** of action j is the loss L_{ij} for each combination of event i and action j times P_i, the probability of occurrence of the event i, summed over all events [see Equation (17.2)].

EXPECTED OPPORTUNITY LOSS

$$EOL(j) = \sum_{i=1}^{N} L_{ij}P_i \tag{17.2}$$

where L_{ij} = opportunity loss that occurs when course of action j is selected and event i occurs

P_i = probability of occurrence of event i

Criterion: Select the course of action with the smallest *EOL*. Selecting the course of action with the smallest *EOL* is equivalent to selecting the course of action with the largest *EMV*. See Equation (17.1) on page 731.

Example 17.5 illustrates the application of opportunity loss for the television set marketing example.

EXAMPLE 17.5

COMPUTING THE EXPECTED OPPORTUNITY LOSS (*EOL*) FOR THE TELEVISION SET MARKETING EXAMPLE

Referring to the opportunity loss table given in Table 17.3 on page 729, and assuming that the probability is 0.40 that the television set will be successful, compute the expected opportunity loss for each alternative course of action. (See Table 17.7.) Determine whether or not to market the television set.

SOLUTION

TABLE 17.7

Expected Opportunity Loss in Millions of Dollars for Each Alternative for the Television Set Marketing Example

		Alternative Courses of Action			
Event E_i	**P_i**	**Market, A_1**	**$L_{ij}P_i$**	**Do Not Market, A_2**	**$L_{ij}P_i$**
Successful E_1	0.40	0	0(0.4) = 0	\$48	\$48(0.4) =\$19.2
Unsuccessful E_2	0.60	\$33	\$33(0.6) = \$19.8	0	0(0.6) = 0
			$EOL(A_1)$ = \$19.8		$EOL(A_2)$ = \$19.2

The expected opportunity loss is lower for not marketing the television set (\$19.2 million) than for marketing the television (\$19.8 million). Therefore, using the *EOL* criterion, the optimal decision is not to market the television set. This outcome is expected since the equivalent *EMV* criterion produced the same optimal strategy. Note once again that if the probability the television set is successful, P_1 is slightly greater than the assumed value of 0.40, a different decision is made. Specifically, if $P_1 = 0.41$, then $EOL(A_1)$ = \$19.47 million, $EOL(A_2)$ = \$19.68 million, and the best decision is to market the television set.

The expected opportunity loss from the best decision is called the **expected value of perfect information (*EVPI*)**. Equation (17.3) defines the *EVPI*.

EXPECTED VALUE OF PERFECT INFORMATION

$$EVPI = \text{expected profit under certainty} - \text{expected monetary value of the best alternative} \quad \textbf{(17.3)}$$

The **expected profit under certainty** represents the expected profit that you could make if you have perfect information about which event will occur.

Example 17.6 illustrates the expected value of perfect information.

EXAMPLE 17.6

COMPUTING THE EXPECTED VALUE OF PERFECT INFORMATION (*EVPI*) IN THE TELEVISION SET MARKETING EXAMPLE

Referring to Example 17.5, compute the expected profit under certainty and the expected value of perfect information.

SOLUTION As the marketing manager of the consumer electronics company, if you could always predict the future, a profit of \$45 million would be made for the 40% of the television sets that are successful, and a loss of \$3 million would be incurred for the 60% of the television sets that are not successful. Thus,

$$\begin{aligned}\text{Expected profit under certainty} &= 0.40(\$45) + 0.60(-\$3)\\ &= \$18 - \$1.80\\ &= \$16.2\end{aligned}$$

This value, \$16.2 million, represents the profit you could make if you knew with *certainty* whether or not the television set was going to be successful. Use Equation (17.3) to compute the expected value of perfect information:

$$\begin{aligned}EVPI &= \text{expected profit under certainty} - \text{expected monetary value of the best alternative}\\ &= \$16.2 - (-\$3) = \$19.2\end{aligned}$$

This $EVPI$ value of \$19.2 million represents the maximum amount that you should be willing to pay for perfect information. Of course, you can never have perfect information and you should never pay the entire $EVPI$ for more information. Rather, the $EVPI$ provides a guideline for an upper bound on how much you might consider paying for better information. The $EVPI$ is also the expected opportunity loss of not marketing the television set.

Return to the "Using Statistics" scenario and the opportunity loss table presented in Table 17.4 on page 729. Table 17.8 presents the computations to determine the expected opportunity loss for Stock A and Stock B.

TABLE 17.8

Expected Opportunity Loss for Each Alternative for the Stock Selection Example

		Alternative Courses of Action			
Event, E_i	P_i	A	$L_{ij}P_i$	B	$L_{ij}P_i$
Recession	0.10	0	0(0.1) = 0	80	80(0.1) = 8
Stable economy	0.40	0	0(0.4) = 0	40	40(0.4) = 16
Moderate growth	0.30	150	150(0.3) = 45	0	0(0.3) = 0
Boom	0.20	250	250(0.2) = 50	0	0(0.2) = 0
			$EOL(A) = 95$		$EOL(B) = EVPI = 24$

The expected opportunity loss is lower for stock B than for stock A. Your optimal decision is to choose stock B, which is consistent with the decision made using expected monetary value. The expected value of perfect information is \$24 (per \$1,000 invested), meaning that you should be willing to pay up to \$24 for perfect information.

Return-to-Risk Ratio

Unfortunately, neither the expected monetary value nor the expected opportunity loss criterion takes into account the *variability* of the payoffs for the alternative courses of action under different events. From Table 17.2 on page 728, you see that the return for stock A varies from \$30 in a recession to \$150 in an economic boom, while stock B (the one chosen according to the expected monetary value and expected opportunity loss criteria) varies from a loss of \$50 in a recession to a profit of \$400 in an economic boom.

To take into account the variability of the events (in this case, the different economic conditions), you can compute the variance and standard deviation of each stock using Equations

(5.2) and (5.3) on page 160. Using the information presented in Table 17.6 on page 732, for stock A, $\mu_A = \$91$ and the variance is equal to

$$\sigma_A^2 = \sum_{i=1}^{N} (X_i - \mu)^2 P(X_i)$$

$$= (30 - 91)^2(0.1) + (70 - 91)^2(0.4) + (100 - 91)^2(0.3) + (150 - 91)^2(0.2)$$

$$= 1{,}269$$

and the standard deviation of stock A $(\sigma_A) = \sqrt{1{,}269} = \$35.62$.

For stock B, $\mu_B = \$162$ and the variance is

$$\sigma_B^2 = \sum_{i=1}^{N} (X_i - \mu)^2 P(X_i)$$

$$= (-50 - 162)^2(0.1) + (30 - 162)^2(0.4) + (250 - 162)^2(0.3) + (400 - 162)^2(0.2)$$

$$= 25{,}116$$

and the standard deviation of stock B $(\sigma_B) = \sqrt{25{,}116} = \$158.48$.

Because you are comparing two sets of data with vastly different means, you should evaluate the relative risk associated with each stock. Once you compute the standard deviation of the return from each stock, you compute the coefficient of variation discussed in section 3.2. Substituting σ for S and EMV for $\bar{X}$ in Equation (3.11) on page 86, you find that the coefficient of variation for stock A is equal to

$$CV_A = \left(\frac{\sigma_A}{EMV_A}\right)100\%$$

$$= \left(\frac{35.62}{91}\right)100\% = 39.1\%$$

while the coefficient of variation for stock B is equal to

$$CV_B = \left(\frac{\sigma_B}{EMV_B}\right)100\%$$

$$= \left(\frac{158.48}{162}\right)100\% = 97.8\%$$

Thus, there is much more variation in the return for stock B than for stock A.

Since the coefficient of variation shows the relative size of the variation compared to the mean (or expected monetary value), a criterion other than EMV or EOL is needed to express the relationship between the return (as expressed by the EMV) and the risk (as expressed by the standard deviation). Equation (17.4) defines the **return-to-risk ratio** as the expected monetary value of action j divided by the standard deviation of action j.

RETURN-TO-RISK RATIO

$$RTRR(j) = \frac{EMV(j)}{\sigma_j} \qquad \textbf{(17.4)}$$

where $EMV(j)$ = expected monetary value for alternative course of action j

σ_j = standard deviation for alternative course of action j

Criterion: Select the course of action with the largest $RTRR$.

For each of the two stocks discussed previously, you compute the return-to-risk ratios as follows. For stock A, the return-to-risk ratio is equal to

$$RTRR(A) = \frac{91}{35.62} = 2.55$$

For stock B, the return-to-risk ratio is equal to

$$RTRR(B) = \frac{162}{158.48} = 1.02$$

Thus, relative to the risk as expressed by the standard deviation, the expected return is much higher for stock A than for stock B. Stock A has a smaller expected monetary value than stock B, but also has a much smaller risk than stock B. The return-to-risk ratio shows A to be preferable to B. Figure 17.4 represents Microsoft Excel output.

FIGURE 17.4
Microsoft Excel Expected Value and Standard Deviation Worksheet

	A	B	C	D	E	F	G	H
1	**Stock Selection Analysis**							
2								
3	**Probabilities & Payoffs Table:**							
4		**P**	**Stock A**	**Stock B**				
5	**Recession**	**0.1**	**30**	**-50**				
6	**Stable economy**	**0.4**	**70**	**30**		Calculations Area		
7	**Moderate growth**	**0.3**	**100**	**250**		For variance and std. deviation		
8	**Boom**	**0.2**	**150**	**400**		Stock A	Stock B	
9						3721	44944	
10	**Statistics for:**		**Stock A**	**Stock B**		441	17424	
11	**Expected Monetary Value**		**91**	**162**		81	7744	
12	**Variance**		**1269**	**25116**		3481	56644	
13	**Standard Deviation**		**35.6230**	**158.4803**				
14	**Coefficient of Variation**		**0.3915**	**0.9783**				
15	**Return to Risk Ratio**		**2.5545**	**1.0222**				
16								
17	Opportunity Loss Table:							
18		Optimum	Optimum	Alternatives				
19		Action	Profit	Stock A	Stock B			
20	Recession	Stock A	30	0	80			
21	Stable economy	Stock A	70	0	40			
22	Moderate growth	Stock B	250	150	0			
23	Boom	Stock B	400	250	0			
24				**Stock A**	**Stock B**			
25	**Expected Opportunity Loss**			**95**	**24**			
26					**EVPI**			

PROBLEMS FOR SECTION 17.2

Learning the Basics

PH Grade ASSIST **17.6** For the following payoff table:

	Action	
Event	***A***	***B***
1	50	100
2	200	125

The probability of event 1 is 0.5 and the probability of event 2 is also 0.5.

a. Compute the expected monetary value (EMV) for actions A and B.

b. Compute the expected opportunity loss (EOL) for actions A and B.

c. Explain the meaning of the expected value of perfect information ($EVPI$) in this problem.

d. Based on the results of (a) or (b), which action would you choose? Why?

e. Compute the coefficient of variation for each action.

f. Compute the return-to-risk ratio ($RTRR$) for each action.

g. Based on (e) and (f), what action would you choose? Why?

h. Compare the results of (d) and (g) and explain any differences.

PH Grade ASSIST **17.7** For the following payoff table:

	Action	
Event	***A***	***B***
1	50	10
2	300	100
3	500	200

The probability of event 1 is 0.8, the probability of event 2 is 0.1, and the probability of event 3 is 0.1.

a. Compute the expected monetary value (*EMV*) for actions *A* and *B*.
b. Compute the expected opportunity loss (*EOL*) for actions *A* and *B*.
c. Explain the meaning of the expected value of perfect information (*EVPI*) in this problem.
d. Based on the results of (a) or (b), which action would you choose? Why?
e. Compute the coefficient of variation for each action.
f. Compute the return-to-risk (*RTRR*) ratio for each action.
g. Based on (e) and (f), what action would you choose? Why?
h. Compare the results of (d) and (g) and explain any differences.
i. Would your answers to (d) and (g) be different if the probabilities for the three events were 0.1, 0.1, and 0.8, respectively? Discuss.

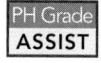

PH Grade ASSIST **17.8** For a potential investment of $1,000, if a stock has an *EMV* of $100 and a standard deviation of $25, what is the

a. rate of return?
b. coefficient of variation?
c. return-to-risk ratio?

PH Grade ASSIST **17.9** A stock has the following predicted returns under the following economic conditions:

Economic Condition	Probability	Return
Recession	0.30	$50
Stable economy	0.30	100
Moderate growth	0.30	120
Boom	0.10	200

Compute the

a. expected monetary value.
b. standard deviation.
c. coefficient of variation.
d. return-to-risk ratio.

PH Grade ASSIST **17.10** The following are the results for two stocks:

	A	*B*
Expected monetary value	$90	$60
Standard deviation	10	10

Which stock would you choose and why?

PH Grade ASSIST **17.11** The following are the results for two stocks:

	A	*B*
Expected monetary value	$60	$60
Standard deviation	20	10

Which stock would you choose and why?

Applying the Concepts

SELF Test **17.12** A vendor at a local baseball stadium must determine whether to sell ice cream or soft drinks at today's game. The vendor believes that the profit made will depend on the weather. The payoff table is as follows:

	Action	
Event	**Sell Soft Drinks**	**Sell Ice Cream**
Cool weather	$50	$30
Warm weather	60	90

Based on her past experience at this time of year, the vendor estimates the probability of warm weather as 0.60.

a. Compute the expected monetary value (*EMV*) for selling soft drinks and selling ice cream.
b. Compute the expected opportunity loss (*EOL*) for selling soft drinks and selling ice cream.
c. Explain the meaning of the expected value of perfect information (*EVPI*) in this problem.
d. Based on the results of (a) or (b), would you choose to sell soft drinks or ice cream? Why?
e. Compute the coefficient of variation for selling soft drinks and for selling ice cream.
f. Compute the return-to-risk ratio (*RTRR*) for selling soft drinks and selling ice cream.
g. Based on (e) and (f), would you choose to sell soft drinks or ice cream? Why?
h. Compare the results of (d) and (g) and explain any differences.

17.13 The Islander Fishing Company purchases clams for $1.50 per pound from fishermen and sells them to various restaurants for $2.50 per pound. Any clams not sold to the restaurants by the end of the week can be sold to a local soup company for $0.50 per pound. The company can purchase

500, 1,000, or 2,000 pounds. The probabilities of various levels of demand are as follows:

Demand (Pounds)	Probability
500	0.2
1,000	0.4
2,000	0.4

a. For each possible purchase level (500, 1,000, or 2,000 pounds), compute the profit (or loss) for each level of demand.
b. Using the expected monetary value (*EMV*) criterion, determine the optimal number of pounds of clams the company should purchase from the fishermen. Discuss.
c. Compute the standard deviation for each possible purchase level.
d. Compute the expected opportunity loss (*EOL*) for purchasing 500, 1,000, and 2,000 pounds of clams.
e. Explain the meaning of the expected value of perfect information (*EVPI*) in this problem.
f. Compute the coefficient of variation for purchasing 500, 1,000, and 2,000 pounds of clams. Discuss.
g. Compute the return-to-risk ratio (*RTRR*) for purchasing 500, 1,000, and 2,000 pounds of clams. Discuss.
h. Based on (b) and (d), what would you choose to sell, 500, 1,000, or 2,000 pounds of clams? Why?
i. Compare the results of (b), (d), (f), and (g) and explain any differences.
j. Suppose that clams can be sold to restaurants for $3 per pound. Repeat (a) through (h) with this selling price for clams and compare the results with those in (i).
k. What would be the effect on the results in (a) through (i) if the probability of the demand for 500, 1,000, and 2,000 clams were 0.4, 0.4, and 0.2, respectively?

PH Grade ASSIST **17.14** An investor has a certain amount of money available to invest now. Three alternative investments are available. The estimated profits of each investment under each economic condition are indicated in the following payoff table:

	Investment Selection		
Event	*A*	*B*	*C*
Economy declines	$500	–$2,000	–$7,000
No change	$1,000	$2,000	–$1,000
Economy expands	$2,000	$5,000	$20,000

Based on his own past experience, the investor assigns the following probabilities to each economic condition:

$$P(\text{economy declines}) = 0.30$$

$$P(\text{no change}) = 0.50$$

$$P(\text{economy expands}) = 0.20$$

a. Determine the best investment according to the expected monetary value (*EMV*) criterion. Discuss.
b. Compute the standard deviation for investments *A*, *B*, and *C*.
c. Compute the expected opportunity loss (*EOL*) for investments *A*, *B*, and *C*.
d. Explain the meaning of the expected value of perfect information (*EVPI*) in this problem.
e. Compute the coefficient of variation for investments *A*, *B*, and *C*.
f. Compute the return-to-risk ratio (*RTRR*) for investments *A*, *B*, and *C*.
g. Based on (e) and (f), what would you choose, investments *A*, *B*, or *C*? Why?
h. Compare the results of (a) and (g) and explain any differences.
i. Suppose the probabilities of the different economic conditions are as follows:
(1) 0.1, 0.6, and 0.3
(2) 0.1, 0.3, and 0.6
(3) 0.4, 0.4, and 0.2
(4) 0.6, 0.3, and 0.1
Repeat (a) through (h) with each of these sets of probabilities and compare the results with those originally computed in (h). Discuss.

17.15 In problem 17.3 on page 730, you developed a payoff table for building a small factory and a large factory for manufacturing designer jeans. Given the results of that problem, suppose that the probabilities of the demand are as follows:

Demand	Probability
10,000	0.1
20,000	0.4
50,000	0.2
100,000	0.3

a. Compute the expected monetary value (*EMV*) for building a small factory and building a large factory.
b. Compute the expected opportunity loss (*EOL*) for building a small factory and building a large factory.
c. Explain the meaning of the expected value of perfect information (*EVPI*) in this problem.
d. Based on the results of (a) or (b), would you choose to build a small factory or a large factory? Why?
e. Compute the coefficient of variation for building a small factory and building a large factory.
f. Compute the return-to-risk ratio (*RTRR*) for building a small factory and building a large factory.
g. Based on (e) and (f), would you choose to build a small factory or a large factory? Why?
h. Compare the results of (d) and (g) and explain any differences.
i. Suppose that the probabilities of demand are 0.4, 0.2, 0.2, and 0.2, respectively. Repeat (a) through (h) with

these probabilities and compare the results with those of (a) through (h).

PH Grade ASSIST **17.16** In problem 17.4 on page 730, you developed a payoff table to assist an author in choosing between signing with company *A* or with company *B*. Given the results computed in that problem, suppose that the probabilities of the levels of demand for the novel are as follows:

Demand	Probability
1,000	0.45
2,000	0.20
5,000	0.15
10,000	0.10
50,000	0.10

a. Compute the expected monetary value (*EMV*) for signing with company *A* and with company *B*.
b. Compute the expected opportunity loss (*EOL*) for signing with company *A* and with company *B*.
c. Explain the meaning of the expected value of perfect information (*EVPI*) in this problem.
d. Based on the results of (a) or (b), if you were the author, which company would you choose to sign with, company *A* or company *B*? Why?
e. Compute the coefficient of variation for signing with company *A* and signing with company *B*.
f. Compute the return-to-risk ratio (*RTRR*) for signing with company *A* and signing with company *B*.
g. Based on (e) and (f), which company would you choose to sign with, company *A* or company *B*? Why?
h. Compare the results of (d) and (g) and explain any differences.
i. Suppose that the probabilities of demand are 0.3, 0.2, 0.2, 0.1, and 0.2, respectively. Repeat (a) through (h) with these probabilities and compare the results with those for (a) through (h).

17.17 In problem 17.5 on page 730, you developed a payoff table for whether to purchase 100, 200, 500, or 1,000 Christmas trees. Given the results of that problem, suppose that the probabilities of the demand for the different number of trees are as follows:

Demand (Number of Trees)	Probability
100	0.20
200	0.50
500	0.20
1,000	0.10

a. Compute the expected monetary value (*EMV*) for purchasing 100, 200, 500, and 1,000 trees.
b. Compute the expected opportunity loss (*EOL*) for purchasing 100, 200, 500, and 1,000 trees.
c. Explain the meaning of the expected value of perfect information (*EVPI*) in this problem.
d. Based on the results of (a) or (b), would you choose to purchase 100, 200, 500, or 1,000 trees? Why?
e. Compute the coefficient of variation for purchasing 100, 200, 500, and 1,000 trees.
f. Compute the return-to-risk ratio (*RTRR*) for purchasing 100, 200, 500, and 1,000 trees.
g. Based on (e) and (f), would you choose to purchase 100, 200, 500, or 1,000 trees? Why?
h. Compare the results of (d) and (g) and explain any differences.
i. Suppose that the probabilities of demand are 0.4, 0.2, 0.2, and 0.2, respectively. Repeat (a) through (h) with these probabilities and compare the results with those for (a) through (h).

17.3 DECISION MAKING WITH SAMPLE INFORMATION

In Sections 17.1 and 17.2 you learned the framework for making decisions when there are several alternative courses of action. You then studied three different criteria for choosing between alternatives. For each criterion, you assigned the probabilities of the various events using the past experience and/or the subjective judgment of the decision-maker. This section introduces decision making when sample information is available to estimate probabilities. Example 17.7 illustrates decision making with sample information.

EXAMPLE 17.7 DECISION MAKING USING SAMPLE INFORMATION FOR THE TELEVISION SET MARKETING EXAMPLE

In section 4.3 on page 144, you found that the probability of a successful television set, given that the company receives a favorable report is 0.64. Thus, the probability of an unsuccessful television set, given that the company receives a favorable report is 1 – 0.64 = 0.36.

Given that the company receives a favorable report, compute the expected monetary value for each alternative course of action, and determine whether or not to market the television set.

SOLUTION You need to use the revised probabilities, not the original subjective probabilities, to compute the expected monetary value of each alternative. Table 17.9 illustrates the computations.

TABLE 17.9
Expected Monetary Value in Millions of Dollars Using Revised Probabilities for Each Alternative in the Television Set Marketing Example

		Alternative Courses of Action			
Event E_i	P_i	**Market, A_1**	$x_{ij}P_i$	**Do Not Market, A_2**	$x_{ij}P_i$
Successful E_1	0.64	+\$45	\$45(0.64) = \$28.8	–\$3	–\$3(0.64) = –\$1.92
Unsuccessful E_2	0.36	–\$36	–\$36(0.36) = –\$12.96	–\$3	–\$3(0.36) = –\$1.08
			$EMV(A_1)$ = \$15.84		$EMV(A_2)$ = –\$3

In this case, the optimal decision is to market the product, because a profit of \$15.84 million is expected as compared to a loss of \$3 million if the television set is not marketed. This decision is different from the one considered optimal prior to the collection of the sample information in the form of the market research report. The favorable recommendation contained in the report greatly increases the probability that the television set will be successful.

Because the relative desirability of the two stocks under consideration in the "Using Statistics" scenario is directly affected by economic conditions, you should use a forecast of the economic conditions in the upcoming year. You can then use Bayes' theorem, introduced in section 4.3, to revise the probabilities associated with the different economic conditions. Suppose that such a forecast can predict either an expanding economy (F_1) or a declining or stagnant economy (F_2). Past experience indicates that, when there is a recession, prior forecasts predicted an expanding economy 20% of the time. When there is a stable economy, prior forecasts predicted an expanding economy 40% of the time. When there is moderate growth, prior forecasts predicted an expanding economy 70% of the time. Finally, when there is a boom economy, prior forecasts predicted an expanding economy 90% of the time.

If the forecast is for an expanding economy, you can revise the probabilities of economic conditions using Bayes' theorem, Equation (4.9) on page 144. Let

event E_1 = recession
event E_2 = stable economy
event E_3 = moderate growth
event E_4 = boom economy

event F_1 = expanding economy is predicted
event F_2 = declining or stagnant economy is predicted

and

$$P(E_1) = 0.10 \quad P(F_1|E_1) = 0.20$$
$$P(E_2) = 0.40 \quad P(F_1|E_2) = 0.40$$
$$P(E_3) = 0.30 \quad P(F_1|E_3) = 0.70$$
$$P(E_4) = 0.20 \quad P(F_1|E_4) = 0.90$$

Then, using Bayes' theorem defined by Equation (4.9) on page 144

$$P(E_1|F_1) = \frac{P(F_1|E_1)P(E_1)}{P(F_1|E_1)P(E_1) + P(F_1|E_2)P(E_2) + P(F_1|E_3)P(E_3) + P(F_1|E_4)P(E_4)}$$

$$= \frac{(0.20)(0.10)}{(0.20)(0.10) + (0.40)(0.40) + (0.70)(0.30) + (0.90)(0.20)}$$

$$= \frac{0.02}{0.57} = 0.035$$

$$P(E_2|F_1) = \frac{P(F_1|E_2)P(E_2)}{P(F_1|E_1)P(E_1) + P(F_1|E_2)P(E_2) + P(F_1|E_3)P(E_3) + P(F_1|E_4)P(E_4)}$$

$$= \frac{(0.40)(0.40)}{(0.20)(0.10) + (0.40)(0.40) + (0.70)(0.30) + (0.90)(0.20)}$$

$$= \frac{0.16}{0.57} = 0.281$$

$$P(E_3|F_1) = \frac{P(F_1|E_3)P(E_3)}{P(F_1|E_1)P(E_1) + P(F_1|E_2)P(E_2) + P(F_1|E_3)P(E_3) + P(F_1|E_4)P(E_4)}$$

$$= \frac{(0.70)(0.30)}{(0.20)(0.10) + (0.40)(0.40) + (0.70)(0.30) + (0.90)(0.20)}$$

$$= \frac{0.21}{0.57} = 0.368$$

$$P(E_4|F_1) = \frac{P(F_1|E_4)P(E_4)}{P(F_1|E_1)P(E_1) + P(F_1|E_2)P(E_2) + P(F_1|E_3)P(E_3) + P(F_1|E_4)P(E_4)}$$

$$= \frac{(0.90)(0.20)}{(0.20)(0.10) + (0.40)(0.40) + (0.70)(0.30) + (0.90)(0.20)}$$

$$= \frac{0.18}{0.57} = 0.316$$

Table 17.10 summarizes the computation of these probabilities. Figure 17.5 displays the joint probabilities in a decision tree. You need to use the revised probabilities, not the original subjective probabilities to compute the expected monetary value. Table 17.11 shows these computations.

TABLE 17.10

Bayes' Theorem Calculations for the Stock Selection Example

Event, E_i	Prior Probability, $P(E_i)$	Conditional Probability, $P(F_1\|E_i)$	Joint Probability, $P(F_1\|E_i)P(E_i)$	Revised Probability, $P(E_i\|F_1)$
E_1 = Recession	0.10	0.20	0.02	$0.02/0.57 = 0.035 = P(E_1\|F_1)$
E_2 = Stable economy	0.40	0.40	0.16	$0.16/0.57 = 0.281 = P(E_2\|F_1)$
E_3 = Moderate growth	0.30	0.70	0.21	$0.21/0.57 = 0.368 = P(E_3\|F_1)$
E_4 = Boom	0.20	0.90	0.18	$0.18/0.57 = 0.316 = P(E_4\|F_1)$
			0.57	

FIGURE 17.5

Decision Tree with Joint Probabilities for Stock Selection Example

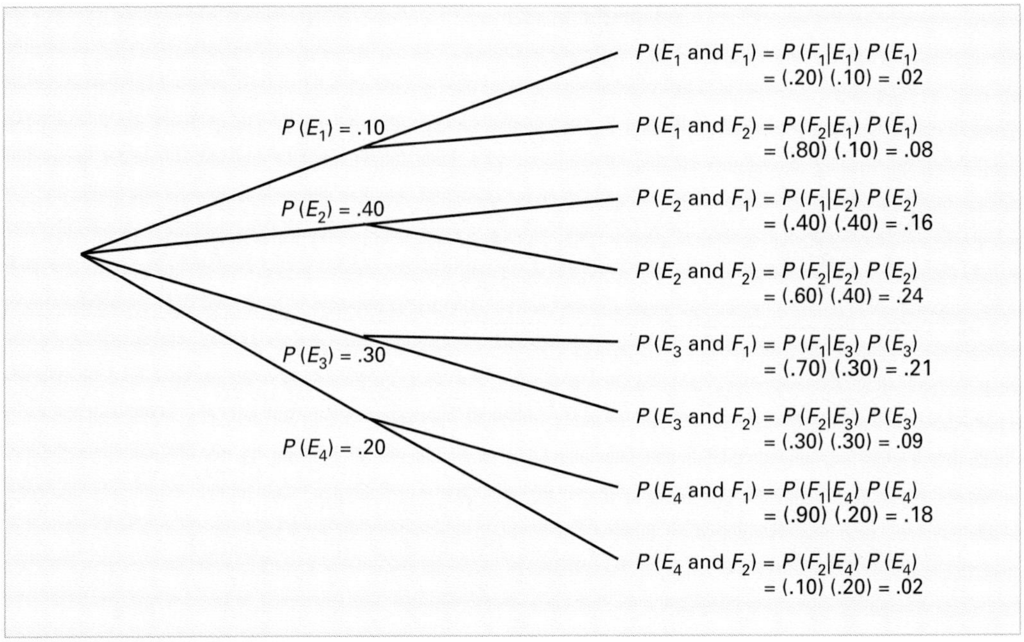

TABLE 17.11

Expected Monetary Value Using Revised Probabilities for Each Alternative for Each of Two Stocks under Four Economic Conditions

		Alternative Courses of Action			
Event, E_i	P_i	A	$x_{ij}P_i$	B	$x_{ij}P_i$
Recession	0.035	30	30(0.035) = 1.05	−\$50	−50(0.035) = −1.75
Stable economy	0.281	70	70(0.281) = 19.67	30	30(0.281) = 8.43
Moderate growth	0.368	100	100(0.368) = 36.80	250	250(0.368) = 92.00
Boom	0.316	150	150(0.316) = 47.40	400	400(0.316) = 126.40
			$EMV(A)$ = \$104.92		$EMV(B)$ = \$225.08

Thus, the expected monetary value or profit for stock A is \$104.92, and the expected monetary value or profit for stock B is \$225.08. Using this criterion, you should once again choose stock B, since the expected monetary value is much higher for this stock. However, you should reexamine the return-to-risk ratios in light of these revised probabilities. Using Equations (5.2) and (5.3) on page 160, for stock A since $\mu_A = \$104.92$,

$$\sigma_A^2 = \sum_{i=1}^{N}(X_i - \mu)^2 P(X_i)$$

$$= (30 - 104.92)^2(0.035) + (70 - 104.92)^2(0.281) + (100 - 104.92)^2(0.368) + (150 - 104.92)^2(0.316)$$

$$= 1{,}190.194$$

The standard deviation of stock A $(\sigma_A) = \sqrt{1{,}190.194} = \$34.50$.

For stock B since $\mu_B = \$225.08$,

$$\sigma_B^2 = \sum_{i=1}^{N}(X_i - \mu)^2 P(X_i)$$

$$= (-50 - 225.08)^2(0.035) + (30 - 225.08)^2(0.281) + (250 - 225.08)^2(0.368) + (400 - 225.08)^2(0.316)$$

$$= 23{,}239.39$$

The standard deviation of stock B $(\sigma_B) = \sqrt{23{,}239.39} = \$152.445$.

To compute the coefficient of variation, substitute σ for S and EMV for $\overline{X}$ in Equation (3.11) on page 86,

$$CV_A = \left(\frac{\sigma_A}{EMV_A}\right)100\%$$

$$= \left(\frac{34.50}{104.92}\right)100\% = 32.88\%$$

and

$$CV_B = \left(\frac{\sigma_B}{EMV_B}\right)100\%$$

$$= \left(\frac{152.445}{225.08}\right)100\% = 67.73\%$$

Thus, there is still much more variation in the returns from stock B than from stock A.

For each of these two stocks, you calculate the return-to-risk ratios as follows. For stock A, the return-to-risk ratio is equal to

$$RTRR(A) = \frac{104.92}{34.50} = 3.041$$

For stock B, the return-to-risk ratio is equal to

$$RTRR(B) = \frac{225.08}{152.445} = 1.476$$

Thus, using the return-to-risk ratio, you should select stock A. This decision is different from the one you reached when using expected monetary value (or the equivalent expected opportunity loss). What stock should you buy? Your final decision will depend on whether you believe that it is more important to maximize the expected return on investment (select stock B) or to control the relative risk (select stock A).

PROBLEMS FOR SECTION 17.3

Learning the Basics

PH Grade ASSIST **17.18** Consider the following payoff table.

	Action	
Event	***A***	***B***
1	50	100
2	200	125

For this problem, $P(E_1) = 0.5$, $P(E_2) = 0.5$, $P(F \mid E_1) = 0.6$, and $P(F \mid E_2) = 0.4$. Suppose that you are informed that event F occurs.

a. Revise the probabilities $P(E_1)$ and $P(E_2)$ now that you know event F has occurred.

Based on these revised probabilities, answer (b) through (i).

b. Compute the expected monetary value of action A and action B.

c. Compute the expected opportunity loss of action A and action B.

d. Explain the meaning of the expected value of perfect information ($EVPI$) in this problem.

e. On the basis of (b) or (c), which action should you choose? Why?

f. Compute the coefficient of variation for each action.

g. Compute the return-to-risk ratio ($RTRR$) for each action.

h. On the basis of (f) and (g), which action should you choose? Why?

i. Compare the results of (e) and (h), and explain any differences.

PH Grade ASSIST **17.19** Consider the following payoff table:

	Action	
Event	***A***	***B***
1	50	10
2	300	100
3	500	200

For this problem $P(E_1) = 0.8$, $P(E_2) = 0.1$, and $P(E_3) = 0.1$, $P(F | E_1) = 0.2$, $P(F | E_2) = 0.4$, and $P(F | E_3) = 0.4$. Suppose you are informed that event F occurs.

a. Revise the probabilities $P(E_1)$, $P(E_2)$, and $P(E_3)$ now that you know event F has occurred.

Based on these revised probabilities, answer (b) through (i).

b. Compute the expected monetary value of action A and action B.

c. Compute the expected opportunity loss of action A and action B.

d. Explain the meaning of the expected value of perfect information (*EVPI*) in this problem.

e. On the basis of (b) and (c), which action should you choose? Why?

f. Compute the coefficient of variation for each action.

g. Compute the return-to-risk ratio (*RTRR*) for each action.

h. On the basis of (f) and (g), which action should you choose? Why?

i. Compare the results of (e) and (h) and explain any differences.

Applying the Concepts

SELF Test **17.20** In problem 17.12 on page 737, a vendor at a baseball stadium was deciding on whether to sell ice cream or soft drinks at today's game. Prior to making her decision, she decides to listen to the local weather forecast. In the past, when it has been cool, the weather reporter has forecast cool weather 80% of the time. When it has been warm, the weather reporter has forecast warm weather 70% of the time. The local weather forecast is for cool weather.

a. Revise the prior probabilities now that you know the weather forecast is for cool weather.

b. Use these revised probabilities to do problem 17.12.

c. Compare the results in (b) to those of problem 17.12.

PH Grade ASSIST **17.21** In problem 17.14 on page 738, an investor is trying to determine the optimal investment decision among three investment opportunities. Prior to making his investment decision, the investor decides to consult with his stockbroker. In the past when the economy has declined, the stockbroker has given a rosy forecast 20% of the time (with a gloomy forecast 80% of the time). When there has been no change in the economy, the stockbroker has given a rosy forecast 40% of the time. When there has been an expanding economy, the stockbroker has given a rosy forecast 70% of the time. The stockbroker in this case gives a gloomy forecast for the economy.

a. Revise the probabilities of the investor based on this economic forecast by the stockbroker.

b. Use these revised probabilities to do problem 17.14.

c. Compare the results in (b) to those of problem 17.14.

17.22 In problem 17.16 on page 739, an author is deciding which of two competing publishing companies to select to publish her new novel. Prior to making a final decision, the author decides to have an experienced reviewer examine her novel. This reviewer has an outstanding reputation for predicting the success of a novel. In the past, for novels that sold 1,000 copies, only 1% received favorable reviews. Of novels that sold 5,000 copies, 25% received favorable reviews. Of novels that sold 10,000 copies, 60% received favorable reviews. Of novels that sold 50,000 copies, 99% received favorable reviews. After examining the author's novel, the reviewer gives it an unfavorable review.

a. Revise the probabilities of the number of books sold in light of the reviewer's unfavorable review.

b. Use these revised probabilities to do problem 17.16.

c. Compare the results in (b) to those of problem 17.16.

17.4 UTILITY

The methods used in sections 17.1 through 17.3 assume that each *incremental* amount of profit or loss has the same value as the previous amounts of profits attained or losses incurred. In fact, under many circumstances in the business world, this assumption of incremental changes is not valid. Most companies, as well as most individuals, make special efforts to avoid large losses. At the same time, many companies, as well as most individuals, place less value on extremely large profits as compared to initial profits. Such differential evaluation of incremental profits or losses is referred to as **utility**, a concept first discussed by Daniel Bernoulli in the eighteenth century (see reference 1). To illustrate this concept, suppose that you are faced with the following two choices:

Choice 1 A fair coin is to be tossed—if it lands on heads, you will receive \$0.60; if it lands on tails, you will pay \$0.40.
Choice 2 Do not play the game.

What decision should you choose? The expected value of playing this game is $(0.60)(0.50) + (-0.40)(0.50) = +\$0.10$, and the expected value of not playing the game is 0.

Most people will decide to play the game, since the expected value is positive and only small amounts of money are involved. Suppose, however, that the game is formulated with a payoff of \$600,000 when the coin lands on heads and a loss of \$400,000 when the coin lands on tails. The expected value of playing the game is now +\$100,000. With these payoffs, even though the expected value is positive, most individuals will not play the game because of the severe negative consequences of losing \$400,000. Each additional dollar amount of either profit or loss does not have the same utility as the previous amount. Large negative amounts for most individuals have severely negative utility; conversely, the extra value of each incremental dollar of profit decreases, when high enough profit levels are reached.

An important part of the decision-making problem, which is beyond the scope of this text (see references 2 and 3), is to develop a utility curve for the decision-maker that represents the utility of each specified dollar amount.

Figure 17.6 illustrates three types of utility curves: those of the risk averter, the risk seeker, and the risk-neutral person.

FIGURE 17.6

Three Types of Utility Curves: Panel A Risk Averter, Panel B Risk Seeker, Panel C Risk Neutral.

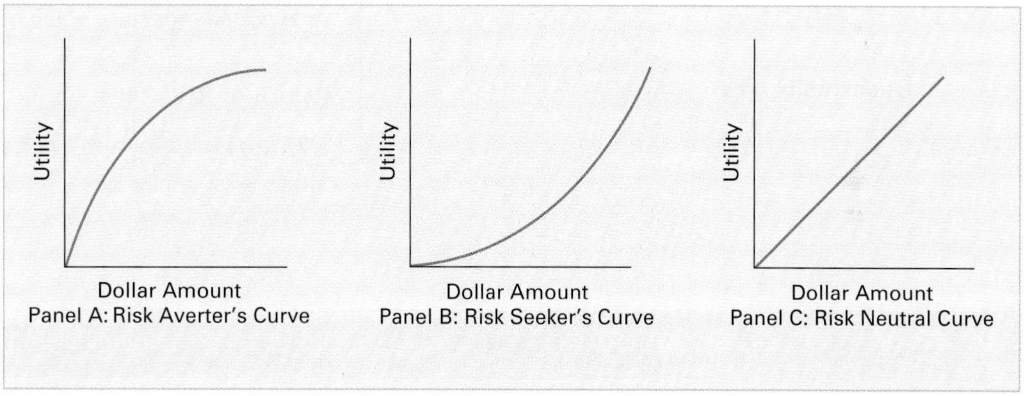

The **risk averter's curve** shows a rapid increase in utility for initial amounts of money followed by a gradual leveling off for increasing dollar amounts. This curve is appropriate for most individuals or businesses, because the value of each additional dollar is not as great after large amounts of money have already been earned.

The **risk seeker's curve** represents the utility of someone who enjoys taking risks. The utility is greater for large dollar amounts. This curve represents an individual who is interested only in "striking it rich" and is willing to take large risks for the opportunity of making large profits.

The **risk-neutral curve** represents the expected monetary value approach. Each additional dollar of profit has the same value as the previous dollar.

After a utility curve is developed in a specific situation, you convert the dollar amounts to utilities. Then you compute the utility of each alternative course of action and apply the decision criteria of expected utility value, expected opportunity loss, and return-to-risk ratio to make a decision.

PROBLEMS FOR SECTION 17.4

Applying the Concepts

17.23 Do you consider yourself a risk seeker, a risk averter, or a risk-neutral person? Explain.

17.24 Refer to problems 17.3–17.5 and 17.12–17.14 on pages 730 and 737–738, respectively. In which problems do you think the expected monetary value (risk-neutral) criteria is inappropriate? Why?

SUMMARY

In this chapter, you learned how to develop payoff tables and decision trees, to use various criteria to choose between alternative courses of action, and to revise probabilities in light of sample information using Bayes' theorem. In the "Using Statistics" scenario, you learned how a manager of a mutual fund could use these tools to decide whether to purchase stock A or stock B. You found that stock B had a higher expected monetary value, a lower expected opportunity loss, but a lower return-to-risk ratio.

KEY FORMULAS

Expected Monetary Value

$$EMV(j) = \sum_{i=1}^{N} x_{ij}P_i \quad \textbf{(17.1)}$$

Expected Opportunity Loss

$$EOL(j) = \sum_{i=1}^{N} L_{ij}P_i \quad \textbf{(17.2)}$$

Expected Value of Perfect Information

$EVPI$ = expected profit under certainty – expected monetary value of the best alternative **(17.3)**

Return-to-Risk Ratio

$$RTRR(j) = \frac{EMV(j)}{\sigma_j} \quad \textbf{(17.4)}$$

KEY TERMS

CHAPTER REVIEW PROBLEMS

Checking Your Understanding

17.25 What is the difference between an event and an alternative course of action?

17.26 What are the advantages and disadvantages of a payoff table as compared to a decision tree?

17.27 How are opportunity losses computed from payoffs?

17.28 Why can't an opportunity loss be negative?

17.29 How does expected monetary value (EMV) differ from expected opportunity loss (EOL)?

17.30 What is the meaning of the expected value of perfect information ($EVPI$)?

17.31 How does the expected value of perfect information differ from the expected profit under certainty?

17.32 What are the advantages and disadvantages of using expected monetary value (EMV) as compared to the return-to-risk ratio ($RTRR$)?

17.33 How is Bayes' theorem used to revise probabilities in light of the sample information?

17.34 What is the difference between a risk averter and a risk seeker?

17.35 Why should you use utilities instead of payoffs in certain circumstances?

Applying the Concepts

17.36 A supermarket chain purchases large quantities of white bread for sale during a week. The stores purchase the bread for \$0.75 per loaf and sell it for \$1.10 per loaf. Any loaves not sold by the end of the week can be sold to a local thrift shop for \$0.40. Based on past demand, the probability of various levels of demand is as follows:

Demand (Loaves)	Probability
6,000	0.10
8,000	0.50
10,000	0.30
12,000	0.10

a. Construct the payoff table indicating the events and alternative courses of action.
b. Construct the decision tree.
c. Compute the expected monetary value (*EMV*) for purchasing 6,000, 8,000, 10,000, and 12,000 loaves.
d. Compute the expected opportunity loss (*EOL*) for purchasing 6,000, 8,000, 10,000, and 12,000 loaves.
e. Explain the meaning of the expected value of perfect information (*EVPI*) in this problem.
f. Based on the results of (c) or (d), how many loaves would you purchase? Why?
g. Compute the coefficient of variation for each purchase level.
h. Compute the return-to-risk ratio (*RTRR*) for each purchase level.
i. Based on (g) and (h), what action would you choose? Why?
j. Compare the results of (f) and (i) and explain any differences.
k. Suppose that new information changes the probabilities associated with the demand level as follows:

Demand (Loaves)	Probability
6,000	0.30
8,000	0.40
10,000	0.20
12,000	0.10

Repeat (c) through (j) of this problem with these new probabilities. Compare the results with these new probabilities to those of (c) through (j).

17.37 The owner of a company that supplies home heating oil would like to determine whether to offer a solar heating installation service to its customers. The owner of the company has determined that a startup cost of \$150,000 would be necessary, but a profit of \$2,000 can be made on each solar heating system installed. The owner estimates the probability of various demand levels as follows:

Number of Units Installed	Probability
50	0.40
100	0.30
200	0.30

a. Construct the payoff table indicating the events and alternative courses of action.
b. Construct the decision tree.
c. Construct the opportunity loss table.
d. Compute the expected monetary value (*EMV*) for offering this solar heating system installation service.
e. Compute the expected opportunity loss (*EOL*) for offering this solar heating system installation service.
f. Explain the meaning of the expected value of perfect information (*EVPI*) in this problem.
g. Compute the return-to-risk ratio (*RTRR*) for offering this solar heating system installation service.
h. Based on the results of (d) or (e) and (g), should the company offer this solar heating system installation service? Why?
i. How would your answers to (a) through (h) be affected if the startup cost were \$200,000?

17.38 The manufacturer of a nationally distributed brand of potato chips wants to determine the feasibility of changing the product package from a cellophane bag to an unbreakable container. The product manager believes that there are three possible national market responses to a change in product package: weak, moderate, and strong. The projected payoffs, in millions of dollars, in increased or decreased profit compared to the current package are as follows:

	Strategy	
Event	Use New Package	Keep Old Package
Weak national response	–\$4	0
Moderate national response	1	0
Strong national response	5	0

Based on past experience, the product manager assigns the following probabilities to the different levels of national response:

$$P(\text{Weak national response}) = 0.30$$

$$P(\text{Moderate national response}) = 0.60$$

$$P(\text{Strong national response}) = 0.10$$

a. Construct the decision tree.
b. Construct the opportunity loss table.
c. Compute the expected monetary value (*EMV*) for offering this new product package.
d. Compute the expected opportunity loss (*EOL*) for offering this new product package.

e. Explain the meaning of the expected value of perfect information (*EVPI*) in this problem.
f. Compute the return-to-risk ratio (*RTRR*) for offering this new product package.
g. Based on the results of (c) or (d) and (f), should the company offer this new product package? Why?
h. What are your answers to parts (c) through (g) if the probabilities were 0.6, 0.3, and 0.1, respectively?
i. What are your answers to parts (c) through (g) if the probabilities were 0.1, 0.3, and 0.6, respectively.

Before making a final decision, the product manager would like to test market the new package in a selected city by substituting the new package for the old package. A determination is then made about whether sales have increased, decreased, or stayed the same. In previous test marketing of other products, when there was a subsequent weak national response, sales in the test city decreased 60% of the time, stayed the same 30% of the time, and increased 10% of the time. Where there was a moderate national response, sales in the test city decreased 20% of the time, stayed the same 40% of the time, and increased 40% of the time. When there was a strong national response, sales in the test city decreased 5% of the time, stayed the same 35% of the time, and increased 60% of the time.

j. If sales in the test city stayed the same, revise the original probabilities in light of this new information.
k. Use the revised probabilities in (j) to repeat (c) through (g).
l. If sales in the test city decreased, revise the original probabilities in light of this new information.
m. Use the revised probabilities in (l) to repeat (c) through (g).

17.39 An entrepreneur wants to determine whether it would be profitable to establish a gardening service in a local suburb. The entrepreneur believes that there are four possible levels of demand for this gardening service:

1. Very low demand—1% of the households would use the service.
2. Low demand—5% of the households would use the service.
3. Moderate demand—10% of the households would use the service.
4. High demand—25% of the households would use the service.

Based on past experiences in other suburbs, the entrepreneur assigns the following probabilities to the various demand levels:

$$P(\text{very low demand}) = 0.20$$

$$P(\text{low demand}) = 0.50$$

$$P(\text{moderate demand}) = 0.20$$

$$P(\text{high demand}) = 0.10$$

The entrepreneur has calculated the following profits or losses of this garden service for each demand level (over a period of 1 year):

	Action	
Demand	**Provide Garden Service ($)**	**Do Not Provide Garden Service**
Very low	–$50,000	0
Low	$60,000	0
Moderate	$130,000	0
High	$300,000	0

a. Construct a decision tree.
b. Construct an opportunity loss table.
c. Compute the expected monetary value (*EMV*) for offering this garden service.
d. Compute the expected opportunity loss (*EOL*) for offering this garden service.
e. Explain the meaning of the expected value of perfect information (*EVPI*) in this problem.
f. Compute the return-to-risk ratio (*RTRR*) for offering this garden service.
g. Based on the results of (c) or (d) and (f), should the entrepreneur offer this garden service? Why?

Before making a final decision, the entrepreneur conducts a survey to determine demand for the gardening service. A random sample of 20 households is selected and 3 indicate that they would use this gardening service.

h. Revise the prior probabilities in light of this sample information. (*Hint:* Use the binomial distribution to determine the probability of the outcome that occurred given a particular level of demand.)
i. Use the revised probabilities in (h) to repeat (c) through (g).

17.40 A manufacturer of a brand of inexpensive felt tip pens maintains a production process that produces 10,000 pens per day. In order to maintain the highest quality of this product, the manufacturer guarantees free replacement of any defective pen sold. Each defective pen produced costs 20 cents for the manufacturer to replace. Based on past experience, four rates of producing defective pens are possible:

Very low—1% of the pens manufactured will be defective.
Low—5% of the pens manufactured will be defective.
Moderate—10% of the pens manufactured will be defective.
High—20% of the pens manufactured will be defective.

The manufacturer can reduce the rate of defective pens produced by having a mechanic fix the machines at the end of each day. This mechanic can reduce the rate to 1%, but his services will cost $80.

A payoff table based on the daily production of 10,000 pens, indicating the replacement costs for each of the two alternatives (calling in the mechanic and not calling in the mechanic), is as follows:

	Action	
Defective Rate	**Do Not Call Mechanic**	**Call Mechanic**
Very low (1%)	$20	$100
Low (5%)	$100	$100
Moderate (10%)	$200	$100
High (20%)	$400	$100

Based on past experience, each defective rate is assumed to be equally likely to occur.

a. Construct a decision tree.
b. Construct an opportunity loss table.
c. Compute the expected monetary value (*EMV*) for calling and for not calling the mechanic.
d. Compute the expected opportunity loss (*EOL*) for calling and for not calling the mechanic.
e. Explain the meaning of the expected value of perfect information (*EVPI*) in this problem.
f. Compute the return-to-risk ratio (*RTRR*) for not calling the mechanic.
g. Based on the results of (c) or (d) and (f), should the company call the mechanic? Why?
h. At the end of a day's production, a sample of 15 pens is selected, and 2 are defective. Revise the prior probabilities in light of this sample information. (*Hint:* Use the binomial distribution to determine the probability of the outcome that occurred given a particular defective rate.)
i. Use the revised probabilities in (h) to repeat (c) through (g).

WEB CASE

Apply your knowledge of decision-making techniques in this Web Case that extends the StockTout Web case from earlier chapters.

StraightArrow Banking & Investments is a Tri-Cities competitor of StockTout that is currently advertising its StraightArrow StraightDeal fund.

Visit the StraightArrow Web site **www.prenhall.com/Springville/SA_Home.htm** and examine the claims and supporting data for the fund. Compare those claims and data to the supporting data found on the StockTout Happy Bull and Worried Bear Funds Web page **www.prenhall.com/Springville/ST_BullsandBears.htm** first discussed in the Web case for Chapter 5. Then answer the following:

1. Is the StraightArrow StraightDeal fund a better investment than either of the StockTout funds? Support your answer by performing and summarizing an appropriate analysis.
2. Before making a decision about which fund makes a better investment, you decide that you need a reliable forecast for the direction of the economy in the next year. After further investigation, you find that the consensus of leading economists is that the economy will be expanding in the next year. You also find out that in the past, when there has been a recession, the consensus of leading economists predicted an expanding economy 10% of the time. When there was a stable economy, they predicted an expanding economy 50% of the time, and when there was an expanding economy, they predicted an expanding economy 75% of the time. When there was a rapidly expanding economy, they predicted an expanding economy 90% of the time.

   Does this information change your answer to question 1? Why or why not?

REFERENCES

1. Bernstein, P. L. *Against the Gods: The Remarkable Story of Risk* (New York: John Wiley, 1996).
2. Render, B., R. M. Stair, and M. Hanna. *Quantitative Analysis for Management*, 8th ed. (Upper Saddle River, NJ: Prentice Hall, 2003).
3. Tversky, A., and D. Kahneman, "Rationale Choice and the Framing of Decisions," *Journal of Business*, 59 (1986) 251–278.

Appendix 17 Using Software for Decision Making

A17.1 MICROSOFT EXCEL

For Opportunity Loss

See section G.39 (**Opportunity Loss**) if you want PHStat2 to produce an opportunity loss analysis worksheet. (There are no Microsoft Excel commands that directly perform opportunity loss analysis.)

For Expected Monetary Value

See section G.40 (**Expected Monetary Value**) if you want PHStat2 to produce an expected monetary value analysis worksheet. (There are no Microsoft Excel commands that directly perform expected monetary value analysis.)

CHAPTER 18

Statistical Applications in Quality and Productivity Management

USING STATISTICS: Service Quality at the Beachcomber Hotel

LEARNING OBJECTIVES

In this chapter, you learn:

- The basic themes of quality management and Deming's 14 points
- The basic aspects of Six Sigma management
- How to construct various control charts
- Which control chart to use for a particular type of data
- How to measure the capability of a process

USING STATISTICS

Service Quality at the Beachcomber Hotel

In the Chapter 12 "Using Statistics" scenario, you were the manager of T. C. Resort Properties. For this scenario, consider that you manage only the Beachcomber Hotel. As the hotel manager, you would want to continually improve the quality of service that your guests receive so that overall guest satisfaction increases. To help you achieve this improvement, T. C. Resort Properties has provided its managers with training in Six Sigma management. In order to meet the business objective of increasing the return rate of guests at your hotel, you have decided to focus on the critical first impressions of the service that your hotel provides. Is the assigned hotel room ready when a guest checks in? Are all expected amenities such as extra towels and a complimentary guest basket in the room when the guest first walks in? Are the video-entertainment center and high-speed Internet access working properly? And, do guests receive their luggage in a reasonable amount of time?

To study these guest satisfaction issues, you have embarked on an improvement project that measures two critical-to-quality (CTQ) measurements, the readiness of the room and the time it takes to deliver luggage. You would like to learn the following:

- Are the proportion of rooms ready and the time required to deliver luggage to the rooms acceptable?
- Are the proportion of rooms ready and the luggage delivery time consistent from day to day, or are they increasing or decreasing?
- On the days when the proportion of rooms that are not ready or the time to deliver luggage is greater than normal, is this due to a chance occurrence or is there a fundamental flaw in the process used to make rooms ready and deliver luggage?

In this chapter the focus is on quality and productivity management. Companies manufacturing products, as well as those providing services, such as the Beachcomber Hotel in the "Using Statistics" scenario, realize that quality and productivity are essential for survival in the global economy. Among the areas in which quality has an impact on our everyday work and personal lives are:

- The design, production, and subsequent reliability of our automobiles
- The services provided by hotels, banks, schools, retailing operations, and mail-order companies
- The continuous improvement in computer chips that makes for faster and more powerful computers
- The ever-expanding capability of communication devices such as data transmission lines, paging devices, facsimile machines, and cellular telephones
- The availability of new technology and equipment that has led to improved diagnosis of illnesses and the improved delivery of health care services

18.1 TOTAL QUALITY MANAGEMENT

During the past twenty-five years, the renewed interest in quality and productivity in the United States followed as a reaction to improvements of Japanese industry that had began as early as 1950. Individuals such as W. Edwards Deming, Joseph Juran, and Kaoru Ishikawa developed an approach that focuses on continuous improvement of products and services through an increased emphasis on statistics, process improvement, and optimization of the total system. This approach, widely known as **total quality management (TQM)**, is characterized by these themes:

- The primary focus is on process improvement.
- Most of the variation in a process is due to the system and not the individual.
- Teamwork is an integral part of a quality management organization.
- Customer satisfaction is a primary organizational goal.
- Organizational transformation must occur in order to implement quality management.
- Fear must be removed from organizations.
- Higher quality costs less not more, but requires an investment in training.

In the 1980s the federal government of the United States increased its efforts to improve quality in American business. Congress passed the Malcolm Baldrige National Improvement Act of 1987 and began awarding the Malcolm Baldrige Award to companies making the greatest strides in improving quality and customer satisfaction. W. Edwards Deming became a prominent consultant to many Fortune 500 companies including Ford, General Motors, and Proctor and Gamble. Through four-day seminars, Deming widely promoted his **"14 points for management"** listed below, and many companies adopted some or all of them.

1. Create constancy of purpose for improvement of product and service.
2. Adopt the new philosophy.
3. Cease dependence on inspection to achieve quality.
4. End the practice of awarding business on the basis of price tag alone. Instead, minimize total cost by working with a single supplier.
5. Improve constantly and forever every process for planning, production, and service.
6. Institute training on the job.
7. Adopt and institute leadership.
8. Drive out fear.
9. Break down barriers between staff areas.
10. Eliminate slogans, exhortations, and targets for the workforce.
11. Eliminate numerical quotas for the workforce and numerical goals for management.
12. Remove barriers that rob people of pride of workmanship. Eliminate the annual rating or merit system.
13. Institute a vigorous program of education and self-improvement for everyone.
14. Put everyone in the company to work to accomplish the transformation.

Point 1, create constancy of purpose, refers to how an organization deals with problems that arise both at present and in the future. The focus is on the constant improvement of a product or service. This improvement process is illustrated by the **Shewhart-Deming cycle** shown in Figure 18.1. The Shewhart-Deming cycle represents a continuous cycle of "plan, do, study, and act." The first step, planning, represents the initial design phase for planning a change in a manufacturing or service process. This step involves teamwork among individuals from different areas within an organization. The second step, doing, involves implementing the change, preferably on a small scale. The third step, studying, involves an analysis of the results using statistical tools to determine what was learned. The fourth step, acting, involves the acceptance of the change, its abandonment, or further study of the change under different conditions.

FIGURE 18.1
Shewhart-Deming Cycle

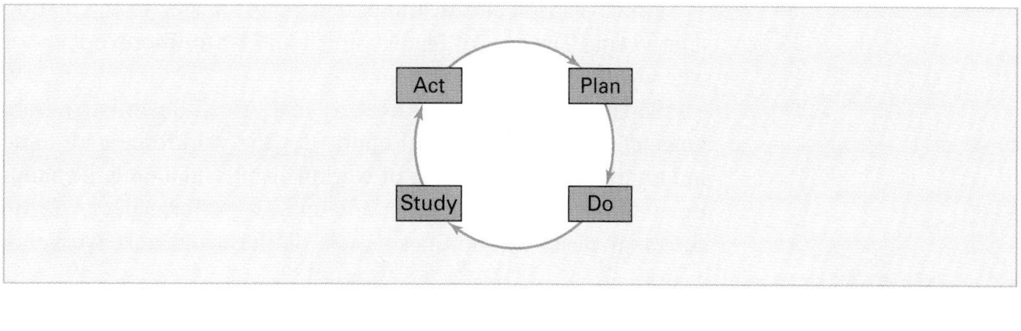

Point 2, adopt the new philosophy, refers to the urgency with which companies need to realize that there is a new economic age of global competition. It is better to be proactive and change before a crisis occurs than to react to some negative experiences that may have occurred. Rather than take the approach "if it's not broke, don't fix it," it is better to continually work on improvement and avoid expensive fixes.

Point 3, cease dependence on inspection to achieve quality, implies that any inspection whose purpose is to improve quality is too late because the quality is already built into the product. It is better to focus on making it right the first time. Among the difficulties involved in inspection (besides high costs) are the failure of inspectors to agree on the operational definitions for nonconforming items, and the problem of separating good and bad items. The following example illustrates the difficulties inspectors face.

Suppose your job involves proofreading the sentence in Figure 18.2 with the objective of counting the number of occurrences of the letter "F." Perform this task and record the number of occurrences of the letter F that you discover.

FIGURE 18.2

An Example of a Proofreading Process

Source: W. W. Scherkenbach, The Deming Route to Quality and Productivity: Road Maps and Roadblocks *(Washington, DC: CEEP Press, 1986).*

FINISHED FILES ARE THE RESULT
OF YEARS OF SCIENTIFIC
STUDY COMBINED WITH THE
EXPERIENCE OF MANY YEARS

People usually see either three Fs or six Fs. The correct number is six Fs. The number you see depends on the method you use to examine the sentence. You are likely to find three Fs if you read the sentence phonetically and six Fs if you count the number of Fs carefully. The point of the exercise is to show that if such a simple process as counting Fs leads to inconsistency of inspectors' results, what will happen when a much more complicated process fails to contain a clear operational definition of nonconforming? Certainly, in such situations, a large amount of variability occurs from inspector to inspector.

Point 4, ending the practice of awarding business on the basis of price tag alone, represents the antithesis of lowest-bidder awards. There is no real long-term meaning to price without knowledge of the quality of the product.

Point 5—improve constantly and forever every process for planning, production and service—reinforces the importance of the continuous focus of the Shewhart-Deming cycle and the belief that quality needs to be built in at the design stage. Attaining quality is a never-ending process in which reduction in variation translates into a reduction in the financial losses resulting from products and services experiencing large fluctuations in quality.

Point 6, institute training, reflects the needs of all employees, including production workers, engineers, and managers. It is critically important for management to understand the differences between special causes and common causes of variation (see section 18.3) so that proper action is taken in each circumstance.

Point 7, adopt and institute leadership, relates to the distinction between leadership and supervision. The aim of leadership should be to improve the system and achieve greater consistency of performance.

Points 8 through 12—drive out fear, break down barriers between staff areas, eliminate slogans, eliminate numerical quotas for the workforce and numerical goals for management, and remove barriers to pride of workmanship (including the annual rating and merit system)—are all related to the evaluation of employee performance. An emphasis on targets and exhortations may place an improper burden on the workforce. Workers cannot produce beyond what the system will allow (this is clearly illustrated in section 18.5). It is management's job to *improve* the system, not to raise the expectations on workers beyond the system's capability.

Point 13, encourage education and self-improvement for everyone, reflects the notion that the most important resource of any organization is its people. Efforts that improve the knowledge of people in the organization also serve to increase the assets of the organization. Education and self-improvement can lead to reduced turnover within an organization.

Point 14, take action to accomplish the transformation, again reflects the approach of management as a process in which one continuously strives toward improvement in a never-ending cycle.

Although Deming's points were thought-provoking, some criticized his approach for lacking a formal, objective accountability. Many managers of large organizations, used to seeing financial analyses of policy changes, needed a more prescriptive approach.

18.2 SIX SIGMA MANAGEMENT

Six Sigma management is a quality improvement system originally developed by Motorola in the mid-1980s. Six Sigma offers a more prescriptive and systematic approach to process improvement than TQM, and places a higher emphasis on accountability and bottom line results. Many companies all over the world are using Six Sigma management to improve efficiency, cut costs, eliminate defects and reduce product variation.

The name Six Sigma comes from the fact that it is a managerial approach designed to create processes that result in no more than 3.4 defects per million.[1] One of the aspects that distinguishes Six Sigma from other approaches is a clear focus on achieving bottom-line results in a relatively short three- to six-month period of time. After seeing the huge financial successes at Motorola, GE, and other early adopters of Six Sigma, many companies worldwide have now instituted Six Sigma programs (see references 1, 8, 9, and 15).

[1]The Six Sigma approach assumes that the process may shift as much as 1.5 standard deviations over the long term. Six standard deviations minus a 1.5 standard deviation shift produces a 4.5 standard deviation goal. The area under the normal curve outside 4.5 standard deviations is approximately 3.4 out of a million (0.0000034).

To guide managers in their task of improving short- and long-term results, Six Sigma uses a five-step process known as the **DMAIC model**—named for the five steps in the process: **D**efine, **M**easure, **A**nalyze, **I**mprove, and **C**ontrol.

- ***Define*** The problem is defined along with the costs, benefits, and the impact on the customer.
- ***Measure*** Operational definitions for each **critical-to-quality (CTQ)** characteristic are developed. In addition, the measurement procedure is verified so that it is consistent over repeated measurements.
- ***Analyze*** The root causes of *why* defects occur are determined, and variables in the process causing the defects are identified. Data are collected to determine benchmark values for each process variable. This analysis often uses control charts (to be discussed in sections 18.3 through 18.7).
- ***Improve*** The importance of each process variable on the CTQ characteristic is studied using designed experiments (see Chapter 11). The objective is to determine the best level for each variable.
- ***Control*** The objective is to maintain the benefits for the long term by avoiding potential problems that can occur when a process is changed.

Implementation of Six Sigma management requires a data-oriented approach that is heavily based on using statistical tools such as control charts and designed experiments. It also involves training everyone in the organization in the DMAIC model.

18.3 THE THEORY OF CONTROL CHARTS

Both total quality management and Six Sigma management make use of a wide array of statistical tools. One tool widely used in each approach to analyze process data collected sequentially over time is the control chart.

The **control chart** monitors variation in a characteristic of a product or service over time. You can use a control chart to study past performance, to evaluate present conditions, or to predict future outcomes (see reference 7). Information gained from analyzing a control chart forms the basis for process improvement. Different types of control charts allow you to analyze different types of critical-to-quality (CTQ) variables—for categorical variables such as the proportion of hotel rooms that are nonconforming in terms of the availability of amenities and the working order of all appliances in the room, discrete variables such as the number of hotel guests registering a complaint during a week, and continuous variables such as the length of time required for delivering luggage to the room. In addition to providing a visual display of data representing a process, a principal focus of the control chart is the attempt to separate special causes of variation from common causes of variation.

Special causes of variation represent large fluctuations or patterns in the data that are not inherent to a process. These fluctuations are often caused by changes in the process that represent either problems to correct or opportunities to exploit. Some organizations refer to special causes of variation as **assignable causes of variation.**
Common causes of variation represent the inherent variability that exists in a process. These fluctuations consist of the numerous small causes of variability that operate randomly or by chance. Some organizations refer to common causes of variation as **chance causes of variation.**

The distinction between the two causes of variation is crucial because special causes of variation are not part of a process and are correctable or exploitable without changing the system. Common causes of variation, however, can be reduced only by changing the system. Such systemic changes are the responsibility of management.

Control charts allow you to monitor a process and identify the presence or absence of special causes. By doing so, control charts help prevent two types of errors. The first type of error involves the belief that an observed value represents special-cause variation when it is due to the common cause variation of the system. Treating common-cause variation as special-cause variation often results in overadjusting a process. This overadjustment, known as **tampering**, increases the variation in the process. The second type of error involves treating special-cause variation as common-cause variation. This error results in not taking immediate corrective action when necessary. Although both of these types of errors can occur even when using a control chart, they are far less likely.

To construct a control chart, you collect samples from the output of a process over time. The samples used for constructing control charts are known as **subgroups**. For each subgroup (i.e., sample), you calculate the value of a statistic associated with a CTQ variable. Commonly used statistics include the fraction nonconforming (see section 18.4), the number of nonconformities (see section 18.6), and the mean and range of a numerical variable (see section 18.7). You then plot the values versus time and add control limits to the chart. The most typical form of a control chart sets control limits that are within ±3 standard deviations[2] of the statistical measure of interest. Equation (18.1) defines, in general, the upper and lower control limits for control charts.

[2] *Recall from section 6.2 that in the normal distribution, $\mu \pm 3\sigma$ includes almost all (99.73%) of the observations in the population.*

CONSTRUCTING CONTROL LIMITS

$$\text{Process mean} \pm 3 \text{ standard deviations} \qquad \textbf{(18.1)}$$

so that

Upper control limit (UCL) = process mean +3 standard deviations
Lower control limit (LCL) = process mean −3 standard deviations

When these control limits are set, you evaluate the control chart by trying to find any pattern that might exist in the values over time and by determining whether any points fall outside the control limits. Figure 18.3 illustrates three different situations.

FIGURE 18.3
Three Control Chart Patterns

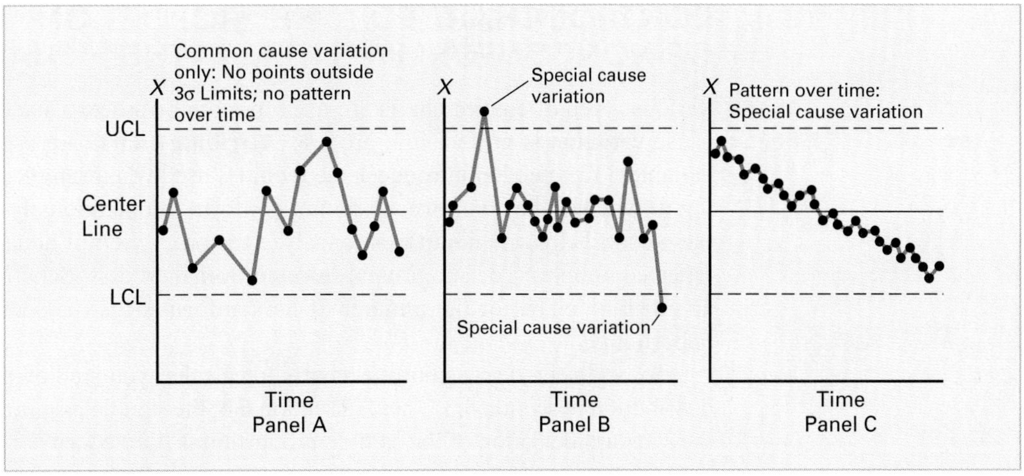

In panel A of Figure 18.3, there is no apparent pattern in the values over time and there are no points that fall outside the 3 standard deviation control limits. The process appears stable and contains only common-cause variation. Panel B, on the contrary, contains two points that fall outside the 3 standard deviation control limits. You should investigate these points to try to determine the special causes that led to their occurrence. Although panel C does not have any points outside the control limits, it has a series of consecutive points above the mean value (the centerline) as well as a series of consecutive points below the mean value. In addition, a long-term overall downward trend is clearly visible. You should investigate the situation to try to determine what may have caused this pattern.

The detection of a trend is not always so obvious. Two other simple rules[3] (see references 7 and 8) that allow you to detect a shift in the mean level of a process are:

[3]*Minitab uses different rules—see reference 13.*

- Eight or more *consecutive points* that lie above the center line or eight or more *consecutive points* that lie below the center line.
- Eight or more *consecutive points* move upward in value or eight or more *consecutive points* move downward in value.

A process whose control chart indicates an out-of-control condition (a point outside the control limits or exhibiting a trend) is said to be out of control. An **out-of-control process** contains both common causes of variation and special causes of variation. Because special causes of variation are not part of the process design, an out-of-control process is unpredictable. Once you determine a process is out of control, you must identify the special causes of variation that are producing the out-of-control conditions. If the special causes are detrimental to the quality of the product or service, you need to implement plans to eliminate this source of variation. When a special cause increases quality, you should change the process so that the special cause is incorporated into the process design. Thus, this beneficial special cause now becomes a common-cause source of variation and the process is improved.

A process whose control chart is not indicating any out-of-control conditions is said to be in control. An **in-control process** contains only common causes of variation. Because these sources of variation are inherent to the process itself, an in-control process is predictable. In-control processes are sometimes said to be in a **state of statistical control**. When a process is in control, you must determine whether the amount of common-cause variation in the process is small enough to satisfy the customers of the products or services. (In section 18.8, you will learn statistical methods that allow you to compare common-cause variation to customer expectations.) If the common-cause variation is small enough to consistently satisfy the customer, you then use control charts to monitor the process on a continuing basis to make sure that the process remains in control. If the common-cause variation is too large, you need to alter the process itself.

18.4 CONTROL CHART FOR THE PROPORTION OF NONCONFORMING ITEMS—THE *p* CHART

Various types of control charts are used to monitor processes and determine whether special-cause variation is present in a process. **Attribute charts** are used for categorical or discrete variables. This section introduces the ***p* chart**, used when sampled items are classified according to whether they conform or do not conform to operationally defined requirements. Thus, the p chart helps you monitor and analyze the proportion of nonconforming items there are in repeated samples (i.e., subgroups) selected from a process. Section 18.6 introduces the c chart, an attribute chart for the number of nonconformances (or occurrences) in a given area of opportunity.

To begin the discussion of p charts, recall that you studied proportions and the binomial distribution in section 5.3. Then, in section 6.4, the sample proportion is defined as $p = X/n$, and the standard deviation of the sample proportion is defined as

$$\sigma_p = \sqrt{\frac{\pi(1-\pi)}{n}}$$

Using Equation (18.1) on page 756, control limits for the proportion of nonconforming[4] items from the sample data are established in Equation (18.2).

[4] In this chapter the term nonconforming items is used, while in earlier chapters the term success was used.

CONTROL LIMITS FOR THE *p* CHART

$$\bar{p} \pm 3\sqrt{\frac{\bar{p}(1-\bar{p})}{\bar{n}}} \qquad \textbf{(18.2)}$$

$$\text{UCL} = \bar{p} + 3\sqrt{\frac{\bar{p}(1-\bar{p})}{\bar{n}}}$$

$$\text{LCL} = \bar{p} - 3\sqrt{\frac{\bar{p}(1-\bar{p})}{\bar{n}}}$$

For equal n_i,

$$\bar{n} = n_i \text{ and } \bar{p} = \frac{\sum_{i=1}^{k} p_i}{k}$$

or in general,

$$\bar{n} = \frac{\sum_{i=1}^{k} n_i}{k} \text{ and } \bar{p} = \frac{\sum_{i=1}^{k} X_i}{\sum_{i=1}^{k} n_i}$$

where X_i = number of nonconforming items in subgroup i

n_i = sample (or subgroup) size for subgroup i

$p_i = X_i/n_i$ = proportion of nonconforming items in subgroup i

k = number of subgroups selected

$\bar{n}$ = mean subgroup size

$\bar{p}$ = estimated proportion of nonconforming items

Any negative value for the lower control limit means that the lower control limit does not exist. To show the application of the *p* chart, return to the "Using Statistics" scenario concerning the Beachcomber Hotel.

During the Measure phase of the Six Sigma DMAIC model, a nonconformance was operationally defined as the absence of an amenity in the room or a room appliance not in working order upon check-in. During the Analyze phase of the Six Sigma DMAIC model, data on the nonconformances were collected daily from a sample of 200 rooms. Table 18.1 lists the number and proportion of nonconforming rooms for each day in the 4-week period. **HOTEL1**.

TABLE 18.1

Nonconforming Hotel Rooms at Check-In over a 28 Day Period

Day	Rooms Studied	Rooms Not Ready	Proportion	Day	Rooms Studied	Rooms Not Ready	Proportion
1	200	16	0.080	15	200	18	0.090
2	200	7	0.035	16	200	13	0.065
3	200	21	0.105	17	200	15	0.075
4	200	17	0.085	18	200	10	0.050
5	200	25	0.125	19	200	14	0.070
6	200	19	0.095	20	200	25	0.125
7	200	16	0.080	21	200	19	0.095
8	200	15	0.075	22	200	12	0.060
9	200	11	0.055	23	200	6	0.030
10	200	12	0.060	24	200	12	0.060
11	200	22	0.110	25	200	18	0.090
12	200	20	0.100	26	200	15	0.075
13	200	17	0.085	27	200	20	0.100
14	200	26	0.130	28	200	22	0.110

For these data, $k = 28$, $\sum_{i=1}^{k} p_i = 2.315$ and because the n_i are equal $n_i = \bar{n} = 200$. Thus,

$$\bar{p} = \frac{\sum_{i=1}^{k} p_i}{k} = \frac{2.315}{28} = 0.0827$$

Using Equation (18.2),

$$0.0827 \pm 3\sqrt{\frac{(0.0827)(0.9173)}{200}}$$

so that

$$\text{UCL} = 0.0827 + 0.0584 = 0.1411$$

and

$$\text{LCL} = 0.0827 - 0.0584 = 0.0243$$

Figure 18.4 displays the Microsoft Excel control chart for the data of Table 18.1. Figure 18.5 provides Minitab output. Figure 18.4 and 18.5 indicate a process in a state of statistical control, with the individual points distributed around $\bar{p}$ without any pattern and all the points within the control limits. Thus, any improvement in the process of making rooms ready for guests in the Improve phase of the DMAIC model must come from the reduction of common-cause variation. Such reductions require a change in the process. These changes are the responsibility of management. Remember that improvements cannot occur until improvements to the process itself are successfully implemented.

FIGURE 18.4
Microsoft Excel *p* Chart for the Nonconforming Hotel Room Data

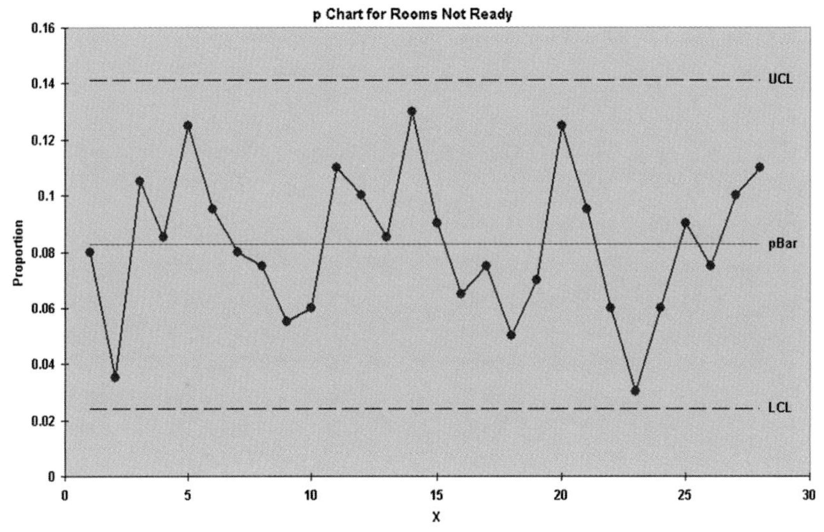

FIGURE 18.5
Minitab *p* Chart for the Nonconforming Hotel Room Data

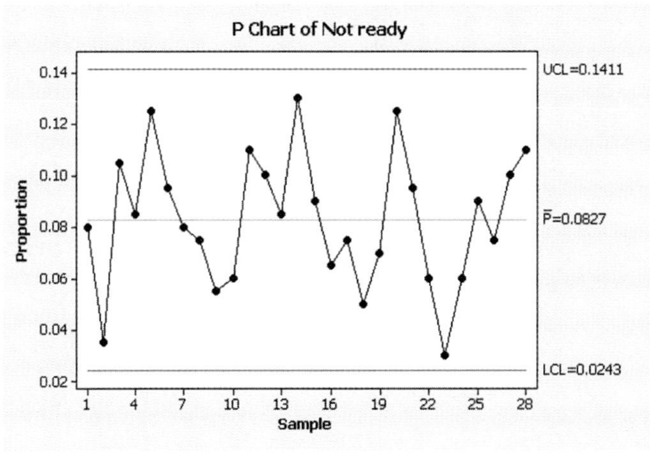

The first example illustrated a situation in which the subgroup size did not vary. As a general rule, as long as none of the subgroup sizes n_i differ from the mean subgroup size $\bar{n}$ by more than ±25% of $\bar{n}$ (see reference 7), you can use Equation (18.2) on page 758 to compute the control limits for the *p* chart. If any subgroup size differs by more than ±25% of $\bar{n}$, you use alternative formulas for calculating the control limits (see reference 7). To illustrate the use of the *p* chart when the subgroup sizes are unequal, Example 18.1 studies the production of gauze sponges.

EXAMPLE 18.1 USING THE *p* CHART FOR UNEQUAL SUBGROUP SIZES

Table 18.2 indicates the number of sponges produced daily and the number of nonconforming sponges for a period of 32 days. **SPONGE** Construct a control chart for these data.

TABLE 18.2
Nonconforming Sponges over a 32-Day Period

Day	Sponges Produced	Non-Conforming Sponges	Proportion	Day	Sponges Produced	Non-Conforming Sponges	Proportion
1	690	21	0.030	17	575	20	0.035
2	580	22	0.038	18	610	16	0.026
3	685	20	0.029	19	596	15	0.025
4	595	21	0.035	20	630	24	0.038
5	665	23	0.035	21	625	25	0.040
6	596	19	0.032	22	615	21	0.034
7	600	18	0.030	23	575	23	0.040
8	620	24	0.039	24	572	20	0.035
9	610	20	0.033	25	645	24	0.037
10	595	22	0.037	26	651	39	0.060
11	645	19	0.029	27	660	21	0.032
12	675	23	0.034	28	685	19	0.028
13	670	22	0.033	29	671	17	0.025
14	590	26	0.044	30	660	22	0.033
15	585	17	0.029	31	595	24	0.040
16	560	16	0.029	32	600	16	0.027

SOLUTION For these data,

$$k = 32, \sum_{i=1}^{k} n_i = 19,926, \text{ and } \sum_{i=1}^{k} X_i = 679$$

Thus, using Equation (18.2) on page 758,

$$\bar{n} = \frac{19,926}{32} = 622.69 \text{ and } \bar{p} = \frac{679}{19,926} = 0.034$$

so that

$$0.034 \pm 3\sqrt{\frac{(0.034)(1-0.034)}{622.69}}$$
$$= 0.034 \pm 0.022$$

Thus,

$$\text{UCL} = 0.034 + 0.022 = 0.056$$

and

$$\text{LCL} = 0.034 - 0.022 = 0.012$$

Figure 18.6 displays the Microsoft Excel control chart for the sponge data. Figure 18.7 shows the Minitab control chart. An examination of either of these figures indicates that day 26, in which there were 39 nonconforming sponges produced out of 651 sampled, is above the upper control limit. Management needs to determine the reason (i.e., root cause) for this special cause variation and take corrective action. Once actions are taken, you can remove the data from day 26 and then construct and analyze a new control chart.

Notice that the UCL and LCL on Figure 18.7 are represented by a jagged line. Minitab is calculating separate control limits for each day depending on the subgroup size for that day.

FIGURE 18.6
Microsoft Excel *p* Chart for the Proportion of Nonconforming Sponges

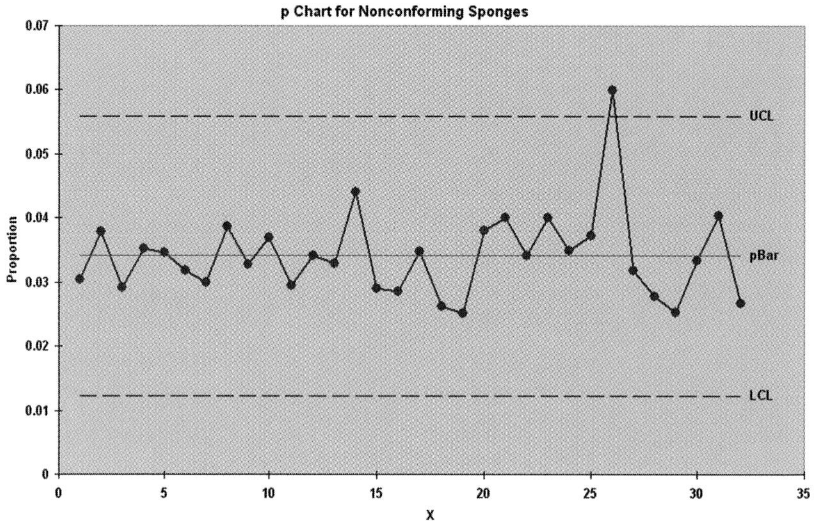

FIGURE 18.7
Minitab *p* Chart for the Proportion of Nonconforming Sponges

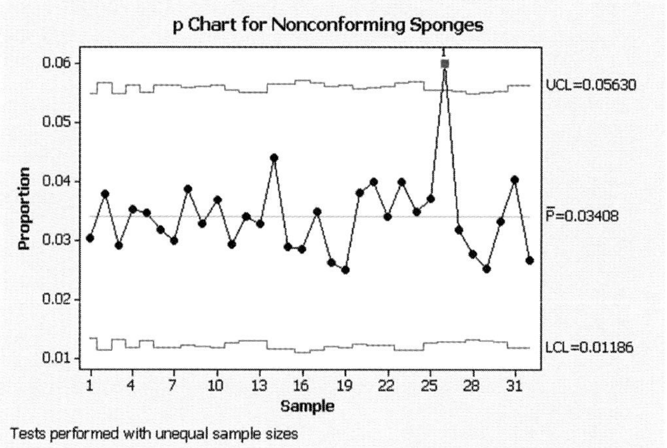

PROBLEMS FOR SECTION 18.4

Learning the Basics

PH Grade ASSIST **18.1** The following data were collected on nonconformances for a period of 10 days.

Day	Sample Size	Nonconformances
1	100	12
2	100	14
3	100	10
4	100	18
5	100	22
6	100	14
7	100	15
8	100	13
9	100	14
10	100	16

a. On what day is the proportion of nonconformances largest? smallest?

b. What are the LCL and UCL?

c. Are there any special causes of variation?

PH Grade ASSIST **18.2** The following data were collected on nonconformances for a period of 10 days.

Day	Sample Size	Nonconformances
1	111	12
2	93	14
3	105	10
4	92	18
5	117	22
6	88	14
7	117	15
8	87	13
9	119	14
10	107	16

a. On what day is the proportion of nonconformances largest? smallest?
b. What are the LCL and UCL?
c. Are there any special causes of variation?

Applying the Concepts

You can solve problems 18.3–18.8 manually or by using Microsoft Excel, Minitab, or SPSS.

18.3 A medical transcription service enters medical data on patient files for hospitals. The service studied ways to improve the turnaround time (defined as the time between receiving data and the time the client receives completed files). After studying the process, it was determined that turnaround time was increased by transmission errors. A transmission error was defined as data transmitted that did not go through as planned, and needed to be retransmitted. Each day a sample of 125 record transmissions were randomly selected and evaluated for errors. The table below presents the number and proportion of transmissions with errors. TRANSMIT

Day	Number of Errors	Proportion of Errors	Day	Number of Errors	Proportion of Errors
1	6	0.048	17	4	0.032
2	3	0.024	18	6	0.048
3	4	0.032	19	3	0.024
4	4	0.032	20	5	0.040
5	9	0.072	21	1	0.008
6	0	0.000	22	3	0.024
7	0	0.000	23	14	0.112
8	8	0.064	24	6	0.048
9	4	0.032	25	7	0.056
10	3	0.024	26	3	0.024
11	4	0.032	27	10	0.080
12	1	0.008	28	7	0.056
13	10	0.080	29	5	0.040
14	9	0.072	30	0	0.000
15	3	0.024	31	3	0.024
16	1	0.008			

a. Construct a p chart.
b. Is the process in a state of statistical control? Why?

18.4 The following data represent the findings from a study conducted at a factory that manufactures film canisters. For 32 days, 500 film canisters were sampled and inspected. The following table lists the number of defective film canisters (i.e., nonconforming items) for each day (i.e., subgroup). CANISTER

Day	Number Nonconforming	Day	Number Nonconforming
1	26	6	20
2	25	7	21
3	23	8	27
4	24	9	23
5	26	10	25
11	22	22	24
12	26	23	26
13	25	24	23
14	29	25	27
15	20	26	28
16	19	27	24
17	23	28	22
18	19	29	20
19	18	30	25
20	27	31	27
21	28	32	19

a. Construct a p chart.
b. Is the process in a state of statistical control? Why?

PH Grade ASSIST **18.5** A hospital administrator is concerned with the time to process patients' medical records after discharge. She determined that all records should be processed within 5 days of discharge. Thus, any record not processed within 5 days of a patient's discharge is nonconforming. The administrator recorded the number of patients discharged and the number of records not processed within the 5-day standard for a 30-day period in the file MEDREC.

a. Construct a p chart for these data.
b. Does the process give an out-of-control signal? Explain.
c. If the process is out of control, assume that special causes were subsequently identified and corrective action taken to keep them from happening again. Then eliminate the data causing the out-of-control signals, and recalculate the control limits.

SELF Test **18.6** The bottling division of Sweet Suzy's Sugarless Cola maintains daily records of the occurrences of unacceptable cans flowing from the filling and sealing machine. The following table lists the number of cans filled and the number of nonconforming cans for one month (based on a 5-day workweek). COLASPC

Day	Cans Filled	Unacceptable Cans	Day	Cans Filled	Unacceptable Cans
1	5,043	47	12	5,314	70
2	4,852	51	13	5,097	64
3	4,908	43	14	4,932	59
4	4,756	37	15	5,023	75
5	4,901	78	16	5,117	71
6	4,892	66	17	5,099	68
7	5,354	51	18	5,345	78
8	5,321	66	19	5,456	88
9	5,045	61	20	5,554	83
10	5,113	72	21	5,421	82
11	5,247	63	22	5,555	87

a. Construct a p chart for the proportion of unacceptable cans for the month. Does the process give an out-of-control signal?

b. If you want to develop a process for reducing the proportion of unacceptable cans, how should you proceed?

18.7 The manager of the accounting office of a large hospital is studying the problem of entering incorrect account numbers into the computer system. A subgroup of 200 account numbers is selected from each day's output, and each account number is inspected to determine whether it is a nonconforming item. The results for a period of 39 days are in the file ERRORSPC.

a. Construct a p chart for the proportion of nonconforming items. Does the process give an out-of-control signal?

b. Based on your answer to (a), if you were the manager of the accounting office, what would you do to improve the process of account number entry?

18.8 A regional manager of a telephone company is responsible for processing requests concerning additions, changes, or deletions of telephone service. She forms a service improvement team to look at the corrections in terms of central office equipment and facilities required to process the orders that are issued to service requests. Data collected over a period of 30 days are in the file TELESPC.

a. Construct a p chart for the proportion of corrections. Does the process give an out-of-control signal?

b. What should the regional manager do to improve the processing of requests for changes in telephone service?

18.5 THE RED BEAD EXPERIMENT: UNDERSTANDING PROCESS VARIABILITY

This chapter began with a discussion of total quality management, Deming's 14 points, Six Sigma management, and definitions of common-cause variation and special-cause variation. Now that you have studied the p chart, this section presents a famous parable, the **red bead experiment**, to enhance your understanding of common cause and special cause variation.

The red bead experiment involves the selection of beads from a box that contains 4,000 beads. Unknown to the participants in the experiment, 3,200 (80%) of the beads are white and 800 (20%) are red. You can use several different scenarios for conducting the experiment. The one used here begins with a facilitator (who will play the role of company supervisor) asking members of the audience to volunteer for the jobs of workers (at least four are needed), inspectors (two are needed), chief inspector (one is needed), and recorder (one is needed). A worker's job consists of using a paddle that has five rows of 10 bead-size holes to select 50 beads from the box of beads.

When the participants have been selected, the supervisor explains the jobs to them. The job of the workers is to produce white beads, because red beads are unacceptable to the customers. Strict procedures are to be followed. Work standards call for the daily production of exactly 50 beads by each worker (a strict quota system). Management has established a standard that no more than 2 red beads (4%) per worker are to be produced on any given day. Each worker dips the paddle into the box of beads so that when it is removed, each of the 50 holes contains a bead. The worker carries the paddle to the two inspectors, who independently record the count of red beads. The chief inspector compares their counts and announces the results to the audience. The recorder writes down the number and percentage of red beads next to the name of the worker.

When all the people know their jobs, "production" can begin. Suppose that on the first "day," the number of red beads "produced" by the four workers (call them Alyson, David, Peter, and Sharyn) was 9, 12, 13, and 7, respectively. How should management react to the day's production when the standard says that no more than 2 red beads per worker should be produced? Should all the workers be reprimanded, or should only David and Peter be given a stern warning that they will be fired if they don't improve?

Suppose that production continues for an additional two days. Table 18.3 summarizes the results for all three days.

TABLE 18.3

Red Bead Experiment Results for 4 Workers over 3 Days

	Day			
Name	**1**	**2**	**3**	**All Three Days**
Alyson	9 (18%)	11 (22%)	6 (12%)	26 (17.33%)
David	12 (24%)	12 (24%)	8 (16%)	32 (21.33%)
Peter	13 (26%)	6 (12%)	12 (24%)	31 (20.67%)
Sharyn	7 (14%)	9 (18%)	8 (16%)	24 (16.0%)
All 4 workers	41	38	34	113
Mean	10.25	9.5	8.5	9.42
Percentage	20.5%	19%	17%	18.83%

From Table 18.3, on each day, some of the workers were above the mean and some below the mean. On day 1, Sharyn did best, but on day 2, Peter (who had the worst record on day 1) was best, and on day 3, Alyson was best. How can you explain all this variation? Using Equation (18.2) on page 758 to develop a p chart for these data,

$$k = 4 \text{ workers} \times 3 \text{ days} = 12, \; n = 50, \text{ and } \sum_{i=1}^{k} X_i = 113$$

Thus,

$$\bar{p} = \frac{113}{(50)(12)} = 0.1883$$

so that

$$\bar{p} \pm 3\sqrt{\frac{\bar{p}(1-\bar{p})}{n}}$$

$$= 0.1883 \pm 3\sqrt{\frac{0.1883(1-0.1883)}{50}}$$

$$= 0.1883 \pm 0.1659$$

Thus,

$$\text{UCL} = 0.1883 + 0.1659 = 0.3542$$

and

$$\text{LCL} = 0.1883 - 0.1659 = 0.0224$$

Figure 18.8 represents the p chart for the data of Table 18.3. In Figure 18.8, all of the points are within the control limits and there are no patterns in the results. The differences between the workers merely represent common-cause variation inherent in a stable system.

Four morals to the parable of the red beads are:

- Variation is an inherent part of any process.
- Workers work within a system over which they have little control. It is the system that primarily determines their performance.
- Only management can change the system.
- There will always be some workers above the mean and some workers below the mean.

FIGURE 18.8
p Chart for the Red Bead Experiment

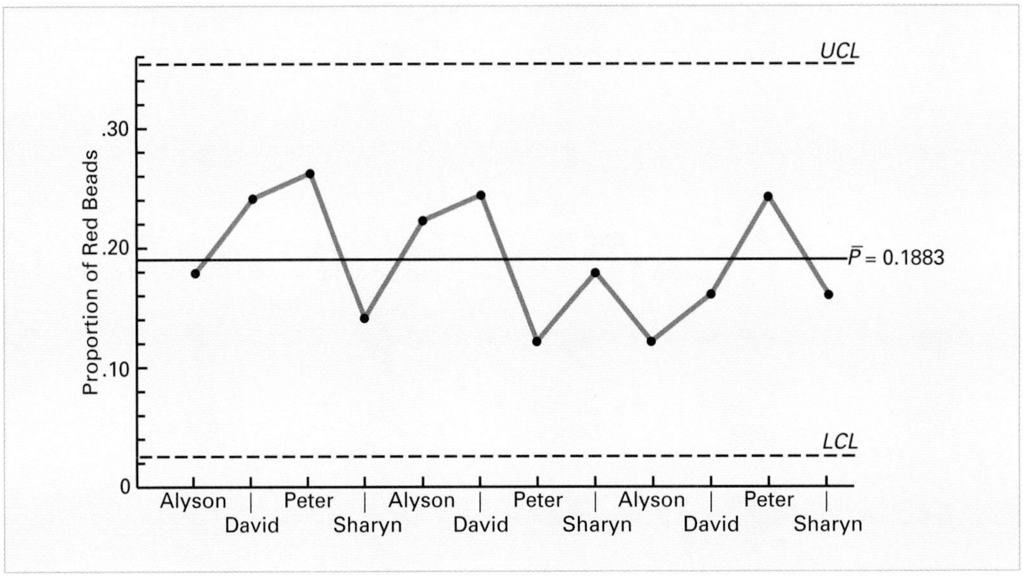

PROBLEMS FOR SECTION 18.5

Applying the Concepts

18.9 In the Red Bead experiment, how do you think many managers would have reacted after day 1? day 2? day 3?

18.10 (**Class Project**) Obtain a version of the red bead experiment for your class.

a. Conduct the experiment in the same way as described in this section.

b. Remove 400 red beads from the bead box before beginning the experiment. How do your results differ from those in (a)? What does this tell you about the effect of the "system" on the workers?

18.6 CONTROL CHART FOR AN AREA OF OPPORTUNITY— THE *c* CHART

In this text, you are introduced to two types of attribute charts, the *p* chart in section 18.4 and the *c* chart in this section. Recall that you use the *p* chart for monitoring and analyzing the proportion of nonconforming items. You use the ***c* chart** for monitoring and analyzing the number of nonconformities in an area of opportunity. **Areas of opportunity** can be individual units of products or services, or units of time, space, or area. Nonconformities are sometimes referred to as defects or flaws. Examples of the number of nonconformities in areas of opportunity are the number of flaws in a square foot of carpet, the number of typographical errors on a printed page, the number of breakdowns per day in an academic computer center, and the number of hotel customers filing a complaint in a given week. This situation differs from the one in which you used the *p* chart; instead of classifying each unit as conforming or nonconforming, you count the number of nonconformities in each area of opportunity.

The above situation fits the assumptions of a Poisson distribution (see section 5.4). For the Poisson distribution, the standard deviation of the number of nonconformities is the square root of the mean number of nonconformities (λ). Assuming that the size of each area of opportunity remains constant,[5] you can compute the control limits for the number of nonconformities per area of opportunity using the observed mean number of nonconformities as an estimate of λ. Equation (18.3) defines the control limits for the *c* chart, which you use to monitor and analyze the number of nonconformities per area of opportunity.

[5] *If the size of the unit varies, you should use the u chart instead of the c chart (see references 7, 8, and 11).*

CONTROL LIMITS FOR THE *c* CHART

$$\bar{c} \pm 3\sqrt{\bar{c}} \qquad (18.3)$$

$$\text{UCL} = \bar{c} + 3\sqrt{\bar{c}}$$

$$\text{LCL} = \bar{c} - 3\sqrt{\bar{c}}$$

where

$$\bar{c} = \frac{\sum_{i=1}^{k} c_i}{k}$$

k = number of units sampled

c_i = number of nonconformities in unit i

To help study the hotel service quality in the "Using Statistics" scenario, you can use a *c* chart to monitor the number of customer complaints filed with the hotel. If guests of the hotel are dissatisfied with any part of their stay, they are asked to file a customer complaint form. At the end of each week, the number of complaints filed is recorded. In this example, a complaint is a nonconformity and the area of opportunity is one week. Table 18.4 lists the number of complaints from the last 50 weeks. **COMPLAINTS**

TABLE 18.4
Number of Complaints in the Last 50 Weeks

Week	Number of Complaints	Week	Number of Complaints	Week	Number of Complaints
1	8	18	7	35	3
2	10	19	10	36	5
3	6	20	11	37	2
4	7	21	8	38	4
5	5	22	7	39	3
6	7	23	8	40	3
7	9	24	6	41	4
8	8	25	7	42	2
9	7	26	7	43	4
10	9	27	5	44	5
11	10	28	8	45	5
12	7	29	6	46	3
13	8	30	7	47	2
14	11	31	5	48	5
15	10	32	5	49	4
16	9	33	4	50	4
17	8	34	4		

For these data,

$$k = 50 \text{ and } \sum_{i=1}^{k} c_i = 312$$

Thus,

$$\bar{c} = \frac{312}{50} = 6.24$$

so that using Equation (18.3) on page 767,

$$\begin{aligned} &\bar{c} \pm 3\sqrt{\bar{c}} \\ &= 6.24 \pm 3\sqrt{6.24} \\ &= 6.24 \pm 7.494 \end{aligned}$$

Thus,

$$\text{UCL} = 6.24 + 7.494 = 13.734$$

$$\text{LCL} = 6.24 - 7.494 < 0$$

Therefore, the LCL does not exist.

Figure 18.9 displays the Minitab control chart for the complaint data of Table 18.4. Figure 18.9 does not indicate any points outside the control limits. However, there is a clear pattern to the number of customer complaints over time. During the first half of the sequence almost all the weeks had more than the mean number of complaints, and almost all the weeks in the second half had less than the mean. There are more than 8 points in a row above the centerline (weeks 6–23) and below the centerline (weeks 31–50), thus signaling a trend. This change, which is an improvement, is due to a special cause of variation. The next step is to investigate the process and determine the special cause that produced this pattern. When identified, you then need to ensure that this becomes a permanent improvement, not a temporary phenomenon. In other words, the source of the special-cause variation must become part of the permanent ongoing process in order for the number of customer complaints not to slip back to the high levels experienced in the first half of the data.

FIGURE 18.9

Minitab *c* Chart for Hotel Complaints

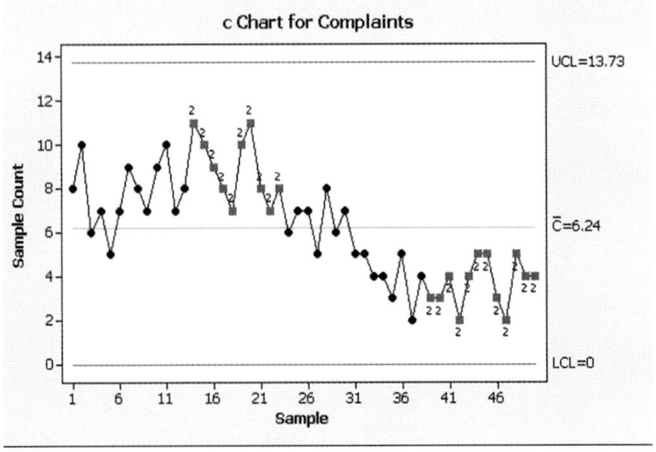

PROBLEMS FOR SECTION 18.6

Learning the Basics

PH Grade ASSIST **18.11** The following data were collected on the number of nonconformities per unit for 10 time periods:

Time	Nonconformities Per Unit	Time	Nonconformities Per Unit
1	7	6	5
2	3	7	3
3	6	8	5
4	3	9	2
5	4	10	0

a. Construct the appropriate control chart and determine the LCL and UCL.
b. Are there any special causes of variation?

PH Grade ASSIST **18.12** The following data were collected on the number of nonconformities per unit for 10 time periods:

Time	Nonconformities Per Unit	Time	Nonconformities Per Unit
1	25	6	15
2	11	7	12
3	10	8	10
4	11	9	9
5	6	10	6

a. Construct the appropriate control chart and determine the LCL and UCL.
b. Are there any special causes of variation?

Applying the Concepts

You can solve problems 18.13 to 18.17 manually or by using Microsoft Excel, Minitab, or SPSS.

PH Grade ASSIST **18.13** To improve service quality, the owner of a dry cleaning business wants to study the number of dry-cleaned items that are returned for rework per day. Records were kept for a 4-week period (the store is open Monday through Saturday) with the results given below. **DRYCLEAN**

Day	Items Returned For Rework	Day	Items Returned For Rework
1	4	4	7
2	6	5	6
3	3	6	8
7	6	16	4
8	4	17	10
9	8	18	9
10	6	19	6
11	5	20	5
12	12	21	8
13	5	22	6
14	8	23	7
15	3	24	9

a. Construct a *c* chart for the number of items per day that are returned for rework. Do you think that the process is in a state of statistical control?
b. Should the owner of the dry cleaning store take action to investigate why 12 items were returned for rework on day 12? Explain. Would your answer change if 20 items were returned for rework on day 12?
c. On the basis of the results in (a), what should the owner of the dry cleaning store do to reduce the number of items per day that are returned for rework?

PH Grade ASSIST SELF Test **18.14** The branch manager of a savings bank has recorded the number of errors that each of 12 tellers has made during the past year. The results were as follows: **TELLER**

Teller	Number of Errors	Teller	Number of Errors
Alice	4	Mitchell	6
Carl	7	Nora	3
Gina	12	Paul	5
Jane	6	Susan	4
Linda	2	Thomas	7
Marla	5	Vera	5

a. Do you think the bank manager will single out Gina for any disciplinary action regarding her performance in the last year?
b. Construct a *c* chart for the number of errors committed by the 12 tellers. Is the number of errors in a state of statistical control?
c. Based on the *c* chart developed in (b), do you now think that Gina should be singled out for disciplinary action regarding her performance? Does your conclusion now agree with what you expected the manager to do?
d. On the basis of the results in (b), what should the branch manager do to reduce the number of errors?

PH Grade ASSIST **18.15** Falls are one source of preventable hospital injury. Although most patients who fall are not hurt, a risk of serious injury is involved. The data in the file PTFALLS represent the number of patient falls per month over a 28-month period in a 19-bed AIDS unit at a major metropolitan hospital.

a. Construct a *c* chart for the number of patient falls per month. Is the process of patient falls per month in a state of statistical control?

b. What effect would it have on your conclusions if you knew that the AIDS unit was started only one month prior to the beginning of data collection?

c. What other factors might contribute to special cause variation in this problem?

18.16 A member of the volunteer fire department for Trenton, Ohio, decided to apply the control chart methodology he learned in his business statistics class to data collected by the fire department. He was interested in determining whether weeks containing more than the mean number of fire runs were due to inherent, chance-cause variation, or if there were special causes of variation such as increased arson, severe drought, or holiday-related activities. The data in the file FIRERUNS contain the number of fire runs made per week (Sunday through Saturday) during 2001. Week one begins on Sunday, December 31, 2000, and week 53 ends on Saturday, January 5, 2002.

Source: Extracted from The City of Trenton *2001* Annual Report, *Trenton, Ohio, February 21, 2002.*

a. What is the mean number of fire runs made per week?

b. Construct a *c* chart for the number of fire runs per week.

c. Is the process in a state of statistical control?

d. Weeks 15 and 41 experienced seven fire runs each. Are these large values explainable by common causes, or does it appear that special causes of variation occurred in these weeks?

e. Explain how the fire department can use these data to chart and monitor future weeks in real-time (i.e., on a week-to-week basis)?

18.17 Rochester-Electro-Medical Inc. is a manufacturing company based in Tampa, Florida, that produces medical products. Recently, management felt the need to improve the safety of the workplace and began a safety sampling study. The following data represent the number of unsafe acts observed by the company safety director over an initial time period in which he made 20 tours of the plant. SAFETY

Tour	Number of UnsafeActs	Tour	Number of Unsafe Acts
1	10	11	2
2	6	12	8
3	6	13	7
4	10	14	6
5	8	15	6
6	12	16	11
7	2	17	13
8	1	18	9
9	23	19	6
10	3	20	9

Source: H. Gitlow, A. R. Berkins, and M. He, "Safety Sampling: A Case Study," Quality Engineering, *14, 2002, 405–419. Reproduced by permission of Taylor & Francis, Inc.,* **taylorandfrancis.com**.

a. Construct a *c* chart for the number of unsafe acts.

b. Based on the results of (a), is the process in a state of statistical control?

c. What should management do next to improve the process?

18.7 CONTROL CHARTS FOR THE RANGE AND THE MEAN

You use **variables control charts** to monitor and analyze a process when you have numerical data. Common numerical variables include time, money, and weight. Because numeric variables provide more information than attribute data such as the proportion of nonconforming items or the number of nonconformities, variables control charts are more sensitive in detecting special-cause variation than the *p* chart or the *c* chart. Variables charts are typically used in pairs. One chart monitors the dispersion (or variability) in a process, and the other monitors the process mean. You must examine the chart that monitors dispersion first because if it indicates the presence of out-of-control conditions, the interpretation of the chart for the mean will be misleading. Although businesses currently use several alternative pairs of charts (see references 7, 8, and 11), this text considers only the control charts for the range and the mean.

The *R* Chart

You can use several different types of control charts to monitor the dispersion (i.e., variability) in a numerically measured characteristic of interest. The simplest and most common is the control chart for the range, the ***R* chart**. You use the range chart only when the sample size is 10 or less. If the sample size is greater than 10, a standard deviation chart is preferable. Because sample sizes of five or less are typically used in many applications, the standard deviation chart is not illustrated in this text. (For a discussion of standard deviation charts, see references 7, 8, or 11.) The *R* chart enables you to determine whether the variability in a process is in control or whether changes in the amount of variability are occurring over time. If the process range is in control, then the amount of variation in the process is consistent over time, and you can use the results of the *R* chart to develop the control limits for the mean.

To develop control limits for the range, you need an estimate of the mean range and the standard deviation of the range. As shown in Equation (18.4), these control limits depend on two constants, the **d_2 factor**, which represents the relationship between the standard deviation and the range for varying sample sizes, and the **d_3 factor**, which represents the relationship between the standard deviation and the standard error of the range for varying sample sizes. Table E.13 contains values for these factors. Equation (18.4) defines the control limits for the *R* chart.

CONTROL LIMITS FOR THE RANGE

$$\bar{R} \pm 3\bar{R}\frac{d_3}{d_2} \quad \textbf{(18.4)}$$

$$\text{UCL} = \bar{R} + 3\bar{R}\frac{d_3}{d_2}$$

$$\text{LCL} = \bar{R} - 3\bar{R}\frac{d_3}{d_2}$$

where

$$\bar{R} = \frac{\sum_{i=1}^{k} R_i}{k}$$

You can simplify the calculations in Equation (18.4) by using the **D_3 factor**, equal to $1 - 3(d_3/d_2)$, and the **D_4 factor**, equal to $1 + 3(d_3/d_2)$, to express the control limits as shown in Equations (18.5a) and (18.5b).

CALCULATING CONTROL LIMITS FOR THE RANGE

$$\text{UCL} = D_4\bar{R} \quad \textbf{(18.5a)}$$

$$\text{LCL} = D_3\bar{R} \quad \textbf{(18.5b)}$$

To illustrate the *R* chart, return to the "Using Statistics" scenario concerning hotel service quality. During the Measure phase of the Six Sigma DMAIC model, the amount of time to deliver luggage was operationally defined as the time from when the guest completes check-in procedures to the time the luggage arrives in the guest's room. During the Analyze phase of the Six Sigma DMAIC model, data were recorded over a 4-week period. Subgroups of five deliveries were selected from the evening shift on each day. Table 18.5 summarizes the results for all 28 days. **HOTEL2**

TABLE 18.5
Luggage Delivery Times and Subgroup Mean and Range for 28 Days

Day	Luggage Delivery Times in Minutes					Mean	Range
1	6.7	11.7	9.7	7.5	7.8	8.68	5.0
2	7.6	11.4	9.0	8.4	9.2	9.12	3.8
3	9.5	8.9	9.9	8.7	10.7	9.54	2.0
4	9.8	13.2	6.9	9.3	9.4	9.72	6.3
5	11.0	9.9	11.3	11.6	8.5	10.46	3.1
6	8.3	8.4	9.7	9.8	7.1	8.66	2.7
7	9.4	9.3	8.2	7.1	6.1	8.02	3.3
8	11.2	9.8	10.5	9.0	9.7	10.04	2.2
9	10.0	10.7	9.0	8.2	11.0	9.78	2.8
10	8.6	5.8	8.7	9.5	11.4	8.80	5.6
11	10.7	8.6	9.1	10.9	8.6	9.58	2.3
12	10.8	8.3	10.6	10.3	10.0	10.00	2.5
13	9.5	10.5	7.0	8.6	10.1	9.14	3.5
14	12.9	8.9	8.1	9.0	7.6	9.30	5.3
15	7.8	9.0	12.2	9.1	11.7	9.96	4.4
16	11.1	9.9	8.8	5.5	9.5	8.96	5.6
17	9.2	9.7	12.3	8.1	8.5	9.56	4.2
18	9.0	8.1	10.2	9.7	8.4	9.08	2.1
19	9.9	10.1	8.9	9.6	7.1	9.12	3.0
20	10.7	9.8	10.2	8.0	10.2	9.78	2.7
21	9.0	10.0	9.6	10.6	9.0	9.64	1.6
22	10.7	9.8	9.4	7.0	8.9	9.16	3.7
23	10.2	10.5	9.5	12.2	9.1	10.30	3.1
24	10.0	11.1	9.5	8.8	9.9	9.86	2.3
25	9.6	8.8	11.4	12.2	9.3	10.26	3.4
26	8.2	7.9	8.4	9.5	9.2	8.64	1.6
27	7.1	11.1	10.8	11.0	10.2	10.04	4.0
28	11.1	6.6	12.0	11.5	9.7	10.18	5.4
					Sums:	265.38	97.5

For the data in Table 18.5,

$$k = 28, \sum_{i=1}^{k} R_i = 97.5, \text{ and } \bar{R} = \frac{\sum_{i=1}^{k} R_i}{k} = \frac{97.5}{28} = 3.482$$

Using Equation (18.4) on page 771 and from Table E.13 for $n = 5$, $d_2 = 2.326$, and $d_3 = 0.864$,

$$3.482 \pm 3(3.482)\left(\frac{0.864}{2.326}\right)$$

$$= 3.482 \pm 3.880$$

Thus,

$$\text{UCL} = 3.482 + 3.880 = 7.362$$

$$\text{LCL} = 3.482 - 3.880 < 0$$

Therefore, the LCL does not exist because it is impossible to get a negative range. Alternatively, using Equation (18.5) on page 771, and $D_3 = 0$ and $D_4 = 2.114$ from Table E.13,

$$\text{UCL} = D_4\bar{R} = (2.114)(3.482) = 7.36$$

and the LCL does not exist.

Figure 18.10 displays the Microsoft Excel R chart for the luggage delivery times. Figure 18.10 does not indicate any individual ranges outside the control limits or any trends.

FIGURE 18.10
Microsoft Excel R Chart for the Luggage Delivery Times

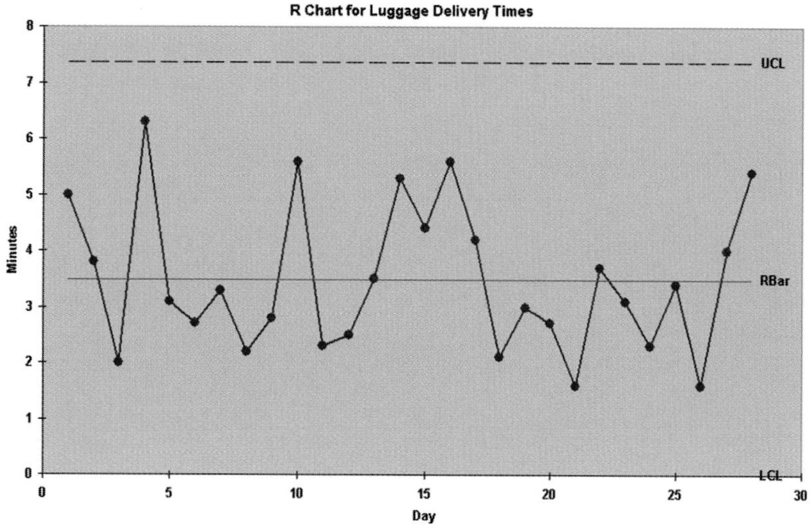

The $\overline{X}$ Chart

Now that you have determined that the control chart shows that the range is in control, you continue by examining the control chart for the process mean, the $\overline{X}$ **chart**.

The control chart for $\overline{X}$ uses subgroups each of size n for k consecutive periods of time. To compute control limits for the mean, you need to compute the mean of the subgroup means (called $\overline{\overline{X}}$) and the standard deviation of the mean (which is called the standard error of the mean $\sigma_{\overline{X}}$ in Chapter 7). The estimate of the standard deviation of the mean is a function of the d_2 factor, which represents the relationship between the standard deviation and the range for varying sample sizes.[6] Equations (18.6) and (18.7) define the control limits for the $\overline{X}$ chart.

[6] *$\overline{R}/d_2$ is used to estimate the standard deviation of the population, and $\overline{R}/d_2\sqrt{n}$ is used to estimate the standard deviation of the mean.*

CONTROL LIMITS FOR THE $\overline{X}$ CHART

$$\overline{\overline{X}} \pm 3\frac{\overline{R}}{d_2\sqrt{n}} \qquad \textbf{(18.6)}$$

$$\text{UCL} = \overline{\overline{X}} + 3\frac{\overline{R}}{d_2\sqrt{n}}$$

$$\text{LCL} = \overline{\overline{X}} - 3\frac{\overline{R}}{d_2\sqrt{n}}$$

where $\overline{\overline{X}} = \dfrac{\sum_{i=1}^{k}\overline{X}_i}{k}$ $\quad \overline{R} = \dfrac{\sum_{i=1}^{k}R_i}{k}$

$\overline{X}_i$ = sample mean of n observations at time i

R_i = range of n observations at time i

k = number of subgroups

You can simplify the calculations in Equation (18.6) by utilizing the A_2 **factor**, equal to $3/\left(d_2\sqrt{n}\right)$. Equations (18.7a) and (18.7b) show the simplified control limits.

CALCULATING CONTROL LIMITS FOR THE MEAN USING THE A_2 FACTOR

$$\text{UCL} = \bar{\bar{X}} + A_2\bar{R} \qquad \textbf{(18.7a)}$$

$$\text{LCL} = \bar{\bar{X}} - A_2\bar{R} \qquad \textbf{(18.7b)}$$

From Table 18.5 on page 772,

$$k = 28, \sum_{i=1}^{k} \bar{X}_i = 265.38, \text{ and } \sum_{i=1}^{k} R_i = 97.5$$

so that

$$\bar{\bar{X}} = \frac{\sum_{i=1}^{k} \bar{X}_i}{k} = \frac{265.38}{28} = 9.478 \text{ and } \bar{R} = \frac{\sum_{i=1}^{k} R_i}{k} = \frac{97.5}{28} = 3.482$$

Using Equation (18.6) on page 773 and Table E.13 for $n = 5$, $d_2 = 2.326$,

$$\bar{\bar{X}} \pm 3\frac{\bar{R}}{d_2\sqrt{n}}$$

$$= 9.478 \pm 3\frac{3.482}{(2.326)\sqrt{5}}$$

$$= 9.478 \pm 2.008$$

Thus,

$$\text{UCL} = 9.478 + 2.008 = 11.486$$

$$\text{LCL} = 9.478 - 2.008 = 7.470$$

Alternatively, using Equations (18.7a) and (18.7b), and $A_2 = 0.577$ from Table E.13,

$$\text{UCL} = 9.478 + (0.577)(3.482) = 9.478 + 2.009 = 11.487$$

$$\text{LCL} = 9.478 - (0.577)(3.482) = 9.478 - 2.009 = 7.469$$

These results are the same as those using Equation (18.6), except for rounding error.

Figure 18.11 displays the Microsoft Excel $\bar{X}$ chart for the luggage delivery time data. Figure 18.12 presents Minitab R and $\bar{X}$ charts. Figures 18.11 and 18.12 do not reveal any points outside the control limits or any trend. Although there is a considerable amount of variability among the 28 subgroup means, since both the R chart and the $\bar{X}$ chart are in control, the luggage delivery process is in a state of statistical control. If you want to reduce the variation or lower the mean delivery time, then you need to change the process.

FIGURE 18.11
Microsoft Excel $\overline{X}$ Chart for the Luggage Delivery Times

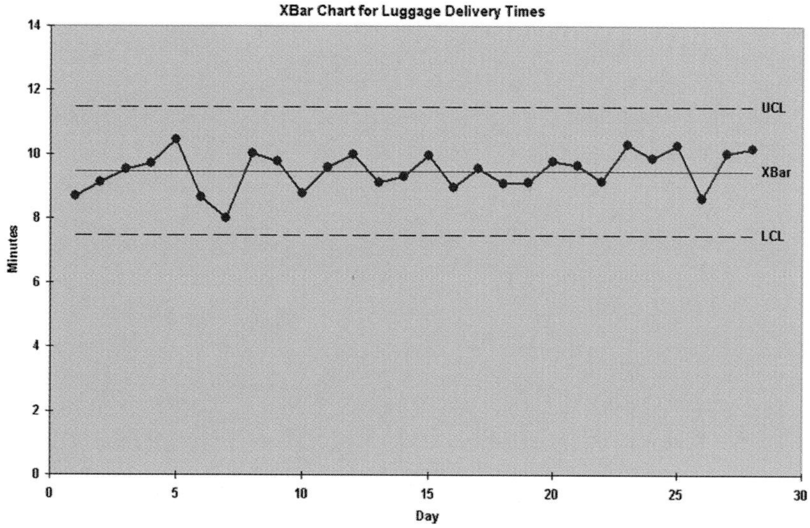

FIGURE 18.12
Minitab $\overline{X}$ and R Charts for the Luggage Delivery Times

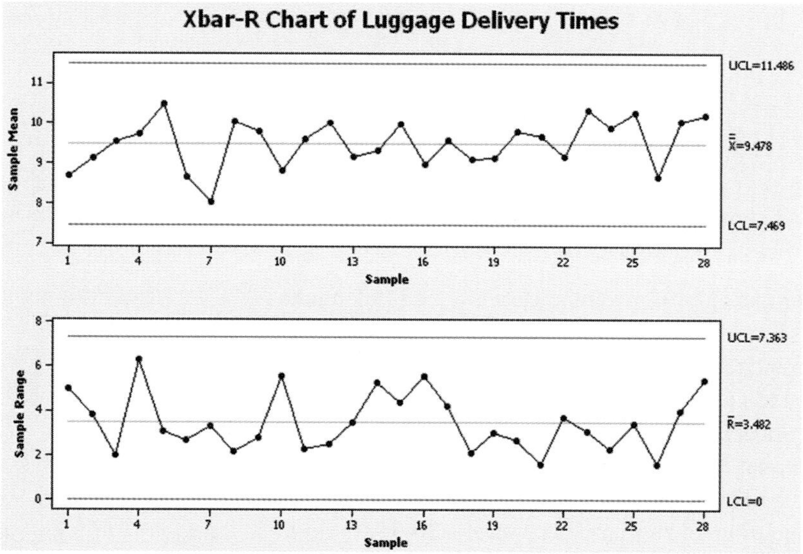

PROBLEMS FOR SECTION 18.7

Learning the Basics

PH Grade ASSIST **18.18** For subgroups of $n = 4$, what is the value of:

a. the d_2 factor?
b. the d_3 factor?
c. the D_3 factor?
d. the D_4 factor?
e. the A_2 factor?

PH Grade ASSIST **18.19** The following summary of data is for subgroups of $n = 4$ for a 10-day period.

Day	Mean	Range	Day	Mean	Range
1	13.6	3.5	6	12.9	4.8
2	14.3	4.1	7	17.3	4.5
3	15.3	5.0	8	13.9	2.9
4	12.6	2.8	9	12.6	3.8
5	11.8	3.7	10	15.2	4.6

a. Compute control limits for the range.
b. Is there evidence of special cause variation in (a)?
c. Compute control limits for the mean.
d. Is there evidence of special-cause variation in (c)?

Applying the Concepts

You should use Microsoft Excel, Minitab, or SPSS to solve problems 18.20–18.26.

SELF Test **18.20** The manager of a branch of a local bank wants to study waiting times of customers for teller service during the 12:00 noon to 1:00 P.M. lunch hour. A subgroup of four customers is selected (one at each 15-minute interval during the hour), and the time in minutes is measured from the point each customer enters the line to when he or she reaches the teller window. The results over a 4-week period are in the data file: **BANKTIME**

Day	Time in Minutes			
1	7.2	8.4	7.9	4.9
2	5.6	8.7	3.3	4.2
3	5.5	7.3	3.2	6.0
4	4.4	8.0	5.4	7.4
5	9.7	4.6	4.8	5.8
6	8.3	8.9	9.1	6.2
7	4.7	6.6	5.3	5.8
8	8.8	5.5	8.4	6.9
9	5.7	4.7	4.1	4.6
10	1.7	4.0	3.0	5.2
11	2.6	3.9	5.2	4.8
12	4.6	2.7	6.3	3.4
13	4.9	6.2	7.8	8.7
14	7.1	6.3	8.2	5.5
15	7.1	5.8	6.9	7.0
16	6.7	6.9	7.0	9.4
17	5.5	6.3	3.2	4.9
18	4.9	5.1	3.2	7.6
19	7.2	8.0	4.1	5.9
20	6.1	3.4	7.2	5.9

a. Construct control charts for the mean and the range.
b. Is the process in control?

18.21 The manager of a warehouse for a local telephone company is involved in a process that receives expensive circuit boards and returns them to central stock so that they can be reused at a later date. Speedy processing of these circuit boards is critical in providing good service to customers and reducing capital expenditures. The data in the file WAREHSE represent the number of circuit boards processed per day by each of a subgroup of five employees over a 30-day period.
a. Construct control charts for the mean and the range.
b. Is the process in control?

18.22 An article in the *Mid-American Journal of Business* presents an analysis for a springwater bottling operation. One of the characteristics of interest is the amount of magnesium, measured in parts per million (ppm), in the water. The data in the following table represent the magnesium levels from 30 subgroups of 4 bottles collected over a 30-hour period. SPWATER

Hour	1	2	3	4
1	19.91	19.62	19.15	19.85
2	20.46	20.44	20.34	19.61
3	20.25	19.73	19.98	20.32
4	20.39	19.43	20.36	19.85
5	20.02	20.02	20.13	20.34
6	19.89	19.77	20.92	20.09
7	19.89	20.45	19.44	19.95
8	20.08	20.13	20.11	19.32
9	20.30	20.42	20.68	19.60
10	20.19	20.00	20.23	20.59
11	19.66	21.24	20.35	20.34
12	20.30	20.11	19.64	20.29
13	19.83	19.75	20.62	20.60
14	20.27	20.88	20.62	20.40
15	19.98	19.02	20.34	20.34
16	20.46	19.97	20.32	20.83
17	19.74	21.02	19.62	19.90
18	19.85	19.26	19.88	20.20
19	20.77	20.58	19.73	19.48
20	20.21	20.82	20.01	19.93
21	20.30	20.09	20.03	20.13
22	20.48	21.06	20.13	20.42
23	20.60	19.74	20.52	19.42
24	20.20	20.08	20.32	19.51
25	19.66	19.67	20.26	20.41
26	20.72	20.58	20.71	19.99
27	19.77	19.40	20.49	19.83
28	19.99	19.65	19.41	19.58
29	19.44	20.15	20.14	20.76
30	20.03	19.96	19.86	19.91

Source: Extracted from Susan K. Humphrey, and Timothy C. Krehbiel, "Managing Process Capability," The Mid-American Journal of Business, *14, Fall 1999, 7–12.*

a. Construct a control chart for the range.
b. Construct a control chart for the mean.
c. Is the process in control?

18.23 The data in the following table are the tensile strengths of bolts of cloth. The data were collected in subgroups of 3 bolts of cloth over a 25-hour period. TENSILE

Hour	1	2	3	Hour	1	2	3
1	15.06	14.62	15.10	14	16.29	14.61	15.67
2	17.58	15.75	16.72	15	15.84	12.16	15.40
3	13.83	14.83	15.61	16	15.12	15.60	13.83
4	17.19	15.75	15.42	17	18.48	16.07	16.31
5	14.56	15.37	15.67	18	17.55	14.73	16.95
6	14.82	17.25	15.73	19	13.57	17.55	15.81
7	17.92	14.76	14.40	20	16.23	16.92	16.45
8	16.53	14.52	17.31	21	14.60	16.83	15.34
9	13.83	14.53	15.32	22	16.73	18.60	16.76
10	16.45	13.85	16.32	23	18.03	14.55	13.87
11	15.20	14.61	18.45	24	16.61	16.45	16.95
12	14.49	16.15	17.80	25	15.86	17.00	18.28
13	15.89	15.04	16.67				

a. Construct a control chart for the range.
b. Construct a control chart for the mean.
c. Is the process in control?

PH Grade ASSIST **18.24** The director of radiology at a large metropolitan hospital is concerned about scheduling in the radiology facilities. On a typical day 250 patients are transported to the radiology department for treatment or diagnostic procedures. If patients do not reach the radiology unit at their scheduled time, backups occur and other patients experience delays. The time it takes to transport patients to the radiology unit is operationally defined as the time between when the transporter is assigned to the patient and the time the patient arrives at the radiology unit. A sample of $n = 4$ patients was selected each day for 20 days and the time to transport each patient (in minutes) was determined, with the results in the data file TRANSPORT.

a. Construct control charts for the mean and the range.
b. Is the process in control?

18.25 A filling machine for a tea bag manufacturer produces approximately 170 tea bags per minute. The process manager monitors the weight of the tea placed in individual bags. A subgroup of $n = 4$ teabags is taken every 15 minutes for 25 consecutive time periods. The results are given below: TEA3

Sample	Weight (in grams)			
1	5.32	5.77	5.50	5.61
2	5.63	5.44	5.54	5.40
3	5.56	5.40	5.67	5.57
4	5.32	5.45	5.50	5.42
5	5.45	5.53	5.46	5.47
6	5.29	5.42	5.50	5.44
7	5.57	5.40	5.52	5.54
8	5.44	5.61	5.49	5.58
9	5.53	5.25	5.67	5.53
10	5.41	5.55	5.51	5.53
11	5.55	5.58	5.58	5.56
12	5.58	5.36	5.45	5.53
13	5.63	5.75	5.46	5.54
14	5.48	5.44	5.45	5.60
15	5.49	5.57	5.43	5.36
16	5.54	5.62	5.66	5.59
17	5.46	5.46	5.38	5.49
18	5.72	5.36	5.59	5.25
19	5.58	5.50	5.36	5.40
20	5.43	5.51	5.37	5.32
21	5.59	5.58	5.60	5.46
22	5.42	5.41	5.40	5.69
23	5.64	5.59	5.42	5.56
24	5.62	5.38	5.75	5.47
25	5.51	5.54	5.73	5.77

a. What are some of the sources of common-cause variation that might be present in this process?
b. What problems might occur that would result in special causes of variation?
c. Construct control charts for the range and the mean.
d. Is the process in control?

18.26 A manufacturing company makes brackets for bookshelves. The brackets provide critical structural support and must have a 90-degree bend ±1 degree. Measurements of the bend of the brackets were taken at 18 different times. Five brackets were sampled at each time. The data are in the file ANGLE.

a. Construct control charts for the range and the mean.
b. Is the process in control?

18.8 PROCESS CAPABILITY

Often it is necessary to analyze the amount of common-cause variation present in an in-control process. Is the common-cause variation small enough to satisfy customers with the product or service? Or is the common-cause variation so large that there are too many dissatisfied customers and a process change is needed?

Analyzing the capability of a process is a way to answer these questions. There are many methods available to analyze and report process capability (see reference 2). This section begins with a relatively simple method to estimate the percentage of products or services that will satisfy the customer. Later in the section, the use of capability indexes is introduced.

Customer Satisfaction and Specification Limits

Quality is defined by the customer. A customer who believes that a product or service has met or exceeded his or her expectations will be satisfied. The management of a company must listen to the customer and translate the customer's needs and expectations into easily measured critical-to-quality (CTQs) variables. Management then sets specification limits for these CTQs.

Specification limits are technical requirements set by management in response to customers' needs and expectations. The **upper specification limit (USL)** is the largest value a CTQ can have and still conform to customer expectations. Likewise, the **lower specification limit (LSL)** is the smallest value a CTQ can have and still conform to customer expectations.

For example, a soap manufacturer understands that customers expect their soap to produce a certain amount of lather. The customer can become dissatisfied if the soap produces too much or too little lather. Product engineers know that the level of free fatty acids in the soap controls the amount of lather. Thus, the process manager, with input from the product engineers, sets both a USL and an LSL for the amount of free fatty acids in the soap.

As an example of a case in which only a single specification limit is involved, consider the "Using Statistics" scenario concerning hotel service quality on page 752. Since customers want their bags delivered as quickly as possible, hotel management sets a USL for the time required for delivery. In this case, there is no LSL. As you can see in both the luggage delivery time and soap examples, specification limits are customer-driven requirements placed on a product or a service. If a process consistently meets these requirements, the process is capable of satisfying the customer.

Process capability is the ability of a process to consistently meet specified customer-driven requirements.

One way to analyze the capability of a process is to estimate the percentage of products or services that are within specifications. To do this, you must have an in-control process because an out-of-control process does not allow you to predict its capability. If you are dealing with an out-of-control process, you must first identify and eliminate the special causes of variation before performing a capability analysis. Out-of-control processes are unpredictable, and therefore, you cannot conclude that such processes are capable of meeting specifications or satisfying customer expectations. In order to estimate the percentage of product or service within specifications, first, you must estimate the mean and standard deviation of the population of all X values, the CTQ variable of interest for the product or service. The estimate for the mean of the population is $\bar{\bar{X}}$, the mean of all the sample means [see Equation (18.6) on page 773]. The estimate of the standard deviation of the population is $\bar{R}$ divided by d_2. You can use the $\bar{\bar{X}}$ and $\bar{R}$ from in-control $\bar{X}$ and R charts, respectively. You need to find the correct d_2 value in Table E.13.

In this text, the population of X values is assumed to be approximately normally distributed. (If your data are not approximately normally distributed, see reference 2 for an alternative approach.) Assuming that the process is in control and X is approximately normal, you can use Equation (18.8) to estimate the probability that a process outcome is within specifications.

ESTIMATING THE CAPABILITY OF A PROCESS

For a CTQ variable with a lower specification limit and an upper specification limit:

$$P(\text{an outcome will be within specifications}) = P(\text{LSL} < X < \text{USL}) \quad \textbf{(18.8a)}$$

$$= P\left(\frac{\text{LSL} - \bar{\bar{X}}}{\frac{\bar{R}}{d_2}} < Z < \frac{\text{USL} - \bar{\bar{X}}}{\frac{\bar{R}}{d_2}}\right)$$

For a CTQ variable with only an upper specification limit:

$$P(\text{an outcome will be within specifications}) = P(X < \text{USL}) \quad \textbf{(18.8b)}$$

$$= P\left(Z < \frac{\text{USL} - \bar{\bar{X}}}{\frac{\bar{R}}{d_2}}\right)$$

For a CTQ variable with only a lower specification limit:

$$P(\text{an outcome will be within specifications}) = P(\text{LSL} < X) \quad \textbf{(18.8c)}$$

$$= P\left(\frac{\text{LSL} - \bar{\bar{X}}}{\frac{\bar{R}}{d_2}} < Z\right)$$

where Z is a standardized normal random variable

In section 18.7, you determined that the luggage delivery process was in control. Suppose that the hotel management has instituted a policy that 99% of all luggage deliveries must be completed in 14 minutes or less. From the summary computations on page 774

$$n = 5 \quad \bar{\bar{X}} = 9.478 \quad \bar{R} = 3.482 \quad \text{and from Table E.13, } d_2 = 2.326$$

Using Equation (18.8b),

$$P(\text{delivery is made within specifications}) = P(X < 14)$$

$$= P\left(Z < \frac{14 - 9.478}{\frac{3.482}{2.326}}\right)$$

$$= P(Z < 3.02)$$

Using Table E.2, $P(Z < 3.02) = 0.99874$

Thus, you estimate that 99.874% of the luggage deliveries will be made within the specified time. The process is capable of meeting the 99% goal set forth by the hotel management.

Capability Indexes

A common approach in business is to use capability indexes to report the capability of a process. A **capability index** is an aggregate measure of a process's ability to meet specification limits. The larger the value of a capability index, the more capable a process is of meeting customer requirements. Equation (18.9) defines C_p, the most commonly used index.

THE C_p INDEX

$$C_p = \frac{\text{USL} - \text{LSL}}{6(\bar{R}/d_2)} \quad \textbf{(18.9)}$$

$$= \frac{\text{specification spread}}{\text{process spread}}$$

The numerator in Equation (18.9) represents the distance between the upper and lower specification limits, referred to as the specification spread. The denominator, $6(\bar{R}/d_2)$, represents a 6 standard deviation spread in the data (the mean ±3 standard deviations), referred to as the process spread. (Recall from Chapter 6 that approximately 99.73% of the values from a normal distribution fall in the interval from the mean ±3 standard deviations.) You want the process spread to be small in comparison to the specification spread in order for the vast majority of the process output to fall within the specification limits. Therefore, the larger the value of C_p, the better is the capability of the process.

C_p is a measure of process potential, not of actual performance, because it does not consider the current process mean. A value of 1 indicates that if the process mean could be centered (i.e., equal to the halfway point between USL and LSL), then approximately 99.73% of the values would be inside the specification limits. A C_p value greater than 1 indicates that a process has the potential of having more than 99.73% of its outcomes within specifications. A C_p value less than 1 indicates that the process is not very capable of meeting customer requirements, for even if the process is perfectly centered, less than 99.73% of the process outcomes will be within specifications. Historically, many companies required a C_p greater than or equal to 1. Now that the global economy has become more quality conscious, many companies are requiring a C_p as large as 1.33, 1.5, and for companies adopting Six Sigma management, 2.0.

To illustrate the calculation and interpretation of the C_p index, suppose a soft-drink producer bottles its beverage into 12-ounce bottles. Each hour four bottles are selected and control charts for the range and the mean are constructed. At the end of 24 hours the capability of the process is studied. The lower specification limit is 11.82 ounces and the upper specification limit is 12.18 ounces. Suppose that the control charts indicate that the process is in control and the following summary calculations were recorded on the control charts:

$$n = 4 \qquad \bar{\bar{X}} = 12.02 \qquad \bar{R} = 0.10$$

To calculate the C_p index assuming that the data are normally distributed, from Table E.13, $d_2 = 2.059$ for $n = 4$. Using Equation (18.9) on page 779,

$$C_p = \frac{USL - LSL}{6(\bar{R}/d_2)}$$

$$= \frac{12.18 - 11.82}{6(0.10/2.059)} = 1.24$$

Because the C_p index is greater than 1, the bottling process has the potential to fill more than 99.73% of the bottles within the specification limits.

In summary, the C_p index is an aggregate measure of process potential. The larger the value of C_p, the more potential the process has of satisfying the customer. In other words, a large C_p indicates that the current amount of common-cause variation is small enough to consistently produce items within specifications. For a process to reach its full potential, the process mean needs to be at or near the center of the specification limits. The following introduces capability indexes that measure actual process performance.

CPL, CPU, and C_{pk}

To measure the capability of a process in terms of actual process performance, the most common indexes are *CPL*, *CPU*, and C_{pk}. Equation (18.10) defines the capability indexes *CPL* and *CPU*.

CPL AND *CPU*

$$CPL = \frac{\bar{\bar{X}} - \text{LSL}}{3(\bar{R}/d_2)} \quad \textbf{(18.10a)}$$

$$CPU = \frac{\text{USL} - \bar{\bar{X}}}{3(\bar{R}/d_2)} \quad \textbf{(18.10b)}$$

Since the process mean is used in the calculation of the *CPL* and *CPU* indices, the value of the index gives a measure of process performance—unlike C_p, which measures only potential. A value of *CPL* (or *CPU*) equal to 1.0 indicates that the process mean is 3 standard deviations away from the lower specification limit (or upper specification limit). For CTQ variables with only an LSL, the *CPL* measures the process performance. For CTQ variables with only a USL, the *CPU* measures the process performance. In either case, the larger the value of the index, the better the capability of the process.

In the "Using Statistics" scenario, the Beachcomber Hotel has a policy that luggage deliveries are to be made in 14 minutes or less. Thus, the CTQ variable delivery time has an upper specification limit of 14, and there is no lower specification limit. Because you previously determined that the luggage delivery process was in control, you can now calculate the *CPU*. From the summary computations on page 774,

$$\bar{\bar{X}} = 9.478 \quad \text{and} \quad \bar{R} = 3.482$$

And, from Table E.13, $d_2 = 2.326$. Then, using Equation (18.10b),

$$CPU = \frac{\text{USL} - \bar{\bar{X}}}{3(\bar{R}/d_2)} = \frac{14 - 9.478}{3(3.482/2.326)} = 1.01$$

The capability index for the luggage delivery CTQ variable is 1.01. Because this value is slightly more than 1, the upper specification limit is slightly more than 3 standard deviations above the mean. To increase *CPU* even farther above 1.00 and therefore increase customer satisfaction, you need to investigate changes in the luggage delivery process. To study a process that has a *CPL* and a *CPU*, see the bottling process in Example 18.2.

EXAMPLE 18.2

CALCULATING *CPL* AND *CPU* FOR THE BOTTLING PROCESS

In the soft-drink bottle filling process described on page 780, the following information was presented:

$$n = 4 \quad \bar{\bar{X}} = 12.02 \quad \bar{R} = 0.10 \quad \text{LSL} = 11.82 \quad \text{USL} = 12.18 \quad d_2 = 2.059$$

Calculate the *CPL* and *CPU* for these data.

SOLUTION You compute the capability indexes using Equations (18.10a) and (18.10b):

$$CPL = \frac{\bar{\bar{X}} - \text{LSL}}{3(\bar{R}/d_2)} = \frac{12.02 - 11.82}{3(0.10/2.059)} = 1.37$$

$$CPU = \frac{\text{USL} - \bar{\bar{X}}}{3(\bar{R}/d_2)} = \frac{12.18 - 12.02}{3(0.10/2.059)} = 1.10$$

Both the *CPL* and *CPU* are greater than 1, indicating that the process mean is more than 3 standard deviations away from both the LSL and USL. Since the *CPU* is less than the *CPL*, you know that the mean is closer to the USL than the LSL.

The capability index C_{pk} measures actual process performance for quality characteristics with two-sided specification limits. C_{pk} is equal to the value of either the *CPL* or *CPU*, whichever is smallest:

C_{pk}

$$C_{pk} = MIN[CPL, CPU] \quad \textbf{(18.11)}$$

A value of 1 for C_{pk} indicates that the process mean is 3 standard deviations away from the closest specification limit. If the characteristic is normally distributed, then a value of 1 indicates that at least 99.73% of the current output is within specifications. Like all capability indices, the larger the value of C_{pk} the better. Example 18.3 illustrates the use of the C_{pk} index.

EXAMPLE 18.3

CALCULATING C_{pk} FOR THE BOTTLING PROCESS

The soft-drink producer in Example 18.2 requires the bottle filling process to have a C_{pk} greater than or equal to 1. Calculate the C_{pk} index.

SOLUTION In Example 18.2, $CPL = 1.37$ and $CPU = 1.10$. Using Equation (18.11):

$$\begin{aligned} C_{pk} &= MIN[CPL, CPU] \\ &= MIN[1.37, 1.10] = 1.10 \end{aligned}$$

The C_{pk} index is greater than 1, indicating that the actual process performance exceeds the company's requirement. More than 99.73% of the bottles contain between 11.82 and 12.18 ounces.

PROBLEMS FOR SECTION 18.8

Learning the Basics

PH Grade ASSIST **18.27** For an in-control process with subgroup data $n = 4$, $\bar{\bar{X}} = 20$ and $\bar{R} = 2$ find the estimate of:

a. the population mean of all X values.
b. the population standard deviation of all X values.

PH Grade ASSIST **18.28** For an in-control process with subgroup data $n = 3$, $\bar{\bar{X}} = 100$ and $\bar{R} = 3.386$ calculate the percentage of outcomes within specifications if:

a. LSL = 98 and USL = 102.
b. LSL = 93 and USL = 107.5.
c. LSL = 93.8 and there is no USL.
d. USL = 110 and there is no LSL.

18.29 For an in-control process with subgroup data $n = 3$, $\bar{\bar{X}} = 100$, and $\bar{R} = 3.386$, calculate the C_p, *CPL*, *CPU*, and C_{pk} if:

a. LSL = 98 and USL = 102.
b. LSL = 93 and USL = 107.5.

Applying the Concepts

SELF Test **18.30** Referring to the data of problem 18.22 **SPWATER** on page 776, the researchers stated, "Some of the benefits of a capable process are increased customer satisfaction, increased operating efficiencies, and reduced costs." To illustrate this point, the authors presented a capability analysis for a springwater bottling operation. One of the critical-to-quality (CTQ) variables is the amount of magnesium, measured in parts per million (ppm), in the water. The LSL and USL for the level of magnesium in a bottle are 18 ppm and 22 ppm, respectively.

a. Estimate the percentage of bottles within specifications.
b. Calculate C_p, *CPL*, *CPU*, and C_{pk}.

18.31 Refer to the data in problem 18.23 on page 776 concerning the tensile strengths of bolts of cloth. There is no upper specification limit for tensile strength, and the lower specification limit is 13. **TENSILE**

a. Estimate the percentage of bolts within specifications.
b. Calculate C_p and *CPL*.

PH Grade ASSIST **18.32** Refer to problem 18.25 on page 777 concerning a filling machine for a tea bag manufacturer. In this problem you should have concluded that the process is in control. The label weight for this product is 5.5 grams, the lower specification limit is 5.2 grams, and the upper specification limit is 5.8 grams. Company policy states that at least 99% of the tea bags produced must be inside the specifications in order for the process to be considered capable. TEA3

a. Estimate the percentage of the tea bags that are inside the specification limits. Is the process capable of meeting the company policy?

b. If management implemented a new policy stating that 99.7% of all tea bags are required to be within specifications, is this process capable of reaching that goal? Explain.

18.33 Refer to problem 18.20 on page 775 concerning waiting time for customers at a bank. (Note: Ignore the fact that in this problem, $\bar{X}$ is out of control). Suppose management has set an upper specification limit of 5 minutes on waiting time and that at least 99% of the waiting times must be less than 5 minutes in order for the process to be considered capable. BANKTIME

a. Estimate the percentage of the waiting times that are inside the specification limits. Is the process capable of meeting the company policy?

b. If management implemented a new policy stating that 99.7% of all waiting times are required to be within specifications, is this process capable of reaching that goal? Explain.

SUMMARY

This chapter has introduced you to quality and productivity including total quality management, Deming's 14 points, and Six Sigma management. You have learned how to use several different types of control charts to distinguish between common causes and special causes of variation. You have learned how to measure the capability of a process by estimating the percentage of items within specifications and by calculating capability indices. By applying these concepts to the services provided by the Beachcomber Hotel, you learned how a manager can identify problems and continually improve service quality.

KEY FORMULAS

Constructing Control Limits

Process mean ±3 standard deviations **(18.1)**

Upper control limit (UCL) =
process mean +3 standard deviations

Lower control limit (LCL) =
process mean −3 standard deviations

Control Limits for the p Chart

$$\bar{p} \pm 3\sqrt{\frac{\bar{p}(1-\bar{p})}{\bar{n}}} \quad \textbf{(18.2)}$$

$$UCL = \bar{p} + 3\sqrt{\frac{\bar{p}(1-\bar{p})}{\bar{n}}}$$

$$LCL = \bar{p} - 3\sqrt{\frac{\bar{p}(1-\bar{p})}{\bar{n}}}$$

Control Limits for the c Chart

$$\bar{c} \pm 3\sqrt{\bar{c}} \quad \textbf{(18.3)}$$

$$UCL = \bar{c} + 3\sqrt{\bar{c}}$$

$$LCL = \bar{c} - 3\sqrt{\bar{c}}$$

Control Limits for the Range

$$\bar{R} \pm 3\bar{R}\frac{d_3}{d_2} \quad \textbf{(18.4)}$$

$$UCL = \bar{R} + 3\bar{R}\frac{d_3}{d_2}$$

$$LCL = \bar{R} - 3\bar{R}\frac{d_3}{d_2}$$

Calculating Control Limits for the Range

$UCL = D_4\bar{R}$ **(18.5a)**

$LCL = D_3\bar{R}$ **(18.5b)**

Control Limits for the $\bar{X}$ Chart

$$\bar{\bar{X}} \pm 3\frac{\bar{R}}{d_2\sqrt{n}} \quad \textbf{(18.6)}$$

$$UCL = \bar{\bar{X}} + 3\frac{\bar{R}}{d_2\sqrt{n}}$$

$$LCL = \bar{\bar{X}} - 3\frac{\bar{R}}{d_2\sqrt{n}}$$

Calculating Control Limits for the Mean Using the A_2 Factor

$$UCL = \bar{\bar{X}} + A_2\bar{R} \quad \textbf{(18.7a)}$$

$$LCL = \bar{\bar{X}} - A_2\bar{R} \quad \textbf{(18.7b)}$$

Estimating the Capability of a Process

For a CTQ variable with a lower specification limit and an upper specification limit:

P(an outcome will be within specifications) **(18.8a)**

$$= P(LSL < X < USL)$$

$$= P\left(\frac{LSL - \bar{\bar{X}}}{\frac{\bar{R}}{d_2}} < Z < \frac{USL - \bar{\bar{X}}}{\frac{\bar{R}}{d_2}}\right)$$

For a CTQ variable with only an upper specification limit:

P(an outcome will be within specifications) **(18.8b)**

$$= P(X < USL)$$

$$= P\left(Z < \frac{USL - \bar{\bar{X}}}{\frac{\bar{R}}{d_2}}\right)$$

For a CTQ variable with only a lower specification limit:

P(an outcome will be within specifications) **(18.8c)**

$$= P(LSL < X)$$

$$= P\left(\frac{LSL - \bar{\bar{X}}}{\frac{\bar{R}}{d_2}} < Z\right)$$

The C_p Index

$$C_p = \frac{USL - LSL}{6(\bar{R}/d_2)} \quad \textbf{(18.9)}$$

$$= \frac{\text{specification spread}}{\text{process spread}}$$

CPL* and *CPU

$$CPL = \frac{\bar{\bar{X}} - LSL}{3(\bar{R}/d_2)} \quad \textbf{(18.10a)}$$

$$CPU = \frac{USL - \bar{\bar{X}}}{3(\bar{R}/d_2)} \quad \textbf{(18.10b)}$$

C_{pk}

$$C_{pk} = MIN[CPL, CPU] \quad \textbf{(18.11)}$$

KEY TERMS

CHAPTER REVIEW PROBLEMS

Checking Your Understanding

18.34 What is the difference between common-cause variation and special-cause variation?

18.35 What should you do to improve a process when special causes of variation are present?

18.36 What should you do to improve a process when only common causes of variation are present?

18.37 Under what circumstances do you use a *p* chart?

18.38 What is the difference between attribute control charts and variables control charts?

18.39 Why are the $\bar{X}$ and R charts used together?

18.40 What principles did you learn from the red bead experiment?

18.41 How do you decide if you should use a p chart or a c chart?

18.42 What is the difference between process potential and process performance?

18.43 A company requires a C_{pk} value of 1 or larger. If a process has a $C_p = 1.5$ and a $C_{pk} = 0.8$, what changes should you make to the process?

18.44 Why is a capability analysis *not* performed on out-of-control processes?

Applying the Concepts

You should use Microsoft Excel, Minitab, or SPSS to solve problems 18.45–18.53.

18.45 A producer of cat food constructed control charts and analyzed several quality characteristics. One characteristic of interest is the weight of the filled cans. The lower specification limit for weight is 2.95 pounds. The data file CATFOOD contains the weights of five cans tested every fifteen minutes during a day's production.

a. Construct a control chart for the range.
b. Construct a control chart for the mean.
c. Is the process in control?
d. If the process is in control, estimate the percentage of the cans whose weight is inside the specification limits.
e. If the process is in control, calculate *CPL*.
f. If the manufacturer requires that 99.7% of all cans be within the specification limits, comment on the capability of the process based on your calculations in (d) and (e).

18.46 Researchers at Miami University in Oxford, Ohio, investigated the use of p charts to monitor the market share of a product and to document the effectiveness of marketing promotions. Market share is defined as the company's proportion of the total number of products sold in a category. If a p chart based on a company's market share indicates an in-control process, then their share in the marketplace is deemed to be stable and consistent over time. In the example given in the article, the RudyBird Diskette Company collected daily sales data from a nationwide retail audit service. The first 30 days of data in the accompanying table indicate the total number of cases of computer diskettes sold and the number of RudyBird diskettes sold. The final 7 days of data were taken after RudyBird launched a major in-store promotion. A control chart was used to see if the in-store promotion would result in special-cause variation in the marketplace.
RUDYBIRD

Cases Sold Before the Promotion

Day	Total	Rudybird	Day	Total	Rudybird
1	154	35	16	177	56
2	153	43	17	143	43
3	200	44	18	200	69
4	197	56	19	134	38
5	194	54	20	192	47
6	172	38	21	155	45
7	190	43	22	135	36
8	209	62	23	189	55
9	173	53	24	184	44
10	171	39	25	170	47
11	173	44	26	178	48
12	168	37	27	167	42
13	184	45	28	204	71
14	211	58	29	183	64
15	179	35	30	169	43

Cases Sold After the Promotion

Day	Total	Rudybird
31	201	92
32	177	76
33	205	85
34	199	90
35	187	77
36	168	79
37	198	97

Source: Extracted from Charles T. Crespy, Timothy C. Krehbiel, and James M. Stearns, "Integrating Analytic Methods into Marketing Research Education: Statistical Control Charts as an Example," Marketing Education Review, *5, Spring 1995, 11–23.*

a. Construct a p chart using data from the first 30 days (prior to the promotion) to monitor the market share for RudyBird Diskettes.
b. Is the market share for RudyBird in control before the start of the in-store promotion?
c. On your control chart, extend the control limits generated in (b) and plot the proportions for days 31 through 37. What effect, if any, did the in-store promotion have on RudyBird's market share?

18.47 The manufacturer of "Boston" and "Vermont" asphalt shingles constructed control charts and analyzed several quality characteristics. One characteristic of interest is the strength of the sealant on the shingle. During each day of production, three shingles are tested for their sealant strength. (Thus, a subgroup is operationally defined as one day of production, and the sample size for each subgroup is 3.) Separate pieces are cut from the upper and lower portions of a shingle, and then reassembled to simulate shingles on a roof. A timed heating process is used to simulate the sealing process. The sealed shingle pieces are pulled apart, and the amount of force (in pounds) required to break the sealant bond is measured and recorded. This variable is called the *sealant strength*. The lower and upper specification limits for sealant strength are 1.0 and 1.5 pounds, respectively. The data file **SEALANT** contains sealant strength measurements on 25 days of production for "Boston" shingles and 19 days for "Vermont" shingles.

For the "Boston" shingles:

a. Construct a control chart for the range.
b. Construct a control chart for the mean.
c. Is the process in control?
d. If the process is in control, estimate the percentage of the shingles whose sealant strength is inside the specification limits.
e. If the process is in control, calculate C_p, CPL, CPU, and C_{pk}.
f. If the manufacturer requires that 99.7% of all shingles be within the specification limits, comment on the capability of the process based on your calculations in (d) and (e).
g. Repeat (a) through (f) using the 19 production days for "Vermont" shingles.

PH Grade ASSIST **18.48** A professional basketball player has embarked on a program to study his ability to shoot foul shots. On each day in which a game is not scheduled, he intends to shoot 100 foul shots. He maintains records over a period of 40 days of practice, with the following results: **FOULSPC**

Day	Foul Shots Made	Day	Foul Shots Made	Day	Foul Shots Made
1	73	15	73	29	76
2	75	16	76	30	80
3	69	17	69	31	78
4	72	18	68	32	83
5	77	19	72	33	84
6	71	20	70	34	81
7	68	21	64	35	86
8	70	22	67	36	85
9	67	23	72	37	86
10	74	24	70	38	87
11	75	25	74	39	85
12	72	26	76	40	85
13	70	27	75		
14	74	28	78		

a. Construct a p chart for the proportion of successful foul shots. Do you think that the player's foul-shooting process is in statistical control? If not, why not?
b. What if you were told that the player used a different method of shooting foul shots for the last twenty days? How might this information change your conclusions in (a)?
c. If you knew the information in (b) prior to doing (a), how might you do the analysis differently?

PH Grade ASSIST **18.49** The funds-transfer department of a bank is concerned with turnaround time for investigations of funds-transfer payments. A payment may involve the bank as a remitter of funds, a beneficiary of funds, or an intermediary in the payment. An investigation is initiated by a payment inquiry or query by a party involved in the payment or any department affected by the flow of funds. When a query is received, an investigator reconstructs the transaction trail of the payment and verifies that the information is correct and the proper payment is transmitted. The investigator then reports the results of the investigation and the transaction is considered closed. It is important that investigations are closed rapidly, preferably within the same day. The number of new investigations and the number and proportion closed on the same day that the inquiry was made are in the file **FUNDTRAN**.

a. Construct a control chart for these data.
b. Is the process in a state of statistical control? Explain.
c. Based on the results of (a) and (b), what should management do next to improve the process?

PH Grade ASSIST **18.50** A branch manager of a brokerage company is concerned with the number of undesirable trades made by her sales staff. A trade is considered undesirable if there is an error on the trade ticket. Trades with errors are canceled and resubmitted. The cost of correcting errors is billed to the brokerage company. The branch manager wants to know whether the proportion of undesirable trades is in a state of statistical control so she can plan the next step in a quality improvement process. Data were collected for a 30-day period with the following results: **TRADE**

Day	Undesirable Trades	Total Trades	Day	Undesirable Trades	Total Trades
1	2	74	16	3	54
2	12	85	17	12	74
3	13	114	18	11	103
4	33	136	19	11	100
5	5	97	20	14	88
6	20	115	21	4	58
7	17	108	22	10	69
8	10	76	23	19	135
9	8	69	24	1	67
10	18	98	25	11	77
11	3	104	26	12	88
12	12	98	27	4	66
13	15	105	28	11	72
14	6	98	29	13	118
15	21	204	30	15	138

a. Construct a control chart for these data.
b. Is the process in control? Explain.
c. Based on the results of (a) and (b), what should the manager do next to improve the process?

18.51 As chief operating officer of a local community hospital, you have just returned from a 3-day seminar on quality and productivity. It is your intention to implement many of the ideas that you learned at the seminar. You have decided to maintain control charts for the upcoming month for the following variables: number of daily admissions, proportion of rework in the laboratory (based on 1,000 daily samples), and time (in hours) between receipt of a specimen at the laboratory and completion of the work (based on a subgroup of 10 specimens per day). The data collected are summarized in the file **HOSPADM**. You are to make a presentation to the chief executive officer of the hospital and the board of directors. Prepare a report that summarizes the conclusions drawn from analyzing control charts for these variables. In addition, recommend additional variables to measure and monitor using control charts.

18.52 On each morning for a period of 4 weeks, record your pulse rate (in beats per minute) just after you get out of bed and also before you go to sleep at night. Set up $\bar{X}$ and R charts and determine whether your pulse rate is in a state of statistical control. Explain.

18.53 (**Class Project**) Use the table of random numbers (Table E.1) to simulate the selection of different colored balls from an urn as follows:

1. Start in the row corresponding to the day of the month you were born plus the year in which you were born. For example, if you were born October 15, 1976, you would start in row 15 + 76 = 91. If your total exceeds 100, subtract 100 from the total.
2. Select two-digit random numbers.
3. If you select a random number from 00 to 94, consider the ball to be white; if the random number is from 95 to 99, consider the ball to be red.

Each student is to select 100 such two-digit random numbers and report the number of "red balls" in the sample. Construct a control chart for the proportion of red balls. What conclusions can you draw about the system of selecting red balls? Are all the students part of the system? Is anyone outside the system? If so, what explanation can you give for someone who has too many red balls? If a bonus were paid to the top 10% of the students (the 10% with the fewest red balls), what effect would that have on the rest of the students? Discuss.

CASE STUDY
THE HARNSWELL SEWING MACHINE COMPANY CASE

Phase 1

For almost 50 years, the Harnswell Sewing Machine Company has manufactured industrial sewing machines. The company specializes in automated machines called pattern tackers that sew repetitive patterns on such mass-produced products as shoes, garments, and seat belts. Aside from the sales of machines, the company sells machine parts. Because the company's products have a reputation for being superior, Harnswell is able to command a price premium for its product line.

Recently, the production manager, Natalie York, purchased several books relating to quality at a local bookstore. After reading them, she considered the feasibility of beginning some type of quality program at the company. At the current time, the company has no formal quality program. Parts are 100% inspected at the time of shipping to a customer or installation in a machine, yet Natalie has always wondered why inventory of certain parts (in particular the half-inch cam roller) invariably falls short before a full year lapses, even though 7,000 pieces have been produced for a demand of 5,000 pieces per year.

After a great deal of reflection and with some apprehension, Natalie has decided that she will approach John Harnswell, the owner of the company, about the possibility of beginning a program to improve quality in the company, starting with a trial project in the machine parts area. As she is walking to Mr. Harnswell's office for the meeting, she has second thoughts about whether this is such a good idea. After all, just last month Mr. Harnswell told her, "Why do you need to go to graduate school for your master's degree in business? That is a waste of your time and will not be of any value to the Harnswell Company. All those professors are just up in their ivory towers and don't know a thing about running a business like I do."

As she enters his office, Mr. Harnswell, ever courteous to her, invites Natalie to sit down across from him. "Well, what do you have on your mind this morning?" Mr. Harnswell asks her in an inquisitive tone. She begins by starting to talk about the books that she has just completed reading and about how she has some interesting ideas for making production even better than it is now and improving profits. Before she can finish, Mr. Harnswell has started to answer. "Look, my dear young lady," he says, "everything has been fine since I started this company in 1955. I have built this company up from nothing to one that employs more than 100 people. Why do you want to make waves? Remember, if it ain't broke, don't fix it." With that he ushers her from his office with the admonishment of, "What am I going to do with you if you keep coming up with these ridiculous ideas?"

EXERCISES

HS.1 Based upon what you have read, which of Deming's 14 points of management are most lacking in the Harnswell Sewing Machine Company? Explain.

HS.2 What changes if any, do you think that Natalie York might be able to institute in the company? Explain.

DO NOT CONTINUE UNTIL YOU HAVE COMPLETED THE PHASE 1 EXERCISES

Phase 2

Natalie slowly walks down the hall after leaving Mr. Harnswell's office, feeling rather downcast. He just won't listen to anyone, she thinks. As she walks, Jim Murante, the shop foreman, comes up beside her. "So," he says, "did you really think that the old man would just listen to you? I've been here more than 25 years. The only way he listens is if he is shown something that worked after it has already been done. Let's see what we can plan out together."

Natalie and Jim decide to begin by investigating the production of the cam rollers that are a precision ground part. The last part of the production process involves the grinding of the outer diameter. After grinding, the part mates with the cam groove of the particular sewing pattern. The half-inch rollers technically have an engineering specification for the outer diameter of the roller of 0.5075 inch (the specifications are actually metric, but in factory floor jargon they are referred to as half-inch), plus a tolerable error of 0.0003 inch on the lower side. Thus, the outer diameter is allowed to be between 0.5072 and 0.5075 inch. Anything larger is reclassified into a different and less costly category, and anything smaller is unusable for anything other than scrap.

The grinding of the cam roller is done on a single machine with a single tool setup and no change in the grinding wheel after initial setup. The operation is done by Dave Martin, the head machinist, who has 30 years of experience in the trade and specific experience producing the cam roller part. Since production occurs in batches, Natalie and Jim sample five parts produced from each batch. Table HS.1 presents data collected over 30 batches. HARNSWELL

TABLE HS.1

Diameter of Cam Rollers (in inches)

	Cam Roller				
Batch	**1**	**2**	**3**	**4**	**5**
1	.5076	.5076	.5075	.5077	.5075
2	.5075	.5077	.5076	.5076	.5075
3	.5075	.5075	.5075	.5075	.5076
4	.5075	.5076	.5074	.5076	.5073
5	.5075	.5074	.5076	.5073	.5076
6	.5076	.5075	.5076	.5075	.5075
7	.5076	.5076	.5076	.5075	.5075
8	.5075	.5076	.5076	.5075	.5074
9	.5074	.5076	.5075	.5075	.5076
10	.5076	.5077	.5075	.5075	.5075
11	.5075	.5075	.5075	.5076	.5075
12	.5075	.5076	.5075	.5077	.5075
13	.5076	.5076	.5073	.5076	.5074
14	.5075	.5076	.5074	.5076	.5075
15	.5075	.5075	.5076	.5074	.5073
16	.5075	.5074	.5076	.5075	.5075
17	.5075	.5074	.5075	.5074	.5072
18	.5075	.5075	.5076	.5075	.5076
19	.5076	.5076	.5075	.5075	.5076
20	.5075	.5074	.5077	.5076	.5074
21	.5075	.5074	.5075	.5075	.5075
22	.5076	.5076	.5075	.5076	.5074
23	.5076	.5076	.5075	.5075	.5076
24	.5075	.5076	.5075	.5076	.5075
25	.5075	.5075	.5075	.5075	.5074
26	.5077	.5076	.5076	.5074	.5075
27	.5075	.5075	.5074	.5076	.5075
28	.5077	.5076	.5075	.5075	.5076
29	.5075	.5075	.5074	.5075	.5075
30	.5076	.5075	.5075	.5076	.5075

EXERCISE

HS.3 **a.** Is the process in control? Why?

b. What recommendations do you have for improving the process?

DO NOT CONTINUE UNTIL YOU HAVE COMPLETED THE PHASE 2 EXERCISE

Phase 3

Natalie examined the $\bar{X}$ and R charts developed from the data presented in Table HS.1. The R chart indicated that the process is in control, but the $\bar{X}$ chart revealed that the mean for batch 17 was outside the lower control limit. This immediately gave her cause for concern because low values for the roller diameter could mean that parts had to be scrapped. Natalie went down to see Jim Murante, the shop foreman, to try to find out what had happened on batch 17. Jim looked up the production records to determine when this batch was produced. "Aha," He exclaims, "I think I've got the answer! This batch was produced on that really cold morning we had last month. I've been after Mr. Harnswell for a long time to let us install an automatic thermostat here in the shop so that the place doesn't feel so cold when we get here in the morning. All he ever tells me is that people aren't as tough as they used to be and if I want to see real cold, I should have been back in that foxhole during the Korean winter of 1952."

Natalie stood there almost in shock. What she realized had happened is that, rather than standing idle until the environment and the equipment warmed to acceptable temperatures, the machinist opted to manufacture parts that might have to be scrapped. In fact, Natalie recalled that a major problem had occurred on that same day when several other expensive parts had to be scrapped. Natalie said to Jim, "We just have to do something. We can't let this go on now that we know what problems it is potentially causing." Natalie and Jim decided to take enough money out of petty cash to get the thermostat without having to fill out a requisition requiring Mr. Harnswell's signature. They installed the thermostat and set the heating control so that the heat would turn on a half hour before the shop opened each morning.

EXERCISES

HS.4 What should Natalie now do concerning the cam roller data? Explain.

HS.5 Explain how the actions of Natalie and Jim to avoid this particular problem in the future has resulted in quality improvement.

DO NOT CONTINUE UNTIL YOU HAVE COMPLETED THE PHASE 3 EXERCISES

Phase 4

Because corrective action was taken to eliminate the special cause of variation, the data for batch 17 were removed from the analysis. The control charts for the remaining days indicate a stable system with only common causes of variation operating on the system. Thus, Natalie and Jim sat down with Dave Martin and several other machinists to try to determine all the possible causes for the existence of oversized and scrapped rollers. Natalie was still troubled by the data. After all, she wanted to find out whether the process is giving oversizes (which are downgraded) and undersizes (which are scrapped). She thought about which tables and charts would be most helpful.

EXERCISE

HS.6 **a.** Construct a frequency distribution and a stem-and-leaf display of the cam roller diameters. Which one do you prefer?

b. Based on your results in (a), construct all appropriate graphs of the cam roller diameters.

c. Write a report expressing your conclusions concerning the cam roller diameters. Be sure to discuss the diameters as they relate to the specifications.

DO NOT CONTINUE UNTIL YOU HAVE COMPLETED THE PHASE 4 EXERCISE

Phase 5

Natalie noticed immediately that the overall mean diameter with batch 17 eliminated is 0.507527, which is higher than the specification value. Thus, the mean diameter of the rollers produced is so high that they would be downgraded in value. In fact, 55 of the 150 rollers sampled (36.67%) were above the specification value. If this percentage is extrapolated to the full year's production, 36.67% of the 7,000 pieces manufactured, or 2,567, could not be sold as half-inch rollers, leaving only 4,433 available for sale. "No wonder we often have shortages that require costly emergency runs," she thought. She also notes that not one diameter is below the lower specification of 0.5072, so not one of the rollers had to be scrapped.

Natalie realized that there had to be a reason for all this. Along with Jim Murante, she decided to show the results to Dave Martin, the head machinist. Dave said that the results didn't surprise him that much. "You know," he says "there is only 0.0003 inch in diameter that I'm allowed in variation. If I aim for exactly halfway between 0.5072 and 0.5075, I'm afraid that I'll make a lot of short pieces that will have to be scrapped. I know from way back when I first started here that Mr. Harnswell and everybody else will come down on my head if they start seeing too many of those scraps. I figure that if I aim at 0.5075, the worst thing that will happen will be a bunch of downgrades, but I won't make any pieces that have to be scrapped."

EXERCISES

HS.7 What approach do you think the machinist should take in terms of the diameter he should aim for? Explain.

HS.8 What do you think that Natalie should do next? Explain.

RUNNING CASE MANAGING THE *SPRINGVILLE HERALD*

Phase 1

An advertising production team is charged with reducing the number and dollar amount of the advertising errors, with initial focus on the ran-in-error category. The team collected data including the number of ads with errors on a Monday to Saturday basis. Table SH18.1 includes the total number of ads and the number containing errors for a period of one month. (Sundays are excluded because a special type of production is used for that day.) SH18-1

TABLE SH18.1

Number of Ads with Errors and Daily Number of Display Ads

Day	Number of Ads with Errors	Number of Ads	Day	Number of Ads with Errors	Number of Ads
1	4	228	14	5	245
2	6	273	15	7	266
3	5	239	16	2	197
4	3	197	17	4	228
5	6	259	18	5	236
6	7	203	19	4	208
7	8	289	20	3	214
8	14	241	21	8	258
9	9	263	22	10	267
10	5	199	23	4	217
11	6	275	24	9	277
12	4	212	25	7	258
13	3	207			

EXERCISES

SH18.1 What is the first thing that the team from the advertising production department should do to reduce the number of errors? Explain.

SH18.2 **a.** Construct the appropriate control chart for these data.
b. Is the process in a state of statistical control? Explain.
c. What should the team recommend as the next step to improve the process?

DO NOT CONTINUE UNTIL YOU HAVE COMPLETED THE PHASE 1 EXERCISES

Phase 2

The advertising production team examined the p chart developed from the data of Table SH18.1. Using the rules for determining out-of-control points, they observed that day 8 is above the upper control limit. Upon investigation, it was determined that on that day there was an employee from another work area assigned to the processing of the ads because several employees were out ill. The group brainstormed ways of avoiding the problem in the future and recommended that a team of people from other work areas receive training on the work done by this area. Members of this team could then cover the processing of the ads by rotating in one- or two-hour shifts.

EXERCISES

SH18.3 What should the advertising production team now do concerning the data of Table SH18.1? Explain.

SH18.4 Explain how the actions of the team to avoid this particular problem in the future has resulted in quality improvement.

SH18.5 In addition to the number of ads with errors, what other information concerning errors on a daily basis should the team collect?

DO NOT CONTINUE UNTIL YOU HAVE COMPLETED THE PHASE 2 EXERCISES

Phase 3

A print production team also is charged with improving the quality of the *Herald*. The team has chosen as its first project the *blackness* of the print of the newspaper. Each day the print production team determines how "black" the newspaper is printed. Blackness is measured on a densimometer that records the results on a standard scale. Five spots on the first newspaper printed each day are randomly selected and the blackness of each spot is measured. Table SH18.2 presents the results for 25 days. **SH18-2**

TABLE SH18.2

Newsprint Blackness for 25 Consecutive Days

	Spot				
Day	**1**	**2**	**3**	**4**	**5**
1	0.96	1.01	1.12	1.07	0.97
2	1.06	1.00	1.02	1.16	0.96
3	1.00	0.90	0.98	1.18	0.96
4	0.92	0.89	1.01	1.16	0.90
5	1.02	1.16	1.03	0.89	1.00
6	0.88	0.92	1.03	1.16	0.91
7	1.05	1.13	1.01	0.93	1.03
8	0.95	0.86	1.14	0.90	0.95
9	0.99	0.89	1.00	1.15	0.92
10	0.89	1.18	1.03	0.96	1.04
11	0.97	1.13	0.95	0.86	1.06
12	1.00	0.87	1.02	0.98	1.13
13	0.96	0.79	1.17	0.97	0.95
14	1.03	0.89	1.03	1.12	1.03
15	0.96	1.12	0.95	0.88	0.99
16	1.01	0.87	0.99	1.04	1.16
17	0.98	0.85	0.99	1.04	1.16
18	1.03	0.82	1.21	0.98	1.08
19	1.02	0.84	1.15	0.94	1.08
20	0.90	1.02	1.10	1.04	1.08
21	0.96	1.05	1.01	0.93	1.01
22	0.89	1.04	0.97	0.99	0.95
23	0.96	1.00	0.97	1.04	0.95
24	1.01	0.98	1.04	1.01	0.92
25	1.01	1.00	0.92	0.90	1.11

EXERCISE

SH18.6 **a.** Construct the appropriate control charts for these data.
b. Is the process in a state of statistical control? Explain.
c. What should the team recommend as the next step to improve the process?

REFERENCES

1. Arndt, M., "Quality Isn't Just for Widgets," *Business Week*, July 22, 2002, 72–73.
2. Bothe, D. R., *Measuring Process Capability* (New York: McGraw-Hill, 1997).
3. Deming, W. E., *Out of the Crisis* (Cambridge, MA: MIT Center for Advanced Engineering Study, 1986).
4. Deming, W. E., *The New Economics for Business, Industry, and Government* (Cambridge, MA: MIT Center for Advanced Engineering Study, 1993).
5. Friedman, T. L., *The Lexus and the Olive Tree: Understanding Globalization* (New York: Farrar, Straus and Giroux, 1999).
6. Gabor, A., *The Man Who Discovered Quality* (New York: Time Books, 1990).
7. Gitlow, H., A. Oppenheim, R. Oppenheim, and D. Levine, *Tools and Methods for the Improvement of Quality*, 3rd ed. (Homewood, IL: Irwin, 2005).
8. Gitlow, H. and D. Levine, *Six Sigma for Green Belts and Champions* (Upper Saddle River, NJ: Financial Times-Prentice-Hall, 2005).
9. Hahn, G. J., N. Doganaksoy, and R. Hoerl, "The Evolution of Six Sigma," *Quality Engineering*, 12 (2000): 317–326.
10. Halberstam, D., *The Reckoning* (New York: Morrow, 1986).
11. Levine, D. M., P. P. Ramsey, and R. K. Smidt, *Applied Statistics for Engineers and Scientists Using Microsoft Excel and Minitab* (Upper Saddle River, NJ: Prentice Hall, 2001).
12. *Microsoft Excel* 2003 (Redmond, WA: Microsoft Corp., 2002).
13. *Minitab for Windows Version 14* (State College, PA: Minitab, Inc., 2004).
14. Scherkenbach, W. W., *The Deming Route to Quality and Productivity: Road Maps and Roadblocks* (Washington, DC: CEEP Press, 1987).
15. Snee, R. D., "Impact of Six Sigma on Quality," *Quality Engineering*, 12 (2000): ix–xiv.
16. Walton, M., *The Deming Management Method* (New York: Perigee Books, 1986).

Appendix 18 Using Software for Control Charts

A18.1 MICROSOFT EXCEL

For *p* Chart

See section G.41 (**p Chart**) if you want PHStat2 to produce a *p* chart as a Microsoft Excel chart. (There are no Microsoft Excel commands that directly produce this chart.)

For *R* and $\overline{X}$ Chart

See section G.42 (**R & XBar Chart**) if you want PHStat2 to produce an *R* or $\overline{X}$ chart as a Microsoft Excel chart. (There are no Microsoft Excel commands that directly produce these charts.)

A18.2 MINITAB

Using Minitab for the *p* Chart

To illustrate how to construct a *p* chart, refer to the data of Table 18.1 on page 759 concerning the number of rooms not ready. Open the **HOTEL1.MTW** worksheet.

1. Select **Stat → Control Charts → Attribute Charts → P**. In the P Chart dialog box (see Figure A18.1) enter **C3** or **'Not ready'** in the Variables: edit box. Since the subgroup sizes are equal, enter **200** in the Subgroup sizes: edit box. Click the **P Chart Options** button.

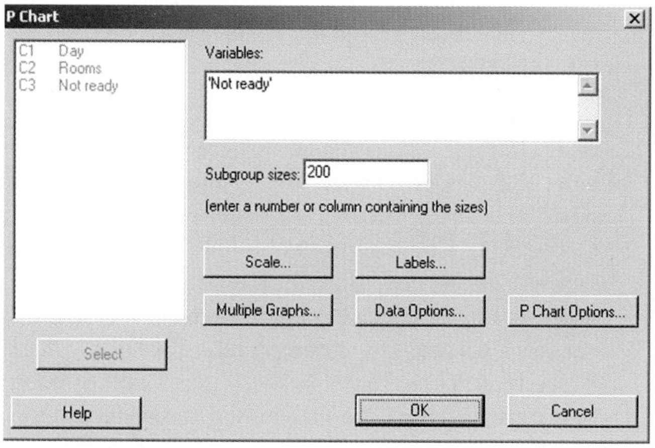

FIGURE A18.1 Minitab P Chart Dialog Box

2. In the P Chart - Options dialog box, click the **Tests** tab (see Figure A18.2). In the drop-down list box, select **Perform all tests for special causes**. Click the **OK** button to return to the P Chart dialog box. (These values will stay intact until Minitab is restarted.) Click the **OK** button to get the *p* chart.

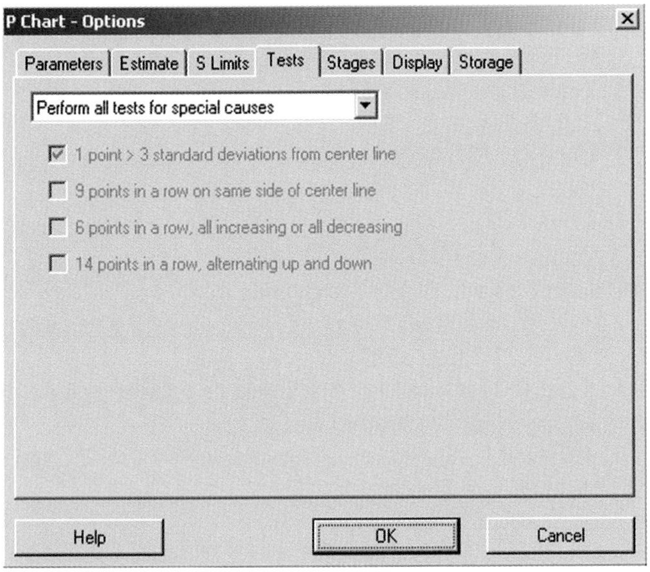

FIGURE A18.2 Minitab *p* Chart - Options Dialog Box, Tests Tab

3. If there are points you want to omit when estimating the center line and control limits, click the **Estimate** tab in the P Chart - Options dialog box. Enter the points to omit in the edit box shown. Click the **OK** button to return to the P Chart dialog box. In the P Chart dialog box, click the **OK** button to get the *p* chart.

Using Minitab for the *c* Chart

To illustrate how to construct a *c* chart, refer to the data of Table 18.4 on page 767 concerning the number of complaints at the hotel. Open the **COMPLAINTS.MTW** worksheet.

1. Select **Stat → Control Charts → Attribute Charts → C**. In the C Chart dialog box (see Figure A18.3), enter **C2** or **COMPLAINTS** in the Variables: edit box.
2. Click the **C Chart Options** button. In the C Chart-Options dialog box, click the **Tests** tab. In the drop-down list box, select **Perform all tests for special causes**. Click the **OK** button to return to the C Chart dialog box. (These values will stay intact until Minitab is restarted.) Click the **OK** button to produce the *c* chart.

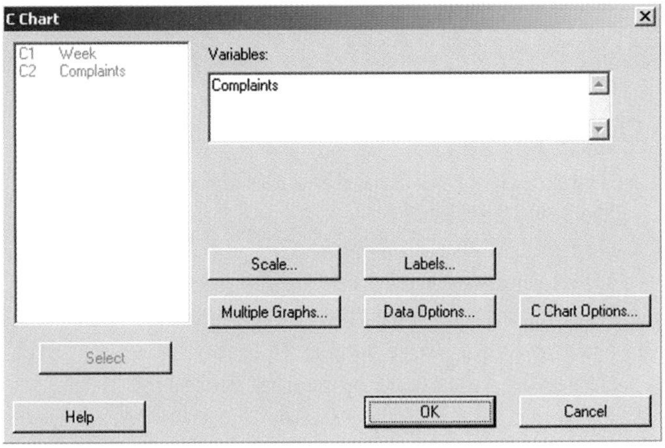

FIGURE A18.3 Minitab *c* Chart Dialog Box

3. If there are points you want to omit when estimating the center line and control limits, click the **Estimate** tab in the C Chart - Options dialog box. Enter the points to omit in the edit box shown. Click the **OK** button to return to the C Chart dialog box. Click the **OK** button to get the *c* Chart.

Using Minitab for the R and $\bar{X}$ Charts

To construct *R* and $\bar{X}$ charts using Minitab, select **Stat → Control Charts → Variable Charts for Subgroups → Xbar-R** from the menu bar. The format for entering the variable name is different, depending on whether the data are stacked down a single column or unstacked across a set of columns with the data for each time period located in a single row. If the data for the variable of interest are stacked down a single column, choose **All observations for a chart are in one column** in the drop-down list box and enter the variable name in the edit box below. If the subgroups are unstacked with each row representing the data for a single time period, choose **Observations for a subgroup are in one row of columns** in the drop-down list box and enter the variable names for the data in the edit box below.

To illustrate how to construct *R* and $\bar{X}$ charts, refer to the data of Table 18.5 on page 772 concerning the luggage delivery times. Open the **HOTEL2.MTW** worksheet.

1. Select **Stat → Control Charts → Variable Charts for Subgroups → Xbar-R**. Since the data are unstacked, select **Observations for a subgroup are in one row of columns** in the drop-down list box. In the Xbar-R Chart dialog box (see Figure A18.4) enter **C2** or **'Time 1,' C3** or **'Time 2,' C4** or **'Time 3,' C5** or **'Time 4,'** and **C6** or **'Time 5'** in the edit box. Click the **Xbar-R Options** button.

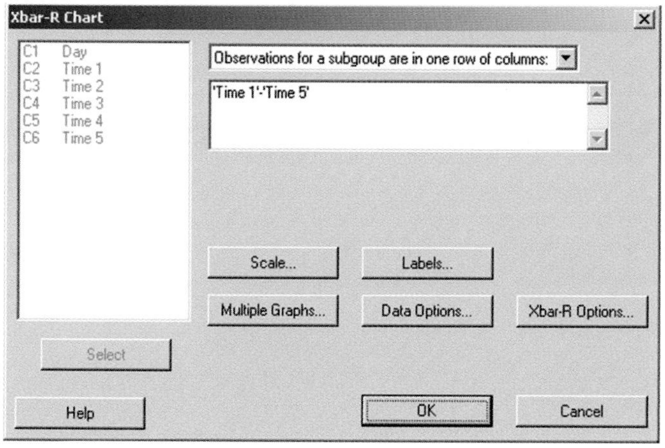

FIGURE A18.4 Minitab Xbar-R Chart Dialog Box

2. In the Xbar-R Chart - Options dialog box, click the **Tests** tab. In the drop-down list box, select **Perform all tests for special causes**. (These values will stay intact until Minitab is restarted.)
3. Click the **Estimate** tab in the Xbar-R Chart Options dialog box (see Figure A18.5). Click the **Rbar** option button. If there are points you want to omit when estimating the center line and control limits, enter the points to omit in the edit box shown. Click the **OK** button to return to the Xbar-R Chart dialog box. (Note: When creating more than one set of R and $\overline{X}$ charts in the same session, reset the points to omit before creating new charts. These values will stay intact until Minitab is restarted.)
4. In the Xbar-R Chart dialog box, click the **OK** button to produce the R and $\overline{X}$ charts.

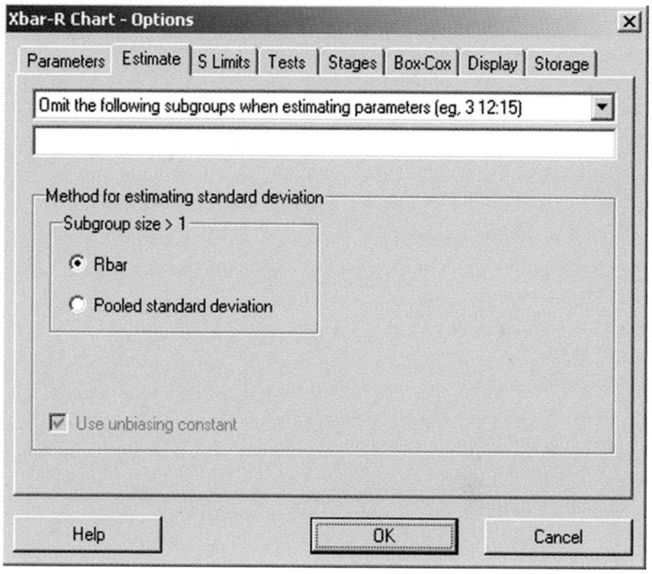

FIGURE A18.5 Minitab Xbar-R Chart Options Dialog Box, Estimate Tab

APPENDICES

A. REVIEW OF ARITHMETIC, ALGEBRA, AND LOGARITHMS

A.1 RULES FOR ARITHMETIC OPERATIONS

RULE	EXAMPLE
1. $a + b = c$ and $b + a = c$	$2 + 1 = 3$ and $1 + 2 = 3$
2. $a + (b + c) = (a + b) + c$	$5 + (7 + 4) = (5 + 7) + 4 = 16$
3. $a - b = c$ but $b - a \neq c$	$9 - 7 = 2$ but $7 - 9 = -2$
4. $a \times b = b \times a$	$7 \times 6 = 6 \times 7 = 42$
5. $a \times (b + c) = (a \times b) + (a \times c)$	$2 \times (3 + 5) = (2 \times 3) + (2 \times 5) = 16$
6. $a \div b \neq b \div a$	$12 \div 3 \neq 3 \div 12$
7. $\frac{a+b}{c} = \frac{a}{c} + \frac{b}{c}$	$\frac{7+3}{2} = \frac{7}{2} + \frac{3}{2} = 5$
8. $\frac{a}{b+c} \neq \frac{a}{b} + \frac{a}{c}$	$\frac{3}{4+5} \neq \frac{3}{4} + \frac{3}{5}$
9. $\frac{1}{a} + \frac{1}{b} = \frac{b+a}{ab}$	$\frac{1}{3} + \frac{1}{5} = \frac{5+3}{(3)(5)} = \frac{8}{15}$
10. $\frac{a}{b} \times \frac{c}{d} = \frac{a \times c}{b \times d}$	$\frac{2}{3} \times \frac{6}{7} = \frac{2 \times 6}{3 \times 7} = \frac{12}{21}$
11. $\frac{a}{b} \div \frac{c}{d} = \frac{a \times d}{b \times c}$	$\frac{5}{8} \div \frac{3}{7} = \frac{5 \times 7}{8 \times 3} = \frac{35}{24}$

A.2 RULES FOR ALGEBRA: EXPONENTS AND SQUARE ROOTS

RULE	EXAMPLE
1. $X^a \cdot X^b = X^{a+b}$	$4^2 \cdot 4^3 = 4^5$
2. $(X^a)^b = X^{ab}$	$(2^2)^3 = 2^6$
3. $(X^a/X^b) = X^{a-b}$	$\frac{3^5}{3^3} = 3^2$
4. $\frac{X^a}{X^a} = X^0 = 1$	$\frac{3^4}{3^4} = 3^0 = 1$
5. $\sqrt{XY} = \sqrt{X}\sqrt{Y}$	$\sqrt{(25)(4)} = \sqrt{25}\sqrt{4} = 10$
6. $\sqrt{\frac{X}{Y}} = \frac{\sqrt{X}}{\sqrt{Y}}$	$\sqrt{\frac{16}{100}} = \frac{\sqrt{16}}{\sqrt{100}} = 0.40$

A.3 RULES FOR LOGARITHMS

Base-10

LOG is the symbol used for base-10 logarithms:

RULE	EXAMPLE
1. $\text{LOG}(10^A) = A$	$\text{LOG}(100) = \text{LOG}(10^2) = 2$
2. If $\text{LOG}(A) = B$, then $A = 10^B$	If $\text{LOG}(A) = 2$, then $A = 10^2 = 100$
3. $\text{LOG}(A \times B) = \text{LOG}(A) + \text{LOG}(B)$	$\text{LOG}(100) = \text{LOG}(10 \times 10)$ $= \text{LOG}(10) + \text{LOG}(10) = 1 + 1 = 2$
4. $\text{LOG}(A^B) = \text{B} \times \text{LOG}(A)$	$\text{LOG}(1000) = \text{LOG}(10^3) = 3 \times \text{LOG}(10)$ $= 3 \times 1 = 3$
5. $\text{LOG}(A/B) = \text{LOG}(A) - \text{LOG}(B)$	$\text{LOG}(100) = \text{LOG}(1000/10)$ $= \text{LOG}(1000) - \text{LOG}(10) = 3 - 1 = 2$

EXAMPLE

Take the base-10 logarithm of each side of the following equation:

$$Y = \beta_0\beta_1^X\varepsilon$$

SOLUTION Apply rules 3 and 4:

$$\begin{aligned}\text{LOG}(Y) &= \text{LOG}(\beta_0\beta_1^X\varepsilon)\\ &= \text{LOG}(\beta_0) + \text{LOG}(\beta_1^X) + \text{LOG}(\varepsilon)\\ &= \text{LOG}(\beta_0) + X \times \text{LOG}(\beta_1) + \text{LOG}(\varepsilon)\end{aligned}$$

Base-e

LN is the symbol used for base-*e* logarithms, commonly referred to as natural logarithms. *e* is Euler's number and $e \cong 2.718282$:

RULE	EXAMPLE
1. $\text{LN}(e^A) = A$	$\text{LN}(7.389056) = \text{LN}(e^2) = 2$
2. If $\text{LN}(A) = B$, then $A = e^B$	If $\text{LN}(A) = 2$, then $A = e^2 = 7.389056$
3. $\text{LN}(A \times B) = \text{LN}(A) + \text{LN}(B)$	$\text{LN}(100) = \text{LN}(10 \times 10)$ $= \text{LN}(10) + \text{LN}(10)$ $= 2.302585 + 2.302585 = 4.605170$
4. $\text{LN}(A^B) = B \times \text{LN}(A)$	$\text{LN}(1000) = \text{LN}(10^3) = 3 \times \text{LN}(10)$ $= 3 \times 2.302585 = 6.907755$
5. $\text{LN}(A/B) = \text{LN}(A) - \text{LN}(B)$	$\text{LN}(100) = \text{LN}(1000/10)$ $= \text{LN}(1000) - \text{LN}(10)$ $= 6.907755 - 2.302585 = 4.605170$

EXAMPLE

Take the base-*e* logarithm of each side of the following equation:

$$Y = \beta_0\beta_1^X\varepsilon$$

SOLUTION Apply rules 3 and 4:

$$\begin{aligned}\text{LN}(Y) &= \text{LN}(\beta_0\beta_1^X\varepsilon)\\ &= \text{LN}(\beta_0) + \text{LN}(\beta_1^X) + \text{LN}(\varepsilon)\\ &= \text{LN}(\beta_0) + X \times \text{LN}(\beta_1) + \text{LN}(\varepsilon)\end{aligned}$$

B. SUMMATION NOTATION

The symbol Σ, the Greek capital letter sigma, is used to denote "taking the sum of." Consider a set of n values for variable X. The expression $\sum_{i=1}^{n} X_i$ means that these n values are to be added together. Thus:

$$\sum_{i=1}^{n} X_i = X_1 + X_2 + X_3 + \cdots + X_n$$

The following problem illustrates the use of the summation notation. Consider five values of a variable X: $X_1 = 2$, $X_2 = 0$, $X_3 = -1$, $X_4 = 5$, and $X_5 = 7$. Thus:

$$\sum_{i=1}^{5} X_i = X_1 + X_2 + X_3 + X_4 + X_5 = 2 + 0 + (-1) + 5 + 7 = 13$$

In statistics, the squared values of a variable are often summed. Thus:

$$\sum_{i=1}^{n} X_i^2 = X_1^2 + X_2^2 + X_3^2 + \cdots + X_n^2$$

and, in the example above:

$$\begin{aligned}\sum_{i=1}^{5} X_i^2 &= X_1^2 + X_2^2 + X_3^2 + X_4^2 + X_5^2 \\ &= 2^2 + 0^2 + (-1)^2 + 5^2 + 7^2 \\ &= 4 + 0 + 1 + 25 + 49 \\ &= 79\end{aligned}$$

$\sum_{i=1}^{n} X_i^2$, the summation of the squares, is *not* the same as $\left(\sum_{i=1}^{n} X_i\right)^2$, the square of the sum.

$$\sum_{i=1}^{n} X_i^2 \neq \left(\sum_{i=1}^{n} X_i\right)^2$$

In the example given earlier, the summation of squares is equal to 79. This is not equal to the square of the sum, which is $13^2 = 169$.

Another frequently used operation involves the summation of the product. Consider two variables, X and Y, each having n values. Then:

$$\sum_{i=1}^{n} X_i Y_i = X_1 Y_1 + X_2 Y_2 + X_3 Y_3 + \cdots + X_n Y_n$$

Continuing with the previous example, suppose there is a second variable, Y, whose five values are $Y_1 = 1$, $Y_2 = 3$, $Y_3 = -2$, $Y_4 = 4$, and $Y_5 = 3$. Then,

$$\begin{aligned}\sum_{i=1}^{5} X_i Y_i &= X_1 Y_1 + X_2 Y_2 + X_3 Y_3 + X_4 Y_4 + X_5 Y_5 \\ &= (2)(1) + (0)(3) + (-1)(-2) + (5)(4) + (7)(3) \\ &= 2 + 0 + 2 + 20 + 21 \\ &= 45\end{aligned}$$

In computing $\sum_{i=1}^{n} X_iY_i$ realize that the first value of X is multiplied by the first value of Y, the second value of X is multiplied by the second value of Y, and so on. These cross products are then summed in order to compute the desired result. However, the summation of products is *not* equal to the product of the individual sums.

$$\sum_{i=1}^{n} X_iY_i \neq \left(\sum_{i=1}^{n} X_i\right)\left(\sum_{i=1}^{n} Y_i\right)$$

In this example, $\sum_{i=1}^{5} X_i = 13$ and $\sum_{i=1}^{5} Y_i = 1 + 3 + (-2) + 4 + 3 = 9$ so that $\left(\sum_{i=1}^{5} X_i\right)\left(\sum_{i=1}^{5} Y_i\right) = (13)(9) = 117$. However $\sum_{i=1}^{5} X_iY_i = 45$. The following table summarizes these results.

VALUE	X_i	Y_i	X_iY_i
1	2	1	2
2	0	3	0
3	−1	−2	2
4	5	4	20
5	7	3	21
	$\sum_{i=1}^{5} X_i = 13$	$\sum_{i=1}^{5} Y_i = 9$	$\sum_{i=1}^{5} X_iY_i = 45$

RULE 1 The summation of the values of two variables is equal to the sum of the values of each summed variable.

$$\sum_{i=1}^{n} (X_i + Y_i) = \sum_{i=1}^{n} X_i + \sum_{i=1}^{n} Y_i$$

Thus,

$$\begin{aligned}\sum_{i=1}^{5} (X_i + Y_i) &= (2 + 1) + (0 + 3) + (-1 + (-2)) + (5 + 4) + (7 + 3)\\ &= 3 + 3 + (-3) + 9 + 10\\ &= 22\end{aligned}$$

$$\sum_{i=1}^{5} X_i + \sum_{i=1}^{5} Y_i = 13 + 9 = 22$$

RULE 2 The summation of a difference between the values of two variables is equal to the difference between the summed values of the variables.

$$\sum_{i=1}^{n} (X_i - Y_i) = \sum_{i=1}^{n} X_i - \sum_{i=1}^{n} Y_i$$

Thus,

$$\sum_{i=1}^{5}(X_i - Y_i) = (2-1) + (0-3) + (-1-(-2)) + (5-4) + (7-3)$$
$$= 1 + (-3) + 1 + 1 + 4$$
$$= 4$$
$$\sum_{i=1}^{5} X_i - \sum_{i=1}^{5} Y_i = 13 - 9 = 4$$

RULE 3 The summation of a constant times a variable is equal to that constant times the summation of the values of the variable.

$$\sum_{i=1}^{n} cX_i = c\sum_{i=1}^{n} X_i$$

where c is a constant.

Thus, if $c = 2$,

$$\sum_{i=1}^{5} cX_i = \sum_{i=1}^{5} 2X_i = (2)(2) + (2)(0) + (2)(-1) + (2)(5) + (2)(7)$$
$$= 4 + 0 + (-2) + 10 + 14$$
$$= 26$$
$$c\sum_{i=1}^{5} X_i = 2\sum_{i=1}^{5} X_i = (2)(13) = 26$$

RULE 4 A constant summed n times will be equal to n times the value of the constant.

$$\sum_{i=1}^{n} c = nc$$

where c is a constant. Thus, if the constant $c = 2$ is summed 5 times,

$$\sum_{i=1}^{5} c = 2 + 2 + 2 + 2 + 2 = 10$$
$$nc = (5)(2) = 10$$

Problem

Suppose there are six values for the variables X and Y such that $X_1 = 2$, $X_2 = 1$, $X_3 = 5$, $X_4 = -3$, $X_5 = 1$, $X_6 = -2$, and $Y_1 = 4$, $Y_2 = 0$, $Y_3 = -1$, $Y_4 = 2$, $Y_5 = 7$, and $Y_6 = -3$. Compute each of the following:

(a) $\sum_{i=1}^{6} X_i$

(b) $\sum_{i=1}^{6} Y_i$

(c) $\sum_{i=1}^{6} X_i^2$

(d) $\sum_{i=1}^{6} Y_i^2$

(e) $\sum_{i=1}^{6} X_iY_i$

(f) $\sum_{i=1}^{6} (X_i + Y_i)$

(g) $\sum_{i=1}^{6}(X_i - Y_i)$

(h) $\sum_{i=1}^{6}(X_i - 3Y_i + 2X_i^2)$

(i) $\sum_{i=1}^{6}(cX_i)$, where $c = -1$

(j) $\sum_{i=1}^{6}(X_i - 3Y_i + c)$, where $c = +3$

ANSWER

(a) 4 (b) 9 (c) 44 (d) 79 (e) 10 (f) 13 (g) −5 (h) 65 (i) −4 (j) −5

References

1. Bashaw, W. L., *Mathematics for Statistics* (New York: Wiley, 1969).
2. Lanzer, P., *Video Review of Arithmetic* (Hicksville, NY: Video Aided Instruction, 1990).
3. Levine, D., *The MBA Primer: Business Statistics* (Cincinnati, OH: Southwestern Publishing, 2000).
4. Levine, D., *Video Review of Statistics* (Hicksville, NY: Video Aided Instruction, 1989).
5. Shane, H., *Video Review of Elementary Algebra* (Hicksville, NY: Video Aided Instruction, 1990).

C. STATISTICAL SYMBOLS AND GREEK ALPHABET

C.1 STATISTICAL SYMBOLS

+ add	× multiply
− subtract	÷ divide
= equal to	≠ not equal to
≅ approximately equal to	
> greater than	< less than
≥ greater than or equal to	≤ less than or equal to

C.2 GREEK ALPHABET

GREEK LETTER	LETTER NAME	ENGLISH EQUIVALENT	GREEK LETTER	LETTER NAME	ENGLISH EQUIVALENT
Α α	Alpha	a	Ν ν	Nu	n
Β β	Beta	b	Ξ ξ	Xi	x
Γ γ	Gamma	g	Ο ο	Omicron	ŏ
Δ δ	Delta	d	Π π	Pi	p
Ε ε	Epsilon	ĕ	Ρ ρ	Rho	r
Ζ ζ	Zeta	z	Σ σ	Sigma	s
Η η	Eta	ē	Τ τ	Tau	t
Θ θ	Theta	th	Υ υ	Upsilon	u
Ι ι	Iota	i	Φ φ	Phi	ph
Κ κ	Kappa	k	Χ χ	Chi	ch
Λ λ	Lambda	l	Ψ ψ	Psi	ps
Μ μ	Mu	m	Ω ω	Omega	ō

D. CD-ROM CONTENTS

D.1 CD-ROM OVERVIEW

The CD-ROM packaged with this text contains program and data files that support your learning of statistics. This CD-ROM includes the following folders:

PHStat2

Contains the setup program and files for PHStat2 version 2.5. You must run the setup program successfully before you can use PHStat2 inside Microsoft Excel. (Be sure to read the instructions in Appendix F and the contents of the PHStat2 readme file on the CD-ROM, before you run the setup program.)

Excel Data Files

Contains the Microsoft Excel workbook files (with the extension .xls) used in the textbook. A detailed list of the files found in this folder starts at the bottom of this page.

Worksheet Template Examples

Contains the Microsoft Excel workbook files used in the Microsoft Excel appendices. Copies of these files also appear in the Excel Data Files folder.

Minitab

Contains the Minitab worksheet files (with the extension .mtw) used in the textbook. A detailed list of the files found in this folder starts at the bottom of this page.

SPSS

Contains the SPSS data files (with the extension .sav) used in the textbook. A detailed list of the files found in this folder starts at the bottom of this page.

Visual Explorations in Statistics

Contains the files necessary to use the Visual Explorations in Statistics macro workbook. If you are using Microsoft Excel 2000 SR-1 or any later version of Excel, you must first make sure that your Microsoft Office security setting is not **High**. (To check your security setting, select **Tools → Macro → Security** and, if necessary, select the **Medium** option in the Security Level tab and click **OK**. When you finish using Visual Explorations, you can go back and reset the security level to **High**, if desired.)

To use this workbook, you can open the **Visual Explorations.xla** file directly from the CD-ROM in Microsoft Excel. If you prefer to use Visual Explorations without always inserting the CD-ROM, copy the **Visual Explorations.xla** file as well as the **Veshelp.hlp** file, containing the orientation and help files, to the hard disk folder of your choice.

CD-ROM Topics

Contains supplemental textbook sections in Adobe PDF file format. You will need the Adobe Acrobat reader software (available on the CD-ROM) in order to read these sections.

Most likely, you will be using files in the Excel, Minitab, or SPSS folder on a regular basis. You can retrieve those files directly from the CD-ROM or copy them first to a hard disk folder. When you copy the files, the files will likely have read-only status. If you wish, you can change the status of the files to read-write by doing the following:

- Open Windows Explorer to the hard disk folder containing the files.
- Select the files to make read-write. (Select **Edit → Select All** to select every file in list.)
- Select **Files → Properties** and in the Properties dialog box, clear (uncheck) the **Read-only** attribute and click **OK**.

D.2 DATA FILE DESCRIPTIONS

The following is an alphabetical listing and description of the data files stored in Excel, Minitab, and SPSS format. In this text, these names appear in smallcaps (such as MUTUALFUNDS2004). For each data file listed in the text, you will find a corresponding file in .xls, .mtw, and .sav format in the appropriate folder.

ACCESS Coded access read times (in msec), file size, programmer group, and buffer size. (Chapter 11)
ACCRES Processing time in seconds and type of computer jobs (research = 0, accounting = 1). (Chapter 10)
ADVERTISE Sales (in thousands of dollars), radio ads (in thousands of dollars), and newspaper ads (in thousands of dollars) for 22 cities. (Chapters 14, 15)
AIRCLEANERS Name, price, energy cost, and filter cost. (Chapters 2, 3)
ALLOY Lifetime of four different alloys. (Chapters 11, 12)
AMPHRS Capacity of batteries. (Chapter 12)
ANGLE Subgroup number and angle. (Chapter 18)
ANSCOMBE Data sets A, B, C, and D—each with 11 pairs of X and Y values. (Chapter 13)
ASSETS Assets of bond funds (Chapter 3)
AUTO Miles per gallon, horsepower, and weight for a sample of 50 car models. (Chapters 14, 15)

AUTO2002 Name, sports utility vehicle (Yes or No), drive type, horsepower, fuel type, miles per gallon, length, width, weight, cargo volume, and turning circle. (Chapters 2, 3, 10, 14, 15)
BANK1 Waiting time (in minutes) spent by a sample of 15 customers at a bank located in a commercial district. (Chapters 3, 9, 10, 12)
BANK2 Waiting time (in minutes) spent by a sample of 15 customers at a bank located in a residential area. (Chapters 3, 10, 12)
BANKCOST1 Bank name, minimum deposit to open, bounced check fee, foreign ATM fee, and online access. (Chapters 2, 3, 8)
BANKCOST2 Bank name, minimum deposit to open, minimum balance to avoid fees, monthly service charge, bounced check fee, foreign ATM fee, and online access. (Chapters 2, 3, 8)
BANKRETURN Year, one-year certificate of deposit return, thirty-month certificate of deposit return, and money market return. (Chapter 3)
BANKTIME Waiting times of four bank customers per day for 20 days. (Chapter 18)
BASEBALL Team, attendance, high temperature on game day, winning percentage of home team, opponent's winning percentage, game played on Friday, Saturday, or Sunday (0 = no, 1 = yes), promotion held (0 = no, 1 = yes). (Chapter 15)
BASKET Year, price of bread, beef, eggs, and lettuce. (Chapter 16)
BATFAIL Times to failure for low, normal, high, and very high. (Chapter 12)
BATTERIES Time to failure (in hours) for 13 flashlight batteries. (Chapters 3, 9)
BATTERIES2 Name, price, cold-cranking amps (CCA). (Chapters 2, 13)
BB2001 Team, league (0 = American, 1 = National), wins, earned run average, runs scored, hits allowed, walks allowed, saves, errors, average ticket prices, fan cost index, regular season gate receipts, local television radio and cable revenues, other local operating revenue, player compensation and benefits, national and other local expenses, income from baseball operations. (Chapters 2, 13, 14, 15)
BB2003 Team, league (0 = American, 1 = National), wins, earned run average, runs scored, hits allowed, walks allowed, saves, and errors. (Chapters 13, 14, 15)
BBREVENUE Team, revenue, value. (Chapter 13)
BBSALARY Year, mean salary, median salary, minimum salary. (Chapter 16)
BEARING Outer ring osculation (low = 0, high = 1), heat treatment (low = 0, high = 1), life of roller bearings. (Chapter 11)
BEER Brand, price in dollars, calories, percentage of alcoholic content, type (craft lager = 1, craft ale = 2, imported lager = 3, regular and ice beer = 4, light and no alcohol beer = 5), and country of origin (U.S. = 1, imported = 0). (Chapters 3, 10, 11)
BREAKSTW Breaking strength for operators (rows) and machines (columns). (Chapter 11)
BUBBLEGUM Bubble diameters for four brands for six students. (Chapters 11, 12)
BULBS Length of life of 40 lightbulbs from manufacturer A (= 1) and 40 lightbulbs from manufacturer B (= 2). (Chapters 2, 10)
CABOT Year and revenue for Cabot Corporation. (Chapter 16)
CAMERA Prices of cameras. (Chapter 3)
CANISTER Day and number of nonconforming film canisters. (Chapter 18)
CATFOOD Time period and weight of cat food (Chapter 18)
CELLPHONE Name, type (CDMA or TDMA), price, talk time, battery capacity. (Chapters 2, 3, 13)
CEREALS Name, cost, calories, fiber, and sugar. (Chapter 3)
CHANGE2004 Mutual fund, change in dollars (Chapter 9)
CHEMICAL Viscosity of batches of chemicals. (Chapters 2, 9)
CIRCUITS Thickness of semiconductor wafers by batch and position. (Chapters 11, 12)
CIRCULATION Magazine, reported newsstand sales, audited newsstand sales. (Chapter 13)
COCACOLA Year, coded year, and operating revenues (in billions of dollars) at Coca-Cola Company. (Chapter 16)
COFFEE Rating of coffees by expert and brand. (Chapters 11, 12)
COFFEEDRINK Product, calories, and fat in coffee drinks. (Chapter 3)
COFFEEPRICE Year and price per pound of coffee in the United States. (Chapter 16)
COLA Sales for normal and end-aisle locations. (Chapters 10, 12)
COLASPC Day, total number of cans filled, and number of unacceptable cans (over a 22-day period). (Chapter 18)
COLLEGECOST University and change in college cost from 2001–2002 to 2002–2003 (Chapter 3)
COLLEGES2002 School, type (0 = public, 1 = private), first quartile SAT score, third quartile SAT score, room and board, total cost (out-of-state cost for public colleges), and average indebtedness at graduation. (Chapters 10, 14, 15)
COMPLAINTS Day and number of complaints. (Chapter 18)
COMPTIME Completion time with current market leader and completion time with new software package. (Chapters 10, 12)

COMPUTERS Download time of three brands of computers. (Chapter 11)
COMPUTERS2 Download time, brand, and browser. (Chapter 11)
CONCRETE1 Compressive strength after two days and seven days. (Chapters 10, 12)
CONCRETE2 Compressive strength after 2 days, 7 days, and 28 days. (Chapters 11, 12)
CONTEST2001 Returns for experts, readers, and dart throwers. (Chapter 11)
CPI-U Year, coded year, and value of CPI-U, the Consumer Price Index. (Chapter 16)
CRACK Type of crack and crack size. (Chapters 10, 12)
CREDIT Month, coded month, credit card charges. (Chapter 16)
CREDITSCORE City and credit score (Chapter 6)
CURRENCY Year, coded year, and mean annual exchange rates (against the U.S. dollar) for the Canadian dollar, Japanese yen, and English pound. (Chapter 16)
CUSTSALE Week number, number of customers, and sales (in thousands of dollars) over a period of 15 consecutive weeks. (Chapter 13)
DELIVERY Customer number, number of cases, and delivery time. (Chapter 13)
DENTAL Annual family dental expenses for 10 employees. (Chapter 8)
DIFFTEST Differences in the sales invoices and actual amounts from a sample of 50 vouchers. (Chapter 8)
DISCOUNT The amount of discount taken from 150 invoices. (Chapter 8)
DISPRAZ Price, price squared, and sales of disposable razors in 15 stores. (Chapter 15)
DJIA Year, coded year, and Dow Jones Industrial Average at the end of the year. (Chapter 16)
DOWRETURN Company, ticker symbol, ten-year return. (Chapter 3)
DRILL Time to drill additional five feet, depth, and type of hole. (Chapter 14)
DRINK Amount of soft drink filled in a subgroup of 50 consecutive 2-liter bottles. (Chapters 2, 9)
DRYCLEAN Day and items returned for rework. (Chapter 18)
ELECTRICITY Year and cost of electricity. (Chapter 16)
ELECUSE Electricity consumption (in kilowatts) and mean temperature (in degrees Fahrenheit) over a consecutive 24-month period. (Chapter 13)
ENERGY State and per capita kilowatt hour use. (Chapter 3)
ENERGY2 Year, price of electricity, natural gas, and fuel oil. (Chapter 16)
ERRORSPC Number of nonconforming items and number of accounts processed over 39 days. (Chapter 18)
ESPRESSO Tamp (the distance in inches between the espresso grounds and the top of the portafilter) and time (the number of seconds the heart, body, and crema are separated). (Chapter 13)
EXPIMP Country, exports, imports. (Chapter 3)
FASTFOOD Product, type (burger vs. chicken), price, size, total fat, saturated fat, calories, and sodium. (Chapters 2, 3)
FASTFOODSALES Year and U.S. sales in billions of dollars. (Chapter 16)
FEDRECPT Year, coded year, and federal receipts (in billions of current dollars). (Chapter 16)
FFCHAIN Raters and restaurant ratings. (Chapters 11, 12)
FIFO Historical cost (in dollars) and audited value (in dollars) for a sample of 120 inventory items. (Chapter 8)
FIRERUNS Week and number of fire runs. (Chapter 18)
FLYASH Fly ash percentage, fly ash percentage squared, and strength. (Chapter 15)
FORCE Force required to break insulator. (Chapters 2, 3, 8, 9)
FORD-REV Quarter, coded quarter, revenue, and three dummy variables for quarters. (Chapter 16)
FOULSPC Number of foul shots made and number taken over 40 days. (Chapter 18)
FREEPORT Address, appraised value, property size (acres), house size, age, number of rooms, number of bathrooms, and number of cars that can be parked in the garage located in Freeport, N.Y. (Chapter 15)
FRUIT Fruit and year, price, and quantity. (Chapter 16)
FUNDTRAN Day, number of new investigations, and number closed over a 30-day period. (Chapter 18)
FURNITURE Days between receipt and resolution of a sample of 50 complaints regarding purchased furniture. (Chapters 2, 3, 8, 9)
GASOLINE Year, gasoline price, 1980 price index, 1995 index. (Chapter 16)
GASPRICE Gasoline price in Manhattan, Bronx, Queens, Brooklyn, Nassau, and Suffolk counties. (Chapter 11)
GCFREEROSLYN Address, appraised value, location, property size (acres), house size, age, number of rooms, number of bathrooms, and number of cars that can be parked in the garage in Glen Cove, Freeport, and Roslyn, NY. (Chapter 15)
GCROSLYN Address, appraised value, location, property size (acres), house size, age, number of rooms, number of bathrooms, and number of cars that can be parked in the garage in Glen Cove and Roslyn, NY. (Chapter 15)
GDP Year and real gross domestic product (in billions of constant 1996 dollars). (Chapter 16)
GEAR Tooth size, part positioning, and gear distortion. (Chapter 11)

GLENCOVE Address, appraised value, property size (acres), house size, age, number of rooms, number of bathrooms, and number of cars that can be parked in the garage in Glen Cove, NY. (Chapters 14, 15)
GOLFBALL Distance for designs 1, 2, 3, and 4. (Chapter 11)
GPIGMAT GMAT scores and GPI for 20 students. (Chapter 13)
GRANULE Granule loss in Boston and Vermont shingles. (Chapters 3, 8, 9, 10)
GROSSREV Year, coded year, and real annual gross revenues. (Chapter 16)
HARDNESS Tensile strength and hardness of aluminum specimens. (Chapter 13)
HARNSWELL Day and diameter of cam rollers (in inches) for samples of five parts produced in each of 30 batches. (Chapter 18)
HOMES Price, location, condition, bedrooms, bathrooms, and other rooms. (Chapter 15)
HOSPADM Day, number of admissions, mean processing time (in hours), range of processing times, and proportion of laboratory rework (over a 30-day period). (Chapter 18)
HOTEL1 Day, number of rooms, number of nonconforming rooms per day over a 28-day period, and proportion of nonconforming items. (Chapter 18)
HOTEL2 Day and delivery time for subgroups of five luggage deliveries per day over a 28-day period. (Chapter 18)
HOTEL-CAR City, hotel cost, rental car cost. (Chapters 2, 3, 8)
HOTEL-PRICE City and hotel price. (Chapter 6)
HOTELPRICE2 City, price in 2004, price in 2002. (Chapter 10)
HOUSE1 Selling price (in thousands of dollars), assessed value (in thousands of dollars), type (new = 0, old = 1), and time period of sale for 30 houses. (Chapters 13, 14, 15)
HOUSE2 Assessed value (in thousands of dollars), size (in thousands of square feet), and age (in years) for 15 houses. (Chapters 13, 14)
HOUSE3 Assessed value (in thousands of dollars), size (in thousands of square feet), and presence of a fireplace for 15 houses. (Chapters 14, 15)
HOUSESNY Year, median price, mortgage rate, mortgage payment, mortgage payment in 2002 dollars. (Chapter 2)
HTNGOIL Monthly consumption of heating oil (in gallons), temperature (in degrees Fahrenheit), attic insulation (in inches), and style (0 = not ranch, 1 = ranch). (Chapters 14, 15)
ICECREAM Daily temperature (in degrees Fahrenheit) and sales (in thousands of dollars) for 21 days. (Chapter 13)
INDPSYCH Reaction time using different assembly methods. (Chapter 12)
INSURANCE Processing time of insurance policies. (Chapters 3, 8, 9)
INTAGLIO Surface hardness of untreated and treated steel plates. (Chapter 10)
INVOICE Number of invoices processed and amount of time (in hours) for 30 days. (Chapter 13)
INVOICES Amount recorded (in dollars) from a sample of 12 sales invoices. (Chapter 9)
ITEMERR Amount of error (in dollars) from a sample of 200 items. (Chapter 8)
JAPANCPI Year and Consumer Price Index for Japan. (Chapter 16)
KEYBOARD Cause, frequency, and percentage. (Chapter 2)
LARGESTBONDS Five-year return of bond funds. (Chapter 3)
LAUNDRY Dirt (in pounds) removed for detergent brands (rows) and cycle times (columns). (Chapter 11)
LAWN Lawn service (0 = no, 1 = yes), family income, lawn size, attitude toward activities (0 = unfavorable, 1 = favorable), number of teenagers, and age of head of household. (Chapter 14)
LOCATE Sales volume (in thousands of dollars) for front, middle, and rear locations. (Chapters 11, 12)
LOGPURCH Annual spending, purchase behavior (0 = no, 1 = yes), and possession of additional credit cards (0 = no, 1 = yes). (Chapter 14)
MAIL Weight of mail and orders. (Chapters 2, 13)
MANAGERS Sales (ratio of yearly sales divided by the target sales value for that region), score from the Wonderlic Personnel test, score on the Strong-Campbell Interest Inventory test, number of years of selling experience prior to becoming a sales manager, and whether the sales manager has a degree in electrical engineering (0 = no, 1 = yes). (Chapter 15)
MBA Success in program (0 = no, 1 = yes), GPA, and GMAT score. (Chapter 14)
MCDONALD Year, coded year, and annual total revenues (in billions of dollars) at McDonald's Corporation. (Chapter 16)
MEASUREMENT Sample, in-line measurement, and analytical lab measurement. (Chapters 10, 12)
MEDICARE Difference in amount reimbursed and amount that should have been reimbursed for office visits. (Chapter 8)
MEDREC Day, number of discharged patients, and number of records not processed for a 30-day period. (Chapter 18)
METALRETURN Year, platinum return, gold return, silver return. (Chapter 3)
MILK Year and price of milk. (Chapter 16)

MOISTURE Moisture content of Boston shingles and Vermont shingles. (Chapter 9)
MOVING Labor hours and cubic feet of packaged items transported in a sample of 36 clients moved by a particular company. (Chapter 13)
MUSICONLINE Album/artist, price at ITunes, Wal-Mart, MusicNow, Musicmatch, Napster. (Chapter 11)
MUTUALFUNDS2004 Fund, category, objective, assets, fees, expense ratio, 2003 Return, three-year return, five-year return, risk, best quarter, and worst quarter. (Chapters 2, 3, 6, 8, 10, 11, 12, 15)
MYELOMA Patient, measurement before transplant, measurement after transplant. (Chapter 10)
NASDAQWEEKLY Week and NASDAQ value. (Chapter 16)
NATURALGAS Coded quarter, price, and three dummy variables for quarters. (Chapter 16)
NBA2004 Team, number of wins, points per game (for team, opponent, and the difference between team and opponent), field goal (shots made), percentage (for team, opponent, and the difference between team and opponent), turnovers (losing the ball before a shot is taken) per game (for team, opponent, and the difference between team and opponent), offensive rebound percentage, and defensive rebound percentage. (Chapters 14, 15)
NEIGHBOR Selling price (in thousands of dollars), number of rooms, and neighborhood location (east = 0, west = 1) for 20 houses. (Chapter 14)
O-RING Flight number, temperature, and O-ring damage index. (Chapter 13)
OIL-GAS Date, price of gasoline in cents per gallon, price of crude oil in dollars per barrel. (Chapter 13)
OMNI Bars sold, price, and promotion expenses. (Chapters 14, 15)
ONLINEBANKING Year and number of households. (Chapter 2)
ONLINESHOPPING Reason, percentage. (Chapter 2)
P&G Year, coded year, and adjusted stock price. (Chapter 16)
PAIN-RELIEF Temperature, brand of pain relief tablet, and time to dissolve. (Chapter 11)
PALLET Weight of Boston and weight of Vermont shingles. (Chapters 2, 8, 9, 10)
PARACHUTE Tensile strength of parachutes from suppliers 1, 2, 3, and 4. (Chapters 11, 12)
PARACHUTE2 Tensile strength for looms (rows) and suppliers (columns). (Chapter 11)
PASTA Weight for type of pasta (rows) and cooking time (columns). (Chapter 11)
PEN Gender, ad, and product rating. (Chapter 11)
PERFORM Performance rating before and after motivational training. (Chapters 10, 12)
PETFOOD Shelf space (in feet), weekly sales (in hundreds of dollars), and aisle location (back = 0, front = 1). (Chapters 13, 14)
PETFOOD2 Pet, type of food, cost per serving, cups per can, protein in grams, and fat in grams. (Chapters 3, 10)
PHONE Time (in minutes) to clear telephone line problems and location (I and II) for samples of 20 customer problems reported to the two office locations. (Chapters 3, 6, 10, 12)
PHOTO Density for developer strength (rows) and development time (columns). (Chapter 11)
PIZZA Product, type (cheese, pepperoni, or chain), cost, calories, and fat. (Chapters 2, 3)
PIZZAHUT Gender (0 = female, 1 = male), price, and purchase behavior (0 = no, 1 = yes). (Chapter 14)
PIZZATIME Time period, delivery time for local restaurant, delivery times for national chain. (Chapter 10)
PLUMBINV Differences (in dollars) between actual amounts recorded on sales invoices and the amounts entered into the accounting system for a sample of 100 invoices. (Chapter 8)
POLIO Year and annual incidence rates per 100,000 persons of reported poliomyelitis. (Chapter 16)
POTATO Percentage of solids content in filter cake, acidity (in pH), lower pressure, upper pressure, cake thickness, varidrive speed, and drum speed setting for 54 measurements. (Chapter 15)
PRINTERS Name, price, text speed, text cost, color photo time, and color photo cost. (Chapters 2, 3, 15)
PROTEIN Calories (in grams), protein, percentage of calories from fat, percentage of calories from saturated fat, and cholesterol (in mg) for 25 popular protein foods. (Chapter 2)
PTFALLS Month and patient falls. (Chapter 18)
PUMPKIN Circumference and weight of pumpkins. (Chapter 13)
RADON Solar radiation, soil temperature, vapor pressure, wind speed, relative humidity, dew point, ambient temperature, and radon concentration. (Chapter 15)
RAISINS Weight of packages of raisins. (Chapter 12)
REDWOOD Height, diameter, and bark thickness. (Chapter 14)
REFRIGERATOR Name, price, and energy cost. (Chapters 2, 13)
RENT Monthly rental cost (in dollars) and apartment size (in square footage) for a sample of 25 apartments. (Chapter 13)
RESTRATE Location, food rating, décor rating, service rating, summated rating, coded location (0 = New York City, 1 = Long Island), and price of restaurants. (Chapters 3, 10, 13, 14, 15)
RETURNS Week and stock price of Microsoft, stock price of General Motors, stock price of Ford, and stock price of International Aluminum. (Chapter 13)

REUSE Percentage of code reused. (Chapter 3)
ROSLYN Address, appraised value, property size (acres), house size, age, number of rooms, number of bathrooms, and number of cars that can be parked in the garage in Roslyn, NY. (Chapter 15)
ROYALS Game, attendance, and whether there was a promotion at Kansas City Royals games. (Chapter 2)
RUBBER Weight of rubber edges. (Chapter 6)
RUDYBIRD Day, total cases sold, and cases of Rudybird sold. (Chapter 18)
S&PSTKIN Coded quarters, end-of-quarter values of the quarterly Standard & Poor's Composite Stock Price Index, and three quarterly dummy variables. (Chapter 16)
SAFETY Tour and number of unsafe acts. (Chapter 18)
SALARIES Job titles, salary for men, and salary for women. (Chapter 10)
SCRUBBER Airflow, water flow, recirculating water flow, orifice diameter, and NTU. (Chapter 15)
SEALANT Sample number, sealant strength for Boston shingles, and sealant strength for Vermont shingles. (Chapter 18)
SEARS Year, coded year, and annual total revenues (in billions of dollars) at Sears, Roebuck & Company. (Chapter 16)
SECURITY City, turnover rate, and violations per million passengers. (Chapters 2, 3, 13)
SH2 Day and number of calls received at the help desk. (Chapters 2, 3)
SH9 Blackness of newsprint. (Chapter 9)
SH10 Length of early calls (in seconds), length of late calls (in seconds), and difference (in seconds). (Chapter 10)
SH11-1 Call, presentation plan (structured = 1, semi-structured = 2, unstructured = 3), and length of call (in seconds). (Chapter 11)
SH11-2 Gender of caller, type of greeting, and length of call. (Chapter 11)
SH13 Hours per month spent telemarketing and number of new subscriptions per month over a 24-month period. (Chapter 13)
SH14 Hours per week spent telemarketing, number of new subscriptions, and type of presentation. (Chapter 14)
SH16 Month and number of home-delivery subscriptions over the most recent 24-month period. (Chapter 16)
SH18-1 Day, number of ads with errors, number of ads, and number of errors over a 25-day period. (Chapter 18)
SH18-2 Day and newsprint blackness measures for each of five spots made over 25 consecutive weekdays. (Chapter 18)
SITE Store number, square footage in thousands of square feet, and sales (in millions of dollars) in 14 stores. (Chapter 13)
SP500 Week, weekly change in the S&P 500, weekly change in the price of Sears, weekly change in the price of Target, and weekly change in the price of Sara Lee. (Chapter 13)
SPEED Miles per gallon and speed (in miles per hour) on automobile test trials. (Chapter 15)
SPONGE Day, number of sponges produced, number of sponges nonconforming over a 32-day period, and proportion of nonconforming sponges. (Chapter 18)
SPORTING Sales, age, annual population growth, income, percentage with high school diploma, and percentage with college diploma. (Chapters 13, 15)
SPORTSTV Year, mean number of viewers in millions for National Football League (NFL), National Basketball Association (NBA), Major League Baseball (MLB), and National Hockey League (NHL). (Chapters 2, 16)
SPWATER Sample number and amount of magnesium. (Chapter 18)
STANDBY Standby hours, staff, remote hours, Dubner hours, and labor hours for 26 weeks. (Chapters 14, 15)
STATES State, commuting time, percentage of homes with more than 8 rooms, median income, percentage of housing that costs more than 30% of family income. (Chapters 2, 3)
STEEL Error in actual length and specified length. (Chapters 2, 6, 8, 9)
STOCKRETURN Year, rate of return of DJIA, S&P 500, Russell 2000, and Wilshire 5000.(Chapter 3)
STOCKS&BONDS Date, closing price of Vanguard Long-Term Bond Index Fund, closing price of the Dow Jones Industrial Average. (Chapter 13)
STOCKS2003 Week, closing weekly stock price for S&P, Sears, Target, Sara Lee. (Chapter 2)
STRATEGIC Year and number of barrels in U.S. strategic oil reserve. (Chapter 16)
TAX Quarterly sales tax receipts (in thousands of dollars) for all 50 business establishments. (Chapter 3)
TAXES County taxes (in dollars) and age of house (in years) for 19 single-family houses. (Chapter 15)
TEA3 Sample number and weight of tea bags. (Chapter 18)
TEABAGS Weight of tea bags. (Chapters 3, 8, 9)
TELESPC Number of orders and number of corrections over 30 days. (Chapter 18)
TELLER Number of errors by tellers. (Chapter 18)
TENSILE Sample number and strength. (Chapter 18)
TESTRANK Rank scores for 10 people trained by a "traditional" method (Method = 0) and 10 people trained by an "experimental" method (Method = 1). (Chapter 12)
TEXTBOOK Textbook, book store price, and Amazon price. (Chapters 10, 12)
TIMES Times to get ready. (Chapter 3)

TOMATOES Year and price per pound in the United States. (Chapter 16)
TOMYLD2 Amount of fertilizer (in pounds per 100 square feet) and yield (in pounds) for 12 plots of land. (Chapter 15)
TOYS-REV Quarter, coded quarter, revenue, and three dummy variables for quarters. (Chapter 16)
TRADE Days, number of undesirable trades, and number of total trades made over a 30-day period. (Chapter 18)
TRADES Day, number of incoming calls, and number of trade executions per day over a 35-day period. (Chapter 13)
TRAIN2 Time (in minutes) early or late with respect to scheduled arrival for 10 LIRR trains (Railroad = 0) and 12 NJT trains (Railroad = 1). (Chapter 12)
TRAINING Assembly time and training program (team-based = 0, individual-based = 1). (Chapters 10, 12)
TRANSMIT Day and number of errors in transmission. (Chapter 18)
TRANSPORT Days and patient transport times (in minutes) for samples of four patients per day over a 30-day period. (Chapter 18)
TRASHBAGS Weight required to break four brands of trashbags. (Chapters 11, 12)
TREASURY Year and interest rate. (Chapter 16)
TROUGH Width of trough. (Chapters 2, 3, 8)
TRSNYC Year, unit value of variable A, and unit value of variable B. (Chapter 16)
TSMODEL1 Years, coded years, and three time series (I, II, III). (Chapter 16)
TSMODEL2 Years, coded years, and two time series (I, II). (Chapter 16)
TUITION School, tuition in state 2002–03, tuition in state 2003–04, difference in tuition in state 2003–04 and 2002–03, tuition out of state 2002–03, tuition out of state 2003–04, difference in tuition out of state 2003–04 and 2002–03. (Chapter 3)
UERATE Year, month, and monthly unemployment rates. (Chapters 2, 16)
UKCPI Year and Consumer Price Index for United Kingdom. (Chapter 16)
UNDERWRITING Score on proficiency exam, score on end of training exam, and training method. (Chapter 14)
UNLEADED Year, month, and price. (Chapter 16)
UTILITY Utility charges for 50 one-bedroom apartments. (Chapters 2, 6)
VB Time (in minutes) for nine students to write and run a Visual Basic program. (Chapter 10)
WALMART Quarter and quarterly revenues. (Chapter 16)
WARECOST Distribution cost (in thousands of dollars), sales (in thousands of dollars), and number of orders for 24 months. (Chapters 13, 14, 15)
WAREHSE Number of units handled per day, and employee number. (Chapter 18)
WHOLEFOODS1 Item, price at Whole Foods, price at Fairway. (Chapters 10, 12)
WHOLEFOODS2 Item, price at Whole Foods, price at Gristede's, price at Fairway, price at Stop & Shop. (Chapters 11, 12)
WINE Summated ratings of different wines by experts. (Chapter 11)
WIP Processing times for samples of 20 books at each of two plants (A = 1, B = 2). (Chapters 3, 6, 10)
WORKFORCE Year, population, and size of the workforce. (Chapter 16)
WRIGLEY Year, coded year, actual revenue, consumer price index, and real revenue. (Chapter 16)
YARN Breaking strength score, pressure (30, 40, or 50 psi), yarn sample, and side-to-side aspect (nozzle = 1, opposite = 2). (Chapter 11)
YIELD Cleansing step, etching step, and yield. (Chapter 11)

E. Tables

E. TABLES

TABLE E.1

Table of Random Numbers

	Column							
Row	**00000 12345**	**00001 67890**	**11111 12345**	**11112 67890**	**22222 12345**	**22223 67890**	**33333 12345**	**33334 67890**
01	49280	88924	35779	00283	81163	07275	89863	02348
02	61870	41657	07468	08612	98083	97349	20775	45091
03	43898	65923	25078	86129	78496	97653	91550	08078
04	62993	93912	30454	84598	56095	20664	12872	64647
05	33850	58555	51438	85507	71865	79488	76783	31708
06	97340	03364	88472	04334	63919	36394	11095	92470
07	70543	29776	10087	10072	55980	64688	68239	20461
08	89382	93809	00796	95945	34101	81277	66090	88872
09	37818	72142	67140	50785	22380	16703	53362	44940
10	60430	22834	14130	96593	23298	56203	92671	15925
11	82975	66158	84731	19436	55790	69229	28661	13675
12	30987	71938	40355	54324	08401	26299	49420	59208
13	55700	24586	93247	32596	11865	63397	44251	43189
14	14756	23997	78643	75912	83832	32768	18928	57070
15	32166	53251	70654	92827	63491	04233	33825	69662
16	23236	73751	31888	81718	06546	83246	47651	04877
17	45794	26926	15130	82455	78305	55058	52551	47182
18	09893	20505	14225	68514	47427	56788	96297	78822
19	54382	74598	91499	14523	68479	27686	46162	83554
20	94750	89923	37089	20048	80336	94598	26940	36858
21	70297	34135	53140	33340	42050	82341	44104	82949
22	85157	47954	32979	26575	57600	40881	12250	73742
23	11100	02340	12860	74697	96644	89439	28707	25815
24	36871	50775	30592	57143	17381	68856	25853	35041
25	23913	48357	63308	16090	51690	54607	72407	55538
26	79348	36085	27973	65157	07456	22255	25626	57054
27	92074	54641	53673	54421	18130	60103	69593	49464
28	06873	21440	75593	41373	49502	17972	82578	16364
29	12478	37622	99659	31065	83613	69889	58869	29571
30	57175	55564	65411	42547	70457	03426	72937	83792
31	91616	11075	80103	07831	59309	13276	26710	73000
32	78025	73539	14621	39044	47450	03197	12787	47709
33	27587	67228	80145	10175	12822	86687	65530	49325
34	16690	20427	04251	64477	73709	73945	92396	68263
35	70183	58065	65489	31833	82093	16747	10386	59293
36	90730	35385	15679	99742	50866	78028	75573	67257
37	10934	93242	13431	24590	02770	48582	00906	58595
38	82462	30166	79613	47416	13389	80268	05085	96666
39	27463	10433	07606	16285	93699	60912	94532	95632
40	02979	52997	09079	92709	90110	47506	53693	49892
41	46888	69929	75233	52507	32097	37594	10067	67327
42	53638	83161	08289	12639	08141	12640	28437	09268
43	82433	61427	17239	89160	19666	08814	37841	12847
44	35766	31672	50082	22795	66948	65581	84393	15890
45	10853	42581	08792	13257	61973	24450	52351	16602
46	20341	27398	72906	63955	17276	10646	74692	48438
47	54458	90542	77563	51839	52901	53355	83281	19177
48	26337	66530	16687	35179	46560	00123	44546	79896
49	34314	23729	85264	05575	96855	23820	11091	79821
50	28603	10708	68933	34189	92166	15181	66628	58599

continued

TABLE E.1
Table of Random Numbers (*Continued*)

	Column							
Row	**00000 12345**	**00001 67890**	**11111 12345**	**11112 67890**	**22222 12345**	**22223 67890**	**33333 12345**	**33334 67890**
51	66194	28926	99547	16625	45515	67953	12108	57846
52	78240	43195	24837	32511	70880	22070	52622	61881
53	00833	88000	67299	68215	11274	55624	32991	17436
54	12111	86683	61270	58036	64192	90611	15145	01748
55	47189	99951	05755	03834	43782	90599	40282	51417
56	76396	72486	62423	27618	84184	78922	73561	52818
57	46409	17469	32483	09083	76175	19985	26309	91536
58	74626	22111	87286	46772	42243	68046	44250	42439
59	34450	81974	93723	49023	58432	67083	36876	93391
60	36327	72135	33005	28701	34710	49359	50693	89311
61	74185	77536	84825	09934	99103	09325	67389	45869
62	12296	41623	62873	37943	25584	09609	63360	47270
63	90822	60280	88925	99610	42772	60561	76873	04117
64	72121	79152	96591	90305	10189	79778	68016	13747
65	95268	41377	25684	08151	61816	58555	54305	86189
66	92603	09091	75884	93424	72586	88903	30061	14457
67	18813	90291	05275	01223	79607	95426	34900	09778
68	38840	26903	28624	67157	51986	42865	14508	49315
69	05959	33836	53758	16562	41081	38012	41230	20528
70	85141	21155	99212	32685	51403	31926	69813	58781
71	75047	59643	31074	38172	03718	32119	69506	67143
72	30752	95260	68032	62871	58781	34143	68790	69766
73	22986	82575	42187	62295	84295	30634	66562	31442
74	99439	86692	90348	66036	48399	73451	26698	39437
75	20389	93029	11881	71685	65452	89047	63669	02656
76	39249	05173	68256	36359	20250	68686	05947	09335
77	96777	33605	29481	20063	09398	01843	35139	61344
78	04860	32918	10798	50492	52655	33359	94713	28393
79	41613	42375	00403	03656	77580	87772	86877	57085
80	17930	00794	53836	53692	67135	98102	61912	11246
81	24649	31845	25736	75231	83808	98917	93829	99430
82	79899	34061	54308	59358	56462	58166	97302	86828
83	76801	49594	81002	30397	52728	15101	72070	33706
84	36239	63636	38140	65731	39788	06872	38971	53363
85	07392	64449	17886	63632	53995	17574	22247	62607
86	67133	04181	33874	98835	67453	59734	76381	63455
87	77759	31504	32832	70861	15152	29733	75371	39174
88	85992	72268	42920	20810	29361	51423	90306	73574
89	79553	75952	54116	65553	47139	60579	09165	85490
90	41101	17336	48951	53674	17880	45260	08575	49321
91	36191	17095	32123	91576	84221	78902	82010	30847
92	62329	63898	23268	74283	26091	68409	69704	82267
93	14751	13151	93115	01437	56945	89661	67680	79790
94	48462	59278	44185	29616	76537	19589	83139	28454
95	29435	88105	59651	44391	74588	55114	80834	85686
96	28340	29285	12965	14821	80425	16602	44653	70467
97	02167	58940	27149	80242	10587	79786	34959	75339
98	17864	00991	39557	54981	23588	81914	37609	13128
99	79675	80605	60059	35862	00254	36546	21545	78179
00	72335	82037	92003	34100	29879	46613	89720	13274

Source: Partially extracted from the Rand Corporation, A Million Random Digits with 100,000 Normal Deviates *(Glencoe, IL, The Free Press, 1955).*

TABLE E.2

The Cumulative Standardized Normal Distribution

Entry represents area under the cumulative standardized normal distribution from −∞ to Z

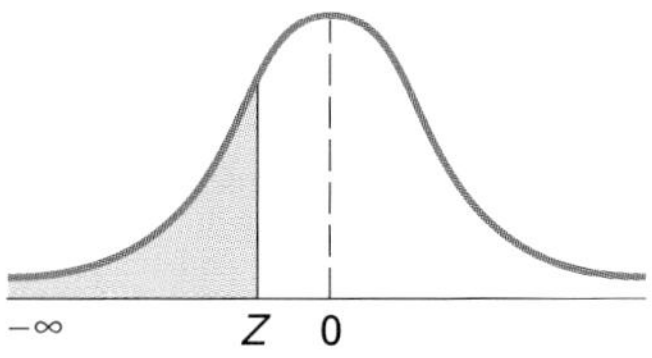

Z	0.00	0.01	0.02	0.03	0.04	0.05	0.06	0.07	0.08	0.09
−6.0	0.000000001									
−5.5	0.000000019									
−5.0	0.000000287									
−4.5	0.000003398									
−4.0	0.000031671									
−3.9	0.00005	0.00005	0.00004	0.00004	0.00004	0.00004	0.00004	0.00004	0.00003	0.00003
−3.8	0.00007	0.00007	0.00007	0.00006	0.00006	0.00006	0.00006	0.00005	0.00005	0.00005
−3.7	0.00011	0.00010	0.00010	0.00010	0.00009	0.00009	0.00008	0.00008	0.00008	0.00008
−3.6	0.00016	0.00015	0.00015	0.00014	0.00014	0.00013	0.00013	0.00012	0.00012	0.00011
−3.5	0.00023	0.00022	0.00022	0.00021	0.00020	0.00019	0.00019	0.00018	0.00017	0.00017
−3.4	0.00034	0.00032	0.00031	0.00030	0.00029	0.00028	0.00027	0.00026	0.00025	0.00024
−3.3	0.00048	0.00047	0.00045	0.00043	0.00042	0.00040	0.00039	0.00038	0.00036	0.00035
−3.2	0.00069	0.00066	0.00064	0.00062	0.00060	0.00058	0.00056	0.00054	0.00052	0.00050
−3.1	0.00097	0.00094	0.00090	0.00087	0.00084	0.00082	0.00079	0.00076	0.00074	0.00071
−3.0	0.00135	0.00131	0.00126	0.00122	0.00118	0.00114	0.00111	0.00107	0.00103	0.00100
−2.9	0.0019	0.0018	0.0018	0.0017	0.0016	0.0016	0.0015	0.0015	0.0014	0.0014
−2.8	0.0026	0.0025	0.0024	0.0023	0.0023	0.0022	0.0021	0.0021	0.0020	0.0019
−2.7	0.0035	0.0034	0.0033	0.0032	0.0031	0.0030	0.0029	0.0028	0.0027	0.0026
−2.6	0.0047	0.0045	0.0044	0.0043	0.0041	0.0040	0.0039	0.0038	0.0037	0.0036
−2.5	0.0062	0.0060	0.0059	0.0057	0.0055	0.0054	0.0052	0.0051	0.0049	0.0048
−2.4	0.0082	0.0080	0.0078	0.0075	0.0073	0.0071	0.0069	0.0068	0.0066	0.0064
−2.3	0.0107	0.0104	0.0102	0.0099	0.0096	0.0094	0.0091	0.0089	0.0087	0.0084
−2.2	0.0139	0.0136	0.0132	0.0129	0.0125	0.0122	0.0119	0.0116	0.0113	0.0110
−2.1	0.0179	0.0174	0.0170	0.0166	0.0162	0.0158	0.0154	0.0150	0.0146	0.0143
−2.0	0.0228	0.0222	0.0217	0.0212	0.0207	0.0202	0.0197	0.0192	0.0188	0.0183
−1.9	0.0287	0.0281	0.0274	0.0268	0.0262	0.0256	0.0250	0.0244	0.0239	0.0233
−1.8	0.0359	0.0351	0.0344	0.0336	0.0329	0.0322	0.0314	0.0307	0.0301	0.0294
−1.7	0.0446	0.0436	0.0427	0.0418	0.0409	0.0401	0.0392	0.0384	0.0375	0.0367
−1.6	0.0548	0.0537	0.0526	0.0516	0.0505	0.0495	0.0485	0.0475	0.0465	0.0455
−1.5	0.0668	0.0655	0.0643	0.0630	0.0618	0.0606	0.0594	0.0582	0.0571	0.0559
−1.4	0.0808	0.0793	0.0778	0.0764	0.0749	0.0735	0.0721	0.0708	0.0694	0.0681
−1.3	0.0968	0.0951	0.0934	0.0918	0.0901	0.0885	0.0869	0.0853	0.0838	0.0823
−1.2	0.1151	0.1131	0.1112	0.1093	0.1075	0.1056	0.1038	0.1020	0.1003	0.0985
−1.1	0.1357	0.1335	0.1314	0.1292	0.1271	0.1251	0.1230	0.1210	0.1190	0.1170
−1.0	0.1587	0.1562	0.1539	0.1515	0.1492	0.1469	0.1446	0.1423	0.1401	0.1379
−0.9	0.1841	0.1814	0.1788	0.1762	0.1736	0.1711	0.1685	0.1660	0.1635	0.1611
−0.8	0.2119	0.2090	0.2061	0.2033	0.2005	0.1977	0.1949	0.1922	0.1894	0.1867
−0.7	0.2420	0.2388	0.2358	0.2327	0.2296	0.2266	0.2236	0.2206	0.2177	0.2148
−0.6	0.2743	0.2709	0.2676	0.2643	0.2611	0.2578	0.2546	0.2514	0.2482	0.2451
−0.5	0.3085	0.3050	0.3015	0.2981	0.2946	0.2912	0.2877	0.2843	0.2810	0.2776
−0.4	0.3446	0.3409	0.3372	0.3336	0.3300	0.3264	0.3228	0.3192	0.3156	0.3121
−0.3	0.3821	0.3783	0.3745	0.3707	0.3669	0.3632	0.3594	0.3557	0.3520	0.3483
−0.2	0.4207	0.4168	0.4129	0.4090	0.4052	0.4013	0.3974	0.3936	0.3897	0.3859
−0.1	0.4602	0.4562	0.4522	0.4483	0.4443	0.4404	0.4364	0.4325	0.4286	0.4247
−0.0	0.5000	0.4960	0.4920	0.4880	0.4840	0.4801	0.4761	0.4721	0.4681	0.4641

continued

TABLE E.2

The Cumulative Standardized Normal Distribution (*Continued*)

Entry represents area under the cumulative standardized normal distribution from $-\infty$ to Z

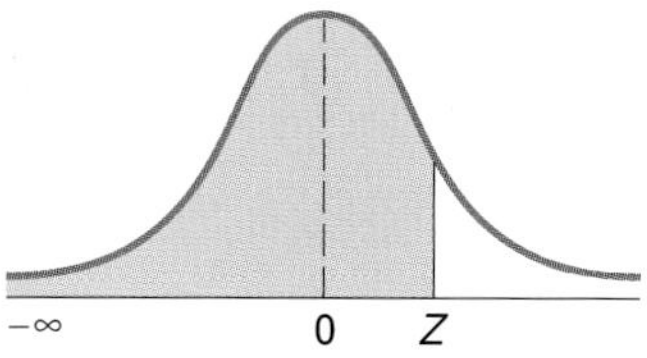

Z	0.00	0.01	0.02	0.03	0.04	0.05	0.06	0.07	0.08	0.09
0.0	0.5000	0.5040	0.5080	0.5120	0.5160	0.5199	0.5239	0.5279	0.5319	0.5359
0.1	0.5398	0.5438	0.5478	0.5517	0.5557	0.5596	0.5636	0.5675	0.5714	0.5753
0.2	0.5793	0.5832	0.5871	0.5910	0.5948	0.5987	0.6026	0.6064	0.6103	0.6141
0.3	0.6179	0.6217	0.6255	0.6293	0.6331	0.6368	0.6406	0.6443	0.6480	0.6517
0.4	0.6554	0.6591	0.6628	0.6664	0.6700	0.6736	0.6772	0.6808	0.6844	0.6879
0.5	0.6915	0.6950	0.6985	0.7019	0.7054	0.7088	0.7123	0.7157	0.7190	0.7224
0.6	0.7257	0.7291	0.7324	0.7357	0.7389	0.7422	0.7454	0.7486	0.7518	0.7549
0.7	0.7580	0.7612	0.7642	0.7673	0.7704	0.7734	0.7764	0.7794	0.7823	0.7852
0.8	0.7881	0.7910	0.7939	0.7967	0.7995	0.8023	0.8051	0.8078	0.8106	0.8133
0.9	0.8159	0.8186	0.8212	0.8238	0.8264	0.8289	0.8315	0.8340	0.8365	0.8389
1.0	0.8413	0.8438	0.8461	0.8485	0.8508	0.8531	0.8554	0.8577	0.8599	0.8621
1.1	0.8643	0.8665	0.8686	0.8708	0.8729	0.8749	0.8770	0.8790	0.8810	0.8830
1.2	0.8849	0.8869	0.8888	0.8907	0.8925	0.8944	0.8962	0.8980	0.8997	0.9015
1.3	0.9032	0.9049	0.9066	0.9082	0.9099	0.9115	0.9131	0.9147	0.9162	0.9177
1.4	0.9192	0.9207	0.9222	0.9236	0.9251	0.9265	0.9279	0.9292	0.9306	0.9319
1.5	0.9332	0.9345	0.9357	0.9370	0.9382	0.9394	0.9406	0.9418	0.9429	0.9441
1.6	0.9452	0.9463	0.9474	0.9484	0.9495	0.9505	0.9515	0.9525	0.9535	0.9545
1.7	0.9554	0.9564	0.9573	0.9582	0.9591	0.9599	0.9608	0.9616	0.9625	0.9633
1.8	0.9641	0.9649	0.9656	0.9664	0.9671	0.9678	0.9686	0.9693	0.9699	0.9706
1.9	0.9713	0.9719	0.9726	0.9732	0.9738	0.9744	0.9750	0.9756	0.9761	0.9767
2.0	0.9772	0.9778	0.9783	0.9788	0.9793	0.9798	0.9803	0.9808	0.9812	0.9817
2.1	0.9821	0.9826	0.9830	0.9834	0.9838	0.9842	0.9846	0.9850	0.9854	0.9857
2.2	0.9861	0.9864	0.9868	0.9871	0.9875	0.9878	0.9881	0.9884	0.9887	0.9890
2.3	0.9893	0.9896	0.9898	0.9901	0.9904	0.9906	0.9909	0.9911	0.9913	0.9916
2.4	0.9918	0.9920	0.9922	0.9925	0.9927	0.9929	0.9931	0.9932	0.9934	0.9936
2.5	0.9938	0.9940	0.9941	0.9943	0.9945	0.9946	0.9948	0.9949	0.9951	0.9952
2.6	0.9953	0.9955	0.9956	0.9957	0.9959	0.9960	0.9961	0.9962	0.9963	0.9964
2.7	0.9965	0.9966	0.9967	0.9968	0.9969	0.9970	0.9971	0.9972	0.9973	0.9974
2.8	0.9974	0.9975	0.9976	0.9977	0.9977	0.9978	0.9979	0.9979	0.9980	0.9981
2.9	0.9981	0.9982	0.9982	0.9983	0.9984	0.9984	0.9985	0.9985	0.9986	0.9986
3.0	0.99865	0.99869	0.99874	0.99878	0.99882	0.99886	0.99889	0.99893	0.99897	0.99900
3.1	0.99903	0.99906	0.99910	0.99913	0.99916	0.99918	0.99921	0.99924	0.99926	0.99929
3.2	0.99931	0.99934	0.99936	0.99938	0.99940	0.99942	0.99944	0.99946	0.99948	0.99950
3.3	0.99952	0.99953	0.99955	0.99957	0.99958	0.99960	0.99961	0.99962	0.99964	0.99965
3.4	0.99966	0.99968	0.99969	0.99970	0.99971	0.99972	0.99973	0.99974	0.99975	0.99976
3.5	0.99977	0.99978	0.99978	0.99979	0.99980	0.99981	0.99981	0.99982	0.99983	0.99983
3.6	0.99984	0.99985	0.99985	0.99986	0.99986	0.99987	0.99987	0.99988	0.99988	0.99989
3.7	0.99989	0.99990	0.99990	0.99990	0.99991	0.99991	0.99992	0.99992	0.99992	0.99992
3.8	0.99993	0.99993	0.99993	0.99994	0.99994	0.99994	0.99994	0.99995	0.99995	0.99995
3.9	0.99995	0.99995	0.99996	0.99996	0.99996	0.99996	0.99996	0.99996	0.99997	0.99997
4.0	0.999968329									
4.5	0.999996602									
5.0	0.999999713									
5.5	0.999999981									
6.0	0.999999999									

TABLE E.3

Critical Values of *t*

For a particular number of degrees of freedom, entry represents the critical value of *t* corresponding to a specified upper-tail area (α).

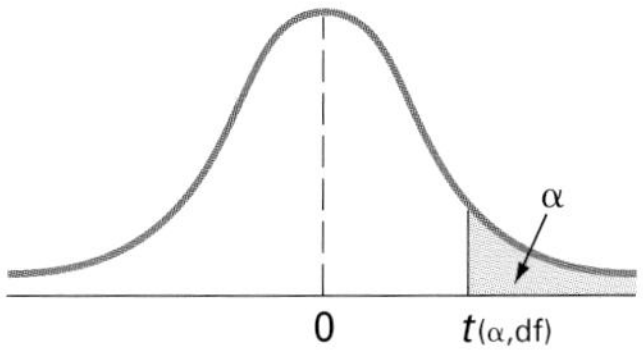

Degrees of Freedom	Upper-Tail Areas 0.25	0.10	0.05	0.025	0.01	0.005
1	1.0000	3.0777	6.3138	12.7062	31.8207	63.6574
2	0.8165	1.8856	2.9200	4.3027	6.9646	9.9248
3	0.7649	1.6377	2.3534	3.1824	4.5407	5.8409
4	0.7407	1.5332	2.1318	2.7764	3.7469	4.6041
5	0.7267	1.4759	2.0150	2.5706	3.3649	4.0322
6	0.7176	1.4398	1.9432	2.4469	3.1427	3.7074
7	0.7111	1.4149	1.8946	2.3646	2.9980	3.4995
8	0.7064	1.3968	1.8595	2.3060	2.8965	3.3554
9	0.7027	1.3830	1.8331	2.2622	2.8214	3.2498
10	0.6998	1.3722	1.8125	2.2281	2.7638	3.1693
11	0.6974	1.3634	1.7959	2.2010	2.7181	3.1058
12	0.6955	1.3562	1.7823	2.1788	2.6810	3.0545
13	0.6938	1.3502	1.7709	2.1604	2.6503	3.0123
14	0.6924	1.3450	1.7613	2.1448	2.6245	2.9768
15	0.6912	1.3406	1.7531	2.1315	2.6025	2.9467
16	0.6901	1.3368	1.7459	2.1199	2.5835	2.9208
17	0.6892	1.3334	1.7396	2.1098	2.5669	2.8982
18	0.6884	1.3304	1.7341	2.1009	2.5524	2.8784
19	0.6876	1.3277	1.7291	2.0930	2.5395	2.8609
20	0.6870	1.3253	1.7247	2.0860	2.5280	2.8453
21	0.6864	1.3232	1.7207	2.0796	2.5177	2.8314
22	0.6858	1.3212	1.7171	2.0739	2.5083	2.8188
23	0.6853	1.3195	1.7139	2.0687	2.4999	2.8073
24	0.6848	1.3178	1.7109	2.0639	2.4922	2.7969
25	0.6844	1.3163	1.7081	2.0595	2.4851	2.7874
26	0.6840	1.3150	1.7056	2.0555	2.4786	2.7787
27	0.6837	1.3137	1.7033	2.0518	2.4727	2.7707
28	0.6834	1.3125	1.7011	2.0484	2.4671	2.7633
29	0.6830	1.3114	1.6991	2.0452	2.4620	2.7564
30	0.6828	1.3104	1.6973	2.0423	2.4573	2.7500
31	0.6825	1.3095	1.6955	2.0395	2.4528	2.7740
32	0.6822	1.3086	1.6939	2.0369	2.4487	2.7385
33	0.6820	1.3077	1.6924	2.0345	2.4448	2.7333
34	0.6818	1.3070	1.6909	2.0322	2.4411	2.7284
35	0.6816	1.3062	1.6896	2.0301	2.4377	2.7238
36	0.6814	1.3055	1.6883	2.0281	2.4345	2.7195
37	0.6812	1.3049	1.6871	2.0262	2.4314	2.7154
38	0.6810	1.3042	1.6860	2.0244	2.4286	2.7116
39	0.6808	1.3036	1.6849	2.0227	2.4258	2.7079
40	0.6807	1.3031	1.6839	2.0211	2.4233	2.7045
41	0.6805	1.3025	1.6829	2.0195	2.4208	2.7012
42	0.6804	1.3020	1.6820	2.0181	2.4185	2.6981
43	0.6802	1.3016	1.6811	2.0167	2.4163	2.6951
44	0.6801	1.3011	1.6802	2.0154	2.4141	2.6923
45	0.6800	1.3006	1.6794	2.0141	2.4121	2.6896
46	0.6799	1.3022	1.6787	2.0129	2.4102	2.6870
47	0.6797	1.2998	1.6779	2.0117	2.4083	2.6846
48	0.6796	1.2994	1.6772	2.0106	2.4066	2.6822

continued

TABLE E.3
Critical Values of t (*Continued*)

	Upper-Tail Areas					
Degrees of Freedom	**0.25**	**0.10**	**0.05**	**0.025**	**0.01**	**0.005**
49	0.6795	1.2991	1.6766	2.0096	2.4049	2.6800
50	0.6794	1.2987	1.6759	2.0086	2.4033	2.6778
51	0.6793	1.2984	1.6753	2.0076	2.4017	2.6757
52	0.6792	1.2980	1.6747	2.0066	2.4002	2.6737
53	0.6791	1.2977	1.6741	2.0057	2.3988	2.6718
54	0.6791	1.2974	1.6736	2.0049	2.3974	2.6700
55	0.6790	1.2971	1.6730	2.0040	2.3961	2.6682
56	0.6789	1.2969	1.6725	2.0032	2.3948	2.6665
57	0.6788	1.2966	1.6720	2.0025	2.3936	2.6649
58	0.6787	1.2963	1.6716	2.0017	2.3924	2.6633
59	0.6787	1.2961	1.6711	2.0010	2.3912	2.6618
60	0.6786	1.2958	1.6706	2.0003	2.3901	2.6603
61	0.6785	1.2956	1.6702	1.9996	2.3890	2.6589
62	0.6785	1.2954	1.6698	1.9990	2.3880	2.6575
63	0.6784	1.2951	1.6694	1.9983	2.3870	2.6561
64	0.6783	1.2949	1.6690	1.9977	2.3860	2.6549
65	0.6783	1.2947	1.6686	1.9971	2.3851	2.6536
66	0.6782	1.2945	1.6683	1.9966	2.3842	2.6524
67	0.6782	1.2943	1.6679	1.9960	2.3833	2.6512
68	0.6781	1.2941	1.6676	1.9955	2.3824	2.6501
69	0.6781	1.2939	1.6672	1.9949	2.3816	2.6490
70	0.6780	1.2938	1.6669	1.9944	2.3808	2.6479
71	0.6780	1.2936	1.6666	1.9939	2.3800	2.6469
72	0.6779	1.2934	1.6663	1.9935	2.3793	2.6459
73	0.6779	1.2933	1.6660	1.9930	2.3785	2.6449
74	0.6778	1.2931	1.6657	1.9925	2.3778	2.6439
75	0.6778	1.2929	1.6654	1.9921	2.3771	2.6430
76	0.6777	1.2928	1.6652	1.9917	2.3764	2.6421
77	0.6777	1.2926	1.6649	1.9913	2.3758	2.6412
78	0.6776	1.2925	1.6646	1.9908	2.3751	2.6403
79	0.6776	1.2924	1.6644	1.9905	2.3745	2.6395
80	0.6776	1.2922	1.6641	1.9901	2.3739	2.6387
81	0.6775	1.2921	1.6639	1.9897	2.3733	2.6379
82	0.6775	1.2920	1.6636	1.9893	2.3727	2.6371
83	0.6775	1.2918	1.6634	1.9890	2.3721	2.6364
84	0.6774	1.2917	1.6632	1.9886	2.3716	2.6356
85	0.6774	1.2916	1.6630	1.9883	2.3710	2.6349
86	0.6774	1.2915	1.6628	1.9879	2.3705	2.6342
87	0.6773	1.2914	1.6626	1.9876	2.3700	2.6335
88	0.6773	1.2912	1.6624	1.9873	2.3695	2.6329
89	0.6773	1.2911	1.6622	1.9870	2.3690	2.6322
90	0.6772	1.2910	1.6620	1.9867	2.3685	2.6316
91	0.6772	1.2909	1.6618	1.9864	2.3680	2.6309
92	0.6772	1.2908	1.6616	1.9861	2.3676	2.6303
93	0.6771	1.2907	1.6614	1.9858	2.3671	2.6297
94	0.6771	1.2906	1.6612	1.9855	2.3667	2.6291
95	0.6771	1.2905	1.6611	1.9853	2.3662	2.6286
96	0.6771	1.2904	1.6609	1.9850	2.3658	2.6280
97	0.6770	1.2903	1.6607	1.9847	2.3654	2.6275
98	0.6770	1.2902	1.6606	1.9845	2.3650	2.6269
99	0.6770	1.2902	1.6604	1.9842	2.3646	2.6264
100	0.6770	1.2901	1.6602	1.9840	2.3642	2.6259
110	0.6767	1.2893	1.6588	1.9818	2.3607	2.6213
120	0.6765	1.2886	1.6577	1.9799	2.3578	2.6174
∞	0.6745	1.2816	1.6449	1.9600	2.3263	2.5758

TABLE E.4

Critical Values of χ^2

For a particular number of degrees of freedom, entry represents the critical value of χ^2 corresponding to a specified upper-tail area (α).

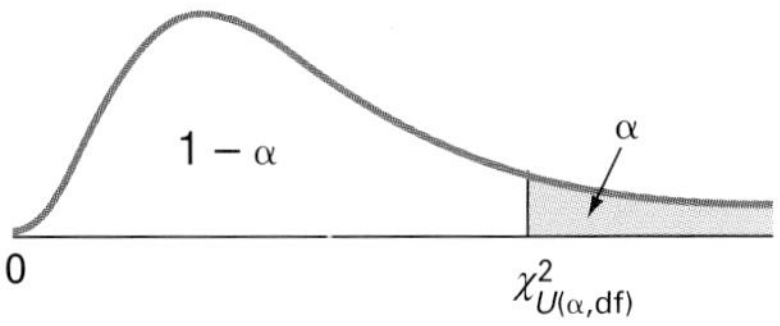

	Upper Tail Areas (α)											
Degrees of Freedom	**0.995**	**0.99**	**0.975**	**0.95**	**0.90**	**0.75**	**0.25**	**0.10**	**0.05**	**0.025**	**0.01**	**0.005**
1			0.001	0.004	0.016	0.102	1.323	2.706	3.841	5.024	6.635	7.879
2	0.010	0.020	0.051	0.103	0.211	0.575	2.773	4.605	5.991	7.378	9.210	10.597
3	0.072	0.115	0.216	0.352	0.584	1.213	4.108	6.251	7.815	9.348	11.345	12.838
4	0.207	0.297	0.484	0.711	1.064	1.923	5.385	7.779	9.488	11.143	13.277	14.860
5	0.412	0.554	0.831	1.145	1.610	2.675	6.626	9.236	11.071	12.833	15.086	16.750
6	0.676	0.872	1.237	1.635	2.204	3.455	7.841	10.645	12.592	14.449	16.812	18.458
7	0.989	1.239	1.690	2.167	2.833	4.255	9.037	12.017	14.067	16.013	18.475	20.278
8	1.344	1.646	2.180	2.733	3.490	5.071	10.219	13.362	15.507	17.535	20.090	21.955
9	1.735	2.088	2.700	3.325	4.168	5.899	11.389	14.684	16.919	19.023	21.666	23.589
10	2.156	2.558	3.247	3.940	4.865	6.737	12.549	15.987	18.307	20.483	23.209	25.188
11	2.603	3.053	3.816	4.575	5.578	7.584	13.701	17.275	19.675	21.920	24.725	26.757
12	3.074	3.571	4.404	5.226	6.304	8.438	14.845	18.549	21.026	23.337	26.217	28.299
13	3.565	4.107	5.009	5.892	7.042	9.299	15.984	19.812	22.362	24.736	27.688	29.819
14	4.075	4.660	5.629	6.571	7.790	10.165	17.117	21.064	23.685	26.119	29.141	31.319
15	4.601	5.229	6.262	7.261	8.547	11.037	18.245	22.307	24.996	27.488	30.578	32.801
16	5.142	5.812	6.908	7.962	9.312	11.912	19.369	23.542	26.296	28.845	32.000	34.267
17	5.697	6.408	7.564	8.672	10.085	12.792	20.489	24.769	27.587	30.191	33.409	35.718
18	6.265	7.015	8.231	9.390	10.865	13.675	21.605	25.989	28.869	31.526	34.805	37.156
19	6.844	7.633	8.907	10.117	11.651	14.562	22.718	27.204	30.144	32.852	36.191	38.582
20	7.434	8.260	9.591	10.851	12.443	15.452	23.828	28.412	31.410	34.170	37.566	39.997
21	8.034	8.897	10.283	11.591	13.240	16.344	24.935	29.615	32.671	35.479	38.932	41.401
22	8.643	9.542	10.982	12.338	14.042	17.240	26.039	30.813	33.924	36.781	40.289	42.796
23	9.260	10.196	11.689	13.091	14.848	18.137	27.141	32.007	35.172	38.076	41.638	44.181
24	9.886	10.856	12.401	13.848	15.659	19.037	28.241	33.196	36.415	39.364	42.980	45.559
25	10.520	11.524	13.120	14.611	16.473	19.939	29.339	34.382	37.652	40.646	44.314	46.928
26	11.160	12.198	13.844	15.379	17.292	20.843	30.435	35.563	38.885	41.923	45.642	48.290
27	11.808	12.879	14.573	16.151	18.114	21.749	31.528	36.741	40.113	43.194	46.963	49.645
28	12.461	13.565	15.308	16.928	18.939	22.657	32.620	37.916	41.337	44.461	48.278	50.993
29	13.121	14.257	16.047	17.708	19.768	23.567	33.711	39.087	42.557	45.722	49.588	52.336
30	13.787	14.954	16.791	18.493	20.599	24.478	34.800	40.256	43.773	46.979	50.892	53.672

For larger values of freedom (*df*) the expression $Z = \sqrt{2\chi^2} - \sqrt{2(df) - 1}$ may be used and the resulting upper-tail area can be found from the cumulative standardized normal distribution (Table E.2).

TABLE E.5

Critical Values of F

For a particular combination of numerator and denominator degrees of freedom, entry represents the critical values of F corresponding to a specified upper-tail area (α).

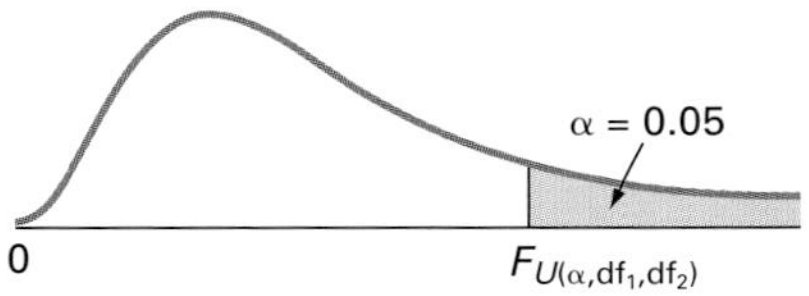

Denominator df_2	Numerator, df_1: 1	2	3	4	5	6	7	8	9	10	12	15	20	24	30	40	60	120	∞
1	161.40	199.50	215.70	224.60	230.20	234.00	236.80	238.90	240.50	241.90	243.90	245.90	248.00	249.10	250.10	251.10	252.20	253.30	254.30
2	18.51	19.00	19.16	19.25	19.30	19.33	19.35	19.37	19.38	19.40	19.41	19.43	19.45	19.45	19.46	19.47	19.48	19.49	19.50
3	10.13	9.55	9.28	9.12	9.01	8.94	8.89	8.85	8.81	8.79	8.74	8.70	8.66	8.64	8.62	8.59	8.57	8.55	8.53
4	7.71	6.94	6.59	6.39	6.26	6.16	6.09	6.04	6.00	5.96	5.91	5.86	5.80	5.77	5.75	5.72	5.69	5.66	5.63
5	6.61	5.79	5.41	5.19	5.05	4.95	4.88	4.82	4.77	4.74	4.68	4.62	4.56	4.53	4.50	4.46	4.43	4.40	4.36
6	5.99	5.14	4.76	4.53	4.39	4.28	4.21	4.15	4.10	4.06	4.00	3.94	3.87	3.84	3.81	3.77	3.74	3.70	3.67
7	5.59	4.74	4.35	4.12	3.97	3.87	3.79	3.73	3.68	3.64	3.57	3.51	3.44	3.41	3.38	3.34	3.30	3.27	3.23
8	5.32	4.46	4.07	3.84	3.69	3.58	3.50	3.44	3.39	3.35	3.28	3.22	3.15	3.12	3.08	3.04	3.01	2.97	2.93
9	5.12	4.26	3.86	3.63	3.48	3.37	3.29	3.23	3.18	3.14	3.07	3.01	2.94	2.90	2.86	2.83	2.79	2.75	2.71
10	4.96	4.10	3.71	3.48	3.33	3.22	3.14	3.07	3.02	2.98	2.91	2.85	2.77	2.74	2.70	2.66	2.62	2.58	2.54
11	4.84	3.98	3.59	3.36	3.20	3.09	3.01	2.95	2.90	2.85	2.79	2.72	2.65	2.61	2.57	2.53	2.49	2.45	2.40
12	4.75	3.89	3.49	3.26	3.11	3.00	2.91	2.85	2.80	2.75	2.69	2.62	2.54	2.51	2.47	2.43	2.38	2.34	2.30
13	4.67	3.81	3.41	3.18	3.03	2.92	2.83	2.77	2.71	2.67	2.60	2.53	2.46	2.42	2.38	2.34	2.30	2.25	2.21
14	4.60	3.74	3.34	3.11	2.96	2.85	2.76	2.70	2.65	2.60	2.53	2.46	2.39	2.35	2.31	2.27	2.22	2.18	2.13
15	4.54	3.68	3.29	3.06	2.90	2.79	2.71	2.64	2.59	2.54	2.48	2.40	2.33	2.29	2.25	2.20	2.16	2.11	2.07
16	4.49	3.63	3.24	3.01	2.85	2.74	2.66	2.59	2.54	2.49	2.42	2.35	2.28	2.24	2.19	2.15	2.11	2.06	2.01
17	4.45	3.59	3.20	2.96	2.81	2.70	2.61	2.55	2.49	2.45	2.38	2.31	2.23	2.19	2.15	2.10	2.06	2.01	1.96
18	4.41	3.55	3.16	2.93	2.77	2.66	2.58	2.51	2.46	2.41	2.34	2.27	2.19	2.15	2.11	2.06	2.02	1.97	1.92
19	4.38	3.52	3.13	2.90	2.74	2.63	2.54	2.48	2.42	2.38	2.31	2.23	2.16	2.11	2.07	2.03	1.98	1.93	1.88
20	4.35	3.49	3.10	2.87	2.71	2.60	2.51	2.45	2.39	2.35	2.28	2.20	2.12	2.08	2.04	1.99	1.95	1.90	1.84
21	4.32	3.47	3.07	2.84	2.68	2.57	2.49	2.42	2.37	2.32	2.25	2.18	2.10	2.05	2.01	1.96	1.92	1.87	1.81
22	4.30	3.44	3.05	2.82	2.66	2.55	2.46	2.40	2.34	2.30	2.23	2.15	2.07	2.03	1.98	1.91	1.89	1.84	1.78
23	4.28	3.42	3.03	2.80	2.64	2.53	2.44	2.37	2.32	2.27	2.20	2.13	2.05	2.01	1.96	1.91	1.86	1.81	1.76
24	4.26	3.40	3.01	2.78	2.62	2.51	2.42	2.36	2.30	2.25	2.18	2.11	2.03	1.98	1.94	1.89	1.84	1.79	1.73
25	4.24	3.39	2.99	2.76	2.60	2.49	2.40	2.34	2.28	2.24	2.16	2.09	2.01	1.96	1.92	1.87	1.82	1.77	1.71
26	4.23	3.37	2.98	2.74	2.59	2.47	2.39	2.32	2.27	2.22	2.15	2.07	1.99	1.95	1.90	1.85	1.80	1.75	1.69
27	4.21	3.35	2.96	2.73	2.57	2.46	2.37	2.31	2.25	2.20	2.13	2.06	1.97	1.93	1.88	1.84	1.79	1.73	1.67
28	4.20	3.34	2.95	2.71	2.56	2.45	2.36	2.29	2.24	2.19	2.12	2.04	1.96	1.91	1.87	1.82	1.77	1.71	1.65
29	4.18	3.33	2.93	2.70	2.55	2.43	2.35	2.28	2.22	2.18	2.10	2.03	1.94	1.90	1.85	1.81	1.75	1.70	1.64
30	4.17	3.32	2.92	2.69	2.53	2.42	2.33	2.27	2.21	2.16	2.09	2.01	1.93	1.89	1.84	1.79	1.74	1.68	1.62
40	4.08	3.23	2.84	2.61	2.45	2.34	2.25	2.18	2.12	2.08	2.00	1.92	1.84	1.79	1.74	1.69	1.64	1.58	1.51
60	4.00	3.15	2.76	2.53	2.37	2.25	2.17	2.10	2.04	1.99	1.92	1.84	1.75	1.70	1.65	1.59	1.53	1.47	1.39
120	3.92	3.07	2.68	2.45	2.29	2.17	2.09	2.02	1.96	1.91	1.83	1.75	1.66	1.61	1.55	1.50	1.43	1.35	1.25
∞	3.84	3.00	2.60	2.37	2.21	2.10	2.01	1.94	1.88	1.83	1.75	1.67	1.57	1.52	1.46	1.39	1.32	1.22	1.00

continued

TABLE E.5

Critical Values of *F* (*Continued*)

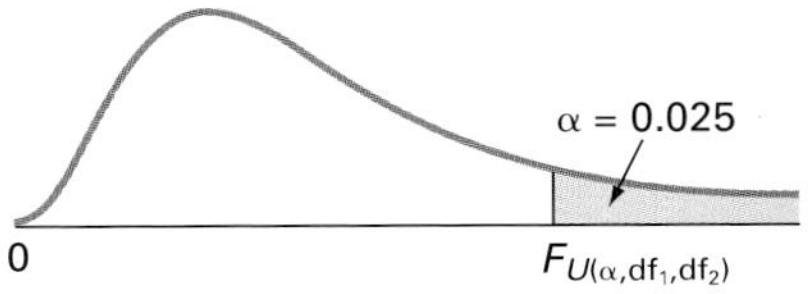

Denominator df_2	Numerator, df_1 1	2	3	4	5	6	7	8	9	10	12	15	20	24	30	40	60	120	∞
1	647.80	799.50	864.20	899.60	921.80	937.10	948.20	956.70	963.30	968.60	976.70	984.90	993.10	997.20	1,001.00	1,006.00	1,010.00	1,014.00	1,018.00
2	38.51	39.00	39.17	39.25	39.30	39.33	39.36	39.39	39.39	39.40	39.41	39.43	39.45	39.46	39.46	39.47	39.48	39.49	39.50
3	17.44	16.04	15.44	15.10	14.88	14.73	14.62	14.54	14.47	14.42	14.34	14.25	14.17	14.12	14.08	14.04	13.99	13.95	13.90
4	12.22	10.65	9.98	9.60	9.36	9.20	9.07	8.98	8.90	8.84	8.75	8.66	8.56	8.51	8.46	8.41	8.36	8.31	8.26
5	10.01	8.43	7.76	7.39	7.15	6.98	6.85	6.76	6.68	6.62	6.52	6.43	6.33	6.28	6.23	6.18	6.12	6.07	6.02
6	8.81	7.26	6.60	6.23	5.99	5.82	5.70	5.60	5.52	5.46	5.37	5.27	5.17	5.12	5.07	5.01	4.96	4.90	4.85
7	8.07	6.54	5.89	5.52	5.29	5.12	4.99	4.90	4.82	4.76	4.67	4.57	4.47	4.42	4.36	4.31	4.25	4.20	4.14
8	7.57	6.06	5.42	5.05	4.82	4.65	4.53	4.43	4.36	4.30	4.20	4.10	4.00	3.95	3.89	3.84	3.78	3.73	3.67
9	7.21	5.71	5.08	4.72	4.48	4.32	4.20	4.10	4.03	3.96	3.87	3.77	3.67	3.61	3.56	3.51	3.45	3.39	3.33
10	6.94	5.46	4.83	4.47	4.24	4.07	3.95	3.85	3.78	3.72	3.62	3.52	3.42	3.37	3.31	3.26	3.20	3.14	3.08
11	6.72	5.26	4.63	4.28	4.04	3.88	3.76	3.66	3.59	3.53	3.43	3.33	3.23	3.17	3.12	3.06	3.00	2.94	2.88
12	6.55	5.10	4.47	4.12	3.89	3.73	3.61	3.51	3.44	3.37	3.28	3.18	3.07	3.02	2.96	2.91	2.85	2.79	2.72
13	6.41	4.97	4.35	4.00	3.77	3.60	3.48	3.39	3.31	3.25	3.15	3.05	2.95	2.89	2.84	2.78	2.72	2.66	2.60
14	6.30	4.86	4.24	3.89	3.66	3.50	3.38	3.29	3.21	3.15	3.05	2.95	2.84	2.79	2.73	2.67	2.61	2.55	2.49
15	6.20	4.77	4.15	3.80	3.58	3.41	3.29	3.20	3.12	3.06	2.96	2.86	2.76	2.70	2.64	2.59	2.52	2.46	2.40
16	6.12	4.69	4.08	3.73	3.50	3.34	3.22	3.12	3.05	2.99	2.89	2.79	2.68	2.63	2.57	2.51	2.45	2.38	2.32
17	6.04	4.62	4.01	3.66	3.44	3.28	3.16	3.06	2.98	2.92	2.82	2.72	2.62	2.56	2.50	2.44	2.38	2.32	2.25
18	5.98	4.56	3.95	3.61	3.38	3.22	3.10	3.01	2.93	2.87	2.77	2.67	2.56	2.50	2.44	2.38	2.32	2.26	2.19
19	5.92	4.51	3.90	3.56	3.33	3.17	3.05	2.96	2.88	2.82	2.72	2.62	2.51	2.45	2.39	2.33	2.27	2.20	2.13
20	5.87	4.46	3.86	3.51	3.29	3.13	3.01	2.91	2.84	2.77	2.68	2.57	2.46	2.41	2.35	2.29	2.22	2.16	2.09
21	5.83	4.42	3.82	3.48	3.25	3.09	2.97	2.87	2.80	2.73	2.64	2.53	2.42	2.37	2.31	2.25	2.18	2.11	2.04
22	5.79	4.38	3.78	3.44	3.22	3.05	2.93	2.84	2.76	2.70	2.60	2.50	2.39	2.33	2.27	2.21	2.14	2.08	2.00
23	5.75	4.35	3.75	3.41	3.18	3.02	2.90	2.81	2.73	2.67	2.57	2.47	2.36	2.30	2.24	2.18	2.11	2.04	1.97
24	5.72	4.32	3.72	3.38	3.15	2.99	2.87	2.78	2.70	2.64	2.54	2.44	2.33	2.27	2.21	2.15	2.08	2.01	1.94
25	5.69	4.29	3.69	3.35	3.13	2.97	2.85	2.75	2.68	2.61	2.51	2.41	2.30	2.24	2.18	2.12	2.05	1.98	1.91
26	5.66	4.27	3.67	3.33	3.10	2.94	2.82	2.73	2.65	2.59	2.49	2.39	2.28	2.22	2.16	2.09	2.03	1.95	1.88
27	5.63	4.24	3.65	3.31	3.08	2.92	2.80	2.71	2.63	2.57	2.47	2.36	2.25	2.19	2.13	2.07	2.00	1.93	1.85
28	5.61	4.22	3.63	3.29	3.06	2.90	2.78	2.69	2.61	2.55	2.45	2.34	2.23	2.17	2.11	2.05	1.98	1.91	1.83
29	5.59	4.20	3.61	3.27	3.04	2.88	2.76	2.67	2.59	2.53	2.43	2.32	2.21	2.15	2.09	2.03	1.96	1.89	1.81
30	5.57	4.18	3.59	3.25	3.03	2.87	2.75	2.65	2.57	2.51	2.41	2.31	2.20	2.14	2.07	2.01	1.94	1.87	1.79
40	5.42	4.05	3.46	3.13	2.90	2.74	2.62	2.53	2.45	2.39	2.29	2.18	2.07	2.01	1.94	1.88	1.80	1.72	1.64
60	5.29	3.93	3.34	3.01	2.79	2.63	2.51	2.41	2.33	2.27	2.17	2.06	1.94	1.88	1.82	1.74	1.67	1.58	1.48
120	5.15	3.80	3.23	2.89	2.67	2.52	2.39	2.30	2.22	2.16	2.05	1.94	1.82	1.76	1.69	1.61	1.53	1.43	1.31
∞	5.02	3.69	3.12	2.79	2.57	2.41	2.29	2.19	2.11	2.05	1.94	1.83	1.71	1.64	1.57	1.48	1.39	1.27	1.00

continued

TABLE E.5

Critical Values of *F* (*Continued*)

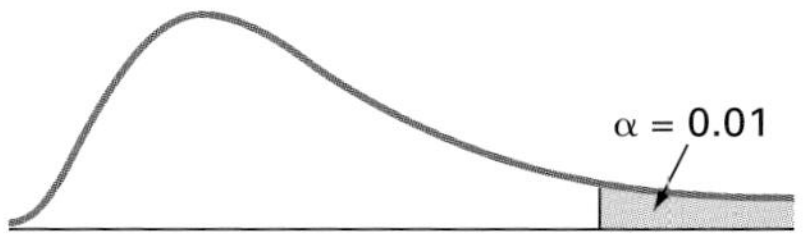

Denominator df_2	Numerator, df_1: 1	2	3	4	5	6	7	8	9	10	12	15	20	24	30	40	60	120	∞
1	4,052.00	4,999.50	5,403.00	5,625.00	5,764.00	5,859.00	5,928.00	5,982.00	6,022.00	6,056.00	6,106.00	6,157.00	6,209.00	6,235.00	6,261.00	6,287.00	6,313.00	6,339.00	6,366.00
2	98.50	99.00	99.17	99.25	99.30	99.33	99.36	99.37	99.39	99.40	99.42	99.43	44.45	99.46	99.47	99.47	99.48	99.49	99.50
3	34.12	30.82	29.46	28.71	28.24	27.91	27.67	27.49	27.35	27.23	27.05	26.87	26.69	26.60	26.50	26.41	26.32	26.22	26.13
4	21.20	18.00	16.69	15.98	15.52	15.21	14.98	14.80	14.66	14.55	14.37	14.20	14.02	13.93	13.84	13.75	13.65	13.56	13.46
5	16.26	13.27	12.06	11.39	10.97	10.67	10.46	10.29	10.16	10.05	9.89	9.72	9.55	9.47	9.38	9.29	9.20	9.11	9.02
6	13.75	10.92	9.78	9.15	8.75	8.47	8.26	8.10	7.98	7.87	7.72	7.56	7.40	7.31	7.23	7.14	7.06	6.97	6.88
7	12.25	9.55	8.45	7.85	7.46	7.19	6.99	6.84	6.72	6.62	6.47	6.31	6.16	6.07	5.99	5.91	5.82	5.74	5.65
8	11.26	8.65	7.59	7.01	6.63	6.37	6.18	6.03	5.91	5.81	5.67	5.52	5.36	5.28	5.20	5.12	5.03	4.95	4.86
9	10.56	8.02	6.99	6.42	6.06	5.80	5.61	5.47	5.35	5.26	5.11	4.96	4.81	4.73	4.65	4.57	4.48	4.40	4.31
10	10.04	7.56	6.55	5.99	5.64	5.39	5.20	5.06	4.94	4.85	4.71	4.56	4.41	4.33	4.25	4.17	4.08	4.00	3.91
11	9.65	7.21	6.22	5.67	5.32	5.07	4.89	4.74	4.63	4.54	4.40	4.25	4.10	4.02	3.94	3.86	3.78	3.69	3.60
12	9.33	6.93	5.95	5.41	5.06	4.82	4.64	4.50	4.39	4.30	4.16	4.01	3.86	3.78	3.70	3.62	3.54	3.45	3.36
13	9.07	6.70	5.74	5.21	4.86	4.62	4.44	4.30	4.19	4.10	3.96	3.82	3.66	3.59	3.51	3.43	3.34	3.25	3.17
14	8.86	6.51	5.56	5.04	4.69	4.46	4.28	4.14	4.03	3.94	3.80	3.66	3.51	3.43	3.35	3.27	3.18	3.09	3.00
15	8.68	6.36	5.42	4.89	4.56	4.32	4.14	4.00	3.89	3.80	3.67	3.52	3.37	3.29	3.21	3.13	3.05	2.96	2.87
16	8.53	6.23	5.29	4.77	4.44	4.20	4.03	3.89	3.78	3.69	3.55	3.41	3.26	3.18	3.10	3.02	2.93	2.81	2.75
17	8.40	6.11	5.18	4.67	4.34	4.10	3.93	3.79	3.68	3.59	3.46	3.31	3.16	3.08	3.00	2.92	2.83	2.75	2.65
18	8.29	6.01	5.09	4.58	4.25	4.01	3.84	3.71	3.60	3.51	3.37	3.23	3.08	3.00	2.92	2.84	2.75	2.66	2.57
19	8.18	5.93	5.01	4.50	4.17	3.94	3.77	3.63	3.52	3.43	3.30	3.15	3.00	2.92	2.84	2.76	2.67	2.58	2.49
20	8.10	5.85	4.94	4.43	4.10	3.87	3.70	3.56	3.46	3.37	3.23	3.09	2.94	2.86	2.78	2.69	2.61	2.52	2.42
21	8.02	5.78	4.87	4.37	4.04	3.81	3.64	3.51	3.40	3.31	3.17	3.03	2.88	2.80	2.72	2.64	2.55	2.46	2.36
22	7.95	5.72	4.82	4.31	3.99	3.76	3.59	3.45	3.35	3.26	3.12	2.98	2.83	2.75	2.67	2.58	2.50	2.40	2.31
23	7.88	5.66	4.76	4.26	3.94	3.71	3.54	3.41	3.30	3.21	3.07	2.93	2.78	2.70	2.62	2.54	2.45	2.35	2.26
24	7.82	5.61	4.72	4.22	3.90	3.67	3.50	3.36	3.26	3.17	3.03	2.89	2.74	2.66	2.58	2.49	2.40	2.31	2.21
25	7.77	5.57	4.68	4.18	3.85	3.63	3.46	3.32	3.22	3.13	2.99	2.85	2.70	2.62	2.54	2.45	2.36	2.27	2.17
26	7.72	5.53	4.64	4.14	3.82	3.59	3.42	3.29	3.18	3.09	2.96	2.81	2.66	2.58	2.50	2.42	2.33	2.23	2.13
27	7.68	5.49	4.60	4.11	3.78	3.56	3.39	3.26	3.15	3.06	2.93	2.78	2.63	2.55	2.47	2.38	2.29	2.20	2.10
28	7.64	5.45	4.57	4.07	3.75	3.53	3.36	3.23	3.12	3.03	2.90	2.75	2.60	2.52	2.44	2.35	2.26	2.17	2.06
29	7.60	5.42	4.54	4.04	3.73	3.50	3.33	3.20	3.09	3.00	2.87	2.73	2.57	2.49	2.41	2.33	2.23	2.14	2.03
30	7.56	5.39	4.51	4.02	3.70	3.47	3.30	3.17	3.07	2.98	2.84	2.70	2.55	2.47	2.39	2.30	2.21	2.11	2.01
40	7.31	5.18	4.31	3.83	3.51	3.29	3.12	2.99	2.89	2.80	2.66	2.52	2.37	2.29	2.20	2.11	2.02	1.92	1.80
60	7.08	4.98	4.13	3.65	3.34	3.12	2.95	2.82	2.72	2.63	2.50	2.35	2.20	2.12	2.03	1.94	1.84	1.73	1.60
120	6.85	4.79	3.95	3.48	3.17	2.96	2.79	2.66	2.56	2.47	2.34	2.19	2.03	1.95	1.86	1.76	1.66	1.53	1.38
∞	6.63	4.61	3.78	3.32	3.02	2.80	2.64	2.51	2.41	2.32	2.18	2.04	1.88	1.79	1.70	1.59	1.47	1.32	1.00

continued

TABLE E.5

Critical Values of F (*Continued*)

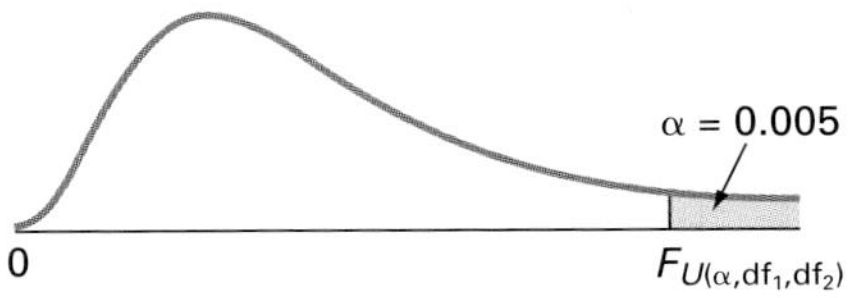

Denominator df_2	Numerator, df_1: 1	2	3	4	5	6	7	8	9	10	12	15	20	24	30	40	60	120	∞
1	16,211.00	20,000.00	21,615.00	22,500.00	23,056.00	23,437.00	23,715.00	23,925.00	24,091.00	24,224.00	24,426.00	24,630.00	24,836.00	24,910.00	25,044.00	25,148.00	25,253.00	25,359.00	25,465.00
2	198.50	199.00	199.20	199.20	199.30	199.30	199.40	199.40	199.40	199.40	199.40	199.40	199.40	199.50	199.50	199.50	199.50	199.50	199.50
3	55.55	49.80	47.47	46.19	45.39	44.84	44.43	44.13	43.88	43.69	43.39	43.08	42.78	42.62	42.47	42.31	42.15	41.99	41.83
4	31.33	26.28	24.26	23.15	22.46	21.97	21.62	21.35	21.14	20.97	20.70	20.44	20.17	20.03	19.89	19.75	19.61	19.47	19.32
5	22.78	18.31	16.53	15.56	14.94	14.51	14.20	13.96	13.77	13.62	13.38	13.15	12.90	12.78	12.66	12.53	12.40	12.27	12.11
6	18.63	14.54	12.92	12.03	11.46	11.07	10.79	10.57	10.39	10.25	10.03	9.81	9.59	9.47	9.36	9.24	9.12	9.00	8.88
7	16.24	12.40	10.88	10.05	9.52	9.16	8.89	8.68	8.51	8.38	8.18	7.97	7.75	7.65	7.53	7.42	7.31	7.19	7.08
8	14.69	11.04	9.60	8.81	8.30	7.95	7.69	7.50	7.34	7.21	7.01	6.81	6.61	6.50	6.40	6.29	6.18	6.06	5.95
9	13.61	10.11	8.72	7.96	7.47	7.13	6.88	6.69	6.54	6.42	6.23	6.03	5.83	5.73	5.62	5.52	5.41	5.30	5.19
10	12.83	9.43	8.08	7.34	6.87	6.54	6.30	6.12	5.97	5.85	5.66	5.47	5.27	5.17	5.07	4.97	4.86	4.75	1.61
11	12.23	8.91	7.60	6.88	6.42	6.10	5.86	5.68	5.54	5.42	5.24	5.05	4.86	4.75	4.65	4.55	4.44	4.34	4.23
12	11.75	8.51	7.23	6.52	6.07	5.76	5.52	5.35	5.20	5.09	4.91	4.72	4.53	4.43	4.33	4.23	4.12	4.01	3.90
13	11.37	8.19	6.93	6.23	5.79	5.48	5.25	5.08	4.94	4.82	4.64	4.46	4.27	4.17	4.07	3.97	3.87	3.76	3.65
14	11.06	7.92	6.68	6.00	5.56	5.26	5.03	4.86	4.72	4.60	4.43	4.25	4.06	3.96	3.86	3.76	3.66	3.55	3.41
15	10.80	7.70	6.48	5.80	5.37	5.07	4.85	4.67	4.54	4.42	4.25	4.07	3.88	3.79	3.69	3.58	3.48	3.37	3.26
16	10.58	7.51	6.30	5.64	5.21	4.91	4.69	4.52	4.38	4.27	4.10	3.92	3.73	3.64	3.54	3.44	3.33	3.22	3.11
17	10.38	7.35	6.16	5.50	5.07	4.78	4.56	4.39	4.25	4.14	3.97	3.79	3.61	3.51	3.41	3.31	3.21	3.10	2.98
18	10.22	7.21	6.03	5.37	4.96	4.66	4.44	4.28	4.14	4.03	3.86	3.68	3.50	3.40	3.30	3.20	3.10	2.89	2.87
19	10.07	7.09	5.92	5.27	4.85	4.56	4.34	4.18	4.04	3.93	3.76	3.59	3.40	3.31	3.21	3.11	3.00	2.89	2.78
20	9.94	6.99	5.82	5.17	4.76	4.47	4.26	4.09	3.96	3.85	3.68	3.50	3.32	3.22	3.12	3.02	2.92	2.81	2.69
21	9.83	6.89	5.73	5.09	4.68	4.39	4.18	4.02	3.88	3.77	3.60	3.43	3.24	3.15	3.05	2.95	2.84	2.73	2.61
22	9.73	6.81	5.65	5.02	4.61	4.32	4.11	3.94	3.81	3.70	3.54	3.36	3.18	3.08	2.98	2.88	2.77	2.66	2.55
23	9.63	6.73	5.58	4.95	4.54	4.26	4.05	3.88	3.75	3.64	3.47	3.30	3.12	3.02	2.92	2.82	2.71	2.60	2.48
24	9.55	6.66	5.52	4.89	4.49	4.20	3.99	3.83	3.69	3.59	3.42	3.25	3.06	2.97	2.87	2.77	2.66	2.55	2.43
25	9.48	6.60	5.46	4.84	4.43	4.15	3.94	3.78	3.64	3.54	3.37	3.20	3.01	2.92	2.82	2.72	2.61	2.50	2.38
26	9.41	6.54	5.41	4.79	4.38	4.10	3.89	3.73	3.60	3.49	3.33	3.15	2.97	2.87	2.77	2.67	2.56	2.45	2.33
27	9.34	6.49	5.36	4.74	4.34	4.06	3.85	3.69	3.56	3.45	3.28	3.11	2.93	2.83	2.73	2.63	2.52	2.41	2.29
28	9.28	6.44	5.32	4.70	4.30	4.02	3.81	3.65	3.52	3.41	3.25	3.07	2.89	2.79	2.69	2.59	2.48	2.37	2.25
29	9.23	6.40	5.28	4.66	4.26	3.98	3.77	3.61	3.48	3.38	3.21	3.04	2.86	2.76	2.66	2.56	2.45	2.33	2.21
30	9.18	6.35	5.24	4.62	4.23	3.95	3.74	3.58	3.45	3.34	3.18	3.01	2.82	2.73	2.63	2.52	2.42	2.30	2.18
40	8.83	6.07	4.98	4.37	3.99	3.71	3.51	3.35	3.22	3.12	2.95	2.78	2.60	2.50	2.40	2.30	2.18	2.06	1.93
60	8.49	5.79	4.73	4.14	3.76	3.49	3.29	3.13	3.01	2.90	2.74	2.57	2.39	2.29	2.19	2.08	1.96	1.83	1.69
120	8.18	5.54	4.50	3.92	3.55	3.28	3.09	2.93	2.81	2.71	2.54	2.37	2.19	2.09	1.98	1.87	1.75	1.61	1.43
∞	7.88	5.30	4.28	3.72	3.35	3.09	2.90	2.74	2.62	2.52	2.36	2.19	2.00	1.90	1.79	1.67	1.53	1.36	1.00

Source: Reprinted from E. S. Pearson and H. O. Hartley, eds., Biometrika Tables for Statisticians, *3rd ed., 1966, by permission of the* Biometrika *Trustees.*

Table E.6

Table of Binomial Probabilities (begins on the following page)

TABLE E.6

Table of Binomial Probabilities

For a given combination of n and p, entry indicates the probability of obtaining a specified value of X. To locate entry: **when $p \leq .50$**, read p across the top heading and both n and X down the left margin; **when $p \geq .50$**, read p across the bottom heading and both n and X up the right margin.

		p																			
n	*X*	**0.01**	**0.02**	**0.03**	**0.04**	**0.05**	**0.06**	**0.07**	**0.08**	**0.09**	**0.10**	**0.15**	**0.20**	**0.25**	**0.30**	**0.35**	**0.40**	**0.45**	**0.50**	*X*	*n*
2	0	0.9801	0.9604	0.9409	0.9216	0.9025	0.8836	0.8649	0.8464	0.8281	0.8100	0.7225	0.6400	0.5625	0.4900	0.4225	0.3600	0.3025	0.2500	2	
	1	0.0198	0.0392	0.0582	0.0768	0.0950	0.1128	0.1302	0.1472	0.1638	0.1800	0.2550	0.3200	0.3750	0.4200	0.4550	0.4800	0.4950	0.5000	1	
	2	0.0001	0.0004	0.0009	0.0016	0.0025	0.0036	0.0049	0.0064	0.0081	0.0100	0.0225	0.0400	0.0625	0.0900	0.1225	0.1600	0.2025	0.2500	0	2
3	0	0.9703	0.9412	0.9127	0.8847	0.8574	0.8306	0.8044	0.7787	0.7536	0.7290	0.6141	0.5120	0.4219	0.3430	0.2746	0.2160	0.1664	0.1250	3	
	1	0.0294	0.0576	0.0847	0.1106	0.1354	0.1590	0.1816	0.2031	0.2236	0.2430	0.3251	0.3840	0.4219	0.4410	0.4436	0.4320	0.4084	0.3750	2	
	2	0.0003	0.0012	0.0026	0.0046	0.0071	0.0102	0.0137	0.0177	0.0221	0.0270	0.0574	0.0960	0.1406	0.1890	0.2389	0.2880	0.3341	0.3750	1	
	3	0.0000	0.0000	0.0000	0.0001	0.0001	0.0002	0.0003	0.0005	0.0007	0.0010	0.0034	0.0080	0.0156	0.0270	0.0429	0.0640	0.0911	0.1250	0	3
4	0	0.9606	0.9224	0.8853	0.8493	0.8145	0.7807	0.7481	0.7164	0.6857	0.6561	0.5220	0.4096	0.3164	0.2401	0.1785	0.1296	0.0915	0.0625	4	
	1	0.0388	0.0753	0.1095	0.1416	0.1715	0.1993	0.2252	0.2492	0.2713	0.2916	0.3685	0.4096	0.4219	0.4116	0.3845	0.3456	0.2995	0.2500	3	
	2	0.0006	0.0023	0.0051	0.0088	0.0135	0.0191	0.0254	0.0325	0.0402	0.0486	0.0975	0.1536	0.2109	0.2646	0.3105	0.3456	0.3675	0.3750	2	
	3	0.0000	0.0000	0.0001	0.0002	0.0005	0.0008	0.0013	0.0019	0.0027	0.0036	0.0115	0.0256	0.0469	0.0756	0.1115	0.1536	0.2005	0.2500	1	
	4	—	—	0.0000	0.0000	0.0000	0.0000	0.0000	0.0000	0.0001	0.0001	0.0005	0.0016	0.0039	0.0081	0.0150	0.0256	0.0410	0.0625	0	4
5	0	0.9510	0.9039	0.8587	0.8154	0.7738	0.7339	0.6957	0.6591	0.6240	0.5905	0.4437	0.3277	0.2373	0.1681	0.1160	0.0778	0.0503	0.0312	5	
	1	0.0480	0.0922	0.1328	0.1699	0.2036	0.2342	0.2618	0.2866	0.3086	0.3280	0.3915	0.4096	0.3955	0.3601	0.3124	0.2592	0.2059	0.1562	4	
	2	0.0010	0.0038	0.0082	0.0142	0.0214	0.0299	0.0394	0.0498	0.0610	0.0729	0.1382	0.2048	0.2637	0.3087	0.3364	0.3456	0.3369	0.3125	3	
	3	0.0000	0.0001	0.0003	0.0006	0.0011	0.0019	0.0030	0.0043	0.0060	0.0081	0.0244	0.0512	0.0879	0.1323	0.1811	0.2304	0.2757	0.3125	2	
	4	—	0.0000	0.0000	0.0000	0.0000	0.0001	0.0001	0.0002	0.0003	0.0004	0.0022	0.0064	0.0146	0.0283	0.0488	0.0768	0.1128	0.1562	1	
	5	—	—	—	—	—	0.0000	0.0000	0.0000	0.0000	0.0000	0.0001	0.0003	0.0010	0.0024	0.0053	0.0102	0.0185	0.0312	0	5
6	0	0.9415	0.8858	0.8330	0.7828	0.7351	0.6899	0.6470	0.6064	0.5679	0.5314	0.3771	0.2621	0.1780	0.1176	0.0754	0.0467	0.0277	0.0156	6	
	1	0.0571	0.1085	0.1546	0.1957	0.2321	0.2642	0.2922	0.3164	0.3370	0.3543	0.3993	0.3932	0.3560	0.3025	0.2437	0.1866	0.1359	0.0937	5	
	2	0.0014	0.0055	0.0120	0.0204	0.0305	0.0422	0.0550	0.0688	0.0833	0.0984	0.1762	0.2458	0.2966	0.3241	0.3280	0.3110	0.2780	0.2344	4	
	3	0.0000	0.0002	0.0005	0.0011	0.0021	0.0036	0.0055	0.0080	0.0110	0.0146	0.0415	0.0819	0.1318	0.1852	0.2355	0.2765	0.3032	0.3125	3	
	4	—	0.0000	0.0000	0.0000	0.0001	0.0002	0.0003	0.0005	0.0008	0.0012	0.0055	0.0154	0.0330	0.0595	0.0951	0.1372	0.1861	0.2344	2	
	5	—	—	—	—	0.0000	0.0000	0.0000	0.0000	0.0000	0.0001	0.0004	0.0015	0.0044	0.0102	0.0205	0.0369	0.0609	0.0937	1	
	6	—	—	—	—	—	—	—	—	—	0.0000	0.0000	0.0001	0.0002	0.0007	0.0018	0.0041	0.0083	0.0156	0	6

n	X	0.99	0.98	0.97	0.96	0.95	0.94	0.93	0.92	0.91	0.90	0.85	0.80	0.75	0.70	0.65	0.60	0.55	0.50	X	n
7	0	0.9321	0.8681	0.8080	0.7514	0.6983	0.6485	0.6017	0.5578	0.5168	0.4783	0.3206	0.2097	0.1335	0.0824	0.0490	0.0280	0.0152	0.0078	7	
	1	0.0659	0.1240	0.1749	0.2192	0.2573	0.2897	0.3170	0.3396	0.3578	0.3720	0.3960	0.3670	0.3115	0.2471	0.1848	0.1306	0.0872	0.0547	6	
	2	0.0020	0.0076	0.0162	0.0274	0.0406	0.0555	0.0716	0.0886	0.1061	0.1240	0.2097	0.2753	0.3115	0.3177	0.2985	0.2613	0.2140	0.1641	5	
	3	0.0000	0.0003	0.0008	0.0019	0.0036	0.0059	0.0090	0.0128	0.0175	0.0230	0.0617	0.1147	0.1730	0.2269	0.2679	0.2903	0.2918	0.2734	4	
	4	—	0.0000	0.0000	0.0001	0.0002	0.0004	0.0007	0.0011	0.0017	0.0026	0.0109	0.0287	0.0577	0.0972	0.1442	0.1935	0.2388	0.2734	3	
	5	—	—	—	0.0000	0.0000	0.0000	0.0000	0.0001	0.0001	0.0002	0.0012	0.0043	0.0115	0.0250	0.0466	0.0774	0.1172	0.1641	2	
	6	—	—	—	—	—	—	—	0.0000	0.0000	0.0000	0.0001	0.0004	0.0013	0.0036	0.0084	0.0172	0.0320	0.0547	1	
	7	—	—	—	—	—	—	—	—	—	—	0.0000	0.0000	0.0001	0.0002	0.0006	0.0016	0.0037	0.0078	0	7
8	0	0.9227	0.8508	0.7837	0.7214	0.6634	0.6096	0.5596	0.5132	0.4703	0.4305	0.2725	0.1678	0.1001	0.0576	0.0319	0.0168	0.0084	0.0039	8	
	1	0.0746	0.1389	0.1939	0.2405	0.2793	0.3113	0.3370	0.3570	0.3721	0.3826	0.3847	0.3355	0.2670	0.1977	0.1373	0.0896	0.0548	0.0312	7	
	2	0.0026	0.0099	0.0210	0.0351	0.0515	0.0695	0.0888	0.1087	0.1288	0.1488	0.2376	0.2936	0.3115	0.2965	0.2587	0.2090	0.1569	0.1094	6	
	3	0.0001	0.0004	0.0013	0.0029	0.0054	0.0089	0.0134	0.0189	0.0255	0.0331	0.0839	0.1468	0.2076	0.2541	0.2786	0.2787	0.2568	0.2187	5	
	4	0.0000	0.0000	0.0001	0.0002	0.0004	0.0007	0.0013	0.0021	0.0031	0.0046	0.0185	0.0459	0.0865	0.1361	0.1875	0.2322	0.2627	0.2734	4	
	5	—	—	0.0000	0.0000	0.0000	0.0000	0.0001	0.0001	0.0002	0.0004	0.0026	0.0092	0.0231	0.0467	0.0808	0.1239	0.1719	0.2187	3	
	6	—	—	—	—	—	—	0.0000	0.0000	0.0000	0.0000	0.0002	0.0011	0.0038	0.0100	0.0217	0.0413	0.0703	0.1094	2	
	7	—	—	—	—	—	—	—	—	—	—	0.0000	0.0001	0.0004	0.0012	0.0033	0.0079	0.0164	0.0312	1	
	8	—	—	—	—	—	—	—	—	—	—	—	0.0000	0.0000	0.0001	0.0002	0.0007	0.0017	0.0039	0	8
9	0	0.9135	0.8337	0.7602	0.6925	0.6302	0.5730	0.5204	0.4722	0.4279	0.3874	0.2316	0.1342	0.0751	0.0404	0.0207	0.0101	0.0046	0.0020	9	
	1	0.0830	0.1531	0.2116	0.2597	0.2985	0.3292	0.3525	0.3695	0.3809	0.3874	0.3679	0.3020	0.2253	0.1556	0.1004	0.0605	0.0339	0.0176	8	
	2	0.0034	0.0125	0.0262	0.0433	0.0629	0.0840	0.1061	0.1285	0.1507	0.1722	0.2597	0.3020	0.3003	0.2668	0.2162	0.1612	0.1110	0.0703	7	
	3	0.0001	0.0006	0.0019	0.0042	0.0077	0.0125	0.0186	0.0261	0.0348	0.0446	0.1069	0.1762	0.2336	0.2668	0.2716	0.2508	0.2119	0.1641	6	
	4	0.0000	0.0000	0.0001	0.0003	0.0006	0.0012	0.0021	0.0034	0.0052	0.0074	0.0283	0.0661	0.1168	0.1715	0.2194	0.2508	0.2600	0.2461	5	
	5	—	—	0.0000	0.0000	0.0000	0.0001	0.0002	0.0003	0.0005	0.0008	0.0050	0.0165	0.0390	0.0735	0.1181	0.1672	0.2128	0.2461	4	
	6	—	—	—	—	—	0.0000	0.0000	0.0000	0.0000	0.0001	0.0006	0.0028	0.0087	0.0210	0.0424	0.0743	0.1160	0.1641	3	
	7	—	—	—	—	—	—	—	—	—	0.0000	0.0000	0.0003	0.0012	0.0039	0.0098	0.0212	0.0407	0.0703	2	
	8	—	—	—	—	—	—	—	—	—	—	—	0.0000	0.0001	0.0004	0.0013	0.0035	0.0083	0.0176	1	
	9	—	—	—	—	—	—	—	—	—	—	—	—	0.0000	0.0000	0.0001	0.0003	0.0008	0.0020	0	9
10	0	0.9044	0.8171	0.7374	0.6648	0.5987	0.5386	0.4840	0.4344	0.3894	0.3487	0.1969	0.1074	0.0563	0.0282	0.0135	0.0060	0.0025	0.0010	10	
	1	0.0914	0.1667	0.2281	0.2770	0.3151	0.3438	0.3643	0.3777	0.3851	0.3874	0.3474	0.2684	0.1877	0.1211	0.0725	0.0403	0.0207	0.0098	9	
	2	0.0042	0.0153	0.0317	0.0519	0.0746	0.0988	0.1234	0.1478	0.1714	0.1937	0.2759	0.3020	0.2816	0.2335	0.1757	0.1209	0.0763	0.0439	8	
	3	0.0001	0.0008	0.0026	0.0058	0.0105	0.0168	0.0248	0.0343	0.0452	0.0574	0.1298	0.2013	0.2503	0.2668	0.2522	0.2150	0.1665	0.1172	7	
	4	0.0000	0.0000	0.0001	0.0004	0.0010	0.0019	0.0033	0.0052	0.0078	0.0112	0.0401	0.0881	0.1460	0.2001	0.2377	0.2508	0.2384	0.2051	6	
	5	—	—	0.0000	0.0000	0.0001	0.0001	0.0003	0.0005	0.0009	0.0015	0.0085	0.0264	0.0584	0.1029	0.1536	0.2007	0.2340	0.2461	5	
	6	—	—	—	—	0.0000	0.0000	0.0000	0.0000	0.0001	0.0001	0.0012	0.0055	0.0162	0.0368	0.0689	0.1115	0.1596	0.2051	4	
	7	—	—	—	—	—	—	—	—	0.0000	0.0000	0.0001	0.0008	0.0031	0.0090	0.0212	0.0425	0.0746	0.1172	3	
	8	—	—	—	—	—	—	—	—	—	—	0.0000	0.0001	0.0004	0.0014	0.0043	0.0106	0.0229	0.0439	2	
	9	—	—	—	—	—	—	—	—	—	—	—	0.0000	0.0000	0.0001	0.0005	0.0016	0.0042	0.0098	1	
	10	—	—	—	—	—	—	—	—	—	—	—	—	—	0.0000	0.0000	0.0001	0.0003	0.0010	0	10

continued

TABLE E.6

Table of Binomial Probabilities (*Continued*)

		p																			
n	*X*	**0.01**	**0.02**	**0.03**	**0.04**	**0.05**	**0.06**	**0.07**	**0.08**	**0.09**	**0.10**	**0.15**	**0.20**	**0.25**	**0.30**	**0.35**	**0.40**	**0.45**	**0.50**	*X*	*n*
20	0	0.8179	0.6676	0.5438	0.4420	0.3585	0.2901	0.2342	0.1887	0.1516	0.1216	0.0388	0.0115	0.0032	0.0008	0.0002	0.0000	0.0000	—	20	
	1	0.1652	0.2725	0.3364	0.3683	0.3774	0.3703	0.3526	0.3282	0.3000	0.2702	0.1368	0.0576	0.0211	0.0068	0.0020	0.0005	0.0001	0.0000	19	
	2	0.0159	0.0528	0.0988	0.1458	0.1887	0.2246	0.2521	0.2711	0.2818	0.2852	0.2293	0.1369	0.0699	0.0278	0.0100	0.0031	0.0008	0.0002	18	
	3	0.0010	0.0065	0.0183	0.0364	0.0596	0.0860	0.1139	0.1414	0.1672	0.1901	0.2428	0.2054	0.1339	0.0716	0.0323	0.0123	0.0040	0.0011	17	
	4	0.0000	0.0006	0.0024	0.0065	0.0133	0.0233	0.0364	0.0523	0.0703	0.0898	0.1821	0.2182	0.1897	0.1304	0.0738	0.0350	0.0139	0.0046	16	
	5	—	0.0000	0.0002	0.0009	0.0022	0.0048	0.0088	0.0145	0.0222	0.0319	0.1028	0.1746	0.2023	0.1789	0.1272	0.0746	0.0365	0.0148	15	
	6	—	—	0.0000	0.0001	0.0003	0.0008	0.0017	0.0032	0.0055	0.0089	0.0454	0.1091	0.1686	0.1916	0.1712	0.1244	0.0746	0.0370	14	
	7	—	—	—	0.0000	0.0000	0.0001	0.0002	0.0005	0.0011	0.0020	0.0160	0.0545	0.1124	0.1643	0.1844	0.1659	0.1221	0.0739	13	
	8	—	—	—	—	—	0.0000	0.0000	0.0001	0.0002	0.0004	0.0046	0.0222	0.0609	0.1144	0.1614	0.1797	0.1623	0.1201	12	
	9	—	—	—	—	—	—	—	0.0000	0.0000	0.0001	0.0011	0.0074	0.0271	0.0654	0.1158	0.1597	0.1771	0.1602	11	
	10	—	—	—	—	—	—	—	—	—	0.0000	0.0002	0.0020	0.0099	0.0308	0.0686	0.1171	0.1593	0.1762	10	
	11	—	—	—	—	—	—	—	—	—	—	0.0000	0.0005	0.0030	0.0120	0.0336	0.0710	0.1185	0.1602	9	
	12	—	—	—	—	—	—	—	—	—	—	—	0.0001	0.0008	0.0039	0.0136	0.0355	0.0727	0.1201	8	
	13	—	—	—	—	—	—	—	—	—	—	—	0.0000	0.0002	0.0010	0.0045	0.0146	0.0366	0.0739	7	
	14	—	—	—	—	—	—	—	—	—	—	—	—	0.0000	0.0002	0.0012	0.0049	0.0150	0.0370	6	
	15	—	—	—	—	—	—	—	—	—	—	—	—	—	0.0000	0.0003	0.0013	0.0049	0.0148	5	
	16	—	—	—	—	—	—	—	—	—	—	—	—	—	—	0.0000	0.0003	0.0013	0.0046	4	
	17	—	—	—	—	—	—	—	—	—	—	—	—	—	—	—	0.0000	0.0002	0.0011	3	
	18	—	—	—	—	—	—	—	—	—	—	—	—	—	—	—	—	0.0000	0.0002	2	
	19	—	—	—	—	—	—	—	—	—	—	—	—	—	—	—	—	—	0.0000	1	
	20	—	—	—	—	—	—	—	—	—	—	—	—	—	—	—	—	—	—	0	20
n	*X*	0.99	0.98	0.97	0.96	0.95	0.94	0.93	0.92	0.91	0.90	0.85	0.80	0.75	0.70	0.65	0.60	0.55	0.50	*X*	*n*

TABLE E.7

Table of Poisson Probabilities

For a given value of λ, entry indicates the probability of a specified value of X.

	λ									
X	**0.1**	**0.2**	**0.3**	**0.4**	**0.5**	**0.6**	**0.7**	**0.8**	**0.9**	**1.0**
0	0.9048	0.8187	0.7408	0.6703	0.6065	0.5488	0.4966	0.4493	0.4066	0.3679
1	0.0905	0.1637	0.2222	0.2681	0.3033	0.3293	0.3476	0.3595	0.3659	0.3679
2	0.0045	0.0164	0.0333	0.0536	0.0758	0.0988	0.1217	0.1438	0.1647	0.1839
3	0.0002	0.0011	0.0033	0.0072	0.0126	0.0198	0.0284	0.0383	0.0494	0.0613
4	0.0000	0.0001	0.0003	0.0007	0.0016	0.0030	0.0050	0.0077	0.0111	0.0153
5	0.0000	0.0000	0.0000	0.0001	0.0002	0.0004	0.0007	0.0012	0.0020	0.0031
6	0.0000	0.0000	0.0000	0.0000	0.0000	0.0000	0.0001	0.0002	0.0003	0.0005
7	0.0000	0.0000	0.0000	0.0000	0.0000	0.0000	0.0000	0.0000	0.0000	0.0001

	λ									
X	**1.1**	**1.2**	**1.3**	**1.4**	**1.5**	**1.6**	**1.7**	**1.8**	**1.9**	**2.0**
0	0.3329	0.3012	0.2725	0.2466	0.2231	0.2019	0.1827	0.1653	0.1496	0.1353
1	0.3662	0.3614	0.3543	0.3452	0.3347	0.3230	0.3106	0.2975	0.2842	0.2707
2	0.2014	0.2169	0.2303	0.2417	0.2510	0.2584	0.2640	0.2678	0.2700	0.2707
3	0.0738	0.0867	0.0998	0.1128	0.1255	0.1378	0.1496	0.1607	0.1710	0.1804
4	0.0203	0.0260	0.0324	0.0395	0.0471	0.0551	0.636	0.0723	0.0812	0.0902
5	0.0045	0.0062	0.0084	0.0111	0.0141	0.0176	0.0216	0.0260	0.0309	0.0361
6	0.0008	0.0012	0.0018	0.0026	0.0035	0.0047	0.0061	0.0078	0.0098	0.0120
7	0.0001	0.0002	0.0003	0.0005	0.0008	0.0011	0.0015	0.0020	0.0027	0.0034
8	0.0000	0.0000	0.0001	0.0001	0.0001	0.0002	0.0003	0.0005	0.0006	0.0009
9	0.0000	0.0000	0.0000	0.0000	0.0000	0.0000	0.0001	0.0001	0.0001	0.0002

	λ									
X	**2.1**	**2.2**	**2.3**	**2.4**	**2.5**	**2.6**	**2.7**	**2.8**	**2.9**	**3.0**
0	0.1225	0.1108	0.1003	0.0907	0.0821	0.0743	0.0672	0.0608	0.0550	0.0498
1	0.2572	0.2438	0.2306	0.2177	0.2052	0.1931	0.1815	0.1703	0.1596	0.1494
2	0.2700	0.2681	0.2652	0.2613	0.2565	0.2510	0.2450	0.2384	0.2314	0.2240
3	0.1890	0.1966	0.2033	0.2090	0.2138	0.2176	0.2205	0.2225	0.2237	0.2240
4	0.0992	0.1082	0.1169	0.1254	0.1336	0.1414	0.1488	0.1557	0.1622	0.1680
5	0.0417	0.0476	0.0538	0.0602	0.0668	0.0735	0.0804	0.0872	0.0940	0.1008
6	0.0146	0.0174	0.0206	0.0241	0.0278	0.0319	0.0362	0.0407	0.0455	0.0504
7	0.0044	0.0055	0.0068	0.0083	0.0099	0.0118	0.0139	0.0163	0.0188	0.0216
8	0.0011	0.0015	0.0019	0.0025	0.0031	0.0038	0.0047	0.0057	0.0068	0.0081
9	0.0003	0.0004	0.0005	0.0007	0.0009	0.0011	0.0014	0.0018	0.0022	0.0027
10	0.0001	0.0001	0.0001	0.0002	0.0002	0.0003	0.0004	0.0005	0.0006	0.0008
11	0.0000	0.0000	0.0000	0.0000	0.0000	0.0001	0.0001	0.0001	0.0002	0.0002
12	0.0000	0.0000	0.0000	0.0000	0.0000	0.0000	0.0000	0.0000	0.0000	0.0001

	λ									
X	**3.1**	**3.2**	**3.3**	**3.4**	**3.5**	**3.6**	**3.7**	**3.8**	**3.9**	**4.0**
0	0.0450	0.0408	0.0369	0.0334	0.0302	0.0273	0.0247	0.0224	0.0202	0.0183
1	0.1397	0.1340	0.1217	0.1135	0.1057	0.0984	0.0915	0.0850	0.0789	0.0733
2	0.2165	0.2087	0.2008	0.1929	0.1850	0.1771	0.1692	0.1615	0.1539	0.1465
3	0.2237	0.2226	0.2209	0.2186	0.2158	0.2125	0.2087	0.2046	0.2001	0.1954
4	0.1734	0.1781	0.1823	0.1858	0.1888	0.1912	0.1931	0.1944	0.1951	0.1954
5	0.1075	0.1140	0.1203	0.1264	0.1322	0.1377	0.1429	0.1477	0.1522	0.1563
6	0.0555	0.0608	0.0662	0.0716	0.0771	0.0826	0.0881	0.0936	0.0989	0.1042
7	0.0246	0.0278	0.0312	0.0348	0.0385	0.0425	0.0466	0.0508	0.0551	0.0595
8	0.0095	0.0111	0.0129	0.0148	0.0169	0.0191	0.0215	0.0241	0.0269	0.0298
9	0.0033	0.0040	0.0047	0.0056	0.0066	0.0076	0.0089	0.0102	0.0116	0.0132
10	0.0010	0.0013	0.0016	0.0019	0.0023	0.0028	0.0033	0.0039	0.0045	0.0053
11	0.0003	0.0004	0.0005	0.0006	0.0007	0.0009	0.0011	0.0013	0.0016	0.0019
12	0.0001	0.0001	0.0001	0.0002	0.0002	0.0003	0.0003	0.0004	0.0005	0.0006
13	0.0000	0.0000	0.0000	0.0000	0.0001	0.0001	0.0001	0.0001	0.0002	0.0002
14	0.0000	0.0000	0.0000	0.0000	0.0000	0.0000	0.0000	0.0000	0.0000	0.0001

continued

TABLE E.7

Table of Poisson Probabilities (*Continued*)

	λ									
X	**4.1**	**4.2**	**4.3**	**4.4**	**4.5**	**4.6**	**4.7**	**4.8**	**4.9**	**5.0**
0	0.0166	0.0150	0.0136	0.0123	0.0111	0.0101	0.0091	0.0082	0.0074	0.0067
1	0.0679	0.0630	0.0583	0.0540	0.0500	0.0462	0.0427	0.0395	0.0365	0.0337
2	0.1393	0.1323	0.1254	0.1188	0.1125	0.1063	0.1005	0.0948	0.0894	0.0842
3	0.1904	0.1852	0.1798	0.1743	0.1687	0.1631	0.1574	0.1517	0.1460	0.1404
4	0.1951	0.1944	0.1933	0.1917	0.1898	0.1875	0.1849	0.1820	0.1789	0.1755
5	0.1600	0.1633	0.1662	0.1687	0.1708	0.1725	0.1738	0.1747	0.1753	0.1755
6	0.1093	0.1143	0.1191	0.1237	0.1281	0.1323	0.1362	0.1398	0.1432	0.1462
7	0.0640	0.0686	0.0732	0.0778	0.0824	0.0869	0.0914	0.0959	0.1002	0.1044
8	0.0328	0.0360	0.0393	0.0428	0.0463	0.0500	0.0537	0.0575	0.0614	0.0653
9	0.0150	0.0168	0.0188	0.0209	0.0232	0.0255	0.0280	0.0307	0.0334	0.0363
10	0.0061	0.0071	0.0081	0.0092	0.0104	0.0118	0.0132	0.0147	0.0164	0.0181
11	0.0023	0.0027	0.0032	0.0037	0.0043	0.0049	0.0056	0.0064	0.0073	0.0082
12	0.0008	0.0009	0.0011	0.0014	0.0016	0.0019	0.0022	0.0026	0.0030	0.0034
13	0.0002	0.0003	0.0004	0.0005	0.0006	0.0007	0.0008	0.0009	0.0011	0.0013
14	0.0001	0.0001	0.0001	0.0001	0.0002	0.0002	0.0003	0.0003	0.0004	0.0005
15	0.0000	0.0000	0.0000	0.0000	0.0001	0.0001	0.0001	0.0001	0.0001	0.0002

	λ									
X	**5.1**	**5.2**	**5.3**	**5.4**	**5.5**	**5.6**	**5.7**	**5.8**	**5.9**	**6.0**
0	0.0061	0.0055	0.0050	0.0045	0.0041	0.0037	0.0033	0.0030	0.0027	0.0025
1	0.0311	0.0287	0.0265	0.0244	0.0225	0.0207	0.0191	0.0176	0.0162	0.0149
2	0.0793	0.0746	0.0701	0.0659	0.0618	0.0580	0.0544	0.0509	0.0477	0.0446
3	0.1348	0.1293	0.1239	0.1185	0.1133	0.1082	0.1033	0.0985	0.0938	0.0892
4	0.1719	0.1681	0.1641	0.1600	0.1558	0.1515	0.1472	0.1428	0.1383	0.1339
5	0.1753	0.1748	0.1740	0.1728	0.1714	0.1697	0.1678	0.1656	0.1632	0.1606
6	0.1490	0.1515	0.1537	0.1555	0.1571	0.1584	0.1594	0.1601	0.1605	0.1606
7	0.1086	0.1125	0.1163	0.1200	0.1234	0.1267	0.1298	0.1326	0.1353	0.1377
8	0.0692	0.0731	0.0771	0.0810	0.0849	0.0887	0.0925	0.0962	0.0998	0.1033
9	0.0392	0.0423	0.0454	0.0486	0.0519	0.0552	0.0586	0.0620	0.0654	0.0688
10	0.0200	0.0220	0.0241	0.0262	0.0285	0.0309	0.0334	0.0359	0.0386	0.0413
11	0.0093	0.0104	0.0116	0.0129	0.0143	0.0157	0.0173	0.0190	0.0207	0.0225
12	0.0039	0.0045	0.0051	0.0058	0.0065	0.0073	0.0082	0.0092	0.0102	0.0113
13	0.0015	0.0018	0.0021	0.0024	0.0028	0.0032	0.0036	0.0041	0.0046	0.0052
14	0.0006	0.0007	0.0008	0.0009	0.0011	0.0013	0.0015	0.0017	0.0019	0.0022
15	0.0002	0.0002	0.0003	0.0003	0.0004	0.0005	0.0006	0.0007	0.0008	0.0009
16	0.0001	0.0001	0.0001	0.0001	0.0001	0.0002	0.0002	0.0002	0.0003	0.0003
17	0.0000	0.0000	0.0000	0.0000	0.0000	0.0000	0.0001	0.0001	0.0001	0.0001

	λ									
X	**6.1**	**6.2**	**6.3**	**6.4**	**6.5**	**6.6**	**6.7**	**6.8**	**6.9**	**7.0**
0	0.0022	0.0020	0.0018	0.0017	0.0015	0.0014	0.0012	0.0011	0.0010	0.0009
1	0.0137	0.0126	0.0116	0.0106	0.0098	0.0090	0.0082	0.0076	0.0070	0.0064
2	0.0417	0.0390	0.0364	0.0340	0.0318	0.0296	0.0276	0.0258	0.0240	0.0223
3	0.0848	0.0806	0.0765	0.0726	0.0688	0.0652	0.0617	0.0584	0.0552	0.0521
4	0.1294	0.1249	0.1205	0.1162	0.1118	0.1076	0.1034	0.0992	0.0952	0.0912
5	0.1579	0.1549	0.1519	0.1487	0.1454	0.1420	0.1385	0.1349	0.1314	0.1277
6	0.1605	0.1601	0.1595	0.1586	0.1575	0.1562	0.1546	0.1529	0.1511	0.1490
7	0.1399	0.1418	0.1435	0.1450	0.1462	0.1472	0.1480	0.1486	0.1489	0.1490
8	0.1066	0.1099	0.1130	0.1160	0.1188	0.1215	0.1240	0.1263	0.1284	0.1304
9	0.0723	0.0757	0.0791	0.0825	0.0858	0.0891	0.0923	0.0954	0.0985	0.1014
10	0.0441	0.0469	0.0498	0.0528	0.0558	0.0588	0.0618	0.0649	0.0679	0.0710
11	0.0245	0.0265	0.0285	0.0307	0.0330	0.0353	0.0377	0.0401	0.0426	0.0452
12	0.0124	0.0137	0.0150	0.0164	0.0179	0.0194	0.0210	0.0277	0.0245	0.0264
13	0.0058	0.0065	0.0073	0.0081	0.0089	0.0098	0.0108	0.0119	0.0130	0.0142
14	0.0025	0.0029	0.0033	0.0037	0.0041	0.0046	0.0052	0.0058	0.0064	0.0071

continued

TABLE E.7
Table of Poisson Probabilities (*Continued*)

	λ									
X	**6.1**	**6.2**	**6.3**	**6.4**	**6.5**	**6.6**	**6.7**	**6.8**	**6.9**	**7.0**
15	0.0010	0.0012	0.0014	0.0016	0.0018	0.0020	0.0023	0.0026	0.0029	0.0033
16	0.0004	0.0005	0.0005	0.0006	0.0007	0.0008	0.0010	0.0011	0.0013	0.0014
17	0.0001	0.0002	0.0002	0.0002	0.0003	0.0003	0.0004	0.0004	0.0005	0.0006
18	0.0000	0.0001	0.0001	0.0001	0.0001	0.0001	0.0001	0.0002	0.0002	0.0002
19	0.0000	0.0000	0.0000	0.0000	0.0000	0.0000	0.0000	0.0001	0.0001	0.0001

	λ									
X	**7.1**	**7.2**	**7.3**	**7.4**	**7.5**	**7.6**	**7.7**	**7.8**	**7.9**	**8.0**
0	0.0008	0.0007	0.0007	0.0006	0.0006	0.0005	0.0005	0.0004	0.0004	0.0003
1	0.0059	0.0054	0.0049	0.0045	0.0041	0.0038	0.0035	0.0032	0.0029	0.0027
2	0.0208	0.0194	0.0180	0.0167	0.0156	0.0145	0.0134	0.0125	0.0116	0.0107
3	0.0492	0.0464	0.0438	0.0413	0.0389	0.0366	0.0345	0.0324	0.0305	0.0286
4	0.0874	0.0836	0.0799	0.0764	0.0729	0.0696	0.0663	0.0632	0.0602	0.0573
5	0.1241	0.1204	0.1167	0.1130	0.1094	0.1057	0.1021	0.0986	0.0951	0.0916
6	0.1468	0.1445	0.1420	0.1394	0.1367	0.1339	0.1311	0.1282	0.1252	0.1221
7	0.1489	0.1486	0.1481	0.1474	0.1465	0.1454	0.1442	0.1428	0.1413	0.1396
8	0.1321	0.1337	0.1351	0.1363	0.1373	0.1382	0.1388	0.1392	0.1395	0.1396
9	0.1042	0.1070	0.1096	0.1121	0.1144	0.1167	0.1187	0.1207	0.1224	0.1241
10	0.0740	0.0770	0.0800	0.0829	0.0858	0.0887	0.0914	0.0941	0.0967	0.0993
11	0.0478	0.0504	0.0531	0.0558	0.0585	0.0613	0.0640	0.0667	0.0695	0.0722
12	0.0283	0.0303	0.0323	0.0344	0.0366	0.0388	0.0411	0.0434	0.0457	0.0481
13	0.0154	0.0168	0.0181	0.0196	0.0211	0.0227	0.0243	0.0260	0.0278	0.0296
14	0.0078	0.0086	0.0095	0.0104	0.0113	0.0123	0.0134	0.0145	0.0157	0.0169
15	0.0037	0.0041	0.0046	0.0051	0.0057	0.0062	0.0069	0.0075	0.0083	0.0090
16	0.0016	0.0019	0.0021	0.0024	0.0026	0.0030	0.0033	0.0037	0.0041	0.0045
17	0.0007	0.0008	0.0009	0.0010	0.0012	0.0013	0.0015	0.0017	0.0019	0.0021
18	0.0003	0.0003	0.0004	0.0004	0.0005	0.0006	0.0006	0.0007	0.0008	0.0009
19	0.0001	0.0001	0.0001	0.0002	0.0002	0.0002	0.0003	0.0003	0.0003	0.0004
20	0.0000	0.0000	0.0001	0.0001	0.0001	0.0001	0.0001	0.0001	0.0001	0.0002
21	0.0000	0.0000	0.0000	0.0000	0.0000	0.0000	0.0000	0.0000	0.0001	0.0001

	λ									
X	**8.1**	**8.2**	**8.3**	**8.4**	**8.5**	**8.6**	**8.7**	**8.8**	**8.9**	**9.0**
0	0.0003	0.0003	0.0002	0.0002	0.0002	0.0002	0.0002	0.0002	0.0001	0.0001
1	0.0025	0.0023	0.0021	0.0019	0.0017	0.0016	0.0014	0.0013	0.0012	0.0011
2	0.0100	0.0092	0.0086	0.0079	0.0074	0.0068	0.0063	0.0058	0.0054	0.0050
3	0.0269	0.0252	0.0237	0.0222	0.0208	0.0195	0.0183	0.0171	0.0160	0.0150
4	0.0544	0.0517	0.0491	0.0466	0.0443	0.0420	0.0398	0.0377	0.0357	0.0337
5	0.0882	0.0849	0.0816	0.0784	0.0752	0.0722	0.0692	0.0663	0.0635	0.0607
6	0.1191	0.1160	0.1128	0.1097	0.1066	0.1034	0.1003	0.0972	0.0941	0.0911
7	0.1378	0.1358	0.1338	0.1317	0.1294	0.1271	0.1247	0.1222	0.1197	0.1171
8	0.1395	0.1392	0.1388	0.1382	0.1375	0.1366	0.1356	0.1344	0.1332	0.1318
9	0.1256	0.1269	0.1280	0.1290	0.1299	0.1306	0.1311	0.1315	0.1317	0.1318
10	0.1017	0.1040	0.1063	0.1084	0.1104	0.1123	0.1140	0.1157	0.1172	0.1186
11	0.0749	0.0776	0.0802	0.0828	0.0853	0.0878	0.0902	0.0925	0.0948	0.0970
12	0.0505	0.0530	0.0555	0.0579	0.0604	0.0629	0.0654	0.0679	0.0703	0.0728
13	0.0315	0.0334	0.0354	0.0374	0.0395	0.0416	0.0438	0.0459	0.0481	0.0504
14	0.0182	0.0196	0.0210	0.0225	0.0240	0.0256	0.0272	0.0289	0.0306	0.0324
15	0.0098	0.0107	0.0116	0.0126	0.0136	0.0147	0.0158	0.0169	0.0182	0.0194
16	0.0050	0.0055	0.0060	0.0066	0.0072	0.0079	0.0086	0.0093	0.0101	0.0109
17	0.0024	0.0026	0.0029	0.0033	0.0036	0.0040	0.0044	0.0048	0.0053	0.0058
18	0.0011	0.0012	0.0014	0.0015	0.0017	0.0019	0.0021	0.0024	0.0026	0.0029
19	0.0005	0.0005	0.0006	0.0007	0.0008	0.0009	0.0010	0.0011	0.0012	0.0014
20	0.0002	0.0002	0.0002	0.0003	0.0003	0.0004	0.0004	0.0005	0.0005	0.0006
21	0.0001	0.0001	0.0001	0.0001	0.0001	0.0002	0.0002	0.0002	0.0002	0.0003
22	0.0000	0.0000	0.0000	0.0000	0.0001	0.0001	0.0001	0.0001	0.0001	0.0001

continued

TABLE E.7

Table of Poisson Probabilities (*Continued*)

	λ									
X	**9.1**	**9.2**	**9.3**	**9.4**	**9.5**	**9.6**	**9.7**	**9.8**	**9.9**	**10**
0	0.0001	0.0001	0.0001	0.0001	0.0001	0.0001	0.0001	0.0001	0.0001	0.0000
1	0.0010	0.0009	0.0009	0.0008	0.0007	0.0007	0.0006	0.0005	0.0005	0.0005
2	0.0046	0.0043	0.0040	0.0037	0.0034	0.0031	0.0029	0.0027	0.0025	0.0023
3	0.0140	0.0131	0.0123	0.0115	0.0107	0.0100	0.0093	0.0087	0.0081	0.0076
4	0.0319	0.0302	0.0285	0.0269	0.0254	0.0240	0.0226	0.0213	0.0201	0.0189
5	0.0581	0.0555	0.0530	0.0506	0.0483	0.0460	0.0439	0.0418	0.0398	0.0378
6	0.0881	0.0851	0.0822	0.0793	0.0764	0.0736	0.0709	0.0682	0.0656	0.0631
7	0.1145	0.1118	0.1091	0.1064	0.1037	0.1010	0.0982	0.0955	0.0928	0.0901
8	0.1302	0.1286	0.1269	0.1251	0.1232	0.1212	0.1191	0.1170	0.1148	0.1126
9	0.1317	0.1315	0.1311	0.1306	0.1300	0.1293	0.1284	0.1274	0.1263	0.1251
10	0.1198	0.1210	0.1219	0.1228	0.1235	0.1241	0.1245	0.1249	0.1250	0.1251
11	0.0991	0.1012	0.1031	0.1049	0.1067	0.1083	0.1098	0.1112	0.1125	0.1137
12	0.0752	0.0776	0.0799	0.0822	0.0844	0.0866	0.0888	0.0908	0.0928	0.0948
13	0.0526	0.0549	0.0572	0.0594	0.0617	0.0640	0.0662	0.0685	0.0707	0.0729
14	0.0342	0.0361	0.0380	0.0399	0.0419	0.0439	0.0459	0.0479	0.0500	0.0521
15	0.0208	0.0221	0.0235	0.0250	0.0265	0.0281	0.0297	0.0313	0.0330	0.0347
16	0.0118	0.0127	0.0137	0.0147	0.0157	0.0168	0.0180	0.0192	0.0204	0.0217
17	0.0063	0.0069	0.0075	0.0081	0.0088	0.0095	0.0103	0.0111	0.0119	0.0128
18	0.0032	0.0035	0.0039	0.0042	0.0046	0.0051	0.0055	0.0060	0.0065	0.0071
19	0.0015	0.0017	0.0019	0.0021	0.0023	0.0026	0.0028	0.0031	0.0034	0.0037
20	0.0007	0.0008	0.0009	0.0010	0.0011	0.0012	0.0014	0.0015	0.0017	0.0019
21	0.0003	0.0003	0.0004	0.0004	0.0005	0.0006	0.0006	0.0007	0.0008	0.0009
22	0.0001	0.0001	0.0002	0.0002	0.0002	0.0002	0.0003	0.0003	0.0004	0.0004
23	0.0000	0.0001	0.0001	0.0001	0.0001	0.0001	0.0001	0.0001	0.0002	0.0002
24	0.0000	0.0000	0.0000	0.0000	0.0000	0.0000	0.0000	0.0001	0.0001	0.0001

X	**λ = 20**	*X*	**λ = 20**	*X*	**λ = 20**	*X*	**λ = 20**
0	0.0000	10	0.0058	20	0.0888	30	0.0083
1	0.0000	11	0.0106	21	0.0846	31	0.0054
2	0.0000	12	0.0176	22	0.0769	32	0.0034
3	0.0000	13	0.0271	23	0.0669	33	0.0020
4	0.0000	14	0.0387	24	0.0557	34	0.0012
5	0.0001	15	0.0516	25	0.0446	35	0.0007
6	0.0002	16	0.0646	26	0.0343	36	0.0004
7	0.0005	17	0.0760	27	0.0254	37	0.0002
8	0.0013	18	0.0844	28	0.0181	38	0.0001
9	0.0029	19	0.0888	29	0.0125	39	0.0001

TABLE E.8
Lower and Upper Critical Values T_1 of Wilcoxon Rank Sum Test

	α		n_1						
n_2	**One-tail**	**Two-tail**	**4**	**5**	**6**	**7**	**8**	**9**	**10**
4	0.05	0.10	11,25						
	0.025	0.05	10,26						
	0.01	0.02	—,—						
	0.005	0.01	—,—						
5	0.05	0.10	12,28	19,36					
	0.025	0.05	11,29	17,38					
	0.01	0.02	10,30	16,39					
	0.005	0.01	—,—	15,40					
6	0.05	0.10	13,31	20,40	28,50				
	0.025	0.05	12,32	18,42	26,52				
	0.01	0.02	11,33	17,43	24,54				
	0.005	0.01	10,34	16,44	23,55				
7	0.05	0.10	14,34	21,44	29,55	39,66			
	0.025	0.05	13,35	20,45	27,57	36,69			
	0.01	0.02	11,37	18,47	25,59	34,71			
	0.005	0.01	10,38	16,49	24,60	32,73			
8	0.05	0.10	15,37	23,47	31,59	41,71	51,85		
	0.025	0.05	14,38	21,49	29,61	38,74	49,87		
	0.01	0.02	12,40	19,51	27,63	35,77	45,91		
	0.005	0.01	11,41	17,53	25,65	34,78	43,93		
9	0.05	0.10	16,40	24,51	33,63	43,76	54,90	66,105	
	0.025	0.05	14,42	22,53	31,65	40,79	51,93	62,109	
	0.01	0.02	13,43	20,55	28,68	37,82	47,97	59,112	
	0.005	0.01	11,45	18,57	26,70	35,84	45,99	56,115	
10	0.05	0.10	17,43	26,54	35,67	45,81	56,96	69,111	82,128
	0.025	0.05	15,45	23,57	32,70	42,84	53,99	65,115	78,132
	0.01	0.02	13,47	21,59	29,73	39,87	49,103	61,119	74,136
	0.005	0.01	12,48	19,61	27,75	37,89	47,105	58,122	71,139

Source: Adapted from Table 1 of F. Wilcoxon and R. A. Wilcox, Some Rapid Approximate Statistical Procedures *(Pearl River, NY: Lederle Laboratories, 1964), with permission of the American Cyanamid Company.*

TABLE E.9

Lower and Upper Critical Values *W* of Wilcoxon Signed Ranks Test

One-Tail Two-Tail	α = .05 α = .10	α = .025 α = .05	α = .01 α = .02	α = .005 α = .01
n		(Lower, Upper)		
5	0,15	—,—	—,—	—,—
6	2,19	0,21	—,—	—,—
7	3,25	2,26	0,28	—,—
8	5,31	3,33	1,35	0,36
9	8,37	5,40	3,42	1,44
10	10,45	8,47	5,50	3,52
11	13,53	10,56	7,59	5,61
12	17,61	13,65	10,68	7,71
13	21,70	17,74	12,79	10,81
14	25,80	21,84	16,89	13,92
15	30,90	25,95	19,101	16,104
16	35,101	29,107	23,113	19,117
17	41,112	34,119	27,126	23,130
18	47,124	40,131	32,139	27,144
19	53,137	46,144	37,153	32,158
20	60,150	52,158	43,167	37,173

Source: Adapted from Table 2 of F. Wilcoxon and R. A. Wilcox, Some Rapid Approximate Statistical Procedures *(Pearl River, NY: Lederle Laboratories, 1964), with permission of the American Cyanamid Company.*

TABLE E.10

Critical Values of the Studentized Range Q

Upper 5% Points ($\alpha = 0.05$)																			
Denominator Degrees of Freedom	Numerator Degrees of Freedom																		
	2	3	4	5	6	7	8	9	10	11	12	13	14	15	16	17	18	19	20
1	18.00	27.00	32.80	37.10	40.40	43.10	45.40	47.40	49.10	50.60	52.00	53.20	54.30	55.40	56.30	57.20	58.00	58.80	59.60
2	6.09	8.30	9.80	10.90	11.70	12.40	13.00	13.50	14.00	14.40	14.70	15.10	15.40	15.70	15.90	16.10	16.40	16.60	16.80
3	4.50	5.91	6.82	7.50	8.04	8.48	8.85	9.18	9.46	9.72	9.95	10.15	10.35	10.52	10.69	10.84	10.98	11.11	11.24
4	3.93	5.04	5.76	6.29	6.71	7.05	7.35	7.60	7.83	8.03	8.21	8.37	8.52	8.66	8.79	8.91	9.03	9.13	9.23
5	3.64	4.60	5.22	5.67	6.03	6.33	6.58	6.80	6.99	7.17	7.32	7.47	7.60	7.72	7.83	7.93	8.03	8.12	8.21
6	3.46	4.34	4.90	5.31	5.63	5.89	6.12	6.32	6.49	6.65	6.79	6.92	7.03	7.14	7.24	7.34	7.43	7.51	7.59
7	3.34	4.16	4.68	5.06	5.36	5.61	5.82	6.00	6.16	6.30	6.43	6.55	6.66	6.76	6.85	6.94	7.02	7.09	7.17
8	3.26	4.04	4.53	4.89	5.17	5.40	5.60	5.77	5.92	6.05	6.18	6.29	6.39	6.48	6.57	6.65	6.73	6.80	6.87
9	3.20	3.95	4.42	4.76	5.02	5.24	5.43	5.60	5.74	5.87	5.98	6.09	6.19	6.28	6.36	6.44	6.51	6.58	6.64
10	3.15	3.88	4.33	4.65	4.91	5.12	5.30	5.46	5.60	5.72	5.83	5.93	6.03	6.11	6.20	6.27	6.34	6.40	6.47
11	3.11	3.82	4.26	4.57	4.82	5.03	5.20	5.35	5.49	5.61	5.71	5.81	5.90	5.99	6.06	6.14	6.20	6.26	6.33
12	3.08	3.77	4.20	4.51	4.75	4.95	5.12	5.27	5.40	5.51	5.62	5.71	5.80	5.88	5.95	6.03	6.09	6.15	6.21
13	3.06	3.73	4.15	4.45	4.69	4.88	5.05	5.19	5.32	5.43	5.53	5.63	5.71	5.79	5.86	5.93	6.00	6.05	6.11
14	3.03	3.70	4.11	4.41	4.64	4.83	4.99	5.13	5.25	5.36	5.46	5.55	5.64	5.72	5.79	5.85	5.92	5.97	6.03
15	3.01	3.67	4.08	4.37	4.60	4.78	4.94	5.08	5.20	5.31	5.40	5.49	5.58	5.65	5.72	5.79	5.85	5.90	5.96
16	3.00	3.65	4.05	4.33	4.56	4.74	4.90	5.03	5.15	5.26	5.35	5.44	5.52	5.59	5.66	5.72	5.79	5.84	5.90
17	2.98	3.63	4.02	4.30	4.52	4.71	4.86	4.99	5.11	5.21	5.31	5.39	5.47	5.55	5.61	5.68	5.74	5.79	5.84
18	2.97	3.61	4.00	4.28	4.49	4.67	4.82	4.96	5.07	5.17	5.27	5.35	5.43	5.50	5.57	5.63	5.69	5.74	5.79
19	2.96	3.59	3.98	4.25	4.47	4.65	4.79	4.92	5.04	5.14	5.23	5.32	5.39	5.46	5.53	5.59	5.65	5.70	5.75
20	2.95	3.58	3.96	4.23	4.45	4.62	4.77	4.90	5.01	5.11	5.20	5.28	5.36	5.43	5.49	5.55	5.61	5.66	5.71
24	2.92	3.53	3.90	4.17	4.37	4.54	4.68	4.81	4.92	5.01	5.10	5.18	5.25	5.32	5.38	5.44	5.50	5.54	5.59
30	2.89	3.49	3.84	4.10	4.30	4.46	4.60	4.72	4.83	4.92	5.00	5.08	5.15	5.21	5.27	5.33	5.38	5.43	5.48
40	2.86	3.44	3.79	4.04	4.23	4.39	4.52	4.63	4.74	4.82	4.91	4.98	5.05	5.11	5.16	5.22	5.27	5.31	5.36
60	2.83	3.40	3.74	3.98	4.16	4.31	4.44	4.55	4.65	4.73	4.81	4.88	4.94	5.00	5.06	5.11	5.16	5.20	5.24
120	2.80	3.36	3.69	3.92	4.10	4.24	4.36	4.48	4.56	4.64	4.72	4.78	4.84	4.90	4.95	5.00	5.05	5.09	5.13
∞	2.77	3.31	3.63	3.86	4.03	4.17	4.29	4.39	4.47	4.55	4.62	4.68	4.74	4.80	4.85	4.89	4.93	4.97	5.01

continued

TABLE E.10

Critical Values of the Studentized Range Q (*Continued*)

Upper 1% Points ($\alpha = 0.01$)

Denominator Degrees of Freedom	Numerator Degrees of Freedom 2	3	4	5	6	7	8	9	10	11	12	13	14	15	16	17	18	19	20
1	90.00	135.00	164.00	186.00	202.00	216.00	227.00	237.00	246.00	253.00	260.00	266.00	272.00	277.00	282.00	286.00	290.00	294.00	298.00
2	14.00	19.00	22.30	24.70	26.60	28.20	29.50	30.70	31.70	32.60	33.40	34.10	34.80	35.40	36.00	36.50	37.00	37.50	37.90
3	8.26	10.60	12.20	13.30	14.20	15.00	15.60	16.20	16.70	17.10	17.50	17.90	18.20	18.50	18.80	19.10	19.30	19.50	19.80
4	6.51	8.12	9.17	9.96	10.60	11.10	11.50	11.90	12.30	12.60	12.80	13.10	13.30	13.50	13.70	13.90	14.10	14.20	14.40
5	5.70	6.97	7.80	8.42	8.91	9.32	9.67	9.97	10.24	10.48	10.70	10.89	11.08	11.24	11.40	11.55	11.68	11.81	11.93
6	5.24	6.33	7.03	7.56	7.97	8.32	8.61	8.87	9.10	9.30	9.49	9.65	9.81	9.95	10.08	10.21	10.32	10.43	10.54
7	4.95	5.92	6.54	7.01	7.37	7.68	7.94	8.17	8.37	8.55	8.71	8.86	9.00	9.12	9.24	9.35	9.46	9.55	9.65
8	4.74	5.63	6.20	6.63	6.96	7.24	7.47	7.68	7.87	8.03	8.18	8.31	8.44	8.55	8.66	8.76	8.85	8.94	9.03
9	4.60	5.43	5.96	6.35	6.66	6.91	7.13	7.32	7.49	7.65	7.78	7.91	8.03	8.13	8.23	8.32	8.41	8.49	8.57
10	4.48	5.27	5.77	6.14	6.43	6.67	6.87	7.05	7.21	7.36	7.48	7.60	7.71	7.81	7.91	7.99	8.07	8.15	8.22
11	4.39	5.14	5.62	5.97	6.26	6.48	6.67	6.84	6.99	7.13	7.25	7.36	7.46	7.56	7.65	7.73	7.81	7.88	7.95
12	4.32	5.04	5.50	5.84	6.10	6.32	6.51	6.67	6.81	6.94	7.06	7.17	7.26	7.36	7.44	7.52	7.59	7.66	7.73
13	4.26	4.96	5.40	5.73	5.98	6.19	6.37	6.53	6.67	6.79	6.90	7.01	7.10	7.19	7.27	7.34	7.42	7.48	7.55
14	4.21	4.89	5.32	5.63	5.88	6.08	6.26	6.41	6.54	6.66	6.77	6.87	6.96	7.05	7.12	7.20	7.27	7.33	7.39
15	4.17	4.83	5.25	5.56	5.80	5.99	6.16	6.31	6.44	6.55	6.66	6.76	6.84	6.93	7.00	7.07	7.14	7.20	7.26
16	4.13	4.78	5.19	5.49	5.72	5.92	6.08	6.22	6.35	6.46	6.56	6.66	6.74	6.82	6.90	6.97	7.03	7.09	7.15
17	4.10	4.74	5.14	5.43	5.66	5.85	6.01	6.15	6.27	6.38	6.48	6.57	6.66	6.73	6.80	6.87	6.94	7.00	7.05
18	4.07	4.70	5.09	5.38	5.60	5.79	5.94	6.08	6.20	6.31	6.41	6.50	6.58	6.65	6.72	6.79	6.85	6.91	6.96
19	4.05	4.67	5.05	5.33	5.55	5.73	5.89	6.02	6.14	6.25	6.34	6.43	6.51	6.58	6.65	6.72	6.78	6.84	6.89
20	4.02	4.64	5.02	5.29	5.51	5.69	5.84	5.97	6.09	6.19	6.29	6.37	6.45	6.52	6.59	6.65	6.71	6.76	6.82
24	3.96	4.54	4.91	5.17	5.37	5.54	5.69	5.81	5.92	6.02	6.11	6.19	6.26	6.33	6.39	6.45	6.51	5.56	6.61
30	3.89	4.45	4.80	5.05	5.24	5.40	5.54	5.65	5.76	5.85	5.93	6.01	6.08	6.14	6.20	6.26	6.31	6.36	6.41
40	3.82	4.37	4.70	4.93	5.11	5.27	5.39	5.50	5.60	5.69	5.77	5.84	5.90	5.96	6.02	6.07	6.12	6.17	6.21
60	3.76	4.28	4.60	4.82	4.99	5.13	5.25	5.36	5.45	5.53	5.60	5.67	5.73	5.79	5.84	5.89	5.93	5.98	6.02
120	3.70	4.20	4.50	4.71	4.87	5.01	5.12	5.21	5.30	5.38	5.44	5.51	5.56	5.61	5.66	5.71	5.75	5.79	5.83
∞	3.64	4.12	4.40	4.60	4.76	4.88	4.99	5.08	5.16	5.23	5.29	5.35	5.40	5.45	5.49	5.54	5.57	5.61	5.65

Source: Reprinted from E. S. Pearson and H. O. Hartley, eds., Table 29 of Biometrika Tables for Statisticians, Vol. 1, *3rd ed., 1966, by permission of the* Biometrika *Trustees, London.*

TABLE E.11

Critical Values d_L and d_U of the Durbin-Watson Statistic D (Critical Values are One-Sided)[a]

	α = 0.05										α = 0.01									
	k = 1		k = 2		k = 3		k = 4		k = 5		k = 1		k = 2		k = 3		k = 4		k = 5	
n	d_L	d_U	d_L	d_U	d_L	d_U	d_L	d_U	d_L	d_U	d_L	d_U	d_L	d_U	d_L	d_U	d_L	d_U	d_L	d_U
15	1.08	1.36	.95	1.54	.82	1.75	.69	1.97	.56	2.21	.81	1.07	.70	1.25	.59	1.46	.49	1.70	.39	1.96
16	1.10	1.37	.98	1.54	.86	1.73	.74	1.93	.62	2.15	.84	1.09	.74	1.25	.63	1.44	.53	1.66	.44	1.90
17	1.13	1.38	1.02	1.54	.90	1.71	.78	1.90	.67	2.10	.87	1.10	.77	1.25	.67	1.43	.57	1.63	.48	1.85
18	1.16	1.39	1.05	1.53	.93	1.69	.82	1.87	.71	2.06	.90	1.12	.80	1.26	.71	1.42	.61	1.60	.52	1.80
19	1.18	1.40	1.08	1.53	.97	1.68	.86	1.85	.75	2.02	.93	1.13	.83	1.26	.74	1.41	.65	1.58	.56	1.77
20	1.20	1.41	1.10	1.54	1.00	1.68	.90	1.83	.79	1.99	.95	1.15	.86	1.27	.77	1.41	.68	1.57	.60	1.74
21	1.22	1.42	1.13	1.54	1.03	1.67	.93	1.81	.83	1.96	.97	1.16	.89	1.27	.80	1.41	.72	1.55	.63	1.71
22	1.24	1.43	1.15	1.54	1.05	1.66	.96	1.80	.86	1.94	1.00	1.17	.91	1.28	.83	1.40	.75	1.54	.66	1.69
23	1.26	1.44	1.17	1.54	1.08	1.66	.99	1.79	.90	1.92	1.02	1.19	.94	1.29	.86	1.40	.77	1.53	.70	1.67
24	1.27	1.45	1.19	1.55	1.10	1.66	1.01	1.78	.93	1.90	1.04	1.20	.96	1.30	.88	1.41	.80	1.53	.72	1.66
25	1.29	1.45	1.21	1.55	1.12	1.66	1.04	1.77	.95	1.89	1.05	1.21	.98	1.30	.90	1.41	.83	1.52	.75	1.65
26	1.30	1.46	1.22	1.55	1.14	1.65	1.06	1.76	.98	1.88	1.07	1.22	1.00	1.31	.93	1.41	.85	1.52	.78	1.64
27	1.32	1.47	1.24	1.56	1.16	1.65	1.08	1.76	1.01	1.86	1.09	1.23	1.02	1.32	.95	1.41	.88	1.51	.81	1.63
28	1.33	1.48	1.26	1.56	1.18	1.65	1.10	1.75	1.03	1.85	1.10	1.24	1.04	1.32	.97	1.41	.90	1.51	.83	1.62
29	1.34	1.48	1.27	1.56	1.20	1.65	1.12	1.74	1.05	1.84	1.12	1.25	1.05	1.33	.99	1.42	.92	1.51	.85	1.61
30	1.35	1.49	1.28	1.57	1.21	1.65	1.14	1.74	1.07	1.83	1.13	1.26	1.07	1.34	1.01	1.42	.94	1.51	.88	1.61
31	1.36	1.50	1.30	1.57	1.23	1.65	1.16	1.74	1.09	1.83	1.15	1.27	1.08	1.34	1.02	1.42	.96	1.51	.90	1.60
32	1.37	1.50	1.31	1.57	1.24	1.65	1.18	1.73	1.11	1.82	1.16	1.28	1.10	1.35	1.04	1.43	.98	1.51	.92	1.60
33	1.38	1.51	1.32	1.58	1.26	1.65	1.19	1.73	1.13	1.81	1.17	1.29	1.11	1.36	1.05	1.43	1.00	1.51	.94	1.59
34	1.39	1.51	1.33	1.58	1.27	1.65	1.21	1.73	1.15	1.81	1.18	1.30	1.13	1.36	1.07	1.43	1.01	1.51	.95	1.59
35	1.40	1.52	1.34	1.58	1.28	1.65	1.22	1.73	1.16	1.80	1.19	1.31	1.14	1.37	1.08	1.44	1.03	1.51	.97	1.59
36	1.41	1.52	1.35	1.59	1.29	1.65	1.24	1.73	1.18	1.80	1.21	1.32	1.15	1.38	1.10	1.44	1.04	1.51	.99	1.59
37	1.42	1.53	1.36	1.59	1.31	1.66	1.25	1.72	1.19	1.80	1.22	1.32	1.16	1.38	1.11	1.45	1.06	1.51	1.00	1.59
38	1.43	1.54	1.37	1.59	1.32	1.66	1.26	1.72	1.21	1.79	1.23	1.33	1.18	1.39	1.12	1.45	1.07	1.52	1.02	1.58
39	1.43	1.54	1.38	1.60	1.33	1.66	1.27	1.72	1.22	1.79	1.24	1.34	1.19	1.39	1.14	1.45	1.09	1.52	1.03	1.58
40	1.44	1.54	1.39	1.60	1.34	1.66	1.29	1.72	1.23	1.79	1.25	1.34	1.20	1.40	1.15	1.46	1.10	1.52	1.05	1.58
45	1.48	1.57	1.43	1.62	1.38	1.67	1.34	1.72	1.29	1.78	1.29	1.38	1.24	1.42	1.20	1.48	1.16	1.53	1.11	1.58
50	1.50	1.59	1.46	1.63	1.42	1.67	1.38	1.72	1.34	1.77	1.32	1.40	1.28	1.45	1.24	1.49	1.20	1.54	1.16	1.59
55	1.53	1.60	1.49	1.64	1.45	1.68	1.41	1.72	1.38	1.77	1.36	1.43	1.32	1.47	1.28	1.51	1.25	1.55	1.21	1.59
60	1.55	1.62	1.51	1.65	1.48	1.69	1.44	1.73	1.41	1.77	1.38	1.45	1.35	1.48	1.32	1.52	1.28	1.56	1.25	1.60
65	1.57	1.63	1.54	1.66	1.50	1.70	1.47	1.73	1.44	1.77	1.41	1.47	1.38	1.50	1.35	1.53	1.31	1.57	1.28	1.61
70	1.58	1.64	1.55	1.67	1.52	1.70	1.49	1.74	1.46	1.77	1.43	1.49	1.40	1.52	1.37	1.55	1.34	1.58	1.31	1.61
75	1.60	1.65	1.57	1.68	1.54	1.71	1.51	1.74	1.49	1.77	1.45	1.50	1.42	1.53	1.39	1.56	1.37	1.59	1.34	1.62
80	1.61	1.66	1.59	1.69	1.56	1.72	1.53	1.74	1.51	1.77	1.47	1.52	1.44	1.54	1.42	1.57	1.39	1.60	1.36	1.62
85	1.62	1.67	1.60	1.70	1.57	1.72	1.55	1.75	1.52	1.77	1.48	1.53	1.46	1.55	1.43	1.58	1.41	1.60	1.39	1.63
90	1.63	1.68	1.61	1.70	1.59	1.73	1.57	1.75	1.54	1.78	1.50	1.54	1.47	1.56	1.45	1.59	1.43	1.61	1.41	1.64
95	1.64	1.69	1.62	1.71	1.60	1.73	1.58	1.75	1.56	1.78	1.51	1.55	1.49	1.57	1.47	1.60	1.45	1.62	1.42	1.64
100	1.65	1.69	1.63	1.72	1.61	1.74	1.59	1.76	1.57	1.78	1.52	1.56	1.50	1.58	1.48	1.60	1.46	1.63	1.44	1.65

[a]*n = number of observations; k = number of independent variables.*

Source: This table is reproduced from Biometrika, *41 (1951): 173 and 175, with the permission of the* Biometrika *Trustees.*

TABLE E.12
Selected Critical Values of F for Cook's D_i Statistic

	$\alpha = 0.50$											
	Numerator $df = k + 1$											
Denominator $df = n - k - 1$	**2**	**3**	**4**	**5**	**6**	**7**	**8**	**9**	**10**	**12**	**15**	**20**
10	.743	.845	.899	.932	.954	.971	.983	.992	1.00	1.01	1.02	1.03
11	.739	.840	.893	.926	.948	.964	.977	.986	.994	1.01	1.02	1.03
12	.735	.835	.888	.921	.943	.959	.972	.981	.989	1.00	1.01	1.02
15	.726	.826	.878	.911	.933	.949	.960	.970	.977	.989	1.00	1.01
20	.718	.816	.868	.900	.922	.938	.950	.959	.966	.977	.989	1.00
24	.714	.812	.863	.895	.917	.932	.944	.953	.961	.972	.983	.994
30	.709	.807	.858	.890	.912	.927	.939	.948	.955	.966	.978	.989
40	.705	.802	.854	.885	.907	.922	.934	.943	.950	.961	.972	.983
60	.701	.798	.849	.880	.901	.917	.928	.937	.945	.956	.967	.978
120	.697	.793	.844	.875	.896	.912	.923	.932	.939	.950	.961	.972
∞	.693	.789	.839	.870	.891	.907	.918	.927	.934	.945	.956	.967

Source: Extracted from E. S. Pearson and H. O. Hartley, eds., Biometrika Tables for Statisticians, *3rd ed., 1966, by permission of the* Biometrika *Trustees.*

TABLE E.13
Control Chart Factors

Number of Observations in Sample	d_2	d_3	D_3	D_4	A_2
2	1.128	0.853	0	3.267	1.880
3	1.693	0.888	0	2.575	1.023
4	2.059	0.880	0	2.282	0.729
5	2.326	0.864	0	2.114	0.577
6	2.534	0.848	0	2.004	0.483
7	2.704	0.833	0.076	1.924	0.419
8	2.847	0.820	0.136	1.864	0.373
9	2.970	0.808	0.184	1.816	0.337
10	3.078	0.797	0.223	1.777	0.308
11	3.173	0.787	0.256	1.744	0.285
12	3.258	0.778	0.283	1.717	0.266
13	3.336	0.770	0.307	1.693	0.249
14	3.407	0.763	0.328	1.672	0.235
15	3.472	0.756	0.347	1.653	0.223
16	3.532	0.750	0.363	1.637	0.212
17	3.588	0.744	0.378	1.622	0.203
18	3.640	0.739	0.391	1.609	0.194
19	3.689	0.733	0.404	1.596	0.187
20	3.735	0.729	0.415	1.585	0.180
21	3.778	0.724	0.425	1.575	0.173
22	3.819	0.720	0.435	1.565	0.167
23	3.858	0.716	0.443	1.557	0.162
24	3.895	0.712	0.452	1.548	0.157
25	3.931	0.708	0.459	1.541	0.153

Source: Reprinted from ASTM-STP 15D by kind permission of the American Society for Testing and Materials.

TABLE E.14

The Standardized Normal Distribution

Entry represents area under the standardized normal distribution from the mean to Z

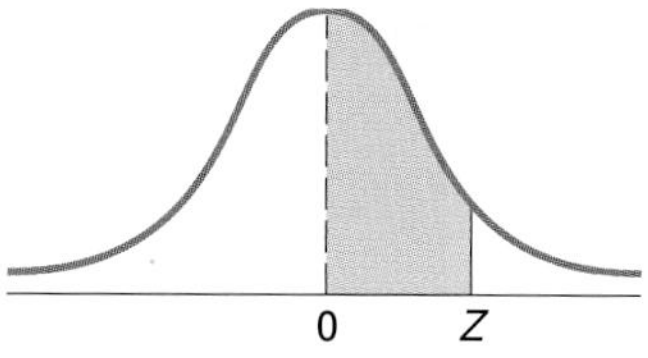

Z	.00	.01	.02	.03	.04	.05	.06	.07	.08	.09
0.0	.0000	.0040	.0080	.0120	.0160	.0199	.0239	.0279	.0319	.0359
0.1	.0398	.0438	.0478	.0517	.0557	.0596	.0636	.0675	.0714	.0753
0.2	.0793	.0832	.0871	.0910	.0948	.0987	.1026	.1064	.1103	.1141
0.3	.1179	.1217	.1255	.1293	.1331	.1368	.1406	.1443	.1480	.1517
0.4	.1554	.1591	.1628	.1664	.1700	.1736	.1772	.1808	.1844	.1879
0.5	.1915	.1950	.1985	.2019	.2054	.2088	.2123	.2157	.2190	.2224
0.6	.2257	.2291	.2324	.2357	.2389	.2422	.2454	.2486	.2518	.2549
0.7	.2580	.2612	.2642	.2673	.2704	.2734	.2764	.2794	.2823	.2852
0.8	.2881	.2910	.2939	.2967	.2995	.3023	.3051	.3078	.3106	.3133
0.9	.3159	.3186	.3212	.3238	.3264	.3289	.3315	.3340	.3365	.3389
1.0	.3413	.3438	.3461	.3485	.3508	.3531	.3554	.3577	.3599	.3621
1.1	.3643	.3665	.3686	.3708	.3729	.3749	.3770	.3790	.3810	.3830
1.2	.3849	.3869	.3888	.3907	.3925	.3944	.3962	.3980	.3997	.4015
1.3	.4032	.4049	.4066	.4082	.4099	.4115	.4131	.4147	.4162	.4177
1.4	.4192	.4207	.4222	.4236	.4251	.4265	.4279	.4292	.4306	.4319
1.5	.4332	.4345	.4357	.4370	.4382	.4394	.4406	.4418	.4429	.4441
1.6	.4452	.4463	.4474	.4484	.4495	.4505	.4515	.4525	.4535	.4545
1.7	.4554	.4564	.4573	.4582	.4591	.4599	.4608	.4616	.4625	.4633
1.8	.4641	.4649	.4656	.4664	.4671	.4678	.4686	.4693	.4699	.4706
1.9	.4713	.4719	.4726	.4732	.4738	.4744	.4750	.4756	.4761	.4767
2.0	.4772	.4778	.4783	.4788	.4793	.4798	.4803	.4808	.4812	.4817
2.1	.4821	.4826	.4830	.4834	.4838	.4842	.4846	.4850	.4854	.4857
2.2	.4861	.4864	.4868	.4871	.4875	.4878	.4881	.4884	.4887	.4890
2.3	.4893	.4896	.4898	.4901	.4904	.4906	.4909	.4911	.4913	.4916
2.4	.4918	.4920	.4922	.4925	.4927	.4929	.4931	.4932	.4934	.4936
2.5	.4938	.4940	.4941	.4943	.4945	.4946	.4948	.4949	.4951	.4952
2.6	.4953	.4955	.4956	.4957	.4959	.4960	.4961	.4962	.4963	.4964
2.7	.4965	.4966	.4967	.4968	.4969	.4970	.4971	.4972	.4973	.4974
2.8	.4974	.4975	.4976	.4977	.4977	.4978	.4979	.4979	.4980	.4981
2.9	.4981	.4982	.4982	.4983	.4984	.4984	.4985	.4985	.4986	.4986
3.0	.49865	.49869	.49874	.49878	.49882	.49886	.49889	.49893	.49897	.49900
3.1	.49903	.49906	.49910	.49913	.49916	.49918	.49921	.49924	.49926	.49929
3.2	.49931	.49934	.49936	.49938	.49940	.49942	.49944	.49946	.49948	.49950
3.3	.49952	.49953	.49955	.49957	.49958	.49960	.49961	.49962	.49964	.49965
3.4	.49966	.49968	.49969	.49970	.49971	.49972	.49973	.49974	.49975	.49976
3.5	.49977	.49978	.49978	.49979	.49980	.49981	.49981	.49982	.49983	.49983
3.6	.49984	.49985	.49985	.49986	.49986	.49987	.49987	.49988	.49988	.49989
3.7	.49989	.49990	.49990	.49990	.49991	.49991	.49992	.49992	.49992	.49992
3.8	.49993	.49993	.49993	.49994	.49994	.49994	.49994	.49995	.49995	.49995
3.9	.49995	.49995	.49996	.49996	.49996	.49996	.49996	.49996	.49997	.49997

F. USING MICROSOFT EXCEL WITH THIS TEXT

F.1 CONFIGURING MICROSOFT EXCEL

Complete the following two steps to ensure that the copy of Microsoft Office or Microsoft Excel that you are using is properly configured.

Step 1: **Verify your copy of Microsoft Excel.** Open Microsoft Excel and select **Help → About Microsoft Excel** to display a dialog box that in its first line states the version number, build number, and any service release (SR) or service packs (SP) numbers. To use Microsoft Excel with this text, you should have Microsoft Excel version 97 or a later version, such as Excel 2000, Excel 2002 (Excel XP), or Excel 2003. Locate and find the original CD-ROM that contains the version of Microsoft Office or Microsoft Excel installed on your system for possible later use. If you maintain your own computer system, you should also visit the Microsoft Office Web site **http://office.microsoft.com** and determine if there are any new updates, service releases, or service packs to be downloaded and installed.

Step 2: **Verify installation of the Microsoft Excel Analysis ToolPak add-ins.** The Microsoft Excel Analysis ToolPak add-in may not have been installed on your system when Microsoft Office or Microsoft Excel was installed. Open Microsoft Excel and select **Tools → Add-Ins** to display the Add-Ins dialog box. Make sure that the **Analysis ToolPak** and **Analysis ToolPak - VBA** check boxes appear and are checked in the list of add-ins available. (See illustration below, the dialog box for Microsoft Excel 2002 and 2003; the dialog box for other versions of Excel are similar.) Click the **OK** button when you are done.

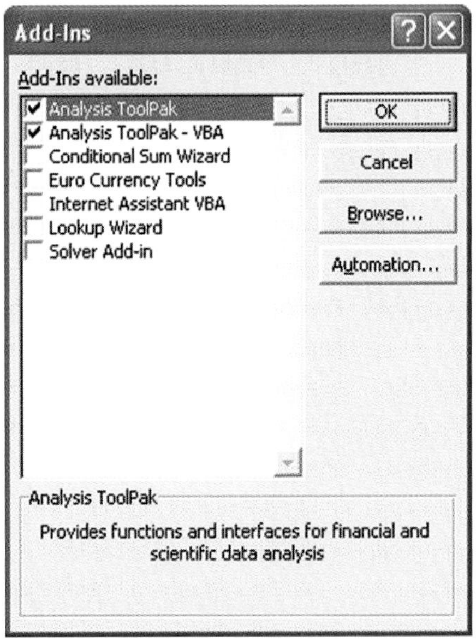

Should the two check boxes not appear in the list of available add-ins, exit Microsoft Excel and rerun the Microsoft Office (or Excel) setup program and have the original program CD-ROM ready to use. You should select the add-or-remove or modify option of the setup program. If you subsequently see a **Choose advanced customization of applications** check box, select it. Select the Analysis ToolPak add-ins for installation, and if given a choice of "run from my computer" or "install on first use," choose "run from my computer." Follow all subsequent onscreen instructions to complete the installation.

F.2 USING THE DATA ANALYSIS TOOLS

The Data Analysis Tools are a set of statistical procedures included with Microsoft Excel. To use the Data Analysis tools, first verify that they are properly installed (see previous section). Then select **Tools → Data Analysis** to display the Data Analysis dialog box (shown below). In the **Analysis Tools** list, select a procedure and click the **OK** button. For most procedures, a second dialog box will appear in which you make entries and selections.

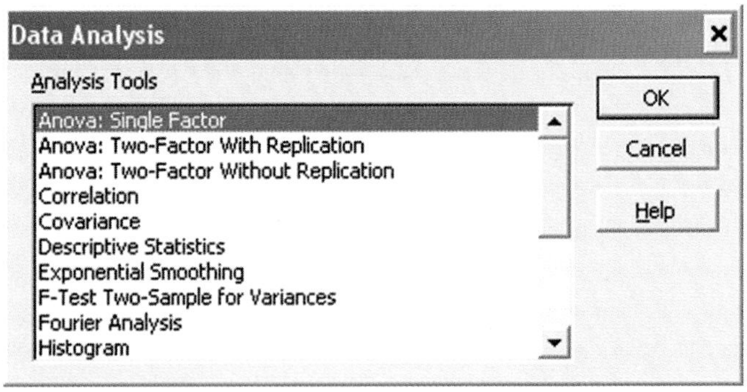

F.3 USING THE PIVOTTABLE WIZARD

You use the PivotTable Wizard to generate PivotTables, summary tables that update themselves automatically as the data on which they are based changes. When reading this book, you can use the wizard to generate one-way and two-way frequency distribution tables for categorical data (see Chapter 2).

To use the PivotTable wizard, you first select **Data → PivotTable and PivotChart Report** (**Data → PivotTable Report** if you are using Microsoft Excel 97). Then you enter information about the design of the table as you step through a series of dialog boxes by clicking a Next button. You click the **Finish** button in the last dialog box to create the table. At any point, you can cancel the operation of the wizard by clicking **Cancel** or move to a previous dialog box by clicking **Back**. In the Step 3 dialog box, you must also click the **Layout** and **Options** buttons in order to complete the process of creating a PivotTable.

The PivotTable Wizards of the various versions of Microsoft Excel differ slightly. For Microsoft Excel 2003, the three-step wizard (see Figure F.1) requires you to do the following:

Step 1: Select the source for the data for the PivotTable and the type of report to be produced in the Step 1 dialog box. In this text, you will always select **Microsoft Excel list or database** as the source and **PivotTable** as the report type. (You do not select a report type in Microsoft Excel 97; PivotTable is assumed.)

Step 2: Enter the cell range of the data that will be summarized in the PivotTable. The first row of this cell range should contain column headings which the wizard will later use as the variable name(s).

Step 3: Choose the location of the PivotTable. In this text, you will always select the **New worksheet** option. Click **Layout** to display the supplemental Layout dialog box. In the Layout dialog box, you drag name labels from a list of variables that appears on the right side (obscured in Figure F.2) into a template that contains page, row, column, and data areas. When you drag a label into the DATA area, the label changes to **Count of *variable*** to indicate that the PivotTable will automatically tally the variable. When you have finished dragging labels, you click **OK** to return to the main Step 3 dialog box. Then click **Options** to display the PivotTable Options dialog box.

In the PivotTable Options dialog box, you enter a self-descriptive table name in the **Name** box and usually enter **0** in the **For empty cells, show** box and leave all other settings as is. You then click **OK** to return to the main Step 3 dialog box and then click **Finish** there to produce the PivotTable.

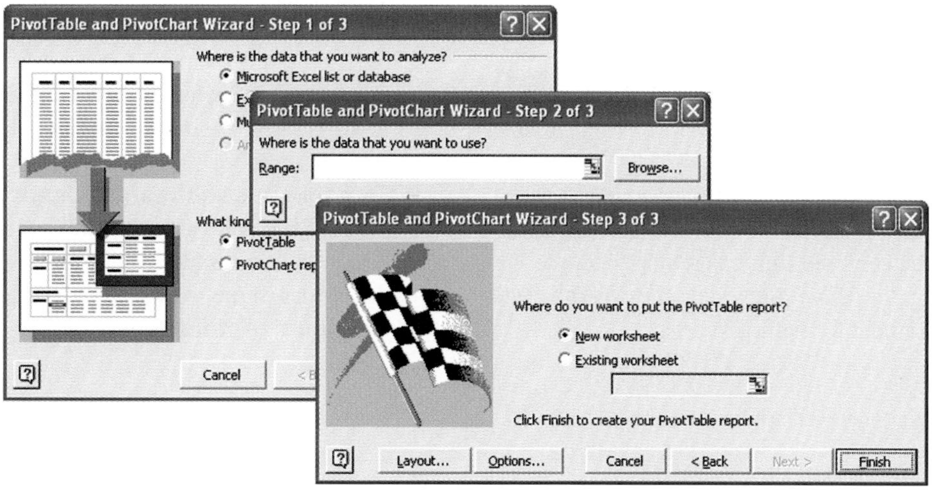

FIGURE F.1 PivotTable Wizard

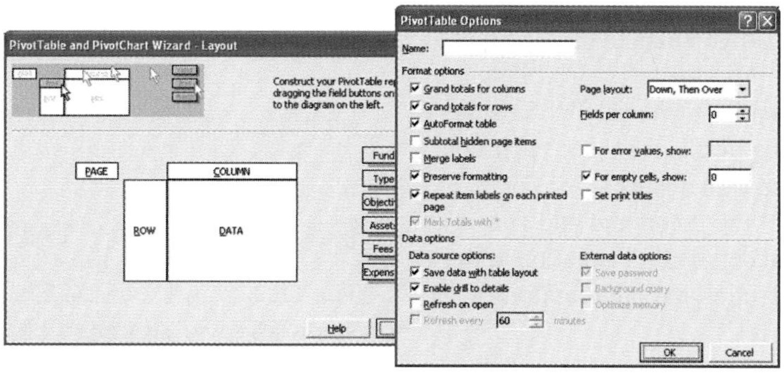

FIGURE F.2 PivotTable Layout (slightly obscured) and Options Dialog Boxes

F.4 ENHANCING THE APPEARANCE OF WORKSHEETS

You can use any number of formatting commands to enhance the appearance of worksheets, regardless of whether those worksheets were produced manually or with the help of a wizard or an add-in. You can use the Excel formatting toolbar button icons to gain easy access to the most commonly used formatting commands, or, if you prefer, you can select **Format → Cells** and make choices in a Format Cells dialog box. Figure F.3 shows the Excel formatting toolbar. Some of the commonly used buttons are labeled and explained below.

FIGURE F.3 Formatting Toolbar

To display cell values using boldface type, select the cell (or cell range) containing the values to be boldfaced and click the **Boldface** button on the formatting toolbar.

To center a cell value in its cell, select the cell (or cell range) containing the values to be centered and click the **Center** button on the formatting toolbar. (The **Align Left** and **Align Right** buttons to either side of the Center button similarly left justify or right justify values in a column.)

To center a long cell entry, such as a title, over a range of columns, select the cell row range over which the entry is to be centered (this range must include the cell containing the title) and click the **Merge and Center** button on the formatting toolbar.

To display numeric values as percentages, select the cell range containing the numeric entries to be displayed as percentages and click the **Percent** button on the formatting toolbar.

To align the decimal point in a series of numeric entries, select the cell range containing the numeric entries to be aligned and click either the **Increase Decimal** or **Decrease Decimal** button on the formatting toolbar until the desired decimal alignment occurs.

To change the background color, select the cell range containing the cells to be changed and click the **Fill Color** drop-down list button. In the Fill Color palette dialog box that appears, select the new background color. (Worksheet examples in this text use the Windows color "Light Turquoise" to tint areas that contain user-changeable data values and the color "Light Yellow" to tint areas that contain results. Light Turquoise and Light Yellow are the fifth and third choices, respectively, on the bottom row of the standard palette.)

To change the cell border effect, select the cell range containing the cells whose borders are to be changed and click the **Borders** drop-down list button. In the Borders palette dialog box that appears, select the new border effect. (Many of the worksheets illustrated in this text use a variety of border effects, including those identified by their tool tips as All Borders, Outside Borders, and Top Border.)

To adjust the width of a column to completely display all the cell values in that column, select the column to be formatted by clicking its column heading in the border of the worksheet, and then select **Format → Column → AutoFit Selection**.

G. PHStat2 USER'S GUIDE

About This Appendix

You should read this appendix if you plan to use PHStat2 to perform statistical analyses in Microsoft Excel while learning from this text. This appendix presents PHStat2 commands in order of appearance in this text. For each command presented, this appendix explains what the command does and shows how to select the command and fill in its dialog box. (If you need more detailed information about a particular command, you can click the Help button of the dialog box.)

This appendix does not discuss how to install PHStat2. For help with this task or for program technical requirements, read the PHStat2 readme file on the CD-ROM packaged with this text. You should also visit the PHStat2 Web site **www.prenhall.com/phstat** to get late-breaking PHStat2 news and updates or to contact technical support.

G.1 ONE-WAY TABLES & CHARTS

The One-Way Tables & Charts command simplifies the preparation of summary charts, bar and pie charts, and Pareto diagrams. This command accepts either unsummarized data (**Raw Categorical Data**) or a **Table of**

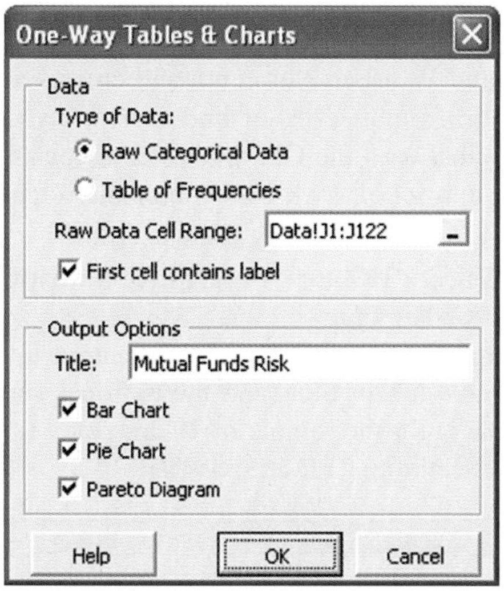

FIGURE G.1

Frequencies in which categories have already been tallied. In either case, the command invisibly uses the PivotTable and Chart Wizards to construct a summary table and charts, or the Pareto diagram, and performs additional formatting operations in order to create a correctly formatted chart.

Use: Select **PHStat ➔ Descriptive Statistics ➔ One-Way Tables & Charts**.
Example: Figure G.1 contains the entries to produce a summary table and charts for the **Risk** variable in the **Data** sheet of the **MUTUALFUNDS2004.xls** file.

G.2 STEM-AND-LEAF DISPLAY

The Stem-and-Leaf Display command creates a stem-and-leaf display as a series of worksheet text entries. The command invisibly sorts the data, decides the stem unit (if you select **Autocalculate stem unit**), and formats cell entries for the stems and leaves. If you select **Summary Statistics**, the command also includes those statistics on the worksheet produced.

The **Set stem unit as** option should be used sparingly and if you use this option, the stem unit you specify must be a power of ten.

Use: Select **PHStat ➔ Descriptive Statistics ➔Stem-and-Leaf Display**.
Example: Figure G.2 contains the entries to produce a stem-and-leaf display for the **5-Yr Return** variable in the **Data** sheet of the **MUTUALFUNDS2004.xls** file.

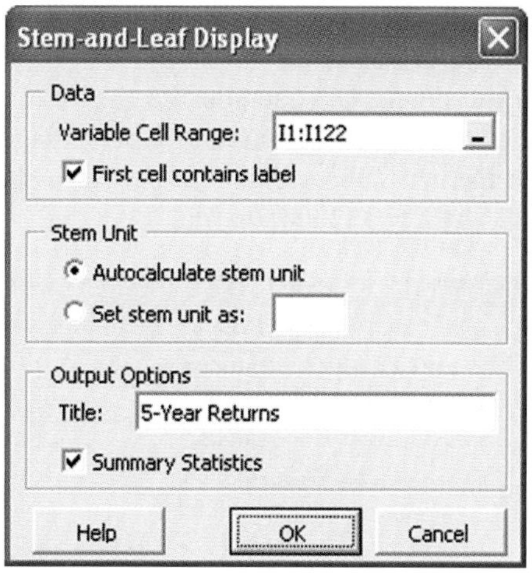

FIGURE G.2

G.3 HISTOGRAM & POLYGONS

The Histogram & Polygons command uses the Data Analysis Histogram procedure and the Chart Wizard to produce frequency distributions, histograms, and frequency, percentage, and cumulative percentage polygons.

This command accepts data either for a single group or multiple groups, and as either unstacked data (each group's data in its own column) or stacked data (data for every group in a single column). If you use the **Multiple Groups – Stacked** option, you will need to enter the **Grouping Variable Cell Range** as well. You control which charts this command creates by selecting the check boxes in **Output Options**.

This command corrects the formatting errors that the Data Analysis Histogram procedure commits when it produces frequency distributions and histograms. Among other things, the command eliminates the extra "More" row in the frequency table and adjusts the gap widths between the bars of the histogram.

Because the command invisibly uses the Data Analysis Histogram procedure, the command uses "bins" and not class groupings to create a frequency table. A bin is a number that specifies the maximum value for a class. Bins are entered as an ordered, ascending list of values in a contiguous cell range (the **Bins Cell Range** of this command or the **Bin Range** of the Data Analysis procedure). A bin range containing values 9.99, 19.99, 29.99 would approximate these three classes: all values less than 10, values greater than or equal to 10 but less than 20, and values greater than or equal to 20 but less than 30.

Because the first class will always be open-ended towards negative infinity, this class will never have a true midpoint. Therefore, the command expects that your **Midpoints Cell Range** will have one cell smaller than your **Bins Cell Range** and will assign the first midpoint to the second class.

Use: Select **PHStat → Descriptive Statistics → Histogram & Polygons**.
Example: Figure G.3 contains the entries to produce a frequency table, histogram, and a percentage polygon for the **Return 2003** variable in the **Data** sheet of the **MUTUALFUNDS2004.xls** file.

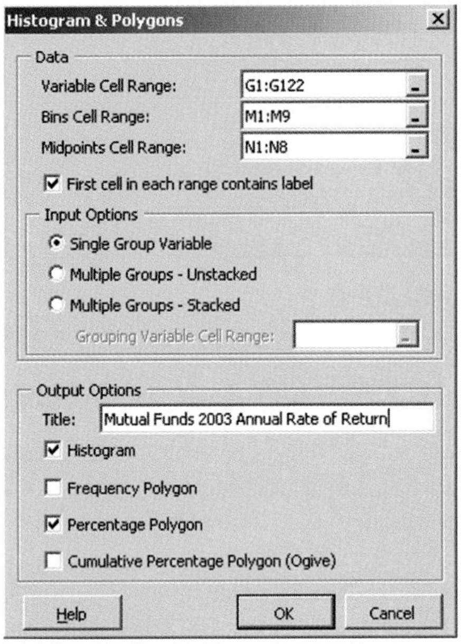

FIGURE G.3

G.4 TWO-WAY TABLES & CHARTS

The Two-Way Tables & Charts command simplifies the preparation of two-way cross-classification tables and side-by-side charts. This command invisibly uses the PivotTable and Chart Wizards to construct a cross-classification table and chart.

Use: Select **PHStat → Descriptive Statistics → Two-Way Tables & Chart**.
Example: Figure G.4 contains the entries to produce a cross-classification table and side-by-side chart for the **Risk** and **Objective** variables in the **Data** sheet of the **MUTUALFUNDS2004.xls** file.

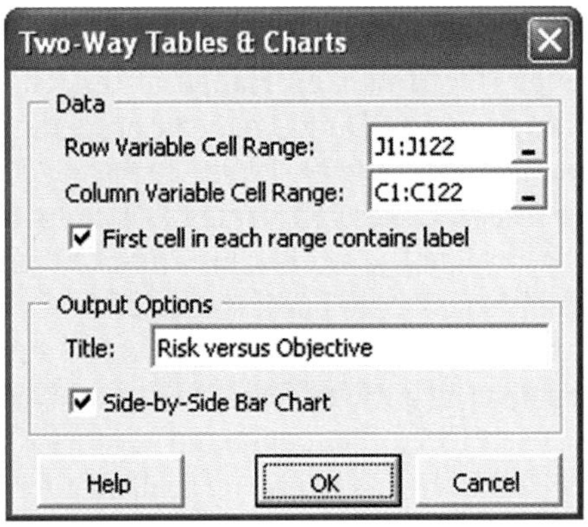

FIGURE G.4

G.5 BOX-AND-WHISKER PLOT

The Box-and-Whisker Plot command creates a box-and-whisker plot from a series of line chart plots. This command invisibly uses the Chart Wizard to construct a plot from a custom set of worksheet entries also created by the command.

Use: Select **PHStat → Descriptive Statistics → Box-and-Whisker Plot**.
Example: Figure G.5 contains the entries to produce a box-and-whisker plot (and the optional five-number summary) for the sample of 10 get-ready times in the **Data** sheet of the **TIMES.xls** file.

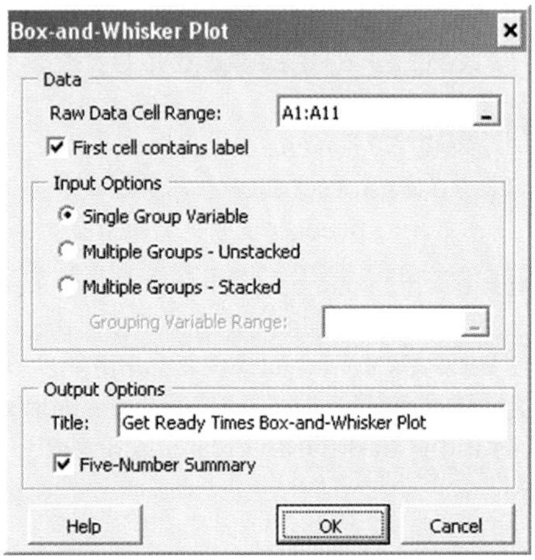

FIGURE G.5

G.6 COVARIANCE AND PORTFOLIO ANALYSIS

The Covariance and Portfolio Analysis command creates a covariance (and portfolio analysis) worksheet into which you add the probabilities and outcomes. Select the **Portfolio Management Analysis** check box to include the portfolio analysis section on the worksheet.

Use: Select **PHStat → Decision-Making → Covariance and Portfolio Analysis**.
Example: Figure G.6 contains the entries to produce a worksheet suitable for the investment data of Table 5.4 on page 163.

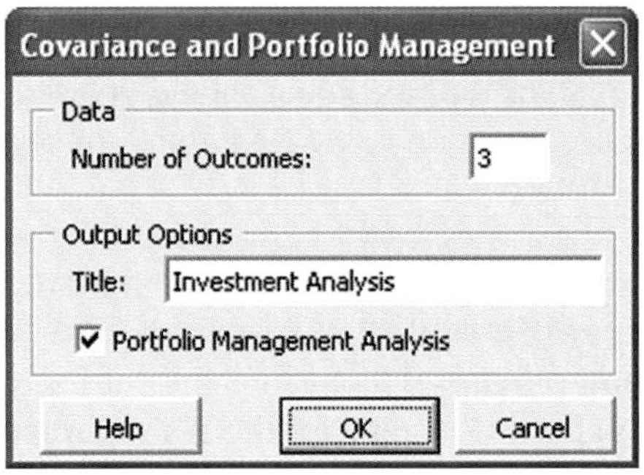

FIGURE G.6

G.7 BINOMIAL PROBABILITIES

The Binomial command creates a binomial probabilities worksheet based on a sample size, probability of success, and an outcomes range that you specify. If you select the **Cumulative Probabilities** check box, the binomial worksheet will include additional columns for $P(<=X)$, $P(<X)$, $P(>X)$, and $P(>=X)$. If you select the **Histogram** check box, the command will invisibly use the Chart Wizard to create a histogram on a separate sheet.

The worksheet created uses the BINOMDIST function to calculate binomial probabilities. The template for this function is **BINOMDIST(*X, n, p, cumulative*)** where X is the number of successes, n is the sample size, p is the probability of success, and *cumulative* is a True or False value that determines whether the function computes the probability of X or fewer successes (True) or computes the probability of exactly X successes (False). For the tagged orders example used in section 5.3, BINOMDIST (3, 4, .1, False) calculates the probability of getting exactly three tagged order forms from a sample of four, whereas BINOMDIST (3, 4, .1, True) calculates the probability of three or fewer tagged order forms.

Use: Select **PHStat → Probability & Prob. Distributions → Binomial**.

Example: Figure G.7 contains the entries to produce a worksheet for the tagged orders example used in section 5.3.

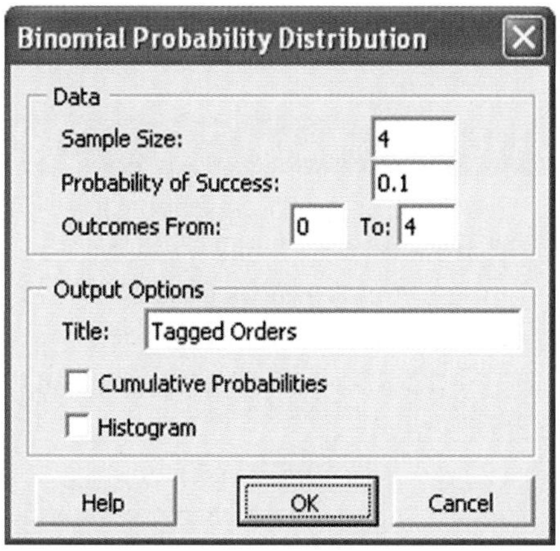

FIGURE G.7

G.8 POISSON PROBABILITIES

The Poisson command creates a Poisson probabilities worksheet based on a sample size, probability of success, and an outcomes range that you specify. If you select the **Cumulative Probabilities** check box, the Poisson worksheet will include additional columns for $P(<=X)$, $P(<X)$, $P(>X)$, and $P(>=X)$. If you select the **Histogram** check box, the command will invisibly use the Chart Wizard to create a histogram on a separate sheet.

The worksheet created uses the POISSON function to calculate Poisson probabilities. The template for this function is **POISSON(*X, lambda, cumulative*)** where X is the number of successes, *lambda* is the mean number of successes, and *cumulative* is a True or False value that determines whether the function computes the probability of X or fewer successes (True) or computes the probability of exactly X successes (False). For the customer arrivals analysis example used in section 5.4, POISSON(2, 3, False) calculates the probability that exactly two customers arrive when the mean number of arrivals is 3.0. Changing the False value to True would calculate the probability of two or fewer customers arriving.

Use: Select **PHStat → Probability & Prob. Distributions → Poisson**.

Example: Figure G.8 contains the entries to produce a worksheet for the customer arrivals analysis example used in section 5.4.

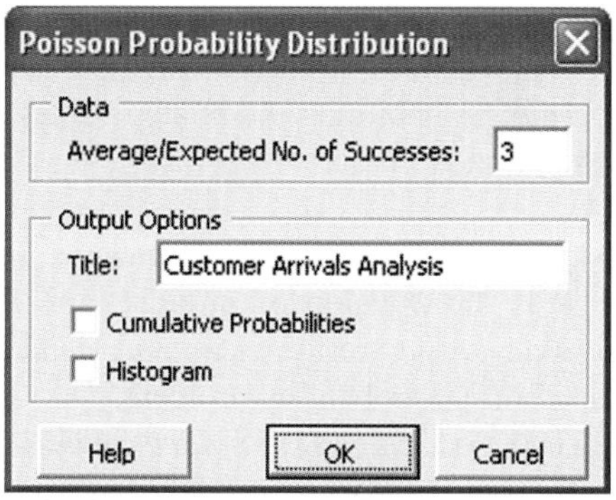

FIGURE G.8

G.9 HYPERGEOMETRIC PROBABILITIES

The Hypergeometric command creates a hypergeometric probabilities worksheet based on the sample size, probability of success, and an outcomes range that you specify. If you select the **Histogram** check box, the command will invisibly use the Chart Wizard to create a histogram on a separate sheet.

The worksheet created uses the HYPGEOMDIST function to calculate hypergeometric probabilities. The template for this function is **HYPGEOMDIST(*X, n, A, N*)** where X is the number of successes, n is the sample size, A is the number of successes in the population, and N is the population size.

Use: Select **PHStat → Probability & Prob. Distributions → Hypergeometric**.

Example: Figure G.9 contains the entries to produce a worksheet for the team formation analysis example used in section 5.5.

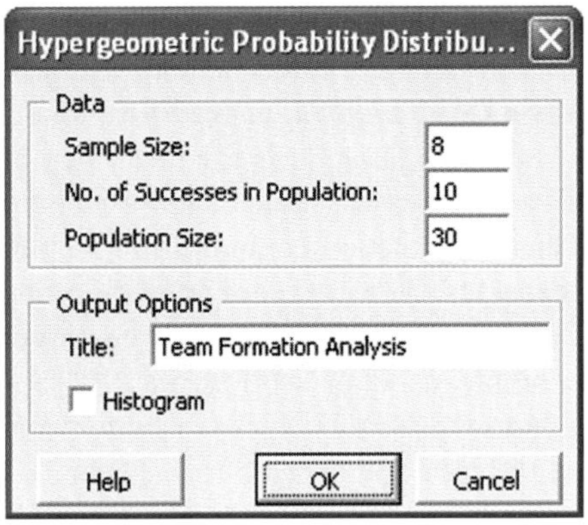

FIGURE G.9

G.10 NORMAL

The Normal command creates a normal probabilities worksheet that solves one or more types of normal probability problems based on the Input Options you select.

The worksheet created uses the STANDARDIZE, NORMDIST, NORMSINV, and NORMINV functions to calculate normal probabilities and related values. The templates for these functions are:

- **STANDARDIZE (*X, mean, standard deviation*)** where *X* is the *X* value of interest, and *mean* and *standard deviation* are the mean and standard deviation for a normal probability problem.
- **NORMDIST (*X, mean, standard deviation*, True)** where *X, mean*, and *standard deviation* are as in the STANDARDIZE function and True indicates that a cumulative probability is to be calculated.
- **NORMSINV(*$P<Z$*)** where $P<Z$ is the area under the curve that is less than *Z*.
- **NORMINV(*$P<X$, mean, standard deviation*)** where $P<X$ is the area under the curve that is less than *X*, and *mean* and *standard deviation* are as in the STANDARDIZE function.

The STANDARDIZE function returns the *Z* value for a particular *X* value, mean, and standard deviation. The NORMDIST returns the area or probability of less than a given *X* value. The NORMSINV function returns the *Z* value corresponding to a given probability. The NORMINV function returns the *X* value for a given probability, mean, and standard deviation. Figure 6.18 on page 206 illustrates how the Normal command uses these functions to create a worksheet solution.

Use: Select **PHStat → Probability & Prob. Distributions → Normal**.
Example: Figure G.10 contains the entries to solve the Example 6.5 and Example 6.6 problems.

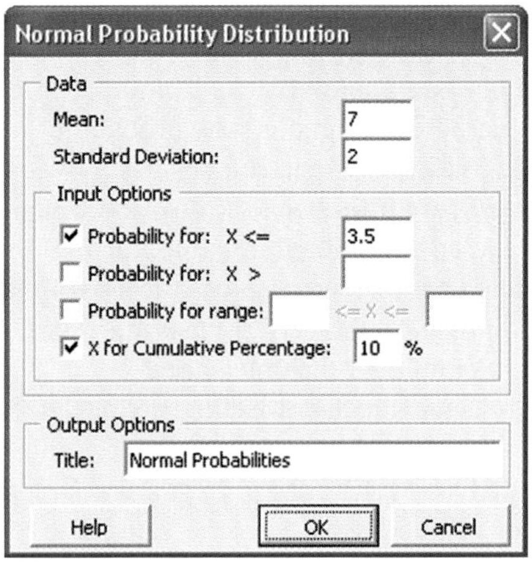

FIGURE G.10

G.11 NORMAL PROBABILITY PLOT

The Normal Probability Plot command creates a normal probability plot as a chart. This command invisibly uses the Chart Wizard to construct a plot from a custom set of worksheet entries also created by the command.

The command creates a **Plot** worksheet on which the *Z* values to be plotted are calculated using the NORMSINV function (see section G.10).

Use: Select **PHStat → Probability & Prob. Distributions → Normal Probability Plot**.
Example: Figure G.11 contains the entries to produce the normal probability plot in Figure 6.23 on page 210.

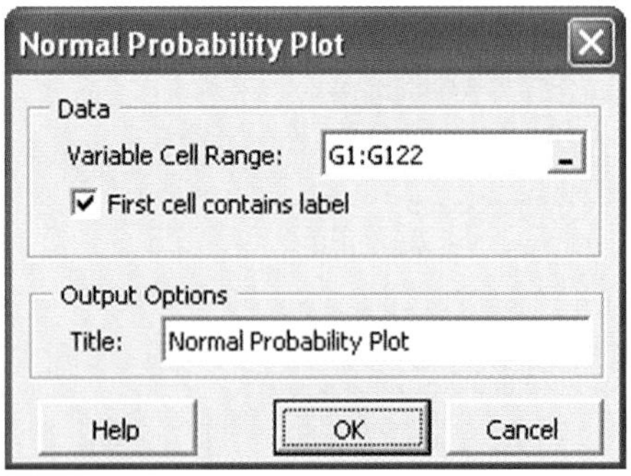

FIGURE G.11

G.12 EXPONENTIAL

The Exponential command creates an exponential probability worksheet based on a mean per unit (*lambda*) and *X* value that you specify.

The worksheet created uses the EXPONDIST function to calculate the exponential probability. The template for this function is **EXPONDIST (*X, mean, True*)** where *X* is the *X* value of interest, *mean* is the population mean λ and True indicates a cumulative probability is to be calculated.

Use: Select **PHStat → Probability & Prob. Distributions → Exponential**.
Example: Figure G.12 contains the entries to produce a worksheet for the bank ATM problem of section 6.5.

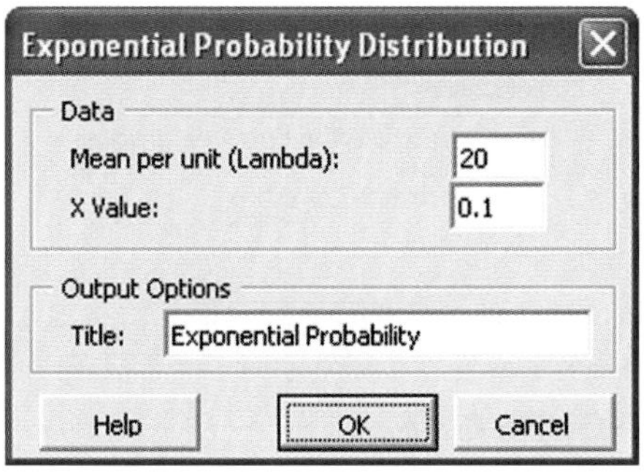

FIGURE G.12

G.13 SAMPLING DISTRIBUTIONS SIMULATION

The Sampling Distributions Simulation command creates a sampling distributions simulation worksheet based on the number of samples, sample size, and type of distribution that you specify.

The command silently uses the Data Analysis Random Number Generation procedure to create the results of the simulation on a new worksheet. The command then adds the sample means, the overall mean, and the standard error of the mean to this worksheet. If you check the Histogram box, the command uses the Data Analysis Histogram procedure to create a histogram of the simulation.

Use: Select **PHStat → Sampling → Sampling Distributions Simulation**.
Example: Figure G.13 contains the entries for generating a simulation of 100 samples of sample size 30 from a uniformly distributed population.

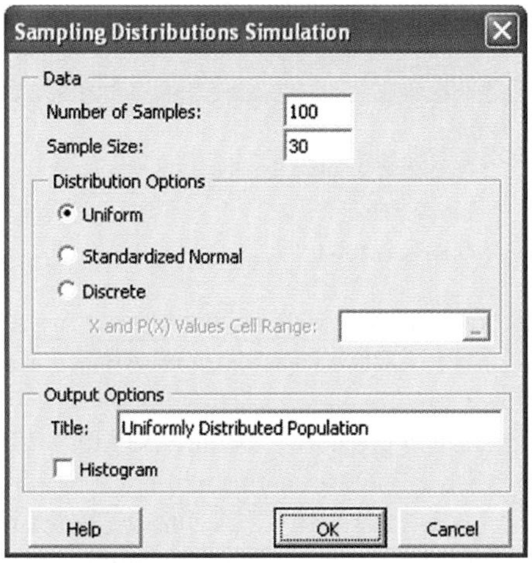

FIGURE G.13

G.14 CONFIDENCE INTERVAL ESTIMATE FOR THE MEAN, SIGMA KNOWN

The Estimate for the Mean, sigma known command creates a confidence interval estimate worksheet based on the values of the population standard deviation, sample mean, sample size, and confidence level that you specify. If you have unsummarized data, you can select the **Sample Statistics Unknown** option and have the command calculate the sample statistics for you.

The worksheet created uses the NORMSINV and CONFIDENCE functions to determine the Z value and calculate the interval half width, respectively. The templates for these functions are:

- **NORMSINV(*P*<*Z*)** where *P*<*Z* is the area under the curve that is less than *Z*.
- **CONFIDENCE(*1 - confidence level, population standard deviation, sample size*)**.

Use: Select **PHStat → Confidence Intervals → Estimate for the Mean, sigma known**.
Example: Figure G.14 contains the entries for solving Example 8.1 on page 264.

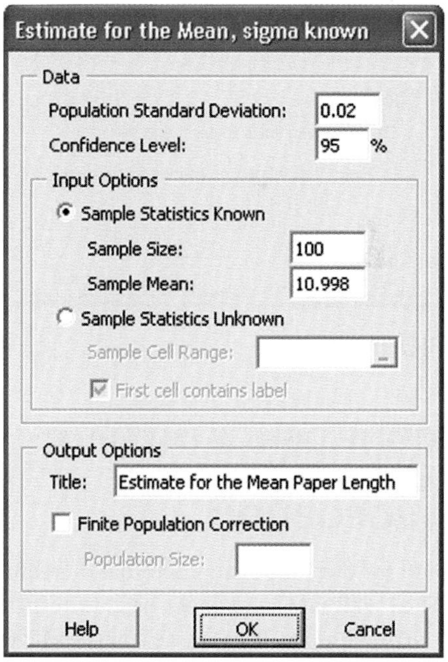

FIGURE G.14

G.15 CONFIDENCE INTERVAL ESTIMATE FOR THE MEAN, SIGMA UNKNOWN

The Estimate for the Mean, sigma unknown command creates a confidence interval estimate worksheet based on sample statistics and a confidence level value that you specify. If you have unsummarized data, you can select the **Sample Statistics Unknown** option and have the command calculate the sample statistics for you.

The worksheet created uses the TINV function to determine the critical value of the *t* distribution. The template for this function is **TINV (1 - *confidence level, degrees of freedom*)**.

Use: Select **PHStat → Confidence Intervals → Estimate for the Mean, sigma unknown**.
Example: Figure G.15 contains the entries for estimating the mean amount of sales invoices (see page 268).

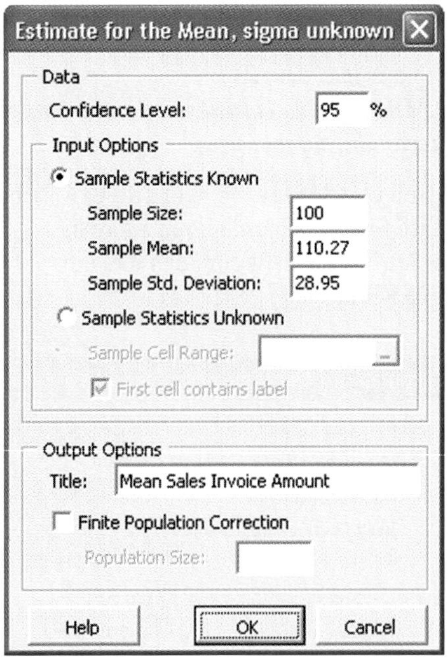

FIGURE G.15

G.16 CONFIDENCE INTERVAL ESTIMATE FOR THE PROPORTION

The Estimate for the Proportion command creates a confidence interval estimate worksheet based on the values of the sample size, number of successes, and confidence level that you specify.

The worksheet created uses the NORMSINV function, the template of which is **NORMSINV($P<Z$)** where $P<Z$ is the area under the curve that is less than *Z*, to determine the *Z* value.

Use: Select **PHStat → Confidence Intervals → Estimate for the Proportion**.
Example: Figure G.16 contains the entries for solving the section 8.3 example.

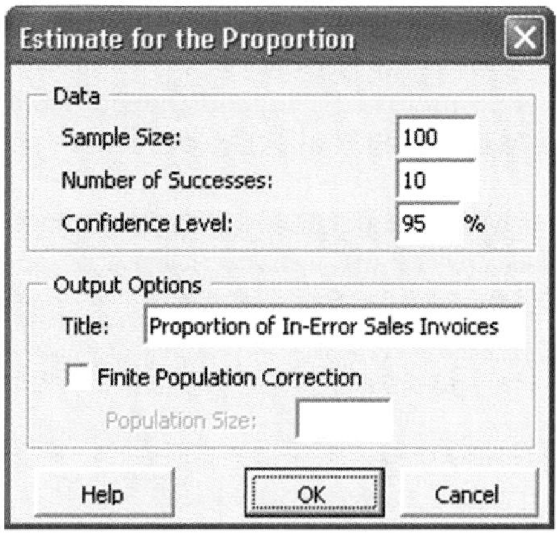

FIGURE G.16

G.17 SAMPLE SIZE DETERMINATION FOR THE MEAN

The Determination for the Mean command creates a worksheet based on the values of the population standard deviation, sampling error, and confidence level that you specify.

The worksheet created uses the NORMSINV function, the template of which is **NORMSINV($P<Z$)** where $P<Z$ is the area under the curve that is less than *Z*, to determine the *Z* value. The worksheet also uses the ROUNDUP function to round the result of the sample size calculation up to the next integer.

Use: Select **PHStat → Sample Size → Determination for the Mean**.
Example: Figure G.17 contains the entries for solving the section 8.4 example.

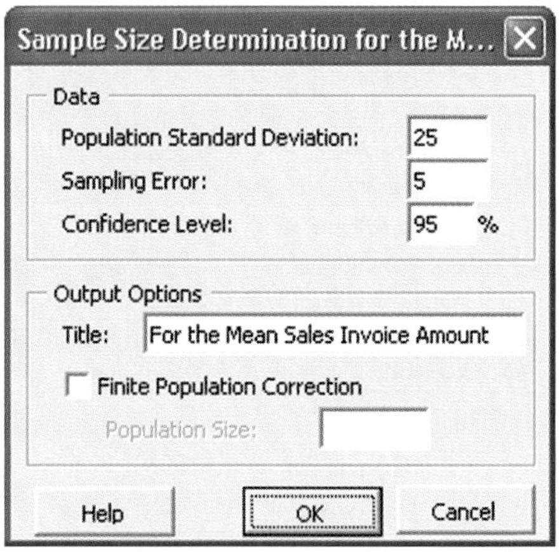

FIGURE G.17

G.18 SAMPLE SIZE DETERMINATION FOR THE PROPORTION

The Determination for the Proportion command creates a worksheet based on the values of the sample size, number of successes, and confidence level that you specify.

The worksheet created uses the NORMSINV function, the template of which is **NORMSINV(*P*<*Z*)** where *P*<*Z* is the area under the curve that is less than *Z*, to determine the *Z* value. The worksheet also uses the ROUNDUP function to round the result of the sample size calculation up to the next integer.

Use: Select **PHStat → Sample Size → Determination for the Proportion**.
Example: Figure G.18 contains the entries for solving the section 8.4 example.

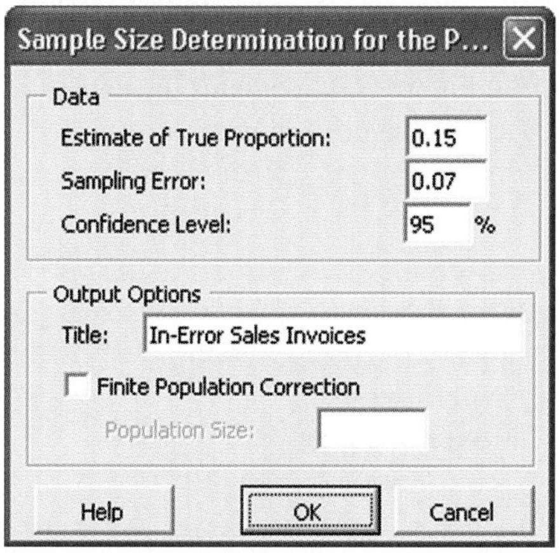

FIGURE G.18

G.19 CONFIDENCE INTERVAL ESTIMATE FOR THE POPULATION TOTAL

The Estimate for the Population Total command creates a confidence interval estimate worksheet based on the values of the population size, sample mean, sample size, sample standard deviation, and confidence level that you specify. If you have unsummarized data, you can select the **Sample Statistics Unknown** option and have the command calculate the sample statistics for you.

The worksheet created uses the TINV function, the template of which is **TINV(1 - *confidence level, degrees of freedom*)**, to determine the critical value from the *t* distribution.

Use: Select **PHStat → Confidence Intervals → Estimate for the Population Total**.
Example: Figure G.19 contains the entries for solving the section 8.5 example.

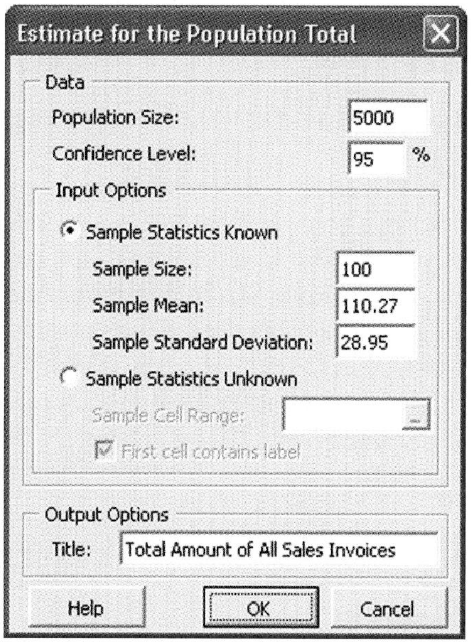

FIGURE G.19

G.20 CONFIDENCE INTERVAL ESTIMATE FOR THE TOTAL DIFFERENCE

The Estimate for the Total Difference command creates a confidence interval estimate worksheet based on the values of the population size, sample size, and confidence level that you specify. To use this command, first create a column of differences starting in row 1 on a new worksheet. (You can compare column A of the **Data** worksheet of the **CIE Total Difference.xls** file as a guide.) With your workbook open to these data, use the command.

Use: Select **PHStat → Confidence Intervals → Estimate for the Total Difference**.
Example: Figure G.20 contains the entries for solving the section 8.5 example. The cell range entry assumes that the differences data are located on a worksheet named **Data**.

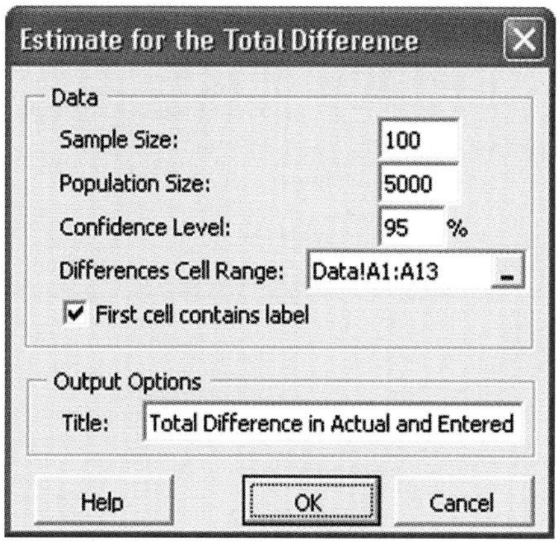

FIGURE G.20

G.21 *Z* TEST FOR THE MEAN, SIGMA KNOWN

The Z Test for the Mean, sigma known command creates a hypothesis-testing worksheet based on the values of the null hypothesis, level of significance, population standard deviation, sample size, and sample mean, and the test option that you specify. If you have unsummarized data, you can select the **Sample Statistics Unknown** option and have the command calculate the sample statistics for you.

The worksheet created uses the NORMSINV and NORMSDIST functions to determine the critical values and *p*-values, respectively. The templates for these functions are:

- **NORMSINV(*P*<*Z*)** where *P*<*Z* is the area under the curve that is less than *Z*.
- **NORMSDIST(*Z value*)**.

Use: Select **PHStat ➔ One-Sample Tests ➔ Z Test for the Mean, sigma known**.
Example: Figure G.21 contains the entries for the section 9.2 example.

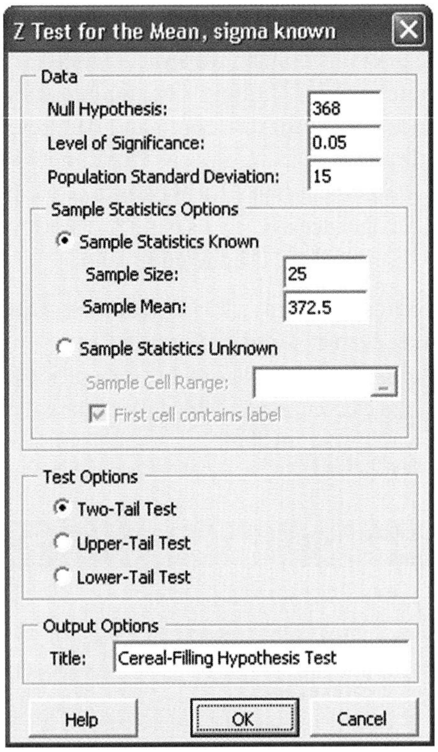

FIGURE G.21

G.22 *t* TEST FOR THE MEAN, SIGMA UNKNOWN

The t Test for the Mean, sigma unknown command creates a hypothesis-testing worksheet based on the values of the null hypothesis, level of significance, sample standard deviation, sample size, and sample mean, and the test option that you specify. If you have unsummarized data, you can select the **Sample Statistics Unknown** option and have the command calculate the sample statistics for you.

The worksheet created uses the TINV and TDIST functions to determine the critical values and *p* -values, respectively. The templates for these functions are:

- **TINV(1 - *confidence level, degrees of freedom*)**
- **TDIST(ABS(*t*), *degrees of freedoms, tails*)**, where *tails* = 2 indicates a two-tail test.

Use: Select **PHStat ➔ One-Sample Tests ➔ t Test for the Mean, sigma unknown**.
Example: Figure G.22 contains the entries for the section 9.4 sales invoices example.

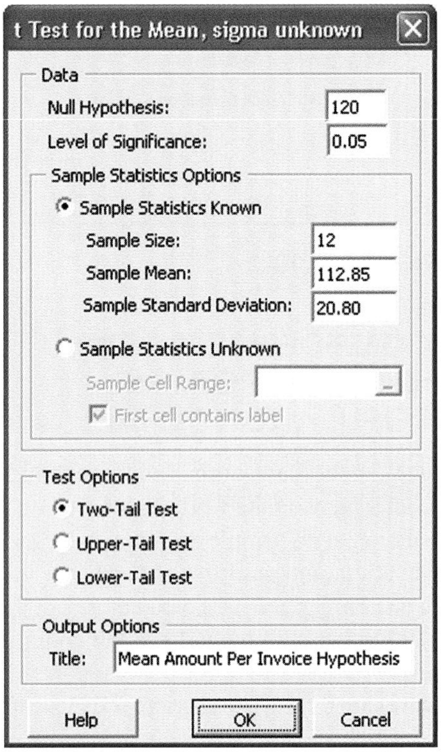

FIGURE G.22

G.23 Z TEST FOR THE PROPORTION

The Z Test for the Proportion command creates a hypothesis-testing worksheet based on the values of the null hypothesis, level of significance, number of successes, and sample size, and the test option that you specify.

The worksheet created uses the NORMSINV and NORMSDIST functions to determine the critical values and *p*-values, respectively. The templates for these functions are:

- **NORMSINV(*$P<Z$*)** where $P<Z$ is the area under the curve that is less than *Z*.
- **NORMSDIST(*Z value*)**.

Use: Select **PHStat → One-Sample Tests → Z Test for the Proportion**.
Example: Figure G.23 contains the entries for the section 9.5 business ownership example.

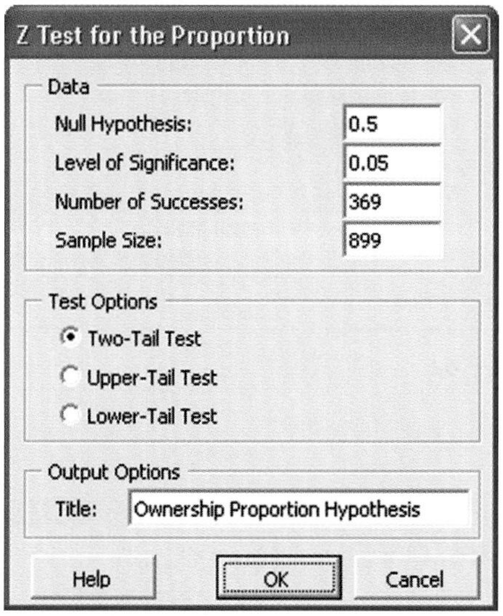

FIGURE G.23

G.24 Z TEST FOR DIFFERENCES IN TWO MEANS

The Z Test for Differences in Two Means command creates a hypothesis-testing worksheet based on the values of the hypothesized difference, level of significance, population standard deviations and sample statistics, and the test option that you specify.

The worksheet created uses the NORMSINV and NORMSDIST functions to determine the critical values and *p*-values, respectively. The templates for these functions are:

- **NORMSINV(*$P<Z$*)** where $P<Z$ is the area under the curve that is less than Z.
- **NORMSDIST(*Z value*)**.

Use: Select **PHStat → Two-Sample Tests → Z Test for Differences in Two Means**.
Example: Figure G.24 contains the entries for problem 10.1 on page 355.

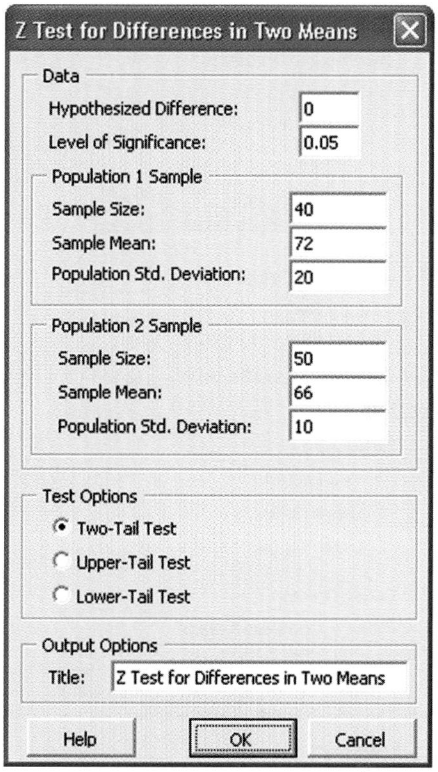

FIGURE G.24

G.25 *t* TEST FOR DIFFERENCES IN TWO MEANS

The t Test for Differences in Two Means command creates a hypothesis-testing worksheet based on the values of the hypothesized difference, level of significance, and the sample statistics, and the test option that you specify.

The worksheet created uses the TINV and TDIST functions to determine the critical values and *p*-values, respectively. The templates for these functions are:

- **TINV(1 - *confidence level, degrees of freedom*)**
- **TDIST(ABS(*t*), *degrees of freedoms, tails*)**, where *tails* = 2 indicates a two-tail test.

Use: Select **PHStat → Two-Sample Tests → t Test for Differences in Two Means**.
Example: Figure G.25 contains the entries for the section 10.1 cola sales example.

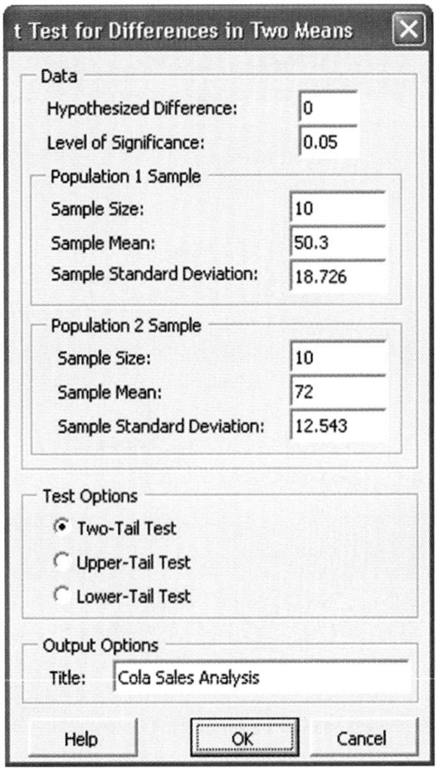

FIGURE G.25

G.26 *Z* TEST FOR THE DIFFERENCES IN TWO PROPORTIONS

The Z Test for the Differences in Two Proportions command creates a hypothesis-testing worksheet based on the hypothesized difference, level of significance, number of successes, and sample size, and the test option that you specify.

The worksheet created uses the NORMSINV and NORMSDIST functions to determine the critical values and *p*-values, respectively. The templates for these functions are:

- **NORMSINV(*P*<*Z*)** where *P*<*Z* is the area under the curve that is less than Z.
- **NORMSDIST(*Z value*)**.

Use: Select **PHStat → Two-Sample Tests → Z Test for the Differences in Two Proportions**.
Example: Figure G.26 contains the entries for the section 10.3 guest satisfaction example.

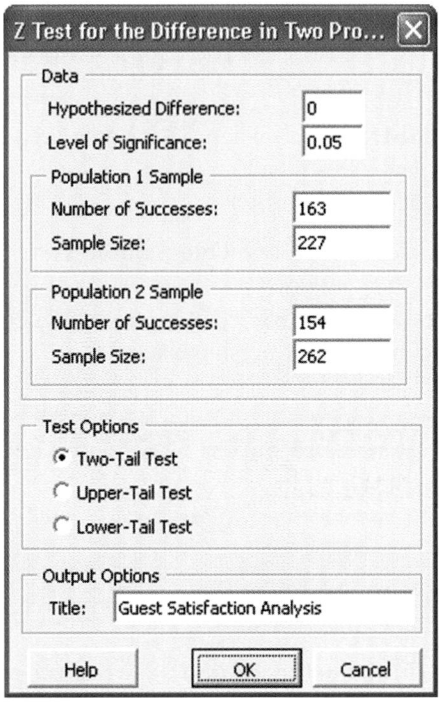

FIGURE G.26

G.27 *F* TEST FOR DIFFERENCES IN TWO VARIANCES

The F Test for Differences in Two Variances command creates a hypothesis-testing worksheet based on the values of the level of significance and sample statistics, and the test option that you specify.

The worksheet created uses the FINV and FDIST functions to determine the critical values and to calculate the *p*-value, respectively. The templates for these functions are:

- **FINV(*upper-tailed p-value, numerator degrees of freedom, denominator degrees of freedom*)**.
- **FDIST(F-*test statistic, numerator degrees of freedom, denominator degrees of freedom*)**.

Use: Select **PHStat → Two-Sample Tests → F Test for the Differences in Two Variances**.
Example: Figure G.27 contains the entries for the section 10.4 cola sales example.

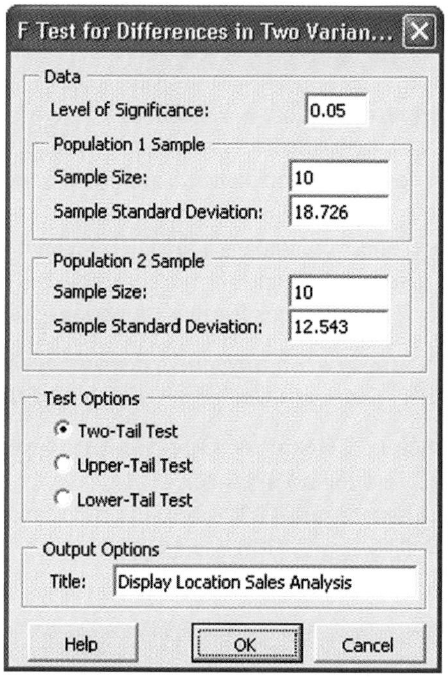

FIGURE G.27

G.28 TUKEY-KRAMER PROCEDURE

The Tukey-Kramer Procedure command runs the Data Analysis Anova: Single Factor procedure and creates a multiple comparisons worksheet in which you enter a Q statistic value. The multiple comparisons worksheet created uses arithmetic formulas to analyze the results.

Use: Select **PHStat → Multiple-Sample Tests → Tukey-Kramer Procedure**.
Example: Figure G.28 contains the entries for the section 11.1 parachute example and assumes that the **Parachute.xls** file is open to the **Data** worksheet.

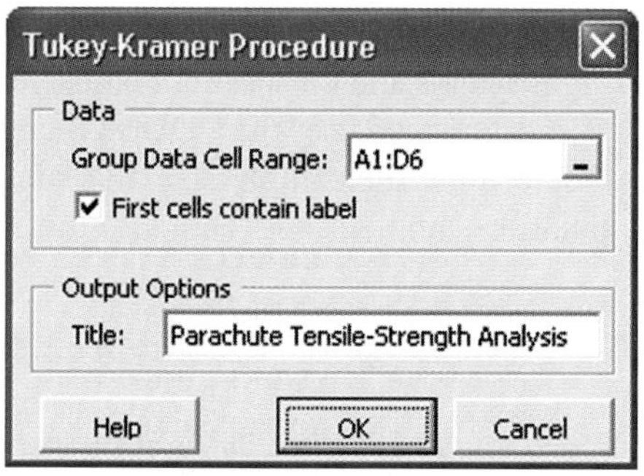

FIGURE G.28

G.29 LEVENE'S TEST

The Levene's Test command runs the Data Analysis Anova:Single Factor procedure using a worksheet of absolute differences from the median that the command creates from the cell range that you specify.

Use: Select **PHStat → Multiple-Sample Tests → Levene's Test**.
Example: Figure G.29 contains the entries for the section 11.1 parachute data example.

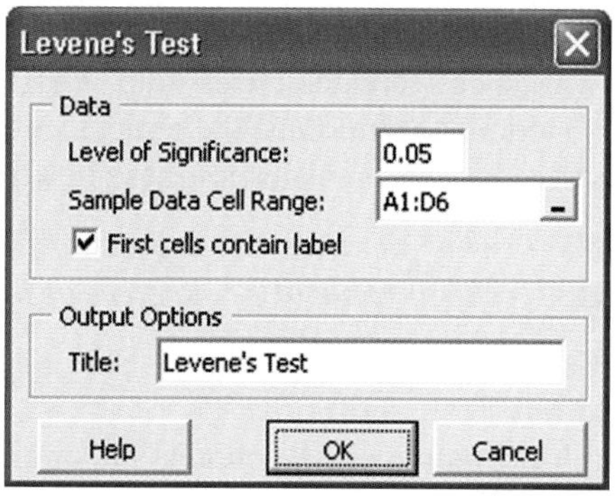

FIGURE G.29

G.30 CHI-SQUARE TEST FOR DIFFERENCES IN TWO PROPORTIONS

The Chi-Square Test for Differences in Two Proportions creates a hypothesis-testing worksheet into which you enter observed frequency table data.

The worksheet created uses the CHIINV and CHIDIST functions to determine the critical values and *p*-value, respectively. The templates for these functions are:

- **CHIINV(*level of significance, degrees of freedom*).**
- **CHIDIST(*critical value of* χ^2, *degrees of freedom*).**

Use: Select **PHStat → Two-Sample Tests → Chi-Square Test for Differences in Two Proportions**.
Example: Figure G.30 contains the entries for the section 12.1 guest satisfaction example.

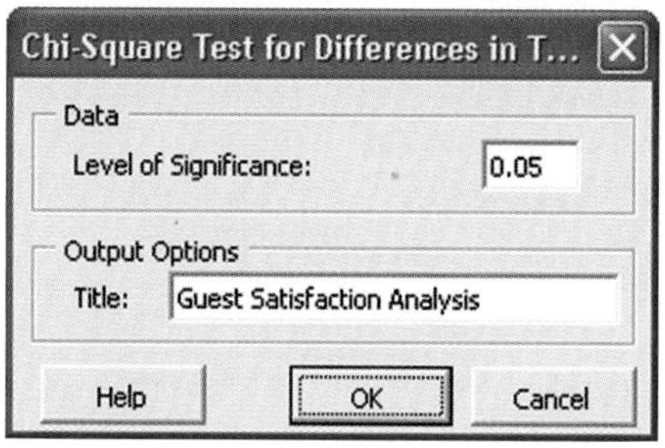

FIGURE G.30

G.31 CHI-SQUARE TEST

The Chi-Square Test command creates a hypothesis-testing worksheet into which you enter the observed frequency table data based on the number of rows and number of columns that you specify. You can use the worksheet produced for either the χ^2 test for the differences in more than two proportions or the χ^2 test of independence. If the number of rows entered is 2, you can also perform the Marascuilo procedure by selecting the **Marascuilo Procedure** output option (this option is enabled only if you enter 2 as the number of rows as was done in Figure G.31A below).

The worksheet created uses the CHIINV and CHIDIST functions to determine the critical values and p-value, respectively. The templates for these functions are:

- **CHIINV(*level of significance, degrees of freedom*).**
- **CHIDIST(*critical value of* χ^2*, degrees of freedom*).**

Use: Select **PHStat → Multiple-Sample Tests → Chi-Square Test**.
Examples: Figure G.31A contains the entries for the section 12.2 guest satisfaction example (with the optional Marascuilo Procedure selected) and Figure G.31B contains the entries for the section 12.3 guest satisfaction survey example.

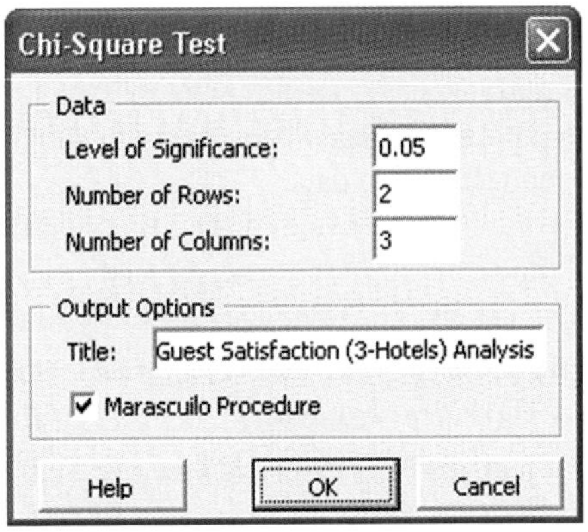

FIGURE G.31A

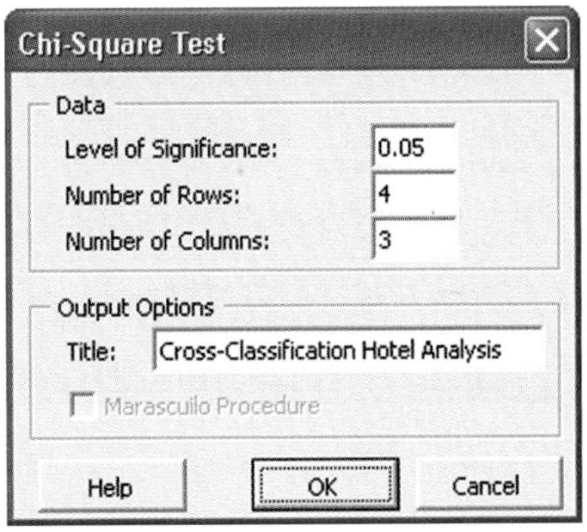

FIGURE G.31B

G.32 CHI-SQUARE TEST FOR A VARIANCE

The Chi-Square Test for a Variance command creates a hypothesis-testing worksheet based on the values of the null hypothesis, level of significance, sample size, and sample standard deviation, and the test option that you specify.

The worksheet created uses the CHIINV and CHIDIST functions to determine the critical values and p-values, respectively. The templates for these functions are:

- **CHIINV(*level of significance, degrees of freedom*).**
- **CHIDIST(*critical value of* χ^2 *, degrees of freedom*).**

Use: Select **PHStat → One-Sample Tests → Chi-Square Test for a Variance**.
Examples: Figure G.32 contains the entries for the section 12.5 cereal-filling example.

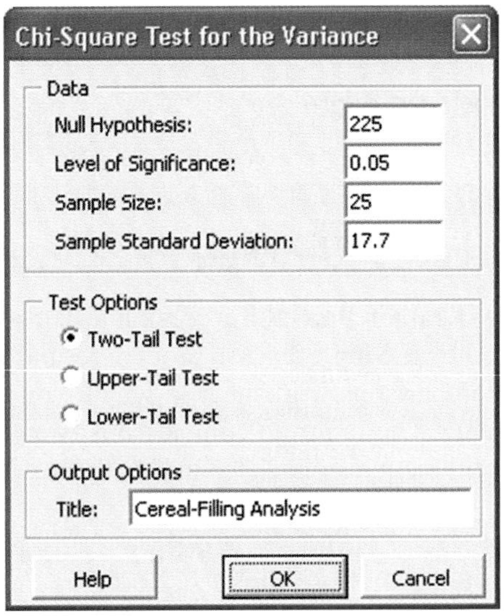

FIGURE G.32

G.33 WILCOXON RANK SUM TEST

The Wilcoxon Rank Sum Test command calculates the rank sums and creates a hypothesis-testing worksheet based on the level of significance and the test option that you specify.

The worksheet created uses the COUNTIF and SUMIF functions to determine the sample size and sum of ranks for each population sampled. The templates for these functions are:

- **COUNTIF(*cell range in which to look for value, value*).**
- **SUMIF(*cell range in which to look for value, value, cell range to be used for sum*).**

The worksheet also uses the NORMSINV and NORMSDIST functions to determine the critical values and p-values, respectively. The templates for these functions are:

- **NORMSINV(*P<Z*)** where *P<Z* is the area under the curve that is less than *Z*.
- **NORMSDIST(*Z value*)**.

Use: Select **PHStat → Two-Sample Tests → Wilcoxon Rank Sum Test**.

Examples: Figure G.33 contains the entries for the section 12.7 display location analysis example and assumes that the **Cola.xls** file is open to the **Data** worksheet.

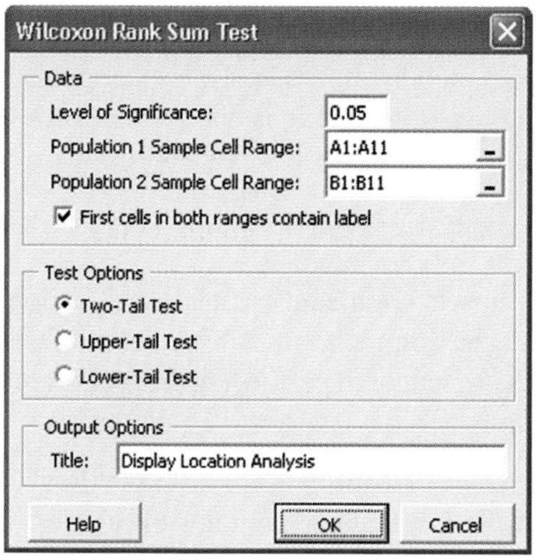

FIGURE G.33

G.34 KRUSKAL-WALLIS RANK TEST

The Kruskal-Wallis Rank Test command calculates the rank sums and creates a hypothesis-testing worksheet based on the level of significance and the test option that you specify.

The worksheet created uses the CHIINV and CHIDIST functions to determine the critical values and *p*-values, respectively. The templates for these functions are:

- **CHIINV(*level of significance, degrees of freedom*).**
- **CHIDIST(*critical value of* χ^2*, degrees of freedom*)**.

Use: Select **PHStat → Multiple-Sample Tests → Kruskal-Wallis Rank Test**.

Examples: Figure G.34 contains the entries for the section 12.9 tensile-strength example and assumes that the **Parachute.xls** file is open to the **Data** worksheet.

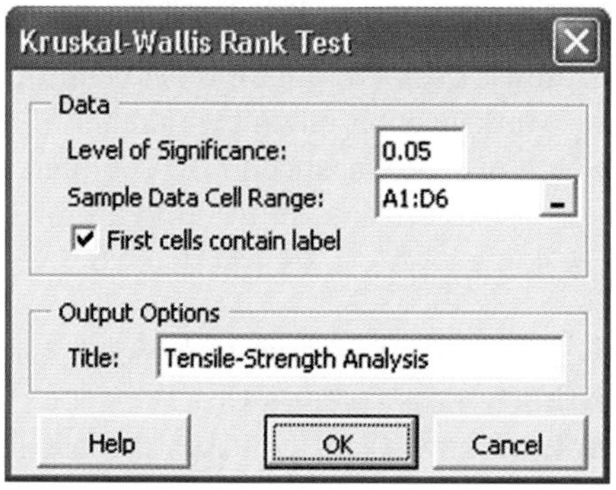

FIGURE G.34

G.35 SIMPLE LINEAR REGRESSION

The Simple Linear Regression command creates a regression results worksheet based on sample data and a confidence level that you specify. The command invisibly uses the Data Analysis Regression procedure to perform the regression and enhances that procedure by offering output options that create a table of residuals, a residual analysis plot, a scatter diagram, a Durbin-Watson statistic worksheet, and a confidence and prediction interval worksheet that is based on an *X* value and confidence level value that you specify.

Use: Select **PHStat → Regression → Simple Linear Regression**.

Example: Figure G.35 contains the entries for the section 13.2 site selection example and assumes that the **Site.xls** file is open to the **Data** worksheet.

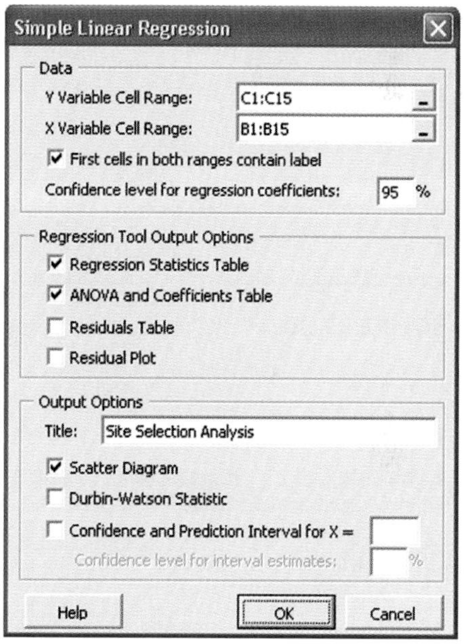

FIGURE G.35

G.36 MULTIPLE REGRESSION

The Multiple Regression command creates a regression results worksheet based on sample data and a confidence level that you specify. The command invisibly uses the Data Analysis Regression procedure to perform the regression and enhances that procedure by offering output options that create a table of residuals, a residual analysis plot, a regression scatter diagram, Durbin-Watson Statistic and Variance Inflationary Factor worksheets as well as a confidence interval estimate and prediction interval worksheet.

Use: Select **PHStat → Regression → Multiple Regression**.

Example: Figure G.36 contains the entries for the section 14.1 OmniPower bar sales example and assumes that the **Omni.xls** file is open to the **Data** worksheet.

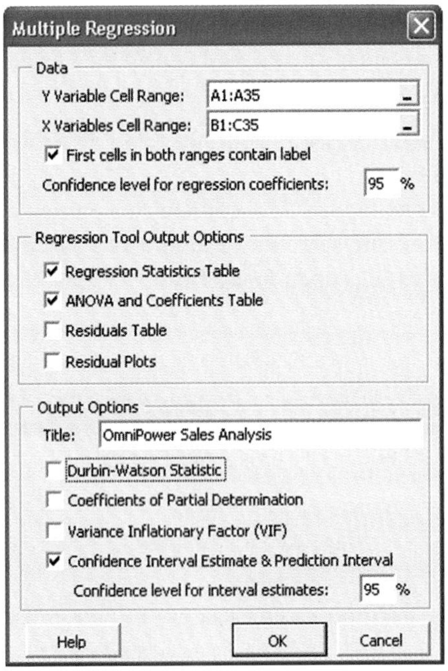

FIGURE G.36

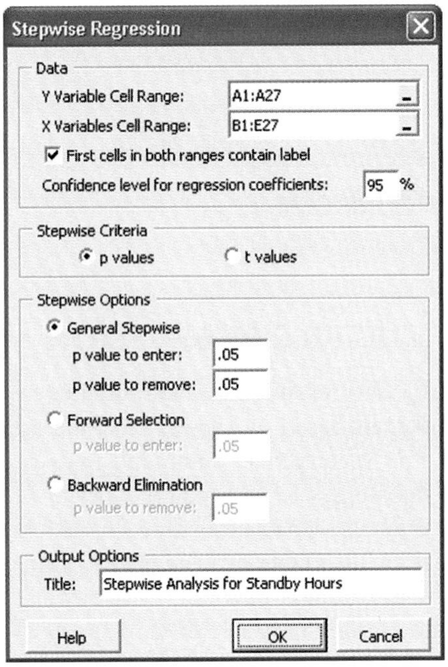

FIGURE G.37

G.37 STEPWISE REGRESSION

The Stepwise Regression command creates a stepwise analysis worksheet based on stepwise criteria and options and a confidence level that you specify. The command invisibly uses the Data Analysis Regression procedure to perform a series of regressions that are then summarized on the stepwise analysis worksheet.

Use: Select **PHStat → Regression → Stepwise Regression**.

Example: Figure G.37 contains the entries for the section 15.5 standby hours analysis example and assumes that the **Standby.xls** file is open to the **Data** worksheet.

G.38 BEST SUBSETS

The Best Subsets command creates a best-subsets analysis worksheet. The command invisibly uses the Data Analysis Regression procedure to perform a series of regressions that are then summarized on the best-subsets analysis worksheet.

Use: Select **PHStat → Regression → Best Subsets**.

Example: Figure G.38 contains the entries for the section 15.5 standby hours analysis example and assumes that the **Standby.xls** file is open to the **Data** worksheet.

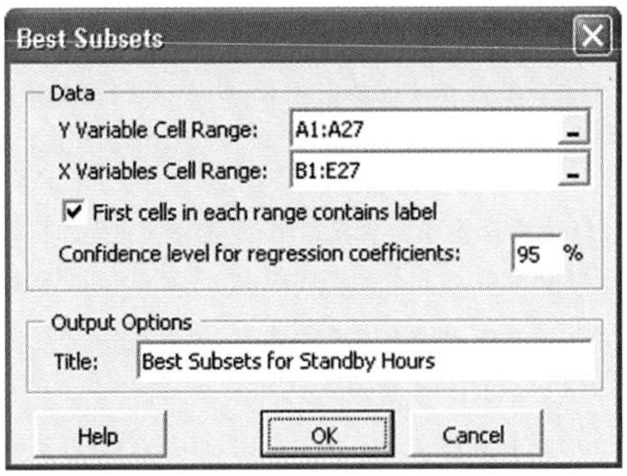

FIGURE G.38

G.39 OPPORTUNITY LOSS

The Opportunity Loss command creates an opportunity loss analysis worksheet based on the number of events and alternative actions that you specify. To complete the worksheet, enter the payoff data in the tinted table cells that begin in row 4.

Use: Select **PHStat → Decision-Making → Opportunity Loss**.

Example: Figure G.39 contains the entries for the Table 17.1 television set marketing example on page 727.

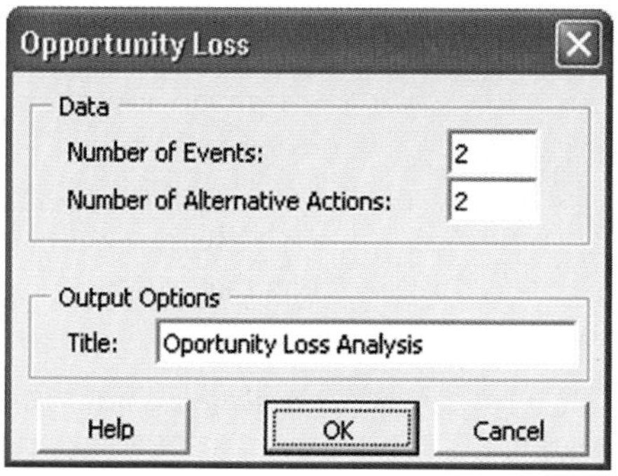

FIGURE G.39

G.40 EXPECTED MONETARY VALUE

The Expected Monetary Value command creates an expected monetary value worksheet based on the number of events and alternative actions that you specify. Select the **Expected Opportunity Loss** and **Measures of Variation** output options to include these additional statistics. To complete the worksheet, enter the probabilities and payoff data in the tinted table cells that begin in row 4.

Use: Select **PHStat → Decision-Making → Expected Monetary Value**.
Example: Figure G.40 contains the entries for the Table 17.7 television set marketing data on page 733.

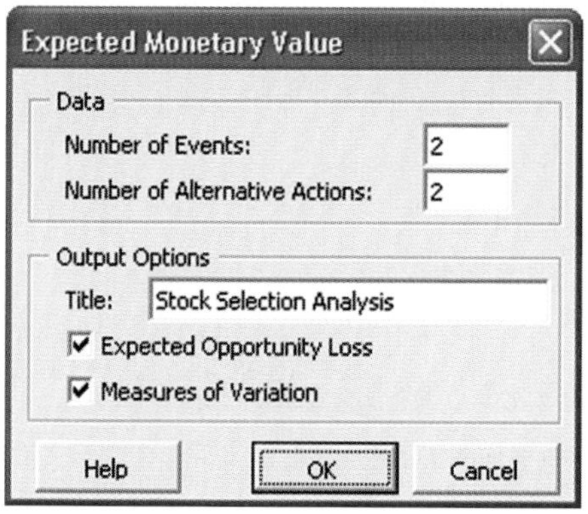

FIGURE G.40

G.41 *p* CHART

The p Chart command creates a *p* chart as a Microsoft Excel chart. (The command also creates two supporting worksheets that contain the data for the chart.) Select the **Size varies** option if the sample (subgroup) size varies and enter the cell range of the sample (subgroup) size variable.

Use: Select **PHStat → Control Charts → p Chart**.
Example: Figure G.41 contains the entries for the section 18.4 hotel rooms study example and assumes that the **Hotel1.xls** file is open to the **Data** worksheet.

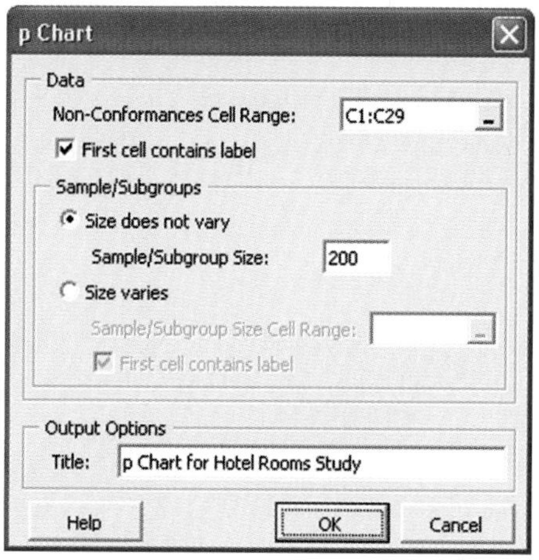

FIGURE G.41

G.42 R AND $\overline{X}$ CHART

The R and XBar Chart command creates *R* and $\overline{X}$ charts as Microsoft Excel charts. Select the **R Chart Only** Chart Option if you do not want an $\overline{X}$ chart. (The command also creates two supporting worksheets that contain data for the charts.)

Use: Select **PHStat → Control Charts → R and XBar Chart**.
Example: Figure G.42 contains the entries for the section 18.7 luggage delivery study example and assumes that the **Hotel2.xls** file is open to the **Data** worksheet.

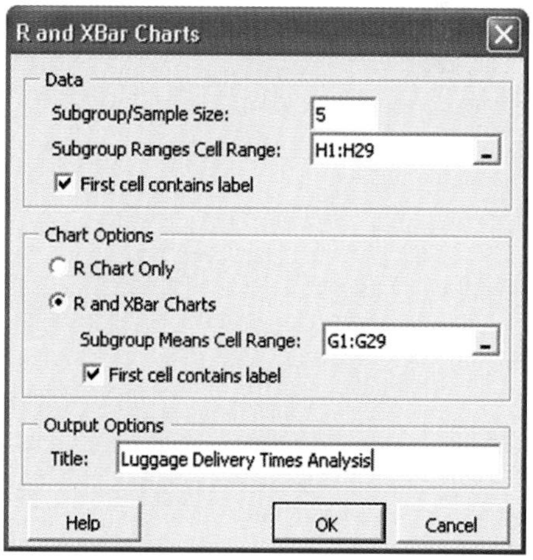

FIGURE G.42

Self-Test Solutions and Answers to Selected Even-Numbered Problems

The following represent worked-out solutions to Self-Test Problems and brief answers to most of the even-numbered problems in the text. For more detailed solutions including explanations, interpretations, and Excel and Minitab output, see the *Student Solutions Manual.*

CHAPTER 1

1.2 Small, medium and large sizes imply order, but do not specify how much more soft drink is added at increasing levels.

1.4 (a) The number of telephones is a numerical variable that is discrete because the variable is counted. It is ratio-scaled because it has a true zero point. **(b)** The length of the longest long-distance call is a numerical variable that is continuous since any value within a range of values can occur. It is ratio-scaled because it has a true zero point. **(c)** Whether there is a telephone line connected to a computer modem in the household is a categorical variable because the answer can only be yes or no. This also makes it a nominal scaled variable. **(d)** Same answer as in (c).

1.6 (a) categorical, nominal scale; **(b)** numerical, continuous, ratio scale; **(c)** numerical, discrete, ratio scale; **(d)** numerical, discrete, ratio scale.

1.8 (a) numerical, continuous, ratio scale; **(b)** numerical, discrete, ratio scale; **(c)** numerical, continuous, ratio scale; **(d)** categorical, nominal scale.

1.10 The underlying variable, ability of the students, may be continuous but the measuring device, the test, does not have enough precision to distinguish between the two students.

1.26 (a) data collected by another organization; **(b)** percentage of flights on-time; **(c)** number of complaints.

1.28 (a) Cat owner households in the United States. **(b)** 1. categorical 2. categorical 3. numerical, discrete 4. categorical.

CHAPTER 2

2.3 (b) The Pareto diagram portrays the data best because it allows you to focus on the categories that have the highest percentage of reasons. **(c)** Try to avoid the following mistakes: little or no knowledge of the company, unprepared to discuss career plans, and limited enthusiasm.

2.4 (b) The Pareto diagram is better than the pie chart to portray these data because it not only sorts the frequencies in descending order, it also provides the cumulative polygon on the same scale. **(c)** From the Pareto diagram, it is obvious that "Google" has the largest market share of 32% followed by Yahoo at 25%.

2.6 (b) 88%; **(d)** The Pareto diagram allows you to see which sources account for most of the electricity.

2.8 (b) The bar chart allows you to see that the "have software for all users" category dominates the company use of anti-spam software.

2.10 (b) Rooms dirty, rooms not stocked, and rooms need maintenance have the largest number of complaints, so focusing on these categories can reduce the number of complaints the most.

2.12 Stem-and-leaf of Finance Scores

5	34
6	9
7	4
8	0
9	38
	$n = 7$

2.14 50 74 74 76 81 89 92

2.16 (a) Ordered array: \$15 \$15 \$18 \$18 \$20 \$20 \$20 \$20 \$20 \$21 \$22 \$22 \$25 \$25 \$25 \$25 \$25 \$26 \$28 \$29 \$30 \$30 \$30
(b) Stem-and-Leaf Display

1	5 5 8 8
2	0 0 0 0 0 1 2 2 5 5 5 5 5 6 8 9
3	0 0 0

(c) The stem-and-leaf display provides more information because it not only orders values from the smallest to the largest into stems and leaves, it also conveys information on how the values distribute and cluster in the data set. **(d)** The bounced check fees seem to be concentrated around \$20 and \$25 since there are five occurences of each of the two values in the sample of 23 banks.

2.18 (a) Ordered array for chicken: 7, 9, 15, 16, 16, 18, 22, 25, 27, 33, 39
Ordered array for burger: 19, 31, 34, 35, 39, 39, 43
(b) Stem-and-leaf display for burgers

1	9
2	
3	14599
4	3

Stem-and-leaf display for chicken

0	79
1	5668
2	257
3	39

(c) The stem-and-leaf display provides more information because it not only orders values from the smallest to the largest into stems and leaves, it also conveys information on how the values distribute and cluster in the data set. **(d)** There seems to be higher fat content for burgers because 6 values in the sample of 7 have fat content higher than 30 as compared to only 2 values in the sample of 11 chicken items. Also, there is only 1 value with a fat content lower than 20 for burgers as compared to 6 values in the sample of 11 chicken items.

2.20 (a) 10 but less than 20, 20 but less than 30, 30 but less than 40, 40 but less than 50, 50 but less than 60, 60 but less than 70, 70 but less than 80, 80 but less than 90, 90 but less than 100. **(b)** 10; **(c)** 15, 25, 35, 45, 55, 65, 75, 85, 95

2.22 (a)

Electricity Costs	Frequency	Percentage
$80 up to $100	4	8%
$100 up to $120	7	14
$120 up to $140	9	18
$140 up to $160	13	26
$160 up to $180	9	18
$180 up to $200	5	10
$200 up to $220	3	6

(c)

Electricity Costs	Frequency	Percentage	Cumulative %
$ 99	4	8.00%	8.00%
$119	7	14.00%	22.00%
$139	9	18.00%	40.00%
$159	13	26.00%	66.00%
$179	9	18.00%	84.00%
$199	5	10.00%	94.00%
$219	3	6.00%	100.00%

(d) The majority of utility charges are clustered between $120 and $180.

2.23 (a)

Error	Frequency	Cumulative %	Percentage
−0.00350—−0.00201	13	13%	13%
−0.00200—−0.00051	26	39%	26%
−0.00050— 0.00099	32	71%	32%
0.00100— 0.00249	20	91%	20%
0.00250— 0.00399	8	99%	8%
0.00400— 0.00549	1	100%	1%

(d) Yes, the steel mill is doing a good job at meeting the requirement as there is only one steel part out of a sample of 100 that is as much as 0.005 inches longer than the specified requirement.

2.24 (a)

Width	Frequency	Percentage
8.310—8.329	3	6.12%
8.330—8.349	2	4.08%
8.350—8.369	1	2.04%
8.370—8.389	4	8.16%
8.390—8.409	5	10.20%
8.410—8.429	16	31.65%
8.430—8.449	5	10.20%
8.450—8.469	5	10.20%
8.470—8.489	6	12.24%
8.490—8.509	2	4.08%

(d) All the troughs will meet the company's requirements of between 8.31 and 8.61 inches wide.

2.26 (a)

Bulb Life (hrs)	Percentage, Mfgr A	Percentage, Mfgr B
650— 749	7.5%	0.0%
750— 849	12.5	5.0
850— 949	50.0	20.0
950—1049	22.5	40.0
1050—1149	7.5	22.5
1150—1249	0.0	12.5

(c)

Bulb Life (hrs)	Percentage Less Than, Mfgr A	Percentage Less Than, Mfgr B
650— 749	7.5%	0.0%
750— 849	20.0	5.0
850— 949	70.0	25.0
950—1049	92.5	65.0
1050—1149	100.0	87.5
1150—1249	100.0	100.0

(d) Manufacturer B produces bulbs with longer lives than Manufacturer A. The cumulative percentage for Manufacturer B shows 65% of their bulbs lasted 1,049 hours or less contrasted with 70% of Manufacturer A's bulbs, which lasted 949 hours or less. None of Manufacturer A's bulbs lasted more than 1,149 hours, but 12.5% of Manufacturer B's bulbs lasted between 1,150 and 1,249 hours. At the same time, 7.5% of Manufacturer A's bulbs lasted less than 750 hours, while all of Manufacturer B's bulbs lasted at least 750 hours.

2.28 (a) Table of frequencies for all student responses

	STUDENT MAJOR CATEGORIES			
GENDER	A	C	M	Totals
Male	14	9	2	25
Female	6	6	3	15
Totals	20	15	5	40

(b) Table of percentages based on overall student responses

	STUDENT MAJOR CATEGORIES			
GENDER	A	C	M	Totals
Male	35.0%	22.5%	5.0%	62.5%
Female	15.0%	15.0%	7.5%	37.5%
Totals	50.0%	37.5%	12.5%	100.0%

(c) Table based on row percentages

	STUDENT MAJOR CATEGORIES			
GENDER	A	C	M	Totals
Male	56.0%	36.0%	8.0%	100.0%
Female	40.0%	40.0%	20.0%	100.0%
Totals	50.0%	37.5%	12.5%	100.0%

(d) Table based on column percentages

	STUDENT MAJOR CATEGORIES			
GENDER	A	C	M	Totals
Male	70.0%	60.0%	40.0%	62.5%
Female	30.0%	40.0%	60.0%	37.5%
Totals	100.0%	100.0%	100.0%	100.0%

2.30 (a)

Contingency Table

	CONDITION OF DIE		
QUALITY	No Particles	Particles	Totals
Good	320	14	334
Bad	80	36	116
Totals	400	50	450

Table of Total Percentages

	CONDITION OF DIE		
QUALITY	No Particles	Particles	Totals
Good	71%	3%	74%
Bad	18%	8%	26%
Totals	89%	11%	100%

Table of Row Percentages

	CONDITION OF DIE		
QUALITY	No Particles	Particles	Totals
Good	96%	4%	100%
Bad	69%	31%	100%
Totals	89%	11%	100%

Table of Column Percentages

	CONDITION OF DIE		
QUALITY	No Particles	Particles	Totals
Good	80%	28%	74%
Bad	20%	72%	26%
Totals	100%	100%	100%

(c) The data suggest that there is some association between condition of the die and the quality of wafer because more good wafers are produced when no particles are found in the die and more bad wafers are produced when there are particles found in the die.

2.32 (a)

Table of row percentages

	GENDER		
ENJOY SHOPPING FOR CLOTHING	Male	Female	Total
Yes	38%	62%	100%
No	74%	26%	100%
Total	48%	52%	100%

Table of column percentages

	GENDER		
ENJOY SHOPPING FOR CLOTHING	Male	Female	Total
Yes	57%	86%	72%
No	43%	14%	28%
Total	100%	100%	100%

Table of total percentages

	GENDER		
ENJOY SHOPPING FOR CLOTHING	Male	Female	Total
Yes	27%	45%	72%
No	21%	7%	28%
Total	48%	52%	100%

(c) The percentage that enjoys shopping for clothing is higher among females as compared to males.

2.34 (b) All five brands increased the amount of rebates from 2001 to 2003. Ford, Chevrolet, and Buick almost doubled the amount of rebates.

2.36 (b) Yes, there is a strong positive relationship between X and Y. As X increases so does Y.

2.38 (b) There does not appear to be a positive relationship between price and energy cost. **(c)** The data do not seem to indicate that higher-priced refrigerators have greater energy efficiency.

2.40 (b) There does not appear to be a relationship between battery capacity and talk time. **(c)** In general, this expectation does not appear to be borne out.

2.42 (b) The unemployment rate followed a downward trend from January of 1998 to September of 2000 and changed to an upward trend afterward.

2.44 (b) There is an upward trend in the number of households using online banking and/or online bill payment. **(c)** The number of U.S. households actively using online banking and/or online bill payment in 2004 will be around 36 million.

2.62 (c) The publisher gets the largest portion (64.8%) of the revenue. About half (32.2%) of the revenue received by the publisher covers manufacturing costs. Publisher's marketing and promotion account for the next larger share of the revenue at 15.4%. Author, bookstore employee salaries and benefits, and publisher administrative costs and taxes each accounts for around 10% of the revenue while the publisher after-tax profit, bookstore operations, bookstore pretax profit and freight constitute the "trivial few" allocations of the revenue.

2.64 (b) From 1999 to 2003, payment by cash and check had declined while payment by debit and other type of payment had increased. The percentage of payment by credit had remained more or less constant.

2.66 (a) The Pareto diagram is most appropriate because it not only sorts the frequencies in descending order, it also provides the cumulative polygon on the same scale. From the Pareto diagram, USA and Brazil make up more than half of the coffee consumption in major markets in 2000. **(b)** The Pareto diagram is most appropriate because it not only sorts the frequencies in descending order, it also provides the cumulative polygon on the same scale. From the Pareto diagram, no single major corporation dominates the coffee market in Brazil. The corporation that owns the largest share of the market, Sara Lee owned brands, captures less than 30% of the market share.

2.68 (a) There is no particular pattern to the deaths due to terrorism on U.S. soil between 1990 and 2001. There are exceptionally high death counts in 1995 and 2001 due to the Oklahoma City and New York City bombings. **(c)** The Pareto diagram is best to portray these data because it not only sorts the frequencies in descending order, it also provides the cumulative polygon on the same scale. The labels in the pie chart become cluttered because there are too many categories in the causes of death.

(d) The major causes of death in the U.S. in 2000 were heart diseases followed by cancer. These two accounted for more than 70% of the total deaths.

2.70 (a)

DESSERT ORDERED	GENDER Male	Female	Total
Yes	71%	29%	100%
No	48%	52%	100%
Total	53%	47%	100%

DESSERT ORDERED	GENDER Male	Female	Total
Yes	30%	14%	23%
No	70%	86%	77%
Total	100%	100%	100%

DESSERT ORDERED	BEEF ENTRÉE Yes	No	Total
Yes	52%	48%	100%
No	25%	75%	100%
Total	31%	69%	100%

DESSERT ORDERED	BEEF ENTRÉE Yes	No	Total
Yes	38%	16%	23%
No	62%	84%	77%
Total	100%	100%	100%

DESSERT ORDERED	BEEF ENTRÉE Yes	No	Total
Yes	12%	11%	23%
No	19%	58%	77%
Total	31%	69%	100%

(b) If the owner is interested in finding out the percentage of males and females who order dessert or the percentage of those who order a beef entrée and a dessert among all patrons, the table of total percentages is most informative. If the owner is interested in the effect of gender on ordering of dessert or the effect of ordering a beef entrée on the ordering of dessert, the table of column percentages will be most informative. Since dessert will usually be ordered after the main entree and the owner has no direct control over the gender of patrons, the table of row percentages is not very useful here. **(c)** 30% of the men ordered desserts compared to 14% of the women. Men are more than twice as likely to order desserts as women. Almost 38% of the patrons ordering a beef entree ordered dessert compared to 16% of patrons ordering all other entrees. Patrons ordering beef are more than 2.3 times as likely to order dessert as patrons ordering any other entree.

2.72 (a) 23575R15 accounts for over 80% of the warranty claims. **(b)** Tread separation accounts for the majority (70%) of the warranty claims. **(c)** Tread separation accounts for more than 70% of the warranty claims among the ATX model. **(d)** The number of claims is evenly distributed among the three incidents; other/unknown incidents account for almost 40% of the claims, tread separation accounts for about 35% of the claims while blow out accounts for about 25% of the claims.

2.74

Cost	Frequency	Percentage	Cumulative Percentage
0.50 but less than 0.75	4	11.11%	11.11%
0.75 but less than 1	16	44.44%	55.56%
1 but less than 1.25	3	8.33%	63.89%
1.25 but less than 1.5	8	22.22%	86.11%
1.5 but less than 1.75	4	11.11%	97.22%
1.75 but less than 2	1	2.78%	100.00%

Calories	Frequency	Percentage	Cumulative Percentage
280 but less than 310	5	13.89%	13.89%
310 but less than 340	9	25.00%	38.89%
340 but less than 370	10	27.78%	66.67%
370 but less than 400	8	22.22%	88.89%
400 but less than 430	4	11.11%	100.00%

Fat	Frequency	Percentage	Cumulative Percentage
Less than 5	1	2.78%	2.78%
5 but less than 10	4	11.11%	13.89%
10 but less than 15	13	36.11%	50.00%
15 but less than 20	9	25.00%	75.00%
20 but less than 25	7	19.44%	94.44%
25 but less than 30	2	5.56%	100.00%

(e) The typical cost for a slice of pizza is between $0.75 and $1.00, since that is both the most frequently occurring interval and more than 50% of the sample is less than or equal to $1.00. The typical caloric content for a slice of pizza is between 310 and 400 calories, since more than 80% of the sample falls in that range. More than 73% of the pizzas have between 10 to 20 grams of fat. Based on the results of the scatter diagrams, calories and fat seem to be related. Other variables do not show any particular pattern in the scatter plot, but the graph of calories and fat has a positive slope because it rises from left to right, showing that as the value of one variable increases, the other also tends to increase.

2.76 (a)

Frequencies (Boston)		
Weight (Boston)	Frequency	Percentage
3015 but less than 3050	2	0.54%
3050 but less than 3085	44	11.96%
3085 but less than 3120	122	33.15%
3120 but less than 3155	131	35.60%
3155 but less than 3190	58	15.76%
3190 but less than 3225	7	1.90%
3225 but less than 3260	3	0.82%
3260 but less than 3295	1	0.27%

(b)

Frequencies (Vermont)		
Weight (Vermont)	Frequency	Percentage
3550 but less than 3600	4	1.21%
3600 but less than 3650	31	9.39%

Frequencies (Vermont)		
Weight (Vermont)	Frequency	Percentage
3650 but less than 3700	115	34.85%
3700 but less than 3750	131	39.70%
3750 but less than 3800	36	10.91%
3800 but less than 3850	12	3.64%
3850 but less than 3900	1	0.30%

(d) 0.54% of the "Boston" shingles pallets are underweight while 0.27% are overweight. 1.21% of the "Vermont" shingles pallets are underweight while 3.94% are overweight.

2.78 (a), (c)

Calories	Frequency	Percentage	Percentage Less Than
50 but less than 100	3	12%	12%
100 but less than 150	3	12	24
150 but less than 200	9	36	60
200 but less than 250	6	24	84
250 but less than 300	3	12	96
300 but less than 350	0	0	96
350 but less than 400	1	4	100

Protein	Frequency	Percentage	Percentage Less Than
16 but less than 20	1	4	4
20 but less than 24	5	20	24
24 but less than 28	8	32	56
28 but less than 32	9	36	92
32 but less than 36	2	8	100

Calories from Fat	Frequency	Percentage	Percentage Less Than
0% but less than 10%	3	12	12
10% but less than 20%	4	16	28
20% but less than 30%	2	8	36
30% but less than 40%	5	20	56
40% but less than 50%	3	12	68
50% but less than 60%	5	20	88
60% but less than 70%	2	8	96
70% but less than 80%	1	4	100

Calories from Saturated Fat	Frequency	Percentage	Percentage Less Than
0% but less than 5%	6	24	24
5% but less than 10%	2	8	32
10% but less than 15%	5	20	52
15% but less than 20%	5	20	72
20% but less than 25%	5	20	92
25% but less than 30%	2	8	100

Cholesterol	Frequency	Percentage	Percentage Less Than
0 but less than 50	2	8	8%
50 but less than 100	17	68	76
100 but less than 150	4	16	92
150 but less than 200	1	4	96
200 but less than 250	0	0	96
250 but less than 300	0	0	96
300 but less than 350	0	0	96
350 but less than 400	0	0	96
400 but less than 450	0	0	96
450 but less than 500	1	4	100

(d) The sampled fresh red meats, poultry, and fish vary from 98 to 397 calories per serving with the highest concentration between 150 to 200 calories. One protein source, spareribs with 397 calories, was over 100 calories beyond the next highest caloric food. The protein content of the sampled foods varies from 16 to 33 grams with 68% of the data values falling between 24 and 32 grams. Spareribs and fried liver are both very different from other foods sampled, the former on calories and the latter on cholesterol content.

2.80 (a)

Count of Drive Type	Fuel Type			
Drive Type	Diesel	Premium	Regular	Grand Total
AWD	0	5	2	7
Front	1	18	63	82
Front, AWD	0	0	1	1
Permanent 4WD	0	3	0	3
Rear	0	11	17	28
Grand Total	1	37	83	121

(c) Based on the results of (a) and (b), the percentage of front-wheel drive cars that use regular gasoline appears to be higher than that for rear-wheel drive cars.

2.82 (a), (c)

Average Ticket$	Frequency	Percentage	Cumulative %
6 but less than 12	3	10.00%	10.00%
12 but less than 18	12	40.00%	50.00%
18 but less than 24	11	36.67%	86.67%
24 but less than 30	3	10.00%	96.67%
30 but less than 36	0	0.00%	96.67%
36 but less than 42	1	3.33%	100.00%

Fan Cost Index	Frequency	Percentage	Cumulative %
80 but less than 105	2	6.67%	6.67%
105 but less than 130	7	23.33%	30.00%
130 but less than 155	10	33.33%	63.33%
155 but less than 180	9	30.00%	93.33%
180 but less than 205	1	3.33%	96.67%
205 but less than 230	1	3.33%	100.00%

Regular season game receipts ($millions)	Frequency	Percentage	Cumulative %
5 but less than 20	5	16.67%	16.67%
20 but less than 35	7	23.33%	40.00%
35 but less than 50	5	16.67%	56.67%
50 but less than 65	6	20.00%	76.67%
65 but less than 80	5	16.67%	93.33%
80 but less than 95	1	3.33%	96.67%
95 but less than 110	1	3.33%	100.00%

Local TV, radio and cable ($millions)	Frequency	Percentage	Cumulative %
0 but less than 10	7	23.33%	23.33%
10 but less than 20	12	40.00%	63.33%
20 but less than 30	6	20.00%	83.33%
30 but less than 40	3	10.00%	93.33%
40 but less than 50	1	3.33%	96.67%
50 but less than 60	1	3.33%	100.00%

Other Local Operating Revenue ($millions)	Frequency	Percentage	Cumulative %
0 but less than 10	6	20.00%	20.00%
10 but less than 20	3	10.00%	30.00%
20 but less than 30	8	26.67%	56.67%
30 but less than 40	8	26.67%	83.33%
40 but less than 50	3	10.00%	93.33%
50 but less than 60	1	3.33%	96.67%
60 but less than 70	1	3.33%	100.00%

Player compensation and benefits ($millions)	Frequency	Percentage	Cumulative %
30 but less than 45	5	16.67%	16.67%
45 but less than 60	8	26.67%	43.33%
60 but less than 75	4	13.33%	56.67%
75 but less than 90	5	16.67%	73.33%
90 but less than 105	5	16.67%	90.00%
105 but less than 120	3	10.00%	100.00%

National and other local Expenses ($millions)	Frequency	Percentage	Cumulative %
30 but less than 40	4	13.33%	13.33%
40 but less than 50	10	33.33%	46.67%
50 but less than 60	9	30.00%	76.67%
60 but less than 70	2	6.67%	83.33%
70 but less than 80	3	10.00%	93.33%
80 but less than 90	2	6.67%	100.00%

Income from Baseball Operations ($millions)	Frequency	Percentage	Cumulative %
–60 but less than –45	2	6.67%	6.67%
–45 but less than –30	2	6.67%	13.33%
–30 but less than –15	8	26.67%	40.00%
–15 but less than 0	8	26.67%	66.67%
0 but less than 15	7	23.33%	90.00%
15 but less than 30	1	3.33%	93.33%
30 but less than 45	2	6.67%	100.00%

(d) There appears to be a weak positive linear relationship between number of wins and player compensation and benefits.

2.84 (b) The only variable that appears to be useful in predicting printer price is text cost. There appears to be a negative relationship between price and text cost. Typically the higher the text cost, the lower is the printer price.

Chapter 3

3.2 (a) mean = 7, median = 7, mode = 7; **(b)** range = 9, interquartile range = 5, $S^2 = 10.8$, $S = 3.286$, $CV = 46.943\%$; **(c)** Z scores: 0, –0.913, 0.609, 0, –1.217, 1.521. None of the Z scores is larger than 3.0 or smaller than –3.0. There is no outlier. **(d)** symmetric since mean = median

3.4 (a) mean = 2, median = 7, mode = 7; **(b)** range = 17, interquartile range = 14.5, $S^2 = 62$, $S = 7.874$, $CV = 393.7\%$; **(c)** left skewed since mean < median

3.6 (a)

	Grade *X*	Grade *Y*
Mean	575	575.4
Median	575	575
Standard deviation	6.40	2.07

(b) If quality is measured by central tendency, Grade *X* tires provide slightly better quality because *X*'s mean and median are both equal to the expected value, 575 mm. If, however, quality is measured by consistency, Grade *Y* provides better quality because, even though *Y*'s mean is only slightly larger than the mean for Grade *X*, *Y*'s standard deviation is much smaller. The range in values for Grade *Y* is 5 mm compared to the range in values for Grade *X*, which is 16 mm.

(c)

	Grade *X*	Grade *Y*, Altered
Mean	575	577.4
Median	575	575
Standard deviation	6.40	6.11

When the fifth *Y* tire measures 588 mm rather than 578 mm, *Y*'s mean inner diameter becomes 577.4 mm, which is larger than *X*'s mean inner diameter, and *Y*'s standard deviation increases from 2.07 mm to 6.11 mm. In this case, *X*'s tires are providing better quality in terms of the mean inner diameter with only slightly more variation among the tires than *Y*'s.

3.7 (a) For the burgers: $\bar{X} = \frac{240}{7} = 34.2857$ Median = (7+1)/2 = 4th ranked value = 35 $Q_1 = (7+1)/4 = 2^{nd}$ ranked value = 31 $Q_3 = 3(7+1)/4 = 6^{th}$ ranked value = 39

For chicken items: $\bar{X} = \frac{227}{11} = 20.6364$ Median = (11+1)/2 = 6th ranked value = 18 $Q_1 = (11+1)/4 = 3^{rd}$ ranked value = 15 $Q_3 = 3(11+1)/4 = 9^{th}$ ranked value = 27

(b) For the burgers: Range = 43 – 19 = 24, Interquartile range = 39 – 31 = 8

Total Fat(X)	X – Mean	(X – Mean)²
19	–15.2857	233.653061
31	–3.28571	10.7959184
34	–0.28571	0.08163265
35	0.714286	0.51020408
39	4.714286	22.2244898
39	4.714286	22.2244898
43	8.714286	75.9387755
34.285714 Mean		Sum: 365.428571

$$S^2 = \frac{365.428571}{6} = 60.904761;\ S = \sqrt{60.904761} = 7.804$$

$$\text{C.V.} = \frac{7.804}{34.28571} \times 100\% = 22.761\%$$

For the chicken items: Range = 39 – 7 = 32; Interquartile range = 27 – 15 = 12

Total Fat(X)	(X – Mean)	(X – Mean)^2
7	–13.6364	185.950413
9	–11.6364	135.404959
15	–5.63636	31.768595
16	–4.63636	21.4958678
16	–4.63636	21.4958678
18	–2.63636	6.95041322
22	1.363636	1.85950413
25	4.363636	19.0413223
27	6.363636	40.4958678
33	12.36364	152.859504
39	18.36364	337.22314
20.636364 Mean		Sum: 954.545455

$S^2 = \frac{954.54545}{10} = 95.454545; S = \sqrt{95.454545} = 9.77$

$\text{C.V.} = \frac{9.77}{20.636} \times 100\% = 47.344\%$

(c) The data for chicken items are skewed to the right and the data for burgers are skewed to the left. **(d)** In general, burgers have more total fat than chicken items. The lowest total fat among burgers is still higher than 50% of the total fat among chicken items. About 25% of the burgers have higher total fat than the highest total fat among the chicken items.

3.8 (a) The distribution of family incomes will most likely be skewed to the right due to the presence of a few millionaires and billionaires. As a result, the median income is a better measure of central tendency than the mean income. **(b)** The article reports the median home price and not the mean home price because the median is a better measure of central tendency in the presence of some extremely expensive homes that will drive the mean home price upward.

3.10 (a) Calories: mean = 380, median = 350, 1st quartile = 260, 3rd quartile = 510. Fat: mean = 15.79, median = 19, 1st quartile = 8, 3rd quartile = 22 **(b)** Calories: variance = 12,800, standard deviation = 113.14, range = 290, interquartile range = 250, CV = 29.77%. None of the Z scores are less than –3 or greater than 3. There is no outlier in calories. Fat: variance = 52.82, standard deviation = 7.27, range = 18.5, Interquartile range = 14, CV = 46.04%. None of the Z scores are less than –3 or greater than 3. There is no outlier in fat. **(c)** Calories are slightly right-skewed while fat is slightly left-skewed. **(d)** The mean calories are 380 while the middle ranked calorie is 350. The average scatter of calories around the mean is 113.14. The middle 50% of the calories are scattered over 250 while the difference between the highest and the lowest calories is 290. The mean fat is 15.79 grams while the middle ranked fat is 19 grams. The average scatter of fat around the mean is 7.27 grams. 50% of the values are scattered over 14 grams while the difference between the highest and the lowest fat is 18.5 grams.

3.12 (a) mean = \$347.86, median = \$340, 1st quartile = \$290, 3rd quartile = \$400. **(b)** variance = 4,910.44, standard deviation = \$70.07, range = \$230, interquartile range = \$110, CV = 20.14%. None of the Z scores are less than –3 or greater than 3. There is no outlier in the price. **(c)** The price of 3-megapixel cameras is symmetrical. **(d)** The mean price is \$347.86 while the middle ranked price is \$340. The average scatter of price around the mean is \$70.07. The middle 50% of the prices are scattered over \$110 while the difference between the highest and the lowest price is \$230.

3.14 (a) Mean = 473.46, Median = 451. There is no mode. The median seems to be a better descriptive measure, since the data are not symmetric. **(b)** Range = 785, Variance = 44,422.44, Standard deviation = 210.77. **(c)** From the manufacturer's viewpoint, the worst measure would be to compute the proportion of batteries that last over 400 hours (8/13 = 0.61). The median (451) and the mean (473.5) are both over 400, and would be better measures for the manufacturer to use in advertisements. **(d)** Mean = 550.38, Median = 492, Mode = none, Range = 1,078, Variance = 99,435.26, Standard deviation =315.33. From the manufacturer's viewpoint, the worst measure remains the proportion of batteries that last over 400 hours (9/13 = 0.69). The median (492) and the mean (550.38) are both well over 400, and would be better measures for the manufacturer to use in advertisements. The shape of the distribution of the altered data set is right-skewed, since its mean is larger than its median.

3.16 (a) Mean = 7.11, Median = 6.68, Q_1 = 5.64, Q_3 = 8.73. **(b)** Variance = 4.336, Standard deviation = 2.082, Range = 6.67, Interquartile range = 3.09, Coefficient of variation = 29.27%. **(c)** Since the mean is greater than the median, the distribution is right-skewed. **(d)** The mean and median are both more than 5 minutes. The distribution is right-skewed, meaning that there are more unusually high values than low values. Further, 13 of the 15 bank customers sampled (or 86.7%) had waiting times in excess of 5 minutes. So, the customer is more likely to experience a waiting time in excess of 5 minutes. The manager overstated the bank's service record in responding that the customer would "almost certainly" not wait longer than 5 minutes for service.

3.17 $\overline{R}_G = [(1 + 0.61)(1 + 0.55)]^{1/2} - 1 = 57.97\%$

3.18 (a)

Year	DJIA	SP500	Russell2000	Wilshire5000
2003	25.30	26.40	45.40	29.40
2002	–15.01	–22.10	–21.58	–20.90
2001	–5.44	–11.90	–1.03	–10.97
2000	–6.20	–9.10	–3.02	–10.89
Geometric mean	–1.42%	–5.77%	2.28%	–5.07%

(b) The rate of return of SP500 is the worst at –5.77% followed by Wilshire 5000 at –5.07% and DJIA –1.42%. Russell 2000 is the only index among the four that has a positive rate of return at 2.28% over the four-year period. **(c)** In general, investments in the metal market achieved a higher rate of return than investments in the certificate of deposit market from 2000 to 2003. Investments in the stock market had the worst rate of return.

3.20 (a)

Year	Platinum	Gold	Silver
2003	34.2	19.5	24.0
2002	24.5	24.5	5.5
2001	–21.3	1.2	–3.0
2000	–23.3	1.8	–5.9
Geometric Mean	0.21%	11.27%	4.53%

(b) All three metals achieved positive rate of returns over the four-year period with gold yielding the highest rate of return at 11.27%, followed by silver at 4.53% and platinum at 0.21%. **(c)** In general, investments in the metal market achieved a higher rate of return than investments in the certificate of deposit market from 2000 to 2003. Investments in the stock market had the worst rate of return.

3.22 (a) Population Mean $\mu = 6$ **(b)** Population Standard deviation, σ = 1.673, Population Variance, $\sigma^2 = 2.8$

3.23 (a) $\mu = \frac{514}{50} = 10.28, \sigma^2 = \frac{204.92}{50} = 4.0984, \sigma = \sqrt{4.0984} = 2.02445$

(b) 64%, 94%, 100% **(c)** These percentages are lower than the empirical rule would suggest.

3.24 (a) 68%; **(b)** 95%; **(c)** not calculable, 75%, 88.89%. **(d)** $\mu - 4\sigma$ to $\mu + 4\sigma$ or −2.8 to 19.2

3.26 (a) mean = 12,999.2158, variance = 14,959,700.52, std. dev. = 3,867.7772. **(b)** 64.71%, 98.04% and 100% of these states have mean per capita energy consumption within 1, 2, and 3 standard deviation of the mean, respectively. **(c)** This is consistent with the 68%, 95%, and 99.7% according to the empirical rule. **(d)** (a)mean = 12,857.7402, variance = 14,238,110.67, std. dev. = 3,773.3421. **(b)** 66%, 98%, and 100% of these states have a mean per capita energy consumption within 1, 2, and 3 standard deviations of the mean, respectively. **(c)** This is consistent with the 68%, 95% and 99.7% according to the empirical rule.

3.28 (a) 25; **(b)** 11.01

3.30 (a) March: mean = 4,720, April: Mean = 4,400. **(b)** March: S = 2,250.0794, April: S = 2,657.2965. **(c)** The mean has declined by $320 while the standard deviation has increased by $407.2171.

3.32 (a) *A* mean = 40.435, *B* mean = 36.831; **(b)** *A S* = 11.097, *B S* = 8.334; **(c)** Yes, division *B* appears to have younger employees who are less variable in age than in division *A*.

3.34 (a) 3,4,7,9,12; **(b)** The distances between the median and the extremes are close, 4 and 5, but the differences in the sizes of the whiskers are different (1 on the left and 3 on the right) so this distribution is slightly right-skewed. **(c)** In 3.2(c) since the mean = median, the distribution was said to be symmetric. The box part of the graph is symmetric, but the whiskers show right skewness.

3.36 (a) −8, −6.5, 7, 8, 9; **(b)** The shape is left-skewed. **(c)** This is consistent with the answer in 3.4(c).

3.38 (a) Five-number summary: 309 593 895.5 1,425 1,720. **(b)** right-skewed

3.40 (a) Bounced-check fee: Five-number summary: 15 20 22 26 30. Monthly service fee: Five-number summary: 0 5 7 10 12. **(b)** The distribution of bounced-check fees is skewed slightly to the right. The distribution of monthly service charges is skewed to the left. **(c)** The central tendency of the bounced-check fee is substantially higher than that of monthly service fees. While the distribution of the bounced-check fees is symmetrical, the distribution of monthly service fees is skewed more to the left with a few banks charging very low or no monthly service fee.

3.42 (a) Commercial district: Five-number summary: 0.38 3.2 4.5 5.55 6.46. Residential area: Five-number summary: 3.82 5.64 6.68 8.73 10.49. **(b)** Commercial district: The distribution is skewed to the left. Residential area: The distribution is skewed slightly to the right. **(c)** The central tendency of the waiting times for the bank branch located in the commercial district of a city is lower than that of the branch located in the residential area. There are a few longer than normal waiting times for the branch located in the residential area whereas there are a few exceptionally short waiting times for the branch located in the commercial area.

3.44 (a) You can say that there is a strong positive linear relationship between the return on investment of U.S. stocks and the International Large Cap stocks, U.S. stocks and Emerging market stocks, a moderate positive linear relationship between U.S. stocks and International Small Cap stocks, U.S. stocks and Emerging market debt stocks, and a very weak positive linear relationship between U.S. stocks and International Bonds. **(b)** In general, there is a positive linear relationship between the return on investment of U.S. stocks and international stocks, U.S. bonds and international bonds, U.S. stocks and Emerging market debt, and a very weak negative linear relationship, if any, between the return on investment of U.S. bonds and international stocks.

3.46 (a) cov (X, Y) = 591.667 **(b)**, r = 0.7196 **(c)** The correlation coefficient is more valuable for expressing the relationship between calories and fat because it does not depend on the units used to measure calories and fat. **(d)** There is a strong positive linear relationship between calories and fat.

3.48 (a) cov (X, Y) = −336.958. **(b)** $r = \dfrac{-336.958}{(105.3617)(7.967)} = -0.4014$ **(c)** The coefficient of correlation between turnover rate and security violations indicates that there is a weak negative linear relationship between the two.

3.62 (a) mean = 43.89, median = 45, 1st quartile = 18, 3rd quartile = 63. **(b)** range = 76, interquartile range = 45, variance = 639.2564, standard deviation = 25.28, coefficient of variation = 57.61%. **(c)** The distribution is skewed to the right because there are a few policies that require an exceptionally long period to be approved. **(d)** The mean approval process takes 43.89 days with 50% of the policies being approved in less than 45 days. 50% of the applications are approved between 18 and 63 days. About 67% of the applications are approved between 18.6 to 69.2 days.

3.64 (a) mean = 8.421, median = 8.42, range = 0.186, S = 0.0461. The mean and median width are both 8.42 inches. The range of the widths is 0.186 inches, and the average scatter around the mean is 0.0461 inches. **(b)** 8.312, 8.404, 8.42, 8.459, 8.498; **(c)** Even though the mean = median, the left whisker is longer so the distribution is left-skewed. **(d)** All the troughs in this sample meet the specifications.

3.66 (a) Office I: mean = 2.214, median = 1.54; Office II: mean = 2.011, median = 1.505; Office I: Q_1 = 0.93, Q_3 = 3.93; Office II: Q_1 = 0.6, Q_3 = 3.75; **(b)** Office I: Range = 5.80, IQR = 3.00, S^2 = 2.952, S = 1.718, CV = 77.597%; Office II: Range = 7.47, IQR = 3.15, S^2 = 3.579, S = 1.892, CV = 94.04%; **(c)** Yes, they are both right-skewed. **(d)** Office II has more variability in times to clear problems, with a wider range and a larger standard deviation. Office II has a lower mean time to clear problems.

3.68 (a) Cost: mean = 0.171; median = 0.17; Calories: mean = 165.758, median = 190; Fiber: mean = 6.909, median = 6; Sugar: mean = 11.394, median = 11; Cost: Q_1 = 0.13, Q_3 = 0.20; Calories: Q_1 = 135, Q_3 = 200; Fiber: Q_1 = 5, Q_3 = 8; Sugar: Q_1 = 6, Q_3 = 17.5; **(b)** Cost: Range = 0.17, IQR = 0.07, S^2 = 0.00220, S = 0.0469, CV = 27.49%; Calories: Range = 160, IQR = 65, S^2 = 2,681.439, S = 51.783, CV = 31.240%; Fiber: Range = 8, IQR = 3, S^2 = 5.773, S = 2.403, CV = 34.781%; Sugar: Range = 23, IQR = 11.50, S^2 = 44.246, S = 6.652, CV = 58.382%; **(c)** Cost: right-skewed, calories: left skewed, fiber: right skewed, sugar: approximately symmetric; **(d)** Cost: The mean cost is about 17 cents per ounce. Most cereals cluster around this cost with a few high-priced cereals. The average scatter around the mean is about 5 cents per ounce. Calories: The mean calories is about 166 and a middle value of 190, with an average scatter around the mean of about 52. Since the data are left-skewed, most of the calories are clustered at the high end with a few lower-calorie cereals. Fiber: The mean amount of fiber is about 6.9 grams with a middle value of 6 grams, with an average scatter around the mean of about 2.4 grams. Since the fiber is right-skewed, the cereals tend to cluster around the lower fiber numbers. Three cereals increase the fiber mean. Sugar: The mean amount of sugar is 11.4 grams with a middle value of 11 grams. The average scatter around the middle is 6.65 grams. The data are fairly symmetric.

3.70 (a) With Promotion: mean = 20,748.93, standard deviation = 8,109.50; Without Promotion: mean = 13,935.70, standard deviation = 4,437.92. **(b)** With Promotion: minimum = 10,470, 1st quartile = 14,905, median = 19,775, 3rd quartile = 24,456, maximum = 40,605. Without

Promotion: minimum = 9,555, 1st quartile = 11,779, median = 12,952, 3rd quartile = 14,367, maximum = 28,834. **(d)** The mean attendance is 6,813 more when there is a promotion than when there is not, and the variation in attendance when there is a promotion is larger than that when there is no promotion. There are many factors that can cause variation in the paid attendance. Some of them are weather condition, time and day of the game, the visiting team, etc.

3.72 (a) Boston: 0.04, 0.17, 0.23, 0.32, 0.98; Vermont: 0.02, 0.13, 0.20, 0.28, 0.83. **(b)** Both distributions are right-skewed. **(c)** Both sets of shingles did quite well in achieving a granule loss of 0.8 grams or less. The Boston shingles had only 2 data points greater than 0.8 grams. The next lowest to these was 0.6 grams. These 2 data points can be considered outliers. Only 1.176% of the shingles failed the specification. In the Vermont shingles only 1 data point was greater than 0.8 grams. The next point was 0.58 grams. Thus only 0.714% of the shingles failed to meet the specification.

3.74 (a), (b)

	Mean	Median	Q_1	Q_3
Average Ticket $	18.1333	17.83	15.20	20.84
Fan Cost Index	144.5737	143.475	124.25	160.76
Reg.Season $	46.1367	47.55	30.20	62.10
Local TV, Radio, Cable	19.0467	16.35	10.90	23.60
Other Local Revenue	27.5933	29.05	13.90	37.00
Player Comp.	71.3567	70.80	49.40	92.80
National and local expenses	54.6467	50.50	46.90	58.50
Income from baseball operations	−8.3733	−8.35	−18.50	1.90

	Variance	Standard deviation	Range	Interquartile range	C.V.
Average Ticket $	35.9797	5.9983	33.07	5.64	33.08%
Fan Cost Index	843.4552	29.0423	143.84	36.51	20.09%
Reg.Season $	512.5445	22.6394	91.60	31.90	49.07%
Local TV, Radio, Cable	151.0184	12.2890	56.30	12.70	64.52%
Other Local Revenue	234.6186	15.3173	58.70	23.10	55.51%
Player Comp.	663.8405	25.7651	88.00	43.40	36.11%
National and local expenses	176.4081	13.2819	49.20	11.60	24.3%
Income from baseball operations	428.1531	20.6919	93.80	20.40	−247.12%

(c) Average ticket prices, local TV, radio and cable receipts, national and other local expenses are skewed to the right; fan cost index is slightly skewed to the right; all other variables are approximately symmetrical. **(d)** $r = 0.3985$. There is a moderate positive linear relationship between the number of wins and player compensation and benefits.

3.76 (a) $r = -0.384$; $r = -0.512$; $r = -0.544$; $r = -0.261$; **(b)** Color photo time has the strongest relationship with price of the four variables, so it would be the most helpful in predicting price. As the price increases the color photo time tends to decrease. All four variables have a negative (inverse) relationship with price.

3.78 Not SUVs:

(a), (b)

	MPG	Length	Width
Minimum	17	155	65
First Quartile	19	178	68
Median	21	189	71
Third Quartile	23	198	73
Maximum	41	215	79
Mean	22.1556	187.9778	71.0000
Variance	18.7396	161.4377	9.8652
Std. Dev	4.3289	12.7058	3.1409
Range	24	60	14
Interquartile Range	4	20	5
Coefficient of Variation	19.54%	6.76%	4.42%

	Cargo Volume	Turning Circle	Weight
Minimum	5	33	2,150
First Quartile	13	38	3,095
Median	15	40	3,427.5
Third Quartile	19	41	3750
Maximum	75.5	45	4315
Mean	22.3944	39.7000	3,391.7222
Variance	323.7837	6.3247	232,210.7647
Std. Dev	17.9940	2.5149	481.8825
Range	70.5	12	2,165
Interquartile Range	6	3	655
Coefficient of Variation	80.35%	6.33%	14.21%

SUVs:

	MPG	Length	Width
Minimum	10	163	67
First Quartile	15	175	70
Median	16	183	72
Third Quartile	18	190	74
Maximum	22	227	80
Mean	16.4839	184.9032	72.3226
Variance	7.2581	209.5570	11.0925
Std. Dev	2.6941	14.4761	3.3305
Range	12	64	13
Interquartile Range	3	15	4
Coefficient of Variation	16.34%	7.83%	4.61%

	Cargo Volume	Turning Circle	Weight
Minimum	28	37	3,055
First Quartile	34.5	39	3,590
Median	37.5	40	4,135
Third Quartile	45.5	41	4,715
Maximum	84	52	7,270
Mean	42.3548	40.5806	4,267.4194
Variance	183.7532	11.3183	783,086.4516
Std. Dev	13.5556	3.3643	884.9217
Range	56	15	4,215
Interquartile Range	11	2	1,125
Coefficient of Variation	32.00%	8.29%	20.74%

(c) Not SUVs: Miles per gallon, and luggage capacity are skewed to the right; length and weight are skewed to the left; width is slightly skewed to the right and turning circle requirement is quite symmetrical. **SUVs:** All variables are skewed to the right except miles per gallon is only slightly skewed to the right.

CHAPTER 4

4.2 (a) Simple events include selecting a red ball. **(b)** Selecting a white ball

4.4 (a) 60/100 = 3/5 = 0.6 **(b)** 10/100 = 1/10 = 0.1 **(c)** 35/100 = 7/20 = 0.35 **(d)** 9/10 = 0.9

4.6 (a) Mutually exclusive, not collectively exhaustive **(b)** Not mutually exclusive, not collectively exhaustive **(c)** Mutually exclusive, not collectively exhaustive **(d)** Mutually exclusive, collectively exhaustive

4.8 (a) "Is a homeowner." **(b)** "A homeowner who drives himself/herself to work." **(c)** "Does not drive self to work." **(d)** "A person can drive himself/herself to work and is also a homeowner."

4.10 (a) "A wafer is good." **(b)** "A wafer is good and no particle was found on the die." **(c)** "bad wafer." **(d)** A wafer can be a "good wafer" and was produced by a die "with particles."

4.12 (a) 83/369 = 0.2249; **(b)** 137/369 = 0.3713; **(c)** 220/369 = 0.5962; **(d)** The probability of "small-to-midsized or offered stock options" includes the probability of "small-to-midsized and offered stock options," the probability of "small-to-midsized but did not offer stock options" and the probability of "large and offered stock options."

4.14 (a) 360/500 = 18/25 = 0.72; **(b)** 224/500 = 56/125 = 0.448; **(c)** 396/500 = 99/125 = 0.792 **(d)** 500/500 = 1.00

4.16 (a) 10/30 = 1/3 = 0.33; **(b)** 20/60 = 1/3 = 0.33; **(c)** 40/60 = 2/3 = 0.67; **(d)** Since $P(A \mid B) = P(A) = 1/3$, events A and B are statistically independent.

4.18 $\frac{1}{2} = 0.5$

4.20 Since $P(A \text{ and } B) = 0.20$ and $P(A)\,P(B) = 0.12$, events A and B are not statistically independent.

4.21 (a) P (a homeowner | drives to work) = 824/1505 = 0.5475; **(b)** P (drives to work | a homeowner) = 824/1000 = 0.8240 **(c)** The conditional events are reversed. **(d)** Since P (a homeowner) = 1000/2000 = 0.50 is not equal to P (a homeowner | drives to work) = 824/1505 = 0.5475, driving to work and whether the respondent is a homeowner or a renter are not statistically independent.

4.22 (a) 36/116 = 0.3103; **(b)** 14/334 = 0.0419; **(c)** 320/334 = 0.9581 P (no particles) = 400/450 = 0.8889. Since P (no particles | good) ≠ P (no particles), "a good wafer" and "a die with no particle" are not statistically independent.

4.24 (a) 29/56 = 0.5179; **(b)** 29/155 = 0.1871; **(c)** The conditional events are reversed. **(d)** Since P (white | claim bias) = 0.1871 is not equal to P (white) = 0.1210, being white and claiming bias are not statistically independent.

4.26 (a) 0.025/0.6 = 0.0417; **(b)** 0.015/0.4 = 0.0375; **(c)** Since P (needs warranty repair | manufacturer based in U.S.) = 0.0417 and P (needs warranty repair) = 0.04, the two events are not statistically independent.

4.28 (a) 0.0045; **(b)** 0.012; **(c)** 0.0059; **(d)** 0.0483

4.30 0.095

4.32 (a) 0.736; **(b)** 0.997

4.33 (a) H = husband watching; W = wife watching

$$P(H|W) = \frac{P(W|H)\cdot P(H)}{P(W|H)\cdot P(H) + P(W|H')\cdot P(H')}$$

$$= \frac{(0.4)(0.6)}{(0.4)(0.6) + (0.3)(0.4)} = \frac{0.24}{0.36}$$

$$= \frac{2}{3} = 0.667$$

(b) $P(W) = 0.24 + 0.12 = 0.36$

4.34 (a) 0.4615; **(b)** 0.325

4.36 (a) P (huge success | favorable review) = 0.099/0.459 = 0.2157; P (moderate success | favorable review) = 0.14/0.459 = 0.3050; P (break even | favorable review) = 0.16/0.459 = 0.3486; P (loser | favorable review) = 0.06/0.459 = 0.1307; **(b)** P (favorable review) = 0.459.

4.38 $3^{10} = 59{,}049$

4.40 (a) $2^7 = 128$; **(b)** $6^7 = 279{,}936$
(c) There are two mutually exclusive and collectively exhaustive outcomes in (a) and six in (b).

4.41 (7)(3)(3) = 63

4.42 (8)(4)(3)(3) = 288

4.43 $n! = 4! = (4)(3)(2)(1) = 24$

4.44 5! = (5)(4)(3)(2)(1) = 120. Not all these orders are equally likely because the players are different in each team.

4.46 $n! = 6! = 720$

4.47 $\dfrac{n!}{(n-X)!} = \dfrac{12!}{9!} = (12)(11)(10) = 1{,}320$

4.48 28

4.49 $\dfrac{n!}{X!(n-X)!} = \dfrac{7!}{4!(3!)} = \dfrac{(7)(6)(5)}{(3)(2)(1)} = 35$

4.50 4,950

4.62 (a) 0.035, **(b)** 0.49, **(c)** 0.975, **(d)** 0.02, **(e)** 0.2857, **(f)** The conditions are switched. Part (d) answers $P(A \mid B)$ and part (e) answers $P(B \mid A)$.

4.64 (a) A simple event can be "a firm that has a transactional public web site" and a joint event can be "a firm that has a transactional public web site and has sales greater than $10 billion." **(b)** 0.3469, **(c)** 0.1449, **(d)** Since P (transactional public web site) · P (sales in excess of $10 billion) ≠ P (transactional public web site and sales in excess of $10 billion), the two events, "sales in excess of ten billion dollars" and "has a transactional public web site" are not independent.

4.66 (a) 0.0225; **(b)** 3,937.5 ≅ 3,938 can be expected to read the advertisement and place an order. **(c)** 0.03; **(d)** 5,250 can be expected to read the advertisement and place an order.

4.68 (a) 0.4712; **(b)** Since the probability that a fatality involved a rollover given that the fatality involved an SUV, van, or pickup is 0.4712, which is almost twice the probability that a fatality involved a rollover with any vehicle type at 0.24, SUV's, vans, or pickups are generally more prone to rollover accidents.

CHAPTER 5

5.2 (a) C: $\mu = 2$, D: $\mu = 2$; **(b)** C: $\sigma = 1.414$, D: $\sigma = 1.095$; **(c)** Distribution C is uniform and symmetric; distribution D is symmetric and has a single mode.

5.4 (a) $\mu = 2$; **(b)** $\sigma = 1.183$

5.5 (a)–(b)

X	$P(x)$	$X*P(X)$	$(X-\mu)^2$	$(X-\mu)^2*P(X)$
0	0.32	0	1.6129	0.516128
1	0.35	0.35	0.0729	0.025515
2	0.18	0.36	0.5329	0.095922
3	0.08	0.24	2.9929	0.239432
4	0.04	0.16	7.4529	0.298116
5	0.02	0.10	13.9129	0.278258
6	0.01	0.06	22.3729	0.223729
	(a) Mean =	1.27	variance =	1.6771
			(b) Stdev =	1.29503

5.6 (a)

X	$P(X)$
\$ − 1	21/36
\$ + 1	15/36

(b)

X	$P(X)$
\$ − 1	21/36
\$ + 1	15/36

(c)

X	$P(X)$
\$ − 1	30/36
\$ + 4	6/36

(d) \$ − 0.167 for each method of play

5.8 (a) 90; 30; **(b)** 126.10; 10.95; **(c)** –1,300; **(d)** 120

5.10 (a) 9.5 minutes; **(b)** 1.9209 minutes

5.12

$X*P(X)$	$Y*P(Y)$	$(X-\mu_X)^2*P(X)$	$(Y-\mu_Y)^2*P(Y)$	$(X-\mu_X)(Y-\mu_Y)*P(XY)$
−10	5	2,528.1	129.6	−572.4
0	45	1,044.3	5,548.8	−2,407.2
24	−6	132.3	346.8	−214.2
45	−30	2,484.3	3,898.8	−3,112.2

(a) $E(X) = \mu_X = \sum_{i-1}^{N} X_i P(X_i) = 59,\ E(Y) = \mu_Y = \sum_{i-1}^{N} Y_i P(Y_i) = 14$

(b) $\sigma_X = \sqrt{\sum_{i-1}^{N} [X_i - E(X)]^2 P(X_i)} = 78.6702$

$\sigma_Y = \sqrt{\sum_{i=1}^{N} [Y_i - E(Y)]^2 P(Y_i)} = 99.62$

(c) $\sigma_{XY} = \sum_{i=1}^{N} [X_i - E(X)][Y_i - E(Y)]P(X_i Y_i) = 6{,}306$

(d) Stock X gives the investor a lower standard deviation while yielding a higher expected return, so the investor should select stock X.

5.14 (a) \$71; \$97; **(b)** 61.88; 84.27; **(c)** 5,113; **(d)** Risk-averse investors would invest in stock X while risk takers would invest in stock Y.

5.16 (a) $E(X) = \$77$, $E(Y) = \$97$; **(b)** $\sigma_X = 39.76$, $\sigma_Y = 108.95$; **(c)** $\sigma_{XY} = 4161$ **(d)** Common stock fund gives the investor a higher expected return than corporate bond fund, but also has a standard deviation more than 2.5 times higher than that for corporate bond fund. An investor should carefully weigh the increased risk.

5.18 (a) 0.5997; **(b)** 0.0016; **(c)** 0.0439; **(d)** 0.4018

5.20

	Mean	Standard Deviation
(a)	0.40	0.600
(b)	1.60	0.980
(c)	4.00	0.894
(d)	1.50	0.866

5.22 (a) 0.0778; **(b)** 0.6826; **(c)** 0.0870; **(d)(a)** $P(X = 5) = 0.3277$ **(b)** $P(X \geq 3) = 0.9421$; **(c)** $P(X < 2) = 0.0067$

5.24 Given $p = 0.90$ and $n = 3$,

(a) $P(X = 3) = \frac{n!}{X!(n-X)!} p^X (1-p)^{n-X} = \frac{3!}{3!0!}(0.9)^3(0.1)^0 = 0.729$

(b) $P(X = 0) = \frac{n!}{X!(n-X)!} p^X (1-p)^{n-X} = \frac{3!}{0!3!}(0.9)^0(0.1)^3 = 0.001$

(c) $P(X \geq 2) = P(X = 2) + P(X = 3) = \frac{3!}{2!1!}(0.9)^2(0.1)^1 + \frac{3!}{3!0!}(0.9)^3(0.1)^0 =$ 0.972

(d) $E(X) = np = 3(0.9) = 2.7$ $\sigma_X = \sqrt{np(1-p)} = \sqrt{3(0.9)(0.1)} = 0.5196$

5.26 (a) $P(X = 0)$ = approximately 0; **(b)** $P(X = 1)$ = approximately 0; **(c)** $P(X \leq 2) = 0.000000374$; **(d)** $P(X \geq 3) = 1.0$

5.28 (a) Since 68% and 24% come from the survey results conducted by the networks, they are best classified as empirical classical probability. **(b)** 0.000014; **(c)** 0.9721; **(d)** 0.000447

5.30 (a) 0.2565; **(b)** 0.1396; **(c)** 0.3033; **(d)** 0.0247

5.32 (a) 0.0337; **(b)** 0.0067; **(c)** 0.9596; **(d)** 0.0404

5.34 (a)
$$\begin{aligned}P(X<5) &= P(X=0) + P(X=1) + P(X=2) + P(X=3) + P(X=4)\\ &= \frac{e^{-6}(6)^0}{0!} + \frac{e^{-6}(6)^1}{1!} + \frac{e^{-6}(6)^2}{2!} + \frac{e^{-6}(6)^3}{3!} + \frac{e^{-6}(6)^4}{4!}\\ &= 0.002479 + 0.014873 + 0.044618 + 0.089235 + 0.133853 = 0.2851\end{aligned}$$

(b) $P(X = 5) = \frac{e^{-6}(6)^5}{5!} = 0.1606$

(c) $P(X \geq 5) = 1 - P(X < 5) = 1 - 0.2851 = 0.7149$

(d) $P(X = 4 \text{ or } X = 5) = P(X = 4) + P(X = 5) = \frac{e^{-6}(6)^4}{4!} + \frac{e^{-6}(6)^5}{5!} =$ 0.2945

5.36 (a) 0.0404; **(b)** 0.9596; **(c)** 0.8301; **(d)** Because Delta has a higher mean rate of mishandled bags per 1,000 passengers than Jet Blue, its probability of mishandling at least a certain number of bags is higher than that of Jet Blue.

5.38 (a) 0.0176; **(b)** 0.9093; **(c)** 0.9220

5.40 (a) 0.0062; **(b)** 0.1173; **(c)** Because Kia has a higher mean rate of problems per car, the probability that a randomly selected Kia will have no more than 2 problems is lower than that of a randomly chosen Lexus. Likewise, the probability that a randomly selected Kia will have zero problems is lower than that of a randomly chosen Lexus.

5.42 (a) 0.2165; **(b)** 0.8013; **(c)** Because Kia has a lower mean rate of problems per car in 2004 compared to 2003, the probability that a randomly selected Kia has zero problems and the probability of no more than 2 problems are both higher than their values in 2003.

5.44 (a) 0.238; **(b)** 0.2; **(c)** 0.1591; **(d)** 0.0083

5.46 (a) If $n = 6$, $A = 25$, and $N = 100$,

$$P(X \geq 2) = 1 - [P(X = 0) + P(X = 1)]$$
$$= 1 - \left[\frac{\binom{25}{0}\binom{100-25}{6-0}}{\binom{100}{6}} + \frac{\binom{25}{1}\binom{100-25}{6-1}}{\binom{100}{6}} \right]$$
$$= 1 - [0.1689 + 0.3620] = 0.4691$$

(b) If $n = 6$, $A = 30$, and $N = 100$,

$$P(X \geq 2) = 1 - [P(X = 0) + P(X = 1)]$$
$$= 1 - \left[\frac{\binom{30}{0}\binom{100-30}{6-0}}{\binom{100}{6}} + \frac{\binom{30}{1}\binom{100-30}{6-1}}{\binom{100}{6}} \right]$$
$$= 1 - [0.1100 + 0.3046] = 0.5854$$

(c) If $n = 6$, $A = 5$, and $N = 100$,

$$P(X \geq 2) = 1 - [P(X = 0) + P(X = 1)]$$
$$= 1 - \left[\frac{\binom{5}{0}\binom{100-5}{6-0}}{\binom{100}{6}} + \frac{\binom{5}{1}\binom{100-5}{6-1}}{\binom{100}{6}} \right]$$
$$= 1 - [0.7291 + 0.2430] = 0.0279$$

(d) If $n = 6$, $A = 10$, and $N = 100$,

$$P(X \geq 2) = 1 - [P(X = 0) + P(X = 1)]$$
$$= 1 - \left[\frac{\binom{10}{0}\binom{100-10}{6-0}}{\binom{100}{6}} + \frac{\binom{10}{1}\binom{100-10}{6-1}}{\binom{100}{6}} \right]$$
$$= 1 - [0.5223 + 0.3687] = 0.1090$$

(e) The probability that the entire group will be audited is very sensitive to the true number of improper returns in the population. If the true number is very low ($A = 5$), the probability is very low (0.0279). When the true number is increased by a factor of six ($A = 30$), the probability the group will be audited increases by a factor of almost 21 (0.5854).

5.48 (a) 0.00000131; **(b)** 0.0802; **(c)** 0.9198; **(d)** 0; 0.3128; 0.6872

5.50 (a) 0.43956; **(b)** 0.8462; **(c)** 0.923; **(d)** 1.33

5.56 (a) 0.74; **(b)** 0.74; **(c)** 0.3898; **(d)** 0.0012; **(e)** The assumption of independence may not be true.

5.58 (a) 0.0547; **(b)** 0.3828; **(c)** 0.9298; **(d)** If the indicator is a random event, the probability that it will make a correct prediction in 8 or more times out of 10 is virtually zero. If one is willing to accept the argument that the amount of campaign expenditures spent during an election year exerts some multiplying impact on the stock market, the probability that the Dow Jones Industrial Average will increase in a U.S. presidential election year is likely to be near 0.90 based on the result of (a)–(c).

5.60 (a) 0.018228; **(b)** 0.089782; **(c)** 0.89199; **(d)** mean = 3.3, standard deviation = 1.486943

5.62 (a) 0.0000; **(b)** 0.04924; **(c)** 0.909646; **(d)** 0.49578

5.64 (a) 0.0003; **(b)** 0.2289; **(c)** 0.4696; **(d)** 0.5304; **(e)** 0.469581; **(f)** 4.4 so about 4 people on average will refuse to participate

5.66 (a) $\mu = 17.6$, **(b)** $\sigma = 1.453$, **(c)** 0.0776, **(d)** 0.5631, **(e)** 0.9740

5.68 (a) 0.0000192791; **(b)** 0.0334; **(c)** 0.8815; **(d)** Based on the results in (a)–(c), the probability that the Standard & Poor's 500 index will increase if there is an early gain in the first five trading days of the year is very likely to be close to 0.90 because that yields a probability of 88.15% that at least 29 of the 34 years the Standard & Poor's 500 index will increase the entire year. However, you should be aware that a high correlation between two events does not always imply a causal relationship.

5.70 (a) The assumptions needed are (i) the probability that a golfer loses a golf ball in a given interval is constant, (ii) the probability that a golfer loses more than one golf ball approaches 0 as the interval gets smaller, (iii) the probability that a golfer loses a golf ball is independent from interval to interval. **(b)** 0.0111; **(c)** 0.70293; **(d)** 0.29707

5.72 (a) virtually zero, **(b)** 0.00000037737, **(c)** 0.00000173886, **(d)** 0.000168669, **(e)** 0.0011998, **(f)** 0.00407937, **(g)** 0.006598978, **(h)** 0.0113502, **(i)** 0.976601

CHAPTER 6

6.2 (a) 0.9089, **(b)** 0.0911, **(c)** + 1.96, **(d)** – 1.00 and + 1.00.

6.4 (a) 0.1401, **(b)** 0.4168, **(c)** 0.3918, **(d)** + 1.00

6.6 (a) 0.9599, **(b)** 0.0228, **(c)** 43.42, **(d)** 46.64 and 53.36

6.8 (a) $P(34 < X < 50) = P(-1.33 < Z < 0) = 0.4082$; **(b)** $P(X < 30) + P(X > 60) = P(Z < -1.67) + P(Z > 0.83) = 0.0475 + (1.0 - 0.7967) = 0.2508$;
(c) $P(Z < -0.84) \cong 0.20$, $Z = -0.84 = \dfrac{X - 50}{12}$
$X = 50 - 0.84(12) = 39.92$ thousand miles or 39,920 miles
(d) The smaller standard deviation makes the Z-values larger.
(a) $P(34 < X < 50) = P(-1.60 < Z < 0) = 0.4452$
(b) $P(X < 30) + P(X > 60) = P(Z < -2.00) + P(Z > 1.00) = 0.0228 + (1.0 - 0.8413) = 0.1815$
(c) $X = 50 - 0.84(10) = 41.6$ thousand miles or 41,600 miles

6.10 (a) 0.9878; **(b)** 0.8185; **(c)** 86.16%; **(d)** Option 1: Since your score of 81% on this exam represents a Z-score of 1.00, which is below the minimum Z-score of 1.28, you will not earn an "A" grade on the exam under this grading option. Option 2: Since your score of 68% on this exam represents a Z-score of 2.00, which is well above the minimum Z-score of 1.28, you will earn an "A" grade on the exam under this grading option. You should prefer Option 2.

6.12 (a) 0.9772; **(b)** 0.1587, **(c)** 0.0038; **(d)** 0.9962

6.14 With 39 values, the smallest of the standard normal quantile values covers an area under the normal curve of 0.025. The corresponding Z-value is – 1.96. The largest of the standard normal quantile values covers an area under the normal curve of 0.975 and its corresponding Z-value is +1.96.

6.16 (a) mean = 99.662, median = 95.78, range = 104.55, $6 \cdot S_X = 149.3072$, interquartile range = 43.105, $1.33 \cdot S_X = 33.0964$. The mean is greater than the median; the range is smaller than 6 times the standard deviation and the interquartile range is larger than 1.33 times the standard deviation. The data do not appear to follow a normal distribution. **(b)** The normal probability plot suggests that the data are right skewed.

6.18 (a) Plant A: $\bar{X} = 9.382$ $S = 3.998$
Five-number summary: 4.42 7.29 8.515 11.42 21.62
The distribution is right-skewed since the mean is greater than the median.
Plant B: $\bar{X} = 11.354$ $S = 5.126$
Five-number summary: 2.33 6.25 11.96 14.25 25.75
Although the results are inconsistent due to an extreme value in the sample, since the mean is less than the median, we can say that the data for Plant B is left-skewed. **(b)** The normal probability plot for Plant *A* is right skewed. Except for the extreme value, the normal probability plot for Plant *B* is left skewed.

6.20 (a) Interquartile range = 0.0025; $S_X = 0.0017$; Range = 0.008; 1.33 $(S_X) = 0.0023$; 6 $(S_X) = 0.0102$ Since the interquartile range is close to 1.33 (S_X) and the range is also close to 6 (S_X), the data appear to be approximately normally distributed. **(b)** The normal probability plot suggests that the data appear to be approximately normally distributed.

6.22 (a) Five-number summary: 82 127 148.5 168 213; mean = 147.06; mode = 130; range = 131; interquartile range = 41; standard deviation = 31.69. The mean is very close to the median. The five-number summary suggests that the distribution is approximately symmetrical around the median. The interquartile range is very close to 1.33 times the standard deviation. The range is about $50 below 6 times the standard deviation. In general, the distribution of the data appears to closely resemble a normal distribution. **(b)** The normal probability plot confirms that the data appear to be approximately normally distributed.

6.24 (a) 0.1667; **(b)** 0.1667; **(c)** 0.7083; **(d)** mean = 60, standard deviation = 34.641

6.26 (a) $P(5{:}55 \text{ a.m.} < X < 7{:}38 \text{ p.m.}) = P(355 < X < 1{,}178) = (1178 - 355)/(1440) = 0.5715$; **(b)** $P(10 \text{ p.m.} < X < 5 \text{ a.m.}) = P(1{,}320 < X < 1{,}440) + P(0 < X < 300) = (1{,}440 - 1{,}320)/1{,}440 + (300)/1{,}440 = 0.2917$; **(c)** Let X be duration between the occurrence of a failure and its detection. $a = 0$, $b = 60$, $P(0 < X < 10) = 10/60 = 0.1667$ **(d)** $P(40 < X < 60) = (60 - 40)/60 = 0.3333$

6.28 (a) 0.6321; **(b)** 0.3679; **(c)** 0.2326 **(d)** 0.7674

6.30 (a) approximately 1.0; **(b)** 0.0003; **(c)** 0.00029; **(d)** 0.99971

6.31 (a) P (arrival time $\le 0.05) = 1 - e^{-(50)(0.05)} = 0.9179$; **(b)** P (arrival time $\le 0.0167) = 1 - 0.4339 = 0.5661$; **(c)** If $\lambda = 60$, P (arrival time $\le 0.05) = 0.9502$, P (arrival time $\le 0.0167) = 0.6329$; **(d)** If $\lambda = 30$, P (arrival time $\le 0.05) = 0.7769$ P (arrival time $\le 0.0167) = 0.3941$

6.32 (a) 0.864665, **(b)** 0.99996, **(c)** 0.6321, 0.9933

6.34 (a) 0.6321, **(b)** 0.3935, **(c)** 0.0952

6.36 (a) 0.8647, **(b)** 0.3297, **(c)** 0.9765, 0.5276

6.38 (a) 0.0457, **(b)** 0.0846, **(c)** 0.9154, **(d)** 0.8697

6.40 (a) 0.2051, **(b)** 0.8281, **(c)** 0.7734, **(d)(a)** 0.2045, **(b)** 0.8286, **(c)** 0.7717

6.42 (a) 0.7864, **(b)** 0.0852, **(c)** 0.1932

6.52 (a) 0.4772, **(b)** 0.9544, **(c)** 0.0456, **(d)** 1.8835, **(e)** 1.8710 and 2.1290

6.54 (a) 0.2734, **(b)** 0.2038, **(c)** 4.404 ounces, **(d)** 4.188 ounces and 5.212 ounces

6.56 (a) 0.7273, **(b)** 0.2884, **(c)** 0.0426, **(d)** 0.0386, **(e)** The common stocks have higher mean annual returns than the long-term government bonds. But they also have higher volatility as reflected by their larger standard deviation. This is the usual trade-off between high return and high volatility in an investment instrument.
Note: The above answers are computed using Microsoft Excel. They may be slightly different when Table E.2 is used.

6.58 (a) 0.8413; **(b)** 0.9330; **(c)** 0.9332; **(d)** 0.3347; (e) 0.4080 and 1.1920

CHAPTER 7

7.2 (a) virtually zero; **(b)** 0.1587; **(c)** 0.0139; **(d)** 50.195

7.4 (a) Both means are equal to 6. This property is called unbiasedness. **(c)** The distribution for $n = 3$ has less variability. The larger sample size has resulted in sample means being closer to μ.

7.6 (a) Since $\bar{X} >$ median, the shape of the sampling distribution of $\bar{X}$ for samples of size 2 will be right skewed. **(b)** When the sample size is 100, the sampling distribution of $\bar{X}$ will be very close to a normal distribution as a result of the central limit theorem. **(c)** 0.1333

7.8 (a) $P(\bar{X} > 3) = P(Z > -1.00) = 1.0 - 0.1587 = 0.8413$
(b) $P(Z < 1.04) = 0.85$ $\bar{X} = 3.10 + 1.04(0.1) = 3.204$
(c) To be able to use the standardized normal distribution as an approximation for the area under the curve, you must assume that the population is approximately symmetrical. **(d)** $P(Z < 1.04) = 0.85$
$\bar{X} = 3.10 + 1.04(0.05) = 3.152$

7.10 (a) 0.9969, **(b)** 0.0142, **(c)** 2.3830 and 2.6170, **(d)** 2.6170. *Note:* These answers are computed using Microsoft Excel. They may be slightly different when Table E.2 is used.

7.12 (a) 0.30, **(b)** 0.0693

7.14 (a) $\mu = 0.501$, $\sigma = \sqrt{\frac{\pi(1-\pi)}{n}} = \sqrt{\frac{0.501(1-0.501)}{100}} = 0.05$

$P(p > 0.55) = P(Z > 0.98) = 1.0 - 0.8365 = 0.1635$

(b) $\mu = 0.60$, $\sigma = \sqrt{\frac{\pi(1-\pi)}{n}} = \sqrt{\frac{0.6(1-0.6)}{100}} = 0.04899$

$P(p > 0.55) = P(Z > -1.021) = 1.0 - 0.1539 = 0.8461$

(c) $\mu = 0.49$, $\sigma_p = \sqrt{\frac{\pi(1-\pi)}{n}} = \sqrt{\frac{0.49(1-0.49)}{100}} = 0.05$

$P(p > 0.55) = P(Z > 1.20) = 1.0 - 0.8849 = 0.1151$
(d) Increasing the sample size by a factor of 4 decreases the standard error by a factor of 2.
(a) $P(p > 0.55) = P(Z > 1.96) = 1.0 - 0.9750 = 0.0250$
(b) $P(p > 0.55) = P(Z > -2.04) = 1.0 - 0.0207 = 0.9793$
(c) $P(p > 0.55) = P(Z > 2.40) = 1.0 - 0.9918 = 0.0082$

7.16 (a) 0.50, **(b)** 0.5717, **(c)** 0.9523, **(d)** **(a)** 0.50, **(b)** 0.4246, **(c)** 0.8386

7.18 (a) 0.5926, **(b)** 0.7211 and 0.8189, **(c)** 0.7117 and 0.8283

7.20 (a) 0.6314, **(b)** 0.0041, **(c)** $P(p > .35) = P(Z > 1.3223) = 0.0930$. If the population proportion is 29%, the proportion of the samples with 35% or more who do not intend to work for pay at all is 9.3%, an unlikely occurrence. Hence, the population estimate of 29% is likely to be an underestimation. **(d)** When the sample size is smaller in (c) compared to (b), the standard error of the sampling distribution of sample proportion is larger.

7.22 (a) 0.3626, **(b)** 0.9816, **(c)** 0.0092. *Note:* These answers are computed using Microsoft Excel. They may be slightly different when Table E.2 is used.

7.24 Sample without replacement: Read from left to right in 3-digit sequences and continue unfinished sequences from end of row to beginning of next row.

Row 05: 338 505 855 551 438 855 077 186 579 488 767 833 170
Rows 05–06: 897
Row 06: 340 033 648 847 204 334 639 193 639 411 095 924
Rows 06–07: 707
Row 07: 054 329 776 100 871 007 255 980 646 886 823 920 461
Row 08: 893 829 380 900 796 959 453 410 181 277 660 908 887
Rows 08–09: 237
Row 09: 818 721 426 714 050 785 223 801 670 353 362 449
Rows 09–10: 406

Note: All sequences above 902 are discarded.

7.26 A simple random sample would be less practical for personal interviews because of travel costs (unless interviewees are paid to attend a central interviewing location).

7.28 Here all members of the population are equally likely to be selected and the sample selection mechanism is based on chance. But selection of two elements is not independent; for example, if *A* is in the sample, we know that *B* is also, and that *C* and *D* are not.

7.29 (a) Since a complete roster of full-time students exists, a simple random sample of 200 students could be taken. If student satisfaction with the quality of campus life randomly fluctuates across the student body, a systematic 1-in-20 sample could also be taken from the population frame. If student satisfaction with the quality of life may differ by gender and by experience/class level, a stratified sample using eight strata, female freshmen through female seniors and male freshmen through male seniors, could be selected. If student satisfaction with the quality of life is thought to fluctuate as much within clusters as between them, a cluster sample could be taken. **(b)** A simple random sample is one of the simplest to select. The population frame is the registrar's file of 4,000 student names. **(c)** A systematic sample is easier to select from the registrar's records than a simple random sample, since an initial person is selected at random and then every 20th person thereafter would be sampled. The systematic sample would have the additional benefit that the alphabetic distribution of sampled students' names would be more comparable to the alphabetic distribution of student names in the campus population. **(d)** If rosters by gender and class designations are readily available, a stratified sample should be taken. Since student satisfaction with the quality of life may indeed differ by gender and class level, the use of a stratified sampling design will not only ensure all strata are represented in the sample, it will generate a more representative sample and produce estimates of the population parameter that have greater precision. **(e)** If all 4,000 full-time students reside in one of 20 on-campus residence halls, which fully integrate students by gender and by class, a cluster sample should be taken. A cluster could be defined as an entire residence hall, and the students of a single randomly selected residence hall could be sampled. Since the dormitories are fully integrated by floor, a cluster could alternatively be defined as one floor of one of the 20 dormitories. Four floors could be randomly sampled to produce the required 200 student sample. Selection of an entire dormitory may make distribution and collection of the survey easier to accomplish. In contrast, if there is some variable other than gender or class that differs across dormitories, sampling by floor may produce a more representative sample.

7.30 (a) Row 16: 2323 6737 5131 8888 1718 0654 6832 4647 6510 4877
Row 17: 4579 4269 2615 1308 2455 7830 5550 5852 5514 7182
Row 18: 0989 3205 0514 2256 8514 4642 7567 8896 2977 8822
Row 19: 5438 2745 9891 4991 4523 6847 9276 8646 1628 3554
Row 20: 9475 0899 2337 0892 0048 8033 6945 9826 9403 6858
Row 21: 7029 7341 3553 1403 3340 4205 0823 4144 1048 2949
Row 22: 8515 7479 5432 9792 6575 5760 0408 8112 2507 3742
Row 23: 1110 0023 4012 8607 4697 9664 4894 3928 7072 5815
Row 24: 3687 1507 7530 5925 7143 1738 1688 5625 8533 5041
Row 25: 2391 3483 5763 3081 6090 5169 0546

Note: All sequences above 5,000 are discarded. There were no repeating sequences.

(b) 089 189 289 389 489 589 689 789 889 989
1089 1189 1289 1389 1489 1589 1689 1789 1889 1989
2089 2189 2289 2389 2489 2589 2689 2789 2889 2989
3089 3189 3289 3389 3489 3589 3689 3789 3889 3989
4089 4189 4289 4389 4489 4589 4689 4789 4889 4989

(c) With the single exception of invoice #0989, the invoices selected in the simple random sample are not the same as those selected in the systematic sample. It would be highly unlikely that a simple random sample would select the same units as a systematic sample.

7.50 (a) 0.4999; **(b)** 0.00009; **(c)** 0; **(d)** 0; **(e)** 0.7518

7.52 (a) 0.8944; **(b)** 4.617, 4.783; **(c)** 4.641

7.54 (a) 0.0092; **(b)** 0.9823; **(c)** 0.0000

7.56 Even though Internet polling is less expensive, faster, and offers higher response rates than telephone surveys, it may lead to more coverage error since a greater proportion of the population may have telephones than have Internet access. It may also lead to nonresponse bias since a certain class and/or age group of people may not use the Internet or may use the Internet less frequently. Due to these errors, the data collected are not appropriate for making inferences about the general population.

7.58 (a) With a response rate of only 15.5%, nonresponse error should be the major cause of concern in this study. Measurement error is a possibility also. **(b)** The researchers should follow up with the nonrespondents. **(c)** The step mentioned in (b) could have been followed to increase the response rate to the survey, thus increasing its worthiness.

7.60 (a) What was the comparison group of "other workers"? Were they another sample? Where did they come from? Were they truly comparable? What was the sampling scheme? What was the population from which the sample was selected? How was salary measured? What was the mode of response? What was the response rate? **(b)** Various answers are possible.

CHAPTER 8

8.2 $114.68 \le \mu \le 135.32$

8.4 In order to have 100% certainty, the entire population would have to be sampled.

8.6 Yes, it is true since 5% of intervals will not include the true mean.

8.8 (a) $\bar{X} \pm Z \cdot \frac{\sigma}{\sqrt{n}} = 350 \pm 1.96 \cdot \frac{100}{\sqrt{64}}$; $325.50 \le \mu \le 374.50$. **(b)** No.

The manufacturer cannot support a claim that the bulbs have a mean of 400 hours. Based on the data from the sample, a mean of 400 hours would represent a distance of 4 standard deviations above the sample mean of 350 hours. **(c)** No. Since σ is known and $n = 64$, from the Central Limit Theorem, you know that the sampling distribution of $\bar{X}$ is approximately normal. **(d)** The confidence interval is narrower based on a population standard deviation of 80 hours rather than the original standard deviation of 100 hours.

(a) $\bar{X} \pm Z \cdot \frac{\sigma}{\sqrt{n}} = 350 \pm 1.96 \cdot \frac{80}{\sqrt{64}}$, $330.4 \le \mu \le 369.6$. **(b)** Based on the smaller standard deviation, a mean of 400 hours would represent a distance of 5 standard deviations above the sample mean of 350 hours. No, the manufacturer cannot support a claim that the bulbs have a mean life of 400 hours.

8.10 (a) 2.2622; **(b)** 3.2498; **(c)** 2.0395; **(d)** 1.9977; **(e)** 1.7531

8.12 $38.95 \le \mu \le 61.05$

8.14 $-0.12 \le \mu \le 11.84$, $2.00 \le \mu \le 6.00$. The presence of the outlier increases the sample mean and greatly inflates the sample standard deviation.

8.15 (a) $\bar{X} \pm t\left(\frac{S}{\sqrt{n}}\right) = 1.67 \pm 2.0930\left(\frac{0.32}{\sqrt{20}}\right)$, $\$1.52 \le \mu \le \$1.82$. **(b)** The store owner can have 95% confidence that the population mean retail value of greeting cards that it has in its inventory is between \$1.52 and \$1.82. The store owner could multiply the ends of the confidence interval by the number of cards to estimate the total value of her inventory.

8.16 (a) $29.44 \le \mu \le 34.56$. **(b)** The quality improvement team can be 95% confident that the population mean turnaround time is between 29.44 hours and 34.56 hours. **(c)** The project was a success because the initial turnaround time of 68 hours does not fall into the interval.

8.18 (a) $\$21.01 \le \mu \le \$24.99$. **(b)** You can be 95% confident that the population mean bounced-check fee is between \$21.01 and \$24.99.

8.20 (a) $31.12 \le \mu \le 54.96$; **(b)** The number of days is approximately normally distributed. **(c)** Yes, the outliers skew the data. **(d)** Since the sample size is fairly large at $n = 50$, the use of the t distribution is appropriate.

8.22 (a) $\bar{X} \pm t\left(\frac{S}{\sqrt{n}}\right) = 182.4 \pm 2.0930\left(\frac{44.2700}{\sqrt{20}}\right)$, $\$161.68 \le \mu \le \$203.12$

(b) $\bar{X} \pm t\left(\frac{S}{\sqrt{n}}\right) = 45 \pm 2.0930\left(\frac{10.0263}{\sqrt{20}}\right)$, $\$40.31 \le \mu \le \$49.69$

(c) The population distribution needs to be normally distributed. **(d)** Both the normal probability plot and the box-and-whisker plot show that the population distributions for hotel cost and car rental are not normally distributed and are skewed to the right.

8.24 $0.19 \le \pi \le 0.31$

8.26 (a) $p = \frac{X}{n} = \frac{135}{500} = 0.27$

$p \pm Z \cdot \sqrt{\frac{p(1-p)}{n}} = 0.27 \pm 2.58\sqrt{\frac{0.27(0.73)}{500}}$ $\quad 0.2189 \le \pi \le 0.3211$

(b) The manager in charge of promotional programs concerning residential customers can infer that the proportion of households that would purchase an additional telephone line if it were made available at a substantially reduced installation cost is somewhere between 0.22 and 0.32 with 99% confidence.

8.28 (a) $0.74 \le \pi \le 0.80$; **(b)** $0.75 \le \pi \le 0.79$; **(c)** The 95% confidence interval is wider. The loss in precision results in a wider confidence interval is the price you pay to achieve a higher level of confidence.

8.30 (a) $0.416 \le \pi \le 0.504$; **(b)** $0.074 \le \pi \le 0.126$

8.32 (a) $0.419 \le \pi \le 0.481$; **(b)** You estimate with 95% confidence that between 41.9% and 48.1% of all working women in North America believe that companies should hold positions for those on maternity leave for more than 6 months.

8.34 $n = 35$

8.36 $n = 1{,}041$

8.38 (a) $n = \frac{Z^2\sigma^2}{e^2} = \frac{(1.96^2)(400^2)}{50^2} = 245.86$ Use $n = 246$

(b) $n = \frac{Z^2\sigma^2}{e^2} = \frac{(1.96^2)(400^2)}{25^2} = 983.41$ Use $n = 984$

8.40 $n = 97$

8.42 (a) $n = 167$; **(b)** $n = 97$

8.44 $n = 62$

8.46 (a) $n = 2{,}377$; **(b)** $n = 1{,}978$; **(c)** The sample sizes differ because the estimated population proportions are different. **(d)** Since purchasing groceries at wholesale clubs and purchasing groceries at convenience stores are not necessary mutually exclusive events, it is appropriate to use one sample and ask the respondents both questions.

8.48 (a) $0.9017 \le \pi \le 0.9572$; **(b)** You are 95% confident that the population proportion of business men and women who have their presentations disturbed by cell phones is between 0.9017 and 0.9572. **(c)** $n = 158$; **(d)** $n = 273$

8.50 $\$10{,}721.5 \le \text{Total} \le \$14{,}978.5$

8.52 (a) 0.054; **(b)** 0.058; **(c)** 0.066

8.54 $\$543{,}176 \le \text{Total} \le \$1{,}025{,}224$

8.56 $\$5{,}443 \le \text{Total Difference} \le \$54{,}229$

8.58 (a) 0.0542; **(b)** Since the upper bound is higher than the tolerable exception rate of 0.04, the auditor should request a larger sample.

8.66 (a) People visiting the *Redbook* Web site. **(b)** no; **(c)** no

8.68 (a) $0.512 \le \pi \le 0.648$; **(b)** $0.431 \le \pi \le 0.569$; **(c)** $0.163 \le \pi \le 0.277$; **(d)** $0.136 \le \pi \le 0.244$; **(e)** $n = 2{,}401$

8.70 (a) $14.085 \le \mu \le 16.515$; **(b)** $0.530 \le \pi \le 0.820$; **(c)** $n = 25$; **(d)** $n = 784$; **(e)** If a single sample were to be selected for both purposes, the larger of the two sample sizes ($n = 784$) should be used.

8.72 (a) $8.049 \le \mu \le 11.351$; **(b)** $0.284 \le \pi \le 0.676$; **(c)** $n = 35$; **(d)** $n = 121$; **(e)** If a single sample were to be selected for both purposes, the larger of the two sample sizes ($n = 121$) should be used.

8.74 (a) $\$25.80 \le \mu \le \$31.24$; **(b)** $0.3037 \le \pi \le 0.4963$; **(c)** $n = 97$; **(d)** $n = 423$; **(e)** If a single sample were to be selected for both purposes, the larger of the two sample sizes ($n = 423$) should be used.

8.76 (a) $\$36.66 \le \mu \le \$40.42$; **(b)** $0.2027 \le \pi \le 0.3973$; **(c)** $n = 110$; **(d)** $n = 423$; **(e)** If a single sample were to be selected for both purposes, the larger of the two sample sizes ($n = 423$) should be used.

8.78 (a) $\pi \le 0.2013$; **(b)** Since the upper bound is higher than the tolerable exception rate of 0.15, the auditor should request a larger sample.

8.80 (a) $n = 27$; **(b)** \$402,652.53 $\le$ Population Total $\le$ \$450,950.79

8.82 (a) $8.41 \le \mu \le 8.43$; **(b)** With 95% confidence, the population mean width of troughs is somewhere between 8.41 and 8.43 inches.

8.84 (a) $0.2425 \le \mu \le 0.2856$; **(b)** $0.1975 \le \mu \le 0.2385$; **(c)** The amount of granule loss for both brands are skewed to the right. **(d)** Since the two confidence intervals do not overlap, you can conclude that the mean granule loss of Boston shingles is higher than that of Vermont shingles.

CHAPTER 9

9.2 H_1 denotes the alternative hypothesis.

9.4 β

9.6 α is the probability of making a Type I error.

9.8 The power of the test is $1 - \beta$.

9.10 It is possible to not reject a null hypothesis when it is false, since it is possible for a sample mean to fall in the nonrejection region even if the null hypothesis is false.

9.12 All else being equal, the closer the population mean is to the hypothesized mean, the larger β will be.

9.14 H_0: defendant is guilty, H_1: defendant is innocent. A Type I error would be not convicting a guilty person. A Type II error would be convicting an innocent person.

9.16 H_0: $\mu = 20$ minutes. 20 minutes is adequate travel time between classes.
H_1: $\mu \ne 20$ minutes. 20 minutes is not adequate travel time between classes.

9.18 H_0: $\mu = 1.00$. The mean amount of paint per one-gallon can is one gallon.
H_1: $\mu \ne 1.00$. The mean amount of paint per one-gallon can differs from one gallon.

9.20 Since $Z_{calc} = +2.21 > 1.96$, reject H_0.

9.22 Reject H_0 if $Z_{calc} < -2.58$ or if $Z_{calc} > 2.58$.

9.24 p-value = 0.0456

9.26 p-value = 0.1676

9.28 (a) H_0: $\mu = 70$ pounds. H_1: $\mu \ne 70$ pounds.
Decision rule: Reject H_0 if $Z < -1.96$ or $Z > +1.96$.

Test statistic: $Z = \dfrac{\bar{X} - \mu}{\sigma/\sqrt{n}} = \dfrac{69.1 - 70}{3.5/\sqrt{49}} = -1.80$

Decision: Since $-1.96 < Z_{calc} = -1.80 < 1.96$, do not reject H_0. There is insufficient evidence to conclude that the cloth has a mean breaking strength that differs from 70 pounds. **(b)** p-value = 2(0.0359) = 0.0718. Interpretation: The probability of getting a sample of 49 pieces that yield a mean strength that is farther away from the hypothesized population mean than this sample is 0.0718 or 7.18%.
(c) Decision rule: Reject H_0 if $Z < -1.96$ or $Z > +1.96$.

Test statistic: $Z = \dfrac{\bar{X} - \mu}{\sigma/\sqrt{n}} = \dfrac{69.1 - 70}{1.75/\sqrt{49}} = -3.60$. Decision: Since $Z_{calc} = -3.60 < -1.96$, reject H_0. There is enough evidence to conclude that the cloth has a mean breaking strength that differs from 70 pounds.
(d) Decision rule: Reject H_0 if $Z < -1.96$ or $Z > +1.96$. Test statistic:

$Z = \dfrac{\bar{X} - \mu}{\sigma/\sqrt{n}} = \dfrac{69 - 70}{3.5/\sqrt{49}} = -2.00$. Decision: Since $Z_{calc} = -2.00 < -1.96$, reject H_0. There is enough evidence to conclude that the cloth has a mean breaking strength that differs from 70 pounds.

9.30 (a) Since $Z_{calc} = -2.00 < -1.96$, reject H_0. **(b)** p-value = 0.0456; **(c)** $325.5 \le \mu \le 374.5$; **(d)** The conclusions are the same.

9.32 (a) Since $-1.96 < Z_{calc} = -0.80 < 1.96$, do not reject H_0; **(b)** p-value = 0.4238; **(c)** Since $Z_{calc} = -2.40 < -1.96$, reject H_0; **(d)** Since $Z_{calc} = -2.26 < -1.96$, reject H_0.

9.34 $Z = +2.33$

9.36 $Z = -2.33$

9.38 p-value = 0.0228

9.40 p-value = 0.0838

9.42 p-value = 0.9162

9.44 (a) Since $Z_{calc} = -1.75 < -1.645$, reject H_0; **(b)** p-value = 0.0401 < 0.05, reject H_0; **(c)** The probability of getting a sample mean of 2.73 feet or less if the population mean is 2.8 feet is 0.0401. **(d)** They are the same.

9.46 (a) H_0: $\mu \le 5$ H_1: $\mu > 5$; **(b)** A Type I error occurs when you conclude that children take a mean of more than 5 trips a week to the store, when in fact they take a mean of no more than 5 trips a week to the store. A Type II error occurs when you conclude that children take a mean of no more than 5 trips a week to the store, when in fact they take a mean of more than 5 trips a week to the store. **(c)** Since $Z_{calc} = 2.9375 > 2.3263$ or the p-value of 0.0017 is less than 0.01, reject H_0. There is enough evidence to conclude the population mean number of trips to the store is greater than 5 per week. **(d)** The probability that the sample mean is 5.47 trips or more when the null hypothesis is true is 0.0017.

9.48 $t = 2.00$

9.50 (a) $t_{crit} = \pm 2.1315$; **(b)** $t_{crit} = +1.7531$

9.52 No, you should not use a t test, since the original population is left-skewed, and the sample size is not large enough for the t to be influenced by the Central Limit Theorem.

9.54 (a) H_0: $\mu \le \$300$. H_1: $\mu > \$300$.
Decision rule: $df = 99$. If $t > 1.2902$, reject H_0.

Test statistic: $t = \dfrac{\bar{X} - \mu}{S/\sqrt{n}} = \dfrac{\$315.40 - \$300.00}{\$43.20/\sqrt{100}} = 3.5648$. Decision:

Since $t_{calc} = 3.5648 > 1.2902$, reject H_0. There is enough evidence to conclude that the mean cost of textbooks per semester at a large university is more than \$300.
(b) Decision rule: $df = 99$. If $t > 1.6604$, reject H_0.

Test statistic: $t = \dfrac{\bar{X} - \mu}{S/\sqrt{n}} = \dfrac{\$315.40 - \$300.00}{\$75.00/\sqrt{100}} = 2.0533$

Decision: Since $t_{calc} = 2.0533 > 1.6604$, reject H_0. There is enough evidence to conclude that the mean cost of textbooks per semester at a large university is more than \$300. **(c)** Decision rule: $df = 99$. If $t > 1.2902$, reject H_0.

Test statistic: $t = \dfrac{\bar{X} - \mu}{S/\sqrt{n}} = \dfrac{\$305.11 - \$300.00}{\$43.20/\sqrt{100}} = 1.1829$. Decision:

Since $t_{calc} = 1.1829 < t = 1.2902$, do not reject H_0. There is not enough evidence to conclude that the mean cost of textbooks per semester at a large university is more than \$300.

9.56 Since $t_{calc} = -1.30 > -1.6694$ and the *p*-value of $0.0992 > 0.05$, do not reject H_0. There is not enough evidence to conclude that the mean waiting time is less than 3.7 minutes.

9.58 (a) Since $t = 1.2567 < 1.7823$, do not reject H_0. There is not enough evidence to conclude that the mean life of the batteries is more than 400 hours. **(b)** *p*-value = 0.1164. The probability that a sample of 13 batteries results in a sample mean of 473.46 or more is 11.64% if the population mean life is 400 hours. **(c)** Allowing for only 5% probability of making a Type I error, the manufacturer should not say in advertisements that these batteries should last more than 400 hours.
(d) (a) Since $t = 1.7195 < 1.7823$, do not reject H_0. There is not enough evidence to conclude that the mean life of the batteries is more than 400 hours. **(b)** *p*-value = 0.0556. The probability that a sample of 13 batteries results in a sample mean of 550.38 or more is 5.56% if the population mean life is 400 hours. **(c)** Allowing for only 5% probability of making a Type I error, the manufacturer should not say in advertisements that these batteries should last more than 400 hours. They can make the claim that these batteries should last more than 400 hours only if they are willing to raise the level of significance to more than 0.0556.The extremely large value of 1,342 raises the sample mean and sample standard deviation and, hence, results in a higher measured *t* statistic of 1.7195. This is, however, still not enough to offset the lower hours in the remaining sample to the degree that the null hypothesis can be rejected.

9.60 (a) Since $-2.0096 < t = 0.114 < 2.0096$, do not reject H_0; **(b)** *p*-value = 0.9095; **(c)** Yes, the data appear to have met the normality assumption. **(d)** The amount of fill is decreasing over time. Therefore, the *t* test is invalid.

9.62 (a) Since $t = -5.684 < -2.6178$, reject H_0. There is enough evidence to conclude that the mean viscosity has changed from 15.5. **(b)** The population distribution needs to be normal. **(c)** The normal probability plot indicates that the distribution is slightly skewed to the right.

9.64 (a) Since $-2.68 < t = 0.094 < 2.68$, do not reject H_0; **(b)** $5.462 \le \mu \le 5.542$; **(c)** The conclusions are the same.

9.66 $p = 0.22$

9.68 Do not reject H_0.

9.69 (a) H_0: $\pi \le 0.5$; H_1: $\pi > 0.5$.
Decision rule: If $Z > 1.645$, reject H_0.

Test statistic: $Z = \dfrac{p - \pi}{\sqrt{\dfrac{\pi(1-\pi)}{n}}} = \dfrac{0.5702 - 0.5}{\sqrt{\dfrac{0.5(0.5)}{1040}}} = 4.5273$

Decision: Since $Z_{calc} = 4.5273 > 1.645$, reject H_0. There is enough evidence to show that more than half of all Americans would rather have \$100 than a day off from work. **(b)** *p*-value = 0.00. The probability is approximately 0 of observing a sample of 593 or more out of 1,040 Americans who will rather have the \$100 than a day off from work, if the population proportion is 0.5.

9.70 Since $Z_{calc} = 2.6902 > 1.645$ and the *p*-value of $0.0036 < 0.05$, reject H_0. There is enough evidence to show that the proportion of employers that planned to hire new employees in 2004 is larger than the 2003 proportion of 0.43.

9.72 (a) Since $-1.96 < Z_{calc} = 0.6381 < 1.96$, do not reject H_0 and conclude that there is not enough evidence to show that the percentage of people who trust energy-efficiency ratings differs from 50%. **(b)** *p*-value = 0.5234. Since the *p*-value of $0.5234 > 0.05$, do not reject H_0

9.74 (a) $p = 0.7112$. **(b)** Since $Z_{calc} = 5.7771 > 1.6449$, reject H_0. There is enough evidence to conclude that more than half of all successful women executives have children. **(c)** Since $Z_{calc} = 1.2927 < 1.6449$, do not reject H_0. There is not enough evidence to conclude that more than two-thirds of all successful women executives have children. **(d)** The random sample assumption is not likely to be valid because the criteria used in defining "successful women executives" is very likely to be quite different than those used in defining the "most powerful women in business" who attended the summit.

9.75 H_0: $\mu \ge 7$ vs. H_1: $\mu < 7$, $\alpha = 0.05$, $n = 16$, $\sigma = 0.2$

Lower critical value: $Z_L = -1.6449$,

$$\bar{X}_L = \mu + Z_L\left(\frac{\sigma}{\sqrt{n}}\right) = 7 - 1.6449\left(\frac{.2}{\sqrt{16}}\right) = 6.9178$$

(a) $Z = \dfrac{\bar{X}_L - \mu}{\dfrac{\sigma}{\sqrt{n}}} = \dfrac{6.9178 - 6.9}{\dfrac{.2}{\sqrt{16}}} = 0.3551$

$\text{power} = 1 - \beta = P(\bar{X} < \bar{X}_L) = P(Z < 0.3551) = 0.6388$
$\beta = 1 - 0.6388 = 0.3612$

(b) $Z = \dfrac{\bar{X}_L - \mu}{\dfrac{\sigma}{\sqrt{n}}} = \dfrac{6.9178 - 6.8}{\dfrac{.2}{\sqrt{16}}} = 2.3551$

$\text{power} = 1 - \beta = P(\bar{X} < \bar{X}_L) = P(Z < 2.3551) = 0.9907$
$\beta = 1 - 0.9907 = 0.0093$

9.76 (a) power = 0.3721, $\beta = 0.6279$; **(b)** power = 0.9529, $\beta = 0.0471$; **(c)** Holding everything else constant, the larger the difference between the population mean and the hypothesized mean, the higher is the power of the test and the lower is the probability of committing a Type II error. Holding everything else constant, the lower the level of significance, the lower is the power of the test and the higher is the probability of committing a Type II error.

9.78 (a) power = 0.8873, $\beta = 0.1127$; **(b)** power = 0.0871, $\beta = 0.9129$

9.80 (a) power = 0.4144, $\beta = 0.5856$; **(b)** power = 0.0665, $\beta = 0.9335$; **(c)** Holding everything else constant, the larger the sample size, the higher is the power of the test and the lower is the probability of committing a Type II error.

9.92 (a) buying a site that is not profitable; **(b)** not buying a profitable site; **(c)** Type I; **(d)** If the executives adopt a less stringent rejection criterion by buying sites for which the computer model predicts moderate or large profit, the probability of committing a Type I error will increase. Many more of the sites the computer model predicts that will generate moderate profit may end up not being profitable at all. On the other hand, the less stringent rejection criterion will lower the probability of committing a Type II error, since now, more potentially profitable sites will be purchased.

9.94 (a) Since $t = 3.248 > 2.0010$, reject H_0. **(b)** *p*-value = 0.0019; **(c)** Since $Z = -0.32 > -1.645$, do not reject H_0. **(d)** Since $-2.0010 < t = 0.75 < 2.0010$, do not reject H_0. **(e)** Since $t = -1.61 > -1.645$, do not reject H_0.

9.96 (a) Since $t = -1.69 > -1.7613$, do not reject H_0. **(b)** The data are from a population that is normally distributed.

9.98 (a) Since $t = -1.47 > -1.6896$, do not reject H_0; **(b)** p-value = 0.0748; **(c)** Since $t = -3.10 < -1.6973$, reject H_0; **(d)** p-value = 0.0021; **(e)** The data in the population are assumed to be normally distributed.

9.100 (a) $t = -21.61$, reject H_0; **(b)** p-value = 0.0000; **(c)** $t = -27.19$, reject H_0; **(d)** p-value = 0.0000

Chapter 10

10.2 Since $-2.58 \le Z = 1.73 \le 2.58$, do not reject H_0.

10.4 (a) $t = 3.8959$; **(b)** $df = 21$; **(c)** 2.5177; **(d)** Since 3.8959 > 2.5177, reject H_0.

10.6 $3.73 \le \mu_1 - \mu_2 \le 12.27$

10.8 (a) Since 5.20 > 2.33, reject H_0. **(b)** p-value < 0.00003.

10.10 (a) Since $-2.0117 < t_{calc} = 0.1023 < 2.0117$, do not reject H_0. There is no evidence of a difference in the two means for the Age 8 group. Since $t_{calc} = 3.375 > 1.9908$, reject H_0. There is evidence of a difference in the two means for the Age 12 group. Since $t_{calc} = 3.3349 > 1.9966$, reject H_0. There is evidence of a difference in the two means for the Age 16 group. **(b)** The test results show that children in the United States begin to develop preferences for brand name products as early as Age 12.

10.12 (a) $H_0: \mu_1 = \mu_2$ where Populations: 1 = Males, 2 = Females
$H_1: \mu_1 \ne \mu_2$
Decision rule: $df = 170$. If $t < -1.974$ or $t > 1.974$, reject H_0.
Test statistic:

$$S_p^2 = \frac{(n_1 - 1)(S_1^2) + (n_2 - 1)(S_2^2)}{(n_1 - 1) + (n_2 - 1)} = \frac{(99)(13.35^2) + (71)(9.42^2)}{99 + 71} = 140.8489$$

$$t = \frac{(\bar{X}_1 - \bar{X}_2) - (\mu_1 - \mu_2)}{\sqrt{S_p^2\left(\frac{1}{n_1} + \frac{1}{n_2}\right)}} = \frac{(40.26 - 36.85) - 0}{\sqrt{140.8489\left(\frac{1}{100} + \frac{1}{72}\right)}} = 1.859$$

Decision: Since $-1.974 < t_{calc} = 1.859 < 1.974$, do not reject H_0. There is not enough evidence to conclude that the mean computer anxiety experienced by males and females is different. **(b)** p-value = 0.0648. **(c)** In order to use the pooled-variance t test, you need to assume that the populations are normally distributed with equal variances.

10.14 (a) Since $-4.1343 < -2.0484$, reject H_0. **(b)** p-value = 0.0003; **(c)** The original populations of waiting times are approximately normally distributed. **(d)** $-4.2292 \le \mu_1 - \mu_2 \le -1.4268$

10.16 (a) Since the $-2.024 < t = 0.354 < 2.024$ or p-value = 0.725 > 0.05, do not reject the null hypothesis. There is not enough evidence to conclude that the mean time to clear problems in the two offices is different. **(b)** p-value = 0.725. The probability that a sample will yield a t-test statistic more extreme than 0.3544 is 0.725 if the mean waiting time between Office 1 and Office 2 is the same. **(c)** You need to assume that the two populations are normally distributed. **(d)** $-0.9543 \le \mu_1 - \mu_2 \le 1.3593$

10.18 (a) Since $t_{calc} = 4.10 > 2.024$, reject H_0. There is evidence of a difference in the mean surface hardness between untreated and treated steel plates. **(b)** p-value = 0.0002. The probability that two samples have a mean difference of 9.3634 or more is 0.02% if there is no difference in the mean surface hardness between untreated and treated steel plates. **(c)** You need to assume that the population distribution of hardness of both untreated and treated steel plates is normally distributed. **(d)** $4.7447 \le \mu_1 - \mu_2 \le 13.9821$

10.20 (a) Since $t_{calc} = -2.1522 < -2.0211$, reject H_0. There is enough evidence to conclude that the mean assembly times in seconds are different between employees trained in a computer-assisted, individual-based program and those trained in a team-based program. **(b)** You must assume that each of the two independent populations is normally distributed. **(c)** Since $t = -2.152 < -2.052$ or p-value = 0.041 < 0.05, reject H_0. **(d)** The results in (a) and (c) are the same. **(e)** $-4.52 \le \mu_1 - \mu_2 \le -0.14$. You are 95% confident that the difference between the population means of the two training methods is between −4.52 and −0.14.

10.22 $df = 19$

10.24 (a) Since $t = -6.8672 < -2.1098$, reject H_0. There is a difference in the mean daily hotel rate in March 2004 and June 2002. **(b)** You must assume that the distribution of the differences between the daily hotel rate in March 2004 and June 2002 is approximately normally distributed. **(c)** p-value is virtually zero. The probability that the t statistic for the mean difference in daily hotel rate is 6.8672 or more in either direction is virtually zero, if there is no difference in the mean daily hotel rate in March 2004 and June 2002. **(d)** $-102.86 \le \mu_D \le -54.51$. You are 95% confident that the mean difference in the hotel rate between March 2004 and June 2002 is somewhere between −\$102.86 and −\$54.51.

10.26 (a) $H_0: \mu_D = 0$ $H_1: \mu_D \ne 0$
Decision rule: $df = 14$. If $t < -2.9768$ or $t > 2.9768$, reject H_0.

$$\text{Test statistic: } t = \frac{\bar{D} - \mu_D}{S_D/\sqrt{n}} = \frac{3.5307 - 0}{13.8493/\sqrt{15}} = 0.9874$$

Decision: Since $-2.9768 < t_{calc} = 0.9874 < 2.9768$, do not reject H_0. There is insufficient evidence to conclude that there is a difference in the mean price of textbooks between the local bookstore and Amazon.com. **(b)** You must assume that the distribution of the differences between the mean price of business textbooks between the local bookstore and Amazon.com is approximately normally distributed.

(c) $$\bar{D} \pm t\left(\frac{S_D}{\sqrt{n}}\right) = 3.5307 \pm 2.9768\left(\frac{13.8493}{\sqrt{15}}\right), \quad -7.1141 \le \mu_D \le 14.1755$$

You are 99% confident that the mean difference between the price is somewhere between −7.1141 and 14.1755. **(d)** The results in (a) and (c) are the same. The hypothesized value of 0 for the difference in the mean price for textbooks between the local bookstore and Amazon.com is inside the 99% confidence interval.

10.28 (a) Since $t = 1.8425 < 1.943$, do not reject H_0. There is not enough evidence to conclude that the mean bone marrow microvessel density is higher before the stem cell transplant than after the stem cell transplant. **(b)** p-value = 0.0575. The probability that the t statistic for the mean difference in density is 1.8425 or more is 5.75% if the mean density is not higher before the stem cell transplant than after the stem cell transplant. **(c)** $-28.26 \le \mu_D \le 200.55$. You are 95% confident that the mean difference in bone marrow microvessel density before and after the stem cell transplant is somewhere between −28.26 and 200.55.

10.30 (a) Since $t = -9.3721 < -2.4258$, reject H_0. **(b)** The population of differences in strength is approximately normally distributed. **(c)** $p = 0.000$.

10.32 (a) Since $-2.58 \le Z = -0.58 \le 2.58$, do not reject H_0; **(b)** $-0.273 \le \pi_1 - \pi_2 \le 0.173$

10.34 (a) Since $Z = -13.53 < -1.96$, reject H_0. **(b)** p-value < 0.00003; **(c)** $-0.2841 \le \pi_1 - \pi_2 \le -0.2165$

10.36 (a) Since $Z_{calc} = 5.8019 > 1.645$, reject H_0. There is sufficient evidence to conclude that the proportion of adults online who use the Internet to gather data about products/services is higher in December 2003 than in 2000. **(b)** The p-value is virtually 0. The probability that the difference in two sample proportions is 0.16 or larger is virtually zero when the null hypothesis is true.

10.38 (a) H_0: $\pi_1 \le \pi_2$ $\quad H_1$: $\pi_1 > \pi_2$
Decision rule: If $Z > 1.645$, reject H_0.

$$\text{Test statistic: } \bar{p} = \frac{X_1 + X_2}{n_1 + n_2} = \frac{29 + 126}{56 + 407} = 0.3348$$

$$Z = \frac{(p_1 - p_2) - (\pi_1 - \pi_2)}{\sqrt{(\bar{p})(1-\bar{p})\left(\frac{1}{n_1} + \frac{1}{n_2}\right)}}$$

$$= \frac{(0.5179 - 0.3096) - 0}{\sqrt{(0.3348)(1 - 0.3348)\left(\frac{1}{56} + \frac{1}{407}\right)}}$$

$$= 3.0965$$

Decision: Since $Z_{calc} = 3.0965 > 1.645$, reject H_0. There is sufficient evidence to conclude that white workers are more likely to claim bias than black workers. **(b)** The p-value is 0.00098. The probability that the difference in two sample proportions is 0.20828 or larger is 0.00098 when the null hypothesis is true.

10.42 (a) 0.429; **(b)** 0.362; **(c)** 0.297; **(d)** 0.258

10.44 $df_{\text{numerator}} = 24$, $df_{\text{denominator}} = 24$.

10.46 Since $0.4405 < F = 0.8258 < 2.27$, do not reject H_0.

10.48 (a) Since $0.3378 < F = 1.2995 < 3.18$, do not reject H_0. **(b)** Since $F = 1.2995 < 2.62$, do not reject H_0. **(c)** Since $F = 1.2995 > 0.4032$, do not reject H_0.

10.50 (a) $H_0: \sigma_1^2 = \sigma_2^2 \quad H_1: \sigma_1^2 \ne \sigma_2^2$
Decision rule: If $F > 1.556$ or $F < 0.653$, reject H_0.

$$\text{Test statistic: } F = \frac{S_1^2}{S_2^2} = \frac{(13.35)^2}{(9.42)^2} = 2.008$$

Decision: Since $F_{calc} = 2.008 > 1.556$, reject H_0. There is enough evidence to conclude that the two population variances are different. **(b)** p-value $= 0.0022$. **(c)** The test assumes that each of the two populations are normally distributed. **(d)** Based on (a) and (b), a separate variance t test should be used.

10.52 (a) Since $0.3958 < F = 0.8248 < 2.5265$, do not reject H_0; **(b)** 0.6789; **(c)** The test assumes that the populations of times are approximately normally distributed; **(d)** yes

10.54 Since $F = 5.76 > 2.54$, reject H_0.

10.62 (a) $H_0: \sigma_1^2 \ge \sigma_2^2 \quad H_1: \sigma_1^2 < \sigma_2^2$
(b) Type I error: Rejecting the null hypothesis that price variance on the Internet is no lower than the price variance in the brick-and-mortar market when the price variance on the Internet is no lower than the price variance in the brick-and-mortar market. Type II error: Failing to reject the null hypothesis that price variance on the Internet is no lower than the price variance in the brick-and-mortar market when the price variance on the Internet is lower than the price variance in the brick-and-mortar market. **(c)** An F test for differences in two variances can be used. **(d)** You need to assume that each of the two populations are normally distributed.
(e) (a) H_0: $\mu_1 \ge \mu_2$ $\quad H_1$: $\mu_1 < \mu_2$
(b) Type I error: Rejecting the null hypothesis that the mean price in electronic market is no lower than the mean price in physical market when the mean price in the electronic market is no lower than the mean price in the physical market. Type II error: Failing to reject the null hypothesis that the mean price in the electronic market is no lower than the mean price in the physical market when the mean price in the electronic market is lower than the mean price in the physical market. **(c)** A paired t test for the mean difference can be used. **(d)** You must assume that the distribution of the difference between the mean price in the electronic market and the physical market is approximately normally distributed.

10.64 (a) The researchers can ask the teenagers, after viewing each ad, to rate the dangers of smoking using a scale from 0 to 10 with 10 representing the most dangerous.
(b) H_0: $\mu_T \ge \mu_S$ $\quad H_1$: $\mu_T < \mu_S$; **(c)** Type I error is the error made by concluding that ads produced by the state are more effective than those produced by Philip Morris while it is not true. The risk of Type I error here is that teenagers can miss the opportunity from the better ads produced by Philip Morris to recognize the true dangers of smoking and the additional expenses the state will have to incur to produce and run the ads. Type II error is the error made by concluding that ads produced by Philip Morris is no less effective than those produced by the state while ads produced by the state are more effective. The risk of Type II error here is that more teenagers will miss the opportunity to recognize the true dangers of smoking from the ads produced by the state. **(d)** Since both ads are shown to the same group of teenagers, a paired t test for the mean difference is most appropriate. **(e)** Statistically reliable here means the conclusions drawn from the test are reliable because all the assumptions needed for the test to be valid are fulfilled.

10.66 (a) Since $t_{calc} = 4.5826 > 2.528$, reject H_0. There is sufficient evidence to conclude that the mean monthly electric bill is more than \$80 for single-family homes in County II during the summer season. **(b)** Since $0.3265 < F_{calc} = 2.78 < 3.2220$, do not reject H_0. There is not enough evidence to conclude that County I and II have different population variances for monthly electric bills. **(c)** Since $t = 2.273 < 2.414$ or p-value $= 0.0140 > 0.01$, do not reject H_0. There is not enough evidence to conclude that the mean monthly bill is higher in County I than in County II. **(d)** $-3.1323 \le \mu_1 - \mu_2 \le 37.1323$

10.68 (a) Since $t_{calc} = 3.3282 > 1.8595$, reject H_0. There is enough evidence to conclude that the introductory computer students required more than a mean of 10 minutes to write and run a program in Visual Basic. **(b)** Since $t_{calc} = 1.3636 < 1.8595$, do not reject H_0. There is not enough evidence to conclude that the introductory computer students required more than a mean of 10 minutes to write and run a program in Visual Basic. **(c)** Although the mean time necessary to complete the assignment increased from 12 to 16 minutes as a result of the increase in one data value, the standard deviation went from 1.8 to 13.2, which reduced the t-value. **(d)** Since $0.2328 < F_{calc} = 0.8125 < 3.8549$, do not reject H_0. There is not enough evidence to conclude that the population variances are different for the Introduction to Computers students and computer majors. Hence, the pooled-variance t test is a valid test to determine whether computer majors can write a Visual Basic program in less time than introductory students, assuming that the distributions of the time needed to write a Visual Basic program for both the Introduction to Computers students and the computer majors are approximately normally distributed. Since $t_{calc} = 4.0666 > 1.7341$, reject H_0. There is enough evidence that the mean time is higher for Introduction to Computers students than for computer majors. **(e)** p-value = 0.000362. If the true population mean amount of time needed for Introduction to Computer students to write a Visual Basic program is no more than 10 minutes, the probability for observing a sample mean greater than the 12 minutes in the current sample is 0.0362%. Hence, at a 95% level of confidence, you can conclude that the population mean amount of time needed for Introduction to Computer students to write a Visual Basic program is more than 10 minutes. As illustrated in part (d) in which there is not enough evidence to conclude that the population variances are different for the Introduction to Computers students and computer majors, the pooled-variance t test performed is a valid test to determine whether computer majors can write a Visual Basic program in less time than introductory students, assuming that the distribution of the time needed to write a Visual Basic program for both the Introduction to Computers students and the computer majors are approximately normally distributed.

10.70 (a) Since $t_{calc} = 7.8735 > 2.3598$, reject H_0. There is enough evidence to conclude that the mean salary for men is greater than the mean salary for women.
(b) The p-value is approximately zero.

10.72 From the box-and-whisker plot and the summary statistics, both distributions are approximately normally distributed. $0.5289 < F = 0.947 < 1.89$ There is insufficient evidence to conclude that the two population variances are significantly different at 5% level of significance. $t = -5.084 < -1.99$ At 5% level of significance, there is sufficient evidence to reject the null hypothesis of no difference in the mean life of the bulbs between the two manufacturers. You can conclude that there is significant difference between the mean life of the bulbs between the two manufacturers.

10.78 (a) Since $F = 3.339 > 1.9096$ or p-value = 0.000361 < 0.05, reject H_0. There is evidence of a difference between the variances in the attendance at games with promotions and games without promotions. **(b)** Since the variances cannot be assumed to be equal, you should use a separate-variance t test for the difference in two means. Since $t = 4.745 > 1.996$ or since the p-value is approximately zero, you reject H_0. **(c)** There is evidence that there is a difference in the mean attendance at games with promotions and games without promotions at the 5% level of significance.

10.80 The normal probability plots suggest that the two populations are not normally distributed. An F test is inappropriate for testing the difference in two variances. The sample variances for Boston and Vermont shingles are 0.0203 and 0.015, respectively, which are not very different. It appears that a pooled-variance t test is appropriate for testing the difference in means. Since $t = 3.015 > 1.967$ or the p-value = 0.0028 $< \alpha = 0.05$, reject H_0. There is sufficient evidence to conclude that there is a difference in the mean granule loss of Boston and Vermont shingles.

CHAPTER 11

11.2 (a) $SSW = 150$; **(b)** $MSA = 15$; **(c)** $MSW = 5$; **(d)** $F = 3$

11.4 (a) 2; **(b)** 18; **(c)** 20

11.6 (a) Reject H_0 if $F > 2.95$, otherwise do not reject H_0. **(b)** Since $F = 4 > 2.95$ reject H_0. **(c)** The table does not have 28 degrees of freedom in the denominator so use the next larger critical value, $Q_U = 3.90$. **(d)** Critical range = 6.166

11.8 (a) Since $F = 10.99 > F_{0.05,2,9} = 4.26$, reject H_0. **(b)** Critical range = 40.39. The experts and darts are not different from each other, but they are both different from the readers. **(c)** It is not valid to infer that the dartboard is better than the professionals since their means are not significantly different. **(d)** Since $F = 0.101 < 4.26$, do not reject H_0. There is no evidence of a significant difference in the variation in the return for the three categories.

11.10 (a) H_0: $\mu_A = \mu_B = \mu_C = \mu_D$ H_1: At least one mean is different.

$$MSA = \frac{SSA}{c-1} = \frac{1986.475}{3} = 662.1583$$

$$MSW = \frac{SSW}{n-c} = \frac{495.5}{36} = 13.76389$$

$$F = \frac{MSA}{MSW} = \frac{662.1583}{13.76389} = 48.1084$$

$$F_{\alpha,c-1,n-c} = F_{0.05,3,36} = 2.8663$$

Since the p-value is approximately zero and $F = 48.1084 > 2.8663$, you reject H_0. There is sufficient evidence of a difference in the mean strength of the four brands of trash bags.

(b) Critical range $= Q_u\sqrt{\frac{MSW}{2}\left(\frac{1}{n_j}+\frac{1}{n_{j'}}\right)} = 3.79\sqrt{\frac{13.7639}{2}\left(\frac{1}{10}+\frac{1}{10}\right)}$
= 4.446
From the Tukey-Kramer procedure, there is a difference in mean strength between Kroger and Tuffstuff, Glad and Tuffstuff, and Hefty and Tuffstuff.
(c) ANOVA output for Levene's test for homogeneity of variance:

$$MSA = \frac{SSA}{c-1} = \frac{24.075}{3} = 8.025$$

$$MSW = \frac{SSW}{n-c} = \frac{198.2}{36} = 5.5056$$

$$F = \frac{MSA}{MSW} = \frac{8.025}{5.5056} = 1.4576$$

$$F_{\alpha,c-1,n-c} = F_{0.05,3,36} = 2.8663$$

Since the p-value = 0.2423 > 0.05 and $F = 1.458 < 2.866$, do not reject H_0. There is insufficient evidence to conclude that the variances in strength among the four brands of trash bags are different. **(d)** From the results in (a) and (b), Tuffstuff has the lowest mean strength and should be avoided.

11.12 (a) Since $F = 12.56 > F_{0.05,4,25} = 2.76$, reject H_0. **(b)** Critical range = 4.67. Advertisements A and B are different from Advertisements C and D. Advertisement E is only different from Advertisement D. **(c)** Since $F = 1.927 < 2.76$, do not reject H_0. There is no evidence of a significant difference in the variation in the ratings among the 5 advertisements. **(d)** The advertisement underselling the pen's characteristics had the highest mean ratings and the advertisements overselling the pen's characteristics had the lowest mean ratings. Therefore, use an advertisement that undersells the pen's characteristics and avoid advertisements that oversell the pen's characteristics.

11.14 (a) Since $F = 53.03 > F_{0.05,3,30} = 2.92$, reject H_0. **(b)** Critical range = 5.27 (using 30 degrees of freedom). Designs 3 and 4 are different from designs 1 and 2. Designs 1 and 2 are different from each other. **(c)** The assumptions are that samples are randomly and independently selected (or randomly assigned), the original populations of distances are approximately normally distributed, and the variances are equal. **(d)** Since $F = 2.093 < F_{3,30} = 2.92$, do not reject H_0. There is no evidence of a significant difference in the variation in the distance among the 4 designs. **(e)** The manager should choose design 3 or 4.

11.16 (a) $SSE = 75$; **(b)** $MSA = 15$, $MSBL = 12.5$, $MSE = 3.125$; **(c)** $F = 4.8$; **(d)** $F = 4$

11.18 (a) *df* numerator = 5, *df* denominator = 24; **(b)** $Q_U = 4.17$; **(c)** Critical range = 2.786

11.20 (a) $MSE = 3$ $SSE = 36$; **(b)** $SSBL = 72$; **(c)** $SST = 144$; **(d)** Since $F = 6 < F_{0.01,2,12} = 6.93$, do not reject H_0. Since $F = 4 < F_{0.01,6,12} = 4.82$, do not reject H_0.

11.22 (a) Since $F = 5.185 > 3.07$, reject H_0. **(b)** Since $F = 5 > 2.49$, reject H_0.

11.24 (a) $H_0: \mu_{.1} = \mu_{.2} = \mu_{.3} = \mu_{.4}$ H_1: At least one mean differs.

$$MSA = \frac{SSA}{c-1} = \frac{16.625}{3} = 5.5417$$

$$MSE = \frac{SSE}{(r-1)(c-1)} = \frac{25.9688}{9} = 2.8854$$

$$F = \frac{MSA}{MSE} = \frac{5.5417}{2.8854} = 1.9206$$

Since $F = 1.9206 < 3.86$ or the p-value $= 0.1968 > 0.05$, do not reject H_0. There is no evidence of a difference in the mean diameter of the bubbles produced by the different brands. **(b)** It is inappropriate to perform the Tukey procedure.
(c) $H_0: \mu_{1.} = \mu_{2.} = \mu_{3.} = \mu_{4.}$ H_1: At least one mean differs.

$$MSBL = \frac{SSBL}{r-1} = \frac{9.2813}{3} = 3.0938$$

$$F = \frac{MSBL}{MSE} = \frac{3.0938}{2.8854} = 1.0722$$

Since $F = 1.0722 < 3.86$ or the p-value $= 0.4084 > 0.05$, do not reject H_0. There is no evidence of difference in the mean diameter of the bubbles across the different students. Hence, there is no significant block effect in this experiment. **(d)** From the conclusion in (c), there is insufficient evidence to conclude that Kyle is the best at blowing big bubbles.

11.28 (a) Since $F_{calc} = 268.26 > F = 3.114$, reject H_0. There is enough evidence to conclude that there is a difference in the mean compressive strength after 2, 7, and 28 days. **(b)** Critical range = 0.1651. At the 0.05 level of significance, all of the comparisons are significant. **(c)** RE = 2.558; **(e)** The compressive strength of the concrete increases over the 3 time periods.

11.30 (a) 40; **(b)** 60 and 55; **(c)** 10; **(d)** 10

11.32 (a) Since $F_{calc} = 6.00 > F = 3.35$, reject H_0. **(b)** Since $F_{calc} = 5.50 > F = 3.35$, reject H_0. **(c)** Since $F_{calc} = 1.00 < F = 2.73$, do not reject H_0.

11.34 *df* B = 4, *df* Total = 44, $SSA = 160$, $SSAB = 80$, $SSE = 150$, $SST = 610$, $MSB = 55$, $MSE = 5$, $F_A = 16$, $F_{AB} = 2$

11.36 (a) Since $F = 1.37 < F_{0.05,1,12} = 4.75$, do not reject H_0. **(b)** Since $F = 23.58 > F_{0.05,1,12} = 4.75$, reject H_0. **(c)** Since $F = 0.70 < F_{0.05,1,12} = 4.75$, do not reject H_0.

11.38 (a) H_0: There is no interaction between brand and water temperature. H_1: There is an interaction between brand and water temperature.

$$MSAB = \frac{SSAB}{(r-1)(c-1)} = \frac{506.3104}{(1)(2)} = 253.1552$$

$$F = \frac{MSAB}{MSE} = \frac{253.1552}{12.2199} = 20.7167$$

Since $F = 20.7167 > 3.555$ or the p-value $= 0.0000214 < 0.05$, reject H_0. There is evidence of interaction between brand of pain reliever and temperature of the water. **(b)** Since there is interaction between brand and the temperature of the water, it is inappropriate to analyze the main effect due to brand. **(c)** Since there is interaction between brand and the temperature of the water, it is inappropriate to analyze the main effect due to water temperature. **(e)** The difference in the mean time a tablet took to dissolve in cold and hot water depends on the brand, with Alka-Seltzer having the largest difference and Equate with the smallest difference.

11.40 (a) Since $F = 0.43 < F_{0.05,1,28} = 4.20$, do not reject H_0. **(b)** Since $F = 0.02 < F_{0.05,1,28} = 4.20$, do not reject H_0. **(c)** Since $F = 45.47 > F_{0.05,1,28} = 4.20$, reject H_0.

11.52 (a) Since $F = 1.48 < F_{0.05,9,16} = 2.54$, do not reject H_0. **(b)** Since $F = 0.79 < F_{0.05,3,16} = 3.24$, do not reject H_0. **(c)** Since $F = 52.087 > F_{0.05,3,16} = 3.24$, reject H_0. **(e)** Critical range = 0.0189. Washing cycles for 22 and 24 minutes are not different with respect to dirt removal, but they are both different from 18 and 20 minute cycles. **(f)** 22 minutes (24 minutes was not different, but 22 does just as well and would use less energy). **(g)** The results are the same.

11.54 (a) Since $F = 0.075 < F_{0.05,2,15} = 3.68$, do not reject H_0. **(b)** Since $F = 4.09 > F_{0.05,2,15} = 3.68$, reject H_0. **(c)** Critical range = 1.489. Breaking strength is significantly different between 30 and 50 psi.

11.56 (a) Since $F = 1.97 < F_{0.05,2,12} = 3.89$, do not reject H_0. **(b)** Since $F = 4.87 > F_{0.05,1,12} = 4.75$, reject H_0. **(c)** Since $F = 5.67 > F_{0.05,2,12} = 3.89$, reject H_0. **(e)** Critical range = 1.30. Mean breaking strength under 30 psi is significantly different from 40 psi and 50 psi. **(f)** The mean breaking strength is highest under 30 psi. **(g)** The two-factor experiment gave a more complete set of results than the one-factor experiment. Not only was the side-to-side aspect factor significant, the application of the Tukey procedure on the air-jet pressure factor determined that breaking strength scores are highest under 30 psi.

11.58 (a) Since $F_{calc} = 4.0835 > 3.55$, reject H_0. There is sufficient evidence to conclude that there is an interaction between buffer size and data file size. **(b)** and **(c)** You cannot directly conclude that there is a significant difference in mean read times due to buffer size or file size (main effects) because the mean read times for different buffer sizes are different for different sizes of files. **(e)** Conclusions about the relative speed of the buffers or file sizes cannot be drawn because buffers and file sizes interact. **(f)** In the completely randomized design, there is not enough evidence of a difference between mean access read times for the three file sizes. In the two-factor factorial design, there is significant interaction between buffer size and data file size.

11.62 (a) $F_{calc} = 4.81 > 2.95$, reject H_0. There is sufficient evidence to show a difference in the mean rating scores among the wines. **(b)** The "within wines" term in (a) is *SSW*. It consists of the among blocks (*SSBL*) and "error" term (*SSE*) in (a). **(c)** In this study, the mean rating scores among the eight wines are so significantly different that employing the wrong procedure did not affect the overall conclusion—there is evidence of a difference among the wines. However, by not taking advantage of the actual blocking the experimental error was inflated in *SSW* and this could have led to an erroneous decision. Here the conclusions from the pairwise comparisons are much weaker. Using the Tukey-Kramer procedure, Wine

#2 and Wine #5 are no longer significantly preferred over Wine #8. In addition, Wine #5 now differs from Wine #4 only by chance.

CHAPTER 12

12.2 (a) For $df = 1$ and $\alpha = 0.95$, $\chi^2 = 0.004$. **(b)** For $df = 1$ and $\alpha = 0.975$, $\chi^2 = 0.00098$. **(c)** For $df = 1$ and $\alpha = 0.99$, $\chi^2 = 0.000157$.

12.4 (a) All $f_e = 25$; **(b)** Since $\chi^2 = 4.00 > 3.841$, reject H_0.

12.6 (a) Since $\chi^2 = 1.0 < 3.841$, do not reject H_0. There is not enough evidence to conclude that there is a significant difference between males and females in the proportion who make gas mileage a priority. **(b)** Since $\chi^2 = 10.0 > 3.841$, reject H_0. There is enough evidence to conclude that there is a significant difference between males and females in the proportion who make gas mileage a priority. **(c)** The larger sample size in (b) increases the difference between the observed and expected frequencies and, hence, results in a larger test statistic value.

12.8 (a) H_0: $\pi_1 = \pi_2$ H_1: $\pi_1 \neq \pi_2$

f_o	f_e	$(f_o - f_e)$	$(f_o - f_e)^2/f_e$
370	395	−25	1.58227
130	105	25	5.95238
420	395	25	1.58227
80	105	−25	5.95238
			15.06932

Decision rule: $df = 1$. If $\chi^2 > 3.841$, reject H_0.

Test statistic: $\chi^2 = \sum_{\text{all cells}} \frac{(f_0 - f_e)^2}{f_e} = 15.0693$

Decision: Since $\chi^2_{\text{calc}} = 15.0693 > 3.841$, reject H_0. There is enough evidence to conclude that there is a significant difference in the proportion of African Americans and whites who invest in stocks. **(b)** p-value = 0.0001. The probability of a test statistic as large as 15.0693 or larger when the null hypothesis is true is 0.0001. **(c)** The results of (a) and (b) are exactly the same as those of problem 10.39. The χ^2_{calc} in (a) and the Z_{calc} in problem 10.39 (a) satisfy the relationship that $\chi^2_{\text{calc}} = 15.0693 = (Z_{calc})^2 = (-3.8819)^2$ and the p-value in problem 10.39 (b) is exactly the same as the p-value in (b).

12.10 (a) Since $\chi^2 = 33.333 > 3.841$, reject H_0. (b) p-value < 0.001.

12.12 (a) The expected frequencies for the first row are 20, 30, and 40. The expected frequencies for the second row are 30, 45, and 60. **(b)** Since $\chi^2 = 12.5 > 5.991$, reject H_0. **(c)** A vs. B: $0.20 > 0.196$, therefore A and B are different, A vs. C: $0.30 > 0.185$, therefore A and C are different, B vs. C: $0.10 < 0.185$, therefore B and C are not different.

12.14 Since the calculated test statistic $742.3961 > 9.4877$, reject H_0 and conclude that there is a difference in the proportion of people who eat out at least once a week in the various countries. **(b)** p-value is virtually zero. The probability of a test statistic greater than 742.3961 or more is approximately zero if there is no difference in the proportion of people who eat out at least once a week in the various countries. **(c)** At the 5% level of significance, there is no significant difference between the proportions in Germany and France, while there is significant difference between all the remaining pairs of countries.

12.16 (a) $H_0: \pi_1 = \pi_2 = \pi_3$ H_1: at least one proportion differs

f_o	f_e	$(f_o - f_e)$	$(f_o - f_e)^2/f_e$
48	42.667	5.333	0.667
152	157.333	−5.333	0.181
56	42.667	13.333	4.167
144	157.333	−13.333	1.130
24	42.667	−18.667	8.167
176	157.333	18.667	2.215
			16.5254

Decision rule: $df = (c - 1) = (3 - 1) = 2$. If $\chi^2 > 5.9915$, reject H_0.

Test statistic: $\chi^2 = \sum_{\text{all cells}} \frac{(f_0 - f_e)^2}{f_e} = 16.5254$

Decision: Since $\chi^2_{\text{calc}} = 16.5254 > 5.9915$, reject H_0. There is a significant difference in the age groups with respect to major grocery shopping day. **(b)** p-value = 0.0003. The probability that the test statistic is greater than or equal to 16.5254 is 0.03%, if the null hypothesis is true.

(c) Pairwise Comparisons	**Critical Range**	$\lvert p_{s_j} - p_{s_{j'}} \rvert$
1 to 2	0.1073	0.04
2 to 3	0.0959	0.16*
1 to 3	0.0929	0.12*

There is a significance difference between the 35–54 and over 54 groups, and between the under 35 and over 54 groups. **(d)** The stores can use this information to target their marketing on the specific group of shoppers on Saturday and the days other than Saturday.

12.18 (a) Since $\chi^2_{\text{calc}} = 128.24 > 5.9915$, reject H_0. There is enough evidence to show that there is a significant difference in the proportion of people who object to their medical records being shared with the three organizations. **(b)** There is a significant difference between each pair of organizations.

12.20 Since $\chi^2_{\text{calc}} = 3.50 < 5.991$, do not reject the null hypothesis. There is insufficient evidence to conclude that there is a difference in the proportion of hotels that correctly post minibar charges among the three cities. **(b)** The p-value is 0.174. The probability of a test statistic larger than 3.5 is 0.174 if the null hypothesis is true.

12.22 Since the null hypotheses are not rejected for any of the three items, it is not necessary to perform the Marascuilo procedure.

12.24 $df = (r - 1)(c - 1) = (3 - 1)(4 - 1) = 6$.

12.26 (a) and **(b)** Since the $\chi^2_{\text{calc}} = 20.680 > 12.592$, reject H_0. There is evidence of a relationship between the quarter of the year in which draft-aged men were born and the numbers assigned as their draft eligibilities during the Vietnam War. It appears that the results of the lottery drawing are different from what would be expected if the lottery were random. **(c)** Since $\chi^2_{\text{calc}} = 9.803 < 12.592$, do not reject H_0. There is not enough evidence to conclude there is any relationship between the quarter of the year in which draft-aged men were born and the numbers assigned as their draft eligibilities during the Vietnam War. It appears that the results of the lottery drawing are consistent with what would be expected if the lottery were random.

12.28 (a) H_0: There is no relationship between the commuting time of company employees and the level of stress-related problems observed on the job. H_1: There is a relationship between the commuting time of company employees and the level of stress-related problems observed on the job.

f_o	f_e	$(f_o - f_e)$	$(f_o - f_e)^2/f_e$
9	12.1379	−3.1379	0.8112
17	20.1034	−3.1034	0.4791
18	11.7586	6.2414	3.3129
5	5.2414	−0.2414	0.0111
8	8.6810	−0.6810	0.0534
6	5.0776	0.9224	0.1676
18	14.6207	3.3793	0.7811
28	24.2155	3.7845	0.5915
7	14.1638	−7.1638	3.6233
			9.8311

(a) Decision rule: If $\chi^2 > 13.277$, reject H_0.

Test statistic: $\chi^2 = \sum_{\text{all cells}} \frac{(f_0 - f_e)^2}{f_e} = 9.8311$

Decision: Since the $\chi^2_{\text{calc}} = 9.8311 < 13.277$, do not reject H_0. There is not enough evidence to conclude that there is a relationship between the commuting time of company employees and the level of stress-related problems observed on the job.
(b) Since $\chi^2_{\text{calc}} = 9.831 > 9.488$, reject H_0. There is enough evidence at the 0.05 level to conclude that there is a relationship.

12.30 Since $\chi^2 = 129.520 > 21.026$, reject H_0.

12.32 (a) $H_0: \pi_1 \geq \pi_2 \quad H_1: \pi_1 < \pi_2$ where 1 = beginning, 2 = end
Decision rule: If $Z < -1.645$, reject H_0.

Test statistic: $Z = \frac{B - C}{\sqrt{B + C}} = \frac{9 - 22}{\sqrt{9 + 22}} = -2.3349$

Decision: Since $Z = -2.3349 < -1.645$, reject H_0. There is evidence that the proportion of coffee drinkers who prefer Brand *A* is lower at the beginning of the advertising campaign than at the end of the advertising campaign. **(b)** *p*-value = 0.0098. The probability of a test statistic smaller than −2.3349 is 0.98% if the proportion of coffee drinkers who prefer Brand *A* is not lower at the beginning of the advertising campaign than at the end of the advertising campaign.

12.34 (a) Since $Z = -2.2361 < -1.645$, reject H_0. There is evidence that the proportion who prefer Brand *A* is lower before the advertising than after the advertising.
(b) *p*-value = 0.0127. The probability of a test statistic less than − 2.2361 is 1.27% if the proportion who prefer Brand *A* is not lower before the advertising than after the advertising.

12.36 (a) Since $Z = -3.8996 < -1.645$, reject H_0. There is evidence that the proportion of employees absent less than 5 days was lower in year 1 than in year 2. **(b)** *p*-value is virtually zero. The probability of a test statistic smaller than −3.8996 is essentially zero if the proportion of employees absent less than 5 days was not lower in year 1 than in year 2.

12.38 (a) 9.2604 and 44.1814; **(b)** 8.9065 and 32.8523; **(c)** 7.2609 and 24.9958.

12.40 10.417

12.42 (a) 6.262 and 27.488; **(b)** 7.261.

12.44 You must assume that the data in the population are normally distributed to be able to use the chi-square test of a population variance or standard deviation. If the data selected do not come from an approximately normally distributed population, the accuracy of the test can be seriously affected.

12.46 (a) H_0: $\sigma = \$200$. H_1: $\sigma \neq \$200$.

Decision rule: $df = 24$. If $\chi^2 < 12.401$ or $\chi^2 > 39.364$, reject H_0.

Test statistic: $\chi^2 = \frac{(n-1) \cdot S^2}{\sigma^2} = \frac{24 \cdot 237.52^2}{200^2} = 33.849$

Decision: Since the test statistic of $12.401 < \chi^2_{\text{calc}} = 33.849 < 39.364$, do not reject H_0. There is insufficient evidence to conclude that the standard deviation of the amount of auto repairs is not equal to \$200. **(b)** You must assume that the data in the population are normally distributed to be able to use the chi-square test of a population variance or standard deviation. **(c)** *p*-value = 2(0.0874) = 0.1748. The *p*-value is the probability that a sample is farther away from the hypothesized value of \$200 than this sample value of \$237.52 when the null hypothesis is true is 0.1748.

12.48 (a) Since $\chi^2_{\text{calc}} = 12.245 < 13.848$, reject H_0. There is sufficient evidence to conclude that the standard deviation of the diameter of doorknobs is less than 0.035 inch in the redesigned production process. **(b)** You must assume that the data in the population are normally distributed to be able to use the chi-square test of a population variance or standard deviation. **(c)** *p*-value = (1 − 0.9770) = 0.0230. The probability of a test statistic equal to or more extreme than the result from this sample data is 0.0230 if the population standard deviation is no less than 0.035 inch.

12.50 (a) Since $\chi^2_{\text{calc}} = 21.492 < 30.144$, do not reject H_0. There is insufficient evidence to conclude that the standard deviation in the capacity of a certain type of battery is greater than 2.5 ampere-hours. **(b)** You must assume that the data in the population are normally distributed to be able to use the chi-square test of a population variance or standard deviation. **(c)** *p*-value = 0.3103. The probability of a test statistic equal to or more extreme than the result from this sample data is 0.3103 if the population standard deviation is no greater than 2.5 ampere-hours.

12.52 Since 61.0684 > 16.8119, reject H_0. There is sufficient evidence to conclude that the distribution of service interruptions does not follow a Poisson distribution with a population mean of 1.5 at the 1% level of significance.

12.54 H_0: Battery life follows a normal distribution.
H_1: Battery life does not follow a normal distribution.

Life	$P(X)$	f_e
under 0	0.001938	0.968776928
0–under 1	0.029741	14.87047242
1–under 2	0.172953	86.47668571
2–under 3	0.377089	188.5444078
3–under 4	0.310381	155.1906456
4–under 5	0.096272	48.13607992
5–under 6	0.011144	5.571843365
6 or more	0.0005	0.241088258
Total	1.000000	500

Combine the first two and the last two classes:

Life	f_o	f_e	$(f_o - f_e)^2/f_e$
under 1	12	15.83925	0.930589
1–under 2	94	86.47669	0.654515
2–under 3	170	188.5444	1.823947
3–under 4	188	155.1906	6.936331
4–under 5	28	48.13608	8.423239
5 or more	8	5.812932	0.822867
Total	500	500	19.59149

$$\chi^2_{k-p-1} = \sum_k \frac{(f_0 - f_e)^2}{f_e} = 19.5915$$

$$= \chi^2_{3,0.05} = 7.8147$$

Since 19.5915 > 7.8147, reject H_0. There is sufficient evidence to conclude that the distribution of battery life does not follow a normal distribution.

12.56 **(a)** 31 and 59; **(b)** 29 and 61; **(c)** 25 and 65; **(d)** As the level of significance α gets smaller, the width of the nonrejection region gets wider.

12.58 **(a)** 31; **(b)** 29; **(c)** 27; **(d)** 25; **(e)** As the level of significance α gets smaller, the width of the nonrejection region gets wider.

12.60 40 and 79

12.62 **(a)** The ranks for Sample 1 are 1, 2, 4, 5, and 10. The ranks for Sample 2 are 3, 6.5, 6.5, 8, 9, and 11. **(b)** 22; **(c)** 44.

12.64 Decision: Since $T_1 = 22 > 20$, do not reject H_0.

12.66 **(a)** H_0: $M_1 = M_2$ H_1: $M_1 \neq M_2$
LIRR: Sum of Ranks = 141 NJT: Sum of Ranks = 112

$$\mu_{T_1} = \frac{n_1(n+1)}{2} = \frac{10(22+1)}{2} = 115$$

$$\sigma_{T_1} = \sqrt{\frac{n_1 n_2 (n+1)}{12}}$$

$$= \sqrt{\frac{(10)(12)(22+1)}{12}} = 15.1658$$

$$Z = \frac{141 - 115}{15.1658} = 1.714$$

Decision: Since $-2.58 < Z = 1.714 < 2.58$, do not reject H_0. There is not enough evidence to conclude that there is any difference in their median tendencies to be late. **(b)** You must assume approximately equal variability in the two populations. **(c)** Allowing for a 1% probability of committing a Type I error, there is insufficient evidence to conclude that there is any significant difference in the median tendencies of both railroads to be late. Hence, you cannot conclude that there is any significant difference in the lateness of the two railroads.

12.68 **(a)** Since $Z_{calc} = -4.118 < -1.645$, reject H_0. There is enough evidence to conclude that the median crack size is less for the unflawed sample than for the flawed sample. **(b)** You must assume approximately equal variability in the two populations. **(c)** Using both the pooled-variance *t* test and the separate-variance *t* test allowed you to reject the null hypothesis and conclude in problem 10.21 that the mean crack size is less for the unflawed sample than for the flawed sample. In this test, using the Wilcoxon rank sum test with large-sample *Z*-approximation also allowed you to reject the null hypothesis and conclude that the median crack size is less for the unflawed sample than for the flawed sample.

12.70 **(a)** Since $-1.96 < Z_{calc} = 0.6627 < 1.96$, do not reject H_0. There is not enough evidence to conclude that the median time to clear these problems between the two offices is different. **(b)** You must assume approximately equal variability in the two populations. **(c)** Using both the pooled-variance *t* test and the separate-variance *t* test, you do not reject the null hypothesis and conclude in problem 10.16 that there is not enough evidence to show that the mean waiting time between the two branches is different. In this test, using the Wilcoxon rank sum test with large-sample *Z*-approximation, you also do not reject the null hypothesis and conclude that there is not enough evidence to show that the median waiting time between the two branches is different.

12.72 **(a)** 53; **(b)** 56; **(c)** 59; **(d)** 61

12.74 $W = 50$

12.76 Since $W = 50 > W_U = 47$, reject H_0.

12.78 61

12.80 **(a)** Since the *p*-value is approximately zero, reject H_0. There is sufficient evidence of a difference in the median daily hotel rate in March 2004 and June 2002. **(b)** The results are the same.

12.82 **(a)** Since the *p*-value = 0.784 is greater than the 0.05 level of significance, do not reject H_0. There is insufficient evidence of a difference in the median measurements in-line and from an analytical lab. **(b)** The results are the same.

12.83 **(a)** H_0: $M_D = 0$ H_1: $M_D \neq 0$

D_i	$\lvert D_i \rvert$	R_i	$R_i(+)$
−5.12	5.12	10	
8.27	8.27	11	11
−2.44	2.44	3	
27.56	27.56	14	14
−4.59	4.59	9	
18.03	18.03	13	13
−10.08	10.08	12	
38.80	38.80	15	15
−2.94	2.94	5	
−4.36	4.36	8	
0.06	0.06	1	1
−3.27	3.27	6	
−0.68	0.68	2	
−3.57	3.57	7	
−2.71	2.71	4	

$n' = 15$ $W_L = 16$, $W_U = 104$ $W = \sum_{i=1}^{n} R_i^{(+)} = 54$

Since the *p*-value = 0.755 is greater than the 0.01 level of significance and $16 < W = 54 < 104$, do not reject H_0. There is insufficient evidence of a difference in the median price of textbooks between the local bookstore and Amazon.com. **(b)** The results are the same.

12.84 **(a)** Since the *p*-value = 0.009 is less than the 0.01 level of significance, reject H_0. There is sufficient evidence of a difference in the median performance ratings between the two programs. **(b)** The results are the same.

12.86 15.086

12.88 Since $H = 0.64 < 9.210$, do not reject H_0. There is insufficient evidence to show any real difference in the median reaction times for the three learning methods.

12.90 **(a)** H_0: $M_{Front} = M_{Middle} = M_{Rear}$ H_1: At least one of the medians differs.
Decision rule: If $H > \chi^2_U = 5.991$, reject H_0.

Sample	Value	Rank
Middle	1.4	1
Middle	1.6	2
Middle	1.8	3
Middle	2.0	4
Rear	2.2	5
Middle	2.4	6
Rear	2.8	7.5
Rear	2.8	7.5
Middle	3.2	9
Front	4.0	10.5
Rear	4.0	10.5
Rear	4.6	12
Front	5.0	13
Front	5.4	14
Rear	6.0	15
Front	6.2	16
Front	7.2	17
Front	8.6	18

Decision: Reject H_0 if $H > \chi^2_{.05,2} = 5.991$.
Test statistic:

$$H = \left[\frac{12}{n(n+1)}\sum_{j=1}^{c}\frac{T_j^2}{n_j}\right] - 3(n+1)$$

$$= \left[\frac{12}{18(19)}(1960.5833)\right] - 3(19) = 11.79$$

Decision: Since $H = 11.79 > 5.991$, reject H_0. There is sufficient evidence to show there is a difference in the median sales (in thousands of dollars) of pet toys among the three store aisle locations. **(b)** The results are consistent.

12.92 (a) Since $H = 22.26 > 7.815$ or the p-value is approximately zero, reject H_0. There is sufficient evidence of a difference in the median strength of the four brands of trash bags. **(b)** The results are the same.

12.94 (a) Reject H_0 if $F_R > 9.2363$. **(b)** Since $F_R = 11.56 > 9.2363$, reject H_0.

12.96 (a) $H_0: M_A = M_B = M_C = M_D$ $\quad H_1$: medians are different

	Rank			
	Bazooka	Bubbletape	Bubbleyum	Bubblicious
	2	3	1	4
	3	1	2	4
	4	1	2.5	2.5
	4	3	1.5	1.5
$R_{.j}$	13	8	7	12
$(R_{.j})^2$	169	64	49	144.0

Test statistic:

$$F_R = \frac{12}{rc(c+1)}\sum_{j=1}^{c} R_{.j}^2 - 3r(c+1)$$

$$= \frac{12}{(4)(4)(5)}(426) - 3(4)(5) = 3.90$$

Upper critical value: $\chi^2_U = \chi^2_{0.05,3} = 7.8147$ $\quad p$-value $= 0.272$
Since the p-value $= 0.272 > 0.05$ and $F_R = 3.90 < 7.81$, do not reject H_0 at the 0.05 level of significance. There is insufficient evidence of a difference in the median diameter of the bubbles produced by the different brands. **(b)** The results are the same.

12.108 (a) Since the $\chi^2_{\text{calc}} = 0.412 < 3.841$, do not reject H_0. There is not enough evidence to conclude that there is a relationship between a student's gender and their pizzeria selection. **(b)** Since the $\chi^2_{\text{calc}} = 2.624 < 3.841$, do not reject H_0. There is not enough evidence to conclude that there is a relationship between a student's gender and their pizzeria selection. **(c)** Since the $\chi^2_{\text{calc}} = 4.956 < 5.991$, do not reject H_0. There is not enough evidence to conclude that there is a relationship between price and pizzeria selection. **(d)** p-value = 0.0839. The probability of a sample that gives a test statistic equal to or greater than 4.956 is 8.39% if the null hypothesis of no relationship between price and pizzeria selection is true. **(e)** Since there is no evidence that price and pizzeria selection are related, it is inappropriate to determine which prices are different in terms of pizzeria preference.

12.110 (a) Since $\chi^2 = 3.3084 < 3.8415$, do not reject H_0 and conclude that there is insufficient evidence of a difference between the proportion of boys and girls who worry about having enough money. **(b)** p-value = 0.0689. The probability of a test statistic larger than 3.3084 is 6.89% if there is not a significant difference between the proportion of boys and girls who worry about having enough money.

12.112 (a) Since $\chi^2_{\text{calc}} = 12.026 > 3.841$, reject H_0. There is enough evidence to conclude there is a relationship between the type of user and the seriousness of concern over the first statement. **(b)** The p-value = 0.000525. The probability of a test statistic of 12.026 or larger is 0.000525 if there is no relationship between type of user and the seriousness of concern over the first statement. **(c)** Since $\chi^2_{\text{calc}} = 7.297 > 3.841$, reject H_0. There is enough evidence to conclude there is a relationship between type of user and the seriousness of concern over the second statement. **(d)** The p-value = 0.00691. The probability of a test statistic of 7.297 or larger is 0.00691 if there is no relationship between type of user and the seriousness of concern over the second statement.

12.114 (a) Since $\chi^2_{\text{calc}} = 11.895 < 12.592$, do not reject H_0. There is not enough evidence to conclude that there is a relationship between the attitudes of employees toward the use of self-managed work teams and employee job classification. **(b)** Since $\chi^2_{\text{calc}} = 3.294 < 12.592$, do not reject H_0. There is not enough evidence to conclude that there is a relationship between the attitudes of employees toward vacation time without pay and employee job classification.

12.116 Since $\chi^2_{\text{calc}} = 160.38 > 5.991$, reject H_0. There is enough evidence to conclude that there is a relationship between the type of incident and the type of model.

12.118 (a) Since $Z = -1.7889 < -1.645$, reject H_0. There is enough evidence of a difference in the proportion of respondents who prefer Coca-Cola before and after viewing the ads. **(b)** p-value = 0.0736. The probability of a test statistic that differs from 0 by 1.7889 or more in either direction is 7.36% if there is not a difference in the proportion of respondents who prefer Coca-Cola before and after viewing the ads. **(c)** The frequencies in the second table are computed from the row and column totals of the first table. **(d)** Since the calculated test statistic is $0.6528 < 3.8415$, do not reject H_0 and conclude that there is not a significant difference in preference for Coca-Cola before and after viewing the ads. **(e)** p-value = 0.4191. The probability of a test statistic larger than 0.6528 is 41.91% if there is not a significant difference in preference for Coca-Cola before and after viewing the ads. **(f)** The

McNemar test performed using the information in the first table takes into consideration the fact that the same set of respondents are surveyed before and after viewing the ads while the chi-square test performed using the information in the second table ignores this fact. The McNemar test should be used because of the related samples (before-after comparison).

CHAPTER 13

13.2 (a) yes; **(b)** no; **(c)** no; **(d)** yes.

13.4 (b) $b_1 = \dfrac{SSXY}{SSX} = \dfrac{27.75}{375} = 0.074$

$b_0 = \bar{Y} - b_1\bar{X} = 2.375 - 0.074(12.5) = 1.45$

For each increase in shelf space of an additional foot, weekly sales are estimated to increase by 0.074 hundreds of dollars, or \$7.40.
(c) $\hat{Y} = 1.45 + 0.074X = 1.45 + 0.074(8) = 2.042$, or \$204.20

13.6 (b) $b_0 = -2.37$, $b_1 = 0.0501$; **(c)** For every cubic foot increase in the amount moved, labor hours are estimated to increase by 0.0501. **(d)** 22.67 labor hours

13.8 (b) $b_0 = -246.2599$, $b_1 = 4.1897$ **(c)** For each additional million dollar increase in revenue, the mean value is estimated to increase by 4.1897 million dollars. **(d)** 382.2005 million dollars.

13.10 (b) $b_0 = 6.048$, $b_1 = 2.019$; **(c)** For every one Rockwell E unit increase in hardness, the mean tensile strength is estimated to increase by 2,019 pounds per square inch. **(d)** 147,382 pounds per square inch

13.12 $r^2 = 0.90$. 90% of the variation in the dependent variable can be explained by the variation in the independent variable.

13.14 $r^2 = 0.75$. 75% of the variation in the dependent variable can be explained by the variation in the independent variable.

13.16 (a) $r^2 = \dfrac{SSR}{SST} = \dfrac{2.0535}{3.0025} = 0.684.$

68.4% of the variation in sales can be explained by the variation in shelf space.

(b) $S_{YX} = \sqrt{\dfrac{SSE}{n-2}} = \sqrt{\dfrac{\sum_{i=1}^{n}(Y_i - \hat{Y}_i)^2}{n-2}} = \sqrt{\dfrac{0.949}{10}} = 0.308$ **(c)** Based on (a) and (b), the model should be very useful for predicting sales.

13.18 (a) $r^2 = 0.8892$. 88.92% of the variation in labor hours can be explained by the variation in cubic feet moved. **(b)** $S_{YX} = 5.0314$;
(c) Based on (a) and (b), the model should be very useful for predicting the labor hours.

13.20 (a) $r^2 = 0.9424$. 94.24% of the variation in value of a baseball franchise can be explained by the variation in its annual revenue.
(b) $S_{YX} = 33.7876$ **(c)** Based on (a) and (b), the model should be very useful for predicting value of a baseball franchise.

13.22 (a) $r^2 = 0.4613$. 46.13% of the variation in the tensile strength can be explained by the variation in the hardness. **(b)** $S_{YX} = 9.0616$
(c) Based on (a) and (b), the model is only marginally useful for predicting tensile strength.

13.24 A residual analysis of the data indicates a pattern, with sizeable clusters of consecutive residuals that are either all positive or all negative. This pattern indicates a violation of the assumption of linearity. A quadratic model should be investigated.

13.26 (a) There does not appear to be a pattern in the residual plot.
(b) The assumptions of regression do not appear to be seriously violated.

13.28 (a) Based on the residual plot, there appears to be a nonlinear pattern in the residuals. A quadratic model should be investigated.
(b) The assumptions of normality and equal variance do not appear to be seriously violated.

13.30 (a) Based on the residual plot, there appears to be a nonlinear pattern in the residuals. A quadratic model should be investigated.
(b) The assumptions of normality and equal variance do not appear to be seriously violated.

13.32 (a) An increasing linear relationship exists. **(b)** There is evidence of a strong positive autocorrelation among the residuals.

13.34 (a) No, because the data were not collected over time. **(b)** If a single store had been selected, then studied over a period of time, you would compute the Durbin-Watson statistic.

13.36 (a) $b_1 = \dfrac{SSXY}{SSX} = \dfrac{201399.05}{12495626} = 0.0161$

$b_0 = \bar{Y} - b_1\bar{X} = 71.2621 - 0.0161(4393) = 0.458$ **(b)** $\hat{Y} = 0.458 + 0.0161X = 0.458 + 0.0161(4500) = 72.908$ or \$72,908 **(c)** There is no evidence of a pattern in the residuals over time.

(d) $D = \dfrac{\sum_{i=2}^{n}(e_i - e_{i-1})^2}{\sum_{i=1}^{n} e_i^2} = \dfrac{1243.2244}{599.0683} = 2.08 > 1.45$. There is no evidence of positive autocorrelation among the residuals. **(e)** Based on a residual analysis, the model appears to be adequate.

13.38 (a) $b_0 = -2.535$, $b_1 = .06073$; **(b)** \$2,505.40; **(d)** $D = 1.64 > d_U = 1.42$ so there is no evidence of positive autocorrelation among the residuals. **(e)** The plot does show some nonlinear pattern suggesting that a nonlinear model might be better. Otherwise, the model appears to be adequate.

13.40 (a) 3.00; **(b)** $t_{16} = \pm 2.1199$; **(c)** Reject H_0. There is evidence that the fitted linear regression model is useful. **(d)** $1.32 \le \beta_1 \le 7.68$.

13.42 (a) $t = \dfrac{b_1 - \beta_1}{S_{b_1}} = \dfrac{0.074}{0.0159} = 4.65 > t_{10} = 2.2281$ with 10 degrees of freedom for $\alpha = 0.05$. Reject H_0. There is evidence that the fitted linear regression model is useful. **(b)** $b_1 \pm t_{n-2} S_{b_1} = 0.074 \pm 2.2281(0.0159)$
$0.0386 \le \beta_1 \le 0.1094$

13.44 (a) $t = 16.52 > 2.0322$, reject H_0; **(b)** $0.044 \le \beta_1 \le 0.0562$

13.46 (a) Since the p-value is approximately zero, reject H_0 at 5% level of significance. There is evidence of a linear relationship between annual revenue and franchise value. **(b)** $3.7888 \le \beta_1 \le 4.5906$

13.48 (a) p-value is virtually $0 < 0.05$, reject H_0; **(b)** $1.246 \le \beta_1 \le 2.792$

13.50 (b) If the S&P gains 30% in a year, the ULPIX is expected to gain an estimated 60%. **(c)** If the S&P loses 35% in a year, the ULPIX is expected to lose an estimated 70%.

13.52 (a) –0.401; **(b)** p-value $= 0.0885 > 0.05$, do not reject H_0; **(c)** At the 0.05 level of significance, there is no relationship between turnover rate of preboarding screeners and the security violations detected.

13.54 (a) 0.484; **(b)** p-value $= 0.017 < 0.05$, reject H_0; **(c)** At the 0.05 level of significance, there is a relationship between cold-cranking amps and price. **(d)** Yes, there is a positive linear relationship indicating that as cold-cranking amps increase so does the price.

13.56 (a) $15.95 \le \mu_{Y|X=400} \le 18.05$; **(b)** $14.651 \le Y_{X=400} \le 19.349$

13.58 (a) $\hat{Y}_i \pm t_{n-2} S_{YX} \sqrt{h_i}$

$= 2.042 \pm 2.2281(0.3081)\sqrt{0.1373}$

$1.7876 \le \mu_{Y|X=8} \le 2.2964$

(b) $\hat{Y}_i \pm t_{n-2} S_{YX} \sqrt{1 + h_i}$

$= 2.042 \pm 2.2281(0.3081)\sqrt{1 + 0.1373}$

$1.3100 \le Y_{X=8} \le 2.7740$

(c) Part (b) provides a prediction interval for the individual response given a specific value of the independent variable and part (a) provides an interval estimate for the mean value given a specific value of the independent variable. Since there is much more variation in predicting an individual value than in estimating a mean value, a prediction interval is wider than a confidence interval estimate.

13.60 (a) $20.799 \le \mu_{Y|X=500} \le 24.542$; **(b)** $12.276 \le Y_{X=500} \le 33.065$

13.62 (a) $367.0757 \le \mu_{Y|X=150} \le 397.3254$; **(b)** $311.3562 \le Y_{X=150} \le 453.0448$

13.74 (a) $b_0 = 24.84$, $b_1 = 0.14$ **(b)** For each additional case, the predicted delivery time is estimated to increase by 0.14 minutes. **(c)** 45.84; **(d)** No, 500 is outside the relevant range of the data used to fit the regression equation. **(e)** $r^2 = 0.972$; **(f)** There is no obvious pattern in the residuals so the assumptions of regression are met. The model appears to be adequate. **(g)** $t = 24.88 > 2.1009$, reject H_0; **(h)** $44.88 \le \mu_{Y|X=150} \le 46.80$; **(i)** $41.56 \le Y_{X=150} \le 50.12$; **(j)** $0.128 \le \beta_1 \le 0.152$

13.76 (a) $b_0 = -44.172$, $b_1 = 1.782$; **(b)** For each additional dollar in assessed value, the predicted selling price is estimated to increase by \$1.78. **(c)** \$80,458; **(d)** $r^2 = 0.926$; **(e)** There is no obvious pattern in the residuals so the assumptions of regression are met. The model appears to be adequate. **(f)** $t = 18.66 > 2.0484$, reject H_0; **(g)** $78.707 \le \mu_{Y|X=70} \le 82.388$; **(h)** $73.195 \le Y_{X=70} \le 87.900$; **(i)** $1.586 \le \beta_1 \le 1.977$

13.78 (a) $b_0 = 0.30$, $b_1 = 0.00487$; **(b)** For each additional point on the GMAT score, the predicted GPI is estimated to increase by 0.00487. **(c)** 3.2225; **(d)** $r^2 = 0.798$ **(e)** There is no obvious pattern in the residuals so the assumptions of regression are met. The model appears to be adequate. **(f)** $t = 8.43 > 2.1009$, reject H_0; **(g)** $3.144 \le \mu_{Y|X=600} \le 3.301$; **(h)** $2.886 \le Y_{X=600} \le 3.559$; **(i)**$.00366 \le \beta_1 \le .00608$

13.80 (a) There is no clear relationship shown on the scatterplot. **(c)** Looking at all 23 flights, when the temperature is lower, there is likely to be some O-ring damage, particularly if the temperature is below 60 degrees. **(d)** 31 degrees is outside the relevant range so a prediction should not be made. **(e)** predicted $Y = 18.036 - 0.240X$ where X = temperature and Y = O-ring damage; **(g)** A nonlinear model would be more appropriate. **(h)** The appearance on the residual plot of a nonlinear pattern indicates a nonlinear model would be better.

13.82 (a) $b_0 = -2629.222$, $b_1 = 82.472$; **(b)** For each additional centimeter in circumference, the mean weight is estimated to increase by 82.472 grams. **(c)** 2,319.08 grams; **(e)** $r^2 = 0.937$; **(f)** There appears to be a nonlinear relationship between circumference and weight. **(g)** p-value is virtually $0 < 0.05$, reject H_0; **(h)** $72.787 \le \beta_1 \le 92.156$; **(i)** $2186.959 \le \mu_{Y|X=60} \le 2451.202$; **(j)** $1726.551 \le Y_{X=60} \le 2911.610$

13.84 (b) $\hat{Y} = 931{,}626.16 + 21{,}782.76X$ **(c)** $b_1 = 21{,}782.76$ means that as the median age of the customer base increases by one year, the latest one-month sales total is estimated to increase by \$21,782.76. **(d)** $r^2 = 0.0017$. Only 0.17% of the total variation in the franchise's latest one-month sales total can be explained by using the median age of customer base. **(e)** The residuals are very evenly spread out across different range of median age. **(f)** Since $-2.4926 < t = 0.2482 < 2.4926$, do not reject H_0. There is not enough evidence to conclude that there is a linear relationship between the one-month sales total and the median age of the customer base. **(g)** $-156{,}181.50 \le \beta_1 \le 199{,}747.02$.

13.86 (a) There is a positive linear relationship between total sales and the percentage of customer base with a college diploma. **(b)** $\hat{Y} = 789{,}847.38 + 35{,}854.15X$ **(c)** $b_1 = 35{,}854.15$ means that for each increase of one percent of the customer base having received a college diploma, the latest one-month sales total is estimated to increase by \$35,854.15. **(d)** $r^2 = 0.1036$. So, 10.36% of the total variation in the franchise's latest one-month sales total can be explained by the percentage of the customer base with a college diploma. **(e)** The residuals are quite evenly spread out around zero. **(f)** Since $t = 2.0392 > 2.0281$, reject H_0. There is enough evidence to conclude that there is a linear relationship between one-month sales total and percentage of customer base with a college diploma. **(g)** $b_1 \pm t_{n-2} S_{b_1} = 35{,}854.15 \pm 2.0281(17{,}582.269)$
$195.75 \le \beta_1 \le 71{,}512.60$

13.88 (a) $b_0 = -24.247$, $b_1 = 1.046$; **(b)** For each additional unit increase in summated rating, the price per person is estimated to increase by \$1.05. Since no restaurant will receive a summated rating of 0, it is inappropriate to interpret the Y-intercept. **(c)** \$28.07; **(d)** $r^2 = 0.658$; **(e)** There is no obvious pattern in the residuals so the assumptions of regression are met. The model appears to be adequate. **(f)** p-value is virtually $0 < 0.05$, reject H_0; **(g)** $\$26.55 \le \mu_{Y|X=50} \le \$29.59$; **(h)** $\$16.00 \le Y_{X=50} \le \$40.15$; **(i)** $0.895 \le \beta_1 \le 1.198$

13.90 (a) MSFT and Ford 0.167; MSFT and GM 0.157; MSFT and IAL −0.232; Ford and GM 0.867; Ford and IAL 0.697; and GM and IAL 0.629; **(b)** There is a strongly positive linear relationship between the stock price of Ford and GM, a moderately strong positive linear relationship between the stock price of Ford and IAL and between GM and IAL, a very weak positive linear relationship between the stock price of Ford and Microsoft, between GM and Microsoft, and a rather weak negative linear relationship between Microsoft and IAL. **(c)** It is not a good idea to have all the stocks in an individual's portfolio be strongly, positively correlated among each other because the portfolio risk can be reduced when a pair of stock prices is negatively related in a portfolio.

Chapter 14

14.2 (a) For each one unit increase in X_1, you estimate that Y will decrease 2 units, holding X_2 constant. For each one unit increase in X_2, you estimate that Y will increase 7 units, holding X_1 constant. **(b)** The Y-intercept equal to 50, estimates the predicted value of Y when both X_1 and X_2 are zero.

14.4 (a) $\hat{Y} = -2.72825 + 0.047114X_1 + 0.011947X_2$; **(b)** For a given number of orders, for each increase of \$1,000 in sales, distribution cost is estimated to increase by \$47.114. For a given amount of sales, for each increase of one order, distribution cost is estimated to increase by \$11.95. **(c)** The interpretation of b_0 has no practical meaning here because it would represent the estimated distribution cost when there were no sales and no orders. **(d)** $\hat{Y} = -2.72825 + 0.047114(400) + 0.011947(4500) = 69.878$ or \$69,878; **(e)** $\$66{,}419{,}93 \le \mu_{Y|X} \le \$73{,}337.01$; **(f)** $\$59{,}380.61 \le Y_X \le \$80{,}376.33$

14.6 (a) $\hat{Y} = 156.4 + 13.081X_1 + 16.795X_2$; **(b)** For a given amount of newspaper advertising, each increase by \$1,000 in radio advertising is estimated to result in a mean increase in sales of \$13,081. For a given amount of radio advertising, each increase by \$1,000 in newspaper advertising is estimated to result in a mean increase in sales of \$16,795.

(c) When there is no money spent on radio advertising and newspaper advertising, the estimated mean sales is \$156,430.44. **(d)** $\hat{Y} = 156.4 + 13.081(20) + 16.795(20) = 753.95$ or \$753,950; **(e)** $\$623{,}038.31 \le \mu_{Y|X} \le \$884{,}860.93$; **(f)** $\$396{,}522.63 \le Y_X \le \$1{,}111{,}376.60$

14.8 (a) $\hat{Y} = 400.8057 + 456.4485X_1 - 2.4708X_2$ where X_1 = Land, X_2 = age; **(b)** For a given age, each increase by one acre in land area is estimated to result in a mean increase in appraised value by \$456.45 thousands. For a given acreage, each increase of one year in age is estimated to result in the mean decrease in appraised value by \$2.47 thousands. **(c)** The interpretation of b_0 has no practical meaning here because it would represent the estimated appraised value of a new house that has no land area. **(d)** $\hat{Y} = 400.8057 + 456.4485(0.25) - 2.4708(45) =$ \$403.73 thousands. **(e)** $372.7370 \le \mu_{Y|X} \le 434.7243$; **(f)** $235.1964 \le Y_X \le 572.2649$

14.10 (a) $MSR = 15$, $MSE = 12$; **(b)** 1.25; **(c)** $F = 1.25 < 4.10$, do not reject H_0; **(d)** 0.20; **(e)** 0.04

14.12 (a) $F = 97.69 > F_{U(2,15-2-1)} = 3.89$. Reject H_0. There is evidence of a significant linear relationship with at least one of the independent variables. **(b)** The p-value is 0.0001. **(c)** $r^2 = 0.9421$. 94.21% of the variation in the long-term ability to absorb shock can be explained by variation in forefoot absorbing capability and variation in midsole impact. **(d)** $r^2_{adj} = 0.93245$

14.14 (a) $F = 74.13 > 3.467$, reject H_0; **(b)** p-value = 0; **(c)** $r^2 = 0.8759$. 87.59% of the variation in distribution cost can be explained by variation in sales and variation in number of orders. **(d)** $r^2_{adj} = 0.8641$

14.16 (a) $F = 40.16 > F_{U(2,22-2-1)} = 3.522$. Reject H_0. There is evidence of a significant linear relationship. **(b)** The p-value is less than 0.001. **(c)** $r^2 = 0.8087$. 80.87% of the variation in sales can be explained by variation in radio advertising and variation in newspaper advertising. **(d)** $r^2_{adj} = 0.7886$

14.18 (a) Based upon a residual analysis, the model appears adequate. **(b)** There is no evidence of a pattern in the residuals versus time. **(c)** $D = \dfrac{1{,}077.0956}{477.0430} = 2.26$; **(d)** $D = 2.26 > 1.55$. There is no evidence of positive autocorrelation in the residuals.

14.20 There appears to be a quadratic relationship in the plot of the residuals against both radio and newspaper advertising. Thus, quadratic terms for each of these explanatory variables should be considered for inclusion in the model.

14.22 There is no particular pattern in the residual plots and the model appears to be adequate.

14.24 (a) Variable X_2 has a larger slope in terms of the t statistic of 3.75 than variable X_1, which has a smaller slope in terms of the t statistic of 3.33. **(b)** $1.46824 \le \beta_1 \le 6.53176$; **(c)** For X_1: $t = 4/1.2 = 3.33 > 2.1098$ with 17 degrees of freedom for $\alpha = 0.05$. Reject H_0. There is evidence that X_1 contributes to a model already containing X_2. For X_2: $t = 3/0.8 = 3.75 > 2.1098$ with 17 degrees of freedom for $\alpha = 0.05$. Reject H_0. There is evidence that X_2 contributes to a model already containing X_1. Both X_1 and X_2 should be included in the model.

14.26 (a) 95% confidence interval on β_1: $b_1 \pm t_{n-k-1}S_{b_1}$, $0.0471 \pm 2.0796 \cdot 0.0203$, $0.00488 \le \beta_1 \le 0.08932$; **(b)** For X_1: $t = b_1/S_{b_1} = 0.0471/0.0203 = 2.32 > 2.0796$. Reject H_0. There is evidence that X_1 contributes to a model already containing X_2. For X_2: $t = b_2/S_{b_2} = 0.01195/0.00225 = 5.31 > 2.0796$ with 21 degrees of freedom for $\alpha = 0.05$. Reject H_0. There is evidence that X_2 contributes to a model already containing X_1. Both X_1 (sales) and X_2 (orders) should be included in the model.

14.28 (a) $9.398 \le \beta_1 \le 16.763$; **(b)** For X_1: $t = 7.43 > 2.093$. Reject H_0. There is evidence that X_1 contributes to a model already containing X_2. For X_2: $t = 5.67 > 2.093$. Reject H_0. There is evidence that X_2 contributes to a model already containing X_1. Both X_1 (radio advertising) and X_2 (newspaper advertising) should be included in the model.

14.30 (a) $227.5865 \le \beta_1 \le 685.3104$; **(b)** For X_1: $t = 4.0922$ and p-value = 0.0003. Since p-value < 0.05, reject H_0. There is evidence that X_1 contributes to a model already containing X_2. For X_2: $t = -3.6295$ and p-value = 0.0012. Since p-value < 0.05, reject H_0. There is evidence that X_2 contributes to a model already containing X_1. Both X_1 (land area) and X_2 (age) should be included in the model.

14.32 (a) For X_1: $F = 1.25 < 4.96$, do not reject H_0. For X_2: $F = 0.833 < 4.96$, do not reject H_0; **(b)** 0.1111, 0.0769

14.34 (a) For X_1: $SSR(X_1|X_2) = SSR(X_1 \text{ and } X_2) - SSR(X_2) = 3{,}368.087 - 3{,}246.062 = 122.025$,

$$F = \frac{SSR(X_1 \mid X_2)}{MSE} = \frac{122.025}{477.043/21} = 5.37 > F_{U(1,21)} = 4.325.$$

Reject H_0. There is evidence that X_1 contributes to a model already containing X_2. For X_2: $SSR(X_2|X_1) = SSR(X_1 \text{ and } X_2) - SSR(X_1) = 3{,}368.087 - 2{,}726.822 = 641.265$,

$$F = \frac{SSR(X_2 \mid X_1)}{MSE} = \frac{641.265}{477.043/21} = 28.23 > F_{U(1,21)} = 4.325.$$ Reject H_0.

There is evidence that X_2 contributes to a model already containing X_1. Since both X_1 and X_2 make a significant contribution to the model in the presence of the other variable, both variables should be included in the model. **(b)** $r^2_{Y1.2} = \dfrac{SSR(X_1 \mid X_2)}{SST - SSR(X_1 \text{ and } X_2) + SSR(X_1 \mid X_2)} =$

$$\frac{122.025}{3{,}845.13 - 3{,}368.087 + 122.025} = 0.2037.$$ Holding constant the effect of the number of orders, 20.37% of the variation in distribution cost can be explained by the variation in sales.

$$r^2_{Y2.1} = \frac{SSR(X_2 \mid X_1)}{SST - SSR(X_1 \text{ and } X_2) + SSR(X_2 \mid X_1)} =$$

$$\frac{641.265}{3{,}845.13 - 3{,}368.087 + 641.265} = 0.5734.$$

Holding constant the effect of sales, 57.34% of the variation in distribution cost can be explained by the variation in the number of orders.

14.36 (a) For X_1: $F = 55.28 > 4.381$. Reject H_0. There is evidence that X_1 contributes to a model already containing X_2. For X_2: $F = 32.12 > 4.381$. Reject H_0. There is evidence that X_2 contributes to a model already containing X_1. Since both X_1 and X_2 make a significant contribution to the model in the presence of the other variable, both variables should be included in the model. **(b)** $r^2_{Y1.2} = 0.7442$. Holding constant the effect of newspaper advertising, 74.42% of the variation in sales can be explained by the variation in radio and television advertising. $r^2_{Y2.1} = 0.6283$. Holding constant the effect of radio and television advertising, 62.83% of the variation in sales can be explained by the variation in newspaper advertising.

14.40 (a) predicted Price = 43.737 + 9.219 Rooms + 12.697 west; **(b)** Holding constant the effect of neighborhood, for each additional room, the mean selling price is estimated to increase by 9.219 thousands of dollars, or \$9,219. For a given number of rooms, the mean selling price in the west neighborhood is estimated to be 12.697 thousand dollars (\$12,697) greater than the east neighborhood. **(c)** \$126,710; $\$121{,}470 \le \mu_{Y|X} \le \$131{,}940$; $\$109{,}560 \le Y_X \le \$143{,}860$; **(d)** the model appears adequate; **(e)** $F = 55.39 > 3.59$, reject H_0; **(f)** $t = 8.95 > 2.1098$, reject H_0.

$t = 3.59 > 2.1098$, reject H_0. include both variables; **(g)** $7.0466 \le \beta_1 \le 11.392$; $5.239 \le \beta_2 \le 20.155$; **(h)** 86.7% of the variation in selling price is explained by the variation in number of rooms and location. **(i)** 85.1%; **(j)** 0.825, 0.431; **(k)** The slope of selling price with number of rooms is the same regardless of location. **(l)** p-value = 0.330, do not reject H_0, do not include the interaction term; **(m)** the model developed in part (a) should be used.

14.42 (a) predicted Time = 8.01 + 0.00523 Depth – 2.105 Dry; **(b)** Holding constant the effect of type of drilling, for each foot increase in depth of the hole, the drilling time is estimated to increase by 0.0052 minutes. For a given depth, a dry drilling is estimated to reduce the drilling time over wet drilling by 2.1052 minutes. **(c)** 6.428 minutes, $6.210 \le \mu_{Y|X} \le 6.646$; $4.923 \le Y_X \le 7.932$; **(d)** The model appears adequate. **(e)** $F = 111.11 > 3.09$, reject H_0; **(f)** $t = 5.03 > 1.9847$, reject H_0. $t = -14.03 < -1.9847$, reject H_0. include both variables; **(g)** $0.0032 \le \beta_1 \le 0.0073$; $-2.403 \le \beta_2 \le -1.807$; **(h)** 69.6% of the variation in drill time is explained by the variation of depth and variation in type of drilling. **(i)** 69.0%; **(j)** 0.207, 0.670; **(k)** The slope of the additional drilling time with the depth of the hole is the same regardless of the type of drilling method used. **(l)** The p-value of the interaction term = 0.462 > 0.05, so the term is not significant and should not be included in the model. **(m)** The model in part (a) should be used.

14.44 (a) $\hat{Y} = 31.5594 + 0.0296X_1 + 0.0041X_2 + 0.000017159X_1X_2$. where X_1 = sales, X_2 = orders, the p-value is 0.3249 > 0.05. Do not reject H_0. There is not enough evidence that the interaction term makes a contribution to the model. **(b)** Since there is not enough evidence of any interaction effect between sales and orders, the model in problem 14.4 should be used.

14.46 (a) The p-value of the interaction term = 0.002 < .05, so the term is significant and should be included in the model. **(b)** Use the model developed in this problem.

14.48 (a) For X_1X_2: the p-value is 0.2353 > 0.05. Do not reject H_0. There is not enough evidence that the interaction term makes a contribution to the model. **(b)** Since there is not enough evidence of an interaction effect between total staff present and remote hours, the model in problem 14.7 should be used.

14.50 Holding constant the effect of other variables, the natural logarithm of the estimated odds ratio for the dependent categorical response will increase by 2.2 for each unit increase in the independent variable to which the coefficient corresponds.

14.52 0.4286

14.54 (a) ln (estimated odds ratio) = $-6.94 + 0.13947X_1 + 2.774X_2 = -6.94 + 0.13947(36) + 2.774(0) = -1.91908$
Estimated odds ratio = $e^{-1.91908} = 0.1467$
Estimated Probability of Success = Odds Ratio / (1 + Odds Ratio) = 0.1467/(1 + 0.1467) = 0.1280; **(b)** From the text discussion of the example, 70.16% of the individuals who charge \$36,000 per annum and possess additional cards can be expected to purchase the premium card. Only 12.80% of the individuals who charge \$36,000 per annum and do not possess additional cards can be expected to purchase the premium card. For a given amount of money charged per annum, the likelihood of purchasing a premium card is substantially higher among individuals who already possess additional cards than for those who do not possess additional cards. **(c)** ln (estimated odds ratio) = $-6.94 + 0.13947X_1 + 2.774X_2 = -6.94 + 0.13947(18) + 2.774(0) = -4.42954$
Estimated odds ratio = $e^{-4.42954} = 0.0119$
Estimated Probability of Success = Odds Ratio / (1 + Odds Ratio) = 0.0119/(1 + 0.0119) = 0.01178; **(d)** Among individuals who do not purchase additional cards, the likelihood of purchasing a premium card diminishes dramatically with a substantial decrease in the amount charged per annum.

14.56 (a) ln (estimated odds) = –121.95 + 8.053 GPA + 0.157 GMAT; **(b)** Holding constant the effects of GMAT score, for each increase of one point in GPA, ln(estimated odds) increases by an estimate of 8.053. Holding constant the effects of GPA, for each increase of one point in GMAT score, ln(estimated odds) increases by an estimate of 0.15729. **(c)** 0.197; **(d)** deviance statistic = 8.122 < 40.133, do not reject H_0 so model is adequate; **(e)** For GPA: $Z = 1.60 < 1.96$, do not reject H_0. For GMAT: $Z = 2.07 > 1.96$, reject H_0. **(f)** ln (estimated odds) = –2.765 + 1.02 GPA; **(g)** ln (estimated odds) = –60.15 + 0.099 GMAT; **(h)** Use model in (g)

14.58 (a) ln (estimated odds) = 1.243 – 0.250 Price, yes; **(b)** ln (estimated odds) = 1.220 – .250 price + 0.0377 gender, yes, no; **(c)** model in part (a) is better; **(d)** 0.2675; **(e)** 0.163; **(f)** 0.0946

14.68 (a) predicted Price = –44.988 + 1.751 Value + 0.368 Time; **(c)** \$81,969; **(d)** The model is adequate; **(e)** $F = 223.46 > 3.35$, reject H_0; **(f)** p-value = 0 so the model is significant. **(g)** 94.3% of the variation in selling price is explained by the variation in assessed value and time since reassessment. **(h)** 93.9%; **(i)** Since $t = 20.41 > 2.0518$, reject H_0. Since $t = 2.87 > 2.0518$, reject H_0. Use a model that includes both variables. **(j)** 0.000 and 0.008 < 0.05, so both variables are significant; **(k)** $1.574 \le \beta_1 \le 1.927$; **(l)** 0.939, 0.234

14.70 (a) predicted Value = 63.775 + 10.725 Size – 0.284 Age; **(c)** \$79,702; **(d)** The residual plot against age indicates a potential pattern. **(e)** $F = 28.58 > 3.89$, reject H_0. **(f)** p-value = 0 indicating the model is significant. **(g)** 82.6% of the variation in assessed value is explained by the variation in size and the age of the home. **(h)** 79.8%; **(i)** Since $t = 3.56 > 2.1788$, reject H_0. Since $t = -3.40 < -2.1788$, reject H_0. **(j)** p-values = 0.004 and 0.005 < 0.05, so both variables are significant. **(k)** $4.158 \le \beta_1 \le 17.292$; **(l)** 0.513, 0.491; **(m)** no

14.72 (a) predicted MPG = 40.877 – 0.0121 Length – 0.00495 Weight; **(c)** 23.66 miles per gallon; **(d)** A quadratic model should be fit. There is some evidence of unequal variance. **(e)** $F = 92.93 > 3.07$, reject H_0. **(f)** p-value = 0 indicating a significant model. **(g)** 61.2% of the variation in miles per gallon is explained by the variation in car length and weight. **(h)** 60.5%; **(i)** Since $t = -0.46 > -1.9799$, do not reject H_0. Since $t = -10.24 < -1.9799$, reject H_0. Use only weight in the model. **(j)** Length is not significant since the p-value for length = 0.647. Weight is significant since the p-value for weight = 0; **(k)** $-0.00591 \le \beta_2 \le -0.00400$; **(l)** 0.00186, 0.471

14.74 (a) $\hat{Y} = 152.0316 - 2.4587X_1 - 15.8687X_2$, where X_1 = league dummy (0 = American, 1 = National) and X_2 = ERA. **(b)** Holding constant the effect of ERA, the estimated number of wins for a team in the American league is 2.4587 above that of the National league. Holding constant the effect of league, for each unit increase in ERA, the number of wins is estimated to decrease by 15.8687. **(c)** 81wins. **(d)** Based on the residual analysis, the model appears adequate. **(e)** $F = 11.4784$ and p-value = 0.00025 < 0.05. Reject H_0. There is evidence of a relationship between number of wins and the two independent variables. **(f)** For X_1: $t = -0.6447$. p-value = 0.5246 > 0.05. Do not reject H_0. Which league the team is in does not make a significant contribution and should not be included in the model. For X_2: $t = -4.7763$. p-value is virtually zero < 0.05. Reject H_0. ERA makes a significant contribution and should be included in the model. The model should include ERA but not the league. **(g)** $-10.2842 \le \beta_1 \le 5.3668$, $-22.6856 \le \beta_2 \le -9.0518$; **(h)** $r^2_{Y.12} = 0.4595$. 45.95% of the variation in number of wins can be explained by variation in type of league and variation in ERA. **(i)** $r^2_{adj} = 0.4195$; **(j)** $r^2_{Y1.2} = 0.0152$. Holding constant the effect of ERA, 1.52% of the variation in number of wins can be explained by variation in the league.

$r^2_{Y2.1} = 0.4580$. Holding constant the effect of league, 45.80% of the variation in number of wins can be explained by variation in ERA. **(k)** The slope of number of wins with ERA is the same regardless of whether the team is in the American or National league. **(l)** $\hat{Y} = 162.6777 - 21.8842X_1 - 18.2211X_2 + 4.4027X_1X_2$. For X_1X_2: the p-value is 0.5189. Do not reject H_0. There is no evidence that the interaction term makes a contribution to the model. **(m)** The model with only ERA should be used.

CHAPTER 15

15.2 (b) $\hat{Y} = -7.556 + 1.2717X - 0.0145X^2$; **(c)** $\hat{Y} = -7.556 + 1.2717(55) - 0.0145(55)^2 = 18.52$; **(d)** Based on residual analysis, there are patterns in the residuals vs. highway speed, vs. the quadratic variable (speed squared), and vs. the fitted values. **(e)** $F = 141.46 > F_{2,25} = 3.39$. Reject H_0. The overall model is significant. The p-value < 0.001. **(f)** $t = -16.63 < -2.0595$. Reject H_0. The quadratic effect is significant. The p-value $<$ 0.001. **(g)** $r^2 = 0.919$. So, 91.9% of the variation in miles per gallon can be explained by the quadratic relationship between miles per gallon and highway speed. **(h)** $r^2_{adj} = 0.912$

15.4 (b) predicted Yield = 6.643 + 0.895 AmtFert – 0.00411 AmtFert^2; **(c)** 49.168 pounds; **(d)** the model appears adequate; **(e)** $F = 157.32 > 4.26$, reject H_0; **(f)** p-value $= 0 < .05$, so the model is significant. **(g)** $t = -4.27 < -2.2622$, reject H_0; **(h)** p-value $= 0.002 < .05$, so the quadratic term is significant. **(i)** 97.2%; **(j)** 96.6%

15.6 (a) 211.84; **(b)** For each additional unit of the logarithm of X_1, the logarithm of Y is estimated to increase by 0.9 units, holding all other variables constant. For each additional unit of the logarithm of X_2, the logarithm of Y is estimated to increase by 1.41 units, holding all other variables constant.

15.8 (a) predicted MPG = 9.036 + 0.852 $\sqrt{\text{MPG}}$; **(b)** 15.35 MPG; **(c)** a quadratic pattern exists so the model is not adequate; **(d)** $t = 1.35 < 2.0555$, do not reject H_0; **(e)** 6.6%; **(f)** 3.0%; **(g)** choose model from problem 15.2

15.10 (a) predicted ln(Yield) = 2.475 + 0.0185 AmtFert; **(b)** 32.95 pounds; **(c)** A quadratic pattern exists so the model is not adequate. **(d)** $t = 6.11 > 2.2281$, reject H_0; **(e)** 78.9%; **(f)** 76.8%; **(g)** choose model from problem 15.4

15.12 Observations 14 ($h_i = 0.3568$) and 19 ($h_i = 0.2828$) are influential as they have $h_i > 2(2+1)/24 = 0.25$. Observations 1, 2, and 14 are possible outliers: $|t_1^*| = 2.5783$, $|t_2^*| = 1.8465$, and $|t_{14}^*| = 1.8429$. All of these values $> t_{24-2-2,.05} = 1.7247$. The largest value for Cook's D_i is $0.5637 < F_{.50,2+1,24-2-1} = 0.8149$. Thus, there is insufficient evidence to delete any observations in the model.

15.14 Largest Cook's $D_i = 0.652 < 0.8177$ but this observation (case 2) is also influential as well as case 13.

15.16 Observations 8 ($h_i = 0.2370$), 18 ($h_i = 0.3449$), and 26 ($h_i = 0.2285$) are influential as they have $h_i > 2(2 + 1)/30 = 0.20$. Observations 8, 9, 10, and 14 are possible outliers: $|t_8^*| = 2.06$, $|t_9^*| = 2.09$, $|t_{10}^*| = 3.40$, and $|t_{14}^*| = 2.24$. All of these values $> t_{30-2-2,.05} = 1.7056$. The largest value for Cook's D_i is 0.5475 for Observation 10 $< F_{.50,2+1,30-2-1} = 0.8089$. Using the three criteria, there is insufficient evidence for removal of any observation from the model.

15.18 1.25

15.20 $R_1^2 = 0.64$, $VIF_1 = \dfrac{1}{1-0.64} = 2.778$, $R_2^2 = 0.64$,

$$VIF_2 = \frac{1}{1-0.64} = 2.778$$

There is no evidence of collinearity.

15.22 $VIF = 1.0 < 5$. There is no evidence of collinearity.

15.24 $VIF = 1.0428$. There is no evidence of collinearity.

15.26 (a) 35.04; **(b)** $C_p > 3$, This does not meet the criterion for consideration of a good model.

15.28 Let Y = selling price, X_1 = assessed value, X_2 = time period, and X_3 = whether house was new (0 = no, 1 = yes). Based on a full regression model involving all of the variables, all of the VIF values (1.3, 1.0, and 1.3, respectively) are less than 5. There is no reason to suspect the existence of collinearity.

Based on a best subsets regression and examination of the resulting C_p values, the best models appear to be a model with variables X_1 and X_2, which has $C_p = 2.8$, and the full regression model, which has $C_p = 4.0$. Based on a regression analysis with all original variables, variable X_3 fails to make a significant contribution to the model at the 0.05 level. Thus, the best model is the model using assessed value (X_1) and time (X_2) as the independent variables. A residual analysis shows no strong patterns.

The final model is: $\hat{Y} = -44.9882 + 1.7506X_1 + 0.3680X_2$, $r^2 = 0.9430$, $r^2_{adj} = 0.9388$. Overall significance of the model: $F = 223.4575$, $p < 0.001$. Each independent variable is significant at the 0.05 level. A stepwise regression in Minitab yields the same model.

15.30 VIF for weight $= 5.1 > 5$ indicates collinearity. Therefore, you remove weight from the model. Best subsets regression selects a best model that includes width, length, and SUV. The residual plot indicates quadratic effects for length and width. Both of these were significant. The model is $\hat{Y} = 236.01 - 0.2072X_2 - 2.0024X_3 - 4.0526X_4 - 0.00000026X_2^2 + 0.0051X_3^2$; $r^2 = 0.693$.

15.32 The most appropriate model is $\hat{Y} = 326.8080 - 23.8570X_1 - 10.2344X_3$, $r^2 = 0.452$.

15.38 The most appropriate model includes only gate receipts as an independent variable. $\hat{Y} = 69.0198 + 0.2582X$; $r^2 = 0.202$

15.40 (a) Best Model: predicted Appraised Value = 136.794 + 276.0876 Land + 0.1288 House Size(sq ft) – 1.3989 Age

15.42 (a) predicted Appraised Value = 110.27 + 0.0821 House Size(sq ft)

15.44 Let Y = appraised value, X_1 = land area, X_2 = interior size, X_3 = age, X_4 = number of rooms, X_5 = number of bathrooms, X_6 = garage size, X_7 = 1 if Glen Cove and 0 otherwise, X_8 = 1 if Roslyn and 0 otherwise.
(a) All VIFs are less than 5 in a full regression model involving all of the variables: There is no reason to suspect collinearity between any pair of variables. The following is the multiple regression model that has the smallest C_p (9.0) and the highest adjusted r^2 (0.891):

Appraised Value = 49.4 + 343 Land (acres) + 0.115 House Size (sq ft) - 0.585 Age - 8.24 Rooms + 26.9 Baths + 5.0 Garage + 56.4 Glen Cove + 210 Roslyn

The individual t test for the significance of each independent variable at the 5% level of significance concludes that only X_1, X_2, X_5, X_7, and X_8 are significant individually. This subset, however, is not chosen when the C_p criterion is used.

The following is the multiple regression output for the model chosen by stepwise regression:

Appraised Value = 23.4 + 347 Land (acres) + 0.106 House Size (sq ft) - 0.792 Age + 26.4 Baths + 57.7 Glen Cove + 213 Roslyn

This model has a C_p value of 7.7 and an adjusted r^2 of 89.0. All the variables are significant individually at 5% level of significance. Combining the stepwise regression and the best subset regression results along with the individual t-test results, the most appropriate multiple regression model for predicting the appraised value is:

$$\hat{Y} = 23.40 + 347.02X_1 + 0.10614X_2 - 0.7921X_3 + 26.38X_5 + 57.74X_7 + 213.46X_8$$

(b) The estimated mean appraised value in Glen Cove is 57.74 above Freeport for two otherwise identical properties. The estimated mean appraised value in Roslyn is 213.46 above Freeport for two otherwise identical properties.

15.48 Explanation: remove vapor pressure, air temperature, soil radiation, wind speed due to $VIF > 5$. After removing insignificant variables, the model has the single independent variable, dew point, and a coefficient of determination of only 10%.

15.50 (a) Let X_1 = Temp, X_2 = Win%, X_3 = OpWin%, X_4 = Weekend, X_5 = Promotion **(b)** $\hat{Y} = -3{,}862.481 + 51.703X_1 + 21.108X_2 + 11.345X_3 + 367.538X_4 + 6{,}927.882X_5$ **(c)** Intercept: Since all the non-dummy independent variables cannot have zero values, the intercept cannot be interpreted. Temp: As the high temperature increases by one degree, the mean paid attendance is estimated to increase by 51.70, taking into consideration all the other independent variables included in the model. Win%: As the winning percentage of the team improves by 1%, the mean paid attendance is estimated to increase by 21.11, taking into consideration all the other independent variables included in the model. OpWin%: As the opponent team's winning percentage at the time of the game improves by 1%, the mean paid attendance is estimated to increase by 11.35, taking into consideration all the other independent variables included in the model. Weekend: The mean paid attendance of a game played on a weekend is estimated to be 367.54 higher than when the game is played on a weekday, taking into consideration all the other independent variables included in the model. Promotion: The mean paid attendance on a promotion day is estimated to be 6,927.88 higher than when there is no promotion, taking into consideration all the other independent variables included in the model. **(d)** At a 0.05 level of significance, the independent variable that makes a significant contribution to the regression model individually is the promotion dummy variable. **(e)** Adjusted $r^2 = 0.2538$. 25.38% of the variation in attendance can be explained by the 5 independent variables after adjusting for the number of independent variables and the sample size. **(f)** The residual plots of temperature, team's winning percentage and opponent team's winning percentage reveal potential violation of the equal variance assumption. The normal probability plot also reveals non-normality in the residuals. **(g)** With all the 5 independent variables in the model: None of the VIF is > 5. Based on the smallest C_p value and the highest adjusted r^2, the best model includes Win percentage, opponent's win percentage, and promotion. Since only X_5 makes a significant contribution to the regression model at the 5% level of significance, the more parsimonious model includes only X_5. **(b)** $\hat{Y} = 13{,}935.703 + 6{,}813.228X_5$; **(c)** Intercept: The estimated mean paid attendance on non-promotion days is 13,935.70. Promotion: The estimated mean paid attendance on promotion days will be 6,813.23 higher than when there is no promotion. **(d)** At the 0.05 level of significance, promotion makes a significant contribution to the regression model. **(e)** $r^2 = 0.2101$. So, 21.02% of the variation in attendance can be explained by promotion. **(f)** The residual plot reveals non-normality in the residuals.

15.52 (a) Let X_1 = Temp, X_2 = Win%, X_3 = OpWin%, X_4 = Weekend X_5 = Promotion **(b)** $\hat{Y} = 10{,}682.455 + 82.205X_1 + 26.263X_2 + 7.367X_3 + 3{,}369.908X_4 + 3{,}129.013X_5$ **(c)** Intercept: Since all the non-dummy independent variables cannot have zero values, the intercept cannot be interpreted. Temp: As the high temperature increases by one degree, the mean paid attendance is estimated to increase by 82.21, taking into consideration all the other independent variables included in the model. Win%: As the winning percentage of the team improves by 1%, the mean paid attendance is estimated to increase by 26.26, taking into consideration all the other independent variables included in the model. OpWin%: As the opponent team's winning percentage at the time of the game increases by 1%, the mean paid attendance is estimated to increase by 7.37, taking into consideration all the other independent variables included in the model. Weekend: The mean paid attendance of a game played on a weekend is estimated to be 3,369.91 higher than when the game is played on a weekday, taking into consideration all the other independent variables included in the model. Promotion: The mean paid attendance on a promotion day is estimated to be 3,129.01 higher than when there is no promotion, taking into consideration all the other independent variables included in the model. **(d)** At the 0.05 level of significance, the independent variables that make a significant contribution to the regression model individually are temperature, team's winning percentage, and the weekend and promotion dummy variables. **(e)** Adjusted $r^2 = 0.3504$. So, 35.04% of the variation in attendance can be explained by the 5 independent variables after adjusting for the number of independent variables and the sample size. **(f)** The residual plots of temperature, team's winning percentage and opponent team's winning percentage do not show serious departure from the equal variance assumption. The normal probability plot of the residuals does not show evidence of serious departure from normality. **(g)** With all the 5 independent variables in the model: None of the VIF is > 5. Based on the smallest C_p value and the highest adjusted r^2, the best model is the full model that includes all the independent variables. Since only X_3 does not make significant contribution to the regression model at 5% level of significance, the more parsimonious model includes X_1, X_2, X_4, and X_5. **(b)** $\hat{Y} = 14{,}965.626 + 88.888X_1 + 23.269X_2 + 3{,}562.425X_4 + 3{,}029.087X_5$
(c) Intercept: Since all the non-dummy independent variables cannot have zero values, the intercept cannot be interpreted. Temp: As the high temperature increases by one degree, the mean paid attendance is estimated to increase by 88.89, taking into consideration all the other independent variables included in the model. Win%: As the winning percentage of the team increases by 1%, the mean paid attendance is estimated to increase by 23.27, taking into consideration all the other independent variables included in the model. Weekend: The mean paid attendance of a game played on a weekend is estimated to be 3,562.53 higher than when the game is played on a weekday, taking into consideration all the other independent variables included in the model. Promotion: The mean paid attendance on a promotion day is estimated to be 3,029.09 higher than when there is no promotion, taking into consideration all the other independent variables included in the model. **(d)** At the 0.05 level of significance, temperature, and the weekend and promotion dummy variables make a significant contribution to the regression model individually. **(e)** Adjusted $r^2 = 0.3457$. So, 34.57% of the variation in attendance can be explained by the 4 independent variables after adjusting for the number of independent variables and

the sample size. **(f)** The residual plots of temperature and team's winning percentage do not show serious departure from equal variance assumption The normal probability plot of the residuals also does not show evidence of serious departure from normality.

15.54 The Philadelphia Phillies ran the most effective promotions in 2002. An estimated additional 11,184.54 fans attended on days when a promotion was held.

CHAPTER 16

16.2 (a) 1959; **(b)** first four years and the last four years

16.4 (b), (c), (d)

Year	Price	MA(3)	ES ($W = 0.5$)	ES ($W = 0.25$)
1996	3.03		3.0300	3.0300
1997	3.05	3.0500	3.0400	3.0350
1998	3.07	3.0567	3.0550	3.0438
1999	3.05	3.0433	3.0525	3.0453
2000	3.01	3.0267	3.0313	3.0365
2001	3.02	2.9767	3.0256	3.0324
2002	2.90	2.8800	2.9628	2.9993
2003	2.72	2.8800	2.8414	2.9295
2004	3.02		2.9307	2.9521

(e) The exponentially smoothed series with $W = 0.5$ is generally lower than that with $W = 0.25$ in the more recent years. The exponential smoothing with $W = 0.5$ assigns more weight to the more recent values and is better for forecasting. The exponential smoothing with $W = 0.25$, which assigns less weight to more recent values, is better suited for eliminating unwanted cyclical and irregular variations.

16.6 (b), (c), (d)

Week	Nasdaq	MA(3)	ES ($W = 0.5$)	ES ($W = 0.25$)
2-Jan	2007		2007.00	2007.00
9-Jan	2087	2078.00	2047.00	2027.00
16-Jan	2140	2137.00	2093.50	2055.25
23-Jan	2184	2130.00	2138.75	2087.44
30-Jan	2066	2104.67	2102.38	2082.08
6-Feb	2064	2061.33	2083.19	2077.56
13-Feb	2054	2052.00	2068.59	2071.67
20-Feb	2038	2040.67	2053.30	2063.25
27-Feb	2030	2038.67	2041.65	2054.94
5-Mar	2048	2021.00	2044.82	2053.20
12-Mar	1985	1991.00	2014.91	2036.15
19-Mar	1940	1961.67	1977.46	2012.11
26-Mar	1960	1985.67	1968.73	1999.09
2-Apr	2057	2023.33	2012.86	2013.56
8-Apr	2053	2035.33	2032.93	2023.42
16-Apr	1996	2033.00	2014.47	2016.57
23-Apr	2050	1988.67	2032.23	2024.93
30-Apr	1920	1962.67	1976.12	1998.69
7-May	1918	1914.00	1947.06	1978.52
14-May	1904	1911.33	1925.53	1959.89
21-May	1912	1934.33	1918.76	1947.92
28-May	1987	1959.33	1952.88	1957.69
4-Jun	1979	1988.67	1965.94	1963.02
10-Jun	2000	1992.33	1982.97	1972.26
18-Jun	1998		1990.49	1978.70

(e) There is a general downward trend after January 23, 2004, but recovered after May 21, 2004.

16.8 (b), (c), (e)

Year	Cost	MA(3)	ES ($W = 0.5$)	ES ($W = 0.25$)
1994–1995	8.16		8.16	8.16
1995–1996	8.10	8.14	8.13	8.15
1996–1997	8.17	8.13	8.15	8.15
1997–1998	8.12	8.08	8.14	8.14
1998–1999	7.94	8.01	8.04	8.09
1999–2000	7.98	8.01	8.01	8.06
2000–2001	8.11	8.15	8.06	8.08
2001–2002	8.37	8.23	8.21	8.15
2002–2003	8.20	8.35	8.21	8.16
2003–2004	8.49		8.35	8.24

(d) $W = 0.50$: $\hat{Y}_{2004\text{–}2005} = E_{2003\text{–}2004} = 8.35$ **(e)** $W = 0.25$: $\hat{Y}_{2004\text{–}2005} = E_{2003\text{–}2004} = 8.24$ **(f)** The exponentially smoothed forecast for 2003–2004 with $W = 0.5$ is higher than that with $W = 0.25$.

16.10 (a) The Y-intercept $b_0 = 4.0$ is the fitted-trend value reflecting the real total revenues during the base year 1985. **(b)** The slope $b_1 = 1.5$ indicates that the real total revenues are increasing at a rate of 1.5 million dollars per year. **(c)** \$10 million; **(d)** \$32.5 million; **(e)** \$37 million

16.12 (b) There has been an upward trend in the CPI in the United States over the 39-year period. The rate of increase accelerated in the late 70s and mid-80s, but the rate of increase tapered off in the early 80s and early 90s.

16.14 (b) $\hat{Y} = 296.1348 + 67.0448X$; **(c)** \$1,972.254 billion and \$2,039.299 billion; **(d)** There is an upward trend in federal receipts between 1978 and 2002. The trend appears to be nonlinear.

16.16 (b) Linear trend: $\hat{Y} = 1.1414 + 0.7288X$.
(c) Quadratic trend: $\hat{Y} = 2.4098 + 0.4469X + 0.0101X^2$.
(d) Exponential trend: $\log_{10} \hat{Y} = 0.5319 + 0.0318X$.
(e) Linear trend: \$22.2771 and \$23.0059 billion, Quadratic trend: \$23.8374 and \$24.8783 billion, Exponential trend: \$28.5329 and \$30.7036 billion

16.18 (b) Linear trend: $\hat{Y} = -19.8639 + 2.6699X$
(c) Quadratic trend: $\hat{Y} = 10.4629 - 3.0164X + 0.1723X^2$
(d) Exponential trend: $\log_{10} \hat{Y} = 0.0075 + 0.0592X$
(e) The exponential trend does seem to fit the data better, especially in the early years. **(f)** Using the exponential trend model, $\hat{Y}_{2004} = \$105.07$

16.20 (b) predicted $Y_{\text{Series I}} = 100.082 + 14.975X$, predicted $\log_{10} Y_{\text{Series II}} = 1.999 + 0.0607X$; **(c)** series I: 249.834; series II: 403.709

16.22 (d) predicted $Y = 50.673 + 1.026X$; **(e)** predicted $Y = 50.532 + 1.132X - 0.0117X^2$; **(f)** predicted $\log_{10}Y = 1.706 + .0081X$; **(g)** part (d) 60.933, part (e) 60.628, part (f) 61.235

16.24 25.30, 27.43, 29.56, 31.69, 33.82

16.26 (a) $U = 0.3$, $V = 0.3$
$\hat{Y}_{2004} = E_{2003} + 1T_{2003} = 10{,}955.8956 + (1)(420.0791) = \$11{,}375.9747$ billion
$\hat{Y}_{2005} = E_{2003} + 2T_{2003} = 10{,}955.8956 + (2)(420.0791) = \$11{,}796.0538$ billion.

(b) $U = 0.7$, $V = 0.7$
$\hat{Y}_{2004} = E_{2003} + 1T_{2003} = 10{,}958.4446 + (1)(457.5822) = \$11{,}416.0268$ billion
$\hat{Y}_{2005} = E_{2003} + 2T_{2003} = 10{,}958.4446 + (2)(457.5822) = \$11{,}873.6090$ billion.
(c) $U = 0.3$, $V = 0.7$
$\hat{Y}_{2004} = E_{2003} + 1T_{2003} = 10{,}971.1321 + (1)(431.5698) = \$11{,}402.7019$ billion
$\hat{Y}_{2005} = E_{2003} + 2T_{2003} = 10{,}971.1321 + (2)(431.5698) = \$11{,}834.2718$ billion.
(d) Given the historical movement of the time series, which suggests that there is a cyclical component in addition to the upward trend, a better model would give more weight to the more recent levels and trends. Hence, the forecast in model (a) will be a better choice. **(e)** The forecasts in (a)–(c) are higher than those of problem 16.13 (c). The GDP has been on the expansionary path since 1992. This is being captured and reflected in the Holt-Winters method with $U = 0.3$ and $V = 0.3$, which gives more weight to recent levels and trends of the time series, but not in the linear trend, which is very much constrained by the model specification. The linear trend model will be more appropriate if the forecasts are longer-run than the immediate short-run forecasts.

16.28 (a) $U = 0.3$, $V = 0.3$, 20.9886 and 21.3941 billions of constant 1982–1984 dollars. **(b)** $U = 0.7$, $V = 0.7$, 22.1441 and 22.5256 billions of constant 1982–1984 dollars. **(c)** $U = 0.3$, $V = 0.7$, 21.1180 and 21.4811 billions of constant 1982–1984 dollars. **(d)** Given the historical movement of the time series, which suggests that there is a cyclical component in addition to the upward trend, a better model will be to give more weight to the more recent levels and trends. Hence, the forecast in model (a) will be a better choice. **(e)** The real operating revenue of Coca-Cola has experienced contraction since 1996. This is being captured and reflected in the Holt-Winters method with $U = 0.3$ and $V = 0.3$, which gives more weight to recent levels and trends of the time series, but not in the three trend models in problem 16.15(e). The Holt-Winters method with $U = 0.7$ and $V = 0.7$, and $U = 0.3$ and $V = 0.7$, which gives more weight to past trends behaves more like the trend models.

16.30 (a) $U = 0.3$, $V = 0.3$, 81.6566; **(b)** $U = 0.7$, $V = 0.7$, 107.0478; **(c)** $U = 0.3$, $V = 0.7$, 89.8731; **(d)** Given the historical movement of the time series, which suggests that the stock price increased at an increasing rate up until 1999, experienced a drastic decline in values in 2000, and has since shown signs of recovery, a better model would give more weight to the more recent levels and past trend. Since there is strong evidence from the last two years on a likely recovery back to the level prior to the decline in 1999, a Holt-Winters method with $U = 0.3$ and $V = 0.7$ that assigns more weight to the recent values and past trend should be used. **(e)** The forecasts in (a) and (c) are lower and the forecast in (b) is higher than that in the exponential trend model of problem 16.18 (f), reflecting the fact that the exponential trend model is very much restricted by its model specification while the Holt-Winters method with $U = 0.3$ and $V = 0.3$ is more capable of capturing the more recent drop in price and downward adjustment in trend. The Holt-Winters method with $U = 0.3$ and $V = 0.7$ is more capable of capturing the more recent drop in price but a potential recovery back to the past trend before the downturn in 2000.

16.32 $t = 2.40 > 2.2281$, reject H_0.

16.34 (a) $t = 1.60 < 2.2281$, do not reject H_0.

16.36 (a) Since p-value $= 0.83 > 0.05$, do not reject H_0: $A_3 = 0$. Third-order term can be deleted. **(b)** Since p-value $= 0.14 > 0.05$, do not reject H_0: $A_2 = 0$. Second-order term can be deleted. **(c)** Since p-value is approximately zero, reject H_0: $A_1 = 0$. A first-order autoregressive model is appropriate. $\hat{Y}_i = 0.5905 + 1.0051Y_{i-1}$ **(d)** 21.6973 and 22.3981

16.38 (a) Since p-value $= 0.73 > 0.05$, do not reject H_0: $A_3 = 0$. Third-order term can be deleted. **(b)** Since p-value $= 0.44 > 0.05$, do not reject H_0: $A_2 = 0$. Second-order term can be deleted. **(c)** Since p-value is approximately zero, reject H_0 that $A_1 = 0$. A first-order autoregressive model is appropriate. $\hat{Y}_i = 2.0644 + 1.0249Y_{i-1}$ **(d)** 92.0576 and 96.4107

16.39 (a) Since the p-value $= 0.20 > 0.05$, do not reject H_0: $A_3 = 0$. Third-order term can be deleted. **(b)** Since the p-value $= 0.998 > 0.05$, do not reject H_0: $A_2 = 0$. Second-order term can be deleted. **(c)** Since the p-value is virtually 0, reject H_0: $A_1 = 0$. First-order autoregressive model is the appropriate model. **(d)** $\hat{Y}_i = 4.8248 + 0.9308Y_{i-1}$. Forecasts are 60.2705123, 60.92171152, and 61.52781659.

16.40 (a) 2.121; **(b)** 1.515

16.42 The residuals in the linear trend model show strings of consecutive positive and negative values. **(b)** 256.6194; **(c)** 214.9249; **(d)** The residuals in the linear trend model show strings of consecutive positive and negative values. The linear trend model is inadequate in capturing the nonlinear trend.

16.44 (b) S_{YX} for linear, quadratic, exponential trend models, three Holt-Winters models and first-order autoregressive models are: 1.3948, 1.2565, 1.9068, 0.2211, 0.9346, 0.2342, 0.6834. **(c)** *MAD* for linear quadratic, exponential trend models, three Holt-Winters models and first-order autoregressive models are: 1.1311, 0.8981, 1.1808, 0.1748, 0.7168, 0.1766, and 0.4775. **(d)** The residuals in the three trend models show strings of consecutive positive and negative values. The Holt-Winters method with $U = 0.7$, $V = 0.7$ and $U = 0.3$, $V = 0.7$ also shows consecutive positive and negative values. The Holt-Winters method with $U = 0.3$, $V = 0.3$ and the autoregressive model perform well for the historical data and has a fairly random pattern of residuals. The Holt-Winters method with $U = 0.3$, $V = 0.3$ also has the smallest values in *MAD*. The autoregressive model would probably be the best model for forecasting.

16.46 (a) The residuals in the three trend models show strings of consecutive positive and negative values. The Holt-Winters method with $U = 0.7$, $V = 0.7$ also shows consecutive positive and negative values. The Holt-Winters method with $U = 0.3$ and $V = 0.3$, $U = 0.3$ and $V = 0.7$, and the autoregressive model perform well for the historical data and has a fairly random pattern of residuals.
(b) S_{YX} for the linear trend model is 18.13, for the quadratic trend model is 9.94, for the exponential trend model is 12.39, for the Holt Winters model with $U = 0.3$, $V = 0.3$ is 2.96, $U = 0.7$, $V = 0.7$ is 8.26, $U = 0.3$, $V = 0.7$ is 2.86, and for the first-order autoregressive model is 8.31.
(c) *MAD* for the linear trend model is 15.11, for the quadratic trend model is 6.93, for the exponential trend model is 5.34, for the Holt Winters model with $U = 0.3$, $V = 0.3$ is 1.16, $U = 0.7$, $V = 0.7$ is 4.53, $U = 0.3$, $V = 0.7$ is 1.20, and for the first-order autoregressive model is 4.28.
(d) The Holt-Winters method with $U = 0.3$ and $V = 0.3$ has the smallest values in *MAD* while the Holt-Winters method with $U = 0.3$ and $V = 0.7$ has the smallest S_{YX}. So either the Holt-Winters method with $U = 0.3$ and $V = 0.3$, or $U = 0.3$ and $V = 0.7$ would be a good model for forecasting.

16.47 (b) S_{YX} for linear, quadratic, exponential trend models, and first-order autoregressive models are: 0.3773, 0.3850, 0.4071, and 0.4765.
(c) *MAD* for linear quadratic, exponential trend models, and first-order autoregressive models are: 0.2702, 0.2644, 0.2902, and 0.3308. **(d)** The

residuals in all the three trend models contain consecutive positive and negative values. The first-order autoregressive model has a fairly random pattern of residuals. The quadratic trend model has the smallest *MAD* while the linear trend model has the smallest S_{YX}. The first-order autoregressive model would be the best model for forecasting.

16.48 (a) 100; **(b)** 1.023; **(c)** 1.259

16.50 (a) 1000; **(b)** 1.259; **(c)** 1.585

16.52 (b) $\log_{10}\hat{Y} = 2.7841 + 0.0090X - 0.0128Q_1 - 0.0062Q_2 - 0.0200Q_3$ **(c)** 1,356.0097; **(d)** 1,405.8415; **(e)** 2004: 1,390.2687 and 1,486.5076 2005: 1,473.5496, 1,527.7008, 1,510.7782, and 1,615.3591; **(f)** The estimated quarterly compound growth rate is 2.10%. **(g)** The second quarter values in the time series are estimated to be on average 1.41% below the fourth quarter values.

16.54 (b)

$$\begin{aligned}\log\hat{Y} = {} & 0.656160 + 0.000751X + 0.003547M_1 \\ & + 0.003190M_2 + 0.000033M_3 + 0.000999M_4 \\ & - 0.001073M_5 + 0.000384M_6 - 0.001428M_7 \\ & - 0.002460M_8 - 0.004697M_9 - 0.003683M_{10} \\ & - 0.000496M_{11}\end{aligned}$$

(c) 5.34%; **(d)** 2004:5.39%, 5.40%, 5.37%, 5.39%, 5.37%, 5.40%, 5.39%, 5.38%, 5.36%, 5.39%, 5.44%, and 5.45%. **(e)** The estimated monthly compound growth rate is $(b_1 - 1)$ 100% = 0.1731%. **(f)** The July values in the time series are estimated to be 0.33% below the December values.

16.55 (b) The amount of charges in January and February are among the lowest in a year. The amount of charges in September is usually lower than in August and October. The largest amount of charges in a year appears in December. **(c)** In general, there is an upward trend, and, hence, the dollar amounts charged on the bank's credit cards is increasing. **(d)** The large difference between the December 2002 charges and the February 2003 charges is as expected based on the monthly pattern described in (b).

(e)

$$\begin{aligned}\log\hat{Y} = {} & 1.6803 + 0.006138X - 0.1698M_1 \\ & - 0.2308M_2 - 0.1779M_3 - 0.1652M_4 \\ & - 0.1088M_5 - 0.1178M_6 - 0.08193M_7 \\ & - 0.04657M_8 - 0.09539M_9 - 0.04955M_{10} \\ & - 0.04081M_{11}\end{aligned}$$

(f) $\log b_1 = 0.0061375$. $b_1 = 10^{0.0061375} = 1.01423$. The estimated monthly compound growth rate is $(b_1 - 1)$ 100% = 1.423%. **(g)** $\log b_2 = -0.1697687$. $b_2 = 10^{-0.1697687} = 0.67644$. The January values in the time series are estimated to be 32.36% below the December values. **(h)** March of 2003: $X = 26$, $M_3 = 1$, $\hat{Y}_{27}$ = \$45.9171 millions; **(i)** April of 2003: $X = 27$, $M_4 = 1$, $\hat{Y}_{28}$ = \$47.9538 millions; **(j)** This classical multiplicative time-series model enables the bank to predict more accurately the amount of charges on its credit cards for each of the 12 months of a year. The bank can then plan to allocate its resources more effectively to reflect the seasonal fluctuation.

16.56 (a) The retail industry is heavily subject to seasonal variation due to the holiday season and so are the revenues for Toys Я Us. **(b)** There is an obvious seasonal effect in the time series. **(c)** $\hat{Y}_i = 3007.9022(1.0149)^{X_i} (0.4051)^{Q_1} (0.4008)^{Q_2} (0.4322)^{Q_3}$ **(d)** The estimated quarterly compound growth rate is 1.49%. **(e)** The first quarter values in the time series are estimated to be 59.49% below the fourth quarter values. The second quarter values in the time series are estimated to be 59.92% below the fourth quarter values. The third quarter values in the time series are estimated to be 56.78% below the fourth quarter values. **(f)** Forecasts for 2004: 2,480.3268, 2,490.5146, 2,725.5678, and 6,400.7917.

16.58 The price of the commodity is 75% higher in 2002 than in 1995.

16.60 (a) 186.96; **(b)** 162.16; **(c)** 154.42

16.62 (a), (b)

Year	DJIA	Price Index (base = 1979)	Price Index (base = 1990)
1979	838.7	100.00	31.84
1980	964	114.94	36.60
1981	875	104.33	33.22
1982	1,046.5	124.78	39.73
1983	1,258.6	150.07	47.79
1984	1,211.6	144.46	46.00
1985	1,546.7	184.42	58.73
1986	1,896	226.06	71.99
1987	1,938.8	231.17	73.62
1988	2,168.6	258.57	82.34
1989	2,753.2	328.27	104.54
1990	2,633.7	314.02	100.00
1991	3,168.8	377.82	120.32
1992	3,301.1	393.60	125.34
1993	3,754.1	447.61	142.54
1994	3,834.4	457.18	145.59
1995	5,117.1	610.12	194.29
1996	6,448.3	768.84	244.84
1997	7,908.3	942.92	300.27
1998	9,181.4	1094.72	348.61
1999	11,497.1	1370.82	436.54
2000	10,788	1286.28	409.61
2001	10,021.5	1194.88	380.51

(c) The price index using 1990 as the base year is more useful because it is closer to the present and the DJIA has grown more than 1,000% over the 23-year period.

16.64 (a), (b)

Year	CPI	Price Index (base = 1990)	Price Index (base = 2003)
1990	129.9	100.00	71.61
1991	135.7	104.46	74.81
1992	139.2	107.16	76.74
1993	141.9	109.24	78.22
1994	146	112.39	80.49
1995	150.7	116.01	83.08
1996	154.4	118.86	85.12
1997	160	123.17	88.20
1998	164.4	126.56	90.63
1999	167.3	128.79	92.23
2000	172.2	132.56	94.93
2001	173.4	133.49	95.59
2002	176.3	135.72	97.19
2003	181.4	139.65	100.00

(c) Both price indices are useful. The one using 1990 as the base year conveys a picture of how the CPI has grown since 1990. The one using 2003 as the base year reveals what the CPI in prior years was as a percentage of the current level. Since the price index is usually used to compare the growth of price from some base year in the past, the price using 1990 as the base is more useful. **(d)** The CPI in UK has grown 39.65% from 1990 to 2003 compared to the 1.45% growth in Japan over the same period.

16.66 (a), (c)

Year	Price	Price Index (base = 1980)	Price Index (base = 1990)
1980	0.703	100.00	40.52
1981	0.792	112.66	45.65
1982	0.763	108.53	43.98
1983	0.726	103.27	41.84
1984	0.854	121.48	49.22
1985	0.697	99.15	40.17
1986	1.104	157.04	63.63
1987	0.943	134.14	54.35
1988	0.871	123.90	50.20
1989	0.797	113.37	45.94
1990	1.735	246.80	100.00
1991	0.912	129.73	52.56
1992	0.936	133.14	53.95
1993	1.141	162.30	65.76
1994	1.604	228.17	92.45
1995	1.323	188.19	76.25
1996	1.103	156.90	63.57
1997	1.213	172.55	69.91
1998	1.452	206.54	83.69
1999	1.904	270.84	109.74
2000	1.443	205.26	83.17
2001	1.414	201.14	81.50
2002	1.451	206.40	83.63
2003	1.711	243.39	98.62
2004	1.472	209.39	84.84

(b) The mean price per pound of fresh tomatoes in 2004 in the United States is 109.39% higher than that in 1980. **(d)** The mean price per pound of fresh tomatoes in 2004 in the United States is 15.16% lower than that in 1990. **(e)** There is an upward trend in the cost of fresh tomatoes from 1980 to 2004.

16.82 (b) Linear trend: $\hat{Y} = 175{,}224.1714 + 2{,}266.7662X$. **(c)** 2004: 220,559.4947 thousands 2005: 222,826.2609 thousands **(d)** Linear trend: $\hat{Y} = 114{,}399.0286 + 1{,}695.1759X$. 2004:148,302.5474 thousands 2005: 149,997.7233 thousands

16.84 (b) $\hat{Y} = -0.8821 + 0.5588X$; **(c)** $\hat{Y} = 1.2129 + 0.0933X + 0.0166X^2$; **(d)** $\hat{Y} = 1.3251(1.1022)^X$; **(e)** Test of A_3: p-value $= 0.4197 > 0.05$. Do not reject H_0. Third-order term can be deleted. Test of A_2: p-value $= 0.6322$. Do not reject H_0. Second-order term can be deleted. Test of A_1: p-value is virtually 0. Reject H_0 that $A_1 = 0$. A first-order autoregressive model is appropriate. $\hat{Y}_i = 0.1880 + 1.0588Y_{i-1}$; **(f)** S_{YX} and MAD for linear, quadratic, exponential, and first-order autoregressive models are: 1.1220 and 0.9264, 0.3206 and 0.2237, 1.0041 and 0.5904, 0.3020 and 0.1905; **(h)** The residuals in the first three models contain consecutive positive and negative values. The autoregressive model performs well for the historical data and has a fairly random pattern of residuals. It also has the smallest S_{YX} and MAD. Based on the principle of parsimony, the autoregressive model would probably be the best model for forecasting. **(i)** $18.2940 billions

16.86 (b) Variable A: $\hat{Y} = 8.5765 + 2.8170X$. Variable B: $\hat{Y} = 11.4482 + 0.4732X$. **(c)** Variable A: $\hat{Y} = 8.2424 + 2.9225X - 0.0053X^2$. Variable B: $\hat{Y} = 10.3022 + 0.8351X - 0.0181X^2$. **(d)** Variable A: $\hat{Y} = 13.6166(1.0889)^X$. Variable B: $\hat{Y} = 11.6639(1.0314)^X$. **(e)** Variable A: Test of A_3: p-value $= 0.69 > 0.05$. Do not reject H_0. Third-order term can be deleted. Test of A_2: p-value $= 0.30 > 0.05$. Do not reject H_0. Second-order term can be deleted. Test of A_1: p-value is virtually 0. Reject H_0. A first-order autoregressive model is appropriate. $\hat{Y}_i = 4.7661 + 0.9267Y_{i-1}$ Variable B: Test of A_3: p-value $= 0.04 < 0.05$. Reject H_0. A third-order autoregressive model is appropriate. $\hat{Y}_i = 1.1414 + 1.9239Y_{i-1} - 1.6231Y_{i-2} + 0.6482Y_{i-3}$ **(g)** Variable A: S_{YX} and MAD for linear, quadratic, exponential, and first-order autoregressive models are: 8.5255 and 5.9916, 8.7571 and 6.0414, 10.0067 and 6.1069, 6.6876 and 5.1562. Variable B: S_{YX} and MAD for linear, quadratic, exponential, and first-order autoregressive models are: 0.6297 and 0.4888, 0.1024 and 0.0814, 0.9030 and 0.7164, 0.0844 and 0.0495. **(h)** Variable A: The residuals in the linear and quadratic trend models contain consecutive positive and negative values. There is no apparent pattern in the residuals of the exponential trend and the first-order autoregressive model. The autoregressive model has the smallest standard error of the estimate and MAD. The autoregressive model would probably be the best model for forecasting. Variable B: The residuals in the quadratic and exponential trend models contain consecutive positive and negative values. There is no apparent pattern in the residuals of the linear trend and the third-order autoregressive model. The third-order autoregressive model has the smallest standard error of the estimate and MAD. The autoregressive model would probably be the best model for forecasting. **(i)** Variable A: 56.6564 Variable B: 19.5860

Chapter 17

17.4 (a)–(b) Payoff table:

	Action			
Event	Company A		Company B	
1	$10,000 + $2·1,000 =	$12,000	$2,000 + $4·1,000 =	$6,000
2	$10,000 + $2·2,000 =	$14,000	$2,000 + $4·2,000 =	$10,000
3	$10,000 + $2·5,000 =	$20,000	$2,000 + $4·5,000 =	$22,000
4	$10,000 + $2·10,000 =	$30,000	$2,000 + $4·10,000 =	$42,000
5	$10,000 + $2·50,000 =	$110,000	$2,000 + $4·50,000 =	$202,000

(d) Opportunity loss table:

			Alternative Courses of Action	
Event	Optimum Action	Profit of Optimum Action	*A*	*B*
1	*A*	12,000	0	6,000
2	*A*	14,000	0	4,000
3	*B*	22,000	2,000	0
4	*B*	42,000	12,000	0
5	*B*	202,000	92,000	0

17.6 (a) 125, 112.5; **(b)** 25, 37.5; **(d)** action A; **(e)** 60%, 11.11%; **(f)** 1.667, 9.0; **(g)** action B

17.8 (a) 10%; **(b)** 25%; **(c)** 4.0

17.10 stock A

17.12 (a) *EMV*(Soft drinks) = 50(0.4) + 60(0.6) = 56, *EMV*(Ice cream) = 30(0.4) + 90(0.6) = 66; **(b)** *EOL*(Soft drinks) = 0(0.4) + 30(0.6) = 18, *EOL*(Ice cream) = 20(0.4) + 0(0.6) = 8; **(c)** *EVPI* is the maximum amount of money the vendor is willing to pay for information about which event will occur. **(d)** Based on (a) and (b), choose to sell ice cream because you will earn a higher expected monetary value and incur a lower opportunity loss than choosing to sell soft drinks. **(e)** *CV*(Soft drinks) = $\frac{4.899}{56} \cdot 100\% = 8.748\%$, *CV*(Ice cream) = $\frac{29.394}{66} \cdot 100\% = 44.536\%$; **(f)** Return to risk ratio for soft drinks = $\frac{56}{4.899} = 11.431$, Return to risk ratio for ice cream = $\frac{66}{29.394} = 2.245$

17.14 (a) 1,050, 1,400, 1,400; **(b)** 522.02, 2498.00, 9656.09; **(c)** 4,100, 3,750, 3,750; **(d)** EVPI = 3,750; **(e)** 49.72%, 178.43%, 689.72%; **(f)** 2.01, .56, .14

17.16 (a) 25,200, 32,400; **(b)** 10,700, 3,500; **(c)** *EVPI* = 3,500; **(d)** sign with company *B*; **(e)** 114.25%, 177.73%; **(f)** 0.875, 0.563; **(g)** sign with company *B* if not risk averse

17.18 (a) $P(E_1 \mid F) = .6$, $P(E_2 \mid F) = .4$; **(b)** 110, 110; **(c)** 30, 30; **(d)** *EVPI* = 30; **(e)** choose either one; **(f)** 66.8%, 11.1%; **(g)** 1.497, 8.981; **(h)** action *B*

17.20 *P*(forecast cool | cool weather) = 0.80, *P*(forecast warm | warm weather) = 0.70
(a) Revised probabilities: *P*(cool | forecast cool) = $\frac{0.32}{0.5} = 0.64$ *P*(warm | forecast cool) = $\frac{0.18}{0.5} = 0.36$ **(b)** *EMV*(Soft drinks) = 50(0.64) + 60(0.36) = 53.6, *EMV*(Ice cream) = 30(0.64) + 90(0.36) = 51.6, *EOL*(Soft drinks) = 0(0.64) + 30(0.36) = 10.8, *EOL*(Ice cream) = 20(0.64) + 0(0.36) = 12.8, *EMV* with perfect information = 50(0.64) + 90(0.36) = 64.4 *EVPI* = *EMV*, perfect information – EMV_A = 64.4 – 53.6 = 10.8. The vendor should not be willing to pay more than $10.80 for a perfect forecast of the weather. The vendor should sell soft drinks to maximize value and minimize loss.
CV(Soft drinks) = $\frac{4.8}{53.6} \cdot 100\% = 8.96\%$, *CV*(Ice cream) = $\frac{28.8}{51.6} \cdot 100\% = 55.81\%$ Return to risk ratio for soft drinks = $\frac{53.6}{4.8} = 11.1667$ Return to risk ratio for ice cream = $\frac{51.6}{28.8} = 1.7917$ Based on these revised probabilities, the vendor's decision changes because of the increased likelihood of cool weather given a forecast for cool. Under these conditions, she should sell soft drinks to maximize the expected monetary value and minimize her expected opportunity loss.

17.22 (a) 0.5590, 0.2484, 0.1412, 0.0502, 0.0013; **(b)** *EMV*: 14,658.60, 11,315.4, *EOL*: 1004.4, 4347.6, *CV*: 38.42%, 99.55%, *RTRR*: 2.603, 1.005; **(c)** under new circumstances, sign with company A

17.36 (c) 2,100, 2,660, 2,520, 1,960; **(d)** 980, 420, 560, 1,120; **(e)** *EVPI* = $420; **(f)** buy 8,000 loaves; **(g)** 0%, 15.79%, 35.57%, 57.14%; **(h)** undefined, 6.333, 2.811, 1.750; **(i)** buy 8,000 loaves; **(k) (c)** 2,100, 2,380, 2,100, 1,540 **(d)** 700, 490, 770, 1,330, **(e)** *EVPI* = $490 **(f)** buy 8,000 loaves, **(g)** 0%, 26.96%, 51.64%, 85.76%, **(h)** undefined, 3.7097, 1.9365, 1.1660, **(i)** buy 8,000 loaves

17.38 (d) 1,200,000, 1,100,000; **(e)** *EVPI* = $1,100,000; **(f)** –.0356, undefined; **(g)** continue using old packaging; **(h) (c)** –1,600,0000, 0, **(d)** 2,400,000, 800,000, **(e)** *EVPI* = $800,000, **(f)** –.5101, undefined, **(g)** continue using old packaging; **(j)**.2466, .6575, .0959; **(k) (c)** 150,600, 0, **(d)** 986,400, 1,137,000, **(e)** *EVPI* = $986,400, **(f)** .0570, undefined, **(g)** use new packaging, **(l)**.5902, .3934, .0164, **(m) (c)** –1,885,400, 0, **(d)** 2,360,800, 475,400, **(e)** *EVPI* = $475,400, **(f)** –.7288, undefined, **(g)** use old packaging

17.40 (c) 180, 100; **(d)** 20, 100; **(e)** *EVPI* = $20; **(f)** 1.2665, undefined; **(g)** call the mechanic; **(h)** 0.0143, 0.2100, 0.4159, 0.3598; **(i) (c)** 248.386, 100, **(d)** 1.144, 149.53, **(e)** *EVPI* = 1.144, **(f)** 2.055, undefined, **(g)** call the mechanic

CHAPTER 18

18.2 (a) Day 4, Day 3; **(b)** LCL = 0.0397, UCL = 0.2460; **(c)** No, proportions are within control limits.

18.4 (a) $n = 500$, $\bar{p} = 761/16000 = 0.0476$

$$\text{UCL} = \bar{p} + 3\sqrt{\frac{\bar{p}(1-\bar{p})}{n}} = 0.0476 + 3\sqrt{\frac{0.0476(1-0.0476)}{500}} = 0.0761$$

$$\text{LCL} = \bar{p} - 3\sqrt{\frac{\bar{p}(1-\bar{p})}{n}} = 0.0476 - 3\sqrt{\frac{0.0476(1-0.0476)}{500}} = 0.0190$$

(b) Since the individual points are distributed around $\bar{p}$ without any pattern and all the points are within the control limits, the process is in a state of statistical control.

18.6 (a) UCL = 0.0176, LCL = 0.0082. The proportion of unacceptable cans is below the LCL on Day 4. There is evidence of a pattern over time since the last eight points are all above the mean and most of the earlier points are below the mean. Therefore, this process is out of control.

18.8 (a) UCL = 0.1431, LCL = 0.0752. Days 9, 26, and 30 are above the UCL. Therefore, this process is out of control.

18.12 (a) UCL = 21.6735, LCL = 1.3265; **(b)** Yes, time 1 is above the UCL.

18.14 (a) The twelve errors committed by Gina appear to be much higher than all others, and Gina would need to explain her performance. **(b)** $\bar{c} = 5.5$, UCL = 12.56, LCL does not exist. The number of errors is in a state of statistical control since none of the tellers are outside the UCL. **(c)** Since Gina is within the control limits, she is operating within the system and should not be singled out for further scrutiny. **(d)** The process needs to be studied and potentially changed using principles of Six Sigma management and/or Deming management.

18.16 (a) $\bar{c} = 3.0566$; **(b)** LCL does not exist, UCL = 8.3015; **(c)** There are no weeks outside the control limits. Therefore, this process is in control; **(d)** Since these weeks are within the control limits, the results are explainable by common cause variation.

18.18 (a) $d_2 = 2.2059$; **(b)** $d_3 = 0.880$; **(c)** $D_3 = 0$; **(d)** $D_4 = 2.282$; **(e)** $A_2 = 0.729$

18.20 (a) $\overline{R} = \frac{\sum_{i=1}^{k} R_i}{k} = \frac{66.8}{20} = 3.34,\ \overline{\overline{X}} = \frac{\sum_{i=1}^{k} \overline{X}_i}{k} = \frac{118.325}{20} = 5.916.$

R chart: UCL $= D_4\overline{R} = 2.282(3.34) = 7.6219$ LCL does not exist.
$\overline{X}$ chart:
UCL $= \overline{\overline{X}} + A_2\overline{R} = 5.9163 + 0.729(3.34) = 8.3511$, LCL $= \overline{\overline{X}} - A_2\overline{R} = 5.9163 - 0.729(3.34) = 3.4814$; **(b)** The mean of sample 10 is slightly below the LCL. The process is out-of-control.

18.22 (a) $\overline{R} = 0.8794$, LCL does not exist, UCL $= 2.0068$;

(b) $\overline{\overline{X}} = 20.1065$, LCL $= 19.4654$, UCL $= 20.7475$; **(c)** The process is in control.

18.24 (a) $\overline{R} = 8.145$, LCL does not exist, UCL $= 18.5869$; $\overline{\overline{X}} = 18.12$, UCL $= 24.0577$, LCL $= 12.1823$; **(b)** There are no sample ranges outside the control limits and there does not appear to be a pattern in the range chart. The mean is above the UCL on day 15 and below the LCL on day 16. Therefore, the process is not in control.

18.26 (a) $\overline{R} = 0.3022$, LCL does not exist, UCL $= 0.6389$;

$\overline{\overline{X}} = 90.1312$, UCL $= 90.3060$, LCL $= 89.9573$; **(b)** On days 5 and 6, the sample ranges were above the UCL. The mean chart may be erroneous since the range is out of control. The process is out of control.

18.28 (a) $P(98 < X < 102) = P(-1 < Z < 1) = 0.6826$; **(b)** $P(93 < X < 107.5) = P(-3.5 < Z < 3.75) = 0.99968$; **(c)** $P(93.8 < X) = P(-3.1 < Z) = 0.99903$; **(d)** $P(X < 110) = P(Z < 5) > 0.999999713$

18.30 (a)

$$\begin{aligned} &P(18 < X < 22) \\ &= P\left(\frac{18 - 20.1065}{0.8794 / 2.059} < Z < \frac{22 - 20.1065}{0.8794 / 2.059}\right) \\ &= P(-4.932 < Z < 4.4335) = 0.9999 \end{aligned}$$

(b)

$$\begin{aligned} C_p &= \frac{(USL - LSL)}{6(\overline{R}/d_2)} = \frac{(22 - 18)}{6(0.8794/2.059)} \\ &= 1.56 \end{aligned}$$

$$\begin{aligned} CPL &= \frac{(\overline{\overline{X}} - LSL)}{3(\overline{R}/d_2)} = \frac{(20.1065 - 18)}{3(0.8704/2.059)} \\ &= 1.644 \end{aligned}$$

$$\begin{aligned} CPU &= \frac{(USL - \overline{\overline{X}})}{3(\overline{R}/d_2)} = \frac{(22 - 20.1065)}{3(0.8704/2.059)} \\ &= 1.477 \end{aligned}$$

$$C_{pk} = \min(CPL, CPU) = 1.477$$

18.32 $\overline{R} = 0.2248$, $\overline{\overline{X}} = 5.509$, $n = 4$, $d_2 = 2.059$; **(a)** $P(5.2 < X < 5.8) = P(-2.83 < Z < 2.67) = 0.9962 - 0.0023 = 0.9939$; **(b)** Since only 99.39% of the tea bags are within the specification limits, this process is not capable of meeting the goal of 99.7%.

18.46 (a) $\overline{p} = 0.2702$, LCL $= 0.1700$, UCL $= 0.3703$; **(b)** Yes, RudyBird's market share is in control before the in-store promotion. **(c)** All 7 days of the in-store promotion are above the UCL. The promotion increased market share.

18.48 (a) $\overline{p} = 0.75175$, LCL $= 0.62215$, UCL $= 0.88135$. Although none of the points are outside the control limits there is a clear pattern over time with the last 13 points above the center line. Therefore, this process is not in control. **(b)** Since the increasing trend begins around day 20, this change in method would be the assignable cause. **(c)** The control chart would have been developed using the first 20 days and then, using those limits, the additional proportions could have been plotted.

18.50 (a) $\overline{p} = 0.1198$, LCL $= 0.0205$, UCL $= 0.2191$; **(b)** Day 24 is below the LCL, therefore, the process is out of control. **(c)** Special causes of variation should be investigated to improve the process. Next the process should be improved to decrease the proportion of undesirable trades.

Index

K

L

M

SITE LICENSE AGREEMENT AND LIMITED WARRANTY

READ THIS LICENSE CAREFULLY BEFORE USING THIS PACKAGE. BY USING THIS PACKAGE, YOU ARE AGREEING TO THE TERMS AND CONDITIONS OF THIS LICENSE. IF YOU DO NOT AGREE, DO NOT USE THE PACKAGE. PROMPTLY RETURN THE UNUSED PACKAGE AND ALL ACCOMPANYING ITEMS TO THE PLACE YOU OBTAINED. ***THESE TERMS APPLY TO ALL LICENSED SOFTWARE ON THE DISK EXCEPT THAT THE TERMS FOR USE OF ANY SHAREWARE OR FREEWARE ON THE DISKETTES ARE AS SET FORTH IN THE ELECTRONIC LICENSE LOCATED ON THE DISK:***

1. GRANT OF LICENSE and OWNERSHIP: The enclosed computer programs and data ("Software") are licensed, not sold, to you by Prentice-Hall, Inc. ("We" or the "Company") in consideration of your purchase or adoption of the accompanying Company textbooks and/or other materials, and your agreement to these terms. We reserve any rights not granted to you. You own only the disk(s) but we and/or licensors own the Software itself. This license allows you to install, use, and display the enclosed copy of the Software on individual computers in the computer lab designated for use by any students of a course requiring the accompanying Company textbook and only for the as long as such textbook is a required text for such course, at a single campus or branch or geographic location of an educational institution, for academic use only, so long as you comply with the terms of this Agreement..

2. RESTRICTIONS: You may not transfer or distribute the Software or documentation to anyone else. Except for backup, you may not copy the documentation or the Software. You may not reverse engineer, disassemble, decompile, modify, adapt, translate, or create derivative works based on the Software or the Documentation. You may be held legally responsible for any copying or copyright infringement that is caused by your failure to abide by the terms of these restrictions.

3. TERMINATION: This license is effective until terminated. This license will terminate automatically without notice from the Company if you fail to comply with any provisions or limitations of this license. Upon termination, you shall destroy the Documentation and all copies of the Software. All provisions of this Agreement as to limitation and disclaimer of warranties, limitation of liability, remedies or damages, and our ownership rights shall survive termination.

4. LIMITED WARRANTY AND DISCLAIMER OF WARRANTY: Company warrants that for a period of 60 days from the date you purchase this Software (or purchase or adopt the accompanying textbook), the Software, when properly installed and used in accordance with the Documentation, will operate in substantial conformity with the description of the Software set forth in the Documentation, and that for a period of 30 days the disk(s) on which the Software is delivered shall be free from defects in materials and workmanship under normal use. The Company does not warrant that the Software will meet your requirements or that the operation of the Software will be uninterrupted or error-free. Your only remedy and the Company's only obligation under these limited warranties is, at the Company's option, return of the disk for a refund of any amounts paid for it by you or replacement of the disk. THIS LIMITED WARRANTY IS THE ONLY WARRANTY PROVIDED BY THE COMPANY AND ITS LICENSORS, AND THE COMPANY AND ITS LICENSORS DISCLAIM ALL OTHER WARRANTIES, EXPRESS OR IMPLIED, INCLUDING WITHOUT LIMITATION, THE IMPLIED WARRANTIES OF MERCHANTABILITY AND FITNESS FOR A PARTICULAR PURPOSE. THE COMPANY DOES NOT WARRANT, GUARANTEE OR MAKE ANY REPRESENTATION REGARDING THE ACCURACY, RELIABILITY, CURRENTNESS, USE, OR RESULTS OF USE, OF THE SOFTWARE.

5. LIMITATION OF REMEDIES AND DAMAGES: IN NO EVENT, SHALL THE COMPANY OR ITS EMPLOYEES, AGENTS, LICENSORS, OR CONTRACTORS BE LIABLE FOR ANY INCIDENTAL, INDIRECT, SPECIAL, OR CONSEQUENTIAL DAMAGES ARISING OUT OF OR IN CONNECTION WITH THIS LICENSE OR THE SOFTWARE, INCLUDING FOR LOSS OF USE, LOSS OF DATA, LOSS OF INCOME OR PROFIT, OR OTHER LOSSES, SUSTAINED AS A RESULT OF INJURY TO ANY PERSON, OR LOSS OF OR DAMAGE TO PROPERTY, OR CLAIMS OF THIRD PARTIES, EVEN IF THE COMPANY OR AN AUTHORIZED REPRESENTATIVE OF THE COMPANY HAS BEEN ADVISED OF THE POSSIBILITY OF SUCH DAMAGES. IN NO EVENT SHALL THE LIABILITY OF THE COMPANY FOR DAMAGES WITH RESPECT TO THE SOFTWARE EXCEED THE AMOUNTS ACTUALLY PAID BY YOU, IF ANY, FOR THE SOFTWARE OR THE ACCOMPANYING TEXTBOOK. SOME JURISDICTIONS DO NOT ALLOW THE LIMITATION OF LIABILITY IN CERTAIN CIRCUMSTANCES, THE ABOVE LIMITATIONS MAY NOT ALWAYS APPLY.

6. GENERAL: THIS AGREEMENT SHALL BE CONSTRUED IN ACCORDANCE WITH THE LAWS OF THE UNITED STATES OF AMERICA AND THE STATE OF NEW YORK, APPLICABLE TO CONTRACTS MADE IN NEW YORK, AND SHALL BENEFIT THE COMPANY, ITS AFFILIATES AND ASSIGNEES. This Agreement is the complete and exclusive statement of the agreement between you and the Company and supersedes all proposals, prior agreements, oral or written, and any other communications between you and the company or any of its representatives relating to the subject matter. If you are a U.S. Government user, this Software is licensed with "restricted rights" as set forth in subparagraphs (a)-(d) of the Commercial Computer-Restricted Rights clause at FAR 52.227-19 or in subparagraphs (c)(1)(ii) of the Rights in Technical Data and Computer Software clause at DFARS 252.227-7013, and similar clauses, as applicable.

Should you have any questions concerning this agreement or if you wish to contact the Company for any reason, please contact in writing:

Director, Media Production
Pearson Education
1 Lake Street
Upper Saddle River, NJ 07458